CULTURE
OF ANIMAL CELLS

CULTURE OF ANIMAL CELLS

A MANUAL OF BASIC TECHNIQUE

Fifth Edition

R. Ian Freshney

Cancer Research UK Centre for Oncology and Applied Pharmacology
Cancer Research UK Beatson Laboratories
University of Glasgow

WILEY-LISS

A JOHN WILEY & SONS, INC., PUBLICATION

Library of Congress Cataloging-in-Publication Data:

Freshney, R. Ian.
 Culture of animal cells : a manual of basic techniques/R. Ian Freshney.—5th ed. p. cm.
Includes bibliographical references and index.
ISBN-13 978-0-471-45329-1 (alk. paper)
ISBN-10 0-471-45329-3 (alk. paper)
1. Tissue culture Laboratory manuals. 2. Cell culture Laboratory manuals. I. Title.
QH585.2.F74 1994
571.6′38—dc21 99-23536

Printed in the United States of America.

10 9 8 7 6 5 4 3 2

This book is dedicated to all of the many friends and colleagues whose help and advice over the years has enabled me to extend the scope of this book beyond my own limited experience.

Contents

13. Subculture and Cell Lines, 199

14. Cloning and Selection, 217

15. Cell Separation, 237

16. Characterization, 247

List of Figures

List of Color Plates

Preface

This, the fifth edition of *Culture of Animal Cells*, is structured in a similar style to previous editions, but with some significant changes. A new chapter has been introduced, Chapter 2, Training Programs, designed to enhance the use of this book as a teaching manual in addition to its role as a reference text. Chapter 2 sets out suggested programs for training new staff or students and adds experimental and analytical elements to some of the protocols to make the learning experience more interesting, more informative, and a little more challenging.

The cross-referencing has been extensively revised and updated. Individual sections now have numbers that incorporate the chapter number. A binomial, e.g., 4.1, cross-refers to Chapter 4, Section 1 and a trinomial, e.g., 14.6.2, would refer to Chapter 14, Section 6, Subsection 2, so the first digits, cross-referring to text, tables, or figures, will always refer to the chapter number. In part, this is designed to facilitate hyperlinking in the future electronic version of this book, which should appear on the Wiley website, *www.wiley.com*, shortly after publication of the paper edition.

The number of color plate pages has been doubled and, in combination with an extended Figure 16.2, now provides photographs of around 40 different cell lines, plus primary cultures, equipment and processes. I am greatly indebted to Yvonne Reid and Greg Sykes of ATCC, Peter Thraves of ECACC, and many others for kindly providing new illustrations; I hope that this will encourage readers to look at their cells more carefully and become sensitive to any changes that occur during routine maintenance.

For most of the book, I have retained the emphasis of previous editions and focused on basic techniques with some examples of more specialized cultures and methods. These techniques are presented as detailed step-by-step protocols that should give sufficient information to carry out a procedure without recourse to the prime literature. There is also introductory material to each protocol to explain the background and supplementary information to provide alternative procedures and applications. Some basic biology is explained, but it is assumed that the reader will have a little knowledge of anatomy, histology, biochemistry, and cell and molecular biology. The book is targeted at those with little or no previous experience in tissue culture, including technicians in training, senior undergraduates, graduate students, postdoctoral workers, and clinicians with an interest in laboratory science.

Specialized techniques are now all contained in one chapter, as it does not seem entirely logical to separate molecular techniques from the others as the boundary is not distinct. The molecular techniques that are included are all seen as having direct relevance to cell culture. No attempt has been made to present basic molecular methodology, as these are already available [e.g., Sambrook et al., 1989; Ausubel et al., 1996, 2002]. Similarly, the chapter on scale-up serves as an interface with biotechnology and provides some background on systems for increasing cell yield, but takes no account of full-scale biopharmaceutical production processes.

The suppliers list has been updated and, hopefully, will remain so at the time of publication. However, company names change, mergers occur, and some suppliers disappear, so it is difficult to maintain currency. Again, it is hoped that, in the future, this will be accomplished more effectively on the website.

Abbreviations used in the text are listed separately after the preface. Conventions employed throughout are D-PBSA for Dulbecco's PBS without Ca^{2+} and Mg^{2+} and UPW for ultrapure water, regardless of how it is prepared.

Concentrations are given in molarity wherever possible, and actual weights have been omitted from the media tables on the assumption that very few people will attempt to make up their own media, but will, more likely, want to compare constituents, for which molar equivalents are more useful.

Protocols are identified in the text by boxing and shading of the text. Reagents that are specific to a particular protocol are detailed in the materials sections of the protocols, the recipes for the common reagents, such as Hanks' BSS or trypsin, are given in Appendix I. Details of the sources of equipment and materials are given in Appendix II, and the contact details of suppliers and other resources are given in Appendix.

As always, I owe a great debt of gratitude to the authors who have contributed protocols, and to others who have advised me in areas where my knowledge is imperfect including Robert Auerbach, Bob Brown, Kenneth Calman, Richard Ham, Rob Hay, Stan Kaye, Nicol Keith, Wally McKeehan, Rona McKie, Stephen Merry, Jane Plumb, Peter Vaughan, Paul Workman, Roland Grafström, and the late John Paul. I am fortunate in having had the clinical collaboration of David I. Graham, David G. T. Thomas, and the late John Maxwell Anderson. In the early stages of the preparation of this book I also benefited from discussions with Don Dougall, Peter del Vecchio, Sergey Federoff, Mike Gabridge, Dan Lundin, John Ryan, Jim Smith, and Charity Waymouth. I am eternally grateful to Paul Chapple who first persuaded me that I should write a basic techniques book on tissue culture and who, more recently, suggested the development of this text into a multimedia presentation, now published by Wiley-Liss. Many of the original illustrations were produced by Jane Gillies and Marina LaDuke, although many of these have now been replaced due to the demands of electronic publishing. I am grateful to the Cancer Research UK Beatson Laboratories for allowing me to take photographs of many of the pieces of equipment that I have illustrated in this book. Some of the data presented were generated by those who have worked with me over the years including Sheila Brown, Ian Cunningham, Lynn Evans, Margaret Frame, Elaine Hart, Carol McCormick, Alison Mackie, John McLean, Alistair McNab, Diana Morgan, Alison Murray, Irene Osprey, Mohammad Zareen Khan, and Natasha Yevdokimova.

I have been fortunate to receive excellent advice and support from the staff of John Wiley & Sons. I would also like to acknowledge with sincere gratitude all those who have taken the trouble to write to me or to John Wiley & Sons with advice and constructive criticism on previous editions. It is pleasant and satisfying to hear from those who have found this book beneficial, but even more important to hear from those who have found deficiencies, which I can then attempt to rectify. I can only hope that those of you who use this book retain the same excitement that I feel about the future prospects emerging in the field.

I would like to thank my daughter Gillian and son Norman for all the help they gave me in the preparation of the first edition, many years ago, and for their continued advice and support. Above all, I would like to thank my wife, Mary, for her hours of help in compilation, proofreading, and many other tasks; without her help and support, the original text would never have been written and I would never have completed this revision by the assigned deadline, nor attained the necessary level of technical accuracy that is the keynote of a good tissue culture manual.

Ian Freshney

Abbreviations

ATCC	American Type Culture Collection	EM	electron microscope
BMP	bone morphogenetic protein	FBS	fetal bovine serum
bp	base pairs (in DNA)	FCS	fetal calf serum
BPE	bovine pituitary extract	FGF	fibroblast growth factor
BrUdR	bromodeoxyuridine	G_0	Gap in cell cycle when cell exit cycle
BSA	bovine serum albumin	G_1	gap one (of the cell cycle)
BUdR	bromodeoxyuridine	G_2	gap two (of the cell cycle)
CAM	chorioallantoic membrane	GLP	good laboratory practice
CAM	cell adhesion molecule	H&E	hematoxylin and eosin (stains)
cAMP	Cyclic adenosine monophosphate	HAT	hypoxanthine, aminopterin, and thymidine
CCD	charge-coupled device	HBS	HEPES-buffered saline
CCTV	closed-circuit television	HBSS	Hanks' balanced salt solution
cDNA	complementary DNA	HC	hydrocortisone
CE	cloning efficiency	hCG	human chorionic gonadotropin
CMC	carboxymethylcellulose	HGPRT	hypoxanthine guanosine phosphoribosyl transferase
CMF	calcium- and magnesium-free saline		
CMRL	Connaught Medical Research Laboratories	HITES	hydrocortisone, insulin, transferrin, estradiol, and selenium
D-PBSA	Dulbecco's phosphate-buffered saline lacking Ca^{2+} and Mg^{2+}	HPV	human papilloma virus
		HuS	human serum
D-PBSB	Dulbecco's phosphate-buffered saline, solution B (Ca^{2+} and Mg^{2+})	HS	horse serum
		HSV	herpes simplex virus
DEPC	diethyl pyrocarbonate	HT	hypoxanthine/thymidine
DMEM	Dulbecco's modification of Eagle's medium	ITS	insulin, transferrin, selenium
DNA	deoxyribonucleic acid	IUdR	iododeoxyuridine
DT	population doubling time	KBM	keratinocyte basal medium
EBSS	Earle's balanced salt solution	kbp	kilobase pairs (in DNA)
EBV	Epstein−Barr virus	KGM	keratinocyte growth medium
ECACC	European Collection of Animal Cell Cultures (now European Collection of Cell Cultures)	LI	labeling index
		M	Mitosis (in cell cycle)
		M199	medium 199
ECGF	endothelial cell growth factor	MACs	mammalian artificial chromosomes
EGF	epidermal growth factor		

MACS	magnet-activated cell sorting	PVP	polyvinylpyrrolidone
MEM	Eagle's Minimal Essential Medium	PWM	pokeweed mitogen
mRNA	messenger RNA	RNA	ribonucleic acid
MTT	3-(4,5-dimethylthiazol-2-yl)-2,5-diphenyltetrazolium bromide	RPMI	Roswell Park Memorial Institute
		RT-PCR	reverse transcriptase PCR
NBCS	newborn calf serum	S	DNA synthetic phase of cell cycle
NCI	National Cancer Institute	SD	saturation density
O.D.	optical density	SIT	selenium, insulin, transferrin
PA	plasminogen activator	S-MEM	MEM with low Mg^{2+} and no Ca^{2+}
PBS	phosphate-buffered saline	SSC	sodium citrate/sodium chloride
PBSA	phosphate-buffered saline, solution A (Ca^{2+} and Mg^{2+} free)	SV40	simian virus 40
		SV40LT	SV40 gene for large T-antigen
PBSB	phosphate-buffered saline, solution B (Ca^{2+} and Mg^{2+})	TCA	trichloroacetic acid
		T_D	population doubling time
PCA	perchloric acid	TEB	Tris/EDTA buffer
PCR	polymerase chain reaction	TGF	transforming growth factor
PDGF	platelet-derived growth factor	TK	thymidine kinase
PE	plating efficiency	UPW	ultrapure water
PE	PBSA/EDTA	VEGF	vascular endothelial growth factor
PEG	polyethylene glycol	XTT	2,3-bis(2-methoxy-4-nitro-5-sulfophenyl)-2H-tetrazolium-5-carboxanilide
PHA	phytohemagglutinin		
PMA	phorbol myristate acetate	YACs	yeast artificial chromosomes

CHAPTER 1

Introduction

1.1 HISTORICAL BACKGROUND

Tissue culture was first devised at the beginning of the twentieth century [Harrison, 1907; Carrel, 1912] (Table 1.1) as a method for studying the behavior of animal cells free of systemic variations that might arise *in vivo* both during normal homeostasis and under the stress of an experiment. As the name implies, the technique was elaborated first with undisaggregated fragments of tissue, and growth was restricted to the migration of cells from the tissue fragment, with occasional mitoses in the outgrowth. As culture of cells from such primary explants of tissue dominated the field for more than 50 years [Fischer, 1925; Parker, 1961], it is not surprising that the name "tissue culture" has remained in use as a generic term despite the fact that most of the explosive expansion in this area in the second half of the twentieth century (Fig. 1.1) was made possible by the use of dispersed cell cultures.

Disaggregation of explanted cells and subsequent plating out of the dispersed cells was first demonstrated by Rous [Rous and Jones, 1916], although passage was more often by surgical subdivision of the culture [Fischer, Carrel, and others] to generate what were then termed cell strains. L929 was the first cloned cell strain, isolated by capillary cloning from mouse L-cells [Sanford et al., 1948]. It was not until the 1950s that trypsin became more generally used for subculture, following procedures described by Dulbecco to obtain passaged monolayer cultures for viral plaque assays [Dulbecco, 1952], and the generation of a single cell suspension by trypsinization, which facilitated the further development of single cell cloning. Gey established the first

continuous human cell line, HeLa [Gey et al., 1952]; this was subsequently cloned by Puck [Puck and Marcus, 1955] when the concept of an X-irradiated feeder layer was introduced into cloning. Tissue culture became more widely used at this time because of the introduction of antibiotics, which facilitated long-term cell line propagation although many people were already warning against continuous use and the associated risk of harboring cryptic, or antibiotic-resistant, contaminations [Parker, 1961]. The 1950s were also the years of the development of defined media [Morgan et al., 1950; Parker et al., 1954; Eagle, 1955, 1959; Waymouth, 1959], which led ultimately to the development of serum-free media [Ham, 1963, 1965] (*see* Section 10.6).

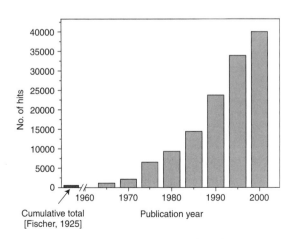

Fig. 1.1. Growth of Tissue Culture. Number of hits in PubMed for "cell culture" from 1965. The pre-1960 figure is derived from the bibliography of Fischer [1925].

TABLE 1.1. Key Events in the Development of Cell and Tissue Culture

Date	Event	Reference
1907	Frog embryo nerve fiber outgrowth *in vitro*	Harrison, 1907
1912	Explants of chick connective tissue; heart muscle contractile for 2–3 months	Carrel, 1912; Burrows, 1912
1916	Trypsinization and subculture of explants	Rous & Jones, 1916
1920s/30s	Subculture of fibroblastic cell lines	Carrel & Ebeling, 1923
1925–1926	Differentiation *in vitro* in organ culture	Strangeways & Fell, 1925, 1926
1940s	Introduction of use of antibiotics in tissue culture	Keilova, 1948; Cruikshank & Lowbury, 1952
1943	Establishment of the L-cell mouse fibroblast; first continuous cell line	Earle et al., 1943
1948	Cloning of the L-cell	Sanford et al., 1948
1949	Growth of virus in cell culture	Enders et al., 1949
1952	Use of trypsin for generation of replicate subcultures	Dulbecco, 1952
	Virus plaque assay	Dulbecco, 1952
1952–1955	Establishment the first human cell line, HeLa, from a cervical carcinoma,	Gey et al., 1952
1952	Nuclear transplantation	*see* Briggs & King, 1960
1954	Fibroblast contact inhibition of cell motility	Abercrombie & Heaysman, 1954
	Salk polio vaccine grown in monkey kidney cells	*see* Griffiths, 1991
1955	Cloning of HeLa on a homologous feeder layer	Puck & Marcus, 1955
	Development of defined media	Eagle, 1955, 1959
	Requirement of defined media for serum growth factors	Sanford et al., 1955; Harris, 1959
Late 1950s	Realization of importance of mycoplasma (PPLO) infection	Coriell et al., 1958; Rothblat & Morton, 1959; Nelson, 1960
1961	Definition of finite life span of normal human cells	Hayflick & Moorhead, 1961
	Cell fusion–somatic cell hybridization	Sorieul & Ephrussi, 1961
1962	Establishment and transformation of BHK21	Macpherson & Stoker, 1962
	Maintenance of differentiation (pituitary & adrenal tumors)	Buonassisi et al., 1962; Yasamura et al., 1966; Sato & Yasumura, 1966
1963	3T3 cells & spontaneous transformation	Todaro & Green, 1963
1964	Pluripotency of embryonal stem cells	Kleinsmith & Pierce, 1964
	Selection of transformed cells in agar	Macpherson & Montagnier, 1964
1964–1969	Rabies, Rubella vaccines in WI-38 human lung fibroblasts	Wiktor et al., 1964; Andzaparidze, 1968
1965	Serum-free cloning of Chinese hamster cells	Ham, 1965
	Heterokaryons—man–mouse hybrids	Harris & Watkins, 1965
1966	Nerve growth factor	Levi-Montalcini, 1966
	Differentiation in rat hepatomas	Thompson et al., 1966
1967	Epidermal growth factor	Hoober & Cohen, 1967
	HeLa cell cross-contamination	Gartler, 1967
	Density limitation of cell proliferation	Stoker & Rubin, 1967
	Lymphoblastoid cell lines	Moore et al., 1967; Gerper et al., 1969; Miller et al., 1971
1968	Retention of differentiation in cultured normal myoblasts	Yaffe, 1968
	Anchorage-independent cell proliferation	Stoker et al., 1968
1969	Colony formation in hematopoietic cells	Metcalf, 1969; *see also* Metcalf, 1990
1970s	Development of laminar-flow cabinets	*see* Kruse et al., 1991; Collins & Kennedy, 1999
1973	DNA transfer, calcium phosphate	Graham & Van der Eb, 1973
1975	Fibroblast growth factor	Gospodarowicz et al., 1975
	Hybridomas—monoclonal antibodies	Kohler & Milstein, 1975
1976	Totipotency of embryonal stem cells	Illmensee & Mintz, 1976
	Growth factor-supplemented serum-free media	Hayashi & Sato, 1976
1977	Confirmation of HeLa cell cross-contamination of many cell lines	Nelson-Rees & Flandermeyer, 1977
	3T3 feeder layer and skin culture	Rheinwald & Green, 1975
1978	MCDB-selective, serum-free media	Ham & McKeehan, 1978
	Matrix interactions	Gospodarowicz et al., 1978b; Reid & Rojkind, 1979
	Cell shape and growth control	Folkman & Moscona, 1978

TABLE 1.1. Key Events in the Development of Cell and Tissue Culture (*Continued*)

Date	Event	Reference
1980s	Regulation of gene expression	*see*, e.g., Darnell, 1982
	Oncogenes, malignancy, and transformation	*see* Weinberg, 1989
1980	Matrix from EHS sarcoma (later Matrigel™)	Hassell et al., 1980
1983	Regulation of cell cycle	Evans et al., 1983; *see also* Nurse, 1990
	Immortalization by SV40	Huschtscha & Holliday, 1983
1980–1987	Development of many specialized cell lines	Peehl & Ham, 1980; Hammond et al., 1984; Knedler & Ham, 1987
1983	Reconstituted skin cultures	Bell et al., 1983
1984	Production of recombinant tissue-type plasminogen activator in mammalian cells	Collen et al., 1984
1990s	Industrial-scale culture of transfected cells for production of biopharmaceuticals	Butler, 1991
1991	Culture of human adult mesenchymal stem cells	Caplan, 1991
1998	Tissue-engineered cartilage	Aigner et al., 1998
1998	Culture of human embryonic stem cells	Thomson et al., 1998
2000+	Human Genome Project: genomics, proteomics, genetic deficiencies and expression errors	Dennis et al., 2001
	Exploitation of tissue engineering	Atala & Lanza, 2002; Vunjak-Novakovic & Freshney, 2004

See also Pollack, 1981.

Throughout this book, the term *tissue culture* is used as a generic term to include organ culture and cell culture. The term *organ culture* will always imply a three-dimensional culture of undisaggregated tissue retaining some or all of the histological features of the tissue *in vivo*. *Cell culture* refers to a culture derived from dispersed cells taken from original tissue, from a primary culture, or from a cell line or cell strain by enzymatic, mechanical, or chemical disaggregation. The term *histotypic culture* implies that cells have been reaggregated or grown to re-create a three-dimensional structure with tissuelike cell density, e.g., by cultivation at high density in a filter well, perfusion and overgrowth of a monolayer in a flask or dish, reaggregation in suspension over agar or in real or simulated zero gravity, or infiltration of a three-dimensional matrix such as collagen gel. *Organotypic* implies the same procedures but recombining cells of different lineages, e.g., epidermal keratinocytes in combined culture with dermal fibroblasts, in an attempt to generate a *tissue equivalent*.

Harrison [1907] chose the frog as his source of tissue, presumably because it was a cold-blooded animal, and consequently, incubation was not required. Furthermore, because tissue regeneration is more common in lower vertebrates, he perhaps felt that growth was more likely to occur than with mammalian tissue. Although his technique initiated a new wave of interest in the cultivation of tissue *in vitro*, few later workers were to follow his example in the selection of species. The stimulus from medical science carried future interest into warm-blooded animals, in which both normal development and pathological development are closer to that found in humans. The accessibility of different tissues, many of which grew well in culture, made the embryonated hen's egg a favorite choice; but the development of experimental animal husbandry, particularly with genetically pure strains of rodents, brought mammals to the forefront as the favorite material. Although chick embryo tissue could provide a diversity of cell types in primary culture, rodent tissue had the advantage of producing continuous cell lines [Earle et al., 1943] and a considerable repertoire of transplantable tumors. The development of transgenic mouse technology [Beddington, 1992; Peat et al., 1992], together with the well-established genetic background of the mouse, has added further impetus to the selection of this animal as a favorite species.

The demonstration that human tumors could also give rise to continuous cell lines [e.g., HeLa; Gey et al., 1952] encouraged interest in human tissue, helped later by the classic studies of Leonard Hayflick on the finite life span of cells in culture [Hayflick & Moorhead, 1961] and the requirement of virologists and molecular geneticists to work with human material. The cultivation of human cells received a further stimulus when a number of different serum-free selective media were developed for specific cell types, such as epidermal keratinocytes, bronchial epithelium, and vascular endothelium (*see* Section 10.2.1). These formulations are now available commercially, although the cost remains high relative to the cost of regular media.

For many years, the lower vertebrates and the invertebrates were largely ignored, although unique aspects of their development (tissue regeneration in amphibians, metamorphosis in insects) make them attractive systems for the study of the molecular basis of development. More recently, the needs of agriculture and pest control have encouraged toxicity and virological studies in insects, and developments in gene technology have suggested that insect cell lines with baculovirus and other vectors may be useful producer cell lines because of the possibility of inserting larger genomic sequences in the viral DNA and a reduced risk of propagating human pathogenic viruses. Furthermore, the economic importance of fish farming and the role of freshwater and marine pollution have stimulated more studies of normal development and pathogenesis in fish. Procedures for handling nonmammalian cells have tended to follow those developed for mammalian cell culture, although a limited number of specialized media are now commercially available for fish and insect cells (*see* Sections 27.7.1, 27.7.2).

The types of investigation that lend themselves particularly to tissue culture are summarized in Fig. 1.2: (1) intracellular activity, e.g., the replication and transcription of deoxyribonucleic acid (DNA), protein synthesis, energy metabolism, and drug metabolism; (2) intracellular flux, e.g., RNA, the translocation of hormone receptor complexes and resultant signal transduction processes, and membrane trafficking; (3) environmental interaction, e.g., nutrition, infection, cytotoxicity, carcinogenesis, drug action, and ligand–receptor interactions; (4) cell–cell interaction, e.g., morphogenesis, paracrine control, cell proliferation kinetics, metabolic cooperation, cell adhesion and motility, matrix interaction, and organotypic models for medical prostheses and invasion; (5) genetics, including genome analysis in normal and pathological conditions, genetic manipulation, transformation, and immortalization; and (6) cell products and secretion, biotechnology, bioreactor design, product harvesting, and downstream processing.

The development of cell culture owed much to the needs of two major branches of medical research: the production of antiviral vaccines and the understanding of neoplasia. The standardization of conditions and cell lines for the production and assay of viruses undoubtedly provided much impetus to the development of modern tissue culture technology, particularly the production of large numbers of cells suitable for biochemical analysis. This and other technical improvements made possible by the commercial supply of reliable media and sera and by the greater control of contamination with antibiotics and clean-air equipment have made tissue culture accessible to a wide range of interests.

An additional force of increasing weight from public opinion has been the expression of concern by many animal-rights groups over the unnecessary use of experimental animals. Although most accept the idea that some requirement for animals will continue for preclinical trials of new pharmaceuticals, there is widespread concern that extensive use of animals for cosmetics development and similar activities may not be morally justifiable. Hence, there is an ever-increasing lobby for more *in vitro* assays, the adoption of which, however, still requires their proper validation and general acceptance. Although this seemed a distant prospect some years ago, the introduction of more sensitive and more readily performed *in vitro* assays, together with a very real prospect of assaying for inflammation *in vitro*, has promoted an unprecedented expansion in *in vitro* testing (*see* Section 22.4).

In addition to cancer research and virology, other areas of research have come to depend heavily on tissue culture techniques. The introduction of cell fusion techniques (*see* Section 27.9) and genetic manipulation [Maniatis et al., 1978; Sambrook et al., 1989; Ausubel et al., 1996] established somatic cell genetics as a major component in the genetic analysis of higher animals, including humans. A wide range of techniques for genetic recombination now includes DNA transfer [Ravid & Freshney, 1998], monochr`msomal transfer

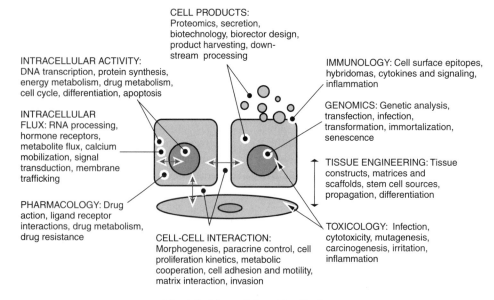

Fig. 1.2. Tissue Culture Applications.

[Newbold & Cuthbert, 1998], and nuclear transfer [Kono, 1997], which have been added to somatic hybridization as tools for genetic analysis and gene manipulation. DNA transfer itself has spawned many techniques for the transfer of DNA into cultured cells, including calcium phosphate coprecipitation, lipofection, electroporation, and retroviral infection (*see* Section 27.11).

In particular, human genetics has progressed under the stimulus of the Human Genome Project [Baltimore, 2001], and the data generated therefrom have recently made feasible the introduction of multigene array expression analysis [Iyer et al., 1999].

Tissue culture has contributed greatly, via the monoclonal antibody technique, to the study of immunology, already dependent on cell culture for assay techniques and the production of hematopoietic cell lines. The insight into the mechanism of action of antibodies and the reciprocal information that this provided about the structure of the epitope, derived from monoclonal antibody techniques [Kohler & Milstein, 1975], was, like the technique of cell fusion itself, a prologue to a whole new field of studies in genetic manipulation. This field has supplied much basic information on the control of gene transcription, and a vast new technology and a multibillion-dollar industry have grown out of the ability to insert exploitable genes into prokaryotic and eukaryotic cells. Cell products such as human growth hormone, insulin, and interferon are now produced routinely by transfected cells, although the absence of posttranscriptional modifications, such as glycosylation, in bacteria suggests that mammalian cells may provide more suitable vehicles [Grampp et al., 1992], particularly in light of developments in immortalization technology (*see* Section 18.4).

Other areas of major interest include the study of cell interactions and intracellular control mechanisms in cell differentiation and development [Jessell and Melton, 1992; Ohmichi et al., 1998; Balkovetz & Lipschutz, 1999] and attempts to analyze nervous function [Richard et al., 1998; Dunn et al., 1998; Haynes, 1999]. Progress in neurological research has not had the benefit, however, of working with propagated cell lines from normal brain or nervous tissue, as the propagation of neurons *in vitro* has not been possible, until now, without resorting to the use of transformed cells (*see* Section 18.4). However, developments with human embryonal stem cell cultures [Thomson et al., 1998; Rathjen et al., 1998; Wolf et al., 1998; Webber & Minger, 2004] suggest that this approach may provide replicating cultures that will differentiate into neurons.

Tissue culture technology has also been adopted into many routine applications in medicine and industry. Chromosomal analysis of cells derived from the womb by amniocentesis (*see* Section 27.6) can reveal genetic disorders in the unborn child, the quality of drinking water can be determined, and the toxic effects of pharmaceutical compounds and potential environmental pollutants can be measured in colony-forming and other *in vitro* assays (*see* Sections 22.3.1, 22.3.2, 22.4).

Further developments in the application of tissue culture to medical problems have followed from the demonstration that cultures of epidermal cells form functionally differentiated sheets [Green et al., 1979] and endothelial cells may form capillaries [Folkman & Haudenschild, 1980], offering possibilities in homografting and reconstructive surgery using an individual's own cells [Tuszynski et al., 1996; Gustafson et al., 1998; Limat et al., 1996], particularly for severe burns [Gobet et al., 1997; Wright et al., 1998; Vunjak-Novakovic & Freshney, 2005] (*see also* Section 25.3.8). With the ability to transfect normal genes into genetically deficient cells, it has become possible to graft such "corrected" cells back into the patient. Transfected cultures of rat bronchial epithelium carrying the β-gal reporter gene have been shown to become incorporated into the rat's bronchial lining when they are introduced as an aerosol into the respiratory tract [Rosenfeld et al., 1992]. Similarly, cultured satellite cells have been shown to be incorporated into wounded rat skeletal muscle, with nuclei from grafted cells appearing in mature, syncytial myotubes [Morgan et al., 1992].

The prospects for implanting normal cells from adult or fetal tissue-matched donors or implanting genetically reconstituted cells from the same patient have generated a whole new branch of culture, that of *tissue engineering* [Atala and Lanza, 2002; Vunjak-Novakovic and Freshney, 2005], encompassing the generation of tissue equivalents by organotypic culture (*see* Section 25.3.8), isolation and differentiation of human embryonal stem (ES) cells and adult totipotent stem cells such as mesenchymal stem cells (MSCs), gene transfer, materials science, utilization of bioreactors, and transplantation technology. The technical barriers are steadily being overcome, bringing the ethical questions to the fore. The technical feasibility of implanting normal fetal neurons into patients with Parkinson disease has been demonstrated; society must now decide to what extent fetal material may be used for this purpose. Where a patient's own cells can be grown and subjected to genetic reconstitution by transfection of the normal gene—e.g., transfecting the normal insulin gene into β-islet cells cultured from diabetics, or even transfecting other cell types such as skeletal muscle progenitors [Morgan et al., 1992]—it would allow the cells to be incorporated into a low-turnover compartment and, potentially, give a long-lasting physiological benefit. Although the ethics of this type of approach seem less contentious, the technical limitations of this approach are still apparent.

In vitro fertilization (IVF), developed from early experiments in embryo culture [*see* review by Edwards, 1996], is now widely used [*see*, e.g., Gardner and Lane, 2003] and has been accepted legally and ethically in many countries. However, another area of development raising significant ethical debate is the generation of gametes *in vitro* from the culture of primordial germ cells isolated from testis and ovary [Dennis, 2003] or from ES cells. Oocytes have been cultured from embryonic mouse ovary and implanted, generating normal mice [Eppig, 1996; Obata et al., 2002], and spermatids have been cultured from newborn bull testes and co-cultured with Sertoli cells [Lee et al., 2001]. Similar work with mouse testes generated spermatids that were used to fertilize mouse eggs, which developed into mature, fertile adults [Marh, et al., 2003].

1.2 ADVANTAGES OF TISSUE CULTURE

1.2.1 Control of the Environment

The two major advantages of tissue culture (Table 1.2) are control of the physiochemical environment (pH, temperature, osmotic pressure, and O_2 and CO_2 tension), which may be controlled very precisely, and the physiological conditions, which may be kept relatively constant, but cannot always be defined. Most cell lines still require supplementation of the medium with serum or other poorly defined constituents. These supplements are prone to batch variation and contain undefined elements such as hormones and other regulatory substances. The identification of some of the essential components of serum (*see* Table 9.5), together with a better understanding of factors regulating cell proliferation (*see* Table 10.3), has made the replacement of serum with defined constituents feasible (*see* Section 10.4). As laboratories seek to express the normal phenotypic properties of cells *in vitro*, the role of the extracellular matrix becomes increasingly important. Currently, that role is similar to the use of serum—that is, the matrix is often necessary, but

TABLE 1.2. Advantages of Tissue Culture

Category	Advantages
Physico-chemical environment	Control of pH, temperature, osmolality, dissolved gases
Physiological conditions	Control of hormone and nutrient concentrations
Microenvironment	Regulation of matrix, cell–cell interaction, gaseous diffusion
Cell line homogeneity	Availability of selective media, cloning
Characterization	Cytology and immunostaining are easily performed
Preservation	Can be stored in liquid nitrogen
Validation & accreditation	Origin, history, purity can be authenticated and recorded
Replicates and variability	Quantitation is easy
Reagent saving	Reduced volumes, direct access to cells, lower cost
Control of C × T	Ability to define dose, concentration (C), and time (T)
Mechanization	Available with microtitration and robotics
Reduction of animal use	Cytotoxicity and screening of pharmaceutics, cosmetics, etc.

TABLE 1.3. Limitations of Tissue Culture

Category	Examples
Necessary expertise	Sterile handling
	Chemical contamination
	Microbial contamination
	Cross-contamination
Environmental control	Workplace
	Incubation, pH control
	Containment and disposal of biohazards
Quantity and cost	Capital equipment for scale-up
	Medium, serum
	Disposable plastics
Genetic instability	Heterogeneity, variability
Phenotypic instability	Dedifferentiation
	Adaptation
	Selective overgrowth
Identification of cell type	Markers not always expressed
	Histology difficult to recreate and atypical
	Geometry and microenvironment change cytology

not always precisely defined, yet it can be regulated and, as cloned matrix constituents become available, may still be fully defined.

1.2.2 Characterization and Homogeneity of Sample

Tissue samples are invariably heterogeneous. Replicates—even from one tissue—vary in their constituent cell types. After one or two passages, cultured cell lines assume a homogeneous (or at least uniform) constitution, as the cells are randomly mixed at each transfer and the selective pressure of the culture conditions tends to produce a homogeneous culture of the most vigorous cell type. Hence, at each subculture, replicate samples are identical to each other, and the characteristics of the line may be perpetuated over several generations, or even indefinitely if the cell line is stored in liquid nitrogen. Because experimental replicates are virtually identical, the need for statistical analysis of variance is reduced.

The availability of stringent tests for cell line identity (Chapter 15) and contamination (Chapter 18) means that preserved stocks may be validated for future research and commercial use.

1.2.3 Economy, Scale, and Mechanization

Cultures may be exposed directly to a reagent at a lower, and defined, concentration and with direct access to the cell. Consequently, less reagent is required than for injection *in vivo*, where 90% is lost by excretion and distribution to tissues other than those under study. Screening tests with many variables and replicates are cheaper, and the legal,

moral, and ethical questions of animal experimentation are avoided. New developments in multiwell plates and robotics also have introduced significant economies in time and scale.

1.2.4 *In Vitro* Modeling of *In Vivo* Conditions

Perfusion techniques allow the delivery of specific experimental compounds to be regulated in concentration, duration of exposure (*see* Table 1.2), and metabolic state. The development of histotypic and organotypic models also increases the accuracy of *in vivo* modeling.

1.3 LIMITATIONS

1.3.1 Expertise

Culture techniques must be carried out under strict aseptic conditions, because animal cells grow much less rapidly than many of the common contaminants, such as bacteria, molds, and yeasts. Furthermore, unlike microorganisms, cells from multicellular animals do not normally exist in isolation and, consequently, are not able to sustain an independent existence without the provision of a complex environment simulating blood plasma or interstitial fluid. These conditions imply a level of skill and understanding on the part of the operator in order to appreciate the requirements of the system and to diagnose problems as they arise (Table 1.3; *see also* Chapter 28). Also, care must be taken to avoid the recurrent problem of cross-contamination and to authenticate stocks. Hence, tissue culture should not be undertaken casually to run one or two experiments.

1.3.2 Quantity

A major limitation of cell culture is the expenditure of effort and materials that goes into the production of relatively little tissue. A realistic maximum per batch for most small laboratories (with two or three people doing tissue culture) might be 1–10 g of cells. With a little more effort and the facilities of a larger laboratory, 10–100 g is possible; above 100 g implies industrial pilot-plant scale, a level that is beyond the reach of most laboratories but is not impossible if special facilities are provided, when kilogram quantities can be generated.

The cost of producing cells in culture is about 10 times that of using animal tissue. Consequently, if large amounts of tissue (>10 g) are required, the reasons for providing them by culture must be very compelling. For smaller amounts of tissue (~10 g), the costs are more readily absorbed into routine expenditure, but it is always worth considering whether assays or preparative procedures can be scaled down. Semimicro- or microscale assays can often be quicker, because of reduced manipulation times, volumes, centrifuge times, etc., and are frequently more readily automated (*see* Sections 21.8, 22.3.5).

1.3.3 Dedifferentiation and Selection

When the first major advances in cell line propagation were achieved in the 1950s, many workers observed the loss of the phenotypic characteristics typical of the tissue from which the cells had been isolated. This effect was blamed on *dedifferentiation*, a process assumed to be the reversal of differentiation, but later shown to be largely due to the overgrowth of undifferentiated cells of the same or a different lineage. The development of serum-free selective media (*see* Section 10.2.1) has now made the isolation of specific lineages quite possible, and it can be seen that, under the right conditions, many of the differentiated properties of these cells may be restored (*see* Section 17.7).

1.3.4 Origin of Cells

If differentiated properties are lost, for whatever reason, it is difficult to relate the cultured cells to functional cells in the tissue from which they were derived. Stable markers are required for characterization of the cells (*see* Section 16.1); in addition, the culture conditions may need to be modified so that these markers are expressed (*see* Sections 3.4.1, 17.7).

1.3.5 Instability

Instability is a major problem with many continuous cell lines, resulting from their unstable aneuploid chromosomal constitution. Even with short-term cultures of untransformed cells, heterogeneity in growth rate and the capacity to differentiate within the population can produce variability from one passage to the next (*see* Section 18.3).

1.4 MAJOR DIFFERENCES *IN VITRO*

Many of the differences in cell behavior between cultured cells and their counterparts *in vivo* stem from the dissociation of cells from a three-dimensional geometry and their propagation on a two-dimensional substrate. Specific cell interactions characteristic of the histology of the tissue are lost, and, as the cells spread out, become mobile, and, in many cases, start to proliferate, so the growth fraction of the cell population increases. When a cell line forms, it may represent only one or two cell types, and many heterotypic cell–cell interactions are lost.

The culture environment also lacks the several systemic components involved in homeostatic regulation *in vivo*, principally those of the nervous and endocrine systems. Without this control, cellular metabolism may be more constant *in vitro* than *in vivo*, but may not be truly representative of the tissue from which the cells were derived. Recognition of this fact has led to the inclusion of a number of different hormones in culture media (*see* Sections 10.4.2, 10.4.3), and it seems likely that this trend will continue.

Energy metabolism *in vitro* occurs largely by glycolysis, and although the citric acid cycle is still functional, it plays a lesser role.

It is not difficult to find many more differences between the environmental conditions of a cell *in vitro* and *in vivo* (*see* Section 22.2), and this disparity has often led to tissue culture being regarded in a rather skeptical light. Still, although the existence of such differences cannot be denied, many specialized functions are expressed in culture, and as long as the limits of the model are appreciated, tissue culture can become a very valuable tool.

1.5 TYPES OF TISSUE CULTURE

There are three main methods of initiating a culture [Schaeffer, 1990; *see* Appendix IV, Fig. 1.3, and Table 1.4]: (1) *Organ culture* implies that the architecture characteristic of the tissue *in vivo* is retained, at least in part, in the culture (*see* Section 25.2). Toward this end, the tissue is cultured at the liquid–gas interface (on a raft, grid, or gel), which

favors the retention of a spherical or three-dimensional shape. (2) In *primary explant culture*, a fragment of tissue is placed at a glass (or plastic)–liquid interface, where, after attachment, migration is promoted in the plane of the solid substrate (*see* Section 12.3.1). (3) *Cell culture* implies that the tissue, or outgrowth from the primary explant, is dispersed (mechanically or enzymatically) into a cell suspension, which may then be cultured as an adherent monolayer on a solid substrate or as a suspension in the culture medium (*see* Sections 12.3, 13.7).

Because of the retention of cell interactions found in the tissue from which the culture was derived, organ cultures tend to retain the differentiated properties of that tissue. They do not grow rapidly (cell proliferation is limited to the periphery of the explant and is restricted mainly to embryonic tissue)

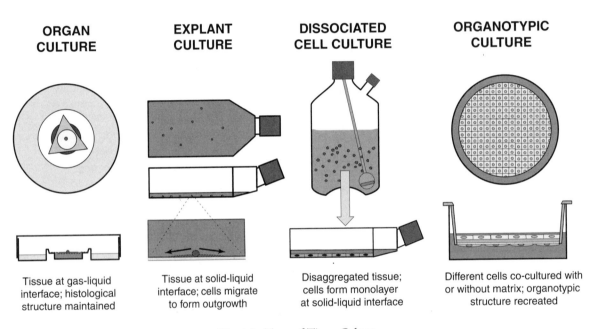

Fig. 1.3. Types of Tissue Culture.

TABLE 1.4. Properties of Different Types of Culture

Category	Organ culture	Explant	Cell culture
Source	Embryonic organs, adult tissue fragments	Tissue fragments	Disaggregated tissue, primary culture, propagated cell line
Effort	High	Moderate	Low
Characterization	Easy, histology	Cytology and markers	Biochemical, molecular, immunological, and cytological assays
Histology	Informative	Difficult	Not applicable
Biochemical differentiation	Possible	Heterogeneous	Lost, but may be reinduced
Propagation	Not possible	Possible from outgrowth	Standard procedure
Replicate sampling, reproducibility, homogeneity	High intersample variation	High intersample variation	Low intersample variation
Quantitation	Difficult	Difficult	Easy; many techniques available

TABLE 1.5. Subculture

Advantages	Disadvantages
Propagation	Trauma of enzymatic or mechanical disaggregation
More cells	Selection of cells adapted to culture
Possibility of cloning	Overgrowth of unspecialized or stromal cells
Increased homogeneity	Genetic instability
Characterization of replicate samples	Loss of differentiated properties (may be inducible)
Frozen storage	Increased risk of misidentification or cross-contamination

and hence cannot be propagated; each experiment requires fresh explantations, which implies greater effort and poorer reproducibility of the sample than is achieved with cell culture. Quantitation is, therefore, more difficult, and the amount of material that may be cultured is limited by the dimensions of the explant ($\sim$1 mm^3) and the effort required for dissection and setting up the culture. However, organ cultures do retain specific histological interactions without which it may be difficult to reproduce the characteristics of the tissue.

Cell cultures may be derived from primary explants or dispersed cell suspensions. Because cell proliferation is often found in such cultures, the propagation of cell lines becomes feasible. A monolayer or cell suspension with a significant growth fraction (*see* Section 21.11.1) may be dispersed by enzymatic treatment or simple dilution and reseeded, or subcultured, into fresh vessels (Table 1.5; *see also* Sections 13.1, 13.7). This constitutes a *passage*, and the daughter cultures so formed are the beginnings of a *cell line*.

The formation of a cell line from a primary culture implies (1) an increase in the total number of cells over several generations and (2) the ultimate predominance of cells or cell lineages with the capacity for high growth, resulting in (3) a degree of uniformity in the cell population (*see* Table 1.5). The line may be characterized, and the characteristics will apply for most of its finite life span. The derivation of *continuous* (or "established," as they were once known) cell lines usually implies a phenotypic change, or *transformation* (*see* Sections 3.8, 18.2).

When cells are selected from a culture, by cloning or by some other method, the subline is known as a *cell strain*. A detailed characterization is then implied. Cell lines or cell strains may be propagated as an adherent monolayer or in suspension. *Monolayer* culture signifies that, given the opportunity, the cells will attach to the substrate and that normally the cells will be propagated in this mode. *Anchorage dependence* means that attachment to (and usually, some degree of spreading onto) the substrate is a prerequisite for cell proliferation. Monolayer culture is the mode of culture common to most normal cells, with the exception of hematopoietic cells. *Suspension* cultures are derived from cells that can survive and proliferate without attachment (*anchorage independent*); this ability is restricted to hematopoietic cells, transformed cell lines, and cells from malignant tumors. It can be shown, however, that a small proportion of cells that are capable of proliferation in suspension exists in many normal tissues (*see* Section 18.5.1). The identity of these cells remains unclear, but a relationship to the stem cell or uncommitted precursor cell compartment has been postulated. This concept implies that some cultured cells represent precursor pools within the tissue of origin (*see* Section 3.10). Cultured cell lines are more representative of precursor cell compartments *in vivo* than of fully differentiated cells, as, normally, most differentiated cells do not divide.

Because they may be propagated as a uniform cell suspension or monolayer, cell cultures have many advantages, in quantitation, characterization, and replicate sampling, but lack the potential for cell–cell interaction and cell–matrix interaction afforded by organ cultures. For this reason, many workers have attempted to reconstitute three-dimensional cellular structures by using aggregates in cell suspension (*see* Section 25.3.3) or perfused high-density cultures on microcapillary bundles or membranes (*see* Section 25.3.2). Such developments have required the introduction, or at least redefinition, of certain terms. *Histotypic* or *histotypic culture*, or *histoculture* (I use *histotypic culture*), has come to mean the high-density, or "tissuelike," culture of one cell type, whereas *organotypic* culture implies the presence of more than one cell type interacting as they might in the organ of origin (or a simulation of such interaction). Organotypic culture has given new prospects for the study of cell interaction among discrete, defined populations of homogeneous and potentially genetically and phenotypically defined cells.

In many ways, some of the most exciting developments in tissue culture arise from recognizing the necessity of specific cell interaction in homogeneous or heterogeneous cell populations in culture. This recognition may mark the transition from an era of fundamental molecular biology, in which many of the regulatory processes have been worked out at the cellular level, to an era of cell or tissue biology, in which that understanding is applied to integrated populations of cells and to a more precise elaboration of the signals transmitted among cells.

CHAPTER 2

Training Programs

2.1 OBJECTIVES

This book has been designed, primarily, as a source of information on procedures in tissue culture, with additional background material provided to place the practical protocols in context and explain the rationale behind some of the procedures used. There is a need, however, to assist those who are engaged in the training of others in tissue culture technique. Whereas an independent worker will access those parts of the book most relevant to his or her requirements, a student or trainee technician with limited practical experience may need to be given a recommended training program, based on his/her previous experience and the requirements of his/her supervisor. This chapter is intended to provide programs at basic and advanced levels for an instructor to use or modify in the training of new personnel.

The programs are presented as a series of exercises in a standard format with cross-referencing to the appropriate protocols and background text. Standard protocol instructions are not repeated in the exercises, as they are provided in detail in later chapters, but suggestions are made for possible experimental modifications to the standard protocol to make each exercise more interesting and to generate data that the trainee can then analyze. Most are described with a minimal number of samples to save manipulation time and complexity, so the trainee should be made aware of the need for a greater number of replicates in a standard experimental situation. The exercises are presented in a sequence, starting from the most basic and progressing toward the more complex, in terms of technical manipulation. They are summarized in Table 2.1, with those that are regarded as indispensable presented in bold

type. The basic and advanced exercises are assumed to be of general application and good general background although available time and current laboratory practices may dictate a degree of selection.

Where more than one protocol is required, the protocol numbers are separated by a semicolon; where there is a choice, the numbers are separated by "or", and the instructor can decide which is more relevant or best suited to the work of the laboratory. It is recommended that all the basic and advanced exercises in Table 2.1 are attempted, and those in bold type are regarded as essential. The instructor may choose to be more selective in the specialized section.

Additional ancillary or related protocols are listed within each exercise. These do not form a part of the exercise but can be included if they are likely to be of particular interest to the laboratory or the student/trainee.

2.2 BASIC EXERCISES

These are the exercises that a trainee or student should attempt first. Most are simple and straightforward to perform, and the protocols and variations on these protocols to make them into interesting experiments are presented in detail in the cross-referenced text. Amounts specified in the Materials sections for each protocol are for the procedure described, but can be scaled up or down as required. Exercises presented in bold font in Table 2.1 are regarded as essential.

A tour of the tissue culture facilities is an essential introduction; this lets the trainee meet other staff, determine their roles and responsibilities, and see the level of preparation

Culture of Animal Cells: A Manual of Basic Technique, Fifth Edition, by R. Ian Freshney
Copyright © 2005 John Wiley & Sons, Inc.

TABLE 2.1. Training Programs

Exercise no.	Procedure	Training objectives	Protocol
Basic:			
1	Pipetting and transfer of fluids.	Familiarization. Handling and accuracy skills.	In exercise
2	**Observation of cultured cells.**	Use of inverted microscope. Appreciation differences in cell morphology within and among cell lines. Use of camera and preparation of reference photographs.	**16.1** 16.6
3	**Aseptic technique: preparing medium for use.**	Aseptic handling. Skill in handling sterile reagents and flasks without contamination. Adding supplements to medium.	**6.1; 11.7**
4	**Feeding a culture.**	Assessing a culture. Changing medium.	**13.1**
5	Washing and sterilizing glassware.	Familiarization with support services. Appreciation of need for clean and nontoxic glass containers.	11.1
6	Preparation and sterilization of water.	Appreciation of need for purity and sterility. Applications and limitations. Sterilization by autoclaving.	11.5
7	Preparation of PBS.	Constitution of salt solutions. Osmolality. Buffering and pH control. Sterilization of heat-stable solutions by autoclaving.	11.6
8	Preparation of pH standards.	Familiarization with use of phenol red as a pH indicator.	9.1
9	Preparation of stock medium and sterilization by filtration.	Technique of filtration and appreciation of range of options.	11.11, 11.12, 11.13, 11.14
10	**Preparation of complete medium from powder or 10× stock.**	Aseptic handling. Constitution of medium. Control of pH.	**11.8, 11.9**
11	**Counting cells by hemocytometer and an electronic counter.**	Quantitative skill. Counting cells and assessment of viability. Evaluation of relative merits of two methods.	**21.1, 21.2**
12	**Subculture of continuous cell line growing in suspension.**	Assessing a culture. Aseptic handling. Cell counting and viability. Selecting reseeding concentration.	**13.3**
13	**Subculture of continuous cell line growing in monolayer.**	Assessing a culture. Aseptic handling. How to disaggregate cells. Technique of trypsinization.	**13.2**
14	Staining a monolayer cell culture with Giemsa.	Cytology of cells. Phase-contrast microscopy. Fixation and staining. Photography.	16.2, 16.6
15	Construction and analysis of growth curve.	Replicate subcultures in multiwell plates. Cell counting. Selecting reseeding concentration.	21.8, 21.1, 21.2
Advanced:			
16	**Cell line characterization.**	Confirmation of cell line identity. Increase awareness of overgrowth, misidentification, and cross-contamination.	**16.7 or 16.8 or 16.9 or 16.10**
17	**Detection of mycoplasma.**	Awareness of importance of mycoplasma screening. Experience in fluorescence method or PCR for routine screening of cell lines for mycoplasma contamination.	**19.2 or 19.3**
18	**Cryopreservation.**	How to freeze cells, prepare cell line and freezer inventory records, stock control.	**20.1; 20.2**
19	**Primary culture.**	Origin and diversity of cultured cells. Variations in primary culture methodology.	**12.2; 12.6 or 12.7**
20	Cloning of monolayer cells.	Technique of dilution cloning. Determination of plating efficiency. Clonal isolation.	14.1, 21.10, 14.6
Specialized:			
21	Cloning in suspension.	Technique of dilution cloning in suspension. Isolation of suspension clones.	14.4 or 14.5; 14.8
22	Selective media.	Demonstration of selective growth of specific cell types.	23.1 or 23.2

TABLE 2.1. Training Programs (*Continued*)

Exercise No.	Procedure	Training objectives	Protocol
23	Cell separation.	Isolation of cell type with desired phenotype by one of several separation methods.	15.1 or 15.2
24	Preparation of feeder layers	How to improve cloning efficiency. Selective effects of feeder layers.	14.3
25	Histotypic culture in filter well inserts.	Familiarization with high-density culture. Potential for differentiation, nutrient transport, and invasion assay.	25.4
26	Cytotoxicity assay.	Familiarization with high-throughput screening methods. Positive and negative effects.	22.4
27	Survival assay. (Can be run as a component of Exercise 20 or as a separate exercise.)	Use of clonal growth to identify positive and negative effects on cell survival and proliferation.	22.3

that is required. The principles of storage should also be explained and attention drawn to the distinctions in location and packaging between sterile and nonsterile stocks, tissue culture grade and non-tissue culture grade plastics, using stocks and backup storage, fluids stored at room temperature versus those stored at 4°C or −20°C. The trainee should know about replacement of stocks: what the shelf life is for various stocks, where replacements are obtained, who to inform if backup stocks are close to running out, and how to rotate stocks so that the oldest is used first.

Exercise 1   Aseptic Technique I: Pipetting and Transfer of Fluids

Purpose of Procedure
To transfer liquid quickly and accurately from one container to another.

Training Objectives
Skill in handling pipettes; appreciation of level of speed, accuracy, and precision required.

Supervision: Continuous initially, then leave trainee to repeat exercise and record accuracy.

Time: 30 min −1 h.

Background Information
Sterile liquid handling (*see* Section 5.2.7); handling bottles and flasks (*see* Section 6.3.4); pipetting (*see* Section 6.3.5).

Demonstration material or operations: Instructor should demonstrate handling pipette, inserting in pipetting aid, and fluid transfer and give some guidance on the compromise required between speed and accuracy. Instructor should also demonstrate fluid withdrawal by vacuum pump (if used in laboratory) and explain the mechanism and safety constraints.

Exercise
Summary of Procedure
Transfer of liquid by pipette from one vessel to another.

Standard Protocol
Aseptic Technique in Vertical Laminar Flow (*see* Protocol 6.1) or Working on the Open Bench (*see* Protocol 6.2). Before embarking on the full protocol, it is useful to have the trainee practice simple manipulations by simply pipetting from a bottle of water into a waste beaker. This gives some familiarization with the manipulations before undertaking aseptic work.

Experimental Variations
1) This exercise is aimed at improving manual dexterity and handling of pipettes and bottles in an aseptic environment. The following additional steps are suggested to add interest and to monitor how the trainee performs:
2) Preweigh the flasks used as receiving vessels.
   a) Add 5 ml to each of 5 flasks.
   b) Using a 5-mL pipette.
3) Using a 25-mL pipette.
4) Record the time taken to complete the pipetting.
5) Weigh the flasks again.
6) Incubate the flasks to see whether any are contaminated.

Data
1) Calculate the mean weight of liquid in each flask.
2) Note the range and
   a) Calculate as a percentage of the volume dispensed, or
   b) Calculate the mean and standard deviation:
      i) Key values into Excel.
      ii) Place the cursor in the cell below the column of figures.
      iii) Press arrow to right of Σ button on standard toolbar, and select average.

iv) Select column of figures, if not already selected, and enter. This will give the average or mean of your data.

v) Place cursor in next cell.

vi) Press arrow to right of Σ button on standard toolbar, and select other functions and then select STDEV.

vii) Select column of figures, if not already selected, and enter. This will give the standard deviation of your data, which you can then calculate as a percentage of the mean to give you an idea of how accurate your pipetting has been.

Analysis

1) Compare the results obtained with each pipette and comment on the differences:
   a) In accuracy
   b) In time
2) When would it be appropriate to use each pipette?
3) What is an acceptable level of error in the precision of pipetting?
4) Which is more important: absolute accuracy or consistency?

Exercise 2   Introduction to Cell Cultures

Purpose of Procedure
Critical examination of cell cultures.

Applications
Checking consistency during routine maintenance; evaluation of status of cultures before feeding, subculture, or cryopreservation; assessment of response to new or experimental conditions; detection of overt contamination.

Training Objectives
Familiarization with appearance of cell cultures of different types and at different densities; use of digital or film camera; distinction between sterile and contaminated, and healthy and unhealthy cultures; assessment of growth phase of culture.

Supervision: Continuous during observation, then intermittent during photography.

Time: 30 min.

Background Information
Morphology, photography (*see* Section 16.4.5).

Demonstration material or operations: Photo examples of cell morphology, phase contrast, fixed and stained, immunostained; types of culture vessel suitable for morphological studies, e.g., Petri dishes (*see* Fig. 8.3), coverslip tubes, chamber slides (*see* Fig. 16.3); cytocentrifuge for suspension cultures (*see* Fig. 16.4).

Safety: No special safety requirements.

Exercise
Summary of Procedure
Examine and photograph a range of cell lines at different cell densities.

Equipment and Materials
❑ Range of flask or Petri dish cell cultures at different densities, preferably with normal and transformed variants of the same cell (e.g., 3T3 and SV3T3, or BHK21-C13 and BHK21-PyY) at densities including mid-log phase (~50% confluent with evidence of mitoses), confluent (100% of growth area covered cells packed but not piling up), and postconfluent (cells multilayering and piling up if transformed). Include suspension cell cultures and low and high concentrations if available.

❑ If possible, include examples of contaminated cultures (preferably not Petri dishes to avoid risk of spread) and unhealthy cultures, e.g., cultures that have gone too long without feeding.

❑ Inverted microscope with 4×, 10×, and 20× phase-contrast objectives and condenser.

❑ Automatic camera, preferably digital with monitor, or film camera with photo-eyepiece.

Standard Protocol
1) Set up microscope and adjust lighting (*see* Protocol 16.1).
2) Bring cultures from incubator. It is best to examine a few flasks at a time, rather than have too many out of the incubator for a prolonged period. Choose a pair, e.g., the same cells at low or high density, or a normal and transformed version of the same cell type.
3) Examine each culture by eye, looking for turbidity of the medium, a fall in pH, or detached cells. Also try to identify monolayer of cells and look for signs of patterning. This can be normal, e.g., swirling patterns of fibroblasts at confluence (*see* Fig 16.2b,h and Plate 5b).
4) Examine at low power (4× objective) by phase-contrast microscopy on inverted microscope, and check cell density and any sign of cell–cell interaction.
5) Examine at medium (10× objective) and high (20× objective) power and check for the healthy status of the cells (*see* Fig. 13.1), signs of rounding up, contraction of the monolayer, or detachment.
6) Check for any sign of contamination (*see* Fig. 19.1a–c).
7) Look for mitoses and estimate, roughly, their frequency.
8) Photograph each culture (*see* Protocol 16.6), noting the culture details (cell type, date form last passage) and cell density and any particular feature that interests you.
9) Return cultures to incubator and repeat with a new set.

Ancillary Protocols: Staining (*see* Protocols 16.2, 16.3; Cytocentrifuge (*see* Protocol 16.4); Indirect Immunofluorescence (*see* Protocol 16.11).

Experimental Variations
1) Look for differences in growth pattern, cell density, and morphology in related cultures.

2) Assess health status of cells.

3) Is there any sign of contamination?

4) Are cells ready for feeding (*see* Section 13.6.2) or passage (*see* Section 13.7.1)?

5) Make a numerical estimate of cell density by calculating the area of the 20× objective field and counting the number of cells per field. This will be easiest if a digital camera and monitor are used where the screen can be overlaid with cling film and each cell ticked with a fine felt-tipped marker.

6) Try to identify and count mitotic cells in these high-power fields.

Data

Qualitative

1) Record your observations on morphology, shape, and patterning for all cultures.

2) Note any contaminations.

3) Confirm healthy status or otherwise.

Quantitative

1) Record cell density (cells/cm^2) for each culture.

2) Record mitotic index for each culture.

Analysis

1) Account for differences in cell density.

2) Account for differences in mitotic index.

3) Compare appearance of cells from normal and transformed cultures and high and low densities and try to explain differences in behavior.

Exercise 3　Aseptic Technique II: Preparing Medium for Use

Purpose of Procedure

To maintain asepsis while handling sterile solutions.

Training Objectives

Aseptic handling: Training in dexterity and sterile manipulation.

Supervision: Continuous.

Background Information

Objectives of aseptic technique (*see* Section 6.1); elements of aseptic environment (*see* Section 6.2); sterile handling (*see* Section 6.3); working in laminar flow (*see* Section 6.4); visible microbial contamination (*see* Section 19.3.1).

Demonstration of materials and operations: Demonstrate how to swab work surface and items brought into hood. Explain the principles of laminar flow and particulate air filtration. Show trainee how to uncap and recap flasks and bottles and how to place the cap on the work surface. Demonstrate holding a pipette, inserting it into a pipetting aid, and using it without touching anything that is not sterile and would contaminate

it, how to transfer solutions aseptically, sloping bottles and flasks during pipetting. Emphasize clearing up and swabbing the hood and checking below the work surface.

Exercise

Summary of Procedure

Add the necessary additions and supplements to 1× stock medium.

Standard Protocol

Preparation of Medium from 1× Stock (*see* Protocol 11.7).

Experimental variations to standard protocol

1) Dispense 50 mL medium into each of 2 sterile bottles.

2) Place one bottle at 4°C.

3) Incubate the other bottle for 1 week and check for signs of contamination (*see* Section 19.3.1).

4) Use these bottles in Exercise 4.

Exercise 4　Feeding a Monolayer Culture

Purpose of Procedure

To replace exhausted medium in a monolayer culture with fresh medium.

Applications

Used to replenish medium between subcultures in rapidly growing cultures, or to change from one type of medium to another.

Training Objectives

Reinforces aseptic manipulation skills. Introduces one of the basic principles of cell maintenance, that of medium replenishment during propagation cycles. Makes trainee observe culture and become aware of signs of medium exhaustion, such as cell density and/or fall in pH, and also looking for contamination. Awareness of risk of cross-contamination. Comparison of preincubated with refrigerated medium.

Supervision: Trainee will require advice in interpreting signs of medium exhaustion and demonstration of medium withdrawal and replenishment.

Time: 30 min.

Background Information

Replacement of medium (*see* Section 13.6.2); monitoring for contamination (*see* Section 19.4); cross-contamination (*see* Section 19.5).

Demonstration material or operations: Exercise requires at least three semiconfluent flasks from a continuous cell line such as HeLa or Vero, with details of number of cells seeded and date seeded. Trainee should also be shown how to bring medium from refrigerator, and it should be stressed that medium is specific to each cell line and not shared among cell lines or operators. Also demonstrate swabbing and laying out hood, use of incubator, retrieving culture from incubator, and observing status of cells by eye and on microscope (freedom from contamination, need to feed, healthy status). Aspirator with pump for medium withdrawal or discard beaker will be required and the process of medium withdrawal and replacement demonstrated, with gassing with 5% CO_2 if necessary.

Safety: If human cells are being handled, a Class II biological safety cabinet must be used and waste medium must be discarded into disinfectant (*see* Section 7.8.5 and Table 7.7).

Exercise
Summary of Procedure
Spent medium is withdrawn and discarded and replaced with fresh medium.

Standard Protocol
Feeding a Monolayer Culture (*see* Protocol 13.1).

Ancillary Protocols: Preparation of Complete Medium (*see* Protocols 11.7, 11.8, or 11.9 and Exercise 3); Preparation of pH Standards (*see* Protocol 9.1); Handling Dishes or Plates (*see* Protocol 6.3).

Experimental Variations
The flasks that are fed with the refrigerated and preincubated medium in this exercise should be used later for cell counting (*see* Exercise 12), and another identical flask should be kept without feeding, to be trypsinized at the same time.

Background: Complete media (*see* Section 9.5); replacement of medium (*see* Section 13.6.2).

Data
Compare appearance and yield (cell counts in Exercise 12) from flasks that have been fed with refrigerated or preincubated medium with yield from the flask that has not been fed.

Routine maintenance should be recorded in a record sheet (*see* Table 12.5) and experimental data tabulated in Exercise 12.

Exercise 5   Washing and Sterilizing Glassware

Purpose of Procedure
To clean and resterilize soiled glassware.

Training Objectives
Appreciation of preparative practices and quality control measures carried on outside aseptic area.

Supervision: Nominated senior member of washup staff should take trainee through standard procedures.

Time: 20–30 min should be adequate for each session, but the time spent will depend on the degree of participation by the trainee in procedures as determined by his/her ultimate role and the discretion of the supervisor.

Background Information
Washup area (*see* Section 4.5.2); washup (*see* Section 5.4.1); glassware washing machine (*see* Section 5.4.11; Fig. 5.21); sterilizer (*see* Section 5.4.4; Fig. 5.18,19); washing and sterilizing apparatus (*see* Section 11.3).

Demonstration material or operations: Trainee should observe all steps in preparation and participate where possible; this may require repeated short visits to see all procedures. Trainee should see all equipment in operation, including stacking, quality control (QC), and safety procedures, although not operating the equipment, unless future duties will include washup and sterilization.

Exercise
Summary of Procedure
Collecting, rinsing, soaking, washing and sterilizing glassware and pipettes.

Equipment and Materials
As in regular use in preparation area (*see* Protocols 11.1–11.3).

Standard Protocols
1) Preparation and Sterilization of Glassware (*see* Protocol 11.1).
2) Preparation and Sterilization of Pipettes (*see* Protocol 11.2).
3) Preparation and Sterilization of Screw Caps (*see* Protocol 11.3).

Ancillary Protocols: Sterilizing Filter Assemblies (*see* Protocol 11.4).

Data
Trainee should become familiar with noting and recording QC data, such as numerical and graphical output from ovens and autoclaves.

Exercise 6   Preparation and Sterilization of Water

Purpose of Procedure
Provision of regular supply of pure, sterile water.

Training Objectives
Appreciation of preparative practices carried on outside aseptic area. Knowledge of need for purity of water and process of preparation.

Supervision: Intermittent.

Time: 30 min

Background Information
Preparation and sterilization of ultra pure water (UPW; *see* Section 11.1.4; Figs. 5.17, 11.10).

Demonstration material or operations: Preparation supervisor should discuss principles and operation of water purification equipment and demonstrate procedures for collection, bottling, sterilization, and QC. Trainee participation at discretion of supervisor and instructor.

Exercise
Summary of Procedure
Purify water, bottle, and sterilize by autoclaving.

Standard Protocol
Preparation and Sterilization of Ultra Pure Water (*see* Protocol 11.5).

Ancillary Protocol: *Preparation of Glassware (see Protocol 11.1)*.

Data

Acquisition
Resistivity (or conductivity) meter on water purifier and total organic carbon (TOC) meter. Automatic printout from autoclave. Sterile tape on bottles. Sterility indicator in center bottle.

Recording
Enter appropriate readings and observations in log book.

Analysis
Review log book at intervals of 1 week, 1 month, and 3 months to detect trends or variability in water quality or sterilizer performance.

Exercise 7 Preparation and Sterilization of Dulbecco's Phosphate-Buffered Saline (D-PBS) without Ca²⁺ and Mg²⁺ (D-PBSA)

Purpose of Procedure
Preparation of isotonic salt solution for use in an atmosphere of air.

Applications
Diluent for concentrates such as 2.5% trypsin, prerinse for trypsinization, washing solution for cell harvesting or changing reagents. As it contains no calcium, magnesium, sodium bicarbonate, or glucose, it is not suitable for prolonged incubations.

Training Objectives
Constitution of simple salt solution. Osmolality. Buffering and pH control. Sterilization of heat stable solutions by autoclaving.

Supervision: Continuous while preparing solution, then intermittent during QC steps. Continuous at start and completion of sterilization and interpretation of QC data.

Time: 2 h.

Background Information
Balanced salt solutions (*see* Section 9.3; Table 9.2); buffering (*see* Section 9.2.3).

Demonstration material or operations: Use of osmometer or conductivity meter. Supervised use of autoclave or bench-top sterilizer.

Safety issues: Steam sterilizers have a high risk of burns and possible risk of explosion (*see* Sections 7.5.2, 7.5.7). Simple bench-top autoclaves can burn dry and, consequently, have a fire risk, unless protected with an automatic, temperature-controlled cut-out.

Exercise
Summary of Procedure
Dissolve premixed powder or tablet in UPW and sterilize by autoclaving.

Standard Protocol
Preparation of D-PBSA (*see* Protocol 11.6).

QC Data

Acquisition: Measure osmolality or conductivity and pH after dissolving constituents.

Recording: Enter details into log book with date and batch number.

Exercise 8 Preparation of pH Standards

Purpose of Procedure
To prepare a series of flasks, similar to those in current use in the laboratory, containing a simple medium or salt solution with phenol red, and adjusted to a pH range embracing the range normally found in culture.

Applications
Assessment of pH during preparation of medium and before feeding or subculturing.

Training Objectives
Familiarization with use of phenol red as a pH indicator. Sterilization with syringe filter.

Supervision: Continuous at start, but minimal thereafter until operation complete.

Time: 2 h.

Background Information
Physicochemical properties, pH (*see* Section 9.2).

Demonstration material or operations: Use of pH meter. Principle, use, and range of syringe filters (*see* Fig. 11.12a,c).

Safety issues: None, as long as no needle is used on outlet.

Exercise
Summary of Procedure
Prepare a range of media at different pHs.

Standard Protocols
Preparation of pH Standards (*see* Protocol 9.1 and Plate 22b), using option of 25-cm² flasks.

Sterile Filtration with Syringe-Tip Filter (*see* Protocol 11.11).

Exercise 9   Preparation of Stock Medium from Powder and Sterilization by Filtration

Purpose of Procedure
Preparation of complex solutions and sterilization of heat-labile reagents and media.

Training Objectives
Technique of filtration and appreciation of range of options. Comparison of positive- and negative-pressure filtration.

Supervision: Instruction on preparation of medium. Constant supervision during setup of filter, intermittent during filtration process, and continuous during sampling for quality control.

Time: 2 h.

Background Information
Preparation of medium from powder (*see* Protocol 11.9); sterile filtration (*see* Section 11.5.2; Protocol 11.12); alternative procedures (*see* Protocols 11.11, 11.13, 11.14).

Demonstration material or operations: Range of disposable filters and reusable filter assemblies, preferably the items themselves but if not, photographs may be used. Emphasize concept of filter size (surface area) and scale. Principles and advantages/disadvantages of positive-/negative-pressure filtration (*see* Section 11.5.2). Handling of filter, filtration, collection, and QC sampling should be demonstrated.

Exercise
Summary of Procedure
Dissolve powder in UPW, filter-sterilize, bottle, and sample for sterility.

Standard Protocol
1) Prepare medium from powder (*see* Protocol 11.9)
2) Sterilize 450 mL by vacuum filtration (*see* Protocol 11.12)
3) Sterilize 550 mL by positive-pressure filtration (*see* Protocol 11.13)

Ancillary Protocols
Preparation of Customized Medium (*see* Protocol 11.10); Autoclavable Media (*see* Section 11.5.1); Reusable Sterilizing Filters (*see* Section 11.3.6); Sterile Filtration with Syringe-Tip Filter (*see* Protocol 11.11); Sterile Filtration with Large

In-Line Filter (*see* Protocol 11.14); Serum (*see* Section 11.5.3; Protocol 11.15).

Experimental Variations
Divide dissolved medium into two lots and sterilize 550 mL by positive-pressure filtration (*see* Protocol 11.13) and 450 mL by negative-pressure filtration (*see* Protocol 11.12).

Background: CO_2 and Bicarbonate (*see* Section 9.2.2); Buffering (*see* Section 9.2.3); Standard Sterilization Protocols (*see* Section 11.5).

Data
1) Note pH before and immediately after filtering.
2) Incubate universal containers or bottles at 37°C for 1 week, and check for contamination.

Analysis
1) Explain the difference in pH between vacuum-filtered versus positive-pressure-filtered medium.
2) Does the pH recover on storage?
3) When would you use one rather than the other?
4) What filters would you use
   a) For 5 mL of a crystalline solution?
   b) For 10 L of medium?
   c) For 1 L of serum?

Exercise 10   Preparation of Complete Medium from 10× Stock

Purpose of Procedure
Addition of unstable components and supplements to stock medium to produce a complete medium designed for a specific task.

Applications
Production of growth medium that will allow the cells to proliferate, maintenance medium that simply maintains cell viability, or differentiation medium that allows cells to differentiate in the presence of the appropriate inducers.

Training Objectives
Further experience in aseptic handling. Increased understanding of the constitution of medium and its supplementation. Stability of components. Control of pH with sodium bicarbonate.

Supervision: A trainee who has responded well to Exercise 6 should need minimum supervision but will require some clarification of the need to add components or supplements before use.

Time: 30 min.

Background Information
Media (*see* Sections 11.4.3, 11.4.4)

Demonstration material or operations: Set of pH standards (*see* Protocol 9.1). Range of bottles available for medium preparation.

Safety: No major safety implications unless a toxic (e.g., cholera toxin or cytotoxic drug) or radioactive constituent is being added.

Exercise

Summary of Procedure

Sterile components or supplements are added to presterilized stock medium to make it ready for use.

Standard Protocol

Preparation of Complete Medium from $10\times$ Concentrate (*see* Protocol 11.8). One or all of the options may be selected from Protocol 11.8.

Ancillary Protocols: Customized Medium (*see* Protocol 11.10); Preparation of Stock Medium from Powder and Sterilization by Filtration (*see* Protocol 11.9); Preparation of pH Standards (*see* Protocol 9.1).

Experimental Variations

Venting flasks

1) Prepare medium according to Protocol 11.8A without HEPES.
2) Pipette 10 mL into each of four 25-cm^2 flasks.
3) Add 20 μL 1M HEPES to each of two flasks.
4) Seal two flasks, one with HEPES and one without, and slacken the caps on the other two.
5) Incubate at 37°C without CO_2 overnight.

Data

Record pH and tabulate against incubation condition.

Analysis

1) Check pH and account for differences.
2) Which condition is correct for this low-bicarbonate medium?
3) What effect has HEPES on the stability of pH?
4) When would venting be appropriate?

Bicarbonate concentration (see Section 9.2.2)

1) Omit the bicarbonate from the standard procedure in Protocol 11.8B and add varying amounts of sodium bicarbonate as follows:
   a) Prepare and label 5 aliquots of 10 mL bicarbonate-free medium in 25-cm^2 flasks.
   b) Add 200 μL, 250 μL, 300 μL, 350 μL, 400 μL of 7.5% $NaHCO_3$ to separate flasks.
2) Leave the cap slack (only just engaging on the thread), or use a filter cap, on the flasks and place at 37°C in a 5% CO_2 incubator.
3) Leave overnight and check pH against pH standards (*see* Protocol 9.1).

Data

Record pH and tabulate against volume of $NaHCO_3$ added.

Analysis

1) Explain what is happening to change the pH.
2) Calculate the final concentration of bicarbonate in each case and determine the correct amount of $NaHCO_3$ to use.
3) If none are correct, what would you do to attain the correct pH?

Exercise 11   Preparation of Complete Medium from Powder

See Protocol 11.9

Exercise 12   Counting Cells by Hemocytometer and Electronic Counter

There are several options in the organization of this exercise. It could be used as an exercise either in the use of the hemocytometer or in the use of an electronic cell counter, or it could be arranged as a joint exercise utilizing both techniques and comparing the outcomes. Alternatively, any of these options could be combined with Exercise 11, to save time. However, as the initial training in cell counting can make the actual counting process quite slow, it is recommended that cell counting is run as a stand–alone exercise, utilizing cultures set up previously (e.g., from Exercise 9), and not as a preliminary to another exercise. The combined use of both counting methods will be incorporated in the following description.

Purpose of Procedure

To quantify the concentration of cells in a suspension.

Applications

Standardization of cell concentrations at routine subculture; analysis of quantitative growth experiments and cell production via growth curves and cell yields.

Training Objectives

Quantitative skill. Counting cells and assessment of viability. Evaluation of relative merits of hemocytometer and electronic counting.

Supervision: Required during preparation and examination of sample and setting up both counting procedures. Counting samples can proceed unsupervised, although the trainee may require help in analyzing results.

Time: 45 min.

TABLE 2.2. Data Record from Exercise 12, Cell Counting

Cells per flask at seeding	Hemocytometer or electronic count at harvest	Dilution or sampling fraction*	Cell/mL of trypsinate or suspension	Cells harvested per flask	Yield: Cells harvested ÷ cells seeded

*Electronic counter dilution of 50× (e.g., 0.4 mL cell suspension in 20 mL counting fluid), with counter sample set at 0.5 mL, would give a factor of 100. Hemocytometer chamber (Improved Neubauer) counts usually sample $1\ mm^2 \times 0.1$ mm deep, i.e., $0.1\ mm^3$, so a factor of 1×10^4 will give cells/mL (*see* Section 21.1.1).

Background Information

Cell counting, hemocytometer (*see* Section 21.1.1); electronic counting (*see* Section 21.1.2).

Demonstration material or operations: Cell cultures used for counting should be derived from Exercise 9. The use of the hemocytometer and electronic counter will require demonstration, with appropriate advice on completing calculations at the end. The principles of operation of the electronic counter should also be explained.

Safety: When human cells are used, handling should be in a Class II microbiological safety cabinet. All plastics and glassware, including the hemocytometer slide and coverslip, should be placed in disinfectant after use, and counting cups and fluid from electronic counting should be disposed of into disinfectant (*see* Section 7.8.5).

Exercise
Summary of Procedure

Cells growing in suspension or detached from monolayer culture by trypsin are counted directly in an optically correct counting chamber, or diluted in D-PBSA and counted in an electronic counter. Cells may be stained beforehand with a viability stain before counting in a hemocytometer.

Standard Protocol

Exercise should be performed first by electronic cell counting (*see* Protocol 21.2) with a diluted cell suspension and then with the concentrated cell suspension, using cell counting by hemocytometer (*see* Protocol 21.1), then repeated with the same concentrated cell suspension, using estimation of viability by dye exclusion (*see* Protocol 22.1).

Ancillary Protocols: Staining with Crystal Violet (*see* Protocol 16.3); DNA Content (*see* Section 16.6); Microtitration Assays (*see* Section 22.3.5).

Experimental Variations

1) Repeat counts 5–10 times with hemocytometer and electronic cell counter and calculate the mean and standard deviation (*see* Exercise 1).
2) Compare fed and unfed flasks from Exercise 9.

Data

Cell counts, with viability where appropriate, calculated per flask, by each counting method.

Details of routine maintenance should be recorded in a record sheet (*see* Table 13.7) and experimental data in a separate table (Table 2.2).

Analysis

1) Calculate viable cell yield relative to cells seeded into flasks.
2) Has changing the medium made any difference?

Exercise 13   Subculture of Cells Growing in Suspension

Purpose of Procedure

Reduction in cell concentration in proportion to growth rate to allow cells to remain in exponential growth.

Applications

Routine passage of unattached cells such as myeloma or ascites-derived cultures; expansion of culture for increased cell production; setting up replicate cultures for experimental purposes.

Training Objectives

Familiarization with suspension mode of growth; cell counting and viability estimation.

Supervision: Initial supervision required to explain principles, but manipulations are simple and, given that the trainee has already performed at least one method of counting in Exercise 10, should not require continuous supervision, other than intermittent checks on aseptic technique.

Time: 30 min.

Background Information

Propagation in suspension, subculture of suspension culture (*see* Sections 13.7.4, 13.7.5); viability (*see* Section 22.3.1).

Demonstration material or operations: Trainee will require two suspension cultures, one in late log phase and one in

TABLE 2.3. Record of Exercise 13, Subculture of Cells Growing in Suspension

Sample	Volume of cell susp. in culture flask	Cell count from hemocytometer or electronic counter	Dilution or sampling fraction[1]	Cells/mL in flask	Viability (ratio of unstained cells to total)[2]	Dilution factor for viability stain	Viable cells/mL	Cells/flask
	v	c	d	$c \times d$	r	f	$c \times d \times r \times f$	$c \times d \times r \times f \times v$
e.g., cell counter without viability stain	20	15321	100	1532100	1	1	1532100	30642000
e.g., hemocytometer with viability stain	20	76	10000	760000	0.85	2	1292000	25840000

[1]For electronic counting, 0.4 mL cell suspension in 20 mL counting fluid is a ×50 dilution, and ×2 as the counter counts a sample of 0.5 mL of the diluted suspension, giving a factor of ×100. For hemocytometer counting, if the center 1 mm^2 is counted, the factor is 1×10^4; if all 9 fields of 1 mm^2 are counted (because the count was low) then the factor is $1 \times 10^4 \div 9$. In practice, it is better to count 5 fields of 1 mm^2 on each side of the slide, whereupon the factor becomes $1 \times 10^4 \div 10$, or 1×10^3.

[2](Total cell count—stained cells) ÷ Total cell count

plateau, with details of seeding date and cell concentration, and should be shown how to add viability determination into hemocytometer counting (*see* Sections 21.1.1, 22.3.1, Protocol 22.1). If stirred culture is to be used rather than static flasks, the preparation of the flasks and the use of the stirrer platform will need to be demonstrated.

Safety: Where human cells are used, handling should be in a Class II microbiological safety cabinet and all materials must be disposed of into disinfectant (*see* Exercise 12).

Exercise
Summary of Procedure
A sample of cell suspension is removed from the culture, counted electronically or by hemocytometer, diluted in medium, and reseeded.

Standard Protocol
Subculture in Suspension (*see* Protocol 13.3); Viability (*see* Protocol 22.1).

Ancillary Protocol: Scale-up in Suspension (*see* Protocol 26.1).

Experimental Variations
Comparison of subculture from log-phase and plateau-phase cells.

Background: Cell concentration at subculture (*see* Section 13.7.3).

Data
Determine cell concentration and viability in subcultures from log-phase and plateau-phase cells 72 h after subculture. Recording is best done in a table (Table 2.3) and transferred to a spreadsheet.

Analysis
1) Calculate the cell yield as described in Table 2.4.
2) Compare the yield from cells seeded from log and plateau phase.

TABLE 2.4. Analysis of Exercise 13

Sample	Cells/flask at seeding	Cells/flask at next subculture	Yield: Cells harvested ÷ cells seeded
	n	$c \times d \times v \times r \times f$	$\dfrac{c \times d \times v \times r \times f}{n}$
Example	200,000	30642000	153.21
Log-phase cells			
Plateau-phase cells			

Exercise 14 Subculture of Continuous Cell Line Growing in Monolayer

Purpose of Procedure

Propagating a culture by transferring the cells of a culture to a new culture vessel. This may involve dilution to reseed the same size of culture vessel, or increasing the size of vessel if expansion is required.

Training Objectives

Assessment of culture: This exercise requires the trainee to examine and assess the status of a culture. The trainee should note the general appearance, condition, freedom from contamination, pH of the medium, and density of the cells.

Aseptic handling: Reinforces skills learned in Exercises 5, 7, and 8.

Subculture or passage: This exercise introduces the principle of transferring the culture from one flask to another with dilution appropriate to the expected growth rate. It shows the trainee how to disaggregate cells by the technique of trypsinization, and how to count cells and assess viability. The trainee is then required to determine the cell concentration and select the correct concentration for reseeding, instilling a concept of quantitation in cell culture and enhancing numeracy skills.

Supervision: Provided that the trainee has shown competence in aseptic technique, continuous direct supervision should not be necessary, but the instructor should be on hand for intermittent supervision and to answer questions.

Background Information

Standard Protocols

Subculture, Criteria for Subculture (*see* Section 13.7.1; Figs. 13.2, 13.3, 13.4); Growth Cycle and Split Ratios (*see* Section 13.7.2), Cell Concentration at Subculture (*see* Section 13.7.3; Fig. 13.4); Choice of Culture Vessel (*see* Section 8.2); CO_2 and Bicarbonate (*see* Section 9.2.2, Table 9.1).

Experimental variations

1) Cell concentration at subculture (*see* Section 13.7.3)
2) Growth cycle (*see* Section 21.9.2)
3) Effect of cell density (*see* Section 25.1.1).

Demonstration of materials and operations: The trainee should be shown different types of culture vessel (*see* Table 8.1; Figs. 8.1–8.8) and photographs of cells, healthy (*see* Fig. 16.2, Plates 5, 6), unhealthy (*see* Fig. 13.1), contaminated (*see* Fig. 19. a–c), and at different densities (*see* Fig. 16.2; Plates 5, 6). Instruction should be given in examining cells by phase-contrast microscopy. A demonstration of trypsinization (*see* Protocol 13.2) will be required.

Exercise

Summary of Procedure

A cell monolayer is disaggregated in trypsin, diluted, and reseeded.

Equipment and Materials

See materials for standard subculture (*see* Protocol 13.2).

Standard Protocol

Subculture of Monolayer (*see* Protocol 13.2).

Experimental variations

Apply in Protocol 13.2 at Step 11:

a) Seed six flasks at 2×10^4 cells/mL.
b) Feed three flasks after 4 days.
c) Determine cell counts after 7 days in two flasks that have been fed and two that have not:
  i) Remove medium and discard.
  ii) Wash cells gently with 2 mL D-PBSA, remove completely, and discard.
  iii) Add 1 mL trypsin to each flask.
  iv) Incubate for 10 min.
  v) Add 1 mL medium to trypsin and disperse cells by pipetting vigorously to give a single cell suspension.
  vi) Count cells by hemocytometer or electronic cell counter.
  vii) Calculate number of cells per flask, cells/mL culture medium, and cells/cm² at time of trypsinization.
d) Fix and stain cells in other flasks (*see* Protocol 16.2).

Note: Pipettors should only be used for counting cells from isolated samples and not for dispensing cells for subculture, unless plugged tips are used.

Ancillary Protocols: Using an Inverted microscope (*see* Protocol 16.1); Cell Counting (*see* Protocols 21.1, 21.2); Preparation of Media (*see* Protocols 11.7, 10.8, 10.9); Staining with Giemsa (*see* Protocol 16.2).

Data

1) Cell counts at start and in one set of flasks after 1 week.
2) Examine and photograph stained flasks in Exercise 13. Best resolution is obtained if examined before cell layer dries.

Analysis

1) Calculate fold yield (number of cells recovered ÷ number of cells seeded; Table 2.6) and explain the differences (*see* Sections 18.5.2, 21.9.3).
2) Is an intermediate feed required for these cells?
3) Comment on differences in cell morphology of fed and unfed cultures in Exercise 13.

TABLE 2.5. Record of Exercise 14

Sample	Volume of trypsinate	Cell count from hemocytometer or electronic counter	Dilution or sampling fraction	Cells/mL in trypsinate	Cells/flask
	t	c	d	c × d	c×d×t

TABLE 2.6. Analysis of Exercise 14

Sample	Cells per flask at seeding	Cells harvested per flask[1]	Yield: Cells harvested ÷ cells seeded
	n	c×d×t (see last column of record)	$\frac{c \times d \times t}{n}$
Not fed	100,000		
	100,000		
Fed	100,000		
	100,000		

[1]Viability has not been taken into account in this instance as trypsinization, or at least the prewashes before trypsinization, tend to remove most of the nonviable cells when handling a continuous cell line. This is not necessarily the case with an early passage or primary culture, when viability may need to be taken into account (*see* Recording, Exercise 20).

Exercise 15   Staining a Monolayer Cell Culture with Giemsa

Purpose of Procedure

Staining with a polychromatic stain like Giemsa reveals the morphology characteristic of the fixed cell and can allow analysis of the status and origin of the cells.

Applications

Monitoring cell morphology, usually in conjunction with phase-contrast observations, during routine passage or under experimental conditions. Preparation of permanent record of appearance of the cells for reference purposes. Identification of cell types present in a primary culture.

Training Objectives

Emphasizes need for observation of cells during and after culture and alerts the user to the morphology of the cell.

Supervision: Minimal.

Time: 30 min.

Background Information

Morphology (*see* Section 16.4); staining (*see* Section 16.4.2).

Demonstration material or operations: Preparation of stain and staining procedure should be demonstrated and examples provided of previously stained material.

Safety: Precautions for human cells as in previous exercises until material is fixed.

Exercise
Summary of Procedure

Remove medium, rinse cell monolayer with PBSA, fix in ethanol, and stain with Giemsa.

Standard Protocol

Staining with Giemsa (*see* Protocol 16.2).

Ancillary Protocols: Staining with Crystal Violet (*see* Protocol 16.3); Using an Inverted Microscope (*see* Protocol 16.1); Digital Photography on a Microscope (*see* Protocol 16.6).

Experimental Variations

1) Use flasks from Exercise 12 and compare cell morphology of fed and unfed cultures.
2) Photograph before (phase-contrast illumination) and after (normal bright-field illumination) staining.

Exercise 16   Construction and Analysis of Growth Curve

Purpose of Procedure

Familiarization with the pattern of regrowth following subculture; demonstration of the growth cycle in routine subculture and as an analytical tool.

Applications

Growth curves, cell proliferation assays, cytotoxicity assays, growth stimulation assays, testing media and sera.

Training Objectives

Setting up experimental replicates. Cell counting and viability. Plotting and analyzing a growth curve. Awareness of differences in doubling times and saturation densities. Selecting reseeding concentration.

Supervision: Trypsinization and counting should not need supervision, given satisfactory progress in Exercise 12, but some supervision will be necessary while setting up plates.

Time: 1 h on day 0; 30 min each day thereafter up to day 10.

Background Information

Choice of culture vessel (*see* Section 8.2); handling dishes or plates (*see* Protocol 6.3); replicate sampling (*see* Section 21.8); growth cycle (*see* Section 21.9.2); microtitration assays (*see* Section 22.3.5).

Demonstration material or operations: Trainee should be shown the range of multiwell plates available (*see* Table 8.1 and Fig. 8.2) and given some indication of their applications. Setting up plates, with the handling precautions to prevent contamination, will need to be demonstrated (*see* Section 6.5.1).

Safety: Care should be taken when handling human cells (*see* Exercise 10 and Section 7.8). Particular care is required in handling open plates and dishes because of the increased risk of spillage (*see* Section 6.6.2 and Fig. 6.10).

Exercise
Summary of Procedure

There are two options for this exercise: a simple growth curve of one cell line using flasks as for regular subculture, or using multiwell plates to analyze differences in growth at different densities, between two different cell lines, or under any other selected set of conditions. The cell monolayer is trypsinized, counted, and diluted in sufficient medium to seed the requisite number of culture flasks or multiwell plates.

Standard Protocols

Growth Curve, Monolayer (*see* Protocol 21.7 for growth curve in flasks to define conditions for routine maintenance and Protocol 21.8 in multiwell plates to analyze growth at different seeding densities and/or to compare two different cell lines).

Ancillary Protocols: Handling Dishes or Plates (*see* Protocol 6.3); MTT-based Cytotoxicity Assay (*see* Protocol 22.4).

Experimental Variations
Cell concentration at seeding

1) Seed flasks at 2×10^4 cells/mL for a rapidly growing continuous cell line or at 1×10^5 cells/mL for a slower-growing finite cell line. Repeat to optimize seeding concentration.

2) One cell line can be set up conveniently at three different cell concentrations on a 12-well plate, with each concentration in triplicate and one well left over for staining (*see* Fig. 21.7a). Separate plates should be set up for each day's sampling for a total of 10 days.

Differences between normal and transformed cell lines: In this case, use a 24-well plate and seed two cell lines on each plate (*see* Fig. 21.7b).

Background: Growth cycle (*see* Section 21.9.2); volume, depth, and surface area (*see* Section 13.6.2); cell concentration at subculture (*see* Section 13.7.3); phases of the growth cycle (*see* Section 21.9.6); derivatives of the growth cycle (*see* Section 21.9.7); contact inhibition (*see* Section 18.5.2); density limitation of cell proliferation (*see* Protocol 18.3).

Data

Cell counts per well per day.

Analysis

1) Calculate cells/mL medium in the wells and cells/cm² from the cells/well and plot mean and standard error (standard deviation divided by square root of the number of samples) for each day on a log scale against days from seeding on a linear scale (*see* Fig. 21.6).
2) Derive lag time, doubling time, and saturation density.
3) Which cell concentration would be best for routine subculture?
4) Account for differences between normal and transformed cell lines (if both have been used).

2.3   ADVANCED EXERCISES

These exercises are dependent on satisfactory progress in the Basic Exercises and should not be attempted until that is achieved. Although advanced, they are still of general application and would be required for anyone claiming general expertise in cell culture. It is, however, possible, if there are time constraints, to defer these exercises if others in the laboratory are already carrying them out and the new trainee or student will not be called upon to perform them. It should be realized, however, if this alternative is adopted, the training cannot be regarded as complete to a reasonable all-round standard until the advanced exercises are performed.

As some basic knowledge is now assumed, variations to the standard protocol will not be presented in the same detail as for the basic exercises, as it is assumed that a greater degree of experimental planning will be beneficial and a significant part of the training objectives. Whereas the basic exercises are presented in the sequence in which they should be performed, the advanced exercises need not be performed in a specific sequence, and, with a class of students, could be performed in rotation.

Exercise 17 Cryopreservation of Cultured Cells

Purpose of Procedure
To provide a secure cell stock to protect against accidental loss and genetic and phenotypic instability.

Applications
Protection of new and existing cell lines; cell banking for archiving and distribution; provision of working cell bank for the lifetime of a project or program; storage of irradiated or mitomycin C–treated feeder cells.

Training Objectives
Familiarization with cell freezing and thawing procedures. Indication of possible variations to improve procedure for difficult cell lines. Comparison or dye exclusion viability with actual cell survival.

Supervision: Basic procedures, such as trypsinization, counting, adding preservative, and filling ampoules should not require supervision, but supervision will be required for accessing the liquid nitrogen storage inventory control system, for freezing and transfer of the ampoules to the nitrogen freezer, and for recovery and thawing.

Time: 1 h on day 1; 15 min on day 2; 30 min on day 3; 1 h on day 4.

Background Information
Rationale for freezing (*see* Section 20.1); cooling rate, cryofreezers, and freezer records (*see* Section 20.3.6); genetic instability (*see* Section 18.3); evolution of cell lines (*see* Section 3.8); control of senescence (*see* Section 18.4.1); serial replacement (*see* Section 20.4.2); cell banks (*see* Section 20.5).

Demonstration material or operations: Trainee should be shown types of ampoules in regular use, freezing devices (*see* Figs. 20.3, 20.4, 20.5) and types of freezer (*see* Section 20.7), and be introduced to the use and upkeep of the freezer inventory control system and record of cell lines.

Safety: Care in the use of human cell lines, as previously. Risk of frostbite, asphyxiation, and, where ampoules are stored submerged in liquid nitrogen, explosion (*see* Section 7.5.6). It is strongly recommended that, for the purposes of this exercise, ampoules are not submerged in liquid nitrogen but are stored in the vapor phase or in a perfused wall freezer.

Exercise
Summary of Procedure
Cells at a high concentration in medium with preservative are cooled slowly, frozen slowly, and placed in a liquid nitrogen freezer. They are then thawed rapidly and reseeded.

Standard Protocol
Freezing Cells (*see* Protocol 20.1); Thawing Frozen Cells (*see* Protocol 20.2).

Ancillary Protocols: Subculture of Monolayer (*see* Protocol 13.2); Subculture in Suspension (*see* Protocol 13.3); Cell Counting by Hemocytometer (*see* Protocol 21.1); Electronic Cell Counting (*see* Protocol 20.2); Estimation of Viability by Dye Exclusion (*see* Protocol 22.1).

Experimental Variations
There are several experimental variables that can be introduced into this exercise:

1) Comparison of DMSO and glycerol as preservative.
2) Alterations in the freezing rate.
3) Holding at room temperature or at 4°C before freezing.
4) Removal of preservative by replacing medium next day or by centrifugation after thawing. If this is selected, then it would be interesting to compare cells from suspension (e.g., L5178Y lymphoma, a hybridoma, or HL60) and from attached monolayers (e.g., HeLa, A549, Vero, or NRK).
5) Rapid or slow dilution after thawing.

A scheme is suggested incorporating a comparison of DMSO and glycerol (step 1) with variations in pretreatment (step 3) before freezing (Fig. 2.1).

Background: Cryoprotectants (*see* Section 20.3.3).

Data
1) Routine records should be completed as for standard freezing (*see* Tables 20.2, 20.3).
2) Cell viability should be performed on the dregs of each ampoule.
3) The thawed cultures should be trypsinized and counted the day after thawing (only for generating survival data; not as a routine).
4) The results can be tabulated as in Exercises 13 and 14 and the recovery of viable cells calculated on the day of thawing and the recovery of attached cells (for monolayer cultures only) calculated on the day after thawing.

Analysis
1) Which preservative is best for your cells?
2) Does dye exclusion viability agree with recovery after 24 h? If not, why not?
3) Is a delay before freezing harmful? Does chilling the cells after adding preservative help?

Exercise 18 Detection of Mycoplasma

Purpose of Procedure
Validation of cell line by proving it to be free of mycoplasmal contamination.

Applications
Routine cell line maintenance; quality control of cell lines before freezing; checking imported cell lines, tissue, and biopsies while in quarantine.

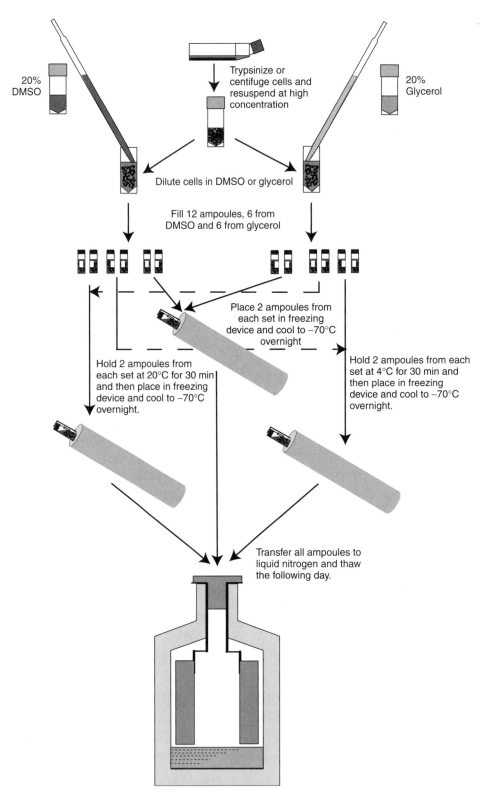

Fig. 2.1. Freezing Exercise. Suggested experimental protocol to compare the cryoprotective effect of DMSO and glycerol, with and without holding the ampoules at 4°C or room temperature before freezing.

Awareness of importance of mycoplasma screening. Experience in fluorescence method or PCR for routine screening of cell lines for mycoplasma contamination.

Supervision: Setting up cultures and infecting feeder layers should not require supervision, provided a real mycoplasma contamination is not suspected, in which case the procedure should be carried out in quarantine under strict supervision. Intermittent supervision will be required during mycoplasma staining or DNA extraction and PCR, depending on the experience of the trainee in these areas. Continuous supervision will be required during interpretation of results.

Time: 30 min 5 days before start to set up or refeed test culture; 1 h on day 0 to set up indicator cultures; 30 min on day 1 to transfer medium from test culture; 2–4 h on day 5 to stain or PCR the cultures and a further 30 min to examine, then or later.

Background Information

Mycoplasma (*see* Section 19.3.2); validation (*see* Section 7.10).

Demonstration material or operations: Quarantine procedures (*see* Section 19.1.8); use of fluorescence microscope or PCR machine; provision of fixed positive cultures.

Safety: No special procedures other than standard precautions for human cells (*see* Section 7.8.3). Trainee should be made aware of the severe risks attached to unshrouded UV sources and the risk attached to removing working light source from fluorescence microscope.

Exercise
Summary of Procedure

A test culture is fed with antibiotic-free medium for 5 days, a sample of the medium is transferred to an indicator cell line, known to support mycoplasma growth, and mycoplasma is assayed in the indicator cells by fluorescent DNA staining or PCR.

Standard Protocol

Fluorescence Detection of Mycoplasma (*see* Protocol 19.2) or Detection of Mycoplasma Contamination by PCR (*see* Protocol 19.3).

Ancillary Protocols: Digital Photography (*see* Section 16.4.5).

Experimental Variations

It is difficult to add an experimental element to this exercise except by exploring potential routes of contamination with infected cultures. It is unlikely that any laboratory would wish to undertake this rather hazardous course of action unless special facilities were available.

Data

1) Results are scored as positive or negative against a fixed positive control (fluorescence) or mycoplasma DNA (PCR).

2) Records should be kept of all assays and outcomes in a written log or by updating the cell line database.

Analysis

1) Mycoplasma-positive specimens will show punctuate or filamentous staining over the cytoplasm (*see* Plate 10a,b).
2) Alternatively, electrophoretic migration of PCR product DNA can be compared with incorporated controls (*see* Fig. 19.2).

Exercise 19   Cell Line Characterization

Together with mycoplasma detection (*see* Exercise 18), cell line characterization is one of the most important technical requirements in the cell culturist's repertoire. Some form of characterization is essential in order to confirm the identity of cell lines in use, but the techniques selected will be determined by the methodology currently in use in the laboratory. If DNA fingerprinting or profiling is available in the laboratory this single parameter will usually be sufficient to identify individual lines if previous comparable data are available for that line; otherwise more than one technique will be required. Cell lines currently in use will probably have a characteristic already monitored related to the use of the line, e.g., expression of a particular receptor, expression of a specific product, or resistance to a drug, and it may only be necessary to add one other parameter, e.g., chromosomal or isoenzyme analysis. If DNA fingerprinting or profiling is not available it is unlikely that anyone would want to get involved in setting them up for the sake of a training exercise and the choice is more likely to be to send the cells to a commercial laboratory for analysis, which has the added advantage that the commercial laboratory will have reference material with which to compare the results. Nevertheless, it is advisable to insert this exercise in the training program to impress upon the student or trainee the importance of cell line authentication, given the widespread use of misidentified cell lines, and the possible consequences (*see* Sections 16.3, 19.5, 20.2).

Isoenzyme electrophoresis, which is easily and cheaply performed with a commercially available kit, is suggested as an easily conducted experiment for this exercise.

Purpose of Procedure

To confirm the identity of a cell line.

Applications

Checking for accidental cross-contamination; quality control of cell lines before freezing and/or initiating a project or program; confirming identity of imported cell lines.

Training Objectives

Confirmation of cell line identity. Increase awareness of overgrowth, misidentification, and cross-contamination.

Supervision: Preparation of samples and electrophoresis will need supervision, although probably not continuous.

Time: 2 h.

Background Information

Need for characterization (*see* section 16.1); morphology (*see* Section 16.4); isoenzymes (*see* Section 16.8.1); chromosome content (*see* Section 16.5); DNA fingerprinting and profiling (*see* Sections 16.6.2, 16.6.3); antigenic markers (*see* Section 16.9); authentication (*see* Section 16.3).

Demonstration material or operations: Use of Authentikit™ electrophoresis apparatus (*see* Fig. 16.12); examples of DNA fingerprints and/or profiles (*see* Fig. 16.9, 16.11); examples of karyotypes (*see* Figs. 16.6, 16.7, 16.8).

Safety: Other than precautions, as before, in the handling of human cell lines, there are no special safety requirements for this exercise. However, if DNA fingerprinting is attempted, care will be required in the handling of radioactively labeled probes (*see* Section 7.7).

Exercise
Summary of Procedure
Cell extracts are prepared, electrophoresed on agarose gels, and developed with chromogenic substrates.

Standard Protocol
Isoenzyme Analysis (*see* Protocol 16.10).

Ancillary or Related Protocols: Chromosome Preparations (*see* Protocol 16.7); Multilocus DNA Fingerprinting of Cell Lines (*see* Protocol 16.8); DNA Profiling (*see* Protocol 16.9); Indirect Immunofluorescence (*see* Protocol 16.11).

Experimental Variations
Six different cell lines should be examined, chosen from those available in the laboratory, or from the following list: HeLa; KB or Hep-2; Vero; L929, 3T3 or 3T6; BHK-21-C13, CHO-K1. Most cell lines from different species can be distinguished by using four isoenzymes: nucleoside phosphorylase, glucose-6-phosphate dehydrogenase, lactate dehydrogenase, and malate dehydrogenase.

Background: Isoenzymes (*see* Section 16.8.1, 16.8.2 and reviews [Hay et al., 2000; Steube et al., 1995]).

Data
Once the gels have been developed with the appropriate chromogenic substrates, they should be photographed or scanned. The gels can be kept.

Analysis
1) Compare results among different cell lines for each enzyme.
2) Comment on the significance of KB or Hep-2 having the same G-6-PD isoenzyme as HeLa.
3) How would you improve resolution among cell lines?

Exercise 20 Primary Culture

Purpose of Procedure
The isolation of cells from living tissue to create a cell culture.

Applications
Initiation of primary cultures for vaccine production; isolation of specialized cell types for study; chromosomal analysis; development of selective media; provision of short-term cell lines for tissue engineering.

Training Objectives
Awareness of origin and diversity of cultured cells. Appreciation of variations in primary culture methodology.

Supervision: Intermittent.

Time: 2–4 h day 0; 1 h day 1; 2 h day 3.

Background Information
Primary culture (*see* Chapter 12).

Demonstration material or operations: Dissection of chick embryos (or alternative tissue source).

Safety: Minimal requirements if chick embryo material is used.

Exercise
Chick embryos have been selected for this exercise for a variety of reasons. They are readily available with minimal animal care backup and can be dissected without restrictions if less than half-term; full term is 21 days, so 10-day embryos are suitable, if a little small. They are larger, at a given stage, than mouse embryos and give a high yield of cells either from the whole chopped embryo or from isolated organs.

Summary of Procedure
Embryos are removed from the egg and dissected, and primary explant and disaggregated cell cultures are set up.

Standard Protocol
1) Isolate embryos from fertile eggs (*see* Protocol 12.2).
2) Disaggregate one embryo in warm trypsin (*see* Protocol 12.5).
3) Disaggregate a similar-sized embryo in cold trypsin (*see* Protocol 12.6).
4) Dissect a third embryo and isolate individual organs, e.g., brain, liver, heart, gut, lungs, and thigh muscle, and disaggregate each tissue separately by the cold trypsin method (*see* Protocol 12.7).
5) Collect some of the tissue remaining from step 4) and set up primary explants (*see* Protocol 12.4).

Ancillary Protocols: Disaggregation in collagenase (*see* Protocol 12.8); Mechanical Disaggregation by Sieving (*see* Protocol 12.9); Enrichment of Viable Cells (*see* Protocol 12.10).

Experimental Variations

1) Cells isolated from different tissues.
2) Recovery after warm and cold trypsinization.

Background Information

Enzymatic disaggregation (*see* Section 12.3.2); trypsinization after cold preexposure (*see* Section 12.3.4).

Data

1) Examine living cultures from organ rudiments after 3–5 days and check for morphological differences and signs of contraction in the heart cells.
2) Count and perform viability stain on cells recovered from warm and cold trypsinization.
3) Trypsinize and count cultures derived from warm and cold trypsin 3 days after seeding.
4) Record data as for Primary Culture (*see* Table 12.2).

Analysis

1) Try to identify different cell types present in primary cultures from organ rudiments.
2) How would you propagate these cultures to retain specific cell types?
3) Calculate and tabulate the recovery of total cells per embryo, cells/g, viable cells/embryo, and viable cells/g.
4) From the number of cells recovered at the first subculture (3 days in this case), calculate the yield of cells per embryo, and as a ratio of the total cells seeded, and viable cells seeded.
5) Was the dye exclusion staining a good predictor of recovery?

Exercise 21   Cloning of Monolayer Cells

Purpose of Procedure

To dilute cells such that they grow as isolated colonies derived from single cells.

Applications

Isolation of genetic or phenotypic variants; survival assay; growth assay.

Training Objectives

Introduction to technique of dilution cloning. Determination of plating efficiency as a growth or survival parameter. Clonal isolation of selected cell types.

Supervision: Initial supervision only on day 0, and help later (days 10–14) with identifying clones.

Time: 1 h.

Background Information

Cloning (*see* Protocols 14.1–14.4); plating efficiency (*see* Section 21.10).

Demonstration material or operations: Previous cloned and stained cultures; options for cloning, e.g., Petri dishes versus microtitration plates; cloning rings for isolation.

Safety: Minimal if a nonhuman cell line is used, e.g., CHO-K1. Use of a growth-arrested feeder layer would, however, require attention to toxicity of mitomycin C or irradiation risk of source, depending on which is used.

Exercise
Summary of Procedure

Monolayer culture is trypsinized in middle to late log phase, and the cells are diluted serially and seeded at a low concentration into Petri dishes or microtitration plates.

Standard Protocols

Use Dilution Cloning (*see* Protocol 14.1) for a simple exercise with a cell line of known plating efficiency. For a more advanced exercise, using a cell line of unknown plating efficiency, and with the option of adding a further experimental variable, use Determination of Plating Efficiency (*see* Protocol 21.10).

Ancillary Protocols: Preparation of Conditioned Medium (*see* Protocol 14.2); Preparation of Feeder Layers (*see* Protocol 14.3); Cloning in Agar (*see* Protocol 14.4); Cloning in Methocel (*see* Protocol 14.5); Clonogenic Assay (*see* Protocol 22.3).

Experimental Variations

1) *Seeding concentration*: Seed at 10, 20, 50, 100, 200 cells per mL (*see* Protocol 21.10).
2) *Serum concentration*: Seed in different serum concentrations from 0–20%. This is a good assay for serum quality (*see* Protocol 14.1). Dilute cells in serum-free medium down to 1000 cells/mL and then dilute 200 μL to 20 mL (1:100) to give 10 cells/mL in separate containers with medium containing the appropriate serum concentration (e.g., 0, 0.5, 1, 2, 5, 10, 20%).
3) *Cytotoxicity*: A simple variable to add into this exercise is the addition of a cytotoxic drug to the cells for 24 h before cloning (*see* Protocol 22.3). If mitomycin C is used, it becomes a useful preliminary to preparing a feeder layer (*see* Protocol 14.3). An exponential range of concentrations between 0 and 50 μg/mL would be suitable, e.g., 0, 0.1, 0.2, 0.5, 1.0, 2.0, 5, 10, 20, 50 μg/mL.
4) *Feeder layer*: Repeat step 1) with and without feeder layer (*see* Protocol 14.3).
5) *Isolation of clones*: Isolate and compare morphology of cloned strains.

Background Information

Cell cloning (*see* Section 14.1); isolation of clones (*see* Sections 14.6, 14.7, 14.8); plating efficiency (*see* Section 21.10); survival (*see* Section 22.3.2).

Data

1) Stain dishes with Crystal Violet (*see* Protocol 16.3) after colonies are visible by naked eye.
2) Count the number of colonies per dish.

Analysis

1) Calculate the plating efficiency (colonies formed ÷ cells seeded ×100) (*see* Protocol 21.10).
2) If cell concentration was varied, plot number of colonies per dish against number of cells seeded per dish (*see* Protocol 21.10). This should give a linear plot.
3) If serum concentration varied, plot plating efficiency against serum concentration (*see* Section 11.6.3).
   a) Why is this a good test for serum?
   b) How would you compare several serum batches?
4) If cytotoxin was used, plot the ratio of colonies per dish at each drug concentration to colonies per dish of control.

This give the surviving fraction, which should be plotted on a log scale against the drug concentration, also on a log scale in this case, although it can be on a linear scale for a narrower range of drug concentrations.

2.4 SPECIALIZED EXERCISES

It is assumed that anyone progressing to the specialized exercises (21–27) will have a specific objective in mind and will select protocols accordingly. The parameters of variability will also be determined by the objectives, so these exercises are not detailed and it is assumed that the student/trainee will, by now, have the skills necessary to design his/her own experiments. They are included in Table 2.1, however, as they are thought to have enough general interest to belong to an extended training program, albeit at a more advanced level.

CHAPTER 3

Biology of Cultured Cells

3.1 THE CULTURE ENVIRONMENT

The validity of the cultured cell as a model of physiological function *in vivo* has frequently been criticized. Often, the cell does not express the correct *in vivo* phenotype because the cell's microenvironment has changed. Cell–cell and cell–matrix interactions are reduced because the cells lack the heterogeneity and three-dimensional architecture found *in vivo*, and many hormonal and nutritional stimuli are absent. This creates an environment that favors the spreading, migration, and proliferation of unspecialized progenitor cells, rather than the expression of differentiated functions. The influence of the environment on the culture is expressed via four routes: (1) the nature of the substrate on or in which the cells grow—solid, as on plastic or other rigid matrix, semisolid, as in a gel such as collagen or agar, or liquid, as in a suspension culture; (2) the degree of contact with other cells; (3) the physicochemical and physiological constitution of the medium; (4) the constitution of the gas phase; and (5) the incubation temperature. The provision of the appropriate environment, including substrate adhesion, nutrient and hormone or growth factor concentration, and cell interaction, is fundamental to the expression of specialized functions (*see* Sections 17.1, 17.7 and Alberts et al., 2002).

3.2 CELL ADHESION

Most cells from solid tissues grow as adherent monolayers, and, unless they have transformed and become anchorage independent (*see* Section 18.5.1), after tissue disaggregation or subculture they will need to attach and spread out on the substrate before they will start to proliferate (*see* Sections 13.7, 21.9.2). Originally, it was found that cells would attach to, and spread on, glass that had a slight net negative charge. Subsequently, it was shown that cells would attach to some plastics, such as polystyrene, if the plastic was appropriately treated with an electric ion discharge or high-energy ionizing radiation. We now know that cell adhesion is mediated by specific cell surface receptors for molecules in the extracellular matrix (*see also* Sections 8.4, 17.7.3), so it seems likely that spreading may be preceded by the secretion of extracellular matrix proteins and proteoglycans by the cells. The matrix adheres to the charged substrate, and the cells then bind to the matrix via specific receptors. Hence, glass or plastic that has been conditioned by previous cell growth can often provide a better surface for attachment, and substrates pretreated with matrix constituents, such as fibronectin or collagen, or derivatives, such as gelatin, will help the more fastidious cells to attach and proliferate.

With fibroblast-like cells, the main requirement is for substrate attachment and spreading and the cells migrate·individually at low densities. Epithelial cells may also require cell–cell adhesion for optimum survival and growth and, consequently, they tend to grow in patches.

3.2.1 Cell Adhesion Molecules

Three major classes of transmembrane proteins have been shown to be involved in cell–cell and cell–substrate adhesion (Fig. 3.1). Cell–cell adhesion molecules, CAMs (Ca^{2+} independent), and *cadherins* (Ca^{2+} dependent) are involved primarily in interactions between homologous cells. These

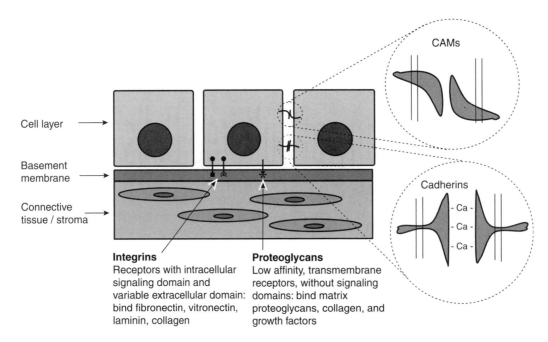

Integrins
Receptors with intracellular signaling domain and variable extracellular domain: bind fibronectin, vitronectin, laminin, collagen

Proteoglycans
Low affinity, transmembrane receptors, without signaling domains: bind matrix proteoglycans, collagen, and growth factors

Fig. 3.1. Cell Adhesion. Diagrammatic representation of a layer of epithelial cells above connective tissue containing fibrocytes and separated from it by a basal lamina. CAMs and cadherins are depicted between like cells, integrins and proteoglycans between the epithelial layer and the matrix of the basal lamina.

proteins are self-interactive; that is, homologous molecules in opposing cells interact with each other [Rosenman & Gallatin, 1991; Alberts et al., 2002], and the cell–cell recognition that this generates has a signaling role in cell behavior [Cavallaro & Christofori, 2004]. Cell–substrate interactions are mediated primarily by *integrins*, receptors for matrix molecules such as fibronectin, entactin, laminin, and collagen, which bind to them via a specific motif usually containing the arginine–glycine–aspartic acid (RGD) sequence [Yamada & Geiger, 1997]. Each integrin comprises one α and one β subunit, the extracellular domains of which are highly polymorphic, thus generating considerable diversity among the integrins. Both integrins and cadherins interact with vinculin, a step in signaling to the nucleus [Bakolitsa et al., 2004].

The third group of cell adhesion molecules is the transmembrane proteoglycans, also interacting with matrix constituents such as other proteoglycans or collagen, but not via the RGD motif. Some transmembrane and soluble proteoglycans also act as low-affinity growth factor receptors [Subramanian et al., 1997; Yevdokimova & Freshney, 1997] and may stabilize, activate, and/or translocate the growth factor to the high-affinity receptor, participating in its dimerization [Schlessinger et al., 1995].

Disaggregation of the tissue, or an attached monolayer culture, with protease will digest some of the extracellular matrix and may even degrade some of the extracellular domains of transmembrane proteins, allowing cells to become dissociated from each other. Epithelial cells are generally more resistant to disaggregation, as they tend to have tighter junctional complexes (desmosomes, adherens junctions, and tight junctions) holding them together, whereas mesenchymal cells, which are more dependent on matrix interactions for intercellular bonding, are more easily dissociated. Endothelial cells may also express tight junctions in culture, especially if left at confluence for prolonged periods on a preformed matrix, and can be difficult to dissociate. In each case, the cells must resynthesize matrix proteins before they attach or must be provided with a matrix-coated substrate.

3.2.2 Intercellular Junctions

Although some cell adhesion molecules are diffusely arranged in the plasma membrane, others are organized into intercellular junctions. The role of the junctions varies between mechanical, such as the desmosomes and adherens junctions, which hold epithelial cells together, tight junctions, which seal the space between cells, e.g. between secretory cells in an acinus or duct or between endothelial cells in a blood vessel, and gap junctions, which allow ions, nutrients, and small signaling molecules such as cyclic adenosine monophosphate (cAMP) to pass between cells in contact [*see* Alberts et al., 2002]. Although desmosomes may be distributed throughout the area of plasma membranes in contact (Fig. 3.2a), they are often associated with tight and adherens junctions at the apical end of lateral cell contacts (Fig. 3.2b).

As epithelial cells differentiate in confluent cultures they can form an increasing number of desmosomes and, if

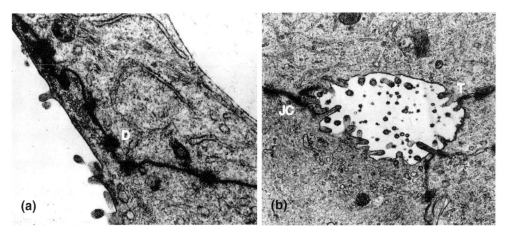

Fig. 3.2. Intercellular Junctions. Electron micrograph of culture of CA-KD cells, an early-passage culture from an adenocarcinoma secondary in brain (primary site unknown). Cells grown on Petriperm dish (Vivascience). (a) Desmosomes (D) between two cells in contact; mag. 28,000×. (b) Canaliculus showing tight junctions (T) and junctional complex (JC); mag. 18,500×. (Courtesy of Carolyn MacDonald).

some morphological organization occurs, can form complete junctional complexes. This is one reason why epithelial cells, if left at confluence for too long, can be difficult to disaggregate. As many of the adhesion molecules within these junctions depend on Ca^{2+} ions, a chelating agent, such as EDTA, is often added to the trypsin during or before disaggregation.

3.2.3 Extracellular Matrix

Intercellular spaces in tissues are filled with extracellular matrix (ECM), the constitution of which is determined by the cell type, e.g., fibrocytes secrete type I collagen and fibronectin into the matrix, whereas epithelial cells produce laminin. Where adjacent cell types are different, e.g., at the boundary of the dermis (fibrocytes) and epidermis (epithelial keratinocytes), both cell types will contribute to the composition of the ECM, often producing a *basal lamina*. The complexity of the ECM is a significant component in the phenotypic expression of the cells attached to it, so a dynamic equilibrium exists in which the cells attached to the ECM control its composition and, in turn, the composition of the ECM regulates the cell phenotype [Kleinman et al., 2003; Zoubiane et al., 2003; Fata et al., 2004]. Hence a proliferating, migratory fibroblast will require a different ECM from a differentiating epithelial cell or neuron. Mostly, cultured cell lines are allowed to generate their own ECM, but primary culture and propagation of some specialized cells, and the induction of their differentiation, may require exogenous provision of ECM.

ECM is comprised variously of collagen, laminin, fibronectin, hyaluronan, proteoglycans, and bound growth factors or cytokines [Alberts et al., 1997, 2002]. It can be prepared by mixing purified constituents, such as collagen and fibronectin, by using cells to generate ECM and washing the producer cells off before reseeding with

the cells under study (*see* Protocol 8.1), or by using a preformed matrix generated by the Engelberth-Holm-Swarm (EHS) mouse sarcoma, available commercially as Matrigel (*see* Section 8.4.1). Matrigel is often used to encourage differentiation and morphogenesis in culture and frequently generates a latticelike network with epithelial (Fig 3.3; Plate 12c) or endothelial cells.

At least two components of interaction with the substrate may be recognized: (1) adhesion, to allow the attachment and spreading that are necessary for cell proliferation [Folkman & Moscona, 1978], and (2) specific interactions, reminiscent of the interaction of an epithelial cell with basement membrane, with other ECM constituents, or with adjacent tissue cells, and required for the expression of some specialized functions (*see* Sections 3.4.1 and 17.7.3). Rojkind et al. [1980], Vlodavsky et al. [1980], and others explored the growth of cells on other natural substrates related to basement membrane. Natural matrices and defined-matrix macromolecules such as Matrigel, Natrigel, collagen, laminin, and vitronectin (B-D Biosciences, Invitrogen) are now available for controlled studies on matrix interaction.

The use of ECM constituents can be highly beneficial in enhancing cell survival, proliferation, or differentiation, but, unless recombinant molecules are used [*see*, e.g., Ido et al., 2004] there is a significant risk of the introduction of adventitious agents from the originating animal (*see* Section 10.1). Fibronectin and laminin fragments are now available commercially (*see* Appendix II).

3.2.4 Cytoskeleton

Cell adhesion molecules are attached to elements of the cytoskeleton. The attachment of integrins to actin microfilaments via linker proteins is associated with reciprocal signaling between the cell surface and the nucleus [Fata et al., 2004]. Cadherins can also link to the actin cytoskeleton

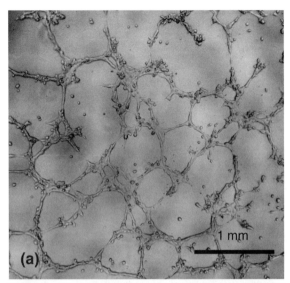

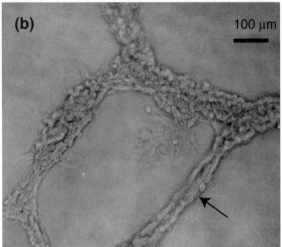

Fig. 3.3. A549 Cells Growing on Matrigel. Cultures of A549 adenocarcinoma cells growing on Matrigel. (a) Low-power shot showing lattice formation 24 h after seeding at 1×10^5 cells/mL. (b) Higher power, 3 days after seeding at 1×10^5 cells/mL. Arrow indicates possible tubular formation. (Courtesy of Jane Sinclair; *see also* Plate 12c.)

in adherens junctions, mediating changes in cell shape. Desmosomes, which also employ cadherins, link to the intermediate filaments—in this case, cytokeratins—via an intracellular plaque, but it is not yet clear whether this linkage is a purely structural feature or also has a signaling capacity. Intermediate filaments are specific to cell lineages and can be used to characterize them (*see* Section 16.3.2; Plate 11a–c). The microtubules are the third component of the cytoskeleton; their role appears to be related mainly to cell motility and intracellular trafficking of micro-organelles, such as the mitochondria and the chromatids at cell division.

3.2.5 Cell Motility

Time-lapse recording (*see* Section 27.3) demonstrates that cultured cells are capable of movement on a substrate.

The most motile are fibroblasts at a low cell density (when cells are not in contact), and the least motile are dense epithelial monolayers. Fibroblasts migrate as individual cells with a recognizable polarity of movement. A lamellipodium, generated by polymerization of actin [Pollard & Borisy, 2003], extends in the direction of travel and adheres to the substrate, and the plasma membrane at the opposite side of the cell retracts, causing the cell to undergo directional movement. If the cell encounters another cell, the polarity reverses, and migration proceeds in the opposite direction. Migration proceeds in erratic tracks, as revealed by colloidal gold tracking [Scott et al., 2000], until the cell density reaches confluence, whereupon directional migration ceases. The cessation of movement at confluence, which is accompanied by a reduction in plasma membrane ruffling, is known as *contact inhibition* (*see* Section 18.5.2) and leads eventually to withdrawal of the cell from the division cycle. Myoblasts and endothelial cells migrate in a similar fashion and, like fibroblasts, may differentiate when they reach confluence, depending on the microenvironment.

Epithelial cells, unless transformed, tend not to display random migration as polarized single cells. When seeded at a low density, they will migrate until they make contact with another cell and the migration stops. Eventually, cells accumulate in patches and the whole patch may show signs of coordinated movement [Casanova, 2002].

3.3 CELL PROLIFERATION

3.3.1 Cell Cycle

The cell cycle is made up of four phases (Fig. 3.4). In the *M phase* (M = mitosis), the chromatin condenses into chromosomes, and the two individual chromatids, which make up the chromosome, segregate to each daughter cell. In the G_1 (Gap 1) *phase*, the cell either progresses toward DNA

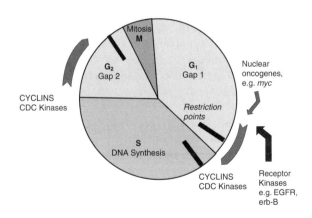

Fig. 3.4. Cell Cycle. The cell cycle is divided into four phases: G_1, S, G_2, and M. Progression round the cycle is driven by cyclins interacting with CDC kinases and stimulated by nuclear oncogenes and cytoplasmic signals initiated by receptor kinase interaction with ligand. The cell cycle is arrested at restriction points by cell cycle inhibitors such as Rb and p53.

synthesis and another division cycle or exits the cell cycle reversibly (G₀) or irreversibly to commit to differentiation. It is during G₁ that the cell is particularly susceptible to control of cell cycle progression at a number of restriction points, which determine whether the cell will re-enter the cycle, withdraw from it, or withdraw and differentiate. G₁ is followed by the *S phase* (DNA synthesis), in which the DNA replicates. S in turn is followed by the G₂ (Gap 2) *phase* in which the cell prepares for reentry into mitosis. Checkpoints at the beginning of DNA synthesis and in G₂ determine the integrity of the DNA and will halt the cell cycle to allow DNA repair or entry into apoptosis if repair is impossible. Apoptosis, or programmed cell death [al-Rubeai & Singh, 1998], is a regulated physiological process whereby a cell can be removed from a population. Marked by DNA fragmentation, nuclear blebbing, and cell shrinkage (*see* Plate 17c,d), apoptosis can also be detected by a number of marker enzymes with kits such as Apotag (Oncor) or the COMET assay [Maskell & Green, 1995].

3.3.2 Control of Cell Proliferation

Entry into the cell cycle is regulated by signals from the environment. Low cell density leaves cells with free edges and renders them capable of spreading, which permits their entry into the cycle in the presence of mitogenic growth factors, such as epidermal growth factor (EGF), fibroblast growth factors (FGFs), or platelet-derived growth factor (PDGF) (*see* Sections 9.5.2, 10.4.3 and Table 10.3), interacting with cell surface receptors. High cell density inhibits the proliferation of normal cells (though not transformed cells) (*see* Section 18.5.2). Inhibition of proliferation is initiated by cell contact and is accentuated by crowding and the resultant change in the shape of the cell and reduced spreading.

Intracellular control is mediated by positive-acting factors, such as the cyclins [Planas-Silva & Weinberg, 1997; Reed, 2003] (*see* Fig. 3.2), which are upregulated by signal transduction cascades activated by phosphorylation of the intracellular domain of the receptor when it is bound to growth factor. Negative-acting factors such as p53 [Sager, 1992; McIlwrath et al., 1994], p16 [Russo et al., 1998], or the Rb gene product [Sager, 1992] block cell cycle progression at restriction points or checkpoints (Fig. 3.5). The link between the extracellular control elements (both positive-acting, e.g., PDGF, and negative-acting, e.g., TGF-β) and intracellular effectors is made by cell membrane receptors and signal transduction pathways, often involving protein phosphorylation and second messengers such as cAMP, Ca²⁺, and diacylglycerol [Alberts et al., 2002]. Much of the evidence for the existence of these steps in the control of cell proliferation has emerged from studies of oncogene and suppressor gene expression in tumor cells, with the ultimate objective of the therapeutic regulation of uncontrolled cell proliferation in cancer. The immediate benefit, however, has been a better understanding of the factors required to regulate cell proliferation in culture [Jenkins, 1992]. These studies

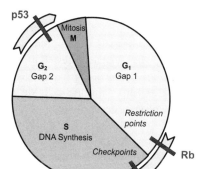

(a) CELL CYCLE ARREST

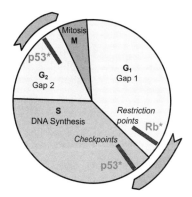

(b) CELL CYCLE PROGRESSION

Fig. 3.5. Cell Cycle Inhibition and Progression. The cell cycle is arrested at restriction points or checkpoints by the action of Rb, p53, and other cell cycle inhibitors (a). When these are inactivated, usually by phosphorylation, cells proceed round the cycle (b).

have had other benefits as well, including the identification of genes that enhance cell proliferation, some of which can be used to immortalize finite cell lines (*see* Section 18.4).

3.4 DIFFERENTIATION

As stated earlier (*see* Section 1.3.3), the expression of differentiated properties in cell culture is often limited by the promotion of cell proliferation, which is necessary for the propagation of the cell line and the expansion of stocks. The conditions required for the induction of differentiation—a high cell density, enhanced cell–cell and cell–matrix interaction, and the presence of various differentiation factors (*see* Sections 17.1.1, 17.7)—may often be antagonistic to cell proliferation and vice versa. So if differentiation is required, it may be necessary to define two distinct sets of conditions—one to optimize cell proliferation and one to optimize cell differentiation.

3.4.1 Maintenance of Differentiation

It has been recognized for many years that specific functions are retained longer when the three-dimensional structure of the tissue is retained, as in organ culture (*see* Section 25.2). Unfortunately, organ cultures cannot be propagated, must be prepared *de novo* for each experiment, and are more difficult to quantify than cell cultures. Re-creating three-dimensional structures by perfusing monolayer cultures (*see* Sections 25.3, 26.2.6) and culturing cells on or in special matrices, such as collagen gel, cellulose, or gelatin sponge, or other matrices (*see* Sections 3.2.3, 8.4.1, 8.4.3, 17.7.3) may be a better option. A number of commercial products, the best known of which is Matrigel (BD Biosciences), reproduce the characteristics of extracellular matrix, but are undefined, although a growth factor-depleted version is also available (GFR Matrigel). These techniques present some limitations, but with their provision of homotypic cell interactions and cell–matrix interactions, and with the possibility of introducing heterotypic cell interactions, they hold considerable promise for the examination of tissue-specific functions, particularly when interactions may be regulated by growing cultures in filter-well inserts (*see* Section 25.3.6). Expression of the differentiated phenotype may also require maintenance in the appropriate selective medium (*see* Section 10.2.1), with appropriate soluble inducers, such as hydrocortisone, retinoids, or planar polar compounds (*see* Sections 17.7.1, 17.7.2), and usually in the absence of serum.

The development of normal tissue functions in culture would facilitate the investigation of pathological behavior such as demyelination and malignant invasion. However,

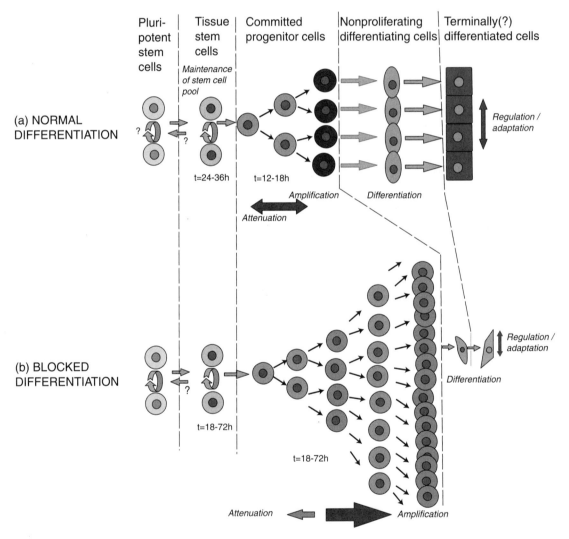

Fig. 3.6. *Differentiation from Stem Cells.* (a) *In vivo*, a small stem cell pool gives rise to a proliferating progenitor compartment that produces the differentiated cell pool. (b) *In vitro*, differentiation is limited by the need to proliferate, and the population becomes predominantly progenitor cells, although stem cells may also be present. Pluripotent stem cells (far left) have also been cultured from some tissues, but their relationship to the tissue stem cells is as yet unclear. Culture conditions select mainly for the proliferating progenitor cell compartment of the tissue or induce cells that are partially differentiated to revert to a progenitor status.

from a fundamental viewpoint, it is only when cells *in vitro* express their normal functions that any attempt can be made to relate them to their tissue of origin. The expression of the differentiated phenotype need not be complete, because the demonstration of a single type-specific surface antigen may be sufficient to place a cell in the correct lineage. More complete functional expression may be required, however, to place a cell in its correct *position* in the lineage and to reproduce a valid model of its function *in vivo*.

3.4.2 Dedifferentiation

Historically, the inability of cell lines to express the characteristic *in vivo* phenotype was blamed on *dedifferentiation*. According to this concept, differentiated cells lose their specialized properties *in vitro*, but it is often unclear whether (1) the wrong lineage of cells is selected *in vitro*, (2) undifferentiated cells of the same lineage (Fig. 3.6) overgrow terminally differentiated cells of reduced proliferative capacity, or (3) the absence of the appropriate

TABLE 3.1. Cell Lines with Differentiated Properties

Cell type	Origin	Cell line	N*	Species	Marker	Reference
Endocrine	Adrenal cortex	Y-1	T	Mouse	Adrenal steroids	Yasamura et al., 1966
Endocrine	Pituitary tumor	GH3	T	Rat	Growth hormone	Buonassisi et al., 1962
Endocrine	Hypothalamus	C7	N	Mouse	Neurophysin; vasopressin	De Vitry et al., 1974
Endothelium	Dermis	HDMEC		Human	Factor VIII, CD36	Gupta et al., 1997
Endothelium	Pulmonary artery	CPAE	C	Cow	Factor VIII, angiotensin II-converting enzyme	Del Vecchio & Smith, 1981
Endothelium	Hepatoma	SK/HEP-1	T	Human	Factor VIII	Heffelfinger et al., 1992
Epithelium	Prostate	PPEC	N	Human	PSA	Robertson & Robertson, 1995
Epithelium	Kidney	MDCK	C	Dog	Domes, transport	Gaush et al., 1966; Rindler et al., 1979
Epithelium	Kidney	LLC–PKI	C	Pig	Na$^+$-dependent glucose uptake	Hull et al., 1976; Saier, 1984
Epithelium	Breast	MCF-7	T	Human	Domes, α-lactalbumin	Soule et al., 1973
Glia	Glioma	MOG–G-CCM	T	Human	Glutamyl synthetase	Balmforth et al., 1986
Glia	Glioma	C6	T	Rat	Glial fibrillary acidic protein, GPDH	Benda et al., 1968
Hepatocytes	Hepatoma	H4–11–E–C3	T	Rat	Tyrosine aminotransferase	Pitot et al., 1964
Hepatocytes	Liver		T	Mouse	Aminotransferase	Yeoh et al., 1990
Keratinocytes	Epidermis	HaCaT	C	Human	Cornification	Boukamp et al., 1988
Leukemia	Spleen	Friend	T	Mouse	Hemoglobin	Scher et al., 1971
Melanocytes	Melanoma	B16	T	Mouse	Melanin	Nilos & Makarski, 1978
Myeloid	Leukemia	K562	T	Human	Hemoglobin	Andersson et al., 1979a, b
Myeloid	Myeloma	Various	T	Mouse	Immunoglobulin	Horibata & Harris, 1970
Myeloid	Marrow	WEHI–3B D+	T	Mouse	Morphology	Nicola, 1987
Myeloid	Leukemia	HL60	T	Human	Phagocytosis; Neotetrazolium Blue reduction	Olsson & Ologsson, 1981
Myocytes	Skeletal muscle	C2	C	Mouse	Myotubes	Morgan et al., 1992
		L6	C	Rat	Myotubes	Richler & Yaffe, 1970
Neuroendocrine	Pheochromocytoma	PC12	T	Rat	Catecholamines; dopamine; norepinephrine	Greene & Tischler, 1976
Neurons	Neuroblastoma	C1300	T	Rat	Neurites	Lieberman & Sachs, 1978
Type II pneumocyte or Clara cell	Lung carcinoma	A549	T	Human	Surfactant	Giard et al., 1972
		NCI-H441	T	Human	Surfactant	Brower et al., 1986
Type II pneumocyte	Lung carcinoma		I	Mouse	Surfactant	Wilkenheiser et al., 1991
Various	Embryonal teratocarcinoma	F9	T	Mouse	PA, laminin, type IV collagen	Bernstine et al., 1973

*Normal (N), continuous (C), immortalized (I), transformed (T).

inducers (hormones: cell or matrix interaction) causes an adaptive, and potentially reversible, loss of differentiated properties (*see* Section 17.1.1). In practice, all of these may contribute to loss of differentiation; even in the correct lineage-selective conditions, continuous proliferation will favor undifferentiated precursors, which, in the absence of the correct inductive environment, do not differentiate.

An important distinction should be made between dedifferentiation, deadaptation, and selection. Dedifferentiation implies that the specialized properties of the cell are lost by conversion to a more primitive phenotype. For example, a hepatocyte would lose its characteristic enzymes (arginase, aminotransferases, etc.) and could not store glycogen or secrete serum proteins, because of reversion or conversion to a precursor cell [Kondo & Raff, 2004]. Deadaptation, on the other hand, implies that the synthesis of specific products or other aspects of specialized function are under regulatory control by hormones, cell–cell interaction, cell–matrix interaction, etc., and can be reinduced if the correct conditions can be re-created. For instance, the presence of matrix as a floating collagen raft [Michalopoulos & Pitot, 1975], Matrigel [Bissell et al., 1987], or dimethyl sulfoxide (DMSO) [Cable & Isom, 1997] allows retention of differentiated properties in hepatocytes. It is now clear that, given the correct culture conditions, differentiated functions can be reexpressed (Table 3.1; *see also* Section 17.5).

For induction to occur, the appropriate cells must be present. In early attempts at liver cell culture, the failure of cells to express hepatocyte properties was due partly to overgrowth of the culture by connective tissue fibroblasts or endothelium from blood vessels or sinusoids. With the correct disaggregation technique and the correct culture conditions [Guguen-Guillouzo, 2002] (*see also* Protocol 23.6), hepatocytes can be selected preferentially. Similarly, epidermal cells can be grown by using either a confluent feeder layer [Rheinwald & Green, 1975] or a selective medium [Peehl & Ham, 1980; Tsao et al., 1982] (*see* Protocol 23.1). Selective media also have been used for many other types of epithelium [Freshney, 2002]. These and other examples [e.g., selective feeder layers (*see* Protocols 23.1, 23.4, 24.1), D-valine for the isolation of kidney epithelium, and the use of cytotoxic antibodies (*see* Section 14.6)] clearly demonstrate that the selective culture of specialized cells is achievable. Many selective media, based mainly on supplemented Ham's F12:DMEM or modifications of the MCDB series (*see* Section 10.2.1), have been devised [Cartwright & Shah, 1994; Mather, 1998], and many are now available commercially (*see* Appendix II), often with specialized cultures.

3.5 CELL SIGNALING

Cell proliferation, migration, differentiation, and apoptosis *in vivo* are regulated by cell–cell interaction, cell–matrix interaction, and nutritional and hormonal signals, as discussed above (*see* Section 3.4.1). Some signaling is contact-mediated via cell adhesion molecules (*see* Section 3.2), but signaling can also result from soluble, diffusible factors. Signals that reach the cell from another tissue via the systemic vasculature are called *endocrine*, and those that diffuse from adjacent cells without entering the bloodstream are called *paracrine*. It is useful to recognize that some soluble signals arise in, and interact with, the same type of cell. I will call this *homotypic paracrine*, or *homocrine*, signaling (Fig. 3.7). Signals that arise in a cell type different from the responding cells are

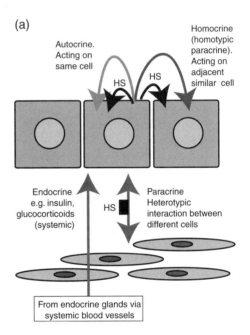

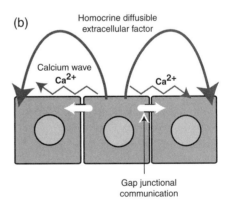

Fig. 3.7. Cell Interaction and Signaling. Routes of interaction among cells. (a) Factors influencing the behavior of a cell include endocrine hormones from the vasculature, paracrine factors from the stroma, homocrine factors from adjacent similar cells, and autocrine factors from the cell itself. Matrix, soluble, and cell-associated heparan sulfate (HS) may help the activation, stabilization, and/or translocation of paracrine factors. (b) Uniformity of response in target tissue is improved by gap junctional communication, by calcium signaling, and, possibly, by homocrine factors from the stimulated cell.

heterotypic paracrine and will be referred to simply as *paracrine* in any subsequent discussion. A cell can also generate its own signaling factors that bind to its own receptors, and this is called *autocrine* signaling.

Although all of these forms of signaling occur *in vivo*, under normal conditions with basal media *in vitro*, only autocrine and homocrine signaling will occur. The failure of many cultures to plate with a high efficiency at low cell densities may be due, in part, to the dilution of one or more autocrine and homocrine factors, and this is part of the rationale in using conditioned medium (*see* Protocol 14.2) or feeder layers (*see* Protocol 14.3) to enhance plating efficiency. As the maintenance and proliferation of specialized cells, and the induction of their differentiation, may depend on paracrine and endocrine factors, these must be identified and added to differentiation medium (*see* Sections 17.7.1, 17.7.2). However, their action may be quite complex as not only may two or more factors be required to act in synergy [*see*, e.g., McCormick and Freshney, 2000], but, in trying to simulate cell–cell interaction by supplying exogenous paracrine factors, it is necessary to take into account that the phenotype of interacting cells, and hence the factors that they produce and the time frame in which they are produced, will change as a result of the interaction. Heterotypic combinations of cells may be, initially at least, a simpler way of providing the correct factors in the correct matrix microenvironment, and analysis of this interaction may then be possible with blocking antibodies or antisense RNA.

3.6 ENERGY METABOLISM

Most culture media contain 4–20 mM glucose, which is used mainly as a carbon source for glycolysis, generating lactic acid as an end product. Under normal culture conditions (atmospheric oxygen and a submerged culture), oxygen is in relatively short supply. In the absence of an appropriate carrier, such as hemoglobin, raising the O_2 tension will generate free radical species that are toxic to the cell, so O_2 is usually maintained at atmospheric levels. This results in anaerobic conditions and the use of glycolysis for energy

metabolism. However, the citric acid cycle remains active, and it has become apparent that amino acids—particularly glutamine—can be utilized as a carbon source by oxidation to glutamate by glutaminase and entry into the citric acid cycle by transamination to 2-oxoglutarate [Reitzer et al., 1979; Butler & Christie, 1994]. Deamination of the glutamine tends to produce ammonia, which is toxic and can limit cell growth, but the use of dipeptides, such as glutamyl-alanine or glutamyl-glycine, appears to minimize the production of ammonia and has the additional advantage of being more stable in the medium (e.g., Glutamax, Invitrogen).

3.7 INITIATION OF THE CULTURE

Primary culture techniques are described in detail later (*see* Chapter 12). Briefly, a culture is derived either by the outgrowth of migrating cells from a fragment of tissue or by enzymatic or mechanical dispersal of the tissue. Regardless of the method employed, primary culture is the first in a series of selective processes (Table 3.2) that may ultimately give rise to a relatively uniform cell line. In primary explantation (*see* Section 12.3.1), selection occurs by virtue of the cells' capacity to migrate from the explant, whereas with dispersed cells, only those cells that both survive the disaggregation technique and adhere to the substrate or survive in suspension will form the basis of a primary culture. If the primary culture is maintained for more than a few hours, a further selection step will occur. Cells that are capable of proliferation will increase, some cell types will survive but not increase, and yet others will be unable to survive under the particular conditions of the culture. Hence, the relative proportion of each cell type will change and will continue to do so until, in the case of monolayer cultures, all the available culture substrate is occupied. It should be realized that primary cultures, although suitable for some studies such as cytogenetic analysis, may be unsuitable for other studies because of their instability. Both cell population changes and adaptive modifications within the cells are occurring continuously throughout the culture, making it difficult

TABLE 3.2. Selection in Cell Line Development

Stage	Primary explant	Enzymatic disaggregation
	Factors influencing selection	
Isolation	Mechanical damage	Enzymatic damage
Primary culture	Adhesion of explant; outgrowth (migration), cell proliferation	Cell adhesion and spreading, cell proliferation
First subculture	Trypsin sensitivity; nutrient, hormone, and substrate limitations; proliferative ability	
Propagation as a cell line	Relative growth rates of different cells; selective overgrowth of one lineage Nutrient, hormone, and substrate limitations Effect of cell density on predominance of normal or transformed phenotype	
Senescence; transformation	Normal cells die out; transformed cells overgrow	

to select a period when the culture may be regarded as homogeneous or stable.

After *confluence* is reached (i.e., all the available growth area is utilized and the cells make close contact with one another), cells whose growth is sensitive to contact inhibition and density limitation of cell proliferation (*see* Section 18.5.2) will stop dividing, while any transformed cells, which are insensitive to density limitation, will tend to overgrow. Keeping the cell density low (e.g., by frequent subculture) helps to preserve the normal phenotype in cultures such as mouse fibroblasts, in which spontaneous transformants tend to overgrow at high cell densities [Todaro & Green, 1963].

Some aspects of specialized function are expressed more strongly in primary culture, particularly when the culture becomes confluent. At this stage, the culture will show its closest morphological resemblance to the parent tissue and retain some diversity in cell type.

3.8 EVOLUTION OF CELL LINES

After the first subculture, or passage (Fig. 3.8), the primary culture becomes known as a *cell line* and may be propagated and subcultured several times. With each successive subculture, the component of the population with the ability to proliferate most rapidly will gradually predominate, and nonproliferating or slowly proliferating cells will be diluted out. This is most strikingly apparent

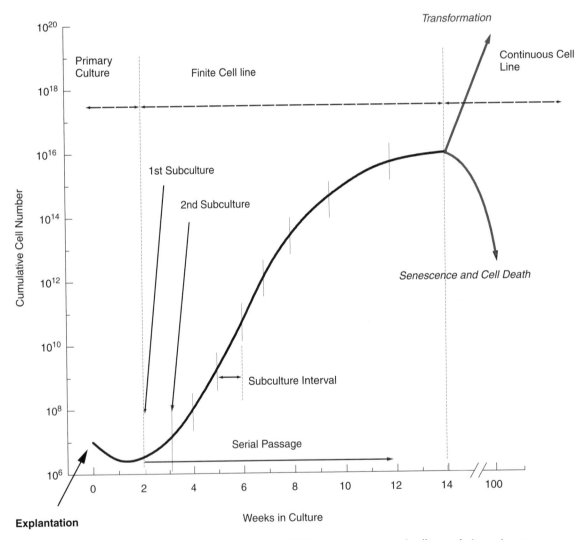

Fig. 3.8. Evolution of a Cell Line. The vertical (*Y*) axis represents total cell growth (assuming no reduction at passage) for a hypothetical cell culture. Total cell number (cell yield) is represented on this axis on a log scale, and the time in culture is shown on the *X*-axis on a linear scale. Although a continuous cell line is depicted as arising at 14 weeks, with different cells it could arise at any time. Likewise, senescence may occur at any time, but for human diploid fibroblasts it is most likely to occur between 30 and 60 cell doublings, or 10 to 20 weeks, depending on the doubling time. Terms and definitions used are as in the glossary. (After Hayflick and Moorhead [1961].)

after the first subculture, in which differences in proliferative capacity are compounded with varying abilities to withstand the trauma of trypsinization and transfer (*see* Section 13.1).

Although some selection and phenotypic drift will continue, by the third passage the culture becomes more stable and is typified by a rather hardy, rapidly proliferating cell. In the presence of serum and without specific selection conditions, mesenchymal cells derived from connective tissue fibroblasts or vascular elements frequently overgrow the culture. Although this has given rise to some very useful cell lines (e.g., WI-38 human embryonic lung fibroblasts [Hayflick and Moorhead, 1961], BHK21 baby hamster kidney fibroblasts [Macpherson and Stoker, 1962], COS cells [Gluzman, 1981], CHO cells [Puck et al., 1958] (*see* Table 13.1), and perhaps the most famous of all, the L-cell, a mouse subcutaneous fibroblast treated with methylcholanthrene [Earle et al., 1943; Sanford et al., 1948]), this overgrowth represents one of the major challenges of tissue culture since its inception—namely, how to prevent the overgrowth of the more fragile or slower-growing specialized cells such as hepatic parenchyma or epidermal keratinocytes. Inadequacy of the culture conditions is largely to blame for this problem, and considerable progress has now been made in the use of selective media and substrates for the maintenance of many specialized cell lines (*see* Section 10.2.1, Chapter 23).

3.8.1 Senescence

Normal cells can divide a limited number of times; hence, cell lines derived from normal tissue will die out after a fixed number of population doublings. This is a genetically determined event involving several different genes and is known as *senescence*. It is thought to be determined, in part, by the inability of terminal sequences of the DNA in the telomeres to replicate at each cell division. The result is a progressive shortening of the telomeres until, finally, the cell is unable to divide further [Bodnar et al., 1998]. Exceptions to this rule are germ cells, stem cells, and transformed cells, which often express the enzyme telomerase, which is capable of replicating the terminal sequences of DNA in the telomere and extending the life span of the cells, infinitely in the case of germ cells and some tumor cells (*see also* Section 18.4.1).

3.9 THE DEVELOPMENT OF CONTINUOUS CELL LINES

Some cell lines may give rise to continuous cell lines (*see* Fig. 3.7). The ability of a cell line to grow continuously probably reflects its capacity for genetic variation, allowing subsequent selection. Genetic variation often involves the deletion or mutation of the p53 gene, which would normally arrest cell cycle progression, if DNA were to become mutated, and overexpression of the telomerase gene. Human fibroblasts remain predominantly euploid throughout their life span in culture and never give rise to continuous cell lines

[Hayflick and Moorhead, 1961], whereas mouse fibroblasts and cell cultures from a variety of human and animal tumors often become aneuploid in culture and frequently give rise to continuous cultures. Possibly the condition that predisposes most to the development of a continuous cell line is inherent genetic variation, so it is not surprising to find genetic instability perpetuated in continuous cell lines. A common feature of many human continuous cell lines is the development of a subtetraploid chromosome number (Fig. 3.9). The alteration in a culture that gives rise to a continuous cell line is commonly called *in vitro transformation* (*see* Chapter 18) and may occur spontaneously or be chemically or virally induced (*see* Sections 18.2, 18.4). The word *transformation* is used rather loosely and can mean different things to different people. In this volume, *immortalization* means the acquisition of an infinite life span and *transformation* implies an additional alteration in growth characteristics (anchorage independence, loss of contact inhibition and density limitation of growth) that will often, but not necessarily, correlate with tumorigenicity.

Continuous cell lines are usually *aneuploid* and often have a chromosome number between the diploid and tetraploid values (*see* Fig. 3.9). There is also considerable variation in chromosome number and constitution among cells in the population (*heteroploidy*) (*see also* Section 18.3.) It is not clear whether the cells that give rise to continuous lines are present at explantation in very small numbers or arise later as a result of the transformation of one or more cells. The second alternative would seem to be more probable on cell kinetic grounds, as continuous cell lines can appear quite late in a

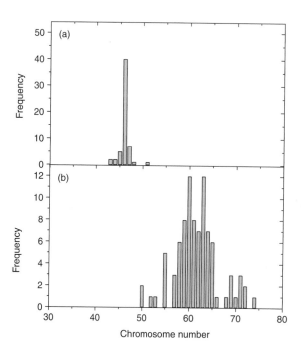

Fig. 3.9. Chromosome Numbers of Finite and Continuous Cell Lines. (a) A normal human glial cell line. (b) A continuous cell line from human metastatic melanoma.

culture's life history, long after the time it would have taken for even one preexisting cell to overgrow. The possibility remains, however, that there is a subpopulation in such cultures with a predisposition to transform that is not shared by the rest of the cells.

The term *transformation* has been applied to the process of formation of a continuous cell line partly because the culture undergoes morphological and kinetic alterations, but also because the formation of a continuous cell line is often accompanied by an increase in tumorigenicity. A number of the properties of continuous cell lines, such as a reduced serum requirement, reduced density limitation of growth, growth in semisolid media, aneuploidy (*see also* Table 18.1 and Plate 14), and more, are associated with *malignant* transformations (*see* Section 18.6). Similar morphological and behavioral changes can also be observed in cells that have undergone virally or chemically induced transformation.

Many (if not most) normal cells do not give rise to continuous cell lines. In the classic example, normal human fibroblasts remain euploid throughout their life span and at crisis (usually around 50 generations) will stop dividing, although they may remain viable for up to 18 months thereafter. Human glia [Pontén & Westermark, 1980] and chick fibroblasts [Hay & Strehler, 1967] behave similarly. Epidermal cells, on the other hand, have shown gradually increasing life spans with improvements in culture techniques [Rheinwald & Green, 1977; Green et al., 1979] and may yet be shown capable of giving rise to continuous growth. Such growth may be related to the self-renewal capacity of the tissue *in vivo* and successful propagation of the stem cells *in vitro* (*see* Section 3.10). Continuous culture of lymphoblastoid cells is also possible [Gjerset et al., 1990] by transformation with Epstein–Barr virus.

3.10 ORIGIN OF CULTURED CELLS

Because most people working under standard conditions do so with finite or continuous proliferating cell lines, it is important to consider the cellular composition of the culture. The capacity to express differentiated markers under the influence of inducing conditions may mean either that the cells being cultured are mature and only require induction to continue synthesizing specialized proteins or that the culture is composed of precursor or stem cells that are capable of proliferation but remain undifferentiated until the correct inducing conditions are applied, whereupon some or all of the cells mature and become differentiated. It may be useful to think of a cell culture as being in equilibrium between multipotent stem cells, undifferentiated but committed precursor cells, and mature differentiated cells (*see* Fig. 3.6) and to suppose that the equilibrium may shift according to the environmental conditions. Routine serial passage at relatively low cell densities would promote cell proliferation and constrain differentiation, whereas high cell densities, low serum, and the appropriate hormones would promote differentiation and inhibit cell proliferation.

The source of the culture will also determine which cellular components may be present. Hence, cell lines derived from the embryo may contain more stem cells and precursor cells and be capable of greater self-renewal than cultures from adults. In addition, cultures from tissues undergoing continuous renewal *in vivo* (epidermis, intestinal epithelium, hematopoietic cells) may still contain stem cells, which, under the appropriate conditions, will have a prolonged life span, whereas cultures from tissues that renew solely under stress (fibroblasts, muscle, glia) may contain only committed precursor cells with a limited life span.

Thus the identity of the cultured cell is defined not only by its lineage *in vivo* (hematopoietic, hepatocyte, glial, etc.), but also by its position in that lineage (stem cell, precursor cell, or mature differentiated cell). Although progression down a differentiation pathway has been thought of as irreversible, the concept of commitment is now being questioned [Kondo & Raff, 2004; Le Douarin et al., 2004] and some precursor cells may be able to convert or revert to stem cell status and redifferentiate along the same or a different lineage.

When cells are cultured from a neoplasm, they need not adhere to these rules. Thus a hepatoma from rat may proliferate *in vitro* and still express some differentiated features, but the closer they are to those of the normal phenotype, the more induction of differentiation may inhibit proliferation. Although the relationship between position in the lineage and cell proliferation may become relaxed (though not lost—B16 melanoma cells still produce more pigment at a high cell density and at a low rate of cell proliferation than at a low cell density and a high rate of cell proliferation), transfer between lineages has not been clearly established (*see* Section 17.4).

CHAPTER 4

Laboratory Design and Layout

4.1 PLANNING

The major requirement that distinguishes tissue culture from most other laboratory techniques is the need to maintain asepsis. Although it is usually not economically viable to create large sterile areas, it is important that the tissue culture laboratory be dust free and have no through traffic. The introduction of laminar-flow hoods has greatly simplified the problem and allows the utilization of unspecialized laboratory accommodation, provided that the location is suitable (*see* Sections 4.3.2, 5.2.1, 6.4).

Several considerations need to be taken into account in planning new accommodation. Is a new building to be constructed, or will an existing one be converted? A conversion limits you to the structural confines of the building; modifying ventilation and air-conditioning can be expensive, and structural modifications that involve load-bearing walls can be costly and difficult to make. When a new building is contemplated, there is more scope for integrated and innovative design, and facilities may be positioned for ergonomic and energy-saving reasons, rather than structural ones. The following items should be considered:

(1) **Ventilation**

    a) *Pressure balance.* Ideally, a tissue culture laboratory should be at positive pressure relative to surrounding work areas, to avoid any influx of contaminated air from outside. However, if human material is being used, most safety regulations will require that the tissue culture laboratory be designated as Category I (*see* Section 7.8.1) and this requires that it is at negative pressure relative to the surrounding areas. To satisfy both requirements it may be preferable to have a positive-pressure buffer zone outside the tissue culture laboratory, such as the Preparation Area and Microscope Room (*see* Fig. 4.3) or the corridor (*see* Fig. 4.4).

    b) *Laminar flow hoods.* Consider where air inlets and extracts must be placed. It is preferable to duct laminar-flow hoods to the exterior to improve air circulation and remove excess heat (300–500 W per hood) from the room. This also facilitates decontamination with formaldehyde, should it be required. Venting hoods to the outside will probably provide most of the air extraction required for the room, and it remains only to ensure that the incoming air, from a central plant or an air conditioner, does not interfere with the integrity of the airflow in the hood. Laminar-flow hoods are better left to run continuously, but if they are to be switched off when not in use, then an alternative air extract must be provided and balanced with the extract via the hoods.

(2) **Accommodation**

    a) *Staff numbers.* How many people will work in the facility, how long will they work each week, and what kinds of culture will they perform? These considerations determine how many laminar-flow hoods will be required (based on whether people can share hoods or whether they will require a hood for most of the day) and whether a large area will be needed to handle bioreactors, animal tissue dissections, or large numbers of cultures. As a rough

Culture of Animal Cells: A Manual of Basic Technique, Fifth Edition, by R. Ian Freshney
Copyright © 2005 John Wiley & Sons, Inc.

guide, 12 laminar-flow hoods in a communal facility can accommodate 50 people with different, but not continuous, requirements.

b) *Space.* What space is required for each facility? The largest area should be given to the culture operation, which has to accommodate laminar-flow hoods, cell counters, centrifuges, incubators, microscopes, and some stocks of reagents, media, glassware, and plastics. The second largest is for washup, preparation, and sterilization, third is storage, and fourth is incubation. A reasonable estimate is 4:2:1:1, in the order just presented.

c) *Aseptic area.* Will people require access to the animal facility for animal tissue? If so, ensure that tissue culture is reasonably accessible to, but not contiguous with, the animal facility. Windows can be a disadvantage in a tissue culture laboratory, leading to heat gain, ultraviolet (UV) denaturation of the medium, and the incursion of microorganisms if they are not properly sealed.

d) *Hoods.* The space between hoods should be approximately 500 mm (2 ft), to allow access for maintenance and to minimize interference in airflow between hoods. This space is best filled with a removable cart or trolley, which allows space for bottles, flasks, reagents, and a notebook.

e) *Incubation.* What type of incubation will be required in terms of size, temperature, gas phase, and proximity to the work space? Will regular, nongassed incubators or a hot room suffice, or are CO_2 and a humid atmosphere required? Generally, large numbers of flasks or large-volume flasks that are sealed are best incubated in a hot room, whereas open plates and dishes will require a humid CO_2 incubator.

f) *Preparation area.* Facilities for washing up and for sterilization should be located (i) close to the aseptic area that they service and (ii) on an outside wall to allow for the possibility of heat extraction from ovens and steam vents from autoclaves. Give your washup, sterilization, and preparation staff a reasonable visual outlook; they usually perform fairly repetitive duties, whereas the scientific and technical staff look into a laminar-flow hood and do not need a view.

g) *Servicing aseptic areas.* Will an elevator be required, or will a ramp suffice? If a ramp will do, what will be the gradient and the maximum load that you can expect to be carried up that gradient without mechanical help?

h) *Storage.* What is the scale of the work contemplated and how much storage space will this require for disposable plastics, etc? What proportion of the work will be cell line work, with its requirement for storage in liquid nitrogen?

(3) **Renovations.** If a conversion of existing facilities is contemplated, then there will be significant structural limitations; choose the location carefully, to avoid space constraints and awkward projections into the room that will limit flexibility.

(4) **Access.** Make sure that doorways are both wide enough and high enough and that ceilings have sufficient clearance to allow the installation of equipment such as laminar-flow hoods (which may need additional space for ductwork), incubators, and autoclaves. Make sure that doorways and spacing between equipment provide access for maintenance.

(5) **Quarantine.** What quarantine facilities will be required? Newly introduced cell lines and biopsies need to be screened for mycoplasma before being handled in the same room as general stocks, and some human and primate biopsies and cell lines may carry a biohazard risk that requires containment (*see* Section 7.8.1).

These questions will enable you to decide what size of facility you require and what type of accommodation—one or two small rooms (Figs. 4.1, 4.2), or a suite of rooms incorporating washup, sterilization, one or more aseptic areas, an incubation room, a dark room for fluorescence microscopy and photomicrography, a refrigeration room, and storage (Fig. 4.3). It is better to have a dedicated tissue culture laboratory with an adjacent preparation area, or a number of smaller ones with a common preparation area, rather than to have tissue culture performed alongside regular laboratory work with only a laminar-flow hood for protection. A separate facility gives better contamination protection, allows tissue culture stocks to be kept separate from regular laboratory reagents and glassware, and will, in any case, be required if human or other primate cells are handled, and in some other cases (*see* Section 7.8.1).

4.2 CONSTRUCTION AND SERVICES

The rooms should be supplied with filtered air. If Category I containment is required for the tissue culture laboratory (*see* Section 7.8.1), it will be required to be at negative pressure relative to other areas. In this case adjacent rooms, which lead to the aseptic area, should be regarded as buffer zones and should also receive filtered air but at positive pressure.

The rooms should be designed for easy cleaning. Furniture should fit tightly to the floor or be suspended from the bench, with a space left underneath for cleaning. Cover the floor with a vinyl or other dustproof finish, and allow a slight fall in the level toward a floor drain located outside the door of the room (i.e., well away from the sterile cabinets). This arrangement allows liberal use of water if the floor has to be washed, but, more important, it protects equipment from damaging floods if stills, autoclaves, or sinks overflow.

If possible it is preferable for the tissue culture lab to be separated from the preparation, washup, and sterilization areas, while still remaining adjacent (*see* Figs. 4.3, 4.4). Adequate floor drainage should be provided in the preparation/washup area, with a slight fall in floor level

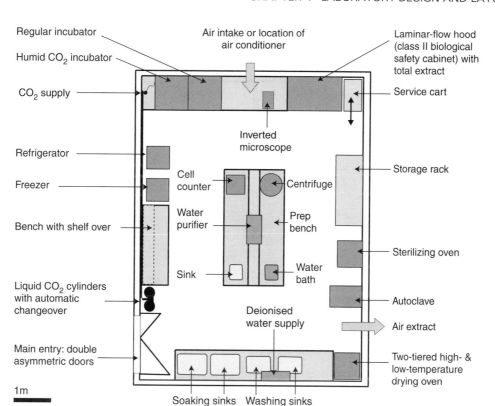

Regular incubator
Humid CO$_2$ incubator
CO$_2$ supply
Air intake or location of air conditioner
Laminar-flow hood (class II biological safety cabinet) with total extract
Service cart
Inverted microscope
Refrigerator
Freezer
Cell counter
Centrifuge
Storage rack
Water purifier
Prep bench
Bench with shelf over
Sterilizing oven
Sink
Water bath
Liquid CO$_2$ cylinders with automatic changeover
Autoclave
Deionised water supply
Air extract
Main entry: double asymmetric doors
Two-tiered high- & low-temperature drying oven
1m
Soaking sinks   Washing sinks

Fig. 4.1. Small Tissue Culture Laboratory. Suggested layout for simple, self-contained tissue culture laboratory for use by two or three persons. Dark-shaded areas represent movable equipment, lighter-shaded areas fixed or movable furniture. Scale 1:100.

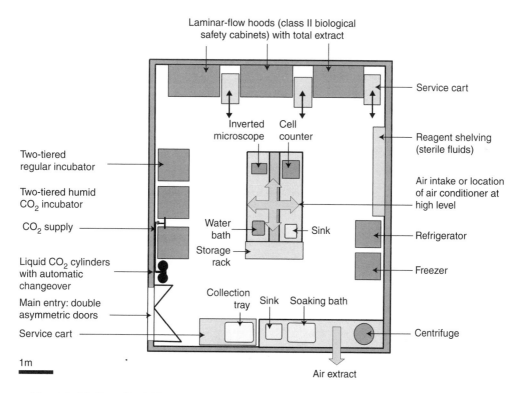

Laminar-flow hoods (class II biological safety cabinets) with total extract
Service cart
Inverted microscope
Cell counter
Reagent shelving (sterile fluids)
Two-tiered regular incubator
Two-tiered humid CO$_2$ incubator
Air intake or location of air conditioner at high level
CO$_2$ supply
Water bath
Sink
Refrigerator
Storage rack
Liquid CO$_2$ cylinders with automatic changeover
Freezer
Main entry: double asymmetric doors
Collection tray
Sink
Soaking bath
Service cart
Centrifuge
1m
Air extract

Fig. 4.2. Medium–Sized Tissue Culture Laboratory. Suitable for five or six persons, with washing up and preparation facility located elsewhere. Dark-shaded areas represent movable equipment, light-shaded areas movable or fixed furniture. Scale 1:100.

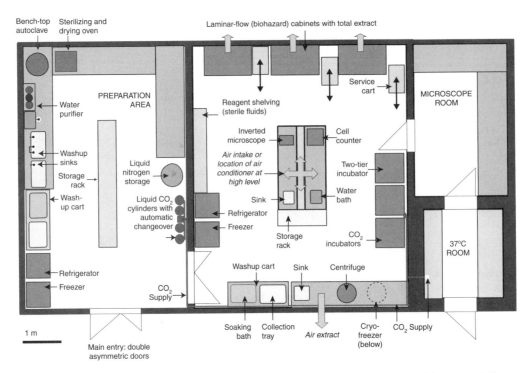

Fig. 4.3. Tissue Culture Lab with Adjacent Prep Room. Medium-sized tissue culture lab (*see* Fig. 4.2), but with attached preparation area, microscope room. and 37°C room. Scale 1:100.

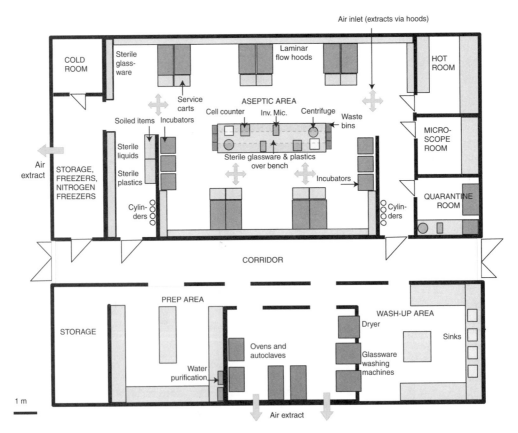

Fig. 4.4. Large Tissue Culture Laboratory. Suitable for 20 to 30 persons. Adjacent washing up, sterilization, and preparation area. Dark-shaded areas represent equipment, light-shaded areas fixed and movable furniture. Scale 1:200.

from the tissue culture lab to the washup. If you have a large tissue culture lab with a separate washup and sterilization facility, it will be convenient to have this on the same floor as, and adjacent to, the laboratory, with no steps to negotiate, so that carts or trolleys may be used. Across a corridor is probably ideal (*see* Fig. 4.4; *see also* Section 4.3).

Try to imagine the flow of traffic—people, reagents, carts, etc.—and arrange for minimum conflict, easy and close access to stores, good access for replenishing stocks without interfering with sterile work, and easy withdrawal of soiled items.

Services that are required include power, combustible gas (domestic methane, propane, etc.), carbon dioxide, compressed air, and vacuum. Power is always underestimated, in terms of both the number of outlets and the amperage per outlet. Assess carefully the equipment that will be required, assume that both the number of appliances and their power consumption will treble within the life of the building in its present form, and try to provide sufficient power, preferably at or near the outlets, but at least at the main distribution board.

Gas is more difficult to judge, as it requires some knowledge of the local provision of power. Electricity is cleaner and generally easier to manage from a safety standpoint, but gas may be cheaper and more reliable. Local conditions will usually determine the need for combustible gas.

If possible, carbon dioxide should be piped into the facility. The installation will pay for itself eventually in the cost of cylinders of mixed gases for gassing cultures, and it provides a better supply, which can be protected (*see* Section 5.3.2), for gassing incubators. Gas flow meters or electronic gas blenders (*see* Appendix II) can be provided at workstations to provide the correct gas mixture.

Compressed air is generally no longer required, as CO_2 incubators regulate the gas mixture from pure CO_2 supplies only, but will be required if a gas mixer is provided at each workstation. Compressed air is also used to expel cotton plugs from glass pipettes before washing and may be required for some types of glassware washing machine (e.g., Scientek 3000).

A vacuum line can be very useful for evacuating culture flasks, but several precautions are needed to run the line successfully. A collection vessel must be present with an additional trap flask, with a hydrophobic filter between the flasks, in order to prevent fluid, vapor, or some contaminant from entering the vacuum line and pump. Also, the vacuum pump must be protected against the line being left open inadvertently; usually this can be accomplished via a pressure-activated foot switch that closes when no longer pressed. In many respects, it is better to provide individual peristaltic pumps at each workstation (*see* Figs. 5.1–5.3), or one pump between two workstations (e.g., Integra Vacusafe, Fig. 5.3).

4.3 LAYOUT OF ASEPTIC ROOM OR SUITE

Six main functions need to be accommodated in the laboratory: sterile handling, incubation, preparation, washup, sterilization, and storage (Table 4.1). If a single room is used, create

TABLE 4.1. Tissue Culture Facilities

Minimum requirements	Desirable features	Useful additions
Sterile area, clean, quiet, and with no through traffic	Filtered air (air-conditioning)	Piped CO_2 and compressed air
Separate from animal house and microbiological labs	Service bench adjacent to culture area	Storeroom for bulk plastics
Preparation area	Separate prep room	Quarantine room
Wash up area (not necessarily within tissue culture laboratory, but at least adjacent to it)	Hot room with temperature recorder	Containment room (could double as quarantine room)
Space for incubator(s)	Separate sterilizing room	Liquid N_2 storage tank (~500 L) and separate storeroom for nitrogen freezers
Storage areas:	Separate cylinder store	Microscope room
Liquids: ambient, 4°C, −20°C		Darkroom
Glassware (shelving)		Vacuum line
Plastics (shelving)		
Small items (drawers)		
Specialized equipment (slow turnover), cupboard(s)		
Chemicals: ambient, 4°C, −20°C; share with liquids, but keep chemicals in sealed container over desiccant		
CO_2 cylinders		
Space for liquid N_2 freezer(s)		
Sink		

a "sterility gradient"; the clean area for sterile handling should be located at one end of the room, farthest from the door, and washup and sterilization facilities should be placed at the other end, with preparation, storage, and incubation in between. The preparation area should be adjacent to the washup and sterilization areas, and storage and incubators should be readily accessible to the sterile working area (*see* Figs. 4.1–4.4).

4.3.1 Sterile Handling Area

Sterile work should be located in a quiet part of the tissue culture laboratory and should be restricted to tissue culture (i.e., not shared with chemical work or with work on other organisms such as bacteria or yeast), and there should be no through traffic or other disturbance that is likely to cause dust or drafts. Use a separate room or cubicle if laminar-flow hoods are not available. The work area, in its simplest form, should be a plastic laminate-topped bench, preferably plain white or neutral gray, to facilitate the observation of cultures, dissection, etc., and to allow an accurate reading of pH when phenol red is used as an indicator. Nothing should be stored on the bench, and any shelving above should be used only in conjunction with sterile work (e.g., for holding pipette cans and instruments). The bench should be either freestanding (away from the wall) or sealed to the wall with a plastic sealing strip or mastic.

4.3.2 Laminar Flow

The introduction of laminar-flow hoods with sterile air blown onto the work surface (*see* Section 5.2.1 and Figs. 5.1, 6.2) affords greater control of sterility at a lower cost than providing a separate sterile room. Individual freestanding hoods are preferable, as they separate operators and can be moved around, but laminar-flow wall or ceiling units in batteries can be used. With individual hoods, only the operator's arms enter the sterile area, whereas with laminar-flow wall or ceiling units, there is no cabinet and the operator is part of the work area. Although this arrangement may give more freedom of movement, particularly with large pieces of apparatus (roller bottles, bioreactors), greater care must be taken by the operator not to disrupt the laminar flow, and it will be necessary to wear sterile caps and gowns to avoid contamination.

Select hoods that suit your accommodation—freestanding or bench top—and allow plenty of legroom underneath with space for pumps, aspirators, and so forth. Freestanding cabinets should be on lockable castors so that they can be moved if necessary. Chairs should be a suitable height, with adjustable seat height and back angle, and able to be drawn up close enough to the front edge of the hood to allow comfortable working well within it. A small cart, trolley, or folding flap (300–500 mm minimum) should be provided beside each hood for materials which may be required but are not in immediate use.

Laminar-flow hoods should have a lateral separation of at least 500 mm (~2 ft) and, if hoods are opposed, there should be a minimum of 2000 mm between the fronts of each hood. Laminar-flow hoods should be installed as part of the construction contract as they will influence ventilation.

4.3.3 Quarantine and Containment

If sufficient space is available, it is worth designating a separate room as a quarantine and/or containment room (*see* Fig. 4.4). This is a separate aseptic room with its own laminar-flow hood (Class II microbiological safety cabinet), incubators, freezer, refrigerator, centrifuge, supplies, and disposal. This room must be separated by a door or air lock from the rest of the suite and be at negative pressure to the rest of the aseptic area. Newly imported cell lines or biopsies can be handled here until they are shown to be free of contamination, particularly mycoplasma (*see* Section 19.3.2 and Protocols 19.2, 19.3), and proscribed pathogens such as HIV or hepatitis B. If local rules will allow, the same room can serve as a Level II containment room at different designated times. If used at a higher level of containment it will also require a biohazard cabinet or pathogen hood with separate extract and pathogen trap (*see* Section 7.8.2).

4.3.4 Service Bench

It may be convenient to position a bench for a cell counter, microscope, etc., close to the sterile handling area and either dividing the area or separating it from the other end of the lab (*see* Figs. 4.1–4.4). The service bench should also provide for the storage of sterile glassware, plastics, pipettes, screw caps, syringes, etc., in drawer units below and open shelves above. The bench may also be used for other accessory equipment, such as a small centrifuge, and should make its contents readily accessible.

4.4 INCUBATION

The requirement for cleanliness is not as stringent as that for sterile handling, but clean air, a low disturbance level, and minimal traffic will give your incubation area a better chance of avoiding dust, spores, and the drafts that carry them.

4.4.1 Incubators

Incubation may be carried out in separate incubators or in a thermostatically controlled hot room (Fig. 4.5). If only one or two are required, incubators are inexpensive and economical in terms of space; but as soon as you require more than two, their cost is more than that of a simple hot room, and their use is less convenient. Incubators also lose more heat when they are opened and are slower to recover than a hot room. As a rough guide, you will need 0.2 m^3 (200 L, 6 ft^3) of incubation space with 0.5 m^2 (6 ft^2) shelf space per person. Extra provision may need to be made for one or more humid incubators with a controlled CO_2 level in the atmosphere (*see* Section 5.3.2).

4.4.2 Hot Room

If you have the space within the laboratory area or have an adjacent room or walk-in cupboard readily available and accessible, it may be possible to convert the area into a hot room (*see* Fig. 4.5). The area need not be specifically constructed as a hot room, but it should be insulated to prevent cold spots being generated on the walls. If insulation is required, line the area with plastic laminate-veneered board, separated from the wall by about 5 cm (2 in.) of fiberglass, mineral wool, or fire-retardant plastic foam. Mark the location of the straps or studs carrying the lining panel in order to identify anchorage points for wall-mounted shelving if that is to be used. Use demountable shelving and space shelf supports at 500–600 mm (21 in.) to support the shelving without sagging. Freestanding shelving units are preferable, as they can be removed for cleaning the rack and the room. Allow 200–300 mm (9 in.) between shelves, and use wider shelves (450 mm, 18 in.) at the bottom and narrower (250–300 mm, 12 in.) ones above eye level. Perforated shelving mounted on adjustable brackets will allow for air circulation. The shelving must be flat and perfectly horizontal, with no bumps or irregularities.

Do not underestimate the space that you will require over the lifetime of the hot room. It costs very little more to equip a large hot room than a small one. Calculate costs on the basis of the amount of shelf space you will require; if you have just started, multiply by 5 or 10; if you have been working for sometime, multiply by 2 or 4.

Wooden furnishings should be avoided as much as possible, as they warp in the heat and can harbor infestations. A small bench, preferably stainless steel or solid plastic laminate, should be provided in some part of the hot room. The bench should accommodate an inverted microscope, the flasks that you wish to examine, and a notebook. If you contemplate doing cell synchrony experiments or having to make any sterile manipulations at 37°C, you should also allow space for a small laminar-flow unit with a 300 × 300 or 450 × 450 mm (12–18 in.) filter size, mounted either on a wall or on a stand over part of the bench. Alternatively, a small laminar-flow hood [not more than 1,000 mm (3 ft) wide] could be located in the room. The fan motor should be specified as for use in the tropics and should not run continuously. If it does run continuously, it will generate heat in the room and the motor may burn out.

Once a hot room is provided, others may wish to use the space for non-tissue-culture incubations, so the area of bench space provided should also take account of possible usage for incubation of tubes, shaker racks, etc. However, ban the use of other microorganisms, such as bacteria or yeast.

Incandescent lighting is preferable to fluorescent, which can cause degradation of the medium. Furthermore, some fluorescent tubes have difficulty lighting up in a hot room.

The temperature of the hot room should be controlled within ±0.5°C at any point and at any time and depends on the sensitivity and accuracy of the control gear, the location of the thermostat sensor, the circulation of air in the room, the nature of the insulation, and the evolution of heat by other apparatus (stirrers, etc.) in the room.

Heaters. Heat is best supplied via a fan heater, domestic or industrial, depending on the size of the room. Approximately 2–3 kW per 20 m^3 (700 ft^3) will be required (or two heaters could be used, each generating 1.0–1.5 kW), depending on the insulation. The fan on the heater should run continuously, and the power to the heating element should come from a proportional controller.

Air circulation. A second fan, positioned on the opposite side of the room and with the airflow opposing that of the fan heater, will ensure maximum circulation. If the room is more than 2 × 2 m (6 × 6 ft), some form of ducting may be necessary. Blocking off the corners (*see* Fig. 4.5a) is often easiest and most economical in terms of space in a square room. In a long, rectangular room, a false wall may be built at either end, but be sure to insulate it from the room and make it strong enough to carry shelving.

Thermostats. Thermostats should be of the "proportional controller" type, acting via a relay to supply heat at a rate proportional to the difference between the room temperature and the set point. When the door opens and the room temperature falls, recovery will be rapid; on the other hand, the temperature will not overshoot its mark, as the closer it approaches the set point, the less heat is supplied.

Ideally, there should be two separate heaters (H1 and H2), each with its own thermostat (HT1 and HT2). One thermostat (HT1) should be located diagonally opposite and behind the opposing fan (F1) and should be set at 37°C. The other thermostat (HT2) should be located diagonally opposite H2 and behind its opposing fan (F2) and should be set at 36°C (*see* Fig. 4.5a). Two safety override cutout thermostats should also be installed, one in series with HT1 and set at 38°C and the other in series with the main supply to both heaters. If the first heater (H1) stays on above the set point, ST1 will cut out, and the second heater (H2) will take over, regulating the temperature on HT2. If the second heater also overheats, ST2 will cut out all power to the heaters (Table 4.2). Warning lights should be installed to indicate when ST1 and ST2 have been activated. The thermostat sensors should be located in an area of rapid airflow, close to the effluent from the second, circulating, fan for greatest sensitivity. A rapid-response, high-thermal-conductivity sensor (thermistor or thermocouple) is preferred over a pressure-bulb type.

Overheating. The problem of unwanted heat gain is often forgotten because so much care is taken to provide heat and minimize loss. It can arise because of (1) a rise in ambient temperature in the laboratory in hot weather or (2) heat produced from within the hot room by apparatus such as stirrer motors, roller racks, laminar-flow units, etc. Try to avoid heat-producing equipment in the hot room. Induction-drive magnetic stirrers produce less heat that mechanically driven magnets, and drive motors for roller

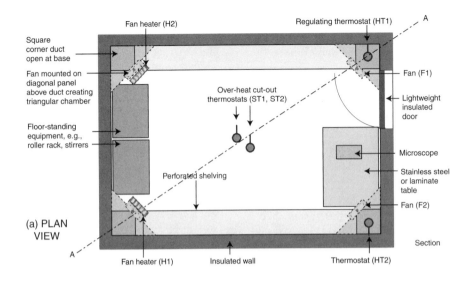

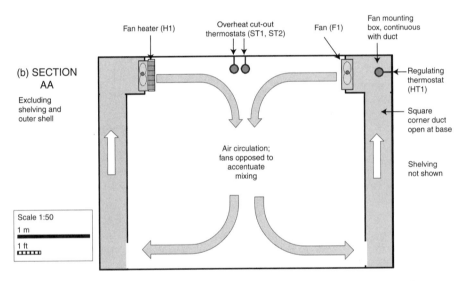

Fig. 4.5. Hot Room. Dual heating circuits and safety thermometers. (a) Plan view. (b) Diagonal section. Arrows represent air circulation. Layout and design were developed in collaboration with M. McLean of Boswell, Mitchell & Johnson (architects) and J. Lindsay of Kenneth Munro & Associates (consulting engineers).

TABLE 4.2. Hot Room Thermostats

Thermostat	37°C	<37°C	≪37°C	>37°C <38°C	>37°C >38°C	>37°C >39°C
HT1	O	I	I	O	O	O
ST1	I	I	I	I	O	O
HT2	O	O	I	O	O	O
ST2	I	I	I	I	I	O
Warning light ST1					♦	♦
Warning light ST2						♦

I = on; **O** = off; ♦ = pilot light illuminated.
HT1, regulating thermostat for heater H1; HT2, regulating thermostat for heater H2; ST1, safety override cut-out thermostat for H1, ST2, safety override cut-out thermostat for common supply to H1 and H2.

racks can sometimes be located outside the hot room. In tropical regions, or where overheating is a frequent problem, it may be necessary to incorporate cooling coils in the duct work of the heaters.

Access. If a proportional controller, good circulation, and adequate heating are provided, an air lock will not be required. The door should still be well insulated (with foam plastic or fiberglass), light, and easily closed—preferably, self-closing. It is also useful to have a hatch leading into the tissue culture area, with a shelf on both sides, so that cultures may be transferred easily into the room. The hatch door should have an insulated core as well. Locating the hatch above the bench will avoid any risk of creating a "cold spot" on the shelving.

Thermometer. A temperature recorder should be installed and should have a chart that is visible to the people working in the tissue culture room. The chart should be changed weekly. If possible, one high-level and one low-level warning light should be placed beside the chart or at a different, but equally obvious, location.

4.5 PREPARATION AREA

4.5.1 Media Preparation

The need for extensive preparation of media in small laboratories can be avoided if there is a proven source of reliable commercial culture media. Although a large enterprise (approximately 50 people doing tissue culture) may still find it more economical to prepare its own media, smaller laboratories may prefer to purchase ready-made media. These laboratories would then need only to prepare reagents, such as salt solutions and ethylenediaminetetraacetic acid (EDTA), bottle these and water, and package screw caps and other small items for sterilization. In that case, although the preparation area should still be clean and quiet, sterile handling is not necessary, as all the items will be sterilized.

If reliable commercial media are difficult to obtain, the preparation area should be large enough to accommodate a coarse and a fine balance, a pH meter, and, if possible, an osmometer. Bench space will be required for dissolving and stirring solutions and for bottling and packaging various materials, and additional ambient and refrigerated shelf space will also be needed. If possible, an extra horizontal laminar-flow hood should be provided in the sterile area for filtering and bottling sterile liquids, and incubator space must be allocated for quality control of sterility (i.e., incubation of samples of media in broth and after plating out).

Heat-stable solutions and equipment can be autoclaved or dry-heat sterilized at the nonsterile end of the preparation area. Both streams then converge on the storage areas (*see* Fig. 4.3).

4.5.2 Washup

Washup and sterilization facilities are best situated outside the tissue culture lab, as the humidity and heat that they produce may be difficult to dissipate without increasing the airflow above desirable limits. Autoclaves, ovens, and distillation apparatus should be located in a separate room if possible (*see* Figs. 4.3, 4.4), with an efficient extraction fan. The washup area should have plenty of space for soaking glassware and space for an automatic washing machine, should you require one. There should also be plenty of bench space for handling baskets of glassware, sorting pipettes, and packaging and sealing packs for sterilization. In addition, you will need space for a pipette washer and dryer. If the sterilization facilities must be located in the tissue culture lab, place them nearest the air extract and farthest from the sterile handling area.

If you are designing a lab from scratch, then you can get sinks built in of the size that you want. Stainless steel or polypropylene are best, the former if you plan to use radioisotopes and the latter for hypochlorite disinfectants.

Sinks should be deep enough (450 mm, 18 in.) to allow manual washing and rinsing of your largest items without having to stoop too far to reach into them. They should measure about 900 mm (3 ft) from floor to rim (Fig. 4.6). It is better to be too high than too low—a short person can always stand on a raised step to reach a high sink, but a tall person will always have to bend down if the sink is too low. A raised edge around the top of the sink will contain spillage and prevent the operator from getting wet when bending over the sink. The raised edge should go around behind the taps at the back.

Each washing sink will require four taps: a single cold-water tap, a combined hot-and-cold mixer, a cold tap for a hose connection for a rinsing device, and a nonmetallic or stainless steel tap for deionized water from a reservoir above the sink (*see* Fig. 4.6). A centralized supply for deionized water should be avoided, as the piping can build up dirt and algae and is difficult to clean.

Trolleys or carts are often useful for collecting dirty glassware and redistributing fresh sterile stocks, but remember to allocate parking space for them.

4.5.3 Storage

Storage must be provided for the following items ensuring sterile and nonsterile are kept separate and clearly labeled:

(1) Sterile liquids, at room temperature (salt solutions, water, etc.), at 4°C (media), and at −20°C or −70°C (serum, trypsin, glutamine, etc.)
(2) Sterile and nonsterile glassware, including media bottles and pipettes
(3) Sterile disposable plastics (e.g., culture flasks and Petri dishes, centrifuge tubes and vials, and syringes)
(4) Screw caps, stoppers, etc., sterile and nonsterile
(5) Apparatus such as filters, sterile and nonsterile
(6) Gloves, disposal bags, etc.
(7) Liquid nitrogen to replenish freezers; the liquid nitrogen should be stored in two ways:

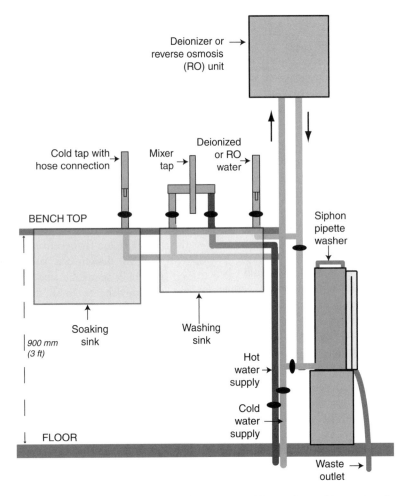

Fig. 4.6. Washing Up Sink and Pipette Washer. Suggested layout for soaking and wash up sinks, with hot, cold, and deionized water supplies. Scale 1:16.

a) in Dewars (25–50 L) under the bench, or

b) in a large storage vessel (100–150 L) on a trolley or in storage tanks (500–1000 L) permanently sited in a room of their own with adequate ventilation or, preferably, outdoors in secure, weatherproof housing (Fig. 4.7).

Note. Liquid-nitrogen storage vessels can build up contamination, so they should be kept in clean areas.

Δ ***Safety Note.*** Adequate ventilation must be provided for the room in which the nitrogen is stored and dispensed, preferably with an alarm to signify when the oxygen tension falls below safe levels. The reason for this safety measure is that filling, dispensing, and manipulating freezer stocks are accompanied by the evaporation of nitrogen, which can replace the air in the room (1 L liquid $N_2 \rightarrow \sim700$ L gaseous N_2).

(8) Cylinder storage for carbon dioxide, in separate cylinders for transferring to the laboratory as required

Δ ***Safety Note.*** The cylinders should be tethered to the wall or bench in a rack (*see* Fig. 7.2).

(9) A piped supply of CO_2 can be taken to work stations, or else the CO_2 supply can be piped from a pressurized tank of CO_2 that is replenished regularly (and must therefore be accessible to delivery vehicles). Which of the two means of storage you actually use will be based on your scale of operation and unit cost. As a rough guide, 2 to 3 people will only require a few cylinders, 10 to 15 will probably benefit from a piped supply from a bank of cylinders, and for more than 15 it will pay to have a storage tank.

Storage areas 1–6 should be within easy reach of the sterile working area. Refrigerators and freezers should be located toward the nonsterile end of the lab, as the doors and compressor fans create dust and drafts and may harbor fungal spores. Also, refrigerators and freezers require maintenance and periodic defrosting, which creates a level and kind of activity best separated from your sterile working area.

The key ideal regarding storage areas is ready access for both withdrawal and replenishment of stocks. Double-sided

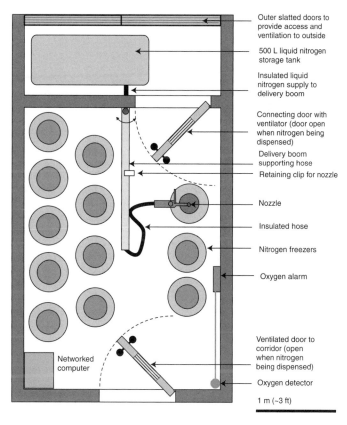

Outer slatted doors to provide access and ventilation to outside

500 L liquid nitrogen storage tank

Insulated liquid nitrogen supply to delivery boom

Connecting door with ventilator (door open when nitrogen being dispensed)

Delivery boom supporting hose

Retaining clip for nozzle

Nozzle

Insulated hose

Nitrogen freezers

Oxygen alarm

Ventilated door to corridor (open when nitrogen being dispensed)

Oxygen detector

Networked computer

1 m (~3 ft)

Fig. 4.7. *Liquid Nitrogen Store and Freezer Store.* The liquid nitrogen store is best located on an outer wall with ventilation to the outside and easy access for deliveries. If the freezer store is adjacent, freezers may be filled directly from an overhead supply line and flexible hose. Doors are left open for ventilation during filling, and a wall-mounted oxygen alarm with a low-mounted detector sounds if the oxygen level falls below a safe level.

units are useful because they may be restocked from one side and used from the other.

△ *Safety Note.* It is essential to have a lip on the edge of both sides of a shelf if the shelf is at a high level and glassware and reagents are stored on it. This prevents items being accidentally dislodged during use and when stocks are replenished.

Remember to allocate sufficient space for storage, as doing so will allow you to make bulk purchases, thereby saving money, and, at the same time, reduce the risk of running out of valuable stocks at times when they cannot be replaced. As a rough guide, you will need 200 L ($\sim$8 ft^3) of 4°C storage and 100 L ($\sim$4 ft^3) of −20°C storage per person. The volume per person increases with fewer people. Thus one person may need a 250-L (10-ft^3) refrigerator and a 150-L (6-ft^3) freezer. Of course, these figures refer to storage space only, and allowance must be made for

access and working space in walk-in cold rooms and deep freezer rooms.

In general, separate −20°C freezers are better than a walk-in −20°C room. They are easier to clean out and maintain, and they provide better backup if one unit fails. You may also wish to consider whether a cold room has any advantage over refrigerators. No doubt, a cold room will give more storage per cubic meter, but the utilization of that space is important—How easy is it to clean and defrost, and how well can space be allocated to individual users? Several independent refrigerators will occupy more space than the equivalent volume of cold room, but may be easier to manage and maintain in the event of failure.

It is also well worth considering budgeting for additional freezer and refrigerator space to allow for routine maintenance and unpredicted breakdowns.

CHAPTER 5

Equipment

5.1 REQUIREMENTS OF A TISSUE CULTURE LABORATORY

Unless unlimited funds are available, it will be necessary to prioritize the specific needs of a tissue culture laboratory: (1) essential—you cannot perform tissue culture reliably without this equipment; (2) beneficial—culture would be done better, more efficiently, quicker, or with less labor; and (3) useful—items that would improve working conditions, reduce fatigue, enable more sophisticated analyses to be made, or generally make your working environment more attractive (Table 5.1). In the following sections, items are presented, under subject headings, in order of priority, as seen by the author, but, clearly, the reader may have different priorities.

The need for a particular piece of equipment is often very subjective—a product of personal aspirations, merchandizing, technical innovation, and peer pressure. The real need is harder to define but is determined objectively by the type of work, the saving in time that the equipment would produce, the greater technical efficiency in terms of asepsis, quality of data, analytical capability, and sample requirements, the saving in time or personnel, the number of people who would use the device, the available budget and potential cost benefit, and the special requirements of your own procedures.

5.2 ASEPTIC AREA

5.2.1 Laminar-Flow Hood

It is quite possible to carry out aseptic procedures without laminar flow if you have appropriate isolated, clean accommodation, with restricted access. However, it is clear that, for most laboratories, which are typically busy and overcrowded, the simplest way to provide aseptic conditions is to use a laminar-flow hood (see Section 6.4). Usually, one hood is sufficient for two to three people. A horizontal flow hood is cheaper and provides the best sterile protection for your cultures, but it is really suitable only for preparing medium and other sterile reagents and for culturing nonprimate cells. It is also particularly suitable for dissecting nonprimate material for primary culture. For potentially hazardous materials (any primate, including human, cell lines, virus-producing cultures, radioisotopes, or carcinogenic or toxic drugs), a Class II or Class III biohazard cabinet should be used (Fig. 5.1; see also Fig. 7.4). In practice, most laboratories now use a Class II microbiological safety cabinet as standard.

Δ *Safety Note.* It is important to familiarize yourself with local and national biohazard regulations before installing equipment, as legal requirements and recommendations vary (see Section 7.8.1).

Choose a hood that is (1) large enough [a minimum working surface of 1200 mm (4 ft) wide ×600 mm (2 ft) deep]; (2) quiet, as noisy hoods are more fatiguing; (3) easily cleaned both inside the working area and below the work surface in the event of spillage; and (4) comfortable to sit at; some cabinets have awkward ducting below the work surface, which leaves no room for your knees, lights or other accessories above that strike your head, or screens that obscure your vision. The front screen should be able to be raised, lowered, or removed completely, to facilitate cleaning and handling bulky culture apparatus. Remember, however, that a biohazard cabinet will not give you, the operator, the

Culture of Animal Cells: A Manual of Basic Technique, Fifth Edition, by R. Ian Freshney
Copyright © 2005 John Wiley & Sons, Inc.

TABLE 5.1. Tissue Culture Equipment

Basic requirements	Nonessential, but beneficial	Useful additions
Laminar-flow hood (biohazard if for human cells)	Cell counter	Glassware washing machine
Incubator (humid CO_2 incubator if using open plates or dishes)	Peristaltic pump	Low-temperature ($\leq -70°C$) freezer
	Pipettor(s)	Conductivity meter
	PH meter	Osmometer
5% CO_2 cylinder (for gassing cultures)	Sterilizing oven	Polyethylene bag sealer (for packaging sterile items for long-term storage)
Liquid CO_2 cylinders, without siphon (for CO_2 incubator)	Hot room	
	Temperature recorders on sterilizing oven and autoclave and in hot room	Computer for freezer records and cell line database
Balance		
Sterilizer (autoclave, pressure cooker)	Phase-contrast, fluorescence microscope	Colony counter
Refrigerator		High-capacity centrifuge (6 × 1 L)
Freezer (for $-20°C$ storage)	Pipette plugger	Digital camera and monitor for inverted microscope(s)
Inverted microscope	Pipette drier	
Soaking bath or sink	Automatic dispenser	Time-lapse video equipment
Deep washing sink	Trolleys or carts	Cell sizer (e.g., Schärfe, Coulter)
Pipette cylinder(s)	Drying oven(s), high and low temperature	Portable temperature recorder for checking hot room or incubators
Pipette washer		
Still or water purifier	Roller racks for roller bottle culture	Plastics shredder/sterilizer
Bench centrifuge	Piped CO_2 supply from cylinder store	Controlled-rate cooler (for cell freezing)
Liquid N_2 freezer (~35 L, 1,500–3,000 ampoules)	Automatic changeover device on CO_2 cylinders	Fluorescence-activated cell sorter
		Confocal microscope
Liquid N_2 storage Dewar (~25 L)		Microtitration plate scintillation counter
Slow-cooling device for cell freezing (see Section 20.3.4)		Centrifugal elutriator centrifuge and rotor
Magnetic stirrer racks for suspension cultures		
Hemocytometer		

required protection if you remove the front screen, and you will lose efficiency in protecting the sterile environment of the cabinet.

The person who will use the hood should sit at it before purchasing in order to simulate normal use. Consider the following questions:

(1) Can you get your knees under the hood while sitting comfortably and close enough to work, with your hands at least halfway into the hood?

(2) Is there a footrest in the correct place?

(3) Are you able to see what you are doing without placing strain on your neck?

(4) Is the work surface perforated, and, if so, will that give you trouble with spillage? (A solid work surface vented at the front and back is preferable).

(5) Is the work surface easy to remove for cleaning?

(6) If the work surfaces are raised, are the edges sharp, or are they rounded so that you will not cut yourself when cleaning out the hood?

(7) Are there crevices in the work surface that might accumulate spillage and contamination? Some cabinets have sectional work surfaces that are easier to remove for cleaning but leave capillary spaces when in place.

(8) Is the lighting convenient and adequate?

(9) Will you be able to get the hood into the tissue culture laboratory?

(10) When in place, can the hood be serviced easily? (Ask the service engineer, not the salesperson!) Hoods require a minimum of 500 mm (~2 ft) of lateral separation.

(11) Will there be sufficient headroom for venting to the room or for ducting to the exterior?

(12) Will the airflow from other cabinets, the room ventilation, or independent air-conditioning units interfere with the integrity of the work space of the hood? That is, will air spill in or aerosols leak out because of turbulence? Meeting this condition will require a minimum of 2000 mm (~7 ft), and preferably 3000 mm (10 ft), of face-to-face separation and correct testing by an engineer with experience in microbiological safety cabinets.

5.2.2 Pipette Cylinders

Pipette cylinders (sometimes known as pipette hods) should be made from polypropylene and should be freestanding and distributed around the lab, one per workstation, with sufficient numbers in reserve to allow full cylinders to stand for 2 h in disinfectant (see Section 7.8.5) before washing.

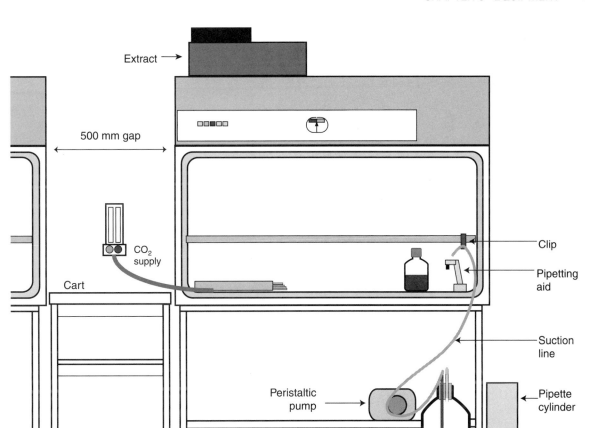

Extract →

500 mm gap

CO₂ supply

Cart

Clip

Pipetting aid

Suction line

Pipette cylinder

Peristaltic pump

Receiver flask

Footswitch →

Fig. 5.1. *Laminar-Flow Hood.* A peristaltic pump, connected to a receiver vessel, is shown on the right side below the hood, with a foot switch to activate the pump. The suction line from the pump leads to the work area, and a delivery tube from a gas mixer provides a supply of CO_2 mixed in air.

5.2.3 Aspiration Pump

Suction from a peristaltic pump may be used to remove spent medium or other reagents from a culture flask (Fig. 5.2a), and the effluent can be collected directly into disinfectant (*see* Section 7.8.5) in a vented container (Fig. 5.2b), with minimal risk of discharging aerosol into the atmosphere if the vent carries a cotton plug or micropore filter. The inlet line should extend further below the stopper than the outlet, by at least 5 cm (2 in.), so that waste does not splash back into the vent. The pump tubing should be checked regularly for wear, and the pump should be operated by a self-canceling foot switch.

A vacuum pump, similar to that supplied for sterile filtration, may be used instead of a peristaltic pump. If necessary, the same pump could serve both purposes; however, a trap will be required to avoid the risk of waste entering the pump. The effluent should be collected in a reservoir (*see* Figs. 5.2b, 5.3) and a disinfectant such as hypochlorite added (*see* Section 7.8.5) when work is finished

and left for at least 2 h before the reservoir is emptied. A hydrophobic micropore filter (e.g., Pall Gelman) and a second trap should be placed in the line to the pump to prevent fluid or aerosol from being carried over. Do not draw air through a pump from a reservoir containing hypochlorite, as the free chlorine will corrode the pump and could be toxic. Also, avoid vacuum lines; if they become contaminated with fluids, they can be very difficult to clean out.

To keep effluent from running back into a flask, always switch on the pump before inserting a pipette in the tubing. (*see* Section 6.5).

5.2.4 Service Carts

It is useful to locate items for use at the laminar flow hoods on moveable carts. These carts can conveniently fill the lower space between adjacent hoods and are easily removed for maintenance of the hoods. They can also be used to carry materials to and from the hoods and basic items restocked by service staff. Larger carts are useful for clearing soiled

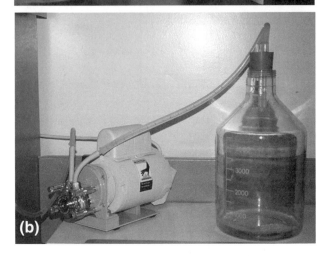

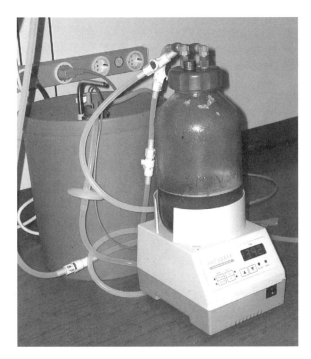

Fig. 5.3. Integra Vacuum Pump. Suction pump and receiver for removing fluids from flasks; available from Integra. (Courtesy of Norbert Fusenig, Deutches Krebsforschungzentrum, Heidelberg).

Fig. 5.2. Withdrawing Medium by Suction. (a) Pipette connected via tube to a peristaltic pump being used to remove medium from a flask. (b) Peristaltic pump on the suction line from the hood leading to a waste receiver.

glassware and used items from the aseptic area to the wash-up. They can be parked at a convenient location (*see* Figs. 4.3, 4.4, 5.1).

5.2.5   Inverted Microscope

It cannot be overemphasized that, despite considerable progress in the quantitative analysis of cultured cells and molecular probes, it is still vital to look at cultures regularly. A morphological change is often the first sign of deterioration in a culture (*see* Section 13.6.1 and Fig. 13.1), and the characteristic pattern of microbiological infection (*see* Section 19.4.1 and Fig. 19.1) is easily recognized.

A simple inverted microscope is essential (Fig. 5.4). Make certain that the stage is large enough to accommodate large roller bottles between it and the condenser (*see* Section 26.2.3) in case you should require such bottles. Many simple and inexpensive inverted microscopes are available on the market, and it is worth getting one with a phototube for digital recording or viewing. If you foresee the need for photographing living cultures, then you should invest in a microscope with good-quality optics and a long-working-distance phase-contrast condenser and objectives,

with provision for a CCD camera and monitor (*see* Fig. 5.14). The increasing use of green fluorescent protein (GFP) for tagging live cells means that fluorescence optics may be necessary.

The Nikon marking ring is a useful accessory to the inverted microscope. This device is inserted in the nosepiece in place of an objective and can be used to mark the underside of a dish in which an interesting colony or patch of cells is located. The colony can then be picked (*see* Section 14.4 and Figs. 14.8, 14.9) or the development of a particularly interesting area in a culture followed.

5.2.6   Centrifuge

Periodically, cell suspensions require centrifugation to increase the concentration of cells or to wash off a reagent. A small bench-top centrifuge, preferably with proportionally controlled braking, is sufficient for most purposes. Cells sediment satisfactorily at 80–100 g; higher gravity may cause damage and promote agglutination of the pellet. A large-capacity refrigerated centrifuge, say, 4 × 1 L or 6 × 1 L, will be required if large-scale suspension cultures (*see* Section 26.1) are contemplated.

5.2.7   Sterile Liquid Handling—Pipetting and Dispensing

Removal of fluids. Fluids can be removed simply by pipette, but the process can be easier and faster if a wide-bore pipette is attached to a vacuum pump or vacuum line,

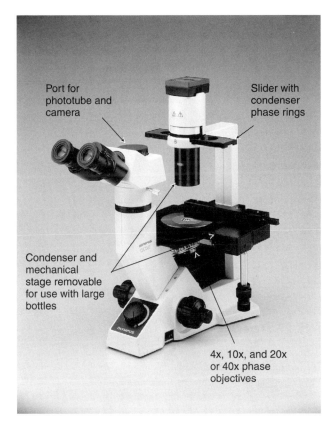

Fig. 5.4. Inverted Microscope. Olympus CKX41 inverted microscope fitted with phase-contrast optics and trinocular head with port for attaching a digital camera. (Photo courtesy of Olympus, UK, Ltd).

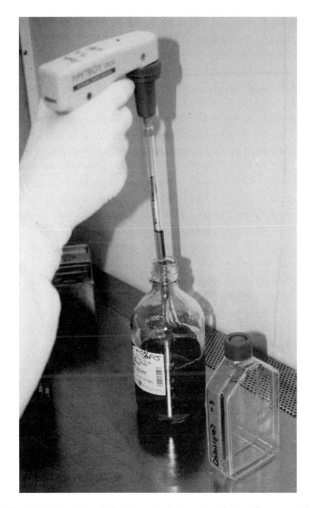

Fig. 5.5. Pipetting Aid. Motorized pipetting device for use with conventional graduated pipettes.

Fig. 5.6. Pipettor. Variable-volume pipetting device. Also available in fixed volume. The pipettor is not itself sterilized but is used with sterilized plastic tips.

with suitable reservoirs to collect the effluent and prevent contamination of the pump, or with the use of a simple peristaltic pump discharging into a reservoir with disinfectant (*see* Section 5.2.3).

Pipetting aids. Simple pipetting is one of the most frequent tasks required in the routine handling of cultures. Although a rubber bulb or other proprietary pipetting device is cheap and simple to use, speed, accuracy, and reproducibility are greatly enhanced by a motorized pipetting aid (Fig. 5.5), which may be obtained with a separate or built-in pump and can be mains operated or rechargeable (*see* Appendix II). The major determinants in choosing a pipetting aid are the weight and feel of the instrument during continuous use, and it is best to try one out before purchasing it. Pipetting aids usually have a filter at the pipette insert to minimize the transfer of contaminants. Some filters are disposable, and some are reusable after resterilization (*see* Fig. 7.2 for the proper method of inserting a pipette into a pipetting device).

Pipettors. These devices originated from micropipettes marketed by Eppendorf and used for dispensing 10–200 μL. As the working range now extends up to 5 mL, the term

"micropipette" is not always appropriate, and the instrument is more commonly called a *pipettor* (Fig. 5.6). Only the tip needs to be sterile, but the length of the tip then limits the size of vessels used. If a sterile fluid is withdrawn from a container with a pipettor, the nonsterile stem must not touch the sides of the container. Reagents of 10–20 mL in volume may be sampled in 5-μL to 1-mL volumes from a universal container or in 5- to 200-μL volumes from a bijou bottle. Eppendorf-type tubes may also be used for volumes of 100 μL to 1 mL, but they should be of the shrouded-cap variety and will require sterilization.

It is assumed that the inside of a pipettor is sterile or does not displace enough air for this to matter. However, it clearly does matter in certain situations. For example, if you are performing serial subculture of a stock cell line (as opposed to a short-term experiment with cells that ultimately will be sampled or discarded, but *not* propagated), the security of the cell line is paramount, and you must use either a plugged regular glass pipette or a disposable pipette with a sterile length that is sufficient to reach into the vessel that you are sampling. If you are using a small enough container to preclude contact from nonsterile parts of a pipettor, then it is permissible to use a pipettor, provided that the tip has a filter that prevents cross-contamination and minimizes microbial contamination. Otherwise, you run the risk of microbial contamination from nonsterile parts of the pipette or, more subtle and potentially more serious, cross-contamination from aerosol or fluid drawn up into the stem of the pipettor.

Routine subculture, which should be rapid and secure from microbial and cross-contamination but need not be very accurate, is best performed with conventional glass or disposable plastic pipettes. Experimental work, which must be accurate but should not involve stock propagation of the cells used, may benefit from using pipettors.

Tips can be bought loose and can be packaged and sterilized in the laboratory, or they can be bought already sterile and mounted in racks ready for use. Loose tips are cheaper, but more labor intensive. Prepacked tips are much more convenient, but considerably more expensive. Some racks can be refilled and resterilized, which presents a reasonable compromise.

Large-volume dispensing. When culture vessels exceed 100 mL in the volume of medium, a different approach to fluid delivery must be adopted. If only a few flasks are involved, a 100-mL pipette (BD Biosciences) or a graduated bottle (Fig. 5.7) or bag (Sigma) may be quite adequate, but if very large volumes (~500 mL) or a large number of high-volume replicates are required, then a peristaltic pump will probably be necessary. Single fluid transfers of large volumes (10–10,000 L) are usually achieved by preparing the medium in a sealed pressure vessel and then displacing it by positive pressure into the culture vessel (Alfa Laval). It is possible to dispense large volumes by pouring, but this should be restricted to a single action with a premeasured volume (*see* Section 6.3.6).

Repetitive dispensing. The traditional repetitive dispenser was the type known as a Cornwall syringe, in which liquid is alternately taken into a syringe via one tube and expelled via another, using a simple two-way valve. The syringe plunger is spring loaded, so the whole procedure is semiautomatic and repetitive. There are numerous variants of this type of dispenser, many of which are still in regular use. The major problems arise from the valves sticking, but

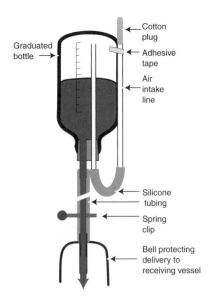

Fig. 5.7. Graduated Bottle Dispenser. Two-hole stopper inserted in the neck of a graduated bottle with a delivery line connected to a dispensing bell, a spring clip on the line, and an inlet line for balancing air. The stopper may be sterilized without the bottle and inserted into any standard bottle containing a medium as required. (From an original design by Dr. John Paul).

this can be minimized by avoiding the drying cycle after autoclaving and flushing the syringe out with medium or a salt solution before and after use. Small-volume repetitive dispensing can also be achieved by incremental movement of the piston in a syringe (Fig. 5.8).

A peristaltic pump can also be used for repetitive serial deliveries and has the advantage that it may be activated via a foot switch, leaving the hands free (Fig. 5.9). Care must be taken setting up such devices to avoid contaminating the tubing at the reservoir and delivery ends. In general, they are worthwhile only if a very large number of flasks is being handled. Automated pipetting provided by a peristaltic pump can be controlled in small increments. In addition, only the delivery tube is autoclaved, and accuracy and reproducibility can be maintained at high levels over a range from 10 to 100 mL. A number of delivery tubes may be sterilized and held in stock, allowing a quick changeover in the event of accidental contamination or change in cell type or reagent.

Automation. Many attempts have been made to automate changing fluids in tissue culture, but few devices or

Fig. 5.8. Repette. Stepping dispenser operated by incremental movement of piston in syringe, activated by thumb button (Jencons).

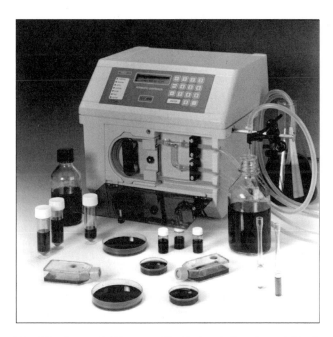

Fig. 5.9. Automatic Dispenser. The Perimatic Premier, suitable for repetitive dispensing and dilution in the 1- to 1000-mL range. If the device is used for sterile operations, only the delivery tube needs to be autoclaved.

systems have the flexibility required for general use. When a standard production system is in use, automatic feeding may be useful, but the time taken in setting it up, the modifications that may be needed if the system changes, and the overriding importance of complete sterility have deterred most laboratories from investing the necessary time and funds. The introduction of microtitration plates (*see* Fig. 8.2) has brought with it many automated dispensers, plate readers, and other accessories (Figs. 5.10, 5.11, 5.12). Transfer devices using perforated trays or multipoint pipettes make it easier to seed from one plate to another, and plate mixers and centrifuge carriers also are available. The range of equipment is so extensive that it cannot be covered here, and the appropriate trade catalogs should be consulted (*see* Appendix II). Two items worthy of note are the Rainin programmable single or multitip pipettor and the Corning Costar Transtar media transfer and replica plating device (Fig. 5.13).

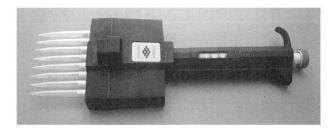

Fig. 5.10. Multipoint Pipettor. Pipettor with manifold to take 8 plastic tips. Also available for 4 and 12.

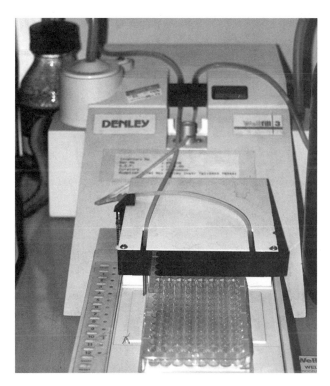

Fig. 5.11. Plate Filler. Automatic filling device for loading microtitration plates. The photo shows a nonsterile application, but the device can be used in sterile applications. (Courtesy of Centre for Oncology and Applied Pharmacology, Glasgow University).

Robotic systems (e.g., from Packard) are now being introduced into tissue culture assays as a natural extension of the microtitration system. Robots provide totally automated procedures but also allow for reprogramming if the analytical approach changes.

Choice of system. Whether a simple manual system or a complex automated one is chosen, the choice is governed mainly by five criteria:

(1) Ease of use and ergonomic efficiency
(2) Cost relative to time saved and increased efficiency
(3) Accuracy and reproducibility in serial or parallel delivery
(4) Ease of sterilization and effect on accuracy and reproducibility
(5) Mechanical, electrical, chemical, biological, and radiological safety

Δ ***Safety Note.*** Most pipetting devices tend to expel fluid at a higher rate than during normal manual operation and consequently have a greater propensity to generate aerosols. This must be kept in mind when using substances that are potentially hazardous.

5.2.8 Cell Counter

A cell counter (*see* Fig. 21.2) is a great advantage when more than two or three cell lines are carried and is essential

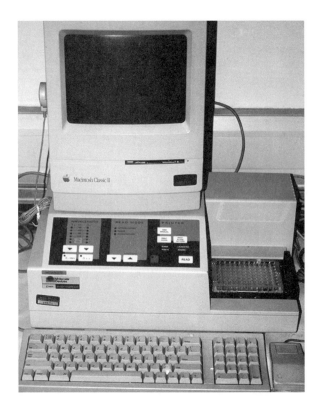

Fig. 5.12. Plate Reader. Densitometer for measuring absorbance of each well; some models also measure fluorescence. (Courtesy of Centre for Oncology and Applied Pharmacology, Glasgow University.)

for precise quantitative growth kinetics. Several companies now market models ranging in sophistication from simple particle counting up to automated cell counting and size analysis. Beckman Coulter and Schärfe market machines suitable for routine counting (*see also* Section 21.1.2). Automatic dispensers (*see also* Section 5.2.7) are useful for delivering nonsterile counting fluid, but, when using bottletop dispensers for cell counting, you should let the fluid settle for a few minutes before counting, as small bubbles are often generated by rapid dispensing, and these will be counted by the cell counter.

Cell sizing. Most midrange or top-of-the-range cell counters (Schärfe, Beckman Coulter) (*see* Fig. 21.2) will provide cell size analysis and the possibility of downloading data to a PC, directly or via a network.

5.2.9 CCD Camera and Monitor

Since the advent of cheap microcircuits, television cameras and monitors have become a valuable aid to the discussion of cultures and the training of new staff or students (Fig. 5.14). Choose a high-resolution, but not high-sensitivity, camera, as the standard camera sensitivity is usually sufficient, and high sensitivity may lead to over-illumination. Black and white usually gives better resolution and is quite adequate for phase-contrast observation of living cultures. Color is preferable for

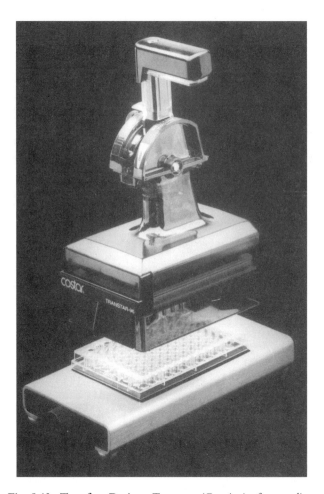

Fig. 5.13. Transfer Device. Transtar (Corning) for seeding, transferring medium, replica plating, and other similar manipulations with microtitration plates, enabling simultaneous handling of all 96 wells. (Reproduced by permission of Corning).

Fig. 5.14. Closed-Circuit Television. CCD camera attached to Zeiss Axiovert inverted microscope. Can be used for direct printing or for time-lapse studies when linked to a video recorder (*see* Section 27.3). Microinjection port on right. (Courtesy of Beatson Institute).

fixed and stained specimens. If you will be discussing cultures with an assistant or one or two colleagues, a 300- or 400-mm (12- or 15-in.) monitor is adequate and gives good definition, but if you are teaching a group of 10 or more students, then go for a 500- or 550-mm (19- or 21-in.) monitor.

High-resolution charge-coupled-device (CCD) video cameras can be used to record and digitize images for subsequent analysis and publication; the addition of a video recorder will allow real-time or time-lapse recordings (*see* Section 27.3).

5.2.10 Dissecting Microscope

Dissection of small pieces of tissue (e.g., embryonic organs or tissue from smaller invertebrates) will require a dissecting microscope (Nikon, Olympus, Leica). A dissecting microscope is also useful for counting monolayer colonies and essential for counting and picking small colonies in agar.

5.3 INCUBATION

5.3.1 Incubator

If a hot room (*see* Section 4.4.2) is not available, it may be necessary to buy an equivalent dry incubator. Even with a hot room, it is sometimes convenient to have another incubator close to the hood for trypsinization. The incubator should be large enough, $\sim$50–200 L (1.5–6 ft^3) per person, and should have forced-air circulation, temperature control to within $\pm$0.2°C, and a safety thermostat that cuts off if the incubator overheats or, better, that regulates the incubator if the first thermostat fails. The incubator should be resistant to corrosion (e.g., stainless steel, although anodized aluminum is acceptable for a dry incubator) and easily cleaned. A double chamber, or two incubators stacked, one above the other, independently regulated, is preferable to one large incubator because it can accommodate more cultures with better temperature control, and if one half fails or needs to be cleaned, the other can still be used.

Many incubators have a heated water jacket to distribute heat evenly around the cabinet, thus avoiding the formation of cold spots. These incubators also hold their temperature longer in the event of a heater failure or cut in power. However, new high-efficiency insulation and diffuse surface heater elements have all but eliminated the need for a water jacket and make moving the incubator much simpler. (A water jacket generally needs to be emptied if the incubator is to be moved.)

Incubator shelving is usually perforated to facilitate the circulation of air. However, the perforations can lead to irregularities in cell distribution in monolayer cultures, with variations in cell density following the pattern of spacing on the shelves. The variations may be due to convection currents generated over points of contact relative to holes in the shelf, or they may be related to areas that cool down more quickly when the door is opened. Although no problem may arise in

routine maintenance, flasks and dishes should be placed on an insulated tile or metal tray in experiments in which uniform density is important.

5.3.2 Humid CO$_2$ Incubator

Although cultures can be incubated in sealed flasks in a regular dry incubator or a hot room, some vessels, e.g., Petri dishes or multiwell plates, require a controlled atmosphere with high humidity and elevated CO$_2$ tension. The cheapest way of controlling the gas phase is to place the cultures in a plastic box, or chamber (Bellco, MP Biomedicals): Gas the container with the correct CO$_2$ mixture and then seal it. If the container is not completely filled with dishes, include an open dish of water to increase the humidity inside the chamber.

CO$_2$ incubators (Fig. 5.15) are more expensive, but their ease of use and superior control of CO$_2$ tension and temperature (anaerobic jars and desiccators take longer to warm up) justify the expenditure. A controlled atmosphere is achieved by using a humidifying tray (Fig 5.16) and controlling the CO$_2$ tension with a CO$_2$-monitoring device, which draws air from the incubator into a sample chamber, determines the concentration of CO$_2$, and injects pure CO$_2$ into the incubator to make up any deficiency. Air is circulated around the incubator by natural convection or by using a fan to keep both the CO$_2$ level and the temperature uniform. It is claimed that fan-circulated incubators recover faster after

Fig. 5.15. CO$_2$ Incubator. CO$_2$ incubator (Forma) with door open showing flasks, multiwell plates and boxed Petri dishes. (*see also* Fig. 5.16).

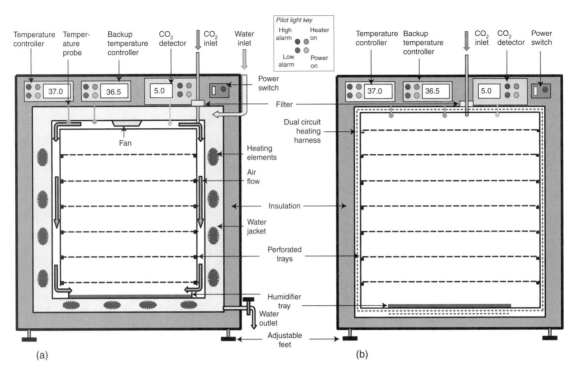

Fig. 5.16. CO₂ Incubator Design. Front view of control panel and section of chamber of two stylized humid CO₂ incubators. (a) Water-jacketed with circulating fan. (b) Dry-walled with no circulating fan (not representative of any particular makes).

opening, although natural convection incubators can still have a quick recovery and greatly reduce the risks of contamination. Dry, heated wall incubators also encourage less fungal contamination on the walls, as the walls tend to remain dry, even at high relative humidity. Some CO₂ controllers need to be calibrated every few months, but the use of gold wire or infrared detectors minimizes drift and many models reset the zero of the CO₂ detector automatically.

The size of incubator required will depend on usage, both the numbers of people using it and the types of cultures. Five people using only microtitration plates could have 1000 plates (~100,000 individual cultures) or 10 experiments each in a modest-sized incubator, while one person doing cell cloning could fill one shelf with one or two experiments. Flask cultures, especially large flasks, are not an economical use of CO₂ incubators. They are better incubated in a regular incubator or hot room. If CO₂ is required, flasks can be gassed from a cylinder or CO₂ supply.

Frequent cleaning of incubators—particularly humidified ones—is essential (*see* Section 19.1.4), so the interior should dismantle readily without leaving inaccessible crevices or corners. Flasks or dishes, or boxes containing them, that are taken from the incubator to the laminar-flow hood should be swabbed with alcohol before being opened (*see* Section 6.3.1).

5.3.3 Temperature Recorder

A recording thermometer with ranges from below −50°C to about +200°C will enable you to monitor frozen storage, the freezing of cells, incubators, and sterilizing ovens with one instrument fitted with a resistance thermometer or thermocouple with a long Teflon-coated lead.

Ovens, incubators, and hot rooms should be monitored regularly for uniformity and stability of temperature control. Recording thermometers should be permanently fixed into the hot room, sterilizing oven, and autoclave, and dated records should be kept to check regularly for abnormal behavior, particularly in the event of a problem arising.

5.3.4 Roller Racks

Roller racks are used to scale up monolayer culture (*see* Section 26.2.3). The choice of apparatus is determined by the scale (i.e., the size and number of bottles to be rolled). The scale may be calculated from the number of cells required, the maximum attainable cell density, and the surface area of the bottles (*see* Table 8.1). A large number of small bottles gives the highest surface area but tends to be more labor intensive in handling, so a usual compromise is bottles around 125 mm (5 in.) in diameter and various lengths from 150 to 500 mm (6–20 in.). The length of the bottle will determine the maximum yield but is limited by the size of the rack; the height of the rack will determine the number of tiers (i.e., rows) of bottles. Although it is cheaper to buy a larger rack than several small ones, the latter alternative (1) allows you to build up your racks gradually (having confirmed that the system works), (2) can be easier to locate in a hot room, and (3) will still allow you to operate if one rack requires maintenance. Bellco or Thermo Electron bench-top models may be satisfactory for smaller scale activities, and Bellco, Integra, and New Brunswick Scientific supply larger racks.

5.3.5 Magnetic Stirrer

Certain specific requirements apply to magnetic stirrers. A rapid stirring action for dissolving chemicals is available with any stirrer, but for the stirrer to be used for enzymatic tissue disaggregation (*see* Section 12.3.3) or suspension culture (*see* Sections 13.7.4, 13.7.5), (1) the motor should not heat the culture (use the rotating-field type of drive or a belt drive from an external motor); (2) the speed must be controlled down to 50 rpm; (3) the torque at low rpm should still be capable of stirring up to 10 L of fluid; (4) the device should be capable of maintaining several cultures simultaneously; (5) each stirrer position should be individually controlled; and (6) a readout of rpm should appear for each position. It is preferable to have a dedicated stirrer for culture work (e.g., Techne or Bellco).

5.4 PREPARATION AND STERILIZATION

5.4.1 Washup

Soaking baths or sinks. Soaking baths or sinks should be deep enough so that all your glassware (except pipettes and the largest bottles) can be totally immersed in detergent during soaking, but not so deep that the weight of the glass is sufficient to break smaller items at the bottom. A sink that is 400 mm (15 in.) wide ×600 mm (24 in.) long ×300 mm (12 in.) deep is about right (*see* Fig. 4.6).

Pipette washer. Reusable glass pipettes are easily washed in a standard siphon-type washer. (*see* Section 11.3.2 and Fig. 11.5). The washer should be placed just above floor level rather than on the bench, to avoid awkward lifting of the pipettes, and should be connected to the deionized water supply, as well as the regular cold water supply, so that the final few rinses can be done in deionized water. If possible, a simple changeover valve should be incorporated into the deionized water feed line (*see* Fig. 4.6).

Pipette drier. If a stainless-steel basket is used in the washer, pipettes may subsequently be transferred directly to an electric drier. Alternatively, pipettes can be dried on a rack or in a regular drying oven.

5.4.2 Water Purifier

Purified water is required for rinsing glassware, dissolving powdered media, and diluting concentrates. The first of these purposes is usually satisfied by deionized or reverse-osmosis water, but the second and third require ultrapure water (UPW), which demands a three- or four-stage process (Fig. 5.17; *see also* Fig. 11.10). The important principle is that

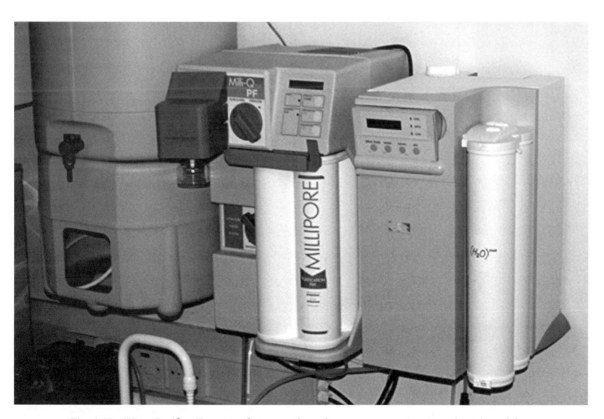

Fig. 5.17. Water Purifier. Tap water first passes through a reverse-osmosis unit on the right and then goes to the storage tank on the left. It then passes through carbon filtration and deionization (center unit) before being collected via a micropore filter (Millipore Milli-Q). (Courtesy of Beatson Institute).

each stage be qualitatively different; reverse osmosis may be followed by charcoal filtration, deionization, and micropore filtration (e.g., via a sterilizing filter; *see* Fig. 11.11), or distillation (with a silica-sheathed element) may be substituted for the first stage. Reverse osmosis is cheaper if you pay the fuel bills; if you do not, distillation is better and more likely to give a sterile product. If reverse osmosis is used, the type of cartridge should be chosen to suit the pH of the water supply, as some membranes can become porous in extreme pH conditions (check with supplier).

The deionizer should have a conductivity meter monitoring the effluent, to indicate when the cartridge must be changed. Other cartridges should be dated and replaced according to the manufacturer's instructions. A total organic carbon (TOC) meter (Millipore) can be used to monitor colloids.

Purified water should not be stored but should be recycled through the apparatus continually to minimize infection with algae or other microorganisms. Any tubing or reservoirs in the system should be checked regularly (every 3 months or so) for algae, cleaned out with hypochlorite and detergent (e.g., Clorox or Chloros), and thoroughly rinsed in purified water before reuse.

Water is the simplest, but probably the most critical, constituent of all media and reagents, particularly serum-free media (*see* Chapter 10). A good quality-control measure is to check the plating efficiency of a sensitive cell line in medium made up with the water (*see* Section 11.6.3 and Protocol 21.10) at regular intervals.

5.4.3 Sterilizing and Drying Ovens

Although all sterilizing can be done in an autoclave, it is preferable to sterilize pipettes and other glassware by dry heat, avoiding the possibility of both chemical contamination from steam condensate and corrosion of pipette cans. Such sterilization, however, will require a high-temperature (160–180°C) fan-powered oven to ensure even heating throughout the load. As with autoclaves, do not get an oven that is too big for the amount or size of glassware that you use. It is better to use two small ovens than one big one; heating is easier, more uniform, quicker, and more economical when only a little glassware is being used. You are also better protected during breakdowns.

5.4.4 Steam Sterilizer (Autoclave)

The simplest and cheapest sterilizer is a domestic pressure cooker that generates 100 kPa (1 atm, 15 lb/in.^{2}) above ambient pressure. Alternatively, a bench-top autoclave (Fig. 5.18) gives automatic programming and safety locking. A larger, freestanding model with a programmable timer, a choice of pre- and poststerilization evacuation, and temperature recording (Fig. 5.19) has a greater capacity, provides more flexibility, and offers the opportunity to comply with good laboratory practice (GLP).

A "wet" cycle (water, salt solutions, etc.) is performed without evacuation of the chamber before or after sterilization. Dry items (instruments, swabs, screw caps, etc.) require that the chamber be evacuated before sterilization or the air replaced by downward displacement, to allow

Fig. 5.18. Bench-Top Autoclave. Simple, top-loading autoclave from Prestige Medical; left with lid closed, right with lid removed for filling. (Courtesy of Beatson Institute.)

Fig. 5.19. Freestanding Autoclave. Medium-sized (300 L; 10 ft³) laboratory autoclave with square chamber for maximum load. The recorder on the top console is connected to a probe in the bottle in the center of the load. (Courtesy of Beatson Institute).

efficient access of hot steam. The chamber should also be evacuated *after* sterilization, to remove steam and promote subsequent drying; otherwise the articles will emerge wet, leaving a trace of contamination from the condensate on drying. To minimize this risk when a "postvac" cycle is not available, always use deionized or reverse-osmosis water to supply the autoclave. If you require a high sterilization capacity (>300 L, 9 ft³), buy two medium-sized autoclaves rather than one large one, so that during routine maintenance and accidental breakdowns you still have one functioning machine. Furthermore, a medium-sized machine will heat up and cool down more quickly and can be used more economically for small loads. Leave sufficient space around the

sterilizer for maintenance and ventilation, provide adequate air extraction to remove heat and steam, and ensure that a suitable drain is available for condensate.

Most small autoclaves come with their own steam generator (calorifier), but larger machines may have a self-contained steam generator, a separate steam generator, or the facility to use a steam line. If high-pressure steam is available on-line, that will be the cheapest and simplest method of heating and pressurizing the autoclave; if not, it is best to purchase a sterilizer complete with its own self-contained steam generator. Such a sterilizer will be cheaper to install and easier to move. With the largest machines, you may not have a choice, as they are frequently offered only with a separate generator. In that case, you will need to allow space for the generator at the planning stage.

5.4.5 Balances

Although most laboratories obtain media that are already prepared, it may be cheaper to prepare some reagents in house. Doing so will require a balance (an electronic one with automatic tare is best) capable of weighing items from around 10 mg up to 100 g or even 1 kg, depending on the scale of the operation. If you are a service provider, it is often preferable to prepare large quantities, sometimes 10× concentrated, so the amounts to be weighed can be quite high. It may prove better to buy two balances, coarse and fine, as the outlay may be similar and the convenience and accuracy are increased.

5.4.6 pH Meter

A simple pH meter for the preparation of media and special reagents is a useful addition to the tissue culture area. Although a phenol red indicator is sufficient for monitoring pH in most solutions, a pH meter will be required when phenol red cannot be used (e.g., in preparing cultures for fluorescence assays and in preparing stock solutions).

5.4.7 Hot Plate Magnetic Stirrer

In addition to the ambient temperature stirrers used for suspension cultures and trypsinization, it may be desirable to have a magnetic stirrer with a hot plate to accelerate the dissolution of some reagents. Placing a solution on a stirrer in the hot room may suffice, but leaving solutions stirring at 37°C for extended periods can lead to microbial growth, so stable solutions are best stirred at a higher temperature for a shorter time.

5.4.8 Automatic Dispensers

Bottle-top dispensers (Fig. 5.20) are suitable for volumes up to about 50 mL; above that level, gravity dispensing from a reservoir, which may be a graduated bottle (*see* Fig. 5.7) or plastic bag, is acceptable if accuracy is not critical. If the volume dispensed is more critical, then a peristaltic pump (*see* Fig. 5.9) is preferable. The duration of dispensing and diameter of tube determine the volume and accuracy of the dispenser. A long dispense cycle with a narrow delivery tube will be more accurate, but a wide-bore tube will be faster.

5.4.9 Conductivity Meter

When solutions are prepared in the laboratory, it is essential to perform quality-control measures to guard against errors (*see* Section 11.6.1). A simple check of ionic concentration can be made with a conductivity meter against a known standard, such as normal saline (0.15 M).

5.4.10 Osmometer

One of the most important physical properties of a culture medium, and one that is often difficult to predict, is the

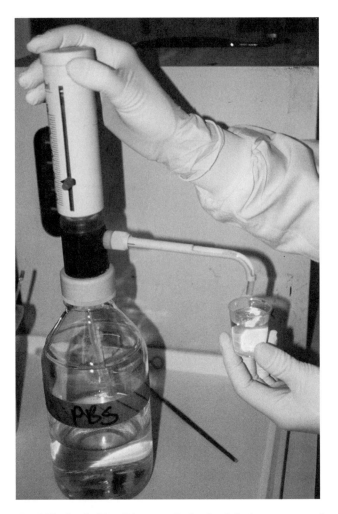

Fig. 5.20. Bottle-Top Dispenser. Spring-loaded piston connected to a two-way valve, alternately drawing a preset volume from the reservoir and dispensing through a side arm. The device is useful for repetitive dispensing of nonsterile solutions such as PBSA for cell counting. Some models can be autoclaved for sterile use.

osmolality. Although the conductivity is controlled by the concentration of ionized molecules, nonionized particles can also contribute to the osmolality. An osmometer (*see* Fig. 9.1) is therefore a useful accessory to check solutions as they are made up, to adjust new formulations, or to compensate for the addition of reagents to the medium. Osmometers usually work by depressing the freezing point of a medium or elevating its vapor pressure. Choose one with a low sample volume (≤1 mL), because you may want to measure a valuable or scarce reagent on occasion, and the accuracy (±10 mosmol/kg) may be less important than the value or scarcity of the reagent.

5.4.11 Glassware Washing Machine

Probably the best way of producing clean glassware is to have a reliable person do your washing up, but when the amount gets to be too great, it may be worth considering the purchase of an automatic washing machine (Fig. 5.21). Several of these

Fig. 5.21. Glassware Washing Machine. Glassware is placed on individual jets, which ensures thorough washing and rinsing (Betterbuilt). (Courtesy of Beatson Institute).

are currently available that are quite satisfactory. Look for the following principles of operation:

(1) A choice of racks with individual spigots over which you can place bottles, flasks, etc. Open vessels such as Petri dishes and beakers will wash satisfactorily in a whirling-arm spray, but narrow-necked vessels need individual jets. The jets should have a cushion or mat at the base to protect the neck of the bottle from chipping.

(2) The pump that forces the water through the jets should have a high delivery pressure, requiring around 2–5 hp, depending on the size of the machine.

(3) Water for washing should be heated to a minimum of 80°C.

(4) There should be a facility for a deionized water rinse at the end of the cycle. This should be heated to 50–60°C; otherwise the glassware may crack after the hot wash and rinse. The rinse should be delivered as a continuous flush, discarded, and not recycled. If recycling is unavoidable, a minimum of three separate deionized rinses will be required.

(5) Preferably, rinse water from the end of the previous wash cycle should be discarded and not retained for the prerinse of the next wash. Discarding the rinse water reduces the risk of chemical carryover when the machine is used for chemical and radioisotope washup.

(6) The machine should be lined with stainless steel and plumbed with stainless steel or nylon piping.

(7) If possible, a glassware drier should be chosen that will accept the same racks as the washer (*see* Fig. 5.21), so that they may be transferred directly, without unloading, via a suitably designed trolley.

5.5 STORAGE

5.5.1 Refrigerators and Freezers

Usually, a domestic refrigerator or freezer is quite efficient and cheaper than special laboratory equipment. Domestic refrigerators are available without a freezer compartment ("larder refrigerators"), giving more space and eliminating the need for defrosting. However, if you require 400 L (12 ft^3) or more storage (*see* Section 4.5.3), a large hospital, blood bank, or catering refrigerator may be better.

If space is available and the number of people using tissue culture is more than three or four, it is worth considering the installation of a cold room, which is more economical in terms of space than several separate refrigerators and is also easier to access. The walls should be smooth and easily cleaned, and the racking should be on castors to facilitate moving for cleaning. Cold rooms should be cleaned out regularly to eliminate old stock, and the walls and shelves should be washed with an antiseptic detergent to minimize fungal contamination.

Similar advice applies to freezers—several inexpensive domestic freezers will be cheaper, and just as effective as a specialized laboratory freezer. Most tissue culture reagents will keep satisfactorily at −20°C, so an ultradeep freeze is not essential. A deep-freeze room is not recommended—it is very difficult and unpleasant to clear out, and it creates severe problems in regard to relocating the contents if extensive maintenance is required.

Although autodefrost freezers may be bad for some reagents (enzymes, antibiotics, etc.), they are quite useful for most tissue culture stocks, whose bulk precludes major temperature fluctuations and whose nature is less sensitive to severe cryogenic damage. Conceivably, serum could deteriorate during oscillations in the temperature of an autodefrost freezer, but in practice it does not. Many of the essential constituents of serum are small proteins, polypeptides, and simpler organic and inorganic compounds that may be insensitive to cryogenic damage, particularly if solutions are stored in volumes ≥100 mL.

5.5.2 Cryostorage Containers

Details of cryostorage containers and advice on selection are given in Chapter 20 (*see* Section 20.3.5). In brief, the choice depends on the size and the type of storage system required. For a small laboratory, a 35-L freezer with a narrow neck and storage in canes and canisters (*see* Figs. 20.6a, 20.7a) or in drawers in a rack system (*see* Figs. 20.6d, 20.7c) should hold about 500–1000 ampoules (*see* Appendix II). Larger freezers will hold >10,000 ampoules and include models with walls perfused with liquid nitrogen, cutting down on nitrogen consumption, and providing safe storage without any liquid nitrogen in the storage chamber itself. It is important to establish, however, if selecting the perfused wall type of freezer, that adequate precautions have been taken to ensure that no particulate material, water, or water vapor can enter the perfusion system, as blockages can be difficult, or even impossible, to clear.

An appropriate storage vessel should also be purchased to enable a backup supply of liquid nitrogen to be held. The size of the vessel depends on (1) the size of the freezer, (2) the frequency and reliability of delivery of liquid nitrogen, and (3) the rate of evaporation of the liquid nitrogen. A 35-L narrow-necked freezer using 5–10 L/wk will only require a 25-L Dewar as long as a regular supply is available. Larger freezers are best supplied on-line from a dedicated storage tank (e.g., a 160-L storage vessel linked to a 320-L freezer with automatic filling and alarm, or a 500-L tank for a larger freezer or for several smaller freezers).

5.5.3 Controlled-Rate Freezer

Although cells may be frozen simply by placing them in an insulated box at −70°C, some cells may require different cooling rates or complex, programmed cooling curves (*see* Section 20.3.4). A programmable freezer (e.g., Cryomed, Planer) enables the cooling rate to be varied by controlling the rate of injecting liquid nitrogen into the freezing chamber, under the control of a preset program (*see* Fig. 20.5). Cheaper alternatives for controlling the cooling rate during cell freezing are the variable-neck plug (Taylor Wharton), a specialized cooling box (Nalge Nunc), a simple polystyrene foam packing container, or foam insulation for water pipes (*see* Figs. 20.2–20.4).

5.6 LABORATORY BACKUP

5.6.1 Computers and Networks

Most laboratories will be equipped with one or more computers, which may be networked. Whether or not a computer or terminal is located in the tissue culture laboratory itself, entering records for cell line maintenance (*see* Section 13.7.8), primary culture (*see* Section 12.3.11), and experiments makes later retrieval and analysis so much easier. Cell line data are also best maintained on a computer database that can also serve as an inventory control for the nitrogen freezer (*see* Section 20.4.1). In larger laboratories stock control of plastics, reagents, and media can also be simplified. There is considerable advantage in networking,

as individual computers can then be backed up centrally on a routine basis and information keyed in at one point can be retrieved elsewhere. For example, photographs recorded digitally in the tissue culture laboratory can be saved to a central server and retrieved in an office or writing area.

5.6.2 Upright Microscope

An upright microscope may be required, in addition to an inverted microscope, for chromosome analysis, mycoplasma detection, and autoradiography. Select a high-grade research microscope (e.g., Leica, Zeiss, or Nikon) with regular bright-field optics up to 100× objective magnification; phase contrast up to at least 40×, and preferably 100×, objective magnification; and fluorescence optics with epi-illumination and 40× and 100× objectives for mycoplasma testing by fluorescence (*see* Protocol 19.2) and fluorescent antibody observation. Leica supplies a 50× water-immersion objective, which is particularly useful for observing routine mycoplasma preparations with Hoechst stain. An automatic digital camera or charge-coupled device (CCD) should also be fitted for photographic records of permanent preparations.

5.6.3 Low-Temperature Freezer

Most tissue culture reagents can be stored at 4°C or −20°C, but occasionally, some drugs, reagents, or products from cultures may require a temperature of −70°C to −90°C, at which point most, if not all, of the water is frozen and most chemical and radiolytic reactions are severely limited. A −70°C to −90°C freezer is also useful for freezing cells within an insulated container (*see* Protocol 20.1). The chest type of freezer is more efficient at maintaining a low temperature with minimum power consumption, but vertical cabinets are much less extravagant in floor space and easier to access. If you do choose a vertical cabinet type, make sure that it has individual compartments, e.g., six to eight in a 400-L (15-ft^3) freezer, with separate close-fitting doors, and expect to pay at least 20% more than for a chest type.

Low-temperature freezers generate a lot of heat, which must be dissipated for them to work efficiently (or at all). Such freezers should be located in a well-ventilated room or one with air-conditioning such that the ambient temperature does not rise above 23°C. If this is not possible, invest in a freezer designed for tropical use; otherwise you will be faced with constant maintenance problems and a shorter working life for the freezer, with all the attendant problems of relocating valuable stocks. One or two failures costing $1000 or more in repairs and the loss of valuable material soon cancel any savings that would be realized in buying a cheap freezer.

5.6.4 Confocal Microscope

Cytological investigations of fluorescently labeled cells often benefit from improved resolution when viewed by confocal microscopy. This technique allows the microscope to view an "optical section" through the specimen presenting the image in one focal plane and avoiding the interference caused by adjacent cells not in the same focal plane. The data are stored digitally and can be processed in a number of ways, including the creation of a vertical section through the sample (a so-called "Z-section"), particularly useful when viewing three-dimensional cultures such as filter wells or spheroids.

5.6.5 PCR Cycler

A number of ancillary techniques in cell line validation, such as mycoplasma detection (*see* Protocol 19.3) and DNA profiling (*see* Protocol 16.9), rely on amplification and detection of specific DNA sequences. If you plan to use these techniques in house, they utilize the polymerase chain reaction (PCR) and require a thermal cycler.

5.7 SPECIALIZED EQUIPMENT

5.7.1 Microinjection Facilities

Micromanipulators can be used to inject directly into a cell, e.g., for nuclear transplantation or dye injection (Fig. 5.22).

5.7.2 Colony Counter

Monolayer colonies are easily counted by eye or on a dissecting microscope with a felt-tip pen to mark off the colonies, but if many plates are to be counted, then an automated counter will help. The simplest uses an electrode-tipped marker pen, which counts when you touch down on a colony. They often have a magnifying lens to help visualize the colonies. From there, a large increase in sophistication and cost takes you to a programmable electronic counter, with a preset program, which counts colonies using image analysis software (e.g., Optomax, Symbiosis). These counters are very rapid, can discriminate between colonies of different

Fig. 5.22. Micromanipulators and Heated Stage. Peltier temperature-controlled stage for time-lapse video or microinjection, with perfusion and sampling capabilities. (Courtesy of Beatson Institute).

diameters, and can even cope with contiguous colonies (*see* Section 21.10.2).

5.7.3 Centrifugal Elutriator

The centrifugal elutriator is a specially adapted centrifuge that is suitable for separating cells of different sizes (*see* Section 15.2.2). The device is costly, but highly effective, particularly for high cell yields.

5.7.4 Flow Cytometer

This instrument can analyze cell populations according to a wide range of parameters, including light scatter, absorbance, and fluorescence (*see* Sections 15.4, 21.7). Multiparametric analysis can be displayed in a two- or three-dimensional format. In the analytic mode, these machines are generally referred to as *flow cytometers* (*see* Section 21.7.2, e.g., BD Biosciences Cytostar), but the signals they generate can also be used in a *fluorescence-activated cell sorter* to isolate individual cell populations with a high degree of resolution (e.g., BD Biosciences FACStar). The cost is high ($100,000–200,000), and the best results are obtained with a skilled operator.

5.8 CONSUMABLE ITEMS

This category includes general items such as pipettes and pipette canisters, culture flasks, ampoules for freezing, centrifuge tubes (10–15 mL, 50 mL, 250 mL; Sterilin, Nalge Nunc, Corning), universal containers (Sterilin, Nalge Nunc), disposable syringes and needles (21–23 G for withdrawing fluid from vials, 19 G for dispensing cells), filters of various sizes (*see* Section 11.5.2) for sterilization of fluids, surgical gloves, and paper towels.

5.8.1 Pipettes

Pipettes should be of the blow-out variety, wide tipped for fast delivery, and graduated to the tip, with the maximum point of the scale at the top rather than the tip. Disposable pipettes can be used, but they are expensive and may need to be reserved for holidays or crises in washup or sterilization. Reusable pipettes are collected in pipette cylinders or hods, one per workstation.

Pasteur pipettes are best regarded as disposable and should be discarded not into pipette cylinders, but into secure glassware waste. Alternatively, disposable plastic Pasteur pipettes can be used (Pastettes).

Pipette cans. Glass pipettes are usually sterilized in aluminum or nickel-plated steel cans. Square-sectioned cans are preferable to round, as they stack more easily and will not roll about the work surface. Versions are available with silicone rubber-lined top and bottom ends to avoid chipping the pipettes during handling (Thermo Electron, Bellco).

Plastic versus glass pipettes. Many laboratories have adopted disposable plastic pipettes, which have the advantage of being prepacked and presterilized and do not have the safety problems associated with chipped or broken glass pipettes. Nor do they have to be washed, which is relatively difficult to do, or plugged, which is tedious. On the downside, they are very expensive and slower to use if singly packed. If, on the other hand, they are bulk packed, there is a high wastage rate unless packs are shared, which is not recommended (*see* Section 6.3.5). Plastic pipettes also add a significant burden to disposal, particularly if they have to be disinfected first.

As a rough guide, plastic pipettes cost about $2000 per person per annum. The number of people will therefore determine whether to employ a person to wash and sterilize glass pipettes or whether to buy plastic pipettes. (Remember that reusable glass pipettes must be purchased at a cost of around $200 per person per annum and require energy for washing and sterilizing).

5.8.2 Hemocytometer

It is essential to have some means of counting cells. A hemocytometer slide is the cheapest option and has the added benefit of allowing cell viability to be determined by dye exclusion (*see* Sections 21.1.1, 22.3.1, Fig. 21.1, Plate 17a, and Protocols 21.1, 22.1).

5.8.3 Culture Vessels

The choice of culture vessels is determined by (1) the yield (number of cells) required (*see* Table 8.1); (2) whether the cell is grown in monolayer or suspension; and (3) the sampling regime (i.e., are the samples to be collected simultaneously or at intervals over a period of time?) (*see* Section 21.8). "Shopping around" will often result in a cheaper price, but do not be tempted to change products too frequently, and always test a new supplier's product before committing yourself to it (*see* Section 11.6.3).

Care should be taken to label sterile, nonsterile, tissue culture, and non-tissue-culture grades of plastics clearly and to store them separately. Glass bottles with flat sides can be used instead of plastic, provided that a suitable washup and sterilization service is available. However, the lower cost tends to be overridden by the optical superiority, sterility, quality assurance, and general convenience of plastic flasks. Nevertheless, disposable plastics can account for approximately 60% of the tissue culture budget—even more than serum.

Petri dishes are much less expensive than flasks, though more prone to contamination and spillage. Depending on the pattern of work and the sterility of the environment, they are worth considering, at least for use in experiments if not for routine propagation of cell lines. Petri dishes are particularly useful for colony-formation assays, in which colonies have to be stained and counted or isolated at the end of an experiment.

5.8.4 Sterile Containers

Petri dishes (9 cm) are required for dissection, 5-mL bijou bottles, 30-mL universal containers, or 50-mL sample pots for storage, 15- and 50-mL centrifuge tubes for centrifugation, and 1.2-mL plastic cryostorage vials (Nalge Nunc) for freezing in liquid nitrogen (*see* Protocol 20.1). Color-coded vials are also available (Alpha Laboratories; *see* Plate 22).

5.8.5 Syringes and Needles

Although it is not recommended that syringes and needles be used extensively in normal handling (for reasons of safety, sterility, and problems with shear stress in the needle when cells are handled), syringes are required for filtration in conjunction with syringe filter adapters, and needles may also be required for extraction of reagents (drugs, antibiotics, or radioisotopes) from sealed vials.

5.8.6 Sterilization Filters

Although permanent apparatus is available for sterile filtration, most laboratories now use disposable filters. It is worthwhile to hold some of the more common sizes in stock, such as 25-mm syringe adapters (Pall Gelman, Millipore) and 47-mm bottle-top adapters or filter flasks (Falcon, Nalge Nunc). It is also wise to keep a small selection of larger sizes on hand (*see* Section 11.5.2).

5.8.7 Paper Towels and Swabs

These should be provided in a central location and beside each workstation or hood.

5.8.8 Disinfectants

It is preferable that all discarded biological material goes into disinfectant to prevent growth of potential contaminating organisms in waste containers. Where human and other primate material is being used, it is obligatory to discard into disinfectant (*see* Section 7.8.5). Chlorine-based disinfectants are used most commonly in liquid concentrate form (Clorox, Chloros) or as tablets (Precept).

CHAPTER 6

Aseptic Technique

6.1 OBJECTIVES OF ASEPTIC TECHNIQUE

Contamination by microorganisms remains a major problem in tissue culture. Bacteria, mycoplasma, yeast, and fungal spores may be introduced via the operator, the atmosphere, work surfaces, solutions, and many other sources (*see* Section 19.1 and Table 19.1). Aseptic technique aims to exclude contamination by establishing a strict code of practice and ensuring that everyone using the facility adheres to it.

Contamination can be minor and confined to one or two cultures, can spread among several cultures and compromise a whole experiment, or can be widespread and wipe out your (or even the whole laboratory's) entire stock. Catastrophes can be minimized if (1) cultures are checked carefully by eye and on a microscope, preferably by phase contrast, every time that they are handled; (2) cultures are maintained without antibiotics, preferably at all times but at least for part of the time, to reveal cryptic contaminations (*see* Section 13.7.7); (3) reagents are checked for sterility (by yourself or the supplier) before use; (4) bottles of media or other reagents are not shared with other people or used for different cell lines; and (5) the standard of sterile technique is kept high at all times.

Mycoplasmal infection, invisible under regular microscopy, presents one of the major threats. Undetected, it can spread to other cultures around the laboratory. It is therefore essential to back up visual checks with a mycoplasma test, particularly if cell growth appears abnormal (*see* Section 19.3.2).

6.1.1 Maintaining Sterility

Correct aseptic technique should provide a barrier between microorganisms in the environment outside the culture and the pure, uncontaminated culture within its flask or dish. Hence, all materials that will come into direct contact with the culture must be sterile and manipulations must be designed such that there is no direct link between the culture and its nonsterile surroundings.

It is recognized that the sterility barrier cannot be absolute without working under conditions that would severely hamper most routine manipulations. As testing the need for individual precautions would be an extensive and lengthy controlled trial, procedures are adopted largely on the basis of common sense and experience. Aseptic technique is a combination of procedures designed to reduce the probability of infection, and the correlation between the omission of a step and subsequent contamination is not always absolute. The operator may abandon several precautions before the probability rises sufficiently that a contamination is likely to occur (Fig. 6.1). By then, the cause is often multifactorial, and consequently, no simple single solution is obvious. If, once established, all precautions are maintained consistently, breakdowns will be rarer and more easily detected.

Although laboratory conditions have improved in some respects (with air-conditioning and filtration, laminar-flow facilities, etc.), the modern laboratory is often more crowded, and facilities may have to be shared. However, with rigid adherence to reasonable precautions, sterility is not difficult to maintain.

Culture of Animal Cells: A Manual of Basic Technique, Fifth Edition, by R. Ian Freshney
Copyright © 2005 John Wiley & Sons, Inc.

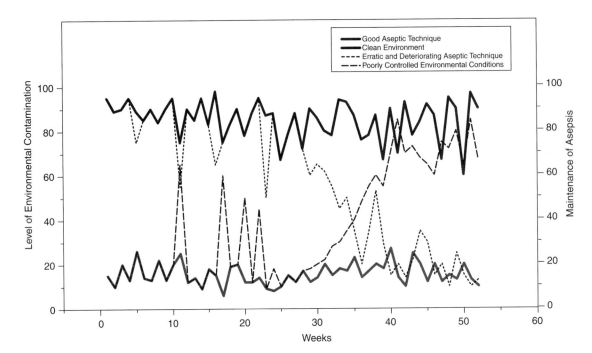

Fig. 6.1. Probability of Contamination. The solid line in the top graph represents variability in technique against a scale of 100, which represents perfect aseptic technique. The solid line in the bottom graph represents fluctuations in environmental contamination, with zero being perfect asepsis. Both lines show fluctuations, the top one representing lapses in technique (forgetting to swab the work surface, handling a pipette too far down the body of the pipette, touching nonsterile surfaces with a pipette, etc.), the bottom one representing crises in environmental contamination (a high spore count, a contaminated incubator, contaminated reagents, etc.). As long as these lapses or crises are minimal in degree and duration, the two graphs do not overlap. When particularly bad lapses in technique (dotted line) coincide with severe environmental crises (dashed line), e.g., at the left-hand side of the chart, where the dashed and dotted lines overlap briefly, the probability of infection increases. If the breakdown in technique is progressive (dotted line sloping down to the right), and the deterioration in the environment is also progressive (dashed line sloping up to the right), then, when the two lines cross, the probability of infection is high, resulting in frequent, multispecific, and multifactorial contamination.

6.2 ELEMENTS OF ASEPTIC ENVIRONMENT

6.2.1 Quiet Area

In the absence of a laminar flow cabinet, a separate sterile room should be used for sterile work. If this is not possible, pick a quiet corner of the laboratory with little or no traffic and no other activity (*see* Section 4.3.1). With laminar flow, an area should be selected that is free from air currents from doors, windows, etc.; the area should also have no through traffic and no equipment that generates air currents (e.g., centrifuges, refrigerators, and freezers; air conditioners should be positioned so that air currents do not compromise the functioning of the hood; *see* Section 4.3). Activity should be restricted to tissue culture, and animals and microbiological cultures should be excluded from the tissue culture area. The area should be kept clean and free of dust and should not contain equipment other than that connected with tissue culture. Nonsterile activities, such as sample processing, staining, or extractions, should be carried out elsewhere.

6.2.2 Work Surface

It is essential to keep the work surface clean and tidy. The following rules should be observed:

(1) Start with a completely clear surface.
(2) Swab the surface liberally with 70% alcohol (*see* Appendix I).
(3) Bring onto the surface only those items you require for a particular procedure.
(4) Remove everything that is not required, and swab the surface down between procedures.
(5) Arrange your work area so that you have (a) easy access to all items without having to reach over one to get at another and (b) a wide, clear space in the center of the bench (not just the front edge!) to work on (Fig. 6.2). If you have too much equipment too close to you, you will inevitably brush the tip of a sterile pipette against a nonsterile surface. Furthermore, the laminar airflow will fail in a hood that is crowded with equipment (Fig. 6.3).

Fig. 6.2. Suggested Layout of Work Area. Laminar-flow hood laid out correctly. Positions may be reversed for left-handed workers.

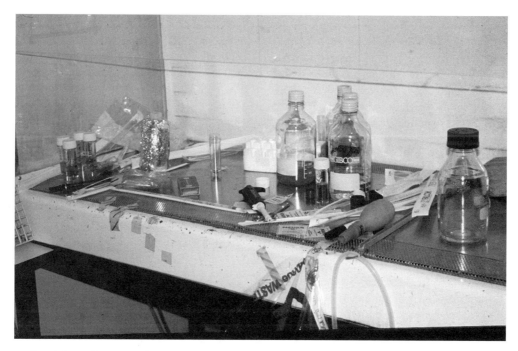

Fig. 6.3. Badly Arranged Work Area. Laminar-flow hood being used incorrectly. The hood is too full, and many items encroach on the air intake at the front, destroying the laminar airflow and compromising both containment and sterility.

(6) Horizontal laminar flow is more forgiving, but you should still work in a clear space with no obstructions between the central work area and the HEPA filter (Fig. 6.4).

(7) Work within your range of vision (e.g., insert a pipette in a bulb or pipetting aid with the tip of the pipette pointing away from you so that it is in your line of sight continuously and not hidden by your arm).

(8) Mop up any spillage immediately and swab the area with 70% alcohol.

(9) Remove everything when you have finished, and swab the work surface down again.

Fig. 6.4. Layout of Horizontal Laminar-Flow Hood. Layout for working properly in a horizontal laminar-flow hood. Positions may be reversed for left-handed workers.

6.2.3 Personal Hygiene

There has been much discussion about whether hand washing encourages or reduces the bacterial count on the skin. Regardless of this debate, washing will moisten the hands and remove dry skin that would otherwise be likely to blow onto your culture. Washing will also reduce loosely adherent microorganisms, which are the greatest risk to your culture. Surgical gloves may be worn and swabbed frequently, but it may be preferable to work without them (where no hazard is involved) and retain the extra sensitivity that this allows.

Caps, gowns, and face masks are required under Good Manufacturing Practice (GMP) [Food and Drug Administration, 1992; Rules and Guidance for Pharmaceutical Manufacturers and Distributors, 1997] conditions but are not necessary under normal conditions, particularly when working with laminar flow. If you have long hair, tie it back. When working aseptically on an open bench, do not talk. Talking is permissible when you are working in a vertical laminar-flow hood, with a barrier between you and the culture, but should still be kept to a minimum. If you have a cold, wear a face mask, or, better still, do not do any tissue culture during the height of the infection.

6.2.4 Reagents and Media

Reagents and media obtained from commercial suppliers will already have undergone strict quality control to ensure that they are sterile, but the outside surface of the bottle they come in is not. Some manufacturers supply bottles wrapped in polyethylene, which keeps them clean and allows them to be placed in a water bath to be warmed or thawed. The wrapping should be removed outside the hood. Unwrapped bottles should be swabbed in 70% alcohol when they come from the refrigerator or from a water bath.

6.2.5 Cultures

Cultures imported from another laboratory carry a high risk, because they may have been contaminated either at the source or in transit. Imported cell lines should always be quarantined (*see* Sections 4.3.3, 19.1.8); i.e., they should be handled separately from the rest of your stocks and kept free of antibiotics until they are shown to be uncontaminated. They may then be incorporated into your main stock. Antibiotics should not be used routinely as they may suppress, but not eliminate, some contaminations (*see* Section 9.4.7).

6.3 STERILE HANDLING

6.3.1 Swabbing

Swab down the work surface with 70% alcohol before and during work, particularly after any spillage, and swab it down again when you have finished. Swab bottles as well, especially those coming from cold storage or a water bath, before using them, and also swab any flasks or boxes from the incubator. Swabbing sometimes removes labels, so use an alcohol-resistant marker.

6.3.2 Capping

Deep screw caps are preferred to stoppers, although care must be taken when washing caps to ensure that all detergent is rinsed from behind rubber liners. Wadless polypropylene caps should be used if possible. The screw cap should be covered with aluminum foil to protect the neck of the bottle from sedimentary dust, although the introduction of deep polypropylene caps (e.g., Duran) has made foil shrouding less necessary.

6.3.3 Flaming

When working on an open bench (Fig. 6.5), flame glass pipettes and the necks of bottles and screw caps before and

Fig. 6.5. Layout of Work Area on Open Bench. Items are arranged in a crescent around the clear work space in the center. The Bunsen burner is located centrally, to be close by for flaming and to create an updraft over the work area.

after opening and closing a bottle, work close to the flame, where there is an updraft due to convection, and do not leave bottles open. Screw caps should be placed with the open side down on a clean surface and flamed before being replaced on the bottle. Alternatively, screw caps may be held in the hand during pipetting, avoiding the need to flame them or lay them down (Fig. 6.6).

Flaming is not advisable when you are working in a laminar-flow hood, as it disrupts the laminar flow, which, in turn, compromises both the sterility of the hood and its containment of biohazardous material. An open flame can also be a fire hazard and can damage the high-efficiency particulate air (HEPA) filter or melt some of the plastic interior fittings.

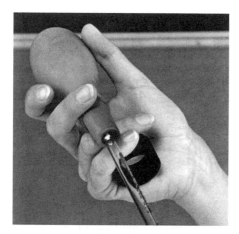

Fig. 6.6. Holding Cap and Bulb. Cap may be unscrewed and held in the crook of the little finger of the hand holding the bulb or pipetting aid.

6.3.4 Handling Bottles and Flasks

When working on an open bench, you should not keep bottles vertical when open; instead, keep them at an angle as shallow as possible without risking spillage. A bottle rack (MP Biomedicals) can be used to keep the bottles or flasks tilted. Culture flasks should be laid down horizontally when open and, like bottles, held at an angle during manipulations. When you are working in laminar flow, bottles can be left open and vertical, but do not let your hands or any other items come between an open vessel or sterile pipette and the HEPA air filter.

6.3.5 Pipetting

Standard glass or disposable plastic pipettes are still the easiest way to manipulate liquids. Syringes are sometimes used but should be discouraged as regular needles are too short to reach into most bottles. Syringing may also produce high shearing forces when you are dispensing cells, and the practice also increases the risk of self-inoculation. Wide-bore cannulae are preferable to needles but still not as rapid to use, except when multiple-stepping dispensers (*see* Fig. 5.8) are used.

Pipettes of a convenient size range should be selected—1 mL, 2 mL, 5 mL, 10 mL, and 25 mL cover most requirements, although 100-mL disposable pipettes are available (BD Biosciences) and are useful for preparing and aliquoting media. Using fast-flow pipettes reduces accuracy slightly but gives considerable benefit in speed. If you are using glass pipettes and require only a few of each, make up mixed cans for sterilization and save space. Disposable plastic pipettes should be double wrapped and removed from their outer wrapping before being placed in a hood. Unused pipettes should be stored in a dust-free container.

Mouth pipetting should be avoided, as it has been shown to be a contributory factor in mycoplasmal contamination and may introduce an element of hazard to the operator (e.g., with virus-infected cell lines and human biopsy or autopsy specimens or other potential biohazards; *see* Section 7.1.7). Inexpensive bulbs (*see* Fig. 6.6) and electric pipetting aids are available (*see* Fig. 5.5); try a selection of these devices to find one that suits you. The instrument you choose should accept all sizes of pipette that you use without forcing them in and without the pipette falling out. Regulation of flow should be easy and rapid but at the same time capable of fine adjustment. You should be able to draw liquid up and down repeatedly (e.g., to disperse cells), and there should be no fear of carryover. The device should fit comfortably in your hand and should be easy to operate with one hand without fatigue.

Pipettors (*see* Fig. 5.6) are particularly useful for small volumes (1 mL and less), although most makes now go up to 5 mL. Because it is difficult to reach down into a larger vessel without touching the inside of the neck of the vessel with the nonsterile stem of the pipettor, pipettors should only be used in conjunction with smaller flasks or by using an intermediary container, such as a universal container. Alternatively, longer tips may be used with larger volumes. Pipettors are particularly useful in dealing with microtitration assays and other multiwell dishes but should not be used for serial propagation unless filter tips are used. Multipoint pipettors (with 4, 8, or 12 points) are available for microtitration plates (*see* Fig. 5.10).

Pipetting in tissue culture is often a compromise between speed and precision; speed is required for minimal damage during manipulations such as subculture, and precision is required for reproducibility under standard culture conditions. However, an error of ±5% is usually acceptable, except under experimental conditions where greater precision may be required. Generally, using the smallest pipette compatible with the maneuver will give the greater precision required of most quantitative experimental work.

It is necessary to insert a cotton plug in the top of a glass pipette before sterilization to keep the pipette sterile during use. The plug prevents contamination from the bulb or pipetting aid entering the pipette and reduces the risk of cross-contamination from pipette contents inadvertently entering the pipetting aid. If the plug becomes wet, discard the pipette into disinfectant for return to the washup. Plugging pipettes for sterile use is a very tedious job, as is the removal of plugs before washing. Automatic pipette pluggers are available (*see* Fig. 11.6); they speed up the process and reduce the tedium.

Plastic pipettes come already plugged. However, they are slower to use if individually wrapped, wasteful if bulk wrapped (some pipettes are always lost, as it is not advisable to share or reuse a pack once opened), and carry a slightly higher risk of contamination, because removing a pipette from a plastic wrapper is not as clean as withdrawing one from a can. They are, however, more likely to be free of chemical and microbial contamination than recycled glass pipettes and reduce washing up requirements.

Automatic pipetting devices and repeating dispensers are discussed in Chapter 5 (*see* Section 5.2.7). Care must be taken to avoid contamination when setting them up, but the increased speed in handling can cut down on fatigue and on the time that vessels are open to contamination.

6.3.6 Pouring

Do not pour from one sterile container into another, unless the bottle you are pouring from is to be used once only to deliver all its contents (premeasured) in one single delivery. The major risk in pouring lies in the generation of a bridge of liquid between the outside of the bottle and the inside, permitting contamination to enter the bottle during storage or incubation.

6.4 LAMINAR FLOW

The major advantage of working in a laminar-flow hood is that the working environment is protected from dust and contamination by a constant, stable flow of filtered air passing over the work surface (Fig. 6.7). There are two main types of flow: (1) *horizontal*, where the airflow blows from the side facing you, parallel to the work surface, and is not recirculated (*see* Fig. 6.7a); and (2) *vertical*, where the air blows down from the top of the hood onto the work surface and is drawn through the work surface and either recirculated or vented (*see* Fig. 6.7b). In most hoods, 20% is vented and made up by drawing in air at the front of the work surface. This configuration is designed to minimize overspill from the work area of the cabinet. Horizontal flow hoods give the most stable airflow and best sterile protection to the culture and reagents; vertical flow hoods give more protection to the operator.

A Class II vertical-flow *biohazard* hood (Microbiological Safety Cabinet, MSC) should be used (*see* Section 7.1.7 and Fig. 7.5a) if potentially hazardous material (human- or primate-derived cultures, virally infected cultures, etc.) is being handled. The best protection from chemical and radiochemical hazards is given by a cytotoxicity hood that is specially designed for the task and that has a carbon filter trap in the recirculating airflow or a hood with all the effluent vented to outside the building (*see* Section 7.5.4). If known human pathogens are handled, a Class III pathogen cabinet with a pathogen trap on the vent is obligatory (*see* Section 7.8.2, Fig. 7.5c).

Laminar-flow hoods depend for their efficiency on a minimum pressure drop across the filter. When resistance builds up, the pressure drop increases, and the flow rate of air in the cabinet falls. Below 0.4 m/s (80 ft/min), the stability of the laminar airflow is lost, and sterility can no longer be maintained. The pressure drop can be monitored with a manometer fitted to the cabinet, but direct measurement of the airflow with an anemometer is preferable.

Routine maintenance checks of the primary filters are required (about every 3 to 6 months). With horizontal-flow

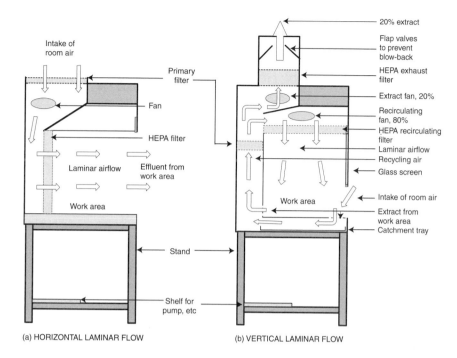

Fig. 6.7. Airflow in Laminar-Flow Hoods. Arrows denote direction of airflow. (a) Horizontal flow. (b) Vertical flow.

hoods, primary filters may be removed (after switching off the fan) and discarded or washed in soap and water. The primary filters in vertical-flow and biohazard hoods are internal and may need to be replaced by an engineer. They should be incinerated or autoclaved and discarded.

Every 6 months the main HEPA filter above the work surface should be monitored for airflow and holes, which are detectable by a locally increased airflow and an increased particle count. Monitoring is best done by professional engineers on a contract basis. Class II biohazard cabinets will have HEPA filters on the exhaust, which will also need to be changed periodically. Again, this should be done by a professional engineer, with proper precautions taken for bagging and disposing of the filters by incineration. If used for biohazardous work, cabinets should be sealed and fumigated before the filters are changed.

Regular weekly checks should be made below the work surface and any spillage mopped up, the tray washed, and the area sterilized with 5% phenolic disinfectant in 70% alcohol. Spillages should, of course, be mopped up when they occur, but occasionally they go unnoticed, so a regular check is imperative. Swabs, tissue wipes, or gloves, if dropped below the work surface during cleaning, can end up on the primary filter and restrict airflow, so take care during cleaning and check the primary filter periodically.

Laminar-flow hoods are best left running continuously, because this keeps the working area clean. Should any spillage occur, either on the filter or below the work surface, it dries fairly rapidly in sterile air, reducing the chance that microorganisms will grow.

Ultraviolet lights are used to sterilize the air and exposed work surfaces in laminar-flow hoods between uses. The effectiveness of the lights is doubtful because they do not reach crevices, which are treated more effectively with alcohol or other sterilizing agents, which will run in by capillarity. Ultraviolet lights present a radiation hazard, particularly to the eyes, and will also lead to crazing of some clear plastic panels (e.g., Perspex) after 6 months to a year if used in conjunction with alcohol.

Δ *Safety Note.* If ultraviolet lights are used, protective goggles must be worn and all exposed skin covered.

6.5 STANDARD PROCEDURE

The essence of good sterile technique embodies many of the principles of standard good laboratory practice (*see* Table 6.1). Keep a clean, clear space to work, and have on it only what you require at one time. Prepare as much as possible in advance, so that cultures are out of the incubator for the shortest possible time and the various manipulations can be carried out quickly, easily, and smoothly. Keep everything in direct line of sight, and develop an awareness of accidental contacts between sterile and nonsterile surfaces. Leave the area clean and tidy when you finish.

The two protocols that follow emphasize aseptic technique. Preparation of media and other manipulations are discussed in more detail under the appropriate headings (*see* Section 11.4). This protocol is designed for use with Exercise 1 (*see* Chapter 2, Exercise 1) but is written in general terms to suit other applications.

TABLE 6.1. Good Aseptic Technique

Subject	Do	Don't
Laminar-flow hoods	Swab down before and after use. Keep minimum amount of apparatus and materials in hood. Work in direct line of sight.	Clutter up the hood. Leave the hood in a mess.
Contamination	Work without antibiotics. Check cultures regularly, by eye and microscope. Box Petri dishes and multiwell plates.	Open contaminated flasks in tissue culture. Carry infected cells. Leave contaminations unclaimed; dispose of them safely.
Mycoplasma	Test cells routinely.	Carry infected cells. Try to decontaminate cultures.
Importing cell lines	Get from reliable source. Quarantine incoming cell lines. Check for mycoplasma. Validate origin. Keep records.	Get from a source far removed from originator.
Exporting cell lines	Check for mycoplasma. Validate origin. Send data sheet. Triple wrap.	Send contaminated cell lines.
Glassware	Keep stocks separate.	Use for non-tissue culture procedures.
Flasks	Pipette with flask slanted. Gas from filtered CO_2 line. Vent briefly if stacked.	Have too many open at once. Gas in CO_2 incubator unless with gas-permeable cap. Stack too high.
Media and reagents	Swab bottles before placing them in hood. Open only in hood.	Share among cell lines. Share with others. Pour.
Pipetting	Use plugged pipettes. Change if contaminated or plug wetted. Use plastic for agar. Discard liquids into waste beaker with funnel (Fig. 6.8).	Use the same pipette for different cell lines. Share with other people. Generate aerosols. Overfill disposal cylinders.

PROTOCOL 6.1. ASEPTIC TECHNIQUE IN VERTICAL LAMINAR FLOW

Outline
Clean and swab down work area, and bring bottles, pipettes, etc. Carry out preparative procedures first (preparation of media and other reagents), followed by culture work. Finally, tidy up and wipe over surface with 70% alcohol.

Materials
Sterile:
- ☐ Eagle's 1×MEM with Hanks' salts and HCO_3, without antibiotics .... 100 mL
- ☐ Pipettes, graduated, and plugged. If glass, an assortment of sizes, 1 mL, 5 mL, 10 mL, 25 mL, in a square pipette can, or, if plastic, individually wrapped and sorted by size on a rack
- ☐ Culture flasks (preweighed if for Exercise 1) .... 25 cm² 10

Nonsterile:
- ☐ Pipetting aid or bulb (*see* Figs. 5.5, 6.6)

- ☐ 70% alcohol in spray bottle (*see* Appendix I)
- ☐ Lint-free swabs or wipes
- ☐ Absorbent paper tissues
- ☐ Pipette cylinder containing water and disinfectant (*see* Section 7.8.5)
- ☐ Scissors
- ☐ Marker pen with alcohol-insoluble ink
- ☐ Notebook, pen, protocols, etc.

Protocol
1. Swab down work surface and all other inside surfaces of laminar-flow hood, including inside of front screen, with 70% alcohol and a lint-free swab or tissue.
2. Bring media, etc., from cold store, water bath, or otherwise thawed from freezer, swab bottles with alcohol, and place those that you will need first in the hood.
3. Collect pipettes and place at one side of the back of work surface in an accessible position (*see* Fig. 6.2)

(a) If glass, open pipette cans and place lids on top or alongside, with the open side down.

(b) If plastic, remove outer packaging and stack individually wrapped pipettes, sorted by size, on a rack or in cans.

4. Collect any other glassware, plastics, instruments, etc., that you will need, and place them close by (e.g., on a cart or an adjacent bench).

5. Slacken, but do not remove, caps of all bottles about to be used.

6. Remove the cap of the bottle into which you are about to pipette, and the bottles that you wish to pipette from, and place the caps open side uppermost on the work surface, at the back of the hood and behind the bottle, so that your hand will not pass over them. Alternatively, if you are handling only one cap at a time, grasp the cap in the crook formed between your little finger and the heel of your hand (*see* Fig. 6.6) and replace it when you have finished pipetting.

7. Select pipette:

(a) If glass:

   i) Take pipette from can, lifting it parallel to the other pipettes in the can and touching them as little as possible, particularly at the tops (if the pipette that you are removing touches the end of any of the pipettes still in the can, discard it),

   ii) Insert the top end into a pipette aid, pointing pipette away from you and holding it well above the graduations, so that the part of the pipette entering the bottle or flask will not be contaminated (Fig. 6.9).

Δ *Safety Note.* As you insert the pipette into the bulb or pipetting aid, take care not to exert too much pressure; pipettes can break if forced (*see* Section 7.5.3 and Fig. 7.2).

(b) If plastic:

   i) Open the pack at the top

   ii) Peel the ends back, turning them outside in

   iii) Insert the end of the pipette into the bulb or pipette aid

   iv) Withdraw the pipette from the wrapping without it touching any part of the outside of the wrapping, or the pipette touching any nonsterile surface (*see* Fig. 6.9)

   v) Discard the wrapping into the waste bin.

8. The pipette in the bulb or pipette aid will now be at right angles to your arm. Take care that the tip of the pipette does not touch the outside of a bottle or the inner surface of the hood (*see* circled areas in Fig. 6.9). Always be aware of where the pipette is. Following this procedure is not easy when you are learning aseptic technique, but it is an essential requirement for success and will come with experience.

9. Tilt the medium bottle toward the pipette so that your hand does not come over the open neck, and, using a 5-mL pipette, withdraw 5 mL of medium and transfer it to a 25-cm² flask, also tilted.

10. Repeat with a further 4 flasks. When you are pipetting into several bottles or flasks, they can be laid down horizontally on their sides. Ensure that the flasks remain well back in the hood and that your hand does not come over open necks. (If performing Exercise 1, record the time taken to add the medium to the 5 flasks).

11. Discard the glass pipette into the pipette cylinder containing disinfectant. Plastic pipettes should be discarded into double-thickness autoclavable biohazard bags.

12. Recap the flasks.

13. Repeat the procedure, using a 25-mL pipette to transfer 5 mL to each of 5 flasks from one filling of the pipette. (If performing Exercise 1, record the time taken to add the medium to the 5 flasks).

14. Replace the cap on the medium bottle and flasks. Bottles may be left open while you complete a particular maneuver, but should always be closed if you leave the hood for any reason.

Note. In vertical laminar flow, do not work immediately above an open vessel. In horizontal laminar flow, do not work behind an open vessel.

15. On completion of the operation, tighten all caps, and remove all solutions and materials no longer required from the work surface.

16. If performing Exercise 1, weigh the flasks to check accuracy of medium dispensing and place at 37°C for 1 week, to check for possible contamination.

Fig. 6.8. Waste Beaker. Filter funnel prevents beaker from splashing back contents.

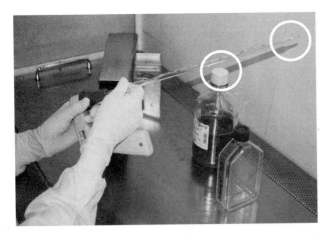

Fig. 6.9. Inserting a Pipette. Pipette being inserted correctly with grip high on the pipette (above the graduations) and the pipette pointing away from the user. Circled areas mark potential risks, i.e., inadvertently touching the bottle or the back of the cabinet.

PROTOCOL 6.2. WORKING ON THE OPEN BENCH

Outline
Clean and swab down work area, and bring bottles, pipettes, etc. (*see* Fig. 6.5). Carry out preparative procedures first. Flame articles as necessary, and keep the work surface clean and clear. Finally, tidy up and wipe over surface with 70% alcohol.

Materials
Sterile or aseptically prepared:
- ❑ Eagle's 1×MEM with Hanks' salts and HCO₃, without antibiotics .... 100 mL
- ❑ Pipettes, graduated, and plugged. If glass, an assortment of sizes, 1 mL, 5 mL, 10 mL, 25 mL, in a square pipette can, or, if plastic, individually wrapped and sorted by size on a rack
- ❑ Culture flasks (preweighed if for Exercise 1) .... 25 cm² 10

Nonsterile:
- ❑ Pipetting aid or bulb (*see* Figs. 5.5, 6.6)
- ❑ 70% alcohol in spray bottle
- ❑ Lint-free swabs or wipes
- ❑ Absorbent paper tissues
- ❑ Pipette cylinder containing water and disinfectant (*see* Section 7.8.5)
- ❑ Bunsen burner (or equivalent) and lighter
- ❑ Scissors
- ❑ Marker pen with alcohol-insoluble ink
- ❑ Notebook, pen, protocols, etc.

Protocol
1. Swab down bench surface with 70% alcohol.

2. Bring media, etc., from cold store, water bath, or otherwise thawed from freezer, swab bottles with alcohol, and place those that you will need first on the bench in the work area, leaving the others at the side.
3. Collect pipettes and place at the side of the work surface in an accessible position (*see* Fig. 6.5).
   (a) If glass, open pipette cans and place lids on top or alongside, open side down.
   (b) If plastic, remove outer packaging and stack individually wrapped pipettes, sorted by size, on a rack or in cans.
4. Collect any other glassware, plastics, instruments, etc., that you will need, and place them close by.
5. Flame necks of bottles, briefly rotating neck in flame, and slacken caps.
6. Select pipette:
   (a) If glass:
      i) Take pipette from can, lifting it parallel to the other pipettes in the can and touching them as little as possible, particularly at the tops (if the pipette that you are removing touches the end of any of the pipettes still in the can, discard it),
      ii) Insert the top end into a pipette aid, pointing pipette away from you and holding it well above the graduations, so that the part of the pipette entering the bottle or flask will not be contaminated (Fig.6.9).
   (b) If plastic:
      i) Open the pack at the top
      ii) Peel the ends back, turning them outside in
      iii) Insert the end of the pipette into the bulb or pipette aid
      iv) Withdraw the pipette from the wrapping without it touching any part of the outside of the wrapping, or the pipette touching any nonsterile surface (*see* Fig. 6.9)
      v) Discard the wrapping into the waste bin.
7. Flame pipette (glass only) by pushing it lengthwise through the flame, rotate 180°, and pull the pipette back through the flame. This should only take 2–3 s, or the pipette will get too hot. You are not attempting to sterilize the pipette; you are merely trying to fix any dust that may have settled on it. If you have touched anything or contaminated the pipette in any other way, discard it into disinfectant for return to the washup facility; do not attempt to resterilize the pipette by flaming.
8. Insert pipette in a bulb or pipette aid, pointing pipette way from you and holding it well

above the graduations, so that the part of the pipette entering the bottle or flask will not be contaminated (*see* Fig. 6.9).

Δ *Safety Note.* Take care not to exert too much pressure, as pipettes can break when being forced into a bulb (*see* Fig. 7.2).

Do not flame plastic pipettes.

9. The pipette in the bulb or pipetting aid will now be at right angles to your arm. Take care that the tip of the pipette does not touch the outside of a bottle or pipette can. Always be aware of where the pipette is. Following this procedure is not easy when you are learning aseptic technique, but it is an essential requirement for success and will come with experience.
10. Holding the pipette still pointing away from you, remove the cap of your first bottle into the crook formed between your little finger and the heel of your hand (*see* Fig. 6.6). If you are pipetting into several bottles or flasks, they can be laid down horizontally on their sides. Work with the bottles tilted so that your hand does not come over the open neck. If you have difficulty holding the cap in your hand while you pipette, place the cap on the bench, open side down. If bottles are to be left open, they should be sloped as close to horizontal as possible in laying them on the bench or on a bottle rest.
11. Flame the neck of the bottle.

Note. If flaming Duran bottles, do not use with pouring ring.

12. Tilt the bottle toward the pipette so that your hand does not come over the open neck.
13. Withdraw the requisite amount of fluid and hold.
14. Flame the neck of the bottle and recap.
15. Remove cap of receiving bottle, flame neck, insert fluid, reflame neck, and replace cap.
16. When finished, tighten caps.
17. On completion of the operation, remove stock solutions from work surface, keeping only the bottles that you will require.

6.5.1 Petri Dishes and Multiwell Plates

Petri dishes and multiwell plates are particularly prone to contamination because of the following factors:

(1) The larger surface area exposed when the dish is open
(2) The risk of touching the rim of the dish when handling an open dish

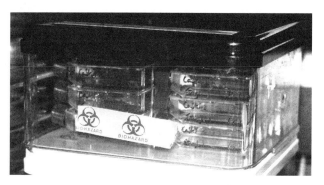

Fig. 6.10. Boxed Dishes. A transparent box, such as a sandwich box or cake box, helps to protect unsealed dishes and plates, and flasks with slackened caps, from contamination in a humid incubator. This type of container should also be used for materials that may be biohazardous, to help contain spillage in the event of an accident.

(3) The risk of carrying contamination from the work surface to the plate via the lid if the lid is laid down
(4) Medium filling the gap between the lid and the dish due to capillarity if the dish is tilted or shaken in transit to the incubator.
(5) The higher risk of contamination in the humid atmosphere of a CO_2 incubator.

The following practices will minimize the risk of contamination:

(1) Do not leave dishes open for an extended period or work over an open dish or lid.
(2) When moving dishes or transporting them to or from the incubator, take care not to tilt them or shake them, to avoid the medium entering the capillary space between the lid and the base.
   a) Using "vented" dishes (*see* Fig. 8.7)
   b) If medium still lodges in this space, discard the lid, blot any medium carefully from the outside of the rim with a sterile tissue dampened with 70% alcohol, and replace the lid with a fresh one. (Make sure that the labeling is on the base!)
(3) Enclose dishes and plates in a transparent plastic box for incubation, and swab the box with alcohol when it is retrieved from the incubator (*see* Fig. 6.10 and *Section* 6.7.1).

The following procedure is recommended for handling Petri dishes or multiwell plates.

PROTOCOL 6.3. HANDLING DISHES OR PLATES

Materials
As for Protocol 6.1 or 6.2, as appropriate

Protocol

To remove medium, etc:

1. Stack dishes or plates on one side of work area.
2. Switch on aspiration pump.
3. Select unplugged pipette and insert in aspiration line.
4. Lift first dish or plate to center of work area.
5. Remove lid and place behind dish, open side up.
6. Grasp the dish as low down on the base as you can, taking care not to touch the rim of the dish or to let your hand come over the open area of the dish or lid.

    With practice, you may be able to open the lid sufficiently and tilt the dish to remove the medium without removing the lid completely. This technique is quicker and safer than the preceding one.
7. Tilt dish and remove medium. Discard into waste beaker with funnel (*see* Fig. 6.8) if aspiration pump not available.
8. Replace lid.
9. Move dish to other side of work area from the untreated dishes in the initial stack.
10. Repeat procedure with remaining dishes or plates.
11. Discard pipette and switch off pump.

To add medium or cells, etc:

1. Position necessary bottles and slacken the cap of the one you are about to use.
2. Bring dish to center of work area.
3. Remove bottle cap and fill pipette from bottle.
4. Remove lid and place behind dish.
5. Add medium to dish, directing the stream gently low down on the side of the base of the dish.
6. Replace lid.
7. Return dish to side where dishes were originally stacked, taking care not to let the medium enter the capillary space between the lid and the base.
8. Repeat with second dish, and so on.
9. Discard pipette.

Again, with practice, you may be able to lift the lid and add the medium without laying the lid down, as you did when you removed the medium.

Note. If you are adding medium to a dish or plate with cells already in it, then the bottle of medium used must be designated for the cell line in use and not used for any other. Preferably, dispense sufficient medium into a separate bottle, and use only for this procedure.

6.6 APPARATUS AND EQUIPMENT

All apparatus used in the tissue culture area should be cleaned regularly to avoid the accumulation of dust and to prevent

microbial growth in accidental spillages. Replacement items, such as gas cylinders, must be cleaned before being introduced to the tissue culture area, and no major movement of equipment should take place while people are working aseptically.

6.6.1 Incubators

Humidified incubators are a major source of contamination. (*see* Section 19.1.4) They should be cleaned out at regular intervals (weekly or monthly, depending on the level of atmospheric contamination and frequency of access) by removing the contents, including all the racks or trays, and washing down the interior and the racks or shelves with a nontoxic detergent such as Decon or Roccall. Traces of detergent should then be removed with 70% alcohol, which should be allowed to evaporate completely before replacing the shelves and cultures.

A fungicide, such as 2% Roccall or 1% copper sulfate, may be placed in the humidifier tray at the bottom of the incubator to retard fungal growth, but the success of such fungicides is limited to the surface that they are in contact with, and there is no real substitute for regular cleaning. Some incubators have high-temperature sterilization cycles, but these are seldom able to generate sufficient heat for long enough to be effective and the length of time that the incubator is out of use can be inconvenient. Some incubators have micropore filtration and laminar airflow to inhibit the circulation of microorganisms (Forma, Jencons), but it may be better to accept a minor increase in recovery time, eliminate the fan, and rely on convection for circulation (*see* Section 5.3.2).

6.6.2 Boxed Cultures

When problems with contamination in humidified incubators recur frequently, it is advantageous to enclose dishes, plates, and flasks with slackened caps, in plastic sandwich boxes (Fig. 6.10). The box should be swabbed before use, inside and outside, and allowed to dry in sterile air. When the box is subsequently removed from the incubator, it should be

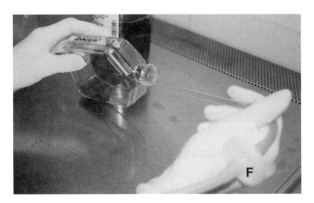

Fig. 6.11. Gassing a Flask. A pipette is inserted into the supply line from the CO_2 source, and the 5% CO_2 is used to flush the air out of the flask without bubbling through the medium. The letter F indicates a micropore filter inserted in the CO_2 line.

swabbed with 70% alcohol before being opened or introduced into your work area. The dishes are then carefully removed and the interior of the box swabbed before reuse.

6.6.3 Gassing with CO_2

It is common practice to place flasks, with the caps slackened, in a humid CO_2 incubator to allow for gaseous equilibration, but doing so does increase the risk of contamination. Several manufacturers provide flasks with gas-permeable caps to allow rapid equilibration in a CO_2 atmosphere without the risk of contamination. Alternatively, purge the flasks from a sterile, premixed gas supply (Fig. 6.11) and then seal them. This avoids the need for a gassed incubator for flasks and gives the most uniform and rapid equilibration.

CHAPTER 7

Safety, Bioethics, and Validation

7.1 LABORATORY SAFETY

In addition to the everyday safety hazards common to any workplace, the cell culture laboratory has a number of particular risks associated with culture work. Despite the scientific background and training of most people who work in this environment, accidents still happen, as familiarity often leads to a casual approach in dealing with regular, biological, and radiological hazards. Furthermore, individuals who service the area often do *not* have a scientific background, and the responsibility lies with those who have to maintain a safe environment for all who work there. It is important to identify potential hazards but, at the same time, not to overemphasize the risks. If a risk is not seen as realistic, then precautions will tend to be disregarded, and the whole safety code will be placed in disrepute.

7.2 RISK ASSESSMENT

Risk assessment is an important principle that has become incorporated into most modern safety legislation. Determining the nature and extent of a particular hazard is only part of the process; the way in which the material or equipment is used, who uses it, the frequency of use, training, and general environmental conditions are all equally important in determining risk (Table 7.1). Such considerations as the amount of a particular material, the degree and frequency of exposure to a hazard, the procedures for handling materials, the type of protective clothing worn, ancillary hazards like exposure to heat, frost, and electric current, and the type of training and experience of the operator all contribute to the risk of a given procedure, although the nature of the hazard itself may remain constant.

A major problem that arises constantly in establishing safe practices in a biomedical laboratory is the disproportionate concern given to the more esoteric and poorly understood hazards, such as those arising from genetic manipulation, relative to the known and proven hazards of toxic and corrosive chemicals, solvents, fire, ionizing radiation, electrical shock, and broken glass. It is important that biohazards be categorized correctly [HHS Publication No. (CDC) 93–8395, 3rd Edition (1993); Health and Safety Commission, 1991a, 1991b, 1992, 1999a, 1999b; Caputo, 1996], neither overemphasized nor underestimated, but the precautions taken should not displace the recognition of everyday safety problems.

7.3 STANDARD OPERATING PROCEDURES

Hazardous substances, equipment, and conditions should not be thought of in isolation but should be taken to be part of a procedure, all the components of which should be assessed. If the procedure is deemed to carry any significant risk beyond the commonplace, then a standard operating procedure (SOP) should be defined, and all who work with the material, equipment, etc., should conform to that procedure. The different stages of the procedure—procurement, storage, operations, and disposal—should be identified, and the possibility must be taken into account that the presence of more than one hazard within a procedure will compound the risk or, at best, complicate the necessary precautions (e.g.,

TABLE 7.1. Elements of Risk Assessment

Category	Items affecting risk
Operator	
Experience	Level
	Relevance
	Background
Training	Previous
	New requirements
Protective clothing	Adequate
	Properly worn (buttoned lab coat)
	Laundered regularly
	Repaired or discarded when damaged
Equipment	
Age	Condition
	Adherence to new legislation
Suitability for task	Access, sample capacity, containment
Mechanical stability	Loading
	Anchorage
	Balance
Electrical safety	Connections
	Leakage to ground (earth)
	Proximity of water
Containment	Aerosols:
	Generation
	Leakage from hood ducting
	Overspill from work area
	Toxic fumes
	Exhaust ductwork:
	Integrity
	Site of effluent and downwind risk
Heat	Generation
	Dissipation
Maintenance	Frequency
	Decontamination required?
Disposal	Route
	Decontamination required?
Physical Risks	
Intense cold	Frostbite
	Numbing
Electric shock	Loss of consciousness
	Cardiac arrest
Fire	General precautions
	Equipment wiring, installation, and maintenance
	Incursion of water near electrical wiring
	Fire drills, procedures, escape routes
	Solvent usage and storage (e.g., do not store ether in refrigerators)
	Flammable mixtures
	Identification of stored biohazards and radiochemicals

TABLE 7.1. Elements of Risk Assessment (*Continued*)

Category	Items affecting risk
Chemicals (including gases and volatile liquids)	
Scale	Amount used
Toxicity	Poisonous
	Carcinogenic
	Teratogenic
	Mutagenic
	Corrosive
	Irritant
	Allergenic
	Asphyxiative
Reaction with water	Heat generation
	Effervescence
Reaction with solvents	Heat generation
	Effervescence
	Generation of explosive mixture
Volatility	Intoxication
	Asphyxiation
Generation of powders and aerosols	Dissemination
	Inhalation
Import, export, and transportation	Breakage, leakage
Location and storage conditions	Access by untrained staff
	Illegal entry
	Weather, incursion of water
	Stability, compression, breakage, leakage
Biohazards	
Pathogenicity	Grade
	Infectivity
	Host specificity
	Stability
Scale	Number of cells
	Amount of DNA
Genetic manipulation	Host specificity
	Vector infectivity
	Disablement
Containment	Room
	Cabinet
	Procedures
Radioisotopes	
Emission	Type
	Energy
	Penetration, shielding
	Interaction, ionization
	Half-life
Volatility	Inhalation
	Dissemination
Localization on ingestion	DNA precursors, such as [^{3}H]thymidine
Disposal	Solid, liquid, gaseous
	Route
	Legal limits

TABLE 7.1. Elements of Risk Assessment (*Continued*)

Category	Items affecting risk
Special Circumstances	
Pregnancy	Immunodeficiency
	Risk to fetus, teratogenicity
Illness	Immunodeficiency
Immunosuppressant drugs	Immunodeficiency
Cuts and abrasions	Increased risk of absorption
Allergy	Powders, e.g., detergents
	Aerosols
	Contact, e.g., rubber gloves
Elements of Procedures	
Scale	Amount of materials used
	Size of equipment & facilities and effect on containment
	Number of staff involved
Complexity	Number of steps or stages
	Number of options
	Interacting systems and procedures
Duration	Process time
	Incubation time
	Storage time
Number of persons involved	Increased risk?
	Diminished risk?
Location	Containment
	Security and access

how does one dispose of broken glass that has been in contact with a human cell line labeled with a radioisotope?).

7.4 SAFETY REGULATIONS

The following recommendations should not be interpreted as a code of practice but rather as advice that might help in compiling safety regulations. The information is designed to provide the reader with some guidelines and suggestions to help construct a local code of practice, in conjunction with regional and national legislation and in full consultation with the local safety committee. These recommendations have no legal standing and should not be quoted as if they do.

General safety regulations should be available from the safety office in the institution or company at which you work. In addition, they are available from the Occupational Safety and Health Administration (OSHA) in the United States [*www.osha-occupational-health-and-safety.com/*]. Europe, including the United Kingdom, comes under new joint regulations [in the UK based on Management of Health and Safety at Work Regulations, 1999]; *www.hmso.*

gov.uk/si/si1999/19993242.htm (Table 7.2). These regulations cover all matters of general safety. The relevant regulations and recommendations for biological safety for the United States are contained in *Biosafety in Microbiological and Biomedical Laboratories* [U.S. Department of Health and Human Services, 1993; *www.orcbs.msu.edu/biological/BMBL/BMBL-1.htm*], a joint document prepared by the Centers for Disease Control and Prevention in Atlanta, Georgia, and the National Institutes of Health in Bethesda, Maryland. For the United Kingdom, the Health Services Advisory Committee of the Health and Safety Commission has published two booklets: *Safe Working and the Prevention of Infection in Clinical Laboratories* [Health and Safety Commission, 1991a] and *Safe Working and the Prevention of Infection in Clinical Laboratories—Model Rules for Staff and Visitors* [Health and Safety Commission, 1991b]. Genetically modified cells are dealt with in *A Guide to the Genetically Modified Organisms (Contained Use) Regulations* [HSE, 1996], also from Her Majesty's Stationery Office. Several of the UK guidelines are under review and may be viewed at *www.hse.gov.uk/hthdir/noframes/bioldex.htm*. The advice given in this chapter is general and should not be construed as satisfying any legal requirement.

7.5 GENERAL SAFETY

Table 7.3 emphasizes those aspects of general safety that are particularly important in a tissue culture laboratory and should be used in conjunction with local safety rules.

7.5.1 Operator

It is the responsibility of the institution to provide the correct training, or to determine that the individual is already trained, in appropriate laboratory procedures and to ensure that new and existing members of staff are and remain familiar with safety regulations. It is the supervisor's responsibility to ensure that procedures are carried out correctly and that the correct protective clothing is worn at the appropriate times.

7.5.2 Equipment

A general supervisor should be appointed to be in charge of all equipment maintenance, electrical safety, and mechanical reliability. A curator should be put in charge of each specific piece of equipment to ensure that the day-to-day operation of the equipment is satisfactory and to train others in its use. Particular risks include the generation of toxic fumes or aerosols from centrifuges and homogenizers, which must be contained either by the design of the equipment or by placing them in a fume cupboard.

The electrical safety of equipment is dealt with in the United States by OHSA [*www.osha-occupational-health-and-safety.com/*], in the United Kingdom by the Safe Use of Work Equipment [1998; *www.hse.gov.uk/pubns/*

TABLE 7.2. Safety Regulations and Guidelines

Topic	US	UK	Europe
General	Occupational Safety and Health Administration (OSHA) [*www.osha-occupational-health-and-safety.com/*].	Management of Health and Safety at Work Regulations, 1999; *www.hmso.gov.uk/si/si1999/ 19993242.htm*	Management of Health and Safety at Work Directive 89/391/EEC
Equipment	OSHA [*www.osha-occupational-health-and-safety.com/*].	Provision and use of work equipment (PUWER) Regulations 1998; SI. 1998 No. 2306	Provision and use of work equipment Directive 89/355/EEC and 95/63/EC
Chemical	OSHA [*www.osha-occupational-health-and-safety.com/*]. National Institutes of Health [*www.niehs.nih.gov/odhsb/*]. National Institute for Occupational Safety and Health [NIOSH, *www.cde.gov/ niosh/homepage.html*]	Control of Substances Hazardous to Health [Health and Safety Commission, 1999; *www.hse.gov.uk/coshh/ index.htm*]	Council Directive 98/24/EC of 7 April 1998 on the protection of the health and safety of workers from the risks related to chemical agents at work (fourteenth individual Directive within the meaning of Article 16 (1) of Directive 89/391/EEC), Official Journal n° L 131 of 05.05.1998, p. 11.
Biological	*Biosafety in Microbiological and Biomedical Laboratories* [U.S. Department of Health and Human Services, 1999; *www.cdc.gov/od/ohs/biosfty/ bmbl4/bmbl4toc.htm*]	*Safe Working and the Prevention of Infection in Clinical Laboratories* [Health and Safety Commission, 1991a] and *Safe Working and the Prevention of Infection in Clinical Laboratories—Model Rules for Staff and Visitors* [Health and Safety Commission, 1991b]. *A Guide to the Genetically Modified Organisms (Contained Use) Regulations* [HSE, 1996; *www.hse.gov.uk/biosafety/*].	EC Directive on the Protection of workers from risks related to exposure to biological agents at work (90/679/EEC)
Radiological	U.S. Nuclear Regulatory Commission. Medical, Industrial, and Academic Uses of Nuclear Materials. *www.nrc. gov/materials/medical.html*	Work with Ionising Radiation, Ionising Radiation Regulations [1999].	

indg291.pdf], and in the European Union by the Provision and Use of Work Equipment [1995].

7.5.3 Glassware and Sharp Items

The most common form of injury in performing tissue culture results from accidental handling of broken glass and syringe needles. Particularly dangerous are broken pipettes in a washup cylinder, which result from too many pipettes being forced into too small a container (Fig. 7.1). Glass Pasteur pipettes should not be inserted into a washup cylinder with other pipettes (*see* below). Needles that have been improperly disposed of together with ordinary waste or forced through the wall of a rigid container when the container is overfilled are also very dangerous.

Accidental inoculation via a discarded needle or broken glass, or because of an accident during routine handling, remains one of the more acute risks associated with handling potentially biohazardous material. It may even carry a risk of transplantation when one handles human tumors [Southam, 1958; Scanlon et al., 1965; Gugel & Sanders, 1986], although reports of this are largely anecdotal.

Pasteur pipettes should be discarded into a sharps bin (*see* Appendix II and *www.cdc.gov/niosh/sharps1.html*) and not into the regular washup as they break very easily and the shards are extremely hazardous. If reused, they should be handled separately and with great care. Disposable plastic Pasteur pipettes (Kimble-Kontes; Alpha Laboratories—Pastettes) are available but tend to have a thicker tip. Avoid using

TABLE 7.3. General Procedures

Category	Action
Regulatory authority	Contact national, regional inspectors
Local Safety Committee	Appoint representatives
	Arrange meetings and discussion
Guidelines	Access local and national
	Generate local guidelines if not already done
Standard operating procedures (SOPs)	Define and make available
Protective clothing	Provide, launder, and ensure that it is worn correctly
Containerization	Specify physical description (e.g., storage and packaging)
Containment levels	Specify chemical, radiological, biological levels
Training	Arrange seminars, supervision
Monitoring	Automatic smoke detectors, oxygen meter
Inspection	Arrange equipment, procedures, laboratory inspections by trained, designated staff
Record keeping	Safety officers and operatives to keep adequate records
Import and export	Regulate and record
Classified waste disposal	Define routes for sharps, radioactive waste (liquid and solid), biohazards, corrosives, solvents, toxins
Access	Limit to trained staff and visitors only
	Exclude children, except in public areas

Fig. 7.1. Overfilled Pipette Cylinder. Pipettes protruding from a pipette cylinder as a result of attempted insertion of pipettes after the cylinder is full; those protruding from the cylinder will not soak properly or be disinfected and are prone to breakage (if glass) when other pipettes are added.

syringes and needles, unless they are needed for loading ampoules (use a blunt cannula) or withdrawing fluid from a capped vial. When disposable needles are discarded, use a rigid plastic or metal container. Do not attempt to bend, manipulate, or resheath the needle. Provide separate hard-walled receptacles for the disposal of sharp items and broken glass, and do not use these receptacles for general waste.

Take care when you are fitting a bulb or pipetting device onto a glass pipette. Choose the correct size of bulb to guard against the risk of the pipette breaking at the neck and lacerating your hand. Check that the neck is sound, hold the pipette as near the end as possible, and apply gentle pressure with the pipette pointing away from your knuckles (Fig. 7.2). Although this is primarily a risk arising from the use of glass pipettes, even plastic pipettes can be damaged and break on insertion, so always check the top of each pipette before use.

7.5.4 Chemical Toxicity

Relatively few major toxic substances are used in tissue culture, but when they are the conventional precautions should be taken, paying particular attention to the distribution of powders and aerosols by laminar-flow hoods (*see* Section 7.8.2). Detergents—particularly those used in automatic washing machines—are usually caustic, and even when they are not, they can cause irritation to the skin, eyes, and lungs. Use liquid-based detergents in a dispensing device whenever possible, wear gloves, and avoid procedures that cause powdered detergent to spread as dust. Liquid detergent concentrates are more easily handled, but often are more expensive. Chemical disinfectants such as hypochlorite should be used cautiously, either in tablet form or as a liquid dispensed from a dispenser. Hypochlorite disinfectants will bleach clothing, cause skin irritations, and even corrode welded stainless steel.

Specific chemicals used in tissue culture that require special attention are (1) dimethyl sulfoxide (DMSO), which is a powerful solvent and skin penetrant and can, therefore,

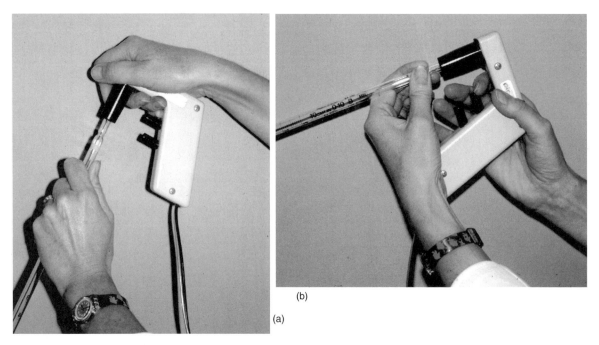

Fig. 7.2. Taking Care in Inserting a Pipette into a Pipetting Device. (a) Wrong position: left hand too far down pipette, risking contamination of the pipette and exerting too much leverage, which might break the pipette; right hand too far over and exposed to end of pipette or splinters should the pipette break at the neck during insertion. Tip of pipette is also obscured by left hand and arm, risking contamination by contact with nonsterile surface. (b) Correct position: left hand farther up pipette with lighter grip; right hand clear of top of pipette; tip of pipette in clear view.

carry many substances through the skin [Horita and Weber, 1964] and even through some protective gloves (e.g., rubber latex or silicone, though little through nitrile), and (2) mutagens, carcinogens, and cytotoxic drugs, which should be handled in a safety cabinet. A Class II laminar-flow hood may be adequate for infrequent handling of small quantities of these substances, but it may be necessary to use a hood designed specifically for cytotoxic chemicals (*see* Fig. 7.5b). Mutagens, carcinogens, and other toxic chemicals are sometimes dissolved in DMSO, increasing the risk of uptake via the skin. Nitrile gloves provide a better barrier but should be tested for the particular agents in use [*see also www.pacifica.com/NitrileGlovesChemicalResistanceBarrierGuide.pdf*].

The handling of chemicals is regulated by the Occupational Safety & Health Administration (OSHA; *www.osha.gov*) and by the Control of Substances Hazardous to Health [Health and Safety Commission 1999a, 1999b] in the United Kingdom. Information and guidelines are also available from the National Institutes of Health (*www.niehs.nih.gov/odhsb/*), and the National Institute for Occupational Safety and Health (NIOSH; *www.cde.gov/niosh/homepage.html*).

7.5.5 Gases

Most gases used in tissue culture (CO_2, O_2, N_2) are not harmful in small amounts but are nevertheless dangerous if handled improperly. They should be contained in pressurized cylinders that are properly secured (Fig. 7.3). If a major leak occurs, there is a risk of asphyxiation from CO_2 and N_2 and of fire from O_2. Evacuation and maximum ventilation are necessary in each case; if there is extensive leakage of O_2, call the fire department. An oxygen monitor should be installed near floor level in rooms where N_2 and CO_2 are stored in bulk, or where there is a piped supply to the room.

If glass ampoules are used, they are sealed in a gas oxygen flame. Great care must be taken both to guard the flame and to prevent inadvertent mixing of the gas and oxygen. A one-way valve should be incorporated into the gas line so that oxygen cannot blow back.

7.5.6 Liquid Nitrogen

Three major risks are associated with liquid nitrogen: frostbite, asphyxiation, and explosion (*see* Protocol 20.1). Because the temperature of liquid nitrogen is $-196°C$, direct contact with the liquid (via splashes, etc.), or with anything—particularly something metallic—that has been submerged in it, presents a serious hazard. Gloves that are thick enough to act as insulation, but flexible enough to allow the manipulation of ampoules, should be worn. When liquid nitrogen boils off during routine use of the freezer, regular ventilation is sufficient to remove excess nitrogen; but when nitrogen is being dispensed, or when a lot of samples are being inserted into the freezer, extra ventilation will be necessary. Remember, 1 L of liquid nitrogen generates nearly 700 L of gas. An oxygen monitor and alarm should be installed (*see* Fig. 4.7)

Fig. 7.3. Cylinder Clamp. Clamps onto edge of bench or rigid shelf and secures gas cylinder with fabric strap. Fits different sizes of cylinder and can be moved from one position to another if necessary; available from most laboratory suppliers.

and linked to the ventilation system so that the number of air changes in the room increases when the alarm goes off.

When an ampoule or vial is submerged in liquid nitrogen, a high pressure difference results between the outside and the inside of the ampoule. If the ampoule is not perfectly sealed, liquid nitrogen may be inspired, causing the ampoule to explode violently when thawed. This problem can be avoided by storing the ampoules in the gas phase or in a perfused jacket freezer (*see* Section 20.3.5) and by ensuring that the ampoules are perfectly sealed. Thawing of ampoules or vials that have been stored submerged in liquid nitrogen should always be performed in a container with a lid, such as a plastic bucket (*see* Protocol 20.2 and Fig. 20.9), and a face shield or goggles must be worn.

7.5.7 Burns
There are three main sources of risk from burns: (1) autoclaves, ovens, and hot plates, (2) handling of items that have just been removed from autoclaves, ovens, and hot plates, and (3) naked flames such as a Bunsen burner, if flaming is being used (*see* Table 7.3). Warning notices should be placed near all hot equipment, including burners, and items that have just been sterilized should be allowed to cool before removal from the sterilizer. Insulated gloves should be provided where hot items are being handled.

7.6 FIRE

Particular fire risks associated with tissue culture stem from the use of Bunsen burners for flaming, together with alcohol

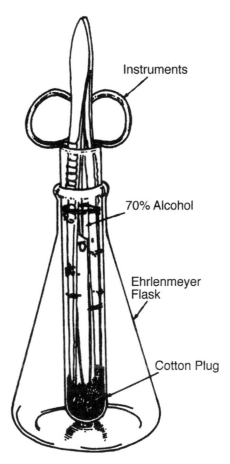

Fig. 7.4. Flask for Alcohol Sterilization of Instruments. The wide base prevents tipping, and the center tube reduces the amount of alcohol required, so that spillage, if it occurs, is minimized. (From an original idea by M. G. Freshney.)

for swabbing or sterilization. Keep the two separate; always ensure that alcohol for sterilizing instruments is kept in minimum volumes in a narrow-necked bottle or flask that is not easily upset (Fig. 7.4). Alcohol for swabbing should be kept in a plastic wash bottle or spray and should not be used in the presence of an open flame. When instruments are sterilized in alcohol and the alcohol is subsequently burnt off, care must be taken not to return the instruments to the alcohol while they are still alight. If you are using this technique, keep a damp cloth nearby to smother the flames if the alcohol ignites.

7.7 IONIZING RADIATION

Three main types of radiation hazard are associated with tissue culture: ingestion, irradiation from labeled reagents, and irradiation from a high-energy source. Guidelines on radiological protection for the United States can be obtained from the Office of Nuclear Regulatory Research (U.S. Nuclear Regulatory Commission, Washington, DC 20555)

TABLE 7.4. Safety Hazards in a Tissue Culture Laboratory

Category	Item	Risk	Precautions
General	Broken glass	Injury, infection	Dispose of carefully in designated bin
	Pipettes	Injury, infection	Check glass for damage and discard; use plastic
	Sharp instruments	Injury, infection	Handle carefully; discard in sharps bin.
	Glass Pasteur pipettes	Injury, infection	Handle carefully; do not use with potentially biohazardous material; use plastic
	Syringe needles	Injury, infection	Minimize or eliminate use; discard into sharps bin
	Cables	Fire, electrocution, snagging, tripping	Check connections sound; clip together and secure in safe place
	Tubing	Leakage, snagging, tripping	Check and replace regularly; clip in place; keep away from passage floors
	Cylinders	Instability, leakage	Secure to bench or wall; check regularly with leak detector
	Liquid nitrogen	Frostbite, asphyxiation, explosion	Wear mask, lab coat, and gloves; do not store ampoules in liquid phase or enclose when thawing
Burns	Autoclaves, ovens, & hot plates	Contact with equipment	Post warning notices
		Handling items just sterilized	Provide gloves
Fire	Bunsen burners; flaming, particularly in association with alcohol	Fire, melting damage, burn risk	Keep out of hoods and do not place under cupboards or shelves; do not return flaming instruments to alcohol
	Manual autoclaves	Can burn dry and contents ignite	Install a timer and a thermostatic cut-out Ensure that a safety valve is present and active
Radiological	Radioisotopes in sterile cabinet	Emission, spillage, aerosols, volatility	Work on absorbent tray in Class II or chemical hazard hood; minimize aerosols
	Irradiation of cultures	Radiation dose	Use monitor, wear personal badge monitor and check regularly
Biological	Importation of cell lines and biopsies	Infection	Do not import from high-risk areas; screen cultures for likely pathogens
	Genetic manipulation	Infection, DNA transfer	Follow genetic manipulation guidelines
	Propagation of viruses	Infection	Observe CDC or ACDP guidelines; work in correct level of containment; minimize aerosols
	Position and maintenance of laminar-flow hoods	Breakdown in containment	Check airflow patterns, pressure drop across filter, and overspill from cabinet regularly

and for the United Kingdom are contained in Work with Ionising Radiation, Ionising Radiation Regulations [1999].

7.7.1 Ingestion

Soluble radiolabeled compounds can be ingested by being splashed on the hands or via aerosols generated by pipetting or the use of a syringe. Tritiated nucleotides, if accidentally ingested, will become incorporated into DNA, and will cause radiolysis within the DNA due to the short path length of the low-energy β-emission from ^3H. Radioactive isotopes of iodine will concentrate in the thyroid and may also cause local damage.

Work in a Class II hood to contain aerosols, and wear gloves. The items that you are working with should be held in a shallow tray lined with paper tissue or Benchcote to contain any accidental spillage. Use the smallest pieces of equipment (e.g., a pipettor with disposable plastic tips, small sample tubes, etc.) compatible with the procedure to generate minimum bulk when they are discarded into a radioactive waste container. Clean up carefully when you are finished, and monitor the area regularly for any spillage.

7.7.2 Disposal of Radioactive Waste

Procedures and routes for disposal of radioactive substances will be defined in local rules governing the laboratory; advice in setting up these rules can be obtained from the authorities given above. Briefly, the amount of radioactivity disposed of over a certain period will have an upper limit, disposal will be limited to certain designated sinks, and the amounts discarded will need to be logged in a record book at the site of disposal. Vessels used for disposal will then need to be decontaminated in an appropriate detergent (*www.oshasafety.com/HazardLab.htm*), and the washings disposed of as radioactive waste. Disposal may need to take account of any biological risk, so items that are to be reused will first have to be biologically decontaminated in hypochlorite and then radioactively decontaminated in Decon or a similar detergent. Both solutions must then be regarded as radioactive waste.

7.7.3 Irradiation from Labeled Reagents

The second type of risk is from irradiation from higher-energy β- and γ-emitters such as ^{32}P, ^{125}I, ^{131}I, and ^{51}Cr. Protection can be obtained by working behind a 2-mm-thick lead shield and storing the concentrated isotope in a lead pot. Perspex screens (5 mm) can be used with ^{32}P at low concentrations for short periods. Work on a tray in a in a Class II hood (*see* Section 7.7.1).

7.7.4 Irradiation from High-Energy Sources

The third type of irradiation risk is from X-ray machines, high-energy sources such as ^{60}Co, or ultraviolet (UV) sources used for sterilizing apparatus or stopping cell proliferation in feeder layers (*see* Section 14.2.3 and Protocols 23.1, 23.4). Because the energy, particularly from X rays or ^{60}Co, is high, these sources are usually located in specially designed accommodations and are subject to strict control. UV sources can cause burns to the skin and damage to the eyes; they should be carefully screened to prevent direct irradiation of the operator, who should wear barrier filter goggles.

Consult your local radiological officer and code of practice before embarking on radioisotopic experiments. Local rules vary, but most places have strict controls on the amount of radioisotopes that can be used, stored, and discarded. Those wishing to use radioisotopes may be required to have a general medical examination, including storage of a blood sample, before starting work.

7.8 BIOHAZARDS

As for radioisotope use, those wishing to use potentially biohazardous material may require a general medical examination, including storage of a blood sample, before starting work. The need for protection against biological hazards [see Caputo, 1996] is defined both by the source of the material and by the nature of the operation being carried out. It is also governed by the conditions under which culture is performed. Using standard microbiological technique on the open bench has the advantage that the techniques in current use have been established as a result of many years of accumulated experience. Problems arise when new techniques are introduced or when the number of people sharing the same area increases. With the introduction of horizontal laminar-flow hoods, the sterility of the culture was protected more effectively, but the exposure of the operator to aerosols was more likely. This led to the development of

vertical laminar-flow hoods with an air curtain at the front (*see* Sections 5.2.1 and 6.4) to minimize overspill from within the cabinet. These are now defined as Class II microbiological safety cabinets (*see* Section 7.8.2).

7.8.1 Levels of Biological Containment

Four biological safety levels have been defined by the National Institutes of Health (NIH) and the Centers for Disease Control and Prevention (CDC) [HHS Publication No. (CDC) 93–8395] (Table 7.5). These concern practices and the facilities and safety procedures that they require. European legislation is still being drafted but is similar to current U.K. guidelines, which also define four levels of biological containment, although there are minor differences from the U.S. classification (Table 7.6). These tables provide summaries only, and anyone undertaking potentially biohazardous work should contact the local safety committee and consult the appropriate website [U.S.: *www.orcbs.msu.edu/biological/BMBL/BMBL-1.htm;* U.K.: *www.hse.gov.uk/biosafety/*]. European Union guidelines are currently under preparation.

7.8.2 Microbiological Safety Cabinets

Within the appropriate level of containment we can define three levels of handling determined by the type of safety cabinet used:

(1) ***Maximum Protection from Known Pathogens.*** A sealed pathogen cabinet with filtered air entering and leaving via a pathogen trap filter (biohazard hood or microbiological safety cabinet, Class III; Fig. 7.4c). The cabinet will generally be housed in a separate room with restricted access and with showering facilities and protection for solid and liquid waste (*see* BSL 4 in Table 7.5 and Level 4 in Table 7.6), depending on the nature of the hazard.

(2) ***Intermediate Level of Protection for Potential Hazards.*** A vertical laminar-flow hood with front protection in the form of an air curtain and a filtered exhaust (biohazard hood or microbiological safety cabinet, Class II; Fig. 7.4a) [National Sanitation Foundation Standard 49, 1983; British Standard BS5726, 1992; European Committee for Standardisation, 1999]. If recognized pathogens are being handled, hoods such as these should be housed in separate rooms, at containment levels 2, 3, or 4, depending on the nature of the pathogen. If there is no reason to suppose that the material is infected, other than by adventitious agents, then hoods can be housed in the main tissue culture facility, which may be categorized as containment level 2, requiring restricted access, control of waste disposal, protective clothing, and no food or drink in the area (*see* Tables 7.4, 7.5). All biohazard hoods must be subject to a strict maintenance program [Osborne et al., 1999], with the filters tested at regular intervals, proper arrangements made for fumigation of the cabinets before changing filters, and disposal of old filters made safe by extracting them into double bags for incineration.

TABLE 7.5. U.S. Biosafety Levels

	BSL 1 (SMP[1])	BSL 2 (SP[2])	BSL 3 (SP)	BSL 4 (SP)
Access	Access limited when work in progress. Controls against insect and rodent infestation.	Restricted. Predisposition assessed; immunization and baseline and periodic serum samples may be required. Hazard warnings posted when appropriate.	As BSL 2 + separate lab with two sets of self-closing doors. Doors closed when work in progress. Immunization or tests and baseline and periodic serum samples required.	As BSL 3 + Only designated personnel. Secure lockable doors. Changing space & shower. Door interlocks. Equipment and materials entry by double-ended autoclave or fumigation. No materials allowed except those required for the work being conducted.
Cleaning	Easy to clean; spaces between cabinets, equipment, etc.; impervious bench surfaces.	As BSL 1 + routine decontamination of work surfaces and equipment.	As BSL 2 + working on plastic-backed absorbent paper recommended. All room surfaces sealed and washable.	As BSL 3 + sealed joints, disinfectant traps on drains, HEPA filters on vents. Minimal surface area for dust.
Personal Hygiene	Lab coats worn. No eating, drinking etc. No mouth pipetting. Wear gloves and protective eyewear (especially with contact lenses). Sink for hand washing. Remove gloves & wash hands on leaving.	As BSL 1 + provision for decontamination and laundering in house. Eyewash facility.	As BSL 2 + goggles and mask or face shield outside BSC. Respiratory protection when aerosol cannot be controlled. Solid-front gowns, removed before leaving lab. Automatic or elbow taps on sink.	As BSL 3 + change of clothing; clothing autoclaved. Shower before leaving.
Airflow and ventilation	Not specified. Windows that open should have screens.	As BSL 1.	Windows closed and sealed. Negative pressure, total extract, exhaust away from occupied areas or air intakes.	Dedicated and alarmed non-recirculating ventilation system. Air exhaust through HEPA filters. Supply and extract interlocked.
Equipment	Not specified.	Routinely decontaminated and particularly before maintenance in house or away.	As BSL 2 + physical containment, e.g., sealed centrifuge cups and rotors. Any exhaust HEPA filtered. Vacuum lines protected by disinfectant traps and HEPA filters. Back-flow prevention devices.	As BSL 3.
Sharps	Not specified.	Restricted to unavoidable use. Used or broken items into containers, decontaminated before disposal.	As BSL 2.	As BSL 2.
MSCs	Not required but aerosol generation minimized.	Class II.	Class II or III exhausting directly via HEPA filter. Class II may recirculate.	Class III or Class II with positive-pressure personnel suit and life support system.

TABLE 7.5. U.S. Biosafety Levels (*Continued*)

	BSL 1 (SMP[1])	BSL 2 (SP[2])	BSL 3 (SP)	BSL 4 (SP)
Disinfection	Disinfectant available. Work surfaces decontaminated at least once per day.	As BSL 1 + procedures specified; autoclave nearby.	As BSL 2 + spills dealt with by trained staff.	As 3 + double-ended exit autoclave required, preferably from Class III cabinet.
Storage and transfer	Not specified.	Leak-proof container.	As BSL 2.	Viable materials leave in double-wrapped, nonbreakable containers via dunk tank or fumigation.
Disposal	Into disinfectant or by sealed container to nearby autoclave.	As BSL 1 + defined decontamination method.	As BSL 2 + decontamination within lab.	As BSL 3 + all effluent, excluding shower and toilet, and other materials disinfected before leaving via double-ended autoclave. Double-ended dunk tank for nonautoclavable waste.
Biosafety manual and training	Not specified.	Training required.	As BSL 2.	As BSL 3 + high proficiency in SMP
Accidents and spills	Not specified.	Written report. Medical evaluation available.	As BSL 2 + spills dealt with by trained staff.	As BSL 3 + monitoring absence, care of illness, and quarantine.
Validation of facility	Not specified.	Not specified.	Not specified.	Safety of effluent

[1] Standard Microbiological Practices.

[2] Special Practices, e.g. handling agents of moderate potential hazard.

(3) ***Minimal Protection.*** Open bench, depending on good microbiological technique. Again, this will normally be conducted in a specially defined area, which may simply be defined as the "tissue culture laboratory," but which will have level 1 conditions applied to it.

Table 7.7 lists common procedures with suggested levels of containment. All those using the facilities, however, should seek the advice of the local safety committees and the appropriate biohazard guidelines (*see* Section 7.2) for legal requirements.

7.8.3 Human Biopsy Material

Issues of biological safety are clearest when known, classified pathogens are being used, because the regulations covering such pathogens are well established both in the United States [U.S. Department of Health and Human Services, 1993] and in the United Kingdom [Advisory Committee on Dangerous Pathogens (ACDP), 1995a,b]. However, in two main areas there is a risk that is not immediately apparent in the nature of the material. One is in the development, by recombinant techniques such as transfection, retroviral infection, and interspecific cell hybridization, of new potentially pathogenic genes. Handling such cultures in facilities such as laminar-flow hoods introduces putative risks for which there are no epidemiological data available for assessment. Transforming viruses, amphitropic viruses, transformed human cell lines, human–mouse hybrids, and cell lines derived from xenografts in immunodeficient mice, for example, should be treated cautiously until there are enough data to show that they carry no risk.

The other area of risk is the inclusion of adventitious agents in human or other primate biopsy or autopsy samples or cell lines [Grizzle & Polt, 1988; Centers for Disease Control, 1988; Wells et al., 1989; Tedder et al., 1995] or in animal products such as serum, particularly if those materials are obtained from parts of the world with a high level of endemic infectious diseases. When infection has been confirmed, the type of organism will determine the degree of containment, but even when infection has not been confirmed, the possibility remains that the sample may yet carry hepatitis B, human immunodeficiency virus (HIV), tuberculosis, or other pathogens as yet undiagnosed. Confidentiality frequently prevents HIV testing without the patient's consent, and, for most of the adventitious infections, the appropriate information will not be available. If possible, biopsy material should be tested for potential adventitious infections before handling. The authority to do so should be written on the consent form that the person donating the tissue will have been asked to sign (*see* example, Table 7.8), but the need to get samples into culture quickly will often

TABLE 7.6. U.K. Biological Containment Levels

	Level 1	Level 2	Level 3	Level 4
Access	Door closed when work in progress.	Restricted.	As 2 + separate lab with observation window. Door locked when lab unoccupied.	As 3 + controls against insect and rodent infestation. Changing space & shower. Door interlocks. Ventilated air lock for equipment. Telephone or intercom.
Space	Not specified.	24 m³/person.	As 2.	As 2.
Cleaning	Easy to clean; impervious bench surfaces.	As 1 + routine decontamination of work surfaces.	As 2.	As 2.
Personal Hygiene	Lab coats (side or back fastening) worn, stored, cleaned, replaced correctly. No eating, drinking, etc. No mouth pipetting.	As 1 + wash basin near exit to decontaminate hands.	As 2 + wear gloves; remove or replace before handling common items such as phone.	As 3 + change of clothing; clothing autoclaved. Shower before leaving.
Airflow	Negative pressure preferable.	Negative pressure required.	As 2.	Negative pressure ≥70 pascals (7 mm H₂O); air exhaust through two HEPA filters in series. Supply and extract interlocked.
Equipment	Not specified.	Not specified.	Should contain own equipment.	Must contain own equipment.
MSCs	Not required but aerosol generation minimized.	MSC or isolator required.	Class I or III (BS5725); Class II (BS5726). Exhaust to via HEPA filter to outside.	Class III.
Disinfection	Disinfectant available.	Procedures specified; autoclave nearby.	As 2 + lab sealable for decontamination. Autoclave preferably within lab.	As 3 + double-ended exit autoclave required, preferably from Class III cabinet.
Storage	Not specified.	Safe storage of biological agents.	As 2.	
Disposal	Into disinfectant	As 1 + waste labeled. Safe collection and disposal.	As 2.	As 3 + all effluent, including shower, disinfected. Double-ended dunk tank for nonautoclavable waste.
Accidents	Report	As 1	As 2.	As 3 + 2ⁿᵈ person present to assist in case of emergency. Respirators available outside.
Validation of facility	Not specified.	Not specified.	Not specified.	Required

mean that you must proceed without this information. Such samples should be handled with caution:

(1) Transport specimens in a double-wrapped container (e.g., a universal container or screw-capped vial within a second screw-top vessel, such as a polypropylene sample jar). This in turn should be enclosed in an opaque plastic or waterproof paper envelope, filled with absorbent tissue packing to contain any leakage, and transported to the lab by a designated carrier.

(2) Enter all specimens into a logbook on receipt, and place the specimens in a secure refrigerator marked with a biohazard label.

(3) Carry out dissection and subsequent culture work in a designated Class II biohazard hood, preferably located in a separate room from that in which routine cell culture is performed. This will minimize the risk of spreading infections, such as mycoplasma, to other cultures and will also reduce the number of people associated with the specimen, should it eventually be found to be infected.

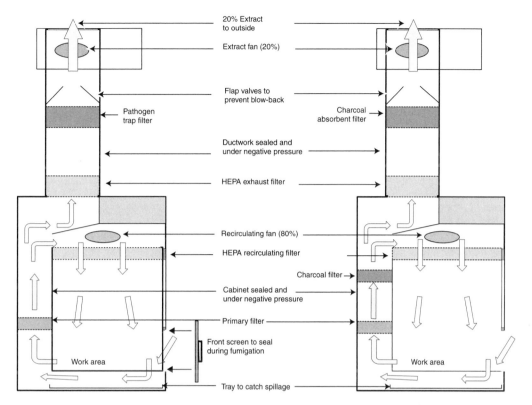

(a) CLASS II SAFETY CABINET   (b) CHEMICAL SAFETY CABINET

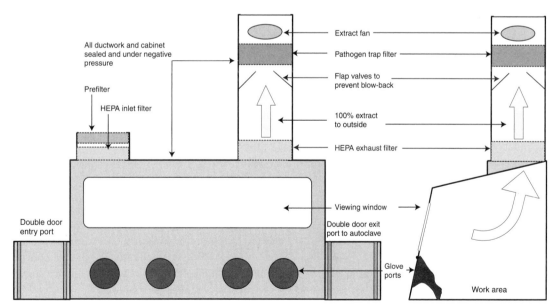

(c) CLASS III SAFETY CABINET: FRONT VIEW   (d) CLASS III SAFETY CABINET: SIDE VIEW

Fig. 7.5. Microbiological Safety Cabinets. (a) Class II vertical laminar flow, recirculating 80% of the air and exhausting 20% of the air via a filter and ducted out of the room through an optional pathogen trap. Air is taken in at the front of the cabinet to make up the recirculating volume and prevent overspill from the work area. (b) Class II chemical safety cabinet with charcoal filters on extract and recirculating air. (c) Class III non-recirculating, sealed cabinet with glove pockets; works at negative pressure and with air lock for entry of equipment and direct access to autoclave, either connected or adjacent. (d) Side view of Class III cabinet.

TABLE 7.7. Biological Procedures and Suggested Levels of Containment*

Procedure	Containment level	Work space
Preparation of media	GLP	Open bench with standard microbiological practice, or horizontal or vertical laminar flow
Primary cultures and cell lines other than human and other primates	1	Open bench with standard microbiological practice, or horizontal or vertical laminar flow
Primary cultures and cell lines, other than human and other primates, that have been infected or transfected	1	Class II laminar-flow hood
Primary culture and serial passage of human and other primate cells	2	Class II laminar-flow hood
Interspecific hybrids or other recombinants, transfected cells, human cells, and animal tumor cells	2	Class II laminar-flow hood
Human cells infected with retroviral constructs	3	Class II laminar-flow hood
Virus-producing human cell lines and cell lines infected with amphitropic virus	3	Class II laminar-flow hood
Tissue samples and cultures carrying known human pathogens	4	Class III pathogen cabinet with glove pockets, filtered air, and pathogen trap on vented air

*These are suggested procedures only and have no legal basis. Consult national legal requirements and local regulations before formulating proper guidelines.

(4) Avoid the use of sharp instruments (e.g., syringes, scalpels, glass Pasteur pipettes) in handling specimens. Clearly, this rule may need to be compromised when a dissection is required, but that should proceed with extra caution.

(5) Put all cultures in a plastic box with tape or labels identifying the cultures as biohazardous and with the name of the person responsible and the date on them (*see* Fig. 6.10).

(6) Discard all glassware, pipettes, instruments, etc., into disinfectant or into biohazard bags for autoclaving.

If appropriate clinical diagnostic tests show that the material is uninfected, and when it has been shown to be sterile and free of mycoplasma, the material may then be cultured with other stocks. However, if more than 1×10^9 cells are to be generated or if pure DNA is to be prepared, the advice of the local safety committee should be sought.

If a specimen is found to be infected, it should be discarded into double biohazard bags together with all reagents used with it and the bags should then be autoclaved or incinerated. Instruments and other hardware should be placed in a container of disinfectant, soaked for at least 2 h, and then autoclaved. If it is necessary to carry on working with the material, the level of containment must increase, according to the category of the pathogen [CDC Office of Health & Safety, 1999 (*www.cdc.gov/od/ohs/biosfty/bmbl4/bmbl4toc.htm*); Advisory Committee on Dangerous Pathogens, 2003; *www.hse.gov.uk/aboutus/meetings/acdp/*].

7.8.4 Genetic Manipulation

Any procedure that involves altering the genetic constitution of the cells or cell line that you are working with by transfer of nucleic acid will need to be authorized by the local biological safety committee. The current regulations may be obtained from the Environmental Protection Branch (NCI, NIH, Bldg 13, Room 2 W64, Bethesda, MD 20892) for the United States (information available on the Internet at *www.nci.nih.gov/intra/resource/biosafe.htm*) and from the Health and Safety Commission [1992] for the United Kingdom; *www.hse.gov.uk/biosafety/gmo/index.htm*.

7.8.5 Disposal of Biohazardous Waste

Potentially biohazardous materials must be sterilized before disposal [National Research Council, 1989; Health Services Advisory Committee, 1992]. They may be placed in unsealed autoclavable sacks and autoclaved, or they may be immersed in a sterilizing agent such as hypochlorite. Various proprietary preparations are available, e.g., Clorox or Chloros liquid concentrates and Precept Tablets (*see* Appendix II). Recommended concentrations vary according to local rules, but a rough guide can be obtained from the manufacturer's instructions. Hypochlorite is often used at 300 ppm of available chlorine, but some authorities demand 2500 ppm (a 1:20 dilution of Chloros), as recommended in the Howie Report [1978]. Hypochlorite is effective and easily washed off those items that are to be reused, but is highly corrosive, particularly in alkaline solutions. It will bleach clothing and even corrode stainless steel, so gloves and a lab coat or apron should be worn when handling hypochlorite, and soaking baths and cylinders should be made of polypropylene.

7.8.6 Fumigation

Some procedures may require that the microbiological safety cabinet be sterilized after use, e.g., if high-grade pathogens are being used, or if the cabinet requires servicing. Fumigation is usually carried out with formaldehyde, requiring the cabinet to be switched off and sealed before fumigation is initiated with an electrically heated generator. The hood is switched on briefly to circulate the gas and then left for 1 h. After this time the hood is allowed to run overnight to exhaust the vapor, opening the front after about 10 min. Fumigation of cabinets can also be carried out with hydrogen peroxide (Bioquell), which is more easily dispersed after fumigation is complete.

7.9 BIOETHICS

In addition to potential biohazards, working with human and animal tissue presents a number of ethical problems involving procurement, subsequent handling, and the ultimate use of the material.

7.9.1 Animal Tissue

Most countries involved in biomedical research will now have in place regulations governing the use of experimental or other donor animals for the provision of tissue. These will apply to higher animals assumed to have sufficient brain capacity and organization to feel pain and distress, and generally will not apply to lower vertebrates such as fish, or to invertebrates. Usually, a higher animal is assumed to be sentient any time after halfway through embryonic development and restrictions will apply to the method by which the animal is killed, or operated upon if it is to remain living, such that the animal suffers minimal pain or discomfort. Restrictions will also apply to the way the animal is housed and maintained, either in an animal house under experimental conditions or in a veterinary hospital under clinical conditions. In each case control is usually exercised locally by an animal ethics committee and nationally by a governmentally or professionally appointed body. Legislation varies considerably from country to country, but the appointment of a local animal ethics committee (AEC) is usually the first step to making contact with the appropriate licensing authority [U.S.: *www.bioscience.org/guides/animcare.htm*; U.K.: *www.ebra.org/regulat/uk.html*; *www.mrc.ac.uk/index/ sitemap.htm*: The Use of Animals in Research].

7.9.2 Human Tissue

The requirements with human material are different and rather more complex. Tissue will normally be collected under clinical conditions by an experienced medical practitioner, and the issues are more to do with the justification of taking the tissue and the uses to which it will be put. Again, there is local control through the local hospital ethics committee (HEC), who will decide whether the work is reasonable and justified by the possible outcome. The HEC must be contacted before any experimental work with human tissue is initiated, and this is best done at the planning stage, as most granting authorities will require evidence of ethical consent before awarding funding. Then there is the question of ownership of the tissue, its contents, such as DNA, any cell

lines that are derived from it, and any products or marketable procedures that might ultimately be developed and sold for profit. The following issues need to be addressed:

(1) The patient's and/or relative's informed consent is required before taking tissue for research purposes, over and above any clinical requirement.
(2) A suitable form (e.g., Table 7.8) should be drafted in a style readily understood by the patient or donor, requesting permission and drawing attention to the use that might be made of the tissue.
(3) Permission may be required from a relative if the donor is too unwell to be considered capable of reasonable judgment.
(4) A short summary of your project should be prepared, in lay terms, explaining what you are doing, why, and what the possible outcome will be, particularly if it is seen to be of medical benefit.
(5) Confidentiality of the origin of the tissue must be ensured.
(6) Ownership of cell lines and their derivatives must be established.
(7) Authority may be needed for subsequent genetic modification of the cell lines.
(8) Patent rights from any commercial collaboration will need to be established.
(9) The donor will need to determine whether any genetic information derived from the tissue should be fed back to the patient and/or physician.
(10) The donor will also be required to consent to screening of the tissue for adventitious pathogens and to say whether he or she wishes to be made aware of the outcome of the tests.

By far the easiest approach is to ask the donor and/or relatives to sign a disclaimer statement before the tissue is removed; otherwise, the legal aspects of ownership of the cell lines that might be derived and any future biopharmaceutical exploitation of the cell lines, their genes, and their products becomes exceedingly complex. Feedback of genetic information and evidence of a possible pathological infection such as HIV are more difficult problems; in the case of a patient in a hospital, the feedback is on a par with a diagnostic test and is most likely to be directed to the doctor, but in the case of a donor who is not hospitalized, you must ask the donor whether he or she wishes to know your findings and any implications that they might have. This may be done best via the donor's general practitioner. These factors are best dealt with by getting the donor to sign a consent form. Such a form may already have been prescribed by the HEC; if not, it will be necessary to prepare one (e.g., as in Table 7.8) in collaboration with the HEC and other involved parties, such as clinical collaborators, patient support groups, and funding authorities. Further information can be obtained under Research Ethics in *www.bioethics.gov/topics/other_index.html*, under Human

TABLE 7.8. Donor Consent Form

CONSENT TO REMOVE TISSUE FOR DIAGNOSIS AND RESEARCH

This form requests your permission to take a sample of your blood or one or more small pieces of tissue to be used for medical research. This sample, or cell lines or other products derived from it, may be used by a number of different research organizations, or it may be stored for an extended period awaiting use. It is also possible that it may eventually be used by a commercial company to develop future drugs. We would like you to be aware of this and of the fact that, by signing this form, you give up any claim that you own the tissue or its components, regardless of the use that may be made of it. You should also be aware of, and agree to, the possible testing of the tissue for infectious agents, such as the AIDS virus or hepatitis.

I am willing to have tissue removed for use in medical research and development. I have read and understand, to the best of my ability, the background material that I have been given. (If the donor is too unwell to sign, a close relative should sign on his or her behalf.)

Name of donor . . . . Name of relative . . . .

Signature . . . .

Date . . . .

This material will be coded, and absolute confidence will be maintained. Your name will not be given to anyone other than the person taking the sample.

Do you wish to receive any information from this material that relates to your health?   ***Yes/No***

Signature . . . . Date . . . .

Would you like, or prefer, that this information be given to your doctor?   ***Yes/No***

If yes, name of doctor . . . .

Address of doctor . . . .

. . . . .

. . . . .

Tissue: Ethical and Legal Issues in *www.nuffieldbioethics.org/ home/*, from the Medical Research Council in the U.K., and from WHO [Operational Guidelines for Ethics Committees That Review Biomedical Research]. In the U.K. approval should be requested from the Central Office for Research Ethics Committees (COREC) *www.corec.org.uk/* if a new HEC is to be set up, or an existing one requires advice.

Perhaps the most controversial aspect of the consent process is the need for the donor to know something about what the tissue will be used for. This requires a brief description in lay terms that will neither burden nor confuse the donor. Tissue transplantation has such clear objectives that little explanation of the science is required,

but some procedures, such as the examination of signal transduction anomalies in transformed cells, will require some generalization of the concept. Often, a brief overview in simple terms given orally can be accompanied by a more detailed description, though still in lay terms, emphasizing the potential advantages but also identifying the ethical issues, such as the subsequent genetic modification of the cells or the transplantation of the cells into another individual after tissue engineering.

7.10 VALIDATION

The proper use of cell lines, whether in research or commercial exploitation, requires that they are validated.

In an industrial environment, this will be a legal obligation if the ultimate product is to be accepted by the Federal Drug Administration (FDA) in the United States or the National Institute for Clinical Excellence (NICE) in the United Kingdom. However, in an academic research laboratory the requirement may be less well defined and the obligation left to individual conscience. Nevertheless, use of cell lines that are not properly validated reduces the reliability of the research and the likelihood that anyone will be able to repeat it.

There are three major elements to validation:

(1) Authentication: Is the cell line what it is claimed to be?
(2) Provenance: What has happened to the cell line since its original isolation?
(3) Contamination: Is the cell line free from all known forms of microbial contamination?

7.10.1 Authentication

Several techniques are available to give a specific profile of the cell line (*see* Section 16.6). DNA fingerprinting and profiling are probably the best but require that DNA is available from the donor, or at least from an earlier generation of the cell line known to be authentic at the time it was preserved. Failing this, a compilation of several characteristics will confirm the origin (species, tissue, etc.; *see* Section 16.3) beyond reasonable doubt. It should be remembered that one or more of the criteria that are used can be specific to the laboratory in which the cells are being used, sufficient for the purpose, though not necessarily readily transferable to another laboratory. The important issue is that some steps must be taken to authenticate the cell line before a major expenditure of time, effort, and funds is committed.

7.10.2 Provenance

Part of the validation process requires that there is a record of how a cell line was isolated and what has happened to it since isolation: maintenance regimens, contamination checks, decontamination procedures if used, properties expressed, genetic modification, spontaneous alterations, etc. Some knowledge of the provenance of the cell line, derived from the published literature or by word of mouth from a colleague, will have been the reason for selecting it in the first place but should be independently documented and added to as work progresses with the cell line. This means that proper records should be kept at all times (*see* Sections 12.3.11, 13.7.8, 20.3.6) detailing routine maintenance, significant experimental observations, and cryostorage. This does not have to be laborious, as the use of a spreadsheet or database will allow a new record to be made of a repeated procedure without having to rekey all the data except the date and that which is new or has changed. A cell line with a good provenance gains value like a piece of antique furniture or a painting.

7.10.3 Contamination

However detailed your records or meticulous your experimental technique, the resulting work is devoid of value, or at least heavily compromised, if the cell line is shown to be contaminated with one or more microorganisms. Where the contamination is overt, it is less of a problem, as cultures can be discarded, but often it is cryptic because (1) the cells have been maintained in antibiotics (*see* Section 19.3); (2) routine testing for organisms such as mycoplasma has not been carried out (*see* Section 19.3.2); or (3) there is no routine test available for the organism, e.g., some viruses or prions (*see* Section 19.3.6). Contamination can be avoided (1) by observing proper aseptic technique (*see* Chapter 6); (2) by obtaining cell lines from a properly validated source, e.g., a cell bank; (3) by culturing cells in the absence of antibiotics, even if only for part of the time (*see* Section 13.7.7); (4) by screening regularly for mycoplasma (staining with Hoechst 33258 will detect any DNA-containing organism big enough to be resolved under a fluorescence microscope; *see* Protocol 19.2); or (5) by screening for the most common viruses using PCR or a commercial contract.

Cell lines that have been properly validated should be stored in liquid nitrogen and issued to end users as required. End users may store their own stock for the duration of a project, but as these user stocks are no longer fully validated, they should not be passed on, and new users should revert to the validated stock.

CHAPTER 8

Culture Vessels and Substrates

8.1 THE SUBSTRATE

8.1.1 Attachment and Growth

The majority of vertebrate cells cultured *in vitro* grow as monolayers on an artificial substrate. Hence the substrate must be correctly charged to allow cell adhesion, or at least to allow the adhesion of cell-derived attachment factors, which will, in turn, allow cell adhesion and spreading. Although spontaneous growth in suspension is restricted to hemopoietic cell lines, rodent ascites tumors, and a few other selected cell lines, such as human small-cell lung cancer [Carney et al., 1981], many transformed cell lines can be made to grow in suspension and become independent of the surface charge on the substrate. However, most normal cells need to spread out on a substrate to proliferate [Folkman & Moscona, 1978; Ireland et al., 1989; Danen & Yamada, 2001], and inadequate spreading due to poor adhesion or overcrowding will inhibit proliferation. Cells shown to require attachment for growth are said to be *anchorage dependent*; cells that have undergone transformation frequently become *anchorage independent* (*see* Section 18.5.1) and can grow in suspension (*see* Section 13.7.4) when stirred or held in suspension with semisolid media such as agar.

8.1.2 Substrate Materials

Glass. This was the original substrate because of its optical properties and surface charge, but it has been replaced in most laboratories by synthetic plastic (usually polystyrene), which has greater consistency and superior optical properties. Glass is now rarely used, although it is cheap, is easily washed without losing its growth-supporting properties, can be sterilized readily by dry or moist heat, and is optically clear. Treatment with strong alkali (e.g., NaOH or caustic detergents) renders glass unsatisfactory for culture until it is neutralized by an acid wash (*see* Section 11.3.1).

Disposable plastic. Single-use sterile polystyrene flasks provide a simple, reproducible substrate for culture. They are usually of good optical quality, and the growth surface is flat, providing uniformly distributed and reproducible monolayer cultures. As manufactured, polystyrene is hydrophobic and does not provide a suitable surface for cell attachment, so tissue culture plastics are treated by corona discharge, gas plasma, or γ-irradiation, or chemically, to produce a charged, wettable surface. Because the resulting product can vary in quality from one manufacturer to another, samples from a number of sources should be tested by determining the growth rate and plating efficiency of cells in current use (*see* Protocols 21.7–21.10) in the appropriate medium containing the optimal and half-optimal concentrations of serum or serum free. (High serum concentrations may mask imperfections in the plastic.)

Although polystyrene is by far the most common and cheapest plastic substrate, cells may also be grown on polyvinylchloride (PVC), polycarbonate, polytetrafluorethylene (PTFE; Teflon), Melinex, Thermanox (TPX), and a number of other plastics. To test a new substrate, grow the cells on it as a regular monolayer, with and without pretreating the surface (*see* Section 8.4.1), and then clone cells (*see* Protocol 21.10). PTFE can be used in a charged (hydrophilic) or uncharged (hydrophobic) form [Janssen et al., 2003;

Lehle et al., 2003]; the charged form can be used for regular monolayer cells and organotypic culture (Biopore, Millipore; Transwell, Corning) and the uncharged for macrophages [von Briesen et al., 1990] and some transformed cell lines.

8.2 CHOICE OF CULTURE VESSEL

Some typical culture vessels are listed in Table 8.1. The anticipated yield of HeLa cells is quoted for each vessel; the yield from a finite cell line (e.g., diploid fibroblasts) would be about one-fifth of the HeLa figure. Several factors govern the choice of culture vessel, including (1) the cell mass required, (2) whether the cells grow in suspension or as a monolayer, (3) whether the culture should be vented to the atmosphere or sealed, (4) the frequency of sampling, (5) the type of analysis required, and (6) the cost.

8.2.1 Cell Yield

For monolayer cultures, the cell yield is proportional to the available surface area of the flask (Fig. 8.1). Small volumes and multiple replicates are best performed in multiwell plates (Fig. 8.2), which can have a large number of small wells (e.g., microtitration plates with 96 or 144 wells, 0.1–0.2 mL of medium, and 0.25-cm^2 growth area or 24-well "cluster dishes" with 1–2 mL medium in each well, 1.75-cm^2 growth area) up to 4-well plates with each well 5 cm in diameter and

TABLE 8.1. Culture Vessel Characteristics

Culture vessel	Replicates	mL	cm^2*	Approximate cell yield (HeLa)
Multiwell plates				
Microtitration	96	0.1	0.3	1×10^5
Microtitration	144	0.1	0.3	1×10^5
4-well plate	4	2	2	5×10^5
6-well plate	6	2	10	2×10^6
12-well plate	12	1	3	7.5×10^5
24-well plate	24	1	2	5×10^5
Petri dishes				
3.5-cm diameter	1	2	8	2×10^6
5-cm diameter	1	4	17.5	4×10^6
6-cm diameter	1	5	21	5×10^6
9-cm diameter	1	10	49	1×10^7
Flasks				
10 cm^2 (T10)	1	2	10	2×10^6
25 cm^2 (T25)	1	5	25	5×10^6
75 cm^2 (T75)	1	25	75	2×10^7
175 cm^2 (T175)	1	75	175	5×10^7
225 cm^2 (T225)	1	100	225	6×10^7
Roller bottle	1	200	850	2.5×10^8
Stirrer bottles				
500 mL (unsparged)	1	50		5×10^7
5000 mL (sparged)	1	4000		4×10^9

*These figures are approximate; actual areas will vary by source.

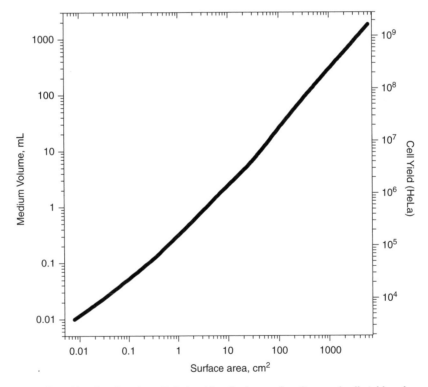

Fig. 8.1. Cell Yield and Surface Area. Relationship of volume of medium and cell yield to the surface area of a culture vessel. The graph is plotted on the basis of the volume of the medium for each size of vessel and is nonlinear, as smaller vessels tend to be used with proportionally more medium than is used with larger vessels. The cell yield is based on the volume of the medium and is approximate.

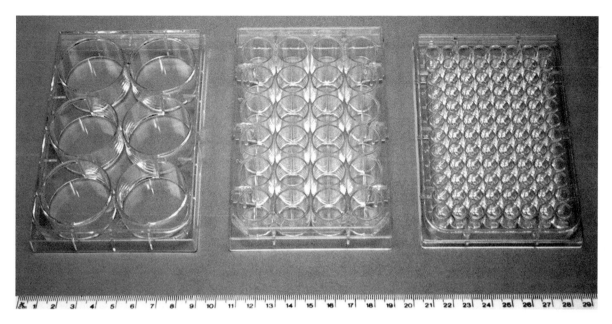

Fig. 8.2. Multiwell Plates. Six-well, 24-well, and 96-well (microtitration) plates. Plates are available with a wide range in the number of wells, from 4 to 144 (*see* Table 8.1 for sizes and capacities).

Fig. 8.3. Petri Dishes. Illustrated are dishes of 3.5-cm, 5-cm, and 9-cm diameter. Square Petri dishes are also available, with dimensions 9 × 9 cm. A grid pattern can be provided to help in scanning the dish—for example, in counting colonies—but can interfere with automatic colony counting.

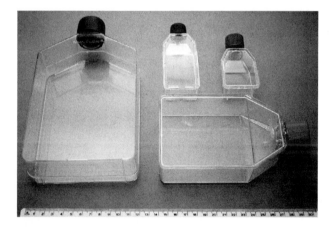

Fig. 8.4. Plastic Flasks. Sizes illustrated are 10 and 25 cm^2 (Falcon, B-D Biosciences), 75 cm^2 (Corning), and 185 cm^2 (Nalge Nunc) (*see* Table 8.1 for representative sizes and capacities).

using 5 mL of culture medium (*see* Table 8.1). The middle of the size range embraces both Petri dishes (Fig. 8.3) and flasks ranging from 10 cm^2 to 225 cm^2 (Fig. 8.4). Flasks are usually designated by their surface area (e.g., 25 cm^2 or 175 cm^2, and sometimes T25 or T175, respectively), whereas Petri dishes are referred to by diameter (e.g., 3.5 cm or 9 cm).

Glass bottles are more variable than plastic because they are usually drawn from standard pharmaceutical supplies. Glass bottles should have (1) one reasonably flat surface, (2) a deep screw cap with a good seal and nontoxic liner, and (3) shallow-sloping shoulders to facilitate harvesting of

monolayer cells after trypsinization and to improve the efficiency of washing.

If you require large cell yields (e.g., ~1 × 10^9 HeLa cervical carcinoma cells or 2 × 10^8 MCR-5 diploid human fibroblasts), then increasing the size and number of conventional bottles becomes cumbersome, and special vessels are required. Flasks with corrugated surfaces (Corning, Becton Dickinson) or multilayered flasks (Corning, Nalge Nunc) offer an intermediate step in increasing the surface area (Fig. 8.5). Cell yields beyond that require multisurface propagators or roller bottles on special racks (*see* Section 26.2.3). Increasing the yield of cells growing in suspension requires only that the volume of the medium be increased, as long as cells in deep

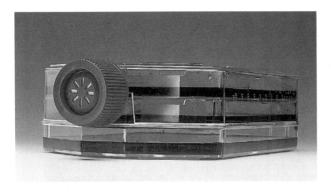

Fig. 8.5. Multisurface Flask. The Nunc Triple-Flask with three 80-cm^2 growth surfaces that are seeded simultaneously. Although the growth surface is 240 cm^2, the shelf space is equivalent to a regular 80-cm^2 flask. As the head space for gas phase is smaller, this flask is best used with a filter cap in a CO_2 incubator. (Photograph courtesy of Nalge Nunc International.)

culture are kept agitated and sparged with 5% CO_2 in air (*see* Section 26.1).

8.2.2 Suspension Culture

Cells that grow in suspension can be grown in any type of flask, plate, or Petri dish that, although sterile, need not be treated for cell attachment. Stirrer bottles are used when agitation is required to keep the cells in suspension. These bottles are available in a wide range of sizes, usually in glass (Bellco, Techne). Agitation is usually by a suspended paddle containing a magnet, whose rotation is driven by a magnetic stirrer (Fig. 8.6; *see also* Figs. 12.7 and 25.1). The rotational

Fig. 8.6. Small Stirrer Flasks. Four small stirrer flasks (Techne), 250-mL capacity, with 50–100 mL medium, on four-place stirrer rack (Techne). Larger flasks, up to 10 L, are available (*see also* Figs. 13.5, 26.1, 26.2).

speed must be kept low, ~60 rpm, to avoid damage from shear stress. Generally, the pendulum design is preferable for minimizing shear. Suspension cultures can be set up as replicates or can be sampled repetitively from a side arm or the flask. They can also be used to maintain a steady-state culture by adding and removing medium continuously (*see* Section 26.1.1).

8.2.3 Venting

Multiwell and Petri dishes, chosen for replicate sampling or cloning, have loose-fitting lids to give easy access to the dish. Consequently, they are not sealed and require a humid atmosphere with the CO_2 concentration controlled (*see* Section 9.2.2). As a thin film of liquid may form around the inside of the lid, partially sealing some dishes, vented lids with molded plastic supports inside should be used (Fig. 8.7). If a perfect seal is required, some multiwell dishes can be sealed with self-adhesive film (*see* Plate sealers in Appendix II).

Flasks may be vented by slackening the caps one full turn, when in a CO_2 incubator, to allow CO_2 to enter or to allow excess CO_2 to escape in excessive acid-producing cell lines. However, caps with permeable filters that permit equilibration with the gas phase (Fig. 8.8) are preferable

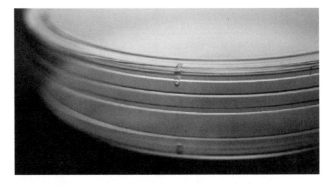

Fig. 8.7. Venting Petri Dishes. Vented dish. Small ridges, 120° apart, raise the lid from the base and prevent a thin film of liquid, e.g., condensate, from sealing the lid and reducing the rate of gas exchange.

Fig. 8.8. Venting Flasks. Gas-permeable cap on 10-cm^2 flask (Falcon, B-D Biosciences).

(although more expensive) as they allow CO_2 diffusion without risk of contamination.

8.2.4 Sampling and Analysis

Multiwell plates are ideal for replicate cultures if all samples are to be removed simultaneously and processed in the same way. If, on the other hand, samples need to be withdrawn at different times and processed immediately, it may be preferable to use separate vessels (*see also* Section 21.8 and Fig. 8.9). Individual wells in microtitration plates can be sampled by cutting and removing only that part of the adhesive plate sealer overlying the wells to be sampled. Alternatively, microtitration plates are available with removable wells for individual processing (Nalge Nunc), although you should ensure that the wells are treated for tissue culture if you wish to use adherent cells.

Low-power microscopic observation is performed easily on flasks, Petri dishes, and multiwell plates with the use of an inverted microscope. In using phase contrast, however, difficulties may be encountered with microtitration plates because of the size of the meniscus; even 24-well plates can only be observed satisfactorily by phase contrast in the center of the well. If microscopy will play a major part in your analysis, it may be advantageous to use a chamber slide (*see* Section 16.4.3 and Fig. 16.3). Large roller bottles give problems with some microscopes; it is usually necessary to remove the condenser, in which case phase contrast will not be available.

If processing of the sample involves extraction in acetone, toluene, ethyl acetate, or certain other organic solvents, then a problem will arise with the solubility of polystyrene. As this problem is often associated with organic solvents used in histological procedures, Lux (Bayer, MP

Biomedicals) supplies solvent-resistant Thermanox (TPX) plastic coverslips, suitable for histology, that fit into regular multiwell dishes (which need not be of tissue culture grade). However, these coverslips are of poor optical quality and should be mounted on slides with cells uppermost and a conventional glass coverslip on top.

Glass vessels are required for procedures such as hot perchloric acid extractions of DNA. Plain-sided test tubes or Erlenmeyer flasks (with no lip), used in conjunction with sealing tape or Oxoid caps, are quick to use and are best kept in a humid CO_2-controlled atmosphere. Regular glass scintillation vials, or "minivials," are also good culture vessels, because they are flat bottomed. Once used with scintillation fluid, however, they should not be reused for culture.

8.2.5 Uneven Growth

Sometimes, cells can be inadvertently distributed nonuniformly across the growth surface. Vibration, caused by opening and closing of the incubator, a faulty fan motor, or vibration from equipment, can perturb the medium, which can result in resonance or standing waves in the flask that, in turn, result in a wave pattern in the monolayer (Fig. 8.10), creating variations in cell density. Eliminating vibration and minimizing entry into the incubator will help reduce uneven growth. Placing a heavy weight in the tray or box with the plates and separating it from the shelf with plastic foam may also help alleviate the problem [Nielsen, 1989], but great care must be taken to wash and sterilize such foam pads, as they will tend to harbor contamination.

8.2.6 Cost

Cost always has to be balanced against convenience; for example, Petri dishes are cheaper than flasks with an

Fig. 8.9. Screw-Cap Vials and Flasks. (a) Glass flasks are suitable for replicate cultures or storage of samples, particularly when plastic may not survive downstream processing. Screw caps are preferable to stoppers, as they are less likely to leak and they protect the neck of the flask from contamination. (b) Scintillation vials are particularly useful for isotope incorporation studies but should not be reused for culture after containing scintillation fluid.

Fig. 8.10. Nonrandom Growth. Examples of ridges seen in cultured monolayers in dishes and flasks, probably due to resonance in the incubator from fan motors or to opening and closing of the incubator doors. (Courtesy of Nalge Nunc.)

equivalent surface area, but require humid, CO_2-controlled conditions and are more prone to infection. They are, however, easier to examine and process.

Cheap soda glass bottles, although not always of good optical quality, are often better for culture than higher-grade Pyrex, or optically clear glass, which usually contains lead. A major disadvantage of glass is that its preparation is labor intensive, because it must be carefully washed and resterilized before it can be reused. Most laboratories now use plastic, because of its convenience, optical clarity, and quality.

8.3 SPECIALIZED SYSTEMS

8.3.1 Permeable Supports

Semipermeable membranes are used as gas-permeable substrates and will also allow the passage of water and small molecules (<500–1000 da), a property that is exploited in some large-scale bioreactors (*see* Sections 25.3.2 and 26.2.6). Growing cells on a water-permeable substrate contributes more than an increased diffusion of oxygen, CO_2, and nutrients. Attachment of cells to a natural substrate such as collagen may control phenotypic expression due to the interaction of integrin receptors on the cell surface with specific sites in the extracellular matrix (*see* Sections 3.2.1, 3.2.3, and 17.7.3). Extensive use has also been made of natural gels, such as collagen. The growth of cells on floating collagen [Michalopoulos and Pitot, 1975; Lillie et al., 1980] has been used to improve the survival of epithelial cells and promote terminal differentiation (*see* Sections 17.7.3 and 25.3.8).

Filter wells. The permeability of the surface to which the cell is anchored may induce polarity in the cell by simulating the basement membrane. Such polarity may be vital to full functional expression in secretory epithelia and many other types of cells [Gumbiner & Simons, 1986; Chambard et al., 1987; Artur: Artursson & Magnusson, 1990; Mullin et al., 1997]. Several manufacturers now provide permeable supports in the form of disposable filter well inserts of many different sizes, materials, and membrane porosities (Costar, B-D Biosciences,

Fig. 8.11. Hollow Fiber Culture. A bundle of hollow fibers of permeable plastic is enclosed in a transparent plastic outer chamber and is accessible via either of the two side arms for seeding cells. During culture, the chamber is perfused down the center of the hollow fibers through connections attached to either end of the chamber (CellMax, Spectrum Labs; *see also* Fig. 26.6).

Millipore, Nunc). Also available are supports precoated with collagen, laminin, or other matrix materials (e.g., Matrigel, B-D Biosciences). All of these supports have been used extensively in studies of cell–cell interaction, cell–matrix interaction, differentiation and polarity, and transepithelial permeability (*see* Section 25.3.6).

Hollow fibers. Knazek et al. [1972] developed a technique for growing cells on the outer surface of bundles of plastic microcapillaries (Fig. 8.11; *see* Section 25.3.2). The plastic allows the diffusion of nutrients and dissolved gases from a medium perfused through the capillaries. Cells will grow up to several cells deep on the outside of the capillaries, and an analogy with whole tissue is implied. Hollow fibers are also used in large-scale bioreactors (*see* Section 26.2.6).

8.4 TREATED SURFACES

8.4.1 Matrix Coating

Conditioning. Cell attachment and growth can be improved by pretreating the substrate [Barnes et al., 1984a]. A well-established piece of tissue culture lore has it that used glassware supports growth better than new. If that is true, it may be due to etching of the surface or minute traces of residue left after culture. The growth of cells in a flask also improves the surface for a second seeding, and this type of conditioning may be due to collagen, fibronectin, or other matrix products [Crouch et al., 1987] released by the cells. The substrate can also be conditioned by treating it with spent medium from another culture [Stampfer et al., 1980], with serum, or with purified fibronectin or collagen (*see* Protocol 23.9).

Polylysine. McKeehan and Ham [1976a] found that it was necessary to coat the surface of plastic dishes with 1 mg/mL of poly-D-lysine before cloning in the absence of serum (*see* Section 14.2.1), so some effects of conditioning may be related to the surface charge.

Collagen and gelatin. Treatment with denatured collagen improves the attachment of many types of cells, such as epithelial cells, and the undenatured gel may be necessary for the expression of differentiated functions (*see* Sections 3.2.3, 17.7.3, 23.2.1). Gelatin coating has been found to be beneficial for the culture of muscle [Richler and Yaffe, 1970] and endothelial cells [Folkman et al., 1979] (*see* Section 23.3.6), and it is necessary for some mouse teratomas. Coating with denatured collagen may be achieved by using rat tail collagen or commercially supplied alternatives (e.g., Vitrogen) and simply pouring the collagen solution over the surface of the dish, draining off the excess, and allowing the residue to dry. Because this procedure sometimes leads to detachment of the collagen layer during culture, a protocol was devised by Macklis et al. [1985] to ensure that the collagen would remain firmly anchored to the substrate, by cross-linking to the plastic with carbodiimide. Collagen can also be used in conjunction with fibronectin (*see* Protocol 23.9).

Collagen may also be applied as an undenatured gel (*see* Section 17.7.3), a type of substrate that has been shown to support neurite outgrowth from chick spinal ganglia [Ebendal, 1976] and morphological differentiation of breast cells [Nicosia and Ottinetti, 1990; Berdichevsky et al., 1992] and hepatocytes [Sattler et al., 1978; Fiorino et al., 1998], and to promote the expression of tissue-specific functions of a number of other cells *in vitro* (e.g., keratinocytes [Maas-Szabowski et al., 2002], *see* Section 23.1.1). Diluting the concentrated collagen 1:10 with culture medium and neutralizing to pH 7.4 causes the collagen to gel, so the dilution and dispensing must be rapid. It is best to add the growth medium to the gel for a further 4–24 h to ensure that the gel equilibrates with the medium before adding cells. At this stage, fibronectin (25–50 μg/mL) or laminin (1–5 μg/mL), or both, may be added to the medium.

Matrigel. Commercially available matrices, such as Matrigel (Becton Dickinson) from the Engelbreth-Holm-Swarm (EHS) sarcoma, contain laminin, fibronectin, and proteoglycans, with laminin predominating (*see also* Section 17.7.3). Other matrix products include Pronectin F (Protein Polymer Technologies), laminin, fibronectin, vitronectin (B-D Biosciences, Biosource International) entactin (US Biological), heparan sulfate, EHS Natrix (B-D Biosciences), ECL (US Biological), and Cell-tak (B-D Biosciences). Some of these products are purified, if not completely chemically defined; others are a mixture of matrix

products that have been poorly characterized and may also contain bound growth factors. If cell adhesion for survival is the main objective, and defined substrates are inadequate, the use of these matrices is acceptable, but if mechanistic studies are being carried out, they can only be an intermediate stage on the road to a completely defined substrate.

Extracellular matrix. Although inert coating of the surface may suffice, it may yet prove necessary to use a monolayer of an appropriate cell type to provide the correct matrix for the maintenance of some specialized cells. Gospodarowicz et al. [1980] were able to grow endothelium on extracellular matrix (ECM) derived from confluent monolayers of 3T3 cells that had been removed with Triton X-100. This residual ECM has also been used to promote differentiation in ovarian granulosa cells [Gospodarowicz et al., 1980] and in studying tumor cell behavior [Vlodavsky et al., 1980].

PROTOCOL 8.1. PREPARATION OF ECM

Outline
Remove a postconfluent monolayer of matrix-forming cells with detergent, wash flask or dish, and seed required cells onto residual matrix.

Materials
- ❏ Mouse fibroblasts, e.g., 3T3, MRC-5 human fibroblasts, or CPAE bovine pulmonary arterial endothelial cells (or any other cell line shown to be suitable for producing extracellular matrix)
- ❏ Sterile, ultrapure water (UPW) (*see* Section 11.4.1)
- ❏ Triton X-100, 1% in sterile UPW

Protocol
1. Set up matrix-producing cultures, and grow to confluence.
2. After 3–5 days at confluence, remove the medium and add an equal volume of sterile 1% Triton X-100 in UPW to the cell monolayer.
3. Incubate for 30 min at 37°C.
4. Remove Triton X-100 solution and wash residue three times with the same volume of sterile UPW.
5. Flasks or dishes may be used directly or may be stored at 4°C for up to 3 wk.

8.4.2 Feeder Layers

Although matrix coating may help attachment, growth, and differentiation, some cultures of more fastidious cells, particularly at low cell densities [Puck & Marcus, 1955], require support from living cells (e.g., mouse embryo fibroblasts; *see* Protocol 13.3). This action is due partly to supplementation of the medium by either metabolite leakage or the secretion of growth factors from the fibroblasts, but may also be due to conditioning of the substrate by cell products. Feeder layers grown as a confluent monolayer may make the surface suitable, or even selective, for attachment for other cells (*see* Sections 14.2.3, 24.7.2 and Protocols 23.1, 23.4). The survival and extension of neurites by central and peripheral neurons can be enhanced by culturing the neurons on a monolayer of glial cells, although in this case the effect may be due to a diffusible factor rather than direct cell contact [Seifert & Müller, 1984].

After a monolayer culture reaches confluence, subsequent proliferation causes cells to detach from the artificial substrate and migrate over the surface of the monolayers. The morphology of the cells may change (Fig. 8.12): The cells may become less well spread, more densely staining, and more highly differentiated. Apparently, and not too surprisingly, the interaction of a cell with a cellular underlay is different from the interaction of the cell with a synthetic substrate. The former can cause a change in morphology and reduce the cell's potential to proliferate.

8.4.3 Three-Dimensional Matrices

It is apparent that many functional and morphological characteristics are lost during serial subculture, due to the loss of tissue architecture and cell-cell interaction (*see* Section 3.4.2). These deficiencies encouraged the exploration of three-dimensional matrices, such as collagen gel [Douglas et al., 1980], cellulose sponge (either alone or coated with collagen) [Leighton et al., 1968], or Gelfoam (*see* Section 25.3.1). Fibrin clots were one of the first media to be used for primary culture and are still used either as crude plasma clots (*see* Section 12.3.1) or as purified fibrinogen mixed with thrombin. Both systems generate a three-dimensional gel in which cells may migrate and grow, either on the solid–gel interface or within the gel [Leighton, 1991].

Three-dimensional matrices are used extensively in tissue engineering [Vunjak-Novakovic and Freshney, 2005] and can be inorganic, such as calcium phosphate, or organic, such as Gelfoam (*see* Section 25.3.8). As well as permitting cell attachment, proliferation, and differentiation, such matrices, or *scaffolds*, are also required to degrade *in vivo* and be replaced by endogenous matrix.

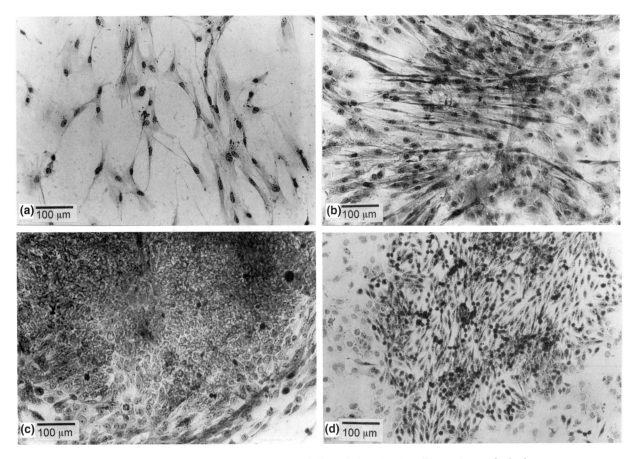

Fig. 8.12. Morphology on Feeder Layers. Morphological alteration in cells growing on feeder layers: Fibroblasts from human breast carcinoma (a) growing on plastic and (b) growing on a confluent feeder layer of fetal human intestinal cells (FHI). Epithelial cells from human breast carcinoma growing (c) on plastic and (d) on same confluent feeder layer as in (b).

Microcarriers made of polystyrene (Nalge Nunc), Sephadex, polyacrylamide, and collagen or gelatin are available in bead form for the propagation of anchorage-dependent cells in suspension (*see* Section 26.2.4 and Appendix II). Although technically three-dimensional, growth on many of these is functionally a two-dimensional monolayer, modified by the radius of curvature of the bead. However, some microcarriers are porous (*see* Section 26.2.4), and porous macrobeads are also used in fluidized-bed bioreactors (*see* Section 26.2.5). Cells grow within the interstices of the matrix in both types of bead. Three-dimensional growth in alginate beads occurs by encapsulation, rather than penetration of the matrix (*see* Section 25.3.5), allows a high level of cell–cell interaction and facilitates differentiation in chondrocytes (*see* Protocol 23.16) and neurons (*see* Protocol 25.3), and has been used to enhance antibody production by hybridomas [Zimmermann et al., 2003].

8.4.4 Metallic Substrates

Cells may be grown on stainless steel disks [Birnie & Simons, 1967] or other metallic surfaces [Litwin, 1973]. Observation of the cells on an opaque substrate requires surface interference microscopy, unless very thin metallic films are used. Westermark [1978] developed a method for the growth of fibroblasts and glia on palladium. Using electron microscopy shadowing equipment, he produced islands of palladium on agarose, which does not allow cell attachment in fluid media. The size and shape of the islands were determined by masks made by photoetching, and the palladium was applied by "shadowing" under vacuum, as used in electron microscopy. Because the layer was very thin, it remained transparent.

8.4.5 Nonadhesive Substrates

Sometimes attachment of the cells is undesirable. The selection of virally transformed colonies, which are anchorage independent, can be achieved by plating cells in agar [Macpherson & Montagnier, 1964], as the untransformed cells do not form colonies readily in this matrix. There are two principles involved in such a system: (1) prevention of attachment at the base of the dish, where spreading and anchorage-dependent growth would occur, and (2) immobilization of the cells such that daughter cells remain associated with the colony, even if they are nonadhesive. Most commonly, agar, agarose, or Methocel (methylcellulose of viscosity 4000 cP) is used (*see* Section 14.3). The first two are gels, and the third is a high-viscosity sol. Because Methocel is a sol, cells will sediment slowly through it. It is therefore commonly used with an underlay of agar (*see* Protocol 14.5). Dishes that are not of tissue culture grade can be used without an agar underlay, but some attachment and spreading may occur.

CHAPTER 9

Defined Media and Supplements

9.1 DEVELOPMENT OF MEDIA

Initial attempts to culture cells were performed in natural media based on tissue extracts and body fluids, such as chick embryo extract, serum, lymph, etc. With the propagation of cell lines, the demand for larger amounts of a medium of more consistent quality led to the introduction of chemically defined media based on analyses of body fluids and nutritional biochemistry. Eagle's Basal Medium [Eagle, 1955] and, subsequently, Eagle's Minimal Essential Medium (MEM) [Eagle, 1959] became widely adopted, variously supplemented with calf, human, or horse serum, protein hydrolysates, and embryo extract. As more continuous cell lines became available (L929 cells, HeLa, etc.), it was apparent that these media were perfectly adequate for the majority of those lines, and most of the succeeding developments were aimed at replacing serum (see Section 10.2), optimizing media for different cell types (e.g., RPMI 1640 for lymphoblastoid cell lines), or modifying for specific conditions (e.g., Leibovitz L15 to eliminate the need for adding CO_2 and $NaHCO_3$) [Leibovitz, 1963].

Isolation and propagation of cells of a specific lineage may require a selective serum-free medium (see Section 10.2.1), whereas cells grown for the formation of products, as hosts for viral propagation, or for non-cell-specific molecular studies rely mainly on Eagle's MEM [Eagle, 1959], Dulbecco's modification of Eagle's medium, DMEM [Dulbecco & Freeman, 1959], or, increasingly, RPMI 1640 [Moore et al., 1967], supplemented with serum. However, many industrial-scale production techniques now use serum-free media, to facilitate downstream processing and reduce the risk of adventitious infectious agents (see Section 10.1). A popular compromise for many laboratories is a mixture of a complex medium, such as Ham's F12 [Ham, 1965], with one with higher amino acid and vitamin concentrations, such as DMEM. This alternative will horrify the purist, but it does generate a useful, all-purpose medium for primary culture as well as cell line propagation.

This chapter concentrates on the general principles of medium composition, using the widely used serum-supplemented media as examples, and Chapter 10 will focus on the design and use of serum-free media.

9.2 PHYSICOCHEMICAL PROPERTIES

9.2.1 pH

Most cell lines grow well at pH 7.4. Although the optimum pH for cell growth varies relatively little among different cell strains, some normal fibroblast lines perform best at pH 7.4–7.7, and transformed cells may do better at pH 7.0–7.4 [Eagle, 1973]. It was reported that epidermal cells could be maintained at pH 5.5 [Eisinger et al., 1979], but this level has not been universally adopted. In special cases it may prove advantageous to do a brief growth experiment (see Protocols 21.7 and 21.8), plating efficiency assay (see Section 21.10), or special function analysis (e.g., see Section 17.7) to determine the optimum pH.

Phenol red is commonly used as an indicator. It is red at pH 7.4 and becomes orange at pH 7.0, yellow at pH 6.5, lemon yellow below pH 6.5, more pink at pH 7.6, and purple at pH 7.8 (see Plate 22b). Because the assessment of color is highly subjective, it is useful to make up a set of standards using a sterile balanced salt solution (BSS) with phenol red at the correct concentration, in the same type of bottle, with the same headspace for air, that you normally use for preparing a medium. The following example will use 25-cm^2 culture flasks as the final receptacles and can be used in conjunction with Exercise 8.

PROTOCOL 9.1. PREPARATION OF pH STANDARDS

Materials
Sterile:
❑ Hanks's balanced salt solution (HBSS), or Eagle's MEM, 10× concentrate, without bicarbonate or glucose, with 20 mM HEPES .... 10 mL
❑ Ultrapure water (UPW) .... 90 mL
❑ 1 N NaOH .... 10 mL
❑ Universal containers, 30 mL 7
❑ Culture flasks, 25 cm^2 7
❑ Pipettor tips, 100 μL
Nonsterile:
❑ pH meter
❑ Pipettor, 10–100 μL

Protocol
1. Make up the HBSS or medium at pH 6.5 or lower, and dispense 10 mL into each of seven labeled universal containers.
2. Allow to equilibrate with air for 30 min.
3. Adjust the pH in the separate containers to 6.5, 6.8, 7.0, 7.2, 7.4, 7.6, and 7.8 with 1 N NaOH, using a pH meter.
4. Filter sterilize 5 mL each HBSS into separate labeled flasks (*see* Protocol 11.11), using the first 5 ml to flush out the filter, before collecting the second 5 mL in the flask.
5. Cap the flasks securely.

9.2.2 CO$_2$ and Bicarbonate

Carbon dioxide in the gas phase dissolves in the medium, establishes equilibrium with HCO$_3^-$ ions, and lowers the pH.

Because dissolved CO$_2$, HCO$_3^-$, and pH are all interrelated, it is difficult to determine the major direct effect of CO$_2$. The atmospheric CO$_2$ tension will regulate the concentration of dissolved CO$_2$ directly, as a function of temperature. This regulation in turn produces H$_2$CO$_3$, which dissociates according to the reaction

$$H_2O + CO_2 \Leftrightarrow H_2CO_3 \Leftrightarrow H^+ + HCO_3^- \qquad (1)$$

HCO$_3^-$ has a fairly low dissociation constant with most of the available cations so it tends to reassociate, leaving the medium acid. The net result of increasing atmospheric CO$_2$ is to depress the pH, so the effect of elevated CO$_2$ tension is neutralized by increasing the bicarbonate concentration:

$$NaHCO_3 \Leftrightarrow Na^+ + HCO_3^- \qquad (2)$$

The increased HCO$_3^-$ concentration pushes equation (1) to the left until equilibrium is reached at pH 7.4. If another alkali (e.g., NaOH) is used instead, the net result is the same:

$$NaOH + H_2CO_3 \Leftrightarrow NaHCO_3 + H_2O \Leftrightarrow Na^+$$
$$+ HCO_3^- + H_2O \qquad (3)$$

The equivalent NaHCO$_3$ concentrations commonly used with different CO$_2$ tensions are listed in Tables 9.1, 9.2, and 9.3. Intermediate values of CO$_2$ and HCO$_3^-$ may be used, provided that the concentration of both is varied proportionally. Because many media are made up in acid solution and may incorporate a buffer, it is difficult to predict how much bicarbonate to use when other alkali may also end up as bicarbonate, as in equation (3). When preparing a new medium for the first time, add the specified amount of bicarbonate and then sufficient 1 N NaOH such that the medium equilibrates to the desired pH after incubation in a Petri dish at 37°C, in the correct CO$_2$ concentration, overnight. When dealing with a medium that is already at working strength, vary the amount of HCO$_3^-$ to suit the gas phase (*see* Table 9.1), and leave the medium overnight to equilibrate at 37°C. Each medium has a recommended bicarbonate concentration and CO$_2$ tension for achieving the correct pH and osmolality, but minor variations will occur in different methods of preparation.

With the introduction of Good's buffers (e.g., HEPES, Tricine) [Good et al., 1966] into tissue culture, there was

---

TABLE 9.1. Relationship Between Bicarbonate, Carbon Dioxide, and HEPES

Compound	Eagle's MEM Hanks's salts	Low HCO$_3^-$ + buffer	Eagle's MEM Earle's salts	DMEM
NaHCO$_3$	4 mM	10 mM	26 mM	44 mM
CO$_2$	Atmospheric and evolved from culture	2%	5%	10%
HEPES* (if required)	10 mM	20 mM	50 mM	—

*If HEPES is used, the equivalent molarity of NaCl must be omitted and osmolality must be checked.

some speculation that, as CO_2 was no longer necessary to stabilize the pH, it could be omitted. This proved to be untrue [Itagaki & Kimura, 1974], at least for a large number of cell types, particularly at low cell concentrations. Although 20 mM HEPES can control pH within the physiological range, the absence of atmospheric CO_2 allows equation (1) to move to the left, eventually eliminating dissolved CO_2, and ultimately HCO_3^-, from the medium. This chain of events appears to limit cell growth, although whether the cells require the dissolved CO_2 or the HCO_3^- (or both) is not clear. Recommended HCO_3^-, CO_2, and HEPES concentrations are given in Table 9.1.

The inclusion of pyruvate in the medium enables cells to increase their endogenous production of CO_2, making them independent of exogenous CO_2, as well as HCO_3^-. Leibovitz L15 medium [Leibovitz, 1963] contains a higher concentration of sodium pyruvate (550 mg/L) but lacks $NaHCO_3$ and does not require CO_2 in the gas phase. Buffering is achieved via the relatively high amino acid concentrations. Because it does not require CO_2, L15 is sometimes recommended for the transportation of tissue samples. Sodium β-glycerophosphate can also be used to buffer autoclavable media lacking CO_2 and HCO_3^- [Waymouth, 1979], and Invitrogen markets a CO_2-independent medium. If the elimination of CO_2 is important for cost saving, convenience, or other reasons, it might be worth considering one of these formulations, but only after appropriate testing.

In sum, cultures in open vessels need to be incubated in an atmosphere of CO_2, the concentration of which is in equilibrium with the sodium bicarbonate in the medium (see Tables 9.1, 9.2, and 9.3). Cells at moderately high concentrations ($\geq 1 \times 10^5$ cells/mL) and grown in sealed flasks need not have CO_2 added to the gas phase, provided that the bicarbonate concentration is kept low ($\sim$4 mM), particularly if the cells are high acid producers. At low cell concentrations, however (e.g., during cloning), and with some primary cultures, it is necessary to add CO_2 to the gas phase of sealed flasks. When venting is required, to allow either the equilibration of CO_2 or its escape in high acid producers, it is necessary to leave the cap slack or to use a CO_2-permeable cap (see Fig. 8.8).

9.2.3 Buffering
Culture media must be buffered under two sets of conditions: (1) open dishes, wherein the evolution of CO_2 causes the pH to rise (see Section 9.2.2), and (2) overproduction of CO_2 and lactic acid in transformed cell lines at high cell concentrations, when the pH will fall. A buffer may be incorporated into the medium to stabilize the pH, but in (1) exogenous CO_2 may still be required by some cell lines, particularly at low cell concentrations, to prevent the total loss of dissolved CO_2 and bicarbonate from the medium. Despite its poor buffering capacity at physiological pH bicarbonate buffer is still used more frequently than any other buffer,

because of its low toxicity, low cost, and nutritional benefit to the culture. HEPES is a much stronger buffer in the pH 7.2–7.6 range and is used at 10–20 mM. It has been found that, when HEPES is used with exogenous CO_2, the HEPES concentration must be more than double that of the bicarbonate for adequate buffering (see Table 9.1). A variation of Ham's F12 with 20 mM HEPES, 10 mM bicarbonate, and 2% CO_2 has been used successfully in the author's laboratory for the culture of a number of different cell lines. It allows the handling of microtitration and other multiwell plates outside the incubator without an excessive rise in pH and minimizes the requirement for HEPES, which is both toxic and expensive.

9.2.4 Oxygen
The other major significant constituent of the gas phase is oxygen. Whereas most cells require oxygen for respiration *in vivo*, cultured cells often rely mainly on glycolysis, a high proportion of which, as in transformed cells, may be anaerobic. Although attempts have been made to incorporate O_2 carriers, by analogy with hemoglobin in blood, as yet this practice is not in general use, and cells rely chiefly on dissolved O_2, which can be toxic due to the elevation in the level of free radicals. Providing the correct O_2 tension is, therefore, always a compromise between fulfilling the respiratory requirement and avoiding toxicity. Strategies involving both elevated and reduced O_2 levels have been employed, as has the incorporation of free radical scavengers, such as glutathione, 2-mercaptoethanol (β-mercaptoethanol) or dithiothreitol, into the medium. Most such strategies have been derived empirically.

Cultures vary in their oxygen requirement, the major distinction lying between organ and cell cultures. Although atmospheric or lower oxygen tensions [Cooper et al., 1958; Balin et al., 1976] are preferable for most cell cultures, some organ cultures, particularly from late-stage embryos, newborns, or adults, require up to 95% O_2 in the gas phase [Trowell, 1959; De Ridder & Mareel, 1978]. This requirement for a high level of O_2 may be a problem of diffusion related to the geometry and gaseous penetration of organ cultures (see Section 25.2.1) but may also reflect the difference between differentiated and rapidly proliferating cells. Oxygen diffusion may also become limiting in porous microcarriers [Preissmann et al., 1997] (see also Section 26.2.4).

Most dispersed cell cultures prefer lower oxygen tensions, and some systems (e.g., human tumor cells in clonogenic assay [Courtenay et al., 1978] and human embryonic lung fibroblasts [Balin et al., 1976]) do better in less than the normal level of atmospheric oxygen tension. McKeehan et al. [1976] suggested that the requirement for selenium in medium is related to oxygen toxicity, as selenium is a cofactor in glutathione synthesis. Oxygen tolerance—and selenium as well—may be provided by serum, so the control of O_2 tension is likely to be more critical in serum-free media.

Because the depth of the culture medium can influence the rate of oxygen diffusion to the cells, it is advisable to keep the depth of the medium within the range 2–5 mm ($0.2–0.5$ mL/cm^2) in static culture. Some types of cells, e.g., bronchial epithelium and keratinocytes, grown in filter well inserts, appear to differentiate better when positioned at the air–liquid interface (*see* Section 25.3.6).

9.2.5 Osmolality

Most cultured cells have a fairly wide tolerance for osmotic pressure [Waymouth, 1970]. As the osmolality of human plasma is about 290 mosmol/kg, it is reasonable to assume that this level is the optimum for human cells *in vitro*, although it may be different for other species (e.g., around 310 mosmol/kg for mice [Waymouth, 1970]). In practice, osmolalities between 260 mosmol/kg and 320 mosmol/kg are quite acceptable for most cells but, once selected, should be kept consistent at ±10 mosmol/kg. Slightly hypotonic medium may be better for Petri dish or open-plate culture to compensate for evaporation during incubation.

Osmolality is usually measured by depression of the freezing point (Fig. 9.1), or elevation of the vapor pressure, of the medium. The measurement of osmolality is a useful quality-control step if you are making up the medium yourself, as it helps to guard against errors in weighing, dilution, etc. It is particularly important to monitor osmolality if alterations are made in the constitution of the medium. The addition of

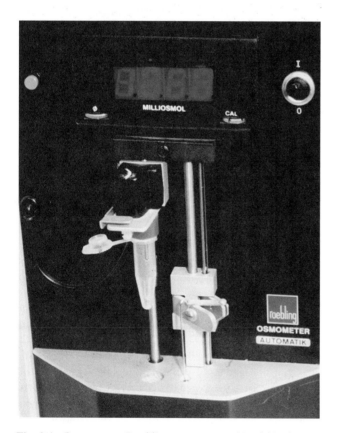

Fig. 9.1. Osmometer. Roebling osmometer (Camlab) showing front panel, sample tube, and sample port. This model accepts samples of 50 μL.

HEPES and drugs dissolved in strong acids and bases and their subsequent neutralization can all markedly affect osmolality.

9.2.6 Temperature

The optimal temperature for cell culture is dependent on (1) the body temperature of the animal from which the cells were obtained, (2) any anatomic variation in temperature (e.g., the temperature of the skin and testis may be lower than that of the rest of the body), and (3) the incorporation of a safety factor to allow for minor errors in regulating the incubator. Thus the temperature recommended for most human and warm-blooded animal cell lines is 37°C, close to body heat, but set a little lower for safety, as overheating is a more serious problem than underheating.

Because of the higher body temperature in birds, avian cells should be maintained at 38.5°C for maximum growth, but they will grow quite satisfactorily, if more slowly, at 37°C.

Cultured mammalian cells will tolerate considerable drops in temperature, can survive several days at 4°C, and can be frozen and cooled to −196°C (*see* Protocol 20.1), but they cannot tolerate more than about 2°C above normal (39.5°C) for more than a few hours and will die quite rapidly at 40°C and over.

Attention must be paid to the consistency of the temperature (within ±0.5°C) to ensure reproducible results. Doors of incubators or hot rooms must not be left open longer than necessary, and large items or volumes of liquid, placed in the warm room to heat, should not be put near any cultures. The spatial distribution of temperature within the incubator or hot room must also be uniform (*see* Sections 4.4 and 5.3); there should be no "cold spots," and air should circulate freely. This means that a large number of flasks should not be stacked together when first placed in the incubator or hot room; space must be allowed between them for air to circulate. As the gas phase within flasks expands at 37°C flasks, particularly when stacked, must be vented by briefly slackening the cap 30 min to 1 h after placing the flasks in the incubator.

Poikilotherms (cold-blooded animals that do not regulate their blood heat within narrow limits) tolerate a wide temperature range, between 15°C and 26°C. Simulating *in vivo* conditions (e.g., for cold-water fish) may require an incubator with cooling as well as heating, to keep the incubator temperature below ambient. If necessary, poikilothermic animal cells can be maintained at room temperature, but the variability of the ambient temperature in laboratories makes this undesirable, and a cooled incubator is preferable.

A number of temperature-sensitive (ts) mutant cell lines have been developed that allow the expression of specific genes below a set temperature, but not above it [Su et al., 1991; Foster & Martin, 1992; Wyllie et al., 1992]. These mutants facilitate studies on cell regulation, but also emphasize the narrow range within which one can operate, as the two discriminating temperatures are usually only about 2–3°C apart. The use of ts mutants usually requires an incubator with cooling as well as heating, to compensate for a warm ambient temperature.

Apart from its direct effect on cell growth, the temperature will also influence pH due to the increased solubility of CO_2 at lower temperatures and, possibly, because of changes in ionization and the pK_a of the buffer. The pH should be adjusted to 0.2 units lower at room temperature than at $37°C$. In preparing a medium for the first time, it is best to make up the complete medium and incubate a sample overnight at $37°C$ under the correct gas tension, in order to check the pH (*see* Section 9.2.2 and Plate 22b).

9.2.7 Viscosity

The viscosity of a culture medium is influenced mainly by the serum content and in most cases will have little effect on cell growth. Viscosity becomes important, however, whenever a cell suspension is agitated (e.g., when a suspension culture is stirred) or when cells are dissociated after trypsinization. Any cell damage that occurs under these conditions may be reduced by increasing the viscosity of the medium with carboxymethylcellulose (CMC) or polyvinylpyrrolidone (PVP) [Cherry & Papoutsakis, 1990; *see also* Appendix I]. This becomes particularly important in low-serum concentrations, in the absence of serum, and in stirred bioreactor cultures (*see* Section 26.1), in which Pluronic F68 is often used, although its effect is probably multifactorial.

9.2.8 Surface Tension and Foaming

The effects of foaming have not been clearly defined, but the rate of protein denaturation may increase, as may the risk of contamination if the foam reaches the neck of the culture vessel. Foaming will also limit gaseous diffusion if a film from a foam or spillage gets into the capillary space between the cap and the bottle, or between the lid and the base of a Petri dish.

Foaming can arise in suspension cultures in stirrer vessels or bioreactors when 5% CO_2 in air is bubbled through medium containing serum. The addition of a silicone antifoam (Dow Chemical) or Pluronic F68 (Sigma), 0.01–0.1%, helps prevent foaming in this situation by reducing surface tension and may also protect cells against shear stress from bubbles. CO_2 should not be bubbled through the medium when gassing a flask as this may generate a foam and can also spread aerosol from the flask.

9.3 BALANCED SALT SOLUTIONS

A balanced salt solution (BSS) is composed of inorganic salts and may include sodium bicarbonate and, in some cases, glucose. The compositions of some common balanced salt solutions are given in Table 9.2. HEPES buffer (5–20 mM) may be added to these solutions if necessary and the equivalent amount of NaCl omitted to maintain the correct osmolality. BSS forms the basis of many complete media, and commercial suppliers will provide Eagle's MEM with Hanks's salts [Hanks & Wallace, 1949] or Eagle's MEM with Earle's salts [Earle et al., 1943], indicating which BSS formulation was used; Hanks's salts would imply the use of sealed flasks with a gas phase of air, whereas Earle's salts would imply a higher bicarbonate concentration compatible with growth in 5% CO_2.

TABLE 9.2. Balanced Salt Solutions

Component	M.W.	Earle's BSS g/L	Earle's BSS mM	Dulbecco's PBS — Without Ca^{2+} and Mg^{2+} (D-PBSA) g/L	Dulbecco's PBS — Without Ca^{2+} and Mg^{2+} (D-PBSA) mM	Dulbecco's PBS — With Ca^{2+} and Mg^{2+} g/L	Dulbecco's PBS — With Ca^{2+} and Mg^{2+} mM	Hanks's BSS g/L	Hanks's BSS mM	Spinner salts (as in S-MEM) g/L	Spinner salts (as in S-MEM) mM
Inorganic salts											
$CaCl_2$ (anhydrous)	111	0.02	0.18			0.2	1.80	0.14	1.3		
KCl	74.55	0.4	5.37	0.2	2.68	0.2	2.68	0.4	5.4	0.40	5.37
KH_2PO_4	136.1			0.2	1.47	0.2	1.47	0.06	0.4		
$MgCl_2 \cdot 6H_2O$	203.3							0.1	0.5		
$MgSO_4 \cdot 7H_2O$	246.5	0.2	0.81			0.98	3.98	0.1	0.4	0.20	0.81
NaCl	58.44	6.68	114.3	8	136.9	8	136.9	8	136.9	6.80	116.4
$NaHCO_3$	84.01	2.2	26.19					0.35	4.2	2.20	26.19
$Na_2HPO_4 \cdot 7H_2O$	268.1			2.2	8.06	2.16	8.06	0.09	0.3		
$NaH_2PO_4 \cdot H_2O$	138	0.14	1.01							1.40	10.14
Total salt			147.9		149.1		154.00		149.4		158.9
Other components											
D-glucose	180.2	1	5.55					1	5.5	1.00	5.55
Phenol red	354.4	0.01	0.03					0.01	0.0	0.01	0.03
Gas phase		5% CO_2				Air		Air		5% CO_2	

BSS is also used as a diluent for concentrates of amino acids and vitamins to make complete media, as an isotonic wash or dissection medium, and for short incubations up to about 4 h (usually with glucose present). BSS recipes are often modified—for instance, by omitting glucose or phenol red from Hanks's BSS or by leaving out Ca^{2+} or Mg^{2+} ions from Dulbecco's PBS [Dulbecco & Vogt, 1954]. PBS without Ca^{2+} and Mg^{2+} is known as PBS Solution A, and the convention D-PBSA will be used throughout this book to indicate the absence of these divalent cations. One should always check for modifications when purchasing BSS and should quote any modifications to the published formula in reports and publications.

The choice of BSS is dependent on both the CO_2 tension (*see* Section 9.2.2 and Tables 9.1 and 9.2) and the intended use of the solution for tissue disaggregation or monolayer dispersal; in these cases Ca^{2+} and Mg^{2+} are usually omitted, as in Moscona's [1952] calcium- and magnesium-free saline (CMF) or D-PBSA (*see* Table 9.2). The choice of BSS also is dependent on whether the solution will be used for suspension culture of adherent cells. S-MEM, based on Eagle's Spinner salt solution, is a variant of Eagle's [1959] minimum essential medium that is deficient in Ca^{2+} in order to reduce cell aggregation and attachment (*see* Table 9.2).

HBSS, EBSS, and PBS rely on the relatively weak buffering of phosphate, which is not at its most effective at physiological pH. Paul [1975] constructed a Tris-buffered BSS that is more effective, but for which the cells sometimes require a period of adaptation. HEPES (10–20 mM) is currently the most effective buffer in the pH 7.2–7.8 range, and Tricine in the pH 7.4–8.0 range, although both tend to be expensive if used in large quantities.

9.4 COMPLETE MEDIA

The term *complete medium* implies a medium that has had all its constituents and supplements added and is sufficient for the use specified. It is usually made up of a defined medium component, some of the constituents of which, such as glutamine, may be added just before use, and various supplements, such as serum, growth factors, or hormones. Defined media range in complexity from the relatively simple Eagle's MEM [Eagle, 1959], which contains essential amino acids, vitamins, and salts, to complex media such as medium 199 (M199) [Morgan et al., 1950], CMRL 1066 [Parker et al., 1957], MB 752/1 [Waymouth, 1959], RPMI 1640 [Moore et al., 1967], and F12 [Ham, 1965] (Table 9.3) and a wide range of serum-free formulations (*see* Tables 10.1 and 10.2). The complex media contain a larger number of different amino acids, including nonessential amino acids and additional vitamins, and are often supplemented with extra metabolites (e.g., nucleosides, tricarboxylic acid cycle intermediates, and lipids) and minerals. Nutrient concentrations are, on the whole, low in F12 (which was optimized by cloning) and high in Dulbecco's modification

of Eagle's MEM (DMEM) [Dulbecco & Freeman, 1959; Morton, 1970], optimized at higher cell densities for viral propagation. Barnes and Sato [1980] used a 1:1 mixture of DMEM and F12 as the basis for their serum-free formulations to combine the richness of F12 and the higher nutrient concentration of DMEM. Although not always entirely rational, this combination has provided an empirical formula that is suitable as a basic medium for supplementation with special additives for many different cell types.

9.4.1 Amino Acids

The essential amino acids (i.e., those that are not synthesized in the body) are required by cultured cells, plus cystine and/or cysteine, arginine, glutamine, and tyrosine, although individual requirements for amino acids will vary from one cell type to another. Other nonessential amino acids are often added as well, to compensate either for a particular cell type's incapacity to make them or because they are made, but lost by leakage into the medium. The concentration of amino acids usually limits the maximum cell concentration attainable, and the balance may influence cell survival and growth rate. Glutamine is required by most cells, although some cell lines will utilize glutamate; evidence suggests that glutamine is also used by cultured cells as a source of energy and carbon [Butler & Christie, 1994; *see also* Section 3.6]. Glutamax (Invitrogen) is a alanyl-glutamine dipeptide which is more stable than glutamine.

9.4.2 Vitamins

Eagle's MEM contains only the water-soluble vitamins (the B group, plus choline, folic acid, inositol, and nicotinamide, but excluding biotin; *see* Table 9.3); other requirements presumably are derived from the serum. Biotin is present in most of the more complex media, including the serum-free recipes, and *p*-aminobenzoic acid (PABA) is present in M199, CMRL 1066 (which was derived from M199), and RPMI 1640. All the fat-soluble vitamins (A, D, E, and K) are present only in M199, whereas vitamin A is present in LHC-9 and vitamin E in MCDB 110 (*see* Table 10.1). Some vitamins (e.g., choline and nicotinamide) have increased concentrations in serum-free media. Vitamin limitation—for example, by precipitation of folate from concentrated stock solutions—is usually expressed in terms of reduced cell survival and growth rates rather than maximum cell density. Like those of the amino acids, vitamin requirements have been derived empirically and often relate to the cell line originally used in their development; e.g., Fischer's medium has a high folate concentration because of the folate dependence of L5178Y, which was used in the development of the medium [Fischer & Sartorelli, 1964].

9.4.3 Salts

The salts are chiefly those of Na^+, K^+, Mg^{2+}, Ca^{2+}, Cl^-, SO_4^{2-}, PO_4^{3-}, and HCO_3^- and are the major components contributing to the osmolality of the medium. Most media derived their salt concentrations originally from Earle's (high

TABLE 9.3. Frequently Used Media

Component	MEM	DMEM	F12	DMEM/F12	αMEM	CMRL 1066	RPMI 1640	M199	L15	McCoy's 5A	Fischer	MB 752/1
Amino acids												
L-alanine			1.0E-04	5.0E-05	2.8E-04	2.8E-04		2.8E-04	2.5E-03	1.5E-04		
L-arginine	6.0E-04	4.0E-04	1.0E-03	7.0E-04	6.0E-04	3.3E-04	1.1E-03	3.3E-04	2.9E-03	2.0E-04	7.1E-05	3.6E-04
L-asparagine			1.0E-04	5.0E-05	3.3E-04	3.8E-04	1.7E-03	3.0E-04	7.6E-05			
L-aspartic acid			1.0E-04	5.0E-05	2.3E-04	2.3E-04	1.5E-04	2.3E-04		1.5E-04		4.5E-04
L-cysteine			2.0E-04	1.0E-04	5.7E-04	1.5E-03		5.6E-07	9.9E-04	2.0E-04		5.0E-04
L-cystine	1.0E-04	2.0E-04		1.0E-04	1.0E-04	8.3E-05	2.1E-04	9.9E-05			9.9E-05	6.3E-05
L-glutamic acid			1.0E-04	5.0E-05	5.1E-04	5.1E-04	1.4E-04	4.5E-04		1.5E-04		1.0E-03
L-glutamine	2.0E-03	4.0E-03	1.0E-03	2.5E-03	2.0E-03	6.8E-04	2.1E-03	6.8E-04	2.1E-03	1.5E-03	1.4E-03	2.4E-03
Glycine		4.0E-04	1.0E-04	2.5E-04	6.7E-04	6.7E-04	1.3E-04	6.7E-04	2.7E-03	1.0E-04		6.7E-04
L-histidine	2.0E-04	2.0E-04	1.0E-04	1.5E-04	2.0E-04	9.5E-05	9.7E-05	1.0E-04	1.6E-03	1.0E-04	3.9E-04	8.3E-04
L-hydroxy-proline						7.6E-05	1.5E-04	7.6E-05		1.5E-04		
L-isoleucine	4.0E-04	8.0E-04	3.0E-05	4.2E-04	4.0E-04	1.5E-04	3.8E-04	1.5E-04	9.5E-04	3.0E-04	5.7E-04	1.9E-04
L-leucine	4.0E-04	8.0E-04	1.0E-04	4.5E-04	4.0E-04	4.6E-04	3.8E-04	4.6E-04	9.5E-04	3.0E-04	2.3E-04	3.8E-04
L-lysine HCl	4.0E-04	8.0E-04	2.0E-04	5.0E-04	4.0E-04	3.8E-04	2.2E-04	3.8E-04	5.1E-04	2.0E-04	2.7E-04	1.3E-03
L-methionine	1.0E-04	2.0E-04	3.0E-05	1.2E-04	1.0E-04	1.0E-04	1.0E-04	1.0E-04	5.0E-04	1.0E-04	6.7E-04	3.4E-04
L-phenylalanine	2.0E-04	4.0E-04	3.0E-05	2.2E-04	1.9E-04	1.5E-04	9.1E-05	1.5E-04	7.6E-04	1.0E-04	4.1E-04	3.0E-04
L-proline			3.0E-04	1.5E-04	3.5E-04	3.5E-04	1.7E-04	3.5E-04		1.5E-04		4.3E-04
L-serine		4.0E-04	1.0E-04	2.5E-04	2.4E-04	2.4E-04	2.9E-04	2.4E-04	1.9E-03	2.5E-04	1.4E-04	
L-threonine	4.0E-04	8.0E-04	1.0E-04	4.5E-04	4.0E-04	2.5E-04	1.7E-04	2.5E-04	2.5E-03	1.5E-04	3.4E-04	6.3E-04
L-tryptophan	4.9E-05	7.8E-05	1.0E-05	4.4E-05	4.9E-05	4.9E-05	2.5E-05	4.9E-05	9.8E-05	1.5E-05	4.9E-05	2.0E-04
L-tyrosine	2.0E-04	4.0E-04	3.0E-05	2.1E-04	2.3E-04	2.2E-04	1.1E-04	2.2E-04	1.7E-03	1.2E-04	3.3E-04	2.2E-04
L-valine	4.0E-04	8.0E-04	1.0E-04	4.5E-04	3.9E-04	2.1E-04	1.7E-04	2.1E-04	8.5E-04	1.5E-04	6.0E-04	5.6E-04
Vitamins												
p-Aminobenzoic acid						3.6E-07	7.3E-06	3.6E-07		7.3E-06		
L-Ascorbic acid					2.5E-04	2.8E-04		2.8E-07		3.2E-06		9.9E-05
Biotin			3.0E-08	1.5E-08	4.1E-07	4.1E-08	8.2E-07	4.1E-08		8.2E-07	4.1E-08	8.2E-08
Calciferol								2.5E-07				
Choline chloride	7.1E-06	2.9E-05	1.0E-04	6.4E-05	7.1E-06	3.6E-06	2.1E-05	3.6E-06	7.1E-06	3.6E-05	1.1E-05	1.8E-03
Folic acid	2.3E-06	9.1E-06	2.9E-06	6.0E-06	2.3E-06	2.3E-08	2.3E-06	2.3E-08	2.3E-06	2.3E-05	2.3E-05	9.1E-07
myo-Inositol	1.1E-05	4.0E-05	1.0E-04	7.0E-05	1.1E-05	2.8E-07	1.9E-04	2.8E-07	1.1E-05	2.0E-04	8.3E-06	5.6E-06
Menadione								6.9E-08				
Nicotinamide	8.2E-06	3.3E-05	3.3E-07	1.7E-05	8.2E-06	2.0E-07	8.2E-06	2.0E-07	8.2E-06	4.1E-06	4.1E-06	8.2E-06
Nicotinic acid								2.0E-07		4.1E-06		
D-Ca pantothenate	4.2E-06	1.7E-05	2.0E-06	9.4E-06	4.2E-06	4.2E-08	1.1E-06	4.2E-08	4.2E-06	8.4E-07	2.1E-06	4.2E-06
Pyridoxal HCl	4.9E-06	2.0E-05	3.0E-07	1.0E-05	4.9E-06	1.2E-07		1.2E-07		2.5E-06	2.5E-06	
Pyridoxine HCl			3.0E-07	1.5E-07		1.2E-07	4.9E-06	1.2E-07		2.4E-06		4.9E-06
Riboflavin	2.7E-07	1.1E-06	1.0E-07	5.8E-07	2.7E-07	2.7E-08	5.3E-07	2.7E-08	1.9E-07	5.3E-07	1.3E-06	2.7E-06
Thiamin	3.0E-06	1.2E-05	1.0E-06	6.4E-06	3.0E-06	3.0E-08	3.0E-06	3.0E-08	2.4E-06	5.9E-07	3.0E-06	3.0E-05
Thiamin mono PO4										4.8E-06		
α-Tocopherol								2.3E-08				
Retinol acetate								3.5E-07				
Vitamin B12			1.0E-06	5.0E-07	1.0E-06		3.7E-09					1.5E-07
Antioxidants												
Glutathione						3.0E-05	3.0E-06	1.5E-07		1.5E-06		4.5E-05
Inorganic salts												
CaCl2	1.8E-03	1.8E-03	3.0E-04	1.1E-03	1.8E-03	1.8E-03		1.3E-03	1.3E-03	9.0E-04	6.2E-04	8.2E-04
KCl	5.3E-03	5.3E-03	3.0E-03	4.2E-03	5.3E-04	5.3E-03	5.3E-03	5.3E-03	5.3E-03	5.3E-03	5.3E-03	2.0E-03
KH2PO4								4.4E-04	4.4E-04			5.9E-04
MgCl2					1.2E-01							1.2E-03
MgSO4	8.1E-04	8.1E-04		4.0E-04	8.1E-04	8.1E-04	4.0E-04	8.1E-04	1.6E-03	8.1E-04	4.9E-04	8.1E-04
NaCl	1.2E-01	1.1E-01	1.3E-01	1.2E-01		1.2E-01	1.0E-01	1.4E-01	1.4E-01	1.1E-01	1.4E-01	1.0E-01
NaHCO3	2.6E-02	4.4E-02	1.4E-02	2.9E-02	2.6E-02	2.6E-02	4.2E-03		2.6E-02	1.3E-02	2.7E-02	
NaH2PO4	1.0E-03	9.1E-04		4.5E-04		1.0E-03				4.2E-03	5.7E-04	
Na2HPO4			1.0E-03	5.0E-04			5.6E-03	4.0E-04	1.6E-03		5.0E-04	2.1E-03

(*Continued overleaf*)

TABLE 9.3. Frequently Used Media (*Continued*)

Component	MEM	DMEM	F12	DMEM/F12	αMEM	CMRL 1066	RPMI 1640	M199	L15	McCoy's 5A	Fischer	MB 752/1
Trace elements												
$CuSO_4 \cdot 5H_2O$			1.6E-08	7.8E-09								
$Fe(NO_3)_3 \cdot 9H_2O$		2.5E-07		1.2E-07								
$FeSO_4 \cdot 7H_2O$			3.0E-06	1.5E-06								
$ZnSO_4 \cdot 7H_2O$			3.0E-06	1.5E-06								
Bases, nucleosides, etc.												
Adenine SO_4								5.4E-05				
Adenosine					3.7E-05							
AMP								5.8E-07				
ATP								1.8E-05				
Cytidine					4.1E-05							
Deoxyadenosine					4.0E-05	4.0E-05						
Deoxycytidine					4.2E-05	3.8E-05						
Deoxyguanosine					3.7E-05	3.7E-05						
2-Deoxyribose								3.7E-06				
DPN						9.5E-06						
FAD						1.2E-06						
Glucuronate, Na						1.9E-05						
Guanine								1.6E-06				
Guanosine					3.5E-05							
Hypoxanthine			3.0E-05	1.5E-05				2.2E-06				
5-Me-deoxycytidine						4.1E-07						
D-Ribose								3.3E-06				
Thymidine			3.0E-06	1.5E-06	4.1E-05	4.1E-05						
Thymine								2.4E-06				
TPN						1.3E-06						
Uracil								2.7E-06				
Uridine					4.1E-05							
UTP						1.8E-06						
Xanthine								2.0E-06				
Energy metabolism												
Cocarboxylase						2.2E-06						
Coenzyme A						3.3E-06						
D-galactose									5.0E-02			
D-glucose	5.6E-03	2.5E-02	1.0E-02	1.8E-02	5.6E-03	5.6E-03	1.1E-02	5.6E-03		1.7E-02	5.6E-03	2.8E-02
Sodium acetate						6.1E-04		4.5E-04				
Sodium pyruvate		1.0E-03	1.0E-03	1.0E-03	1.0E-03				5.0E-03			
Lipids and precursors												
Cholesterol						5.2E-07		5.2E-07				
Ethanol (solvent)						3.5E-04						
Linoleic acid		3.0E-07		1.5E-07								8.9E-05
Lipoic acid			1.0E-06	5.1E-07	9.7E-07							
Tween 80						1.8E-05		1.8E-05				
Other components												
Peptone, mg/mL										0.6		
Phenol red	2.7E-05	4.0E-05	3.2E-05	3.6E-05	2.9E-05	5.3E-05	1.3E-05	4.5E-05	2.7E-05	2.9E-05	1.3E-05	2.7E-05
Putrescine			1.0E-06	5.0E-07								
Gas phase												
CO_2	5%	10%	2%	7%	5%	5%	5%	5%	Air	5%	2%	5%

All concentrations are molar, and computer-style notation is used (e.g., 3.0E-2 = 3.0×10^{-2} = 30 mM). Molecular weights are given for root compounds; although some recipes use salts or hydrated forms, molarities will, of course, remain the same. *Synonyms and abbreviations:* AMP, adenosine monophosphate; ATP, adenosine triphosphate; biotin = vitamin H; calciferol = vitamin D_2; FAD, flavine adenine dinucleotide; lipoic acid = thioctic acid; menadione = vitamin K_3; *myo*-inositol; = L-inositol; nicotinamide = niacinamide; nicotinic acid = niacin; pyridoxine HCl = vitamin B_6; thiamin = vitamin B_1; α-tocopherol = vitamin E; retinol = vitamin A_1; TPN, triphosphopyridine nucleotide; UTP, uridine triphosphate; vitamin B_{12} = cobalamin. See text for references.

bicarbonate; gas phase, 5% CO_2) or Hanks's (low bicarbonate; gas phase, air) BSS. Divalent cations, particularly Ca^{2+}, are required by some cell adhesion molecules, such as the cadherins [Yamada & Geiger, 1997]. Ca^{2+} also acts as an intermediary in signal transduction [Alberts et al., 2002], and the concentration of Ca^{2+} in the medium can influence whether cells will proliferate or differentiate (*see* Sections 17.2.2, 23.2.1). Na^+, K^+, and Cl^- regulate membrane potential, whereas SO_4^{2-}, PO_4^{3-}, and HCO_3^- have roles as anions required by the matrix and nutritional precursors for macromolecules, as well as regulators of intracellular charge.

Calcium is reduced in suspension cultures in order to minimize cell aggregation and attachment (*see* Section 9.3). The sodium bicarbonate concentration is determined by the concentration of CO_2 in the gas phase (*see* Section 9.2.2) and has a significant nutritional role in addition to its buffering capability.

9.4.4 Glucose

Glucose is included in most media as a source of energy. It is metabolized principally by glycolysis to form pyruvate, which may be converted to lactate or acetoacetate and may enter the citric acid cycle and is oxidized to form CO_2 and water. The accumulation of lactic acid in the medium, particularly evident in embryonic and transformed cells, implies that the citric acid cycle may not function entirely as it does *in vivo*, and recent data have shown that much of its carbon is derived from glutamine rather than glucose. This finding may explain the exceptionally high requirement of some cultured cells for glutamine or glutamate.

9.4.5 Organic Supplements

A variety of other compounds, including proteins, peptides, nucleosides, citric acid cycle intermediates, pyruvate, and lipids, appear in complex media. Again, these constituents have been found to be necessary when the serum concentration is reduced, and they may help in cloning and in maintaining certain specialized cells, even in the presence of serum.

9.4.6 Hormones and Growth Factors

Hormones and growth factors are not specified in the formulas of most regular media, although they are frequently added to serum-free media (*see* Sections 9.5.2, 9.5.3, 10.4.2, 10.4.3).

9.4.7 Antibiotics

Antibiotics were originally introduced into culture media to reduce the frequency of contamination. However, the use of laminar-flow hoods, coupled with strict aseptic technique, makes antibiotics unnecessary. Indeed, antibiotics have a number of significant disadvantages:

(1) They encourage the development of antibiotic-resistant organisms.

(2) They hide the presence of low-level, cryptic contaminants that can become fully operative if the antibiotics are removed, the culture conditions change, or resistant strains develop.
(3) They may hide mycoplasma infections.
(4) They have antimetabolic effects that can cross-react with mammalian cells.
(5) They encourage poor aseptic technique.

Hence it is strongly recommended that routine culture be performed in the absence of antibiotics and that their use be restricted to primary culture or large-scale labor-intensive experiments with a high cost of consumables. If conditions demand the use of antibiotics, then they should be removed as soon as possible, or, if they are used over the long term, parallel cultures should be maintained free of antibiotics (*see* Section 13.7.7).

A number of antibiotics used in tissue culture are moderately effective in controlling bacterial infections (Table 9.4). However, a significant number of bacterial strains are resistant to antibiotics, either naturally or by selection, so the control that they provide is never absolute. Fungal and yeast contaminations are particularly hard to control with antibiotics; they may be held in check, but are seldom eliminated (*see* Section 19.4).

9.5 SERUM

Serum contains growth factors, which promote cell proliferation, and adhesion factors and antitrypsin activity, which promote cell attachment. Serum is also a source of minerals, lipids, and hormones, many of which may be bound to protein (Table 9.5). The sera used most in tissue culture are bovine calf, fetal bovine, adult horse, and human serum. Calf (CS) and fetal bovine (FBS) serum are the most widely used, the latter particularly for more demanding cell lines and for cloning. Human serum is sometimes used in conjunction with some human cell lines, but it needs to be screened for viruses, such as HIV and hepatitis B. Horse serum is preferred to calf serum by some workers, as it can be obtained from a closed donor herd and is often more consistent from batch to batch. Horse serum may also be less likely to metabolize polyamines, due to lower levels of polyamine oxidase; polyamines are mitogenic for some cells [Hyvonen et al., 1988; Kaminska et al., 1990].

9.5.1 Protein

Although proteins are a major component of serum, the functions of many proteins *in vitro* remain obscure; it may be that relatively few proteins are required other than as carriers for minerals, fatty acids, and hormones. Those proteins for which requirements have been found are albumin [Iscove & Melchers, 1978; Barnes & Sato, 1980], which may be important as a carrier of lipids, minerals, and globulins [Tozer

TABLE 9.4. Antibiotics Used in Tissue Culture

| Antibiotic | Concentration, μg/mL (unless otherwise stated) | | Activity against |
	Working	Cytotoxic	
Amphotericin B (Fungizone)	2.5	30	Fungi, yeasts
Ampicillin	2.5		Bacteria, gram positive and gram negative
Ciprofloxacin	100		Mycoplasma
Erythromycin	50	300	Mycoplasma
Gentamycin Gentamicin	50	>300	Bacteria, gram positive and gram negative; mycoplasma
Kanamycin	100	10 mg/mL	Bacteria, gram positive and gram negative; mycoplasma
MRA (ICN)	0.5		Mycoplasma
Neomycin	50	3,000	Bacteria, gram positive and gram negative
Nystatin	50		Fungi, yeasts
Penicillin-G	100 U/mL	10,000 U/mL	Bacteria, gram positive
Polymixin B	50	1 mg/mL	Bacteria, gram negative
Streptomycin SO$_4$	100	20 mg/mL	Bacteria, gram positive and gram negative
Tetracyclin	10	35	Bacteria, gram positive and gram negative
Tylosin	10	300	Mycoplasma

After Paul, 1975.

& Pirt, 1964]; fibronectin (cold-insoluble globulin), which promotes cell attachment [Yamada & Geiger, 1997; Hynes, 1992], although probably not as effectively as cell-derived fibronectin; and α_2-macroglobulin, which inhibits trypsin [de Vonne & Mouray, 1978]. Fetuin in fetal serum enhances cell attachment [Fisher et al., 1958], and transferrin [Guilbert & Iscove, 1976] binds iron, making it less toxic and bioavailable. Other proteins, as yet uncharacterized, may be essential for cell attachment and growth.

Protein also increases the viscosity of the medium, reducing shear stress during pipetting and stirring, and may add to the medium's buffering capacity.

9.5.2 Growth Factors

Natural clot serum stimulates cell proliferation more than serum from which the cells have been removed physically (e.g., by centrifugation). This increased stimulation appears to be due to the release of platelet-derived growth factor (PDGF) from the platelets during clotting. PDGF [Antoniades et al., 1979; Heldin et al., 1979] is one of a family of polypeptides with mitogenic activity and is probably the major growth factor in serum. PDGF stimulates growth in fibroblasts and glia, but other platelet-derived factors, such as TGF-β, may inhibit growth or promote differentiation in epithelial cells [Lechner et al., 1981].

Other growth factors (*see* Table 10.3), such as fibroblast growth factors (FGFs) [Gospodarowicz, 1974], epidermal growth factor (EGF) [Cohen, 1962; Carpenter & Cohen, 1977; Gospodarowicz et al., 1978a], endothelial cell growth factors such as vascular endothelial growth factor (VEGF) and angiogenin [Hu et al., 1997; Folkman & d'Amore, 1996; Joukov et al., 1997; Folkman et al., 1979; Maciag et al., 1979], and insulin-like growth factors IGF-I and IGF-II [le Roith & Raizada, 1989], which have been isolated from whole tissue

or released into the medium by cells in culture, have varying degrees of specificity [Hollenberg & Cuatrecasas, 1973] and are probably present in serum in small amounts [Gospodarowicz & Moran, 1974]. Many of these growth factors are available commercially (*see* Appendix II) as recombinant proteins, some of which also are available in long-form analogs (Sigma) with increased mitogenic activity and stability.

9.5.3 Hormones

Insulin promotes the uptake of glucose and amino acids [Kelley et al., 1978; Stryer, 1995] and may owe its mitogenic effect to this property or to activity via the IGF-I receptor. IGF-I and IGF-II bind to the insulin receptor, but also have their own specific receptors, to which insulin may bind with lower affinity. IGF-II also stimulates glucose uptake [Sinha et al., 1990]. Growth hormone may be present in serum—particularly fetal serum—and, in conjunction with the somatomedins (IGFs), may have a mitogenic effect. Hydrocortisone is also present in serum—particularly fetal bovine serum—in varying amounts and it can promote cell attachment [Ballard & Tomkins, 1969; Fredin et al., 1979] and cell proliferation [Guner et al., 1977; McLean et al., 1986; *see also* Sections 23.1.1, 23.1.3, 23.1.4], but under certain conditions (e.g., at high cell density) may be cytostatic [Freshney et al., 1980a,b] and can induce cell differentiation [Moscona & Piddington, 1966; Ballard, 1979; McLean et al., 1986; Speirs et al., 1991; McCormick et al., 1995, 2000].

9.5.4 Nutrients and Metabolites

Serum may also contain amino acids, glucose, oxo (keto) acids, nucleosides, and a number of other nutrients and intermediary metabolites. These may be important in simple media but less so in complex media, particularly those with higher amino acid concentrations and other defined supplements.

TABLE 9.5. Constituents of Serum

Constituent	Range of concentration[a]
Proteins and Polypeptides	40–80 mg/mL
Albumin	20–50 mg/mL
Fetuin[b]	10–20 mg/mL
Fibronectin	1–10 μg/mL
Globulins	1–15 mg/mL
Protease inhibitors: α_1-antitrypsin, α_2-macroglobulin	0.5–2.5 mg/mL
Transferrin	2–4 mg/mL
Growth factors:	
EGF, PDGF, IGF-I and -II, FGF, IL-1, IL-6	1–100 ng/mL
Amino acids	0.01–1.0 μM
Lipids	2–10 mg/mL
Cholesterol	10 μM
Fatty acids	0.1–1.0 μM
Linoleic acid	0.01–0.1 μM
Phospholipids	0.7–3.0 mg/mL
Carbohydrates	1.0–2.0 mg/mL
Glucose	0.6–1.2 mg/mL
Hexosamine[c]	6–1.2 mg/mL
Lactic acid[d]	0.5–2.0 mg/mL
Pyruvic acid	2–10 μg/mL
Polyamines:	
Putrescine, spermidine	0.1–1.0 μM
Urea	170–300 μg/mL
Inorganics:	0.14–0.16 M
Calcium	4–7 mM
Chlorides	100 μM
Iron	10–50 μM
Potassium	5–15 mM
Phosphate	2–5 mM
Selenium	0.01 μM
Sodium	135–155 mM
Zinc	0.1–1.0 μM
Hormones:	0.1–200 nM
Hydrocortisone	10–200 nM
Insulin	1–100 ng/mL
Triiodothyronine	20 nM
Thyroxine	100 nM
Vitamins:	10 ng–10 μg/mL
Vitamin A	10–100 ng/mL
Folate	5–20 ng/mL

[a]The range of concentrations is very approximate and is intended to convey only the order of magnitude. Data are from Evans and Sanford [1978], and Cartwright and Shah [1994].

[b]In fetal serum only.

[c]Highest in human serum.

[d]Highest in fetal serum.

9.5.5 Lipids

Linoleic acid, oleic acid, ethanolamine, and phospho-ethanolamine are present in serum in small amounts, usually bound to proteins such as albumin.

9.5.6 Minerals

Serum replacement experiments [Ham & McKeehan, 1978] have also suggested that trace elements and iron, copper, and zinc may be bound to serum protein. McKeehan et al. [1976] demonstrated a requirement for selenium, which probably helps to detoxify free radicals as a cofactor for GSH synthetase.

9.5.7 Inhibitors

Serum may contain substances that inhibit cell proliferation [Harrington & Godman, 1980; Liu et al., 1992; Varga Weisz & Barnes, 1993]. Some of these may be artifacts of preparation (e.g., bacterial toxins from contamination before filtration, or antibodies, contained in the γ-globulin fraction, that cross-react with surface epitopes on the cultured cells), but others may be physiological negative growth regulators, such as TGF-β [Massague et al., 1992]. Heat inactivation removes complement from the serum and reduces the cytotoxic action of immunoglobulins without damaging polypeptide growth factors, but it may also remove some more labile constituents and is not always as satisfactory as untreated serum.

9.6 SELECTION OF MEDIUM AND SERUM

All 12 media described in Table 9.3 were developed to support particular cell lines or conditions. Many were developed with L929 mouse fibroblasts or HeLa cervical carcinoma cells, and Ham's F12 was designed for Chinese hamster ovary (CHO) cells; all now have more general applications and have become classic formulations. Among them, data from suppliers would indicate that RPMI 1640, DMEM, and MEM are the most popular, making up about 75% of sales. Other formulations seldom account for more than 5% of the total; most constitute 2–3%, although blended DMEM/F12 comes closer, with over 4%.

Eagle's Minimal Essential Medium (MEM) was developed from Eagle's Basal Medium (BME) by increasing the range and concentration of the constituents. For many years, Eagle's MEM had the most general use of all media. Dulbecco's modification of BME (DMEM) was developed for mouse fibroblasts for transformation and virus propagation studies. It has twice the amino acid concentrations of MEM, has four times the vitamin concentrations, and uses twice the HCO_3^- and CO_2 concentrations to achieve better buffering. αMEM [Stanners et al., 1971] has additional amino acids and vitamins, as well as nucleosides and lipoic acid; it has been used for a wide range of cell types, including hematopoietic cells. Ham's F12 was developed to clone CHO cells in low-serum medium; it is also used widely, particularly for clonogenic assays (*see* Protocol 22.3.2) and primary culture (*see* Protocol 12.7).

CMRL 1066, M199, and Waymouth's media were all developed to grow L929 cells serum-free but have been used alone or in combination with other media, such as DMEM or F12, for a variety of more demanding conditions. RPMI

1640 and Fischer's media were developed for lymphoid cells—Fischer's specifically for L5178Y lymphoma, which has a high folate requirement. RPMI 1640 in particular has quite widespread use, often for attached cells, despite being designed for suspension culture and lacking calcium. L15 medium was developed specifically to provide buffering in the absence of HCO_3^- and CO_2. It is often used as a transport and primary culture medium for this reason, but its value was diminished by the introduction of HEPES and the demonstration that HCO_3^- and CO_2 are often essential for optimal cell growth, regardless of the requirement for buffering.

Information regarding the selection of the appropriate medium for a given type of cell is usually available in the literature in articles on the origin of the cell line or the culture of similar cells. Information may also be obtained from the source of the cells. Cell banks, such as ATCC and ECACC, provide information on media used for currently available cell lines, and data sheets can be accessed from their websites (*see* Appendix III; *see also* Table 9.6). Failing this, the choice is made either empirically or by comparative testing of several media, as for selection of serum (*see* Section 9.6.2).

Many continuous cell lines (e.g., HeLa, L929, BHK21), primary cultures of human, rodent, and avian fibroblasts, and cell lines derived from them can be maintained on a relatively simple medium such as Eagle's MEM, supplemented with calf serum. More complex media may be required when a specialized function is being expressed (*see* Section 17.7) or when cells are subcultured at low seeding density ($<1 \times 10^3$/mL), as in cloning (*see* Section 14.2). Frequently, the more demanding culture conditions that require complex media also require fetal bovine serum rather than calf or horse serum, unless the formulation specifically allows for the omission of serum.

Some suggestions for the choice of medium and several examples of cell types and the media used for them are given in Table 9.6 [*see also* Mather, 1998]. If information is not available, a simple cell growth experiment with commercially available media and multiwell plates (*see* Protocols 21.7–21.9) can be carried out in about two weeks. Assaying for clonal growth (*see* Protocol 21.10) and measuring the expression of specialized functions may narrow the choice further. You may be surprised to find that your best conditions do not agree with those mentioned in the literature; reproducing the conditions found in another laboratory may be difficult, because of variations in preparation or supplier, the impurities present in reagents and water, and differences between batches of serum. It is to be hoped that as serum requirements are reduced and the purity of reagents increases, the standardization of media will improve.

Finally, you may have to compromise in your choice of medium or serum because of cost. Autoclavable media are available from commercial suppliers (*see* Appendix II). They are simple to prepare from powder and are suitable for many

TABLE 9.6. Selecting a Suitable Medium

Cells or cell line	Medium	Serum
3T3 cells	MEM, DMEM	CS
Chick embryo fibroblasts	Eagle's MEM	CS
Chinese hamster ovary (CHO)	Eagle's MEM, Ham's F12	CS
Chondrocytes	Ham's F12	FB
Continuous cell lines	Eagle's MEM, DMEM	CS
Endothelium	DMEM, M199, MEM	CS
Fibroblasts	Eagle's MEM	CS
Glial cells	MEM, DMEM/F12	FB
Glioma	MEM, DMEM/F12	FB
HeLa cells	Eagle's MEM	CS
Hematopoietic cells	RPMI 1640, Fischer's, αMEM	FB
Human diploid fibroblasts	Eagle's MEM	CS
Human leukemia	RPMI 1640	FB
Human tumors	L15, RPMI 1640, DMEM/F12	FB
Keratinocytes	αMEM	FB
L cells (L929, LS)	Eagle's MEM	CS
Lymphoblastoid cell lines (human)	RPMI 1640	FB
Mammary epithelium	RPMI 1640, DMEM/F12	FB
MDCK dog kidney epithelium	DMEM, DMEM/F12	FB
Melanocytes	M199	FB
Melanoma	MEM, DMEM/F12	FB
Mouse embryo fibroblasts	Eagle's MEM	CS
Mouse leukemia	Fisher's, RPMI 1640	FB, HoS
Mouse erythroleukemia	DMEM/F12, RPMI 1640	FB, HoS
Mouse myeloma	DMEM, RPMI 1640	FB
Mouse neuroblastoma	DMEM, DMEM/F12	FB
Neurons	DMEM	FB
NRK rat kidney fibroblasts	MEM, DMEM	CS
Rat minimal-deviation hepatoma (HTC, MDH)	Swim's S77, DMEM/F12	FB
Skeletal muscle	DMEM, F12	FB, HoS
Syrian hamster fibroblasts (e.g., BHK 21)	MEM, GMEM, DMEM	CS

Abbreviations: CS, calf serum; FB, fetal bovine serum; HoS, horse serum. SF12 is Ham's F12 plus Eagle's essential amino acids and nonessential amino acids as in DMEM (available as 100× stock, *see* Appendix II). Further recommendations on the choice of medium can be found in McKeehan [1977], Barnes et al. [1984(a–d)], and Mather [1998].

continuous cell strains. They may need to be supplemented with glutamine for most cells and usually require serum. The cost of serum should be calculated on the basis of the volume of the medium when cell yield is not important, but if the objective is to produce large quantities of cells, one should calculate serum costs on a per-cell basis. Thus, if a culture grows to 1×10^6/mL in serum A and 2×10^6 mL in serum B, serum B becomes the less expensive by a factor of

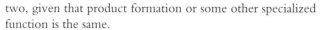

two, given that product formation or some other specialized function is the same.

If fetal bovine serum seems essential, try mixing it with calf serum. This may allow you to reduce the concentration of the more expensive fetal serum. If you can, leave out serum altogether, or reduce the concentration, and use a serum-free formulation (*see* Section 10.5).

9.6.1 Batch Reservation

Considerable variation may be anticipated between batches of serum. Such variation results from differing methods of preparation and sterilization, different ages and storage conditions, and variations in animal stocks from which the serum was derived, including different strains and disparities in pasture, climate, and other environmental conditions. It is important to select a batch, use it for as long as possible, and replace it, eventually, with one as similar to it as possible.

Serum standardization is difficult, as batches vary considerably, and one batch will last only about six months to a year, stored at −20°C. Select the type of serum that is most appropriate for your purposes, and request batches to test from a number of suppliers. Most serum suppliers will reserve a batch until a customer can select the most suitable one (provided that this does not take longer than three weeks or so). When a suitable batch has been selected, the supplier is requested to hold the appropriate volume for up to one year, to be dispatched on demand. Other suppliers should also be informed, so that they may return the rejected batches to their stocks.

9.6.2 Testing Serum

The quality of a given serum is assured by the supplier, but the firm's quality control is usually performed with one of a number of continuous cell lines. If your requirements are more demanding, then you will need to do your own testing. There are four main parameters for testing serum.

Plating efficiency. During cloning, the cells are at a low density and hence are at their most sensitive, making this a very stringent test. Plate the cells out at 10 to 100 cells/mL, and look for colonies after 10 days to two weeks. Stain and count the colonies (*see* Protocol 21.10), and look for differences in plating efficiency (survival) and colony size (cell proliferation). Each serum should be tested at a range of concentrations from 2% to 20%. This approach will reveal whether one serum is equally effective at a lower concentration, thereby saving money and prolonging the life of the batch, and will show up any toxicity at a high serum concentration.

Growth curve. A growth curve should be plotted for cell growth in each serum (*see* Protocols 21.7−21.9), so that the lag period, doubling time, and saturation density (cell density at "plateau") can be determined. A long lag implies that the culture has to adapt to the serum; short doubling times

are preferable if you want a lot of cells quickly; and a high saturation density will provide more cells for a given amount of serum and will be more economical.

Preservation of cell culture characteristics. Clearly, the cells must do what you require of them in the new serum, whether they are acting as host to a given virus, producing a certain cell product, differentiating, or expressing a characteristic sensitivity to a given drug.

Sterility. Serum from a reputable supplier will have been tested and shown to be free of microorganisms. However, in the unlikely event that a sample of serum is contaminated but has escaped quality control, the fact that it is contaminated should show up in mycoplasma screening (*see* Section 19.3.2).

9.6.3 Heat Inactivation

Serum is heat inactivated by incubating it for 30 min at 56°C. It may then be dispensed into aliquots and stored at −20°C. Originally, heating was used to inactivate complement for immunoassays, but it may achieve other effects not yet documented. Often, heat-inactivated serum is used because of the adoption of a previous protocol, without any concrete evidence that it is beneficial. Claims that heat inactivation removes mycoplasma are probably unfounded, although heat treatment may reduce the titer for some mycoplasma.

9.7 OTHER SUPPLEMENTS

In addition to serum, tissue extracts and digests have traditionally been used as supplements to tissue culture media.

9.7.1 Amino Acid Hydrolysates

Many such supplements are derived from microbiological culture techniques and autoclavable broths. Bactopeptone, tryptose, and lactalbumin hydrolysate (Difco—B-D Biosciences) are proteolytic digests of beef heart or lactalbumin and contain mainly amino acids and small peptides. Bactopeptone and tryptose may also contain nucleosides and other heat-stable tissue constituents, such as fatty acids and carbohydrates.

9.7.2 Embryo Extract

Embryo extract is a crude homogenate of 10-day-old chick embryo that is clarified by centrifugation (*see* Appendix I). The crude extract was fractionated by Coon and Cahn [1966] to give fractions of either high or low molecular weight. The low-molecular-weight fraction promoted cell proliferation, whereas the high-molecular-weight fraction promoted pigment and cartilage cell differentiation. Although Coon and Cahn did not fully characterize these fractions, more recent evidence would suggest that the low-molecular-weight fraction probably contains peptide growth factors and the high-molecular-weight fraction proteoglycans and other matrix constituents.

Embryo extract was originally used as a component of plasma clots (*see* Sections 3.7, 12.3.1) to promote cell migration from the explant and has been retained in some organ culture techniques (*see* Section 25.2).

9.7.3 Conditioned Medium

Puck and Marcus [1955] found that the survival of low-density cultures could be improved by growing the cells in the presence of feeder layers (*see* Section 14.2.3). This effect is probably due to a combination of effects including conditioning of the substrate and conditioning of the medium by the release into it of small molecular metabolites and growth factors [Takahashi & Okada, 1970]. Hauschka and Konigsberg [1966] showed that the conditioning of culture medium that was necessary for the growth and differentiation of myoblasts was due to collagen released by the feeder cells. Using feeder layers and conditioning the medium with embryonic fibroblasts or other cell lines remains a valuable method of culturing difficult cells [*see*, e.g., Stampfer et al., 1980].

Attempts have been made to isolate active fractions from conditioned medium, and the original supposition is still probably close to the correct interpretation. Conditioned medium contains both substrate-modifying matrix constituents, like collagen, fibronectin, and proteoglycans, and growth factors, such as those of the heparin-binding group (FGF, etc.), insulin-like growth factors (IGF-I and -II), PDGF, and several others (*see* Table 10.3), in addition to the intermediary metabolites previously proposed. However, conditioned medium adds undefined components to medium and should be eliminated after the active constituents are determined.

CHAPTER 10

Serum-Free Media

Although many cell lines are still propagated in medium supplemented with serum, in many instances cultures may now be propagated in serum-free media (Tables 10.1, 10.2). The need to (1) standardize media among laboratories, (2) provide specialized media for specific cell type, and (3) eliminate variable natural products, led to the development of more complex media, such as M199 of Morgan et al. [1950], CMRL 1066 of Parker et al. [1957], NCTC109 [Evans et al., 1956], Waymouth's MB 572/1 [1959], NCTC135 [Evans & Bryant, 1965], and Birch and Pirt [1971] for L929 mouse fibroblast cells, and Ham's F10 [1963] and F12 [1965] clonal growth media for Chinese hamster ovary (CHO) cells. Serum-free media were also developed for HeLa human cervical carcinoma cells [Blaker et al., 1971; Higuchi, 1977].

Although a degree of cell selection may have been involved in the adaptation of continuous cell lines to serum-free conditions, the MCDB series of media [Ham & McKeehan, 1978; see also Table 10.1], Sato's DMEM/F12-based media [Barnes & Sato, 1980], and others based on RPMI 1640 [Carney et al., 1981; Brower et al., 1986; see also Table 10.2], demonstrated that serum could be reduced or omitted without apparent cell selection if appropriate nutritional and hormonal modifications were made to the media [Barnes et al., 1984a−d; Cartwright & Shah, 1994; Mather, 1998]. These also provided selective conditions for primary culture of particular cell types. Specific formulations (e.g., MCDB 110 [Bettger et al., 1981]) were derived to culture human fibroblasts [Ham, 1984], many normal and neoplastic murine and human cells [Barnes & Sato, 1980], lymphoblasts [Iscove & Melchers, 1978], and several different primary cultures [Mather & Sato, 1979a,b; Sundqvist et al., 1991; Gupta et al., 1997; Keen & Rapson, 1995; Vonen et al., 1992] in the absence of serum, with, in several cases, some protein added [Tsao et al., 1982; Benders et al., 1991]. This list now covers a wide range of cell types, and many of the media are available commercially (see Appendix II). In addition, the need to remove animal proteins from the in vitro production of biopharmaceuticals has generated a number of formulations for continuous cell lines such as CHO and hybridomas [Froud, 1999; Ikonomou et al., 2003; Shah, 1999], not least for some of the safety issues involved [Merten, 1999].

10.1 DISADVANTAGES OF SERUM

Using serum in a medium has a number of disadvantages:

(1) **Physiological Variability.** The major constituents of serum, such as albumin and transferrin, are known, but serum also contains a wide range of minor components that may have a considerable effect on cell growth (see Table 9.5). These components include nutrients (amino acids, nucleosides, sugars, etc.), peptide growth factors, hormones, minerals, and lipids, the concentrations and actions of which have not been fully determined.

(2) **Shelf Life and Consistency.** Serum varies from batch to batch, and at best a batch will last one year, perhaps deteriorating during that time. It must then be replaced with another batch that may be selected as similar, but will never be identical, to the first batch.

(3) **Quality Control.** Changing serum batches ·requires extensive testing to ensure that the replacement is as

TABLE 10.1. Examples of Serum-Free Media Formulations

Medium		MCDB 110	MCDB 131	MCDB 170	MCDB 202	MCDB 302	MCDB 402	WAJC 404	MCDB 153	Iscove's	LHC-9
Cell type		Human lung fibroblasts	Human vascular endothelium	Mammary epithelium	Chick embryo fibroblasts	CHO cells	3T3 cells	Prostatic epithelium[1]	Keratinocytes	Lymphoid cells	Bronchial epithelium
Reference		Bettger et al., 1981	Knedler & Ham, 1987; Gupta et al., 1997	Hammond et al., 1984	McKeehan & Ham, 1976b	Hamilton & Ham, 1977	Shipley & Ham, 1983	McKeehan et al., 1984; Chaproniere & McKeehan, 1986	Peehl & Ham, 1980	Iscove & Melchers, 1978	Lechner & LaVeck, 1985
Component	Mol. wt.										
Amino acids											
L-alanine	89	1.0E-04	3.0E-05	1.0E-04	1.0E-04		1.0E-04	1.0E-04	2.8E-04	1.0E-04	
L-arginine	211	1.0E-03	3.0E-04	3.0E-04	3.0E-04	1.0E-03	3.0E-04	1.0E-03	1.0E-03	4.0E-04	2.0E-03
L-asparagine	132	1.0E-04	1.0E-04	1.0E-03	1.0E-03	1.1E-04	1.0E-04	1.0E-04	1.0E-04	1.9E-04	1.0E-04
L-aspartic acid	133	1.0E-04	1.0E-04	1.0E-04	1.0E-04	1.0E-04	1.0E-05	3.0E-05	3.0E-05	2.3E-04	3.0E-05
L-cysteine	176	5.0E-05	2.0E-04	7.0E-05	2.0E-04	1.0E-04			2.4E-04		2.4E-04
L-cystine	240			2.0E-04	2.0E-04		4.0E-04	2.4E-04		2.9E-04	
L-glutamic acid	147	1.0E-04	3.0E-05	1.0E-04	1.0E-04	1.0E-04	1.0E-05	1.0E-04	1.0E-04	5.1E-04	1.0E-04
L-glutamine	146	2.5E-03	1.0E-02	2.0E-03	1.0E-03	3.0E-03	5.0E-03	6.0E-03	6.0E-03	4.0E-03	6.0E-03
Glycine	75	3.0E-04	3.0E-05	1.0E-04	1.0E-04	1.0E-04	1.0E-04	1.0E-04	1.0E-04	4.0E-04	1.0E-04
L-histidine	210	1.0E-04	2.0E-04	1.0E-04	1.0E-04	1.0E-04	2.0E-03	8.0E-05	8.0E-05	2.0E-04	1.6E-04
L-isoleucine	131	3.0E-05	5.0E-04	1.0E-04	1.0E-04	3.0E-05	1.0E-03	1.5E-05	1.5E-05	8.0E-04	3.0E-05
L-leucine	131	1.0E-04	1.0E-03	3.0E-04	3.0E-04	1.0E-04	2.0E-03	5.0E-04	5.0E-04	8.0E-04	1.0E-03
L-lysine HCl	183	2.0E-04	1.0E-03	2.0E-04	2.0E-04	2.0E-04	8.0E-04	1.0E-04	1.0E-04	8.0E-04	2.0E-04
L-methionine	149	3.0E-05	1.0E-04	3.0E-05	3.0E-05	3.0E-05	2.0E-04	3.0E-05	3.0E-05	2.0E-04	6.0E-05
L-phenylalanine	165	3.0E-05	2.0E-04	3.0E-05	3.0E-05	3.0E-05	3.0E-04	3.0E-05	3.0E-05	4.0E-04	6.0E-05
L-proline	115	3.0E-04	1.0E-04	5.0E-05	5.0E-05	3.0E-04		3.0E-04	3.0E-04	3.5E-04	3.0E-04
L-serine	105	1.0E-04	3.0E-04	3.0E-04	3.0E-04	1.0E-04	1.0E-04	6.0E-04	6.0E-04	4.0E-04	1.2E-03
L-threonine	119	1.0E-04	1.0E-04	3.0E-04	3.0E-04	1.0E-04	5.0E-04	1.0E-04	1.0E-04	8.0E-04	2.0E-04
L-tryptophan	204	1.0E-05	2.0E-05	3.0E-05	3.0E-05	1.0E-05	1.0E-05	1.5E-05	1.5E-05	7.8E-05	3.0E-05
L-tyrosine	181	3.5E-05	1.0E-04	5.0E-05	5.0E-05	4.4E-05	2.0E-04	1.5E-05	1.5E-05	4.6E-04	3.0E-05
L-valine	117	1.0E-04	1.0E-03	3.0E-04	3.0E-04	1.0E-04	2.0E-03	3.0E-04	3.0E-04	8.0E-04	6.0E-04
Vitamins											
Biotin	244	3.0E-08	3.0E-08	3.0E-08	3.0E-08	3.0E-08	3.0E-08	6.0E-08	6.0E-08	5.3E-08	6.0E-08
Choline chloride	140	1.0E-04		1.0E-04	1.0E-04	1.0E-04	1.0E-04	1.0E-04	1.0E-04	2.0E-05	2.0E-04
Folic acid	441		1.0E-04			3.0E-06		1.8E-06	1.8E-06	9.1E-06	1.8E-06
Folinic acid	512	1.0E-09	1.0E-06	1.0E-08	1.0E-08		1.0E-06				
myo-Inositol	180	1.0E-04	4.0E-05	1.0E-04	1.0E-04	1.0E-04	4.0E-05	1.0E-04	1.0E-04	4.0E-05	1.0E-04
Nicotinamide	122	5.0E-05	5.0E-05	5.0E-05	5.0E-05	3.0E-07	5.0E-05	3.0E-07	3.0E-07	3.3E-05	3.0E-07
Pantothenate	238	1.0E-06	5.0E-05	1.0E-06	1.0E-06	1.0E-06	5.0E-05	1.0E-06	1.0E-06	1.7E-05	1.0E-06
Pyridoxal HCl	204									2.0E-05	
Pyridoxine HCl	206	3.0E-07	1.0E-05	3.0E-07	3.0E-07	3.0E-07	1.0E-04	3.0E-07	3.0E-07		3.0E-07
Riboflavin	376	3.0E-07	1.0E-08	3.0E-07	3.0E-07	1.0E-07	1.0E-06	1.0E-07	1.0E-07	1.1E-06	1.0E-07
Thiamin HCl	337	1.0E-06	1.0E-05	1.0E-06	1.0E-06	1.0E-06	1.0E-04	1.0E-06	1.0E-06	1.2E-05	1.0E-06
α-Tocopherol	430	1.4E-07									
Retinoic acid	300										3.3E-07
Retinol acetate	329	4.2E-07									
Vitamin B$_{12}$	1355	1.0E-07	1.0E-08	1.0E-07	1.0E-07	1.0E-07	1.0E-08	3.0E-07	3.0E-07	9.6E-09	3.0E-07
Antioxidants											
Dithiothreitol	154	6.5E-06									
Glutathione	307	6.5E-07									
Inorganic salts											
$CaCl_2$	147	1.0E-03	1.6E-03	2.0E-03	2.0E-03	6.0E-04	1.6E-03	1.3E-04	3.0E-05	1.5E-03	1.1E-04
KCl	75	5.0E-03	4.0E-03	3.0E-03	3.0E-03	3.0E-03	4.0E-03	1.5E-03	1.5E-03	4.4E-03	1.5E-03

TABLE 10.1. Examples of Serum-Free Media Formulations (*Continued*)

Medium		MCDB 110	MCDB 131	MCDB 170	MCDB 202	MCDB 302	MCDB 402	WAJC 404	MCDB 153	Iscove's	LHC-9
Cell type		Human lung fibroblasts	Human vascular endothelium	Mammary epithelium	Chick embryo fibroblasts	CHO cells	3T3 cells	Prostatic epithelium[1]	Keratinocytes	Lymphoid cells	Bronchial epithelium
Reference		Bettger et al., 1981	Knedler & Ham, 1987; Gupta et al., 1997	Hammond et al., 1984	McKeehan & Ham, 1976b	Hamilton & Ham, 1977	Shipley & Ham, 1983	McKeehan et al., 1984; Chaproniere & McKeehan, 1986	Peehl & Ham, 1980	Iscove & Melchers, 1978	Lechner & LaVeck, 1985
Component	Mol. wt.										
KNO$_3$	160					1.6E-08				7.5E-07	
MgCl$_2$	203					6.0E-04		6.0E-04	6.0E-04		2.2E-02
MgSO$_4$	247	1.0E-03	1.0E-02	1.5E-03	1.5E-03	6.1E-10	8.0E-04			8.1E-04	
NaCl	58	1.1E-01	1.1E-01	1.2E-01	1.2E-01	1.3E-01	1.2E-01	1.2E-01	1.2E-01	7.7E-02	1.0E-01
NaHCO$_3$	84		1.4E-02			1.4E-02	1.4E-02	1.4E-02	1.4E-02	3.6E-02	1.2E-02
NA$_2$HPO$_4$	120	3.0E-03	5.9E-04	5.0E-04	5.0E-04	1.2E-03	5.0E-04	2.0E-03	2.0E-03	1.0E-03	2.0E-03
Trace elements											
CuSO$_4$·5H$_2$O	160	1.0E-09	7.5E-09	1.0E-09	1.0E-09		5.0E-09	1.0E-09	1.1E-08		1.0E-08
FeSO$_4$	278	5.0E-06	1.0E-06	5.0E-06	5.0E-06	3.0E-06	1.0E-06	5.0E-06	5.0E-06		5.4E-04
MnSO$_4$·H$_2$O	169	1.0E-09	1.2E-09	5.0E-10	5.0E-10		1.0E-09	1.0E-09	1.0E-09		1.0E-09
(NH$_4$)$_6$Mo$_7$O$_{24}$	1236	1.0E-09	3.0E-09	1.0E-09	1.0E-09	1.0E-08	3.0E-09	1.0E-09	1.0E-09		1.0E-09
NiCl$_2$	238	5.0E-10	4.2E-10	5.0E-12	5.0E-12		3.0E-10	5.0E-10	5.0E-10		5.0E-10
H$_2$SeO$_3$	129	3.0E-08		3.0E-08	3.0E-08	1.3E-08	1.0E-08	3.0E-08	3.0E-08	1.0E-07	3.0E-08
Na$_2$SiO$_3$	122	5.0E-07	2.3E-05	5.0E-07	5.0E-07		1.0E-05	5.0E-07	5.0E-07		5.0E-07
SnCl$_2$	190	5.0E-10		5.0E-12	5.0E-12			5.0E-10	5.0E-10		5.0E-10
NH$_4$VO$_3$	117	5.0E-09	5.1E-09	5.0E-09	5.0E-09	1.0E-08	5.0E-09	5.0E-09	5.0E-09		5.0E-09
ZnSO$_4$·7H$_2$O	288	5.0E-07	1.0E-09	1.0E-07	1.0E-07	3.0E-06	1.0E-06	5.0E-07	5.0E-07		4.8E-07
Lipids and precursors											
Cholesterol	387	7.6E-06									
Ethanolamine	61			1.0E-04					1.0E-04		
Linoleic acid	280			2.0E-07	2.0E-07	3.0E-07	3.0E-07				
Lipoic acid	206	1.0E-08	1.0E-08	1.0E-08	1.0E-08	9.8E-07	1.0E-08	1.0E-06	1.0E-06		1.0E-06
Phosphoethanolamine	141			1.0E-04							5.0E-07
Soya lecithin, µg/mL		6									
Soybean lipid, µg/mL										50	
Sphingomyelin, µg/mL		1									
Hormones and growth factors											
EGF, ng/mL			30	10				25	25		5
Epinephrine	183										2.7E-06
Hydrocortisone[2]	362	5.0E-07		1.4E-07				5.0E-07	1.4E-07		2.0E-07
Insulin. µg/mL			1	5				10	5		5
Prolactin, µg/mL			1	5							
PGE$_1$	355	2.5E-08							2.5E-08		
Triiodothyronine	673										1.0E-08
Nucleosides, etc.											
Adenine SO$_4$	184	1.0E-05	1.0E-06	1.0E-06	1.0E-06		1.0E-06	1.8E-04	1.8E-04		1.8E-04
Hypoxanthine	136					3.0E-05					
Thymidine	242	3.0E-07	1.0E-07	3.0E-07	3.0E-07		1.0E-06	3.0E-06	3.0E-06		3.0E-06
Energy metabolism											
D-glucose	180	4.0E-03	5.6E-03	8.0E-03	8.0E-03	1.0E-02	5.5E-03	6.0E-03	6.0E-03	2.5E-02	6.0E-03
Phosphoenolpyruvate	190	1.0E-05									

(*Continued overleaf*)

TABLE 10.1. Examples of Serum-Free Media Formulations (*Continued*)

Medium		MCDB 110	MCDB 131	MCDB 170	MCDB 202	MCDB 302	MCDB 402	WAJC 404	MCDB 153	Iscove's	LHC-9
Cell type		Human lung fibroblasts	Human vascular endothelium	Mammary epithelium	Chick embryo fibroblasts	CHO cells	3T3 cells	Prostatic epithelium[1]	Keratinocytes	Lymphoid cells	Bronchial epithelium
Reference		Bettger et al., 1981	Knedler & Ham, 1987; Gupta et al., 1997	Hammond et al., 1984	McKeehan & Ham, 1976b	Hamilton & Ham, 1977	Shipley & Ham, 1983	McKeehan et al., 1984; Chaproniere & McKeehan, 1986	Peehl & Ham, 1980	Iscove & Melchers, 1978	Lechner & LaVeck, 1985
Component	Mol. wt.										
Sodium acetate $3H_2O$	136					1.0E		3.7E-03	3.7E-03		3.7E-03
Sodium pyruvate	110	1.0E-03	1.0E-03	1.0E-03	5.0E-04	1.1E+02		5.0E-04	5.0E-04	1.0E-03	5.0E-04
Other components											
Cholera toxin	~90,000							2.0E-10			
HEPES, Na salt	260	3.0E-02		3.0E-02	3.0E-02			2.8E-02	2.8E-02	2.5E-02	2.3E-02
Phenol red	376	3.3E-06	3.3E-05	3.3E-06	3.3E-06	3.3E-06	3.3E-05	3.3E-05	3.3E-05	4.0E-05	3.3E-06
Putrescine 2HCl	161	1.0E-09	1.2E-09	1.0E-09	1.0E-09	1.0E-06	1.0E-09	1.0E-06	1.0E-06		1.0E-06
Protein supplements											
BPE, µgP/mL[3]		70						25			35
BSA, mg/mL										0.5–10	
Dialyzed FBS, µgP/mL									1		
Transferrin, Fe^{3+} saturated, µg/mL				5						30–300	10
Gas phase											
CO_2	44	2%	5%	2%	2%	5%	5%	5%	5%	10%	5%

[1]See also complete PFMR-4A [Peehl, 2002]. [2]Soluble analogs of hydrocortisone, such as dexamethasone, can be used. [3]Ovine prolactin can be substituted for BPE. Most concentrations are molar, and computer-style notation is used, e.g., 3.0E-2 = 3.0×10^{-2} = 30 mM. Molecular weights are given for root compounds; although some recipes use salts or hydrated forms, molarities will, of course, remain the same. The units are given in the component column where molarity is not used. *Abbreviations:* BPE, bovine pituitary extract; BSA, bovine serum albumin; EGF, epidermal growth factor; FBS, fetal bovine serum. See text for references.

close as possible to the previous batch. This can involve several tests (for growth, plating efficiency, and special functions; *see* Section 9.7.2) and a number of different cell lines.

(4) **Specificity.** If more than one cell type is used, each type may require a different batch of serum, so that several batches must be held on reserve simultaneously. Coculturing different cell types will present an even greater problem.

(5) **Availability.** Periodically, the supply of serum is restricted because of drought in the cattle-rearing areas, the spread of disease among the cattle, or economic or political reasons. This can create problems at any time, restricting the amount of serum available and the number of batches to choose from, but can be particularly acute at times of high demand. Today, demand is increasing, and it will probably exceed supply unless the majority of commercial users are able to adopt serum-free media. Although an average research laboratory may reserve 100–200 L of serum per year, a commercial biotechnology laboratory can use that amount or more in a week.

(6) **Downstream Processing.** The presence of serum creates a major obstacle to product purification and may even limit the pharmaceutical acceptance of the product.

(7) **Contamination.** Serum is frequently contaminated with viruses, many of which may be harmless to cell culture but represent an additional unknown factor outside the operator's control [Merten, 1999]. Fortunately, improvements in serum sterilization techniques have virtually eliminated the risk of mycoplasma infection from sera from most reputable suppliers, but viral infection remains a problem, despite claims that some filters may remove viruses (Pall Gelman). Because of the risk of spreading bovine spongiform encephalitis among cattle, cell cultures and serum shipped to the United States or

TABLE 10.2. Examples of Serum-Free Media; Supplemented Basal Media

Medium	HITES	ACL-3		N3	G3	K-1	K-2			
Reference	Carney et al., 1981	Brower et al., 1986	Masui et al., 1986b	Bottenstein, 1984	Michler-Stuke & Bottenstein, 1982	Taub, 1984	Taub, 1984	Chopra & Xue-Hu, 1993	Robertson & Robertson, 1995	Naeyaert et al., 1991
Cell type	Human small-cell lung carcinoma	Human non-small-cell lung carcinoma	Human lung adeno-carcinoma	Human neuro-blastoma, LA-N-1	Rat glial cells	MDCK (dog kidney)	LLC-PK, (pig kidney)	Human parotid	Human prostate	Human melanocytes
Basal medium	RPMI 1640	RPMI 1640	DMEM/F12	DMEM/F12	DMEM	DMEM/F12	DMEM/F12	MCDB 153	αMEM/F12	M199
Supplements										
Arg VP, μU/mL							10			
BPE, μgP/mL								25	25	
BSA, mg/mL		5.0			0.3					
Cholera toxin										1.0E-09
Cholesterol							1.0E-08			
EGF, ng/mL		10						10	10	1
Epinephrine										
Estradiol	1.0E-08									
Ethanolamine			5.0E-07						1.0E-03	
FGF-2 (basic FGF), ng/mL										10
Glucagon, μg/mL			0.2							
Glutamine (additional)		2.0E-03							2.0E-03	
Hydrocortisone	1.0E-08	5.0E-08				5.0E-08	2.0E-07	1.4E-06	1.0E-06	1.4E-06
Insulin, μg/mL	5.0	20.0	5.0	5.0	0.5	5.0	25.0	5.0	5.0	10.0
Na pyruvate (additional)		5.0E-04								
Na$_2$SeO$_3$	3.0E-08	2.5E-08	2.5E-08	3.0E-08						
Phosphoethanol-amine									1.0E-04	
Progesterone				2.0E-08	2.0E-07					
Prolactin										
Prostaglandin E$_1$						7.0E-08				
Putrescine				1.0E-04	1.0E-07					
Transferrin, Fe^{3+} saturated, μg/mL	100	10		50	100	5	10		5	10
Triiodothyronine		1.0E-10	5.0E-10		4.9E-07	5.0E-12	1.0E-09			1.0E-09
Thyroxine					4.5E-07					

Most concentrations are molar, and computer-style notation is used, e.g., 3.0E-2 = 3.0 × 10^{-2} = 30 mM. Molecular weights are given for root compounds; although some recipes use salts or hydrated forms, molarities will, of course, remain the same. The units are given in the component column where molarity is not used. Soluble analogs of hydrocortisone, such as dexamethasone, can be used. *Abbreviations:* Arg VP, arginine vasopressin; BPE, bovine pituitary extract; BSA, bovine serum albumin; EGF, epidermal growth factor; FBS, fetal bovine serum; FGF, fibroblast growth factor.

Australia require information on the country of origin and the batch number of the serum. Serum derived from cattle in New Zealand probably has the lowest endogenous viral contamination, as many of the viruses found in European and North American cattle are not found in New Zealand.

(8) **Cost.** Cost is often cited as a disadvantage of serum supplementation. Certainly, serum constitutes the major part of the cost of a bottle of medium (more than 10 times the cost of the chemical constituents), but if it is replaced by defined constituents, the cost of these may be as high as that of the serum. However, as the demand for such items as transferrin, selenium, insulin, etc., rises, the cost is likely to come down with increasing market size, and serum-free media will become relatively cheaper. The availability of recombinant growth factors, coupled with market demand, may help to reduce their intrinsic cost.

(9) **Growth Inhibitors.** As well as its growth-promoting activity, serum contains growth-inhibiting activity, and although stimulation usually predominates, the net effect of the serum is an unpredictable combination of both inhibition and stimulation of growth. Although substances such as PDGF may be mitogenic to fibroblasts, other constituents of serum can be cytostatic. Hydrocortisone, present at around 1×10^{-8} M in fetal serum, is cytostatic to many cell types, such as glia [Guner et al., 1977] and lung epithelium [McLean et al., 1986], at high cell densities (although it may be mitogenic at low cell densities), and TGF-β, released from platelets, is cytostatic to many epithelial cells.

(10) **Standardization.** Standardization of experimental and production protocols is difficult, both at different times and among different laboratories, because of batch-to-batch variations in serum.

10.2 ADVANTAGES OF SERUM-FREE MEDIA

All of the above problems can be eliminated by removal of serum and other animal products. Serum-free media have, in addition, two major positive benefits.

10.2.1 Selective Media

One of the major advantages of the control over growth-promoting activity afforded by serum-free media is the ability to make a medium selective for a particular cell type (*see* Tables 10.1 and 10.2). Fibroblastic overgrowth can be inhibited in breast and skin cultures by using MCDB 170 [Hammond et al., 1984] and 153 [Peehl and Ham, 1980], melanocytes can be cultivated in the absence of fibroblasts and keratinocytes [Naeyaert et al., 1991], and separate lineages and even stages of development may be selected in hematopoietic cells by choosing the correct growth factor or group of growth factors (*see* Section 23.4). Many of these selective media are now available commercially along with cultures of selected cell types (*see* Table 10.4).

10.2.2 Regulation of Proliferation and Differentiation

Add to the ability to select for a specific cell type the possibility of switching from a growth-enhancing medium for propagation to a differentiation-inducing medium by altering the concentration and types of growth factors and other inducers.

10.3 DISADVANTAGES OF SERUM-FREE MEDIA

Serum-free media are not without disadvantages:

(1) **Multiplicity of Media.** Each cell type appears to require a different recipe, and cultures from malignant tumors may vary in requirements from tumor to tumor, even within one class of tumors. Although this degree of specificity may be an advantage to those isolating specific cell types, it presents a problem for laboratories maintaining cell lines of several different origins.

(2) **Selectivity.** Unfortunately, the transition to serum-free conditions, however desirable, is not as straightforward as it seems. Some media may select a sublineage that is not typical of the whole population, and even in continuous cell lines, some degree of selection may still be required. Cells at different stages of development (e.g., stem cells vs. committed precursor cells) may require different formulations, particularly in the growth factor and cytokine components.

(3) **Reagent Purity.** The removal of serum also requires that the degree of purity of reagents and water and the degree of cleanliness of all apparatus be extremely high, as the removal of serum also removes the protective, detoxifying action that some serum proteins may have. Although removing this action is no doubt desirable, it may not always be achievable, depending on resources.

(4) **Cell Proliferation.** Growth is often slower in serum-free media, and fewer generations are achieved with finite cell lines.

(5) **Availability.** Although improving steadily, the availability of properly quality-controlled serum-free media is quite limited, and the products are often more expensive than conventional media.

10.4 REPLACEMENT OF SERUM

The essential factors in serum have been described (*see* Section 9.6) and include (1) adhesion factors such as fibronectin; (2) peptides, such as insulin, PDGF, and TGF-β, that regulate growth and differentiation; (3) essential nutrients, such as minerals, vitamins, fatty acids, and intermediary metabolites; and (4) hormones, such as insulin, hydrocortisone, estrogen, and triiodothyronine, that regulate membrane transport, phenotypic status, and the constitution of the cell surface. Although some of these are catered for in the formulation of serum-free media, others are not and may require addition and optimization.

10.4.1 Serum-Free Subculture

Adhesion factors. When serum is removed, it may be necessary to treat the plastic growth surface with fibronectin (25–50 µg/mL) or laminin (1–5 µg/mL), added directly to the medium [Barnes et al., 1984a; *see* Protocol 23.9]. Pretreating the plastic with poly-L-lysine (1 mg/mL) was shown to enhance the survival of human diploid fibroblasts [McKeehan & Ham, 1976a; *see also* Section 8.4 and Barnes et al., 1984a].

Protease inhibitors. After trypsin-mediated subculture, the addition of serum inhibits any residual proteolytic activity. Consequently, protease inhibitors such as soya bean trypsin inhibitor or 0.1 mg/mL aprotinin (Sigma) must be added to serum-free media after subculture. Furthermore, because crude trypsin is a complex mixture of proteases, some of which may require different inhibitors, it is preferable to use pure trypsin (e.g., Sigma Gr. III) followed by a trypsin inhibitor. Alternatively, one may wash cells by centrifugation to remove trypsin.

Trypsin and other proteases. Special care may be required when trypsinizing cells from serum-free media, as the cells are more fragile and may need purified crystalline trypsin, and to be chilled to 4°C, to reduce damage [McKeehan, 1977]. Alternative sources of proteases are available to avoid animal products coming in contact with cells. Purified porcine trypsin may be replaced with recombinant trypsin (TrypLE™, Invitrogen; TrypZean, Sigma). Nonmammalian proteases are also available (e.g., Accutase™ or Accumax™, Sigma). Pronase, Dispase, and collagenase are bacterial proteases not neutralized by trypsin inhibitors and may require removal by centrifugation. It is possible that Pronase can be inactivated by dilution without subsequent neutralization in serum-free conditions [McKeehan, personal communication]. Pronase is very effective but can be toxic to some cells; Dispase and collagenase will not give a single cell suspension if there are epithelial cells present.

10.4.2 Hormones

Hormones that have been used to replace serum include growth hormone (somatotropin) at 50 ng/mL, insulin at 1–10 U/mL, which enhances plating efficiency in a number of different cell types, and hydrocortisone, which improves the cloning efficiency of glia and fibroblasts (*see* Tables 10.1 and 10.2 and Section 14.2.1) and has been found necessary for the maintenance of epidermal keratinocytes and some other epithelial cells (*see* Section 23.1.1). Barnes and Sato [1980] described 10 pM triiodothyronine (T_3) as a necessary supplement for MDCK (dog kidney) cells, and it has also been used for lung epithelium [Lechner & LaVeck, 1985; Masui et al., 1986b]. Various combinations of estrogen, androgen, or progesterone with hydrocortisone and prolactin at around 10 nM can be shown to be necessary for the maintenance of mammary epithelium [Klevjer-Anderson & Buehring, 1980; Hammond et al., 1984; Strange et al., 1991; Lee et al., 1996].

Other hormones with activities not usually associated with the cells they were tested on were found to be effective in replacing serum, e.g., follicle-stimulating hormone (FSH) with B16 murine melanoma [Barnes and Sato, 1980]. It is possible that sequence homologies exist between some growth-stimulating polypeptides and well-established peptide hormones. Alternatively, the processing of some of the large proteins or polypeptides may release active peptide sequences with quite different functions.

10.4.3 Growth Factors

The family of polypeptides that has been found to be mitogenic *in vitro* is now quite extensive (Table 10.3) and includes the heparin-binding growth factors (including the FGF family), EGF, PDGF [Barnes et al., 1984a, c], IGF-I and -II, and the interleukins [Thomson, 1991] that are active in the 1–10 ng/mL range. Growth factors and cytokines tend to have a wide-ranging specificity, except for some that are active in the hematopoietic system [Barnes et al., 1984d; *see also* Section 23.4). Keratinocyte growth factor (KGF) [Aaronson et al., 1991], besides showing activity with epidermal keratinocytes, will also induce proliferation and differentiation in prostatic epithelium [Planz et al., 1998; Thomson et al., 1997]. Hepatocyte growth factor (HGF) [Kenworthy et al., 1992] is mitogenic for hepatocytes but is also morphogenic for kidney tubules [Furue & Saito, 1997; Montesano et al., 1997; Balkovetz & Lipschutz, 1999]. Growth factors and cytokines acquire their specificity by virtue of the fact that their production is localized and that they have a limited range. Most act as paracrine factors (they are active on adjacent cells) and not by systemic distribution in the blood.

Growth factors may act synergistically or additively with each other or with other hormones and paracrine factors, such as prostaglandin $F_{2\alpha}$ and hydrocortisone [Westermark & Wasteson, 1975; Gospodarowicz, 1974]. For example, the action of interleukin 6 (IL-6) and oncostatin M on A549 cells is dependent on dexamethasone, a synthetic hydrocortisone analog [McCormick et al., 1995, 2000]. The action is due to the production of a heparan sulfate proteoglycan (HSPG) [Yevdokimova & Freshney, 1997]. The requirement for heparin or HSPG was first observed with FGF [Klagsbrun & Baird, 1991], but may be a more general phenomenon, e.g., β-glycan has been shown to be involved in the cellular response to TGF-β [Lopez-Casillas et al., 1993]. Some growth factors are dependent on the activity of a second growth factor before they act [Phillips and Christofalo, 1988]; e.g., bombesin alone is not mitogenic in normal cells but requires the simultaneous or prior action of insulin or one of the IGFs [Aaronson et al., 1991].

10.4.4 Nutrients in Serum

Iron, copper, and a number of minerals have been included in serum-free recipes, although evidence that some of the rarer minerals are required is still lacking. Selenium (Na_2SeO_3), at around 20 nM, is found in most formulas, and there appears to be some requirement for lipids or lipid precursors such as choline, linoleic acid, ethanolamine, or phosphoethanolamine.

10.4.5 Proteins and Polyamines

The inclusion in medium of proteins such as bovine serum albumin (BSA) or tissue extracts often increases cell growth and survival but adds undefined constituents to the medium and retains the problem of adventitious infectious agents.

TABLE 10.3. Growth Factors and Mitogens

Name and synonyms	Abbreviation	Mol. mass. (KDa)	Source*	Function
Acidic fibroblast gf; aFGF; heparin binding gf 1, HBGF-1; endothelial cell gf (ECGF); myoblast gf (MGF)	FGF-1	13 h	Bovine brain; pituitary	Mitogen for endothelial cells
Activin; TGF-β family		g	Gonads	Morphogen; stimulates FSH secretion
Angiogenin		16	Fibroblasts, lymphocytes, colonic epithelial; cells	Angiogenic; endothelial mitogen
Astroglial growth factor-1; member of acidic FGFs	AGF-1	14	Brain	Mitogen for astroglia
Astroglial growth factor-2; member of basic FGFs	AGF-2	14	Brain	Mitogen for astroglia
Basic fibroblast gf; bFGF; HBGF-2; prostatropin	FGF-2	13 h	Bovine brain; pituitary	Mitogen for many mesodermal and neuroectodermal cells; adipocyte and ovarian granulosa cell differentiation
Brain-derived neurotrophic factor	BDNF	28	Brain	Neuronal viability
Cachectin	TNF-α	17	Monocytes	Catabolic; cachexia; shock
Cholera toxin	CT	80–90	Cholera bacillus	Mitogen for some normal epithelia
Ciliary neurotrophic factor; member of IL-6 group	CNTF		Eye	
Endothelial cell growth factor; acidic FGF family	ECGF	h	Recombinant	Endothelial mitogenesis
Endothelial growth supplement; mixture of endothelial mitogens	ECGS		Bovine pituitary	Endothelial mitogenesis
Endotoxin			Bacteria	Stimulates TNF production
Epidermal growth factor, Urogastrone	EGF	6	Submaxillary salivary gland (mouse) human urine; guinea pig prostate	Active transport; DNA, RNA, protein, synthesis; mitogen for epithelial and fibroblastic cells; synergizes with IGF-1 and TGF-β
Erythropoietin	EPO	34–39 g	Juxtaglomerular cells of kidney	Erythroid progenitor proliferation and differentiation
Eye-derived growth factor-1; member of basic FGFs	EDGF-1	14	Eye	
Eye-derived growth factor-2; member of acidic FGFs	EDGF-2	14	Eye	
Fibroblast gf-3; product of int-2 oncogene	FGF-3	14 h	Mammary tumors	Mitogen; morphogen; angiogenic
Fibroblast gf-4; product of hst/KS3 oncogene	FGF-4	14 h	Embryo; tumors	Mitogen; morphogen; angiogenic
Fibroblast gf-5	FGF-5	14 h	Fibroblasts; epithelial cells; tumors	Mitogen; morphogen; angiogenic
Fibroblast gf-6; product of hst-2 oncogene	FGF-6	14 h	Testis; heart; muscle	Mitogenic for fibroblasts; morphogen
Fibroblast gf-9	FGF-9	14 h	Recombinant	
Fibroblast gf-10	FGF-10	14 h		Trophoblast invasion; stimulate uPA and PAI-1; alveolar epithelial mitogen
Granulocyte colony-stimulating factor; pluripoietin; CFS-β	G-CSF	18–22		Granulocyte progenitor proliferation and differentiation

TABLE 10.3. Growth Factors and Mitogens (*Continued*)

Name and synonyms	Abbreviation	Mol. mass. (KDa)	Source*	Function
Granulocyte/macrophage colony-stimulating factor CSA; human CSFα	GM-CSF	14–35 g		Granulocyte/macrophage progenitor proliferation
Heparin-binding EGF-like factor	HB-EGF		Recombinant	
Hepatocyte gf, HBGF-8; Scatter factor	HGF	h	Fibroblasts	Epithelial morphogenesis; hepatocyte proliferation
Heregulin; *erb*B2 ligand	HRG	70	Breast cancer cells; recombinant	Mammary and other epithelial cell mitogen
Immune interferon; macrophage-activating factor (MAF)	IFN-γ	20–25	Activated lymphocytes	Antiviral; activates macrophages
Inhibin; TGF-β family		31 g	Ovary	Morphogen; inhibits FSH secretion
Insulin	Ins	6	βIslet cells of pancreas	Glucose uptake and oxidation; amino acid uptake; glyconeogenesis
Insulin-like gf 1; somatomedin-C; NSILA-1	IGF-1			Mediates effect of growth hormone on cartilage sulfation; insulin-like activity
Insulin-like gf2; MSA in rat	IGF-2	7	BRL-3A cell-conditioned medium	Mediates effect of growth hormone on cartilage sulfation; insulin-like activity
Interferon-α1; leukocyte interferon	IFN-α1	18–20	Macrophages	Antiviral; differentiation inducer; anticancer
Interferon-α2; leukocyte interferon	IFN-α2	18–20	Macrophages	Antiviral; differentiation inducer; anticancer
Interferon-β; fibroblast interferon	IFN-β1	22–27 g	Fibroblasts	Antiviral; differentiation inducer; anticancer
Interferon β2; fibroblast interferon; IL-6, BSF-2 (*see also* IL-6)	IFN-β2	22–27 g	Activated T-cells; fibroblasts; tumor cells	Keratinocyte differentiation; PC12 differentiation (*see also* IL-6)
Interferon γ; immune interferon	IFNγ	Activated lymphocytes	Antiviral, macrophage activator; antiproliferative on transformed cells	
Interleukin-1; lymphocyte-activating factor (LAF); B-cell-activating factor (BAF); hematopoietin-1	IL-1	12–18	Activated macrophages	Induces IL-2 release
Interleukin-2; T-cell gf (TCGF)	IL-2	15	CD4+ve lymphocytes (NK); murine LBRM-5A4 and human Jurkat FHCRC cell lines	Supports growth of activated T-cells; stimulates LAK cells
Interleukin-3; multipotential colony-stimulating factor; mast cell growth factor	IL-3	14–28 g	Activated T-cells; WEHI-3b myelomonocytic cell lines	Granulocyte/macrophage production and differentiation
Interleukin-4; B-cell gf; BCGF-1; BSF-1	IL-4	15–20	Activated CD4+ve lymphocytes	Competence factor for resulting B-cells; mast cell maturation (with IL-3)
Interleukin-5; T-cell-replacing factor (TRF); eosinophil-differentiating factor (EDF) BCGF-2	IL-5	12–18 g	T-lymphocytes	Eosinophil differentiation; progression factor for competent B-cells

(Continued overleaf)

TABLE 10.3. Growth Factors and Mitogens (*Continued*)

Name and synonyms	Abbreviation	Mol. mass. (KDa)	Source*	Function
Interleukin-6; Interferon β-2; B-cell-stimulating factor (BSF-2); hepatocyte-stimulating factor; hybridoma-plasmacytoma gf	IL-6	22–27 g	Activated T-cells macrophage/monocytes; fibroblasts; tumor cells	Acute phase response; B-cell differentiation; keratinocyte differentiation; PC12 differentiation
Interleukin-7; hematopoietic growth factor; lymphopoietin 1	IL-7	15–17 g	Bone marrow stroma	Pre- and pro-B-cell growth factor
Interleukin-8; monocyte-derived neutrophil chemotactic factor (MDNCF); T-cell chemotactic factor; neutrophil-activating protein (NAP-1)	IL-8	8–10 h	LPS monocytes; PHA lymphocytes; endothelial cells; IL-1- and TNF-stimulated fibroblasts and keratinocytes	Chemotactic factor for neutrophils, basophils, and T-cells
Interleukin-9; human P-40; mouse T-helper gf; mast-cell-enhancing activity (MEA)	IL-9	30–40	CD4+ve T-cells; stimulated by anti-CD4 antibody PHA or PMA	Growth factor for T-helper, megakaryocytes, mast cells (with IL-3)
Interleukin-10; cytokine synthesis inhibitory factor (CSIF)	IL-10	20		Immune suppressor
Interleukin-11; adipogenesis inhibitory factor (AGIF)	IL-11	21		Stimulates plasmacytoma proliferation and T-cell-dependent development of Ig-producing B-cells
Interleukin-12; cytotoxic lymphocyte maturation factor	IL-12	40, 35 subunits		Activated T-cell and NK cell growth factor; induces IFN-γ
Keratinocyte gf, FGF-7	KGF	14 h	Fibroblasts	Keratinocyte proliferation and differentiation; prostate epithelial proliferation and differentiation
Leukemia inhibitory factor; HILDA; member of IL-6 group	LIF	24	SCO cells	Inhibits differentiation in embryonal stem cells
Lipopolysaccharide	LPS	10	Gram-positive bacteria	Lymphocyte activation
Lymphotoxin	TNF-β	20–25	Lymphocytes	Cytotoxic for tumor cells
Macrophage inflammatory protein-1α	MIP–1α	10	Macrophages	Hematopoietic stem cell inhibitor
Monocyte/macrophage colony-stimulating factor CSF-1	M-CSF	47–74	B- and T-cells, monocytes, mast cells, fibroblasts	Macrophage progenitor proliferation and differentiation
Müllerian inhibition factor	MIF		Testis	Inhibition of Müllerian duct; inhibition of ovarian carcinoma
Nerve gf, β	βNGF	27	Male mouse submaxillary salivary gland	Trophic factor; chemotactic factor; differentiation factor; neurite outgrowth in peripheral nerve
Oncostatin M; member of IL-6 group	OSM	28	Activated T-cells and PMA-treated monocytes	Differentiation inducer (with glucocorticoid); fibroblast mitogen
Phytohemagglutinin	PHA	30	Red kidney bean (*Phaseolus vulgaris*)	Lymphocyte activation
Platelet-derived endothelial cell growth factor; similar to gliostatin	PD-ECGF	~70	Blood platelets, fibroblasts, smooth muscle	Angiogenesis; endothelial cell mitogen; neuronal viability; glial cytostasis

TABLE 10.3. Growth Factors and Mitogens (*Continued*)

Name and synonyms	Abbrevi- ation	Mol. mass. (KDa)	Source*	Function
Platelet-derived growth factor	PDGF	30	Blood platelets	Mitogen for mesodermal and neuroectodermal cells; wound repair; synergizes with EGF and IGF-1
Phorbol myristate acetate; TPA; phorbol ester	PMA	0.617	Croton oil	Tumor promoter; mitogen for some epithelial cells and melanocytes; differentiation factor for HL-60 and squamous epithelium
Pokeweed mitogen	PWM		Roots of pokeweed (*Phytolacca americana*)	Monocyte activation
Stem cell factor; mast cell growth factor; steel factor; *c-kit* ligand	SCF	31 g	Endothelial cells, fibroblasts, bone marrow, Sertoli cells	Promotes first maturation division of pluripotent hematopoietic stem cell
Transferrin	Tfn	78	Liver	Iron transport; mitogen
Transforming growth factor α	TGF-α	6		Induces anchorage-independent growth and loss of contact inhibition
Transforming growth factor β (six species)	TGF-β1–6	23–25 dimer	Blood platelets	Epithelial cell proliferation inhibitor; squamous differentiation inducer

*Sources described are some of the original tissues from which the natural product was isolated. In many cases the natural product has been replaced by cloned recombinant material that is available commercially (*see* Appendix II). *Abbreviation* (other than in column 2): gf, growth factor; g, glycosylated; h, heparin binding; HBGF, heparin-binding growth factor. Some of the information in this table was taken from Barnes et al. [1984a], Lange et al. [1991], Jenkins [1992], and Smith et al. [1997].

Tissue extracts include bovine pituitary extract used in conjunction with keratinocyte serum-free media, but it may be possible to replace this with defined recombinant growth factors. BSA, fatty acid free, is used at 1–10 mg/mL. Transferrin, at around 10 ng/mL, is required as a carrier for iron and may also have a mitogenic role. Putrescine has been used at 100 nM.

10.4.6 Matrix

One of the properties of serum is to provide a number of proteins, such as fibronectin, that coat the plastic and make it more adhesive (*see* Sections 10.4.1, 8.4.1). In the absence of serum, the plastic substrate may need to be coated with fibronectin (*see* Protocol 23.9) or polylysine [McKeehan & Ham, 1976a] (*see* Tables 10.1 and 10.2 and Section 14.2.1).

10.5 SELECTION OF SERUM-FREE MEDIUM

If the reason for using a serum-free medium is to promote the selective growth of a particular type of cell, then that reason will determine the choice of medium (e.g., MCDB 153 for epidermal keratinocytes, LHC-9 for bronchial epithelium, HITES for small-cell lung cancer, MCDB 130 for endothelium, etc.; *see* Tables 10.1 and 10.2). If the reason is simply to avoid using serum with continuous cell lines, such as CHO cells or hybridomas, in order to reduce the likelihood of viral or serum proteins in the cell product, then the choice will be wider, and there will be several commercial sources to choose from (Table 10.4). When a cell line is obtained from the originator or a reputable cell bank, the supplier will recommend the appropriate medium, and the only reason to change will be if the medium is unavailable or is incompatible with other stocks. If possible, it is best to stay with the originator's recommendation, as this may be the only way to ensure that the line exhibits its specific properties. Table 10.4 summarizes the availability and selection of the serum-free media listed in Tables 10.1 and 10.2 and makes a few additional suggestions. (*See also* [Mather, 1998], [Barnes et al., 1984a–d], and [Ham & McKeehan, 1979].)

10.5.1 Commercially Available Serum-Free Media

Several suppliers (*see* Table 10.4 and Appendix II) now make serum-free media. Some are defined formulations, such as MCDB 131 (Sigma) for endothelial cells and LHC-9 (Biosource International) for bronchial epithelium, whereas others are proprietary formulations, such as CHO-S-SFM for CHO-K1 cells and Opti-MEM for hematopoietic cells (Invitrogen). Many are designed primarily for culture of hybridomas, when the formation of a product that is free of serum proteins is clearly important, but others are applicable to other cell types. Although evidence exists in the literature for the use of proprietary media with specific cell types (*see* Table 10.4), commercial recipes are often a trade secret, and you can only rely on the supplier's advice or, better, screen a number of media over several subcultures with your own cells. The latter can be an extensive exercise but is justified if you are planning long-term work with the cells.

TABLE 10.4. Selecting a Serum-Free Medium

Cells or cell line	Serum-free medium (*see* Table 10.1)	Refs. (*see* Tables 10.1 and 10.2)	Commercial suppliers (of specified media or alternatives)
Adipocytes		Rodriguez et al., 2004	PromoCell
BHK 21	HyQ PF CHO; HyG PF CHO MPS; PC-1	Pardee et al., 1984	Cambrex
Bronchial epithelium	LHC-9	*See* Protocol 22.9	Biosource; Cambrex; PromoCell
Chick embryo fibroblasts	MCDB 201, 202	McKeehan & Ham, 1976b	Sigma
Chinese hamster ovary (CHO)	MCDB 302; PC-1	Hamilton & Ham, 1977	Cambrex; Sigma; Invitrogen; JRH Biosciences; PromoCell
Chondrocytes	Supplemented DMEM/F12	Adolphe, 1984; *see* Protocol 23.13	PromoCell
Continuous cell lines	Eagle's MEM, M199, MB752/1, CMRL 1066, MCDB media, DMEM:F12 + supplements	Waymouth, 1984	Cambrex; Invitrogen; JRH Biosciences; ICN; Sigma
Corneal epithelial cells	MCDB 153	*See* Protocol 23.2	Cambrex, KGM; Cascade
COS-1,7		Doering et al., 2002	Cambrex
Endothelium	MCDB 130, 131	Knedler & Ham, 1987; Gupta et al., 1997; Hoheisel et al., 1998	Cambrex; Cascade; PromoCell; Sigma
Fibroblasts	MCDB 110, 202, 402	Bettger et al., 1981; Shipley & Ham, 1983	Cambrex; Cascade; PromoCell; Sigma
Glial cells	Michler-Stuke	Michler-Stuke & Bottenstein, 1982	Cambrex; Invitrogen
Glioma	SF12 (Ham's F12 with extra essential and nonessential amino acids)	Frame et al., 1980; Freshney, 1980	Cambrex; Invitrogen
HeLa cells		Blaker et al., 1971; Bertheussen, 1993	Cambrex
Hematopoietic cells	αMEM; Iscove's	Stanners et al., 1971; Iscove & Melchers, 1978	Sigma; Roche
HL-60		Li et al., 1997	Cambrex
HT-29		Oh et al., 2001	Cambrex
Human diploid fibroblasts	MCDB 110, 202; PC-1	Bettger et al., 1981; Ham, 1984	Cascade; Cambrex; PromoCell
Human leukemia and normal leukocytes	Iscove's	Breitman et al., 1984	Cambrex; Hyclone; MP Biomedicals Invitrogen; JRH Biosciences; Sigma
Human tumors	Brower; HITES; Masui; Bottenstein N3	Brower et al., 1986; Carney et al., 1981; Masui et al., 1986b; Bottenstein, 1984; Chopra et al., 1996	
Hepatocytes, liver epithelium	Williams E, L15	Williams & Gunn, 1974; Mitaka et al., 1993]	Cambrex; Sigma
Hybridomas	Iscove's	Iscove & Melchers, 1978; Murakami, 1984	Sigma; Invitrogen; Cambrex; JRH Biosciences; MP Biomedicals; Roche; Irvine Scientific; Metachem; PromoCell
Insect cells		Ikonomou et al., 2003	JRH Biosciences; Cell Gro

TABLE 10.4. Selecting a Serum-Free Medium (*Continued*)

Cells or cell line	Serum-free medium (*see* Table 10.1)	Refs. (*see* Tables 10.1 and 10.2)	Commercial suppliers (of specified media or alternatives)
Keratinocytes	MCDB 153	Peehl & Ham, 1980; Tsao et al., 1982; Boyce & Ham, 1983; *see* Protocol 23.1	Invitrogen; Cascade; Cambrex; PromoCell; Sigma
L cells (L929, LS)	NCTC109; NCTC135	Birch & Pirt, 1970, 1971; Higuchi, 1977	Sigma
Lymphoblastoid cell lines (human)	Iscove's	Iscove & Melchers, 1978	JRH Biosciences; CellGenix; Amersham
Mammary epithelium	MCDB 170	Hammond et al., 1984; *see* Protocol 22.3	Cascade; Cambrex
MDCK dog kidney epithelium	K-1; PC-1	Taub, 1984	Cambrex
LLC-PK, pig kidney	K-2	Taub, 1984	
Melanocytes	Gilchrest	*See* Protocol 22.18	Cambrex; Cascade; PromoCell
Melanoma	Gilchrest	*See* Protocol 22.18; Halaban, 2004	Cambrex; Cascade; PromoCell
Mouse embryo fibroblasts; 3T3 cells	MCDB 402	Shipley & Ham, 1983; Ham, 1984	Biosource
Mouse leukemia		Murakami, 1984	
Mouse erythroleukemia	SF12 (Ham's F12 with extra essential and nonessential amino acids); Iscove's	Frame et al., 1980; Freshney, 1980; Iscove & Melchers, 1978	MP Biomedicals; Invitrogen; Sigma
Mouse myeloma		Murakami, 1984	Sigma, Invitrogen
Mouse neuroblastoma	MCDB 411; DMEM:F12/N1	Agy et al., 1981; Bottenstein, 1984	
Neurons	DMEM:F12/N3; B27/Neurobasal	Bottenstein, 1984; Brewer, 1995	N2, Invitrogen
Osteoblasts		Shiga et al., 2003	OGM, Cambrex; PromoCell
Prostate	WJAC 404;	*See* Table 10.11; Peehl, 2002	Cambrex
Renal			REGM (epithelial) MsGM (mesangial), Cambrex
Smooth muscle cells			Cascade; Cambrex; PromoCell
Skeletal myoblasts		Goto et al., 1999	PromoCell
Urothelium	MCDB 153; KSFMc	Southgate et al., 2002	KGM-2, Cambrex
Matrix-coating products			B-D Biosciences; MP Biomedicals; Invitrogen; Biosource; Sigma

Further recommendations on the choice of medium can be found in McKeehan [1977], Barnes et al. [1984a–d], and Mather [1998]; *see also* Tables 10.1 and 10.2.

It is important to determine the quality control performed by commercial suppliers of serum-free media. The ideal situation is for the medium to have been tested against the cells that you wish to grow (e.g., keratinocyte growth medium should have been tested on keratinocytes). Some suppliers, such as Cambrex, Cascade, and PromoCell will supply the appropriate cells with the medium, ensuring the correct quality control, but others may have performed quality control with routine cell lines, in which case you will have to do your own quality control before purchasing the medium.

10.5.2 Serum Substitutes
A number of products have been developed commercially to replace all or part of the serum in conventional media. Some of these products are Ex-cyte (Bayer), Ventrex

(JRH Biosciences), CPSR-1,2,3 (Sigma), Nu-serum (Becton Dickinson), and Biotain-MPS (Cambrex). Although they may offer a degree of consistency not obtainable with regular sera, variations in batches can still occur, and the constitution of the products is not fully defined. They may be useful as an *ad hoc* measure or for purposes of economy but are not a replacement for serum-free media. Nutridoma (BCL), ITS Premix, TCM, TCH (ICN), SIT (Sigma), and Excell-900 (JRH Biosciences) are defined supplements aimed at replacing serum, partially or completely [Cartwright & Shah, 1994]. Sigma markets the MegaCell series of media, which are MEM, DMEM, RPMI 1640, and MEM or DMEM/F12 mixtures supplemented with growth factors and additional amino acids, which enable a substantial reduction in the concentration of serum.

The choice of serum-free media and serum replacements is now so large and diverse that it is not possible to make individual recommendations for specific tasks. The best approach is to check the literature, contact the suppliers, obtain samples of those products that seem most relevant from previous reports, and screen the products in your own assays (e.g., for growth, survival, or special functions).

Amino acid hydrolysates (*see* Section 9.7.1) are used to help reduce or eliminate serum in large-scale cultures for biotechnology applications. They have the advantage that they can be sterilized by autoclaving, but have an undefined constitution.

10.5.3 Adaptation to Serum-Free Media

Many continuous cell lines, such as HeLa, CHO-K1, or mouse myelomas, may be adapted to growth in the absence of serum. This often involves a prolonged period of selection of a minority component of the cell population, so it is important to ensure that the properties of the cell line are not lost during this period of selection. If a myeloma is to be used to generate a hybridoma, or a CHO cell used for transfection, the selection of a serum-free line should be done before fusion or transfection, to minimize the risk of loss of properties during selection.

Adaptation to serum-free medium is usually carried out over several serial subcultures, with the serum concentration being reduced gradually. Once stable cell proliferation is established at one concentration, subculture cells into a lower concentration until stable growth is reestablished and then dilute the serum again. In suspension cultures, this is done by monitoring the viable cell count and diluting the cell suspension accordingly, perhaps keeping the minimal cell concentration higher than for normal subculture. For monolayer cultures, reduce the serum concentration a few days before subculture, and then subculture into the new low serum concentration. During the adaptation process it may be necessary to supplement the medium with factors known to replace serum (*see* Sections 10.4, 10.6).

10.6 DEVELOPMENT OF SERUM-FREE MEDIUM

There are two general approaches to the development of a serum-free medium for a particular cell line or primary culture. The first is to take a known recipe for a related cell type, with or without 10–20% dialyzed serum, and alter the constituents individually or in groups, while reducing the serum, until the medium is optimized for your own particular requirement. This was the approach adopted by Ham and co-workers [Ham, 1984] and generally will provide optimal conditions. If a group of compounds is found to be effective in reducing serum supplementation, the active constituents may be identified by the systematic omission of single components and then the concentrations of the essential components optimized [Ham, 1984]. However, this is a very time-consuming and laborious process, involving growth curves and clonal growth assays at each stage, and it is not unreasonable to expect to spend at least three years developing a new medium for a new type of cell.

The time-consuming nature of the first approach has led to the second approach: supplementing existing media such as RPMI 1640 [Carney et al., 1981] or combining media such as Ham's F12 with DMEM [Barnes & Sato, 1980] and restricting the manipulation of the constituents to a shorter list of substances. Among the latter are selenium, transferrin, albumin, insulin, hydrocortisone, estrogen, triiodothyronine, ethanolamine, phosphoethanolamine, growth factors (EGF, FGF, PDGF, endothelial growth supplement, etc.), prostaglandins (PGE_1, $PGF_{2\alpha}$), and any other substances that may have special relevance. (*see* Section 10.4; Table 10.3.) Selenium, transferrin, and insulin will usually be found to be essential for most cells, whereas the requirements for the other constituents will be more variable.

10.7 PREPARATION OF SERUM-FREE MEDIUM

A number of recipes for serum-free media are now available—some commercially (*see* Appendix II)—for particular cell types [Cartwright & Shah, 1994; Mather, 1998; *see also* Tables 10.1 and 10.2 and Chapter 23]. The procedure for making up serum-free recipes is similar to that for preparing regular media (*see* Protocol 11.9; *see also* [Waymouth, 1984]). Ultrapure reagents and water should be used and care taken with solutions of Ca^{2+} and Fe^{2+} or Fe^{3+} to avoid precipitation. Metal salts tend to precipitate in alkaline pH in the presence of phosphate, particularly when the medium or salt solution is autoclaved, so cations in stock solutions should be kept at a low pH (below 6.5) and maintained phosphate free. They should be sterilized by autoclaving or filtration (*see* Section 11.3). It is often recommended that cations be added last, immediately before using the medium. Otherwise the constituents are generally made up as a series of stock solutions, minerals and vitamins at 1000×, tyrosine, tryptophan, and phenylalanine in 0.1 N HCl at 50×, essential amino acids at 100× in water, salts at 10× in water, and any other special cofactors, lipids, etc., at 1000× in the appropriate solvents. These are combined in the correct proportions and diluted to the final concentration, and then the pH and osmolality are checked (*see* Section 11.2.5).

Growth factors, hormones, and cell adhesion factors are best added separately just before the medium is used, as they may need to be adjusted to suit particular experimental conditions.

10.8 PROTEIN-FREE MEDIA

There is increasing pressure from regulatory authorities to remove all animal products from contact with cultured cells used in the production of biopharmaceuticals. Trypsin can be replaced with recombinant trypsin or a nonvertebrate protease (*see* Section 10.4.1) and growth factors with recombinant growth factors. BSA can often be replaced by supplementation with those factors, such as lipids, hormones, minerals, and growth factors (*see* Section 10.4), normally bound to BSA in serum, as it has yet to be established that

BSA has a role in itself. Adaptation to protein-free medium may require further selection.

10.9 CONCLUSIONS

However desirable serum-free conditions may be, there is no doubt that the relative simplicity of retaining serum, the specialized techniques required for the use of some serum-free media, the considerable investment in time, effort, and resources that go into preparing new recipes or even adapting existing ones, and the multiplicity of media required if more than one cell type is being handled all act as considerable deterrents to most laboratories to enter the serum-free arena. There is also no doubt, however, that the need for consistent and defined conditions for the investigation of regulatory processes governing growth and differentiation, the pressure from biotechnology to make the purification of products easier, and the need to eliminate all sources of potential infection will eventually force the adoption of serum-free media on a more general scale. But first, recipes must be found that are less "temperamental" than some current recipes and that can be used with equal facility and effectiveness in different laboratories.

CHAPTER 11

Preparation and Sterilization

11.1 PREPARATION OF REAGENTS AND MATERIALS

All stocks of chemicals and glassware used in tissue culture should be reserved for that purpose alone. Traces of heavy metals or other toxic substances can be difficult to remove and are detectable only by a gradual deterioration in the culture. It follows that separate stocks imply separate washing of glassware. The requirements of tissue culture washing are higher than for general glassware, although the level of soil may be less; a special detergent may be necessary (*see* Section 11.3.4), and chemical contamination from regular laboratory glassware must be avoided. The almost universal adoption of single-use disposable plasticware has removed the problem of maintaining glass flasks suitable for culture vessels; nevertheless, whenever glass is used, either for storage or for culture, the problem of chemical contaminants leaching out into media or reagents remains, and absolute cleanliness is therefore essential.

11.2 STERILIZATION OF APPARATUS AND LIQUIDS

All apparatus and liquids that come in contact with cultures or other reagents must be sterilized. The choice of method depends largely on the stability of the item at high temperatures (Table 11.1). In general, items that have a high resistance to heat, e.g., metals, glass, and thermostable plastics, such as PTFE, are sterilized by dry heat. This is one of the simplest and most effective methods of sterilization, provided all parts of the load reach the correct temperature for the required period. Sterilization by moist heat, or autoclaving, is also highly effective and can be applied to heat-stable liquids, such as water, salt solutions, and some specially formulated media. Irradiation can be used to sterilize heat-labile materials, such as plastics, with γ-irradiation, usually with ^{60}Co or ^{135}Cs, being the most effective. As the dose required is quite high (25 kGy) this is usually done at a central facility, which can also often provide electron beam sterilization. Ethylene oxide can also be used for heat-labile plastics but tends to be adsorbed onto the plastic and may take up to several days to be released. Other chemical sterilants include sodium hypochlorite and formaldehyde, but these are more often used as decontamination agents (*see* Section 7.8.5) for liquid and plastics disposal. Formaldehyde is also used for fumigation (*see* Section 7.8.6).

Heat-labile liquids are sterilized by micropore filtration (*see* Section 11.5.2), which may be by positive pressure, through filters ranging in size from syringe tip to in-line disks of up to 293-mm diameter or pleated cartridge filters of up to 1000 cm^2 (*see* Table 11.4), or by negative pressure using a vacuum flask with filter attached (*see* Protocol 11.12). Where there is a suitable vacuum line or pump this is a simple and effective method of filtration, as the filtrate goes straight into the vessel in which it will be stored, but negative-pressure filtration will tend to increase the pH as it draws off dissolved CO$_2$ from the medium.

The preferred methods of sterilization of specific items (Tables 11.2, 11.3) are described in the following sections.

TABLE 11.1. Methods of Sterilization

Method	Conditions	Materials	Limitations
Dry heat	160°C, 1 h	Heat stable: metals, glass, PTFE.	Some charring may occur, e.g., of indicating tape and cotton plugs.
Moist heat	121°C, 15–20 min	Heat-stable liquids: water, salt solutions, autoclavable media. Moderately heat-stable plastics: silicones, polycarbonate, nylon, polypropylene.	Steam penetration requires steam-permeable packaging. Large fluid loads need time to heat up.
Irradiation:			
γ-Irradiation	25 kGy	Plastics, organic scaffolds, heat-sensitive reagents and pharmaceuticals.	Chemical alteration of plastics can occur. Macromolecular degradation.
Electron beam	25 kGy	Plastics, organic scaffolds, heat-sensitive reagents and pharmaceuticals.	Needs high-energy source. Not suitable for average laboratory installation.
Microwave	5 min full power	Aqueous solutions and gels such as agar.	Only useful for small volumes; usually just for melting agar
Short-wave UV	254 nM, 50–100 W, 30 min	Flat surfaces, circulating air.	Will not reach shadow areas. Spores resistant.
Chemical:			
Ethylene oxide	1 h	Heat-labile plastics.	Items must be ventilated for 24–48 h; leaves toxic residue.
Hypochlorite	300–2500 ppm 30 min	Contaminated solutions. Plastics.	Needs extensive washing. May leave residue.
70% Alcohol	Soak for 1 h	Dissecting instruments (combined with flaming). Some plastics.	Does not kill spores. Fire risk with flaming. Precast Perspex or Lucite may shatter if immersed in alcohol.
Filtration	0.1- to 0.2-μm porosity	All aqueous solutions; particularly suitable for heat-labile reagents and media. Specify low protein binding for growth factors, etc.	Not suitable for some solvents, e.g., DMSO. Slow with viscous solutions.

11.3 APPARATUS

11.3.1 Glassware

If the glass surface is to be used for cell propagation, it must not only be clean but also carry the correct charge. Caustic alkaline detergents render the surface of the glass unsuitable for cell attachment and require subsequent neutralization with dilute HCl or H_2SO_4, but neutral detergents do not alter the glass surface and can be removed more easily. The most effective washing procedure is as follows:

(1) Do not let soiled glassware dry out. A sterilizing agent, such as sodium hypochlorite, should be included in the water used to collect soiled glassware
   a) to remove any potential biohazard, and
   b) to prevent microbial growth in the water
(2) Select a detergent that is effective in the water of your area, that rinses off easily, and that is nontoxic (*see* Section 11.3.4).
(3) Before drying the glassware, make sure that it has been thoroughly rinsed in tap water followed by deionized or distilled water.
(4) Dry glassware inverted so that it drains readily.
(5) Sterilize the glassware by dry heat to minimize the risk of depositing toxic residues from steam sterilization.

Sterilization procedures are designed not just to kill replicating microorganisms but also to eliminate the more resistant spores. Moist heat is more effective than dry heat; however, it does carry a risk of leaving a residue. Dry heat is preferable, but a minimum temperature of 160°C maintained for 1 h is required. Moist heat (for fluids and perishable items) should be maintained at 121°C for 15–20 min (*see* Tables 11.1, 11.2, 11.3). For moist heat to be effective, steam penetration must be ensured, which means that the sterilization chamber must be evacuated before steam injection or the air must be completely replaced with steam by downward displacement.

Inserting Thermalog indicators (*see* Appendix II) and a temperature probe from a recording thermometer into a sample of the load, centrally located, monitors both temperature and humidity during sterilization. The temperature of the chamber effluent is often used, but this does not accurately reflect the temperature of the load. Recording thermometers have the advantage that they will create a permanent record that can be archived. Indicators (e.g., Thermalog), on the other hand, provide a visual confirmation of temperature and humidity and can be used to monitor several parts of the load simultaneously. Both are recommended.

TABLE 11.2. Sterilization of Equipment and Apparatus

Item	Sterilization
Ampoules for freezer, glass	Dry heat[1]
Ampoules for freezer, plastic	Autoclave[2] (usually bought sterile)
Apparatus containing glass and silicone tubing	Autoclave
Disposable tips for micropipettes	Autoclave in autoclavable trays or nylon bags
Filters, reusable	Autoclave; do not use prevacuum or postvacuum
Glassware	Dry heat
Glass bottles with screw caps	Autoclave with cap slack
Glass coverslips	Dry heat
Glass slides	Dry heat
Glass syringes	Autoclave (separate piston if PTFE)
Instruments	Dry heat
Magnetic stirrer bars	Autoclave
Pasteur pipettes, glass	Dry heat
Pipettes, glass	Dry heat
Plexiglas, Perspex, Lucite	70% EtOH (*see* text)
Polycarbonate	Autoclave
Repeating pipettes or syringes	Autoclave (separate PTFE pistons from glass barrels)
Screw caps	Autoclave
Silicone grease (for isolating clones)	Autoclave in glass Petri dish
Silicone tubing	Autoclave
Stoppers, rubber and silicone	Autoclave
Test tubes	Dry heat

[1] Dry heat, 160°C/L h.

[2] Autoclave, 100 kPa (1 bar, 15 lb/in.[2]), 121°C for 20 min.

PROTOCOL 11.1. PREPARATION AND STERILIZATION OF GLASSWARE

Materials

- ❏ Disinfectant: hypochlorite, 300 ppm available chlorine, minimum, when diluted in detergent (e.g., Precept tablets, Clorox or Chloros; *see* Appendix II)
- ❏ Detergent (e.g., 7X™ or Decon®)
- ❏ Soaking baths
- ❏ Bottle brushes
- ❏ Stainless steel baskets (to collect washed and rinsed glassware for drying)
- ❏ Aluminum foil
- ❏ Sterility indicators (Alpha Medical for dry heat or steam sterilization; Thermalog indicators for steam sterilization only)

- ❏ Sterile-indicating tape or tabs (Bennett). This is different from the sterile-indicating tape used in autoclaves, as the sterilizing temperature is higher in an oven. Most autoclave tapes tend to char and release traces of volatile material from the adhesive, which can leave a deposit on the oven or even the glassware.
- ❏ Sterilizing oven, fan assisted, capable of reaching 160°C, and preferably with recording thermometer and flexible probe

Protocol

Collection and Washing of Glassware (Fig. 11.1):

1. Immediately after use, collect glassware into detergent containing disinfectant. It is important that glassware does not dry before soaking, or cleaning will be much more difficult.
2. Soak overnight in detergent.
3. Rinsing:
   (a) Brush glassware by hand or with a mechanized bottle brush (Fig. 11.2) the following morning, and rinse thoroughly in four complete changes of tap water followed by three changes of deionized water. A sink-rinsing spray is a useful accessory; otherwise bottles must be emptied and filled completely each time. Clipping bottles in a basket will help to speed up this stage.
   (b) Machine rinses should be done without detergent. If done on a spigot header (*see* Section 5.4.11), this can be reduced to two rinses with tap water and one with deionized or reverse-osmosis water.
4. After rinsing thoroughly, invert bottles, etc., in stainless steel wire baskets and dry upside down.
5. Cap bottles with aluminum foil when cool, and store.

Sterilization of Glassware:

1. Attach a small square of sterile-indicating tape or other indicator label to glassware, and date.
2. Place glassware in an oven with fan-circulated air and temperature set to 160°C.
3. To ensure that the center of the load reaches 160°C:
   (a) Place a sterility indicator in a bottle or typical item in the middle of the load.
   (b) If using a recording thermometer, place the sensor in a bottle or typical item in the middle of the load.
   (c) Do not pack the load too tightly; leave room for circulation of hot air.
4. Close the oven, check that the temperature returns to 160°C, seal the oven with a strip of tape with the time recorded on it (or use automatic locking and recorder), and leave for 1 h.

TABLE 11.3. Sterilization of Liquids

Solution	Sterilization	Storage
Agar	Autoclave[1] or boil	Room temperature
Amino acids	Filter[2]	4°C
Antibiotics	Filter	−20°C
Bacto-peptone	Autoclave	Room temperature
Bovine serum albumin	Filter (use stacked filters)	4°C
Carboxylmethyl cellulose	Steam, 30 min[3]	4°C
Collagenase	Filter	−20°C
DMSO	Self-sterilizing; dispense into aliquots in sterile tubes	Room temperature; keep dark, avoid contact with rubber or plastics (except polypropylene)
Drugs	Filter (check for binding; use low- binding filter, e.g., Millex-GV, if necessary)	−20°C
EDTA	Autoclave	Room temperature
Glucose, 20%	Autoclave	Room temperature
Glucose, 1–2%	Filter (low concentrations; caramelizes if autoclaved)	Room temperature
Glutamine	Filter	−20°C
Glycerol	Autoclave	Room temperature
Growth factors	Filter (low protein binding)	−20°C
HEPES	Autoclave	Room temperature
HCl, 1 M	Filter	Room temperature
Lactalbumin hydrolysate	Autoclave	Room temperature
Methocel	Autoclave	4°C
NaHCO$_3$	Filter	Room temperature
NaOH, 1 M	Filter	Room temperature
Phenol red	Autoclave	Room temperature
Salt solutions (without glucose)	Autoclave	Room temperature
Serum	Filter; use stacked filters	−20°C
Sodium pyruvate, 100 mM	Filter	−20°C
Transferrin	Filter	−20°C
Tryptose	Autoclave	Room temperature
Trypsin	Filter	−20°C
Vitamins	Filter	−20°C
Water	Autoclave	Room temperature

[1] Autoclave, 100 kPa (15 lb/in.^{2}), 121°C for 20 min.
[2] Filter, 0.2-μm pore size.
[3] Steam, 100°C for 30 min.

5. After 1 h, switch off the oven and allow it to cool with the door closed. It is convenient to put the oven on an automatic timer so that it can be left to switch off on its own overnight and be accessed in the morning. This precaution allows for cooling in a sterile environment and also minimizes the heat generated during the day, when it is hardest to deal with.
6. Use glassware within 24–48 h.

Keep organic matter out of the oven. Do not use paper tape or packaging material, unless you are sure that it will not release volatile products on heating. Such products will eventually build up on the inside of the oven, making it smell when hot, and some deposition may occur inside the glassware being sterilized.

Alternatively, bottles may be loosely capped with screw caps and foil, tagged with autoclave tape, and autoclaved for 20 min at 121°C with a prevacuum and a postvacuum cycle (*see* Section 5.4.4). Then the caps may be tightened when the bottles have cooled down. Caps must be very slack (loosened one complete turn) during autoclaving, so as to allow steam to enter the bottle and to prevent the liner (if one is used) from being sucked out of the cap and sealing the bottle. If a bottle becomes sealed during sterilization in an autoclave, sterilization will not be complete (Fig. 11.3 and Plate 22a). Unfortunately, misting often occurs when bottles are autoclaved, and a slight residue may be left when the mist evaporates. Also, the bottles risk becoming

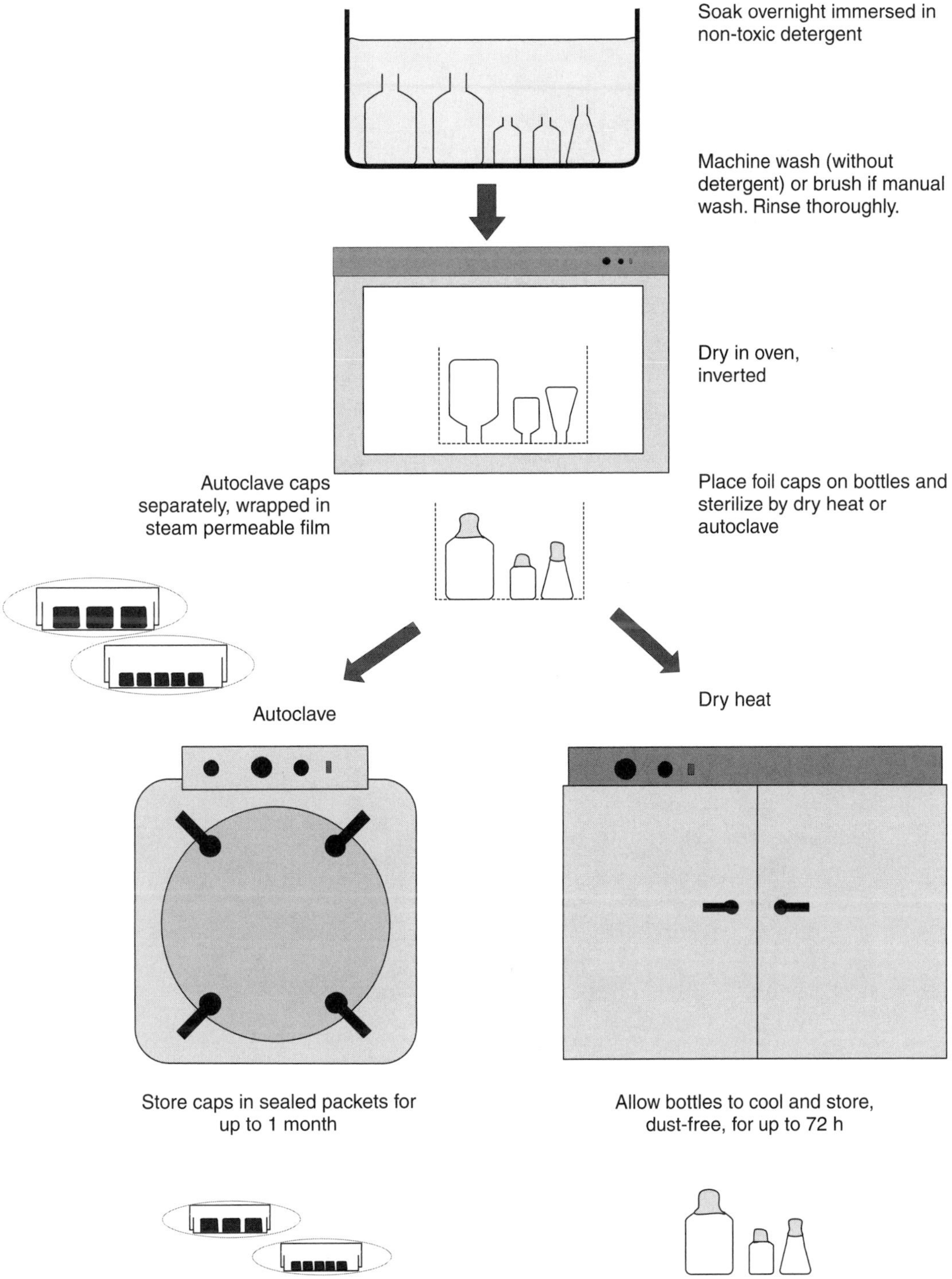

Fig. 11.1. *Washing and Sterilizing Glassware.* Procedure as in Protocol 11.1. Sterilization conditions: autoclave at 121°C for 15 min; oven, 160°C for 1 h. Caps are sterilized separately from bottles to avoid condensation forming in bottles if autoclaved with caps in place (*see* Fig. 11.3).

contaminated as they cool, by drawing in nonsterile air before they are sealed. Dry-heat sterilization is better (autoclaving the caps separately; *see* Section 11.3.1), because it allows the bottles to cool down within the oven before they are removed.

11.3.2 Glass Pipettes

Both glass and plastic pipettes are used in tissue culture. Plastic pipettes have the advantage that they are single use and disposable, avoiding the need for washup and sterilization. Glass pipettes are significantly cheaper but have to be

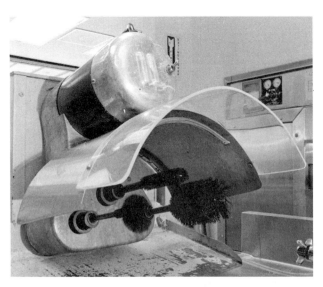

Fig. 11.2. Motorized Bottle Brushes. Rotating brushes located below a protective Plexiglas screen. Care must be taken not to press down too hard on the brush, lest the bottle break.

deplugged and replugged each time they are used. Glass pipettes must also be washed carefully in the pipette washer, so that they will not retain soil and become vulnerable to subsequent blockage.

Δ **Safety Note.** Glass pipettes are prone to damage, and chipped ends present a severe hazard to both users and the washup staff. Discard or repair them.

PROTOCOL 11.2. PREPARATION AND STERILIZATION OF GLASS PIPETTES

Materials
- Pipette cylinders (to collect used pipettes)
- Disinfectant: hypochlorite, 300 ppm available chlorine, minimum, when diluted in detergent (e.g., Precept tablets, Clorox, or Chloros)
- Detergent (e.g., 7X or Decon®)
- Stainless steel baskets (to collect washed and rinsed pipettes for drying)
- Sterility indicators (see Appendix II)
- Pipette cans (square aluminum or stainless steel with silicone cushions at either end; square cans do not roll on the bench) (Thermo Electron)
- Sterile-indicating tape or tabs (see Protocol 11.1)
- Sterilizing oven, fan assisted, capable of reaching 160°C, and preferably with recording thermometer and flexible probe

---

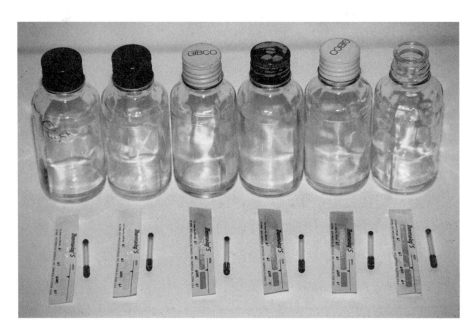

Fig. 11.3. Sterilizing Bottles. These bottles were autoclaved with Thermalog sterility indicators inside. Thermalog turns blue with high temperature and steam, and the blue area moves along the strip with time at the required sterilization conditions. The cap on the leftmost bottle was tight, and each succeeding cap was gradually slacker, until, finally, no cap was used on the bottle furthest to the right. The leftmost bottle is not sterile, because no steam entered it. The second bottle is not sterile either, because the liner drew back onto the neck and sealed it. The next three bottles are all sterile, but the brown stain on the indicator shows that there was fluid in them at the end of the cycle. Only the bottle at the far right is sterile and dry. The glass indicators (Browne's tubes) all implied that their respective bottles were sterile. (*See also* Plate 22a).

Protocol

Collection and Washing:

1. Place water with detergent and a disinfectant in pipette cylinder.
2. Discard pipettes, tip first, into cylinder immediately after use (Fig. 11.5).
   (a) Do not accumulate pipettes in the hood or allow pipettes to dry out.
   (b) Do not put pipettes that have been used with agar or silicones (water repellent, antifoam, etc.) in the same cylinder as regular pipettes. Use disposable pipettes for silicones, and either rinse agar pipettes after use in hot tap water or use disposable pipettes.
3. Soak pipettes overnight or for a minimum of 2 h. If usage of pipettes is heavy, replace cylinders at intervals when full, and soak for 2 h before entering rinse cycle.
4. After soaking, remove plugs with compressed air.
5. Transfer pipettes to pipette washer (Figs. 11.4, 11.5; *see also* Fig. 4.6), tips uppermost.
6. Rinse in tap water by siphoning action of pipette washer for a minimum of 4 h or in an automatic washing machine with a pipette adapter, but without detergent.
7. Turn valve to setting for deionized (DW) or reverse-osmosis (ROW) water (*see* Fig. 4.6), or wait until last of the tap water finally runs out, turn off the tap water, and empty and fill washer three times with DW or ROW. (Use automatic deionized rinse cycle in glassware washing machine.)
8. Transfer pipettes to pipette dryer or drying oven, and dry with tips uppermost.
9. Plug with cotton (Fig. 11.6).
10. Sort pipettes by size and store dust free.

Sterilization:

1. Place pipettes in pipette cans. (It is useful to have both ends of the cans labeled with the size of the pipettes.)
2. Fill each can with one size of pipette.
3. Fill a few cans with an assortment of 1-mL, 2-mL, 10-mL, and 25-mL pipettes (e.g., four of each size).
4. Attach sterile-indicating tape, bridging the cap to the can, and stamp date on tape.
5. Sterilize by dry heat for 1 h at 160°C. Use the smallest amount of tape possible, or replace with temperature indicators (Alpha Medical; Bennett), which are small and are made of less volatile material. The temperature should be measured in the center of the load, to ensure that this, the most difficult part to reach, attains the minimum sterilizing conditions. Leave spaces between cans when loading the oven, to allow for circulation of hot air (Fig. 11.7).
6. Remove pipettes from oven and allow to cool, and transfer cans to tissue culture laboratory. If you anticipate that pipettes will lie idle for more than 48 h, seal cans around the cap with adhesive tape.

11.3.3 Screw Caps

There are two main types of caps that are in common use for glass bottles: (1) aluminum or phenolic plastic caps with synthetic rubber or silicone liners and (2) wadless polypropylene caps that are reusable (Duran); these caps are deeply shrouded and have ring inserts for better sealing and to improve pouring (although pouring is not recommended in sterile work). The following precautions should be observed:

(1) Polypropylene caps will seal only if screwed down tightly on a bottle with no chips or imperfections on the lip of the opening. Discard bottles with chipped necks.

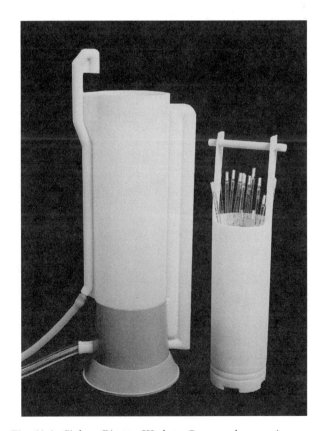

Fig. 11.4. Siphon Pipette Washer. Connected to main water supply. A slow fill establishes a siphoning action and drains the main chamber, which then refills. This process is repeated many times over 4–6 h. The pipettes shown here are as collected from use. Before the pipette basket is placed in the washer, the plugs are blown out and the pipettes inverted, as in Fig. 11.5.

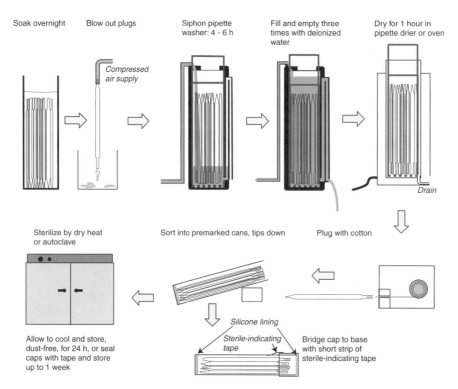

Fig. 11.5. Washing and Sterilizing Pipettes.

Fig. 11.6. Semiautomatic Pipette Plugger (Bellco). A thick strand of cotton is fed through a selected loading aperture, and a section is cut off when the appropriate length is inserted in the end of the pipette. Different thicknesses of cotton are required for different sizes of pipette.

(2) Do not leave aluminum caps or any other aluminum items in alkaline detergents for more than 30 min, as they will corrode.

(3) Do not put glassware together with caps in the same detergent bath, or the aluminum may contaminate the glass.

Fig. 11.7. Sterilizing Oven. Pipette cans are stacked with spaces between to allow circulation of hot air. Brown staining on front of oven shows evidence of volatile material from sterile-indicating tape.

(4) Avoid detergents that are made for machine washing, as they are highly caustic.

PROTOCOL 11.3. PREPARATION AND STERILIZATION OF SCREW CAPS

Materials

❑ Disinfectant: hypochlorite, 300 ppm available chlorine (e.g., Precept tablets)
❑ Detergent (e.g., 7X™ or Decon®)
❑ Soaking baths
❑ Stainless steel baskets (to collect caps for washing and drying)
❑ Sterility indicators (e.g., *see also* Appendix II)
❑ Glass Petri dishes (for packaging)
❑ Autoclavable plastic film (*see* Appendix II) or paper sterilization bags
❑ Sterile-indicating autoclave tape
❑ Autoclave, with recording thermometer with flexible probe that can be inserted in load

Protocol

Collection and Washing:

Metal or phenolic caps with liners:

1. Soak 30 min (maximum) in detergent.
2. Rinse thoroughly for 2 h. (Make sure all caps are submerged.) Liners should be removed and replaced after rinsing, which may be carried out in either of two ways:
   (a) In a beaker (or pail) with running tap water led by a tube to the bottom. Stir the caps by hand every 15 min.
   (b) In a basket or, better, in a pipette-washing attachment. Rinse in an automatic washing machine, but do not use detergent in the machine.

Polypropylene caps:

These may be washed and rinsed by hand as just described (extending the detergent soak if necessary). Because these caps may float, they must be weighted down during soaking and rinsing. For automatic washers, after soaking in detergent, use pipette-washing attachment and normal cycle without machine detergent.

Stoppers:

Shrouded caps are preferred to stoppers, but if the latter are required, use silicone or heavy metal-free white rubber stoppers in preference to those made of natural rubber. Wash and sterilize as for caps. (There will be no problem with flotation in washing and rinsing.)

Sterilization:

3. Place caps in a glass Petri dish with the open side down.

4. Wrap Petri dish containing caps in cartridge paper or steam-permeable nylon film, and seal with autoclave tape (Fig. 11.8).
5. Prepare similar package with sterility indicator enclosed and insert in the middle of the load.
6. Autoclave for 20 min at 121°C and 100 kPa (15 p.s.i.) (Fig. 11.9).

11.3.4 Selection of Detergent

When most culture work was done on glass, the quality (charge, chemical residue) of the glass surface was critical.

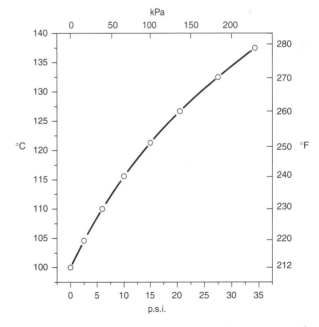

Fig. 11.8. Packaging Screw Caps for Sterilization. The caps are enclosed in a glass Petri dish, which is then sealed in an autoclavable nylon bag.

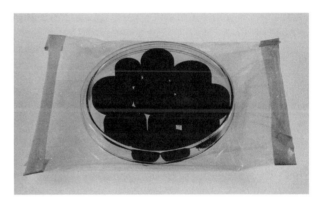

Fig. 11.9. Relationship Between Pressure and Temperature. 121°C and 100 kPa or 1 bar (250°F, 15 lb/in.²) for 15–20 min are the conditions usually recommended.

As most cell culture is now carried out on disposable plastic, the major requirements for cleaning glassware are that (a) the detergent be effective in removing residue from the glass and (b) no toxic residue be left behind to leach out into the medium or other reagents. Some tissue culture suppliers will provide a suitable detergent that has been tested with tissue culture (e.g., 7X, MP Biomedicals or Decon for manual washing), but often, machine detergents will come from a general laboratory supplier or the supplier of the machine. The washing efficiency of a detergent can be determined by washing heavily soiled glassware (e.g., a bottle of serum or a medium containing serum that has been autoclaved). One simply uses the normal washup procedure; a visual check will then show which detergents have been effective.

The presence of a toxic residue is best determined by cloning cells (*see* Protocol 21.10), e.g., on a glass Petri dish that has been washed in the detergent, rinsed as previously indicated (*see* Protocol 11.1), and then sterilized by dry heat. A plastic Petri dish should be used as a control. This technique can also be used if you are anxious about residue left on glassware after autoclaving.

11.3.5 Miscellaneous Equipment

Cleaning. All new apparatus and materials (silicone tubing, filter holders, instruments, etc.) should be soaked in detergent overnight, thoroughly rinsed, and dried. Anything that will corrode in the detergent—mild steel, aluminum, copper, brass, etc.—should be washed directly by hand without soaking (or with soaking for 30 min only, using detergent if necessary) and then rinsed and dried.

Used items should be rinsed in tap water and immersed in detergent immediately after use. Allow them to soak overnight, and then rinse and dry them. Again, do not expose materials that might corrode to detergent for longer than 30 min. Aluminum centrifuge buckets and rotors must never be allowed to soak in detergent.

Particular care must be taken with items treated with silicone grease or silicone fluids. These items must be treated separately and the silicone removed, if necessary, with carbon tetrachloride. Silicones are very difficult to remove if they are allowed to spread to other apparatus, particularly glassware.

Packaging. Ideally, all apparatus used for sterilization should be wrapped in a covering that will allow steam to penetrate but will be impermeable to dust, microorganisms, and mites. Proprietary bags, bearing sterile-indicating marks that show up after sterilization, are available from clinical sterile-supply services (e.g., Polysciences, Bio-Medical Products). Semipermeable transparent nylon film (Portex Plastics, Applied Scientific, KNF) is sold in rolls of flat tubes of different diameters and can be made up into bags with sterile-indicating tape. Although expensive, such film can be reused several times before becoming brittle.

Tubes and orifices should be covered with tape and paper or nylon film before packaging, and needles or other sharp points should be shrouded with a glass test tube or other appropriate guard.

Sterilization. The type of sterilization used will depend on the material (*see* Table 11.1). Metallic items are best sterilized by dry heat. Silicone rubber (which should be used in preference to natural rubber), PTFE, polycarbonate, cellulose acetate, and cellulose nitrate filters, etc., should be autoclaved for 20 min at 121°C and 100 kPa (1 bar, 15 lb/in.^{2}) with preevacuation and postevacuation steps in the cycle, except when filters are sterilized in a filter assembly (*see* Protocol 11.4). In small bench-top autoclaves and pressure cookers, make sure that the autoclave boils vigorously for 10–15 min before pressurizing to displace all the air. (Take care that enough water is put in at the start to allow for evaporation.) After sterilization, the steam is released and the items are removed to dry off in an oven or rack.

Δ *Safety Note.* To avoid burns, take care in releasing steam and handling hot items. Wear elbow–length insulated gloves, and keep your face well clear of escaping steam when you open doors, lids, etc. Use safety locks on autoclaves.

11.3.6 Reusable Sterilizing Filters

Although most laboratories now use disposable filter assemblies, some large-scale users may prefer to use stainless steel filter housings. Filter assemblies should be made up and sterilized in accordance with Protocol 11.4.

PROTOCOL 11.4. STERILIZING FILTER ASSEMBLIES

Materials
Nonsterile:
❑ Filter holder (Table 11.4).
❑ Micropore filters to fit holder
❑ Prefilter, glass fiber, if required
❑ Steam-permeable nylon film
❑ Sterile-indicating autoclave tape
❑ Autoclave

Protocol
1. After thorough washing in detergent (*see* Section 11.3.5), rinse assembly in water, followed by deionized water, and dry.
2. Insert support grid in filter and place filter membrane on grid. If membrane is made of polycarbonate, apply wet to counteract static electricity.
3. Place prefilters (glass fiber and others as required; *see* Section 11.5.3) on top of filter.
4. Reassemble filter holder, but do not tighten up completely. (Leave about one whole turn on bolts.)
5. Cover inlet and outlet of filter with aluminum foil.

6. Pack filter assembly in sterilizing paper or steam-permeable nylon film, and close assembly with sterile-indicating tape.

7. Autoclave at 121°C and 100 kPa (1 bar, 15 lb/in.^{2}) for 20 min with no preevacuation or postevacuation. (Use "liquids cycle" in automatic autoclaves.)

8. Remove and allow to cool.

9. Do not tighten filter holder completely until the filter is wetted at the beginning of filtration (*see* Protocol 11.14.6).

Alternative methods of sterilization. Many plastics cannot be exposed to the temperature required for autoclaving or dry-heat sterilization. To sterilize such items, immerse them in 70% alcohol for 30 min and dry them off under UV light in a laminar-flow hood. Care must be taken with some plastics (e.g., Plexiglas, Perspex, Lucite), as they will depolymerize in alcohol or when they are exposed to UV light. Ethylene oxide may be used to sterilize plastics, but two to three weeks are required to clear it completely from the plastic surface after sterilization. The best method for sterilizing plastics is γ-irradiation, at a level of 25 kGy. Items should be packaged and sealed; polythene may be used and sealed by heat welding.

11.4 REAGENTS AND MEDIA

The ultimate objective in preparing reagents and media is to produce them in a pure form (1) to avoid the accidental inclusion of inhibitors and toxic substances, (2) to enable the reagent to be totally defined and the functions of its constituents to be fully understood, and (3) to reduce the risk of microbial contamination.

Most reagents and media can be sterilized either by autoclaving, if they are heat stable—e.g., water, salt solutions, amino acid hydrolysates—or by membrane filtration, if they are heat labile. For autoclaving, solutions should be dispensed into borosilicate glass or polycarbonate and kept sealed to avoid evaporation and chemical pollution from the autoclave. If soda glass bottles are used, they are better left with the caps slack to minimize breakage. The evolution of vapor will help to prevent steam entering from the autoclave, but the level of the liquid will need to be restored with sterile UPW later. As with sterilizing apparatus, sterile indicators (e.g., Thermalog) and the probe from a recording thermometer should be placed in a mock sample in the center of the load.

Media and reagents supplied on-line to large-scale culture vessels and industrial or semiindustrial bioreactors can be sterilized on-line by short-duration ultrahigh-temperature treatment (Alfa-Laval). Adapting this process to media production might allow increased automation and ultimately reduce costs.

11.4.1 Water

Water used in tissue culture must be of a very high purity, particularly with serum-free media (*see also* Section 5.4.2). Because water supplies vary greatly, the degree of purification required may vary. Hard water will need a conventional ion-exchange water softener on the supply line before entering the purification system, but this will not be necessary with soft water.

There are four main approaches to water purification: reverse osmosis, distillation, deionization, and carbon filtration. For ultrapure water (UPW), the first stage is usually reverse osmosis, but it can be replaced by distillation (Fig. 11.10; *see also* Fig. 5.17). Distillation has the advantage that the water is heat sterilized, but it is more expensive because of power consumption and the boiler needs to be cleaned out regularly. If glass distillation is used for the first stage, the still should be electric and automatically controlled, and the heating elements should be made of borosilicate glass or silica sheathed. Reverse osmosis depends on the integrity of the filtration membrane; hence, the effluent must be monitored. The type of reverse-osmosis cartridge used is determined by the pH of the water supply (*see* manufacturer's specification for details). If the costs of both power for distillation and replacement membranes are deducted directly from your budget, reverse osmosis will probably work out cheaper, but if power is supplied free or is costed independently of usage, then distillation will be cheaper.

The second stage is carbon filtration, which will remove both organic and inorganic colloids. The third stage is high-grade mixed-bed deionization to remove ionized inorganic material, and the final stage is micropore filtration to remove any microorganisms acquired from the system and to trap any resin that may have escaped from the deionizer. To minimize pollution during storage, the water should be collected directly from the final-stage micropore filter without being stored. Water stored after the first stage is completed can be used as rinsing water. If the water is recycled continuously from the micropore filter to the reservoir (*see* Fig. 11.10) with the supply from the first stage turned off (e.g., overnight), the stored water gradually "polishes" (i.e., increases in purity). In such a system, water should be used first thing in the morning for preparing media.

The quality of the deionized water should be monitored by its conductivity (the inverse of resistivity) at regular intervals, and the cartridge should be changed when an increase in conductivity is observed. The ISO 3696 standard sets resistivity of type I water at a level ≥ 10 MΩ/cm (conductivity < 0.1 μS/cm) at 25°C. The total organic carbon (TOC) should be ≤ 10 parts per billion (ppb). Conductivity meters are usually supplied with water purification systems; a TOC meter will need to be purchased separately (*see* Appendix II). Distillation can precede deionization, providing sterile water to the deionization stage. An ultrafiltration stage can be inserted between deionization and micropore filtration to produce pyrogen-free water.

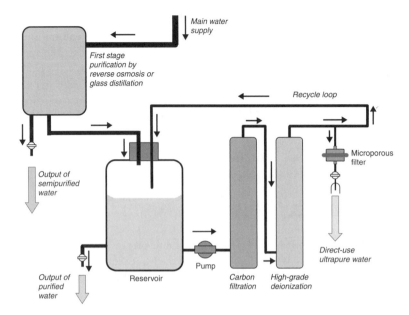

Fig. 11.10. Water Purification. Tap water is fed to a storage container via reverse osmosis or glass distillation. This semipurified water is then recycled back to the storage container via carbon filtration, deionization, and micropore filtration. Reagent-quality water is available at all times from the storage reservoir; media-quality water is available from the micropore filter supply (at right of diagram). If the apparatus recycles continuously, then water of the highest purity will be collected first thing in the morning for the preparation of medium (*see also* Fig. 5.17).

Sterilize water by autoclaving at 121°C and 100 kPa (15 lb/in.^{2}, 1 bar) for 20 min and dispense in suitable aliquots (e.g., for media preparation from concentrates). The bottles should be sealed during sterilization, which will require borosilicate glass (Pyrex) or polycarbonate (Nalge Nunc) bottles. If bottles are unsealed (e.g., if soda glass is used, it may break if sealed), allow 10% extra volume per bottle to allow for evaporation.

The amounts quoted in Protocol 11.5 are designed for use in Exercise 6 (*see* Chapter 2), but may be varied according to requirements.

PROTOCOL 11.5. PREPARATION AND STERILIZATION OF ULTRAPURE WATER (UPW)

Equipment and Materials
Nonsterile:
- ❏ Graduated glass borosilicate (e.g., Pyrex) or autoclavable plastic (polycarbonate) screw-cap bottles (ensure there is sufficient head space for later additions), 500 mL 21
- ❏ Screw caps to fit 21
- ❏ Sterility indicator strips 2
- ❏ Sterile-indicating tape
- ❏ Marker pen or preprinted labels

- ❏ Water purification equipment
- ❏ Autoclave
- ❏ Log book (record of preparation and sterilization)

Standard Protocol
1. Label bottles with date, contents, and batch number.
2. Early in the morning after overnight recycling, run about 50 mL water to waste from purifier, check conductivity (or resistivity) and total organic carbon (TOC) on respective meters, and enter in log book.
3. If water is within specified limits (resistivity ≥10 MΩcm at 25°C, TOC ≤10 ppb), collect ultrapure water directly into labeled bottles.
4. Fill to the specified mark, e.g., 430–450 mL if to be used for diluting 10× concentrated medium (*see* Protocol 11.8). If to be autoclaved open, add 10% extra.
5. Place sterility indicator in one bottle (to be discarded when checked after autoclaving).
6. Seal bottles with screw caps. If using soda glass, i.e., if bottle is liable to breakage, then leave caps slack.
7. Place bottles in autoclave with bottle containing sterility indicator in center of load.

8. Close autoclave and check settings: 121°C, 100 kPa (15 lb/in.^{2}, 1 bar), for 20 min with postvac deselected.
9. Start sterilization cycle.
10. On completion of cycle, check printout to confirm that correct conditions have been attained for the correct duration and enter in log book.
11. Allow load to cool to below 50°C.
12. Open autoclave and retrieve bottles. If caps were slack during autoclaving, tighten when bottles reach room temperature.

Note. Sealing bottles, which have been open during autoclaving, when they are still warm can cause the liner in some caps to be drawn into the bottle as it cools and the contents contract. Also, there will be a rapid intake of air into the bottle when it is first opened, and this can cause contamination. Bottles that have been open during autoclaving should be allowed to cool to room temperature in a sterile atmosphere, e.g., in the autoclave or in horizontal laminar flow, before the caps are tightened.

Check sterility indicator to confirm that sterilization conditions have been achieved and enter in log book.
13. Place bottles in short-term storage at room temperature.
14. Replenish culture room stocks as required, confirming by examination of sterile-indicating tape that the bottles have been through sterilization cycle.

11.4.2 Balanced Salt Solutions

The formulation of BSS has been discussed previously (*see* Section 9.3). The formula for Hanks's BSS [after Paul, 1975] contains magnesium chloride in place of some of the sulfate originally recommended (*see* Table 9.2); it should be autoclaved below pH 6.5 to prevent calcium and magnesium phosphates from precipitating and should be neutralized just before use. Similarly, Dulbecco's phosphate-buffered saline (D-PBS) is made up without calcium and magnesium (D-PBSA), which are made up separately (D-PBSB) and added just before use if required. DPBS is often used without the addition of the Ca^{2+} and Mg^{2+} component, and in that form it should be referred to as (D-PBSA) or (D-PBS) without Ca^{2+} and Mg^{2+}; the "D-PBSA" convention is used throughout this book. "D-PBS" is also used to distinguish Dulbecco's formulation from simple phosphate-buffered saline (PBS), which lacks potassium chloride and is just isotonic sodium chloride with phosphate buffer and should not be regarded as a balanced salt solution. Which formulation is being used should always be made clear in reports and publications.

Most balanced salt solutions contain glucose, which, because it may caramelize on autoclaving, is best omitted during preparation and sterilization and added later. If glucose is prepared as a 100× concentrate (200 g/L), caramelization during autoclaving is reduced, and it can be used at 5–25 mL/L BSS to give 1–5 g/L. Alternatively, complete balanced salt solutions can be prepared as for complete defined media (*see* Protocol 11.8) and sterilized by filtration (*see* Protocols 11.11–11.14). Protocol 11.6 is given for DPBS-A, commonly used as a rinsing solution and solvent for trypsin or ethylenediaminetetraacetic acid (EDTA), but can also be used for any BSS lacking glucose and bicarbonate. The amounts correspond to Exercise 7 (*see* Chapter 2), but can be scaled up as required.

PROTOCOL 11.6. PREPARATION AND STERILIZATION OF D-PBSA

Outline
Dissolve powder with constant mixing, make up to final volume, check pH and conductivity, dispense into aliquots, and autoclave.

Materials
Nonsterile:
- D-PBS powder (Solution A, lacking Ca^{2+} and Mg^{2+}, e.g., Sigma D5652), 1-L pack .... 1 or tablets (Oxoid Br 14a) 10
- Ultrapure water (UPW; *see* Section 11.4.1) .... 1 L
- Container: Clear glass or clear plastic aspirator with tap outlet at base, 1 L .... 1 or Erlenmeyer flask or bottle, peristaltic metering pump and tubing, 1 L 1
- Magnetic stirrer and PTFE-coated follower 1
- Bottles for storage, graduated; borosilicate glass (Pyrex, Schott), 100 mL 10
- Conductivity meter or osmometer
- pH meter
- Autoclave tape or sterile-indicating tabs
- Autoclave

Protocol
1. Add 1 L UPW to container.
2. Place container on magnetic stirrer and set to around 200 rpm.
3. Open packet of D-PBSA powder, or count out 10 tablets, and add slowly to container while mixing.
4. Stir until completely dissolved.

5. Check pH and conductivity of a sample and enter in record.
   (a) pH should not vary more than 0.1 pH unit (pH may vary with different formulations).
   (b) Conductivity should not vary more than 5% from 150 μScm⁻¹. Osmolality can be used as an alternative to conductivity (or in addition), and should show similar consistency between batches.

Note. It is important to check these parameters for consistency, as a quality control measure, to ensure that there has been no mistake in preparation. Any adjustments, e.g., to the osmolality, should be made after the quality control checks have been made.

   (c) Discard sample; do not add back to main stock.
6. Dispense contents of container into graduated bottles.
7. Cap and seal bottles. It is useful to cover the caps with paper secured with an elastic band. The paper can be dated and labeled with the contents, using a stamp; this also avoids having to remove labels from used bottles.
8. Attach a small piece of autoclave tape or sterile-indicating tab and date.
9. Sterilize by autoclaving for 20 min at 121°C and 100 kPa (1 bar, 15 lb/in.²) in sealed bottles (*see* Fig. 11.9).
10. Store at room temperature.

11.4.3 Preparation and Sterilization of Media

During the preparation of complex solutions, care must be taken to ensure that all of the constituents dissolve and do not get filtered out during sterilization and that they remain in solution after autoclaving or storage. Concentrated media are often prepared at a low pH (between 3.5 and 5.0) to keep all the constituents in solution, but even then, some precipitation may occur. If the constituents are properly resuspended, they will usually redissolve on dilution; but if the precipitate has been formed by degradation of some of the constituents of the medium, then the quality of the medium may be reduced. If a precipitate forms, the performance of the medium should be checked by cell growth and cloning and an appropriate assay of special functions (*see* Section 9.6.2).

Commercial media are supplied as (1) working–strength solutions (1×) with or without glutamine; (2) 10× concentrates, usually without NaHCO₃ and glutamine, which are available as separate concentrates; or (3) powdered media, with or without NaHCO₃ and glutamine. Powdered media are the cheapest and not a great deal more expensive than making up medium from your own chemical constituents if you include time for preparation, sterilization, and quality control, the cost of raw materials of high purity, and the cost of overheads such as power and wages. Powdered media are quality controlled by the manufacturer for their growth-promoting properties but not, of course, for sterility. They are mixed very efficiently by ball milling, so, in theory, a pack may be subdivided for use at different times. However, in practice, it is better to match the size of the pack to the volume that you intend to prepare, because once the pack is opened, the contents may deteriorate and some of the constituents may settle.

Tenfold concentrates cost two to three times as much per liter of working-strength medium as powdered media but save on sterilization costs. Buying media at working strength is the most expensive (4–5 times the cost of a 10× concentrate) but is the most convenient, as no further preparation is required other than the addition of serum, if that is required. Protocol 11.7 describes the preparation of complete medium from 1× stock; the amounts refer to the use of this protocol in Exercise 3 (*see* Chapter 2). The volumes are the minimum required, although they will presumably be taken from stock bottles of larger volume.

PROTOCOL 11.7. PREPARATION OF MEDIUM FROM 1× STOCK

Outline
Check the formulation; if complete, it may be used directly, after adding serum if that is required (*see* Sections 9.6, 13.6.2). If the formulation is incomplete (e.g., lacking glutamine), add the appropriate stock concentrate.

Note. A supplement (e.g., serum or antibiotics) is a constituent that is added to the medium and is not in the original formulation. It needs to be indicated in any publication. Other additions (e.g., glutamine or NaHCO₃) are part of the formulation and are not supplements. They need not be indicated in publications unless their concentrations are changed.

Materials
❑ Pipettes, etc., as listed for aseptic technique (*see* Protocol 6.1, 6.2), plus sufficient of the following to provide the volumes stipulated:
❑ Medium stock, e.g., Eagle's MEM with Hanks's salts .... 100 mL
❑ Glutamine, 200 mM (will need to be thawed) .... 1 mL
❑ Serum (will need to be thawed), newborn or fetal bovine .... 10 mL
❑ Antibiotics (if required; **not** recommended for routine use):

Penicillin in BSS or D-PBSA, 10,000
U/mL .... 0.5 mL
Streptomycin in BSS or D-PBSA,
10 mg/mL .... 0.5 mL

Protocol
1. Check formulation of medium, and determine what additions are required, e.g. glutamine.
2. Take medium to hood with any other supplement or addition that is required.
3. Unwrap bottles if they are wrapped in polythene, and swab with 70% alcohol.
4. Uncap bottles.
5. Transfer the appropriate volume of each addition to the stock bottle to make the correct dilution; for 100 mL, use the following ingredients and amounts:

   Glutamine, 200 mM .... 1 mL
   Antibiotics (if used) .... 0.5 mL each
   Serum .... 10 mL (for 10%)
   (a) Use a different pipette for each addition.
   (b) Move each new stock to the opposite side of the hood after it has been added, so that you will know that it has been used.
   (c) Remove all additives or supplements from the hood when the medium is complete.
6. This protocol specifies MEM with Hanks's salts, which has a low bicarbonate concentration (4 mM). If MEM with Earle's salts, or another high-bicarbonate (23 mM) medium, is used then gas the air space with 5% CO_2. Do not bubble gas through any medium containing serum, as the medium will froth out through the neck, risking contamination.
7. Recap bottles.
8. Alter labeling to record additions, date and initial.
9. Return medium to 4°C or use directly (*see* Chapter 2, Exercise 4).
10. If using a new medium for the first time, pipette an aliquot into a flask or Petri dish, and incubate for at least 1 h (preferably overnight) under your standard conditions, to ensure that the pH equilibrates at the correct value. If it does not, readjust the pH of the medium and repeat, or else alter the CO_2 concentration.

Note. Changing the CO_2 concentration of the incubator will affect all other culture media in the same incubator. Changing the CO_2 concentration should be regarded as a one-time adjustment and should not be used for batch-to-batch variations, which are better controlled by adding sterile acid or alkali.

Preparing medium from 10× concentrate represents a good compromise between the economy of preparing medium from powder (*see* Protocol 11.9) and the ease of using a 1× preparation (*see* Protocol 11.7). Protocol 11.8 is designed to be used in conjunction with Exercise 10 (*see* Chapter 2).

PROTOCOL 11.8. PREPARATION OF MEDIUM FROM 10× CONCENTRATE

Outline
Sterilize aliquots of deionized distilled water of such a size that one aliquot, when made up to full-strength medium, will last from one to three weeks. Add concentrated medium and other constituents, adjust the pH, and use the solution or return it to the refrigerator.

Materials
❑ Pipettes, etc., as listed for aseptic technique (*see* Protocols 6.1, 6.2), plus sufficient of the following to provide the volumes stipulated:

A. *For sealed culture flask with gas phase of air, low HCO_3^- concentration, atmospheric CO_2 concentration, and low buffering capacity.*
Sterile solutions for 500 mL of medium:
❑ Premeasured aliquot of UPW .... 443 mL
❑ Medium, 10× concentrate,
  e.g., MEM with Hanks's salts .... 50 mL
❑ Glutamine (will need to be thawed),
  to give 2 mM final, 200 mM.................... 5 mL
❑ $NaHCO_3$, to give 4 mM final concentration,
  7.5% (0.89 M)..................................5 mL
❑ Bovine serum (will need to be thawed),
  to give 10% final, newborn or fetal......... 50 mL
❑ Antibiotics if required (**not** recommended for routine use):
  Penicillin, to give 50 U/mL final,
  10,000 U/mL .... 2.5 mL
  Streptomycin, to give 50 μg/mL final,
  10 mg/mL .... 2.5 mL
❑ HEPES (if required), 1.0 M .... 5–10 mL
❑ NaOH, 1 M .... as required

Protocol
1. Thaw serum and glutamine and bring to the hood.
2. Swab any bottles that have been in a water bath before placing in the hood.
3. Add constituents as listed in sterile solutions section. HEPES may be added to increase the buffering capacity, and the flask may be vented

to atmosphere for some cell lines at a high cell density if a lot of acid is produced.

4. Add 1 M NaOH to give pH 7.2 at 20°C. When incubated, the medium will rise to pH 7.4 at 37°C, but this figure may need to be checked by a trial titration the first time the recipe is used (*see* below, Protocol 11.7, and Chapter 2, Exercise 10).

B. For cultures in open vessels in a CO_2 incubator, or under CO_2 in sealed flasks, with 5% CO_2 and a high bicarbonate concentration.

Sterile solutions for 500 mL of medium:
- ☐ Premeasured aliquot of UPW .... 443 mL
- ☐ Medium 10× concentrate, e.g., MEM with Earle's salts.................. 50 mL
- ☐ Glutamine (will need to be thawed), to give 2 mM final, 200 mM....................5 mL
- ☐ $NaHCO_3$, to give 26 mM final concentration, 7.5% (0.89 M)............................. 14.5 mL
- ☐ Serum, newborn or fetal bovine .... 50 mL
- ☐ Antibiotics if required (**not** recommended for routine use):
  Penicillin, 10,000 U/mL .... 2.5 mL
  Streptomycin, 10 mg/mL .... 2.5 mL
- ☐ NaOH, 1 M .... as required

Protocol
1. Thaw serum and glutamine and bring to the hood.
2. Swab any bottles that have been in a water bath before placing in the hood.
3. Add constituents as listed in sterile solutions section.
4. Add 1 M NaOH to give pH 7.2 at 20°C. When incubated, the medium will rise to pH 7.4 at 37°C, but this figure may need to be checked by a trial titration the first time the recipe is used (*see* Adjusting pH, below):
   (a) Dispense 5 mL into each of 5 Petri dishes.
   (b) Add varying amounts of 1N NaOH from 5 to 50 µL to each Petri dish.
   (c) Incubate under 5% CO_2 for a minimum of 2 h and, preferably, overnight.

C. For cultures in open vessels in a CO_2 incubator, or under CO_2 in sealed flasks, with 2% CO_2 and an intermediate bicarbonate concentration.

Sterile solutions for 500 mL of medium:
- ☐ Premeasured aliquot of UPW .... 443 mL
- ☐ Medium, 10× concentrate, e.g., Ham's F12 .... 50 mL
- ☐ Glutamine (will need to be thawed), to give 2 mM final, 200 mM .... 5 mL
- ☐ $NaHCO_3$, to give 8 mM final concentration, 7.5% (0.89 M) .... 9 mL

- ☐ HEPES, to give 20 mM final concentration, 1 M .... 10 mL
- ☐ Serum, newborn or fetal bovine .... 50 mL
- ☐ Antibiotics if required (**not** recommended for routine use):
  Penicillin, 10,000 U/mL .... 2.5 mL
  Streptomycin, 10 mg/mL .... 2.5 mL
- ☐ NaOH, 1 N .... as required

Protocol
1. Thaw serum and glutamine and bring to the hood.
2. Swab any bottles that have been in a water bath before placing in the hood.
3. Add constituents as listed in sterile solutions section.
4. Add 1 M NaOH to give pH 7.2 at 20°C. When incubated, the medium will rise to pH 7.4 at 37°C, but this figure may need to be checked by a trial titration the first time the recipe is used (*see* Step B4, above, and Adjusting pH, below).

If any other constituents are required it will be necessary to remove an equivalent volume from the amount of water already present in the bottle before adding the main constituents, if these additional constituents are made up in water. If the additional constituents are in isotonic salt, they can be added to the final volume of medium. Because it is isotonic, serum can be added to the final volume of medium, although doing so will dilute the nutrients from the medium.

Always equilibrate and check the pH at 37°C, as the solubility of CO_2 decreases with increased temperature and the pK_a of the HEPES will change.

Adjusting pH. The amount of alkali needed to neutralize 10× concentrated medium (which is made up in acid to maintain the solubility of the constituents) may vary from batch to batch and from one medium to another, and in practice, titrating the medium to pH 7.4 at 37°C can sometimes be a little difficult. When making up a new medium for the first time, add the stipulated amount of $NaHCO_3$ and allow samples with varying amounts of alkali to equilibrate overnight at 37°C in the appropriate gas phase. Check the pH the following morning, select the correct amount of alkali, and prepare the rest of the medium accordingly.

Sodium bicarbonate. The bicarbonate concentration is important in establishing a stable equilibrium with atmospheric CO_2, but regardless of the amount of bicarbonate used, if the medium is at pH 7.4 and 37°C, the bicarbonate concentration at each concentration of CO_2 will be as previously described (*see* Section 9.2.2 and Table 9.1). Some media are designed for use with a high bicarbonate concentration and elevated CO_2 in the atmosphere (e.g., Eagle's MEM with Earle's salts), whereas others have a low bicarbonate concentration for use with a gas phase of air (e.g., Eagle's MEM with Hanks's salts; *see* Tables 8.1 and 8.2). If a medium is changed and its bicarbonate concentration altered, it is important to make sure that the osmolality is still within an acceptable range. The osmolality should always be checked (*see* Section 9.2.5) when any significant alterations are made to a medium that are not in the original formulation.

Additions to medium. If your consumption of medium is fairly high (>200 L/year) and you are buying the medium ready made, then it may be better to get extra constituents included in the formulation, as this practice will work out to be cheaper. HEPES in particular is very expensive to buy separately. Glutamine is often supplied separately, as it is unstable; it is best to buy it separately and store it frozen. The half-life of glutamine in medium at 4°C is about 3 wk and at 37°C about 1 wk. Some dipeptides of glutamine have increased stability, while retaining the bioavailability of the glutamine. One such is Glutamax, available from Invitrogen.

Precipitation. Care should be taken with 10× concentrates to ensure that all of the constituents are in solution, or at least evenly suspended, before dilution. Some constituents (e.g., folic acid or tyrosine) can precipitate and be missed at dilution. Incubation at 37°C for several hours may overcome this problem.

Quality control. Once a batch of medium has been tested for its growth-promoting and other properties, if it is found to be satisfactory, it need not be tested each time it is made up to working strength. The sterility of a medium made by diluting a 10× concentrate should be checked, however, by incubating an aliquot of the complete medium at 37°C for 48 h before use (*see* Section 11.6.2).

11.4.4 Powdered Media

Instructions for the preparation of powdered media are supplied with each pack. Choose a size that you can make up all at once and use the medium within three months. Select a formulation lacking glutamine. If other unstable constituents are present, they also should be omitted and added later as a sterile concentrate just before use.

Although BSS can be sterilized by autoclaving (*see* Protocol 11.6), this requires the omission of glucose and bicarbonate. If BSS is to be sterilized complete, follow the procedure for powdered medium (*see* Protocol 11.9).

PROTOCOL 11.9. PREPARATION OF MEDIUM FROM POWDER

Outline
Dissolve the entire contents of the pack in the correct volume of UPW, using a magnetic stirrer and adding the powder gradually with constant mixing. When all the constituents have dissolved completely, the medium should be filtered immediately and not allowed to stand, in case any of the constituents precipitate or microbial contamination appears. The pH should be adjusted after the final constituents (e.g., glutamine, $NaHCO_3$, or serum) have been added to the medium.

Equipment and Materials
Sterile:
- ❑ Graduated bottles for medium, 100 mL 10
- ❑ Caps for bottles 10
- ❑ Universal containers for contamination control sampling 3

Nonsterile:
- ❑ Ultrapure water .... 1 L
- ❑ Powdered medium, e.g., MEM with Earle's salts, without glutamine.......... 1-L pack
- ❑ Graduated Erlenmeyer flask or bottle, capacity plus head space, 1 L 1
- ❑ Magnetic stirrer and PTFE-coated follower, 50 mm 1
- ❑ Conductivity meter

Protocol
1. Add appropriate volume of ultrapure water to container.
2. Add magnetic follower.
3. Place container on magnetic stirrer and set to around 200 rpm.
4. Open packet of powder and add contents slowly to container while mixing.
5. Stir until powder is completely dissolved.
6. Check pH and conductivity of a sample and enter in record:
   (a) pH should be within 0.1 unit of expected level for particular medium.
   (b) Conductivity should be within 2% of expected value for particular medium.
   (c) Discard sample; do not add back to main stock.
7. Sterilize by filtration (*see* Section 11.5.2, Protocols 11.11–11.14).
8. Cap and seal bottles.
9. Store at 4°C.
10. Add serum and glutamine, and correct pH to 7.4 as required, just before use.

For people using smaller amounts (<1.0 L/wk) or several different types of medium, smaller volumes may be prepared, complete with glutamine, and filtered directly into storage bottles with a bottle-top filter sterilizer or filter flasks (*see* Section 11.5.2 and Protocol 11.12). With this and other negative-pressure filtration systems, some dissolved CO_2 may be lost during filtration, and the pH may rise. Provided that the correct amount of $NaHCO_3$ is in the medium to suit the gas phase (*see* Section 9.2.2), the medium will reequilibrate in the incubator, but this should be confirmed the first time the medium is used.

For large-scale requirements (>10 L/wk), medium can be prepared in a pressure vessel, checked at intervals with a large pipette to determine whether solution is complete, and sterilized by positive pressure through an in-line disposable or reusable filter into a receiver vessel (*see* Section 11.5.2, Protocol 11.14).

11.4.5 Customized Medium

If one intends to explore different formulations, or if the medium is to be made up in house from individual constituents, it is convenient to make up a number of concentrated stocks—amino acids at 50× or 100×, vitamins at 1000×, and tyrosine and tryptophan at 50× in 0.1 M HCl, glucose at 200 g/L, and single-strength BSS. The requisite amount of each concentrate is then mixed, filtered through a sterilizing filter of 0.2-μm porosity, and diluted with high- or low-bicarbonate BSS (*see* Section 11.4.2, Figs. 11.11, 11.12, and Table 9.2).

PROTOCOL 11.10. PREPARATION OF CUSTOMIZED MEDIUM

Outline
Prepare stock concentrates (derived from Tables 9.3 or 10.1) and store frozen. Thaw and blend as required. Sterilize by filtration and store until required.

Materials
❑ Amino acid concentrate, 100× in water, stored frozen
❑ Tyrosine and tryptophan, 50× in 0.1 M HCl
❑ Vitamins, 1000× in water
❑ Glucose, 100× (200 g/L in BSS)
❑ Additional solutions (e.g., trace elements and nucleosides; not lipids, hormones, or growth factors, which should be added just before using the medium)
❑ Storage bottles, selected for optimum aliquot size for subsequent dilution (*see* Step 5 of Protocol)

Protocol
1. Thaw solutions and ensure that all solutes have redissolved.

---

2. Blend constituents in correct proportions:
   (a) amino acid concentrate 100 mL
   (b) tyrosine and tryptophan 200 mL
   (c) vitamins 10 mL
   (d) glucose 100 mL
3. Mix and sterilize by filtration (*see* Protocols 11.12, 11.13).
4. Store frozen.
5. For use, dilute 41 mL of concentrate mixture with 959 mL sterile 1× BSS.
6. Adjust to pH 7.4.
7. Store at 4°C.
8. Add serum or other supplements, such as hormones, growth factors, and lipids, just before using medium. If metals are used as trace elements, it is also better to add these just before using the medium, as they can precipitate in the presence of phosphate in concentrated stocks.

The advantage of this type of recipe is that it can be varied; extra nutrients (oxo-acids, nucleosides, minerals, etc.) can be added or the major stock solutions altered to suit requirements, but, in practice, this procedure is so laborious and time consuming that few laboratories make up their own media from basic constituents, unless they wish to alter individual constituents regularly. The reliability of commercial media depends entirely on the application of appropriate quality-control measures. Any laboratory carrying out its own preparation must make sure that appropriate quality-control measures are used (*see* Section 11.6.1).

There are now many reputable suppliers of standard formulations (*see* Appendix II), many of whom will supply specialized, serum-free formulations and custom media. It is important to ensure that the quality control these suppliers employ is relevant to the medium and the cells you wish to propagate. You might buy MCDB 153, which is tested on HeLa cell colony formation, but this test is of little relevance if you wish to grow primary keratinocytes.

11.5 STERILIZATION OF MEDIA

11.5.1 Autoclavable Media

Some commercial suppliers offer autoclavable versions of Eagle's MEM and other media. Autoclaving is much less labor intensive, is less expensive, and has a much lower failure rate than filtration. The procedure to follow is supplied in the manufacturer's instructions and is similar to that described earlier for D-PBSA (*see* Protocol 11.6). The medium is buffered to pH 4.25 with succinate, in order to stabilize the B vitamins during autoclaving, and is subsequently neutralized. Glutamine is replaced by glutamate or glutamyl dipeptides, or is added sterile after autoclaving.

BSS is autoclavable without glucose and glutamine, and various amino acid hydrolysates, such as tryptose phosphate broth and other microbiological media, are also sterilizable by autoclaving.

11.5.2 Sterile Filtration

Filtration through 0.1- to 0.2-μm microporous filters is the method of choice for sterilizing heat-labile solutions (Fig. 11.11). Numerous kinds of filters are available, made from many different materials, including polyethersulfone (PES), nylon, polycarbonate, cellulose acetate, cellulose nitrate, PTFE, and ceramics, and in sizes from syringe-fitting filters (Fig. 11.12a,c) through small and intermediate in-line filters (Fig. 11.12b,c,f) to multidisk and cartridge filters (Fig. 11.13). Low-protein-binding filters are available from most suppliers (e.g., Durapore, Millipore); PES filters are generally found to be faster flowing. Polycarbonate filter membranes are absolute filters with an array of holes of a uniform porosity; the number of holes per unit area increases as the size diminishes, to maintain a uniform flow rate. Most other filters are of the mesh variety and filter by entrapment; they generally have a faster flow rate and reduced clogging, but will compress at high pressures.

Filtration may be carried out by positive pressure (*see* Fig. 11.11a, Fig. 11.12b,c,f), from a pressurized container (*see* Fig. 11.14) or with a peristaltic pump (Fig. 11.15). High-pressure filtration from a pressure vessel is faster than a peristaltic pump but will require a collecting receiver as the flow cannot be regulated easily, whereas a peristaltic pump can be switched on and off during collection. High pressure also tends to compact the filter and is unsuitable for viscous solutions such as serum. Negative-pressure filtration is often simpler (*see* Figs. 11.11b, 11.12d,e), particularly for small-scale operations, and will collect directly into storage vessels. It may cause an elevation of the pH, however, because of evolution of CO_2.

Disposable filters. Disposable filter holder designs include simple disk filters, hollow-fiber units, and cartridges (*see* Figs. 11.11–11.13). Syringe-tip filters are generally used for low-volume filtration (for 2–20 mL) and vary in size from 13- to 50-mm diameter (*see* Fig. 11.12a). Intermediate-sized filters (for 50 ml–5 L) can be used in-line with a peristaltic pump (Fig. 11.14), or as bottle-top filters used with a vacuum line and a regular medium bottle (Fig. 11.12d). Intermediate-sized filters can also be purchased as complete filter units for attaching to a vacuum line, with an upper chamber for the nonsterile solution and a lower chamber to receive the sterile liquid and to use for storage (Fig. 11.12e). Large-capacity cartridge filters (for 20–500 L) are usually operated in-line under positive pressure from a reservoir (Figs. 11.13, 11.15). Although disposable filters are more expensive than reusable filters, they are less time-consuming to use and give fewer failures.

Reusable filters. Reusable filter holders may use membranes (Fig. 11.16) or cartridges (Fig. 11.13b) and are sterilized by autoclaving (*see* Protocol 11.4). They are usually connected to a pressure reservoir (*see* Fig. 11.15) or a peristaltic pump and operate under positive pressure.

Protocols 11.11–11.14 feature disposable filters in small sizes and a reusable filter for larger volumes.

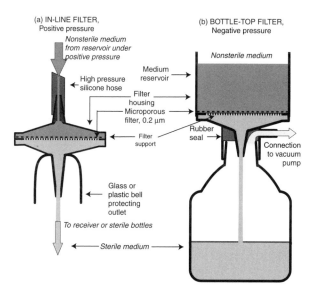

Fig. 11.11. Sterile Filtration. (a) In-line filter. Nonsterile medium from pump or pressure vessel. (b) Bottle-top filter or filter flask (designs are similar). Medium added to upper chamber and collected in lower. Lower chamber can be used for storage.

PROTOCOL 11.11. STERILE FILTRATION WITH SYRINGE-TIP FILTER

Materials
Sterile:
- ❑ Plastic syringe, 10- to 50-mL capacity
- ❑ Syringe-tip filter (e.g., Pall Gelman Acrodisc or Millipore Millex)
- ❑ Receiver vessel (e.g., a universal container)

Nonsterile:
- ❑ Solution for sterilization (5–100 mL)

Protocol
1. Swab down hood and assemble materials.
2. Fill syringe with solution to be sterilized.
3. Uncap receiver vessel.
4. Unpack filter and attach to tip of syringe, holding the sterile filter within the bottom half of the packaging while attaching to syringe.
5. Expel solution through filter into receiver vessel. Only moderate pressure is required.
6. Cap receiver vessel.
7. Discard syringe and filter.

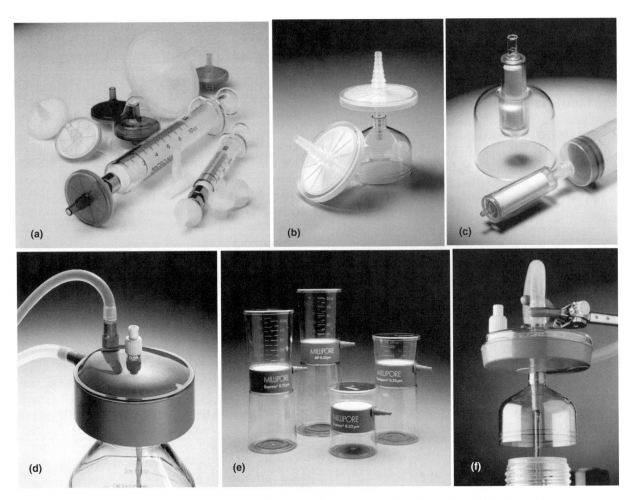

Fig. 11.12. Disposable Sterilizing Filters. (a) Millex 25-mm disk syringe filter. (b) In-line filter with bell. (c) Sterivex with and without bell. (d) Bottle-top filter. (e) Filter cup and storage vessels. (f) Large in-line filter with bell. (a,b,c,f) Positive pressure. (d,e) Negative pressure (*see also* Fig. 11.11). Photographs courtesy of Millipore (UK), Ltd.

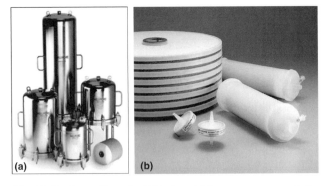

Fig. 11.13. Large-Capacity Filtration. (a) Large pressure vessels for positive-pressure filtration. (b) Stacked disk and pleated cartridge filters for large-scale filtration. (Courtesy of Millipore (UK) Ltd).

The syringe may be refilled several times by returning the filter to the lower half of the sterile packaging, detaching it from the syringe, refilling the syringe, and reattaching the filter. If the back-pressure increases, take a new filter.

If a vacuum line or pump is available, volumes between 50 and 500 mL can be filtered conveniently by negative pressure into an integral filter flask (*see* Figs. 11.11b, 11.12e) or with a bottle-top filter and regular medium bottle (*see* Fig. 11.12d). Protocol 11.12 is presented for use in conjunction with Exercise 9 (*see* Chapter 2), but other volumes and filter sizes may be used (*see* Table 11.4).

PROTOCOL 11.12. STERILE FILTRATION WITH VACUUM FILTER FLASK

Materials
Sterile:
- ❑ Filter flask 500 mL (*see* Fig. 11.11b; *see*, e.g., Fig. 11.12e) 1
- ❑ Cap for lower chamber (if chamber is used for storage) 1

❑ Sample tube or universal container
   for sterility test 1
Nonsterile:
❑ Medium for sterilization .... 450 mL
❑ Vacuum pump or vacuum line
❑ Thick-walled connector tubing from pump or line,
   to fit filter flask inlet

Protocol

1. Connect side arm of filter flask to vacuum pump.
2. Remove cap from bottle and lid from top chamber of filter flask.
3. Pour nonsterile medium into top chamber.
4. Switch on pump.
5. Unpack cap for lower chamber, ready for use.
6. When liquid has all been drawn into lower chamber, switch off pump and detach filter housing and top chamber.
7. Transfer 10 mL from lower flask to universal container or equivalent tube.
8. Gas the head space in the lower chamber and the universal container with 5% CO_2.
9. Cap the lower chamber and universal.
10. Label lower chamber and universal with name of medium, date, and your initials.

11. Store lower chamber at 4°C until required.
12. Incubate the 10-mL sample in the universal at 37°C for 1 week and check for contamination. Do not release medium for use until shown to be free of contamination.

Generally, this method of filtration has a low risk of contamination, so sampling for a contamination check may be omitted in due course.

Filter flasks are available with from 150- to 1000-mL receiver capacity. If a larger volume is to be filtered and dispensed into aliquots, it may be better to use one of several bottle-top filters, which may be set up to filter material directly into standard medium bottles (*see* Fig. 11.12d). Alternatively, solutions may be filtered by positive pressure, either from a pressurized reservoir (*see* Fig. 11.15) or via a peristaltic pump (*see* Fig. 11.14), passing through a sterile in-line membrane or cartridge filter equipped with a bell to protect the receiver vessel from contamination (*see* Figs. 11.11, 11.12, 11.14, 11.15). Protocol 11.13 uses a small in-line filter and can be used in conjunction with Exercise 9.

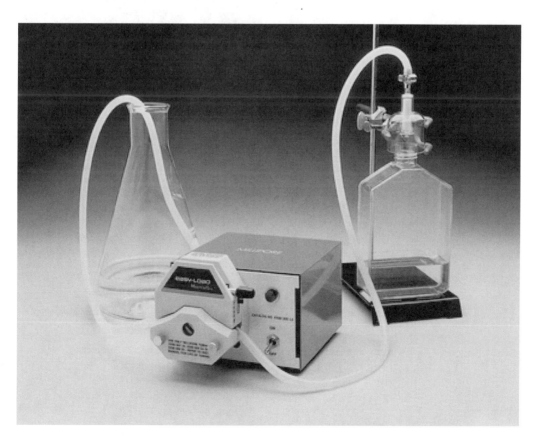

Fig. 11.14. Peristaltic Pump Filtration. Sterile filtration with peristaltic pump between nonsterile reservoir and sterilizing filter (Millipak, Millipore; courtesy of Millipore [U.K.], Ltd).

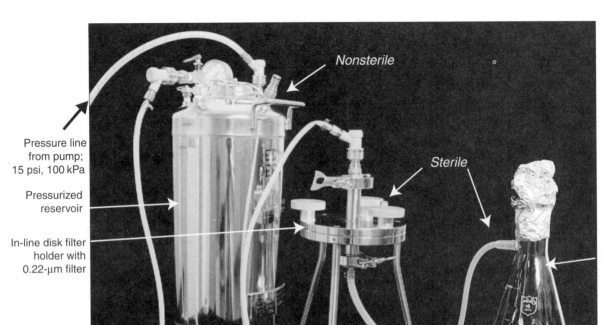

Fig. 11.15. *Large-Scale In-Line Filter Assembly.* In-line filter assembly supplied from a pressurized reservoir (left) and connected to receiver flask (right). Only the filter assembly and the receiver flask need be sterilized. Normally, the glass bell would be covered in protective foil; this is left off here for purposes of illustration.

TABLE 11.4. Filter Size and Fluid Volume

Filter size or designation	Disposable (D) or reusable (R)	Approximate volume that may be filtered	
		Crystalloid	Colloid
25 mm, Millex	D	1–100 mL	1–20 mL
47 mm or Sterivex cartridge	R, D	0.1–1 L	100–250 mL
90 mm	R	1–10 L	0.2–2 L
Millipak-20	D	2–10 L	200 mL–2 L
Millipak-40	D	10–20 L	2–5 L
Millipak-60	D	20–30 L	5–7 L
Millipak-100	D	30–75 L	7–10 L
Millipak-200	D	75–150 L	10–30 L
Millidisk	D	30–300 L	5–50 L
142 mm	R	10–50 L	1–5 L
293 mm	R	50–500 L	5–20 L

Examples in the table are quoted from Millipore catalog. Similar products are available from Pall Gelman, Nalge Nunc, Sartorius, and a number of other suppliers. (*See* Appendix II.)

PROTOCOL 11.13. STERILE FILTRATION WITH SMALL IN-LINE FILTER

Materials
Sterile:
❑ In-line filter with bell (*see* Figs. 11.11a, 11.12f) .... 47 mm
❑ Graduated medium bottles, foil capped, sterilized by dry heat (*see* Protocol 11.1) ... 7
❑ Caps, autoclaved (*see* Protocol 11.3) 6
❑ Sample tube or universal container for sterility test 1
Nonsterile:
❑ Medium for sterilization (from Exercise 9 and Protocol 11.9) .... 550 mL
❑ Peristaltic pump (Fig. 11.14) preferably with foot switch 1

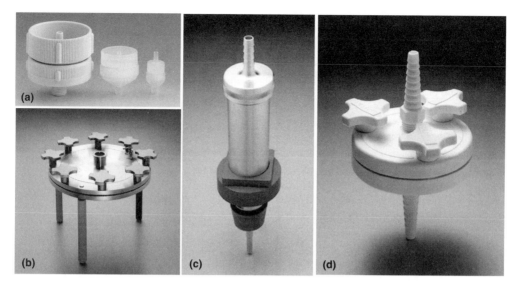

Fig. 11.16.   Reusable Filters. (a) Polypropylene, in-line, Luer fitting. (b) Stainless steel housing, high-capacity disk-type filter. (c) 47-mm filter with reservoir. (d) 90 mm in line with hose connections. (Courtesy of Millipore, U.K., Ltd).

❑ Silicone tubing to fit pump and inlet to filter .... 50 cm
❑ Clamp stand to hold filter 1

Protocol
1. Feed tubing through peristaltic pump.
2. Insert upstream end into medium to be sterilized.
3. Unpack filter and connect to outlet from peristaltic pump.
4. Clamp filter in clamp stand at a suitable height such that bottles for receiving medium can be positioned below the filter with the neck shrouded and removed easily when ready to be filled. Use one of the sterile medium bottles to set up the filter if necessary, but do not use this bottle, ultimately, for sterile collection.
5. Switch on pump and collect ~20 mL into bottle used for setup.
6. Remove setup bottle, cap it and number it "1".
7. Remove foil from first medium bottle and place under filter bell.
8. Switch on pump.
9. Fill bottle to 100 mL mark.
10. Switch off pump.
11. Remove bottle, cap it, number it, and replace with fresh sterile medium bottle.
12. Repeat steps 7–11, filling 5 bottles to 100 mL, and collecting the remainder in the last bottle.
13. Gas the head space in each bottle with 5% CO_2.
14. Replace the aluminum foil over the cap and neck of each bottle to keep it free from dust during storage.

15. Label, date, and initial bottles.
16. Place bottles containing 100 mL medium at 4°C for storage (*see* Section 11.6.4).
17. Place the first and last bottles, sealed, at 37°C and incubate for 1 week (*see* Section 11.6.2).
18. Store medium at 4°C until required; do not release for use until sterility test is complete. If any contamination is found in the test samples, refilter the whole batch or discard it.

Filtration with positive pressure can be scaled up by increasing the filter size and the size of the medium reservoir. The peristaltic pump is often replaced by adding positive pressure to the medium reservoir in a pressure vessel (*see* Fig. 11.13). Protocol 11.14 is suitable for up to 50 L, but the volume can be increased up to industrial scale by selecting the appropriate pressure vessel and filter (often a pleated cartridge or other multisurface filter).

PROTOCOL 11.14. STERILE FILTRATION WITH LARGE IN-LINE FILTER

Materials
(*see* Figs. 11.13, 11.15, and 10.17)
Sterile:
❑ Filter (e.g., 90-mm membrane and reusable filter holder; *see* Protocol 11.4; Table 11.4; Fig. 11.15)

or disposable disk (e.g., Millipore Millex) or cartridge (e.g., Pall Gelman Capsule).

❑ Receiver for filtrate with outlet at the base
❑ Silicone pressure tubing from filter to sterile receiver
❑ Silicone tubing with glass bell and spring clip on receiver outlet (see Fig. 11.15)
❑ Medium bottles, foil capped, dry heat sterilized (see Protocol 11.1)
❑ Caps for medium bottles, autoclaved (see Protocol 11.3)

Nonsterile:
❑ Pressure vessel, 5–50 L.
❑ Pump, 100 kPa (15 p.s.i.)
❑ Clamp stand and clamps to secure filter (unless filter holder has legs) and outlet bell
❑ Silicone pressure tubing from pressure vessel to filter

Protocol

1. Secure the filter holder in position.
2. Connect the outlet to the receiver vessel.
3. Secure the outlet from the receiver at a suitable height such that medium bottles can be positioned below it with the neck shrouded by the bell and removed easily when the bottle filled. Use one of the sterile medium bottles to set up if necessary, but do not use this bottle, ultimately, for sterile collection.
4. Connect the pressure vessel to the filter inlet.
5. Decant the medium into the pressure vessel.
6. For reusable filters, turn on the pump just long enough to wet the filter. Stop the pump and tighten up the filter holder.
7. Switch on the pump to deliver 100 kPa (15 lb/in.^{2}). When the receiver starts to fill, draw off aliquots into medium stock bottles of the desired volume.
8. Cap the bottles as each one is taken from the filter bell.
9. Replace the foil over the cap.
10. Label, date, and initial bottles.
11. Depending on the number and size of bottles filled, remove sample bottles from the beginning, middle, and end of the run, and incubate at 37°C for 1 week to check for contamination.
12. Store medium at 4°C. Do not release for general use until quality control has been performed.

Positive pressure is recommended for optimum performance of the filter and to avoid the removal of CO_2, which results from negative-pressure filtration. Positive pressure may also be applied by using a peristaltic pump (Fig. 11.14) in line between the nonsterile reservoir and a disposable in-line filter, such as the Millipak (Millipore), which may be used instead of the reusable assembly. The only preparation and sterilization required is for the medium bottles as the disposable filter is bought sterile and no receiver is necessary (as flow can be interrupted easily by switching off the peristaltic pump).

11.5.3 Serum

Preparing serum is one of the more difficult procedures in tissue culture, because of variations in the quality and consistency of the raw materials and because of the difficulties encountered in sterile filtration due to particulate material and viscosity. Moreover, serum is also one of the most costly constituents of tissue culture, accounting for 20–30% of the total budget if it is bought from a commercial supplier. Buying sterile serum is certainly the best approach from the point of view of consistency and quality control, but Protocol 11.15 is suggested if serum has to be prepared in the laboratory. The underlying principle is that a graded series of filters, which need not be sterile, are used to remove particulate material before the serum is passed through a sterilizing filter of 0.1 μm at a low pressure (Fig. 11.17).

PROTOCOL 11.15. COLLECTION AND STERILIZATION OF SERUM

Outline

Collect blood, allow it to clot, and separate the serum. Sterilize serum through filters of gradually reducing porosity. Bottle and freeze filtered serum.

Collection

Arrangements may be made to collect whole blood from a slaughterhouse. The blood should be collected directly from the bleeding carcass and not allowed to lie around after collection. Alternatively, blood may be withdrawn from live animals under proper veterinary supervision. The latter alternative, if performed consistently on the same group of animals, gives a more reproducible serum but a lower volume for a greater expenditure of effort. If the procedure is done carefully, blood may be collected aseptically.

Clotting

Allow the blood to clot by having it stand overnight in a covered container at 4°C. This so-called natural-clot serum is superior to serum that is physically separated from the blood cells by centrifugation and defibrination, as platelets release growth factors into the serum during clotting. Separate the serum from

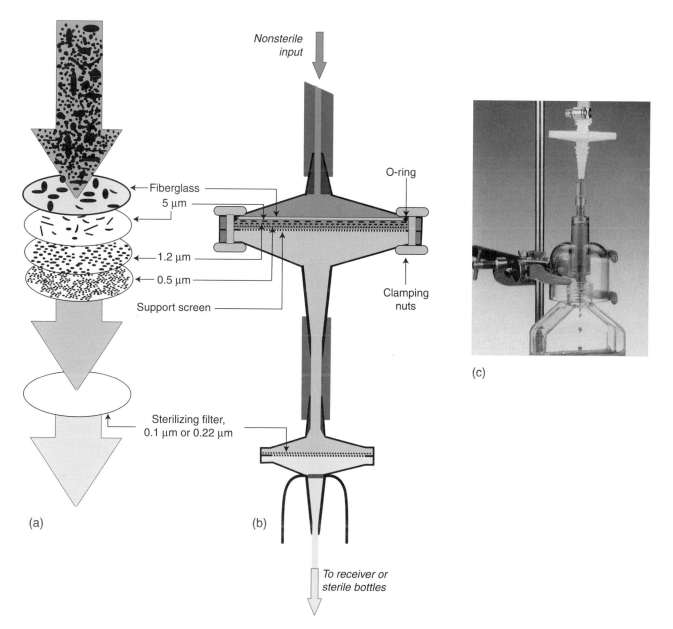

Nonsterile input

Fiberglass
5 μm

1.2 μm

0.5 μm

Support screen

O-ring

Clamping nuts

Sterilizing filter, 0.1 μm or 0.22 μm

(a)

(b)

(c)

To receiver or sterile bottles

Fig. 11.17. Prefilter. Prefiltration for filtering colloidal solutions (e.g., serum) or solutions with high particulate content. Several prefilters can be connected in series, and only the final filter need be sterile. (a) Diagrammatic representation of principle; medium passing through filters of gradually reducing porosity. (b) Filter series stacked in a nonsterile reusable filter holder leading to a sterile disposable 0.1- or 0.22-μm in-line filter as in (c). (c) One disposable prefilter (nonsterile or sterile) inserted upstream of a final 0.22-μm or 0.1-μm sterilizing filter.

the clot, and centrifuge the serum at 2,000 g for 1 h to remove sediment.

Sterilization

Serum is usually sterilized by filtration through a sterilizing filter of 0.1-μm porosity, but because of its viscosity and high particulate content, the serum should be passed through a graded series of fiberglass or other prefilters before passing through the final sterilizing filter (Fig. 11.17). Only the last filter, a 142- to 350-mm in-line disk filter or equivalent disposable filter (e.g., Millipak 200), need be sterile. The prefilter assemblies may be stainless steel with replaceable cartridges, disk filter units, or a single bonded unit (Pall Gelman). The last is easiest to use, but more difficult to clean and reuse.

For Sterilizing Volumes of 5–20 L:

Materials
Sterile:
❏ Sterilizing filter: 200 mm, 0.1-μm-porosity (e.g., Millipak, Millipore). A porosity of 0.2 μm is sufficient for antibacterial and fungal sterilization, but 0.1 μm is required to remove mycoplasma.
❏ Sterile receiving vessel with outlet at base
❏ Sterile bottles with caps and foil
❏ Universal containers for sterility testing
Nonsterile:
❏ Peristaltic pump and tubing
❏ Clamp stand and clamps
❏ Nonsterile prefilters:
   Fiberglass disposable filter or 142-mm reusable filter (e.g., Pall Gelman)
   5-μm-porosity disposable or reusable filter (e.g., Pall Gelman Versapore)
   1.2-μm porosity disposable or reusable filter (e.g., Millipore Opticap)
   0.45-μm porosity disposable or reusable filter (e.g., Millipore Millipak)

Note. The preceding filters are suggestions only; contact your supplier and request a series of filters and prefilters that will suit your serum requirements. If it is a once-only activity, choose disposable filters; if collection will be repeated regularly, it will be more economical to employ reusable filter holders for the prefilter stages, while still using a disposable filter for sterilization, for added security.

Protocol
1. Insert appropriate nonsterile filters into nonsterile prefilter holders (if reusable holders are being used).
2. Connect one or more prefilters in line and upstream from a sterile disposable or reusable filter holder (Fig. 11.17) that contains a 0.1-μm-porosity filter and is connected to a sterile receiver via the peristaltic pump.
3. Place the intake of the pump into the serum container.
4. Switch on the pump, and check any reusable filter holders for leakage as they are wetted; switch off the pump, and tighten filter holders as necessary.
5. Restart the pump and continue filtering, checking for leaks or blockages. Increasing the flow rate will increase the rate of filtration but may cause the filters to become packed or clogged.
6. Collect aliquots in sterile bottles, leaving at least 20% headspace to allow for expansion on freezing.

7. Collect samples at the beginning, middle, and end of the run to check sterility.
8. Cap, label, and number the bottles.
9. Replace the foil over the cap and neck for storage.
10. Store serum at −20°C until quality control is completed (*see* Section 11.6.1).

Small-scale serum processing. If small amounts (<1 L) of serum are required, then the process is similar to Protocol 11.14, but can be scaled down. After clot retraction (*see* Protocol 11.15: Clotting), small volumes of serum may be centrifuged (5–10,000 g) and then filtered through a series of disposable filters (e.g., 50-mm Millipore Millex or Pall Gelman Acrodisc) and, finally, through a 50-mm, 0.1-μm-porosity sterile disposable filter.

Centrifuge very small volumes (10–20 mL) at 10,000 g, and filter the serum directly through a graded series of syringe-tip disposable 25-mm filters (e.g., Acrodisc, Pall Gelman; Millex, Millipore), finishing with a 0.1-μm sterilizing filter (e.g., Millex).

Storage. Bottle the serum in sizes that will be used up within two to three weeks after thawing. Freeze the serum as rapidly as possible, and if it is thawed, do not refreeze it unless further prolonged storage is required.

Serum is best used within 6–12 months of preparation if it is stored at −20°C, but more prolonged storage may be possible at −70°C; usually, however, the bulk of serum stocks makes this impractical. Polycarbonate or high-density polypropylene bottles will eliminate the risk of breakage if storage at −70°C is desired. Regardless of the temperature of the freezer or the nature of the bottles, do not fill them completely; allow for the expansion of water during freezing.

Human serum. Pooled outdated human blood or plasma from a blood bank can be used instead of or in addition to bovine or equine blood. It should be sterile and not require filtration. Titrate out the heparin or citrate anticoagulant with Ca^{2+}, allow the blood to clot overnight, and then separate and freeze the serum.

Δ ***Safety Note.*** Care must be taken with human donor serum to ensure that it is screened for hepatitis, HIV, tuberculosis, and other adventitious infections.

Quality control. Use the same procedures as for medium (*see* Section 11.6).

A major problem that is emerging with the use of serum is the possibility of viral infection. When the possibility of bacterial infection was first appreciated, it was relatively easy to devise filtration procedures to filter out anything above 1.0 μm, and eventually a porosity of 0.45 μm became standard. Subsequently, it was learned that mycoplasma would pass through filters as low as 0.2 μm, and commercial suppliers of serum lowered the exclusion limits of their filters to 0.1 μm. This reduction in size appears to have virtually eliminated mycoplasma from serum batches used in culture, but the problem of viral contamination remains. Filtering out virus would seem to be a much more significant task, but some companies (e.g., Pall Gelman) claim that it may be possible. Collecting serum from areas of low indigenous infection and screening before processing are better alternatives.

Dialysis. For certain studies, the presence of constituents of low molecular weight (amino acids, glucose, nucleosides, etc.) may be undesirable. These constituents may be removed by dialysis through conventional dialysis tubing.

PROTOCOL 11.16. DIALYSIS OF SERUM

Materials
Sterile:
❏ Bottles and caps
❏ Sterilizing filter, 0.1 μm
❏ Prefilters: 5.0-μm fiberglass, 1.2-, 0.45-, and 0.22-μm membrane filters(e.g., Durapore, Millipore; *see* Protocol 11.14)
Nonsterile:
❏ Dialysis tubing
❏ Beaker with ultrapure water
❏ Bunsen burner
❏ Tripod with wire gauze
❏ Serum to be dialyzed
❏ HBSS at 4°C
❏ Measuring cylinder

Protocol
1. Boil five pieces of 30-mm × 500-mm dialysis tubing in three changes of distilled water.
2. Transfer tubing to Hanks's balanced salt solution (HBSS), and allow to cool.
3. Tie double knots at one end of each tube.
4. Half-fill each dialysis tube with serum (20 mL).
5. Express air and knot the open end of the tube, leaving a space of about half the tube between the serum and the knot.
6. Place the tubing in 5 L of HBSS, and stir on a magnetic stirrer overnight at 4°C.
7. Change HBSS and repeat step 6 twice.
8. Collect serum into a measuring cylinder and note the volume collected. (If the volume is less than the starting volume of the serum, add HBSS to return to the starting volume. If the volume is greater than the starting volume of the

---

serum, make due allowance when adding to the medium later.)
9. Sterilize serum through a graded series of filters (*see* Protocol 11.15).
10. Bottle and freeze the serum.

11.5.4 Preparation and Sterilization of Other Reagents

Individual recipes and procedures are given in Appendix I. On the whole, most reagents are sterilized by filtration if they are heat labile and by autoclaving if they are heat stable (*see* Table 11.2). Filters with low binding properties (e.g., Millex-GV) are available for sterilizing of proteins and peptides.

11.6 CONTROL, TESTING, AND STORAGE OF MEDIA

11.6.1 Quality Control

A medium that is prepared in the laboratory needs to be tested before use. If the medium is purchased ready made as a 1× working-strength solution, then it should be possible to rely on the quality control carried out by the supplier, other than any special requirements that you have of the medium. Likewise, if a 10× concentrate is used, the growth and sterility testing will have been done, and the only variable will be the water used for dilution. Provided that the conductivity and level of total organic carbon fall within specifications (*see* Section 11.4.1) and no major changes have been made in the supply of water, most laboratories will accept this compliance as adequate quality control.

However, if medium is prepared from powder, it will have been sterilized in the laboratory, and quality control will be required to confirm sterility, although, given that all the constituents have dissolved, you may be prepared to accept the quality control of the supplier regarding the medium's growth-promoting activity. Media made up from basic constituents will require complete quality control, involving both sterility testing and culture testing.

11.6.2 Sterility Testing

Bubble point. When positive-pressure filtration is complete and all the liquid has passed through the filter, raise the pump pressure until bubbles form in the effluent from the filter. This is the *bubble point* and should occur at more than twice the pressure used for filtration (*see* manufacturer's instructions). If the filter bubbles at the sterilizing pressure (100 kPa, 15 p.s.i.) or lower, then it is perforated and should be discarded. In that case, any filtrate that has been collected should be regarded as nonsterile and refiltered. Single-use, disposable filters rarely fail the bubble point test, which is very quick and easy to perform. Reusable filters can fail, so they should be checked after every filtration run.

Incubation. Collect samples at the beginning, middle, and end of the run. If the bottles are small, this can be done by removing bottles. If the bottles are large, rather than wasting medium, collect samples into smaller containers at intervals during filtration. Remember, however, if you are bottling in 1000-mL sizes, taking 1-mL samples will reduce the sensitivity by a factor of 10^3; if you are not prepared to sacrifice whole bottles at this size, then you should at least sample 100 mL. It is best not to withdraw samples from individual bottles, as this will both increase the risk of contaminating the bottles and reduce the volume in the bottle used for sampling, thereby altering the dilutions of subsequent additions to that bottle.

Incubate samples according to either of the following procedures:

(1) Incubate samples of medium at 37°C for 1 week. If any of the samples become cloudy, discard them and resterilize the batch. If there are signs of contamination in the other stored bottles, the whole batch should be discarded.
(2) For a more thorough test, and when the solution being filtered does not have its own nutrients, take samples, as described in this section, and dilute one-third of each into nutrient broths (e.g., L-broth, beef heart hydrolysate, and thioglycollate). Divide each sample in two, and incubate one at 37°C and one at 20°C for 10 days, with uninoculated controls. If there is any doubt after this incubation, mix and plate out aliquots on nutrient agar and incubate at 37°C and 20°C.

Downstream secondary filtration. Place a demountable 0.45-μm sterile filter in the effluent line from the main sterilizing filter. Any contamination that passes because of failure in the first filter will be trapped in the second. At the end of the run, remove the second filter and place the filter on nutrient agar. If colonies grow, discard or refilter the medium. This method has the advantage that it monitors the entire filtrate, and not just a small fraction of it, although it does not avoid risks of contamination during bottling and capping.

Autoclaved solutions. Sterility testing of autoclaved stocks is much less essential, provided that proper monitoring (of the temperature and the time spent at the sterilizing temperature) of the center of the load of the autoclave is carried out (*see* Section 11.4).

11.6.3 Culture Testing
Media that have been produced commercially will have been tested for their capability of sustaining the growth of one or more cell lines. (If they have not, then you should change your supplier!) However, under certain circumstances, you may wish to test your own media for quality: (1) if it has been made up in the laboratory from basic constituents; (2) if any additions or alterations are made to the medium; (3) if the medium is for a special purpose that the commercial supplier is not able to test; and (4) if the medium is made up from powder and there is a risk of losing constituents during filtration.

The medium can become contaminated with toxic substances during filtration. For example, some filters are treated with traces of detergent to facilitate wetting, and the detergent may leach out into the medium as it is being filtered. Such filters should be washed by passing PBS or BSS through them before use or by discarding the first aliquot of filtrate. Polycarbonate filters (e.g., Nucleopore) are wettable without detergents and are preferred by some workers, particularly when the serum concentration in the medium is low.

There are three main types of culture test: (1) plating efficiency; (2) growth curve at regular passage densities and up to saturation density; and (3) the expression of a special function (e.g., differentiation in the presence of an inducer, viral propagation, the formation of a specific product, or the expression of a specific antigen). All of these tests should be performed on the new batch of medium with your regular medium as a control.

Plating efficiency. The plating efficiency test (*see* Protocol 21.10) is the most sensitive culture test, detecting minor deficiencies and low concentrations of toxins that are not apparent at higher cell densities. Ideally, it should be performed with a limiting concentration of serum, which may otherwise mask deficiencies in the medium. To determine this concentration, do an initial plating efficiency test in different concentrations of serum and select a concentration such that the plating efficiency is about half that of the usual concentration but still gives countable colonies.

Growth curve. A clonal growth assay will not always detect insufficiencies in the amount of particular constituents. For example, if the concentration of one or more amino acids is low, it may not affect clonal growth but could influence the maximum cell concentration attainable.

A growth curve (*see* Protocols 21.7–21.9) gives three parameters of measurement: (1) the lag phase before cell proliferation is initiated after subculture, indicating whether the cells are having to adapt to different conditions; (2) the doubling time in the middle of the exponential growth phase, indicating the growth-promoting capacity of the medium; and (3) the terminal cell density. In cell lines whose growth is not sensitive to density (e.g., continuous cell lines; *see* Section 18.5.2), the terminal cell density indicates the total yield possible and usually reflects the total amino acid or glucose concentration. Remember that a medium that gives half the terminal cell density costs twice as much per cell produced.

Special functions. If you are testing special functions, a standard test from the experimental system you are using (e.g., a virus titer in the medium after a set number of days) should be performed on the new medium alongside the old one.

A major implication of these tests is that they should be initiated well in advance of the exhaustion of the current stock of medium so that proper comparisons may be made and there is time to have fresh medium prepared if the medium fails any of the tests.

Records. All quality control tests should be recorded in a log book or computer database, along with a batch number of the solution being tested and the name of the tester. The person supervising preparation and sterilization should review these records and determine failure rates and trends, to check for the need to alter procedures.

11.6.4 Storage

Opinions differ as to the shelf life of different media. As a rough guide, media made up without glutamine should last 6–9 months at 4°C. Once glutamine, serum, or antibiotics are added, the storage time is reduced to 2–3 weeks. Hence, media that contain labile constituents should either be used within 3 weeks of preparation or be stored at −20°C.

Some forms of fluorescent lighting will cause riboflavin and tryptophan or tyrosine to deteriorate into toxic by-products, mainly peroxides [Wang & Nixon, 1978; Edwards et al., 1994]. Thus incandescent lighting should be used in cold rooms where media are stored and in hot rooms where cells are cultured, and the light should be extinguished when the room is not occupied. Bottles of medium should not be exposed to fluorescent lighting for longer than a few hours; a dark freezer is recommended for long-term storage.

CHAPTER 12

Primary Culture

12.1 TYPES OF PRIMARY CELL CULTURE

A primary culture is that stage of the culture after isolation of the cells but before the first subculture. There are four stages to consider: (1) acquisition of the sample, (2) isolation of the tissue, (3) dissection and/or disaggregation, and (4) culture after seeding into the culture vessel. After isolation, a primary cell culture may be obtained either by allowing cells to migrate out from fragments of tissue adhering to a suitable substrate or by disaggregating the tissue mechanically or enzymatically to produce a suspension of cells, some of which will ultimately attach to the substrate. It appears to be essential for most normal untransformed cells, with the exception of hematopoietic cells, to attach to a flat surface in order to survive and proliferate with maximum efficiency. Transformed cells (see Section 18.5.1), on the other hand, particularly cells from transplantable animal tumors, are often able to proliferate in suspension.

The enzymes used most frequently for tissue disaggregation are crude preparations of trypsin, collagenase, elastase, pronase, dispase, DNase, and hyaluronidase, alone or in various combinations, e.g., elastase and DNase for type II alveolar cell isolation [Dobbs & Gonzalez, 2002], collagenase with Dispase [Booth & O'Shea, 2002], and collagenase with hyaluronidase [Berry & Friend, 1969; Seglen, 1975]. There are other, nonmammalian enzymes, such as Trypzean (Sigma), a recombinant, maize-derived, trypsin, TrypLE (Invitrogen), recombinant microbial, and Accutase and Accumax (Innovative Cell Technologies), also available for primary disaggregation. Crude preparations are often more successful than purified enzyme preparations, because the former contain other proteases as contaminants, although the latter are generally less toxic and more specific in their action. Trypsin and pronase give the most complete disaggregation, but may damage the cells. Collagenase and dispase, on the other hand, give incomplete disaggregation, but are less harmful. Hyaluronidase can be used in conjunction with collagenase to digest the intracellular matrix, and DNase is used to disperse DNA released from lysed cells; DNA tends to impair proteolysis and promote reaggregation (see Table 13.4).

Although each tissue may require a different set of conditions, certain requirements are shared by most of them:

(1) Fat and necrotic tissue are best removed during dissection.
(2) The tissue should be chopped finely with sharp instruments to cause minimum damage.
(3) Enzymes used for disaggregation should be removed subsequently by gentle centrifugation.
(4) The concentration of cells in the primary culture should be much higher than that normally used for subculture, because the proportion of cells from the tissue that survives in primary culture may be quite low.
(5) A rich medium, such as Ham's F12, is preferable to a simple medium, such as Eagle's MEM, and, if serum is required, fetal bovine often gives better survival than does calf or horse. Isolation of specific cell types will probably require selective media (see Section 10.2.1 and Chapter 23).
(6) Embryonic tissue disaggregates more readily, yields more viable cells, and proliferates more rapidly in primary culture than does adult tissue.

12.2   ISOLATION OF THE TISSUE

Before attempting to work with human or animal tissue, make sure that your work fits within medical ethical rules or current legislation on experimentation with animals (*see* Section 7.9.1). For example, in the United Kingdom, the use of embryos or fetuses beyond 50% gestation or incubation is regulated under the Animal Experiments (Scientific Procedures) Act of 1986. Work with human biopsies or fetal material usually requires the consent of the local ethical committee and the patient and/or his or her relatives (*see* Section 7.9.2).

Δ *Safety Note.* Work with human tissue should be carried out at Containment Level 2 in a Class II biological safety cabinet (*see* Section 7.8.3).

An attempt should be made to sterilize the site of the resection with 70% alcohol if the site is likely to be contaminated (e.g., skin). Remove the tissue aseptically and transfer it to the tissue culture laboratory in dissection BSS (DBSS) or transport medium (*see* Appendix I) as soon as possible. Do not dissect animals in the tissue culture laboratory, as the animals may carry microbial contamination. If a delay in transferring the tissue is unavoidable, it can be held at 4°C for up to 72 h, although a better yield will usually result from a quicker transfer.

12.2.1   Mouse Embryo

Mouse embryos are a convenient source of cells for undifferentiated fibroblastic cultures. They are often used as feeder layers (*see* Fig. 14.2.3).

PROTOCOL 12.1. ISOLATION OF MOUSE EMBRYOS

Outline
Remove uterus aseptically from a timed pregnant mouse and dissect out embryos.

Materials
Sterile:
- ❏ DBSS: Dissection BSS (BSS with a high concentration of antibiotics; *see* Appendix I) in 25- to 50-mL screw-capped tube or universal container
- ❏ BSS, 50 mL in a sterile beaker (used to cool instruments after flaming)
- ❏ Petri dishes, 9 cm
- ❏ Pointed forceps
- ❏ Pointed scissors
Nonsterile:
- ❏ Small laminar-flow hood
- ❏ Timed pregnant mice (*see* Step 1 of this protocol)
- ❏ Alcohol, 70%, in wash bottle
- ❏ Alcohol, 70%, to sterilize instruments (*see* Fig. 7.4)
- ❏ Bunsen burner

Δ *Safety Note.* When sterilizing instruments by dipping them in alcohol and flaming them, take care not to return the instruments to alcohol while they are still alight!

Protocol

1. *Induction of estrus.* If males and females are housed separately, then put together for mating, estrus will be induced in the female 3 days later, when the maximum number of successful matings will occur. This process enables the planned production of embryos at the appropriate time. The timing of successful matings may be determined by examining the vaginas each morning for a hard, mucous plug.

2. *Dating the embryos.* The day of detection of a vaginal plug, or the "plug date," is noted as day zero, and the development of the embryos is timed from this date. Full term is about 19–21 days. The optimal age for preparing cultures from a whole disaggregated embryo is around 13 days, when the embryo is relatively large (*Figs.* 12.1, 12.2) but still contains a high proportion of undifferentiated mesenchyme, which is the main source of the culture. However, isolation and handling embryos beyond 50% full-term may require a license (e.g., in the United Kingdom) so 9- or 10-day embryos may be preferable. Although the amount of tissue recovered from these embryos will be substantially less, a higher proportion of the cells will grow. Most individual organs, with the exception of the brain and the heart, begin to form at about the 9th day of gestation, but are difficult to isolate until about the 11th day. Dissection of individual organs is easier at 13–14 days, and most of the organs are completely formed by the 18th day.

3. Kill the mouse by cervical dislocation (U.K. Schedule I procedure), and swab the ventral surface liberally with 70% alcohol (Fig. 12.3a).

4. Tear the ventral skin transversely at the median line just over the diaphragm (Fig. 12.3b), and, grasping the skin on both sides of the tear, pull in opposite directions to expose the untouched ventral surface of the abdominal wall (Fig. 12.3c).

5. Cut longitudinally along the median line of the exposed abdomen with sterile scissors, revealing the viscera (Fig. 12.3d). At this stage, the uterus, filled with embryos, is obvious in the posterior abdominal cavity (Fig. 12.3e).

6. Dissect out the uteri into a 25-mL or 50-mL screw-capped vial containing 10 or 20 mL DBSS (Fig. 12.3f).

Note. All of the preceding steps should be done outside the tissue culture laboratory; a small laminar-flow hood

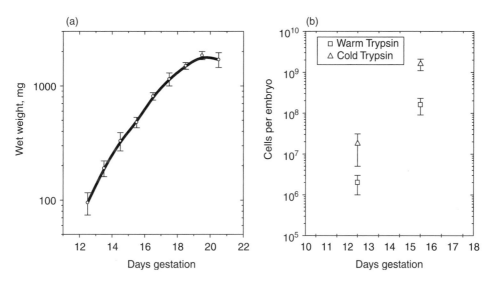

Fig. 12.1. Total Wet Weight and Yield of Cells per Mouse Embryo. (a) Total wet weight of embryo without placenta or membranes, mean ± standard deviation [from Paul et al., 1969]. (b) Cell yield per embryo after incubation in 0.25% trypsin at 37°C for 4 h with no intermediate harvesting (squares) or after soaking in 0.25% trypsin at 4°C for 5 h and incubation at 37°C for 30 min (triangles; Protocol 12.6).

and rapid technique will help to maintain sterility. Do not take live animals into the tissue culture laboratory, as the animals may carry contamination. If an animal carcass must be handled in the tissue culture area, make sure that the carcass is immersed in alcohol briefly, or thoroughly swabbed, and disposed of quickly after use.

7. Take the intact uteri to the tissue culture laboratory, and transfer them to a fresh Petri dish of sterile DBSS (Fig. 12.3g).
8. Dissect out the embryos:
   (a) Tear the uterus with two pairs of sterile forceps, keeping the points of the forceps close together to avoid distorting the uterus and bringing too much pressure to bear on the embryos (Fig. 12.3g,h).
   (b) Free the embryos from the membranes (Fig. 12.3i) and placenta and place them to one side of the dish to bleed.
9. Transfer the embryos to a fresh Petri dish. If a large number of embryos is required (i.e., more than four or five litters), it may be helpful to place the dish on ice (for subsequent dissection and culture; see Protocols 12.4–12.8).

12.2.2 Chick Embryo

Chick embryos are easier to dissect, as they are larger than mouse embryos at the equivalent stage of development. Like mouse embryos, chick embryos are used to provide predominantly mesenchymal cell primary cultures for cell proliferation analysis, to provide feeder layers, and as a substrate for viral propagation. Because of their larger size, it is easier to dissect out individual organs to generate specific cell types, such as hepatocytes, cardiac muscle, and lung epithelium. As with mouse embryos, the use of chick embryos may be subject to animal legislation (e.g., in the United Kingdom) and working with embryos that are more than half-term may require a license.

PROTOCOL 12.2. ISOLATION OF CHICK EMBRYOS

Outline
Remove embryo aseptically from the egg and transfer to dish.

Materials
Sterile:
☐ DBSS: Dissection BSS (BSS with a high concentration of antibiotics; *see* Appendix I) in 25- to 50-mL screw-capped tube or universal container
☐ BSS, 50 mL in a sterile beaker (used to cool instruments after flaming)
☐ Small beaker, 20–50 mL or egg cup
☐ Forceps, straight and curved
☐ Petri dishes, 9 cm
Nonsterile:
☐ Embryonated eggs, 10th day of incubation
☐ Alcohol, 70%

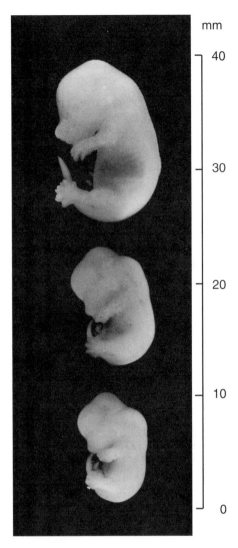

mm

— 40

— 30

— 20

— 10

— 0

Fig. 12.2. Mouse Embryos. Embryos from the 12th, 13th, and 14th days of gestation. The 12-day embryo (bottom) came from a small litter (three) and is larger than would normally be found at this stage.

❑ Swabs
❑ Humid incubator (no additional CO$_2$ above atmospheric level)

Protocol
1. Incubate the eggs at 38.5°C in a humid atmosphere, and turn the eggs through 180° daily. Although hens' eggs hatch at around 20 to 21 days, the lengths of their developmental stages are different from those of mouse embryos. For a culture of dispersed cells from the whole embryo, the egg should be taken at about 8 day, and for isolated-organ rudiments, at about 10–13 day. (10 days is the maximum in the United Kingdom without a license.)

2. Swab the egg with 70% alcohol, and place it with its blunt end facing up in a small beaker (Fig. 12.4a).
3. Crack the top of the shell (Fig. 12.4b), and peel the shell off to the edge of the air sac with sterile forceps (Fig. 12.4c).
4. Resterilize the forceps (i.e., dip them in alcohol, burn off the alcohol, and cool the forceps in sterile BSS), and then use the forceps to peel off the white shell membrane to reveal the chorioallantoic membrane (CAM) below, with its blood vessels (Fig. 12.4d,e).
5. Pierce the CAM with sterile curved forceps (Fig. 12.4f), and lift out the embryo by grasping it gently under the head (Fig. 12.4g,h). Do not close the forceps completely, or else the neck will sever; place the middle digit under the forceps and use the finger pad to restrict the pressure of the forefinger (*see* Fig. 12.4g).
6. Transfer the embryo to a 9-cm Petri dish containing 20 mL DBSS (Fig. 12.4i). (For subsequent dissection and culture, *see* Protocol 12.7.)

12.2.3 Human Biopsy Material

Handling human biopsy material presents certain problems that are not encountered with animal tissue. It usually is necessary to obtain consent (1) from the hospital ethical committee, (2) from the attending physician or surgeon, and (3) from the donor or patient or the patient's relatives (*see* Section 7.9.2). Furthermore, biopsy sampling is usually performed for diagnostic purposes, and hence the needs of the pathologist must be met first. This factor is less of a problem if extensive surgical resection or nonpathological tissue (e.g., placenta or umbilical cord) is involved.

The operation is often performed by one of the resident staff at a time that is not always convenient to the tissue culture laboratory, so some formal collection or storage system must be employed for times when you or someone on your staff cannot be there. If delivery to your lab is arranged, then there must be a system for receiving specimens, recording details of the source, tissue of origin, pathology, etc. (*see* Section 12.3.11), and alerting the person who will perform the culture that the specimens have arrived; otherwise, valuable material may be lost or spoiled.

△ *Safety Note.* Human biopsy material carries a risk of infection (*see* Section 7.8.3), so it should be handled under Containment Level 2 in a Class II biohazard cabinet, and all media and apparatus must be disinfected after use by autoclaving or immersion in a suitable disinfectant (*see* Section 7.8.5). The tissue should be screened for adventitious infections such as hepatitis, HIV, and tuberculosis [in the United States, U.S. Department of Health and Human

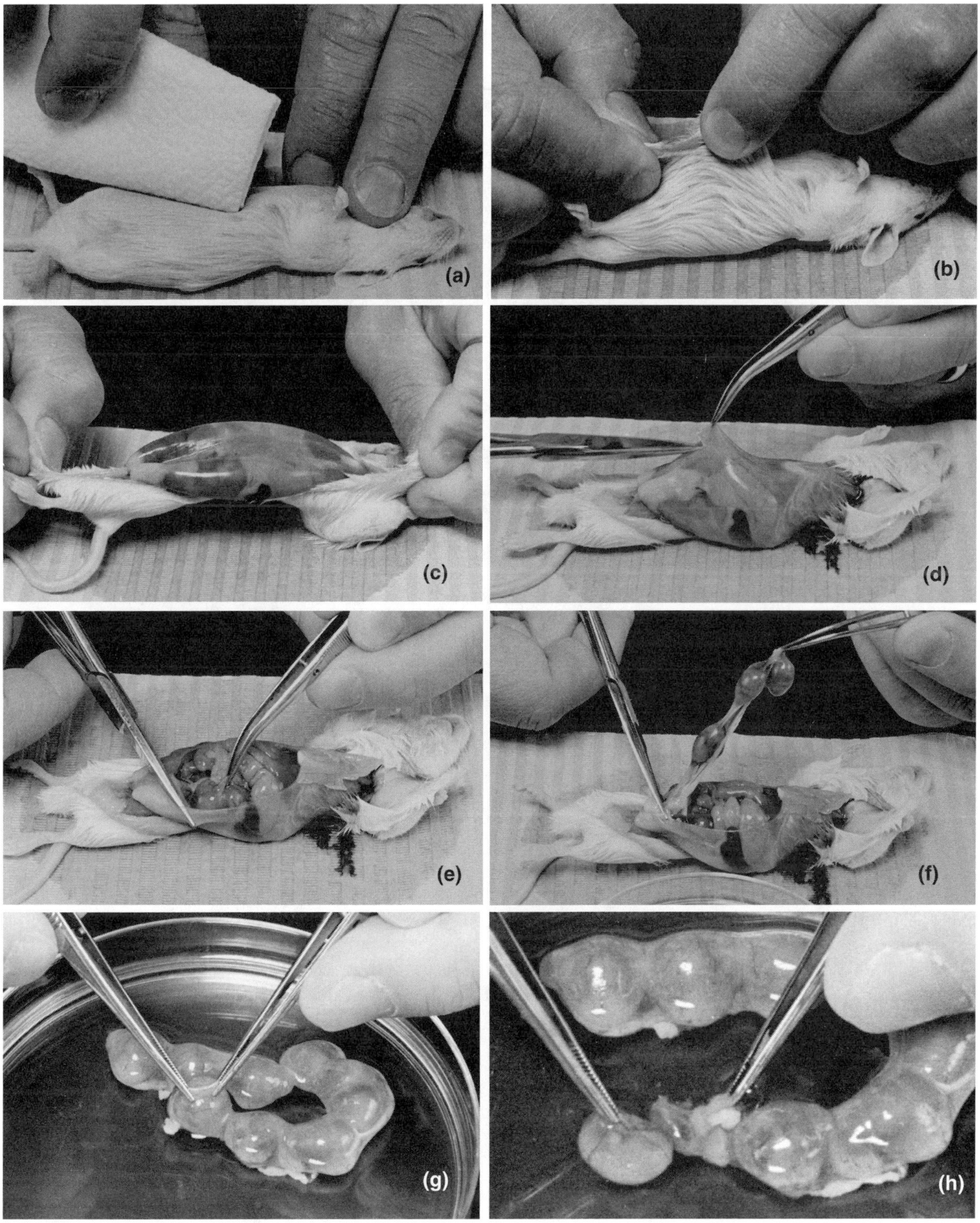

Fig. 12.3. Mouse Dissection. Stages in dissection of a pregnant mouse for the collection of embryos (*see* Protocol 12.1). (a) Swabbing the abdomen. (b), (c) Tearing the skin to expose the abdominal wall. (d) Opening the abdomen. (e) Revealing the uterus *in situ*. (f) Removing the uterus. (g), (h) Dissecting the embryos from the uterus. (i) Removing the membranes. (j) Removing the head (optional). (k) Chopping the embryos. (l) Transferring pieces to trypsinization flask (for warm trypsinization; *see* Protocol 12.5). (m) Transferring the pieces to a small Erlenmeyer flask (for cold trypsinization; *see* Protocol 12.6). (n) Flask on ice.

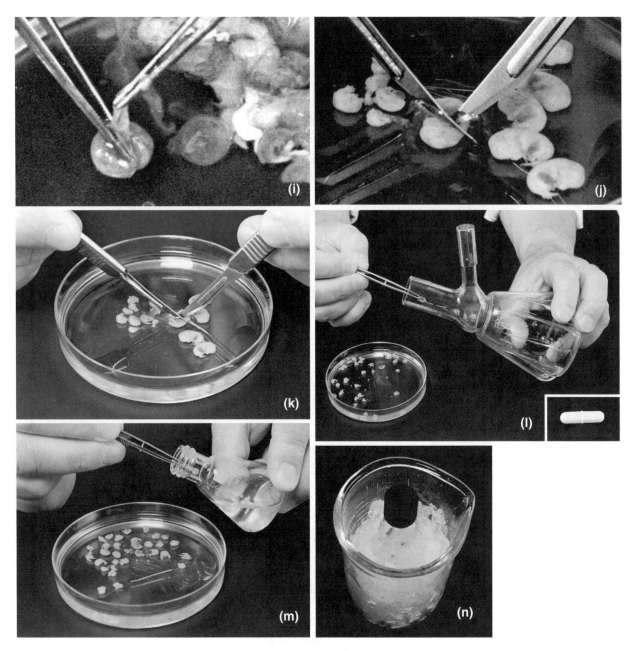

Fig. 12.3. (*Continued*)

Services, 1993; in the UK, Advisory Committee on Dangerous Pathogens, 1995b] unless the patient has already been tested for these infections.

PROTOCOL 12.3. HUMAN BIOPSIES

Outline
Consult with hospital staff, provide labeled container(s) of medium, and arrange for collection of samples from operating room or pathologist.

Materials
❑ Specimen tubes (15–30 mL) with leakproof caps about one-half full with culture medium containing antibiotics (*see* Appendix I: Collection Medium) and labeled with your name, address, and telephone number.

Protocol
1. Provide containers of collection medium, clearly labeled, to the anteroom of the operating theater or to the pathology laboratory.

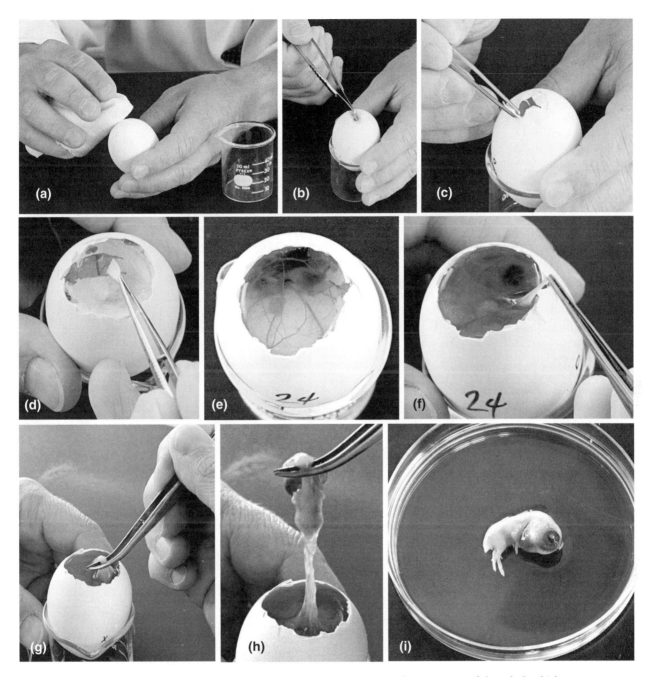

Fig. 12.4. Removing a Chick Embryo from an Egg. Stages in the extraction of the whole chick embryo from an egg. (a) Swabbing the egg with alcohol. (b) Cracking the shell. (c) Peeling off the shell. (d) Peeling off the shell membrane. (e) Chorioallantoic membrane (CAM) and vasculature revealed. (f) Removing CAM with forceps. (g) Grasping the embryo round the neck. (h) Withdrawing the embryo from the egg. (i) Isolated 10-day embryo in Petri dish.

2. Make arrangements to be alerted when the material is ready for collection.
3. Collect the containers after surgery, or have someone send them to you immediately after collection and inform you when they have been dispatched.

4. Transfer the sample to the tissue culture laboratory. The sample should be triple wrapped (e.g., in a sealed tube within a sealed plastic bag full of absorbent tissue, in case of leakage, within a padded envelope with your name, address, and telephone number on it; *see* Section 20.4).

Usually, if kept at 4°C, biopsy samples survive for at least 24 h and even up to 3 or 4 days, although the longer the time from surgery to culture, the more the samples are likely to deteriorate.

5. Log receipt of sample as a numbered entry in a hand-written record book for subsequent transfer to a computerized database, or key into database directly on receipt.

6. *Decontamination.* Although most surgical specimens are sterile when removed, problems may arise with subsequent handling. Superficial specimens (e.g., skin biopsies, melanomas, etc.) and gastrointestinal tract specimens are particularly prone to contamination even when a disinfectant wash is given before skin biopsy and a parenteral antibiotic is given before gastrointestinal surgery. It may be advantageous to consult a medical microbiologist to determine which flora to expect in a given tissue and then choose your antibiotics for collection and dissection accordingly. If the surgical sample is large enough (i.e., 200 mg or more), then a brief dip (i.e., 30 s–1 min) in 70% alcohol will help to reduce superficial contamination without harming the center of the tissue sample.

12.3 PRIMARY CULTURE

Several techniques have been devised for the disaggregation of tissue isolated for primary culture. These techniques can be divided into (1) purely mechanical techniques, involving dissection with or without some form of maceration, and (2) techniques utilizing enzymatic disaggregation (Fig. 12.5). Primary explants are suitable for very small amounts of tissue; enzymatic disaggregation gives a better yield when more tissue is available, and mechanical disaggregation works well with soft tissues, and some firmer tissues when the size of the viable yield is not important.

12.3.1 Primary Explant

The primary explant technique was the original method developed by Harrison [1907], Carrel [1912], and others for initiating a tissue culture. As originally performed, a fragment of tissue was embedded in blood plasma or lymph, mixed with heterologous serum and embryo extract, and placed on a coverslip that was inverted over a concavity slide. The clotted plasma held the tissue in place, and the explant could be examined with a conventional microscope. The heterologous serum induced clotting of the plasma, and the embryo extract and serum, together with the plasma, supplied nutrients and growth factors and stimulated cell migration from the explant. This technique is still used but has been largely replaced by the simplified method described in Protocol 12.4.

PROTOCOL 12.4. PRIMARY EXPLANTS

Outline
The tissue is chopped finely and rinsed, and the pieces are seeded onto the surface of a culture flask or Petri dish in a small volume of medium with a high concentration (i.e., 40–50%) of serum, such that surface tension holds the pieces in place until they adhere spontaneously to the surface (Fig. 12.6a). Once this is achieved, outgrowth of cells usually follows (Fig. 12.6b; Plates 1a, 2b).

Materials
Sterile:
- ☐ Growth medium (e.g., 50:50 DMEM:F12 with 20% fetal bovine serum)
- ☐ 100 mL DBSS
- ☐ Petri dishes, 9 cm, non-tissue-culture grade
- ☐ Forceps
- ☐ Scalpels
- ☐ Pipettes, 10 mL with wide tips
- ☐ Centrifuge tubes, 15 or 20 mL, or universal containers
- ☐ Culture flasks, 25 cm², or tissue-culture-grade Petri dishes, 5–6 cm. The size of flasks and volume of growth medium depend on the amount of tissue: roughly five 25-cm² flasks per 100 mg of tissue.

Protocol

1. Transfer tissue to fresh, sterile DBSS, and rinse.
2. Transfer the tissue to a second dish; dissect off unwanted tissue, such as fat or necrotic material, and transfer to a third dish.
3. Chop finely with crossed scalpels (*see* Fig. 12.6a, top) into about 1-mm cubes.
4. Transfer by pipette (10–20 mL, with wide tip) to a 15- or 50-mL sterile centrifuge tube or universal container. (Wet the inside of the pipette first with BSS or medium, or else the pieces will stick.)
5. Allow the pieces to settle.
6. Wash by resuspending the pieces in DBSS, allowing the pieces to settle, and removing the supernatant fluid. Repeat this step two more times.
7. Transfer the pieces (remember to wet the pipette) to a culture flask, with about 20–30 pieces per 25-cm² flask.
8. Remove most of the fluid, and add 1 mL growth medium per 25-cm² growth surface. Tilt the flask gently to spread the pieces evenly over the growth surface.
9. Cap the flask, and place it in an incubator or hot room at 37°C for 18–24 h.

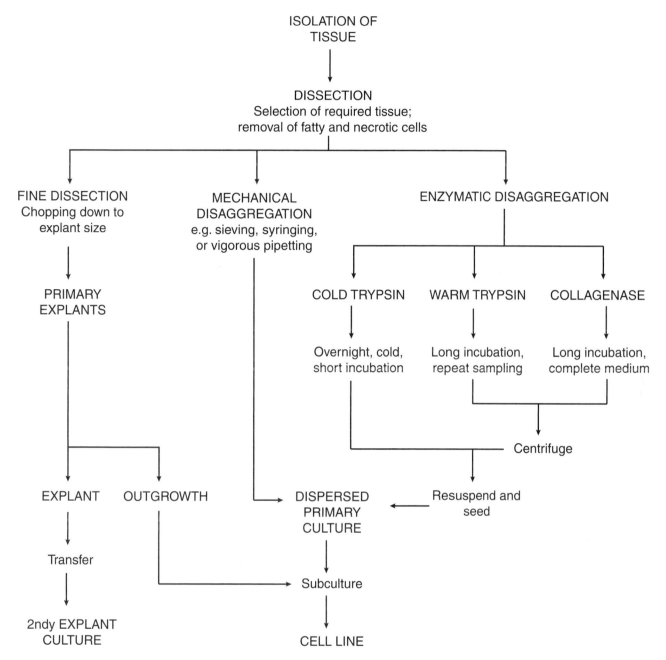

*Fig. 12.5. **Options for Primary Culture.*** Multiple paths to obtaining a cell line; center and left, by mechanical disaggregation, right, by enzymatic disaggregation. An explant may be transferred to allow further outgrowth to form, while the outgrowth from the explant may be subcultured to form a cell line.

10. If the pieces have adhered, then the medium volume may be made up gradually over the next 3–5 days to 5 mL per 25 cm² and then changed weekly until a substantial outgrowth of cells is observed (*see* Fig. 12.6b).

11. Once an outgrowth has formed, the remaining explant may be picked off with a scalpel

(Fig. 12.6c) and transferred by prewetted pipette to a fresh culture vessel. (Then return to step 7.)

12. Replace the medium in the first flask until the outgrowth has spread to cover at least 50% of the growth surface, at which point the cells may be subcultured (*see* Protocol 13.2).

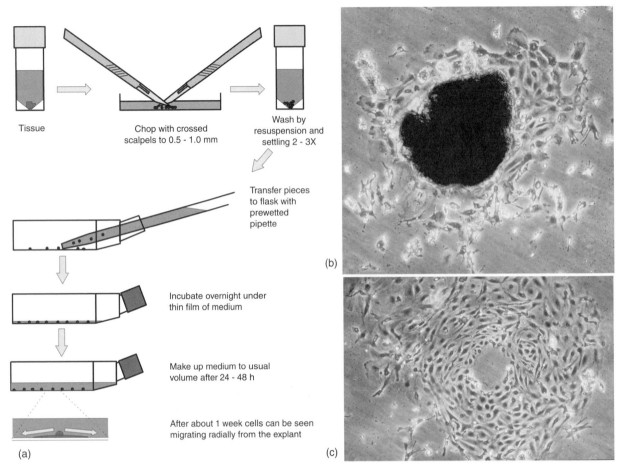

Fig. 12.6. Primary Explant Culture. (a) Schematic diagram of stages in dissection and seeding primary explants. (b) Primary explant culture from mouse squamous skin carcinoma; explant and early stage of outgrowth about 3 days after explantation (*see also* Plate 2b). (c) Outgrowth after removal of explant, about 7 days after explantation. 10× objective.

This technique is particularly useful for small amounts of tissue, such as skin biopsies, for which there is a risk of losing cells during mechanical or enzymatic disaggregation. Its disadvantages lie in the poor adhesiveness of some tissues and the selection of cells in the outgrowth. In practice, however, most cells, particularly embryonic, migrate out successfully.

Attaching explants. Both adherence and migration may be stimulated by placing a glass coverslip on top of the explant, with the explant near the edge of the coverslip, or the plastic dish may be scratched through the explant to attach the tissue to the flask [Elliget & Lechner, 1992] (*see* Protocol 23.9). Attachment may also be promoted by treating the plastic with polylysine or fibronectin (*see* Sections 8.4.11, 10.4.5, 14.2.1), extracellular matrix (*see* Protocol 8.1), or feeder layers (*see* Protocol 14.3). Historically, plasma clots have been used to promote attachment. Place a drop of plasma on the plastic surface, and embed the explant in it. This should induce the plasma to clot in a few minutes, whereupon medium can be added. Alternatively, purified fibrinogen and thrombin can be used [Nicosia & Ottinetti, 1990].

12.3.2 Enzymatic Disaggregation

Cell−cell adhesion in tissues is mediated by a variety of homotypic interacting glycopeptides (cell adhesion molecules, or CAMs) (*see* Section 3.2.1), some of which are calcium dependent (cadherins) and hence are sensitive to chelating agents such as EDTA or EGTA. Integrins, which bind to the arginine-glycine-aspartic acid (RGD) motif in extracellular matrix, also have Ca^{2+}-binding domains and are affected by Ca^{2+} depletion. Intercellular matrix and basement membranes contain other glycoproteins, such as fibronectin and laminin, which are protease sensitive, and proteoglycans, which are less so but can sometimes be degraded by glycanases, such as hyaluronidase or heparinase. The easiest approach is to proceed from a simple disaggregation solution to a more complex solution (*see* Table 13.4) with trypsin alone or trypsin/EDTA as a starting point, adding other proteases to improve disaggregation, and deleting trypsin if necessary to increase viability. In general, increasing the purity of an enzyme will give better control and less toxicity with increased specificity but may result in less disaggregation activity.

Mechanical and enzymatic disaggregation of the tissue avoids problems of selection by migration and yields a higher number of cells that are more representative of the whole tissue in a shorter time. However, just as the primary explant technique selects on the basis of cell migration, dissociation techniques will select protease- and mechanical stress-resistant cells.

Embryonic tissue disperses more readily and gives a higher yield of proliferating cells than newborn or adult tissue. The increasing difficulty in obtaining viable proliferating cells with increasing age is due to several factors, including the onset of differentiation, an increase in fibrous connective tissue and extracellular matrix, and a reduction of the undifferentiated proliferating cell pool. When procedures of greater severity are required to disaggregate the tissue (e.g., longer trypsinization or increased agitation), the more fragile components of the tissue may be destroyed. In fibrous tumors, for example, it is very difficult to obtain complete dissociation with trypsin while still retaining viable carcinoma cells.

The choice of which trypsin grade to use has always been difficult, as there are two opposing trends: (1) The purer the trypsin, the less toxic it becomes, and the more predictable its action; (2) the cruder the trypsin, the more effective it may be, because of other proteases. In practice, a preliminary test experiment may be necessary to determine the optimum grade for viable cell yield, as the balance between sensitivity to toxic effects and disaggregation ability may be difficult to predict.

Crude trypsin is by far the most common enzyme used in tissue disaggregation [Waymouth, 1974], as it is tolerated quite well by many cells and is effective for many tissues. Residual activity left after washing is neutralized by the serum of the culture medium, or by a trypsin inhibitor (e.g., soya bean trypsin inhibitor) when serum-free medium is used.

12.3.3 Warm Trypsin

It is important to minimize the exposure of cells to active trypsin in order to preserve maximum viability. Hence, when whole tissue is being trypsinized at 37°C, dissociated cells should be collected every half hour, and the trypsin should be removed by centrifugation and neutralized with serum in medium.

PROTOCOL 12.5. TISSUE DISAGGREGATION IN WARM TRYPSIN

Outline
The tissue is chopped and stirred in trypsin for a few hours. The dissociated cells are collected every half hour, centrifuged, and pooled in medium containing serum (Fig. 12.7).

Materials
Sterile or aseptically prepared:
- ❑ Tissue, 1–5 g
- ❑ DBSS, 50 mL (*see* Appendix I)
- ❑ Trypsin (crude), 2.5% in D-PBSA or normal saline
- ❑ D-PBSA, 200 mL
- ❑ Growth medium with serum (e.g., DMEM/F12 with 10% fetal bovine serum)
- ❑ Culture flasks, 5–10 flasks per g tissue (varies depending on cellularity of tissue)
- ❑ Petri dishes, 9 cm, non-tissue-culture grade
- ❑ Preweighed vials, 50-mL centrifuge tubes, or universal containers, 2
- ❑ Trypsinization flask: 250-mL Erlenmeyer flask (preferably indented as in Fig. 12.9) or stirrer flask (*see* Fig. 13.5)
- ❑ Magnetic follower, autoclaved in a test tube
- ❑ Curved forceps
- ❑ Pipettes (Pasteur, 2 mL, 10 mL)

Nonsterile:
- ❑ Magnetic stirrer
- ❑ Hemocytometer or cell counter

Protocol
1. Transfer the tissue to fresh, sterile DBSS in 9-cm Petri dish, and rinse.
2. Transfer the tissue to a second dish; dissect off unwanted tissue, such as fat or necrotic material, and transfer to a third dish.
3. Chop with crossed scalpels (*see* Fig. 12.7) into about 3-mm cubes.
4. Transfer the tissue with curved forceps to the preweighed vial or tube.
5. Allow the pieces to settle.
6. Wash the tissue by resuspending the pieces in DBSS, allowing the pieces to settle, and removing the supernatant fluid. Repeat this step two more times.
7. Drain the vial or tube and reweigh.
8. Transfer all the pieces to the empty trypsinization flask, flushing the vial or tube with DBSS.
9. Remove most of the residual fluid, and add 180 mL of D-PBSA.
10. Add 20 mL of 2.5% trypsin. (Other enzymes, e.g., collagenase, hyaluronidase, or DNase, may be added at this stage as well, if required.)
11. Add the magnetic follower to the flask.
12. Cap the flask, and place it on the magnetic stirrer in an incubator or hot room at 37°C.
13. Stir at about 100 rpm for 30 min at 37°C.
14. After 30 min, collect disaggregated cells as follows:
    (a) Allow the pieces to settle.

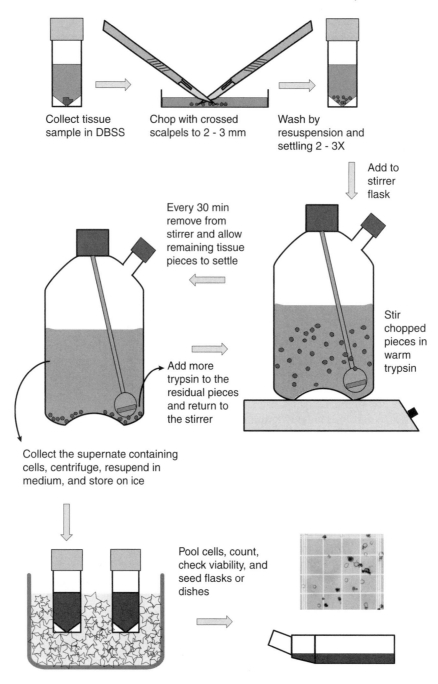

Collect tissue
sample in DBSS

Chop with crossed
scalpels to 2 - 3 mm

Wash by
resuspension and
settling 2 - 3X

Add to
stirrer
flask

Stir
chopped
pieces in
warm
trypsin

Every 30 min
remove from
stirrer and allow
remaining tissue
pieces to settle

Add more
trypsin to the
residual pieces
and return to
the stirrer

Collect the supernate containing
cells, centrifuge, resupend in
medium, and store on ice

Pool cells, count,
check viability, and
seed flasks or
dishes

Fig. 12.7. Warm Trypsin Disaggregation.

(b) Pour off the supernatant into a centrifuge tube and place it on ice.

(c) Add fresh trypsin to the pieces remaining in the flask, and continue to stir and incubate for a further 30 min.

(d) Centrifuge the harvested cells from step 11(b) at approximately 500 *g* for 5 min.

(e) Resuspend the resulting pellet in 10 mL of medium with serum, and store the suspension on ice.

15. Repeat step 11 until complete disaggregation occurs or until no further disaggregation is apparent (usually 3–4 h).

16. Collect and pool chilled cell suspensions, count the cells by hemocytometer or electronic cell

counter (*see* Section 21.1), and check viability (*see* Section 22.3.1).

17. As the cell population will be very heterogeneous, electronic cell counting will require confirmation with a hemocytometer, because calibration can be difficult.

18. Remove any large remaining aggregates by filtering through sterile muslin or a proprietary sieve (*see*, e.g., Fig. 12.8).

19. Dilute the cell suspension to 1×10^6/mL in growth medium, and seed as many flasks as are required, with approximately 2×10^5 cells/cm². When the survival rate is unknown or unpredictable, a cell count is of little value (e.g., in tumor biopsies, in which the proportion of necrotic cells may be quite high). In this case, set up a range of concentrations from about 5 to 25 mg of tissue per mL.

20. Change the medium at regular intervals (2–4 days as dictated by depression of pH). Check the supernate for viable cells before discarding it, as some cells can be slow to attach or may even prefer to proliferate in suspension.

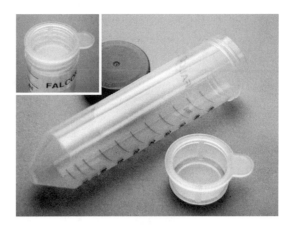

Fig. 12.8. Cell Strainer. Disposable polypropylene filter and tube for straining aggregates from primary suspensions (BD Biosciences). Can also be used for disaggregating soft tissues (*see also* Fig. 12.13).

cells after 30-min incubations in the warm trypsin method rather than have them exposed for the full time (i.e., 3–4 h) required to disaggregate the whole tissue. A simple method of minimizing damage to the cells during exposure is to soak the tissue in trypsin at 4°C for 6–18 h to allow penetration of the enzyme with little tryptic activity (Table 12.1). Following this procedure, the tissue will only require 20–30 min at 37°C for disaggregation [Cole & Paul, 1966].

This technique is useful for the disaggregation of large amounts of tissue in a relatively short time, particularly for whole mouse embryos or chick embryos. It does not work as well with adult tissue, in which there is a lot of fibrous connective tissue, and mechanical agitation can be damaging to some of the more sensitive cell types, such as epithelium. If reaggregation is found after centrifugation and resuspension, incubate in DNase, 10–20 μg/mL, for 10–20 min, and recentrifuge.

12.3.4 Trypsinization with Cold Preexposure

One of the disadvantages of using trypsin to disaggregate tissue is the damage that may result from prolonged exposure to the tissue to trypsin at 37°C; hence the need to harvest

---

PROTOCOL 12.6. TISSUE DISAGGREGATION IN COLD TRYPSIN

Outline
Chop tissue and place in trypsin at 4°C for 6–18 h. Incubate after removing the trypsin, and disperse the cells in warm medium (Fig. 12.9).

Materials
Sterile or aseptically prepared:
- ☐ Tissue, 1–5 g, preweighed
- ☐ Growth medium (e.g., DMEM/F12 with 10% FBS)
- ☐ DBSS

TABLE 12.1. Cell Yield by Warm and Cold Trypsinization

Duration and temperature of trypsinization		After trypsinization			After 24 h in culture	
		Cells recovered per embryo $\times 10^{-7}$	% Viability by dye exclusion (Trypan blue)	Total no. of viable cells $\times 10^{-7}$	Recovered, % of total seeded	Viability, % of viable cells seeded
4°C	37°C					
0 h	4 h	1.69	86	1.45	47.2	54.9
5.5 h	0.5 h	3.32	60	1.99	74.5	124
24 h	0.5 h	3.40	75	2.55	60.3	80.2

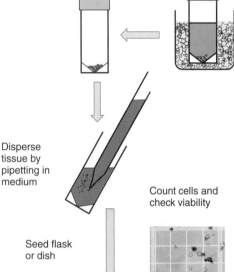

Collect tissue sample in DBSS

Chop with crossed scalpels to 2 - 3 mm

Wash by resuspension and settling 2 - 3X

Remove trypsin and incubate for 20 - 30 min

Replace BSS with trypsin and place on ice overnight

Disperse tissue by pipetting in medium

Count cells and check viability

Seed flask or dish

Fig. 12.9. Cold Trypsin Disaggregation. See also Plate 2a, d, e and Plate 3.

- ❑ 0.25% crude trypsin in serum-free RPMI 1640 or MEM/Stirrer Salts (S-MEM)
- ❑ Petri dishes, 9 cm, non-tissue-culture grade
- ❑ Forceps, straight and curved
- ❑ Scalpels
- ❑ Erlenmeyer flask, 25 or 50 mL, screw capped, preweighed (or glass vial or universal container)
- ❑ Culture flasks, 25 or 75 cm²
- ❑ Pipettes (Pasteur, 2 mL, 10 mL)

Nonsterile:
- ❑ Ice bath

Protocol

1. Transfer the tissue to fresh, sterile DBSS in a 9-cm Petri dish, and rinse.

2. Transfer the tissue to a second dish and dissect off unwanted tissue, such as fat or necrotic material.

3. Transfer to a third dish and chop with crossed scalpels (*see* Fig. 12.9) into about 3-mm cubes. Embryonic organs, if they do not exceed this size, are better left whole.

4. Transfer the tissue with curved forceps to a 15- or 50-mL preweighed sterile vial.

5. Allow the pieces to settle.

6. Wash the tissue by resuspending the pieces in DBSS, allowing the pieces to settle, and removing the supernatant fluid. Repeat this step two more times.

7. Carefully remove the residual fluid and reweigh the vial.

8. Add 10 mL/g of tissue of 0.25% trypsin in RPMI 1640 or S-MEM at 4°C.

9. Place the mixture at 4°C for 6–18 h.

10. Remove and discard the trypsin carefully, leaving the tissue with only the residual trypsin. (Other enzymes, e.g., collagenase, hyaluronidase, or DNase, may be added in 1- to 2-mL amounts at this stage, if required.)

11. Place the tube at 37°C for 20–30 min.

12. Add warm medium, approximately 1 mL for every 100 mg of original tissue, and gently pipette the mixture up and down until the tissue is completely dispersed.

13. If some tissue does not disperse, then the cell suspension may be filtered through sterile muslin or stainless steel mesh (100–200 μm), or a disposable plastic mesh strainer (Fig. 12.10), or the larger pieces may simply be allowed to settle. When there is a lot of tissue, increasing the volume of suspending medium to 20 mL for each gram of tissue will facilitate settling and subsequent collection of supernatant fluid. Two to three minutes should be sufficient to get rid of most of the larger pieces.

14. Determine the cell concentration in the suspension by hemocytometer or electronic cell counter (*see* Section 21.1), and check viability (*see* Protocol 22.1).

15. The cell population will be very heterogeneous; electronic cell counting will initially require confirmation with a hemocytometer, as calibration can be difficult.

16. Dilute the cell suspension to 1×10^6/mL in growth medium, and seed as many flasks as are required, with approximately 2×10^5 cells/cm². When the survival rate is unknown or unpredictable, a cell count is of little value (e.g., in tumor biopsies, for which the proportion of necrotic cells may be quite high). In this case,

17. Change the medium at regular intervals (2–4 days as dictated by depression of pH). Check the supernate for viable cells before discarding it, as some cells can be slow to attach or may even prefer to proliferate in suspension.

set up a range of concentrations from about 5 to 25 mg of tissue per mL.

The cold trypsin method usually gives a higher yield of viable cells, with improved survival after 24 h culture (*see* Figs. 12.1, 12.10 and Table 12.1), and preserves more different cell types than the warm method (*see* Plates 2d,e and 3). Cultures from mouse embryos contain more epithelial cells when prepared by the cold method, and erythroid cultures from 13-day fetal mouse liver respond to erythropoietin after this treatment, but not after the warm trypsin method or mechanical disaggregation [Cole & Paul, 1966; Conkie, personal communication]. The cold trypsin method is also convenient, as no stirring or centrifugation is required and the incubation at 4°C may be done overnight.

12.3.5 Chick Embryo Organ Rudiments

The cold trypsin method is particularly suitable for small amounts of tissue, such as embryonic organs. Protocol 12.7

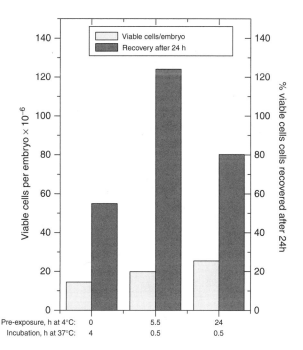

Fig. 12.10. Warm and Cold Trypsinization. Yield of viable cells per embryo by warm and cold trypsinization methods. Yield of viable cells per embryo increases by cold trypsinization up to 24 h at 4°C, but recovery after 24 h culture is greatest with shorter cold trypsinization (>100%, implying cell proliferation), perhaps because some of the cells released by longer cold trypsinization are not proliferative.

gives good reproducible cultures from 10- to 13-day chick embryos with evidence of several different cell types characteristic of the tissue of origin. This protocol forms a good exercise for teaching purposes.

PROTOCOL 12.7. CHICK EMBRYO ORGAN RUDIMENTS

Outline

Dissect out individual organs or tissues, and place them, preferably whole, in cold trypsin overnight. Remove the trypsin, incubate the organs or tissue briefly, and disperse them in culture medium. Dilute and seed the cultures.

Materials
Sterile:
- DBSS
- Crude trypsin (Difco 1:250 or equivalent) 0.25% in RPMI 1640 or S-MEM on ice; lower concentrations may be used with purer grades of trypsin, e.g., 0.05–0.1% Sigma crystalline or Worthington Grade IV
- Culture medium (e.g., DMEM/F12 with 10% FBS), minimum of 12 mL per tissue
- Petri dishes, 9 cm, non-tissue-culture grade
- Culture flasks, 25 cm² (2 per tissue)
- Scalpels (No. 11 blade for most steps)
- Iridectomy knives for fine dissection
- Curved and straight fine forceps
- Pipettes (Pasteur, 2 mL, 10 mL)
- Test tubes, preferably glass, 10–15 mL, with screw caps

Nonsterile:
- Embryonated hen's eggs, 10- to 13-days incubation (>10 days requires license in the United Kingdom)
- Ice bath
- Binocular dissecting microscope

Protocol
1. Remove the embryo from the egg as described previously (*see* Protocol 12.2), and place it in sterile DBSS.
2. Remove the head (Fig. 12.11a,b).
3. Remove an eye and open it carefully, releasing the lens and aqueous and vitreous humors (Fig. 12.11c,d).
4. Grasp the retina in two pairs of fine forceps and gently peel the pigmented retina off the neural retina and connective tissue (Fig. 12.11e). (This step requires a dissection microscope for 10-day embryos. A brief exposure to 0.25% trypsin in

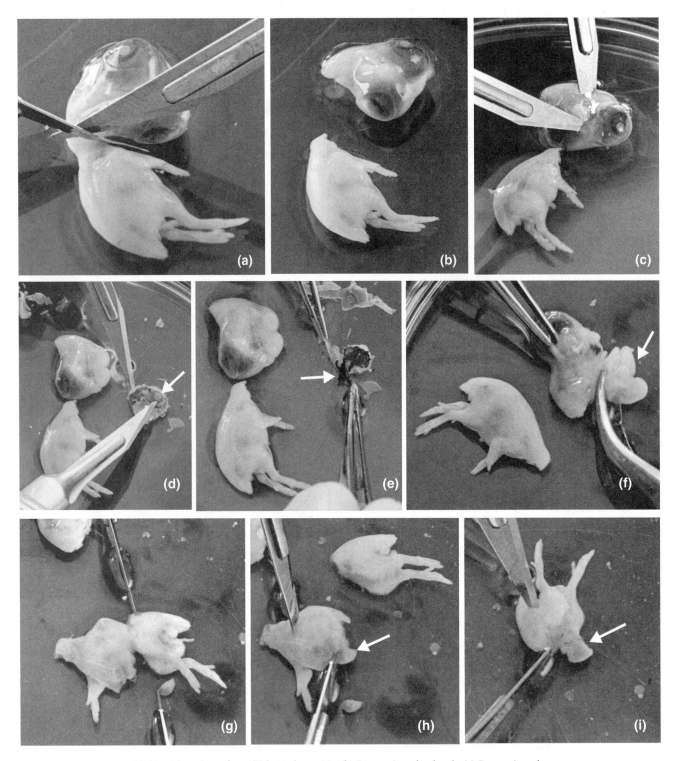

Fig. 12.11. Dissection of a Chick Embryo. (a), (b) Removing the head. (c) Removing the eye. (d) Dissecting out the lens. (e) Peeling off the retina. (f) Scooping out the brain. (g) Halving the trunk. (h) Teasing out the heart and lungs from the anterior half. (i) Teasing out the liver and gut from the posterior half. (j) Inserting the tip of the scalpel between the left kidney and the dorsal body wall. (k) Squeezing out the spinal cord. (l) Peeling the skin off the back of the trunk and hind leg. (m) Slicing muscle from the thigh. (n) Organ rudiments arranged around the periphery of the dish. From the right, clockwise, we have the following organs: brain, heart, lungs, liver, gizzard, kidneys, spinal cord, skin, and muscle.

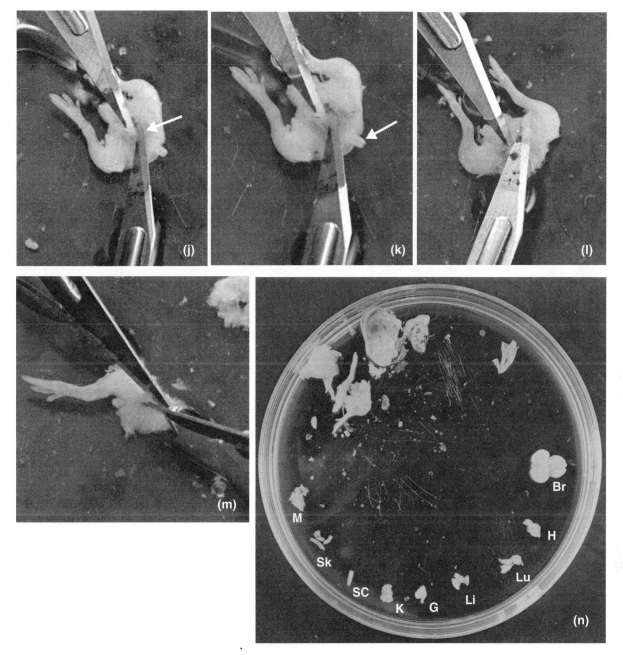

Fig. 12.11. (*Continued*)

4. 1 mM EDTA will allow the two tissues to separate more easily.) Put the tissue to one side.
5. Pierce the top of the head with curved forceps, and scoop out the brain (Fig. 12.11f). Place the brain with the retina at the side of the dish.
6. Halve the trunk transversely where the pink color of the liver shows through the ventral skin (Fig. 12.11g). If the incision is made on the line of the diaphragm, then it will pass between the heart and the liver; but sometimes the liver will go to the anterior instead of the posterior half.

7. Gently probe into the cut surface of the anterior half, and draw out the heart and lungs (Fig. 12.11h; tease the organs out, and do not cut until you have identified them). Separate the heart and lungs and place at the side of the dish.
8. Probe the posterior half, and draw out the liver, with the folds of the gut enclosed in between the lobes (Fig. 12.11i). Separate the liver from the gut and place each at the side of the dish.
9. Fold back the body wall to expose the inside of the dorsal surface of the body cavity in the

posterior half. The elongated lobulated kidneys should be visible parallel to and on either side of the midline.

10. Gently slide the tip of the scalpel under each kidney and tease the kidneys away from the dorsal body wall (Fig. 12.11j). (This step requires a dissection microscope for 10-day embryos.) Carefully cut the kidneys free, and place them on one side.

11. Place the tips of the scalpels together on the midline at the posterior end, and, advancing the tips forward, one over the other, express the spinal cord as you would express toothpaste from a tube (Fig. 12.11k). (This step may be difficult with 10-day embryos.)

12. Turn the posterior trunk of the embryo over, and strip the skin off the back and upper part of the legs (Fig. 12.11l). Collect and place this skin on one side.

13. Dissect off muscle from each thigh, and collect this muscle together (Fig. 12.11m).

14. Transfer all of these tissues, and any others you may want, to separate test tubes containing 1 mL of 0.25% trypsin, and place these tubes on ice. Make sure that the tissue slides right down the tube into the trypsin.

15. Leave the test tubes for 6–18 h at 4°C.

16. Carefully remove the trypsin from the test tubes without disturbing the tissue; tilting and rolling the tube slowly will help.

17. Incubate the tissue in the residual trypsin for 15–20 min at 37°C.

18. Add 4 mL of medium to each of two 25-cm^2 flasks for each tissue to be cultured.

19. Add 2 mL of medium to tubes containing tissues and residual trypsin, and pipette up and down gently to disperse the tissue.

20. Allow any large pieces of tissue to settle.

21. Pipette off the supernatant fluid into the first flask, mix, and transfer 1 mL of diluted suspension to the second flask. This procedure gives two flasks at different cell concentrations and avoids the need to count the cells. Experience will determine the appropriate cell concentration to use in subsequent attempts.

22. Change the medium as required (e.g., for brain, it may need to be changed after 24 h, but pigmented retina will probably last 5–7 days), and check for characteristic morphology and function.

Analysis. After 3–5 days, contracting cells may be seen in the heart cultures, colonies of pigmented cells in the

pigmented retina culture, and the beginning of myotubes in skeletal muscle cultures. Culture may be fixed and stained (*see* Protocol 16.2) for future examination.

12.3.6 Other Enzymatic Procedures

Disaggregation in trypsin can be damaging (e.g., to some epithelial cells) or ineffective (e.g., for very fibrous tissue, such as fibrous connective tissue), so attempts have been made to utilize other enzymes. Because the extracellular matrix often contains collagen, particularly in connective tissue and muscle, collagenase has been the obvious choice [Freshney, 1972 (colon carcinoma); Speirs et al., 1996 (breast carcinoma); Chen et al., 1989 (kidney); Booth & O'Shea, 2002 (gut); Heald et al., 1991 (pancreatic islet cells)] (*see* Sections 12.3.6, 23.2.6–23.2.8). Other bacterial proteases, such as pronase [Schaffer et al., 1997; Glavin et al., 1996] and dispase (Boehringer-Mannheim) [Compton et al., 1998; Inamatsu et al., 1998], have also been used with varying degrees of success. The participation of carbohydrate in intracellular adhesion has led to the use of hyaluronidase [Berry & Friend, 1969] and neuraminidase in conjunction with collagenase. Other proteases continue to appear on the market (*see* Section 12.1). With the selection now available, screening available samples is the only option if trypsin, collagenase, dispase, pronase, hyaluronidase, and DNase, alone and in combinations, do not prove to be successful.

12.3.7 Collagenase

This technique is very simple and effective for many tissues: embryonic, adult, normal, and malignant. It is of greatest benefit when the tissue is either too fibrous or too sensitive to allow the successful use of trypsin. Crude collagenase is often used and may depend, for some of its action, on contamination with other nonspecific proteases. More highly purified grades are available if nonspecific proteolytic activity is undesirable, but they may not be as effective as crude collagenase.

PROTOCOL 12.8. TISSUE DISAGGREGATION IN COLLAGENASE

Outline
Place finely chopped tissue in complete medium containing collagenase and incubate. When tissue is disaggregated, remove collagenase by centrifugation, seed cells at a high concentration, and culture (Fig. 12.12).

Materials
Sterile:
❏ Collagenase (2000 units/mL), Worthington CLS or Sigma 1A

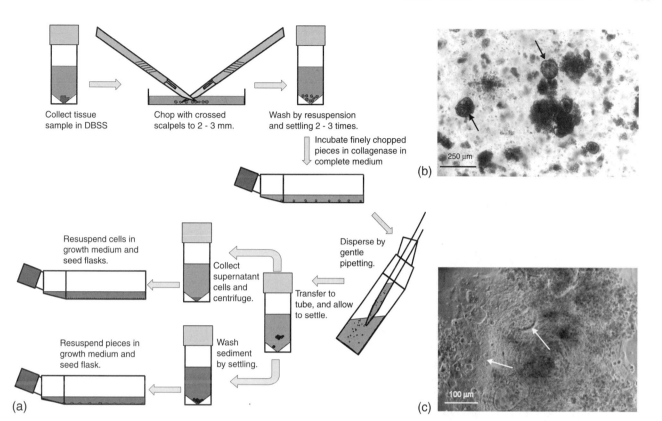

Fig. 12.12. Tissue Disaggregation by Collagenase. (a) Schematic diagram of dissection followed by disaggregation in collagenase. (b) Cell clusters from human colonic carcinoma after 48 h dissociation in crude collagenase (Worthington CLS grade); before removal of collagenase. (c) As (b), but after removal of collagenase, further disaggregation by pipetting, and culture for 48 h. The clearly defined rounded clusters (black arrows) in (b) form epithelium-like sheets (white arrows) in (c), some still three-dimensional, some spreading as a sheet, and the more irregularly shaped clusters produce fibroblasts. (*See also* Plate 2b).

- ☐ Culture medium, e.g., DMEM/F12 with 10% FBS
- ☐ DBSS
- ☐ Pipettes, 1 mL, 10 mL
- ☐ Petri dishes, 9 cm, non-tissue-culture grade
- ☐ Culture flasks, 25 cm^2
- ☐ Centrifuge tubes or universal containers, 15–50 mL, depending on the amount of tissue being processed
- ☐ Scalpels

Nonsterile:
- ☐ Centrifuge

Protocol

1. Transfer the tissue to fresh, sterile DBSS, and rinse.
2. Transfer the tissue to a second dish and dissect off unwanted tissue, such as fat or necrotic material.
3. Transfer to a third dish and chop finely with crossed scalpels (*see* Fig. 12.12) into about 1-mm cubes.
4. Transfer the tissue by pipette (10–20 mL, with wide tip) to a 15- or 50-mL sterile centrifuge tube or universal container. (Wet the inside of the pipette first with DBSS, or else the pieces will stick.)
5. Allow the pieces to settle.
6. Wash the tissue by resuspending the pieces in DBSS, allowing the pieces to settle, and removing the supernatant fluid. Repeat this step two more times.
7. Transfer 20–30 pieces to one 25-cm^2 flask and 100–200 pieces to a second flask.
8. Drain off the DBSS, and add 4.5 mL of growth medium with serum to each flask.
9. Add 0.5 mL of crude collagenase, 2000 units/mL, to give a final concentration of 200 units/mL collagenase.
10. Incubate at 37°C for 4–48 h without agitation. Tumor tissue may be left up to 5 days or more if disaggregation is slow (e.g., in scirrhous carcinomas of the breast or the colon), although it may be necessary to centrifuge the tissue and resuspend it in fresh medium and collagenase

before that amount of time has passed if an excessive drop in pH is observed (i.e., <pH 6.5).

11. Check for effective disaggregation by gently moving the flask; the pieces of tissue will "smear" on the bottom of the flask and, with gentle pipetting, will break up into single cells and small clusters (Fig. 12.13).

12. With some tissues (e.g., lung, kidney, and colon or breast carcinoma), small clusters of epithelial cells can be seen to resist the collagenase and may be separated from the rest by allowing them to settle for about 2 min. If these clusters are further washed with DBSS by resuspension and settling and the sediment is resuspended in medium and seeded, then they will form islands of epithelial cells. Epithelial cells generally survive better if they are not completely dissociated.

13. When complete disaggregation has occurred, or when the supernatant cells are collected after removing clusters by settling, centrifuge the cell suspension from the disaggregate and any washings at 50–100 g for 3 min.

14. Discard the supernatant DBSS or medium, resuspend and combine the pellets in 5 mL of medium, and seed in a 25-cm² flask. If the pH fell during collagenase treatment (to pH 6.5 or less by 48 h), then dilute the suspension two- to threefold in medium after removing the collagenase.

15. Replace the medium after 48 h.

Some cells, particularly macrophages, may adhere to the first flask during the collagenase incubation. Transferring the cells to a fresh flask after collagenase treatment (and subsequent removal of the collagenase) removes many of the macrophages from the culture. The first flask may be cultured as well, if required. Light trypsinization will remove any adherent cells other than macrophages.

Disaggregation in collagenase has proved particularly suitable for the culture of human tumors [Pfragner & Freshney, 2004], mouse kidney, human adult and fetal brain, liver (*see* Protocol 23.6), lung, and many other tissues, particularly epithelium [Freshney & Freshney, 2002]. The process is gentle and requires no mechanical agitation or special equipment. With more than 1 g of tissue, however, it becomes tedious at the dissection stage and can be expensive, because of the amount of collagenase required. It will also release most of the connective tissue cells, accentuating the problem of fibroblastic outgrowth, so it may need to be followed by selective culture (*see* Section 10.2.1) or cell separation (*see* Chapter 15).

The discrete clusters of epithelial cells produced by disaggregation in collagenase (*see* Step 9 of Protocol 12.8) and

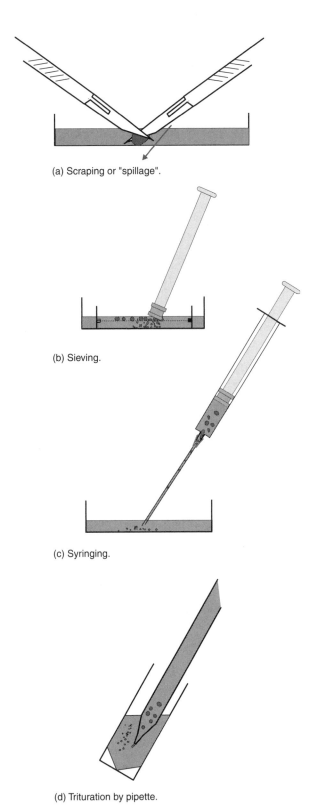

(a) Scraping or "spillage".

(b) Sieving.

(c) Syringing.

(d) Trituration by pipette.

Fig. 12.13. Mechanical Disaggregation. (a) Scraping or "spillage". Cutting action, or abrasion of cut surface, releases cells. (b) Sieving. Forcing tissue through sieve with syringe piston. (Falcon Cell Strainer can be used; *see* Fig. 12.8.) (c) Syringing. Drawing tissue into syringe through wide bore needle or canula and expressing. (d) Trituration by pipette. Pippetting tissue fragments up and down through wide bore pipette.

by the cold trypsin method (*see* Plates 2, 3) can be selected under a dissection microscope and transferred to individual wells in a microtitration plate, alone or with irradiated or mitomycin C-treated feeder cells (*see* Sections 14.2.3, 23.1.1, 23.1.4).

12.3.8 Mechanical Disaggregation

The outgrowth of cells from primary explants is a relatively slow process and can be highly selective. Enzymatic digestion is rather more labor intensive, although, potentially, it gives a culture that is more representative of the tissue. As there is a risk of proteolytic damage to cells during enzymatic digestion, many people have chosen to use the alternative of mechanical disaggregation, e.g., collecting the cells that spill out when the tissue is carefully sliced [Lasfargues, 1973], pressing the dissected tissue through a series of sieves for which the mesh is gradually reduced in size, or, alternatively forcing the tissue fragments through a syringe (with or without a wide-gauge needle) [Zaroff et al., 1961] or simply pipetting it repeatedly (*see* Fig. 12.13). This procedure gives a cell suspension more quickly than does enzymatic digestion but may cause mechanical damage. Scraping ("spillage") (Fig. 12.13a) and sieving (Fig. 12.13b) are probably the gentlest mechanical methods while pipetting (Fig. 12.13d) and, particularly, syringing (Fig. 12.13c), are most likely to generate shear. Protocol 12.9 is one method of mechanical disaggregation that has been found to be moderately successful with soft tissues, such as brain.

PROTOCOL 12.9. MECHANICAL DISAGGREGATION BY SIEVING

Outline
The tissue in culture medium is forced through a series of sieves for which the mesh is gradually reduced in size until a reasonable suspension of single cells and small aggregates is obtained. The suspension is then diluted and cultured directly.

Materials
Sterile:
❑ Growth medium, e.g., DMEM/F12 with 10% FBS
❑ Forceps
❑ Sieve (Fig. 12.13b), or graded series of sieves from 100 μm down to 20 μm, or Falcon "Cell Strainer" (*see* Fig. 12.8)
❑ Petri dishes, 9 cm
❑ Scalpels
❑ Disposable plastic syringes (2 mL or 5 mL)
❑ Culture flasks

Protocol
1. After washing and preliminary dissection of the tissue (*see* Steps 1 and 2 of Protocol 12.5), chop

the tissue into pieces about 3–5 mm across, and place a few pieces at a time into a stainless steel or polypropylene sieve of 1-mm mesh in a 9-cm Petri dish (*see* Fig. 12.13b) or centrifuge tube (*see* Fig. 12.8).
2. Force the tissue through the mesh into medium by applying gentle pressure with the piston of a disposable plastic syringe. Pipette more medium through the sieve to wash the cells through it.
3. Pipette the partially disaggregated tissue from the Petri dish into a sieve of finer porosity, perhaps 100-μm mesh, and repeat Step 2.
4. The suspension may be diluted and cultured at this stage, or it may be sieved further through 20-μm mesh if it is important to produce a single-cell suspension. In general, the more highly dispersed the cell suspension, the higher the sheer stress required and the lower the resulting viability.
5. Seed the culture flasks at 2×10^5, 1×10^6, and 2×10^6 cells/mL by diluting the cell suspension in medium.

Only soft tissues, such as spleen, embryonic liver, embryonic and adult brain, and some human and animal soft tumors, respond well to this technique. Even with brain, for which fairly complete disaggregation can be obtained easily, the viability of the resulting suspension is lower than that achieved with enzymatic digestion, although the time taken may be very much less. When the availability of tissue is not a limitation and the efficiency of the yield is unimportant, it may be possible to produce, in a shorter amount of time, as many viable cells with mechanical disaggregation as with enzymatic digestion, but at the expense of very much more tissue.

12.3.9 Separation of Viable and Nonviable Cells

When an adherent primary culture is prepared from dissociated cells, nonviable cells are removed at the first change of medium. With primary cultures maintained in suspension, nonviable cells are gradually diluted out when cell proliferation starts. If necessary, however, nonviable cells may be removed from the primary disaggregate by centrifuging the cells on a mixture of Ficoll and sodium metrizoate (e.g., Hypaque or Triosil) [Vries et al., 1973]. This technique is similar to the preparation of lymphocytes from peripheral blood (*see* Protocol 27.1). The viable cells collect at the interface between the medium and the Ficoll/metrizoate, and the dead cells form a pellet at the bottom of the tube.

TABLE 12.2. Data Record for Primary Culture

Date . . . . Time . . . . Operator . . . .

		Record
Origin of tissue	Species	
	Race or strain	
	Age	
	Sex	
	Path. no. or animal tag no.	
	Tissue	
	Site	
	Stored tissue/DNA location	
Pathology		
Disaggregation agent	Trypsin, collagenase, etc.	
	Concentration	
	Duration	
	Diluent	
Cell count	Concentration after resuspension (C_I)	
	Volume (V_I)	
	Yield ($Y = C_I \times V_I$)	
	Yield per g (wet weight of tissue)	
Seeding	Number (N) and type of vessel (flask, dish, or plate wells)	
	Final concentration (C_F)	
	Volume per flask, dish, or well (V_F)	
Medium	Type	
	Batch no.	
	Serum type and concentration	
	Batch no.	
	Other additives	
	CO_2 concentration	
Matrix coating	e.g., fibronectin, Matrigel, collagen	
Subculture	Recovery at 1st subculture, cell/flask	
	% (cells recovered ÷ cells seeded)	
	Cell line designation	

PROTOCOL 12.10. ENRICHMENT OF VIABLE CELLS

Outline

Up to 2×10^7 cells in 9 mL of medium may be layered on top of 6 mL of Ficoll-Hypaque in a 25-mL screw-capped centrifuge bottle. The mixture is then centrifuged, and viable cells are collected from the interface.

Materials

Sterile:
- ❏ Cell suspension with as few aggregates as possible
- ❏ Clear centrifuge tubes or universal containers
- ❏ D-PBSA
- ❏ Ficoll Hypaque or equivalent (*see* Appendix II) (Ficoll/metrizoate, adjusted to 1.077 g/cc)
- ❏ Growth medium
- ❏ Syringe with blunt cannula or square-cut needle, Pasteur pipettes, or pipettor

Nonsterile:
- ❏ Hemocytometer or cell counter
- ❏ Centrifuge

Protocol

1. Allow major aggregates in the cell suspension to settle.
2. Layer 9 mL of the cell suspension onto 6 mL of the Ficoll-paque mixture. This step should be done in a wide, transparent centrifuge tube with a cap, such as the 25-mL Sterilin or Nunclon universal container, or in the clear plastic Corning 50-mL tube, using double the aforementioned volumes.
3. Centrifuge the mixture for 15 min at 400 g (measured at the center of the interface).
4. Carefully remove the top layer without disturbing the interface.
5. Collect the interface carefully with a syringe, Pasteur pipette, or pipettor.

Δ *Safety Note.* If you are using human or other primate material, do not use a sharp needle or glass Pasteur pipette.

6. Dilute the mixture to 20 mL in medium (e.g., DMEM/F12/10FB).
7. Centrifuge the mixture at 70 g for 10 min.
8. Discard the supernatant fluid, and resuspend the pellet in 5 mL of growth medium.
9. Repeat Steps 7 and 8 in order to wash cells free of density medium.
10. Count the cells with a hemocytometer or an electronic counter.
11. Seed the culture flask(s).

This procedure can be scaled up or down and works with lower ratios of density medium to cell suspension (e.g., 5 mL of cell suspension over 1 mL of density medium).

12.3.10 Primary Culture in Summary

The disaggregation of tissue and preparation of the primary culture make up the first, and perhaps most vital, stage in the culture of cells with specific functions. If the required cells are lost at this stage, then the loss is irrevocable. Many different cell types may be cultured by choosing the correct techniques (*see* Section 10.2.1 and Chapter 23). In general, trypsin is more severe than collagenase, but is sometimes more effective in creating a single-cell suspension. Collagenase does not dissociate epithelial cells readily, but this characteristic can be an advantage for separating the epithelial cells from stromal cells. Mechanical disaggregation is much quicker than the procedure using collagenase, but damages more cells. The best approach is to try out the techniques described in Protocols 12.4–12.9 and select the method that works best in your system. If none of those methods is successful, try using additional enzymes, such as pronase, dispase, Accutase, and DNase, and consult the literature for examples of previous work with the tissue in which you are interested.

12.3.11 Primary Records

Regardless of the technique used to produce a culture, it is important to keep proper records of the culture's origin and derivation, including the species, sex, and tissue from which it was derived, any relevant pathology, and the procedures used for disaggregation and primary culture (Table 12.2). If you are working under good laboratory practice (GLP; *www.oecd.org/department/0,2688,en_26 49_34381_1_1_1_1_1,00.html*) conditions, then such records are not just desirable, but obligatory [Food and Drug Administration, 1992; Department of Health and Social Security, 1986; Organisation for Economic Co-operation and Development, 2004]. Records can be kept in a notebook or another hard copy file of record sheets, but it is best at this stage to initiate a record in a computer database; this record then becomes the first step in maintaining the *provenance* of the cell line. The database record may never proceed beyond the primary culture stage, but, looking at the opposite extreme, it could be the first stage in creating an accurate record of what will become a valuable cell line. If the record becomes irrelevant, it can always be deleted, but it cannot, with any accuracy, be created later if the cell line assumes some importance.

As it may be necessary to authenticate the origin of a cell line at a later stage in its life history, particularly if cross-contamination is suspected, it is important to save a sample of tissue, blood, or DNA from the same individual at the time of isolation of tissue for the primary culture. This sample can then be used as reference material for DNA fingerprinting (*see* Section 16.6.2) or DNA profiling (*see* Section 16.6.3).

CHAPTER 13

Subculture and Cell Lines

13.1 SUBCULTURE AND PROPAGATION

The first *subculture* represents an important transition for a culture. The need to subculture implies that the primary culture has increased to occupy all of the available substrate. Hence, cell proliferation has become an important feature. Although the primary culture may have a variable growth fraction (*see* Section 21.11.1), depending on the type of cells present in the culture, after the first subculture, the growth fraction is usually high (80% or more). From a very heterogeneous primary culture, containing many of the cell types present in the original tissue, a more homogeneous cell line emerges. In addition to its biological significance, this process has considerable practical importance, as the culture can now be propagated, characterized, and stored, and the potential increase in cell number and the uniformity of the cells open up a much wider range of experimental possibilities (*see* Table 1.5).

13.2 TERMINOLOGY

Once a primary culture is subcultured (or *passaged*), it becomes known as a *cell line*. This term implies the presence of several cell lineages of either similar or distinct phenotypes. If one cell lineage is selected, by cloning (*see* Protocol 14.1), by physical cell separation (*see* Chapter 15), or by any other selection technique, to have certain specific properties that have been identified in the bulk of the cells in the culture, this cell line becomes known as a *cell strain* (*see* Appendix IV). Some commonly used cell lines and cell strains are listed in Table 13.1 (*see also* Table 3.1). If a cell line transforms *in vitro*, it gives rise to a *continuous cell line* (*see* Sections 3.8, 18.4), and if selected or cloned and characterized, it is known as a *continuous cell strain*. It is vital at this stage to confirm the identity of the cell lines and exclude the possibility of cross-contamination; many cell lines in common use are not, in fact, what they are claimed to be, but have been cross-contaminated with HeLa or some other vigorously-growing cell line (Table 13.2). However, continuous cell lines have a number of advantages; the relative advantages and disadvantages of finite cell lines and continuous cell lines are listed in Table 13.3.

The first subculture gives rise to a *secondary* culture, the secondary to a *tertiary*, and so on, although in practice, this nomenclature is seldom used beyond the tertiary culture. In Hayflick's work and others with human diploid fibroblasts [Hayflick & Moorhead, 1961], each subculture divided the culture in half (i.e., the *split ratio* was 1:2), so passage number was the same as generation number. However, they need not be the same. The *passage number* is the number of times that the culture has been subcultured, whereas the *generation number* is the number of doublings that the cell population has undergone, given that the number of doublings in the primary culture is very approximate. When the split ratio is 1:2, as in Hayflick's experiments, the passage number is approximately equal to the generation number. However, if subculture is performed at split ratios greater than 1:2 the generation number, which is the significant indicator of culture age, will increase faster than the passage number based on the number of doublings that the cell population has undergone since the previous

Culture of Animal Cells: A Manual of Basic Technique, Fifth Edition, by R. Ian Freshney
Copyright © 2005 John Wiley & Sons, Inc.

TABLE 13.1. Commonly Used Cell Lines

Cell line	Morphology	Origin	Species	Age	Ploidy	Characteristics	Reference
Finite, from Normal Tissue							
IMR-90	Fibroblast	Lung	Human	Embryonic	Diploid	Susceptible to human viral infection; contact inhibited	Nichols et al., 1977
MRC-5	Fibroblast	Lung	Human	Embryonic	Diploid	Susceptible to human viral infection; contact inhibited	Jacobs, 1970
MRC-9	Fibroblast	Lung	Human	Embryonic	Diploid	Susceptible to human viral infection; contact inhibited	Jacobs, et al., 1979
WI-38	Fibroblast	Lung	Human	Embryonic	Diploid	Susceptible to human viral infection	Hayflick & Moorhead, 1961
Continuous, from Normal Tissue							
293	Epithelial	Kidney	Human	Embryonic	Aneuploid	Readily transfected.	Graham et al., 1977
3T3-A31	Fibroblast		Mouse BALB/c	Embryonic	Aneuploid	Contact inhibited; readily transformed	Aaronson & Todaro, 1968
3T3-L1	Fibroblast		Mouse Swiss	Embryonic	Aneuploid	Adipose differentiation	Green & Kehinde, 1974
BEAS-2B	Epithelial	Lung	Human	Adult			Reddel et al., 1988
BHK21-C13	Fibroblast	Kidney	Syrian hamster	Newborn	Aneuploid	Transformable by polyoma	Macpherson & Stoker, 1962
BRL 3A	Epithelial	Liver	Rat	Newborn		Produce IGF-2	Coon, 1968
C2	Fibroblastoid	Skeletal muscle	Mouse	Embryonic		Myotubes	Morgan et al., 1992
C7	Epithelioid	Hypothalamus	Mouse			Neurophysin; vasopressin	De Vitry et al., 1974
CHO-K1	Fibroblast	Ovary	Chinese hamster	Adult	Diploid	Simple karyotype	Puck et al., 1958
COS-1, COS-7	Epithelioid	Kidney	Pig	Adult		Good hosts for DNA transfection	Gluzman, 1981
CPAE	Endothelial	Pulmonary-artery endothelium	Cow	Adult	Diploid	Factor VIII, Angiotensin II converting enzyme	Del Vecchio & Smith, 1981
HaCaT	Epithelial	Keratinocytes	Human	Adult	Diploid	Cornification	Boukamp et al., 1988
L6	Fibroblastoid	Skeletal muscle	Rat	Embryonic		Myotubes	Richler & Yaffe, 1970
LLC-PKI	Epithelial	Kidney	Pig	Adult	Diploid	Na$^+$-dependent glucose uptake	Hull et al., 1976; Saier, 1984
MDCK	Epithelial	Kidney	Dog	Adult	Diploid	Domes, transport	Gaush et al., 1966; Rindler et al., 1979
NRK49F	Fibroblast	Kidney	Rat	Adult	Aneuploid	Induction of suspension growth by TGF-α,β	De Larco & Todaro, 1978
STO	Fibroblast		Mouse	Embryonic	Aneuploid	Used as feeder layer for embryonal stem cells	Bernstein, 1975
Vero	Fibroblast	Kidney	Monkey	Adult	Aneuploid	Viral substrate and assay	Hopps et al., 1963
Continuous, from Neoplastic Tissue							
A2780	Epithelial	Ovary	Human	Adult	Aneuploid	Chemosensitive with resistant variants	Tsuruo et al., 1986
A549	Epithelial	Lung	Human	Adult	Aneuploid	Synthesizes surfactant	Giard et al., 1972
A9	Fibroblast	Subcutaneous	Mouse	Adult	Aneuploid	Derived from L929; Lacks HGPRT.	Littlefield, 1964b
B16	Fibroblastoid	Melanoma	Mouse	Adult	Aneuploid	Melanin	Nilos & Makarski, 1978
C1300	Neuronal	Neuroblastoma	Rat	Adult	Aneuploid	Neurites	Liebermann & Sachs, 1978
C6	Fibroblastoid	Glioma	Rat	Newborn	Aneuploid	Glial fibrillary acidic protein, GPDH	Benda et al., 1968

TABLE 13.1. Commonly Used Cell Lines (*Continued*)

Cell line	Morphology	Origin	Species	Age	Ploidy	Characteristics	Reference
Caco-2	Epithelial	Colon	Human	Adult	Aneuploid	Transports ions and amino acids	Fogh, 1977
EB-3	Lymphocytic	Peripheral blood	Human	Juvenile	Diploid	EB virus +ve	Epstein & Barr, 1964
Friend	Suspension	Spleen	Mouse	Adult	Aneuploid	Hemoglobin	Scher et al., 1971
GH1, GH2, GH3	Epithelioid	Pituitary tumor	Rat	Adult		Growth hormone	Buonassisi et al., 1962; Yasamura et al., 1966
H4-11-E-C3	Epithelial	Hepatoma	Rat	Adult	Aneuploid	Tyrosine aminotransferase	Pitot et al., 1964
HeLa	Epithelial	Cervix	Human	Adult	Aneuploid	G6PD Type A	Gey et al., 1952
HeLa-S3	Epithelial	Cervix	Human	Adult	Aneuploid	High plating efficiency; will grow well in suspension	Puck & Marcus, 1955
HEP-G2	Epithelioid	Hepatoma	Human	Adult	Aneuploid	Retains some microsomal metabolizing enzymes	Knowles et al., 1980
HL-60	Suspension	Myeloid leukemia	Human	Adult	Aneuploid	Phagocytosis; Neotetrazolium Blue reduction	Olsson & Ologsson, 1981
HT-29	Epithelial	Colon	Human	Adult	Aneuploid	Differentiation inducible with NaBt	Fogh & Trempe, 1975
K-562	Suspension	Myeloid leukemia	Human	Adult	Aneuploid	Hemoglobin	Andersson et al., 1979a,b
L1210	Lymphocytic		Mouse	Adult	Aneuploid	Rapidly growing; suspension	Law et al., 1949
L929	Fibroblast		Mouse	Adult	Aneuploid	Clone of L-cell	Sanford et al., 1948
LS	Fibroblast		Mouse	Adult	Aneuploid	Grow in suspension; derived from L929	Paul & Struthers, personal communication
MCF-7	Epithelial	Pleural effusion from breast tumor	Human	Adult	Aneuploid	Estrogen receptor +ve, domes, α-lactalbumin	Soule et al., 1973
MCF-10	Epithelial	Fibrocystic mammary tissue	Human	Adult	Near diploid	Dome formation	Soule et al., 1990
MOG-G-CCM	Epithelioid	Glioma	Human	Adult	Aneuploid	Glutamyl synthetase	Balmforth et al., 1986
P388D1	Lymphocytic		Mouse	Adult	Aneuploid	Grow in suspension	Dawe & Potter, 1957; Koren et al., 1975
S180	Fibroblast		Mouse	Adult	Aneuploid	Cancer chemotherapy screening	Dunham & Stewart, 1953
SK/HEP-1	Endothelial	Hepatoma, endothelium	Human	Adult	Aneuploid	Factor VIII	Heffelfinger et al., 1992
WEHI-3B D+	Suspension	Marrow	Mouse	Adult	Aneuploid	IL-3 production	Nicola, 1987
ZR-75-1	Epithelial	Ascites fluid from breast tumor	Human	Adult	Aneuploid	ER-ve, EGFr+ve	Engel et al., 1978

subculture (*see* Section 13.7.2). None of these approximations takes account of cell loss through necrosis, apoptosis, or differentiation or premature aging and withdrawal from cycle, which probably take place at every growth cycle between each subculture.

13.3 CULTURE AGE

Cell lines with limited culture life spans are known as *finite* cell lines and behave in a fairly reproducible fashion (*see* Section 3.8.1). They grow through a limited number of

TABLE 13.2. Cross-Contaminated Cell Lines

Cell line	Species	Cell type	Contaminant	Species	Cell type	Source of Data
207	Human	Pre-B leukemia	REH	Human	pre-B leukemia	DSMZ
2474/90	Human	Gastric carcinoma	HT-29	Human	Colorectal carcinoma	DSMZ
2957/90	Human	Gastric carcinoma	HT-29	Human	Colorectal carcinoma	DSMZ
3051/80	Human	Gastric carcinoma	HT-29	Human	Colorectal carcinoma	DSMZ
ADLC-5M2	Human	Lung carcinoma	HELA/-S3	Human	Cervical adenocarcinoma	DSMZ
AV3	Human	Amnion	HeLa	Human	Cervical adenocarcinoma	ATCC
BCC-1/KMC	Human	Basal cell carcinoma	HELA/-S3	Human	Cervical adenocarcinoma	DSMZ
BM-1604	Human	Prostate carcinoma	DU-145	Human	Prostate carcinoma	DSMZ
C16	Human	Fetal lung fibroblast (MRC-5 clone)	HeLa	Human	Cervical adenocarcinoma	ECACC
CHANG liver	Human	Embryonic liver epithelium	HeLa	Human	Cervical adenocarcinoma	ATCC; JCRB
COLO-818	Human	Melanoma	COLO-800	Human	Melanoma	DSMZ
DAMI	Human	Megakaryocytic	HEL	Human	Erythroleukemia	DSMZ
ECV304	Human	Endothelium	T24	Human	Bladder carcinoma	ATCC
ECV304	Human	Normal endothelial	T24	Human	Bladder carcinoma	DSMZ
EJ	Human	Bladder carcinoma	T24	Human	Bladder carcinoma	ATCC; JCRB
EPLC3-2M1	Human	Lung carcinoma	HELA/-S3	Human	Cervical adenocarcinoma	DSMZ; JCRB
EPLC-65	Human	Lung carcinoma	HELA/-S3	Human	Cervical adenocarcinoma	DSMZ
F2-4E5	Human	Thymic epithelium	SK-HEP-1	Human	Hepatoma	DSMZ
F2-5B6	Human	Thymic epithelium	SK-HEP-1	Human	Hepatoma	DSMZ
FL	Human	Amnion	HeLa	Human	Cervical adenocarcinoma	ATCC
GHE	Human	Astrocytoma	T-24	Human	Bladder carcinoma	DSMZ
Girardi Heart	Pig	Adult heart	HeLa	Human	Cervical adenocarcinoma	ATCC
HAG	Human	Adenomatous goitre	T-24	Human	Bladder carcinoma	DSMZ
HEp-2	Human	Adult laryngeal epithelium	HeLa	Human	Cervical adenocarcinoma	ATCC
HMV-1	Human	Melanoma	HeLa-S3	Human	Cervical adenocarcinoma	DSMZ
HuL-1			HeLa	Human	Cervical adenocarcinoma	JCRB
IMC-2	Human	Maxillary carcinoma	HeLa-S3	Human	Cervical adenocarcinoma	DSMZ
Intestine 407	Human	Intestine epithelium	HeLa	Human	Cervical adenocarcinoma	ATCC
J-111			HeLa	Human	Cervical adenocarcinoma	JCRB
JOSK-I	Human	Monocytic leukemia	U-937	Human	Histiocytic lymphoma	DSMZ
JOSK-K	Human	Monocytic leukemia	U-937	Human	Histiocytic lymphoma	DSMZ
JOSK-M	Human	Monocytic leukemia	U-937	Human	Histiocytic lymphoma	DSMZ
JOSK-S	Human	Monocytic leukemia	U-937	Human	histolytic lymphoma	DSMZ
JTC-17			HeLa	Human	Cervical adenocarcinoma	Yamakage (JCRB)
KB	Human	Adult oral cavity epithelium	HeLa	Human	Cervical adenocarcinoma	ATCC; JCRB
KO51			K562	Human	Myeloid leukemia	DSMZ; JCRB
KOSC-3			Ca9-22			JCRB

TABLE 13.2. Cross-Contaminated Cell Lines (*Continued*)

Cell line	Species	Cell type	Contaminant	Species	Cell type	Source of Data
L132	Human	Embryonic lung epithelium	HeLa	Human	Cervical adenocarcinoma	ATCC
LR10.6	Human	Pre-B cell leukemia	NALM-6	Human	Pre-B cell leukemia	DSMZ
MaTu	Human	Breast carcinoma	HELA/-S3	Human	Cervical adenocarcinoma	DSMZ
MC-4000	Human	Breast carcinoma	HELA/-S3	Human	Cervical adenocarcinoma	DSMZ
MKB-1	Human	T-cell leukemia	CCRF-CEM	Human	T-cell leukemia	DSMZ
MKN28			MKN74			JCRB
MOLT-15	Human	T-cell leukemia	CTV-1	Human	Monocytic leukemia	DSMZ
MT-1	Human	Breast carcinoma	HELA/-S3	Human	Cervical adenocarcinoma	DSMZ
NCC16			PHK16-0b			JCRB
P1-1A3	Human	Thymic epithelium	SK-HEP-1	Human	Hepatoma	DSMZ
P1-4D6	Human	Thymic epithelium	SK-HEP-1	Human	Hepatoma	DSMZ
P39 TSU	Human		HL60	Human	Myeloid leukemia	JCRB
PBEI	Human	Pre-B cell leukemia	NALM-6	Human	Pre-B cell leukemia	DSMZ
PSV811	Human	Fibroblast	WI38	Human	Fibroblast	JCRB
RAMAK-1	Human	Muscle synovium	T-24	Human	Bladder carcinoma	DSMZ
SBC-2	Human	Bladder carcinoma	HELA/-S3	Human	Cervical adenocarcinoma	DSMZ
SBC-7	Human	Bladder carcinoma	HELA/-S3	Human	Cervical adenocarcinoma	DSMZ
SCLC-16H	Human	SCLC	SCLC-21/22H	Human	SCLC	DSMZ
SCLC-24H	Human	SCLC	SCLC-21/22H	Human	SCLC	DSMZ
SNB-19			U251MG	Human	Glioma	ATCC
SPI-801	Human	T-cell leukemia	K-562	Human	Myeloid leukemia	DSMZ
SPI-802	Human	T-cell leukemia	K-562	Human	Myeloid leukemia	DSMZ
TK-1	Human?		U251MG	Human	Glioma	JCRB
TMH-1			IHH-4			JCRB
WISH	Human	Newborn amnion epithelium	HeLa	Human	Cervical adenocarcinoma	ATCC

TABLE 13.3. Properties of Finite and Continuous Cell Lines

Properties	Finite	Continuous (transformed)
Ploidy	Euploid, diploid	Aneuploid, heteroploid
Transformation	Normal	Immortal, growth control altered, and tumorigenic
Anchorage dependence	Yes	No
Contact inhibition	Yes	No
Density limitation of cell proliferation	Yes	Reduced or lost
Mode of growth	Monolayer	Monolayer or suspension
Maintenance	Cyclic	Steady state possible
Serum requirement	High	Low
Cloning efficiency	Low	High
Markers	Tissue specific	Chromosomal, enzymic, antigenic
Special functions (e.g., virus susceptibility, differentiation)	May be retained	Often lost
Growth rate	Slow (T_D of 24–96 h)	Rapid (T_D of 12–24 h)
Yield	Low	High
Control parameters	Generation no.; tissue-specific markers	Stain characteristics

cell generations, usually between 20 and 80 cell population doublings, before extinction. The actual number of doublings depends on species and cell lineage differences, clonal variation, and culture conditions, but it is consistent for one cell line grown under the same conditions. It is therefore important that reference to a cell line should express the approximate generation number or number of doublings since explantation; I say "approximate" because the number of generations that have elapsed in the primary culture is difficult to assess.

Continuous cell lines (*see* Table 13.1) have escaped from senescence control, so the generation number becomes less important and the number of passages since last thawed from storage becomes more important (*see* Section 13.7.2). In addition, because of the increased cell proliferation rate and saturation density (*see* Section 18.5), split ratios become much greater (1:20–1:100) and cell concentration at subculture becomes much more critical (*see* Section 13.7.3).

13.4 CELL LINE DESIGNATIONS

New cell lines should be given a code or designation [e.g., normal human brain (NHB)]; a cell strain or cell line number (if several cell lines were derived from the same source; e.g., NHB1, NHB2, etc.); and, if cloned, a clone number (e.g., NHB2-1, NHB2-2, etc.). It is useful to keep a log book or computer database file where the receipt of biopsies or specimens is recorded before initiation of a culture. The accession number in the log book or database file, perhaps linked to an identifier letter code, can then be used to establish the cell line designation; for example, LT156 would be lung tumor biopsy number 156. This method is less likely to generate ambiguities, such as the same letter code being used for two different cell lines, and gives automatic reference to the record of accession of the line. Rules of confidentiality preclude the use of a donor's initials in naming a cell line.

For finite cell lines, the number of population doublings should be estimated and indicated after a forward slash, e.g., NHB2/2, and increases by one for a split ratio of 1:2 (e.g., NHB2/2, NHB2/3, etc.), by two for a split ratio of 1:4 (e.g., NHB2/2, NHB2/4, etc.), and so on. When dealing with a continuous cell line a "p" number at the end is often used to indicate the number the number of passages since the last thaw from the freezer (*see* Section 20.4.2), e.g., HeLa-S3/p4.

When referenced in publications or reports, it is helpful to prefix the cell line designation with a code indicating the laboratory in which it was derived (e.g., WI for Wistar Institute, NCI for National Cancer Institute, SK for Sloan-Kettering) [Federoff, 1975]. In publications or reports, the cell line should be given its full designation the first time it is mentioned and in the Materials and Methods section, and the abbreviated version can then be used thereafter.

It is essential that the cell line designation is unique, or else confusion will arise in subsequent reports in the literature.

Cell banks deal with this problem by giving each cell line an accession number; when reporting on cell lines acquired from a cell bank, you should give this accession number in the Materials and Methods section. Punctuation can also give rise to problems when one is searching for a cell line in a database, so always adhere to a standard syntax, and do not use apostrophes or spaces.

13.5 CHOOSING A CELL LINE

Apart from specific functional requirements, there are a number of general parameters to consider in selecting a cell line:

(1) **Finite vs. Continuous.** Is there a continuous cell line that expresses the right functions? A continuous cell line generally is easier to maintain, grows faster, clones more easily, produces a higher cell yield per flask, and is more readily adapted to serum-free medium (*see* Table 13.3).

(2) **Normal or Transformed.** Is it important whether the line is malignantly transformed or not? If it is, then it might be possible to obtain an immortal line that is not tumorigenic, e.g., 3T3 cells or BKK21-C13.

(3) **Species.** Is species important? Nonhuman cell lines have fewer biohazard restrictions and have the advantage that the original tissue may be more accessible.

(4) **Growth Characteristics.** What do you require in terms of growth rate, yield, plating efficiency, and ease of harvesting? You will need to consider the following parameters:
   a) Population-doubling time (*see* Section 21.9.7)
   b) Saturation density (yield per flask; *see* Section 21.9.5)
   c) Plating efficiency (*see* Section 21.10)
   d) Growth fraction (*see* Section 21.11.1)
   e) Ability to grow in suspension (*see* Section 18.5.1, Table 13.5)

(5) **Availability.** If you have to use a finite cell line, are there sufficient stocks available, or will you have to generate your own line(s)? If you choose a continuous cell line, are authenticated stocks available?

(6) **Validation.** How well characterized is the line (*see* Section 7.10), if it exists already, or, if not, can you do the necessary characterization (*see* Chapter 16)? Is the line authentic (*see* Section 16.3)? It is vital to eliminate the possibility of cross-contamination before embarking on a program of work with a cell line, as so many cross-contaminations have been reported (*see* Table 13.2).

(7) **Phenotypic Expression.** Can the line be made to express the right characteristics (*see* Section 17.7)?

(8) **Control Cell Line.** If you are using a mutant, transfected, transformed, or abnormal cell line, is there a normal equivalent available, should it be required?

(9) **Stability.** How stable is the cell line (*see* Section 18.3 and Plate 7)? Has it been cloned? If not, can you clone it,

and how long would this cloning process take to generate sufficient frozen and usable stocks?

13.6 ROUTINE MAINTENANCE

Once a culture is initiated, whether it is a primary culture or a subculture of a cell line, it will need a periodic medium change, or "feeding," followed eventually by subculture if the cells are proliferating. In nonproliferating cultures, the medium will still need to be changed periodically, as the cells will still metabolize and some constituents of the medium will become exhausted or will degrade spontaneously. Intervals between medium changes and between subcultures vary from one cell line to another, depending on the rate of growth and metabolism; rapidly growing transformed cell lines, such as HeLa, are usually subcultured once per week, and the medium should be changed four days later. More slowly growing, particularly nontransformed, cell lines may need to be subcultured only every two, three, or even four weeks, and the medium should be changed weekly between subcultures (*see also* Sections 13.6.2, 13.7.1, 21.9.2).

13.6.1 Significance of Cell Morphology

Whatever procedure is undertaken, it is vital that the culture be examined carefully to confirm the absence of contamination (*see* Section 19.3.1 and Fig. 19.1). The cells should also be checked for any signs of deterioration, such as granularity around the nucleus, cytoplasmic vacuolation, and rounding up of the cells with detachment from the substrate (Fig. 13.1). Such signs may imply that the culture requires a medium change, or may indicate a more serious problem, e.g., inadequate or toxic medium or serum, microbial contamination, or senescence of the cell line. Medium

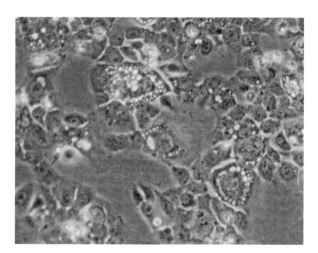

Fig. 13.1. Unhealthy Cells. Vacuolation and granulation in bronchial epithelial cells (BEAS-2B) due, in this case, to medium inadequacy. The cytoplasm of the cells becomes granular, particularly around the nucleus, and vacuolation occurs. The cells may become more refractile at the edge if cell spreading is impaired.

deficiencies can also initiate apoptosis (*see* Section 3.3.1 and Plate 17c,d). During routine maintenance, the medium change or subculture frequency should aim to prevent such deterioration, as it is often difficult to reverse.

Familiarity with the cell's morphology may also give the first indication of cross-contamination or misidentification. It is useful to have a series of photographs of cell types in regular use (*see* Fig. 16.2 and Plates 9 and 10), taken at different cell densities (preferably known cell densities, e.g., by counting the number of cells/cm^2) to refer to when handling cultures, particularly when a new member of staff is being introduced to culture work.

13.6.2 Replacement of Medium

Four factors indicate the need for the replacement of culture medium:

(1) **A Drop in pH.** The rate of fall and absolute level should be considered. Most cells stop growing as the pH falls from pH 7.0 to pH 6.5 and start to lose viability between pH 6.5 and pH 6.0, so if the medium goes from red through orange to yellow, the medium should be changed. Try to estimate the rate of fall; a culture at pH 7.0 that falls 0.1 pH units in one day will not come to harm if left a day or two longer before feeding, but a culture that falls 0.4 pH units in one day will need to be fed within 24–48 h and cannot be left over a weekend without feeding.

(2) **Cell Concentration.** Cultures at a high cell concentration exhaust the medium faster than those at a low concentration. This factor is usually evident in the rate of change of pH, but not always.

(3) **Cell Type.** Normal cells (e.g., diploid fibroblasts) usually stop dividing at a high cell density (*see* Section 18.5.2), because of cell crowding, growth factor depletion, and other reasons. The cells block in the G$_1$ phase of the cell cycle and deteriorate very little, even if left for two to three weeks or longer. Transformed cells, continuous cell lines, and some embryonic cells, however, deteriorate rapidly at high cell densities unless the medium is changed daily or they are subcultured.

(4) **Morphological Deterioration.** This factor must be anticipated by regular examination and familiarity with the cell line (*see* Section 13.6.1). If deterioration is allowed to progress too far, it will be irreversible, as the cells will tend to enter apoptosis (*see* Section 3.3.1).

Volume, depth, and surface area. The usual ratio of medium volume to surface area is 0.2–0.5 mL/cm^2 (*see also* Section 21.9.3). The upper limit is set by gaseous diffusion through the liquid layer, and the optimum ratio depends on the oxygen requirement of the cells. Cells with a high O$_2$ requirement do better in shallow medium (e.g., 2 mm), and those with a low requirement may do better in deep medium (e.g., 5 mm). If the depth of the medium is greater than

5 mm, then gaseous diffusion may become limiting. With monolayer cultures, this problem can be overcome by rolling the bottle (*see* Section 26.2.3) or perfusing the culture with medium and arranging for gas exchange in an intermediate reservoir (*see* Section 25.3.2, 26.2.5).

Holding medium. A *holding medium* may be used when stimulation of mitosis, which usually accompanies a medium change, even at high cell densities, is undesirable. Holding media are usually regular media with the serum concentration reduced to 0.5% or 2% or eliminated completely. For serum-free media, growth factors and other mitogens are omitted. This omission inhibits mitosis in most untransformed cells. Transformed cell lines are unsuitable for this procedure, as either they may continue to divide successfully or the culture may deteriorate, because transformed cells do not block in a regulated fashion in G_1 of the cell cycle (*see* Section 3.3.1).

Holding media are used to maintain cell lines with a finite life span without using up the limited number of cell generations available to them (*see* Section 3.8.1). Reduction of serum and cessation of cell proliferation also promote expression of the differentiated phenotype in some cells [Maltese & Volpe, 1979; Schousboe et al., 1979]. Media used for the collection of biopsy samples can also be referred to as holding media.

Standard feeding protocol. Protocol 13.1 is designed to accompany Exercise 4, using medium prepared from Exercise 3 (*see* Chapter 2). The cells and media are specified, but can easily be changed to suit individual requirements.

PROTOCOL 13.1. FEEDING A MONOLAYER CULTURE

Outline
Examine the culture by eye and on an inverted microscope. If indicated, e.g., by a fall in pH, remove the old medium and add fresh medium. Return the culture to the incubator.

Materials
Sterile:
- ❑ Cell cultures: A549 cells 4 days after seeding at 2×10^4 cells/mL, 25 cm² flasks 4
- ❑ Growth medium .... 100 mL
  e.g., Eagle's 1×MEM with Hanks's salts and 4 mM HCO₃, without antibiotics.
  If the training program is being followed, use the two media prepared in Exercise 3, one of which has been stored at 4°C and one at 37°C, for one week.
- ❑ Pipettes, graduated, and plugged. If glass, an assortment of sizes, 1 mL, 5 mL, 10 mL, 25 mL,

in a square pipette can, or, if plastic, individually wrapped and sorted by size on a rack
- ❑ Unplugged pipettes for aspirating medium if pump or vacuum line is available

Nonsterile:
- ❑ Pipetting aid or bulb (*see* Figs. 5.5, 6.6)
- ❑ Tubing to receiver connected to vacuum line or to receiver via peristaltic pump (*see* Figs. 5.1–5.3)
- ❑ Alcohol, 70%, in spray bottle
- ❑ Lint-free swabs or wipes
- ❑ Absorbent paper tissues
- ❑ Pipette cylinder containing water and disinfectant (*see* Sections 5.8.8, 7.8.5)
- ❑ Marker pen with alcohol-insoluble ink
- ❑ Notebook, pen, protocols, etc.

Protocol
1. Prepare the hood by ensuring that it is clear and swabbing it with 70% alcohol.
2. Bring the reagents and materials necessary for the procedure, swab bottles with 70% alcohol and place items required immediately in the hood (*see* Protocol 6.1)
3. Examine the culture carefully for signs of contamination or deterioration (*see* Figs. 13.1, 19.1).
4. Check the previously described criteria—pH and cell density or concentration—and, based on your knowledge of the behavior of the culture, decide whether or not to replace the medium. If feeding is required, proceed as follows.
5. Take the culture to the sterile work area.
6. Uncap the flask.
7. Take sterile pipette and insert into bulb or pipetting aid, or, selecting an unplugged pipette, connect to vacuum line or pump.
8. Withdraw the medium, and discard into waste beaker (*see* Fig. 6.8). Or, preferably, aspirate medium via a suction line in the hood connected to an external pump (*see* Figs. 5.2, 5.3).
9. Discard pipette.
10. Uncap the medium bottle.
11. Take a fresh pipette and add the same volume of fresh medium as was removed, prewarmed to 37°C if it is important that there be no check in cell growth, and recap the bottle.
12. Discard the pipette.
13. Recap the flask and the medium bottle.
14. Return the culture to the incubator.
15. Complete record of observations and feeding on record sheet or lab book.
16. Clear away all pipettes, glassware, etc., and swab down the work surface.

Note. When a culture is at a low density and growing slowly, it may be preferable to half-feed it—i.e., to remove only half of the medium at Step 8 and replace it in Step 11 with the same volume as was removed.

13.7 SUBCULTURE

When a cell line is subcultured the regrowth of the cells to a point ready for the next subculture usually follows a standard pattern (Fig. 13.2). A *lag period* after seeding is followed by a period of exponential growth, called the *log phase*. When the *cell density* (cells/cm^2 substrate) reaches a level such that all of the available substrate is occupied, or when the *cell concentration* (cells/mL medium) exceeds the capacity of the medium, growth ceases or is greatly reduced (*see* Fig. 16.2b,d,f,h,j,l and Plate 4d). Then either the medium must be changed more frequently or the culture must be divided. For an adherent cell line, dividing a culture, or *subculture* as it is called, usually involves removal of the medium and dissociation of the cells in the monolayer with trypsin, although some loosely adherent cells (e.g., HeLa-S$_3$) may be subcultured by shaking the bottle, collecting the cells in the medium, and diluting as appropriate in fresh medium in new bottles. Exceptionally, some cell monolayers cannot be dissociated in trypsin and require the action of alternative proteases, such as pronase, dispase, and collagenase (Table 13.4). Of these proteases, pronase is the most effective but can be toxic to some cells. Dispase and collagenase are generally less toxic than trypsin but may not give complete dissociation of epithelial cells.

Other proteases, such as Accutase, Accumax (invertebrate proteases), and Trypzean or TrypLE (recombinant trypsins), are available, and their efficacy should be tested where either there is a problem with standard disaggregation protocols or there is a need to avoid mammalian (e.g., porcine trypsin) or bacterial (e.g., Pronase) proteases. The severity of the treatment required depends on the cell type, as does the sensitivity of the cells to proteolysis, and a protocol should be selected with the least severity that is compatible with the generation of a single-cell suspension of high viability.

The attachment of cells to each other and to the culture substrate is mediated by cell surface glycoproteins and Ca^{2+} (*see* Section 3.2). Other proteins, and proteoglycans, derived from the cells and from the serum, become associated with the cell surface and the surface of the substrate and facilitate cell adhesion. Subculture usually requires chelation of Ca^{2+} and degradation of extracellular matrix and, potentially, the extracellular domains of some cell adhesion molecules.

13.7.1 Criteria for Subculture

The need to subculture a monolayer is determined by the following criteria:

(1) **Density of Culture.** Normal cells should be subcultured as soon as they reach confluence. If left more than 24 h, they will withdraw from the cycle and take longer to recover when reseeded. Transformed cells should also be subcultured on reaching confluence; although they will continue to proliferate beyond confluence, they will start

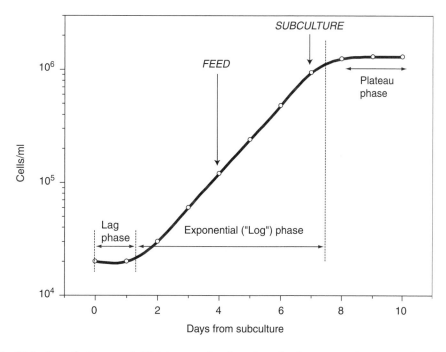

Fig. 13.2. Growth Curve and Maintenance. Semilog plot of cell concentration versus time from subculture, showing the lag phase, exponential phase, and plateau, and indicating times at which subculture and feeding should be performed (*see also* Section 21.9.2 and Fig. 21.6).

TABLE 13.4. Cell Dissociation Procedures

Procedure	Pretreatment	Dissociation agent	Medium	Applicable to
Shake-off	None	Gentle mechanical shaking, rocking, or vigorous pipetting	Culture medium	Mitotic or other loosely adherent cells
Scraping	None	Cell scraper	Culture medium	Cell lines for which proteases are to be avoided (e.g., receptor or cell surface protein analysis); can damage some cells and rarely gives a single-cell suspension
Trypsin* alone	Remove medium completely	0.01–0.5% Crude trypsin; usually 0.25%	D-PBSA, CMF, or saline citrate	Most continuous cell lines
Prewash + trypsin	D-PBSA	0.25% Crude trypsin	D-PBSA	Some strongly adherent continuous cell lines and many early-passage cells
Prewash + trypsin	1 mM EDTA in D-PBSA	0.25% Crude trypsin	D-PBSA	Strongly adherent early-passage cell lines
Prewash + trypsin	1 mM EDTA in D-PBSA	0.25% Crude trypsin	D-PBSA + 1 mM EDTA	Many epithelial cells, but some can be sensitive to EDTA; EGTA can be used
Trypsin + collagenase	1 mM EDTA in D-PBSA	0.25% Crude trypsin; 200 U/mL crude collagenase	D-PBSA + 1 mM EDTA	Dense cultures and multilayers, particularly with fibroblasts
Dispase	None	0.1–1.0 mg/mL Dispase	Culture medium	Removal of epithelium in sheets (does not dissociate epithelium)
Pronase	None	0.1–1.0 mg/mL Pronase	Culture medium	Provision of good single-cell suspensions, but may be harmful to some cells
DNase	D-PBSA or 1 mM EDTA in D-PBSA	2–10 µg/mL crystalline DNase	Culture medium	Use of other dissociation agents which damage cells and release DNA

*Digestive enzymes are available (Difco, Worthington, Roche, Sigma) in varying degrees of purity. Crude preparations—e.g., Difco trypsin, 1:250, or Worthington CLS-grade collagenase—contain other proteases that may be helpful in dissociating some cells, but may be toxic to other cells. Start with a crude preparation, and progress to purer grades if necessary. Purer grades are often used at a lower concentration (µg/mL), as their specific activities (enzyme units/g) are higher. Purified trypsin at 4°C has been recommended for cells grown in low-serum concentrations or in the absence of serum [McKeehan, 1977] and is generally found to be more consistent. Batch testing and reservation, as for serum, may be necessary for some applications.

to deteriorate after about two doublings, and reseeding efficiency will decline.

(2) **Exhaustion of Medium.** Exhaustion of the medium (*see* Section 13.6.2) usually indicates that the medium requires replacement, but if a fall in pH occurs so rapidly that the medium must be changed more frequently, then subculture may be required. Usually, a drop in pH is accompanied by an increase in cell density, which is the prime indicator of the need to subculture. Note that a sudden drop in pH can also result from contamination, so be sure to check (*see* Section 19.3).

(3) **Time Since Last Subculture.** Routine subculture is best performed according to a strict schedule, so that reproducible behavior is achieved and monitored. If cells have not reached a high-enough density (i.e., they are not confluent) by the appropriate time, then increase the seeding density, or if they reach confluence too soon, then reduce the seeding density. Once this routine is established, the recurrent growth should be consistent

in duration and cell yield from a given seeding density. Deviations from this pattern then signify a departure from normal conditions or indicate deterioration of the cells. Ideally, a cell concentration should be found that allows for the cells to be subcultured after 7 days, with the medium being changed after 3–4 days.

(4) **Requirements for Other Procedures.** When cells are required for purposes other than routine propagation, they also have to be subcultured, in order to increase the stock or to change the type of culture vessel or medium. Ideally, this procedure should be done at the regular subculture time, when it will be known that the culture is performing routinely, what the reseeding conditions should be, and what outcome can be expected. However, demands for cells do not always fit the established routine for maintenance, and compromises have to be made, but (1) cells should not be subcultured while still within the lag period, and (2) cells should always be taken between the middle of the log phase and the time before

which they have entered the plateau phase of a previous subculture (unless there is a specific requirement for plateau-phase cells, in which case they will need frequent feeding or continuous perfusion).

Handling different cell lines. Different cell lines should be handled separately, with a separate set of media and reagents. If they are all handled at the same time, there is a significant risk of cross-contamination, particularly if a rapidly growing line, such as HeLa, is maintained alongside a slower-growing line (*see* Section 19.5).

Typical subculture protocol for cells grown as a monolayer. Protocol 13.2 describes trypsinization of a monolayer (Fig. 13.3; Plates 7–12) after an EDTA prewash to remove traces of medium, divalent cations, and serum (if used). This procedure can be carried out without the prewash, or with only D-PBSA as a prewash, and with 1 mM EDTA in the trypsin if required, depending on the type of cell (*see* Table 13.4).

The amounts of materials specified are designed for use with Exercise 13 in Chapter 2, but can be varied as required.

PROTOCOL 13.2. SUBCULTURE OF MONOLAYER CELLS

Outline
Remove the medium. Expose the cells briefly to trypsin. Incubate the cells. Disperse the cells in medium. Count the cells. Dilute and reseed the subculture.

Materials
Sterile:
- A549 cells, 7 days and 14 days after seeding at 2×10^4 cells/mL, 25-cm² flasks .... 1 each
- WI-38, MRC-5, or an equivalent normal diploid fibroblast culture, 7 days and 14 days after seeding at 2×10^4 cells/mL, 25-cm² flasks .... 1 each
- Growth medium, e.g., 1×MEM with Earle's salts, 23 mM HCO₃, without antibiotics .. 100 mL
- Trypsin, 0.25% in D-PBSA (*see* Table 13.4) .... 10 mL
- D-PBSA with 1 mM EDTA .... 20 mL
- Pipettes, graduated, and plugged. If glass, an assortment of sizes, 1 mL, 5 mL, 10 mL, 25 mL, in a square pipette can, or, if plastic, individually wrapped and sorted by size on a rack

- Unplugged pipettes for aspirating medium if pump or vacuum line is available
- Universal containers or 50-mL centrifuge tubes 4
- Culture flasks, 25 cm² 8

Nonsterile:
- Pipetting aid or bulb (*see* Figs. 5.5, 6.6)
- Tubing to receiver connected to vacuum line or to receiver via peristaltic pump (*see* Figs. 5.1–5.3)
- Alcohol, 70%, in spray bottle
- Lint-free swabs or wipes
- Absorbent paper tissues
- Pipette cylinder containing water and disinfectant (*see* Sections 5.8.8, 7.8.5)
- Marker pen with alcohol-insoluble ink
- Notebook, pen, protocols, etc.
- Hemocytometer or electronic cell counter

Protocol
1. Prepare the hood, and bring the reagents and materials to the hood to begin the procedure (*see* Section 6.5).
2. Examine the cultures carefully for signs of deterioration or contamination (*see* Figs. 13.1, 19.1).
3. Check the criteria (*see* Section 13.7.1), and, based on your knowledge of the behavior of the culture, decide whether or not to subculture. (Those following Exercise 13 should make particular note of the density and condition of the cells, e.g., evidence of mitoses, multilayering, cellular deterioration.) If subculture is required, proceed as follows.
4. Take the culture flasks to a sterile work area, and remove and discard the medium (*see* Protocol 13.1, Steps 7–9). Handle each cell line separately, repeating this procedure from this step for each cell line handled.
5. Add D-PBSA/EDTA prewash (0.2 mL/cm²) to the side of the flasks opposite the cells so as to avoid dislodging cells, rinse the prewash over the cells, and discard. This step is designed to remove traces of serum that would inhibit the action of the trypsin and deplete the divalent cations, necessary for cell adhesion.
6. Add trypsin (0.1 mL/cm²) to the side of the flasks opposite the cells. Turn the flasks over and lay them down. Ensure that the monolayer is completely covered. Leave the flasks stationary for 15–30 s.
7. Raise the flasks to remove the trypsin from the monolayer and quickly check that the monolayer is not detaching. Using trypsin at 4°C helps to prevent premature detachment, if this turns out to be a problem.

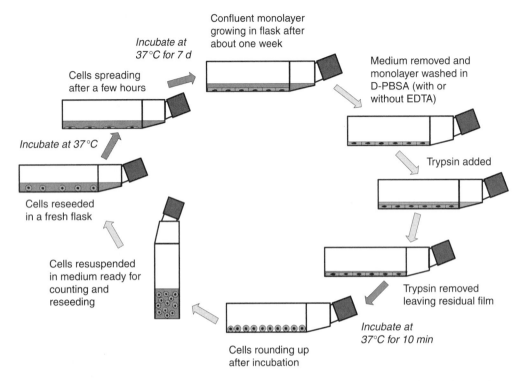

Fig. 13.3. Subculture of Monolayer. Stages in the subculture and growth cycle of monolayer cells after trypsinization (*see also* Plates 4, 5).

8. Withdraw all but a few drops of the trypsin.
9. Incubate, with the flasks lying flat, until the cells round up (Figs. 13.3, 13.4, Plate 5); when the bottle is tilted, the monolayer should slide down the surface. (This usually occurs after 5–15 min.) Do not leave the flasks longer than necessary, but on the other hand, do not force the cells to detach before they are ready to do so, or else clumping may result.

Note. In each case, the main dissociating agent, be it trypsin or EDTA, is present only briefly, and the incubation is performed in the residue after most of the dissociating agent has been removed. If you encounter difficulty in getting cells to detach and, subsequently, in preparing a single-cell suspension, you may employ alternative procedures (*see* Table 13.4).

10. Add medium (0.1–0.2 mL/cm²), and disperse the cells by repeated pipetting over the surface bearing the monolayer.
11. Finally, pipette the cell suspension up and down a few times, with the tip of the pipette resting on the bottom corner of bottle, taking care not to create a foam. The degree of pipetting required will vary from one cell line to another; some cell lines disperse easily, whereas others require vigorous pipetting in order to disperse them. Almost all cells incur mechanical damage from shearing forces if pipetted too vigorously. Primary suspensions and early-passage cultures are particularly prone to damage, partly because of their greater fragility and partly because of their larger size, but continuous cell lines are usually more resilient and require vigorous pipetting for complete disaggregation. Pipette the suspension up and down sufficiently to disperse the cells into a single-cell suspension. If this step is difficult, apply a more aggressive dissociating agent (*see* Table 13.4).

A single-cell suspension is desirable at subculture to ensure an accurate cell count and uniform growth on reseeding. It is essential if quantitative estimates of cell proliferation or of plating efficiency are being made and if cells are to be isolated as clones.

12. Count the cells with a hemocytometer or an electronic particle counter (*see* Section 21.1), and record the cell counts.
13. Dilute the cell suspensions to the appropriate seeding concentration:
   (a) By adding the appropriate volume of cell suspension to a premeasured volume of medium in a culture flask

or

(b) By diluting the cells to the total volume required and distributing that volume among several flasks.

Procedure (a) is useful for routine subculture when only a few flasks are used and precise cell counts and reproducibility are not critical, but procedure (b) is preferable when setting up several replicates, because the total number of manipulations is reduced and the concentrations of cells in each flask will be identical.

For Exercise 13, use procedure (b): dilute the cells from each flask to 2×10^4 cells per mL in 20 mL medium, and seed 5 mL from each suspension into two flasks.

14. If the cells are grown in elevated CO_2 (as indicated by the use of Eagle's MEM with Earle's salts and 23 mM $NaHCO_3$, as listed in Materials above), gas the flask by blowing the correct gas mixture (in this case 5% CO_2) from a premixed cylinder, or a gas blender, through a filtered line into the flask above the medium (*see* Fig. 6.11). Do not bubble gas through the medium, as doing so will generate bubbles, which can denature some constituents of the medium and increase the risk of contamination. If the normal gas phase is air, as with Eagle's MEM with Hanks' salts (*see* Protocol 13.1), this step may be omitted.

15. Cap the flasks, and return them to the incubator. Check the pH after about 1 h. If the pH rises in a medium with a gas phase of air, then return the flasks to the aseptic area and gas the culture briefly (1–2 s) with 5% CO_2. As each culture will behave predictably in the same medium, you eventually will know which cells to gas when they are reseeded, without having to incubate them first. If the pH rises in medium that already has a 5% CO_2 gas phase (as in this case), either increase the CO_2 to 7% or 10% or add sterile 0.1 N of HCl.

16. Repeat this procedure from Step 4 for the second cell line.

Note. The procedure in Step 15 should not become a long-term solution to the problem of high pH after subculture. If the problem persists, then reduce the pH of the medium at the time it is made up, and check the pH of the medium in the incubator or in a gassed flask.

As the expansion of air inside plastic flasks causes larger flasks to swell and prevents them from lying flat, the pressure should be released by briefly slackening the cap 30 min after placing the flask in the incubator. Alternatively, this problem may be prevented by compressing the top and bottom of large flasks before sealing them (care must be taken not to exert too much pressure and crack the flasks). Incubation restores the correct shape as the gas phase expands.

13.7.2 Growth Cycle and Split Ratios

Routine passage leads to the repetition of a standard growth cycle (Fig. 13.4a; *see also* Fig. 13.2). It is essential to become familiar with this cycle for each cell line that is handled, as it controls the seeding concentration, the duration of growth before subculture, the duration of experiments, and the appropriate times for sampling to give greatest consistency. Cells at different phases of the growth cycle behave differently with respect to cell proliferation, enzyme activity, glycolysis and respiration, synthesis of specialized products, and many other properties (*see* Sections 21.9.4, 25.1.1).

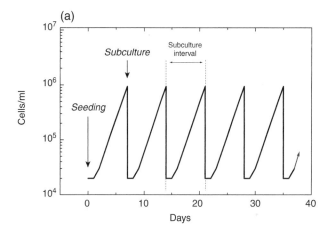

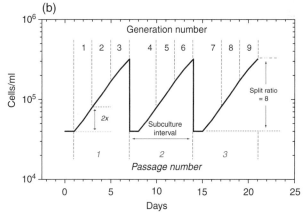

Fig. 13.4. Serial Subculture. (a) Repetition of the standard growth cycle during propagation of a cell line: If the cells are growing correctly, then they should reach the same concentration (peaks) after the same time in each cycle, given that the seeding concentration (troughs) and subculture interval remain constant. (b) Generation number and passage: Each subculture represents one passage, but the generation number (in this case 3 per passage) depends on the split ratio (8 for 3 doublings per passage).

For finite cell lines, it is convenient to reduce the cell concentration at subculture by 2-, 4-, 8-, or 16-fold (i.e., a split ratio of 2, 4, 8, and 16, respectively), making the calculation of the number of population doublings easier (i.e., respectively, a split ratio of 2 corresponds to 1 population doubling, 4 to 2, 8 to 3, and 16 to 4); for example, a culture divided 8-fold requires three doublings to return to the same cell density (Fig. 13.4b). A fragile or slowly growing line should also be split 1:2, whereas a vigorous, rapidly growing normal cell line can be split 1:8 or 1:16 and some continuous cell lines may be split 1:50 or 1:100. Once a cell line becomes continuous (usually taken as beyond 150 or 200 generations), the cell concentration is the main parameter and the culture should be cut back to between 10^4 and 10^5 cells/mL. The split ratio, or dilution, is also chosen to establish a convenient subculture interval, perhaps 1 or 2 weeks, and to ensure that the cells (1) are not diluted below the concentration that permits them to reenter the growth cycle within a reasonable lag period (24 h or less) and (2) do not enter a plateau before the next subculture.

When handling a cell line for the first time, or when using an early-passage culture with which you have little experience, it is good practice to subculture the cell line to a split ratio of 2 or 4 at the first attempt, noting the cell concentrations as you do so. As you gain experience and the cell line seems established in the laboratory, it may be possible to increase the split ratio—i.e., to reduce the cell concentration after subculture—but always keep one flask at a low split ratio when attempting to increase the split ratio of the rest.

Even when a split ratio is used to determine the seeding concentration at subculture, the number of cells per flask should be recorded after trypsinization and at reseeding, so that the growth rate can be estimated at each subculture and the consistency can be monitored (see Protocols 21.7–21.9). Otherwise, minor alterations will not be detected for several passages.

13.7.3 Cell Concentration at Subculture

The ideal method of determining the correct seeding density is to perform a growth curve at different seeding concentrations (see Protocols 21.7–21.9) and thereby determine the minimum concentration that will give a short lag period and early entry into rapid logarithmic growth (i.e., a short population-doubling time) but will reach the top of the exponential phase at a time that is convenient for the next subculture.

As a general rule, most continuous cell lines subculture satisfactorily at a seeding concentration of between 1×10^4 and 5×10^4 cells/mL, finite fibroblast cell lines subculture at about the same concentration, and more fragile cultures, such as endothelium and some early-passage epithelial cells, subculture at around 1×10^5 cells/mL. For a new culture, start at a high seeding concentration and gradually reduce until a convenient growth cycle is achieved without any deterioration in the culture.

13.7.4 Propagation in Suspension

Protocol 13.2 refers to the subculture of monolayers, because the manner in which most primary cultures and cell lines grow. However, cells that grow continuously in suspension, either because they are nonadhesive (e.g., many leukemias and murine ascites tumors) or because they have been kept in suspension mechanically, or selected, may be subcultured like bacteria or yeast. Suspension cultures have a number of advantages (Table 13.5); for example, trypsin treatment is not required so subculture is quicker and less traumatic for the cells, and scale-up is easier (see also Section 26.1). Replacement of the medium (feeding) is not usually carried out with suspension cultures, and instead, the culture is either diluted and expanded, diluted and the excess discarded, or the bulk of the cell suspension is withdrawn and the residue is diluted back to an appropriate seeding concentration. In each case, a growth cycle will result, similar to that for monolayer cells, but usually with a shorter lag period.

Cells that grow spontaneously in suspension can be maintained in regular culture flasks, which need not be tissue culture treated (although they must be sterile, of course). The rules regarding the depth of medium in static cultures are as for monolayers—i.e., 2–5 mm to allow for gas exchange. When the depth of a suspension culture is increased—e.g., if it is expanded—the medium requires agitation, which is best achieved with a suspended rotating magnetic pendulum, with the culture flask placed on a magnetic stirrer (Fig. 13.5; see also Section 26.1). Roller bottles rotating on a rack can also be used to agitate suspension cultures (see Table 8.1 and Fig. 26.11).

13.7.5 Subculture of Cells Growing in Suspension

Protocol 13.3 describes routine subculture of a suspension culture into a fresh vessel. A continuous culture can be

TABLE 13.5. Monolayer vs. Suspension Culture

Monolayer	Suspension
Culture requirements	
Cyclic maintenance	Steady state
Trypsin passage	Dilution
Limited by surface area	Volume (gas exchange)
Growth properties	
Contact inhibition	Homogeneous suspension
Cell interaction	
Diffusion boundary	
Useful for	
Cytology	Bulk production
Mitotic shake-off	Batch harvesting
In situ extractions	
Continuous product harvesting	
Applicable to	
Most cell types, including primaries	Only transformed cells

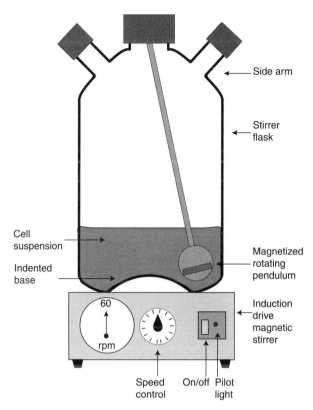

Fig. 13.5. Stirrer Culture. A small stirrer flask, based on the Techne design, with a capacity of 250–1000 mL. The cell suspension is stirred by a pendulum, which rotates in an annular depression in the base of the flask.

maintained in the same vessel, but the probability of contamination gradually increases with any buildup of minor spillage on the neck of the flask during dilution. The criteria for subculture are similar to those for monolayers:

(1) **Cell Concentration,** which should not exceed 1×10^6 cells/mL for most suspension-growing cells.
(2) **pH,** which is linked to cell concentration, and declines as the cell concentration rises.
(3) **Time Since Last Subculture,** which, as for monolayers, should fit a regular schedule.
(4) **Cell Production Requirements** for experimental or production purposes.

Protocol 13.3 is designed for use in conjunction with Exercise 12 in Chapter 2. The cell lines, culture vessels, and other parameters may be adjusted to suit other cell lines as required.

PROTOCOL 13.3. SUBCULTURE IN SUSPENSION

Outline
Withdraw a sample of the cell suspension, count the cells, and seed an appropriate volume of the cell

suspension into fresh medium in a new flask, restoring the cell concentration to the starting level.

Materials
Sterile:
- ☐ Starter culture: HL-60, L1210, or P388, 7 days and 10 days after seeding at 1×10^4 cells/mL, 25-cm² flasks .... 1 each
- ☐ Growth medium, e.g., MEM with Spinner Salts (S-MEM) or RPMI 1640, with 23 mM NaHCO₃, 5% calf serum ...... 200 mL
- ☐ Pipettes, graduated, and plugged .... 1 can of each
  If glass, an assortment of sizes, 1 mL, 5 mL, 10 mL, 25 mL, in a square pipette can, or, if plastic, individually wrapped and sorted by size on a rack
- ☐ Unplugged pipettes for aspirating medium if pump or vacuum line available .... 1 can
- ☐ Universal containers or 50-mL centrifuge tubes 4
- ☐ Stirrer flasks, 500 mL with magnetic pendulum stirrers (Techne, Bellco) 2
- ☐ Culture flasks, 25 cm² 4

Nonsterile:
- ☐ Pipetting aid or bulb (*see* Figs. 5.5, 6.6)
- ☐ Tubing to receiver connected to vacuum line or to receiver via peristaltic pump (*see* Figs. 5.1–5.3)
- ☐ Alcohol, 70%, in spray bottle
- ☐ Lint-free swabs or wipes
- ☐ Absorbent paper tissues
- ☐ Pipette cylinder containing water and disinfectant (*see* Sections 5.8.8, 7.8.5)
- ☐ Marker pen with alcohol-insoluble ink
- ☐ Notebook, pen, protocols, etc.
- ☐ Hemocytometer or electronic cell counter
- ☐ Magnetic stirrer platform

Protocol

1. Prepare the hood, and bring the reagents and materials to the hood to begin the procedure (*see* Section 6.5).
2. Examine the culture carefully for signs of contamination or deterioration. This step is more difficult with suspension cultures than with monolayer cells, but cells that are in poor condition are indicated by shrinkage, an irregular outline, and/or granularity. Healthy cells should look clear and hyaline, with the nucleus visible on phase contrast, and are often found in small clumps in static culture.
3. Take the cultures to the sterile work area, remove a sample from them, and count the cells in the samples.

4. Based on the previously described criteria (*see* Section 13.7.1) and on your knowledge of the behavior of the culture, decide whether or not to subculture. If subculture is required (Exercise 12 will require subculture), proceed as follows.
5. Mix the cell suspension, and disperse any clumps by pipetting the cell suspension up and down.
6. Add 50 mL of medium to each of two stirrer flasks.
7. Add 5 mL of medium to each of four 25-cm^2 flasks.
8. Add a sufficient number of cells to give a final concentration of 1×10^5 cells/mL for slow-growing cells (36–48 h doubling time) or 2×10^4/mL for rapidly growing cells (12–24 h doubling time). For the cells cited in Materials, use a final concentration of 1×10^4 cells/mL.
9. Gas the cultures with 5% CO_2.
10. Cap the flasks, and take to incubator. Lay the flasks flat, as for monolayer cultures.
11. Cap the stirrer flasks and place on magnetic stirrers set at 60–100 rpm, in an incubator or hot room at 37°C. Take care that the stirrer motor does not overheat the culture. Insert a polystyrene foam mat under the bottle if necessary. Induction-driven stirrers generate less heat and have no moving parts.

Suspension cultures have a number of advantages (*see* Table 13.5). First, the production and harvesting of large quantities of cells may be achieved without increasing the surface area of the substrate (*see* Section 26.1). Furthermore, if dilution of the culture is continuous and the cell concentration is kept constant, then a steady state can be achieved; this steady state is not readily achieved in monolayer cultures. Maintenance of monolayer cultures is essentially cyclic, with the result that growth rate and metabolism vary, depending on the phase of the growth cycle.

13.7.6 Standardization of Culture Conditions

Standardization of culture conditions is essential for maintaining phenotypic stability. Although some conditions may alter because of the demands of experimentation, development, and production, routine maintenance should adhere to standard, defined conditions.

Medium. The type of medium used will influence the selection of different cell types and regulate their phenotypic

TABLE 13.6. Data Record, Feeding

Date . . . . Time . . . . Operator . . . .

	Date:				
Cell line	Designation				
	Primary or subculture				
	Generation or pass no.				
Status	Phase of growth cycle				
	Appearance of cells				
	Density of cells				
	pH of medium (approx.)				
	Clarity of medium				
Medium	Type				
	Batch no.				
	Serum type and concentration				
	Batch no.				
	Other additives				
	CO_2 concentration				
Other parameters					

TABLE 13.7. Data Record, Subculture

Date . . . . *Time* . . . . *Operator* . . . .

	Date:				
Cell line	Designation				
	Generation or pass no.				
Status before subculture	Phase of growth cycle				
	Appearance of cells				
	Density of cells				
	pH of medium (approx.)				
	Clarity of medium				
Dissociation agent	Prewash				
	Trypsin				
	EDTA				
	Other				
	Mechanical				
Cell count	Concentration after resuspension (C_I)				
	Volume (V_I)				
	Yield ($Y = C_I \times V_I$)				
	Yield per flask				
Seeding	Number (N) & type of vessel (flask, dish, or plate wells)				
	Final concentration (C_F)				
	Volume per flask, dish, or well (V_F)				
	Split ratio ($Y \div C_F \times V_F \times N$), or number of flasks seeded $\div$ number of flasks trypsinized, where the flasks are of same size				
Medium/serum	Type				
	Batch no.				
	Serum type and concentration				
	Batch no.				
	Other additives				
	CO_2 concentration				
Matrix coating	e.g., fibronectin, Matrigel, collagen				
Other parameters					

expression (*see* Sections 9.6, 10.2.1, 10.2.2, 17.2, 17.7). Consequently, once a medium has been selected, standardize on that medium, and preferably on one supplier, if the medium is being purchased ready made.

Serum. The best method of eliminating serum variation is to convert to a serum-free medium (*see* Section 10.2), although, unfortunately, serum-free formulations are not yet available for all cell types, and the conversion may be costly and time consuming. Serum substitutes (*see* Section 10.5.2) may offer greater consistency and are generally cheaper than serum or growth factor supplementation but do not offer the control over the physiological environment afforded by serum-free medium. If serum is required, select a batch (*see* Section 9.6.1), and use that batch throughout each stage of culture including cryopreservation (*see* Section 20.2).

Plastics. Most of the leading brands of culture flasks and dishes will give similar results, but there may be minor variations due to the treatment of the plastic for tissue culture (*see* Section 8.1.2). Hence it is preferable to adhere to one type of flask or dish and supplier.

Cell line maintenance. Cell lines may alter their characteristics if maintained differently from a standard regime (*see* Section 13.6). The maintenance regime should be optimized (*see* Section 13.7) and then remain consistent throughout the handling of the cell line (*see* Sections 13.7.1, 13.7.2, 13.7.3). Cultures should also be replaced from frozen stocks at regular intervals (*see* Section 20.4.2).

13.7.7 Use of Antibiotics

The continuous use of antibiotics encourages cryptic contaminations, particularly mycoplasma, and the development of antibiotic-resistant organisms (*see* Section 9.4.7). It may also interfere with cellular processes under investigation. However, there may be circumstances for which contamination is particularly prevalent or a particularly valuable cell line is being carried, and in these cases, antibiotics may be used. If they are used, then it is important to maintain some

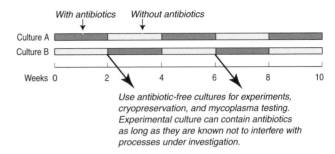

Fig. 13.6. Parallel Cultures and Antibiotics. A suggested scheme for maintaining parallel cell cultures with and without antibiotics, such that each culture always spends part of the time out of antibiotics.

antibiotic-free stocks in order to reveal any cryptic contaminations; these stocks can be maintained in parallel, and stock may be alternated in and out of antibiotics (*see* Fig. 13.6) until antibiotic-free culture is possible. It is not advisable to adopt this procedure as a permanent regime, and, if a chronic contamination is suspected, the cells should be discarded or the contamination eradicated (*see* Sections 19.4.4, 28.6, 28.8.3), and then you may revert to antibiotic-free maintenance.

13.7.8 Maintenance Records

Keep details of routine maintenance, including feeding and subculture (Tables 13.6, 13.7), and deviations or changes should be added to the database record for that cell line. Such records are required for GLP [Food and Drug Administration, 1992; Department of Health and Social Security, 1986; Organisation for Economic Co-operation and Development, 2004], as for primary culture records (*see* 12.3.11), but are also good practice in any laboratory. If standard procedures are defined, then entries need say only "Fed" or "Subcultured," with the date and a note of the cell numbers. Any comments on visual assessment and any deviation from the standard procedure should be recorded as well.

This set of records forms part of the continuing provenance of the cell line, and all data, or at least any major event—e.g., if the medium supplier is changed, the line is cloned, transfected, or changed to serum-free medium—should be entered in the database.

CHAPTER 14

Cloning and Selection

It can be seen from the preceding two chapters that a major recurrent problem in tissue culture is the preservation of a specific cell type and its specialized properties. Although environmental conditions undoubtedly play a significant role in maintaining the differentiated properties of specialized cells in a culture (*see* Sections 17.1, 17.7), the selective overgrowth of unspecialized cells and cells of the wrong lineage remains a major problem.

14.1 CELL CLONING

The traditional microbiological approach to the problem of culture heterogeneity is to isolate pure cell strains by cloning, but, although this technique is relatively easy for continuous cell lines, its success in most primary cultures is limited by poor cloning efficiencies. However, the cloning of primary cultures can be successful, e.g., Sertoli cells [Zwain et al., 1991], juxtaglomerular [Muirhead et al., 1990] and glomerular [Troyer & Kreisberg, 1990] cells from kidney, oval cells from liver [Suh et al., 2003], satellite cells from skeletal muscle [Zeng et al., 2002; McFarland et al., 2003; Hashimoto, 2004], and separation of different lineages from adult stem cell populations [Young et al., 2004].

A further problem of cultures derived from normal tissue is that they may survive only for a limited number of generations (*see* Sections 3.8.1, 18.4.1), and by the time that a clone has produced a usable number of cells, it may already be near senescence (Fig. 14.1). Although cloning of continuous cell lines is more successful than cloning finite cell lines, considerable heterogeneity may still arise within the clone as

it is grown up for use (*see* Section 18.3, Fig. 18.1, and Plate 7). Nevertheless, cloning may help to reduce the heterogeneity of a culture.

Cloning is also used as a survival assay (*see* Sections 21.10, 22.3.3) for optimizing growth conditions (*see* Sections 10.5, 11.6.3) and for determining chemosensitivity and radiosensitivity (*see* Protocol 22.3).

Cloning of attached cells may be carried out in Petri dishes, multiwell plates, or flasks, and it is relatively easy to discern individual colonies. Micromanipulation is the only conclusive method for determining genuine clonality (i.e., that a colony was derived from one cell), but when symmetrical colonies are derived from a single-cell suspension, particularly if colony formation is monitored at the early stages, then it is probable that the colonies are clones.

Cloning can also be carried out in suspension by seeding cells into a gel, such as agar or agarose, or a viscous solution, such as Methocel, with an agar or agarose underlay. The stability of the gel, or viscosity of the Methocel, ensures that daughter cells do not break away from the colony as it forms. Even in monolayer cloning, some cell lines, such as HeLa-S_3 and CHO, are poorly attached, and cells can detach from colonies as they form and generate daughter colonies, which will give an erroneous plating efficiency. This can be minimized by cloning in Methocel without an underlay and allowing the cells to sediment on to the plastic growth surface. Hematopoietic cells are usually cloned in suspension; depending on the cells and growth factors used, the colony generates undifferentiated cells with high repopulation efficiency, *in vivo* or *in vitro*, or may mature into colonies of differentiated hematopoietic cells with very little

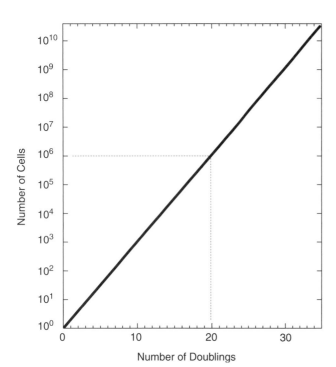

Fig. 14.1. Clonal Cell Yield. The relationship of the cell yield in a clone to the number of population doublings; e.g., 20 doublings are required to produce 10^6 cells.

repopulation efficiency. Cloning then becomes an assay for reproductive potential and stem cell identity.

Continuous cell lines generally have a high plating efficiency in monolayer and in suspension because of their transformed status, whereas normal cells, which may have a moderately high cloning efficiency in monolayer, have a very low cloning efficiency in suspension, because of their need to attach and spread out to enter the cell proliferation cycle. This distinction has allowed suspension cloning to be used as an assay for transformation (*see* Section 18.5.1).

Dilution cloning [Puck & Marcus, 1955] is the technique that is used most widely, based on the observation that cells diluted below a certain density form discrete colonies. The amounts and instruction given in Protocol 14.1 correspond to Exercise 20 but can be varied to suit individual requirements.

PROTOCOL 14.1. DILUTION CLONING

Outline

Seed the cells at low density and incubate until colonies form. Stain the cells (for plating efficiency and survival assays; *see* Plate 6a,e) or isolate them and propagate into a cell strain if they are being used for selection (Fig. 14.2)

Materials

Sterile or aseptically prepared:

- ❏ CHO cells, 25-cm² flask, late log phase 1
- ❏ Medium: Ham's F12, 5% CO_2 equilibrated, 10% FBS .... 250 mL
- ❏ Trypsin, 0.25%, 1:250, or equivalent .... 10 mL
- ❏ Pipettes, 1, 5, 10, and 25 mL .... 10 of each
- ❏ Petri dishes, 6 cm, 1 pack 20
- ❏ Tubes, or universal containers, for dilution .... 4–12 (number depends on final step)

Nonsterile:

- ❏ Hemocytometer or electronic cell counter

Protocol

1. Trypsinize the cells (*see* Protocol 13.2) to produce a single-cell suspension. Undertrypsinizing will produce clumps and overtrypsinizing will reduce the viability of the cells, but it is fundamental to the concept of cloning that the cells be singly suspended. It may be necessary when cloning a new cell line for the first time to try different lengths of trypsinization and different recipes (*see* Table 13.4), to give the optimum plating efficiency from a good single-cell suspension.
2. While the cells are trypsinizing, number the dishes (on the side of the base), and measure out medium for the dilution steps. Up to four dilution steps may be necessary to reduce a regular monolayer accurately to a concentration suitable for cloning.
3. When the cells round up and start to detach, disperse the monolayer in 5 mL medium containing serum or trypsin inhibitor
4. Count the cells, and dilute the cell suspension to 1×10^5 cells/mL, then to a concentration that will give ~50 colonies per Petri dish (*see* Table 14.1). For CHO-K1 this will be 10 cells/mL, and the dilution steps will be:
   (a) Dilute trypsinate to 1×10^5/mL (approximately 1:10 or 1:20, depending on the number of cells in the flask).
   (b) Dilute 200 μL of the 1×10^5/mL suspension to 20 mL (1:100) to give 1×10^3 cells/mL.
   (c) Dilute 200 μL of the 10^3/mL suspension to 20 mL (1:100) to give 10 cells/mL. If you wish to add a variable to the cloning conditions, e.g., a range of serum concentrations, different sera, or a growth factor, prepare a range of tubes at this stage and add 200 μL of the 1×10^3 cells per mL suspension to each of them separately.

If cloning the cells for the first time, choose a range of 10, 20, 50, 100, 200 and 2000 cells/mL (*see* Protocol 21.10) to determine the plating efficiency.

5. Seed 3 Petri dishes each with 5 mL of medium containing cells from the final dilution stage. It is also advisable to seed dishes from the 2000 cells/mL suspension as controls to confirm that cells were present if no clones form at the lower concentration.

6. Place the dishes in a transparent plastic sandwich or cake box.

7. Put the box in a humid CO_2 incubator or gassed sealed container (2–10% CO_2, *see* Section 9.3.2).

8. Leave the cultures untouched for 1 week. If colonies have formed:
   (a) For plating efficiency assay (*see also* Protocol 21.10), stain and count the colonies (*see* Protocol 16.3 and Plate 6).
   (b) For clonal selection, isolate individual colonies (*see* Protocol 14.8).
   If no colonies are visible, replace medium and continue to culture for another week. Feed the dishes again and culture them for a third week if necessary. If no colonies appear by 3 weeks, then it is unlikely that they will appear at all.

Feeding. As the density of cells during cloning is very low, the need to feed the dishes after one week is debatable. Feeding mainly counteracts the loss of nutrients (such as glutamine), which are unstable, and replaces growth factors that have degraded. However, it also increases the risk of contamination, so it is reasonable to leave dishes for two weeks without feeding. If it is necessary to leave the dishes for a third week, then the medium should be replaced, or at least half of it.

The preferential formation of colonies at the center of the plate can be due to incorrect seeding, either from seeding the

TABLE 14.1. Relationship of Seeding Density to Plating Efficiency

Expected plating efficiency (%)	Optimal cell number to be seeded			
	Per mL	Per cm^2	Per dish, 6 cm	Per dish, 9 cm
0.1	1×10^4	2×10^3	40,000	100,000
1.0	1×10^3	200	4000	10,000
10	100	20	400	1000
50	20	4	80	200
100	10	2	40	100

cells into the center of a plate that already contains medium or from swirling the plate such that the cells tend to focus in the center, but it can also be due to resonance in the incubator (*see* Section 8.2.5).

Microtitration plates. If the prime purpose of cloning is to isolate colonies, then seeding into microtitration plates can be an advantage (Fig. 14.3). When the clones grow up, isolation is easy, although the plates have to be monitored at the early stages to mark which wells genuinely have single clones. The statistical probability of a well having a single clone can be increased by reducing the seeding density to a level such that only 1 in 5 or 10 wells would be expected to have a colony; e.g., from Table 14.1, 100 cells/mL at 10% plating efficiency would give 10 colonies/mL, or 1 colony/0.1 mL, as added to a microtitration plate—i.e., 1 colony/well. If the seeding concentration is reduced to 10 cells/mL, then, theoretically, only 1 in 10 wells will contain a colony, and the probability of wells containing more than one colony is very low.

14.2 STIMULATION OF PLATING EFFICIENCY

When cells are plated at low densities, the rate of survival falls in all but a few cell lines. This does not usually present a severe problem with continuous cell lines, for which the plating efficiency seldom drops below 10%, but with primary cultures and finite cell lines, the plating efficiency may be quite low—0.5–5%, or even zero. Numerous attempts have been made to improve plating efficiencies, based on the assumption either that cells require a greater range of nutrients at low densities, because of loss by leakage, or that cell-derived diffusible signals or conditioning factors are present in high-density cultures and are absent or too dilute at low densities. The intracellular metabolic pool of a leaky cell in a dense population will soon reach equilibrium with the surrounding medium, but that of an isolated cell never will. This principle was the basis of the capillary technique of Sanford et al. [1948], by which the L929 clone of L-cells was first produced. The confines of the capillary tube allowed the cell to create a locally enriched environment that mimicked a higher cell concentration. In microdrop techniques developed later, the cells were seeded as a microdrop under liquid paraffin, again maintaining a relatively high cell concentration, keeping one colony separate from another, and facilitating subsequent isolation. As media improved, however, plating efficiencies increased, and Puck and Marcus [1955] were able to show that cloning cells by simple dilution (as described in Protocol 14.1) in association with a feeder layer of irradiated mouse embryo fibroblasts (*see* Protocol 14.3) gave acceptable cloning efficiencies, although subsequent isolation required trypsinization from within a collar placed over each colony (*see* Protocol 14.6).

14.2.1 Conditions that Improve Clonal Growth
(1) **Medium.** Choose a rich medium, such as Ham's F12, or a medium that has been optimized for the cell

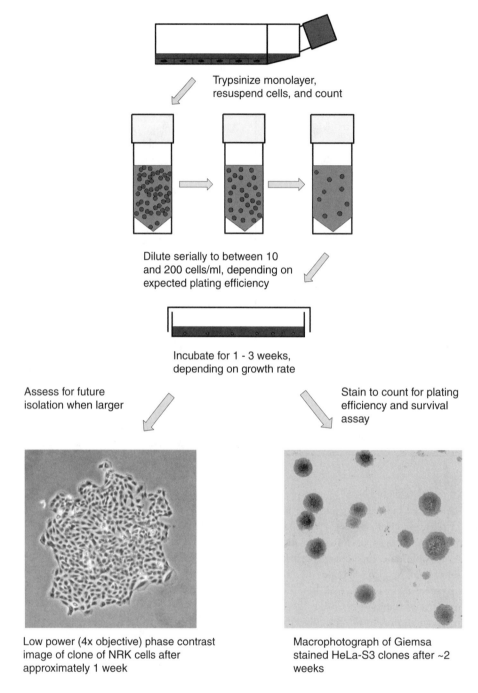

Trypsinize monolayer, resuspend cells, and count

Dilute serially to between 10 and 200 cells/ml, depending on expected plating efficiency

Incubate for 1 - 3 weeks, depending on growth rate

Assess for future isolation when larger

Stain to count for plating efficiency and survival assay

Low power (4x objective) phase contrast image of clone of NRK cells after approximately 1 week

Macrophotograph of Giemsa stained HeLa-S3 clones after ~2 weeks

Fig. 14.2. Dilution Cloning. Cells are from a trypsinized monolayer culture that has been counted and diluted sufficiently to generate isolated colonies. When clones form, they may be isolated (*see* Figs. 14.7; 14.8). If isolation is not required and the cloning is being performed for quantitative assay (*see* Protocols 21.10, 22.3), the colonies are fixed, stained, and counted. (*See also* Plate 6a,e).

type in use (e.g., MCDB 110 [Ham, 1984] for human fibroblasts, Ham's F12 or MCDB 302 for CHO [Ham, 1963; Hamilton & Ham, 1977]) (*see* Sections 9.6, 10.5, Tables 10.1, 10.2 and Chapter 23).

(2) **Serum.** When serum is required, fetal bovine is generally better than calf or horse. Select a batch for cloning

experiments that gives a high plating efficiency during tests (*see* Protocol 21.10).

(3) **Hormones.** Insulin, 1×10^{-10} IU/mL, has been found to increase the plating efficiency of several cell types [Hamilton & Ham, 1977]. Dexamethasone, 2.5×10^{-5} M, 10 μg/mL, a soluble synthetic hydrocortisone

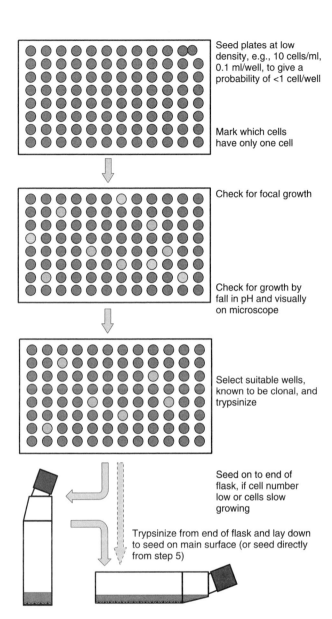

Seed plates at low density, e.g., 10 cells/ml, 0.1 ml/well, to give a probability of <1 cell/well

Mark which cells have only one cell

Check for focal growth

Check for growth by fall in pH and visually on microscope

Select suitable wells, known to be clonal, and trypsinize

Seed on to end of flask, if cell number low or cells slow growing

Trypsinize from end of flask and lay down to seed on main surface (or seed directly from step 5)

Fig. 14.3. Cloning in Microtitration Plates. Cells are seeded at a low enough concentration to give a probability of <1 cell/well, so that some wells will have 1 cell/well, which can be checked by visual examination on the microscope a few hours after plating. Those wells are marked and followed, and when a fall in pH indicates growth, this is confirmed by microscopic observation and the colony is isolated by trypsinization.

analog, improves the plating efficiency of chick myoblasts and human normal glia, glioma, fibroblasts, and melanoma and gives increased clonal growth (colony size) if removed five days after plating [Freshney et al., 1980a, b]. Lower concentrations (e.g., 1×10^{-7} M) have been found to be preferable for epithelial cells (*see* Sections 23.2.1, 23.2.4, 23.2.7).

(4) **Intermediary Metabolites.** Oxo-acids (previously known as keto-acids)—e.g., pyruvate or α-oxoglutarate

(α-ketoglutarate) [Griffiths & Pirt, 1967; McKeehan & McKeehan, 1979] and nucleosides [α-MEM; Stanners et al., 1971]—have been used to supplement media and are already included in the formulation of a rich medium, such as Ham's F12. Pyruvate is also added to Dulbecco's modification of Eagle's MEM [Dulbecco & Freeman, 1959; Morton, 1970].

(5) **Carbon Dioxide.** CO_2 is essential for obtaining maximum cloning efficiency for most cells. Although 5% CO_2 is usually used, 2% is sufficient for many cells and may even be slightly better for human glia and fibroblasts. HEPES (20 mM) may be used with 2% CO_2, protecting the cells against pH fluctuations during feeding and in the event of failure of the CO_2 supply. Using 2% CO_2 also cuts down on the consumption of CO_2. At the other extreme, Dulbecco's modification of Eagle's Basal Medium (DMEM) is normally equilibrated with 10% CO_2 and is frequently used for cloning myeloma hybrids for monoclonal antibody production. The concentration of bicarbonate must be adjusted if the CO_2 tension is altered, so that equilibrium is reached at pH 7.4 (*see* Table 9.1).

(6) **Treatment of Substrate.** Polylysine improves the plating efficiency of human fibroblasts in low serum concentrations [McKeehan & Ham, 1976a] (*see* Section 8.4):

(a) Add 1 mg/mL of poly-D-lysine in UPW to the plates ($\sim$5 mL/25 cm^2).

(b) Remove and wash the plates with 5 mL of D-PBSA per 25 cm^2. The plates may be used immediately or stored for several weeks before use.

Fibronectin also improves the plating of many cells [Barnes and Sato, 1980]. The plates may be pretreated with 5 μg/mL of fibronectin incorporated in the medium.

(7) **Trypsin.** Purified (twice recrystallized) trypsin used at 0.05 μg/mL may be preferable to crude trypsin, but opinions vary. McKeehan [1977] noted a marked improvement in plating efficiency when trypsinization (using pure trypsin) was carried out at 4°C. The introduction of recombinant trypsin, TrypZean™ (Sigma) and TrypLE™ (GIBCO), also gives the opportunity to use a more highly purified trypsin of nonanimal origin; Invitrogen claim that GIBCO TrypLE™ enhances plating efficiency in A549 (with serum) and MDCK cells (serum free).

14.2.2 Conditioned Medium

Medium that has been used for the growth of other cells acquires metabolites, growth factors, and matrix products from these cells. This conditioned medium can improve the plating efficiency of some cells if it is diluted into the regular growth medium.

PROTOCOL 14.2. PREPARATION OF CONDITIONED MEDIUM

Outline
Harvest medium from homologous cells, or a different cell line, from the late log phase. Filter, and dilute with fresh medium as required.

Materials
- Cells for conditioning: same cell line, another cell line (e.g., 3T3 cells), or mouse embryo fibroblasts (*see* Protocols 12.1, 12.5, 12.6)
- Cloning medium: Ham's F12, with 10% FBS, or as appropriate for the cells to be cloned.
- Sterilizing filter: 0.45 μm or 0.22 μm, filter flask

Protocol
1. Grow conditioning cells to 50% of confluence.
2. Change the medium, and incubate for a further 48 h.
3. Collect the medium.
4. Centrifuge the medium at 1000 *g* for 10 min.
5. Filter the medium through a 0.45-μm sterilizing filter. (The medium may need to be clarified first by prefiltration through 5-μm and 1.2-μm filters; *see* Protocol 11.14).
6. Store the medium frozen at −20°C.
7. Thaw the medium before use, and add it to cloning medium in the following proportion: 1 part conditioned medium to 2 parts cloning medium.

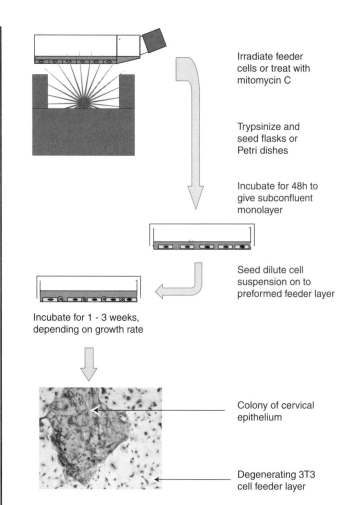

Irradiate feeder cells or treat with mitomycin C

Trypsinize and seed flasks or Petri dishes

Incubate for 48h to give subconfluent monolayer

Seed dilute cell suspension on to preformed feeder layer

Incubate for 1 - 3 weeks, depending on growth rate

Colony of cervical epithelium

Degenerating 3T3 cell feeder layer

Fig. 14.4. Feeder Layers. Cells are irradiated and trypsinized (or may be trypsinized first and then irradiated in suspension, or treated with mitomycin C) and seeded at a low density to enhance cloning efficiency. (Photo courtesy of M. G. Freshney.)

Note. Centrifugation, freezing and thawing, and filtration steps all help to avoid the risk of carrying any cells over from the conditioning cells. If the same cells are used for conditioning as for cloning, then this problem is less important, but better cloning may be obtained by using a different cell line or primary mouse fibroblasts. If a cell strain is derived by this method, then its identity must be confirmed (e.g., *see* Protocols 16.10, 16.11, 16.12) to preclude cross-contamination from the conditioned medium.

14.2.3 Feeder Layers
The reason that some cells do not clone well is related to their inability to survive at low cell densities. One way to maintain cells at clonogenic densities but, at the same time, to mimic high cell densities, is to clone the cells onto a growth-arrested feeder layer (Fig. 14.4). The feeder cells may provide nutrients, growth factors, and matrix constituents that enable the cloned cells to survive more readily.

PROTOCOL 14.3. PREPARATION OF FEEDER LAYERS

Outline
Seed homologous or heterologous cells—e.g., from mouse embryo, rendered nonproliferative by irradiation or drug treatment—at medium density before the cloning of test cells.

Materials
Sterile or aseptically prepared:
- Secondary culture of 13-day mouse embryo fibroblasts (*see* Protocols 12.6 and 13.2)
- Culture medium for cells to be cloned
- Mitomycin C, 100 μg/mL stock, in HBSS or serum-free medium

Nonsterile:
- ☐ X ray or ^{60}Co source capable of delivering 30 Gy in 30 min or less (instead of mitomycin C)

Protocol
1. Trypsinize embryo fibroblasts from the primary culture (*see* Protocols 12.6 and 13.2) and reseed the cells at 1×10^5 cells/mL.
2. Block further proliferation by one of the following four methods:
   (a) *By irradiation*
   i) Expose cultures, in situ in flasks, to 60 Gy from an X-ray machine or source of γ-radiation such as ^{60}Co.
   ii) Expose trypsinized suspension to 60 Gy from an X-ray machine or source of γ-radiation such as ^{60}Co, and reseed at 1×10^5 cells/mL. Alternatively, the irradiated cell suspension may be stored at 4°C for up to 5 days.
   (b) *By treatment with mitomycin C*
   i) Add mitomycin C to subconfluent cells at 0.25 µg/mL, incubate at 37°C overnight (~18 h) [Macpherson & Bryden, 1971], and replace the medium.
   ii) Add at a final concentration of 20 µg/mL to a trypsinized suspension of 1×10^7 cells/mL (giving 2 µg/10^6 cells), incubate for 1 h at 37°C, wash by centrifugation (4 × 10 mL) to remove the mitomycin C, and reseed at 1×10^5 cells/mL [Stanley, 2002] or store at 4°C for up to 5 days.
3. After a further 24–72 h in culture, trypsinize the cells and reseed, or reseed stored cells directly, in fresh medium at 5×10^4 cells/mL (1×10^4 cells/cm²).
4. Incubate the culture for a further 24–48 h, and then seed the cells for cloning.

The feeder cells will remain viable for up to 3 weeks, but will eventually die out and are not carried over if the colonies are isolated. Other cell lines or homologous cells may be used to improve the plating efficiency, but heterologous cells have the advantage that if clones are to be isolated later, chromosome analysis will rule out accidental contamination from the feeder layer. Other cell lines that have been used for feeder cells include 3T3, MRC-5, and STO cells. Early-passage mouse embryo cells probably produce more matrix components than established cell lines, but screening different cells is the only way to be sure which type of cell is best for a particular application.

Cells may vary in their sensitivity to irradiation or mitomycin C so a trial run should be carried out before cloning is attempted, to ensure that none of the feeder cells survives (*see also* Chapter 2, Exercise 21, Experimental Variation 3). Even then, it is advisable to seed two or three feeder layer plates with feeder cells alone, to act as controls, each time cloning is carried out.

14.3 SUSPENSION CLONING

Some cells, particularly hematopoietic stem cells and virally transformed fibroblasts, clone readily in suspension. To hold the colony together and prevent mixing, the cells are suspended in agar or Methocel and plated on an agar underlay or into dishes that need not be treated for tissue culture. Protocol 14.4 has been submitted by Mary Freshney, Cancer Research UK Beatson Institute, Garscube Estate, Switchback Road, Bearsden, Glasgow, G61 1BD, United Kingdom (*see* Fig. 14.5 and Protocol 23.23).

PROTOCOL 14.4. CLONING IN AGAR

Outline
Agar is liquid at high temperatures but gels at 37°C. Cells are suspended in warm agar medium and, when incubated after the agar gels, form discrete colonies that may be isolated easily (Fig. 14.5).

Materials
Sterile:
- ☐ Noble agar, Difco
- ☐ Medium at double strength (i.e., Ham's F12, RPMI 1640, Dulbecco's MEM, or CMRL 1066). Prepare the medium from 10× concentrate to half the recommended final volume, and add twice the normal concentration of serum if required.
- ☐ Fetal bovine serum (if required)
- ☐ Growth medium, 1×, for cell dilutions
- ☐ Sterile ultrapure water (UPW)
- ☐ Sterile conical flask
- ☐ Pipettes, including sterile plastic disposable pipettes for agar solutions
- ☐ Universal containers, bijou bottles, or centrifuge tubes for dilution
- ☐ Petri dishes, 3.5 cm, non-tissue-culture grade

Nonsterile:
- ☐ Bunsen burner and tripod
- ☐ Water bath at 55°C
- ☐ Water bath at 37°C
- ☐ Electronic cell counter or hemocytometer
- ☐ Tray

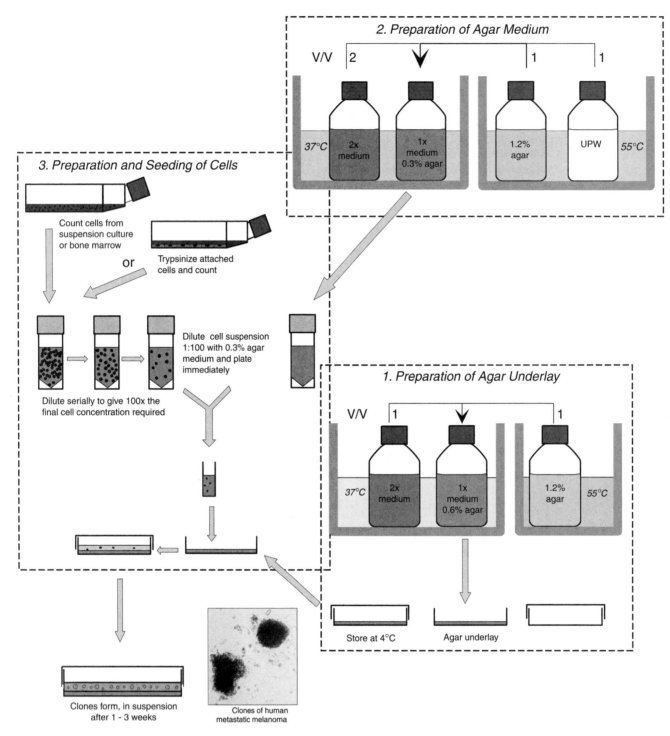

Fig. 14.5. Cloning in Suspension. Cultured cells or primary suspensions from bone marrow or tumors, suspended in agar or low-melting-temperature agarose, which is then allowed to gel, form colonies in suspension. Use of an underlay prevents attachment to the base of the dish. 1. Preparation of agar underlay: Agar, 1.2%, at 55°C is mixed with 2× medium at 37°C and dispensed immediately into dishes, where it is allowed to gel at room temperature or 4°C. 2. Preparation of agar medium: Agar, 1.2%, and UPW are maintained at 55°C and mixed with 2× medium to give 0.3% agar for cloning. The use of low-melting-point agarose allows all solutions to be maintained at 37°C, but this agarose can be more difficult to gel. 3. Cells grown in suspension, derived from bone marrow, or trypsinized from an attached monolayer are counted and diluted serially, and the final product is diluted with agar or agarose and seeded onto an agar underlay.

Note. Before preparing the medium and the cells, work out the cell dilutions and label the Petri dishes. For an assay to measure the cloning efficiency of a cell line, prepare to set up three dishes for each cell dilution. Convenient cell numbers per 3.5-cm dish are 1,000, 333, 111, and 37—i.e., serial one-third dilutions of the cell suspension. If any growth factors, hormones, or other supplements are to be added to the dishes, they should be added to the 0.6% agar underlay.

Protocol

1. Number or label the Petri dishes on the side of the base. It is convenient to place them on a tray.
2. Prepare 2× medium containing 40% FBS, and keep it at 37°C.
3. Weigh out 1.2 g of agar.
4. Measure 100 mL of sterile UPW into a sterile conical flask and another 100 mL into a sterile bottle. Add the 1.2 g of agar to the flask. Cover the flask, and boil the solution for 2 min. Alternatively, the agar may be sterilized in the autoclave in advance, but, if subsequently stored, it will still need to be boiled or microwaved, in order to melt it for use.
5. Transfer the boiled agar and the bottle of sterile UPW to a water bath at 55°C.
6. Prepare a 0.6% agar underlay by combining an equal volume of 2× medium and 1.2% agar (Fig. 14.6). Keep the underlay at 37°C. If any growth factors, hormones, or other supplements are being used, they should be added to the underlay medium at this point.

Note. If a titration of growth factors is being carried out or a selection of different factors is being used, add the required amount to the Petri dishes before the underlay is added.

7. Add 1 mL of 0.6% agar medium to the dishes, mix, and ensure that the medium covers the base of the dish. Leave the dishes at room temperature to set (*see* Fig. 14.5).
8. Prepare the cell suspension, and count the cells.
9. Prepare 0.3% agar medium, and keep it at 37°C. This medium may be prepared by diluting 2× medium at 37°C with 1.2% agar at 55°C and UPW at 55°C in the respective proportions of 2:1:1 (*see* Fig. 14.5).
10. Prepare the following cell dilutions, making the top concentration of cells 1×10^5/mL:
    (a) 1×10^5/mL.
    (b) Dilute 1×10^5/mL by 1/3 to give 3.3×10^4/mL.
    (c) Dilute 3.3×10^4/mL by 1/3 to give 1.1×10^4/mL.
    (d) Dilute 1.1×10^4/mL by 1/3 to give 3.7×10^3/mL.
11. Label four bijou bottles or tubes, one for each dilution, and pipette 40 μL of each cell dilution, including the 1×10^5/mL concentration, into the respective container. Add 4 mL of 0.3% agar medium at 37°C to each container, mix, and pipette 1 mL from each container onto each of three Petri dishes (Fig. 14.5). This will give final concentrations as follows:
    (a) 1×10^3/mL/dish
    (b) 330/mL/dish
    (c) 110/mL/dish
    (d) 37/mL/dish

Note. Always be sure that the agar medium for the top layer has had adequate time to cool to 37°C before adding the cells to it.

12. Allow the solution in the Petri dishes to gel at room temperature.
13. Put the Petri dishes into a clean plastic box with a lid, and incubate them at 37°C in a humid incubator for 10 days.

Agarose, which has a reduced component of sulfated polysaccharides, can be substituted for agar. Some types of agarose have a lower gelling temperature and can be manipulated more easily at 37°C. They are gelled at 4°C and then are returned to 37°C.

Because of the complexity of handling melted agar with cells, and the impurities that may be present in agar, some laboratories prefer to use Methocel, which is a viscous solution and not a gel [Buick et al., 1979]. It has a higher viscosity when warm. Because it is a sol and not a gel, cells will sediment through it slowly. It is, therefore, essential to use an underlay with Methocel. Colonies form at the interface between the Methocel and the agar (or agarose) underlay, placing themselves in the same focal plane and making analysis and photography easier.

PROTOCOL 14.5. CLONING IN METHOCEL

Outline
Suspend the cells in medium containing Methocel, and seed the cells into dishes containing an agar or agarose underlay (Fig. 14.6).

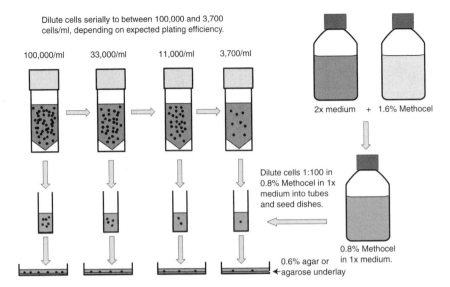

Fig. 14.6. Cloning in Methocel. A series of cell dilutions is prepared as for agar cloning, diluted 1:100 in Methocel medium, and plated into non-tissue-culture-grade dishes or dishes with an agar underlay.

Materials

Sterile:
- ❑ Noble agar, Difco, or agarose
- ❑ Medium at double strength (i.e., Ham's F12, RPMI 1640, Dulbecco's MEM, or CMRL 1066). Prepare the medium from 10× concentrate to half the recommended final volume, and add twice the normal concentration of serum (if required)
- ❑ Fetal bovine serum (if required)
- ❑ Growth medium, 1×, for cell dilutions
- ❑ 1.6% Methocel, 4 Pa-s (4000 centipoises), in UPW; place on ice (*see* Appendix I: Methocel)
- ❑ Sterile ultrapure water (UPW)
- ❑ Sterile conical flask
- ❑ Pipettes, including sterile plastic disposable pipettes for agar solutions
- ❑ Universal containers, bijou bottles, or centrifuge tubes
- ❑ 3.5-cm Petri dishes, non-tissue-culture grade
- ❑ Syringes to dispense Methocel (because of its viscosity Methocel tends to cling to the inside of pipettes, making dispensing inaccurate)

Nonsterile:
- ❑ Bunsen burner and tripod
- ❑ Water bath at 55°C
- ❑ Water bath at 37°C
- ❑ Electronic cell counter or hemocytometer
- ❑ Tray

Protocol

1. Prepare agar underlays as in Protocol 14.4, Steps 1–7.

2. Dilute the Methocel to 0.8% with an equal volume of 2× medium. Mix it well, and keep it on ice.

3. Trypsinize monolayer cells, or collect cells from suspension culture or bone marrow and count them.

4. Prepare the following cell dilutions, making the top concentration of cells 1×10^5/mL:
   (a) 1×10^5/mL.
   (b) Dilute 1×10^5/mL by 1/3 to give 3.3×10^4/mL.
   (c) Dilute 3.3×10^4/mL by 1/3 to give 1.1×10^4/mL.
   (d) Dilute 1.1×10^4/mL by 1/3 to give 3.7×10^3/mL.
   Methocel is viscous, so manipulations are easier to perform with a syringe without a needle.

5. Label four bijou bottles or tubes, one for each dilution, and pipette 40 μL of each cell dilution, including the 1×10^5/mL concentration, into the respective container. Add 4 mL of 0.8% Methocel medium to each container, and mix well with a vortex or, if the cells are known to be particularly fragile, by sucking the solution gently up and down with a syringe several times. Then use a syringe to add 1 mL from each container to each of three Petri dishes (*see* Fig. 14.6). This will give final concentrations as follows:
   (a) 1×10^3/mL/dish
   (b) 330/mL/dish
   (c) 110/mL/dish
   (d) 37/mL/dish

6. Incubate the dishes in a humid incubator until colonies form. Because the colonies form at the interface between the agar and the Methocel, fresh medium may be added, 1 mL per dish or well, after 1 week and then removed and replaced with more fresh medium after 2 weeks without disturbing the colonies.

Many of the recommendations that apply to medium supplementation for monolayer cloning also apply to suspension cloning. In addition, sulfydryl compounds, such as mercaptoethanol (5×10^{-5} M), glutathione (1 mM), and α-thioglycerol (7.5×10^{-5} M) [Iscove et al., 1980], are sometimes used. Macpherson [1973] found that the inclusion of DEAE dextran was beneficial for cloning, a finding that was later confirmed by Hamburger et al. [1978], who also found that macrophages enhanced the cloning of tumor cells, although others have found them to be detrimental. Courtenay et al. [1978] incorporated rat red blood cells into the medium and demonstrated that a low oxygen tension enhanced cloning. The problem may lie with the toxicity of free oxygen; with red blood cells present it will be bound to hemoglobin. It may be possible to mimic this with perfluorocarbons [Lowe et al., 1998].

Most cell types clone in suspension with a lower efficiency than in monolayer, some cells by two or three orders of magnitude. The isolation of colonies is, however, much easier.

14.4 ISOLATION OF CLONES

When cloning is used for the selection of specific cell strains, the colonies that form (*see* Plate 6) need to be isolated for further propagation. If monolayer cells are cloned directly into multiwell plates (*see* Protocol 14.1—Microtitration Plates), then colonies may be isolated by trypsinizing individual wells. It is, however, necessary to confirm the clonal origin of the colony during its formation by regular microscopic observation. If cloning is performed in Petri dishes, there is no physical separation between colonies. This separation must be created by removing the medium and placing a stainless steel or ceramic ring around the colony to be isolated (Fig. 14.7).

PROTOCOL 14.6. ISOLATION OF CLONES WITH CLONING RINGS

Outline

The colony is trypsinized from within a porcelain, glass, PTFE, or stainless steel ring and transferred to

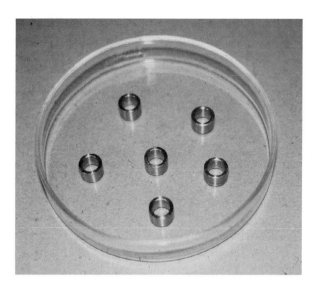

Fig. 14.7. Cloning Rings. Stainless steel rings, cut from tubing, in 9-cm glass Petri dish. Porcelain (Fisher), thick-walled glass (Scientific Laboratory Supplies), or plastic (e.g., cut from nylon, silicone, or Teflon thick-walled tubing) can also be used. Whatever the material, the base must be smooth in order to seal with silicone grease onto the base of the Petri dish, and the internal diameter must be just wide enough to enclose one whole clone without the external diameter overlapping adjacent clones.

---

one of the wells of a 24- or 12-well plate, or directly to a 25-cm² flask (Fig. 14.8).

Materials
Sterile:
- ❏ Cloning rings (*see* Appendix II); sterilize in a glass Petri dish by dry heat or autoclave
- ❏ Silicone grease; sterilize in a glass Petri dish by dry heat, 160°C for 1 h, or autoclave at 121°C for 15 min
- ❏ Yellow tips for pipettor, or Pasteur pipettes with a bent end (Bellco #1273)
- ❏ Trypsin 0.25% in D-PBSA
- ❏ Growth medium
- ❏ Multiwell plate, 24-well, and/or 25-cm² flasks
- ❏ Sterile forceps

Nonsterile:
- ❏ Pipettor, 50–100 µL
- ❏ Felt-tip pen or, preferably, a Nikon ring marker or object marker that fits into the objective nosepiece of a microscope in place of one of the objectives

Protocol
1. Examine the clones, and mark those that you wish to isolate with a felt-tip marker on the underside of the dish, or use a ring marker (Nikon).

2. Remove the medium from the dish, and rinse the clones gently with D-PBSA.
3. Using sterile forceps, take one cloning ring, dip it in silicone grease, and press it down on the dish alongside the silicone grease, to spread the grease around the base of the ring.
4. Place the ring around the desired colony.
5. Repeat Steps 4 and 5 for two or three other colonies in the same dish.
6. Add sufficient 0.25% trypsin to fill the hole in the ring (0.1–0.4 mL, depending on the internal diameter of the ring).
7. Leave the trypsin for 20 s, and then remove it.
8. Close the dish, and incubate it for 15 min at 37°C.
9. Add 0.1–0.4 mL of medium to each ring.
10. Taking each clone in turn, pipette the medium up and down to disperse the cells, and transfer the medium to a well of a 24-well plate or to a 25-cm^2 flask standing on end. Use a separate pipette, or a separate pipettor tip, for each clone.
11. Wash out the ring with another 0.1–0.4 mL of medium, and transfer the medium to the same well or flask.

Note. The dish will dry out if left open for too long. Either limit the number of clones isolated or cover the dish between manipulations.

12. Make up the medium in the wells to 1.0 mL, close the plate, and incubate it. If you are using flasks, then add 1 mL of medium to each flask and incubate the flasks standing on end.
13. When the clone grows to fill the well, subculture to a 25-cm^2 flask, incubated conventionally with 5 mL of medium. If you are using the upended flask technique (*see* Fig. 14.8), then remove the medium when the end of a flask is confluent, trypsinize the cells, resuspend them in 5 mL of medium, and lay the flask down flat. Continue the incubation.

Flasks are available with a removable top film (Nalge Nunc) that may be peeled off to allow harvesting of clones. Alternatively, where an irradiation source is available, clones may be isolated by shielding one with a lead disk and irradiating the rest of the monolayer with 30 Gy (*see* Protocol 14.7).

PROTOCOL 14.7. ISOLATING CELL COLONIES BY IRRADIATION

Outline
Invert the flask under an X-ray machine or ^{60}Co source, screening the desired colony with lead.

Δ *Safety Note.* X-ray machines and ^{60}Co sources must be used under strict supervision and with appropriate monitoring to safeguard your own exposure and that of others (*see* Section 7.7.4). Contact your local radioprotection officer before setting up this type of experiment.

Materials
Sterile:
❑ Growth medium
❑ 0.25% trypsin
❑ D-PBSA
Nonsterile:
❑ X ray or cobalt γ-source
❑ Pieces of lead cut from a 2-mm-thick sheet and of a size from about 2 to 5 mm in diameter

Protocol
1. Select the desired colony, and mark it with a felt-tip pen or a Nikon ring marker.
2. Select a piece of lead of appropriate diameter.
3. Take the flask to the radiation source.
4. Invert the flask under the source.
5. Cover the colony with a 2-mm-thick piece of lead.
6. Irradiate the flask with 30 Gy.
7. Return the flask to the sterile area.
8. Remove the medium, trypsinize the cells, and allow the cells to reestablish in the same bottle, using the irradiated cells as a feeder layer.

If irradiation and trypsinization are carried out when the colony is about 100 cells in size, then the trypsinized cells will reclone, given a reasonably high cloning efficiency. Three serial clonings may be performed within six weeks by this method.

14.4.1 Other Isolation Techniques for Monolayer Clones
(1) Distribute small coverslips or broken fragments of coverslips on the bottom of a Petri dish. When plated out at the correct density, some colonies are found to be singly distributed on a piece of glass and may be transferred to a fresh dish or multiwell plate.
(2) Use the capillary technique of Sanford et al. [1948]. A dilute cell suspension is drawn into a sterile glass

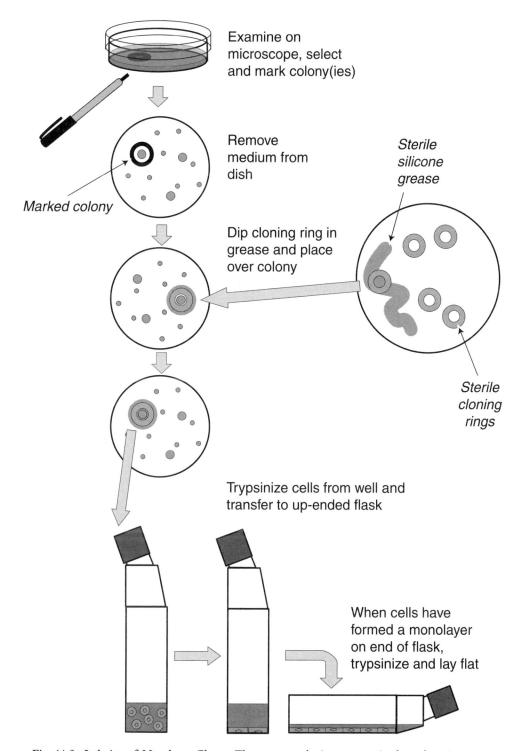

Fig. 14.8. Isolation of Monolayer Clones. The mature colonies are examined on the microscope, and suitable colonies are selected and marked. The medium is removed, and cloning rings, dipped in silicone grease, are placed around each colony, which is then trypsinized from within the ring.

capillary tube (e.g., a 50-µL Drummond Microcap), allowing colonies to form inside the tube. The tube is then carefully broken on either side of a colony and transferred to a fresh plate. This technique was exploited by Echarti and Maurer [1989, 1991] for clonogenic assay

of hematopoietic cells and tumor cells, for which the colony-forming efficiency can be quantified by scanning the capillary in a densitometer.

(3) Cells may be cloned in the OptiCell chamber, which is made up of two opposing growth surfaces of thin flexible

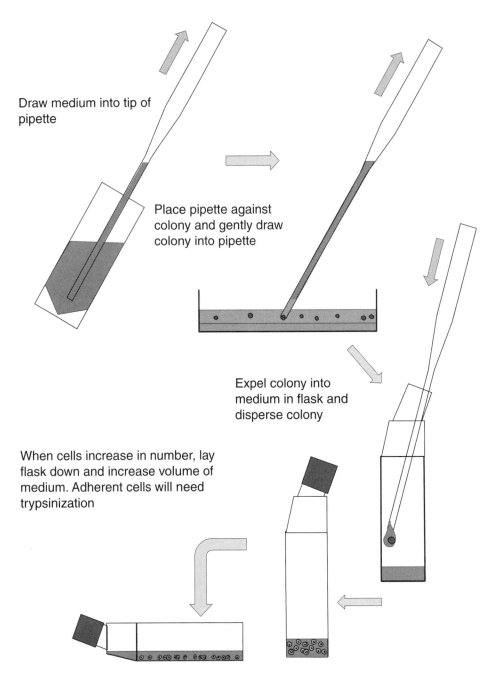

Draw medium into tip of
pipette

Place pipette against
colony and gently draw
colony into pipette

Expel colony into
medium in flask and
disperse colony

When cells increase in number, lay
flask down and increase volume of
medium. Adherent cells will need
trypsinization

Fig. 14.9. Isolation of Suspension Clones. Mark the colony as for monolayer, then draw the colony
into a Pasteur pipette (or pipettor tip). Transfer the colony to a culture flask, disperse it in medium,
and incubate it. Make up medium when cells start to grow.

plastic that may be cut with a scalpel or scissors. Provided
the outer surfaces are kept sterile, this can be used to cut
out segments with colonies, which are then trypsinized
into a multiwell plate or flask.

14.4.2 Suspension Clones
The isolation of colonies growing in suspension is simple but
requires a dissection microscope.

**PROTOCOL 14.8. ISOLATION OF
SUSPENSION CLONES**

Outline
Draw the colony into a pipettor or Pasteur pipette,
and transfer the colony to a flask or the well of a
multiwell plate (Fig. 14.9).

Materials

Sterile:
- ❑ Growth medium in universal container
- ❑ Multiwell plates, 24 well
- ❑ Culture flask, 25 cm²
- ❑ Yellow tips for pipettor or Pasteur pipettes

Nonsterile:
- ❑ Dissecting microscope, 20–50× magnification
- ❑ Pipettor, 100 μL
- ❑ Felt-tip pen or Nikon ring marker

Protocol

1. Examine the dishes on an inverted microscope, and mark the colonies with a felt-tip pen or a Nikon ring marker.
2. Pipette 1 mL of medium into each well of a 24-well plate.
3. Pick the colonies, using a dissecting microscope:
   (a) Use a separate pipettor tip or Pasteur pipette for each colony.
   (b) Set the pipettor to 100 μL.
   (c) Draw approximately 50 μL of medium into the pipette tip, place the tip of the pipette against the colony to be isolated, and gently draw in the colony.
3. Transfer the contents of the pipette to a 24-well dish, and flush out the colony with medium. If the medium is made with Methocel the colony will settle, and if the cells are adherent they will attach and grow out. Cells that normally grow in suspension will settle but, of course, will not attach. If the medium is made from agar, you may need to pipette the colony up and down a few times in the well to disperse the agar. Clones may also be seeded directly into a 25-cm² plastic flask that is standing on end (*see* Protocol 14.6).

14.5 REPLICA PLATING

Bacterial colonies can be replated by pressing a moist pad gently down onto colonies growing on a nutrient agar plate and transferring the pad to a second, fresh agar plate. Various attempts have been made to adapt this technique to cell culture, usually by placing a mesh screen or filter over monolayer clones and transferring it to a fresh dish after a few days [Hornsby et al., 1992]. For clones that have been developed in microtitration plates, there are a number of transfer devices available—e.g., the Corning Transtar (*see* Fig. 5.13), which can be used with suspension cultures directly after agitating the culture or with monolayer cultures after trypsinization and resuspension.

14.6 SELECTIVE INHIBITORS

Manipulating the conditions of a culture by using a selective medium is a standard method for selecting microorganisms. Its application to animal cells in culture is limited, however, by the basic metabolic similarities of most cells isolated from one animal, in terms of their nutritional requirements. The problem is accentuated by the effect of serum, which tends to mask the selective properties of different media. Most selective media that have been shown to be generally successful have been serum-free formulations (*see* Section 10.2.1). A number of metabolic inhibitors, however, have had recurrent success. Gilbert and Migeon [1975, 1977] replaced the L-valine in the culture medium with D-valine and demonstrated that cells possessing D-amino acid oxidase would grow preferentially. Kidney tubular epithelia [Gross et al., 1992], bovine mammary epithelia [Sordillo et al., 1988], endothelial cells from rat brain [Abbott et al., 1992], and Schwann cell cultures [Armati & Bonner, 1990] have been selected in this way. However, this technique appears not to be effective against human fibroblasts [Masson et al., 1993].

Much of the effort in developing selective conditions has been aimed at suppressing fibroblastic overgrowth. Kao and Prockop [1977] used *cis*-OH-proline, although it can prove toxic to other cells. Fry and Bridges [1979] found that phenobarbitone inhibited fibroblastic overgrowth in cultures of hepatocytes, and Braaten et al. [1974] were able to reduce the fibroblastic contamination of neonatal pancreas by treating the culture with sodium ethylmercurithiosalicylate. Fibroblasts also tend to be more sensitive to geneticin (G418) at 100 μg/mL [Levin et al., 1995].

One of the more successful approaches was the development of a monoclonal antibody to the stromal cells of a human breast carcinoma [Edwards et al., 1980]. Used with complement, this antibody proved to be cytotoxic to fibroblasts from several tumors and helped to purify a number of malignant cell lines. Cells may also be killed selectively with drug- or toxin-conjugated antibodies [e.g., Beattie et al., 1990]. However, selective antibodies are used more extensively in "panning" or magnetizable bead separation techniques (*see* Section 15.3.2).

Selective media are also commonly used to isolate hybrid clones from somatic hybridization experiments. HAT medium (*see* Appendix I), a combination of hypoxanthine, aminopterin, and thymidine, selects hybrids with both hypoxanthine guanine phosphoribosyltransferase and thymidine kinase from parental cells deficient in one or the other enzyme (*see* Section 27.9.1) [Littlefield, 1964a].

Transfected cells are also selected by resistance to a number of drugs, such as neomycin, its analog geneticin (G418), hygromycin, and methotrexate, by including a resistance-conferring gene in the construct used for transfection [e.g., *neo* (aminoglycoside phosphotransferase), *hph* (hygromycin B phosphotransferase), or *dhfr* (dihydrofolate reductase); *see* Section 28.5]. Culture in the correct concentration of the selective marker, determined by titration against the transfected and nontransfected controls, selects for stable transfectants. Selection with methotrexate has the additional advantage that increasing the methotrexate concentration leads to amplification of the *dhfr* gene and can coamplify other genes in the construct.

Negative selection is also possible by using the Herpes simplex virus (HSV) *TK* gene, which activates Ganciclovir (Syntex) into a cytotoxic product [Jin et al., 2003]. Transfected cells will be sensitized to the drug.

When a mixture of cells shows different responses to growth factors, it is possible to stimulate one cell type with the appropriate growth factor and then, taking advantage of the increased sensitivity of the more rapidly growing cells, kill the cells selectively with irradiation or cytosine arabinoside (ara-c) (*see* Section 23.4.1). Alternatively, if an inhibitor is known or a growth factor is removed, which will take one population out of the cycle, the remaining cycling cells can be killed with ara-c or irradiation.

14.7 ISOLATION OF GENETIC VARIANTS

Protocol 14.9 for the development of mutant cell lines that amplify the dihydrofolate reductase (DHFR) gene was contributed by June Biedler, Memorial Sloan-Kettering Cancer Center, New York, New York.

PROTOCOL 14.9. METHOTREXATE RESISTANCE AND DHFR AMPLIFICATION

Principle

Cells exposed to gradually increasing concentrations of folic acid antagonists, such as methotrexate (MTX), over a prolonged period of time will develop resistance to the toxic effects of the drug [Biedler et al., 1972]. Resistance resulting from amplification of the DHFR gene generally develops the most rapidly, although other mechanisms—e.g., alteration in antifolate transport and/or mutations affecting enzyme structure or affinity—may confer part or all of the resistant phenotype.

Outline

Expose the cells to a graded series of concentrations of MTX for weeks, periodically replacing the medium with fresh medium containing the same drug concentration. Select for subculturing those flasks in which a small percentage of cells survive and form colonies. Repeatedly subculture such cells in the same and in two- to tenfold higher MTX concentrations until the cells acquire the desired degree of resistance.

Materials
Sterile:
- Chinese hamster cells or rapidly growing human or mouse cell lines
- Methotrexate Sodium Parenteral (Lederle Laboratories)
- 0.15 M NaCl
- Tissue culture flasks
- Pipettes
- Culture medium that does not contain thymidine and hypoxanthine (e.g., Eagle's MEM with 10% fetal bovine serum)

Nonsterile:
- Inverted microscope
- Liquid nitrogen freezer

Protocol
1. Clone the parental cell line to obtain a rapidly growing, genotypically uniform population to be used for selection.
2. Dilute the MTX with sterile 0.15 M (0.85%) NaCl. The drug packaged for use in the clinic is in solution at 2.5 mg/mL.
3. Inoculate 2.5×10^5 cells into replicate 25-cm^2 flasks containing no drug or 0.01, 0.02, 0.05, and 0.1 μg/mL of MTX in complete tissue culture medium. Adjust the pH of each solution to pH 7.4, and incubate the flasks at 37°C for 5–7 days.
4. Observe the cultures with an inverted microscope. Replace the medium with fresh medium containing the same amount of MTX in cultures showing clonal growth of a small proportion of cells amid a background of enlarged, substrate-adherent, and probably dying cells, and reincubate those cultures.
5. Allow the cells to grow for another 5–7 days, changing the growth medium as necessary but continuously exposing the cells to MTX. When the cell density has reached 2–10×10^6 cells/flask, subculture the cells at 2.5×10^5 cells/flask into new flasks containing the same and two- to tenfold higher drug concentrations.
6. After another 5–7 days, observe the new passage flasks as well as the cultures from the previous passage that had been exposed to higher drug concentrations; change the medium and select for viable cells as before.
7. Continue the selection with progressively higher drug concentrations at each subculture step until the desired level of resistance is obtained: 2–3 months for Chinese hamster cells with low to moderate levels of resistance, increase in DHFR activity, and/or transport alteration; 4–6 months or more for high levels of resistance and enzyme

overproduction, for Chinese hamster, mouse, or fast-growing human cells.
8. Periodically freeze samples of the developing lines in liquid nitrogen (*see* Protocol 20.1).

Analysis. Characterize resistant cells for levels of resistance to the drug in a clonal growth assay (*see* Protocol 22.3), for increase in activity or amount of DHFR by biochemical or gel electrophoresis techniques [Albrecht et al., 1972; Melera et al., 1980], and/or for increase in mRNA and copy number of the reductase gene by Northern, Southern, or dot blots [Scotto et al., 1986] with DHFR-specific probes to determine the mechanism(s) of resistance.

Variations. Cell culture media other than Eagle's MEM can be used; the composition of the medium (e.g., folic acid content) can be expected to influence the rate and type of MTX resistance development. Media containing thymidine, hypoxanthine, and glycine (*see* Tables 9.3, 10.1, and 10.2) prevent the development of antifolate resistance and should be avoided. Cells can be treated with chemical mutagens before selection [Thompson & Baker, 1973]; this treatment may also alter the rate and type of mutant selection.

Selection can also be done with cells plated in the drug at low density in 10-cm tissue culture dishes (with the isolation of individual colonies with cloning rings; *see* Protocol 14.6), using single cells in 96-well cluster dishes, or in soft agar, to enable the isolation of one or multiple clonal populations at each or any step during resistance development.

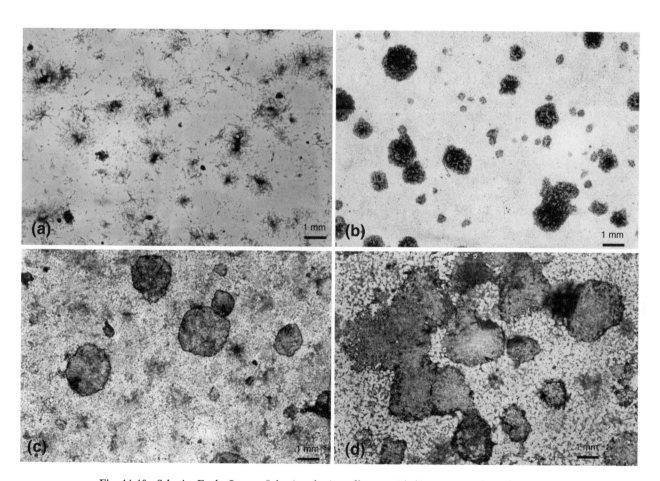

Fig. 14.10. Selective Feeder Layers. Selective cloning of breast epithelium on a confluent feeder layer. (a) Colonies forming on plastic alone after seeding 4000 cells/cm² (2 × 10⁴ cells/mL) from a breast carcinoma culture. Small, dense colonies are epithelial cells, and larger, stellate colonies are fibroblasts. (b) Colonies of cells from the same culture, seeded at 400 cells/cm² (2000 cells/mL) on a confluent feeder layer of FHS74Int cells [Owens et al., 1974]. The epithelial colonies are much larger than those in (a), the plating efficiency is higher, and there are no fibroblastic colonies. (c) Colonies from a different breast carcinoma culture plated onto the same feeder layer. Note the different colony morphology with a lighter-stained center and ring at the point of interaction with the feeder layer. (d) Colonies from normal breast culture seeded onto FHI cells (fetal human intestine; similar to FHS74Int). A few small, fibroblastic colonies are present in (c) and (d). [Smith et al., 1981; A. J. Hackett, personal communication.] (*See also* Plate 6c,d).

Cells can be made to be resistant to a number of other agents, such as antibiotics, other antimetabolites, toxic metals, and so on, by similar techniques; differences in the mechanism of action or degree of toxicity of the agents, however, may require that treatment with the agent be intermittent rather than continuous and may increase the time necessary for selection.

Cell lines of different species or with slower growth rates, such as some human tumor cell lines, may require different (usually lower) initial drug concentrations, longer exposure times at each concentration, and smaller increases in the concentration between selection steps.

Solubilization of MTX other than the Lederle product will require the addition of equimolar amounts of NaOH and sterilization through a 0.2-μm filter.

14.8 INTERACTION WITH SUBSTRATE

14.8.1 Selective Adhesion

Different cell types have different affinities for the culture substrate and attach at different rates. If a primary cell suspension is seeded into one flask and transferred to a second flask after 30 min, a third flask after 1 h, and so on for up to 24 h, then the most adhesive cells will be found in the first flask and the least adhesive in the last. Macrophages will tend to remain in the first flask, fibroblasts in the next few flasks, epithelial cells in the next few flasks, and, finally, hematopoietic cells in the last flask.

If collagenase in complete medium is used for primary disaggregation of the tissue (*see* Protocol 12.8), most of the cells that are released will not attach within 48 h unless the collagenase is removed. However, macrophages migrate out of the fragments of tissue and attach during this period and can be removed from other cells by transferring the disaggregate to a fresh flask after 48–72 h of treatment with collagenase. This technique works well during disaggregation of biopsy specimens from human tumors.

14.8.2 Selective Detachment

Treatment of a heterogeneous monolayer with trypsin or collagenase will remove some cells more rapidly than others. Periodic brief exposure to trypsin removed fibroblasts from cultures of fetal human intestine [Owens et al., 1974] and skin [Milo et al., 1980]. Lasfargues [1973] found that the exposure of cultures of breast tissue to collagenase for a few days at a time removed fibroblasts and left the epithelial cells. EDTA, on the other hand, may release epithelial cells more readily than it will release fibroblasts [Paul, 1975].

Dispase II (Boehringer Mannheim) selectively dislodges sheets of epithelium from human cervical cultures grown on feeder layers of 3T3 cells without dislodging the 3T3 cells (*see* Protocol 24.3). This technique may be effective in subculturing epithelial cells from other sources, selecting against stromal fibroblasts.

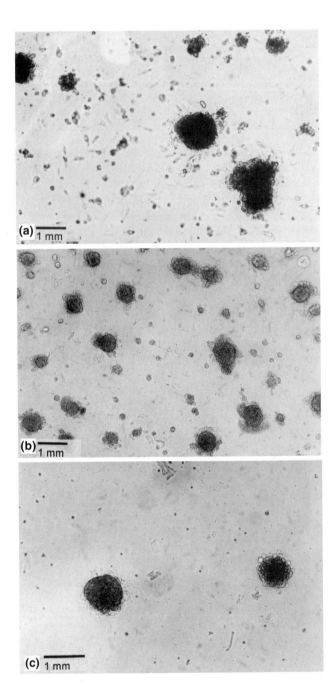

Fig. 14.11. Growth of Melanoma, Fibroblasts, and Glia in Suspension. Cells were plated out at 5×10^5 per 35-mm dish (2.5×10^5 cells/mL) in 1.5% Methocel over a 1.25% agar underlay. Colonies were photographed after 3 wk. (a) Melanoma. (b) Human normal embryonic skin fibroblasts. (c) Human normal adult glia.

14.8.3 Nature of Substrate

Although several sources of ECM are now available (*see* Appendix II), the emphasis so far has been on promoting cell survival or differentiation, and little has been made of the potential for selectivity in "designer" matrices, although collagen has been reported to favor epithelial proliferation [Kibbey et al., 1992; Kinsella et al., 1992] and Matrigel also

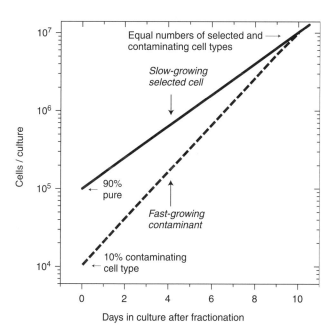

Fig. 14.12. Overgrowth in Mixed Culture. Overgrowth of a slow-growing cell line by a rapidly growing contaminant. This figure portrays a hypothetical example, but it demonstrates that a 10% contamination with a cell population that doubles every 24 h will reach equal proportions with a cell population that doubles every 36 h after only 10 days of growth.

favors epithelial survival and differentiation [Bissell et al., 1987; Ghosh et al., 1991; Kibbey et al., 1992]. Because the constituents are now better understood, mixing various collagens with proteoglycans, laminin, and other matrix proteins could be used to create more selective substrates. The selective effect of substrates on growth may depend on differential rates both of attachment and of growth, or the net result of both. Collagen and fibronectin coating has been used to enhance epithelial cell attachment and growth [Lechner & LaVeck, 1985; Wise, 2002] (*see* Protocols 23.5, 23.9) and to support endothelial cell growth and function [Relou et al., 1998; Martin et al., 2004].

Primaria plastics (BD Biosciences) have a charge on the plastic surface different from that of conventional tissue culture plastics and are designed to enhance epithelial growth relative to fibroblasts. B-D Biosciences and others (*see* Appendix II) also supply plastics coated in natural or synthetic matrices that may facilitate growth of more fastidious cell types, but are probably not selective.

14.8.4 Selective Feeder Layers

As well as conditioning the substrate (*see* Section 14.2.3), feeder layers can also be used for the selective growth of epidermal cells [Rheinwald & Green, 1975] (*see* Protocols 23.1, 23.4) and for repressing stromal overgrowth in cultures of breast (Fig. 14.10; *see also* Plate 6c,d) and colon carcinoma (*see* Protocol 24.3) [Freshney et al., 1982b]. The role of the feeder layer is probably quite complex; it provides not only extracellular matrix for adhesion of the epithelium, but also positively acting growth factors and negative regulators that inactivate TGF-β [Maas-Szabowski & Fusenig, 1996]. Human glioma will grow on confluent feeder layers of normal glia, whereas cells derived from normal brain will not [MacDonald et al., 1985] (*see* Protocol 24.1).

14.8.5 Selection by Semisolid Media

The transformation of many fibroblast cultures reduces the anchorage dependence of cell proliferation (*see* Section 18.5.1) [Macpherson & Montagnier, 1964]. By culturing the cells in agar (*see* Protocol 14.4) after viral transformation, it is possible to isolate colonies of transformed cells and exclude most of the normal cells. Normal cells will not form colonies in suspension with the high efficiency of virally transformed cells, although they will often do so with low plating efficiencies. The difference between transformed and untransformed cells is not as clear with early-passage tumor cell lines, as plating efficiencies can be quite low; normal glia and fetal skin fibroblasts also form colonies in suspension with similar efficiencies (<1%) (Fig. 14.11).

Cell cloning and the use of selective conditions have a significant advantage over physical cell separation techniques (*see* Chapter 15), in that contaminating cells are either eliminated entirely by clonal selection or repressed by constant or repeated application of selective conditions. Even the best physical cell separation techniques still allow some overlap between cell populations, such that overgrowth recurs. A steady state cannot be achieved, and the constitution of the culture changes continuously. From Figure 14.12, it can be seen that a 90%-pure culture of cell line A will be 50% overgrown by a 10% contamination with cell line B in 10 days, given that B grows 50% faster than A. For continued culture, therefore, selective conditions are required in addition to, or in place of, physical separation techniques.

CHAPTER 15

Cell Separation

Although cloning and using selective culture conditions are the preferred methods for purifying a culture (*see* Sections 10.2.1, 14.1, 14.6), there are occasions when cells do not plate with a high enough efficiency to make cloning possible or when appropriate selection conditions are not available. It may then be necessary to resort to a physical or immunological separation technique. Separation techniques have the advantage that they give a high yield more quickly than cloning, although not with the same purity.

The more successful separation techniques depend on differences in (1) cell density (specific gravity), (2) affinity of antibodies to cell surface epitopes, (3) cell size, and (4) light scatter or fluorescent emission as sorted by flow cytometry. The first two techniques involve a relatively low level of technology and are inexpensive, whereas the second two call for high technology with a significant outlay of capital. The most effective separations often employ two or more parameters to obtain a high level of purity [*see*, e.g., Calder et al., 2004; Al-Mufti et al., 2004].

15.1 CELL DENSITY AND ISOPYKNIC SEDIMENTATION

Separation of cells by density can be performed by centrifugation at low *g* with conventional equipment [Pretlow & Pretlow, 1989; Sharpe, 1988; Calder et al., 2004; Al-Mufti et al., 2004]. The cells sediment in a density gradient to an equilibrium position equivalent to their own density (isopyknic sedimentation; Fig. 15.1). The density medium should be nontoxic and nonviscous at high densities (1.10 g/mL) and should exert little osmotic pressure in

solution. Serum albumin [Turner et al., 1967], dextran [Schulman, 1968], Ficoll (Pharmacia) [Sykes et al., 1970], metrizamide (Nygaard) [Munthe-Kaas & Seglen, 1974], and Percoll (Amersham Biosciences) [Pertoft & Laurent, 1982] have all been used successfully (*see also* Protocol 27.1). Percoll (colloidal silica) and the radiopaque iodinated compounds metrizamide and metrizoate are among the more effective media currently used. A gradient may be generated (1) by layering different densities of Percoll with a pipette, syringe, or pump, (2) with a special gradient former (Fig. 15.2d), or (3) by high-speed spin (Fig. 15.3).

PROTOCOL 15.1. CELL SEPARATION BY CENTRIFUGATION ON A DENSITY GRADIENT

Outline
Form a gradient, centrifuge cells through the gradient, collect fractions, dilute in medium, and culture (Fig. 15.1).

Materials
Sterile:
- ❏ Growth medium
- ❏ Growth medium + 20% Percoll
- ❏ Centrifuge tubes, 25 mL
- ❏ D-PBSA
- ❏ Trypsin, 0.25%
- ❏ Syringe or gradient harvester
- ❏ Plates, 24 well or microtitration
Nonsterile:
- ❏ Refractometer or density meter

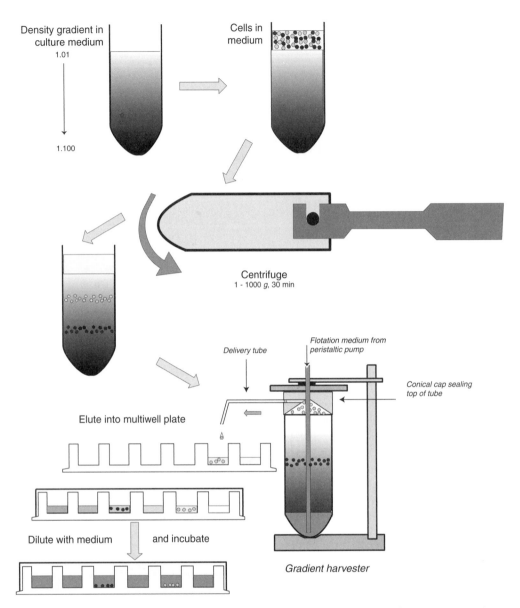

Fig. 15.1. Cell Separation by Density. Cells are layered on to a preformed gradient (*see* Fig. 15.2) and the tube centrifuged. The tube is placed on a gradient harvester, and flotation medium (e.g., Fluorochemical FC43) is pumped down the inlet tube to the bottom of the gradient, displacing the gradient and cells upward and out through the delivery tube into a multiwell plate. The cells are diluted with medium (so that they will sink) and cultured. (Gradient mixer after an original design by Dr. G. D. Birnie).

❑ Hemocytometer or cell counter
❑ Low-speed centrifuge

Protocol
Preparation of the density gradient
(a) Layering:
  i) Adjust the density of the Percoll medium to 1.10 g/cc and its osmotic strength to 290 mosmol/kg.

  ii) Mix the Percoll and regular media in varying proportions to give the desired density range (e.g., 1.020–1.100 g/cc) in 10 or 20 steps.

  iii) Layer one step over another, 1 mL or 2 mL per step, with a pipette, syringe or peristaltic pump, starting with the densest solution and building up a stepwise density gradient in a 25-mL centrifuge tube. It is also possible, and may be preferable, to layer from the bottom, starting with the least dense solution and injecting each layer of progressively higher density below the previous one, using a syringe or peristaltic pump.

Gradients may be used immediately or left overnight.

(b) With a gradient former: A continuous linear gradient may be produced by mixing, for example, 1.020 g/mL with 1.080 g/cc Percoll in a gradient former (Fisons, Pharmacia, Buchler; *see* Fig. 15.2).

(c) Centrifugation:

  i) Place the medium containing Percoll at density 1.085 g/cc in a tube.

  ii) Centrifuge at 20,000 *g* for 1 h.

  iii) Centrifugation generates a sigmoid gradient (*see* Fig. 15.3), the shape of which is determined by the starting concentration of Percoll, the duration and centrifugal force of the centrifugation, the shape of the tube, and the type of rotor.

1. Trypsinize cells and resuspend them in the medium plus serum or a trypsin inhibitor. Check to make sure that the cells are singly suspended.
2. Using a syringe, pipettor, or fine-tipped pipette, layer up to 2×10^7 cells in 2 mL of medium on top of the gradient.
3. The tube may be allowed to stand on the bench for 4 h and will sediment under 1 *g*; or it may be centrifuged for 20 min at 100–1000 *g*. If the latter procedure is used, increase centrifuge speed gradually at start of run and do not apply brake at end of run.
4. Collect fractions with a syringe or a gradient harvester (Fisons; *see* Fig. 15.1). Fractions of 1 mL may be collected into a 24-well plate or of 0.1 mL into a microtitration plate. Samples should be taken at intervals for cell counting and for determining the density (ρ) of the gradient medium. Density may be measured on a refractometer (Hilger) or density meter (Paar).
5. Add an equal volume of medium to each well, and mix to ensure that the cells settle to the bottom of the well. Change the medium to remove the Percoll after 24–48 h incubation.

Variations

Position of cells. Cells may be incorporated into the gradient during its formation by centrifugation. Although

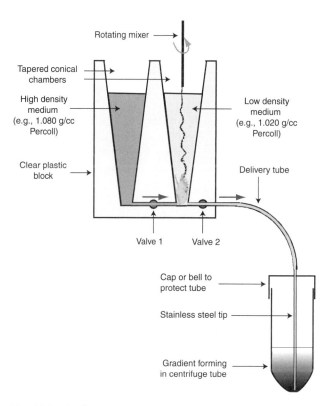

Fig. 15.2. Gradient Mixing Device. Two chambers are cut in a solid transparent plastic block and connected by a thin canal across the bases of the chambers and exiting to the exterior. A delivery tube with a stainless steel tip, long enough to reach the bottom of the centrifuge tube, is inserted in the outlet. With the valves closed, the left-hand chamber is filled with high-density medium and the right-hand chamber with low-density medium. Valve 1 is opened, the stainless steel tip is placed in the bottom of the tube, the stirrer is started, and then valve 2 is opened. As the solution in the left chamber runs into the right, it mixes and gradually increases the density, while the mixture is running out into the tube.

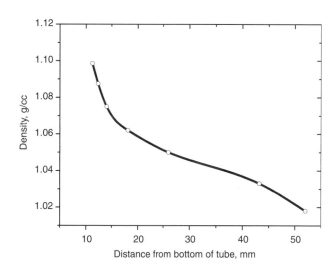

Fig. 15.3. Centrifuge-Derived Gradient. Gradient generated by spinning Percoll at 20,000 *g* for 1 h.

only one spin is required, spinning the cells at such a high-*g* force may damage them. In addition, Percoll [Pertoft & Laurent, 1982], Isopaque [Splinter et al., 1978], and metrizamide [Freshney, 1976] may be taken up by cells, so it is preferable to layer cells on top of a preformed gradient.

Other media. Ficoll is one of the most popular media because, like Percoll, it can be autoclaved. Ficoll is a little more viscous than Percoll at high densities and may cause some cells to agglutinate. Metrizamide (Nycomed), a nonionic derivative of metrizoate, which is a radiopaque iodinated substance used in radiography (Isopaque, Hypaque, Renografin) and in lymphocyte purification (*see* Protocol 27.1), is less viscous than Ficoll at high densities [Rickwood & Birnie, 1975] but may be taken up by some cells (*see* above).

Marker beads. Amersham Biosciences manufactures colored marker beads of standard densities that may be used to determine the density of regions of the gradient.

Isopyknic sedimentation is quicker than velocity sedimentation at unit gravity (*see* Section 15.2) and gives a higher yield of cells for a given volume of gradient. It is ideal when clear differences in density (≥0.02 g/cc) exist between cells. Cell density may be affected by uptake of the medium used to form the gradient, by the position of the cells in the growth cycle (plateau phase cells are denser),

and by serum [Freshney, 1976]. Because high-*g* forces are not required, isopyknic sedimentation can be done on any centrifuge and can even be performed at 1 *g*.

15.2 CELL SIZE AND SEDIMENTATION VELOCITY

Sedimentation of cells is also influenced by cell size (cross-sectional area), which becomes the major determinant of sedimentation velocity at 1 *g* and a significant component at higher sedimentation rates at elevated *g*. The relationship between the particle size and sedimentation rate at 1 *g*, although complex for submicron-sized particles, is fairly simple for cells and can be expressed approximately as

$$v \approx \frac{r^2}{4}$$

[Miller & Phillips, 1969], where *v* is the sedimentation rate in mm/h and *r* is the radius of the cell in μm (*see* Table 21.1).

15.2.1 Unit Gravity Sedimentation
Layering cells over a serum gradient in medium will allow the cells to settle through the medium according to the above equation. However, unit gravity sedimentation is unable to handle large numbers of cells (~1 × 10⁶ cells/cm² of surface

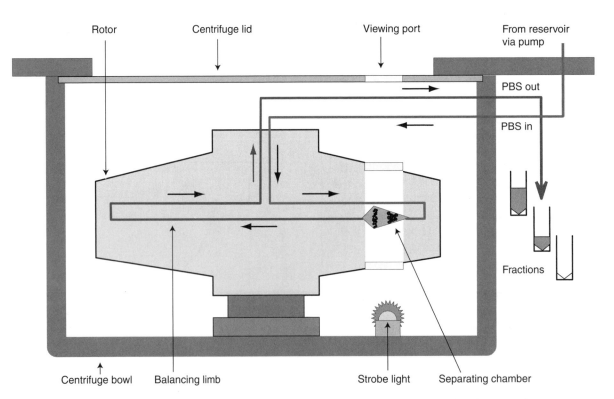

Fig. 15.4. Centrifugal Elutriator Rotor (Beckman). A cell suspension and carrier liquid enter at the center of the rotor and are pumped to the periphery and then into the outer end of the separating chamber. The return loop is via the opposite side of the rotor, to maintain balance.

area at the top of the gradient) and does not give particularly good separations unless the mean cell sizes are very different and the cell populations are homogeneous in size. It is useable where there are major differences in cell size, or when aggregates are being separated from single cells, e.g., after collagenase digestion (*see* Protocol 12.8).

15.2.2   Centrifugal Elutriation

Most cell separations based on cell size use either centrifugal elutriation (giving moderate resolution but a high yield) or a cell sorter (high resolution with a low yield; *see* Section 15.4). The centrifugal elutriator (Beckman Coulter) is a device for increasing the sedimentation rate and improving the yield and resolution of cell separation by performing the separation in a specially designed centrifuge and rotor (Fig. 15.4) [Lutz et al., 1992]. Cells in the suspending medium are pumped into the separation chamber in the rotor while it is turning. While the cells are in the chamber, centrifugal force tends to push the cells to the outer edge of the rotor (Fig. 15.5). Meanwhile, medium is pumped through the chamber such that the centripetal flow rate balances the sedimentation rate of the cells. If the cells were uniform, they would remain stationary, but because they vary in size, density, and cell surface configuration, they tend to sediment at different rates. Because the sedimentation chamber is tapered, the flow rate increases toward the edge of the rotor, and a continuous range of flow rates is generated. Cells of differing sedimentation rates will therefore reach equilibrium at different positions in the chamber. The sedimentation chamber is illuminated by a stroboscopic light and can be observed through a viewing port. When the cells are seen to reach equilibrium, the flow rate is increased and the cells are pumped out into receiving vessels. The separation can be performed in a complete medium and the cells cultured directly afterward.

The procedure comprises four phases: (1) setting up and sterilizing the apparatus, (2) calibration, (3) loading the sample and establishing the equilibrium conditions, and (4) harvesting fractions. The details of the protocol are provided in the operating manual for the elutriator (Beckman Coulter). Equilibrium is reached in a few minutes, and the whole run may take 30 min. On each run, 1×10^8 cells may be separated, and the run may be repeated as often as necessary. The apparatus is, however, fairly expensive, and a considerable amount of experience is required before effective separations may be made. A number of cell types have been separated by this method [Teofili et al., 1996; Lag et al., 1996; Yoshioka et al., 1997], as have cells of different phases of the cell cycle [Breder et al., 1996; Mikulits et al., 1997].

15.3   ANTIBODY-BASED TECHNIQUES

There are a number of techniques that rely on the specific binding of an antibody to the cell surface. These include immune lysis by an antibody against unwanted cells, e.g., fibroblasts in an epithelial population [Edwards et al., 1980], immune targeting of a cytotoxin [Beattie et al., 1990], fluorescence-activated cell sorting (*see* Section 15.4), immune panning (*see* Section 15.3.1), and sorting with antibody-conjugated magnetizable beads (magnetically activated cell sorting, MACS) [Saalbach et al., 1997] (*see* Section 15.3.2). These techniques all depend on the specificity of the selecting antibody and the presentation of the correct epitope on the cell surface of living cells, as confirmed by immune staining (*see* Section 16.11) or flow cytometry (*see* Sections 15.4, 16.9.2, 21.7.2).

15.3.1   Immune Panning

The attachment of cells to dishes coated with antibodies, a process called *immune panning*, has been used successfully with a number of different cell types [Murphy et al., 1992; Fujita et al., 2004]. The vast majority of panning methods are derived from the work of Wysocki and Sata [1978] with lymphocyte subpopulations. A cell-type-specific antibody raised against a cell surface epitope is conjugated to the bottom of a Petri dish, and when the mixed cell population is added to the dish, the cells to which the antibody is directed attach rapidly to the bottom of the dish. The remainder can then be removed. Immune panning can be used positively, to select a specific subset of cells that can be released subsequently by mechanical detachment or light trypsinization, or negatively, to remove unwanted cells.

15.3.2   Magnetic Sorting

Magnetic sorting uses a specific antibody, raised against a cell surface epitope, conjugated to ferritin beads (Dynabeads, from Dynal; Plate 23a–c) or microbeads (Miltenyi; Plate 23d,e). When the cell suspension is mixed with the beads and then placed in a magnetic field (Fig. 15.6), the cells that have attached to Dynabeads are drawn to the side of the separating chamber. The cells and beads are released when the current is switched off, and the cells may be separated from the beads by trypsinization or vigorous pipetting. In the Miltenyi system, cells are immunologically bound to microparamagnetic beads which bind to ferromagnetic spheres in the separation column when placed in an electromagnetic field, and released when the column is removed from the magnetic field (Fig. 15.7). Several cell types have been separated by this method, including stem cell purification by negative sorting [Bertoncello

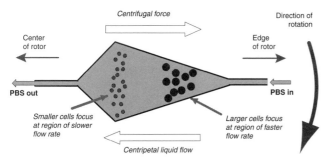

Fig. 15.5. Separation Chamber of Elutriator Rotor.

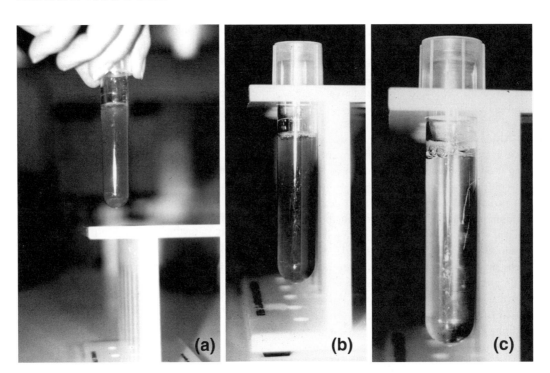

Fig. 15.6. Magnetic Sorting. Negative sort. Committed progenitor cells from bone marrow suspension bound to Dynal paramagnetic beads with antibodies to lineage markers. Lineage-negative (stem) cells are not bound and remain in the suspension ready for sorting by flow cytometry. (a) Inserting the tube into the magnetic holder. (b) Tube immediately after being placed in magnetic holder. (c) Tube 30 s after placement in magnetic holder. (*See also* Plate 23a).

et al., 1991], purging bone marrow of leukemic cells [Trickett et al., 1990], and isolation of kidney tubular epithelium [Pizzonia et al., 1991]. Protocols for the use of Dynabeads are available on the Dynal website (*see* Appendix III).

The use of immunomagnetic microbeads [Gaudernack et al., 1986] (Fig. 15.7; Plate 23d,e) allows the cells to be cultured or processed through further sorting procedures, without the need to remove the beads (Miltenyi). The method is therefore particularly useful for a positive sort. Protocol 15.2 has been abstracted from the Miltenyi instruction sheet (*see also* Protocol 23.18D). Check Miltenyi web site for latest versions of labeling and separation protocols (*www.miltenyibiotec.com/*).

PROTOCOL 15.2. MAGNET-ACTIVATED CELL SORTING (MACS)

Outline

Buffy coat or another mixed-cell suspension is mixed with antibody-conjugated microbeads, diluted, and placed in a magnetic separation column. Cells bound to microbeads bind to ferromagnetic spheres in the column, while unbound cells flow through. Bound cells are released from the column when it is removed

from the separator magnet and are purged from the column with a syringe piston.

Materials
Sterile:
- Buffer: D-PBSA, pH 7.2, with 0.5% BSA and 2 mM EDTA
- Magnetic cell separator: MiniMACS, MidiMACS, VarioMACS, or SuperMACS (Miltenyi)
- RS+ or VS+ column adaptors
- Positive selection column: MS+/RS+ for up to 1×10^7 cells; LS+/VS+ for up to 1×10^8 cells
- Magnetizable microbeads conjugated to antibody raised against cell surface antigen of cells to be collected
- Collection tube to match volume being collected: MS+/RS+ 1 mL; LS+/VS+ 5 mL

Protocol
Labeling:
1. Isolate peripheral blood mononuclear cells by the standard method (*see* Protocol 27.1), or prepare a cell suspension by trypsinization or an alternative procedure (*see* Protocols 12.5, 12.6, 12.8, and 13.2).
2. Remove dead cells with Ficoll-Hypaque (*see* Protocol 12.10).

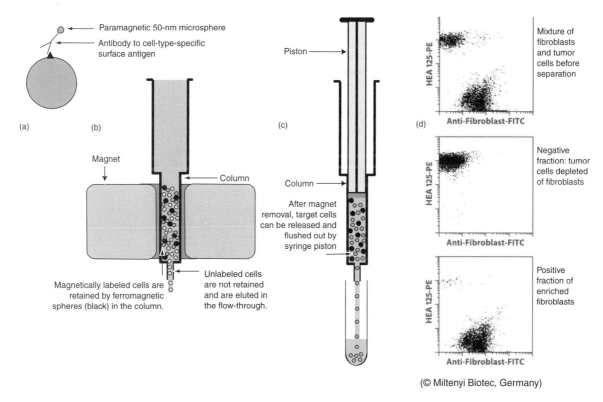

(© Miltenyi Biotec, Germany)

Fig. 15.7. Magnetic Cell Sorting (MACS® Technology). Positive Sort. (a) Cells are preincubated with antibodies raised against a cell type-specific surface antigen and conjugated to paramagnetic MACS MicroBeads. (b) When the cells are introduced into the column, cells bound to MicroBeads are retained in the magnetic field generated by magnet and column matrix; unlabeled cells go straight through. (c) The magnetically bound cells, released when the column is removed from the magnet, are flushed out with the piston. (*See also* plate 23 d,e.) (d) Dot plot from flow cytometry of tumor cell culture containing fibroblasts before MACS sorting (top), flow-through fraction after MACS sorting with Anti-Fibroblast MicroBeads, as in b, enriched fibroblast fraction (bottom) after releasing from magnet. (d. © Miltenyi Biotec, used with permission).

3. Remove clumps by passing cells through 30 μm of nylon mesh or filter (Miltenyi, #414:07). (Wet the filter with a buffer before use.)
4. Wash the cells in the buffer by centrifugation.
5. Resuspend pellet from centrifugation in 80 μL of buffer per 10^7 total cells (80 μL minimum volume, even for $<1 \times 10^7$ cells).
6. Add 20 μL of MACS microbeads per 1×10^7 cells.
7. Mix and incubate suspension for 15 min at 6–12°C. (For fewer cells, use the same volume.)
8. Dilute suspension by adding 10–20 × the labeling volume of buffer.
9. Centrifuge 300 g for 10 min.
10. Remove the supernatant and resuspend cells plus microbeads in 500 μL buffer per 1×10^8 total cells.

Positive magnetic separation:

11. Place the column in the magnetic field of the MACS separator.
12. Wash the column: MS⁺/RS⁺ 500 μL; LS⁺/VS⁺ 3 mL.
13. Apply the cell suspension to the column: MS⁺/RS⁺ 500–1000 μL; LS⁺/VS⁺ 1–10 mL.
14. Allow negative cells to pass through the column.
15. Rinse the column with buffer: MS⁺/RS⁺ 3 × 500 μL; LS⁺/VS⁺ 3 × 3 mL.
16. Remove the column from the separator and place the column on the collection tube.
17. Pipette buffer into the column (MS⁺/RS⁺ 1 mL, LS⁺/VS⁺ 5 mL), and flush out positive cells, using the plunger supplied with the column.
18. Count the cells, adjusting the concentration in the growth medium:
    (a) $1 \times 10^5 - 1 \times 10^6$ cells/mL, and seed culture flasks for primary culture.
    (b) 10–1000 cells per mL, and seed Petri dishes for cloning.

15.4 FLUORESCENCE-ACTIVATED CELL SORTING

Fluorescence-activated cell sorting [Vaughan & Milner, 1989] operates by projecting a single stream of cells through a laser beam in such a way that the light scattered from the cells is detected by one or more photomultipliers and recorded (Figs. 15.8, 15.9). If the cells are pretreated with a fluorescent stain (e.g., propidium iodide or chromomycin A_3 for DNA) or a fluorescent antibody, the fluorescence emission excited by the laser is detected by a second photomultiplier tube. The information obtained is then processed and displayed as a two-(Fig. 15.10) or three-dimensional graph on the monitor.

A *flow cytometer* is an analytical instrument that processes the output of the photomultipliers to analyze the constitution of a cell population (e.g., to determine the proportion of cells in different phases of the cell cycle, measured by a combination of DNA fluorescence and cell size measurements).

A *fluorescence-activated cell sorter* (FACS; e.g., B-D Biosciences FACSAria) is an instrument that uses the emission signals from each cell to sort the cell into one of two sample collection tubes and a waste reservoir. If specific coordinates are set to delineate sections of the display, the cell sorter will divert those cells with properties that would place them within these coordinates (e.g., high or low light scatter, high or low fluorescence) into the appropriate receiver tube, placed below the cell stream. The stream itself is broken up into droplets by a high-frequency vibration applied to the flow chamber, and the droplets containing single cells with specific attributes are charged as they leave the chamber. These droplets are deflected, left or right according to the charge applied, as they pass between two oppositely charged plates. The charge is applied briefly and at a set time after the cell has cut the laser beam such that the droplet containing one specifically marked cell is deflected into the receiver. The concentration in the cell stream must be low enough that the gap between cells is sufficient to prevent two cells from inhabiting one droplet [Vaughan & Milner, 1989].

All cells having similar properties are collected into the same tube. A second set of coordinates may be established and a second group of cells collected simultaneously into another tube by changing the polarity of the cell stream and deflecting the cells in the opposite direction. All remaining cells are collected in a central waste reservoir.

This method may be used to separate cells according to any differences that may be detected by light scatter (e.g., cell

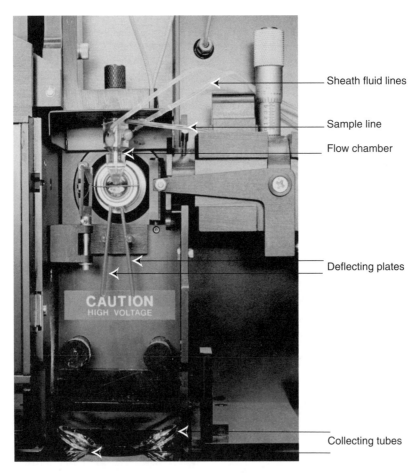

Fig. 15.8. Fluorescence-Activated Cell Sorter (FACS). Close-up of flow chamber and separation compartment of cell sorter version of the FACS IV (Becton Dickinson; *see also* Fig. 15.9).

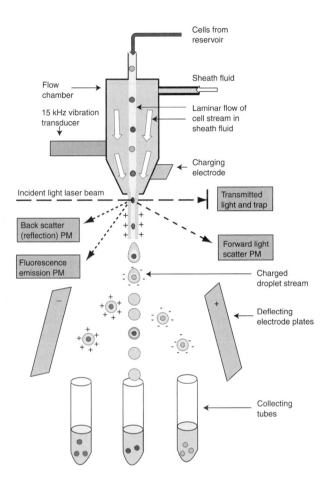

Fig. 15.9. Flow Cytometry. Principle of operation of flow cytophotometer. Cell stream in D-PBSA enters at the top, and sheath liquid, also D-PBSA, is injected around the cell stream to generate a laminar flow within the flow chamber. When the cell stream exits the chamber, it cuts a laser beam, and the signal generated triggers the charging electrode, thereby charging the cell stream. The cell stream then breaks up into droplets, induced by the 15-kHz vibration transducer attached to the flow chamber. The droplets carry the charge briefly applied to the exiting cell stream and are deflected by the electrode plates below the flow chamber. The charge is applied to the plates with a sufficient delay to allow for the transit time from cutting the beam to entering the space between the plates (*see also* Fig. 15.8).

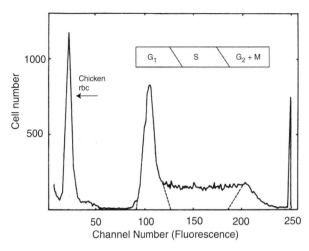

Fig. 15.10. Printout from FACS II. Friend (murine erythroleukemia) cells were fixed in methanol as a single-cell suspension and stained with chromomycin A_3. The suspension was then mixed with chicken erythrocytes (RBC), fixed and stained in the same way, and run through a FACS II. The printout plots cell number on the vertical axis and channel number (fluorescence) on the horizontal axis. As fluorescence is directly proportional to the amount of DNA per cell, the trace gives a distribution analysis of the cell population by DNA content. The lowest DNA content is found in the chicken erythrocytes, included as a standard. The major peak around channel 100 represents those cells as the G_1 phase of the cell cycle. Cells around channel 200 therefore represent G_2 and metaphase cells (with double the amount of DNA per cell), and cells in intermediate channels are in the S (DNA synthetic) phase. Cells accumulated in channel 250 are those for which the DNA value is off the scale or those that have formed aggregates. (Courtesy of Prof. B. D. Young).

size) or fluorescence (e.g., DNA, RNA, or protein content; enzyme activity; specific antigens) and has been applied to a wide range of cell types. It has probably been utilized most extensively for hematopoietic cells [Battye & Shortman, 1991; Pipia & Long, 1997], for which disaggregation into the obligatory single-cell suspension is relatively simple, but has also been used for solid tissues (e.g., lung [Aitken et al., 1991], skin [Swope et al., 1997], and gut [Boxberger et al., 1997; *see also* Protocol 23.17]). It is an extremely powerful tool but is limited by the cell yield (about 1×10^7 cells is a reasonable maximum number of cells that can be processed at one time). Although the more sophisticated machines with cell separation capability have a high capital

cost (approximately $200,000) and require a full-time skilled operator, less expensive bench-top machines are available for analytical use (e.g., Guava Technologies).

15.5   OTHER TECHNIQUES

The many other techniques that have been used successfully to separate cells are too numerous to describe in detail. They are summarized below and listed in Table 15.1.

Electrophoresis is performed either in a Ficoll gradient [Platsoucas et al., 1979] or by curtain electrophoresis; the second technique is probably more effective and has been used to separate kidney tubular epithelium [Kreisberg et al., 1977].

Affinity chromatography uses antibodies [Varon & Manthorpe, 1980; Au & Varon, 1979] or plant lectins [Pereira & Kabat, 1979] that are bound to nylon fiber [Edelman, 1973] or Sephadex (Amersham Pharmacia). This technique appears to be useful for fresh blood cells, but less so for cultured cells.

Countercurrent distribution [Walter, 1975, 1977; Sharpe, 1988] has been utilized to purify murine ascites tumor cells with reasonable viability.

TABLE 15.1. Cell Separation Methods

Method	Basis for separation	Equipment	Comments	Reference
Sedimentation velocity at 1 g	Cell size	Custom-made separating funnel and baffles	Simple technique, but not very high resolution or yield	Miller & Phillips, 1969
Isopyknic sedimentation	Cell density	Centrifuge	Simple and rapid	Pertoft & Laurent, 1982; see Protocol 15.1
MACS	Surface antibodies	Simple magnet and flow chamber	Specific, given a highly specific surface antibody	Gaudernack et al., 1986; see Protocol 15.2
Immune panning	Surface antibodies	Antibody-coated dishes	Simple, low technology with precoated plates available, but also depends on specific surface antibody	Wysocki & Sata, 1978
Centrifugal elutriation	Cell size, density, and surface configuration	Special centrifuge and elutriator rotor	Rapid, high cell yield, but quite complex process	Lutz et al., 1992
FACS	Cell surface area, fluorescent markers, fluorogenic enzyme substrates, multiparameter	Flow cytometer	Complex technology and expensive; very effective; high resolution, but low yield	Vaughan & Milner, 1989
Affinity chromatography	Cell surface antigens, cell surface carbohydrate	Sterilizable chromatography column	Elutriation of cells from columns is difficult, better in free suspension	Edelman, 1973
Countercurrent distribution	Affinity of cell surface constituents for solvent phase	Shaker	Some cells may suffer loss of viability, but method is quite successful for others	Walter, 1977
Electrophoresis in gradient or curtain	Surface charge	Curtain electrophoresis apparatus		Kreisberg et al., 1977; Platsoucas et al., 1979

15.6 BEGINNER'S APPROACH TO CELL SEPARATION

It is best to start with a simple technique such as density gradient centrifugation or, if a specific cell surface phenotype can be predicted, panning on coated dishes or MACS. If the resolution or yield is insufficient, then it may be necessary to employ FACS or centrifugal elutriation. Centrifugal elutriation is useful for the rapid sorting of large numbers of cells, but FACS will probably give the purest cell population, based on the combined application of two or more stringent criteria.

When a high-purity cell suspension is required and selective culture is not an option, it will probably be necessary to employ at least a two-step fractionation, in a manner analogous to the purification of proteins. In many such procedures, density gradient separation is used as a first step, with panning or MACS as a second, and the final purification is performed by FACS, as has been used in the isolation of hematopoietic stem cells [Cooper & Broxmeyer, 1994].

CHAPTER 16

Characterization

16.1   THE NEED FOR CHARACTERIZATION

There are six main requirements for cell line characterization:

(1) Demonstration of the absence of cross-contamination (*see* Sections 7.10.1, 16.3, 16.6.2, 19.5; Table 13.2)
(2) Confirmation of the species of origin
(3) Correlation with the tissue of origin, which comprises the following characteristics:
   a) Identification of the lineage to which the cell belongs
   b) Position of the cells within that lineage (i.e., the stem, precursor, or differentiated status)
(4) Determination of whether the cell line is transformed or not:
   a) Is the cell line finite or continuous (*see* Sections 18.2, 18.4)?
   b) Does it express properties associated with malignancy (*see* Section 18.6)?
(5) Indication of whether the cell line is prone to genetic instability and phenotypic variation (*see* Sections 17.1.1, 18.3)
(6) Identification of specific cell lines within a group from the same origin, selected cell strains, or hybrid cell lines, all of which require demonstration of features unique to that cell line or cell strain.

16.2   RECORD KEEPING AND PROVENANCE

When a new cell line is derived, either from a primary culture or from an existing cell line, it is difficult to assess its future value. Often, it is only after a period of use and dissemination that the true importance of the cell line becomes apparent, and at that point details of its origin are required. However, by that time it is too late to collect information retrospectively. It is therefore vital that adequate records are kept, from the time of isolation of the tissue, or of the receipt of a new cell line, detailing the origin, characteristics, and handling of the cell line. These records form the *provenance* of the cell line, and the more detailed the provenance, the more valuable the cell line (*see* Sections 12.3.11 and 13.7.8).

This aspect of cell culture has become particularly important with the widespread dissemination of cell lines through cell banks and personal contacts to research laboratories and commercial companies far removed from their origin. In particular, if a cell line becomes incorporated into a procedure that requires validation of its components, then the authentication of the cell line becomes crucial. Authentication requires that the cell line be characterized on receipt, and periodically during use, and that these data are compatible with, and added to, the existing provenance.

16.3   AUTHENTICATION

Characterization of a cell line is vital, not only in determining its functionality but also in proving its authenticity; special attention must be paid to the possibility that the cell line has become cross-contaminated with an existing continuous cell line or misidentified because of mislabeling or confusion in handling. The demonstration that the majority of continuous cell lines in use in the United States in the late 1960s had

Culture of Animal Cells: A Manual of Basic Technique, Fifth Edition, by R. Ian Freshney
Copyright © 2005 John Wiley & Sons, Inc.

TABLE 16.1. Characterization of Cell Lines and Cell Strains

Criterion	Method	Reference	Protocol no.
Karyotype	Chromosome spread with banding	Rothfels & Siminovitch, 1958, Rooney & Czepulkowski, 1986	16.9
Isoenzyme analysis	Agar gel electrophoresis	Hay, 2000	16.12
Cell surface antigens	Immunohistochemistry	Hay, 2000, Burchell et al., 1983, 1987	16.13
Cytoskeleton	Immunocytochemistry with antibodies to specific cytokeratins	Lane, 1982, Moll et al., 1982	16.13
DNA fingerprint	Restriction enzyme digest; PAGE; satellite DNA probes	Hay, 2000, Jeffreys et al., 1985	16.10
DNA Profile	PCR of microsatellite repeats	Masters et al., 2001	16.11

become cross-contaminated with HeLa cells [Gartler, 1967; Nelson-Rees & Flandermeyer, 1977; Lavappa, 1978] first brought this serious problem to light, but the continued use of the lines 30 years later indicates that many people are still unaware, or are unwilling to accept, that many lines in common use (e.g., Hep-2, KB, Girardi Heart, WISH, Chang Liver) are not authentic (*see* Sections 16.6.2, 16.6.3, and Table 13.2). The use of cell lines contaminated with HeLa or other continuous cell lines, without proper acknowledgment of the contamination, is still a major problem [Stacey et al., 2000; Drexler et al., 2003] and emphasizes that any vigorously growing cell line can overgrow another, more slowly growing line [Masters et al., 1988; van Helden et al., 1988; Christensen et al., 1993; Gignac et al., 1993; van Bokhoven et al., 2001a,b; Rush et al., 2002]. Although some of the work with these cell lines may remain perfectly valid, e.g., if it is a molecular process of interest regardless of the origin of the cells, any attempt to correlate cell behavior with the tissue or tumor of origin is totally invalidated by cross-contamination. Unfortunately, this fact is frequently ignored by journal editors, granting authorities, referees, and users of the cell lines. Characterization studies, particularly with continuous cell lines, have become a process of authentication that is vital to the validation of the data derived from these cells.

The nature of the technique used for characterization depends on the type of work being carried out; for example, if molecular technology is readily available, then DNA fingerprinting (*see* Protocol 16.8), DNA profiling (*see* Protocol 16.9), or analysis of gene expression (*see* Section 16.7 and Protocol 27.8) are likely to be of most use, whereas a cytology laboratory may prefer to use chromosome analysis (*see* Section 16.5) coupled with FISH and chromosome painting (*see* Protocol 27.9), and a laboratory with immunological expertise may prefer to use MHC analysis (e.g., HLA typing) coupled with lineage-specific markers. Combined with a functional assay related to your own interests, these procedures should provide sufficient data to authenticate a cell line.

Regardless of the intrinsic ability of the laboratory, DNA fingerprinting or DNA profiling and multiple isoenzyme analysis by agarose gels (*see* Protocol 16.10) have now become the major standard procedures for cell line identification; if you are unable or unwilling to carry out these authentication procedures yourself, then seek a commercial service (*see* Appendix II).

Some of the methods in general use for cell line characterization are listed in Table 16.1 [*see also* Hay et al., 2000].

16.3.1 Species Identification

Chromosomal analysis, otherwise known as *karyotyping* (*see* Protocols 16.7, 27.5), is one of the best methods for distinguishing between species. Chromosome banding patterns can be used to distinguish human and mouse chromosomes (*see* Section 16.5), and chromosome painting, i.e., using combinations of specific molecular probes to hybridize to individual chromosomes (*see* Sections 16.5.3 and 27.8.2), adds further resolution and specificity to this technique. These probes identify individual chromosome pairs and are species specific. The availability of probes is limited to a few species at present, and most are either mouse or human, but chromosome painting is a good method for distinguishing between human and mouse chromosomes in potential cross-contaminations and interspecific hybrids. Isoenzyme electrophoresis (*see* Protocol 16.10) is also a good diagnostic test and is quicker than chromosomal analysis. A simple kit is available that makes this technique readily accessible (*see* Protocol 16.10). A combination of the two methods is often used and gives unambiguous results [Hay, 2000].

16.3.2 Lineage or Tissue Markers

Individual organs are comprised of tissues, e.g., skin is made up of an outer epidermis and underlying dermis, and tissues, in turn, are made up of individual lineages, e.g., the dermis contains connective tissue fibrocytes, vascular endothelial cells and smooth muscle cells, and the mesenchymal cells of the dermal papillae, among others. Each cell type can be traced back, via a series of proliferating cell stages, to an originating stem cell (*see* Fig. 3.6), forming a treelike structure. Each "branch" of that "tree" can be regarded as a *lineage*, as in a basal cell of the epidermis following a differentiation

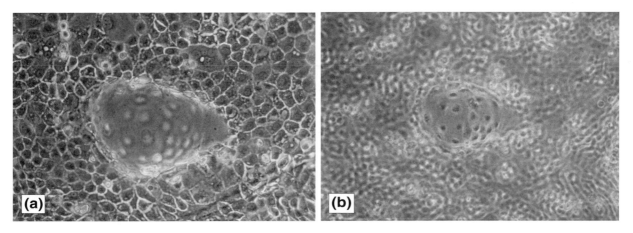

Fig. 16.1. Domes. (a) Dome, or hemicyst, formed in an epithelial monolayer by downward transport of ions and water, lower focus (on monolayer). (b) Upper focus (top of dome). (*See also* Plate 12a,b.)

path to a mature cornified keratinocyte. Some lineages, e.g., the myeloid lineage of hematopoietic differentiation, may branch into sublineages (neutrophilic, eosinophilic, and basophilic), so lineage marker expression is also influenced by *differentiation*, i.e., the position of the cell in the lineage differentiation pathway (*see* Section 16.10). Lineage markers are helpful in establishing the relationship of a particular cell line to its tissue of origin (Table 16.1).

Cell surface antigens. These markers are particularly useful in sorting hematopoietic cells [Visser & De Vries, 1990] and have also been effective in discriminating epithelium from stroma with antibodies such as anti-EMA [Heyderman et al., 1979] and anti-HMFG 1 and 2 [Burchell & Taylor-Papadimitriou, 1989] and neuroectodermally derived cells (e.g., anti-A2B5) [Dickson et al., 1983] from cells derived from other germ layers.

Intermediate filament proteins. These are among the most widely used lineage or tissue markers [Lane, 1982; Ramaekers et al., 1982]. Glial fibrillary acidic protein (GFAP) for astrocytes [Bignami et al., 1980] (*see* Plate 11b) and desmin [Bochaton-Piallat et al., 1992; Brouty-Boyé et al., 1992] for muscle are the most specific, whereas cytokeratin marks epithelial cells and mesothelium [Lane, 1982; Moll et al., 1982] (*see* Plate 11a). Neurofilament protein marks neurons (Wu & de Vellis, 1987; Kondo & Raff, 2000) and some neuroendocrine cells (Bishop et al., 1988). Vimentin (*see* Plate 11c), although usually restricted to mesodermally derived cells *in vivo*, can appear in other cell types *in vitro*.

Differentiated products and functions. Hemoglobin for erythroid cells, myosin or tropomyosin for muscle, melanin for melanocytes, and serum albumin for hepatocytes are among the best examples of specific cell type markers, but, like all differentiation markers, they depend on the complete expression of the differentiated phenotype.

Transport of inorganic ions, and the resultant transfer of water, is characteristic of absorptive and secretory epithelia [Abaza et al., 1974; Lever, 1986]; grown as monolayers, some epithelial cells will produce *domes*, which are hemicysts in the monolayer caused by accumulation of water on the underside of the monolayer [Rabito et al., 1980] (Fig. 16.1; *see* Plate 12a,b). Other specific functions that can be expressed *in vitro* include muscle contraction and depolarization of nerve cell membrane.

Enzymes. Three parameters are available in enzymic characterization: (1) the constitutive level (i.e., in the absence of inducers or repressors); (2) the response to inducers and repressors; and (3) isoenzyme polymorphisms (*see* Sections 16.8.1, 16.8.2). Creatine kinase BB isoenzyme is characteristic of neuronal and neuroendocrine cells, as is neuron-specific enolase; lactic dehydrogenase is present in most tissues, but as different isoenzymes, and a high level of tyrosine aminotransferase, inducible by dexamethasone, is generally regarded as specific to hepatocytes (Table 16.2) [Granner et al., 1968].

Regulation. The level of expression of many differentiated products is under the regulatory control of environmental influences, such as hormones, the matrix, and adjacent cells (*see* Section 17.7). Hence, the measurement of specific lineage markers may require preincubation of the cells in, for example, a hormone such as hydrocortisone, specific growth factors, or growth of the cells on extracellular matrix of the correct type. Maximum expression of both tyrosine aminotransferase in liver cells and glutamine synthetase in glia requires prior induction with dexamethasone. Glutamine synthetase is also repressed by glutamine, so glutamate should be substituted in the medium 48 h before assay [DeMars, 1958].

Lineage fidelity. Although many of the markers described above have been claimed as lineage markers, they are more properly regarded as tissue or cell type markers, as

TABLE 16.2. Enzymic Markers

Enzyme	Cell type	Inducer	Repressor	Reference
Glutamyl synthetase	Astroglia (brain)	Hydrocortisone	Glutamine	Hallermeyer & Hamprecht, 1984
Tyrosine aminotransferase	Hepatocytes	Hydrocortisone		Granner et al., 1968
Sucrase	Enterocytes	NaBt		Pignata et al., 1994; Vachon et al., 1996
Alkaline phosphatase	Type II pneumocyte (in lung alveolus)	Dexamethasone, oncostatin, IL-6	TGF-β	Edelson et al., 1988; McCormick & Freshney, 2000
Alkaline phosphatase	Enterocytes	Dexamethasone, NaBt		Vachon et al., 1996
Nonspecific esterase	Macrophages	PMA, Vitamin D$_3$		Murao et al., 1983
Angiotensin-converting enzyme	Endothelium	Collagen, Matrigel		Del Vecchio & Smith, 1981
Neuron-specific enolase	Neurons, neuroendocrine cells			Hansson et al., 1984
Tyrosinase	Melanocytes	cAMP		Park et al., 1993, 1999
DOPA-decarboxylase	Neurons, SCLC			Dow et al., 1995; Gazdar et al., 1980
Creatine kinase MM	Muscle cells	IGF-II	FGF-1,2,7	Stewart et al., 1996
Creatine kinase BB	Neurons, neuroendocrine cells, SCLC			Gazdar et al., 1981

they are often more characteristic of the function of the cell than its embryonic origin. Cytokeratins occur in mesothelium and kidney epithelium, although both of these tissues derive from the mesoderm. Neuron-specific enolase and creatine kinase BB are expressed in neuroendocrine cells of the lung, although these cells are now recognized to derive from the endoderm and not from neuroectoderm, as one might expect of neuroendocrine-type cells.

16.3.3 Unique Markers

Unique markers include specific chromosomal aberrations (e.g., deletions, translocations, polysomy); major histocompatibility (MHC) group antigens (e.g., HLA in humans), which are highly polymorphic; and DNA fingerprinting (*see* Protocol 16.8) or profiling (*see* Protocol 16.9). Enzymic deficiencies [e.g., thymidine kinase deficiency (TK⁻)] and drug resistance (e.g., vinblastine resistance, usually coupled to the expression of the P-glycoprotein by one of the *mdr* genes, which code for the efflux protein) are not truly unique, but they may be used to distinguish among cell lines from the same tissues but different donors.

16.3.4 Transformation

The transformation status forms a major element in cell line characterization and is dealt with separately (*see* Chapter 17).

16.4 CELL MORPHOLOGY

Observation of morphology is the simplest and most direct technique used to identify cells. It has, however, certain shortcomings that should be recognized. Most of these are related to the plasticity of cellular morphology in response to different culture conditions; for example, epithelial cells growing in the center of a confluent sheet are usually regular, polygonal, and with a clearly defined edge, whereas the same cells growing at the edge of a patch may be more irregular and distended and, if transformed, may break away from the patch and become fibroblast-like in shape.

Subconfluent fibroblasts from hamster kidney or human lung or skin assume multipolar or bipolar shapes (*see* Fig. 16.2a,g and Plate 8a) and are well spread on the culture surface, but at confluence they are bipolar and less well spread (*see* Fig. 16.2b,h and Plates 8b, 10d,e). They also form characteristic parallel arrays and whorls that are visible to the naked eye (Plate 10c). Mouse 3T3 cells (Fig. 16.2s,t) grow like multipolar fibroblasts at low cell density but become epithelium-like at confluence (Fig. 16.2v,w; *see* Plate 10b). Alterations in the substrate [Gospodarowicz et al., 1978b; Freshney, 1980], and the constitution of the medium (*see* Section 17.7.2), can also affect cellular morphology. Hence, comparative observations of cells should always be made at the same stage of growth and cell density in the same medium, and growing on the same substrate.

The terms "fibroblastic" and "epithelial" are used rather loosely in tissue culture and often describe the appearance rather than the origin of the cells. Thus a bipolar or multipolar migratory cell, the length of which is usually more than twice its width, would be called fibroblastic, whereas a monolayer cell that is polygonal, with more regular dimensions, and that grows in a discrete patch along with other cells is usually regarded as epithelial (Fig. 16.2e,f,i,j,m,n,q). However, when the identity of the cells has not been confirmed, the terms "fibroblast-like" (or "fibroblastoid") and "epithelium-like" (or "epithelioid") should be used. Carcinoma-derived cells are often epithelium-like but more variable in morphology (Fig. 16.2m,n,k,l), and some mesenchymal cells like CHO-K1 and endothelium can also assume an epithelium-like morphology at confluence (Fig. 16.2d,r).

Frequent brief observations of living cultures, preferably with phase-contrast optics, are more valuable than infrequent stained preparations that are studied at length. The former give a more general impression of the cell's morphology and its plasticity and also reveal differences in granularity and vacuolation that bear on the health of the culture. Unhealthy cells often become granular and then display vacuolation around the nucleus (*see* Fig. 13.1).

16.4.1 Microscopy

The inverted microscope is one of the most important tools in the tissue culture laboratory, but it is often used incorrectly. As the thickness of the closed culture vessel makes observation difficult from above, because of the long working distance, the culture vessel is placed on the stage, illuminated from above, and observed from below. As the thickness of the wall of the culture vessel still limits the working distance, the maximum objective magnification is usually limited to 40×. The use of phase-contrast optics, where an annular light path is masked by a corresponding dark ring in the objective and only diffracted light is visible, enables unstained cells to be viewed with higher contrast than is available by normal illumination. Because this means that the intensity of the light is increased, an infrared filter should be incorporated for prolonged observation of cells.

PROTOCOL 16.1. USING AN INVERTED MICROSCOPE

Outline
Place the culture on the microscope stage, switch on the power, and select the correct optics. Focus on the specimen and center the condenser and phase ring, if necessary.

Materials
Nonsterile:
- ❏ Inverted microscope (*see* Section 5.2.5) with phase contrast on 10× and 20× or 40× objectives and condenser with appropriate phase rings
- ❏ Lens tissue

Protocol
1. Make sure that the microscope is clean. (Wipe the stage with 70% alcohol, and clean the objectives with lens tissue, if necessary.)
2. Switch on the power, bringing the lamp intensity up from its lowest to the correct intensity with the rheostat, instead of switching the lamp straight to bright illumination.

3. Check the alignment of the condenser and the light source:
   (a) Place a stained slide, flask, or dish on the stage.
   (b) Close down the field aperture.
   (c) Focus on the image of the iris diaphragm.
   (d) Center the image of the iris in the field of view.
4. Center the phase ring:
   (a) If a phase telescope is provided, insert it in place of the standard eyepiece. Then focus on the phase ring, and check that the condenser ring (white on black) is concentric with the objective ring (black on white).
   (b) If no phase telescope is provided, then replace the stained specimen with an unstained culture (living or fixed), refocus, and move the phase ring adjustment until optimum contrast is obtained.
5. Examine the cells. A 10× phase-contrast objective will give sufficient detail for routine examination. A 4× phase-contrast objective will not give sufficient detail, and higher magnifications than 10× restrict the area scanned.
   (a) Note the growth phase (e.g., sparse, subconfluent, confluent, dense).
   (b) Note the state of the cells (e.g., clear and hyaline or granular, vacuolated).
6. Check the clarity of the medium (e.g., absence of debris, floating granules, signs of contamination), selecting a higher-magnification objective as required. (*See* Section 19.3.1, Fig. 19.1a,c,e, and Plate 16a,b,c.)
7. Record your observations.
8. Turn the rheostat down and switch off the power to the microscope.
9. Return the culture to the incubator, or take it to a hood for sterile manipulations.

Culture vessels taken straight from the incubator will develop condensation on the inside of the top or lid. This condensation can be cleared in a flask by inverting the flask, running medium over the inside of the top of the flask (without allowing it to run into the neck), and standing the flask vertically for 10–20 s to allow the liquid to drain down. Clearing the condensation is more difficult for Petri dishes and may require replacement of the lid. If the dish is correctly labeled, i.e., on one side of the base, this should not prejudice identification of the culture. Cultures in microtitration plates are particularly difficult to examine due to the diffraction caused by the meniscus.

It is useful to keep a set of photographs at different cell densities (*see* Fig. 16.2) for each cell line (*see* Protocol 16.6), prepared shortly after acquisition and at intervals thereafter, as a record in case a morphological change is subsequently suspected. Photographs of cell lines in regular use should be displayed above the inverted microscope. Photographic records can be supplemented with photographs of stained preparations and a scanned autoradiograph from a DNA fingerprint or digital output from DNA profiling and stored with the cell line record digitally or in hard copy.

16.4.2 Staining

A polychromatic blood stain, such as Giemsa, provides a convenient method of preparing a stained culture. Giemsa stain is usually combined with May–Grünwald stain when staining blood, but not when staining cultured cells. Alone, it stains the nucleus pink or magenta, the nucleoli dark blue, and the cytoplasm pale gray-blue. It stains cells fixed in alcohol or formaldehyde, but will not work correctly unless the preparation is completely anhydrous.

PROTOCOL 16.2. STAINING WITH GIEMSA

Outline
Fix the culture in methanol, stain it directly with undiluted Giemsa, and then dilute the stain 1:10. Wash the culture, and examine it wet.

Materials
Nonsterile:
❏ D-PBSA
❏ D-PBSA:methanol, 1:1
❏ Undiluted Giemsa stain
❏ Methanol
❏ Deionized water

Protocol
This protocol assumes that a cell monolayer is being used, but fixed cell suspensions (*see* Section 16.4.4) can also be used, starting at Step 6.

1. Remove and discard the medium.
2. Rinse the monolayer with D-PBSA, and discard the rinse.
3. Add 5 mL of D-PBSA/methanol per 25 cm². Leave it for 2 min and then discard the D-PBSA/methanol.
4. Add 5 mL of fresh methanol, and leave it for 10 min.
5. Discard the methanol, and replace it with fresh anhydrous methanol. Rinse the monolayer, and then discard the methanol.
6. At this point, the flask may be dried and stored or stained directly. It is important that staining be done directly from fresh anhydrous methanol, even with a dry flask. If the methanol is poured

off and the flask is left for some time, water will be absorbed by the residual methanol and will inhibit subsequent staining. Even "dry" monolayers can absorb moisture from the air.
7. Add neat Giemsa stain, 2 mL per 25 cm², making sure that the entire monolayer is covered and remains covered.
8. After 2 min, dilute the stain with 8 mL of water, and agitate it gently for a further 2 min.
9. Displace the stain with water so that the scum that forms is floated off and not left behind to coat the cells. Wash the cells gently in running tap water until any pink cloudy background stain (precipitate) is removed, but stain is not leached out of cells (usually about 10–20 s).
10. Pour off the water, rinse the monolayer in deionized water, and examine the cells on the microscope while the monolayer is still wet. Store the cells dry, and rewet them to re-examine.

Note. Giemsa staining is a simple procedure that gives a good high-contrast polychromatic stain, but precipitated stain may give a spotted appearance to the cells. This precipitate forms as a scum at the surface of the staining solution, because of oxidation, and throughout the solution, particularly on the surface of the slide, when water is added. Washing off the stain by upward displacement, rather than pouring it off or removing the slides, is designed to prevent the cells from coming in contact with the scum. Extensive washing at the end of the procedure is designed to remove any precipitate left on the preparation.

Fixed-cell preparations can also be stained with crystal violet. Crystal violet is a monochromatic stain and stains the nucleus dark blue and the cytoplasm light blue. It is not as good as Giemsa for morphological observations, but it is easier to use and can be reused. It is a convenient stain to use for staining clones for counting, as it does not have the precipitation problems of Giemsa and is easier to wash off, giving a clearer background and making automated colony counting easier.

PROTOCOL 16.3. STAINING WITH CRYSTAL VIOLET

Materials
Nonsterile:
❏ D-PBSA
❏ D-PBSA/MeOH: 50% methanol in D-PBSA
❏ Methanol
❏ Crystal Violet

❑ Filter funnel and filter paper of a size appropriate to take 5 mL of stain from each dish
❑ Bottle for recycled stain

Protocol
1. Remove and discard the medium.
2. Rinse the monolayer with D-PBSA, and discard the rinse.
3. Add 5 mL of D-PBSA/MeOH per 25 cm². Leave it for 2 min and then discard the D-PBSA/MeOH.
4. Add 5 mL of fresh methanol, and leave it for 10 min.
5. Discard the methanol. Drain the dishes, and allow them to dry.
6. Add 5 mL of Crystal Violet per 25 cm², and leave for 10 min.
7. Place the filter funnel in the recycle bottle.
8. Discard the stain into the filter funnel.
9. Rinse the dish in tap water and then in deionized water, and allow the dish to dry.

16.4.3 Culture Vessels for Cytology: Monolayer Cultures

The following is a list of culture vessels that have been found suitable for cytology:

(1) Regular 25-cm² flasks or 5-cm Petri dishes.
(2) Coverslips (glass or plastic; *see* Appendix II) in multiwell dishes, Petri dishes, or Leighton tubes (Fig. 16.3a; *see also* Figs. 8.2 and 8.3).
(3) Microscope slides in 9-cm Petri dishes or with attached multiwell chambers (chamber slides; *see* Appendix II; Fig. 16.3b).
(4) The OptiCell double membrane culture chamber (Fig 16.3c).
(5) Petriperm dishes (Heraeus), cellulose acetate or polycarbonate filters, Thermanox, and PTFE-coated coverslips, which have all been used for EM cytology studies (some pretreatment of filters or PTFE may be required—e.g., gelatin, collagen, fibronectin, Matrigel, or serum coating—*see* Section 8.4.1).

16.4.4 Preparation of Suspension Culture for Cytology

Nonadherent cells must be deposited on a glass or plastic surface for cytological observations. The conventional technique for blood cells, the preparation of a blood smear, does not work well because cultured cells tend to rupture during smearing, although this can be reduced by coating the slide with serum and then spreading the cells in serum.

Centrifugation. Cytological centrifuges have become the preferred option for creating monolayers from suspension cells. Centrifuges (usually called *cytocentrifuges*) are available with special slide carriers for spinning cells onto a slide (*see* Appendix II and Fig. 16.4). The carriers have sample compartments leading to an orifice that is placed against a microscope slide and are located within a specially designed, enclosed rotor. Coating the slides with serum helps to prevent the cells from bursting when they hit the slide.

PROTOCOL 16.4. PREPARATION OF SUSPENSION CELLS FOR CYTOLOGY BY CYTOCENTRIFUGE

Outline
Dispense a small volume of concentrated cells into the sample compartment *in situ* and placed against a slide. Centrifuge the cells onto the slide.

Materials
Nonsterile:
❑ Cell suspension, 5×10^5 cells/mL, in 50–100% serum
❑ Serum
❑ Methanol
❑ Cytocentrifuge (*see* Appendix II)
❑ Microscope slides
❑ Slide carriers for centrifuge
❑ Fan

Protocol
1. Coat the slides in serum and drain.
2. Dry the slides with a fan.
3. Label the slides.
4. Place the slides on slide carriers, and insert the carriers in the rotor.
5. Place approximately $1–2 \times 10^5$ cells in 200–400 µL of medium in at least two sample blocks (*see* manufacturer's protocol).
6. Switch on the centrifuge, and spin the cells down onto the slides at 100 *g* for 5 min.
7. Dry off the slides quickly, and fix them in MeOH for 10 min.

Note. Some cytocentrifuges require that fixation be performed *in situ*.

Δ *Safety Note.* The rotor must be sealed during spinning if human or other primate cells are being used.

Drop technique. This procedure is the same as used for chromosome preparation (*see* Protocol 16.7), but without the colcemid and hypotonic treatments. Care must be taken to avoid clumping and the cell concentration must be adjusted to ensure the cells do not pile up. Cells at the edge of the spot may rupture.

Filtration. This technique is sometimes used in exfoliative cytology (*see* the instructions of the following manufacturers for further details: Pall Gelman, Millipore, Sartorius). This technique is suitable for sampling large volumes with a low cell concentration.

PROTOCOL 16.5. FILTRATION CYTOLOGY

Outline
Gently draw a low-concentration cell suspension through a filter by vacuum. Wash the cells and fix it *in situ*.

Materials
Nonsterile:
- D-PBSA, 20 mL
- Methanol, 50 mL
- Giemsa stain
- Mountant (DPX or Permount)
- Transparent filters (e.g., 25-mm TPX or poly-carbonate, 0.5-μm porosity, Millipore, Pall Gelman, Sartorius)

- Filter holder (e.g., Millipore, Pall Gelman, Sartorius)
- Vacuum flask (Millipore)
- Cell suspension (~10^6 cells in 5–10 mL medium with 20% serum)
- Vacuum pump

Protocol
1. Set up the filter assembly (Fig. 16.5) with a 25-mm-diameter, 0.5-μm-porosity transparent filter.
2. Draw the cell suspension onto the filter with a vacuum pump. Do not let all of the medium run through.
3. Gently add 10 mL of D-PBSA when the cell suspension is down to 2 mL.
4. Repeat Step 3 when the D-PBSA is down to 2 mL.
5. Add 10 mL of methanol to the D-PBSA, and keep adding methanol until pure methanol is being drawn through the filter.
6. Switch off the vacuum before all of the methanol runs through.
7. Lift out the filter, and air dry it.
8. Stain the filter in Giemsa, rinse in water, and dry.
9. Mount the filter on a slide in DPX or Permount by pressing the coverslip down with a heavy weight to flatten the filter.

Fig. 16.2. *Examples of Cell Morphology in Culture.* (a) BHK-21 (baby hamster kidney fibroblasts) in log growth. The culture is not confluent, and the cells are well spread and randomly oriented (although some nonrandom orientation is beginning to appear). (b) BHK-21 cells at the end of log phase and entering plateau phase. (c) CHO-K1 cloned line of Chinese hamster ovary; some fibroblast-like, others more epithelioid. (d) CHO-K1 cells at high density; refractile and more elliptical or epithelioid with fewer spindle-shaped cells. (e) Vero cells in log phase; epithelial and forming sheets. (f) High-density population of Vero; postconfluent, dense sheet with smaller cell diameters. (g) Low-density (mid-log phase) MRC-5 human fetal lung fibroblasts; growth random, although some orientation beginning to appear. (h) Confluent population of MRC-5 cells with parallel orientation clearly displayed. (i) HEK 293 human embryonic kidney epithelial cell line, growing in sheets at mid-log phase. (j) High-density HEK 293 cells showing densely packed epithelial cells. (k) LNCaP clone FGC from a lymph node metastasis of prostate carcinoma; medium to high density. (l) High-density LNCaP cells forming aggregate. (m) HeLa cells from human cervical carcinoma. (n) HeLa-S3 clone of HeLa. (o) IMR-32 cells from human neuroblastoma. (p) Cos-7 cells from monkey kidney. (q) MDCK cells, Madin-Darby canine kidney. (r) BAE cells from bovine arterial endothelium. (s) Subconfluent Swiss 3T3 cells (Sw-3T3) from Swiss albino mouse embryo. (t) NIH-3T3 cells from NIH mouse embryo. (u) L-929, clone of mouse L-cells. (v) Sw-3T3 at confluence. Note low-contrast appearance implying very flat monolayer. (w) Postconfluent Sw-3T3. (x) Postconfluent Sw-3T3 showing a transformation focus of refractile cells overgrowing the monolayer, demonstrating why these cells should be subcultured well before they reach confluence. (y) Confluent STO mouse embryo fibroblasts, often used as feeder cells for embryonal stem cell culture. (z) Confluent EMT-6 mouse mammary tumor cell line; forms spheroids readily. (aa) Subconfluent A2780, human ovarian carcinoma. (bb) Caco-2, colorectal carcinoma, used in transepithelial transport studies. (cc) Hep-G2, human hepatoma, sometimes used to metabolize drugs or potential cytotoxins. (dd) HT-29 colorectal carcinoma; differentiation inducible with sodium butyrate. (ee) PC-12, rat adrenal pheochromocytoma; NGF induces neuronal phenotype. (ff) U-373 MG, human glioma; one of an extensive series of malignant and normal glial cell lines developed in Uppsala, Sweden. (Photos a–q, s–v courtesy of ATCC; r, courtesy of Peter Del Vecchio; y,z, aa–ff, courtesy of ECACC; *see also* Plates 7–10).

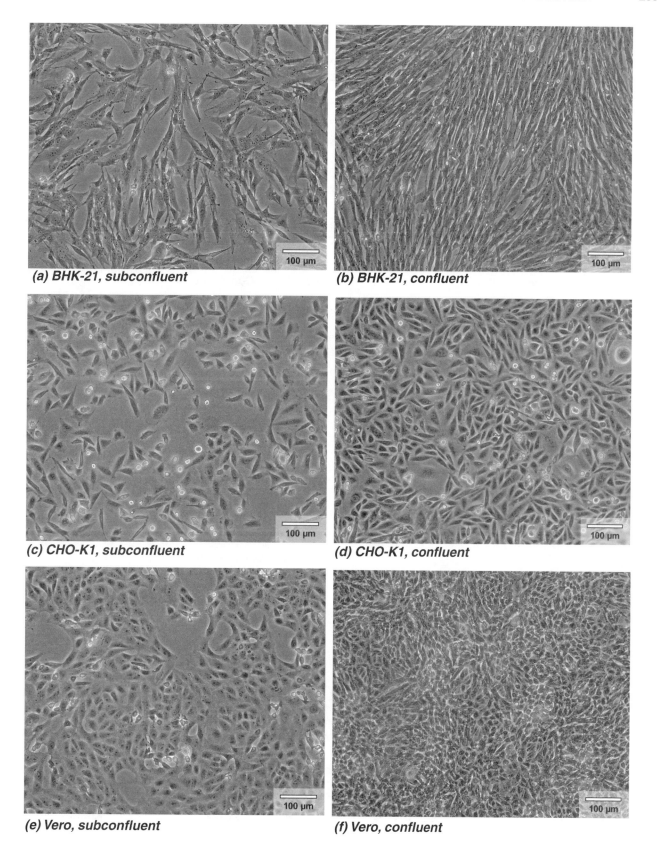

(a) BHK-21, subconfluent

(b) BHK-21, confluent

(c) CHO-K1, subconfluent

(d) CHO-K1, confluent

(e) Vero, subconfluent

(f) Vero, confluent

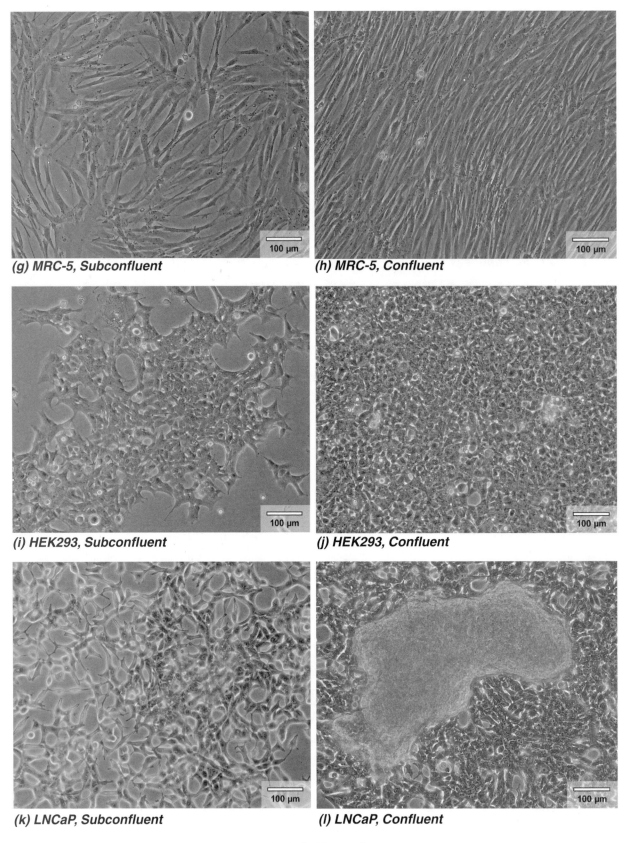

(g) MRC-5, Subconfluent

(h) MRC-5, Confluent

(i) HEK293, Subconfluent

(j) HEK293, Confluent

(k) LNCaP, Subconfluent

(l) LNCaP, Confluent

Fig. 16.2. (*Continued*)

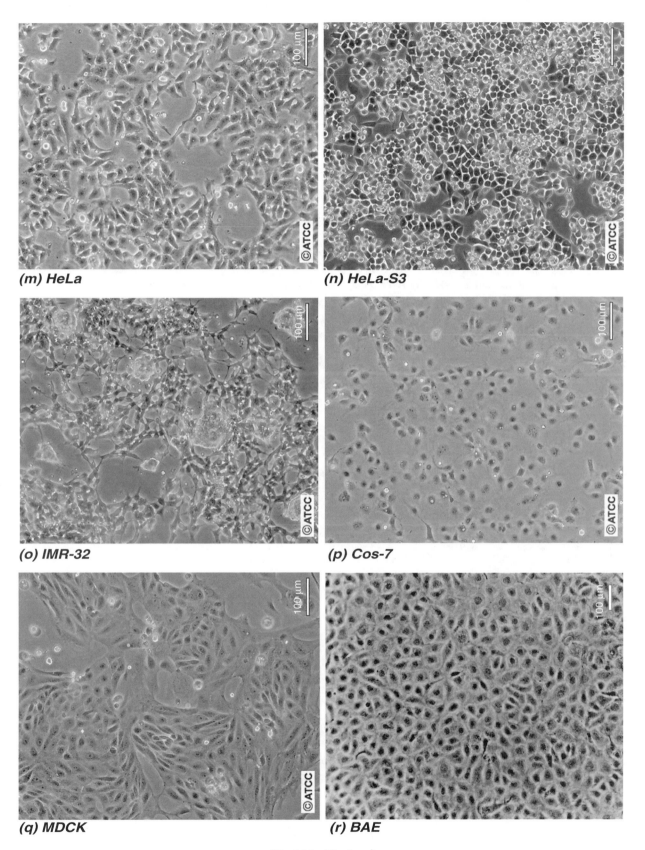

(m) HeLa

(n) HeLa-S3

(o) IMR-32

(p) Cos-7

(q) MDCK

(r) BAE

Fig. 16.2. (Continued)

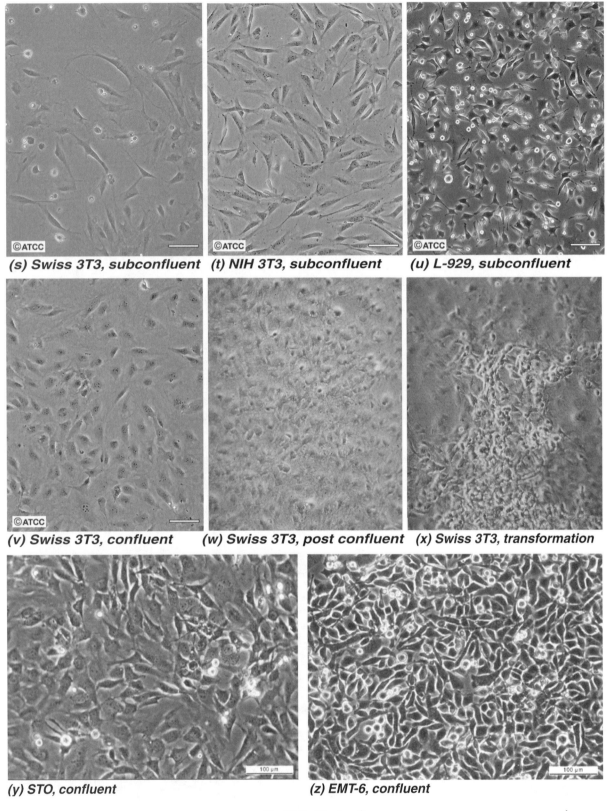

(s) Swiss 3T3, subconfluent   *(t) NIH 3T3, subconfluent*   *(u) L-929, subconfluent*

(v) Swiss 3T3, confluent   *(w) Swiss 3T3, post confluent*   *(x) Swiss 3T3, transformation*

(y) STO, confluent   *(z) EMT-6, confluent*

Fig. 16.2. (*Continued*)

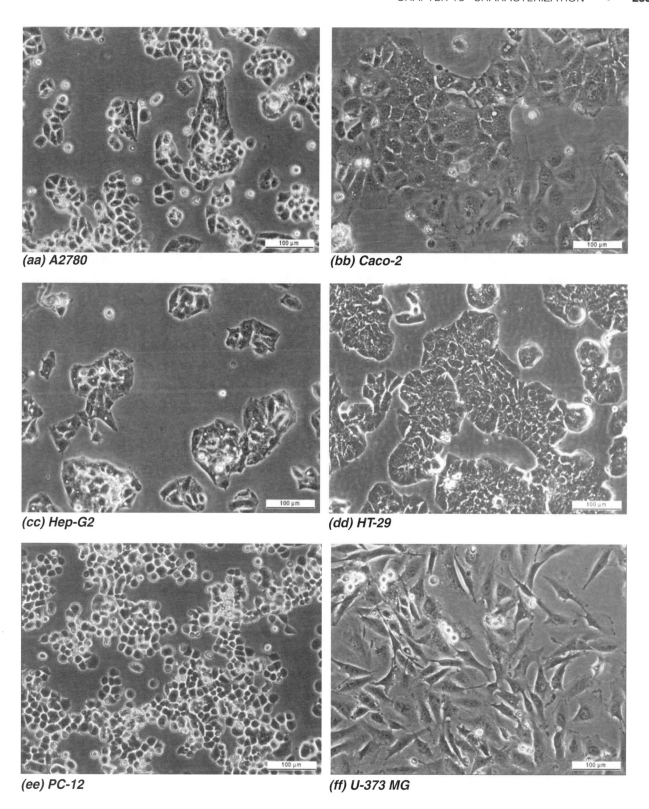

(aa) A2780

(bb) Caco-2

(cc) Hep-G2

(dd) HT-29

(ee) PC-12

(ff) U-373 MG

Fig. 16.2. (*Continued*)

16.4.5 Photomicrography

The simplest way of recording microscope images is digitally via a CCD camera. Choosing at least a 5-megapixel camera gives resolution that is more than adequate for publication.

Electronic storage of images increases the facility of searching, electronic transmission to other sites, formatting, and editing. The image is viewed on-screen at the time it is taken and can be printed instantly if required, in black and white or in color.

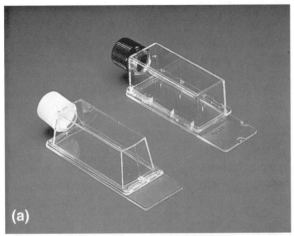

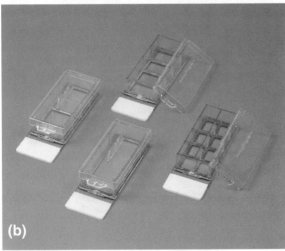

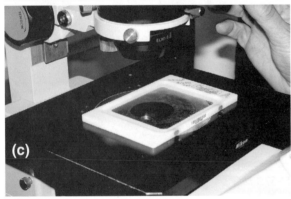

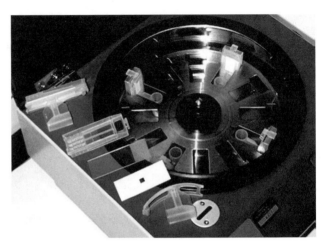

Fig. 16.4. Cytocentrifuge. View of interior of centrifuge showing rotor and polypropylene slide carriers in place with one carrier disassembled on left-hand side (Sakura Finetek).

Fig. 16.3. Culture Vessels for Cytology. (a) Nalge Nunc slide flask; flask is detachable for processing. (b) Lab-Tek chamber slides: detachable plastic chambers on a regular microscope slide. One, 2, 4, 8, and 16 chambers per slide are available. (c) OptiCell culture chamber with upper and lower plastic membranes suitable for cell growth and observation. (a, b courtesy of Nalge Nunc; c courtesy of Dr. Donna Peehl).

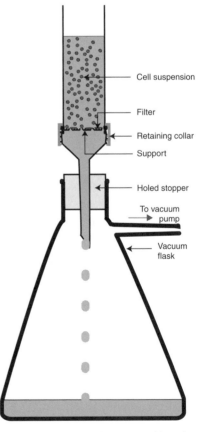

Fig. 16.5. Filter Cytology. Filter assembly for cytological preparation (e.g., Millipore).

PROTOCOL 16.6. DIGITAL PHOTOGRAPHY ON A MICROSCOPE

Outline
Set up the microscope, check the alignment of the optics, and focus on the relevant area of the specimen. Check the connections of the camera to the computer, divert the light beam to the camera, focus, and save the image.

Materials
Sterile or Aseptically Prepared:
❑ Culture for examination (make sure that the culture is free of debris—e.g., change the medium on a primary culture before photography)

Nonsterile:
❑ Inverted microscope (*see* Section 5.2.5) with the following attributes:
  Phase contrast on 10× and 20× or 40× objectives
  Phase condenser with phase rings for 10× and 20× or 40× objectives
  Trinocular head or other port with adapter for CCD camera
  Neutral density filters, 2× and 4×
❑ Lens tissue
❑ Record pad for images
❑ Micrometer slide

Protocol
1. Prepare culture for photography:
   (a) For flask cultures:
      i) Take to a laminar flow hood, loosen the cap on the flask, and allow the culture to cool.
      ii) Tighten the cap on the flask when the contents of the flask are at room temperature, in order to avoid any distortion of the flask as it cools.
      iii) Rinse the medium over the inside of the top of the flask to remove any condensation that may have formed.
      iv) Allow the film of medium to drain by standing the flask vertically.
   (b) For dishes and plates:
      i) Allow the culture to cool to room temperature in a laminar-flow hood.
      ii) Replace the lid of the dish or plate (make sure that the sample is labeled on the base of the dish or plate).
2. Select the field and magnification (4× is best for clones or patterning of a monolayer, 10× for a representative shot, and 20× for cellular detail),

avoiding imperfections or marks on the flask. (Always label the side of the flask or dish, and not the top or bottom.)
3. Check the focus:
   (a) Focus the binocular eyepieces to your own eyes, using the frame or graticule in the eyepiece.
   (b) Focus the microscope.
4. Switch to the monitor, and check the density and contrast of the image on screen, adjusting the light intensity with the rheostat on the microscope for a black-and-white record and with neutral density filters for color (although the image can later be reprocessed electronically if the color temperature is incorrect).
5. Refocus the microscope if necessary.
6. Save the image at a resolution appropriate to the image's ultimate use and the electronic storage available (e.g., 940 × 740 will require 2 Mb storage space and provide reasonable quality if reproduced at 5 × 7 in., 12 × 18 cm.)
7. Turn down the light to avoid overheating the culture.

Note. An infrared filter may be incorporated to minimize overheating.

8. Repeat Steps 3–6 with a micrometer slide at the same magnification to give the scale of magnification. If this image is processed in the same manner as the other shots, it can be used to generate a scale bar.
9. Return the culture to the incubator.
10. Complete a record manually or in a spreadsheet against the filename; otherwise, the images will be difficult to identify.
11. Saved images can be viewed, edited, and formatted with programs such as Adobe PhotoShop and printed by high-quality inkjet, color laser printer, or Kodak ColorEase printer.

With the fall in the cost of digital cameras, it is hard to justify continued use of film. However, if using conventional film, there are two major frame sizes you may wish to consider: 35 mm, which is best for routine color transparencies and high-volume black-and-white photographs, and $3\frac{1}{2} \times 4\frac{1}{2}$ in. (9 × 12 cm) Polaroid with positive/negative film (type 665), which is good for low-volume black-and-white photographs. Instant film, made by Fuji, may also be used. With instant film, you obtain an immediate result and know that you have the record without having to develop and print a film. The unit cost is high, but there is a considerable saving in time. However, some instant films have relatively low contrast, and their color saturation is not always as good as that of conventional film. Instant films also tend to have a shorter shelf life.

16.5 CHROMOSOME CONTENT

Chromosome content or *karyotype* is one of the most characteristic and best-defined criteria for identifying cell lines and relating them to the species and sex from which they were derived. Examples of human and mouse karyotypes are available at *www.pathology.washington.edu/galleries/Cytogallery/*, and the rat ideogram is available at *http://ratmap.gen.gu.se/Idiogram.html*. *See also* Mitelman [1995] for human and Hsu and Benirschke [1967] for other mammals, although the latter predates chromosome banding (*see* Section 16.5.1). Chromosome analysis can also distinguish between normal and transformed cells (*see* Section 18.3.1), because the chromosome number is more stable in normal cells (except in mice, where the chromosome complement of normal cells can change quite rapidly after explantation into culture).

PROTOCOL 16.7. CHROMOSOME PREPARATIONS

Outline
Fix cells arrested in metaphase and swollen in hypotonic medium. Drop the cells on a slide, stain, and examine (Fig. 16.6) [Rothfels & Siminovitch, 1958; Rooney & Czepulkowski, 1986].

Materials
Sterile or Aseptically Prepared:
❑ Culture of cells in log phase
❑ Colcemid, 1×10^{-5} M in D-PBSA
❑ D-PBSA
❑ Trypsin, 0.25% crude
Nonsterile:
❑ Hypotonic solution: 0.04 M KCl, 0.025 M sodium citrate
❑ Acetic methanol fixative: 1 part glacial acetic acid plus 3 parts anhydrous methanol or ethanol, made up fresh and kept on ice
❑ Giemsa stain
❑ DPX or Permount mountant
❑ Ice
❑ Centrifuge tubes
❑ Pasteur pipettes

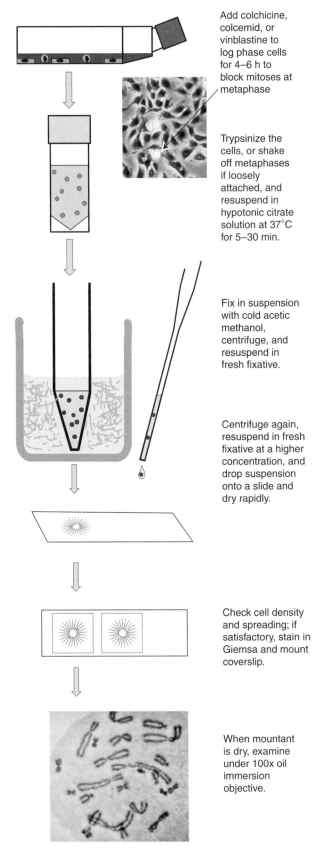

Add colchicine, colcemid, or vinblastine to log phase cells for 4–6 h to block mitoses at metaphase

Trypsinize the cells, or shake off metaphases if loosely attached, and resuspend in hypotonic citrate solution at 37°C for 5–30 min.

Fix in suspension with cold acetic methanol, centrifuge, and resuspend in fresh fixative.

Centrifuge again, resuspend in fresh fixative at a higher concentration, and drop suspension onto a slide and dry rapidly.

Check cell density and spreading; if satisfactory, stain in Giemsa and mount coverslip.

When mountant is dry, examine under 100x oil immersion objective.

Fig. 16.6. Chromosome Preparation. Preparation of chromosome spreads from monolayer cultures by the drop technique.

❑ Slides
❑ Coverslips, #00
❑ Slide dishes
❑ Low-speed centrifuge
❑ Vortex mixer

Protocol

1. Set up a 75-cm² flask culture at between 2×10^4 and 5×10^4 cells/mL (4×10^3 and 1×10^4 cells/cm²) in 20 mL.
2. Approximately 3–5 days later, when the cells are in the log phase of growth, add 0.2 mL 1×10^{-5} M colcemid to the medium already in the flask to give a final concentration of 1×10^{-7} M.
3. After 4–6 h, remove the medium gently, add 5 mL of 0.25% trypsin, and incubate the culture for 10 min.
4. Centrifuge the cells in trypsin, and discard the supernatant trypsin.
5. Resuspend the cells in 5 mL of hypotonic solution, and leave them for 20 min at 37°C.
6. Add an equal volume of freshly prepared, ice-cold acetic methanol, mixing constantly, and then centrifuge the cells at 100 g for 2 min.
7. Discard the supernatant mixture, "buzz" the pellet on a vortex mixer (e.g., hold the bottom of the tube against the edge of the rotating cup), and slowly add fresh acetic methanol with constant mixing.
8. Leave the cells for 10 min on ice.
9. Centrifuge the cells for 2 min at 100 g.
10. Discard the supernatant acetic methanol, and resuspend the pellet by "buzzing" in 0.2 mL of acetic methanol, to give a finely dispersed cell suspension.
11. Draw one drop of the suspension into the tip of a Pasteur pipette, and drop from around 12 in. (30 cm) onto a cold slide. Tilt the slide and let the drop run down the slide as it spreads.
12. Dry off the slide rapidly over a beaker of boiling water, and examine it on the microscope with phase contrast. If the cells are evenly spread and not touching, then prepare more slides at the same cell concentration. If the cells are piled up and overlapping, then dilute the suspension two- to fourfold and make a further drop preparation. If the cells from the diluted suspension are satisfactory, then prepare more slides. If not, then dilute the suspension further and repeat this step.
13. Stain the cells with Giemsa:
    (a) Place the slides on a rack positioned over the sink.
    (b) Cover the cells completely with a few drops of neat Giemsa, and stain for 2 min.
    (c) Flood the slides with approximately 10 volumes of water.
    (d) Leave the slides for a further 2 min.
    (e) Displace the diluted stain with running water.

Note. Do **not** pour the stain off the slides as it will leave a scum of oxidized stain behind; always displace the stain with water. Even when staining in a dish, the stain is never poured off but must be displaced from the bottom with water.

    (f) Finish by running the slides individually under tap water to remove any precipitated stain.
    (g) Check the staining under the microscope. If it is satisfactory, then dry the slide thoroughly and mount with a #00 coverslip in DPX or Permount.

Variations

Metaphase block. (1) Vinblastine, 1×10^{-6} M, may be used instead of colcemid. (2) Duration of the metaphase block may be increased to give more metaphases for chromosome counting, or shortened to reduce chromosome condensation and improve banding (*see* Section 16.5.1).

Collection of mitosis. Some cells, e.g., CHO and HeLa, detach readily when in metaphase if the flask is shaken ("shake-off" technique), eliminating trypsinization. The procedure is as follows: (1) add colcemid, (2) remove the colcemid carefully and replace it with hypotonic citrate/KCl, (3) shake the flask to dislodge cells in metaphase either before or after incubation in hypotonic medium, and (4) fix the cells as previously described.

Hypotonic treatment. Substitute 0.075 M KCl alone or HBSS diluted to 50% with distilled water for the hypotonic citrate used in Protocol 16.9. The duration of the hypotonic treatment may be varied from 5 min to 30 min to reduce lysis or increase spreading.

Spreading. There are perhaps more variations at this stage than any other, all designed to improve the degree and flatness of the spread. They include (1) dropping cells onto a slide from a greater height (clamp the pipette, and mark the position of the slide, using a trial run with fixative alone); (2) flame drying (dry the slide after dropping the cells, by heating the slide over a flame, or actually burn off the fixative by igniting the drop on the slide as it spreads; however, the latter may make subsequent banding more difficult); (3) making the slide ultracold (chill the slide on solid CO_2

before dropping the cells on to it); (4) refrigerating the fixed-cell suspension overnight before dropping the cells onto a slide; (5) dropping cells on a chilled slide (e.g., steep the slide in cold alcohol and dry it off) and then placing the slide over a beaker of boiling water; and (6) tilting the slide or blowing the drop across the slide as the drop spreads (*see also* Protocols 27.5, 27.9).

16.5.1 Chromosome Banding

This group of techniques [*see* Rooney & Czepulkowski, 1986] was devised to enable individual chromosome pairs to be identified when there is little morphological difference between them [Wang & Fedoroff, 1972, 1973]. For Giemsa banding, the chromosomal proteins are partially digested by crude trypsin, producing a banded appearance on subsequent staining. Trypsinization is not required for quinacrine banding. The banding pattern is characteristic for each chromosome pair (Fig. 16.7). The methods for banding include the following: (1) Giemsa banding (use trypsin and EDTA rather than trypsin alone; *see* Fig. 16.7a); (2) Q-banding [Caspersson et al., 1968] [stain the cells in 5% (w/v) quinacrine dihydrochloride in 45% acetic acid, rinse the slide, and mount it in deionized water at pH 4.5 [Lin & Uchida, 1973; Uchida & Lin, 1974]]; (3) C-banding [this technique emphasizes the centromeric regions; the fixed preparations are pretreated for 15 min with 0.2 M of HCl and 2 min with 0.07 M NaOH and then are treated overnight with SSC (either 0.03 M sodium citrate, 0.3 M NaCl or 0.09 M sodium citrate, 0.9 M NaCl) before staining with Giemsa stain [Arrighi & Hsu, 1974]] (*see* Protocol 27.5).

Techniques have been developed for discriminating between human and mouse chromosomes, principally to aid the karyotypic analysis of human and mouse hybrids. These methods include fluorescent staining with Hoechst 33258, which causes mouse centromeres to fluoresce more brightly than human centromeres [Hilwig & Gropp, 1972; Lin et al., 1974], alkaline staining with Giemsa ("Giemsa-11") [Bobrow et al., 1972; Friend et al., 1976] (*see* Fig. 16.7c), and *in situ* hybridization with *chromosome paints* (*see* below and Protocol 27.9).

16.5.2 Chromosome Analysis

The following are methods by which the chromosome complement may be analyzed:

(1) **Chromosome Count.** Count the chromosome number per spread for between 50 and 100 spreads. (The chromosomes need not be banded.) Closed-circuit television or a camera lucida attachment may help. You should attempt to count all of the mitoses that you see and classify them (a) by chromosome number or (b), if counting is impossible, as "near diploid uncountable" or "polyploid uncountable." Plot the results as a histogram (*see* Fig. 3.9).

(2) **Karyotype.** Digitally photograph about 10 or 20 good spreads of banded chromosomes. Using Adobe

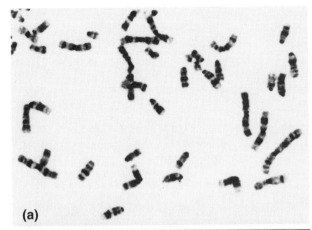

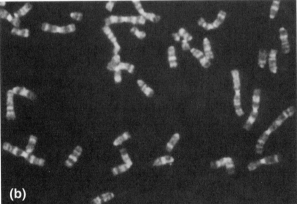

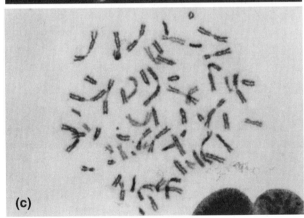

Fig. 16.7. Chromosome Staining. (a) Human chromosomes banded by the standard trypsin-Giemsa technique. (b) The same preparation as in (a), but stained with Hoechst 33258. (c) Human–mouse hybrid stained with Giemsa at pH 11. Human chromosomes are less intensely stained than mouse chromosomes. Several human/mouse chromosomal translocations can be seen. (Courtesy of R. L. Church.)

Photoshop or an equivalent graphics program, cut the individual chromosomes and paste them into a new file where they can be rotated, trimmed, aligned and sorted (Fig. 16.8). Image analysis can be used to sort chromosome images automatically to generate karyotypes (e.g., Leica CW4000).

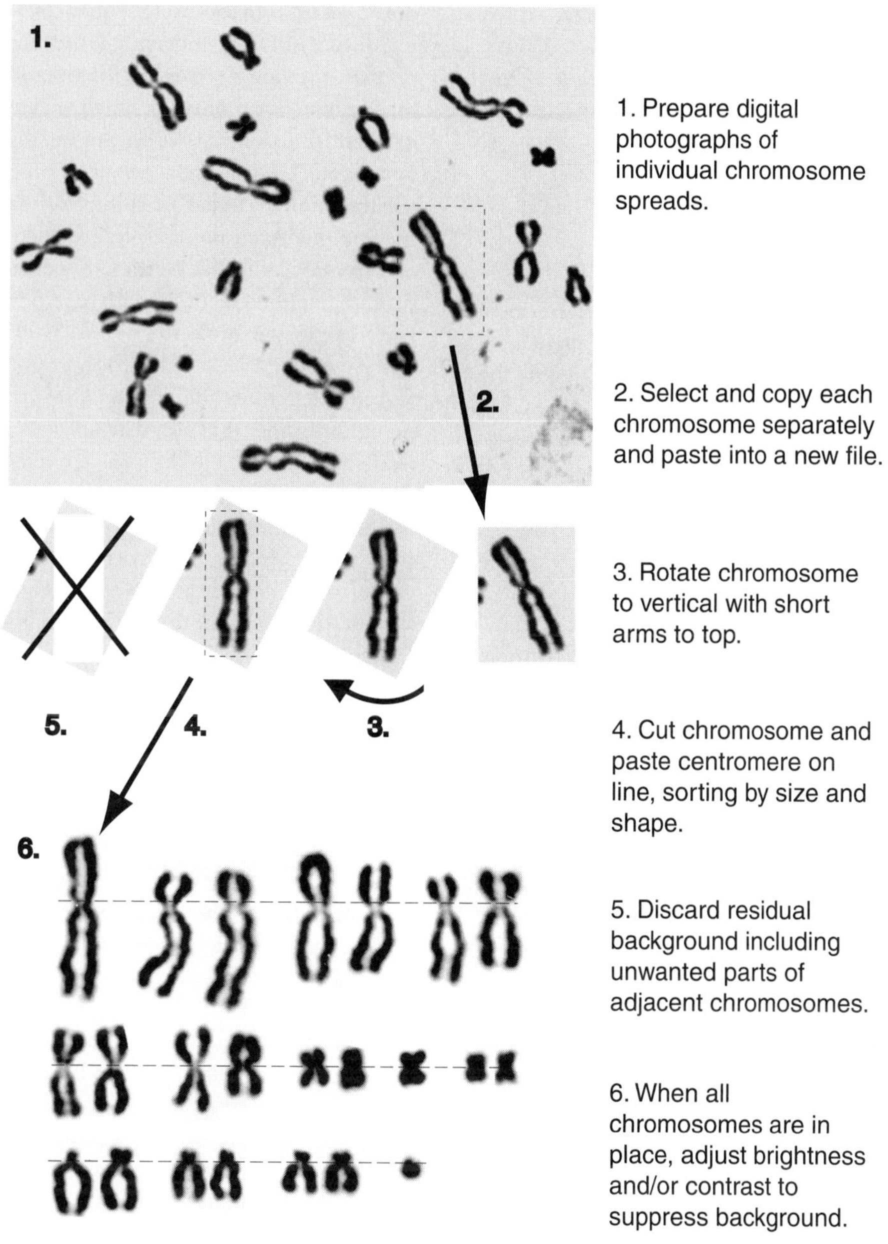

Fig. 16.8. Karyotype Preparation. Steps in the preparation of a karyotype from digital microphotographs of metaphase spreads. Chinese hamster cells recloned from the Y-5 strain of Yerganian and Leonard [1961] (acetic-orcein).

With the assistance of chromosome banding (*see* Protocol 27.5) and chromosome painting (*see* Protocol 27.9) an aggregate image of the typical karyotype can be prepared as a stylized image known as an ideogram. Ideograms for human, mouse, and horse are available at *www.pathology.washington.edu/research/cytopages/* and rat on *ratmap.gen.gu.se/Idiogram.html*.

Chromosome counting and karyotyping allow species identification of the cells and, when banding is used, distinguish cell line variation and marker chromosomes.

However, banding and karyotyping is time-consuming, and chromosome counting with a quick check on gross chromosome morphology may be sufficient to confirm or exclude a suspected cross-contamination.

16.5.3 Chromosome Painting

With the advent of fluorescently labeled probes that bind to specific regions, and even specific genes, on chromosomes, it has become possible to locate genes, identify translocations, and determine the species of origin of chromosomes. This technique employs *in situ* hybridization technology, which can also be used for extrachromosomal and cytoplasmic localization of specific nucleic acid sequences, such as might be used to localize specific messenger RNA species (*see* Protocols 27.8 and 27.9).

16.6 DNA CONTENT

DNA can be measured by propidium iodide fluorescence with a CCD camera or flow cytometry [Vaughan & Milner, 1989] (*see* Sections 15.4 and 21.7.2), although the generation of the necessary single-cell suspension will, of course, destroy the topography of the specimen. Analysis of DNA content is particularly useful in the characterization of transformed cells that are often aneuploid and heteroploid (*see* Section 18.3).

16.6.1 DNA Hybridization

Hybridization of specific molecular probes to unique DNA sequences (Southern blotting) [Sambrook et al., 1989; Ausubel et al., 1996] can provide information about species-specific regions, amplified regions of the DNA, or altered base sequences that are characteristic to that cell line. Thus strain-specific gene amplifications, such as amplification of the dihydrofolate reductase (DHFR) gene, may be detected in cell lines selected for resistance to methotrexate [Biedler et al., 1972]; amplification of the MDR gene in vinblastine-resistant cells [Schoenlein et al., 1992]; overexpression of a specific oncogene, or oncogenes in transformed cell lines [Bishop, 1991; Hames & Glover, 1991; Weinberg, 1989]; or deletion, or loss of heterozygosity in suppressor genes [Witkowski, 1990; Marshall, 1991]. Although this can be detected in restriction digests of extracts of whole DNA, this is limited by the amount of DNA required and it is more common to use the polymerase chain reaction (PCR) with a primer specific to the sequence of interest, enabling detection in relatively small numbers of cells, such as would be available in a laboratory-scale experiment. Alternatively, specific probes can be used to detect specific DNA sequences by *in situ* hybridization (*see* Section 27.8), having the advantage of displaying topographical differences and heterogeneity within a cell population.

It is also possible to label a cell strain for future identification by transfecting in a reporter gene, such as green fluorescent protein (GFP), luciferase, or β-galactosidase [Krotz et al., 2003], and then to detect it by fluorescence, luminescence, or Southern blotting or by chromogenic enzyme assay for the gene product.

16.6.2 DNA Fingerprinting

DNA contains regions known as satellite DNA that are apparently not transcribed. These regions are highly repetitive, and their lengths vary, with minisatellite DNA having 1- to 30-kb repeats and microsatellite DNA having only 2–4 bases in repeating sequences. The functions of these regions are not fully understood; they may be purely structural or may provide a reservoir of potentially codable regions for genetic recombination in further evolution. Regardless of their function, however, these regions are not highly conserved, because they are not transcribed, and they give rise to regions of hypervariability. When the DNA is cut with specific endonucleases, specific sequences may be probed with cDNAs that hybridize to these hypervariable regions or they may be amplified by PCR with specific templates. The probes were originated by Jeffreys et al. [1985] (*see* Protocol 16.8; Stacey et al., 1992). Electrophoresis reveals variations in fragment length in satellite DNA (restriction fragment length polymorphisms, RFLPs) that are specific to the individual from which the DNA was derived. When analyzed by polyacrylamide electrophoresis, each individual's DNA gives a specific hybridization pattern as revealed by autoradiography with radioactive or fluorescent probes. These patterns have come to be known as DNA fingerprints and are cell line specific, except if more than one cell line has been derived from one individual or if highly inbred donor animals have been used.

DNA fingerprints appear to be quite stable in culture, and cell lines from the same origin, but maintained separately in different laboratories for many years, still retain the same or very similar DNA fingerprints. DNA fingerprinting is a very powerful tool in determining the origin of a cell line, if the original cell line, or DNA from it or from the donor individual, has been retained. This emphasizes the need to retain a blood, tissue, or DNA sample when tissue is isolated for primary culture (*see* Section 12.3.11). Furthermore, if a cross-contamination or misidentification is suspected, this can be investigated by fingerprinting the cells and all potential contaminants. DNA fingerprinting has confirmed earlier isoenzyme and karyotypic [Gartler, 1967; Nelson-Rees & Flandermeyer, 1977; Lavappa, 1978; Nelson-Rees et al., 1981] data indicating that many commonly used cell lines are cross-contaminated with HeLa (*see* Sections 16.3, 7.10.1, 19.5; Fig. 16.9 and Table 13.2).

The following protocol for DNA fingerprinting has been provided by Glyn Stacey, NIBSC, South Mimms, Herts EN6 3QG, England, United Kingdom.

Since the first report of multilocus DNA fingerprinting by Jeffreys et al. [1985], many probes have been developed for analysis of polymorphic loci called variable number tandem repeats, or VNTRs. In multilocus DNA fingerprinting, the probe is used under conditions that allow cross-hybridization with different repetitive DNA families. Thus the final result represents polymorphic information from many parts of the genome and is therefore potentially sensitive to change in a cell line's genomic DNA complement. The first protocol given here is based on the original multilocus DNA fingerprinting methods published by Jeffreys et al. (1985) which has been usefully applied to a broad

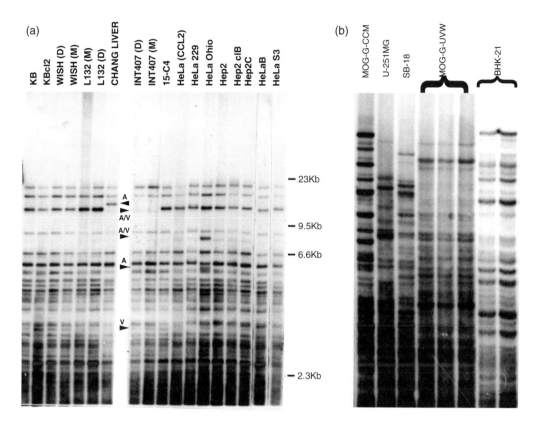

Fig. 16.9. DNA Fingerprints. Southern blots of cell line DNA digested with the *Hin*fI restriction enzyme and hybridized with the minisatellite probe 33.15 [Jeffreys et al., 1985]. (a) DNA fingerprints of HeLa cell-contaminated cell lines and subclones of the HeLa cell line. Banding patterns are identical in cases when master banks (M) and their derivative working or distribution cell banks (D) were analyzed. Fingerprint patterns were generally consistent between the cell lines, but some cases showed additional bands (A) (e.g., HeLa Ohio, INT 407, and Chang Liver). (Photograph courtesy of G. Stacey, NIBSC, UK) (b) Four human glioma cell lines—MOG-G-CCM, U-251 MG, SB-18, and MOG-G-UVW (three separate freezings)—and duplicate lanes of BHK-21 controls. (Fingerprints courtesy of G. Stacey, NIBSC, UK).

range of mammalian, non-mammalian and even plant and microbial species for ecological studies. Whilst this approach still requires a high level of technical expertise in agarose gel electrophoresis and Southern Blotting rapid blot hybridization methods have been developed with a rapid alkaline phosphatase signal system that provides results from Southern membranes within one working day (Tepnel). The use of repetitive sequences in the M13 phage genome as a probe for identity testing has been demonstrated in animals, plants, and bacteria and therefore represents a very wide-ranging authentication technique that has the added advantage that pure probe DNA is available commercially providing a low cost single method for analysis of cell line identity where cell lines from a variety of species are involved. Commercial kits and services available for PCR based analysis provide a rapid solution for the authentication of human cell lines and are discussed below.

PROTOCOL 16.8. MULTILOCUS DNA FINGERPRINTING OF CELL LINES

a) Preparation of Southern Blots of Cell Line DNA

Outline
Extract DNA from cells digested with a restriction enzyme, immobilize the digest on a nylon membrane by the Southern blot technique [Ausubel et al., 1996, 2002], and hybridize the DNA with labeled DNA from the M13 phage.

Materials
Sterile:
❑ D-PBSA: phosphate-buffered saline, without Ca²⁺ and Mg²⁺ (pH 7.2)
❑ *Hin*fI enzyme and buffer (e.g., Life Technologies)

Nonsterile:
- Agarose gels: 0.7% agarose (Type 1A, Sigma) in 1× TBE, 1 mg/L ethidium bromide

Δ *Safety Note.* Ethidium bromide is a potential carcinogen.

- TBE, 5× (stock solution, pH 8.2): 54 g of Tris base (Sigma), 27.5 g of boric acid (Sigma), and 20 mL of 0.5 M EDTA, disodium salt (Sigma)
- Loading buffer, 6× : 0.25% (w/v) bromophenol blue, 0.25% (w/v) xylene cyanol, and 40% (w/v) sucrose in distilled water
- Acid wash: 500 mL of 0.1 M HCl
- Alkaline wash: 500 mL of 0.2 M sodium hydroxide, 0.6 M NaCl
- Neutralizing wash: 500 mL of 0.5 M Tris, pH 7.6, 1.5 M NaCl
- SSC, 2×: 1-in-10 dilution of stock 20× SSC
- SSC, 20× (stock solution): 175.3 g of NaCl (Analar, Merck) and 88.2 g of sodium citrate (Analar, Merck) dissolved in distilled water to a final volume of 1.0 L at final pH 7.2.

Protocol
1. Wash approximately 1×10^7 cells (harvested by scraping or trypsinization) twice in D-PBSA, and recover the cells as a pellet.
2. Extract the high-molecular-weight genomic DNA, and check its integrity by minigel electrophoresis of a small aliquot, as described in Step 6 of this Protocol. A method for DNA extraction that has been used routinely for cell lines and that avoids the use of phenol is described in Stacey et al. [1992].
3. Quantify the DNA by UV spectroscopy using a 1-in-50 dilution of DNA [Sambrook et al., 1989].
4. Mix 5 μg of DNA with 40 U of *Hin*fI enzyme, 3 μL of 10× enzyme buffer, and sterile distilled water to a total volume of 30 μL.
5. Incubate the mixture at 38°C overnight. Remix, and microfuge the digest after 1 h.
6. Test that the digestion is complete by agarose minigel electrophoresis of 1 μL of digest in 9 μL of electrophoresis buffer (1× TBE) and 2 μL of 6× loading buffer. *Hin*fI digests should show a low-molecular-weight smear of DNA fragments (<5 kb) with no residual genomic DNA band (20–30 kb).
7. To the remainder of the digest, add 6 μL of 6× loading mix and load it into an analytical agarose gel (0.7% in 1× TBE, 1 mg/L ethidium bromide at least 20 cm long) in parallel with a *Hin*dIII digest marker and a digest of 5 μg of DNA from

a standard cell line (e.g., HeLa). Electrophorese the mixture at approximately 3 V cm^{-1} until the 2.3-kb *Hin*dIII fragment has run the length of the gel (i.e., after 17–20 h). Photograph the gel against a white ruler over a UV transilluminator as a record of migration distance.

8. Treat the separated DNA fragments in the gel with acid wash (15 min), alkaline wash (30 min), and finally a neutralizing wash (30 min). The agarose gel can then be blotted onto a nylon membrane (Hybond N, Amersham) most simply by capillary action using 20× SSC transfer buffer, as described by Sambrook et al. [1989]. Then rinse the membrane in 2× SSC, dry it, and wrap it in Saran Wrap (Dow) or equivalent.
9. Fix DNA fragments to the membrane by exposing the DNA to ultraviolet light (302 nm) for 5 min using a UV-transilluminator (e.g., TM20, UVP Inc).
10. Fixed membranes should be stored in the dark in dry polythene bags before hybridization.

b) Preparation of Labeled M13 Phage DNA

Outline
Label M13mp9 DNA by random primer extension [Feinberg & Vogelstein, 1983], and purify the DNA on a Sephadex column.

Materials
Sterile:
- Template: M13mp9 ssDNA (0.25 ng/μL; Boehringer Mannheim) diluted 1:4 in distilled water
- HEPES, 1 M, pH 6.6
- DTM reagent: dATP, dCTP, and dTTP, at 100 μM each in 250 mM of Tris·HCl, 25 mM of MgCl$_2$, and 50 mM of 2-mercaptoethanol (pH 8.0)

Δ *Safety Note.* 2-Mercaptoethanol is volatile and toxic.

- OL reagent: 90 O.D. units of oligodeoxyribonucleotides in 1 mM of Tris and 1 mM of EDTA (pH 7.5)
- Labeling buffer: LS buffer (100:100:28 of 1 M HEPES:DTM:OL)
- Bovine serum albumin (20 mg/mL of molecular biology grade; Life Technologies)
- α-[^{32}P]dGTP (3 Ci/mmol; Amersham International plc)

Δ *Safety Note.* ^{32}P is a source of high-energy β particles, exposure to which should be minimized. Always use gloves and 1-cm Perspex screens.

❑ Klenow enzyme (1 μg/μL, sequencing grade; Invitrogen)

❑ Sephadex column (e.g., NICK column, Amersham Pharmacia)

❑ Column buffer: 10 mM Tris·HCl, pH 7.5, 1 mM EDTA

Protocol

1. Mix 2 μL of diluted M13 DNA with 11 μL of distilled water in a sterile microtube.
2. Place the tube in a boiling water bath for 2 min.
3. Transfer the tube directly to ice for 5 min.
4. Add 11.4 μL of LS buffer and 0.5 μL of BSA to the tube.
5. Behind a Perspex screen, add 10 μL of [^{32}P]dGTP and 2 μL of Klenow enzyme to the tube.
6. Mix the contents of the tube, and briefly centrifuge the tube.
7. Incubate the tube at room temperature overnight behind a Perspex screen.
8. Carefully pipette the contents of the tube onto a Sephadex column previously equilibrated with 10 mL of column buffer.
9. Apply 2 × 400 μL of column buffer, collecting the eluate from the second 400 μL in a microtube as the purified probe.

Δ *Safety Note.* Radioactivity in the column will remain high, due to retention of unincorporated nucleotide.

10. Boil the purified probe for 2 min and chill it on ice before mixing it with hybridization solution.

c) Hybridization

Outline

Hybridize labeled M13mp9 DNA to Southern-blotted cell DNA, and, after stringency washing, use autoradiography to visualize the pattern of fragments binding the M13 sequences [Westneat et al., 1988].

Materials

Nonsterile:

❑ Prehybridization/hybridization solutions: 0.263 M disodium hydrogen phosphate, 7% (w/v) sodium dodecyl sulfate, 1 mM EDTA, 1% (w/v) BSA fraction V (Roche)

❑ Stringency wash: SSC, 2×

❑ SSC, 20× (stock solution): 175.3 g of sodium chloride (Analar, Merck) and 88.2 g of sodium citrate (Analar, Merck) dissolved in distilled water to a final volume of 1 L at final pH 7.2.

Protocol

1. Add membranes to 200 mL (for up to three membranes) of prewarmed prehybridization solution at 55°C in a polythene sandwich box.
2. Shake the box at 55°C for 4 h.
3. Transfer the membranes to 150 mL of prewarmed hybridization solution in a second sandwich box at 55°C, with the boiled and chilled probe added.
4. Shake the box at 55°C overnight, for a minimum of 16 h.
5. Transfer up to two membranes to 1 L of stringency wash, and shake the mixture at room temperature for 15 min.
6. Repeat Step 5
7. Replace the wash with 1 L of prewarmed (55°C) stringency wash plus 0.1% SDS, and shake the box at 55°C for 15 min.
8. Rinse the membranes in 1× SSC at room temperature.
9. Allow the membranes to air dry on filter paper until just moist; then wrap them in Saran Wrap and measure their surface counts with a Geiger–Muller monitor. Apply autoradiography at −80°C to permit an estimation of the time for X-ray film exposure. Two films (e.g., Fuji RX) should be applied to the hybridized Southern membrane in an autoradiography cassette.
10. Develop the first film after overnight exposure to enable more accurate estimation of the final exposure time required. Use standard X-ray developing chemicals (e.g., 1:5 v/v of CD15 developer for 5 min and CD40 fixer for 5 min).

Δ *Safety Note.* Procedures involving radioactive materials should be performed according to national and local regulations. Always consult your local biological safety officer before beginning such work.

Discussion. This simple method of fingerprinting using M13 phage DNA (Fig. 16.9) provides useful fingerprint patterns for cell lines from a wide range of animals, including insects. Note that some cultures may show fingerprints with high-molecular-weight bands that are poorly resolved, and it may be helpful in such cases to interpret only those bands in a molecular weight range of about 3–15 kb. The method may be refined by using the 0.78-kb *Cla*I−*Bsm*I M13 DNA fragment, which contains the most useful probe sequences. Nonradioactive probe-labeling methods can also be applied [Moreno, 1990].

Other useful multilocus fingerprint probes include the probes 33.15 and 33.6 [Jeffreys et al., 1985] and oligonucleotide probes such as (GTG)₅ [Ali et al., 1986], both of which are available commercially from Tepnel and Fresenius, respectively. Nevertheless, M13 phage DNA is the most readily available and the cheapest multilocus fingerprinting probe that can be usefully applied to a wide range of species.

A variety of single-locus probes and polymerase chain reaction (PCR) methods for VNTR analysis are also proving very useful for the identification of cell cultures. For human cell lines, multiplex PCR kits readily give profiles that are specific to the individual of origin (e.g., Promega). However, experience with similar systems indicates that STR analysis may not be able to differentiate between clones of common origin (Y. Reid, personal communication, American Type Culture Collection), although this can be achieved with multilocus DNA fingerprinting [Stacey et al., 1993]. Techniques based on random amplified polymorphic DNA analysis using single, short, random sequence primers may prove problematic in terms of reproducibility. However, single-primer methods based on mini- or microsatellite sequences appear to be more promising and can have a very wide range of applications [Meyer et al., 1997]. PCR methods are also available that can determine the species of origin of a cell line [e.g., Stacey et al., 1997]. There are other identification methods that do not require the use of PCR and these include isoenzyme analysis (*see* Section 16.8.2) and agarose electrophoresis of gDNA restriction digests to directly visualize species-specific DNA fragments [Stacey et al., 1997].

For authentication of human cell lines there is an ever increasing number of companies that provide identity testing services (*see* Appendix II). The author cannot vouch for the quality of service provided by any of these organizations and it is important when approaching such companies to ask certain questions to assure yourself that they will provide the service you need. These questions include:

- Do they perform the testing themselves or do they outsource testing services?
- What is the specificity of the methods used for individual identification?
- Are the genetic markers used linked with those used by professional bodies or other expert centers?
- Do they have experience in interpreting cell line data?
- Do they have in house expertise in the methods to assist in interpretation of results?
- Do they have accreditation by an appropriate professional or government body or do they have formal affiliation with an expert group or organization?

It should be remembered that public service collections such as ATCC (*http://www.atcc.org*), ECACC (*hpa.org*), DSMZ (*http://www.dsmz.de*), JCRB (*http://cellbank.nibio.go.jp/*) will also be able to advise on such testing and may provide it as a service.

Multilocus methods have the potential to detect changes in genetic structure at many loci throughout the genome

[Thacker et al., 1988] and are generally more versatile for the analysis of multiple species. The fingerprinting method described in Protocol 16.10 provides a cheap and straightforward route to useful quality control for cell lines from a wide range of species. Such testing enables simultaneous confirmation of identity and exclusion of cross-contamination. As recently reiterated by numerous authors (e.g. Stacey et al., 2000; Masters et al., 2001) these are essential criteria to ensure consistent experimental data and to avoid waste of time and money due to mislabeled or cross-contaminated cultures.

16.6.3 DNA Profiling

Because microsatellite sequences are quite small it has been possible to identify and quantify short tandem repeat (STR) sequences at specific loci. Second-generation multiplex (SGM) examines 7 different areas of the genome and SGM Plus® examines 11, giving a discrimination potential of $1:10^9$. So far this only applies to human cell lines, but there is great potential for the generation of numerical data that would be interchangeable between different laboratories [Masters et al., 2001; Langdon, 2004]. DNA profiling is available as a service from a number of laboratories including LGC and ATCC. The following protocol has been submitted by Amber James of LGC Diagnostics Dept., Queens Road, Teddington, Middlesex, TW11 0NJ, England, UK.

PROTOCOL 16.9. DNA STR PROFILING OF CELL LINES

Background

Cross-contamination between cell lines is a longstanding and frequent cause of scientific misrepresentation. It has been estimated that up to 36% of cell lines are of a different origin than that claimed. STR profiling by the use of SGM Plus™ provides a quick and inexpensive way of identifying the STR profile of a cell line that is reproducible between laboratories [Masters et al., 2001].

Outline

DNA is extracted from cell pellets with Qiagen® QIAamp® mini kits; the extracted DNA is amplified with SGM Plus™ and then run on an acrylamide gel with an ABI 377 DNA sequencer before being analyzed with GeneScan® and Genotyper® software.

a) Extraction of cell pellets

Reagents and Materials

❑ QIAamp® DNA mini kit
❑ Microfuge tubes, 1.5 mL (e.g., Eppendorf Biopure Safelock tubes)

- ❏ Centrifuge (e.g., Eppendorf 5415 D or similar specification)
- ❏ Waterbath or oven capable of maintaining a temperature of 56°C
- ❏ Vortex mixer
- ❏ Absolute ethanol
- ❏ Phosphate buffer, 0.01 M (e.g., sigma PBS tablets, 2 tablets dissolved in 400 mL water)
- ❏ Before starting the procedure the following reagents must be prepared:
  - **(a) Dissolved protease solvent:** Add the appropriate amount of protease solvent to the lyophilized Qiagen® protease. The solution is stable for up to 2 months when stored at 2–8°C.
  - **(b) AW1 and AW2 buffers:** Add the appropriate amount of absolute ethanol to the bottles containing the AW1 and AW2 concentrates. AW1 and AW2 buffers are stable at room temperature for a year.
  - **(c) AL buffer:** Before use, mix thoroughly by inversion.

Protocol

(Kit manufacturer's protocol with minor modifications)

If the cells are not suspended in culture medium, proceed from Step 5.

If the cell pellets have been received in culture medium:

1. Centrifuge the samples at 20,000 g for 3 min.
2. Using a fine-point liquipette, carefully remove the supernate and discard.
3. Resuspend the cells in 200 μL of PBS
4. Repeat Steps 1 and 2 once more.
5. Resuspend the cells in 200 μL of PBS
6. Pipette 20 μL of Qiagen® protease into the bottom of an empty labeled 1.5-mL microfuge tube.
7. Add 200 μL of each sample into its corresponding tube containing the protease.
8. Add 200 μL of AL buffer (lysis buffer) into each tube.
9. Mix each tube by vortexing for approximately 15 s.
10. Incubate the samples at 56°C for 10 min.
11. After this time, settle the contents of each tube by centrifuging for approximately 3 s at 1000 g.
12. Carefully uncap each tube and add 200 μL of absolute ethanol.
13. Mix by vortexing each tube for 15 s, and centrifuge at 1000 g for 3 s to pellet the contents.

14. Carefully transfer the mixture into a correspondingly labeled QIAamp® spin column, which has been placed in a 2-mL collection tube.
15. Close the cap of the column and spin at 6000 g for 1 min.
16. Place the spin column into a clean collection tube and discard the tube containing the filtrate.
17. Carefully open the lid of the spin column and add 500 μL of AW1 buffer.
18. Repeat Steps 15 and 16.
19. Carefully open the lid of the spin column and add 500 μL of AW2 buffer.
20. Close the cap and centrifuge at 20,000 g for 3 min.
21. AW2 buffer can have a detrimental effect on downstream applications, so it is advisable to replace the collection tube and spin the column for a further minute at 20,000 g to ensure that all of the AW2 buffer has been discarded.
22. Place the column into a correspondingly labeled 1.5-mL microfuge tube.
23. Open the lid of the column and add 200 μL of AE buffer (elution buffer).
24. Incubate at 56°C for 1 min, then centrifuge at 6000 g for a further minute.
25. Quantify the DNA in each sample, e.g., with Picogreen (Molecular Probes).
26. If the samples are not to be amplified immediately they should be stored frozen.

b) Amplification with SGM Plus™

Reagents and Materials

- ❏ Components of SGM Plus™ PCR mix (AMPF/STR® PCR reaction mix, AMPF/STR® SGM Plus™ primer set, Amplitaq gold® DNA polymerase)
- ❏ Pure sterile water (e.g., Sigma Tissue culture water)
- ❏ Thermocycler (e.g., ABI 9700)

Protocol

1. With reference to the ABI protocol, prepare sufficient mix for the samples to be amplified. Per sample, this comprises:
   - **(a)** AMPF/STR® PCR reaction mix, 21 μL
   - **(b)** AMPF/STR® SGM Plus™ primer set, 11 μL
   - **(c)** Amplitaq gold® DNA polymerase, 1 μL
2. Per sample, aliquot 30 μL of the mix either into individual PCR tubes or into a thin walled microtitration plate, seal the plate with adhesive foil.
3. Remove the samples from the freezer and allow to defrost.
4. Mix the samples and spin down to settle the contents of each tube.

5. Add approximately 1 ng of DNA to the appropriate aliquot of SGM Plus™ PCR mix and make up the volume added to 20 μL by adding the appropriate volume of pure sterile water.

6. Coamplify a sample of known genotype to act as a control (such as the kit control)

7. Place the samples onto a thermocycler programmed with the appropriate cycling details, as follows, and start the run:
   (a) Hold at 95°C for 11 min ±5 s.
   (b) Hold at 94 ± 1°C for 1 min ± 2 s.
   (c) Hold at 59 ± 1°C for 1 min ± 2 s.
   (d) Hold at 72 ± 1°C for 1 min ± 2 s.

8. Repeat Steps 7(b) to 7(d) 28 times.

9. Hold at 60 ± 1°C for 45 min ± 5 s.

10. Hold at 4°C until required.

b) Gel electrophoresis using the ABI 377 DNA sequencer

Reagents and Materials
- ABI 377 DNA Sequencer (Fig. 16.10)
- *Gel preparation material:* Measuring cylinder, balance, filter unit (e.g., Sartolab 150-mL filtration unit), vacuum pump, sharks tooth comb, 0.2-mm gel spacers, magnetic stirrer, urea (e.g., Fisher), acrylamide stock (e.g., Bio-Rad 40% Acrylamide/BIS 19:1), pure sterile water (e.g., Fisher AR water), mixed bed resin (e.g., Sigma) ammonium persulfate (e.g., Sigma), 10× TBE (e.g., National Diagnostics), TEMED (e.g., Sigma).
- *Gel apparatus materials:* Gel plate and cassette apparatus, buffer chamber apparatus, heat transfer plate.
- Ice or electronic cold plate (e.g., Camlab Ice cube)
- Thermocycler machine (e.g., ABI 9700)
- *SGM Plus™ Gel loading mixture:* Formamide, dextran blue dye (supplied in kit), GeneScan 500 (GS500 ROX) internal size standard (supplied in kit).

Protocol
1. Prepare a 4% polyacrylamide gel by mixing the following components in a beaker:
   (a) Urea, 18 g
   (b) Acrylamide stock, 40%, 5.2 mL
   (c) Water, 27 mL
   (d) Mixed bed resin, 0.5 g

2. Pour the gel.

3. With reference to the technical manual, set up the ABI 377 DNA sequencer for a gel run.

4. Remove the amplified samples from the Thermocycler and prepare the loading mix, per sample, as follows:
   (a) Formamide, 1.2 μL
   (b) GS500 ROX, 0.13 μL
   (c) Dextran Blue loading buffer, 0.2 μL

5. Aliquot 1.5 μL of the loading mix into an appropriate tube or plate. Coprepare loading mix for 2 allelic ladders (supplied in the kit). To the appropriate tube or well add approximately 1 μL of amplified PCR product or allelic ladder. Mix gently.

6. To denature the DNA, heat the gel loading mixture containing the amplified PCR product at approximately 95°C for approximately 3 min before immediately transferring to a tray of ice or an electronic cooling block, set to approximately 0°C.

7. With reference to the ABI technical manual, finalize the setting up of the DNA sequencer and load the gel, starting with the samples to be loaded into the odd lanes.

8. The gel run takes approximately 2.5 h.

c) Analysis of the cell line results

Materials
- GeneScan® and Genotyper® software.
- PowerMac computer

Protocol
This section covers 2 main areas; the analysis of the raw data collected by the DNA sequencer collection software with GeneScan® analysis software and the use of Genotyper® software to assign allelic designations to the profiles previously analyzed with the GeneScan® software.

1. Open the gel image and adjust the contrast to ensure that all the colors can be clearly seen.

2. Set the analysis range to include the 75-bp and 450-bp internal size standard peaks.

3. Regenerate the gel image.

4. Track the gel with the automatic gel tracker.

5. When the tracking is complete, check that it is correct; the data can then be extracted from each lane to form individual sample files. Select extract lanes from the menu.

6. The GeneScan® software uses two main control windows, Analysis control and Results control. Analysis control appears when a project is first opened and is used to define the internal size standard and change the analysis parameters. Results control can be selected when the analysis of the samples has been completed. It controls which results to display and the format of those results.

7. Using the Analysis control window, generate the size standard for each sample analyzed, the peaks

will be sized as follows: 75, 100, 140, 150, 160, 200, 250, 300, 350, 400, 450.

8. Install the appropriate matrix.

9. Click on the colored column headers marked B, G, Y, and R to highlight all the lanes to be analyzed. Click on analyze to start the analysis of the samples. All samples that have been analyzed will display a small triangle in each grid box.

10. The analyzed samples can then be opened in Results control and the ladder peak resolution and acceptability of controls can be checked.

11. From this window print the GeneScan® electropherogram (EPG) for each sample (Fig. 16.11).

12. The Genotyper® software can then be used to assign designations to the peaks seen in GeneScan®.

13. All of the data from a single gel can be processed together as a single Genotyper® file.

14. Import all of the data from the GeneScan® file into the SGM Plus™ Genotyper® template.

15. Label the peaks by pressing (Apple+1).

16. It is important to check that the size standard has been correctly recognized in each sample, and the designations should be reviewed by selecting the red dye sample button and clicking on the plot window button.

17. The allelic ladders should also be reviewed to ensure that all represented alleles have been correctly identified.

18. If the above are acceptable, the labels assigned to each sample can be reviewed.

19. A check should be made that all peaks are correctly assigned, and any designations that are deemed to be incorrect should be removed by clicking on the peak corresponding to that label.

20. Once the analyst is confident of the assigned designations, the Genotyper® can be printed for each sample and attached to the corresponding GeneScan® EPG printout.

21. A comparison can then be made of the analyzed Profile against the published profile for each cell line sample.

Fig. 16.10. DNA Sequencer. ABI 377 DNA sequencer using automated electrophoretic analysis on polyacrylamide gel to separate and quantify amplified STR sequences. (Photo courtesy of A. James, LGC Promochem).

Qualitative analysis of total cell protein reveals differences between cells when whole cells, or cell membrane extracts, are run on two-dimensional gels [O'Farrell, 1975]. This technique produces a characteristic "fingerprint" similar to polypeptide maps of protein hydrolysates, but contains so much information that interpretation can be difficult. It is possible to scan these gels by computerized densitometry (Molecular Dynamics), which attributes quantitative values to each spot and can normalize each fingerprint to set standards and interpret positional differences in spots. Labeling the cells with ^{32}P, [^{35}S]methionine, or a combination of ^{14}C-labeled amino acids, followed by autoradiography, may make analysis easier, but it is not a technique suitable for routine use unless the technology for preparing two-dimensional gels is currently in use in the laboratory.

Analysis of phenotypic expression is now more readily quantified by reverse transcriptase PCR (RT-PCR) of gene transcripts on a large scale by microarray gene expression analysis (Affymetrix) [Kiefer et al., 2004; Staab et al., 2004]. An example is given in Plate 24. The data in this plate compare three different cell types, for which this profiling exercise clearly indicates differences.

16.7 RNA AND PROTEIN EXPRESSION

Cells of a particular characteristic phenotype can be recognized by analysis of gene expression by Northern blotting [Sambrook et al., 1989; Ausubel et al., 1996] using radioactive, fluorescent, or luminescent probes. This procedure can be carried out at the cellular level, as in FISH (*see* Protocol 27.9), or by *in situ* hybridization with radioactive probes (*see* Protocol 27.8).

16.8 ENZYME ACTIVITY

Specialized functions *in vivo* are often expressed in the activity of specific enzymes, some of which may be expressed *in vitro* (*see* Table 16.2). Unfortunately, many enzyme activities are lost *in vitro* (*see* Sections 3.4.2, 17.1.1) and are no longer available as markers of tissue specificity—e.g., liver parenchyma loses arginase activity within a few days of culture. However, some cell lines do express specific enzymes, such as tyrosine aminotransferase in the rat hepatoma

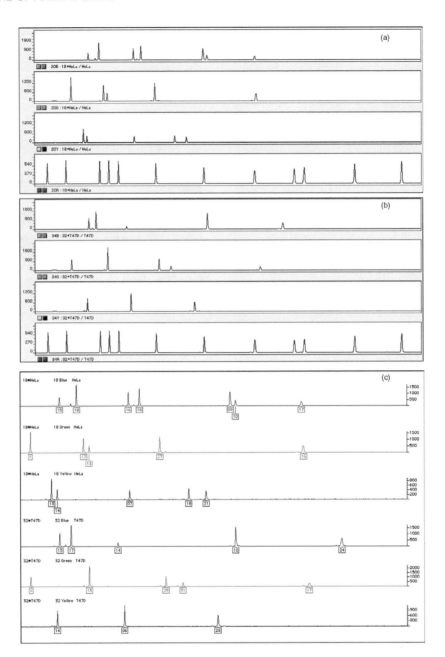

Fig. 16.11. DNA Profiling. (a) & (b) HeLa and T47D cell lines analyzed with Genescan® software to allocate sizes to the peaks detected by the ABI 377 DNA sequencer during gel electrophoresis. (c) HeLa and T47D cell lines analyzed with Genotyper® software to allocate allelic types to the peaks sized during analysis with Genescan® software. (Courtesy of A. James, LGC Promochem).

HTC cell lines [Granner et al., 1968]. When looking for specific marker enzymes, the constitutive (uninduced) level and the induced level should be measured and compared with a number of control cell lines. Glutamyl synthetase activity, for instance, characteristic of astroglia in brain, is increased severalfold when the cells are cultured in the presence of glutamate instead of glutamine [DeMars, 1958; McLean et al., 1986]. Induction of enzyme activity requires specialized conditions for each enzyme, and information on these conditions may be obtained from the literature.

Common inducers are glucocorticoid hormones, such as dexamethasone; polypeptide hormones, such as insulin and glucagon; and alteration in substrate or product concentrations in the medium, as in the aforementioned example with glutamyl synthetase.

16.8.1 Isoenzymes

Enzyme activities can also be compared qualitatively between cell strains using enzyme protein polymorphisms among species and, sometimes, among races, individuals, and tissues

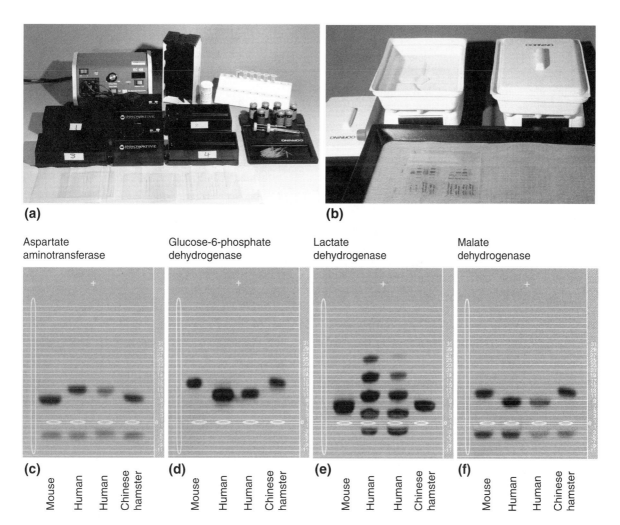

Fig. 16.12. Isoenzyme Electrophoresis. Analysis of four isoenzymes by the Authentik (Innovative Chemistry) gel electrophoresis system. (a) A four-tank setup with power pack at rear, reagents on right, and three precast gels in the foreground. (b) Staining and washing trays (Corning) with developed gel in foreground. (c–f) Images from developed electropherograms. (Photos and electropherograms courtesy of ATCC).

within a species. These so-called *isoenzymes*, or *isozymes*, may be separated chromatographically or electrophoretically, and the distribution patterns (zymograms) may be found to be characteristic of species or tissue (Fig. 16.12). Nims et al. [1998] determined that interspecies cell line cross-contamination can be detected with isoenzyme analysis if the contaminating cells represent at least 10% of the total cell population.

Electrophoresis media include agarose, cellulose acetate, starch, and polyacrylamide. In each case, a crude enzyme extract is applied to one point in the gel, and a potential difference is applied across the gel. The different isoenzymes migrate at different rates and can be detected later by staining with chromogenic substrates. Stained gels can be read directly by eye and photographed, or scanned with a densitometer.

Protocol 16.10 for isoenzyme analysis using an agarose gel system has been contributed by Jane L. Caputo, American

Type Culture Collection, 10801 University Blvd., Manassas, Virginia 20110, USA.

16.8.2 Isoenzyme Electrophoresis with Authentikit

The species of origin of a cell line can be determined with the Authentikit gel electrophoresis system, which can be used to determine the mobility of seven isoenzymes: nucleoside phosphorylase (NP; E.C. 2.4.2.1), glucose-6-phosphate dehydrogenase (G6PD; E.C. 1.1.1.49), malate dehydrogenase (MD, E.C. 1.1.1.37), lactate dehydrogenase (LD; E.C. 1.1.1.27), aspartate aminotransferase (AST, E.C. 2.6.1.1), mannose-6-phosphate isomerase (MPI, E.C. 5.3.1.8), and peptidase B (Pep B, E.C. 3.4.11.4). This system allows easy screening of up to six cell lines per gel for seven genetic markers in less than 3 hours [Hay, 2000]. In most cases, the species of origin can be determined by using only four of the seven isoenzymes listed: nucleoside phosphorylase, glucose-6-phosphate dehydrogenase, malate dehydrogenase, and lactate dehydrogenase. Similarly, interspecies cell line cross-contamination can be detected in most cases by using these four isoenzymes [Nims et al., 1998], although peptidase B was also needed to analyze a contamination of a mouse cell line with a hamster cell line.

PROTOCOL 16.10. ISOENZYME ANALYSIS

Outline
Harvest the cells, wash them in D-PBSA, resuspend them at 5×10^7 cells/mL, and prepare the cell extract. Store the extract at $-70°C$. Prepare the gel apparatus. Apply 1–2 µL of the cell extract, standard, and control to the agarose gels. Fill the chambers with water, place the agarose gels in the chambers, and run electrophoresis for 25 min. Apply enzyme reagents, incubate the gels for 5–20 min, rinse the gels, dry them, and examine the finished gels showing enzyme zones.

Materials
Sterile or Aseptically Prepared:
❑ Cells
❑ Extracts (Innovative Chemistry, Inc.):
   Standard, L929 extract
   Control, HeLa S3 extract
❑ Cell Extraction Buffer (Innovative Chemistry, Inc.) or Triton X-100 Extract Solution: 1:15, v/v, Triton X-100 in 0.9% NaCl, pH 7.1, containing 6.6×10^{-4} M EDTA (store at 4°C)

Nonsterile:
❑ Authentikit apparatus and reagents (all from Innovative Chemistry, Inc.):
   Agarose gel films, SAB 8.6, one for each enzyme tested
   Incubation tray liners, one for each enzyme tested
   Incubation trays
   Stain and wash trays
   Temperature-controlled chamber cover and base
   Power supply (160 V DC) with timer
   Safety interconnector
   Incubation chamber/dryer
   Magnetic stirrer with 1-in.-long stir bar
   Buffer, SAB 8.6
   Enzymatic substrates, 1 vial of each:
      Nucleoside phosphorylase (NP)
      Glucose-6-phosphate dehydrogenase (G6PD)
      Malate dehydrogenase (MD)
      Lactate dehydrogenase (LD)
      Aspartate aminotransferase (AST)
      Mannose-6-phosphate isomerase (MPI)
      Peptidase B (Pep B)
   Sample applicator Teflon tips®
   Gel documentation form
   Enzyme migration data form
   Cell I.D. final analysis form
❑ Microliter syringe
❑ Microcentrifuge (Eppendorf, Brinkmann)
❑ Eppendorf tubes, 1.5 mL
❑ Pipettor, 1.0 mL
❑ Pipettor tips
❑ Pipettes, 5.0 mL
❑ Graduated cylinder, 100 mL
❑ Marking pen
❑ Deionized water
❑ Distilled water
❑ D-PBSA
❑ Protective gloves
❑ Container for disposal of pipettor tips

Protocol

a) Preparation of extract:
1. Grow cells in the conventional manner to give 2×10^7 viable cells.
2. Harvest cells according to the procedure recommended for the particular cell line.
3. Resuspend the cell pellet in D-PBSA, and count the number of viable cells.
4. Centrifuge the suspension at 300 g for 5–10 min to pellet the cells and decant the supernate.
5. Repeat Steps 3 and 4 for a total of three washes.
6. Add 100 µL of extract solution to the cell pellet.
7. Gently draw the cell suspension into a small-bore pipette, and pipette the suspension up and down until all cells are lysed.
8. Transfer the cell lysate to Eppendorf tubes (1.5 mL), and pipette the lysate up and down for several minutes until the cell membranes become cloudy and clump together. Spin the lysate at top speed (~9000 g) for 2 min in a microcentrifuge. Recover the supernate, divide it into aliquots in the desired volume, and store it at $-70°C$.

b) Setup of electrophoresis apparatus:
1. Turn on the incubation chamber.
2. Place one tray liner in each incubator tray.
3. Add 6.0 mL of deionized water to each incubator tray, and place the trays at 37°C for 20 min.
4. Place 95 mL of SAB 8.6 buffer in each chamber of the electrophoresis cell base (a total of 190 mL of buffer per base). This buffer should not be reused, as the pH changes significantly during the electrophoretic run.

5. Connect the filled cell base to the safety interconnector, which is plugged into a voltage-regulated power supply. The power supply should be plugged into a grounded electrical outlet.

6. Fill each temperature-controlled electrophoresis cell cover with 500 mL of cold water (4–10°C). The cell cover must be filled when it is in a vertical position, or else the water will run out. Cool the water with ice, or store a sufficient supply of cold water in a standard refrigerator. Replace this water before each run.

7. Place 500 mL of deionized or distilled water in each stain, and wash the tray with it. This water must not be reused.

c) Electrophoresis procedure:

1. Label each agarose gel to be used. This can be done with labels available from Innovative Chemistry, Inc., or by writing on the back of the gel with a permanent marking pen.

2. Place the gel on the bench, with the plastic nipples down. Orient the gel so that the sample application wells are closest to you.

3. Label each well with the identity of the sample to be tested (e.g., standard, control, unknown 1, unknown 2, etc.). Six unknowns can be tested on each gel.

4. Gently peel the agarose gel from its hard plastic cover. Be careful to handle the gel only by its edges. Discard the hard plastic cover.

5. Add the cell extracts to the sample wells. Use the dispenser with a Teflon tip attached to dispense exactly 1 μL of cell extract into each well. Use a fresh tip for each sample. Dispense the standard extract into track 1, the control into track 2, and the unknowns in tracks 3 to 8. Only the drop of cell extract should touch the well, not the tip, in order to avoid damaging the agarose. If 2 μL of cell extract are required, allow the first 1 μL to diffuse into the agarose before applying the second.

6. Insert each loaded agarose film into a cell cover. The agarose side must face outward. Match the anode (+) side of the agarose film with the anode (+) side of the cell cover. It may be necessary to bend the agarose gel film slightly to insert it into the cell cover.

7. Place each cell cover on a cell base. The black end with the magnet inside must be nearest to the power supply. Turn on the power supply (160 V), and set the timer for 25 min.

8. Remove each reagent substrate vial from the refrigerator about 5 min before the end of the run, and allow it to warm to room temperature.

9. Add 0.5–1 mL of deionized water to each reagent vial immediately before use. Swirl the vial gently to dissolve the reagent.

10. When the electrophoresis is finished, remove the cell cover from the cell base and place it on absorbent paper.

d) Staining procedure:

1. To remove the agarose gel film from the cell cover, grasp the film by its edges, squeeze inward, and remove it from the cover.

2. Place the agarose gel film, gel side up, on absorbent paper on a flat surface, with the wells placed toward you.

3. Carefully blot the residual buffer from both ends of the agarose film with a lint-free tissue.

4. Place a 5-mL pipette along the lower edge of the agarose gel film.

5. Pour the reconstituted substrate evenly onto the agarose film along the leading edge of the pipette.

6. With a single, smooth motion, push the pipette across the surface of the agarose. Drag the pipette back toward you, and push it across the surface one more time. Roll it off the end of the agarose, removing the excess substrate in the process. Be very careful not to damage the agarose. No pressure is necessary to perform this step.

7. Place the agarose gel film, agarose side up, into a prewarmed incubator tray, and place the tray in a 37°C incubator for approximately 5 to 20 min.

8. After incubation, wash the agarose gel film two times in 500 mL of double-distilled or deionized water for 15 min each time, with agitation provided by a magnetic stirring bar. After the first 15 min, remove each gel from the water, discard the water, add 500 mL of fresh water to the dish, and immerse the gel in the water. Cover the gel to protect it from light. Be sure that the gel is immersed and not floating on top of the wash water.

9. Remove the gel from the water, and place it on a drying rack in the drying chamber of the incubator or oven. Dry it for 30 min or until the agarose is dry. Alternatively, agarose gel films will dry at room temperature overnight.

10. To clean up, pour out the buffer from the cell base, and rinse the cell base with distilled water. Pour the water out of the cell cover, rinse the inside of the cell cover, and allow it to dry.

11. Evaluate your results. The bands are permanent, and the films can be kept for future reference. If a background staining develops over time, the gels were not sufficiently washed.

Analysis of electropherogram. Attach the dried gel to
the gel documentation form; line up the sample wells at the
origin, and measure them. The bright yellow background of
the forms are millimeter lined and give maximum contrast
and enhance the purple bands on the dry gel (*see* Fig. 16.12).
Measure the enzyme migration distance by measuring from
the middle of the application zone to the middle of the
enzyme zone. Record measured distances for the standard,
the control, and the unknowns on the enzyme migration data
form. Transfer the enzyme migration data to the cell I.D. final
analysis form to confirm the species identification. Detailed
instructions for identifying unknowns are included with the
forms. Keep these forms in your notebook as a permanent
record of your results. A detailed discussion of each enzyme
can be found in the *Handbook for Cell Authentication and
Identification*, available from Innovative Chemistry, Inc.

Alternative extraction techniques. Cell extracts can be
prepared by ultracentrifugation, freezing and thawing rapidly
three times, or treatment with octyl alcohol [Macy, 1978].
A cell extraction buffer is also available from Innovative
Chemistry, Inc.

The control and standard may be prepared by using the
standard procedure (*see* Protocol 16.10) for preparation of
extract. The standard is prepared from the mouse L929 cell
line (ATCC CCL-1), and the control is prepared from the
human HeLa cell line (ATCC CCL-2). O'Brien et al. [1980]
reported that samples could be stored at −70°C for up to a
year without substantial loss of enzyme activity.

16.9 ANTIGENIC MARKERS

As a result of the abundance of antibodies and kits from
commercial suppliers (*see* Appendix II), immunostaining and
ELISA assays are among the most useful techniques available
for cell line characterization (Table 16.3). Regardless of the
source of the antibody, however, it is essential to be certain
of its specificity by using appropriate control material. This
is true for monoclonal antibodies and polyclonal antisera
alike; a monoclonal antibody is highly specific for a particular
epitope, but the specificity of the expression of the epitope
to a particular cell type must still be demonstrated.

16.9.1 Immunostaining
Antibody localization is determined by fluorescence, wherein
the antibody is conjugated to a fluorochrome, such
as fluorescein or rhodamine, or by the deposition of
a precipitated product from the activity of horseradish
peroxidase or alkaline phosphatase conjugated to the
antibody. Immunological staining may be direct—i.e., the
specific antibody is itself conjugated to the fluorochrome or
enzyme and used to stain the specimen directly. Usually,
however, an indirect method is used in which the primary
(specific) antibody is used in its native form to bind to

the antigen in the specimen, followed by treatment with
a second antibody, raised against the immunoglobulin of
the first antibody. The second antibody may be conjugated
to a fluorochrome [Johnson, 1989] and visualized on a
fluorescence microscope (*see* Plates 11a, 15d,e, and 20c−f),
or to peroxidase and visualized on a conventional microscope,
after development with a chromogenic peroxidase substrate
(*see* Plates 11b−d and 12a,b).

Various methods have been used to enhance the sensitivity
of detection of these methods, particularly the peroxidase-
linked methods. In the peroxidase-anti-peroxidase (PAP)
technique, a further amplifying tier is added by reaction with
peroxidase conjugated to anti-peroxidase antibody from the
same species as the primary antibody [Jones & Gregory, 1989].
Even greater sensitivity has been obtained by using a biotin-
conjugated second antibody with streptavidin conjugated
to peroxidase or alkaline phosphatase (Amersham) or gold-
conjugated second antibody (Janssen) with subsequent silver
intensification (Amersham).

**PROTOCOL 16.11. INDIRECT
IMMUNOFLUORESCENCE**

Outline
Fix the cells, and treat them sequentially with first and
second antibodies. Examine the cells by UV light.

Materials
Sterile or Aseptically Prepared:
☐ Culture grown on glass coverslip, chambered slide,
or polystyrene Petri dish
Nonsterile:
☐ Freshly prepared fixative: 5% acetic acid in ethanol
(place at −20°C)
☐ Primary antibody diluted 1:100−1:1000 in culture
medium with 10% FBS
☐ Second antibody raised against the species of
the first (e.g., if the first antibody was raised in
rabbit, then the second should be from a different
species, such as goat anti-rabbit immunoglobulin);
the second antibody should be conjugated to
fluorescein or rhodamine
☐ Swine serum or another blocking agent
☐ D-PBSA
☐ Mountant: 50% glycerol in D-PBSA and containing
a fluorescence-quenching inhibitor (Vecta)

Protocol
1. Wash the coverslip with cells in D-PBSA, and
place it in a suitable dish—e.g., a 24-well plate
for a 13-mm coverslip.
2. Place the dish at −20°C for 10 min, add cold
fixative to it, and leave it for 20 min.

TABLE 16.3. Antibodies Used in Cell Line Recognition

Cell type	Antibody to	Localization	Specificity	Comments
Anterior pituitary	hGH	Golgi and secreted	High	
Astroglia	GFAP	Intermediate-filament cytoskeleton	High	*See also* Table 22.2
B-cells	CD22	Cell surface		
Breast epithelium	α-Lactalbumin, casein	Golgi and secreted	High	Needs differentiation induction
Colorectal and lung adenocarcinoma	CEA	Cell surface and Golgi	Intermediate	Cell adhesion molecule
Endothelium	Factor VIII	Weibl–Palade bodies	High	Granular appearance diagnostic
	V-CAM	Cell surface	High	Cell adhesion molecule
Epithelium	L-CAM	Cell surface	High	Cell adhesion molecule
Epithelium	Cytokeratin	Intermediate-filament cytoskeleton	High	Some antibodies also stain mesothelium; specific antibodies for different epithelia
Epithelium	EMA	Cell surface	High	Same antigen as HMFG
Epithelium	HMFG I & II	Cell surface	High	Same antigen as EMA
Fetal hepatocytes	AFP	Golgi and secreted	High	Also expressed by hepatomas
Hematopoietic stem cells	CD34	Cell surface	High	Also in precursors but not mature cells
Hepatocytes	Albumin	Golgi and secreted	High	Needs differentiation
Immature B-cells	CD20	Cell surface	High	Not in plasma cells
Mesodermal cells	Vimentin	Intermediate-filament cytoskeleton	Low	Also expressed in epithelial and glial cells in culture
Mesothelium	Mesothelial cell antigen	Cell surface	High	
Monocytes and macrophages	CD14	Cell surface	High	
Myeloid cells	CD13	Cell surface	High	
Myocytes	Desmin	Intermediate-filament cytoskeleton	High	
Neural and neuroendocrine cells	NSE	Cytoplasmic	Intermediate	Can be expressed in SCLC
Neural cells	N-CAM	Cell surface	Intermediate	Also expressed in some SCLC
Neuroendocrine lung and stomach	GRP	Golgi and secreted	Intermediate	Also expressed in some SCLC
Neuronal cells	Neurofilament protein	Intermediate-filament cytoskeleton	High	Also stains some SCLC
Oligodendrocytes	Myelin basic protein	Cell surface	High	Also stains peripheral neurons
	Gal-C	Cell surface	High	*See also* Table 22.2
Placental epithelium	hCG	Golgi and secreted	Low	Also expressed in some lung tumors
Prostatic epithelium	PSA	Golgi and secreted	High	
T-cells	CD3	Cell surface	High	
T-cells and endothelium	ICAM	Cell surface	Intermediate	

Abbreviations: GFAP, glial fibrillary acidic protein; EMA, epithelial membrane antigen; HMFG, human milk fat globule protein; AFP, α-fetoprotein; CEA, carcinoembryonic antigen; SCLC, small cell lung cancer; hCG, human chorionic gonadotropin; ICAM, intercellular cell adhesion molecule; NSE, neuron-specific enolase; PSA, prostate-specific antigen; Gal-C, galactocerebroside; GRP, gastrin-releasing peptide.

3. Remove the fixative, wash the coverslip in D-PBSA, add 1 mL of normal swine serum, and leave the dish at room temperature for 20 min.
4. Rinse the coverslip in D-PBSA, drain it on paper tissue, and place the coverslip, inverted, on a 50-μL drop of diluted primary antibody.
5. Place the coverslip at 37°C for 30 min at room temperature for 1–3 h or overnight at 4°C. For the 4°C incubation the antibody may be diluted 1:1000.
6. Rinse the coverslip in D-PBSA, and transfer it to the second antibody, diluted 1:20, for 20 min at 37°C.
7. Rinse the coverslip in D-PBSA, and mount it on a slide in 50% glycerol in D-PBSA with fluorescence bleaching retardant (Vecta).
8. Examine the slide on a fluorescence microscope.

Variations

Fixation. For the cell surface or, particularly, for fixation-sensitive antigens, treat the cells with antibodies first, and then postfix as in Step 2. When a glass substrate is used, cold acetone may be substituted for acid ethanol.

Indirect peroxidase. Substitute peroxidase-conjugated antibody for the fluorescent antibody at stage 6, and then transfer it to peroxidase substrate. (Diaminobenzidine-stained preparations can be dehydrated and mounted in DPX, but ethyl carbazole must be mounted in glycerol as in Step 7.)

PAP. Use peroxidase-conjugated second antibody at stage 6, and then transfer the coverslip to diluted PAP complex (1:100) (most immunobiological suppliers—e.g., Dako, Vecta) in D-PBSA for 20 min. Rinse it, and add peroxidase substrate as for indirect peroxidase. Incubate, wash, and mount the coverslip.

Cell surface markers. Specific cell surface antigens are usually stained in living cells (at 4°C in the presence of sodium azide to inhibit pinocytosis), whereas intracellular antigens are stained in fixed cells, sometimes requiring

Fig. 16.13. Agilent Immunoanalyzer. Agilent 2100, which can handle the simultaneous analysis of multiple immunofluorescent stained samples. (a) Sample compartment. (b) Sample holder. (Courtesy of PromoCell).

light trypsinization to permit access of the antibody to the antigen.

HLA and blood-group antigens can be demonstrated on many human cell lines and serve as useful characterization tools. HLA polymorphisms are particularly valuable, especially when the donor patient profile is known [Espmark & Ahlqvist-Roth, 1978; Stoner et al., 1981; Pollack et al., 1981].

16.9.2 Immunoanalysis

Assay of cell extracts can be preformed by ELISA assay on a microtitration plate array with antibody–coated wells. Assays are provided in kit form (e.g., Assay Designs; R&D) and allow quantitation of marker and product protein expression. Antigen expression analysis can also be automated and quantitated by flow cytometry (*see* Section 21.7.2) (B-D Biosciences FACSStar or Guava Technologies), by image analysis of multiple cell streams (Agilent 2100; Fig. 16.13), or by antibody microarray analysis.

16.10 DIFFERENTIATION

Many of the characteristics described under antigenic markers or enzyme activities may also be regarded as markers of differentiation, and as such they can help to correlate cell lines with their tissue of origin. Other examples of differentiation, and thus specific markers of cell line identity, are given in Chapter 3 (*see* Table 3.1), and the appropriate assays for these properties may be derived from the references cited in the table (*see also* Section 17.6 and Chapter 23.)

CHAPTER 17

Differentiation

17.1 EXPRESSION OF THE *IN VIVO* PHENOTYPE

The phenotype of cells cultured and propagated as a cell line is often different from that of the predominating cell type in the originating tissue (*see* Section 3.4.2). This is due to several factors that regulate the geometry, growth, and function *in vivo*, but that are absent from the *in vitro* microenvironment (*see* Section 17.1.1). *Differentiation* is the process leading to the expression of phenotypic properties characteristic of the functionally mature cell *in vivo*. This may be irreversible such as the cessation of DNA synthesis in the erythroblast nucleus or mature keratinocyte, or reversible, such as the induction of albumin synthesis in differentiated hepatocytes, which is often lost in culture but can be reinduced.

Differentiation as used in this text describes the combination of *constitutive* (stably expressed without induction) and *adaptive* (subject to positive and negative regulation of expression) properties found in the mature cell. *Commitment*, on the other hand, implies an irreversible transition from a stem cell to a particular defined lineage endowing the cell with the potential to express a limited repertoire of properties, either constitutively or when induced to do so. The concept of commitment has been called into question, however, by the demonstration that some precursor cells can revert or convert to multipotent stem cells (Kondo & Raff, 2000, 2004; Le Douarin et al., 2004).

Terminal differentiation implies that a cell has progressed down a particular lineage to a point at which the mature phenotype is fully expressed and beyond which the cell cannot progress. In principle, this definition need not exclude cells, such as fibrocytes, that can revert to a less differentiated phenotype and resume proliferation, but in practice, the term tends to be reserved for cells like neurons, skeletal muscle, or keratinized squames, for which terminal differentiation is irreversible.

17.1.1 Dedifferentiation

Dedifferentiation has been used to describe the loss of the differentiated properties of a tissue when it becomes malignant or when it is grown in culture. As dedifferentiation comprises complex processes with several contributory factors, including cell death, selective overgrowth, and adaptive responses, the term should be used with caution. When used correctly, dedifferentiation means the loss by a cell of the specific phenotypic properties associated with the mature cell. When dedifferentiation occurs, it may be either an adaptive process, implying that the differentiated phenotype may be regained given the right inducers (*see also* Section 3.4), or a selective process, implying that a progenitor cell has been selected because of its greater proliferative potential. In either case the progenitor cell may be induced to mature to the fully differentiated cell, or even revert to a stem cell (*see* Section 19.1), given the correct environmental conditions.

17.1.2 Lineage Selection

If the wrong lineage has been selected (e.g., stromal fibroblasts from liver instead of hepatocytes), no amount of induction can bring back the required phenotype. In the past, this failure was often erroneously attributed to dedifferentiation but was more likely due to overgrowth of stromal fibroblasts induced by growth factors such as platelet-derived growth factor (PDGF) and inhibition of proliferation in epithelial cells by transforming growth factor β (TGF-β).

17.2 STAGES OF DIFFERENTIATION

There are two main pathways to differentiation in the adult organism. In constantly renewing tissues, like the epidermis, intestinal mucosa, and blood, a small population of totipotent or pluripotent undifferentiated stem cells, capable of self-renewal, gives rise, on demand, to progenitor cells that will proliferate and progress toward terminal differentiation, losing their capacity to divide as they reach the terminal stages (see Fig. 3.6). This process gives rise to mature, differentiated cells that normally will not divide. Proliferation in the progenitor compartment is regulated by feedback to generate the correct size of the differentiated cell pool.

In tissues that do not turn over rapidly, but replenish themselves in response to trauma, the resting tissue shows little proliferation; however, the mature cells may reenter division. In connective tissue, for example, cells such as fibrocytes may respond to a local reduction in cell density and/or the presence of one or more growth factors by losing their differentiated properties (e.g., collagen synthesis) and reentering the cell cycle. When the tissue has regained the appropriate cell density by division, cell proliferation stops and differentiation is reinduced. This type of renewal is rapid, because a relatively large population of cells is recruited.

It is not clear whether the cells that reenter the cell cycle to regenerate the tissue are phenotypically identical to the bulk of the differentiated cell population, or whether they represent a subset of reversibly differentiated cells. In liver, which responds to damage by regeneration, it appears that the mature hepatocyte can reenter the cell cycle, whereas in skeletal muscle, where terminally differentiated cells cannot reenter the cycle, regenerating cells are derived from the satellite cells, which appear to form a quiescent stem cell population (see Section 23.3.3). Mature fibrocytes, blood vessel endothelial cells, and glial cells appear to be able to reenter the cell cycle to regenerate, but the possibility of a regenerative subset within the total population cannot be ruled out.

17.3 PROLIFERATION AND DIFFERENTIATION

As differentiation progresses, cell division is reduced and eventually ceases. In most cell systems, cell proliferation is incompatible with the expression of differentiated properties. Tumor cells can sometimes break this restriction, and in melanoma, for example, melanin continues to be synthesized while the cells are proliferating. Even in these cases, however, synthesis of the differentiated product increases when division stops.

There are severe implications for this relationship in culture, where expansion and propagation are often the main requirements, and it is therefore not surprising to find that the majority of cell lines do not express fully differentiated properties. This fact was noted many years ago by the exponents of organ culture (see Section 25.2), who set out to retain three-dimensional, high-cell-density tissue architecture and to prevent dissociation and selective overgrowth of undifferentiated cells. However, although of considerable value in elucidating cellular interactions regulating differentiation, organ culture has always suffered from the inability to propagate large numbers of identical cultures, particularly if large numbers of cells are required, and the heterogeneity of the sample, assumed to be essential for the maintenance of the tissue phenotype, has in itself made the ultimate biochemical analysis of pure cell populations and of their responses extremely difficult.

Hence, in recent years, there have been many attempts to reinduce the differentiated phenotype in pure populations of cells by recreating the correct environment and, by doing so, to define individual influences exerted on the induction and maintenance of differentiation. This process usually implies the cessation of cell division and the creation of an interactive, high-density cell population, as in histotypic or organotypic culture. Cell–cell interaction has become a key feature in establishing engineered tissues (see Section 25.3.8).

17.4 COMMITMENT AND LINEAGE

Progression from a stem cell to a particular pathway of differentiation traditionally implied a rapid increase in commitment, with advancing stages of progression (see Fig. 3.6). A hematopoietic stem cell, after commitment to lymphocytic differentiation, would not change lineage at a later stage and adopt myeloid or erythrocytic characteristics. Similarly, a primitive neuroectodermal stem cell, once committed to become a neuron, would not change to a glial cell. Commitment was regarded as the point between the stem cell and a particular progenitor stage where a cell or its progeny can no longer transfer to a separate lineage. If such irreversible commitment exists it must occur much later than previously thought, as some progenitor cells can revert to stem cells with multilineage potential (see Section 17.1).

Many claims have been made in the past regarding cells transferring from one lineage to another. Perhaps the most substantiated of these claims is that of the regeneration of the amphibian lens by recruitment of cells from the iris [Clayton et al., 1980; Cioni et al., 1986]. As the iris can be fully differentiated and still regenerate lens, this claim has been proposed as transdifferentiation. It is, however, one of the few examples, and most other claims have been for tumor cell systems for which the origin of the tumor population may not be clear. For example, small-cell carcinoma of the lung has been found to change to squamous or large-cell carcinoma following relapse after the completion of chemotherapy. Whether this implies that one cell type, the Kulchitsky cell [de Leij et al., 1985], presumed to give rise to small-cell lung carcinoma, changed its commitment, or whether it implies that the tumor originally derived from a multipotent stem cell and on recurrence progressed down a different route, is still not clear [Gazdar et al., 1983; Goodwin et al., 1983; Terasaki et al., 1984]. Similarly, advanced anaplastic squamous skin cancers can give rise to cells that appear mesenchymal [Oft et al., 2002].

The K562 cell line was isolated from a myeloid leukemia but subsequently was shown to be capable of erythroid differentiation [Andersson et al., 1979b]. Rather than being a committed myeloid progenitor converting to erythroid, the tumor probably arose in the common stem cell known to give rise to both erythroid and myeloid lineages. For some reason, as yet unknown, continued culture favored erythroid differentiation rather than the myeloid features seen in the original tumor and early culture. In some cases, again in cultures derived from tumors, a mixed phenotype may be generated. For instance, the C$_6$ glioma of rat expresses both astrocytic and oligodendrocytic features, and these features may be demonstrated simultaneously in the same cells (Breen & De Velis, 1974).

In general, however, these cases are unusual and are restricted to tumor cultures. Most cultures from normal tissues, although they may differentiate in different directions, do not alter to a different lineage. This raises the question of the actual status of cell lines derived from normal tissues. Most cultures are derived from (1) stem cells, or early progenitor cells, which may differentiate in one or more different directions (e.g., lung mucosa, which can become squamous or mucin secreting, depending on the stimuli); (2) late progenitor cells, which may stay true to the lineage; or (3) differentiated cells, such as fibrocytes, which may dedifferentiate and proliferate, but still retain lineage fidelity (*see* Section 3.8). Some mouse embryo cultures, loosely called fibroblasts (e.g., the various cell lines designated 3T3), probably more correctly belong to category (1), as they can be induced to become adipocytes, muscle cells, osteocytes, and endothelium, as well as fibrocytes.

Cell lines are perhaps best regarded as a mixed population of stem cells, progenitor cells, and differentiated cells, the balance being determined by soluble or contact mediated signals in the environment. In most propagated cell lines the majority of cells will have the progenitor phenotype, but may retain the plasticity to shift to either a stem cell or differentiated phenotype, given the correct signals.

There are now some well-described examples in which progenitor cells (e.g., the O2A common progenitor of the oligodendrocyte and type 2 astrocyte in the brain, which remains a proliferating progenitor cell in a mixture of PDGF and bFGF) will differentiate into an oligodendrocyte in the absence of growth factors or serum, or into a type 2 astrocyte in fetal bovine serum or a combination of ciliary neurotropic factor (CNTF) and bFGF [Raff et al., 1978; Raff, 1990] (*see* Table 23.2). However, those cells which have become oligodendrocyte precursors can still revert to a stemlike phenotype when exposed to BMPs (Kondo & Raff, 2000, 2004). Cardiac muscle cells remain undifferentiated and proliferative in serum and bFGF but

differentiate in the absence of serum [Goldman & Wurzel, 1992], and primitive embryonal stem cell cultures differentiate spontaneously unless kept in the undifferentiated proliferative phase by bFGF, SCF (stem cell factor, Steel factor, kit ligand), and LIF (lymphocyte inhibitory factor) [Matsui et al., 1992].

Hence, with the advent of more defined media and the identification of more differentiation-inducing factors, it is gradually becoming possible to define the correct inducer environment that will maintain cells in a stemlike, progenitor, or differentiated state.

17.5 STEM CELL PLASTICITY

Conventional stem cell theory predicts that, the more primitive a stem cell, the greater its potency. A *unipotent* stem cell is a cell that will give rise to only one lineage, e.g., stem cells in the basal layer of the epidermis, which give rise to keratinocytes. A *bipotent* stem cell will give rise to two lineages, e.g., a lymphoid stem cell may give rise to T- or B-lymphocytes. A *multipotent* stem cell will generate more than two lineages, e.g., the progenitor cell compartment in the bone marrow, which gives rise to granulocytes, monocytes, megakaryocytes, mast cells, and erythrocytes. A *totipotent* stem cell will give rise to all known cell types. Within the hematopoietic system, this term may be used for the stem cell that gives rise to all blood cell types, but in its more general definition it would imply an embryonal stem cell (ES cell), which, by current dogma, is the only cell that can give rise to every cell lineage. Originating in the inner cell mass of the early embryo, it seems reasonable that ES cells should be totipotent.

The other traditional concept is that the greater the degree of commitment, the greater the likelihood of the stem cell being located in a specific tissue, e.g. hematopoietic stem cells are located in the bone marrow, enterocyte stem cells at the bottom of intestinal crypts, and keratinocyte stem cells in the epidermis. With this commitment and histological localization comes a reduction in potency, and one would expect that, for example, stem cells in the liver will only make liver cells (hepatocytes, bile duct cells). Furthermore, tissues that do not regenerate, e.g., neurons in the central nervous system, would not be expected to possess stem cells. However, this tidy concept of lineage commitment, potency, and tissue localization is currently being questioned [Vescovi et al., 2002] by results that show that (1) tissues that are non-regenerative, such as neurons in the brain, do have stem cells [Pevny& Rao, 2003] and that (2) tissue localization does not necessarily mean lineage commitment and reduced potency, as liver stem cells can generate neurons [Deng et al., 2003], bone marrow stem cells can generate cardiac muscle [Mangi et al., 2003], muscle stem cells can generate hematopoietic cells [Cao et al., 2003], and neural stem cells can generate endothelium [Wurmser et al., 2004]. In some cases differentiation into an unpredicted lineage, e.g., bone marrow cells generating hepatocytes or cardiac muscle or neurons [Alison

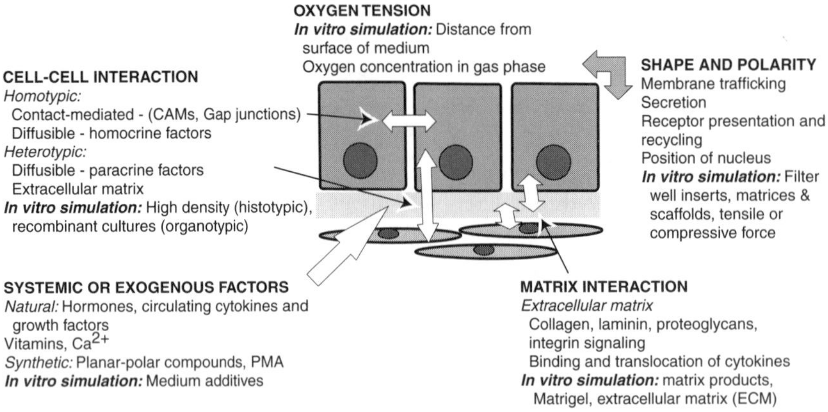

Fig. 17.1. Regulation of Differentiation. Parameters controlling the expression of differentiation *in vitro*.

et al., 2004; Alvarez-Dolado et al., 2003], may be due to cell fusion [Greco & Recht, 2003], but others have claimed that cell fusion is not involved [Wurmser et al., 2004].

These developments have upset conventional thinking on stem cell biology and lineage commitment (*see also* Sections 17.1, 17.4), but they also open up an enormous horizon for the exploitation of somatic stem cell biology in the regulation of gene expression and tissue regeneration. Two problems remain, however: (1) reliable markers for the isolation of such minority cell populations [Zhou et al., 2001; Cai et al., 2004], which by definition are devoid of lineage markers, (2) practical utilization of adult stem cells, which may have fewer replicative generations available, in tissue regeneration, and (3) elaboration of culture conditions for propagation and differentiation which are as effective as those used for embryonal stem cells. A third category of stem cells is also emerging, those found in the umbilical cord. Although superficially hematopoietic, multipotency has been demonstrated [Lee et al., 2004; McGuckin et al., 2004] so the use of this category of stem cells may have fewer ethical limitations than human ES cells, and provide greater longevity in the cell lines generated.

17.6 MARKERS OF DIFFERENTIATION

Markers expressed early and retained throughout subsequent maturation stages are generally regarded as lineage markers—e.g., intermediate filament proteins, such as the cytokeratins (epithelium) [Moll et al., 1982], or glial fibrillary acidic protein (astrocytes) [Eng & Bigbee, 1979; Bignami et al., 1980]. Markers of the mature phenotype representing terminal differentiation are more usually specific cell products or enzymes involved in the synthesis of these products—e.g., hemoglobin in an erythrocyte, serum albumin in a hepatocyte, transglutaminase [Schmidt et al., 1985] or involucrin [Parkinson & Yeudall, 2002] in a differentiating keratinocyte, and glycerol phosphate dehydrogenase in an

oligodendrocyte [Breen & De Vellis, 1974] (*see* Table 3.1). These properties are often expressed late in the lineage and are more likely to be reversible and under adaptive control by hormones, nutrients, matrix constituents, and cell–cell interaction (*see* Fig. 17.1).

As the genes encoding a large number of differentiated products have now been identified and sequenced [International Human Genome Sequencing Consortium, 2001], it is possible to look for the expression of differentiation marker proteins by RT-PCR [Ausubel et al., 1996, 2002] and microarray analysis [Kawasaki, 2004], which identifies the expression of specific mRNAs (*see* Plate 24). This method will not necessarily confirm synthesis of the final product, but, it is possible to distinguish between very low levels (or no expression) and high levels of expression of specific gene transcripts. Expression of many genes can be screened simultaneously, a considerable advantage over other methods of determining product expression.

Differentiation should be regarded as the expression of one, or preferably more than one, marker associated with terminal differentiation. Although lineage markers are helpful in confirming cell identity, the expression of the functional properties of the mature cells is the best criterion for terminal differentiation and confirmation of cellular origin.

17.7 INDUCTION OF DIFFERENTIATION

There are five main parameters that control differentiation (Fig. 17.1): cell–cell interaction, cell–matrix interaction, cell shape and polarity, oxygen tension, and soluble systemic factors.

17.7.1 Cell Interaction

Homotypic. Homologous cell interaction occurs at high cell density. It may involve gap junctional communication [Finbow & Pitts, 1981], in which metabolites, second messengers, such as cyclic AMP, diacylglycerol (DAG), Ca^{2+}, or electrical charge may be communicated between cells. This interaction probably harmonizes the expression of differentiation within a population of similar cells, rather than initiating its expression.

The presence of homotypic cell–cell adhesion molecules, such as CAMs or cadherins, which are calcium-dependent, provides another mechanism by which contacting cells may interact. These adhesion molecules promote interaction primarily between like cells via identical, reciprocally acting, extracellular domains (*see* Section 3.2.1), and they appear to have signal transduction potential via phosphorylation of the intracellular domains [Doherty et al., 1991; Gumbiner, 1995].

Heterotypic. Heterologous cell interaction—e.g., between mesodermally and endodermally or ectodermally derived cells—is responsible for initiating and promoting differentiation. During and immediately after gastrulation in the embryo, and later during organogenesis, mutual interaction between cells originating in different germ layers promotes differentiation [Yamada et al., 1991; Hirai et al., 1992; Hemmati-Brivaniou et al., 1994; Muller et al., 1997]. For example, when endodermal cells form a diverticulum from the gut and proliferate within adjacent mesoderm, the mesoderm induces the formation of alveoli and bronchiolar ducts and is itself induced to become elastic tissue [Hardman et al., 1990; Caniggia et al., 1991].

The extent to which this process is continued in the adult is not clear, but evidence from epidermal maturation suggests that a reciprocal interaction, mediated by growth factors such as KGF and GM-CSF and cytokines such as IL-1α and IL-1β, with the underlying dermis is required for the formation of keratinized squames with fully cross-linked keratin (Fig. 17.2) [Maas-Szabowski et al., 2002].

Paracrine growth factors

Positively acting. If the interacting cells are from different lineages, these are *heterotypic* paracrine factors, e.g., KGF in prostate [Planz et al., 1998], alveolar maturation factor in the lung [Post et al., 1984], and other potential candidates, such as IL-6 and oncostatin M in lung maturation [McCormack et al., 1996, McCormick & Freshney, 2000], glia maturation factor in brain [Keles et al., 1992] (*see* Section 17.7.1; Plates 12d and 13e,f), and interferons [*see*, e.g., Pfeffer & Eisenkraft, 1991].

Some growth factors act as morphogens [Gumbiner, 1992], e.g., epimorphin [Hirai et al., 1992; Radisky et al., 2003] and hepatocyte growth factor/scatter factor (HGF/SF) [Kinoshita & Miyajima, 2002]. HGF is one of the family of heparin-binding growth factors (HBGF) (*see* Table 10.3), shown to be released from fibroblasts, such as the MRC-5 [Kenworthy et al., 1992]. Epimorphin and HGF induce tubule formation

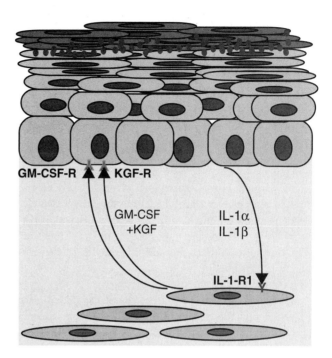

Fig. 17.2. Reciprocal Paracrine Interaction. Schematic illustration of the double paracrine pathways of keratinocyte growth regulation in organotypic cocultures with fibroblasts involving IL-1, KGF, and GM-CSF as well as their receptors [*see also* Maas-Szabowski et al., 2000; Szabowski et al. 2000]. (Figure adapted from Maas-Szabowski et al. [2002]).

in the MDCK continuous cell line from dog kidney [Orellana et al., 1996; Montesano et al., 1997], salivary gland epithelium [Furue & Saito, 1997], and mammary epithelium [Soriano et al., 1995]. KGF, also a paracrine factor, is produced by dermal and prostatic fibroblasts. It influences epidermal differentiation [Aaronson et al., 1991; Gumbiner, 1992; Maas-Szabowski et al., 2002] and regulates differentiation in prostatic epithelium [Planz et al., 1998; Thomson et al., 1997]. Growth factors such as FGF-1, -2, and -3, KGF (FGF-7), TGF-β, and activin are also active in embryonic induction [Jessell & Melton, 1992].

HGF and KGF are both produced only by fibroblasts, bind to receptors found only on other cells (principally epithelium), and are classic examples of paracrine growth factors. The release of paracrine factors may be under the control of systemic hormones in some cases. It has been demonstrated that type II alveolar cells in the lung produce surfactant in response to dexamethasone *in vivo*. *In vitro* experiments have shown that this induction of surfactant synthesis is dependent on the steroid binding to receptors in the stroma, which then releases a peptide to activate the alveolar cells [Post et al., 1984]. Similarly, the response of epithelial cells in the mouse prostate to androgens is mediated by the stroma [Thomson et al., 1997]. KGF has been shown to be at least one component of the interaction in the prostate [Yan et al., 1992].

The differentiation of the intestinal enterocyte, which is stimulated by hydrocortisone, also requires underlying stromal fibroblasts [Kédinger et al., 1987], and, in this case, modification of the extracellular matrix between the two cell types is implicated [Simon-Assmann et al., 1986]. Dexamethasone-dependent matrix modification is also implicated in the response of alveolar Type II cells to paracrine factors (*see* Section 17.7.3), [Yevdokimova & Freshney, 1997]. Differentiation in the hematopoietic system is under the control of several positively acting lineage-specific growth factors, such as IL-1, IL-6, G-CSF, and GM-CSF (*see* Tables 10.3 and 23.3), the last of which is also dependent on the matrix for activation [de Wynter et al., 1993]. Matrix heparan sulfates are also implicated in FGF activation [Klagsbrun & Baird, 1991; Fernig & Gallagher, 1994].

Homocrine factors are *homotypic* paracrine factors arising in one cell and affecting adjacent cells of the same lineage, and *autocrine* factors stimulate the same cell that generates them (*see* Section 3.5 and Fig. 3.7). Homocrine and autocrine factors are likely to be indistinguishable from a practical point of view.

Negatively acting. Factors such as MIP-1α maintain the stem cell phenotype [Graham et al., 1992], and, similarly, PDGF and FGF-2 promote the growth and self-renewal of the O-2A progenitor cell, but inhibit its differentiation [Bögler et al., 1990]. Other negative regulators of differentiation include LIF in ES cell differentiation [Smith et al., 1988] and TGF-β in alveolar type II cell differentiation [Torday & Kourembanas, 1990; McCormick & Freshney, 2000; McCormack et al., 1996].

17.7.2 Systemic or Exogenous Factors

Physiological inducers (Table 17.1). Systemic physiological regulators that induce differentiation include the following. (1) *Hormones* arise from a distant organ or tissue and reach the target tissue via the vasculature *in vivo* (i.e., *endocrine* factors). This category includes hydrocortisone, glucagon, and thyroxin (or triiodothyronine). (2) *Vitamins* such as vitamin D_3 [Jeng et al., 1994; Rattner et al., 1997] and retinoic acid [Saunders et al., 1993; Hafny et al., 1996; Ghigo et al., 1998] are derived from the diet and may be modified by metabolism. (3) Inorganic ions, particularly Ca^{2+}: High Ca^{2+} promotes keratinocyte differentiation [Cho & Bikle, 1997], (*see* Section 23.2.1 for example). This probably relates to the role of calcium in cell interaction (cadherins are calcium-dependent), intracellular signaling, and the membrane flux of calcium in so-called "calcium waves" that propagate signals from one responding cell to adjacent cells of the same lineage (*see* Fig. 3.7). Along with gap junctional communication and, possibly, homocrine factors and heparan sulfate, this helps to generate a coordinated response.

Other physiological inducers are the paracrine factors, which are dealt with above (*see* Section 17.7.1). These are assumed to act directly, cell to cell, without the need for vascular transmission. However, some of these factors (FGF, PDGF, interleukins) are found in the blood, so they may also have a systemic role.

Nonphysiological inducers. Rossi and Friend [1967] observed that mouse erythroleukemia cells treated with dimethyl sulfoxide (DMSO), to induce the production of Friend leukemia virus, turned red because of the production of hemoglobin (*see* Plate 13a,b), and this has been confirmed by an increase in globin gene expression (*see* Plate 13c,d). Subsequently, it was demonstrated that many other cells—e.g., neuroblastoma, myeloma, and mammary carcinoma—also responded to DMSO by differentiating. Many other compounds have now been added to this list of nonphysiological inducers: hexamethylene bisacetamide (HMBA); *N*-methyl acetamide; benzodiazepines, whose action may be related to that of DMSO; and a range of cytotoxic drugs, such as methotrexate, cytosine arabinoside, and mitomycin C (Table 17.2). Sodium butyrate has also been classed with these nonphysiological inducers of differentiation, but there is some evidence that butyrate occurs naturally, e.g. in the gut, and regulates normal behavior.

The action of these compounds is unclear but may be mediated by changes in membrane fluidity (particularly for polar solvents, like DMSO, and anesthetics and tranquillizers); by their influence as lipid intercalators of signal transduction enzymes, such as protein kinase C (PKC) and phospholipase D (PLD), which tend to relocate from the soluble cytoplasm to the endoplasmic reticulum when activated; or by alterations in DNA methylation or histone acetylation. The induction of differentiation by polar solvents, such as DMSO, may be phenotypically normal, but the induction by cytotoxic drugs may also induce gene expression unrelated to differentiation [McLean et al., 1986].

Tumor promoters, such as phorbol myristate acetate (PMA), have been shown to induce squamous differentiation in bronchial mucosa, although not in bronchial carcinoma [Willey et al., 1984; Masui et al., 1986a,b; Saunders et al., 1993]. Although these tumor promoters are not normal regulators *in vivo*, they bind to specific receptors and activate signal transduction, such as by the activation of PKC [Dotto et al., 1985].

17.7.3 Cell–Matrix Interactions

Surrounding the surface of most cells is a complex mixture of glycoproteins and proteoglycans that is highly specific for each tissue, and even for parts of a tissue. Reid [1990] showed that the construction of artificial matrices from different constituents can regulate gene expression. For example, addition of liver-derived matrix material induced expression of the albumin gene in hepatocytes. Furthermore, collagen has been found to be essential for the functional expression

TABLE 17.1. Soluble Inducers of Differentiation: Physiological

Inducer	Cell Type	Reference
Steroid and related:		
Hydrocortisone	Glia, glioma	McLean et al., 1986
	Lung alveolar type II cells	Rooney et al., 1995; McCormick et al., 1995
	Hepatocytes	Granner et al., 1968
	Mammary epithelium	Marte et al., 1994
	Myeloid leukemia	Sachs, 1978
Retinoids	Tracheobronchial epithelium	Kaartinen et al., 1993
	Endothelium	Lechardeur et al., 1995; Hafny et al., 1996
	Enterocytes (Caco-2)	McCormack et al., 1996
	Embryonal carcinoma	Mills et al., 1996
	Melanoma	Lotan & Lotan, 1980; Meyskens and Fuller, 1980
	Myeloid leukemia	Degos, 1997
	Neuroblastoma	Ghigo et al., 1998
Peptide hormones:		
Melanotropin	Melanocytes	Goding and Fisher, 1997
Thyrotropin	Thyroid	Chambard et al., 1983
Erythropoietin	Erythroblasts	Goldwasser, 1975
Prolactin	Mammary epithelium	Takahashi et al., 1991; Rudland, 1992; Marte et al., 1994
Insulin	Mammary epithelium	Marte et al., 1994; Rudland, 1992
Cytokines:		
Nerve growth factor	Neurons	Levi-Montalcini, 1979
Glia maturation factor, CNTF, PDGF, BMP2	Glial cells	Keles et al., 1992; Raff, 1990; Kondo & Raff, 2000, 2004
Epimorphin	Kidney epithelium	Hirai et al., 1992
Fibrocyte-pneumocyte factor	Type II pneumocytes	Post et al., 1984
Interferon-α, β	A549 cells	McCormack et al., 1996
	HL60, myeloid leukemia	Kohlhepp et al., 1987
Interferon-γ	Neuroblastoma	Wuarin et al., 1991
CNTF	Type 2 astrocytes	Raff, 1990
IL-6, OSM	A549 cells	McCormack et al., 1996; McCormick & Freshney, 2000
BMP	10T1/2 cells	Shea et al., 2003
KGF	Keratinocytes	Aaronson et al., 1991
	Prostatic epithelium	Thomson et al., 1997; Yan et al., 1992
HGF	Kidney (MDCK)	Bhargava et al., 1992; Li et al., 1992
	Hepatocytes	Montesano et al., 1991
TGF-β	Bronchial epithelium, melanocytes	Masui et al., 1986a; Fuller and Meyskens, 1981
Endothelin	Melanocytes	Aoki et al., 2005
Vitamins:		
Vitamin E	Neuroblastoma	Prasad et al., 1980
Vitamin D_3	Monocytes (U937)	Yen et al., 1993
	Myeloma	Murao et al., 1983
	Osteoblasts	Vilamitjana-Amedee et al., 1993
	Enterocytes (IEC-6)	Jeng et al., 1994
Vitamin K	Hepatoma	Bouzahzah et al., 1995
	Kidney epithelium	Cancela et al., 1997
Retinoids (see above in this table)		
Minerals:		
Ca^{2+}	Keratinocytes	Boyce & Ham, 1983

of many epithelial cells [Burwen & Pitelka, 1980; Flynn et al., 1982; Berdichevsky et al., 1992] and for endothelium to mature into capillaries [Folkman & Haudenschild, 1980]. The RGD motif (arginine-glycine-aspartic acid) in matrix molecules appears to be the receptor interactive moiety in many cases [Yamada & Geiger, 1997]. Small polypeptides containing this sequence effectively block matrix-induced differentiation, implying that the intact matrix molecule is required [Pignatelli & Bodmer, 1988].

Attempts to mimic matrix effects by use of synthetic macromolecules have been partially successful in using poly-D-lysine to promote neurite extension in neuronal

TABLE 17.2. Soluble Inducers of Differentiation: Nonphysiological

Inducer	Cell type	Reference
Planar–polar compounds:		
DMSO	Murine erythroleukemia	Rossi & Friend, 1967; Dinnen & Ebisuzaki, 1990
	Myeloma	Tarella et al., 1982
	Neuroblastoma	Kimhi et al., 1976
	Mammary epithelium	Rudland, 1992
	Hepatocytes	Mitaka et al., 1993; Hino et al., 1999
Sodium butyrate	Erythroleukemia	Andersson et al., 1979b
	Colon cancer	Velcich et al., 1995
N-methyl acetamide	Glioma	McLean et al., 1986
N-methyl formamide, dimethyl formamide	Colon cancer	Dexter et al., 1979
HMBA	Erythroleukemia	Osborne et al., 1982; Marks et al., 1994
Benzodiazepines	Erythroleukemia	Clarke and Ryan, 1980
Cytotoxic drugs:		
Genistein	Erythroleukemia	Watanabe et al., 1991
Cytosine arabinoside	Myeloid leukemia	Takeda et al., 1982
Mitomycin C; anthracyclines	Melanoma	Raz, 1982
Methotrexate	Colorectal carcinoma	Lesuffleur et al., 1990
Signal transduction modifiers:		
PMA	Bronchial epithelium	Willey et al., 1984; Masui et al., 1986a
	Mammary epithelium	Wada et al., 1994
	Colon (HT29, Caco-2)	Velcich et al., 1995; Pignata et al., 1994
	Monocytes (U937)	Hass et al., 1993
	Erythroleukemia (K562)	Kujoth and Fahl, 1997
	Neuroblastoma	Spinelli et al., 1982

Abbreviations: HMBA, Hexamethylene bisacetamide; PMA (TPA), phorbol myristate acetate.

cultures (*see* Section 23.4.1), but it seems that there is still a great deal to learn about the specificity of matrix interactions. It is unlikely that charge alone is sufficient to mimic the more complex signals demonstrated in many different types of matrix interaction, but charge alterations probably allow cell attachment and spreading, and under these conditions the cells may be capable of producing their own matrix.

It has been shown that endothelial cells [Kinsella et al., 1992; Garrido et al., 1995] and many epithelial cells differentiate more effectively on Matrigel [Kibbey et al., 1992; Darcy et al., 1995; Venkatasubramanian et al., 2000; Portnoy et al., 2004], a matrix material produced by the Engelberth–Holm–Swarm (EHS) sarcoma and made up predominantly of laminin, but also of collagen and proteoglycans. A number of cell lines, e.g., A549 type II or Clara cell adenocarcinoma of lung, show apparent morphogenesis when grown on Matrigel (*see* Plate 12c). This technique is useful but has the problem of introducing another biological variable to the system. Defined matrices are required; however, as yet they must be made in the laboratory and are not commercially available. Although fibronectin, laminin, collagen, and a number of other matrix constituents are available commercially, the specificity probably lies largely in the proteoglycan moiety, within which there is the potential for wide variability, particularly in the number,

type, and distribution of the sulfated glycosaminoglycans, such as heparan sulfate [Fernig & Gallagher, 1994].

The extracellular matrix may also play a role in the modulation of growth factor activity. It has been suggested that the matrix proteoglycans, particularly heparan sulfate proteoglycans (HSPGs), may bind certain growth factors, such as GM-CSF [Damon et al., 1989; Luikart et al., 1990], and make them more available to adjacent cells. Transmembrane HSPGs may also act as low-affinity receptors for growth factors and transport these growth factors to the high-affinity receptors [Klagsbrun & Baird, 1991; Fernig & Gallagher, 1994]. A549 cells require glucocorticoid to respond to differentiation inducers such as OSM, IL-6, and lung fibroblast-conditioned medium. The glucocorticoid, in this case dexamethasone (DX), induces the A549 cells to produce a low-charge density fraction of heparan sulfate, which, when partially purified, was shown to substitute for the DX and activate OSM, IL-6, and fibroblast-conditioned medium [Yevdokimova & Freshney, 1997].

17.7.4 Polarity and Cell Shape

Studies with hepatocytes [Sattler et al., 1978] showed that full maturation required the growth of the cells on collagen gel and the subsequent release of the gel from the bottom of the dish with a spatula or bent Pasteur pipette. This process allowed shrinkage of the gel and an alteration in the shape of the cell from flattened to cuboidal, or even columnar. Accompanying or following the shape change, and also possibly due to access to medium through the gel, the cells developed polarity, visible by electron microscopy; when the nucleus became asymmetrically distributed, nearer to the bottom of the cell, an active Golgi complex formed and secretion toward the apical surface was observed.

A similar establishment of polarity has been demonstrated in thyroid epithelium [Chambard et al., 1983] with a filter well assembly. In this case, the lower (basal) surface generated receptors for thyroid-stimulating hormone (TSH) and secreted triiodothyronine, and the upper (apical) surface released thyroglobulin. Studies with hepatocytes [Guguen-Guillouzo & Guillouzo, 1986] (*see* Section 23.2.6) and bronchial epithelium [Saunders et al., 1993] suggest that floating collagen may not be essential, but the success of filter well culture (*see* Section 24.7.4) confirms that access to medium from below helps to establish polarity.

17.7.5 Oxygen Tension

Expression of fully keratinized squamous differentiation in skin requires the positioning of the epidermal cells in an organotypic construct at the air/liquid interface [Maas-Szabowski et al., 2002]. Likewise, location of alveolar type II cells at the air/liquid interface is necessary for optimal type II differentiation [Dobbs et al., 1997], and tracheal epithelium will only become mucus secreting at the air/liquid interface, becoming squamous if grown on the bottom of the dish [Kaartinen et al., 1993; Paquette et al., 2003]. It is assumed that positioning the cells at the air/liquid interface enhances gas exchange, particularly facilitating oxygen uptake without raising the partial pressure and risking free radical toxicity. However, it is also possible that the thin film above mimics the physical conditions *in vivo* (surface tension, lack of nutrients) as well as oxygenation. The presence of D-PBS on the apical surface may actually be preferable to complete medium [Chambard et al., 1983, 1987].

17.8 DIFFERENTIATION AND MALIGNANCY

It is frequently observed that, with increasing progression of cancer, histology of a tumor indicates poorer differentiation, and from a prognostic standpoint, patients with poorly differentiated tumors generally have a lower survival rate than patients with differentiated tumors. It has also been stated that cancer is principally a failure of cells to differentiate normally. It is therefore surprising to find that many tumors grown in tissue culture can be induced to differentiate (*see* Table 17.2). Indeed, much of the fundamental data on cellular differentiation has been derived from the Friend murine leukemia, mouse and human myeloma, hepatoma, and neuroblastoma. Nevertheless, there appears

to be an inverse relationship between the expression of differentiated properties and the expression of malignancy-associated properties, even to the extent that the induction of differentiation has often been proposed as a mode of therapy [Spremulli & Dexter, 1984; Freshney, 1985].

17.9 PRACTICAL ASPECTS

It is clear that, given the correct environmental conditions, and assuming that the appropriate cells are present, partial, or even complete, differentiation is achievable in cell culture. As a general approach to promoting differentiation, as opposed to cell proliferation and propagation, the following may be suggested:

(1) Select the correct cell type by use of appropriate isolation conditions and a selective medium (*see* Section 10.2.1 and Chapter 23).

(2) Grow the cells to a high cell density ($>1 \times 10^5$ cells/cm^2) on the appropriate matrix. The matrix may be collagen of a type that is appropriate to the site of origin of the cells, with or without fibronectin or laminin, or it may be more complex, tissue-derived or cell-derived (*see* Protocol 8.1), e.g., Matrigel (*see* Section 17.7.3) or a synthetic matrix (e.g., poly-D-lysine for neurons).

(3) Change the cells to a differentiating medium rather than a propagation medium, e.g., for epidermis increase Ca^{2+} to around 3 mM, and for bronchial mucosa increase the serum concentration (*see* Protocol 23.9). For other cell types, this step may require defining the growth factors appropriate to maintaining cell proliferation and those responsible for inducing differentiation.

(4) Add differentiation-inducing agents, such as gluco-corticoids; retinoids; vitamin D$_3$; DMSO; HMBA; prostaglandins; and cytokines, such as bFGF, EGF, KGF, HGF, IL-6, OSM, TGF-β, interferons, NGF, and melanocyte-stimulating hormone (MSH), as appropriate for the type of cell (*see* Tables 17.1 and 17.2).

(5) Add the interacting cell type during the growth phase (Step 2 in this procedure), the induction phase (Steps 3 and 4 in this procedure), or both phases. Selection of the correct cell type is not always obvious, but lung fibroblasts for lung epithelial maturation [Post et al., 1984; Speirs et al., 1991], glial cells for neuronal maturation [Seifert & Müller, 1984], and bone marrow adipocytes for hematopoietic cells (*see* Protocol 23.22) are some of the better-characterized examples.

(6) Elevating the culture in a filter well [*see*, e.g., Chambard et al., 1983] may be advantageous, particularly for certain epithelia, as it provides access for the basal surface to nutrients and ligands, the opportunity to establish polarity, and regulation of the nutrient and oxygen concentration at the apical surface, by adjusting the composition and depth of overlying medium.

Not all of these factors may be required, and the sequence in which they are presented is meant to imply some degree of priority. Scheduling may also be important; for example, the matrix generally turns over slowly, so prolonged exposure to matrix-inducing conditions may be important, whereas some hormones may be effective in relatively short exposures. Furthermore, the response to hormones may depend on the presence of the appropriate extracellular matrix, cell density, or heterologous cell interaction.

CHAPTER 18

Transformation and Immortalization

18.1 ROLE IN CELL LINE CHARACTERIZATION

Transformation is seen as a particular event or series of events that both depends on and promotes genetic instability. It alters many of the cell line's properties, including growth rate, mode of growth (attached or in suspension), specialized product formation, longevity, and tumorigenicity (Table 18.1). It is therefore important that these characteristics are included when a cell line is validated to determine whether it originates from neoplastic cells or has undergone transformation in culture. The transformation status is a vital characteristic that must be known when culturing cells from tumors, in order to confirm that the cells are derived from the neoplastic component of the tumor, rather than from normal infiltrating fibroblasts, blood vessel cells, or inflammatory cells.

More than one criterion is necessary to confirm neoplastic status, as most of the aforementioned characteristics are expressed in normal cells at particular stages of development. The exceptions are gross aneuploidy, heteroploidy, and tumorigenicity, which, taken together, are regarded as conclusive positive indicators of malignant transformation. However, some tumor cell lines can be near euploid and non-tumorigenic, and other criteria are therefore required.

18.2 WHAT IS TRANSFORMATION?

In microbiology, where the term was first used in this context, *transformation* implies a change in phenotype that is dependent on the uptake of new genetic material. Although this process is now achievable artificially in mammalian cells

(*see* Section 27.11), it is called *transfection* or *DNA transfer* in this case to distinguish it from transformation. Transformation of cultured cells implies a spontaneous or induced permanent phenotypic change resulting from a heritable change in DNA and gene expression. Although transformation can arise from infection with a transforming virus, such as polyoma, or from transfection with genes such as mutant *ras*, it can also arise spontaneously or after exposure to ionizing radiation or chemical carcinogens.

Transformation is associated with *genetic instability* and three major classes of phenotypic change, one or all of which may be expressed in one cell strain: (1) *immortalization*, the acquisition of an infinite life span, (2) *aberrant growth control*, the loss of contact inhibition of cell motility, density limitation of cell proliferation, and anchorage dependence, and (3) *malignancy*, as evidenced by the growth of invasive tumors *in vivo*. The term *transformation* is used here to imply all three of these processes. The acquisition of an infinite life span alone is referred to as *immortalization*, because it can be achieved without grossly aberrant growth control and malignancy, which are usually linked.

18.3 GENETIC INSTABILITY

The characteristics of a cell line do not always remain stable. In addition to the phenotypic alterations already described (*see* Sections 3.4.2, 17.1.1) cell lines are also prone to genetic instability. Normal, human finite cell lines are usually genetically stable, but cell lines from other species, particularly the mouse, are genetically unstable and transform quite readily. Continuous cell lines, particularly from tumors

Culture of Animal Cells: A Manual of Basic Technique, Fifth Edition, by R. Ian Freshney
Copyright © 2005 John Wiley & Sons, Inc.

TABLE 18.1. Properties of Transformed Cells

Property	Assay	Protocol, Figure, or Reference
Growth		
Immortal	Grow beyond 100 pd	Kopper & Hajdu 2004; Meeker and De Marzo 2004
Anchorage independent	Clone in agar; may grow in stirred suspension	Protocols 14.4, 14.5, 13.3
Loss of contact inhibition	Microscopic observation; time lapse	Fig. 18.3; Plate 9
Growth on confluent monolayers of homologous cells	Focus formation	Fig. 18.3
Reduced density limitation of growth	High saturation density; high growth fraction at saturation density	Protocols 18.3, 21.10, 21.11
Low serum requirement	Clone in limiting serum	Protocols 21.9, 22.3
Growth factor independent	Clone in limiting serum	Protocols 21.9, 22.3
Production of autocrine growth factors	Immunostaining; clone in limiting serum with conditioned medium; receptor-blocking antibody or peptide inhibitor	Protocols 16.13, 21.9, 22.3
Transforming growth factor production	Suspension cloning of NRK	Protocols 14.4, 14.5
High plating efficiency	Clone in limiting serum	Protocols 21.9, 22.3
Shorter population-doubling time	Growth curve	Protocols 21.7, 21.8
Genetic		
High spontaneous mutation rate	Sister chromatid exchange	Protocol 22.5
Aneuploid	Chromosome content	Protocol 16.9
Heteroploid	Chromosome content	Protocol 16.9
Overexpressed or mutated oncogenes	Southern blot; FISH, immunostaining; microarray analysis	Ausubel et al., 1996, 2002; Protocols 28.2, 16.13; Plate 24
Deleted or mutated suppressor genes	Southern blot; FISH, immunostaining; microarray analysis	Ausubel et al., 1996, 2002; Protocols 28.2, 16.13; Plate 24
Gene and chromosomal translocations	FISH, chromosome paints	Protocol 28.2
Structural		
Modified actin cytoskeleton	Immunostaining	Protocol 16.13
Loss of cell surface-associated fibronectin	Immunostaining	Protocol 16.13
Modified extracellular matrix	Immunostaining; DEAE chromatography	Protocol 16.13; Yevdokimova and Freshney, 1997
Altered expression of cell adhesion molecules (CAMs, cadherins, integrins)	Immunostaining	Protocol 16.13
Disruption in cell polarity	Immunostaining; polarized transport in filter wells	Protocol 16.13; Halleux and Schneider, 1994
Neoplastic		
Tumorigenic	Xenograft in nude or scid mice	Giovanella et al., 1974; Russo et al., 1993
Angiogenic	CAM assay; filter wells, VGEF production	Plate 9; Ment et al., 1997; Buchler et al., 2004
Enhanced protease secretion (e.g., plasminogen activator)	Plasminogen activator assay	Fig. 18.8; Whur et al., 1980; Boxman et al., 1995
Invasive	Organoid confrontation; filter well invasion assay	Figs. 18.6, 18.7; Mareel et al., 1979; Brunton et al., 1997

of all species, are very unstable, not surprisingly, as it was this instability that allowed the necessary mutations to become continuous, and deletion or alteration in DNA surveillance genes, such as p53, are usually implicated. Consequently, continuous cell strains, even after cloning, contain a range of genotypes that are constantly changing.

There are two main causes of genetic heterogeneity: (1) The spontaneous mutation rate appears to be higher *in vitro*, associated, perhaps, with the high rate of cell proliferation, and (2) mutant cells are not eliminated unless their growth capacity is impaired. Minimal-deviation rat hepatoma cells, grown in culture, express tyrosine aminotransferase activity constitutively and may be induced further by dexamethasone [Granner et al., 1968], but subclones of a cloned strain of H4-II-E-C3 [Pitot et al., 1964] differed both in the constitutive level of the enzyme and in its capacity to be induced by dexamethasone (Fig. 18.1).

18.3.1 Chromosomal Aberrations

Evidence of genetic rearrangement can be seen in chromosome counts (*see* Fig. 3.9) and karyotype analysis. Although the mouse karyotype is made up exclusively of small telocentric chromosomes, several metacentrics arise in many continuous murine cell lines due to fusion of the telomeres (Fig. 18.2a; Robertsonian fusion). Furthermore, although virtually every cell in the animal has the normal diploid set of chromosomes, this is more variable in culture. In extreme cases—e.g., continuous cell strains, such as HeLa-S3—less than half of the cells will have exactly the same karyotype; i.e., they are *heteroploid* (*see also* Fig. 3.9).

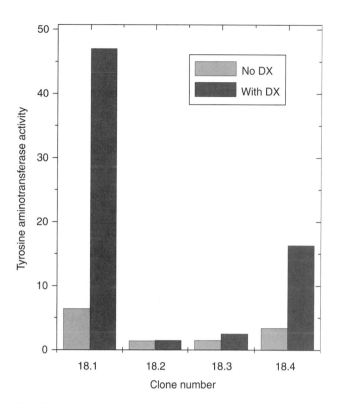

Fig. 18.1. Clonal Variation. Variation in tyrosine aminotransferase activity among four subclones of clone 18 of a rat minimal-deviation hepatoma cell strain, H-4-II-E-C3. Cells were cloned; clone 18 was isolated, grown up, and recloned; and the second-generation clones were assayed for tyrosine aminotransferase activity, with and without pretreatment of the culture with dexamethasone. Light gray bars, basal level; dark gray bars. induced level. (Data J. Somerville).

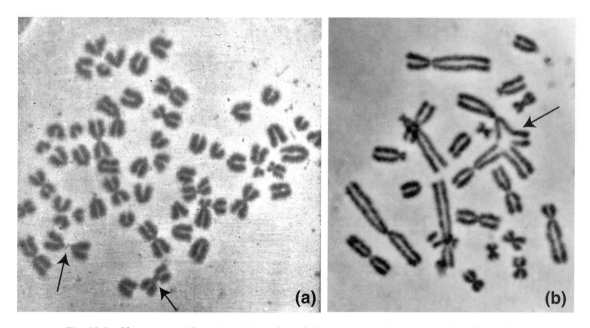

Fig. 18.2. Chromosome Aberrations. Examples of aberrant recombinations. (a) P2 cells, a clone of L929 mouse fibroblasts, showing multiple telomeric fusions, with two marked by arrows. (b) Recombination event between two dissimilar chromosomes of the larger group in Y5 Chinese hamster cells. This cell would be unlikely to survive.

Both variations in ploidy and increases in the frequency of individual chromosomal aberrations (Fig. 18.2b) can be found [Biedler, 1976; Croce, 1991], and variations in chromosome number are found in most tumor cultures [see Fig. 3.8; Protocol 16.9; Sandberg, 1982]. The incidence of genetic instability and frequency of chromosomal rearrangement can be determined by the sister chromatid exchange assay [Venitt, 1984] (see Protocol 22.5 and Plate 17e,f).

Some specific aberrations are associated with particular types of malignancy [Croce, 1991]. The first of these aberrations to be documented was the Philadelphia chromosome in chronic myeloid leukemia (trisomy 13). Subsequently, translocations of the long arms of chromosomes 8 and 14 were found in Burkitt's lymphoma [Lebeau & Rowley, 1984]. Several other leukemias also express other translocations [Mark, 1971]. Meningiomas often have consistent aberrations, and small-cell lung cancer frequently has a 3p2 deletion [Wurster-Hill et al., 1984]. These aberrations constitute tumor-specific markers that can be extremely valuable in cell line characterization and confirmation of neoplasia.

18.3.2 Variations in DNA Content

Flow cytometry [Traganos et al., 1977] shows that the DNA content of tumor cells reflects chromosomal aberrations—i.e., it may vary from the normal somatic cell DNA content and show marked heterogeneity within a population. DNA analysis does not substitute for chromosome analysis, however, as cells with an apparently normal DNA content can still have an aneuploid karyotype. Deletions and polysomy may cancel out, or translocations may occur without net loss of DNA.

18.4 IMMORTALIZATION

Most normal cells have a finite life span of 20–100 generations (see Section 3.8.1), but some cells, notably those from rodents and from most tumors, can produce continuous cell lines with an infinite life span. The rodent cells are karyotypically normal at isolation and appear to go through a crisis after about 12 generations; most of the cells die out in this crisis, but a few survive with an enhanced growth rate and give rise to a continuous cell line.

If continuous cell lines from mouse embryos (e.g., the various 3T3 cell lines) are maintained at a low cell density and are not allowed to remain at confluence for any length of time, they remain sensitive to contact inhibition and density limitation of growth [Todaro & Green, 1963]. If, however, they are allowed to remain at confluence for extended periods, foci of cells appear with reduced contact inhibition, begin to pile up, and will ultimately overgrow (Fig. 18.3).

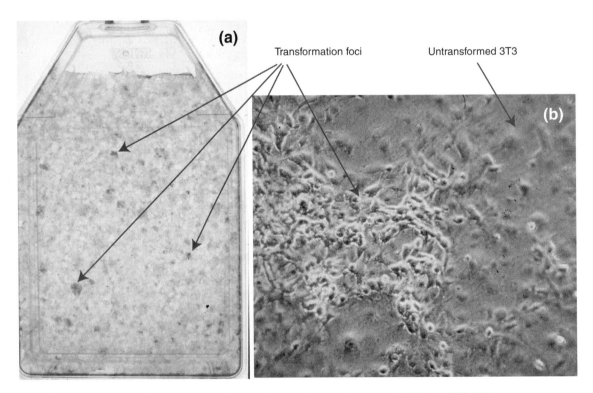

Fig. 18.3. Transformation Foci. A monolayer of normal, contact-inhibited NIH 3T3 mouse fibroblasts left at confluence for 2 weeks. (a) 75-cm² flask stained with Giemsa. (b) Phase-contrast image of focus of transformed cells overgrowing normal monolayer (10× objective). (*See also* Plate 14c).

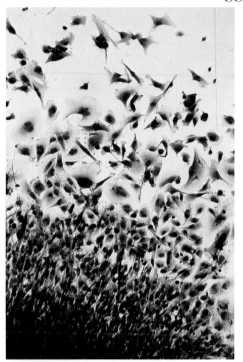

(a) **Human Lung Carcinoma.** *Outgrowth from primary explant from squamous cell carcinoma of human lung, MOG-L-DAN. Giemsa stained. 10× objective.*

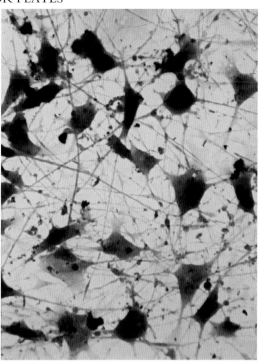

(b) **MOG-GP Astrocytoma.** *Astroglial cells from an anaplastic astrocytoma. Giemsa stained. 40× objective.*

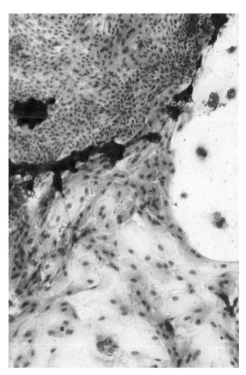

(c) **Human Cervical Intraepithelial Neoplasia (CIN).** *Primary culture from cervical biopsy subcultured onto 3T3 feeder layer of which a few degenerating cells remain, top right. Giemsa stained. 10× objective. (Courtesy of M.G. Freshney.)*

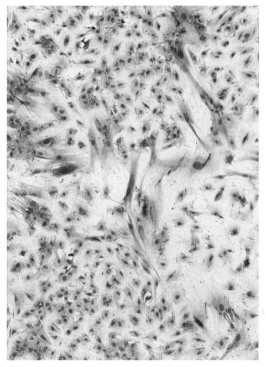

(d) **Human Fibrosarcoma.** *Primary culture subcultured onto fetal intestinal feeder layer. The fibrosarcoma cells are spindle shaped, and the feeder layer is polygonal. Giemsa stained. 10× objective.*

Plate 1. Primary Culture, Human.

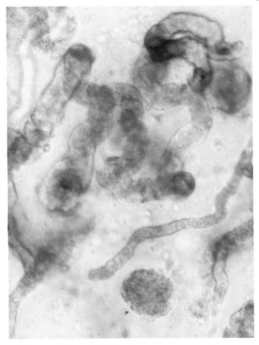

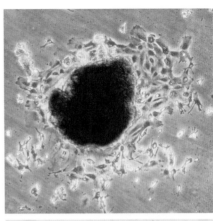

(b) Primary Explant. Mouse squamous cell carcinoma. Phase contrast; 10× objective.

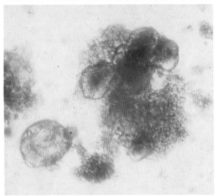

(c) Collagenase Digest. Human colon carcinoma. Bright field; 4× objective.

(a) Mouse Kidney Tubules. Disaggregation of newborn mouse kidney by cold trypsin method. Connective tissue has dissociated, but fragments of tubules and glomeruli remain intact. Also seen with collagenase digestion. Normal bright field illumination; 4× objective.

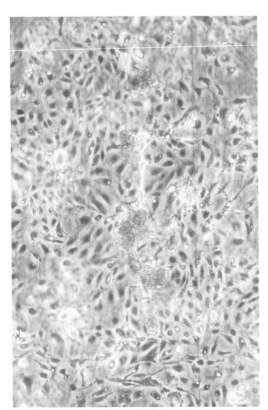

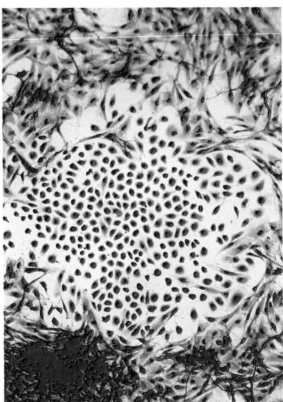

(d) Outgrowth from Kidney Tubules. Outgrowth of cells from attached fragments after disaggregation by cold trysin method. Phase contrast; 10× objective.

(e) Newborn Rat Kidney. Primary culture from trypsinized (cold method) newborn rat kidney. Giemsa stained; 10× objective. (Courtesy of M.G. Freshney).

Plate 2. Primary Culture, Explant, Cold Trypsin and Collagenase.

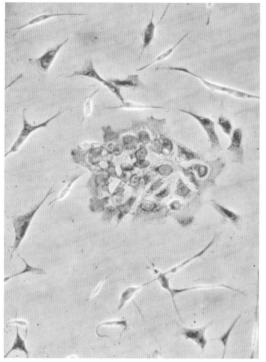

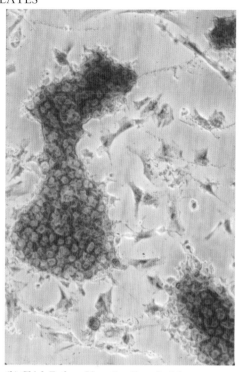

(a) **Chick Embryo Lung.** *Primary culture of 13-day chick embryo lung 48 h after disaggregation by the cold trypsin method. Light Giemsa stain and phase contrast; 10× objective.*

(b) **Chick Embryo Liver.** *Details as for (a).*

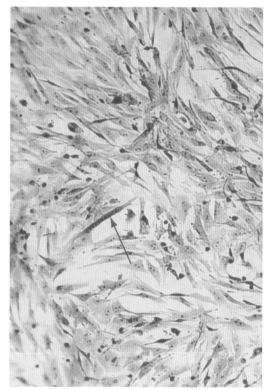

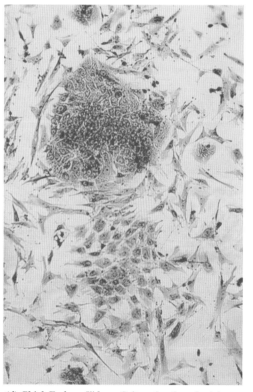

(c) **Chick Embryo Thigh Muscle.** *Culture details as for (a). Note multinucleate myotube (red arrow) resulting from fusion of myocytes (small dark-stained spindles). Giemsa stain, bright field; 10× objective.*

(d) **Chick Embryo Kidney.** *Culture details as for (a). Giemsa stain, bright field; 10× objective.*

Plate 3. Primary Culture, Chick Embryo.

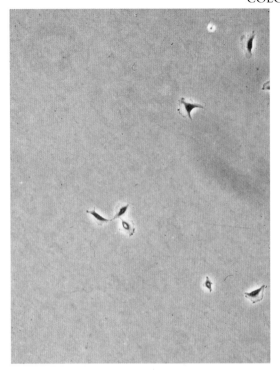

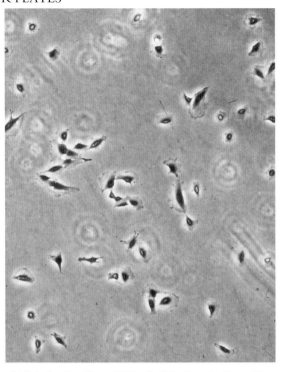

(a) Newly Subcultured Monolayer. NRK cells 24 h after subculture. Phase contrast; 10× objective.

(b) Entering Log Phase. NRK cells 48 h after subculture. Phase contrast; 10× objective.

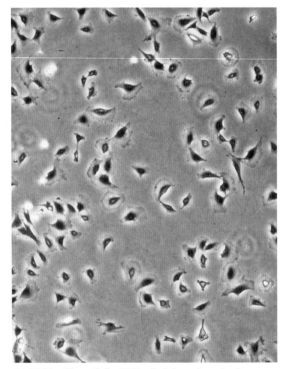

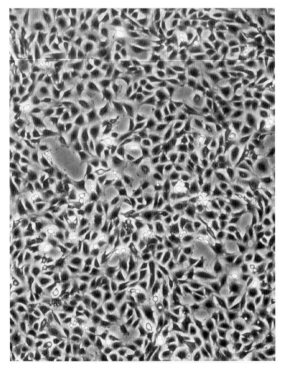

(c) Mid-log Phase Cells. NRK cells 3 days after subculture; ready for medium change. Phase contrast; 10× objective.

(d) Late Log Phase Cells. NRK cells 7 days after subculture and ready for subculture again. Phase contrast; 10× objective.

Plate 4. Phases of the Growth Cycle.

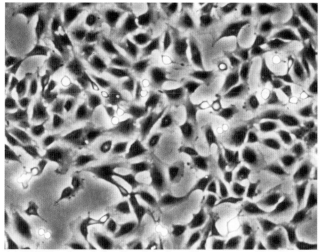

(a) NRK Monolayer before Trypsinization. Phase contrast; 20× objective.

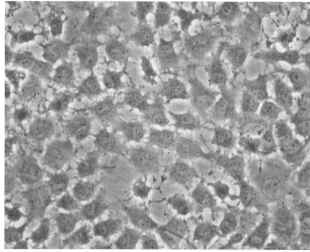

(b) NRK Monlayer after D-PBSA/EDTA Prewash. Phase contrast; 20× objective.

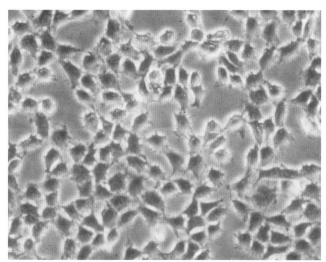

(c) NRK Monolayer Immediately after Trypsin Removal. Phase contrast; 20× objective.

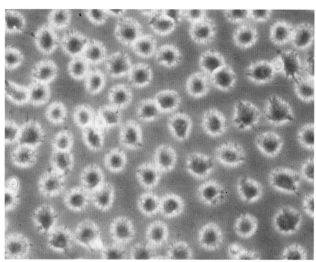

(d) NRK Monolayer 1 min after Trypsin Removal. Phase contrast; 20× objective.

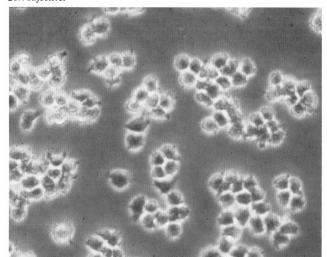

(e) NRK Monolayer 5 min after Trypsin Removal. Phase contrast; 20× objective.

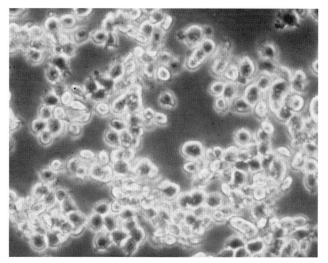

(f) Fully Disaggregated Monolayer. 10 min after removal of trypsin and ready for dispersing and counting. Phase contrast; 20× objective.

Plate 5. Subculture by Trypsinization.

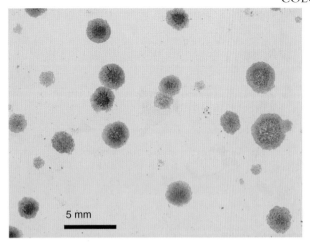

(a) HeLa Clones. *HeLa-S₃ cloned by dilution cloning and stained with Giemsa after 3 weeks. Scale bar 5 mm.*

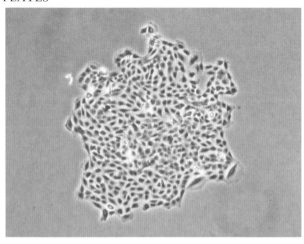

(b) NRK Clone. *Small clone of NRK cells, cloned by dilution cloning. Phase contrast, 4× objective.*

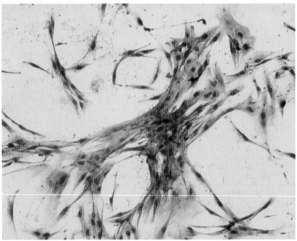

(c) Breast Carcinoma, JUW, Cloned on Plastic. *Microphotograph of secondary culture from human breast carcinoma, 4000 cells/cm², growing on plastic. Mainly fibroblasts. Giemsa stain. 10× objective.*

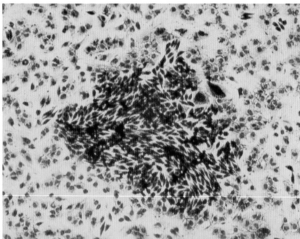

(d) Breast Carcinoma on Feeder Layer. *Microphotograph of epithelial colony from human breast carcinoma cells, 400 cells/cm², growing on confluent feeder layer of FHS74Int human fetal intestinal cells. Giemsa stain. 10× objective.*

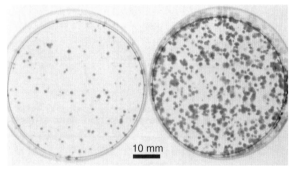

(e) Effect of Serum on Plating Efficiency. *Mv1Lu cells, transfected with myc oncogene, plated with 500 cells per 6-cm Petri dish, and fixed and stained 2 weeks later. 10% FBS (left), 20% FBS (right). Scale bar 10 mm (Courtesy of M. Z. Khan.)*

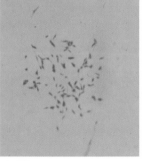

(f) Effect of Glucocorticoid on Plating Efficiency. *Early-passage cell line from human glioma cloned in the absence (left) and presence (right) of 25 μM dexamethasone. Macrophotograph; total magnification approximately 10×.*

Plate 6. Cell Cloning.

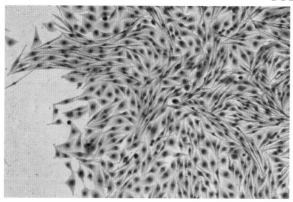

(a) Clone of Continuous Glioma Cell Line, MOG-G-CCM.
Elliptical and spindle-shaped morphology. Giemsa stained. 10×
objective.

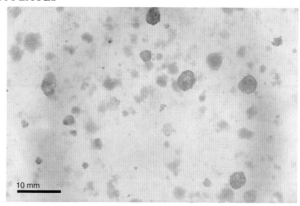

(d) Cloned Culture of Early-Passage Cell Line. *Human non-small cell*
lung carcinoma (NSCLC). Giemsa stained. Scale bar 10 mm.

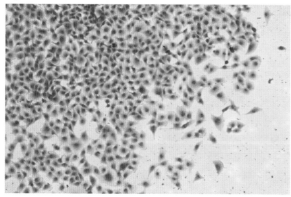

(b) Different Clone of Same Continuous Glioma Cell line as (a),
MOG-G-CCM. *Epithelioid morphology. Giemsa stained. 10×*
objective.

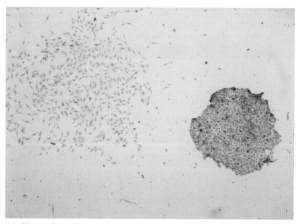

(e) Clones from Human NSCLC. *Giemsa stained. Detail from (d).*
4× objective.

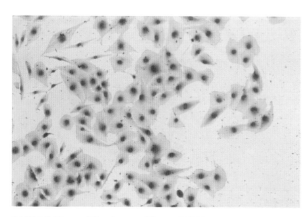

(c) Third Clone of Continuous Glioma Cell Line, MOG-G-CCM.
Squamous epithelioid morphology. Same magnification as (a) and
(b). Giemsa stained. 10× objective.

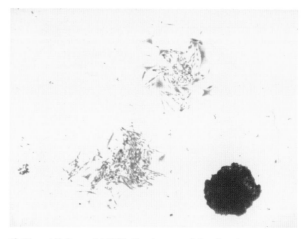

(f) Cloned Culture of A2780. *Continuous cell line from human ovarian*
carcinoma. Giemsa stained. 4× objective.

Plate 7. Cell Cloning, Morphological Diversity.

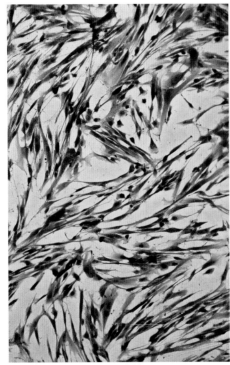

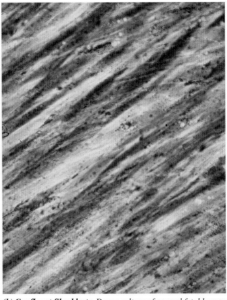

(b) Confluent fibroblasts. Dense culture of normal fetal human lung fibroblasts. Phase contrast; 10x objective.

(a) Normal human fetal lung fibroblasts. Subconfluent normal fetal human lung fibroblasts. Giemsa stained; 10x objective.

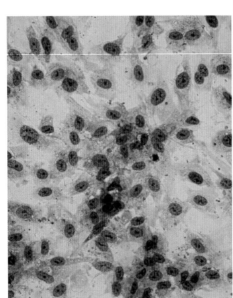

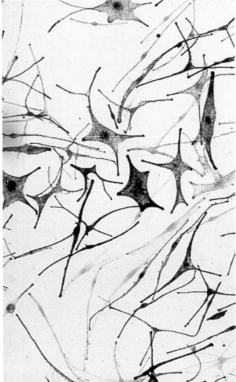

(c) Human Umbilical Cord Endothelium. Cell line from collagenase outwash of human umbilical vein. Goiemsa stained. 20x objective.

(d) Human Melanocytes. Secondary culture of TPA-treated epidermal melanocytes derived from African-American newborn foreskin. Note the dendritic morphology of the cells and their slender spindle shape. In contrast to the melanocytes derived from Caucasian newborn foreskin, these melanocytes display a high level of melanin granules (seen via phase contrast, x320). (Courtesy of H-Y. Park.)

Plate 8. Finite Cell Lines.

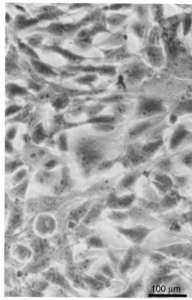

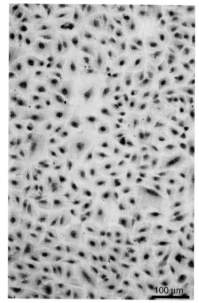

(a) HeLa-S₃. Human cervical carcinoma. Phase contrast. 40× objective.

(b) A549. Human lung adenocarcinoma. Giemsa stained. 10× objective.

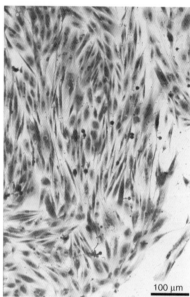

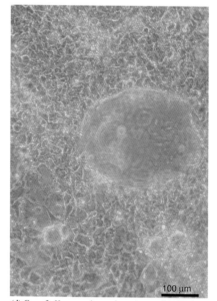

(c) MOG-G-UVW. Human glioma. Giemsa stained. 10× objective

(d) Caco-2. Human colorectal carcinoma showing dome formation. This cell line should be subcultured before it reaches this stage and while it is still subconfluent. Phase contrast; 10× objective.

Plate 9. Continuous Cell Lines from Human Tumors.

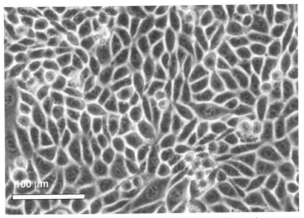

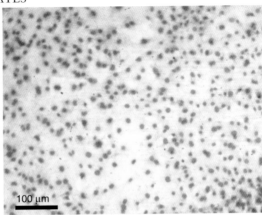

(a) CHO-K1. Clone of CHO, continuous cell line from Chinese hamster ovary. Phase contrast. Scale bar 100 μm.

(b) Swiss 3T3. Fibroblast-like cells from Swiss mouse embryo. Giemsa stained. Scale bar 100 μm.

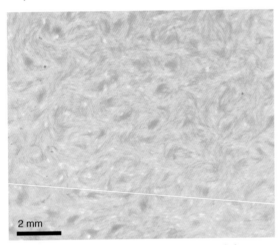

(c) Stained Flask of BHK-21 Clone 13. Baby hamster kidney fibroblasts after confluence showing typical swirling pattern formed by parallel arrays of cells. Giemsa stained. Macrophotograph. Scale bar 2 μm.

(d) BHK-21-C13. Baby hamster kidney fibroblasts approaching confluence and assuming parallel arrays. Giemsa stained. Scale bar 250 μm.

(e) High Density BHK-21 C13. Postconfluent baby hamster kidney fibroblasts forming a second monolayer of more densely stained cells. Giemsa stained. Scale bar 250 μm

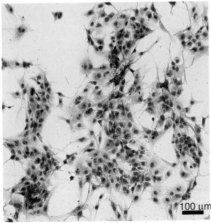

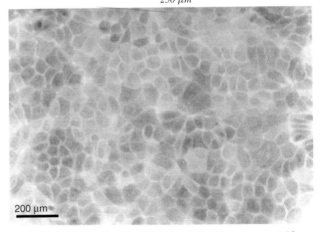

(f) MDCK Madin–Darby Canine Kidney. Epithelial-like cell line from dog kidney. Giemsa stained. Scale bar 100 μm.

(g) Mv1Lu Mink lung Epithelial Cell Line. Stained by immunoperoxidase for cytokeratin (AE3 primary antibody; Photo courtesy of M.Z. Khan.) Scale bar 100 μm.

Plate 10. Continuous Cell Lines from Normal, Nonhuman, Animal Tissue.

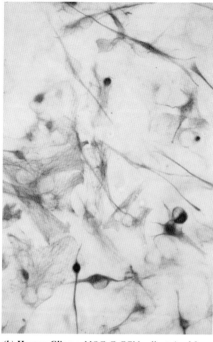

(a) Normal Human Keratinocytes on adult human dermal fibroblast Feeder Layer. Keratinocyte pancytokeratin stained green and vimentin in fibroblasts stained red. 40× objective. (Courtesy of Hans-Jurgen Stark.)

(b) Human Glioma. MOG-G-CCM cells stained for GFAP by immunoperoxidase. 10× objective.

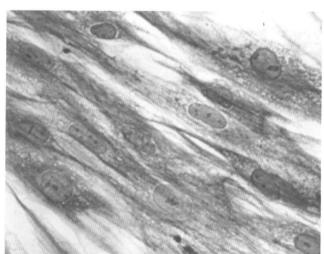

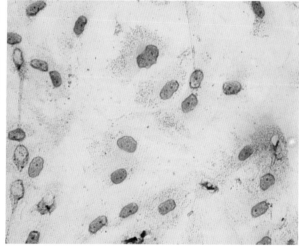

(c) Breast Stromal Fibroblasts. Stained for vimentin (brown) by immunoperoxidase . 40× objective. (Courtesy of Valerie Speirs.)

(d) Human Umbilical Vein Endothelial cells. Factor VIII granular cytoplasmic staining by immunoperoxidase. 40× objective. (Courtesy of R. L. Shaw and M. Frame.)

Plate 11. Immunostaining.

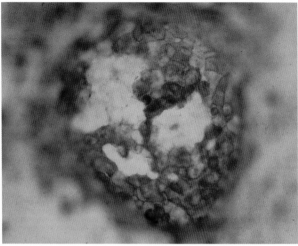

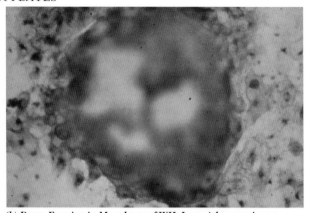

(b) Dome Forming in Monolayer of WIL Lung Adenocarcinoma, Lower Focus. *Mosaic of CEA-positive and CEA-negative cells. Focused on monolayer. Immunoperoxidase stained. 40× objective.*

(a) Dome Forming in Monolayer of WIL Lung Adenocarcinoma. *Mosaic of CEA-positive and CEA-negative cells. Focused on top of dome. Immunoperoxidase stained. 40× objective.*

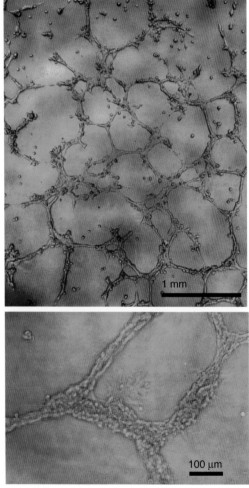

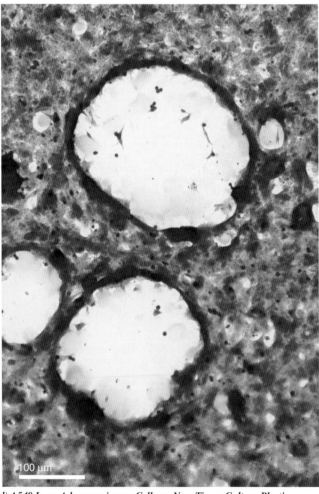

(c) A549 Lung Adenocarcioma Cells Growing on Matrigel. *Upper, 4× objective, lower 10×; bright field, unstained. (Courtesy of Jane Sinclair.)*

(d) A549 Lung Adenocarcinoma Cells on Non-Tissue Culture Plastic. *Growing in non-tissue culture-grade Petri dish, with lung fibroblasts growing on a coverslip in center of dish. 10× objective; bright field, Giemsa stained. (Courtesy of Valerie Speirs.)*

Plate 12. Morphological Differentiation in Epithelial Cells.

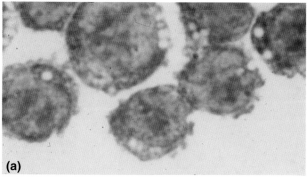

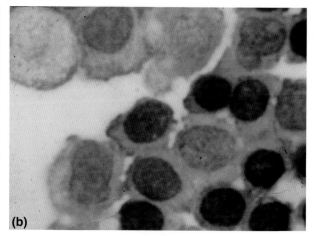

(a, b) Hemoglobin in Friend Cells. *Benzidine staining of hemoglobin in control Friend erythroleukemia cells. (a) control, (b) with 2% DMSO. 100× objective. (Courtesy of David Conkie). Giemsa counterstain.*

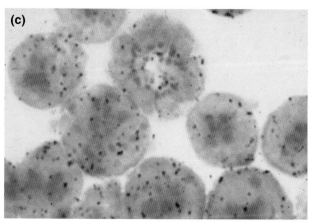

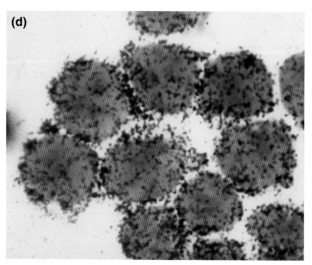

(c,d) In Situ Autoradiographic Labeling of Friend Cells. *Effect of DMSO on induction of mRNA for globin. (c) control, (d) induced with 2% DMSO. Giemsa stained. 100× objective. (Courtesy of David Conkie).*

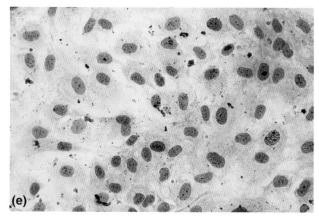

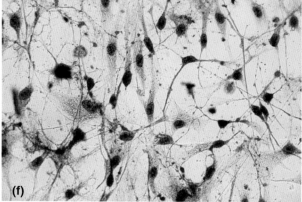

(e) Normal Human Glial Cells. *Undifferentiated glial cells from normal human brain. Giemsa stained. Total magnification 200×. (Courtesy of Margart Frame.)*

(f) Differentiated Human Glial Cells. *Morphological differentiation induced in glial cells from normal human brain by glia maturation factor. Giemsa stained. Total magnification 200×. (Courtesy of Margaret Frame.)*

Plate 13. Differentiation in Friend Cells and Human Glia.

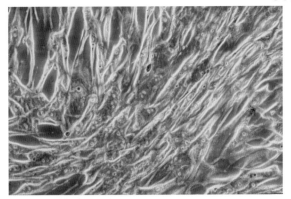

(a) Contact Inhibition. Late log-phase cultures of BHK21-C13. Cells tend to assume a parallel orientation and will not overgrow each other. Olympus CK2 microscope, 10× objective.

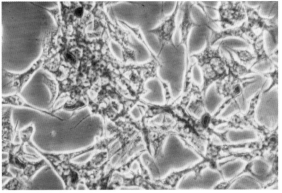

(b) Loss of Contact Inhibition. Polyoma-transformed clone BHK21-PyY cells, which show no recognition of each other and grow randomly over each other. Olympus CK2 microscope 10× objective.

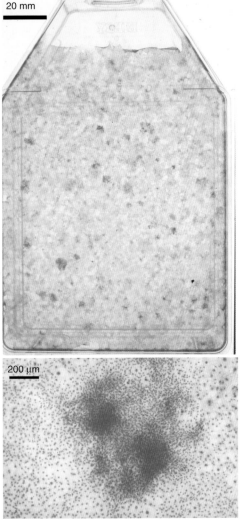

(c) Focus Formation in 3T3 Cells. A monolayer of Sw-3T3 cells, left at confluence for 3 weeks, showing foci of transformed cells escaping contact inhibition. Top, whole 75 cm² flask; bottom 4× objective. Giemsa stained.

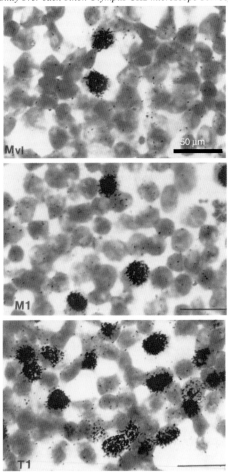

(d) [³H]-thymidine Incorporation in Mv1Lu Mink Lung Cells. Mink lung cells, Mv1Lu, were labeled with [³H]-TdR for 1 h, fixed, coated with autoradiographic emulsion (see Protocol 27.3), and stained with Giemsa. Mv1 (top) is the control cell line Mv1Lu, M1 (middle) is Mv1Lu transfected with the myc oncogene, and T1 (bottom) with mutant ras. T1 cells were tumorigenic and had a statistically significant (p < 0.001) increase in labeling index compared to Mv1Lu, also seen with bromodeoxyuridine labeling [Khan et al., 1991]. Scale bar 50 µm. (Courtesy of M .Z. Khan.)

Plate 14. Transformation.

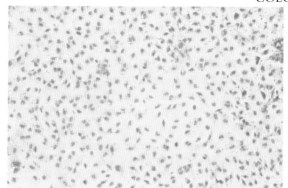

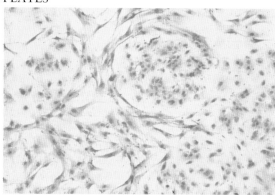

(a) Normal Human Fetal Instestinal Cell Line FHS74Int. *Contact inhibited confluent monolayer. 10× objective; Giemsa.*

(b) Human Glioma Cell Line WLY. *Infiltrating contact inhibited confluent monolayer FHS74Int; normal glial cells do not infiltrate this confluent monolayer (see Plate 15a). 10× objective; Giemsa.*

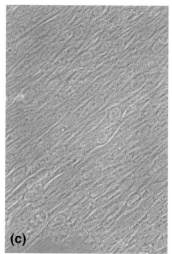

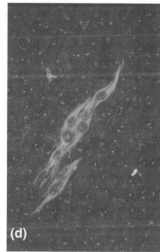

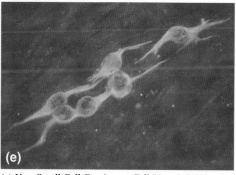

(c) Non-Small Cell Carcinoma Cell Line, L-DAN. *Infiltrating normal fetal lung fibroblasts. Phase contrast, 40× objective.* **(d) Indirect immunofluorescence for Cytokeratin.** *L-DAN migrating in parallel with the fibroblasts. 40× objective.* **(e) L-DAN migrating within fibroblasts.** *100× objective; indirect fluorescence for cytokeratin.*

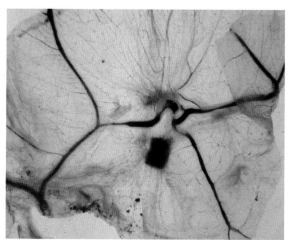

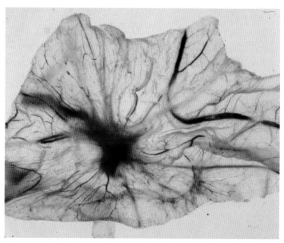

(f) Induction of Angiogenesis, Control. *A crude extract of normal glial cells was placed on the chorioallantoic membrane at 10 days of incubation, and the membrane was removed 2 weeks later. Approximately life-size. (Courtesy of Margaret Frame.)*

(g) Induction of Angiogenesis. *Chick chorioallantoic membrane with a crude extract of Walker 256 carcinoma cells added at 10 days of incubation, and the membrane was removed 2 weeks later. Approximately life-size. (Courtesy of Margaret Frame.)*

Plate 15. Properties of Transformed Cells.

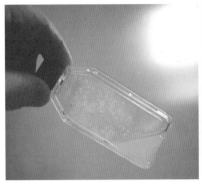

(a) Contaminated Flask. Reduction in pH and cloudiness of medium.

(b) Floculated Contamination. Bacterial, but little drop in pH.

(c) Contaminated Medium Bottle. Cloudy with sediment; yeast.

(d) Samples of Broth. L Broth with test samples of medium. Control clear (left), negative test (center), and positive control (right).

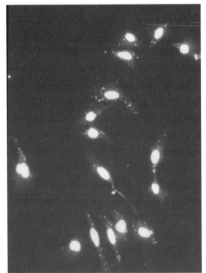

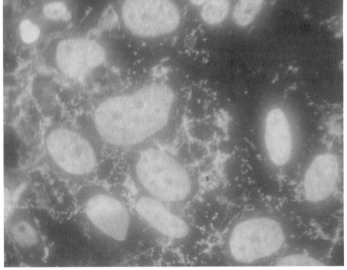

(e) Mycoplasma, Low Power. Mycoplasma infected culture as revealed by Hoechst 33258 staining. 40× objective; total magnification ×180.

(f) Mycoplasma, High Power. Hoechst-stained mycoplasma-infected cells under 100× objective; total magnification ×800.

Plate 16. Examples of Contamination.

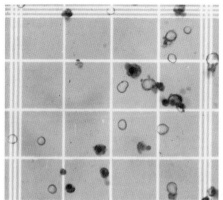

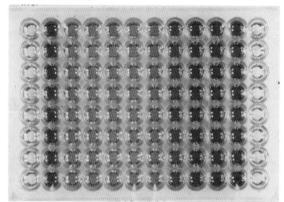

(a) Dye Exclusion, Naphthalene Black.
Hemocytometer slide 200 µm square with viable (unstained) and non-viable (blue stained) cells. 40× objective.

(b) MTT Assay. *Microtitration plate with cells stained with MTT after 24 h in a range of concentrations of VP16. Extreme left- and right-hand columns are blanks, 2nd from left and 2nd from right are untreated controls. Column 3 to column 8, VP16 0–10 µM.*

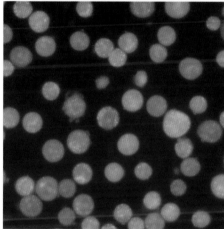

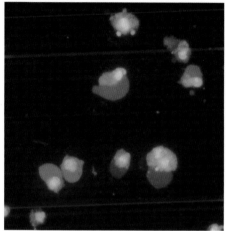

(c) HT29 cells. *HT29 cells treated with 0.25 µg/ml TRAIL (TNF-related apoptosis-inducing ligand) for 16 h and stained with acridine orange. Attached cells showing normal morphology. 40× objective; total magnification ~×300. (Courtesy of Angela Hague.)*

(d) Apoptosis in HT29 cells. *Detached cells showing chromatin condensation and fragmentation characteristic of apoptosis. 40× objective; total magnification ~×300. (Courtesy of Angela Hague.)*

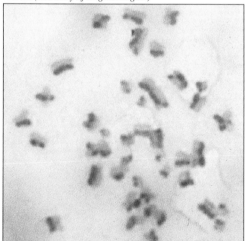

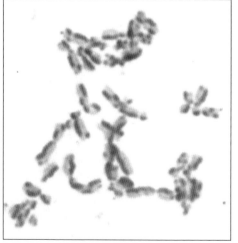

(e) Sister Chromatid Exchange (SCE). *Untreated cells: A2780/Cp70 with an additional human chromosome 2 transferred (A2780/cp70 +chr2). 100× objective. (Courtesy of Robert Brown & Maureen Illand).*

(f) Induced SCE. *A2780/Cp70 cells treated with 10 µM cisplatin for 1 h, showing extensive SCE: dark Giemsa staining alternating between strands within individual chromosomes. 100× objective. (Courtesy of Robert Brown & Maureen Illand).*

Plate 17. Viability and Cytotoxicity.

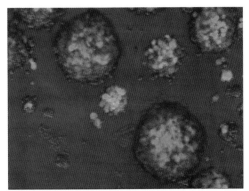

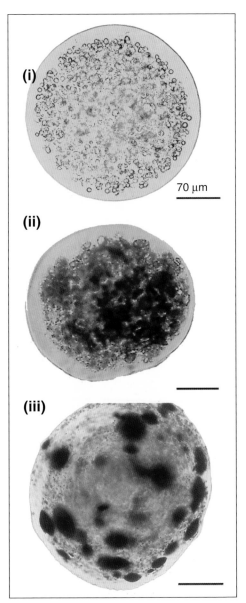

(a) Transfected Mosaic Spheroids. *Derived from the human glioma cell line MOG-G-UVW, the spheroids, ranging in size from 100- to 500-μm diameter, are composed of mixtures of cells transfected with the GFP gene (green) and cells transfected with the NAT gene. 40× objective. (Courtesy of Marie Boyd and Rob Mairs.)*

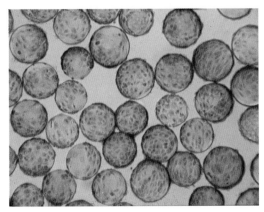

(c) Microcarriers. *Vero cells growing on microcarriers. 10× objective. (Courtesy of MP Biomedicals.)*

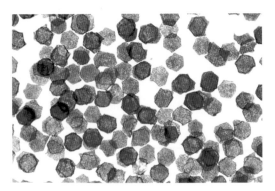

(d) Hexagonal Microcarriers. *Nunc Microhex Beads. Total magnification 20×. (Courtesy of Nalge Nunc.)*

(b) Alginate Encapsulation. *Light microscopic images of cells encapsulated in alginate after 2 hours (i), 3 weeks (ii) and 4 months (iii) in vitro. Within the alginate beads both cell death and cell proliferation will occur, and for many cell lines multicellular spheroids will form inside the beads. Scale bar 70 μm. (Courtesy of Tracy-Ann Read and Rolf Bjerkvig.)*

Plate 18. Spheroids, Encapsulation, and Microcarriers.

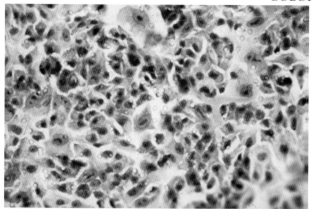

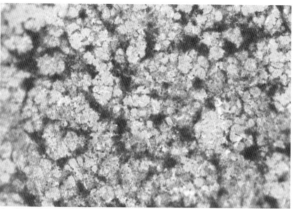

(a) Culture on Filter Well Inserts. *A549 human lung adenocarcinoma grown on polycarbonate filter. Holes in the filter (15 μm) are visible, particularly in top right-hand corner space. 40× objective. Giemsa stained.*

(b) Coculture on Filter Well Inserts. *Human fetal lung epithelial cells. Cells growing on filter with human fetal lung fibroblasts on other side of filter. 4× objective. Giemsa stained.*

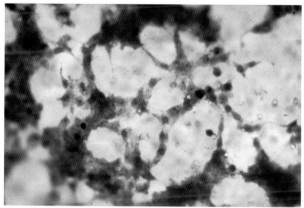

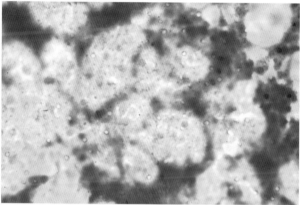

(c) Lung Cell Coculture, Top Focus. *Human fetal lung epithelial cells. As for (b), focused on cells above filter. 40× objective. Giemsa.*

(d) Coculture on Filter Well Inserts. *Human fetal lung epithelial cells. As for (b), focused on cells below filter. 40× objective. Giemsa.*

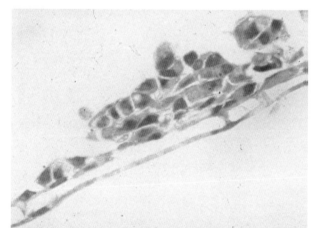

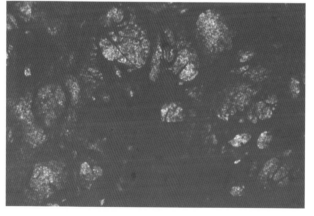

(e) Section of Coculture. *Section through filter with A549 cells and fibroblasts. Both cell types were seeded on top, but fibroblasts migrated through first, although some remained on the top. Three cells are seen travelling through the pores. H&E stain; 40× objective.*

(f) Chick Embryo Lung. *High-density primary culture of cold trypsin-disaggregated 13-day embryonic chick lung. 72 h after seeding at >1×10⁶ cells/mL. 40× objective. Giemsa.*

Plate 19. *Organotypic Culture in Filter Wells.*

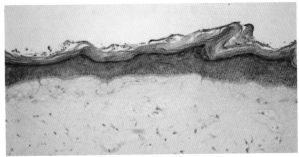

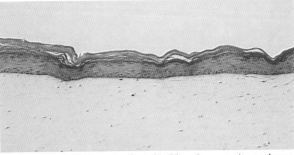

(a) Keratinocyte and Dermal Fibroblast Cocultures. *Organotypic cocultures after 2 weeks in vitro. 40× objective. H&E stain.*

(b) Organotypic Cultures Implanted in Vivo. *Organotypic cocultures in vivo 3 weeks after transplantation onto the nude mouse. 40× objective. H&E-stain.*

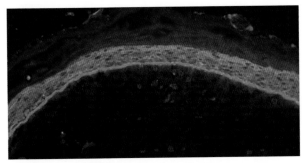

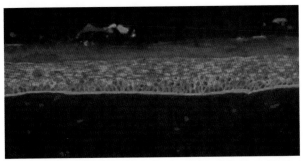

(c) Integrin and cytokeratin in Cocultures in Vitro. *Immunofluorescence for α 6-integrin (green), keratin 10 (red), and DNA fluorescence blue. 40× objective.*

(d) Integrin and Cytokeratin in Organotypic Cocultures Implanted in Vivo. *Immunofluorescent staining for alpha6-integrin (green), keratin 10 (red) with DNA fluorescing blue. 40× objective*

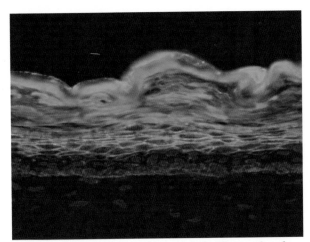

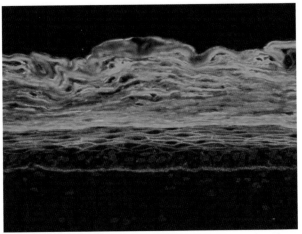

(e) Involucrin and Collagen in Organotypic Cocultures at 2 weeks. *Keratinocyte-fibroblast cocultures with collagen gel stained for involucrin, green, and the basement membrane component collagen VII, red. 40× objective.*

(f) Involucrin and Collagen in Organotypic Cocultures at 3 weeks. *Keratinocyte-fibroblast cocultures with collagen gel stained for involucrin, green, and the basement membrane component collagen VII, red. 40× objective.*

Plate 20. Organotypic Culture of Skin. (Courtesy of Hans-Jürgen Stark).

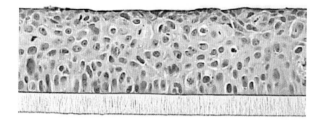

(a) Exposed to control solution for 10 min.

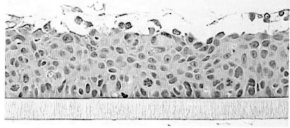

(b) Exposed to 1% SDS for 10 min.

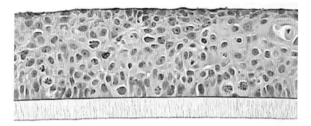

(c) Control solution, 20 min.

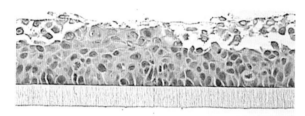

(d) 1% SDS for 20 min.

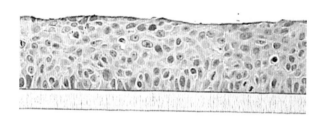

(e) Control solution, 60 min.

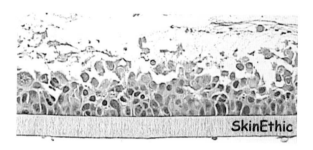

(f) 1% SDS for 60 min.

Human Corneal Epithelium in Filter Well Inserts. *(See also Figs. 22.10, 25.7; these examples and illustrations courtesy of SkinEthic.)*

Plate 21. In Vitro Toxicity in Organotypic Model.

COLOR PLATES

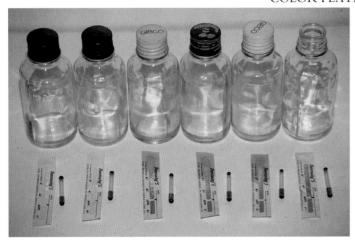

(a) Sterilizing Bottles. *These bottles were autoclaved with Thermalog sterility indicators inside. Thermalog turns blue with high temperature and steam, and the blue area moves along the strip with time at the required sterilization conditions. The cap on the leftmost bottle was tight, and each succeeding cap was gradually slacker, until, finally, no cap was used on the bottle furthest to the right. The leftmost bottle is not sterile, because no steam entered it. The second bottle is not sterile either, because the liner drew back onto the neck and sealed it. The next three bottles are all sterile, but the brown stain on the indicator shows that there was fluid in them at the end of the cycle. Only the bottle at the far right is sterile and dry.*

(b) pH Standards. *Phenol red pH indicator in a standard set of solutions. Far left and far right are unacceptable and need immediate action, i.e. medium change, subculture, or gassing.*

pH6.5   pH7.0   pH7.4   pH7.6

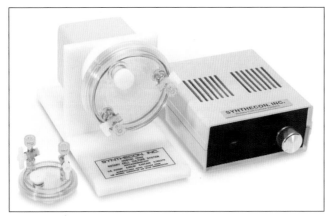

(c) Rotating Cell Culture System. *Chamber rotates to create simulated zero gravity so that cells do not sediment and grow as aggregates within the chamber. (Courtesy of Synthecon.)*

(d) BelloCell Culture System. *Cells are grown on macrobeads in center of chamber, and the bellows medium chamber alternately forces medium up over beads and back down into reservoir. (Courtesy of Metabios.)*

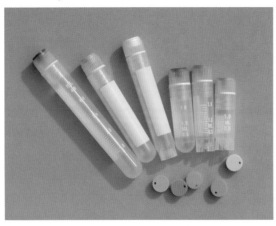

(e) Color-coded Cryovials. *Polypropylene cryostorage vials, 1.0 - 4.6 mL capacity. Colored inserts for caps helps to prevent misidentification. (Alpha Laboratories, Ltd.)*

Plate 22. Medium Preparation, Culture Systems, and Cryovials.

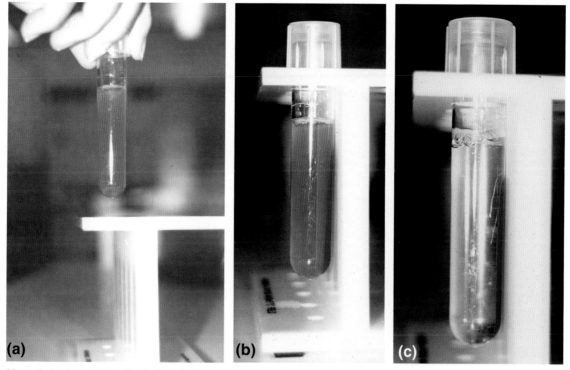

Magnetic Sorting with Dynabeads (Dynal). *Negative Sort; committed progenitor cells from bone marrow suspension bound to Dynal paramagnetic beads with antibodies to lineage markers. Lineage negative (stem) cells are not bound and remain in the suspension ready for sorting by flow cytometry. (a) Inserting the tube into the magnetic holder. (b) Tube immediately after being placed in magnetic holder. (c) Tube 30 s after placement in magnetic holder.*

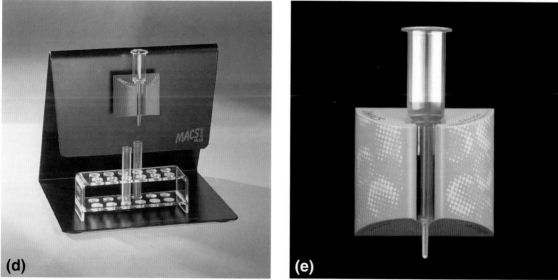

Magnetic cell sorting with the MidiMACS Separator (Miltenyi Biotec). *(d) Column, magnet, rack, and stand. (e) MidiMacs Separator with LD column. (Courtesy of Miltenyi Biotec.)*

Plate 23. *Magnetically Activated Cell Sorting.*

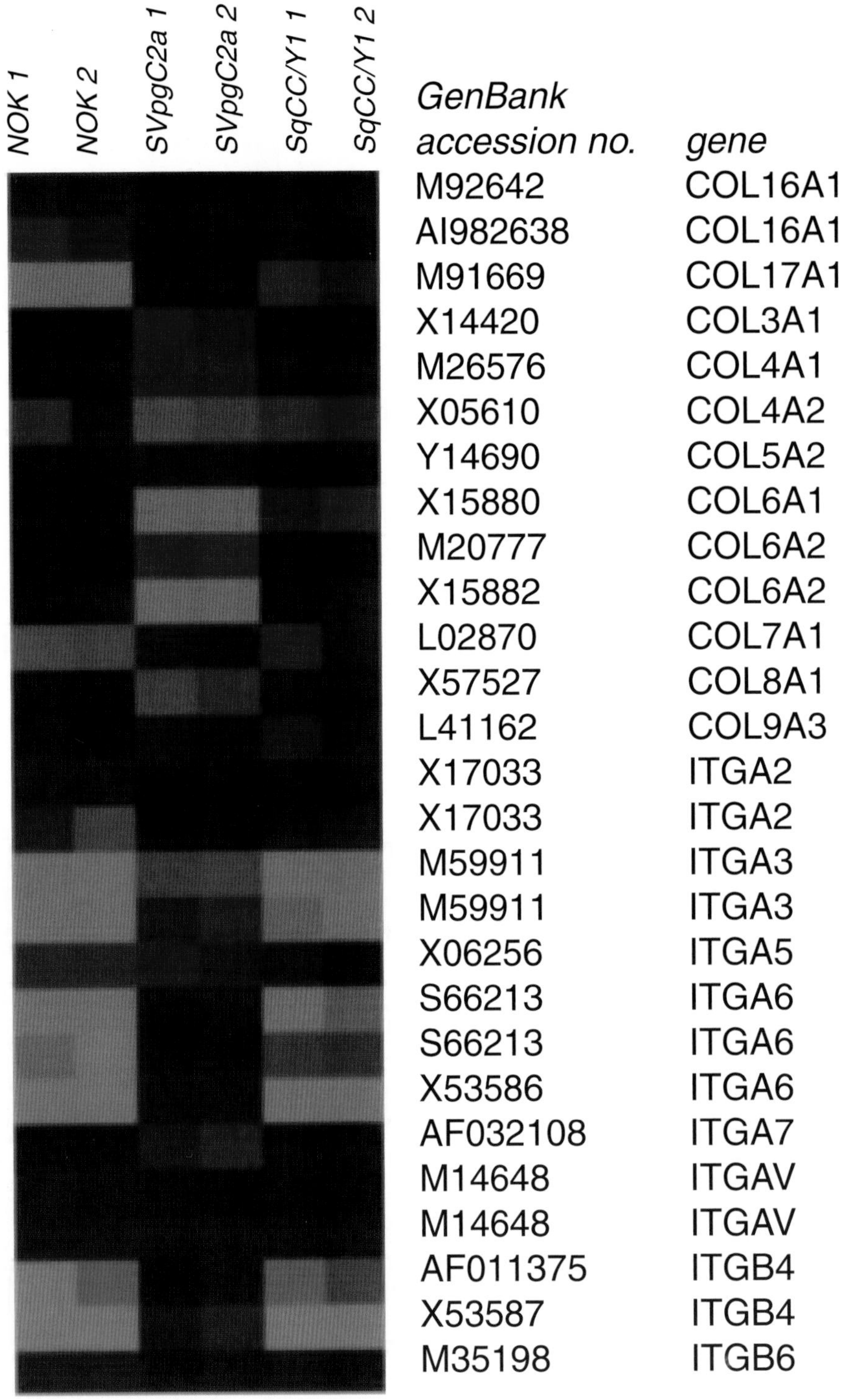

Expression of fibronectin, collagen and integrin transcripts. *Detected by Affymetrix oligonucleotide microarray in normal (NOK), SV40 T-antigen immortalised (SVpgC2a), and malignant (SqCC/Y1) human oral keratinocyte lines. GenBank gene abbreviations: COL16A1 = type XVI collagen 1 chain, etc.; ITGA2 = integrin 2 subunit, etc.; ITGB4 = integrin 4 subunit, etc. (From Sarang et al., 2003, ATLA 31: 575-585 by permission of the publisher and author.)*

Plate 24. *Affymetrix Microarray Analysis of Gene Expression.*

The fact that these cells are not apparent at low densities or when confluence is first reached suggests that they arise *de novo*, by a further transformation event. They appear to have a growth advantage, and subsequent subcultures will rapidly be overgrown by the randomly growing cell. This cell type is often found to be tumorigenic.

18.4.1 Control of Senescence

The finite life span of cells in culture is regulated by a group of 10 or more dominantly acting senescence genes, the products of which negatively regulate cell cycle progression [Goldstein et al., 1989; Sasaki et al., 1996]. Somatic hybridization experiments between finite and immortal cell lines usually generate hybrids with a finite life span, suggesting that the senescence genes are dominant [Pereira-Smith and Smith, 1988]. It is likely that one or more of these genes negatively regulate the expression of telomerase [Holt et al., 1996; Greider & Blackburn, 1996; Smith & de Lange, 1997; Bryan & Reddel, 1997], required for the terminal synthesis of telomeric DNA, which otherwise becomes progressively shorter during a finite life span, until the chromosomal DNA can no longer replicate. Telomerase is expressed in germ cells and has moderate activity in stem cells, but is absent from somatic cells. Deletions and/or mutations within senescence genes, or overexpression or mutation of one or more oncogenes that override the action of the senescence genes, can allow cells to escape from the negative control of the cell cycle and reexpress telomerase.

It has been assumed that immortalization is a multistep process involving the inactivation of a number of cell cycle regulatory genes, such as Rb and p53. The SV40 LT gene is often used to induce immortalization. The product of this gene, T antigen, is known to bind Rb and p53. By doing so, it not only allows an extended proliferative life span but also restricts the DNA surveillance activity of genes like p53, thereby allowing an increase in genomic instability and an increased chance of generating further mutations favorable to immortalization (e.g., the upregulation of telomerase or the downregulation of one of the telomerase inhibitors). Transfection of the telomerase gene with a regulatable promoter is sufficient to immortalize cells [Bodnar et al., 1998; Vaziri & Benchimol, 1998].

Immortalization per se does not imply the development of aberrant growth control and malignancy, as a number of immortal cell lines, such as 3T3 cells and BHK21-C13, retain contact inhibition of cell motility, density limitation of cell proliferation, and anchorage dependence, and are not tumorigenic. It must be assumed, however, that some aspects of growth control are abnormal and that there is a likely increase in genomic instability. Furthermore, immortalized cell lines often lose the ability to differentiate, but there are reports of telomerase-induced immortalization of keratinocytes [Dickson et al., 2000] and skeletal muscle satellite cells [Wootton et al., 2003] without abrogation of p53 activity and retention of the ability to differentiate.

18.4.2 Immortalization with Viral Genes

A number of viral genes have been used to immortalize cells (Table 18.2). It has been recognized for some time that SV40 can be used to immortalize cells, and the gene responsible for

TABLE 18.2. Genes Used in Immortalization

Gene	Insertion	Cell type	Reference
EBV: *ebna, lmp*1	Infection	B-lymphocytes	Bolton and Spurr, 1996; Bourillot et al., 1998; Sugimoto et al., 2004
SV40LT	Lipofection	Keratinocytes	Steinberg, 1996
	Calcium phosphate transfection	Fibroblasts	Mayne et al., 1996
	Calcium phosphate transfection	Astroglial cells	Burke et al., 1996
	Adenovirus infection	Esophageal epithelium	Inokuchi et al., 1995
	Microinjection	Rat brain endothelium	Lechardeur et al., 1995
	Transfection	Prostate epithelium	Rundlett et al., 1992
	Transfection	Mammary epithelium	Shay et al., 1993
	Strontium phosphate transfection	Bronchial epithelium	De Silva et al., 1996
	Strontium phosphate transfection	Mesothelial cells	Duncan et al., 1996
HPV16 E6/E7	Retroviral transfer	Cervical epithelium	Demers et al., 1994
	Transfection	Keratinocytes	Bryan et al., 1995
	Strontium phosphate transfection	Mesothelial cells	De Silva et al., 1994
	Strontium phosphate transfection	Bronchial epithelium	De Silva et al., 1994
	Retroviral infection	Ovarian surface epithelium	Tsao et al., 1995
Ad5 E1a		Epithelial cells	Douglas and Quinlan, 1994
htrt	Transfection	Pigmented retinal epithelium	Bodnar et al., 1998
	Transfection	Foreskin fibroblasts	Bodnar et al., 1998
	Transduction	Bone marrow stem cells	Simonsen et al., 2002
	Transduction	Keratinocytes	Dickson et al., 2000
	Transduction	Myoblasts	Wootton et al., 2003

this appears to be the large T (LT) gene [Mayne et al., 1996]. Other viral genes that have been used to immortalize cells are adenovirus E1a [Seigel, 1996], human papilloma virus (HPV) E6 and E7 [Peters et al., 1996; Le Poole et al., 1997], and Epstein–Barr virus (EBV; usually the whole virus is used) [Bolton & Spurr, 1996]. Most of these genes probably act by blocking the inhibition of cell cycle progression by inhibiting the activity of genes such as CIP-1/WAF-1/p21, Rb, p53, and p16, thus giving an increased life span, reducing DNA surveillance, and giving an enhanced opportunity for further mutations. Those genes that have been used most extensively are EBV for lymphoblastoid cells [Bolton & Spurr, 1996] and SV40LT for adherent cells such as fibroblasts [Mayne et al., 1996], keratinocytes [Steinberg, 1996], and endothelial cells [Punchard et al., 1996], and hTERT for mesenchymal stem cells (see Protocol 18.2) and as number of other cells. Endothelial cells have also been immortalized by irradiation [Punchard et al., 1996].

Typically, cells are transfected or retrovirally infected with the immortalizing gene before they enter senescence. This extends their proliferative life span for another 20–30 population doublings, whereupon the cells cease proliferation and enter *crisis*. After a variable period in crisis (up to several months), a subset of immortal cells overgrows. The fraction of cells that eventually immortalize can be 1×10^{-5} to 1×10^{-9}.

18.4.3 Immortalization of Human Fibroblasts

The following introduction to the immortalization of fibroblasts has been condensed from Mayne et al. [1996].

By far, the most successful and most frequently used method for deriving immortal human fibroblasts is through the expression of SV40 T antigen, which does not lead directly to immortalization but initiates a chain of events that results in an immortalized derivative appearing with a low probability, estimated at about 1 in 10^7 [Shay & Wright, 1989; Huschtcha & Holliday, 1983]. SV40-transfected cells are selected directly under appropriate culture conditions, and the surviving cells are subcultured to give rise to a precrisis SV40-transformed cell population. These cells are cultivated continuously until they reach the end of their proliferative life span, when they inevitably enter crisis. They must then be nurtured with care, and sufficient cells must be cultured, to give a reasonable chance for an immortalized derivative to appear.

The choice of T antigen expression vectors depends on the choice of the dominant selectable marker gene, which is the source of the promoter that drives T antigen expression and alternative forms of the T antigen itself. Although, in our experience [Mayne et al., 1996], selection for *gpt* is effective in human fibroblasts, G418 (*neo*) and hygromycin (*hygB*) are much more effective and easier to use. The majority of human SV40-immortalized fibroblast cell lines have been established with either SV40 virus or constructs, such as pSV3*neo*, that express T antigen from the endogenous promoter. We recommend the use of pSV3*neo* [Southern & Berg, 1982; Mayne et al., 1986] for the constitutive expression of T antigen.

Cells should be used between passages 7 and 15. Trypsinize the cells 24–48 h before transfection, and seed ≤ 2–2.5×10^5 cells per 9-cm dish or ≤ 5.5–6.8×10^5 cells per 175-cm^2 flask. Cells should be 70–80% confluent when transfected, in a final volume of medium of 10 mL/9-cm dish or 30 mL/175-cm^2 flask.

The calcium phosphate precipitation method relies on the formation of a DNA precipitate in the presence of calcium and phosphate ions. The DNA is first sterilized by precipitation in ethanol and resuspension in sterile buffer. It is then mixed carefully with calcium, and the resulting solution is added very slowly, with mixing, to a phosphate solution. When making the precipitate, it is important to note that optimal gene transfer occurs when the final concentration of DNA in the precipitate is 20 μg/mL. The volume of DNA precipitate applied to the cultures should never exceed one-tenth of the total volume. It is necessary to leave the mixture to develop for 30 min before adding it to the cell cultures.

Protocol 18.1 for the immortalization of fibroblasts has been condensed from Mayne et al. [1996].

PROTOCOL 18.1. FIBROBLAST IMMORTALIZATION

Materials
Sterile:
- HEPES buffer: HEPES, 12.5 mM; pH 7.12
- 10× CaHEPES: CaCl$_2$, 1.25 M; HEPES, 125 mM; pH 7.12
- 2× HEPES-buffered phosphate (2× HBP): Na$_2$HPO$_4$, 1.5 mM; NaCl, 280 mM; HEPES, 25 mM; pH 7.12
- NaOAc: NaOAc, 3 M; pH 5.5
- Tris-buffered EDTA (TBE): Tris·HCl, 2 mM; EDTA, 0.1 mM; pH 7.12
- Absolute ethanol
- Eagle's MEM/15% FCS
- G418 (Invitrogen): 20 mg/mL in HEPES buffer, pH 7.5, sterilized by filtration through a 0.2-μm membrane and stored in small aliquots at −20°C
- Hygromycin (Roche Applied Science): 2 mg/mL in UPW, filter sterilized with a 0.2-μm membrane and stored in small aliquots at −20°C
- SV40 T antigen DNA

Protocol
1. Estimate the amount of DNA required for the transfection:
   (a) Use a maximum of 20 μg of your T antigen vector (without carrier DNA) for each plate and 60 μg of the vector DNA per flask.

(b) Include an additional 20 μg of DNA, as it is not always possible to recover the full expected amount after preparation of the precipitate.

2. Prepare a sterile solution of the vector DNA in a microcentrifuge tube:
   (a) Precipitate the DNA with one-tenth of a volume of 3.0 M NaOAc, pH 5.5, and 2.5 volumes of ethanol.
   (b) Mix well, ensuring that the entire inside of the tube has come into contact with the ethanol solution.
   (c) Leave the tube on ice briefly (∼5 min).
   (d) Centrifuge the tube for 15 min at 15,000 rpm in a microcentrifuge to collect the precipitate.
   (e) Gently remove the tube from the centrifuge, and open it in a laminar flow hood.
   (f) Remove the supernatant by aspiration, taking care not to disturb the pellet. Ensure that the ethanol is well drained.
   (g) Allow the pellet to air dry in the cell culture hood until all traces of ethanol have evaporated.
   (h) Resuspend the DNA pellet in TBE to give a final concentration of 0.5 mg/mL. It may be necessary to vortex the tube in order to release the pellet from the side of the tube.
   (i) Incubate the tube at 37°C for 5–10 min, with occasional vortexing to ensure that the pellet is well resuspended.

Note. Do not use higher TBE concentrations for resuspending your DNA, as doing so can interfere with the formation of the DNA precipitate.

3. Prepare the DNA calcium phosphate precipitate:
   (a) Calculate the total volume of precipitate required. The final concentration of the DNA in the precipitate should be 20 μg/mL, and you will need 1 mL for each 9-cm plate and 3 mL for each 175-cm² flask. Remember to make an extra 1 mL of precipitate to ensure recovery of sufficient volume for all of your cultures, as some loss of volume will occur during preparation of the precipitate.
   (b) Dilute the DNA/TBE mix in 12.5 mM HEPES to give 20 μg/mL in the final mix.
   (c) Add 10× CaCl₂, one-tenth of the volume of the final mix.
   (d) Add the solution dropwise, while mixing, to an equal volume of 2× HBP. For example, for nine 9-cm dishes at 1 mL/dish, plus 1 mL to spare (i.e., 10 mL of mix), we have the following:

   i) DNA/TBE...........................0.5 mL
   ii) 12.5 mM HEPES..................3.5 mL
   iii) 10× CaCl₂........................ 1.0 mL
   iv) Add dropwise to 2× HBP.........5.0 mL.

   The final concentrations in the mix are as follows:

DNA	20 μg/mL
CaCl₂	0.125 M
Na₂PO₄	0.75 mM
NaCl	140 mM
HEPES	12.5 mM

   The pH is 7.12

Note. Use plastic pipettes and tubes when preparing DNA calcium phosphate precipitates, as the precipitates stick very firmly to glass. For mixing, we recommend the use of two pipettes in two handheld pipette aids, one for blowing bubbles of sterile air into the mixture and the other for carefully adding the DNA/calcium mix in a dropwise fashion to the phosphate solution; good mixing results in an even precipitate. Use 1- to 5-mL pipettes, depending on the volume of precipitate being made. As the DNA/calcium solution is added to the phosphate solution, a light, even precipitate will begin to form. This precipitate is quite obvious and gives a milky appearance when complete.

   (e) After all of the DNA/calcium solution has been added to the phosphate, replace the lid on the tube, invert the tube gently once or twice, and leave the tube to stand at room temperature for 30 min.
   (f) It is important to make a mock precipitate without DNA. This allows you to assess the effectiveness of your selection conditions. A mock precipitate can be made exactly as described for the regular precipitate, but the DNA for the mock precipitate is replaced with additional HEPES buffer.

4. Add 1 mL of the DNA or mock precipitate to each plate or 3 mL to each flask. Make sure that the volume of precipitate is no more than one-tenth of the total volume of the culture medium already on the cells.

5. Leave the precipitate on the cells for a minimum of 6 h, but not more than overnight (∼16 h). For fibroblasts from some individuals, exposure to calcium and phosphate for more than 6 h may be toxic.

6. Remove the calcium phosphate precipitate by aspiration. There is no need to wash the cells further, or, in our experience, to further treat the cells with either DMSO or glycerol.

7. Add medium to the cultures, and incubate them until 48 h from the start of the experiment;

then add selective agents to the cultures. The agent that you add will depend on the vector used for transfection. Vectors carrying the *neo* gene confer resistance to G418 (Geneticin), and vectors carrying the *hyg b* gene confer resistance to hygromycin B (Roche Applied Science). All fibroblasts, in our experience, require 100–200 μg/mL of G418 or 10–20 μg/mL of hygromycin B to kill the cells gradually over a period of a week.

8. Change the medium on the transfected plates:
    (a) Dispense the total volume of medium required into a suitably sized sterile bottle, and add G418 or hygromycin B from the concentrated stocks to give the correct final concentration.
    (b) Gently swirl the solution to mix it.
    (c) Aspirate the medium from the plates or flasks, and replace it with the selective medium.
    (d) Return the plates or flasks to the incubator.

9. Monitor the effects of the selective medium on a daily basis by examining the culture under the microscope. When a significant number of cells have lifted and died, replace the medium with fresh selective medium. Selection should be maintained at all times.

10. Continue to replace the medium until the background of cells has lifted and died. The mock-transfected plates that have not been transfected with DNA should have no viable cells remaining after 7–10 days. If there are cells remaining, then the selection has not worked adequately, and it may be necessary to raise the concentration of the selective agent.

11. Once the background of cells has died, it is no longer necessary to routinely change the medium on the cells. The cells should then be left undisturbed in the incubator for four to six weeks to allow the transfected cells to grow and form colonies.

12. Once colonies arise, pick out individual colonies by using cloning rings (*see* Protocol 14.6) or bulk the colonies together by trypsinizing the whole dish or flask.

13. Freeze aliquots of cells at the earliest opportunity and regularly thereafter. Once transfectants have been expanded into cultures and ample stocks frozen in liquid nitrogen, it is necessary to keep the culture going for an extended period of time until it reaches crisis. To minimize the risk from fungal contaminants, add amphotericin B (Fungizone, Invitrogen) at 2.5 μg/mL.

14. Subculture the cultures routinely until they reach the end of their *in vitro* life span. As the cells approach crisis, the growth rate often slows. As the cultures begin to degenerate and cell division ceases, it is no longer necessary to subculture the cells. However, if heavy cell debris begins to cling to the remaining viable cells, it may be advisable to trypsinize the cells to remove the debris. Either return all of the cells to the same vessel, or use a smaller vessel to compensate for the cell death that is occurring. In general, the cells grow and survive better if they are not too sparse. With patience, care, and the culture of sufficient cells from your freezer stocks, you should, in most cases, obtain a postcrisis line.

15. When healthy cells begin to emerge, allow the colonies to grow to a reasonable size before subculturing, and then begin to subculture the colonies again. Do not be tempted to put too few cells into a large flask.

16. Freeze an ampoule of cells at the earliest opportunity, and continue to build up a freezer stock before using the culture.

17. To check that your postcrisis line is truly immortal, we recommend selecting and expanding individual clones from the culture. This procedure has the additional benefit of providing a homogeneous culture derived from a single cell.

Posttransfection Care of Cultures

(1) The level of selective agent is chosen to produce a gentle kill over a period of about a week. The majority of cells on the DNA-treated plates should die within seven days, and those cells that remain should be the successful transfectants.

(2) It is advisable to freeze ampoules of cells routinely both to build up a stock of transfected cells and to save time if cultures are lost because of contamination.

(3) In most cases, crisis is a marked event, with the majority of cells showing signs of deterioration and a net loss of viable cells. On average, crisis lasts from 3 to 6 months, and the culture may deteriorate to the point where very few, if any, obviously healthy cells are present.

(4) Once a culture has entered crisis, you can then reliably predict the timing of crisis for parallel cultures stored in liquid nitrogen. This prediction allows one to retrieve ampoules from parallel cultures from the freezer and to build up a number of flasks sitting at the threshold of crisis. As these parallel cultures enter crisis and begin to lose viability, these flasks can be pooled. In some cases, there will be an adequate cover of cells in the flask but high levels of cellular debris. In these cases, replate the cells. The cells should be trypsinized and centrifuged, and the pellet should be returned to the original flask. The

flask may be rinsed several times with trypsin to remove any adhering cell debris before returning the cells to it.

(5) The first sign of a culture emerging from crisis is usually the appearance of one or more foci of apparently healthy, robust cells with the typical appearance of SV40-transformed cells. On subculturing, these foci expand to give a healthy, regenerating culture. In many cases, the early-emerging postcrisis cells grow poorly, but the growth properties improve with further subculturing, and subcloning helps to select individual clones with better growth properties. It is important to freeze an ampoule of your new cell line as early as possible. Our working definition for an immortal line is that the culture has undergone a minimum of 100 population doublings posttransfection and has survived subsequent subcloning.

(6) As the appearance of an immortal derivative within these cultures is a relatively rare event, it is essential that any postcrisis cell lines that emerge are checked to ensure that they were derived from the original starting material and are not the result of cross-contamination from other immortal cell lines in the laboratory (*see* Sections 7.10.1, 16.1, 16.3, 16.6.2, 19.5; Table 13.2; and Protocols 16.8–16.10).

18.4.4 Telomerase-Induced Immortalization

Telomeres play an essential role in chromosome stability and determining cellular life span. Telomerase or terminal transferase is composed of two main subunits, RNA component (hTR) and a protein catalytic subunit (hTERT). The RNA subunit is ubiquitously expressed in both normal and malignant tissues, whereas hTERT is only expressed in cells and tissues such as tumors, germ line cells and activated lymphocytes. The primary cause of senescence appears to be telomeric shortening, followed by telomeric fusion and the formation of dicentric chromosomes and subsequent apoptosis. Transfecting cells with the telomerase gene *htrt* extends the life span of the cell line (*see* Fig. 18.4), and a proportion of these cells become immortal but not malignantly transformed [Bodnar et al., 1998, Simonsen et al., 2002]. As a high proportion of the $htrt^+$ clones become immortal, this appears to be a promising technique for immortalization. Although the functionality of some of these lines has yet to be demonstrated, there are some encouraging reports of uncompromised differentiation, e.g., in keratinocytes [Dickson et al., 2000] and in myocytes [Wootton et al., 2003].

The preceding introduction and Protocol 18.2 were contributed by Nedime Serakinci, Department of Human Genetics, Bartholin Bygningen, Universitetsparken, 8000 Aarhus C, Denmark. This protocol has been used successfully for mesenchymal and neuronal human stem cell lines and primary cultures.

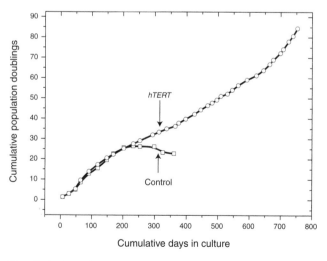

Fig. 18.4. *Cumulative Population Doublings(PD) of hTERT-Immortalized Cells Compared to Senescing Control Cells.* Cumulative PD were calculated for cultures of human mesenchymal stem cells (*see* Protocol 18.2) and plotted against time in culture for cells with ectopic expression of hTERT after retroviral transduction (circles) and control (nontransduced) cells (squares). (Data courtesy of N. Serakinci.)

PROTOCOL 18.2. IMMORTALIZATION OF HUMAN MESENCHYMAL STEM CELLS BY TELOMERASE

Materials
General (sterile):
- ❑ Growth medium: Dulbecco's modified Eagle's medium (DMEM) with high glucose, 4.5 g/L, and L-glutamine, 2 mM, supplemented with 10% fetal bovine serum, 100 U/mL of penicillin, and 100 μg/mL streptomycin
- ❑ Polybrene, 8 mg/mL
- ❑ Universal containers
- ❑ Culture flasks, 25 cm², 75 cm²

Production of retroviral vector
- ❑ Cell lines:
    PG13 [Miller et al., 1991]
    GP+E-86 [Markovitz, 1988]
- ❑ Retroviral vector GCsamhTERT
- ❑ *htrt* DNA for transfection
- ❑ Tx buffer:
    HEPES, 0.5 M, pH 7,1.....................200 μL
    NaCl, 5 M.............................. 100 μL
    Na₂HPO₄/NaH₂PO₄, 1 M................. 3 μL
    UPW...................................1.7 mL
    Total 20 mL
- ❑ Buffer A:
    NaCl (5M).............................. 600 μL
    EDTA (0.5M).............................40 μL
    Tris·HCl pH 7.5 (0.5M)...................400 μL

UPW.. 19 mL

Total 20.04 mL

❑ CaCl, 2.5 M

Transduction of hMSC cells
❑ hMSC cells
❑ Flasks, 75 cm^2
❑ Multiwell plates, 6-well

Posttransduction
❑ Materials for cryopreservation (*see* Protocol 20.1)
❑ Materials for DNA fingerprinting (*see* Protocol 16.8) or DNA profiling (*see* Protocol 16.9) or other authentication procedures (*see* Protocol 16.10 for example)

PCR Reagents
❑ DNA, 100 μg/mL
❑ 10× PCR Buffer (+Mg)L
❑ Sense primer (2 μM)
❑ Antisense primer (2 μM)
❑ dNTP mix (10 mM)
❑ DNA polymerase
❑ UPW

Protocol
Production of retroviral vector
The retroviral vector with the hTERT gene is packaged into the gibbon ape leukemia virus (GALV) packaging cell line PG13 [Miller et al., 1991] by a two-step procedure. First, use 20 μg/mL *htrt* DNA (Geron) to transfect packaging cell line GP+E-86 [Markovitz et al., 1988] and then use the supernate to infect PG13 cells.

1. Seed 6.7×10^5 GP+E-86 cells in a small culture flask (25 cm^2).
2. Transfection of GP+E-86 cells:
   (a) Dispense 280 μL of the Tx-buffer per tube (universal containers).
   (b) Prepare DNA tubes:

Construct Name	DNA conc.	15 μg DNA	buffer A	CaCl, 2.5 M	Total
	Z* μg/μL	X* μL	280-30-X	30	280

*Where X × Z = 15 μg

   (c) Add the Tx buffer drop by drop to the tubes containing the DNA solution. Mix gently.
   (d) Incubate at room temperature for 30 min to allow precipitate formation.
   (e) Add fresh medium on to the GP+E-86 cells, 5 mL per 25-cm^2 flask.
   (f) Gently add the mix solution (Tx+DNA) onto the GP+E-86 cells and incubate 4 to 6 h at 37°C.
   (g) Wash the cells carefully 3 times with D-PBSA in the 25-cm^2 flask.
   (h) Add fresh medium to the cells and incubate at 37°C overnight.
3. Change medium on the cells and add 2 mL fresh medium (instead of 5 ml) for virus production.
4. On the same day as Step 3, seed 1×10^4 PG13 cells in each well of a 6-well plate.
5. On the following day, harvest the supernate from the transfected GP+E-86 cells and add Polybrene to a final concentration of 8 μg/mL (i.e., 1 μL/mL supernate).
6. Pass the supernate through a 0.45-μm filter and add 2 mL filtrate to each well containing PG13 cells.
7. Centrifuge the plates at 32°C at 1000 *g*
8. Incubate at 37°C overnight.
9. Next day, change medium on cells.

Transduction of hMSC cells:
1. Seed 2×10^6 transduced PG13 cells in a 75-cm^2 culture flask.
2. Next day, add 6 mL fresh medium to the PG13 cells.
3. Following day, plate the hMSC cells for transduction in 6-well plates at a concentration of approximately 2.5×10^4–7.5×10^4 cells/mL.
4. Harvest the supernate from the packaging cells, add Polybrene to a final concentration of 8 μg/mL and pass the supernate through a 0.45-μm filter.
5. Add 2 mL of filtered retroviral supernate to each well of the 6-well plates.
6. Centrifuge plates at 32°C at 1000 *g* and incubate at 37°C overnight.
7. Remove the retroviral supernatant and add fresh medium to the cells.

Post-transduction Care of Cultures:
1. The cells that have not received hTERT gene should start to die at about passage 10–12; the remaining cells should be the cells that incorporated the hTERT gene and will be selected with subsequent passage.
2. It is advisable to freeze ampoules of cells routinely to build up a stock of transduced cells at different population doubling levels that match the PD growth curve, and to save time if cultures are lost because of contamination.
3. Use the following formula to derive the PD at each subculture and to establish a PD growth curve:

$$PD = \frac{\ln(N_{finish}/N_{start})}{\ln 2}$$

Where PD is the number of population doublings, ln is the natural logarithm, N_{start} is the number of cells initially seeded, and N_{finish} is the total number of cells recovered at subculture.

Example (for 3.2×10^6 cells recovered from a seeding of 2×10^5):

$$PD = \frac{\ln(3.2 \times 10^6/2 \times 10^5)}{\ln 2}$$

$$PD = \frac{\ln(16)}{\ln 2}$$

$$PD = \frac{2.7726}{0.6931} = 4$$

If the cells are split 1:4 at each subculture then N_{finish} is multiplied by 4 for each time the cells have been subcultured. So, in the above example with 10 subcultures, N_{finish} is $(3.2 \times 10^6) \times 4$, ten times, or 3.4×10^{12}.

Example (for 3.2×10^6 cells recovered from an initial seeding of 2×10^5, after 10 subcultures at 1:4 split):

$$PD = \frac{\ln(3.4 \times 10^{12}/2 \times 10^5)}{\ln 2}$$

$$PD = \frac{\ln(1.67 \times 10^7)}{\ln 2}$$

$$PD = \frac{16.6309}{0.6931} = 23$$

4. It is essential to check the authenticity of the immortalized cell line for a period after it is established to ensure that it was derived from the original starting material and is not the result of cross-contamination from other immortal cell lines in the laboratory (*see* Section 19.5). This can be done, for example, by PCR against the transgene, hTERT (*see* below), DNA fingerprinting (*see* Protocol 16.8), or DNA profiling (*see* Protocol 16.9).

PCR for hTERT:
1. Thaw the PCR reagents and keep on ice
2. Make a master mix and remember to include positive (hTERT positive other sample) and negative control (no DNA template).

    DNA, 100 µg/mL.................. 1 µL (100 ng)
    10× PCR buffer (+Mg)............ 2 µL
    Sense primer (2 µM)............... 2 µL
    Antisense primer (2 µM)......... 2 µL
    dNTP mix (10 mM)............. 0.4 µL
    DNA polymerase............... 0.1 µL
    UPW.......................... 12.5 µL
        Final volume 20 mL (including DNA)
3. Place the tubes in PCR machine and run program: initial denaturation at 94°C for 3 min, denaturation

at 94°C for 30 s, annealing at 59°C for 30 s, extension at 74°C for 1 min for 30 cycles, then 74°C for 10 min.
4. Run PCR products on 1.5–2% agarose gel.

18.4.5 Transgenic Mouse

The transgenic mouse Immortomouse (H-$2K^b$-$tsA58$ SV40 large T) carries the temperature-sensitive SV40LT gene. A number of tissues from this mouse give rise to immortal cell lines, including colonic epithelium [Fenton & Hord, 2004], brain astroglia [Noble & Barnett, 1996], muscle [Ahmed et al., 2004], and retinal endothelium [Su et al., 2003].

18.5 ABERRANT GROWTH CONTROL

Cells cultured from tumors, as well as cultures that have transformed *in vitro*, show aberrations in growth control, such as growth to higher saturation densities [Dulbecco & Elkington, 1973], clonogenicity in agar [Freedman & Shin, 1974], and growth on confluent monolayers of homologous cells [Aaronson et al., 1970]. These cell lines exhibit lower serum or growth factor dependence, usually form clones with a higher efficiency, and are assumed to have acquired some degree of autonomous growth control by overexpression of oncogenes or by deletion of suppressor genes. Growth control is often autocrine—i.e., the cells secrete mitogens for which they possess receptors, or the cells express receptors or stages in signal transduction that are permanently active and unregulated. Although immortalization does not necessarily imply a loss of growth control, many cells progress readily from immortalization to aberrant growth, perhaps because of genetic instability intrinsic to the immortalized genotype.

18.5.1 Anchorage Independence

Many of the properties associated with neoplastic transformation *in vitro* are the result of cell surface modifications [Hynes, 1974; Nicolson, 1976; Bruyneel et al., 1990], e.g., changes in the binding of plant lectins [Laferte & Loh, 1992], in cell surface glycoproteins [Bruyneel et al., 1990; Carraway et al., 1992], and in cell adhesion molecules [Yang et al., 2004], many of which may be correlated with the development of invasion and metastasis *in vivo*. Fibronectin, or large extracellular transformation-sensitive (LETS) protein, is lost from the surface of transformed fibroblasts [Hynes, 1973; Vaheri et al., 1976] due to alterations in integrins. This loss may contribute to a decrease in cell-cell and cell-substrate adhesion [Yamada et al., 1991; Reeves, 1992] and to a decreased requirement for attachment and spreading for the cells to proliferate.

Transformed cells may lack specific CAMs (e.g., L-CAM), which, when transfected back into the cell, regenerate the normal, noninvasive phenotype [Mege et al., 1989], and, as such, they may be recognized as tumor suppressor genes.

Other CAMs may be overexpressed, such as N-CAM in small-cell lung cancer [Patel et al., 1989], when the extracellular domain is subject to alternative splicing [Rygaard et al., 1992]. The expression of and degree of phosphorylation of integrins may also change [Watt, 1991], potentially altering cytoskeletal interactions, the regulation of gene transcription, the substrate adhesion of the cells, and the relationship between cell spreading and cell proliferation [Fata et al., 2004].

In addition, the loss of cell–cell recognition, a product of reduced cell–cell adhesion, leads to a disorganized growth pattern and the loss of contact inhibition of cell motility and density limitation of cell proliferation (*see* Section 18.5.1). Cells can grow detached from the substrate, either in stirred suspension culture or suspended in semisolid media, such as agar or Methocel. There is an obvious analogy between altered cell adhesion in culture and detachment from the tissue in which a tumor arises and the subsequent formation of metastases in foreign sites, but the rational basis for this analogy is not clear; new adhesions are clearly involved.

Suspension cloning. Macpherson and Montagnier [1964] were able to demonstrate that polyoma-transformed BHK21 cells could be grown preferentially in soft agar, whereas untransformed cells cloned very poorly. Subsequently, it has been shown that colony formation in suspension is frequently enhanced after viral transformation. The situation regarding spontaneous tumors is less clear, however, despite the fact that Shin and coworkers demonstrated a close correlation between tumorigenicity and suspension cloning in Methocel [Freedman and Shin, 1974; Kahn & Shin, 1979]. Although Hamburger and Salmon [1977] showed that many human tumors contain a small percentage of cells (<1.0%) that are clonogenic in agar, a number of normal cells will also clone in suspension [Laug et al., 1980; Peehl & Stanbridge 1981; Freshney & Hart, 1982] (*see* Fig. 14.11). Because normal fibroblasts are among these cells, the value of this technique for assaying for the presence of tumor cells in short-term cultures from human tumors is in some doubt. However, it remains a valuable technique for assaying neoplastic transformation *in vitro* by tumor viruses and was used extensively by Styles [1977] to assay for carcinogenesis.

Techniques for cloning in suspension are described in Chapter 14 (*see* Protocols 14.4 and 14.5). Variations with particular relevance to the assay of neoplastic cells lie in the choice of the suspending medium. It has been suggested [Neugut & Weinstein, 1979] that agar may allow only the most highly transformed cells to clone, whereas agarose (which lacks sulfated polysaccharides) is less selective. Montagnier [1968] was able to show that untransformed BHK21 cells, which would grow in agarose but not in agar, could be prevented from growing in agarose by the addition of dextran sulfate.

18.5.2 Contact Inhibition

The loss of contact inhibition may be detected morphologically by the formation of a disoriented monolayer

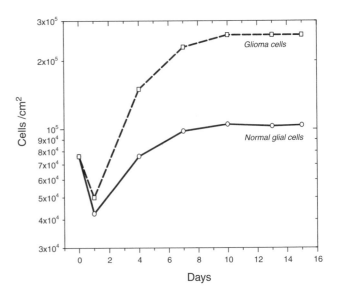

Fig. 18.5. Density Limitation of Cell Proliferation. The difference in plateaus (saturation densities) attained by cultures from normal brain (circles, solid line) and a glioma (squares, broken line). Cells were seeded onto 13-mm coverslips, and 48 h later, the coverslips were transferred to 9-cm Petri dishes containing 20 mL of growth medium, to minimize exhaustion of the medium.

of cells (*see* Plate 14a,b) or rounded cells in foci within the regular pattern of normal surrounding cells (*see* Fig. 18.3 and Plate 14c). Cultures of human glioma show a disorganized growth pattern and exhibit reduced density limitation of growth by growing to a higher saturation density than that of normal glial cell lines (Fig. 18.5) [Freshney et al., 1980a, b]. As variations in cell size influence the saturation density, the increase in the labeling index with [^{3}H]thymidine at saturation density (*see* Protocol 21.11) is a better measure of the reduced density limitation of growth. Human glioma, labeled for 24 h at saturation density with [^{3}H]thymidine, gave a labeling index of 8%, whereas normal glial cells gave 2% [Guner et al., 1977].

PROTOCOL 18.3. DENSITY LIMITATION OF CELL PROLIFERATION

Outline
Grow the culture to saturation density in nonlimiting medium conditions, and determine, autographically, the percentage of cells labeling with [^{3}H]thymidine.

Materials
Sterile or Aseptically Prepared:
❑ Culture of cells ready for subculture
❑ Growth medium

❑ Maintenance medium (no serum or growth factors) containing 37 KBq/mL (1.0 µCi/mL) of [³H]-thymidine, 74 GBq/mmol (2 Ci/mmol)
❑ D-PBSA
❑ Trypsin, 0.25%
❑ Multiwell plates, 24-well, containing coverslips 13 mm in diameter
❑ Petri dishes, 9 cm (bacteriological grade, 1 per coverslip)

Protocol

1. Trypsinize the cells and seed 1×10^5 cells/mL into a 24-well plate, 1 mL/well, each well containing a 13-mm-diameter coverslip.
2. Incubate the cells in a humidified CO_2 incubator for 1–3 days.
3. Transfer the coverslips to 9-cm bacteriological grade Petri dishes, each containing 20 mL of medium, and place the dishes in the CO_2 incubator.
4. Continue culturing, changing the medium every 2 days once the cells become confluent on the coverslips.
5. Trypsinize and count the cells from two coverslips every 3–4 days. As the cells become denser on the coverslip, it may be necessary to add 200–500 units/mL of crude collagenase to the trypsin in order to achieve complete dissociation of the cells for counting.
6. When cell growth ceases, i.e., two sequential counts show no significant increase, add 2.0 mL 37 KBq/mL (1.0 µCi/mL) of [³H]thymidine, 74 GBq/mmol (2 Ci/mmol), and incubate the cells for a further 24 h.

Δ **Safety Note.** Handle [³H]thymidine with care. Although it is a low-energy β-emitter, it localizes to DNA and can induce radiolytic damage. Wear gloves, do not handle [³H]thymidine in a horizontal laminar-flow hood but in a biohazard or cytotoxic drug handling hood (*see* Section 7.5.4), and discard waste liquids and solids by the appropriate route specified in the local rules governing the handling of radioisotopes.

7. Transfer the coverslips back to a 24-well plate, and trypsinize the cells for autoradiography (*see* Section 27.2). The cells may be fixed in suspension and dropped on a slide as for chromosome preparations (*see* Protocol 16.7 without the hypotonic treatment), centrifuged onto a slide with a cytocentrifuge (*see* Protocol 16.4), or trapped on filters by vacuum filtration (*see* Protocol 16.5).

Note. It is necessary to trypsinize high-density cultures for autoradiography because of their thickness and the weak penetration of β-emission from ³H-labeled cells in the underlying layers will not be detected by the radiosensitive emulsion, due to the absorption of the β-particles by the overlying cells (the mean path length of β-particles in water is approximately 1 µm). If the cells remain as a monolayer at saturation density, this step may be omitted, and autoradiographs may be prepared by mounting the coverslips, cells uppermost, on a microscope slide.

Analysis. Count the number of labeled cells as a percentage of the total number of cells. Scan the autoradiographs under the microscope, and count the total number of cells and the proportion of cells labeled in representative parts of the slide (*see* Fig. 21.14).

Variations. Cells in DNA synthesis may also be labeled with bromodeoxyuridine (BUdR) and subsequently detected by antibody to BUdR-labeled DNA (Dako). Human cycling cells can also be labeled with the Ki67 monoclonal antibody (Dako) against DNA polymerase, or with anti-PCNA against proliferating cell nuclear antigen (PCNA). Although there is generally good agreement between [³H]thymidine and BUdR labeling, Ki67 and PCNA will label more cells, as the antigen is present throughout the cycle and is not restricted to the S-phase.

Growth of cells at high density in non-limiting medium can also be achieved by growing the cells in a filter well (B-D Biosciences, Corning Millipore), choosing a filter diameter substantially below that of the dish (e.g., the Corning Costar 8-mm filter in a 24-well plate), and counting the number of cells in plateau, performing an autoradiograph or immunostaining with Ki67 or anti-PCNA.

18.5.3 Serum Dependence

Transformed cells have a lower serum dependence than their normal counterparts [Temin, 1966, Eagle et al., 1970], due, in part, to the secretion of growth factors by tumor cells [Todaro & DeLarco, 1978]. These factors have been collectively described as autocrine growth factors. Implicit in this definition is that (1) the cell produces the factor; (2) the cell has receptors for the factor; and (3) the cell responds to the factor by entering mitosis. Some of these factors may have an apparent transforming activity on normal cells (e.g., TGF-α) binding to the EGF receptor and inducing mitosis [Richmond et al., 1985], although, unlike true transformation (*see* Section 18.2), this type of transformation is reversible. These factors also cause nontransformed cells to adopt a transformed phenotype and grow in suspension [Todaro & DeLarco, 1978]. This effect can be assayed by treating NRK cells with conditioned medium from the test cell and cloning them in suspension (*see* Protocols 14.2, 14.4, and 14.5).

Tumor cells can also produce many hemopoietic growth factors, such as interleukins 1, 2, and 3, along with colony-stimulating factor (CSF) [Fontana et al., 1984; Metcalf, 1990].

It has been proposed [Cuttitta et al., 1985] that some factors, such as gastrin-releasing peptide and vasoactive intestinal peptide (VIP), and human chorionic gonadotropin (hCG), hitherto believed to be ectopic hormones produced by lung carcinomas, may in fact be autocrine growth factors. Autocrine growth factors can be detected by immunostaining (*see* Protocol 16.11), but their value as transformation markers is limited, because many normal cells, e.g., glia, fibroblasts, and endothelial cells, produce autocrine factors when proliferating.

18.5.4 Oncogenes

Autonomous growth control is also achieved in transformed cells by oncogenes, expressed as modified receptors, such as the *erb*-B2 oncogene product, and the modified G protein, such as mutant *ras*, or by the overexpression of genes regulating stages in signal transduction (e.g., *src* kinase) or transcriptional control (e.g., *myc, fos,* and *jun*) [Bishop, 1991]. In many cases, the gene product is permanently active and is unable to be regulated. Overexpression of oncogenes can be detected by immunostaining (*see* Protocol 16.11), *in situ* hybridization (*see* Section 27.8), immunoblotting for the protein product, RT-PCR for mRNA, or microarray analysis (*see* Section 16.6.1 and Plate 24). In some cases, the oncogene product (e.g., *erb*-B2, activated Ha-*ras*) can be distinguished from the normal product (e.g., EGF receptor, normal *ras*, respectively) qualitatively as well as quantitatively, by specific antibodies.

18.6 TUMORIGENICITY

Transformation is a multistep process that often culminates in the production of neoplastic cells [Quintanilla et al., 1986]. However, cell lines derived from malignant tumors, presumably already transformed, can undergo further transformation with an increased growth rate, reduced anchorage dependence, more pronounced aneuploidy, and immortalization. This suggests that a series of steps, not necessarily coordinated or interdependent and not necessarily individually tumorigenic, is required for malignant transformation. Furthermore, all cell lineages present within a tumor need not have the same transformed properties, and the same set of properties need not be expressed in every tumor. Progression may imply the expression of new properties or the deletion of old ones that may induce metastasis or even spontaneous remission.

There are therefore several steps in transformation, the sequence of which may be determined by environmental selective pressure. *In vitro,* where little restriction on growth is imposed, the events need not necessarily follow in the same sequence as *in vivo.*

18.6.1 Malignancy

Malignancy implies that the cells have developed the capacity to generate invasive tumors if implanted *in vivo* into an isologous host or if transplanted as a xenograft into an immune-deprived animal. Although the development of malignancy can be recognized as a discrete phenotypic event, it often accompanies the development of aberrant growth control, suggesting that some of the lesions responsible for aberrant growth control also cause malignancy. An obvious candidate for such lesions is a deficit in cell–cell interaction that deprives the cell of control of proliferation (density limitation of cell proliferation) and of motility control (contact inhibition).

Two approaches have been used to explore malignancy-associated properties: (1) Cells have been cultured from malignant tumors and characterized; and (2) transformation *in vitro* with a virus or a chemical carcinogen, or transfection with oncogenes, has produced cells that were tumorigenic and that could be compared with the untransformed cells. The second approach provides transformed clones of the same lineage, which can be shown to be malignant, and these clones can be compared with untransformed clones, which are not malignant. Unfortunately, many of the characteristics of cells transformed *in vitro* have not been found in cells derived from spontaneous tumors. Ideally, tumor cells and equivalent normal cells should be isolated and characterized. Unfortunately, there have been relatively few instances for which this arrangement has been possible, and even then, although the cells may belong to the same lineage, their position in that lineage is not always clear, and thus comparison is not strictly justified. Although the bulk of cells in normal adult tissue will be differentiated and nonproliferative, those comprising a tumor will tend to be proliferating and undifferentiated, and this status will distinguish the tumor population from the normal, regardless of their malignancy. Furthermore, it would appear that many, if not most, cells in a tumor do not have a prolonged life span in culture, and the population crucial to the advancement of the tumor, and consequently, the main target for chemotherapy, may be quite small and equivalent to a stem cell population in normal tissue. As yet there is little conclusive evidence that these cells can be cultured, but an increasing amount of effort is being directed that way [Petersen et al., 2003; Takahashi et al., 2003; Aarti et al., 2004].

18.6.2 Tumor Transplantation

The only generally accepted sign of malignancy is the demonstration of the formation of invasive or metastasizing tumors *in vivo.* Transplantable tumor cells ($\sim 1 \times 10^6$) injected into isogeneic hosts will produce invasive tumors in a high proportion of cases, whereas 1×10^6 normal cells of similar origin will not. Models have been developed, using immune-suppressed or immune-deficient host animals, to study the tumorigenicity of human tumors. The genetically athymic "nude" mouse [Giovanella et al., 1974] and thymectomized irradiated mice [Bradley et al., 1978a; Selby et al., 1980] have both been used extensively as hosts for xenografts. The take rate of the grafts varies, however, and many clearly defined tumor cell lines and tumor biopsies fail to produce tumors as xenografts; those that do take frequently fail to metastasize, although they may be invasive locally. Take rates can be improved by sublethal irradiation of the host nude mouse (30–60 Gy), by using asplenic athymic (*scid*) mice, or by implanting the cells in Matrigel [Pretlow et al., 1991]. Despite the frequency of false negatives, tumorigenesis remains a good indicator of malignancy.

18.6.3 Invasiveness

Tumorigenesis assays should always be accompanied by histology of the tumor to confirm its histopathological similarity to the original tumor and to demonstrate that it is invasive. However, if the cells are not tumorigenic, or if transplantation facilities are not available or are not considered desirable, then it is possible to utilize a number of *in vitro* assays. Some of these assays also provide models that are more readily quantified than *in vivo* assays.

Chick chorioallantoic membrane. The chorioallantoic membrane (CAM) assay can be performed on chick embryos *in ovo* or on explanted CAM *in vitro*. Easty and Easty [1974] showed that invasion of the CAM could be demonstrated in organ culture, and others [Hart & Fidler, 1978] attempted, with some limited success, to construct a chamber capable of quantifying the penetration of tumor cells across the CAM. An advantage of the CAM assay *in ovo* is that it may also show angiogenesis (*see* Section 18.6.4 and Plate 15f,g), and subsequent histology may reveal whether the tumor cells have penetrated the underlying basement membrane.

Organoid confrontation. Mareel et al. [1979] developed an *in vitro* model for invasion, using chick embryo heart fragments cocultured with reaggregated clusters of tumor cells (Fig. 18.6). Invasion appears to be correlated with the

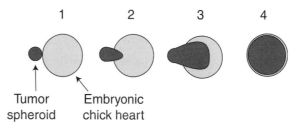

Fig. 18.6. Chick Heart Assay. Tumor spheroids (*see* Protocol 25.2) are cocultured with healed fragments of 8-day embryonic chick heart. (1) The spheroid adheres to the heart after a few hours; (2) after 24–48 h, it starts to penetrate the chick heart. (3) It spreads within the heart fragment; (4) by 8–10 days it has completely replaced the heart tissue. [After Mareel et al., 1979].

malignant origin of the cells, is progressive, and causes destruction of the host tissue. The application of this technique to human tumor cells shows a good correlation between malignancy and invasiveness in the assay [de Ridder & Calliauw, 1990]. This technique has been used extensively, but is difficult to quantify and requires skilled histological interpretation.

Filter wells. A number of filter well techniques have been developed, based on the penetration of filters coated with Matrigel or some other extracellular matrix constituent (Fig. 18.7) [Repesh, 1989; Schlechte et al., 1990; Brunton et al., 1997; Lamb et al., 1997]. The degree of penetration into the gel, or through to the distal side of the filter, is rated as invasiveness and is determined histologically by the number of cells and the distance moved, or by prelabeling the cells with ^{125}I and counting the radioactivity on the distal side of the filter or bottom of the dish. It is more readily quantified, but lacks the presence of normal host cells in the barrier, normally associated with invasion *in vivo*. The penetration

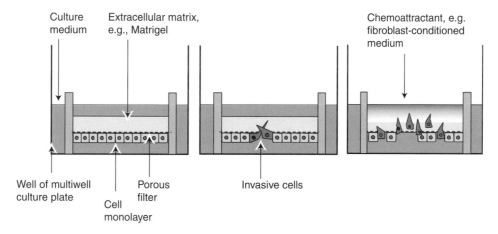

Fig. 18.7. Filter Well Invasion. Cells plated on the underside of the filter migrate through the filter into growth factor-depleted Matrigel in the well of the filter insert, encouraged by the addition of a chemoattractant, such as fibroblast-conditioned medium, to the upper side of the Matrigel [After Brunton et al., 1997].

into Matrigel is likely to be a measure of matrix degradation and reflects the production of proteases or glycosidases by the cells. It is not clear how cells that do not make their own degradative enzymes, but instead rely on the production of proteases induced in the stroma, will perform in these assays.

18.6.4 Angiogenesis

Tumor cells release factors, including VEGF [Joukov et al., 1997], FGF-2 [Thomas et al., 1997], and angiogenin [Hu et al., 1997], that are capable of inducing neovascularization [Folkman, 1992; Skobe & Fusenig, 1998]. Fragments of tumor, pellets of cultured cells, or cell extracts, implanted on the surface of the CAM of a hen's egg, promote an increase in vascularization that is apparent to the naked eye 6–8 days later (Plate 9). Because this assay is not readily quantified, the stimulation of cell migration [Bagley et al., 2003], production of vascular endothelial growth factor (VEGF) [Buchler et al., 2004], or morphogenesis [Chen et al., 1997; Ment et al., 1997; Jain et al., 1997] in filter well cultures of vascular endothelium may provide the basis for more quantitative assays.

18.6.5 Plasminogen Activator

Other products that tend to be increased in transformed cells are proteolytic enzymes [Mahdavi & Hynes, 1979], long since associated with theories of invasive growth [Liotta, 1987]. Because proteolytic activity may be associated with the cell surface of many normal cells and is absent on some tumor cells, an equivalent normal cell must be used as a control when using this criterion. Plasminogen activator (PA) is higher in some cultures from human glioma than in cultures from normal brain [Hince & Roscoe, 1980] (Fig. 18.8), and other cultures have previously shown that PA is associated with many different tumors [de Vries et al., 1995; del Vecchio et al., 1993; Schwartz Albiez et al., 1991]. PA may be measured by clarification of a fibrin clot or by release

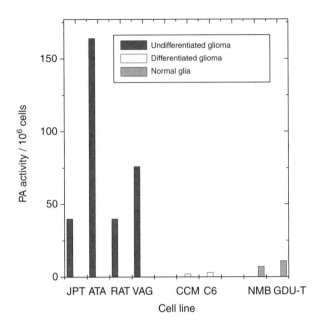

Fig. 18.8. Plasminogen Activator. PA produced by tumor cells *in vitro* (The units are arbitrary). The PA activities of the four gliomas—JPT, ATA, RAT, and VAG—were all higher than cells cultured from normal brain (NMB-C, GDU-T). It was also found that the only cells to produce the differentiated glial marker glial fibrillary acidic protein—CCM and C6—had the lowest PA of all.

of free soluble ^{125}I from [^{125}I]fibrin [Unkless et al., 1974; Strickland & Beers, 1976]. In addition, a simple chromogenic assay has been developed by Whur et al. [1980].

It has been proposed that, for some carcinomas, soluble urokinase-like PA (uPA) is elevated more than tissue-type PA (tPA) [Markus et al., 1980; Duffy et al., 1990], so it is informative to couple the chromogenic assay with zymogram analysis [Davies et al., 1993; Boxman et al., 1995] and then immunoblot to determine the proportion of each type of PA.

CHAPTER 19

Contamination

19.1 SOURCES OF CONTAMINATION

Maintaining asepsis is still one of the most difficult challenges to the newcomer to tissue culture. Awkwardness during early training can be overcome by experience, but, in certain situations, even the most experienced worker will suffer from contamination. There are several potential routes to contamination (Table 19.1) including failure in the sterilization procedures for solutions, glassware and pipettes, turbulence and particulates (dust and spores) in the air in the room, poorly maintained incubators and refrigerators, faulty laminar-flow hoods, the importation of contaminated cell lines or biopsies, and lapses in sterile technique. The last of these is probably the most significant.

19.1.1 Operator Technique

If reagents are sterile and equipment is in proper working order, contamination depends on the interaction of the operator's technique with environmental conditions. If the skill and level of care of the operator is high and the atmosphere is clean, free of dust, and still, contamination as a result of manipulation will be rare. If the environment deteriorates (e.g., as a result of construction work or a seasonal increase in humidity), or if the operator's technique declines (through the omission of one or more apparently unnecessary precautions), the probability of infection increases. If both happen simultaneously or sequentially, the results can be catastrophic.

Let us consider this conjunction of events in graphic form (see Fig. 6.1). The maintenance of good technique may be represented by the top graph, with occasional lapses shown as a downward peak, and the quality of the environment may be depicted by the bottom graph, with occasional sporadic increases in risk, such as a contaminated Petri dish opened accidentally in the area or dust generated from equipment maintenance. Provided that the two curves (of good technique and a high-quality environment) are kept well apart, the coincidence of a lapse in technique and an environmental breakdown will be rare. If, however, there is a progressive decline in technique or in the environment, the frequency of contamination will increase, and if both conditions deteriorate, contamination will be regular and widespread.

19.1.2 Environment

It is fairly obvious that the environment in which tissue culture is carried out must be as clean as possible and free from disturbance and through traffic. Conducting tissue culture in the regular laboratory area should be avoided; a laminar-flow hood will not give sufficient protection from the busy environment of the average laboratory. A clean, traffic-free area should be designated, preferably as an isolated room or suite of rooms (see Sections 4.3, 6.2.1). Equipment brought into the sterile area from storage and air currents from doors, refrigerators, centrifuges, and the movement of operators, all increase the risk of contamination. Maintain a strict cleaning program for surfaces and equipment, and wipe down anything that is brought in.

19.1.3 Use and Maintenance of Laminar-Flow Hood

The commonest example of poor technique is improper use of the laminar-flow hood. If it becomes overcrowded with

TABLE 19.1. Routes to Contamination

Route or cause	Prevention
Technique	
Manipulations, pipetting, dispensing, etc.	
Nonsterile surfaces and equipment	Clear work area of items not in immediate use.
Spillage on necks and outside of bottles and on work surface	Swab regularly with 70% alcohol. Do not pour liquids. Dispense or transfer by pipette, autodispenser, or transfer device. If pouring is unavoidable: (1) do so in one smooth movement, (2) discard the bottle that you pour from, and (3) wipe up any spillage.
Touching or holding pipettes too low down, touching necks of bottles, inside screw caps	Hold pipettes above graduations. Do not work over open vessels.
Splash-back from waste beaker	Discard waste into a beaker with a funnel or, preferably, by drawing off the waste into a reservoir by means of a vacuum pump.
Sedimentary dust or particles of skin settling on the culture or bottle; hands or apparatus held over an open dish or bottle	Do not work over (vertical laminar flow and open bench) or behind and over (horizontal laminar flow) an open bottle or dish.
Work surface	
Dust and spillage	Swab the surface with 70% alcohol before during, and, after work. Mop up spillage immediately.
Operator hair, hands, breath, clothing	
Dust from skin, hair, or clothing dropped or blown into the culture	Wash hands thoroughly or wear gloves. Wear a lint-free lab coat with tight cuffs and gloves overlapping them.
Aerosols from talking, coughing, sneezing, etc.	Keep talking to a minimum, and face away from work when you talk. Avoid working with a cold or throat infection, or wear a mask. Tie back long hair or wear a cap. Wear a lab coat different from the one you wear in the general lab area or animal house.
Materials and reagents	
Solutions	
Nonsterile reagents and media	Filter or autoclave solutions before using them.
Dirty storage conditions	Clean up storage areas and disinfect regularly.
Inadequate sterilization procedures	Monitor the performance of the autoclave with a recording thermometer or sterility indicator (*see* Appendix II and Protocol 11.5). Check the integrity of filters with a bubble-point or microbial assay after using them. Test all solutions after sterilization.
Poor commercial supplier	Test solutions; change suppliers.
Glassware and screw caps	
Dust and spores from storage	Shroud caps with foil. Wipe bottles with 70% alcohol before taking them into the hood. Replace stocks from the back of the shelf. Do not store anything unsealed for more than 24 h.
Ineffective sterilization (e.g., an overfilled oven or sealed bottles, preventing the ingress of steam)	Check the temperature of the load throughout the cycle. In the autoclave, keep caps slack on empty bottles. Stack oven and autoclave correctly (*see* Protocol 11.1).
Instruments, pipettes	
Ineffective sterilization	Sterilize items by dry heat before using them. Monitor the performance of the oven.
Contact with a nonsterile surface or some other material	Resterilize instruments. (Use 70% alcohol; burn and cool off the instruments.) Do not grasp any part of an instrument or pipette that will pass into a culture vessel.

TABLE 19.1. Routes to Contamination (*Continued*)

Route or cause	Prevention
Culture flasks and media bottles in use	
Dust and spores from incubator or refrigerator	Use screw caps instead of stoppers. Swab bottles before placing in hood. Box plates and dishes.
Dirty storage or incubation conditions.	Cover caps and necks of bottles with aluminum foil during storage or incubation.
	Wipe flasks and bottles with 70% alcohol before using them.
	Clean out stores and incubators regularly.
Media under the cap and spreading to the outside of the bottle	Discard all bottles that show spillage on the outside of the neck. Do not pour.
Equipment and Facilities	
Room air	
Drafts, eddies, turbulence, dust, aerosols	Clean filtered air.
	Reduce traffic and extraneous activity.
	Wipe the floor and work surfaces regularly.
Laminar-Flow Hoods	
Perforated filter	Check filters regularly for holes and leaks.
Change of filter needed	Check the pressure drop across the filter.
Spillages, particularly in crevices or below a work surface	Clear around and below the work surface regularly. Let alcohol run into crevices.
Dry incubators	
Growth of molds and bacteria on spillages	Wipe up any spillage with 70% alcohol on a swab.
	Clean out incubators regularly.
CO₂, humidified incubators	
Growth of molds and bacteria on walls and shelves in a humid atmosphere	Clean out with detergent followed by 70% alcohol (*see* Protocol 19.1).
Spores, etc., carried on forced-air circulation	Enclose open dishes in plastic boxes with close-fitting lids (but do not seal the lids).
	Swab incubators with 70% alcohol before opening them.
	Put a fungicide or bacteriocide in humidifying water (but check first for toxicity).
Other equipment	
Dust on cylinders, pumps, etc.	Wipe with 70% alcohol before bring in
Mites, insects, other infestations in wooden furniture, or benches, in incubators, and on mice, etc., taken from the animal house	
Entry of mites, etc., into sterile packages	Seal all sterile packs.
	Avoid wooden furniture if possible; use plastic laminate, one-piece, or stainless steel bench tops.
	If wooden furniture is used, seal it with polyurethane varnish or wax polish and wash it regularly with disinfectant.
	Keep animals out of the tissue culture lab.
Importation of Biological Materials	
Tissue samples	
Infected at source or during dissection	Do not bring animals into the tissue culture lab.
	Incorporate antibiotics into the dissection fluid (*see* Section 12.3).
	Dip all potentially infected large-tissue samples in 70% alcohol for 30 s.
Incoming cell lines	
Contaminated at the source or during transit	Handle these cell lines alone, preferably in quarantine, after all other sterile work is finished. Swab down the bench or hood after use with 2% phenolic disinfectant in 70% alcohol, and do not use it until the next morning.
	Check for contamination by growing a culture for two weeks without antibiotics. (Keep a duplicate culture in antibiotics at the first subculture.)
	Check for contamination visually, by phase-contrast microscopy and Hoechst stain for mycoplasma. Using indicator cells allows screening before first subculture.

Note: No one-to-one relationship between prevention and cause is intended throughout this table; preventative measures are interactive and may relate to more than one cause.

bottles and equipment (Figs. 6.2, 6.3), the laminar airflow is disrupted, and the protective boundary layer between operator and room is lost. This in turn leads to the entry of nonsterile air into the hood and the release of potentially biohazardous materials into the room. In addition, the risk of collision between sterile pipettes and nonsterile surfaces of bottles, etc., increases. One should bring into the hood only those items that are directly involved in the current operation.

Laminar-flow hoods also must be maintained regularly, and the integrity of the filters, ductwork, and cabinets should be checked at least twice a year by a competent engineer. The engineer should also check the containment of the workspace, i.e., that internal air does not spill out and outside air does not enter, both of which are dependent on the internal air velocity and outside turbulence.

19.1.4 Humid Incubators

A major source of contamination stems from the use of humid incubators (*see* Section 6.6.1). High humidity is not required unless open vessels are being used; sealed flasks are better kept in a dry incubator or a hot room (*see* Section 4.4). A low-CO_2 medium (e.g., based on Hanks' salts; *see* Section 9.3.2) can be used to avoid the need to gas flasks with CO_2. If there is a need to gas flasks with CO_2, this is better done from a cylinder or piped supply and the flasks sealed and placed in a normal incubator. Using permeable caps (*see* Section 8.2.3) minimizes the risk of contamination but increases the unit cost and still exposes the flask to a higher-risk atmosphere than in a dry incubator. Permeable caps may be sealed with a secondary rubber cap (B-D Biosciences) for transfer to a nongassed incubator. If flasks are maintained in a CO_2 incubator with slack or permeable caps, it is possible to keep the incubator dry and use a different incubator for open plates. The CO_2 monitoring system will need to be recalibrated if the incubator is used dry, and the flasks will need to be checked for evaporation, and the caps tightened after the pH equilibrates, if necessary.

There are, however, many situations in which a humid incubator must be used. To reduce the risk of contamination, an incubator should be selected with an interior all of which is readily accessible and can be cleaned easily. Cultures placed in the incubator can be enclosed in a plastic box (*see* Section 6.6.2). Fan circulation in a CO_2 incubator shortens recovery time for both CO_2 and temperature, but at the cost of increased risk of contamination (*see* Sections 5.3.2, 6.6.1); open plate cultures are better maintained in static air and frequency of access limited as much as possible. Having a dry, non-CO_2 incubator for sealed flasks and short procedures such as trypsinization will help to limit access to an incubator used for open plates and cloning.

Fungicides. Copper-lined incubators have reduced fungal growth but are usually about 20–30% more expensive than conventional ones. Placing copper foil in the humidifier tray also inhibits the growth of fungus, but only in the tray, and will not protect the walls of the incubator. A number of fungal retardants are in common use, including copper sulfate, riboflavin, sodium dodecyl sulfate (SDS), and Roccall, a proprietary fungicidal cleaner used in a 2% solution. A comparison of colony formation in incubators with and without Roccall shows no toxic effect. Many of these retardants are detergents, so it is important not to have a CO_2 or an air line bubbling through the humidifier tray, or the liquid will foam. Remember, a fungicide will only protect the tray; there is no substitute for regular cleaning!

Cleaning incubators. Cleaning should be carried out regularly with 10% Roccall or an equivalent nontoxic antifungal cleaner. The frequency will depend on where the incubator is located; monthly may suffice for a clean area with filtered room air, but a shorter interval will be required for a rural site, where the spore count is higher, or during construction work or renovation. The frequency of access will also influence the buildup of fungal contamination. When the incubator is in use, any spillage must be mopped up immediately and contaminated cultures removed as soon as they are detected.

PROTOCOL 19.1. CLEANING INCUBATORS

Outline
Remove cultures to an alternative incubator, switch off the empty incubator, wash it out with detergent and alcohol, switch on the heat, and allow the incubator to dry. Replenish the water in the tray and restore the CO_2.

Materials
Nonsterile:
❏ Water containing 2% Roccall or an equivalent fungal inhibitor, to refill the water tray
❏ Alternative CO_2 incubator or plastic box and sealing tape
❏ Detergent: 10% Roccall, or another fungicidal detergent
❏ Alcohol, 70%

Protocol
1. Remove all cultures to another CO_2 incubator, or enclose the cultures in a sealable container (e.g., a desiccator), gas them with CO_2, and place them in a regular incubator or hot room.
2. Switch off the incubator that is to be cleaned.
3. Remove all the shelves, the water tray, and any demountable panels from the incubator.
4. Wash the inside of the incubator with detergent solution; try to reach all corners and crevices.
5. Rinse with water.

6. Wipe the interior of the incubator with 70% alcohol.
7. Restore heat (not CO_2), and leave the door open until the chamber is dry.
8. Wash the shelves and panels in detergent, rinse them in water, and wipe them with 70% alcohol.
9. Return the panels and shelves to the incubator.
10. Replace the water tray and fill it with sterile water containing 2% Roccall.
11. Close the door and restore CO_2.
12. When the temperature and CO_2 have stabilized, return the cultures to the incubator.

If the incubator has a sterilization cycle, it should be run after Step 9.

19.1.5 Cold Stores

Refrigerators and cold rooms also tend to build up fungal contamination on the walls in a humid climate, due to condensation that forms every time the door is opened, admitting moist air. The moist air increases the risk of deposition of spores on stored bottles; hence, they should be swabbed with alcohol before being placed in the hood (see Section 6.5). The cold store should be cleared, and the walls and shelving should be washed down with disinfectant every few months.

19.1.6 Sterile Materials

There should be no risk of contamination from sterile plastics and reagents if the appropriate quality control is carried out, either in house (see Section 11.6) or by the supplier. However, failures in sterilization can occur if, for example, the load is too tightly packed, there is a cold spot in either the sterilizing oven or autoclave, packaging is punctured before or during storage, or a low-level contamination has escaped detection despite quality control. It is difficult to guard against this eventuality, other than by following the correct procedures (see Section 11.6) and checking to make sure that all packing is correctly sealed.

It is critical when new staff are introduced into the tissue culture laboratory that they are made familiar with sterilization procedures (see Chapter 2, Exercises 3–9), even if they will not be called upon to carry out these procedures themselves. They should also be aware of the location of, and distinction between, sterile and nonsterile stocks (see Section 2.2). A simple error by a new recruit can cause severe problems that can last for several days before it is discovered.

19.1.7 Imported Cell Lines and Biopsies

As cell lines and tissue samples brought into the tissue culture laboratory may be contaminated, they should be quarantined (see Section 19.1.8) until shown to be clear of contamination, at which point they, or their derivatives, can join other stocks in general use (see Section 6.2.5). Whenever possible, all cell lines should be acquired via a reputable cell bank, which will have screened for contamination. Cell lines from any other source, as well as biopsies from all animal and human donors, should be regarded as contaminated until shown to be otherwise.

19.1.8 Quarantine

Any culture that is suspected of being contaminated, and any imported material that has not been tested, should be kept in quarantine. Preferably, quarantine should take place in a separate room with its own hood and incubator, but if this is not feasible, one of the hoods that are in general use may be employed. In this case, the hood should be used last thing in the day and should be sprayed and swabbed down with 70% alcohol containing 2% phenolic disinfectant after use. Then it should be withdrawn from service until the following day.

19.2 TYPES OF MICROBIAL CONTAMINATION

Bacteria, yeasts, fungi, molds, mycoplasmas, and occasionally protozoa, can all appear as contaminants in tissue culture. Usually, the species or type of infection is not important, unless it becomes a frequent occurrence. It is only necessary to note the general kind of contaminant (e.g., bacterial rods or cocci, yeast, etc.), how it was detected, the location where the culture was last handled, and the operator's name. If a particular type of infection recurs frequently, it may be beneficial to identify it in order to find its origin. [For more detailed screening procedures for microbial contamination, see European Pharmacopoeia, 1980; United States Pharmacopeia, 1985; Doyle et al., 1990; Doyle & Bolton, 1994; and Hay & Cour, 1997.] In general, rapidly growing organisms are less problematic as they are often overt and readily detected, whereupon the culture can be discarded. Difficulties arise when the contaminant is cryptic, either because it is too small to be seen on the microscope, e.g., mycoplasma, or slow growing such that the level is so low that it escapes detection. Use of antibiotics can be a common cause of cryptic contaminations remaining undetected (see Section 9.4.7).

19.3 MONITORING FOR CONTAMINATION

Potential sources of contamination are listed in Table 19.1, along with the precautions that should be taken to avoid them. Even in the best laboratories, however, contaminations do arise, so the following procedure is recommended:

(1) Check for contamination by eye and with a microscope at each handling of a culture. Check for mycoplasma every month.
(2) If it is suspected, but not obvious, that a culture is contaminated, but the fact cannot be confirmed in situ, clear the hood or bench of everything except your suspected culture and one can of Pasteur pipettes. Because

of the potential risk to other cultures, this is best done after all your other culture work is finished. Remove a sample from the culture and place it on a microscope slide. (Kovaslides are convenient for this, as they do not require a coverslip.) Check the slide with a microscope, preferably by phase contrast. If it is confirmed that the culture is contaminated, discard the pipettes, swab the hood or bench with 70% alcohol containing a phenolic disinfectant, and do not use the hood or bench until the next day.

(3) Record the nature of the contamination.

(4) If the contamination is new and is not widespread, discard the culture, the medium bottle used to feed it, and any other reagent (e.g., trypsin) that has been used in conjunction with the culture. Discard all of these into disinfectant, preferably in a fume hood and outside the tissue culture area.

(5) If the contamination is new and widespread (i.e., in at least two different cultures), discard all media, stock solutions, trypsin, etc.

(6) If the same kind of contamination has occurred before, check stock solutions for contamination (a) by incubation alone or in nutrient broth (*see* Section 11.6.2 and Plate 16d) or (b) by plating out the solution on nutrient agar (Oxoid, Difco). If (a) and (b) prove negative, but contamination is still suspected, incubate 100 mL of solution, filter it through a 0.2-μm filter, and plate out filter on nutrient agar with an uninoculated control.

(7) If the contamination is widespread, multispecific, and repeated, check (a) the laboratory's sterilization procedures (e.g., the temperatures of ovens and autoclaves, particularly in the center of the load, the duration of the sterilization cycle), (b) the packaging and storage practices, (e.g., unsealed glassware should be resterilized every 24 h), and (c) the integrity of the aseptic room and laminar-flow hood filters.

(8) Do not attempt to decontaminate cultures unless they are irreplaceable.

19.3.1 Visible Microbial Contamination

Characteristic features of microbial contamination are as follows:

(1) A sudden change in pH, usually a decrease with most bacterial infections (Plate 16a), very little change with yeast (Plate 16c) until the contamination is heavy, and sometimes an increase in pH with fungal contamination.

(2) Cloudiness in the medium (*see* Plate 16a–c), sometimes with a slight film or scum on the surface or spots on the growth surface that dissipate when the flask is moved (Plate 16b).

(3) Under a low-power microscope (∼×100), spaces between cells will appear granular and may shimmer with bacterial contamination (Fig. 19.1a). Yeasts appear as separate round or ovoid particles that may bud off smaller

particles (Fig. 19.1b). Fungi produce thin filamentous mycelia (Fig. 19.1c) and, sometimes, denser clumps of spores. With toxic infection, some deterioration of the cells will be apparent.

(4) Under high-power microscopy (∼×400), it may be possible to resolve individual bacteria and distinguish between rods and cocci. At this magnification, the shimmering that is visible in some infections will be seen to be caused by mobility of bacteria. Some bacteria form clumps or associate with the cultured cells.

(5) With a slide preparation, the morphology of the bacteria can be resolved at ×1000, but this is not usually necessary. Microbial infection may be confused with precipitates of media constituents (particularly protein) or with cell debris, but can be distinguished by their regular morphology. Precipitates may be crystalline or globular and irregular and are not usually as uniform in size. Clumps of bacteria may be confused with precipitated protein, but, particularly if shaken, many single or strings of bacteria will be seen. If you are in doubt, plate out a sample of medium on nutrient agar (*see* Section 11.6.2).

19.3.2 Mycoplasma

Detection of mycoplasmal infections (Fig. 19.1d–f) is not obvious by routine microscopy, other than through signs of deterioration in the culture, and requires fluorescent staining, PCR, ELISA assay, immunostaining, autoradiography, or microbiological assay. Fluorescent staining of DNA by Hoechst 33258 [Chen, 1977] is the easiest and most reliable method (*see* Protocol 19.2) and reveals mycoplasmal infections as a fine particulate or filamentous staining over the cytoplasm at ×500 magnification (Plate 16e,f). The nuclei of the cultured cells are also brightly stained by this method and thereby act as a positive control for the staining procedure. Most other microbial contaminations will also show up with fluorescence staining, so low levels of contamination, or particularly small organisms such as micrococci, can also be detected.

It is important to appreciate the fact that mycoplasmas do not always reveal their presence by means of macroscopic alterations of the cells or medium. Many mycoplasma contaminants, particularly in continuous cell lines, grow slowly and do not destroy host cells. However, they can alter the metabolism of the culture in many different ways. Because mycoplasmas take up thymidine from the medium, infected cultures show abnormal labeling with [^{3}H]thymidine (*see* Fig. 27.2). Immunological studies can also be totally frustrated by mycoplasmal contamination, as attempts to produce antibodies against the cell surface may raise antimycoplasma antibodies. Mycoplasmas can alter cell behavior and metabolism in many other ways [McGarrity, 1982; Doyle et al., 1990; Izutsu et al., 1996; Dorazio et al., 1996; Giron et al., 1996; Paddenberg et al., 1996], so there is an absolute requirement for routine, periodic assays to detect possible covert contamination of all cell cultures, particularly continuous cell lines.

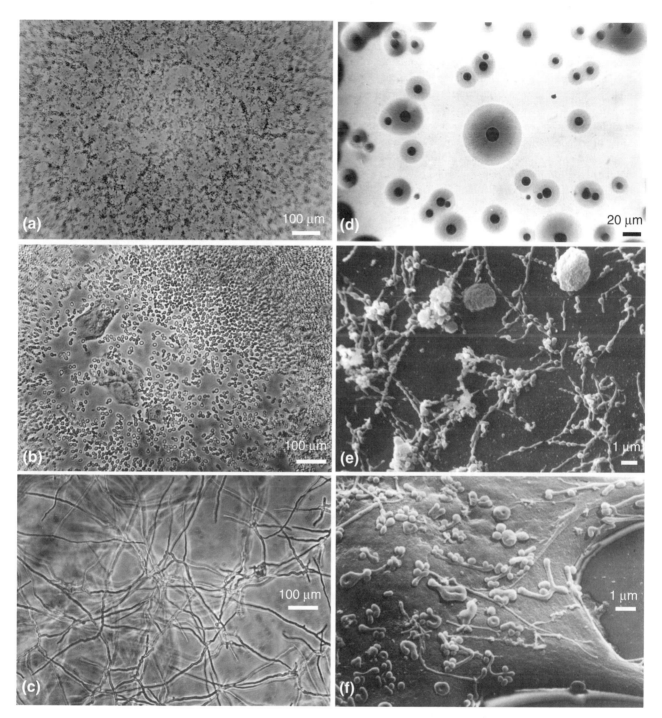

Fig. 19.1. Types of Contamination. Examples of microorganisms found to contaminate cell cultures. (a) Bacteria. (b) Yeast. (c) Mold. (d) Mycoplasma colonies growing on special nutrient agar. (e & f) Scanning electron micrograph of mycoplasma growing on the surface of cultured cells (d–f, courtesy of Dr. M. Gabridge; a–c, 10× objective) (*see* Plate 16e,f for mycoplasma stained with Hoechst 33258).

Monitoring cultures for mycoplasmas. Superficial signs of chronic mycoplasmal infection include a diminished rate of cell proliferation, reduced saturation density [Stanbridge & Doersen, 1978], and agglutination during growth in suspension [Giron et al., 1996]. Acute infection causes total deterioration, with perhaps a few resistant colonies, although

these and any resulting cell lines are not necessarily free of contamination and may carry a chronic infection.

19.3.3 Fluorescence Staining for Mycoplasma

The cultures are stained with Hoechst 33258, a fluorescent dye that binds specifically to DNA [Chen, 1977]. Because mycoplasmas contain DNA, they can be detected readily by their characteristic particulate or filamentous pattern of fluorescence on the cell surface and, if the contamination is heavy, in surrounding areas. Monolayer cell cultures can be fixed and stained directly, but after centrifugation, the medium from cells growing in suspension will need to be added to an indicator cell (i.e., another monolayer) known to be free of mycoplasma but also known to be a good host for mycoplasma and to spread well in culture, with adequate cytoplasm to reveal any adherent mycoplasma. Vero cells, 3T6, NRK, and A549 are all suitable.

The use of an indicator cell is recommended in the following protocol as it also helps to avoid problems with false positives arising from debris when cells are assayed just after thawing or from primary cultures, which you might not want to sacrifice anyway. If there is a lot of debris, the medium can be filtered through a sterile 5-μm filter or centrifuged at 100 g.

PROTOCOL 19.2. FLUORESCENCE DETECTION OF MYCOPLASMA

Outline

Culture cells in the absence of antibiotics for at least one week, transfer the supernatant medium to an early log phase-indicator culture, incubate 3–5 days, fix and stain the cells, and look for fluorescence other than in the nucleus.

Materials

Sterile or Aseptically Prepared:
- ❑ Supernatant medium, free of antibiotics, from 7-day monolayer or centrifuged suspension cell culture
- ❑ Indicator cells (e.g., 3T6, NRK, Vero, A549)
- ❑ Petri dishes, 6 cm

Nonsterile:
- ❑ Hoechst 33258 stain, 50 ng/mL in BSS without phenol red (BSS-PR) or D-PBSA
- ❑ BSS-PR: Hanks' BSS without Phenol Red
- ❑ D-PBSA: Dulbecco's PBS without Ca^{2+} and Mg^{2+}
- ❑ Deionized water
- ❑ Fixative: freshly prepared acetic methanol (1:3, on ice)
- ❑ Buffered glycerol mountant (*see* Appendix I: Mycoplasma Reagents)

Protocol

1. Seed indicator cells into Petri dishes without using antibiotics; seed enough to give 50–60% confluence in 4–5 days (e.g., 2×10^4 NRK or 1×10^5 A549, in 5 mL of medium).
2. Add 1.5 mL of medium from the test culture.
3. Incubate the culture until the cells reach 50–60% confluence.

Note. Cultures must not reach confluence by the end of the assay or staining will be inhibited and the subsequent visualization of mycoplasma will be impaired.

4. Remove the medium and discard it.
5. Rinse the monolayer with BSS-PR or D-PBSA, and discard the rinse.
6. Add fresh BSS-PR or D-PBSA diluted 50:50 with fixative, rinse the monolayer, and discard the rinse.
7. Add pure fixative, rinse, and discard the rinse.
8. Add more fixative ($\sim$0.5 mL/cm^2), and fix for 10 min.
9. Remove and discard the fixative.
10. Dry the monolayer completely if it is to be stored. (Samples may be accumulated at this stage and stained later.)
11. If you are proceeding directly with staining, wash off the fixative with deionized water and discard the wash.
12. Add Hoechst 33258 in BSS-PR or D-PBSA, and stain 10 min at room temperature.
13. Remove and discard the stain.
14. Rinse the monolayer with water and discard the rinse.
15. Mount a coverslip in a drop of buffered glycerol mountant, and blot off any surplus from the edges of the coverslip.
16. Examine the monolayer by epifluorescence with a 330-/380-nm excitation filter and an LP 440-nm barrier filter.

Analysis. Check for extranuclear fluorescence. Mycoplasmas give pinpoints or filaments of fluorescence over the cytoplasm and, sometimes, in intercellular spaces (*see* Plate 16e,f). The pinpoints are close to the limits of resolution with a 50× objective (0.1–1.0 μm) and are usually regular in size and shape. Not all of the cells will necessarily be infected, so as much as possible of the preparation should be scanned before you declare the culture uninfected.

Sometimes a light, uniform staining of the cytoplasm is observed, probably due to RNA. This fluorescence tends to fade on storage of the preparation, and examination the next day (after storing dry and in the dark) usually gives clearer results. This artifact never has the sharp punctuate or filamentous appearance of mycoplasma and can be distinguished fairly readily with further experience in observation.

If there is any doubt regarding the interpretation of the fluorescence test, it should be repeated after generating a further subculture of the test cells in the absence of antibiotics. If results are still equivocal, adopt another assay, such as PCR or ELISA.

19.3.4 PCR for Mycoplasma

The following protocol and introduction have been contributed by Cord C. Uphoff and Hans G. Drexler of DSMZ, Braunschweig, Germany.

The polymerase chain reaction (PCR) provides a very sensitive and specific assay for the direct detection of mycoplasmas in cell cultures with low expenditure of labor, time, and cost, simplicity, objectivity of interpretation, reproducibility, and documentation of results. Several primer sequences are published for both single and nested PCR and with narrow or broad specificity for mycoplasma or eubacteria species. In most cases, the 16S rDNA sequences are used as target sequences, because this gene contains regions with more and less conserved sequences. This gene also offers the opportunity to perform a PCR with the 16S rDNA or an RT-PCR (reverse transcriptase-PCR) with the cDNA of the 16S rRNA.

Here, we describe the use of a mixture of oligonucleotides for the specific detection of mycoplasmas. This approach reduces significantly the generation of false positive results due to possible contamination of the solutions used for sample preparation and the PCR run, and of other materials with airborne bacteria. One of the main problems concerning PCR reactions with samples from cell cultures is the inhibition of the *Taq* polymerase by unspecified substances. To eliminate those inhibitors, we strictly recommend that the sample DNA be extracted and purified by conventional phenol-chloroform extraction or by the more convenient column or matrix binding extraction methods.

To confirm the error-free preparation of the sample and PCR run, appropriate control reactions have to be included in the PCR. These comprise internal control DNA for every sample reaction and in parallel, positive and negative as well as water control reactions. The internal control consists of a DNA fragment with the same primer sequences for amplification, but is of a different size than the amplicon of mycoplasma–contaminated samples. This control DNA is added to the PCR mixture in a previously determined limiting dilution to demonstrate the sensitivity of the PCR reaction.

PROTOCOL 19.3. DETECTION OF MYCOPLASMA BY PCR

Materials

❏ D-PBSA (phosphate-buffered saline): 137 mM NaCl, 2.7 mM KCl, 8.1 mM Na_2HPO_4, 1.5 mM KH_2PO_4, pH 7.2. Autoclave 20 min at 121°C to sterilize the solution

❏ TAE (Tris acetic acid EDTA), 50×: 2 M Tris base, 5.71% glacial acetic (v/v), 100 mM EDTA. Adjust to pH ~8.5

❏ DNA extraction and purification system, e.g., phenol/chloroform extraction and ethanol precipitation, or DNA extraction kits applying DNA binding matrices

❏ Thermal cycler

❏ *Taq* DNA polymerase

❏ Loading buffer, 6×: 0.09% (w/v) bromophenol blue, 0.09% (w/v) xylene cyanol FF, 60% glycerol (v/v), 60 mM EDTA.

❏ Primers (any supplier):

  5′ primers (Myco−5′):

cgc	ctg	agt	agt	acg	t**w**c	gc
tgc	ctg	**r**gt	agt	aca	ttc	gc
cgc	ctg	agt	agt	atg	ctc	gc
cgc	ctg	ggt	agt	aca	ttc	gc

  3′ primers (Myco−3′):

gcg	gtg	tgt	aca	a**r**a	ccc	ga
gcg	gtg	tgt	aca	aac	ccc	ga

  (**r** = mixture of g and a;
  w = mixture of t and a)

Primer stock solutions: 100 μM in dH_2O, stored frozen at −20°C. Working solutions: mix of forward primers at 5 μM each (Myco−5′) and mix of reverse primers at 5 μM each (Myco−3′) in dH_2O, aliquoted in small amounts (i.e., 25- to 50-μL aliquots), and stored frozen at −20°C.

❏ Internal control DNA: The internal control may be prepared as published elsewhere [Uphoff & Drexler, 2002]. A limiting dilution should be determined experimentally by performing PCR with a dilution series of the internal control DNA.

❏ Positive control DNA: a 10-fold dilution of any mycoplasma-positive sample prepared as described below.

❏ Deoxy-nucleotide triphosphate mixture (dNTP-mix): Mixture contains deoxyadenosine triphosphate (dATP), deoxycytidine triphosphate (dCTP), deoxyguanosine triphosphate (dGTP), and deoxythymidine triphosphate (dTTP) at 5 mM in H_2O, and stored as 50-μL aliquots at −20°C.

❏ Agarose, 1.3%-TAE gel.

Protocol

A. Sample Collection and Preparation of DNA

1. Before collecting the samples, the cell line to be tested for mycoplasma contamination should be cultured without any antibiotics for several

days, or for at least two weeks after thawing. This should ensure that the titer of the mycoplasmas in the supernatant medium is within the detection limits of the PCR assay.

2. Collect 1 mL of the supernatant medium of adherent cells, or of cultures with settled suspension cells. Collecting the samples in this way will include some viable or dead cells, an advantage as some mycoplasma strains predominantly adhere to the eukaryotic cells or even invade them. The supernatant medium can be stored at 4°C for a few days or at −20°C for several weeks. After thawing, the samples should be processed immediately.

3. The supernatant medium is centrifuged at 13,000 g for 5 min and the pellet is resuspended in 1 mL D-PBSA by vortexing.

4. The suspension is centrifuged again and washed one more time with D-PBSA as described in Step 3.

5. After centrifugation, the pellet is resuspended in 100 μL D-PBSA by vortexing and then heated to 95°C for 15 min.

6. Immediately after lysing the cells, the DNA is extracted and purified by standard phenol/chloroform extraction and ethanol precipitation or other DNA isolation methods.

B. PCR Reaction

Established rules to avoid DNA carry over should be strictly followed: (i) the places where the DNA is extracted, the PCR reaction is set up, and the gel is run after the PCR should be separated from each other; (ii) all reagents should be stored in small aliquots to provide a constant source of uncontaminated reagents; (iii) avoid reamplifications; (iv) reserve pipettes, tips, and tubes for their use in PCR only and irradiate pipettes frequently by UV light; (v) the succession of the PCR set up described below should be followed strictly; (vi) wear gloves during the whole sample preparation and PCR setup; (vii) include the appropriate control reactions, such as internal, positive, negative, and the water control reaction.

1. Set up two reactions per sample to be tested with the following solutions:
   Sample only: 1 μL dNTPs, 1 μL Myco−5′, 1 μL Myco−3′, 1.5 μL 10 × PCR buffer, 9.5 μL dH$_2$O
   Sample and DNA internal standard: 1 μL dNTPs, 1 μL Myco−5′, 1 μL Myco−3′, 1.5 μL 10 × PCR buffer, 8.5 μL dH$_2$O, 1 μL internal control DNA.

Premix, stocks for multiple samples, three samples for the reaction without internal control DNA (for the positive, negative, and the water controls) and two for the internal control DNA for the positive and the negative controls, plus a surplus for pipetting.

2. Transfer 14 μL of each premixed stock to 0.2-mL PCR reaction tubes and add 1 μL dH$_2$O to the water control reaction.

3. Prepare the *Taq* DNA polymerase mix (10 μL per reaction, plus 1 additional reaction to allow for pipetting) containing 1× PCR buffer and 1 U *Taq* polymerase per reaction.

4. Set aside all reagents used for the preparation of the premixed stocks.

5. Take out the samples of DNA to be tested and the positive control DNA. Do not handle the reagents and samples simultaneously.

6. Add 1 μL per DNA preparation to one reaction tube that contains no internal control DNA and to one tube containing the internal control DNA.

7. To perform a hot start PCR, transfer the reaction mixtures (without *Taq* polymerase) to the thermal cycler and start one thermo cycle with the following parameters: step 1: 7 min at 95°C, step 2: 3 min at 72°C, step 3: 2 min at 65°C, step 4: 5 min at 72°C.

During step 2, open the thermal lid and add 10 μL of the *Taq* polymerase mix to each tube. For many samples, the duration of this step can be prolonged. Open and close each reaction tube separately to prevent evaporation of the samples. Allow at least 30 s after adding the *Taq* polymerase to the last tube and closing the lid of the thermal cycler for equilibration of the temperature within the tubes and removal of condensate from the lid before continuing to the next cycle step.

8. After this initial cycle, perform 32 thermal cycles with the following parameters, step 1: 4 s at 95°C, step 2: 8 s at 65°C, step 3: 16 s at 72°C plus 1 s of extension time during each cycle.

9. Finish the reaction with a final amplification step at 72°C for 10 min and then cool the samples to room temperature.

10. Prepare a 1.3% agarose-TAE gel containing 0.3 μg of ethidium bromide per mL. Submerge the gel in 1× TAE and add 12 μL of the amplification product (10 μL reaction mixtures plus 2 μL of 6 × loading buffer) to each well and run the gel at 10 V/cm.

11. Visualize the specific products on a suitable ultraviolet light screen and document the results.

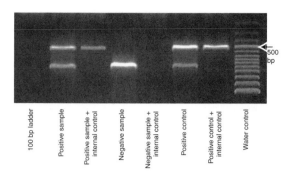

100 bp ladder | Positive sample | Positive sample + internal control | Negative sample | Negative sample + internal control | Positive control | Positive control + internal control | Water control

Fig. 19.2. Mycoplasma Detection by PCR. Ethidium bromide fluorescence of PCR products of infected, uninfected, and control cells, electrophoresed on 1.3% agarose. The 100-bp ladder consists of the following bands: 100, 200, 300, 400, 500 (strongly stained band), 600, 700, 800, 900, 1031, 1200, 1500, 2000, and 3000 bp. The wild-type mycoplasma bands are about 510 bp, and the internal control band is almost 1000 bp. (Courtesy of Cord Uphoff, DSMZ).

Interpretation of Results

(1) Ideally, all samples containing the internal control DNA show a band at 986 bp. If the second run also shows no band for sample and the internal control, the whole procedure should be repeated.

(2) Mycoplasma-positive samples show a band at 502 to 520 bp, depending on the mycoplasma species (Fig. 19.2).

(3) Contaminations of reagents with mycoplasma-specific DNA or PCR product are revealed by a band in the water control and/or in the negative control sample. Weak mycoplasma-specific bands can occur after treatment of infected cell cultures with anti-mycoplasma reagents for the elimination of mycoplasma or when other antibiotics are applied routinely. In these cases the positive reaction might be due either to residual DNA in the culture medium derived from dead mycoplasma cells or to viable mycoplasma cells that are present at a very low titer.

19.3.5 Alternative Methods for Detecting Mycoplasma

Biochemical. Among other methods that have been reported for the detection of mycoplasmal infections are methods that detect mycoplasma-specific enzymes such as arginine deiminase or nucleoside phosphorylase [*see* Schneider & Stanbridge, 1975; Levine & Becker, 1977] and those that detect toxicity with 6-methylpurine deoxyriboside (Mycotect, Invitrogen).

Microbiological culture. This is a very sensitive method, widely employed in quality control and validation procedures. However, it should only be used with isolation facilities and the appropriate background experience in microbiology, as these microorganisms are quite fastidious, and it will be necessary to culture live mycoplasma as a positive control. The cultured cells are seeded into mycoplasma broth [Doyle et al., 1990], grown for 6 days, and plated out onto special nutrient

agar [Hay, 2000]. Colonies form in about 8 days and can be recognized by their size (~200-μm diameter) and their characteristic "fried egg" morphology-dense center with a lighter periphery (Fig. 19.1d). Commercial kits for microbiological detection (Mycotrim) are available from Irvine Scientific.

Although using selective culture conditions and examining the morphology of a colony enables the species of mycoplasma to be identified, the microbiological culture method is much slower and more difficult to perform than the fluorescence technique. Mycoplasma testing by microbiological culture is available (*see* Appendix II). Specific monoclonal antibodies now allow the characterization of mycoplasma contaminations.

Molecular hybridization. Molecular probes specific to mycoplasmal DNA can be used in Southern blot analysis to detect infections by conventional molecular hybridization techniques. A kit is available that uses similar probes and a simple single-step separation of hybrid DNA for subsequent detection by scintillation counting (GenProbe, Fisher Scientific).

[³H]thymidine incorporation. One other method that has been used quite successfully is autoradiography with [³H]thymidine [Nardone et al., 1965]. The culture is incubated overnight with 4 KBq/mL (~0.1 μCi/mL) of [³H]thymidine with high specific activity and an autoradiograph is prepared (*see* Protocol 27.3). Grains over the cytoplasm are indicative of contamination (*see* Fig. 27.2), which can be accompanied by a lack of nuclear labeling because of trapping of the thymidine at the cell surface by the mycoplasma.

19.3.6 Viral Contamination

Incoming cell lines, natural products, such as serum, in media, and enzymes such as trypsin, used for subculture, are all potential sources of viral contamination. A number of reagents are screened by manufacturers against a limited range of viruses, and claims have been made that the larger viruses can be filtered out during processing, but there is no certain way at present to eliminate viral contamination. The best way of avoiding it is to ensure that the products are collected from animals free from known virus infections. For this, you will need to rely on the quality control put in place by the supplier.

Detection of viral contamination. Screening with a panel of antibodies by immunostaining (*see* Protocol 16.11) or ELISA assays is probably the best way of detecting viral infection. Alternatively, one may use PCR with the appropriate viral primers. Some commercial companies (BioReliance) offer viral screening.

19.4 ERADICATION OF CONTAMINATION

19.4.1 Bacteria, Fungi, and Yeasts

The most reliable method of eliminating a microbial contamination is to discard the culture and the medium and reagents used with it, as treating a culture will either be unsuccessful or may lead to the development of an antibiotic-resistant microorganism. Decontamination should be attempted only in extreme situations, under quarantine, and with expert supervision. If unsuccessful, the culture and associated reagents should be autoclaved as soon as failure becomes obvious.

PROTOCOL 19.4. ERADICATION OF MICROBIAL CONTAMINATION

Outline
Wash the culture several times in a high concentration of antibiotics by rinsing the monolayer or by centrifugation and resuspension of nonadherent cells. Then grow the culture for three subcultures with, and three without, antibiotics. Test for contamination after each subculture.

Materials
Sterile:
❑ DBSS (*see* Appendix I: Dissection BSS)
❑ High-antibiotic medium (*see* Appendix I: Collection Medium and Table 9.4; *see* Section 19.4.2 for mycoplasma)
❑ Materials for subculture (*see* Protocol 13.2)
Nonsterile:
❑ Microscope, preferably with 40× and 100× phase-contrast optics
❑ Materials for staining mycoplasma (*see* Protocol 19.2)

Protocol
1. Collect the contaminated medium carefully. If possible, the organism should be tested for sensitivity to a range of individual antibiotics. If not, autoclave the medium or add hypochlorite (*see* Section 7.8.5).
2. Wash the cells in DBSS (dilution can reduce the number of contaminants by two logs with each wash, unless they are adherent to the cells):
    (a) For monolayers, rinse the culture three times with DBSS, trypsinize, and wash the cells twice more in DBSS by centrifugation and resuspension.
    (b) For suspension cultures, wash the culture five times in DBSS by centrifugation and resuspension.
3. Reseed a fresh flask at the lowest reasonable seeding density.
4. Add high-antibiotic medium and change the culture every 2 days.
5. Subculture in a high-antibiotic medium.

---

6. Repeat Steps 1–4 for three subcultures.
7. Remove the antibiotics, and culture the cells without them for a further three subcultures.
8. Check the cultures by phase-contrast microscopy and Hoechst staining (*see* Protocol 19.2).
9. Culture the cells for a further two months without antibiotics, and check to make sure that all contamination has been eliminated (*see* Section 11.6.2).

The general rule should be that contaminated cultures are discarded and that decontamination is not attempted unless it is absolutely vital to retain the cell strain. In any event, complete decontamination is difficult to achieve, particularly with yeast, and attempts to do so may produce hardier, antibiotic-resistant strains.

19.4.2 Eradication of Mycoplasma

If mycoplasma is detected in a culture, the first and overriding rule, as with other forms of contamination, is that the culture should be discarded for autoclaving or incineration. In exceptional cases (e.g., if the contaminated line is irreplaceable), one may attempt to decontaminate the culture. Decontamination should be done, however, only by an experienced operator, and the work must be carried out under conditions of quarantine.

Several agents are active against mycoplasma, including kanamycin, gentamicin, tylosin [Friend et al., 1966], poly-anethol sulfonate [Mardh, 1975], and 5-bromouracil in combination with Hoechst 33258 and UV light [Marcus et al., 1980]. Coculturing with macrophages [Schimmelpfeng et al., 1968], animal passage [Van Diggelen et al., 1977], and cytotoxic antibodies [Pollock & Kenny, 1963] can also be effective in some cases. However, the most successful agents have been tylosin (MP Biomedicals), Mycoplasma Removal Agent [MRA, also from (MP Biomedicals); Drexler et al., 1994], ciprofloxacin (Bayer) [Mowles, 1988; Hlubinova et al., 1994], and BM-Cycline (Roche).

Contaminated cultures should be treated as in Protocol 19.3, using MRA, BM-Cycline, or tylosin at the manufacturer's recommended concentration, in place of the usual antibiotics in DBSS and the collection medium. However, this operation should not be undertaken unless it is absolutely essential, and even then it must be performed in experienced hands and in isolation [Uphoff & Drexler 2004]. It is far safer to discard infected cultures.

19.4.3 Eradication of Viral Contamination

There are no reliable methods for eliminating viruses from a culture at present; disposal or tolerance are the only options.

19.4.4 Persistent Contamination

Many laboratories have suffered from periods of contamination that seems to be refractory to all the remedies suggested in Table 19.1. There is no easy resolution to this problem, other than to follow the previous recommendations in a logical and analytical fashion, paying particular attention to changes in technique, new staff, new suppliers, new equipment, and inadequate maintenance of laminar-flow hoods or other equipment. Typically, an increase in the contamination rate stems from deterioration in aseptic technique, an increased spore count in the atmosphere, poorly maintained incubators, a contaminated cold room or refrigerator, or a fault in a sterilizing oven or autoclave, the way that it is packed, or the monitoring of the sterilization cycle.

The constant use of antibiotics also favors the development of chronic contamination. Many organisms are inhibited, but not killed, by antibiotics. They will persist in the culture, undetected for most of the time, but periodically surfacing when conditions change. It is essential that cultures be maintained in antibiotic-free conditions for at least part of the time (*see* Fig. 13.6), and preferably all the time; otherwise cryptic contaminations will persist, their origins will be difficult to determine, and eliminating them will be impossible.

A slight change in practices, the introduction of new personnel, or an increase in activity as more people use a facility can all contribute to an increase in the rate of contamination. Procedures must remain stringent, even if the reason is not always obvious to the operator, and alterations in routine should not be made casually. If strict practices are maintained, contamination may not be eliminated entirely, but it will be reduced and detected early.

19.5 CROSS-CONTAMINATION

During the development of tissue culture, a number of cell strains have evolved with very short doubling times and high plating efficiencies. Although these properties make such cell lines valuable experimental material, they also make them potentially hazardous for cross-infecting other cell lines [Nelson-Rees & Flandermeyer, 1997; Lavappa, 1978; Nelson-Rees et al., 1981; Kneuchel & Masters, 1999; Dirks et al., 1999; MacLeod et al., 1999; van Bokhoven et al., 2001a,b; Rush et al., 2002; Milanesi et al., 2003; Drexler et al., 2003] (*see also* Table 13.2). The extensive cross-contamination of many cell lines with HeLa and other rapidly rowing cell lines is now clearly established [Stacey et al., 2000; MacLeod et al., 1999], but many operators are still unaware of the seriousness of the risk (*see also* Sections 7.10.1, 16.1, 16.3, 16.6.2; Table 13.2). The responsibility lies with supervisors to impress upon new personnel the severity of the risks, with journal editors and referees to reject manuscripts, and with grant review bodies to reject grant proposals without evidence of proper authentication of cell lines. Without the acceptance of these obligations, the situation can only get worse.

The following practices help avoid cross-contamination:

(1) Obtain cell lines from a reputable cell bank that has performed the appropriate validation of the cell line (*see* Sections 7.10, 20.2; Appendix II: Cell Banks), or perform the necessary authentication yourself as soon as possible (*see* Section 16.3).

(2) Do not have culture flasks of more than one cell line, or media bottles used with them, open simultaneously.

(3) Handle rapidly growing lines, such as HeLa, on their own and after other cultures.

(4) Never use the same pipette for different cell lines.

(5) Never use the same bottle of medium, trypsin, etc., for different cell lines.

(6) Do not put a pipette back into a bottle of medium, trypsin, etc., after it has been in a culture flask containing cells.

(7) Add medium and any other reagents to the flask first, and then add the cells last.

(8) Do not use unplugged pipettes, or pipettors without plugged tips, for routine maintenance.

(9) Check the characteristics of the culture regularly, and suspect any sudden change in morphology, growth rate, or other phenotypic properties. Cross-contamination or its absence may be confirmed by DNA fingerprinting (*see* Protocol 16.8) [Stacey et al., 1992], DNA profiling (*see* Protocol 16.9) karyotype [Nelson-Rees & Flandermeyer, 1977] (*see* Section 16.5; Protocols 16.7, 27.5, 27.9), or isoenzyme analysis [O'Brien et al., 1980] (*see* Protocol 16.10).

It cannot be overemphasized that cross-contaminations can and do occur. It is essential that the preceding precautions be taken and that cell strain characteristics be checked regularly.

19.6 CONCLUSIONS

- Check living cultures regularly for contamination by using normal and phase-contrast microscopy and for mycoplasmas by employing fluorescent staining of fixed preparations or by PCR.

- Do not maintain all cultures routinely in antibiotics. Grow at least one set of cultures of each cell line without antibiotics for a minimum of two weeks at a time, and preferably continuously, in order to allow cryptic contaminations to become overt.

- Do not attempt to decontaminate a culture unless it is irreplaceable, and then do so only under strict quarantine.

- Quarantine all new lines that come into your laboratory until you are sure that they are uncontaminated.

- Do not share media or other solutions among cell lines or among operators, and check cell line characteristics (*see* Section 16.3) periodically to guard against cross-contamination.

- New cell lines should be characterized, preferably by DNA fingerprinting or profiling, as soon after isolation as possible.

CHAPTER 20

Cryopreservation

As cell culture develops within a laboratory a number of cell lines will be developed or acquired, unless the work is exclusively with primary cultures. The use of each cell line adds to its provenance, and each one becomes a valuable resource. If unique, the cell line might be impossible to replace; at best replacement would be expensive and time-consuming. It is, therefore, essential to protect this considerable investment by preserving cell lines.

20.1   RATIONALE FOR FREEZING

Cell lines in continuous culture are prone to variation due to selection in early-passage culture (*see* Section 3.8), senescence in finite cell lines (*see* Sections 3.8.1, 18.4.1), and genetic and phenotypic instability (*see* Section 16.1,18.3) in continuous cell lines. In addition, even the best-run laboratory is prone to equipment failure and contamination. Cross-contamination and misidentification (*see* Sections 7.10.1, 16.1, 16.3, 16.6.2, 19.5; Table 13.2) also continues to occur with an alarming frequency. There are many reasons, therefore, for freezing down a validated stock of cells; these reasons can be summarized as follows:

(1) Genotypic drift due to genetic instability
(2) Senescence and the resultant extinction of the cell line
(3) Transformation of growth characteristics and acquisition of malignancy-associated properties
(4) Phenotypic instability due to selection and dedifferentiation
(5) Contamination by microorganisms

(6) Cross-contamination by other cell lines
(7) Misidentification due to careless handling
(8) Incubator failure
(9) Saving time and materials maintaining lines not in immediate use
(10) Need for distribution to other users

20.2   ACQUISITION OF CELL LINES FOR CRYOPRESERVATION

There are certain requirements that should be met before cell lines are considered for cryopreservation (Table 20.1).

Validation. Cell lines should be shown to be free of contamination (*see* Section 19.3) and authentic (*see* Section 16.3) before cryopreservation. Although a few ampoules of a newly acquired cell line may be frozen before complete validation has been carried out (*see* Section 20.4), proper validation should be carried out before major stocks are frozen (*see* Section 7.10).

When to freeze. If it is a finite cell line, it is grown to around the fifth population doubling in order to create a sufficient number of cells for freezing. Continuous cell lines should be cloned (*see* Sections 14.1, 14.3, 14.4), a characterized clone selected, and sufficient stocks grown for freezing. Continuous cell lines have advantages; they survive indefinitely, grow more rapidly, and can be cloned more easily; but they may be less stable genetically. Finite cell lines are usually diploid or close to it and are stable between

Culture of Animal Cells: A Manual of Basic Technique, Fifth Edition, by R. Ian Freshney
Copyright © 2005 John Wiley & Sons, Inc.

TABLE 20.1. Requirements before Freezing

Acquisition	Finite cell line	Freeze at early passage (<5 subcultures)
	Continuous cell line	Clone, select, and characterize; amplify
Standardization	Medium	Select optimal medium, and adhere to this medium
	Serum (if used)	Select a batch for use at all stages (*see* Section 9.6)
	Substrate	Standardize on one type and supplier, although not necessarily on one size or configuration
Validation	Provenance	Record details of origin, life history, and properties
	Authentication	Check cell line characteristics against origin and provenance (*see* Sections 16.1, 16.3)
	Transformation	Determine transformed status (*see* Table 18.1)
	Contamination	Microbial (*see* Sections 19.3, 11.6.2)
		Mycoplasma (*see* Protocols 19.2, 19.3)
	Cross-contamination and misidentification	Criteria to confirm identity (*see* Section 16.3)

certain passage levels, but they are harder to clone, grow more slowly, and eventually die out or transform.

20.3 PRINCIPLES OF CRYOPRESERVATION

20.3.1 Theoretical Background to Cell Freezing

Optimal freezing of cells for maximum viable recovery on thawing depends on minimizing intracellular ice crystal formation and reducing cryogenic damage from foci of high-concentration solutes formed when intracellular water freezes. This is achieved (a) by freezing slowly to allow water to leave the cell but not so slowly that ice crystal growth is encouraged, (b) by using a hydrophilic cryoprotectant to sequester water, (c) by storing the cells at the lowest possible temperature to minimize the effects of high salt concentrations on protein denaturation in micelles within the ice, and (d) by thawing rapidly to minimize ice crystal growth and generation of solute gradients formed as the residual intracellular ice melts.

20.3.2 Cell Concentration

Cells appear to survive freezing best when frozen at a high cell concentration. This is largely an empirical observation but probably is related partly to the reduced viability on thawing requiring a higher seeding concentration and partly to improved survival at a high cell concentration if cells are leaky because of cryogenic damage. A high concentration at freezing also allows sufficient dilution of the cryoprotectant at reseeding after thawing so that centrifugation is unnecessary (at least for most cells). The number of cells frozen should be sufficient to allow for 1:10 or 1:20 dilution on thawing to dilute out the cryoprotectant but still keep the cell concentration higher than at normal passage; for example, for cells subcultured normally at 1×10^5/mL, 1×10^7 should be frozen in 1 mL of medium, and, after thawing the cells, the whole 1 mL should be diluted to 20 mL of medium, giving 5×10^5 cells/mL (five times the normal seeding concentration). This dilutes the cryoprotectant from 10% to 0.5%, at which concentration it is less likely to be toxic.

Residual cryoprotectant may be diluted out as soon as the cells start to grow (for suspension cultures) or the medium changed as soon as the cells have attached (for monolayers).

20.3.3 Freezing Medium

The cell suspension is frozen in the presence of a cryoprotectant such as glycerol or dimethyl sulfoxide (DMSO) [Lovelock & Bishop, 1959]. Of these two, DMSO appears to be the more effective, possibly because it penetrates the cell better than glycerol. Concentrations of between 5% and 15% have been used, but 7.5% or 10% is more usual. There are situations in which DMSO may be toxic or induce cells to differentiate (*see* Section 17.7.2), e.g., with hematopoietic cell lines such as L5178Y or HL60 (M. Freshney, personal communication), and in these cases it is preferable either to use glycerol or to centrifuge the cells after thawing to remove the cryoprotectant.

It has been claimed that cells should be kept at 4°C after DMSO is added to the freezing medium and before freezing, but preliminary experiments by the author suggested that this did not improve survival after freezing and may even have reduced it [unpublished observations], perhaps by inhibiting intracellular penetration. However, this issue is still not fully resolved and is worth further experimentation (*see* Chapter 2, Exercise 17).

DMSO should be colorless, and it needs to be stored in glass or polypropylene, as it is a powerful solvent and will leach impurities out of rubber and some plastics. Glycerol should be not more than one year old, as it may become toxic after prolonged storage.

Other cryoprotectants have been suggested, such as polyvinylpyrrolidone (PVP) [Suzuki et al., 1995], polyethylene glycol (PEG) [Monroy et al., 1997], and hydroxyethyl starch (HES) [Pasch et al., 2000], but none has had the general acceptance of either DMSO or glycerol, although there may be some improvement with trehalose [Eroglu et al., 2000; Buchanan et al., 2004]. Many laboratories also increase the serum concentration in freezing medium to 40%, 50%, or even 100%.

20.3.4 Cooling Rate

Most cultured cells survive best if they are cooled at 1°C/min [Leibo & Mazur, 1971; Harris & Griffiths, 1977]. This is probably a compromise between fast freezing minimizing ice crystal growth and slow cooling encouraging the extracellular migration of water. The shape of the cooling curve is governed by (a) the ambient temperature, (b) any insulation surrounding the cells, including the ampoule, (c) the specific heat and volume of the ampoule contents, and (d) the latent heat absorption during freezing. When cells are frozen in an insulated container placed in an ultra-deep freeze, as in (1)–(4) below, this results in a curve with a rapid cooling rate at the start, when the temperature differential is greatest, down to a minimum at the eutectic point, a slight rise as freezing commences, followed by a plateau as the latent heat of freezing is absorbed, and then a more rapid fall as freezing is completed, gradually slowing down as the temperature of the freezing chamber is reached (Fig. 20.1).

When cells are transferred to the liquid nitrogen freezer (or the end stage is reached in a programmable freezer) the temperature drops rapidly to between −180 and −196°C.

A controlled cooling rate can be achieved in several ways:

(1) Lay the ampoules on cotton wool in a polystyrene foam box with a wall thickness of ∼15 mm. This box, plus the cotton wool, should provide sufficient insulation such that the ampoules will cool at 1°C/min when the box is placed at −70°C or −90°C in a regular deep freeze or insulated container with solid CO_2.

Fig. 20.2. Ampoules on Cane. Plastic ampoules are clipped onto an aluminum cane (bottom), enclosed in a cardboard tube (middle), and placed inside an insulating foam tube (top). The insulating tube is plugged at either end with cotton or another suitable insulating material.

(2) Insert the canes in tubular foam pipe insulation, with a wall thickness of ∼15 mm (Fig. 20.2), and place the insulation at −70°C or −90°C in a regular deep freeze or insulated container with solid CO_2.

(3) Place the ampoules in a Taylor Wharton freezer neck plug (Fig. 20.3) and insert the plug into the neck of the nitrogen freezer.

(4) Place the ampoules in a Nalge Nunc freezing container (Fig. 20.4) and place at −70°C or −80°C.

(5) Use a controlled-rate freezer programmed to freeze at 1°C/min (Fig. 20.5), with accelerated freezing through the eutectic point (*see* Fig. 20.1).

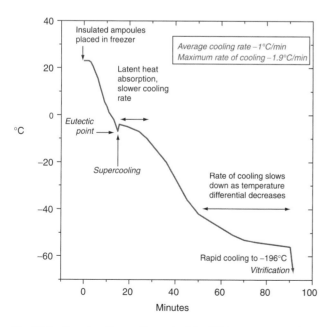

Fig. 20.1. Freezing Curve. Record of the fall in temperature in an ampoule containing medium, clipped with five other ampoules on an aluminum cane, enclosed in a cardboard tube, placed within a polyurea-foam tube (*see* Fig. 20.2), and placed in a freezer at −70°C.

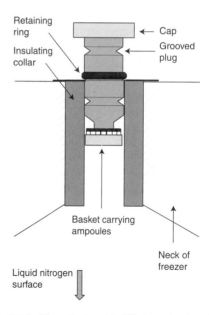

Fig. 20.3. Neck Plug Cooler. Modified neck plug for narrow-necked freezers, allowing controlled cooling at different rates (Taylor Wharton). Shown is the section of the freezer neck with the modified neck plug in place. The "O" ring is used to set the height of the ampoules within the neck of the freezer. The lower the height, the faster the cooling.

Fig. 20.4. Nalge Nunc Cooler. Plastic holder with fluid-filled base. The specific heat of the coolant in the base insulates the container and gives a cooling rate of ∼1°C/min in the ampoules.

With any of the first four methods, e.g., the insulated tube method in (2) (*see* Fig. 20.2), the cooling rate will be an average of varying rates throughout the curve, and no attempt is made to control supercooling at the eutectic point or the duration of the plateau during latent heat absorption, both of which may impair survival. If recovery is low, it is possible to change the average cooling rate (i.e., by use of more or less insulation). Use of a programmable freezer (*see* Fig. 20.5) with a probe, which senses the temperature of the ampoule and adds liquid nitrogen to the freezing chamber at the correct rate to achieve a preprogrammed cooling rate, and which can seed freezing as the cell suspension reaches the eutectic, minimizes the stress of supercooling and can achieve a linear cooling rate throughout the range. Different cooling rates at different phases of the cooling curve, optimized experimentally to suit the cells being frozen, can be programmed into the cooling curve. Programmable coolers are, however, relatively expensive, compared to the simple devices described in (1)−(4) above, and have few advantages unless you wish to vary the cooling rate [e.g., Foreman & Pegg, 1979] or alter the shape of the cooling curve.

With the insulated container methods, the cooling rate is proportional to the difference in temperature between the ampoules and the ambient air. If the ampoules are placed in a freezer at −70°C, they will cool rapidly to around −50°C, but the cooling rate falls off significantly after that (*see* Fig. 20.1). Hence, the time that the ampoules spend in the −70°C freezer needs to be longer than the amount of time projected by a 1°C/min cooling rate, as the bottom of the curve is asymptotic. It is safer to leave the ampoules at −70°C overnight before transferring them to liquid nitrogen. Furthermore, when removed from the freezing device, they will heat up at a rate of ∼10°C/min. It is critical that they do not warm up above −50°C, as they will start to deteriorate, so the transfer to the liquid nitrogen freezer must take significantly less than two minutes.

20.3.5 Cryofreezers

Storage in a liquid nitrogen freezer (Figs. 20.6, 20.7; *see also* Section 5.5.2) is currently the most satisfactory method of

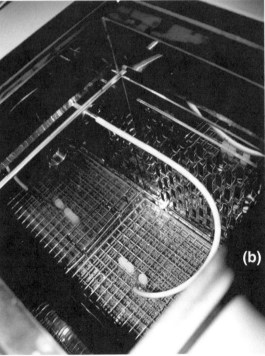

Fig. 20.5. Programmable Freezer. Ampoules are placed in an insulated chamber, and the cooling rate is regulated by injecting liquid nitrogen into the chamber at a rate determined by a sensor on the rack with the ampoules and a preset program in the console unit (Planer Biomed). (a) Control unit and freezing chamber (lid open). (b) Close-up of a freezing chamber with four ampoules, one with a probe in it.

preserving cultured cells [Hay et al., 2000]. The frozen cells are transferred rapidly to the cryofreezer when they are at or below −70°C.

Cryofreezers differ in design depending on size of the access neck, storage system employed, and location of liquid nitrogen (Fig. 20.7).

Neck size. Canister storage systems tend to have narrow necks (Figs. 20.6a,b, 20.7a), which reduces the rate of evaporation of the liquid nitrogen but makes access a little

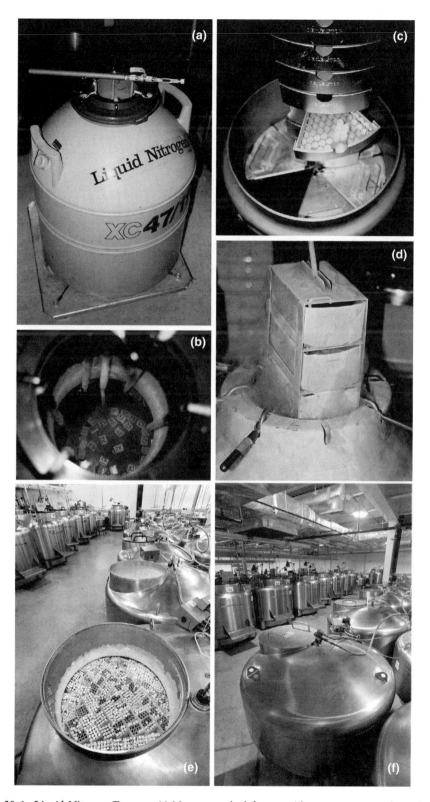

Fig. 20.6. Liquid Nitrogen Freezers. (a) Narrow-necked freezer with storage on canes in canisters. (b) Interior of the narrow-necked freezer, looking down on canes in canister, positioned in center as it would be for withdrawal. Normal storage position is under shoulder of freezer, just visible top right. (c) Wide-necked freezer with storage in triangular drawers. (d) Narrow-necked freezer with storage in square drawers (*see also* Fig. 20.7). (e) High-capacity freezer with offset access port open, revealing canes in canisters. (f) High-capacity freezers in a cell bank. Nearest freezer shows connections for monitoring and automatic filling. (e & f, photos courtesy of ATCC.)

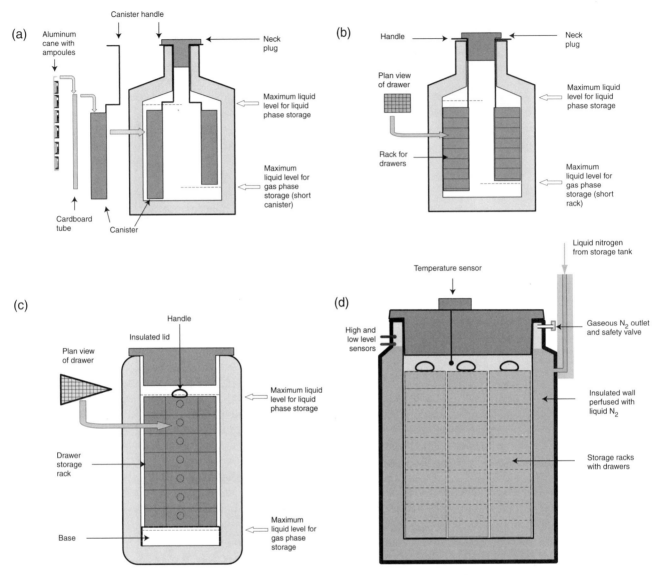

Fig. 20.7. Nitrogen Freezer Design. Four main types of nitrogen freezers: (a) Narrow-necked with ampoules on canes in canisters (high capacity, low boil-off rate). (b) Narrow-necked with ampoules in square racks (moderate capacity, low boil-off rate). (c) Wide-necked with ampoules in triangular racks (high capacity, high boil-off rate). (d) Wide-necked with storage in drawers; piped liquid nitrogen perfused through freezer wall and level controlled automatically by high and low level sensors (top left).

awkward. Wide-necked freezers are chosen for ease of access and maximum capacity, usually with storage in sections within drawers, but tend to have a faster evaporation rate. However, it is possible to select a relatively narrow-necked freezer while still using a tray system for storage. Freezers are available with inventory control based on square-array storage trays; these trays are mounted on racks that are accessed by the same system as the cane and canister of conventional narrow-necked freezers (Figs. 20.6d, 20.7b). These freezers do not have the storage capacity of the cane and canister, but have equivalent holding times and the honeycomb storage array that many people prefer (*see* below).

Storage system. There are two mains types of storage used for 1-mL ampoules for cell culture work. The cane system, based on the storage of sperm in straws, uses ampoules clipped on to an aluminum cane, inserted into a cardboard tube, and placed within cylindrical canisters in the freezer. It has the advantage that ampoules can be handled in multiples of six at a time, with all the ampoules on one cane being from the same cell line, making the transfer from the cooling device to the freezer easier and quicker. The canes can be colored and numbered, making location fairly easy, and ampoules can be withdrawn without exposing all the other ampoules to the warm atmosphere. Storage in rectangular drawers is preferred by some users, who feel that retrieval is easier and individual ampoules can be identified by the drawer number and the coordinates within the drawer. It does mean, however, that the total contents of the drawer, which can be from 20 to 100 ampoules, are exposed at one time when an ampoule is retrieved, and the whole stack must be lifted out if you are accessing one of the lower drawers. Also, loading a large number of ampoules into the drawer must be done one ampoule at a time, risking delay and overheating.

Ampoules. Plastic ampoules are preferred for the average experimental and teaching laboratory, as they are safer and more convenient, but some repositories and cell banks prefer glass ampoules for seed stocks, because the long-term storage properties of glass are well characterized and, when correctly performed, sealing is absolute. Plastic ampoules are usually polypropylene, and 1.2 mL is probably the most popular size (*see* Plate 22e). They may be labeled with a fine-tipped marker, or labels, which should be alcohol resistant and able to withstand the low temperature of the freezer (*see* Appendix II: Cryomarkers and Cryolabels). The label should show the cell strain designation and, preferably, the date and user's initials, although the latter is not always feasible in the available space. Different colored caps also help identification (*see* Plate 22e). Remember, cell cultures stored in liquid nitrogen may well outlive you! They can easily outlive your stay in a particular laboratory. The record therefore should be readily interpreted by others and sufficiently comprehensive so that the cells may be of use to others.

Plastic ampoules require the correct torsion for closing, as they will leak if too slack or too tight (due to distortion of the o-ring). It is worth practicing with a new batch to make sure (a) that they seal correctly and (b) that they withstand the low temperature; occasionally a faulty batch of ampoules may shatter on thawing. Plastic ampoules are of a larger diameter and taller than equivalent glass ampoules, so special canes must be used.

Δ ***Safety Note.*** Inexperienced users should avoid using glass ampoules as they have a serious risk of explosion when thawed. If glass ampoules are used, they must be perfectly and quickly sealed in a gas-oxygen flame. If sealing takes too long, the cells will heat up and die, and the air in the ampoule will expand and blow a hole in the top of the ampoule. If the ampoule is not perfectly sealed, it may inspire liquid nitrogen during storage in the liquid phase of the nitrogen freezer and will subsequently explode violently on thawing.

It is possible to check for leakage by placing ampoules in a dish of 1% methylene blue in 70% alcohol at 4°C for 10 min before freezing. If the ampoules are not properly sealed, the methylene blue will be drawn into the ampoule, and the ampoule should be discarded.

Location of liquid nitrogen. If the liquid nitrogen is located in the main body of the freezer there is a choice of filling the freezer and submerging the ampoules, or only part-filling and storing the ampoules in the vapor phase. Storage

in the liquid phase means that the container can be filled and the liquid nitrogen will therefore last longer, but risks uptake of nitrogen by leaky ampoules, which will then explode violently on thawing. There is also a greater likelihood of transfer of contamination between ampoules and the buildup of contamination from outside, carried in when material is introduced and concentrated by the constant evaporation of the liquid nitrogen in the tank. With the introduction of improved insulation and reduced evaporation, vapor-phase storage is probably preferable. It also eliminates the risk of splashing when the liquid nitrogen boils when something is inserted and reduces evaporation of nitrogen into the room air. There is, however, a gradient in the temperature from the surface of the liquid nitrogen up to the neck of approximately 80°C, from −190°C to around −110°C in gas phase storage [Rowley & Byrne, 1992], although the design and composition of the racking system may help to eliminate this gradient.

Some freezers have the liquid nitrogen located within the wall of the freezer and not in the storage compartment. It is replenished by an automatic feed with high and low level controls (Fig. 20.7d), and evaporated nitrogen is released via a relief valve. This has the advantages of gas-phase storage, with the added advantage of a lower consumption of liquid nitrogen and elimination of the temperature gradient. However, the nitrogen level is not visible and cannot be measured by dipstick, so complete reliance has to be made on electronic monitoring (*see* below). In addition, any blockage of nitrogen flow within the freezer wall can be very difficult or even impossible to eliminate, so it is essential that the liquid nitrogen be filtered and that steps are taken to ensure that no water, or water vapor, enters the system, as ice can also block nitrogen flow.

Δ ***Safety Note.*** Biohazardous material *must* be stored in the gas phase, and teaching and demonstrating should also be done with gas-phase storage. Above all, if liquid-phase storage is used, the user must be made aware of the explosion hazard of both glass and plastic and must wear a face shield or goggles.

Monitoring and replenishing liquid nitrogen. The investment in the contents of a nitrogen freezer can be considerable and must be protected by a strict monitoring regime and electronic liquid level alarms. When the nitrogen is in the storage compartment, the level should be monitored at least once per week with a dipstick, recording the level on a chart. This should be done, even if automatic filling is employed, as these systems can fail. Where the liquid nitrogen is totally enclosed, refilling is automatic, but must still be backed up, preferably with two independent temperature recorders, both of which should sound an alarm, one if the temperature rises above −170°C and one above −150°C.

If liquid nitrogen storage is not available, the cells may be stored in a conventional freezer. The temperature in this freezer should be as low as possible; little deterioration has

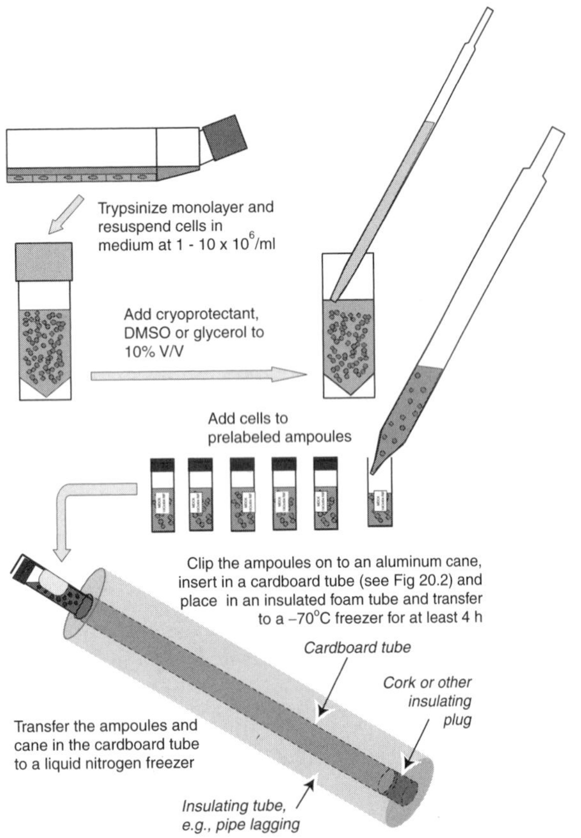
Trypsinize monolayer and resuspend cells in medium at 1 - 10 x 10^6/ml
Add cryoprotectant, DMSO or glycerol to 10% V/V
Add cells to prelabeled ampoules
Clip the ampoules on to an aluminum cane, insert in a cardboard tube (see Fig 20.2) and place in an insulated foam tube and transfer to a –70°C freezer for at least 4 h
Cardboard tube
Cork or other insulating plug
Transfer the ampoules and cane in the cardboard tube to a liquid nitrogen freezer
Insulating tube, e.g., pipe lagging

Fig. 20.8. Freezing Cells. Trypsinized cells, in medium with cryoprotectant, aliquoted into ampoules, which are then clipped on to an aluminum cane, inserted into a cardboard tube, and inserted into an insulated tube. The tube and contents are placed at −70°C or −80°C for 4 h or overnight, before transferring the cardboard tube containing the ampoules to a liquid nitrogen freezer with canister storage (*see* Fig. 20.7a).

been found at −196°C [Green et al., 1967], but significant deterioration (5–10% per annum) may occur at −70°C.

PROTOCOL 20.1. FREEZING CELLS

Outline

Grow the culture to late log phase, prepare a high concentration cell suspension in medium with a cryoprotectant, aliquot into ampoules, and freeze slowly (*see* Fig. 20.8).

Materials

Sterile or Aseptically Prepared:

- ❑ Culture to be frozen
- ❑ If monolayer: PBSA and 0.25% crude trypsin
- ❑ Growth medium (Serum improves survival of the cells after freezing; up to 50%, or even pure, serum has been used. If serum is being used with serum-free cultures, it should be washed off after thawing.)
- ❑ Cryoprotectant, free of impurities (*see* above): DMSO in a glass or polypropylene vial, or glycerol, fresh and in a universal container
- ❑ Syringe, 1–5 mL, for dispensing glycerol (because it is viscous)
- ❑ Plastic ampoules, 1.2 mL, prelabeled with the cell line designation and the date of freezing

Nonsterile:

- ❑ Hemocytometer or electronic cell counter
- ❑ Canes or racks for storage (racks may already be in place in the freezer)
- ❑ Insulated container for freezing: polystyrene box lined with cotton wool or plastic foam insulation tube (*see* Fig. 20.2; or controlled rate freezer if available, *see* Fig. 20.5)
- ❑ Protective gloves, nitrile

Protocol

1. Make sure the culture satisfies the criteria for freezing (*see* Table 20.1) and check by eye and on microscope for:
   (a) Healthy appearance (*see* Section 13.6.1)
   (b) Morphological characteristics (*see* Section 16.4)
   (c) Phase of growth cycle (should be late log phase before entering plateau (*see* Section 13.6.6)
   (d) Freedom from contamination (*see* Section 19.3)
2. Grow the culture up to the late log phase and, if you are using a monolayer, trypsinize and count the cells (*see* Protocol 13.2). If you are using a suspension, count and centrifuge the cells (*see* Protocol 13.3).
3. Resuspend at 2×10^6–2×10^7 cells/mL.
4. Dilute one of the cryoprotectants in growth medium to make freezing medium:
   (a) Add dimethyl sulfoxide (DMSO) to 10–20%

Δ ***Safety Note.*** DMSO can penetrate many synthetic and natural membranes, including *skin* and rubber gloves [Horita & Weber, 1964]. Consequently, any potentially harmful substances in regular use (e.g., carcinogens) may well be carried into the circulation through the skin and even through rubber gloves. DMSO should always be handled with caution, particularly in the presence of any toxic substances.

   or
   (b) Add glycerol to 20–30%.

5. Dilute the cell suspension 1:1 with freezing medium to give approximately 1×10^6–$1 \times$

10^7 cells/mL and 5–10% DMSO (or 10–15% glycerol). It is not advisable to place ampoules on ice in an attempt to minimize deterioration of the cells. A delay of up to 30 min at room temperature is not harmful when using DMSO and is beneficial when using glycerol.

6. Dispense the cell suspensions into prelabeled ampoules, and cap the ampoules with sufficient torsion to seal the ampoule without distorting the gasket.

7. Place the ampoules on canes for canister storage (*see* Figs. 20.2, 20.6a,b, 20.7a), or leave them loose for drawer storage (*see* Figs. 20.6c,d, 20.7b-d).

8. Freeze the ampoules at 1°C/min by one of the methods described above (*see* Section 20.3.4). With the insulated container methods, this will take a minimum of 4–6 h after placing them at −70°C if starting from a 20°C ambient temperature (*see* Fig. 20.1 and Section 20.3.4), but preferably leave the ampoules in the container at −70°C overnight.

9. When the ampoules have reached −70°C or lower, check the freezer record before removing the ampoules from the −70°C freezer or controlled rate freezer, and identify a suitable location for the ampoules.

10. Transfer the ampoules to the liquid N_2 freezer, preferably not submerged in the liquid, placing the cane and tube into the predetermined canister or individual ampoules into the correct spaces in the predetermined drawer. This transfer must be done quickly (<2 min), as the ampoules will reheat at ~10°C/min, and the cells will deteriorate rapidly if the temperature rises above −50°C.

△ *Safety Note.* Protective gloves and a face mask should be used when placing ampoules in liquid nitrogen.

11. When the ampoules are safely located in the freezer, complete the appropriate entries in the freezer index (*see* Tables 20.2 and 20.3).

20.3.6 Freezer Records

Records should provide (a) an inventory showing what is in each part of the freezer, (b) an indication of free storage spaces, and (c) a cell strain index, describing the cell line, its designation, its origin, details of maintenance and freezing procedures, what its special characteristics are, and where it is located. This record may be kept on a conventional card index, but a computerized database will give superior data storage and retrieval. This type of data can be provided by separate tables within the same database used for the

provenance of the cell line (Tables 20.2, 20.3; *see also* Sections 12.3.10, 13.8). Material stored on disks or tape must have back-up copies on disk or tape or must have a hard-copy printout.

Using a computerized database requires that the curator of the freezers manages this database. If entries are to be made by users, user stocks can have both read and write access, whereas seed stock and distribution stock should be accessible only to the curator. Alternatively, the whole file can be read only and updated by the curator from paper entries on cards or a log book.

20.3.7 Thawing Stored Ampoules

When required, cells are thawed and reseeded at a relatively high concentration to optimize recovery. The ampoule should be thawed as rapidly as possible, to minimize intracellular ice crystal growth during the warming process. This can be done in warm water, in a bucket or waterbath, but, if the ampoule has been submerged in liquid nitrogen during storage, the warming bath must be covered in case the ampoule has leaked and inspired liquid nitrogen, when it will explode violently on warming.

The cell suspension should be diluted slowly after thawing as rapid dilution reduces viability. This gradual process is particularly important with DMSO, with which sudden dilution can cause severe osmotic damage and reduce cell survival by half. Most cells do not require centrifugation, as replacing the medium the following day will suffice for a monolayer or dilution for a suspension. However, some cells (often suspension-growing cells) are more sensitive to cryoprotectants, particularly DMSO, and must be centrifuged after thawing but still need to be diluted slowly in medium first.

PROTOCOL 20.2. THAWING FROZEN CELLS

Outline
Thaw the cells rapidly, dilute them slowly, and reseed them at a high cell density (Fig. 20.9).

Materials
Sterile:
❑ Culture flask
❑ Centrifuge tube (if centrifugation is required)
❑ Growth medium
❑ Pipettes, 1 mL, 10 mL
❑ Syringe and 19-g needle (if you are using glass ampoules)
Nonsterile:
❑ Protective gloves and face mask
❑ Sterile water at 37°C, 10 cm deep in a clean, alcohol-swabbed bucket with lid
❑ Forceps

TABLE 20.2. Cell Line Record

Cell line	**Freeze date:**										New card
	Location:										

Species Normal/ neoplastic Adult/ Fetal/NB	**Mycoplasma:** Method Date of test Result	**Freeze instructions:** Rate Cryoprotectant %
Tissue Site Pass./ Gen. No	**Authentication:**	**Thaw instructions:** Thaw rapidly to 37°C
Author Ref. Mode of growth	Date of test Method	Dilute to: 5 mL 10 mL 20 mL 50 mL
	Normal Maintenance:	Centrifuge to remove cryoprotectant? Yes/No
Special characteristics:	Subculture frequency Seeding conc. Agent Medium change frequency Type Serum, etc.. . . . . . %	**Special requirements**
	Gas phase Buffer pH	
Person completing card Date	**Any other special conditions**	**Biohazard precautions:**

- ❑ Alcohol, 70%
- ❑ Swabs
- ❑ Naphthalene Black (Amido Black), 1%, or Trypan Blue, 0.4%.

Protocol

1. Check the index for the location of the ampoule to be thawed.
2. Collect all materials, prepare the medium, and label the culture flask.
3. Retrieve the ampoule from the freezer, check from the label that it is the correct one, and, if it has not been submerged in liquid nitrogen, place it in sterile water at 37°C in a beaker, plastic bucket, or water bath. If possible, avoid getting water up

TABLE 20.3. Freezer Record

Position: Freezer no. . . . . . .   Canister/Section no. . . . . .   Tube/drawer no. . . . .

Cell strain/line: . . . .   Freeze date . . . .   Frozen by . . . .

No. of ampoules frozen . . . .   No. of cells/ampoule . . . .   in . . . . . . mL

Growth medium . . . .   Serum . . . . Conc. . . . . . .   Freeze medium . . . .

Method of cooling . . . .   Cooling rate . . . . . .

Thawing record:

Thaw date	No. of ampoules	No. left	Seeding			Viability			Notes
			Conc.	Vol.	Medium	Dye exclusion	~% attached by 24 h	Cloning efficiency	

to the cap as this will increase the chance of contamination.

Δ *Safety Note.* A closed lab coat and gloves must be worn when removing the ampoule from the freezer. If the ampoule has been stored **submerged** in liquid nitrogen, **a face shield, or protective goggles**, as well as a closed lab coat and gloves, must be worn. Ampoules, including plastic ampoules, stored in the liquid phase may inspire the liquid nitrogen and, on thawing, will explode violently. In this case **a plastic bucket with a lid must be used for thawing** to contain any explosion.

4. When the ampoule has thawed, double-check the label to confirm the identity of the cells; then swab the ampoule thoroughly with 70% alcohol, and open it in a laminar-flow hood.
5. Transfer the contents of the ampoule to a culture flask with a 1-mL pipette.
6. Add medium slowly to the cell suspension: 10 mL over about 2 min added dropwise at the start, and

then a little faster, gradually diluting the cells and cryoprotectant.

For cells that require centrifugation to remove the cryoprotectant:
(a) Dilute the cells slowly, as in Step 6, but in a centrifuge tube or universal container.
(b) Centrifuge them for 2 min at 100 g.
(c) Discard the supernatant medium with the cryoprotectant.
(d) Resuspend the cells in fresh growth medium.
(e) Seed flask for culture.
7. The dregs in the ampoule may be stained with Naphthalene Black or Trypan Blue to determine their viability (*see* Protocol 22.3.1).
8. Check after 24 h:
(a) For attached monolayer cells, confirm attachment and try to estimate percentage survival based on photographs of cells at the expected density (cells/cm^2) with full survival (*see* Sections 13.6.1, 16.4.5; Plate 4).
(b) For suspension-growing cells, check appearance (clear cytoplasm, lack of granularity),

and dilute to regular seeding concentration. This can be made more precise if the cells are counted and an estimate of viability is made (*see* Section 22.3.1), in which case the cells can be diluted to the regular seeding concentration of viable cells.

The dye exclusion viability and the approximate take (e.g., the proportion of cells attached after 24 h) should be recorded on the appropriate record card or file to assist in future thawing. One ampoule should be thawed from each new freezing, to check that the operation was successful.

Freezing flasks. Whole flasks may be frozen by growing the cells to late log phase, adding 5–10% DMSO to the smallest volume of medium that will effectively cover the monolayer, and placing the flask in an expanded polystyrene container of 15-mm wall thickness [Ohno et al., 1991]. The insulated container is placed in a −70°C to −90°C freezer and will freeze at approximately 1°C/min. Survival is good for several months, as long as the flask in its container is not removed from the freezer. Twenty-four-well plates may also be frozen in the same manner [Ure et al., 1992] with about 150 μL freezing medium per well and can be used to store large numbers of clones during evaluation procedures.

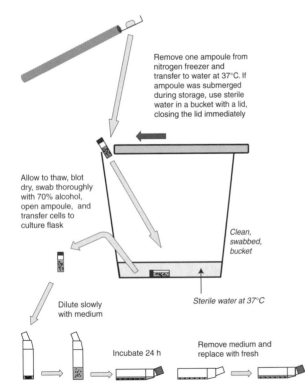

Remove one ampoule from nitrogen freezer and transfer to water at 37°C. If ampoule was submerged during storage, use sterile water in a bucket with a lid, closing the lid immediately

Allow to thaw, blot dry, swab thoroughly with 70% alcohol, open ampoule, and transfer cells to culture flask

Clean, swabbed, bucket

Sterile water at 37°C

Dilute slowly with medium

Incubate 24 h

Remove medium and replace with fresh

Fig. 20.9. Thawing Cells. Ampoules are removed from the freezer and thawed rapidly in warm water; under cover, to avoid the risk of explosion if the ampoules have been stored in the liquid phase.

20.4 DESIGN AND CONTROL OF FREEZER STOCKS

As soon as a small surplus of cells becomes available, from subculturing a primary culture or newly acquired cell line, and shown to be free of contamination, a few ampoules should be frozen as what is called a *token freeze*. When the cell line has been propagated successfully, or a cloned cell strain selected with the desired characteristics, and its identity and freedom from contamination have been confirmed, then a *seed stock* should be stored frozen (Table 20.4).

20.4.1 Freezer Inventory Control

The seed stock should be protected and not be made available for general issue. When an ampoule is thawed to check the viability of the seed stock freezing, and if usage of the cell line is anticipated in the near future, cells can be grown up for a *distribution stock* and frozen; ampoules from this stock are issued to individuals as required. Individual users requiring stocks over a prolonged period should then freeze down their own *user stocks*, which should be discarded when the work is finished. User stocks should never be passed on to another user as they will not have been fully validated; new users should request a culture or ampoule from the distribution stock. When the distribution stock becomes depleted, it may be replenished from the seed stock. When the seed stock falls below five ampoules, it should be replenished before any other ampoules are issued, and with the minimum increase in generation number from the first freezing.

20.4.2 Serial Replacement of Culture Stock

Stock cultures should be replaced from the freezer at regular intervals to minimize the effects of genetic drift and phenotypic variation. After a cell line has been in culture for $2\frac{1}{2}$ months, thaw out another vial, check its characteristics, make sure that it is free from contamination (*see* Section 19.3), and expand it to replace existing stocks. Discard the existing stocks when they have been out of the freezer for 3 months, and move on to the new stock (Fig. 20.10). Repeat this process every 3 months with cells that have a population-doubling time (PDT) of approximately 24 h; cell lines with shorter or longer PDTs may need shorter or longer replacement intervals, respectively.

20.5 CELL BANKS

Several cell banks exist (*see* Appendix II) for the secure storage and distribution of validated cell lines. Because many cell lines may come under patent restrictions, particularly hybridomas and other genetically modified cell lines, it has also been necessary to provide patent repositories with limited access.

As a general rule, it is preferable to obtain your initial seed stock from a reputable cell bank, where the necessary characterization and quality control will have been done.

TABLE 20.4. Acquisition and Storage of Cell Lines

Stage	Source	No. of ampoules	Distribution	Validation
Token Freeze	Originator	1–3	None	Provenance only
Seed Stock	Original stock or token freeze	12	None (replenishment of distribution stock only)	Viability Authentication Transformation Contamination Cross-contamination
Distribution Stock	Test thaw from seed stock	50–100 (or more as required)	Users, including other laboratories	Viability Contamination Cross-contamination
User Stock	Distribution stock	20 (5-year project, culture stock replacement 4× per year)	None	Viability Contamination Cross-contamination

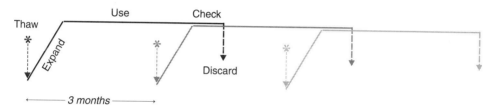

Fig. 20.10. Serial Culture Replacement. An ampoule is thawed and the cells grown up to the desired bulk for regular use and maintained for 3 months. Three months after the first thaw, a fresh ampoule is thawed, grown up, and, after checking characteristics against current stock, used to replace the current stock, which is then discarded. The cycle is repeated every 3 months, so that no culture remains in use for more than 3 months.

Furthermore, it is highly recommended that you submit valuable cultures to a cell bank in addition to maintaining your own frozen stock, as the former will protect you against loss of your own lines and allow their distribution to others. If you feel that your cells should not be distributed, then they can be banked with that restriction placed on them.

Most cell banks also make their catalog information available on-line, where they act as a major information resource. There are also several data banks that can be accessed on-line and that maintain information on cell lines held by subscribers in their own laboratories (*see* Appendix II). This type of data bank provides a vast increase in the amount of material that is potentially available, but it must be remembered that cell lines obtained from other laboratories will vary significantly in the amount of characterization that they have had and in the quality of their maintenance.

20.6 TRANSPORTING CELLS

Cultures may be transferred from one laboratory to another as frozen ampoules or as living cultures. In either case:

(1) Advise the recipient as to when the cells are to be shipped;
(2) Fax or E-mail instructions on the following:

a) what to do on receipt,
b) medium or serum required,
c) any special supplements, and
d) subculture regimen;

(3) Tape the data sheet for the cells and a copy of the instructions to the outside of the package so that the recipient knows what to do before opening it.

20.6.1 Frozen Ampoules

Ship frozen ampoules in solid CO_2, in a thick-walled polystyrene foam container (Fig. 20.11a). Transfer the ampoule from the liquid nitrogen freezer to the solid CO_2 as quickly as possible, as it will warm up at $\sim 10-20°C/min$ and must not rise above $-50°C$. Place a sealable polypropylene centrifuge tube on solid CO_2 in an insulated box. Remove the ampoule from the freezer, quickly wrap it in absorbent paper tissue (in case of leakage), and place it in the polypropylene tube and seal with a tight cap. The box should be about $30 \times 30 \times 30$ cm ($12 \times 12 \times 12$ in.) with a wall thickness of 5 cm (2 in.) and a central space of about $20 \times 20 \times 20$ cm ($8 \times 8 \times 8$ in.). You will need about 5 kg (12 lb) of solid CO_2 to fill the central space. Usually, cells will remain frozen for up to 3 days if properly packed, but if they thaw slowly their viability will decline rapidly. The carrier must be informed when cells are shipped in solid CO_2.

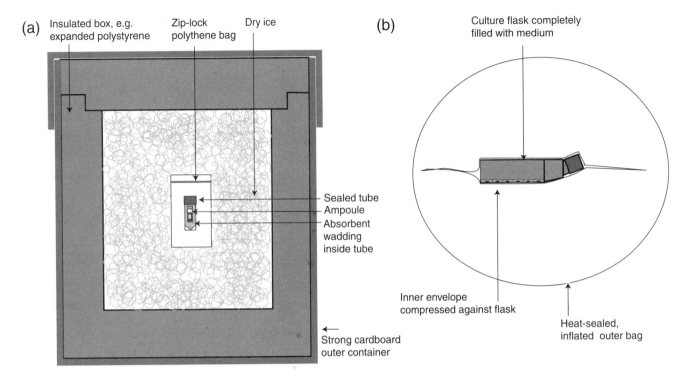

Fig. 20.11. Transportation Containers for Cells. Cells shipped as an ampoule of frozen cells or as a growing culture. (a) Rigid cardboard box with insulating expanded polystyrene lining containing dry ice with the ampoule, sealed in a tube with absorbent wadding and the tube sealed in a zip-lock bag, placed in the center of the dry ice. (b) Double-skin plastic container with flask trapped in inner compartment by inflating the outer chamber (Air Box, Air Packaging Technologies).

On arrival, the ampoule should be thawed and seeded as normal (*see* Protocol 20.2).

20.6.2 Living Cultures

Alternatively, cells may be shipped as a growing culture. The cells should be at the mid- to late log phase; confluent or postconfluent cultures will exhaust the medium more rapidly and may tend to detach in transit. The flask should be filled to the top with medium, taped securely around the neck with a stretch–type waterproof adhesive tape, and sealed in a small polythene bag. The bagged flask is packed within a rigid container filled with absorbent packing. This container is then sealed in a polythene bag and then in plastic foam or bubble wrap. Alternatively, there is an inflatable bag that is ideal for transporting cells (Air Packaging Technologies, Inc). The flask is sealed in an inner envelope (Fig. 20.11b) and cushioned by the inflatable outer jacket. Place on the container a label that says "fragile" and, in large letters: DO NOT FREEZE!

On receipt, the flask is removed from the packing, swabbed thoroughly with 70% alcohol, and opened under sterile conditions. Most of the medium is removed from the flask, leaving only the normal amount for culture—e.g., 5 mL for a 25-cm^2 flask—and the remainder kept for feeding the cells, if required. The culture can be weaned onto new medium when it is ready for the first feed, but keep the original shipping medium in case there are any problems of adaptation to the new stock.

It is usually better to ship cells via a courier, who should be briefed as to the contents of the package and the urgency of delivery. International mailings will require negotiation with customs controls, and a competent nominated agent who is familiar with this type of importation can help to move the package through customs. Cells shipped to the United States are required to be quarantined and tested by the Department of Agriculture, so it is better to arrange this shipment through one of the cell banks, ECACC or ATCC.

CHAPTER 21

Quantitation

Quantitation in cell culture is required for the characterization of the growth properties of different cell lines for experimental analyses and to establish reproducible culture conditions, for the consistency of primary culture and the maintenance of cell lines (*see* Section 13.6). Working under sterile conditions and the need to minimize the time that cultures spend out of the incubator often lead to a degree of compromise between speed and accuracy. Although primary culture and normal maintenance require a quantitative approach to ensure reproducibility, speed of handling is the key to good cell survival and growth, so in these situations reduced accuracy, in pipetting, for example, may be acceptable in pursuit of rapid handling. However, under experimental conditions, accuracy should increase, even if the operations become a little slower. Although an error of ±10% may be acceptable in routine maintenance, this should be ±5% under experimental conditions.

This chapter covers cell counting technology and a number of other assays used in quantifying cell proliferation, as well as other basic assays for determining cell bulk, such as DNA and protein estimations. Assays relating to cytotoxicity and cell survival will be found in Chapter 22.

21.1 CELL COUNTING

Although estimates can be made of the stage of growth of a culture from its appearance under the microscope, the standardization of culture conditions and proper quantitative experiments are difficult to analyze and reproduce unless the cells are counted before and after, and preferably during, each experiment.

21.1.1 Hemocytometer

The concentration of a cell suspension may be determined by placing the cells in an optically flat chamber under a microscope (Fig. 21.1). The cell number within a defined area of known depth (i.e., within a defined volume) is counted, and the cell concentration is derived from the count.

Protocol 21.1 can be used in conjunction with Exercise 12 (*see* Chapter 2).

PROTOCOL 21.1. CELL COUNTING BY HEMOCYTOMETER

Outline
Trypsinize a monolayer culture, or sample a suspension culture, prepare a hemocytometer slide, and add the cells to the counting chamber. Count the cells on a microscope and calculate the cell concentration.

Materials
Sterile:
- D-PBSA
- Crude trypsin, 0.25%
- Growth medium
- Yellow pipettor tips

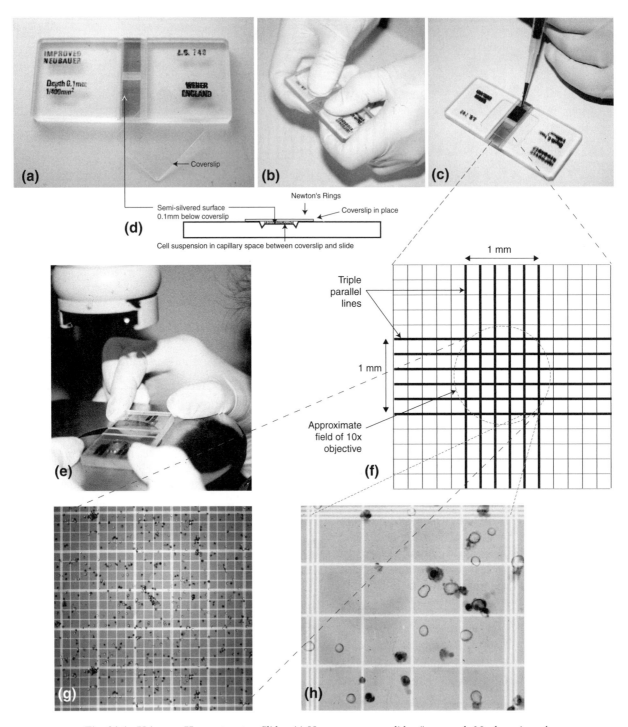

Fig. 21.1. Using a Hemocytometer Slide. (a) Hemocytometer slide (improved Neubauer) and coverslip before use. (b) Pressing coverslip down onto slide. (c) Adding a cell suspension to an assembled slide. (d) Longitudinal section of the slide, showing the position of the cell sample in a 0.1-mm-deep chamber. (e) Viewing slide on microscope. (f) Magnified view of the total area of the grid. The central area enclosed by the dotted circle is that area which would be covered by the average 10× objective. This area covers approximately the central 1 × 1 mm square of the grid. (g) Low-power (10× objective) microphotograph showing the 25 smaller 200 × 200-μm squares of a slide, which make up the central 1 × 1-mm square, loaded with cells pretreated with Naphthalene Black (Amido Black). (h) High-power (40× objective) microphotograph of one of the smaller 200-μm squares, bounded by three parallel lines and containing 16 of the smallest (50 × 50 μm) squares. Viable cells are unstained and clear, with a refractile ring around them; nonviable cells are dark and have no refractile ring.

Nonsterile:

- ❏ Pipettor, 20 µL or adjustable 100 µL
- ❏ Hemocytometer (Improved Neubauer)
- ❏ Tally counter
- ❏ Microscope

Protocol

1. Sample the cells:
   - (a) For a monolayer,
     - i) Trypsinize the monolayer as for routine subculture (*see* Protocol 13.2) and resuspend in medium to give an estimated 1×10^6/mL. Where samples are being counted in a growth experiment, the trypsin need not be removed and the cells can be dispersed in the trypsinate and counted directly, or after diluting 50:50 with medium containing serum if the cells tend to reaggregate.
     - ii) Mix the suspension thoroughly to disperse the cells, and transfer a small sample (~1 mL) to a vial or universal container.
   - (b) For a suspension culture,
     - i) Mix the suspension thoroughly to disperse any clumps;
     - ii) Transfer 1 mL of the suspension to a vial or universal container.
       A minimum of approximately 1×10^6 cells/mL is required for this method, so the suspension may need to be concentrated by centrifuging it (at 100 *g* for 2 min) and resuspending it in a measured smaller volume.

2. Prepare the slide:
   - (a) Clean the surface of the slide with 70% alcohol, taking care not to scratch the semisilvered surface.
   - (b) Clean the coverslip, and, wetting the edges very slightly, press it down over the grooves and semisilvered counting area (*see* Fig. 21.1). The appearance of interference patterns ("Newton's rings"—rainbow colors between the coverslip and the slide, like the rings formed by oil on water) indicates that the coverslip is properly attached, thereby determining the depth of the counting chamber.

3. Mix the cell sample thoroughly, pipetting vigorously to disperse any clumps, and collect 20 µL into the tip of a pipettor.

4. Transfer the cell suspension immediately to the edge of the hemocytometer chamber, and expel the suspension and let it be drawn under the coverslip by capillarity. Do not overfill or underfill the chamber, or else its dimensions may change, due to alterations in the surface tension; the fluid should run only to the edges of the grooves.

5. Mix the cell suspension, reload the pipettor, and fill the second chamber if there is one.

6. Blot off any surplus fluid (without drawing from under the coverslip), and transfer the slide to the microscope stage.

7. Select a 10× objective, and focus on the grid lines in the chamber (*see* Fig. 21.1). If focusing is difficult because of poor contrast, close down the field iris, or make the lighting slightly oblique by offsetting the condenser.

8. Move the slide so that the field you see is the central area of the grid and is the largest area that you can see bounded by three parallel lines. This area is 1 mm². With a standard 10× objective, this area will almost fill the field, or the corners will be slightly outside the field, depending on the field of view (*see* Fig. 21.1).

9. Count the cells lying within this 1-mm² area, using the subdivisions (also bounded by three parallel lines) and single grid lines as an aid for counting. Count cells that lie on the top and left-hand lines of each square, but not those on the bottom or right-hand lines, to avoid counting the same cell twice. For routine subculture, attempt to count between 100 and 300 cells per mm²; the more cells that are counted, the more accurate the count becomes. For more precise quantitative experiments, 500–1000 cells should be counted.
   - (a) If there are very few cells (<100/mm²), count one or more additional squares (each 1 mm²) surrounding the central square.
   - (b) If there are too many cells (>1000/mm²), count only five small squares (each bounded by three parallel lines) across the diagonal of the larger (1 mm²) square.

10. If the slide has two chambers, move to the second chamber and do a second count. If not, rinse the slide and repeat the count with a fresh sample.

Analysis. Calculate the average of the two counts, and derive the concentration of your sample using the formula

$$c = n/v$$

where c is the cell concentration (cells/mL), n is the number of cells counted, and v is the volume counted (mL). For the Improved Neubauer slide, the depth of the chamber is

0.1 mm, and, assuming that only the central 1 mm^2 is used, v is 0.1 mm^3, or 1×10^{-4} mL. The formula then becomes

$$c = n/10^{-4}, \text{ or } c = n \times 10^4$$

If the cell concentration is high and only the five diagonal squares within the central 1 mm^2 were counted (i.e., 1/5 of the total), this equation becomes

$$c = n \times 5 \times 10^4$$

If the cell concentration is low, count ten 1-mm^2 squares, five in each chamber of the slide. The expression then becomes

$$c = \frac{n \times 10^4}{10} \quad \text{or} \quad c = n \times 10^3$$

Hemocytometer counting is cheap and gives you the opportunity to see what you are counting. If the cells were previously mixed with an equal volume of a viability stain (*see* Protocol 22.1; Fig. 21.1g,h), a viability determination may be performed at the same time. (Remember to compensate for the additional dilution with viability stain to obtain an absolute count.) However, the procedure is rather slow and prone to error both in the method of sampling and in the total number of cells counted; it also requires a minimum of 1×10^6 cells/mL.

Most of the errors in this procedure occur by incorrect sampling and transfer of cells to the chamber. Make sure that the cell suspension is properly mixed before you take a sample, and do not allow the cells time to settle or adhere in the tip of the pipette before transferring them to the chamber. Ensure

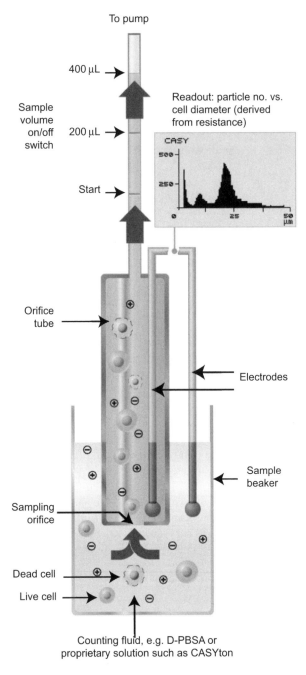

Fig. 21.3. CASY Cell Counter Operation. The electrolyte (e.g., D-PBSA or CASYton) in an orifice tube is connected to a pump and draws a measured volume of the cell sample from a beaker. A cell passing through the orifice alters the flow of current and generates a signal, the amplitude of which is proportional to the volume of the cell. (Courtesy of Schärfe Systems).

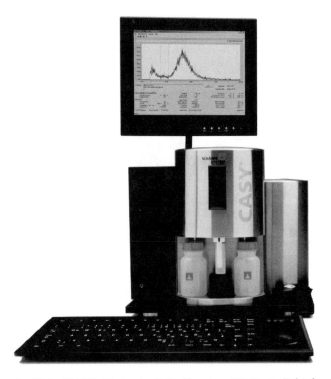

Fig. 21.2. CASY Electronic Cell Counter. CASY 1 (Schärfe Systems) cell counter, also suitable for cell sizing and discrimination between viable and nonviable cells, and single cells and aggregates. (Courtesy of Schärfe Systems).

also that you have a single-cell suspension, as aggregates make counting inaccurate. Larger aggregates may enter the chamber more slowly or not at all. If aggregation cannot be eliminated during preparation of the cell suspension (*see* Table 13.4), lyse the cells in 0.1 M citric acid containing 0.1% crystal violet at 37°C for 1 h and then count the nuclei [Sanford et al., 1951].

21.1.2 Electronic Counting

The main suppliers of electronic cell counters are Beckman Coulter (Coulter Z1 and Z2) and Scharfe Systems (CASY 1; *see* below). Both use the system devised originally by Coulter Electronics (now Beckman Coulter) in which cells, drawn through a fine orifice, increase the electrical resistance to the current flowing through the orifice, in proportion to the volume of the cells, producing a series of pulses that are sorted and counted.

CASY™ electronic cell counter (see Fig. 21.2). There are three main components of this electronic cell counter (*see* Fig. 21.3): (1) an orifice tube, with a 150-μm orifice, connected to a metering pump; (2) an amplifier, pulse-height analyzer, and scaler connected to two electrodes, one in the orifice tube and one in the sample beaker; and (3) an analog and a digital readout showing the cell count and a number of other parameters, such as cell volume and size distribution, depending on the model purchased (Schärfe Systems).

When the count is initiated, a measured volume of cell suspension is drawn through the orifice (*see* Fig. 21.3). As each cell passes through the orifice, it changes the resistance to the current flowing through the orifice by an amount proportional to the volume of the cell. This change in resistance generates a pulse (amp^{-1}) that is amplified and counted. Because the size of the pulse is proportional to the volume of the cell (or any other particle) passing through the orifice, a series of signals of varying pulse height are generated. The lower threshold (in cell diameter) is set to eliminate electronic noise and fine particulate debris, but to retain pulses derived from cells (*see Calibration*, below). The upper threshold is either set to infinity or is adjustable, depending on the model. These thresholds determine which range of particle sizes is counted. The display on the counter shows a histogram depicting the distribution of cell size.

Protocol 21.2 can be used in conjunction with Exercise 12 (*see* Chapter 2).

PROTOCOL 21.2. ELECTRONIC CELL COUNTING BY ELECTRICAL RESISTANCE

Outline
Dilute a sample of cells in electrolyte (physiological saline or D-PBSA), place the diluted sample under the orifice tube, and count the cells by drawing 0.5 mL of the diluted sample through the counter.

Materials
Sterile or Aseptically Prepared:
❑ Cell culture
❑ D-PBSA
❑ 0.25% crude trypsin
❑ Growth medium
Nonsterile:
❑ Counting cups

Protocol
1. Trypsinize the cell monolayer, or collect a sample from the suspension culture. The cells must be well mixed and singly suspended.
2. Dilute the sample of cell suspension to 1:50 in 20 mL of counting fluid in a 25-mL beaker or disposable sample cup. An automatic dispenser (*see* Fig. 5.20) will speed up this dilution and improve reproducibility.

Note. Dispensing counting fluid rapidly can generate air bubbles that will be counted as they pass through the orifice. Consequently, the counting fluid should stand for a few moments before counting. If the fluid is dispensed first and the cells added second, this problem is minimized.

3. Mix the suspension well, and place it under the tip of the orifice tube, ensuring that the orifice is covered and that the external electrode lies submerged in the counting fluid in the sample beaker.
4. Check the program settings:
    (a) Threshold setting(s) (minimum cell size, usually 7.0 μm)
    (b) Volume to be counted (usually 0.5 mL)
    (c) Background subtraction (if used)
    (d) Dilution settings (e.g., 50 if 0.4 mL is counted in 20 mL of D-PBSA).
5. Check the visual analog display
    (a) to ensure that all cells fall within the threshold setting(s);
    (b) to check for viability or cell debris (indicated by a shoulder on the curve or histogram falling below the normal lower threshold setting);
    (c) to check for aggregation (indicated by particles appearing above the normal size range).
6. Initiate the count sequence.
7. When the count cycle is complete, the size distribution will appear on the analog screen (Fig. 21.4). Switching to the digital screen will give the cell count per mL.

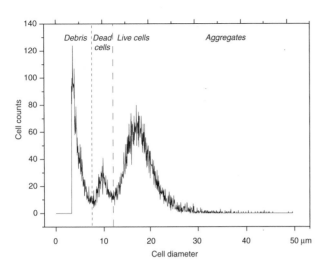

Fig. 21.4. Analog Printout from CASY Electronic Cell Counter. Cell number plotted against cell diameter gives a size distribution analysis that enables the lower threshold (vertical dashed line) to be set. In this display, the upper threshold could be set to 30 μm or set to infinity. The peak below the vertical dashed line and the vertical dotted line represents nonviable cells that can be used with the viable cell count to determine percentage viability. (Data courtesy of Schärfe Systems).

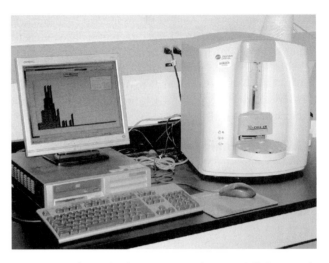

Fig. 21.5. Beckman Coulter Vi-CELL Electronic Cell Counter. An undiluted cell suspension is placed in a cup in the sample holder, sampled by the counter (right), mixed with Trypan Blue, and analyzed in a flow cell. The output appears on the display (left) as a graphic of cell size distribution or a microscope view of the cells with live cells circled in green and dead cells circled in red. (Courtesy of ATCC).

Analysis. If the background and dilution are set correctly, the cell concentration will be that of the starting suspension before dilution in counting fluid. Otherwise, if the counter is set to take 0.5 mL of a 1:50 dilution, the final count on the readout should be multiplied by 100 to give the concentration in cells/mL of the original cell suspension. However, some counters automatically compensate for the volume sampled, so check the instruction manual.

Calibration. Older counters required calibration by counting cells at increasing increments of the lower threshold (upper threshold set to infinity). The plot of these counts generated a plateau, the center of which gave the correct threshold setting. In modern counters, with an analog display, the lower threshold may be set manually on the readout, usually at 7.0 μm, while the upper is set to 30 μm or infinity. Setting the lower threshold to 7.0 μm will include nonviable cells, whereas a higher setting (12 μm in Fig. 21.4) will exclude most of the nonviable cells. Nonviable cells have a smaller apparent diameter because the plasma membrane is leaky, the cytoplasm has the same resistance as the electrolyte, and the resistance signal is generated by the nucleus. Counts at and above 30 μm indicate aggregation; aggregates can be partially excluded by setting an appropriate upper threshold (e.g., 25 μm or 30 μm in Fig. 21.4) or can be allowed for by using a statistical calculation available within the counter's operating program.

Beckman Coulter Vi-CELL electronic cell counter. This counter uses image analysis to scan cells in an optical flow cell and will discriminate between live and dead cells by recognizing Trypan Blue staining. This can be viewed on the display, where viable cells are circled in green and nonviable cells are circled in red. The counter also generates an analog plot of cell size distribution, enabling upper and lower thresholds to be set (Fig. 21.5). The counter must be configured by setting four parameters for each cell type.

Electronic cell counting is rapid and has a low inherent error, because of the high number of cells counted. Although it is prone to misinterpretation, because cell aggregates, dead cells, and particles of debris of the correct size will all be counted, corrections are possible to exclude dead cells and aggregates on both the CASY and the Vi-CELL, and the CASY can make an approximate programmatic correction for aggregation. The cell suspension should still be examined carefully before dilution and counting on the CASY; the Vi-CELL gives a visual display of the cell suspension being counted.

Electronic particle counters are expensive but, if used correctly, are very convenient and give greater speed and accuracy to cell counting.

21.1.3 Stained Monolayers

There are occasions when cells cannot be harvested for counting or are too few to count in suspension (e.g., at low cell concentrations in microtitration plates). In these cases, the cells may be fixed and stained *in situ* and counted by eye with a microscope. Because this procedure is tedious and subject to high operator error, isotopic labeling or the estimation of the total amount of DNA (*see* Protocol 21.3) or protein (*see* Protocol 21.4) is preferable, although these measurements may not correlate directly with the cell number—e.g., if the ploidy of the cell varies. A rough estimate of the cell number per well can also be obtained by staining the cells with crystal violet and measuring the absorption on a densitometer. This method has also been used to calculate the number of cells per colony in clonal growth assays [McKeehan et al., 1977]. Staining cells with Coomassie blue, sulforhodamine B [Boyd, 1989; Skehan et al., 1990], or MTT [Plumb et al., 1989] also gives an estimation of the cell number, given that linearity has been demonstrated previously in a standard plot of absorption against cell number (*see* Protocol 22.4). MTT staining has the advantage that it stains only viable cells.

21.2 CELL WEIGHT

Wet weight is seldom used unless very large cell numbers are involved, because the amount of adherent extracellular liquid gives a large error. As a rough guide, however, there are about 2.5×10^8 HeLa cells ($14-16$ μm in diameter) per gram wet weight, about $8-10 \times 10^8$ cells/g for murine leukemias, e.g., L5178Y murine lymphoma, Friend murine erythroleukemia, myelomas, and hybridomas ($11-12$ μm in diameter), and about 1.8×10^8 cells/g for human diploid fibroblasts ($16-18$ μm in diameter) (Table 21.1). Similarly, dry weight is seldom used, because salt derived from the medium contributes to the weight of unfixed cells, and fixed cells lose some of their low-molecular-weight intracellular constituents and lipids. However, an estimate of dry weight can be derived by interferometry [Brown & Dunn, 1989].

21.3 DNA CONTENT

In practice, besides the cell number, DNA and protein are the two most useful measurements for quantifying the amount of cellular material. DNA may be assayed by several fluorescence methods, including reaction with DAPI [Brunk et al., 1979], PicoGreen (assay kit from Molecular Probes), or Hoechst 33258 [Labarca and Paigen, 1980]. The fluorescence emission of Hoechst 33258 at 458 nm is increased by interaction of the dye with DNA at pH 7.4 and in high salt to dissociate

TABLE 21.1. Relationship Between Cell Size, Volume, and Mass

Cell type	Diameter (μm)	Volume (μm^3)	Cells/g $\times 10^{-6}$ Calculated	Measured
Murine leukemia (e.g., L5178Y or Friend)	11–12	800	1250	1000
HeLa	14–16	1200	800	250
Human diploid fibroblasts	16–18	2500	400	180

the chromatin protein. This method gives a sensitivity of 10 ng/mL, but requires intact double-stranded DNA. DNA can also be measured by its absorbance at 260 nm, where 50 μg/mL has an optical density (O.D.) of 1.0. Because of interference from other cellular constituents, the direct absorbance method is useful only for purified DNA. Protocol 21.3 is a relatively simple and straightforward assay for DNA.

PROTOCOL 21.3. DNA ESTIMATION BY HOECHST 33258

Outline
Homogenize cells or tissue in buffer, and then sonicate the homogenate. Mix aliquots of the culture with H33258, and measure the fluorescence.

Materials
Nonsterile:
- Buffer: 0.05 M NaPO$_4$; 2.0 M NaCl, pH 7.4, containing 2×10^{-3} M EDTA
- Hoechst 33258: in buffer, 1 μg/mL for DNA above 100 ng/mL and 0.1 μg/mL for 10–100 ng/mL

Protocol
1. Homogenize the cells in buffer, 1×10^5 cells/mL for 1 min, using a Potter homogenizer.
2. Sonicate the cells for 30 s.
3. Dilute the cells 1:10 in Hoechst 33258 and buffer.
4. Read fluorescence emission at 492 nm with excitation at 356 nm, using calf thymus DNA as a standard.

21.4 PROTEIN

The protein content of cells is widely used for estimating total cellular material and can be used in growth experiments or as a denominator in expressions of the specific activity of enzymes, the receptor content, or intracellular metabolite concentrations. The amount of protein in solubilized cells can be estimated directly by measuring the absorbance at 280 nm, with minimal interference from nucleic acids and other constituents. The absorbance at 280 nm can detect down to 100 μg of protein, or about 2×10^5 cells.

Colorimetric assays are more sensitive than measurements of absorption, and among these assays, the Bradford reaction with Coomassie blue [Bradford, 1976] is one of the most widely used.

21.4.1 Solubilization of Sample
Because most assays rely on a final colorimetric step, they must be carried out on clear solutions. Cell monolayers and

cell pellets may be dissolved in 0.5–1.0 M NaOH by heating them to 100°C for 30 min or leaving them overnight at room temperature. Alternatively, with 0.3 M NaOH and 1% sodium lauryl sulfate, the solution is complete after 30 min at room temperature.

The Bradford method is not dependent on specific amino acids and is quite sensitive, requiring 50–100,000 cells. Coomassie blue undergoes a spectral change on binding to protein in acidic solution. Color is generated in one step after a short incubation and read within 30 min.

PROTOCOL 21.4. PROTEIN ESTIMATION BY THE BRADFORD METHOD

Outline
Dissolve protein, mix it with a color reagent, and read the O.D. after 10 min.

Materials
- ❑ Sodium lauryl sulfate (SLS, SDS), 3.5 mM in water or 0.3 M NaOH
- ❑ Coomassie Brilliant Blue G-250, 0.12 mM (0.01%), in 4.7% EtOH and 85% (w/v) phosphoric acid: Dissolve 100 mg of Coomassie blue in 50 mL of 95% EtOH; add 100 mL of 85% phosphoric acid; and dilute the solution to 1 L with water
- ❑ Protein standard solution (e.g., BSA, 10 μg/mL); on first setting up the assay and at intervals of 1–2 months, perform a standard curve with BSA or a similar standard protein (1–50 μg/mL)

Protocol
1. Solubilize the protein (1–20 μg) or cells (around 1×10^6) in 100 μL of 3.5 mM sodium dodecyl sulfate in water or 0.3 M NaOH.
2. Add 100 μL of reagent blank (SLS), 100 μL of test protein solution, and 100 μL of BSA standard to separate, triplicate tubes.
3. Add 1.0 mL of Coomassie blue. Mix the solution, and let it stand for 10 min.
4. Read the tests and the BSA standard on a spectrophotometer at 595 nm against the reagent blank.

Variations. Reagents for the Bradford assay are available in a kit from Bio-Rad. Sulforhodamine B [Skehan et al., 1990] and bicinchonic acid (BCA) [Smith et al., 1985] can also be used to measure protein content and the sensitivity of the assays makes them very suitable for microtitration plate assays. A micro-BCA kit available from Pierce

(*see* Appendix III) is very sensitive and suitable for small numbers of cells ($\sim 1 \times 10^3$).

21.5 RATES OF SYNTHESIS

21.5.1 DNA Synthesis

Measurements of DNA synthesis are often taken to be representative of the amount of cell proliferation (*see also* Section 22.3.5). [³H]thymidine ([³H]-TdR) or [³H]deoxy-cytidine is the usual precursor that is employed. Exposure to one of these precursors may be for short periods (0.5–1 h) for rate estimations or for longer periods (24 h or more) to measure accumulated DNA synthesis when the basal rate is low (e.g., in high-density cultures). [³H]-TdR should not be used for incubations longer than 24 h or at high specific activities, as radiolysis of DNA will occur, because of the short path length of β-emission (~ 1 μm) from decaying tritium; the β-emission releases energy within the nucleus and causes DNA strand breaks. If prolonged incubations or high specific activities are required, [¹⁴C]-TdR or ³²P should be used.

PROTOCOL 21.5. ESTIMATION OF DNA SYNTHESIS BY [³H]THYMIDINE INCORPORATION

Outline
Label the cells with [³H]-TdR, extract the DNA, and determine the level of radioactivity by means of a scintillation counter.

Materials
Sterile or Aseptically Prepared:
- ❑ Cells at a suitable stage (grown in glass vials or test tubes if the hot 2 M PCA method is being used; otherwise, grown in conventional plastic)
- ❑ [³H]-TdR, 0.4 MBq/mL (~ 10 μCi/mL), 100 μL for each mL of culture medium

Note. Some media—e.g., Ham's F10 and F12—contain thymidine, which will ultimately determine the specific activity of added [³H]-TdR. Allowance will have to be made for this factor when judging the amount of isotope to add. Although 40 KBq/mL (~ 1.0 Ci/mL) of isotope may be sufficient for most media, 0.2 MBq/mL (~ 5 μCi/mL) should be used with F10 or F12.

Nonsterile:
- ❑ Trichloroacetic acid (TCA), 0.6 M (on ice), 6 mL for each 1 mL of culture
- ❑ HBSS or D-PBSA, ice cold, 2 mL per mL of culture
- ❑ Perchloric acid, 2 M, or 0.3 M NaOH with 35 mM sodium lauryl sulfate (SLS, SDS), 0.5 mL per 1 mL of culture

❑ MeOH, 1 mL per 1 mL of culture
❑ Scintillation vials
❑ Scintillant (10× volume of perchloric acid or NaOH/SDS)

Protocol

1. Grow the culture to the desired density (usually the mid-log phase for maximum DNA synthesis or the plateau for density-limited DNA synthesis; *see* Section 18.5.2).
2. Add [^{3}H]-TdR, 40 KBq/mL (~1.0 µCi/mL), or 2 MBq/mmol (~50 Ci/mol) in HBSS.
3. Incubate the cells for 1–24 h as required.
4. Remove the radioactive medium carefully, and discard it into the proper container for liquid radioactive waste.
5. Wash the cells carefully with 2 mL of HBSS or D-PBSA, and add 2 mL of ice-cold 0.6 M TCA for 10 min. Fix the cells in MeOH first if they are loosely adherent (*see* Protocol 21.6).
6. Repeat the trichloroacetic acid wash twice, for 5 min each time.
7. Add 0.5 mL of 2 M perchloric acid, place the solution on a hot plate at 60°C for 30 min, and allow the solution to cool. Alternatively, add 0.5 mL of SLS in NaOH, and incubate the solution for 30 min at 37°C, or overnight at room temperature.
8. Collect the solubilized pellet, transfer it to the scintillant, and determine the radioactivity on a scintillation counter.

Note. If you are using the perchloric acid method, the residue may be dissolved in alkali for protein determination (*see* Protocol 21.4). Replicate cultures should be set up to provide cell counts to allow calculation of the DNA synthesis related to cell number.

For suspension cultures, spin the cells at 100 *g* for 10 min in Steps 4, 5, and 6. Mix the cells on a vortex mixer to disperse the pellet before each wash (in 0.6 M trichloroacetic acid and 2 M perchloric acid) in Step 7. Spin the cells after Step 7 at 1000 *g* for 10 min to separate the precipitate (for protein estimation, if required) and the supernatant (for scintillation counting).

Δ **Safety Note.** [^{3}H]thymidine represents a particular hazard because it induces radiolytic damage in DNA. Take care to avoid accidental ingestion, injection, or inhalation of aerosols. When using [^{3}H]thymidine, work in a biohazard cabinet or on an open bench, but do not use horizontal laminar flow, or else aerosols will be blown directly at you.

Incubation with an isotopic precursor can provide several different types of data, depending on the incubation conditions and subsequent processing. Incubation followed by a short wash in ice-cold BSS and extraction into 0.6 M TCA will give a measure of the uptake and, if carried out over a few minutes' duration, will give a fair measure of the unidirectional flux. In experiments measuring uptake, the incorporation of precursors into acid–insoluble molecules, such as protein and DNA, is assumed to be minimal, because of the short incubation time, and only the acid–soluble pools are counted by extraction into cold 0.6 M TCA. In longer incubations (i.e., 2–24 h), it is assumed that the precursor pools become saturated. Equilibrium levels may be measured by cold trichloroacetic acid extraction, and the incorporation into polymers may be measured by extraction with hot 2 M perchloric acid (DNA), cold dilute alkali (RNA), or hot 1.0 M NaOH (protein).

21.5.2 Protein Synthesis

Colorimetric assays measure the total amount of protein present at any one time. Sequential observations over a period of time may be used to measure the net protein accumulation or loss (i.e., protein synthesized—protein degraded), while the rate of protein synthesis may be determined by incubating cells with a radioisotopically labeled amino acid, such as [^{3}H]leucine or [^{35}S]methionine, and measuring (e.g., by scintillation counting) the amount of radioactivity incorporated into acid–insoluble material per 10^6 cells or per milligram of protein over a set period of time.

Δ **Safety Note.** Radioisotopes must be handled with care and according to local regulations governing permitted amounts, authorized work areas, handling procedures, and disposal (*see* Section 7.8.5).

PROTOCOL 21.6. PROTEIN SYNTHESIS

Materials
Sterile or aseptically prepared:
❑ Cell cultures: for example, 1×10^4–1×10^6 cells/well, in a 24-well plate
❑ [^{3}H]leucine, 2 MBq/mL (~50 µCi/mL) in culture medium without serum (the specific activity is unimportant, as it will be determined by the leucine concentration in the medium)
Nonsterile:
❑ Sodium lauryl sulfate (SLS or SDS), 1% (35 mM) in 0.3 M NaOH
❑ Trichloroacetic acid (TCA)
❑ Scintillation vials
❑ Eppendorf tubes
❑ Scintillation fluid with a minimum of 10% water tolerance

Protocol
1. Incubate the culture to the required cell density.

2. Remove the culture from the incubator, and add a prewarmed solution of radioisotope in medium or BSS, diluting 1:10 (e.g., 100 μL per 1 mL/well)
3. Return the culture to the incubator as rapidly as possible.
4. Incubate the culture for 4–24 h.

Note. Different proteins turn over at different rates. This protocol is not aimed at any specific subset of proteins, but at the total protein in rapidly proliferating cells. When assaying protein synthesis in a cell line for the first time, check that the rate of synthesis is linear over the chosen incubation time. A lag may be encountered if the amino acid pool is slow to saturate.

5. Remove the culture from the incubator, and withdraw the medium carefully from the wells into a container designated for radioactive liquid waste. (For disposal *see* Section 7.7.2).
6. Wash the cells gently with cold HBSS or D-PBSA.

Note. Some monolayer cultures—particularly some loosely adherent continuous cell lines, such as HeLa-S_3—may detach during washing. In this case, remove the isotope and add methanol to fix the monolayer. Leave the culture for 10 min, carefully remove the methanol, and dry the monolayer.

7. Replace the plates on ice. Add 0.6 M TCA, at 4°C, for 10 min, to remove any unincorporated precursor.
8. Repeat Step 7 twice, but for only 5 min each time.
9. Wash the culture with MeOH, and then dry the plates.
10. Add 0.5 mL of 0.3 M NaOH, containing 1% SLS, and leave for 30 min at room temperature.
11. Mix the contents of each well, and transfer them to separate scintillation vials.
12. Add 5 mL of scintillant, and count on a scintillation counter.

Note. Biodegradable scintillation fluids—e.g., Ecoscint—are preferred over toluene- or xylene-based fluids, as the former are less toxic to handle and can be poured down the sink with excess water provided that the levels of radioactivity fall within the legal limits.

For suspension cultures, spin the cells at 1000 g for 10 min in Step 5 to remove the medium, and at Steps 6, 7, and 8. Also, omit Step 9.
Plates from Step 9 can also be quantified directly on a phosphorimager.

21.6 PREPARATION OF SAMPLES FOR ENZYME ASSAY AND IMMUNOASSAY

As the amount of cellular material available from cultures is often too small for efficient homogenization, other methods of lysis are required to release soluble products and enzymes for assay. It is convenient either to set up cultures of the necessary cell number in sample tubes or multiwell plates (*see* Section 22.3.5 and Protocol 21.8) or to trypsinize a bulk culture and place aliquots of cells into assay tubes. In either case, the cells should be washed in HBSS or D-PBSA to remove the serum, and lysis buffer should be added to the cells. The lysis buffer should be chosen to suit the assay, but, if the particular lysis buffer is unimportant, 0.15 M NaCl or D-PBSA may be used. If the product to be measured is membrane bound, add 1% detergent (Na deoxycholate, Nonidet P40) to the lysis buffer. If the cells are pelleted, resuspend them in the buffer by vortex mixing. Freeze and thaw the preparation three times by placing it in EtOH containing solid CO_2 (~−90°C) for 1 min and then in 37°C water for 2 min (longer for samples greater than 1 mL). Finally, spin the preparation at 10,000 g for 1 min (e.g., in an Eppendorf centrifuge), and collect the supernatant for assay.

Alternatively, the whole extract may be assayed for enzyme activity, and the insoluble material may be removed by centrifugation later if necessary.

21.7 CYTOMETRY

21.7.1 *In Situ* Labeling
Fluorescence labeling, either directly with a fluorescent dye (e.g., Hoechst 33258 for DNA) or with a conjugated antibody for detection of an antigen or molecular probe (*see* Protocols 27.9 and 27.8), can measure the amounts of enzyme, DNA, RNA, protein, or other cellular constituents *in situ* with a CCD camera. This process allows qualitative as well as quantitative analyses to be made, but is slow if large numbers of cells are to be scanned.

21.7.2 Flow Cytometry
Flow cytometry of a cell suspension [Vaughan & Miller, 1989], (*see also* Section 16.9.2; Fig. 15.9), while losing the relationship between cytochemistry and morphology, samples up to 1×10^7 cells, can measure multiple cellular constituents and activities [Kurtz & Wells, 1979; Klingel et al., 1994], and enables correlation of these measurements with other cellular parameters, such as cell size, lineage, DNA content, or viability [Al-Rubeai et al., 1997].

21.8 REPLICATE SAMPLING

Because, in most cases, cultured cells can be prepared in a uniform suspension, the provision of large numbers of replicates for statistical analysis is often unnecessary. Usually, three replicates are sufficient, and for many simple observations (e.g., cell counts), duplicates may be sufficient.

Many types of culture vessel are available for replicate monolayer cultures (*see* Section 8.2), and the choice of which vessel to use is determined (1) by the number of cells required in each sample and (2) by the frequency or type of sampling. For example, if the incubation time is not a variable, replicate sampling is most readily performed in multiwell plates, such as microtitration plates or 24-well plates. If, however, samples are collected over a period of time (e.g., daily for 5 days), then the constant removal of a plate for daily processing may impair growth in the rest of the wells. In this case, the replicates are best prepared in individual tubes or 4-well plates. Plain glass or tissue-culture-treated plastic test tubes may be used, although Leighton tubes are superior, as they provide a flat growth surface. Alternatively, if the optical quality of the tubes is not critical, flat-bottomed glass specimen tubes and even glass scintillation vials may be good containers. If glass vials or tubes are used, they must be washed as tissue culture glassware (*see* Section 11.3.1); they cannot be used for tissue culture after use with scintillant.

Sealing large numbers of vials or tubes can become tedious, so many people seal tubes with vinyl tape rather than screw caps. Such tape can also be color coded to identify different treatments. Adhesive film may be used for sealing microtitration plates (*see* Appendix II: Plate sealers). This reduces evaporation and contamination and gives a more even performance across the plate. It also means that individual wells or rows can be sampled without opening up the rest of the plate.

Handling suspension cultures is generally easier than dealing with monolayer cultures, because the shape of the container and its surface charge are less important. Multiple sampling can also be performed on one culture when using suspension cultures. This sampling is done conveniently by sealing the bottle containing the culture with a silicone rubber membrane closure (Pierce) and then sampling with a syringe and needle. (Remember to replace the volume of culture removed with an equal volume of air.)

21.8.1 Data Acquisition
Analysis of data from cultured cells is not necessarily different from the way that data from any other system are handled. However, the production of large amounts of data in cell culture experiments is relatively easy, particularly when dealing with microtitration plates, with which several hundred data points can be generated without a great deal of effort, or even several thousand by using robotics. Handling tissue culture-derived data will depend on how they are generated, on the scale of the experiment, and the number of parameters. Although cell counting is the accepted method for generating data to construct a growth curve with one cell line under two or three sets of conditions, this does not lend itself to expansion to, say, determining multiple growth curves to

measure the response of several cell lines to combinations of growth factors. In this example, if a colorimetric end point is chosen, then absorbance (e.g., MTT assay; *see* Protocol 22.4) or fluorescence emission (e.g., sulforhodamine [Boyd, 1989]) can be used, the plates analyzed on a plate reader, and data reduction achieved by using the appropriate software.

A radiometric end point (e.g., [^{3}H]thymidine) used in conjunction with microtitration plates can be determined by simultaneous measurement of the whole plate in a microtitration plate scintillation counter (PerkinElmer). Similarly, large numbers of sample tubes can be read automatically by γ or β counting, using robotic systems (Beckman Coulter).

21.8.2 Data Analysis
The ability to generate large amounts of data has been made possible by the creation of multiple replicate analysis systems, such as microtitration. As with ELISA analysis, the rate-limiting step is no longer the generation of the data, but is instead its analysis. It is important, therefore, when choosing a parameter of measurement to suit the culture system, that some thought be given to the amount of data to be generated and how those data will be handled. The easiest approach is to direct the data into a computer, either via a network or to a PC dedicated to that project. A number of companies now market computer programs that display and analyze data from microtitration plate assays (not just ELISA). These programs include titration curves, enzyme kinetics, and binding assays. With the necessary skills, and using a spreadsheet for importation of the data, you may be able to set up this type of program for yourself; alternatively, consultant advice is often available from the suppliers of plate readers (*see* Appendix II).

21.9 CELL PROLIFERATION

Measurements of cell proliferation rates are often used to determine the response of cells to a particular stimulus or toxin (*see* Protocols 21.7 and 21.8). Quantitation of culture growth is also important in routine maintenance, as it is a crucial element for monitoring the consistency of the culture and knowing the best time to subculture (*see* Section 13.7), the optimum dilution, and the estimated plating efficiency at different cell densities. Testing medium, serum, new culture vessels or substrates, and so forth all require quantitative assessment.

21.9.1 Experimental Design
Knowledge of the growth state of a culture, and its kinetic parameters, is critical in the design of cell culture experiments. Cultures vary significantly in many of their properties between the lag phase, the period of exponential growth (log phase), and the stationary phase (plateau). It is therefore important to take account of the status of the culture both at the initiation of an experiment and at the time of sampling,

in order to determine whether is it proliferating or not and, if it is, the duration of the population doubling time (PDT) and the cell cycle time. Cells that have entered the plateau phase have a greatly reduced growth fraction and a different morphology, may be more differentiated, and may become polarized. They generally tend to secrete more extracellular matrix and may be more difficult to disaggregate. Generally, cell cultures are most consistent and uniform in the log phase, and sampling at the end of the log phase gives the highest yield and greatest reproducibility.

It is also important to consider the effects of the duration of an experiment on the transition from one state to another. Adding a drug in the middle of the exponential phase and assaying later may give different results, depending on whether the culture is still in exponential growth when it is harvested or whether it has entered the plateau phase. Microtitration plate assays of cytotoxicity (*see* Section 22.4) are particularly susceptible to error if the culture reaches plateau during an assay; as the cells in those wells that are at the highest density reach plateau, cell proliferation decreases and there is an apparent shift in the 50% inhibitory point (ID_{50}) as more wells reach plateau (*see* Protocol 22.4). Hence, scheduling treatment and sampling requires a detailed knowledge of the parameters of the growth cycle.

21.9.2 Growth Cycle

As described previously (*see* Section 13.7.2), after subculture, cells progress through a characteristic growth pattern of lag phase, exponential, or log phase, and stationary, or plateau phase (Fig. 21.6). The log and plateau phases give vital information about the cell line, the PDT during exponential growth, and the maximum cell density achieved in the plateau phase (i.e., the saturation density). The measurement of the PDT is used to quantify the response of the cells to different inhibitory or stimulatory culture conditions, such as variations in nutrient concentration, hormonal effects, or toxic drugs. It is also a good monitor of the culture during serial passage and enables the calculation of cell yields and the dilution factor required at subculture.

Single time points are unsatisfactory for monitoring growth if the shape of the growth curve is not known. A reduced cell count after, say, 5 days could be caused by a reduced growth rate of some or all of the cells; a longer lag period, implying adaptation or cell loss (it is difficult to distinguish between the two); or a reduction in saturation density (*see* Fig. 21.8). This is not to say that growth curves are of no value. They can be useful for testing media, sera, growth factors and some drugs, and once the response being monitored is fully characterized and the type of response is

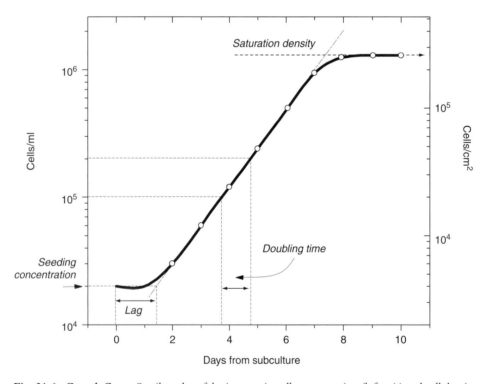

Fig. 21.6. Growth Curve. Semilog plot of the increase in cell concentration (left axis) and cell density (right axis) after subculture. Replicate 25-cm² flask cultures are sampled daily and counted. It should be possible to draw a straight line through the part of the plot that represents the exponential phase and derive the population doubling time (PDT) from the middle region of this best-fit line. The time at the intercept of the line extrapolated from the exponential phase with the seeding concentration is the lag time, and the saturation density is found at the plateau (at least three linear points without an increase in cell concentration) at the top end of the curve.

predictable (e.g., a change in the PDT), then single time point observations can be made at a time point known to be in mid-log phase. Growth curves are particularly useful for the determination of the saturation density, although the amount of growth at saturation density should be assessed by the labeling index with [³H]thymidine (*see* Section 18.5.2 and Protocols 21.11, 21.12).

The PDT derived from a growth curve should not be confused with the cell cycle or generation time. The PDT is an average figure that applies to the whole population, and it describes the net result of a wide range of division rates, including a rate of zero, within the culture. The *cell cycle time* or *generation time* is measured from one point in the cell cycle until the same point is reached again (*see* Section 21.12) and refers only to the dividing cells in the population, whereas the PDT is influenced by nongrowing and dying cells as well. PDTs vary from 12–15 h in rapidly growing mouse leukemias, like the L1210, to 24–36 h in many adherent continuous cell lines, and up to 60 or 72 h in slow-growing finite cell lines.

A new growth cycle begins each time the culture is subcultured and can be analyzed in more detail, as described in Protocol 21.7. Using flasks for a growth curve is more labor intensive, limiting the number of replicates to two per day for 10 days, requiring 20 flasks, with an additional 4 flasks for staining or to act as back-up. A cell concentration of 2×10^4 cells/mL should be chosen for a rapidly growing line and 1×10^5 cells/mL for a slower-growing finite cell line. Repeating the growth curve with higher or lower seeding concentrations should then allow the correct seeding concentration and subculture interval (*see* Section 13.7.2) to be established.

This protocol can be used in conjunction with Exercise 16 (*see* Chapter 2).

PROTOCOL 21.7. GROWTH CURVE WITH A MONOLAYER IN FLASKS

Outline
Set up flasks and count the cells at daily intervals until the culture reaches the plateau phase.

Materials
Sterile or Aseptically Prepared:
- ❏ Monolayer cell culture, A549, Vero, or HeLa-S3, 75-cm² flask, late log phase 1
- ❏ Trypsin, 0.25%, crude, with 10 mM EDTA .... 5 mL
- ❏ Growth medium with 4 mM NaHCO₃ .... 200 mL
- ❏ D-PBSA (prewash and for cell counting) .... 50 mL
- ❏ Flasks, 25 cm² 24

Protocol
1. Trypsinize the cells as for a regular subculture (*see* Protocol 13.2).
2. Dilute the cell suspension to 2×10^4 cells/mL in 150 mL of medium.
3. Seed 24 25-cm² flasks.
4. Seal the flasks and place in an incubator at 37°C.

Note. A low-bicarbonate (4 mM) medium is specified for this protocol for the sake of simplicity, but if a high-bicarbonate medium is used (e.g., 23 mM) then either gas the flasks with 5% CO_2 or place in a CO_2 incubator with slack or gas-permeable caps.

5. After 24 h, remove the first two flasks from the incubator, and count the cells:
   (a) Remove the medium completely.
   (b) Add 2 mL of trypsin/EDTA to each flask.
   (c) Incubate the flasks for 15 min.
   (d) Disperse the cells in the trypsin/EDTA and transfer 0.4 mL of the suspension to 19.6 mL of D-PBSA.
   (e) Count the cells on an electronic cell counter.
6. Repeat sampling at 48 and 72 h, as in Steps 5 and 6.
7. Change the medium at 72 h, or sooner, if indicated by a drop in the pH (*see* Protocol 13.1; Plate 22b).
8. Continue sampling daily for rapidly growing cells (i.e., cells with a PDT of 12–24 h), but reduce the frequency of sampling to every 2 days for slowly growing cells (i.e., cells with a PDT >24 h) until the plateau phase is reached.
9. Keep changing the medium every 1, 2, or 3 days, as indicated by the fall in pH.
10. Stain the cells in one flask at 2, 5, 7, and 10 days (*see* Section 16.4.2).

Note. A hemocytometer may be used to count the cells, but may be difficult to use for the lower cell concentrations at the start of the growth curve. If you use a hemocytometer, reduce the volume of trypsin to 0.5 mL, disperse the cells carefully, using a pipettor without frothing the trypsin, and transfer the cells to the hemocytometer.

This type of assay is useful for comparing different media, supplements, and growth stimulants or inhibitors. However, for a quantitative assay for one or more variables, it is preferable to use multiwell plates. Protocol 21.8 can also be used with 12-well plates in conjunction with Exercise 16 (*see* Chapter 2).

PROTOCOL 21.8. GROWTH CURVE WITH A MONOLAYER IN MULTIWELL PLATES

Outline
Set up a series of multiwell plates with cultures at three different cell concentrations, and count the cells in one plate at daily intervals until the culture reaches the plateau phase.

Materials
Sterile or Aseptically Prepared:
- ☐ Monolayer cell culture, A549, Vero, or HeLa-S3, 75-cm² flask, late log phase 1
- ☐ Trypsin, 0.25%, crude, with 10 mM EDTA .... 5 mL
- ☐ Growth medium, 100 mL, with 26 mM NaHCO₃ .... 300 mL
- ☐ D-PBSA (prewash and for cell counting) .... 50 mL
- ☐ Plates, 12 well 10

Nonsterile:
- ☐ Plastic box to hold the plates
- ☐ CO₂ incubator or CO₂ supply to purge the box with 5% CO₂

Protocol
1. Trypsinize the cells as for a regular subculture (*see* Protocol 13.2).
2. Dilute the cell suspension to 1×10^5 cells/mL, 3×10^4 cells/mL, and 1×10^4 cells/mL, in 25 mL of medium for each concentration.
3. Seed three 12-well plates with 2 mL of the 1×10^4/mL cell suspension to each well of the top 4 wells, 2 mL of the 3×10^4/mL to each well of the second row, and 2 mL of the 1×10^5/mL to each well of the third row (Fig. 21.7). Add the cell suspension slowly from the center of the well, so that it does not swirl around the well. Similarly, do not shake the plate to mix the cells, as the circular movement of the medium will concentrate the cells in the middle of the well.
4. Place the plates in a humid CO₂ incubator or a sealed box gassed with 5% CO₂.
5. After 24 h, remove the first plate from the incubator, and count the cells in three wells at each concentration:
   (a) Remove the medium completely from the three wells containing cells to be counted.
   (b) Add 0.5 mL of trypsin/EDTA to each of the three wells.
   (c) Incubate the plate for 15 min.
   (d) Add 0.5 mL medium with serum, disperse the cells in the trypsin/EDTA/medium, and transfer 0.4 mL of the suspension to 19.6 mL of D-PBSA.
   (e) Count the cells on an electronic cell counter.
6. Stain the cells in the remaining wells at each cell density (*see* Section 16.4.2).

Note. A hemocytometer may be used to count the cells, but may be difficult to use for lower cell concentrations. If you use a hemocytometer, reduce the volume of trypsin to 0.1 mL, and disperse the cells carefully, using a pipettor without frothing the trypsin. Transfer the cells to the hemocytometer.

7. Repeat sampling at 48 and 72 h, as in Steps 5 and 6.
8. Change the medium at 72 h, or sooner, if indicated by a drop in the pH (*see* Protocol 13.1; Plate 22b).
9. Continue sampling daily for rapidly growing cells (i.e., cells with a PDT of 12–24 h), but reduce the frequency of sampling to every 2 days for slowly growing cells (i.e., cells with a PDT >24 h) until the plateau phase is reached.
10. Keep changing the medium every 1, 2, or 3 days, as indicated by the fall in pH.

21.9.3 Analysis of Monolayer Growth Curves
(1) Calculate the number of cells per well, per mL of culture medium (*see* Fig. 21.6), and per cm² of available growth surface in the well as follows:
  a) *Primary count.* The count obtained from electronic counting or hemocytometer is the number of cells/mL of trypsinate.
  b) *Cells per flask or well.* Where 1 mL trypsin has been used for trypsinization, the primary count is the same as the number of cells per flask or well. If 2 mL trypsin was used, double the cell count to give cells per flask or well. If 0.5 mL trypsin was used, divide the primary count by 2 to give cells per flask or well.
  c) *Cells per mL of culture medium (cell concentration).* Divide the number of cells per well or flask by the volume of medium used during culture.
  d) *Cells per cm² of growth surface (cell density).* Divide the number of cells per well or flask by the surface area of the well or flask. As there are small variations among manufacturers, it is best to calculate the surface area for the culture vessel that you use. As a rough guide, however, each well is 2 cm² in a 24-well plate and 3.8 cm² in a 12-well plate. The flasks recommended for Protocol 21.7 have 25-cm² growth area.

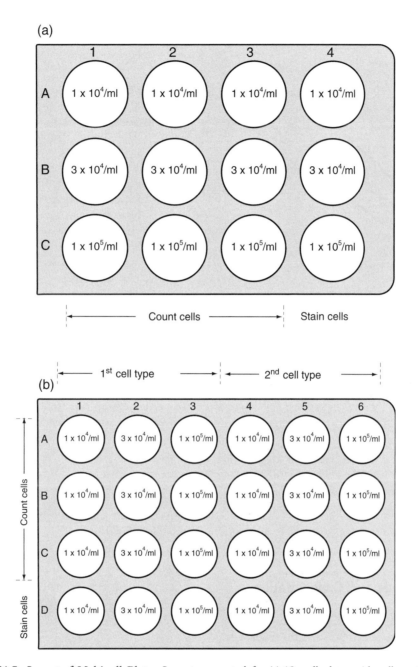

Fig. 21.7. Layout of Multiwell Plates. Layouts suggested for (a) 12-well plates with cells at three different concentrations and wells allocated for counting and staining and (b) 24-well plates with an additional variable, in this case cell type.

(2) Plot the cell density (cells/cm²) and the cell concentration (cells/mL), both on a log scale, against time on a linear scale (*see* Fig. 21.6). The scale on both vertical axes is the same but out of register by a factor that depends on the number of cm² per mL medium. In a 25-cm² flask, there are 5 mL medium covering 25 cm², 0.2 mL/cm² or 5 cm²/mL medium, so the right-hand axis will be out of register with the left by a factor of 5, and 1×10^5 on the left axis will be equivalent to 2×10^4 on the right-hand axis.

(3) Determine the lag time, PDT, and plateau density (*see* Fig. 21.6; *see also* Section 21.11.1).

(4) Establish the appropriate starting density for routine passage from Protocol 21.7. Repeat the growth curve at different cell concentrations if necessary.

(5) Compare growth curves under different conditions (Fig. 21.8), and try to interpret the data (*see* legend for Fig. 21.8).

(6) Examine the stained cells at each density to:

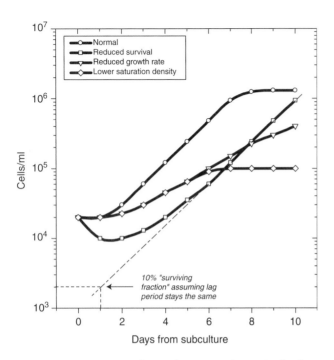

Fig. 21.8. Interpretation of Growth Curves. Changes in the shape of a growth curve can be interpreted in a number of different ways, but the labels in the key of this plot indicate what would normally be deduced from these curves.

a) determine whether the distribution of cells in the flasks or wells is uniform and whether the cells are growing up the sides of the well

b) observe differences in cell morphology as the density increases

c) compare cell–cell interaction in normal and transformed cells.

21.9.4 Medium Volume, Cell Concentration, and Cell Density

It is important when using multiwell plates and comparing data with culture flasks to remember that the volume of medium used in a multiwell plate is often proportionately higher than in a flask for a given surface area and the cell density will be higher for the same cell concentration (*see* Section 13.7.3). If 2×10^4 cells/mL are seeded in 5 mL into a 25-cm^2 flask, the cell density at seeding will be 4000 cells/cm^2 ($20,000 \times 5 \div 25$), whereas if the same cell concentration is seeded in 2 mL in a 12-well plate the cell density will be 10,500 cells/cm^2 ($20,000 \times 2 \div 3.8$). This density is more than twice that of the flask, and the cells will reach plateau at least 1 day earlier. If an exact comparison is intended the ratio of medium to culture surface area must be the same, and a volume of 0.75 mL would be required to achieve the same cell density in a 12-well plate as in a 25-cm^2 flask for a given cell concentration. Unfortunately, such a low volume would cause uneven cellular distribution due to the shape of the meniscus, and cells would tend to concentrate at the edges of the wells. This problem increases as the wells get smaller, because the relative effect of the meniscus increases with a decrease in diameter of the well.

Multiwell plates are suitable for comparing different growth conditions, media, sera, or growth factors or cytotoxins, but if a growth curve is being used to establish conditions for routine maintenance then the growth curve must be performed in the same vessels as being used for routine subculture (although the difference between a 25-cm^2 and a 75-cm^2 flask will be minimal given that the volume of medium per cm^2 remains the same).

It is also possible to perform a growth curve in microtitration plates, e.g., as control plates to monitor cell numbers in a cytotoxicity assay (*see* Section 22.3.3). However, as the growth area per well is very small the cell numbers can be very low, particularly at the start of an experiment, so four or eight wells need to be pooled and counted.

21.9.5 Suspension Cultures

A growth curve can also be generated from cells growing in suspension, usually without replenishing the medium. The objectives are similar as for monolayer cells, i.e., to set conditions for routine maintenance or to assay differences in growth conditions. As trypsinization is not required, several samples can be harvested from the same vessel.

PROTOCOL 21.9. GROWTH CURVE WITH CELLS IN SUSPENSION

Outline
Set up a series of cultures at three different cell concentrations, and count the cells daily until they reach the plateau phase.

Materials
Sterile or Aseptically Prepared:
❑ Suspension cell culture
❑ Growth medium, 100 mL
❑ D-PBSA (for cell counting)
❑ Plates, 24 well
Nonsterile:
❑ Plastic box to hold the plates
❑ CO$_2$ incubator or 5% CO$_2$ supply to purge the box with 5% CO$_2$.

Protocol
1. Add the cell suspension in growth medium to wells at a range of concentrations as for monolayer cultures (*see* Protocol 21.8).

2. Sample 0.4 mL of the culture from triplicate wells at intervals, ensuring that the cells are well mixed and completely disaggregated.
3. Count the samples on electronic cell counter (*see* Section 21.1.2) or hemocytometer (*see* Section 21.1.1).
4. Calculate the cell concentration per sample, and plot on a log scale against time on a linear scale as for monolayer growth (*see* Section 21.9.3). Cell density does not apply to a suspension culture as they are not adherent.

Variations. Seed two 75-cm² flasks with 20 mL of cell suspension in growth medium for each cell concentration, and sample 0.4 mL of culture from each flask daily or as required. Mix the culture well before sampling it, and keep the flasks out of the incubator for the minimum length of time. Do not feed the cultures during the growth curve. Alternatively, set up a stirrer flask (Techne, Integra, Bellco) and sample daily. If a membrane closure is used on the side arm of the stirrer flask, then the flask can be sampled without removing it from the hot room by swabbing the membrane, upending the flask, and sampling via the side arm with a syringe and needle.

21.9.6 Phases of the Growth Cycle

The growth cycle (*see* Fig. 21.6) may be divided into three phases:

The lag phase. This phase is the time after subculture and reseeding during which there is little evidence of an increase in the cell number. It is a period of adaptation during which the cell replaces elements of the cell surface and extracellular matrix lost during trypsinization, attaches to the substrate, and spreads out. During spreading, the cytoskeleton reappears, an integral part of the spreading process. The activity of enzymes, such as DNA polymerase, increases, followed by the synthesis of new DNA and structural proteins. Some specialized cell products may disappear and not reappear until the cessation of cell proliferation at a high cell density.

The log phase. This phase is the period of exponential increase in the cell number following the lag period and terminating one or two population doublings after confluence is reached. The length of the log phase depends on the seeding density, the growth rate of the cells, and the density that inhibits cell proliferation. In the log phase, the growth fraction is high (usually 90–100%), and the culture is in its most reproducible form. It is the optimal time for sampling, because the population is at its most uniform and the viability is high. However, the cells are randomly distributed in the cell cycle and, for some purposes, may need to be synchronized (*see* Section 27.5).

The plateau phase. Toward the end of the log phase, the culture becomes confluent—i.e., all of the available growth surface is occupied and all of the cells are in contact with surrounding cells. After confluence, the growth rate of the culture is reduced, and in some cases, cell proliferation ceases almost completely after one or two further population doublings (*see* Section 18.5.2). At this stage, the culture enters the plateau, or stationary, phase, and the growth fraction falls to between 0 and 10%. The cells may become less motile; some fibroblasts become oriented with respect to one another, forming a typical parallel array of cells. "Ruffling" of the plasma membrane is reduced, and the cell both occupies less surface area of substrate and presents less of its own surface to the medium. There may be a relative increase in the synthesis of specialized versus structural proteins, and the constitution and charge of the cell surface may be changed.

The cessation of motility, membrane ruffling, and growth following contact of cells at confluence was originally described by Abercrombie and Heaysman [1954] and was designated *contact inhibition*. It has since been realized that the reduction in the growth of normal cells after confluence is reached is not due solely to contact, but may also involve reduced cell spreading [Stoker et al., 1968; Folkman & Moscona, 1978], buildup of inhibitors, and depletion of nutrients, particularly growth factors [Dulbecco & Elkington, 1973; Stoker, 1973; Westermark & Wasteson, 1975] in the medium [Holley et al., 1978]. This depletion can be quite local in a static monolayer, generating a diffusion boundary around the cells [Stoker, 1973] that can be overcome by irrigating the monolayer. The term *density limitation (of cell proliferation)* has been used to remove the implication that cell–cell contact is the major limiting factor [Stoker & Rubin, 1967], and the term *contact inhibition* is best reserved for those events resulting directly from cell contact (i.e., reduced cell motility and membrane ruffling, resulting in the formation of a strict monolayer and orientation of the cells with respect to each other).

Cultures of normal simple epithelial and endothelial cells stop growing after reaching confluence and remain as a monolayer. Most cultures, however, with regular replenishment of medium, will continue to proliferate (although at a reduced rate) well beyond confluence, resulting in multilayers of cells. Human embryonic lung and adult skin fibroblasts, which express contact inhibition of movement, will continue to proliferate, laying down layers of collagen between the cell layers until multilayers of six or more cells can be reached under optimal conditions [Kruse et al., 1970]. These fibroblasts still retain an ordered parallel array, however. Therefore, the terms "plateau" and "stationary" are not strictly accurate and should be used with caution.

Cultures that have transformed spontaneously or have been transformed by virus or chemical carcinogens will usually reach a higher cell density in the plateau phase than their normal counterparts [Westermark, 1974] (Fig. 21.9). This higher cell density is accompanied by a higher growth fraction

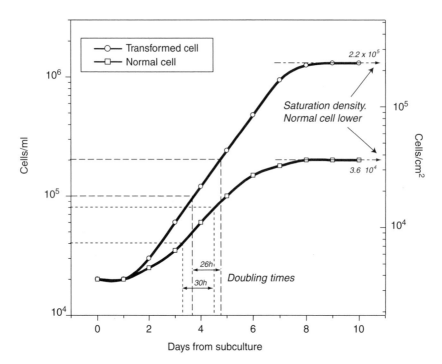

Fig. 21.9. Saturation Density. Transformation produces an increase in the saturation density of transformed cells, relative to that found in the equivalent normal cells. This increase is often accompanied by a shorter PDT. [Data from normal human glia and glioma cell lines; Freshney et al., unpublished observations.]

and the loss of density limitation of cell proliferation. The plateau phase for these cultures is an equilibrium between cell proliferation and cell loss. These cultures are often *anchorage independent* for growth—i.e., they can easily be made to grow in suspension (*see* Section 18.5.1).

21.9.7 Derivatives from the Growth Curve

The construction of a growth curve from cell counts performed at intervals after subculture enables the measurement of a number of parameters that should be found to be characteristic of the cell line under a given set of culture conditions. The first of these parameters is the duration of the lag period, or *lag time*, obtained by extrapolating a line drawn through the points for the exponential phase until it intersects the seeding concentration (*see* Fig. 21.6) and then reading off the elapsed time since seeding equivalent to that intercept. The second parameter is the *population doubling time* (PDT)—i.e., the time taken for the culture to increase twofold in the middle of the exponential, or log, phase of growth. This parameter should not be confused with the generation time or cell cycle time (*see* Section 21.12), which are determined by measuring the transit of a population of cells from one point in the cell cycle until they return to the same point.

The last of the commonly derived measurements from the growth cycle are the *plateau level* and *saturation density*. The plateau level is the cell concentration (cells/mL of medium) in the plateau phase and is dependent on the cell type and the frequency with which the medium is replenished. The

saturation density is the density of the cells (cells/cm^2 of growth surface) in the plateau phase. Saturation density and plateau level are difficult to measure accurately, as a steady state is not easily achieved in the plateau phase. Ideally, the culture should be perfused, to avoid nutrient limitation or growth factor depletion, but a reasonable compromise is to grow the cells on a restricted area, say a 15-mm diameter coverslip or filter well, in a 9-cm-diameter Petri dish with 20 mL of medium that is replaced daily (*see* Protocol 18.3). Under these conditions, the limitation of growth by the medium is minimal, and the cell density exerts the major effect. A count of the cells under these conditions is a more accurate and reproducible measurement than a cell count in plateau under conventional culture conditions. Note that the term "plateau" does not imply the complete cessation of cell proliferation, but instead represents a steady state in which cell division is balanced by cell loss.

Although it is not appropriate to talk of "saturation density" in a suspension culture, nonadherent cells can still enter plateau because of exhaustion of the medium. Frequently, however, suspension cells at plateau phase will enter apoptosis quite quickly and may not even generate a real plateau.

With normal cells, a steady state may be achievable by not replenishing the growth factors in the medium. In this case, the cells are seeded and grown and the plateau reached without changing the medium. Clearly, the conditions used to attain the plateau phase must be carefully defined.

21.10 PLATING EFFICIENCY

Colony formation at low cell density, or *plating efficiency*, is the preferred method for analyzing cell proliferation and survival (*see also* Protocol 22.3). This technique reveals differences in the growth rate within a population and distinguishes between alterations in the growth rate (colony size) and cell survival (colony number). It should be remembered, however, that cells may grow differently as isolated colonies at low cell densities. In this situation fewer cells will survive, even under ideal conditions, and all cell interaction is lost until the colony starts to form. Heterogeneity in clonal growth rates reflects differences in the capacity for cell proliferation between lineages within a population, but these differences are not necessarily expressed in an interacting monolayer at higher densities, when cell communication is possible.

When cells are plated out as a single-cell suspension at low cell densities ($2-50$ cells/cm^2), they grow as discrete colonies (*see* Protocol 14.1 and Plate 6). The number of these colonies can be used to express the plating efficiency:

$$\frac{\text{No. of colonies formed}}{\text{No. of cells seeded}} \times 100 = \text{Plating efficiency.}$$

If it can be confirmed that each colony grew from a single cell, then this term becomes the *cloning efficiency*. Measurements of the plating efficiency are derived by counting the number of colonies over a certain size (usually around 50 cells) growing from a low inoculum of cells, and this term should not be used for the recovery of adherent cells after seeding at higher cell densities. Survival at higher densities is more properly referred to as the *seeding efficiency*:

$$\frac{\text{No. of cells attached}}{\text{No. of cells seeded}} \times 100 = \text{Seeding efficiency.}$$

It should be measured at a time when the maximum number of cells has attached, but before mitosis starts. This time is a difficult point to define, as the window between maximum cell attachment and the initiation of mitosis may be quite narrow, and the events may even overlap; however, it still provides a crude measurement of recovery in, for example, routine cell freezing or primary culture.

Protocol 21.10 for determination of plating efficiency can be used in conjunction with Exercise 21 (*see* Chapter 2), with, for example, varying concentrations of serum (*see* Plate 6e) or with and without a feeder layer (*see* Protocol 14.3).

PROTOCOL 21.10. DETERMINATION OF PLATING EFFICIENCY

Outline
Seed the cells at low density and incubate until colonies form (*see* Protocol 14.1); stain and count the colonies.

Materials
Sterile:
☐ Growth medium .... 400 mL
☐ Trypsin, 0.25%, crude .... 10 mL
☐ Petri dishes, 6 cm 20
☐ Tubes, or universal containers, for dilution 20
Nonsterile:
☐ Hemocytometer or electronic cell counter
☐ Fixative: anhydrous methanol .... 100 mL
☐ D-PBSA .... 200 mL
☐ Stain: crystal violet .... 100 mL
☐ Filter funnel and filter paper (to recycle the stain)

Protocol
1. Trypsinize the cells (*see* Protocol 13.2) to produce a single-cell suspension.
2. While the cells are trypsinizing,
   (a) Number the dishes on the side of the base.
   (b) Measure out medium for the dilution steps (Fig. 21.10). There should be more than enough medium for three replicates at each dilution.
3. When the cells round up and start to detach:
   (a) Disperse the monolayer in medium containing serum or a trypsin inhibitor.
   (b) Count the cells.
   (c) Dilute the cells to:
      i) 2×10^4/mL for two 25-cm^2 flasks for routine maintenance
      ii) 2×10^3 cell/mL as top concentration for subsequent dilutions
      iii) Five further dilutions from (ii) to give 200, 100, 50, 20, and 10 cells/mL
4. Seed the Petri dishes with 5 mL medium containing cells at each of the five concentrations in (iii). Seed two 6-cm Petri dishes at 2×10^3 cells/mL to act as controls in case the cloning is unsuccessful (to prove that there were cells present in the top dilution, at least).
5. Gas the flasks with 5% CO$_2$ and take to incubator.
6. Put the Petri dishes in a transparent plastic box and place in a humid CO$_2$ incubator, preferably one with limited access and reserved for cloning.
7. Incubate the dishes until colonies are visible to the naked eye (1–3 weeks).
8. Stain the colonies with crystal violet:
   (a) Remove the medium from the dishes.
   (b) Rinse the cells with D-PBSA, and discard the rinse.
   (c) Add 5 mL fresh D-PBSA and then add 5 mL methanol with gentle mixing (avoid colonies detaching).

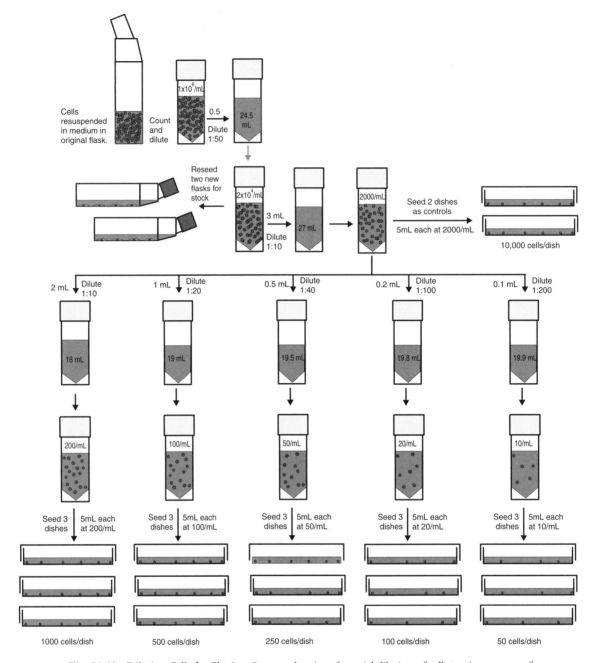

Fig. 21.10. Diluting Cells for Cloning. Suggested regime for serial dilution of cells to give a range of seeding densities, suitable for cloning a cell line for the first time or establishing the linearity of plating efficiency versus seeding concentration (*see* Fig 21.11). Subsequently, when a suitable concentration has been selected, the cells may be diluted, with fewer steps, to the desired concentration.

(d) Replace the 50:50 D-PBSA:methanol mixture with 5 mL fresh methanol and fix the cells for 10 min.

(e) Discard the methanol, and add crystal violet, neat, 2–3 mL per 6-cm dish, making sure that the whole of the growth surface is covered.

(f) Stain for 10 min.

(g) Remove the stain, and return it to the stock bottle of stain via a filter.

(h) Rinse the dishes with water and allow to dry.

9. Count the colonies in each dish, excluding those below 50 cells per colony. Magnifying viewers can make counting the colonies easier.

It will be necessary to define a threshold above which colonies will be counted. If the majority of the colonies are between a hundred and a few thousands, then set the threshold at 50 cells per colony. In practice, this is a fairly natural threshold when counting by eye. However, if the colonies are very small (<100 cells), then set the threshold at 16 cells per colony. Below 16 cells, equivalent to 4 cell consecutive divisions, it would be hard to presume continued cell proliferation.

21.10.1 Analysis of Colony Formation

Calculate the plating efficiency (*see* calculation in introduction to this section) at each seeding density. The plating efficiency should remain constant throughout the range of seeding densities (i.e., a plot of colony number against number of cells seeded should be linear). However, some cells may not plate well at very low densities, and the plating efficiency will fall. This can sometimes be minimized by using a feeder layer (*see* Section 14.2.3). If the plating efficiency falls at the higher concentrations, it implies that the cells are aggregating or colonies are coalescing. A seeding concentration that lies within the linear range of the plating efficiency curve (Fig. 21.11) should be selected for future assays.

The size distribution of the colonies may also be determined (e.g., to assay the growth-promoting ability of a test medium or serum; *see* Section 11.4.3) by counting the number of cells per colony by eye or estimating it by densitometry. To do so, after fixing and staining the colonies with crystal violet, measure absorption on a densitometer

[McKeehan et al., 1977] or size the colonies on an automatic colony counter (*see* Section 21.10.2).

21.10.2 Automatic Colony Counting

If the colonies are uniform in shape and quite discrete, they may be counted on an *automatic colony counter* (*see* Appendix II: Colony Counters), which scans the plate with a CCD camera and analyzes the image to give an instantaneous readout of the number of colonies (Fig. 21.12). A size discriminator gives an analysis of the size of the colonies, based on the average colony diameter, but this is not always proportional to the cell number, as cells may pile up in the center of a colony.

Although expensive, these instruments can save a great deal of time and make colony counting more objective. However, they do not work well with colonies that overlap more than ~20% or have irregular outlines.

21.11 LABELING INDEX

Cells that are synthesizing DNA will incorporate [³H]-TdR (*see* Section 21.5.1). The percentage of labeled cells, determined by autoradiography (*see* Protocol 27.3), is known as the labeling index (LI) [*see*, e.g., Westermark, 1974; Macieira-Coelho, 1973]. Measurement of the LI after a 30-min to 1-h labeling period with [³H]-TdR shows a large difference between exponentially growing cells (LI = 10−20%) and cells at the plateau phase (LI ≈ 1%). Normal Mv1Lu cells were shown to have a lower LI with [³H]-TdR than that of their neoplastic derivative transfected with mutant *ras* neoplastic cells [Khan et al., 1991] (*see* Plate 14d and Fig. 21.13). Because the LI is very low in the plateau phase, the duration of the [³H]-TdR labeling period may have to be increased to 24 h when showing differences at saturation density.

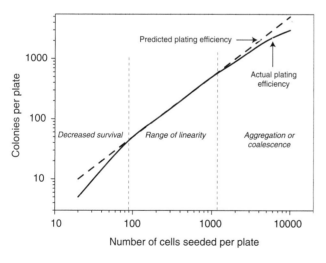

Fig. 21.11. Linearity of Plating Efficiency. If plating efficiency remains constant over a wide range of cell concentrations, the curve is linear (dashed line), whereas if there is poor survival at low densities or aggregation or coalescence at high densities, plating efficiency decreases (solid line).

Fig. 21.12. Automatic Colony Counter. Automatic counter for colonies in plating efficiency and survival assays. (Perceptive Biosystems).

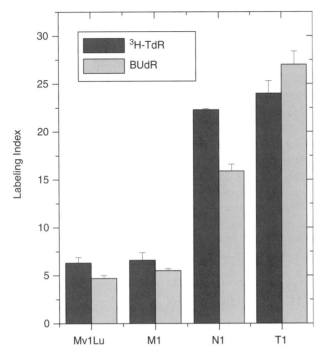

Fig. 21.13. Labeling Index. Mink lung cells, Mv1Lu, and oncogene-transfected derivatives were labeled with [^{3}H]-TdR for 1 h, fixed, and coated with autoradiographic emulsion (*see* Protocol 27.3), stained with Giemsa (*see* Plate 14d), or labeled for 1 h with bromodeoxyuridine (BUdR), fixed, and stained by immunoperoxidase with an antibody directed against BUdR bound to DNA. Labeled nuclei were counted as percentage of the total in each case. Mv1Lu is the control cell line, M1 is Mv1Lu transfected with the *myc* oncogene, N1 is Mv1Lu cells transfected with normal human *ras*, and T1 with mutant human *ras*. T1 cells were tumorigenic and had a statistically significant ($p < 0.001$) increase in labeling index compared to Mv1Lu, also seen with bromodeoxyuridine labeling [Tabular data from Khan et al., 1991].

PROTOCOL 21.11. LABELING INDEX WITH [^{3}H]THYMIDINE

Outline

Grow cells to an appropriate density. Label the cells with [^{3}H]thymidine for 30 min. Wash and fix the cells. Remove any unincorporated precursor from the cells and prepare autoradiographs.

Materials

Sterile or Aseptically Prepared:
❑ Cell culture for assay
❑ Growth medium
❑ Trypsin, 0.25%, crude
❑ D-PBSA
❑ [^{3}H]thymidine, 2.0 MBq/mL (~50 µCi/mL), 75 GBq/mmol (~2 Ci/mmol)

△ *Safety Note.* Handle [^{3}H]thymidine with care, as it is radioactive and genotoxic. Follow local guidelines for its use (*see* Section 7.7).

❑ Multiwell plate(s) containing 13-mm Thermanox coverslips (Nalge Nunc)
Nonsterile:
❑ Hemocytometer or electronic cell counter
❑ D-PBSA
❑ Acetic methanol (1 part glacial acetic acid to 3 parts methanol), ice cold, freshly prepared
❑ Microscope slides
❑ Mountant (e.g., DPX or Permount)
❑ Trichloroacetic acid (TCA), 0.6 M, ice cold
❑ Deionized water
❑ Methanol

Protocol

1. Set up the cultures at 2×10^4 cells/mL–5×10^4 cells/mL in 24-well plates containing coverslips.
2. Allow the cells to attach, start to proliferate (48–72 h), and grow to the desired cell density.
3. Add [^{3}H]thymidine to the medium, 100 KBq/mL (~5 µCi/mL), and incubate the cultures for 30 min.

Note. Some media—e.g., Ham's F10 and F12—contain thymidine (*see* Tables 9.3, 10.1, and 10.2). In these cases, the concentration of radioactive thymidine must be increased to give the same specific activity in the medium.

4. Remove the labeled medium, and discard it into a designated container for radioactive waste.
5. Wash the coverslips three times with D-PBSA. Lift the coverslips off the bottom of the wells (but not right out of them) at each wash to allow removal of the isotope from underneath.
6. Add 1:1 D-PBSA:acetic methanol, 1 mL per well, and then remove it immediately.
7. Add 1 mL of acetic methanol at 4°C to each well, and leave the cultures for 10 min.
8. Remove the coverslips, and dry them with a fan.
9. Mount the coverslip on a microscope slide with the cells uppermost.
10. Leave the mountant to dry overnight.
11. Place the slides in 0.6 M TCA at 4°C in a staining dish, and leave them for 10 min. Replace the TCA twice during this extraction, thereby removing unincorporated precursors.
12. Rinse the slides in deionized water, then in methanol, and dry the slides.
13. Prepare an autoradiograph (*see* Protocol 27.3).

14. When the autoradiograph has been exposed for the appropriate period (usually 1–2 weeks), develop, stain (*see* Protocol 27.3), and examine under microscope with 40× objective.

15. Count the percentage of labeled cells. To cover a representative area, follow the scanning pattern illustrated in Fig. 21.14.

21.11.1 Growth Fraction

If cells are labeled with [³H]thymidine for varying lengths of time up to 48 h, the plot of the LI against time increases rapidly over the first few hours and then flattens out to a very low gradient, almost a plateau (Fig. 21.15). The level of this plateau, read against the vertical axis, is the growth fraction of the culture—i.e., the proportion of the cells in cycle at the time of labeling.

PROTOCOL 21.12. DETERMINATION OF GROWTH FRACTION

Outline
Label the culture continuously for 48 h, sampling at intervals for autoradiography.

Protocol
Follow Protocol 21.10, except that at step 3, incubation should be carried out for 15 min, 30 min, and 1, 2, 4, 8, 24, and 48 h.

Analysis. Count the number of labeled cells as a percentage of the total number of cells, using the scanning pattern from Protocol 21.10. Plot the LI against time (*see* Fig. 21.15).

Note. Autoradiographs with ³H can be prepared only when the cells remain as a monolayer. If they form a multilayer, then they must be trypsinized after labeling, and slides must be prepared by the drop technique (*see* Protocol 16.7, without the hypotonic step) or by cytocentrifugation (*see* Protocol 16.4), as the energy of β-emission from ³H is too low to penetrate an overlying layer of cells.

The LI can also be determined by labeling cells with BUdR, which becomes incorporated into DNA. This effect can be detected subsequently by immunostaining with an anti-BUdR antibody (Dako). Results from this method are generally in agreement with the results of using [³H] thymidine [Khan et al., 1991] (*see also* Fig. 21.13).

21.11.2 Mitotic Index

The mitotic index is the fraction or percentage of cells in mitosis and is determined by counting mitoses in stained

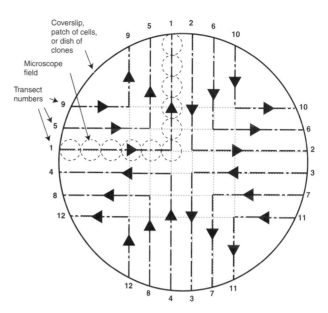

Fig. 21.14. Scanning Slides or Dishes. Scanning pattern for the analysis of cytological preparations on slides or dishes. Each dotted circle represents one microscope field, and the large circle represents the extent of the specimen (e.g., a coverslip, culture dish, or well, or a spot of cells on a slide). Guide lines can be drawn with a nylon-tipped pen with a light, transparent ink.

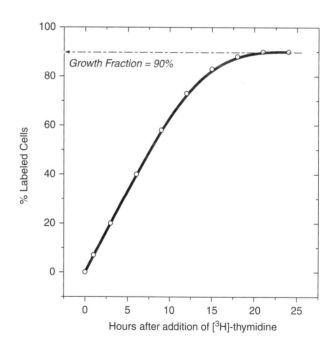

Fig. 21.15. Growth Fraction. To determine the growth fraction, cells are labeled continuously with [³H]thymidine, and the percentage of labeled cells is determined at intervals by autoradiography (*see* text).

cultures as a proportion of the whole population of cells. The scanning pattern should be as for counting labeled nuclei (*see* Fig. 21.14).

21.11.3 Division Index

A number of antibodies, such as Ki67 [Zhu & Joyce, 2004] or anti-PCNA [Katdare et al., 2004], are able to stain cells in the division cycle. These antibodies are raised against proteins expressed during the cell cycle but not expressed in resting cells. Some of the antibodies are directed against DNA polymerase, but the epitopes for others are not known. The cells are stained by immunofluorescence or immunoperoxidase (*see* Protocol 16.11), and the proportion of stained cells is determined cytologically (*see* Fig. 21.14) or by flow cytometry (*see* Section 16.9.2). This figure gives a higher index than that of either mitotic counting or DNA labeling, as cells stain throughout the cell cycle. It also gives a particularly useful indication of the growth fraction.

21.12 CELL CYCLE TIME

To determine the length of the cell cycle (generation time) and its constituent phases, cells are labeled continuously with BUdR, and the incorporation of the label is detected at intervals after labeling by immunofluorescence microscopy or flow cytometry. For flow cytometry, the cells are also stained with propidium iodide and analyzed for BUdR incorporation versus DNA content, to follow the progression of cells around the cycle [Dolbeare & Selden, 1994; Poot et al., 1994].

21.13 CELL MIGRATION

Cells in culture, particularly fibroblasts, are motile and can migrate significant distances across the substrate, dependent on cell density and the presence of stimulants such as growth factors. Motility is evidenced by ruffling of the cell membrane as visualized by time-lapse video. The amount of motility is difficult to quantify, but the quantification of cell migration can be achieved by detailed analysis of time-lapse video sequences (*see* Protocol 27.4) or by image analysis of tracks made by the cell's phagocytosis in dishes coated with colloidal gold [Kawa et al., 1997].

Migration can also be assayed by the movement of cells through a porous membrane, as in chemotaxis assays in a Boyden chamber [Schor, 1994] or invasion assays in a filter well (*see* Section 18.6.3).

CHAPTER 22

Cytotoxicity

22.1 VIABILITY, TOXICITY, AND SURVIVAL

Once a cell is explanted from its normal *in vivo* environment, the question of viability, particularly in the course of experimental manipulations, becomes fundamental. Previous chapters have dealt with the status of the cultured material relative to the tissue of origin and how to quantify changes in growth and phenotypic expression. However, none of these data is acceptable unless the great majority of the cells are shown to be viable. Furthermore, many experiments carried out *in vitro* are for the sole purpose of determining the potential cytotoxicity of the compounds being studied, either because the compounds are being used as pharmaceuticals or cosmetics and must be shown to be nontoxic or because they are designed as anticancer agents and cytotoxicity may be crucial to their action.

New drugs, cosmetics, food additives, and so on go through extensive cytotoxicity testing before they are released for use by the public. This testing usually involves a large number of animal experiments, although in Europe, these experiments will be subject to new legislation, due to be introduced in 2009 for topical application and in 2013 for systemic application. There is much pressure, both humane and economic, to perform at least part of cytotoxicity testing *in vitro*. The introduction of specialized cell lines and interactive organotypic cultures, and the continued use of long-established cultures, may make this a reasonable proposition.

Toxicity is a complex event *in vivo*, where there may be direct cellular damage, as with a cytotoxic anticancer drug, physiological effects, such as membrane transport in the kidney or neurotoxicity in the brain, inflammatory effects, both at the site of application and at other sites, and other systemic effects. Currently, it is difficult to monitor systemic and physiological effects *in vitro*, so most assays determine effects at the cellular level, or *cytotoxicity*. Definitions of cytotoxicity vary, depending on the nature of the study and whether cells are killed or simply have their metabolism altered. Whereas an anticancer agent may be required to kill cells, demonstrating the lack of toxicity of other pharmaceuticals may require a more subtle analysis of specific targets such as an alteration in cell signaling or cell interaction such as might give rise to an inflammatory or allergic response.

All of these assays oversimplify the events that they measure and are employed because they are cheap, easily quantified, and reproducible. However, it has become increasingly apparent that they are inadequate for modern drug development, which requires greater emphasis on molecular target specificity and precise metabolic regulation. Gross tests of cytotoxicity are still required, but there is a growing need to supplement them with more subtle tests of metabolic perturbation. Perhaps the most obvious of these tests is the induction of an inflammatory or allergic response, which need not imply cytotoxicity of the allergen or inflammatory agent and is still one of the hardest results to demonstrate *in vitro*.

It is not within the scope of this text to define all of the requirements of a cytotoxicity assay, many of which may be quite specialized to the needs of the user, so instead I will concentrate on those aspects that influence cell growth or survival. Cell growth is generally taken to be the regenerative potential of cells, as measured by clonal growth

Culture of Animal Cells: A Manual of Basic Technique, Fifth Edition, by R. Ian Freshney
Copyright © 2005 John Wiley & Sons, Inc.

(*see* Protocol 21.10), net change in population size (e.g., in a growth curve; *see* Protocols 21.7 and 21.8), or a change in cell mass (total protein or DNA) or metabolic activity (e.g., DNA, RNA, or protein synthesis; MTT reduction).

22.2 *IN VITRO* LIMITATIONS

It is important that any *in vitro* measurement can be interpreted in terms of the *in vivo* response of the same or similar cells, or at least that the differences that exist between *in vitro* and *in vivo* measurements are clearly understood.

Pharmacokinetics. The measurement of toxicity *in vitro* is generally a cellular event. For example, it would be very difficult to recreate the complex pharmacokinetics of drug exposure *in vitro*, and between *in vitro* and *in vivo* experiments there usually are significant differences in exposure time to and concentration of the drug, rate of change of the concentration, metabolism, tissue penetration, clearance, and excretion. Although it may be possible to simulate these parameters—e.g., using multicellular tumor spheroids for drug penetration or timed perfusion to simulate concentration and time ($C \times T$) effects—most studies concentrate on a direct cellular response, thereby gaining simplicity and reproducibility.

Metabolism. Many nontoxic substances become toxic after being metabolized by the liver; in addition, many substances that are toxic *in vitro* may be detoxified by liver enzymes. For *in vitro* testing to be accepted as an alternative to animal testing, it must be demonstrated that potential toxins reach the cells *in vitro* in the same form as they would *in vivo*. This proof may require additional processing by purified liver microsomal enzyme preparations [McGregor et al., 1988], coculture with activated hepatocytes [Guillouzo, 1989; Frazier, 1992], or genetic modification of the target cells with the introduction of genes for metabolizing enzymes under the control of a regulatable promoter [Macé et al., 1994].

Tissue and systemic responses. The nature of the response must also be considered carefully. A toxic response *in vitro* may be measured by changes in cell survival (*see* Protocol 22.3) or metabolism (*see* Section 22.3.4), whereas the major problem *in vivo* may be a tissue response (e.g., an inflammatory reaction, fibrosis, kidney failure) or a systemic response (e.g., pyrexia, vascular dilatation). For *in vitro* testing to be more effective, models of these responses must be constructed, perhaps utilizing organotypic cultures reassembled from several different cell types and maintained in the appropriate hormonal milieu.

It should not be assumed that complex tissue and even systemic reactions cannot be simulated *in vitro*. Assays for inflammatory responses, teratogenic disorders, and neurological dysfunctions may be feasible *in vitro*, given a proper understanding of cell–cell interaction and the interplay of endocrine hormones with local paracrine and autocrine factors.

22.3 NATURE OF THE ASSAY

The choice of assay will depend on the agent under study, the nature of the response, and the particular target cell. Assays can be divided into five major classes:

(1) *Viability*: An immediate or short-term response, such as an alteration in membrane permeability or a perturbation of a particular metabolic pathway correlated with cell proliferation or survival
(2) *Survival*: The long-term retention of self-renewal capacity (5–10 generations or more)
(3) *Metabolic*: Assays, usually microtitration based, of intermediate duration that can either measure a metabolic response (e.g., dehydrogenase activity; DNA, RNA, or protein synthesis) at the time of, or shortly after, exposure. Making the measurement two or three population doublings after exposure is more likely to reflect cell growth potential and may correlate with survival.
(4) *Transformation*: Survival in an altered state (e.g., a state expressing genetic mutation, alterations in growth control, or malignant transformation)
(5) *Irritancy*: A response analogous to inflammation, allergy, or irritation *in vivo*; as yet difficult to model *in vitro*, but may be possible to assay by monitoring cytokine release in organotypic cultures

22.3.1 Viability

Viability assays are used to measure the proportion of viable cells after a potentially traumatic procedure, such as primary disaggregation, cell separation, or cryostorage.

Most viability tests rely on a breakdown in membrane integrity measured by the uptake of a dye to which the cell is normally impermeable (e.g., Trypan Blue, Erythrosin, or Naphthalene Black) or the release of a dye normally taken up and retained by viable cells (e.g., diacetyl fluorescein or Neutral Red). However, this effect is immediate and does not always predict ultimate survival. Furthermore, dye exclusion tends to overestimate viability—e.g., 90% of cells thawed from liquid nitrogen may exclude Trypan Blue, but only 60% prove to be capable of attachment 24 h later.

Note that routine assessment of viability at subculture can be uninformative regarding trypsinized cells as most of the nonviable cells will be lost in the discarded medium and prewash before trypsinization. An accurate assessment of the viability status at subculture requires that all the cells are recovered from the medium and prewash and combined with the trypsinate. However, viability of *reseeded* cells will be accurately determined without this recovery.

Protocol 22.1 can be used in conjunction with Exercises 12 and 17 (*see* Chapter 2).

PROTOCOL 22.1. ESTIMATION OF VIABILITY BY DYE EXCLUSION

Principle

Viable cells are impermeable to Naphthalene Black, Trypan Blue, and a number of other dyes [Kaltenbach et al., 1958].

Outline

Mix a cell suspension with stain, and examine it by low-power microscopy.

Materials

Sterile or Aseptically Prepared:
- ❑ Cells for testing, e.g., flask for trypsinization, frozen vial to thaw, or primary disaggregate
- ❑ Growth medium appropriate to cell type.... 20 mL
- ❑ Trypsin, 0.25%................................. 5 mL
- ❑ D-PBSA... 10 mL

Nonsterile:
- ❑ Hemocytometer
- ❑ Viability stain
  (e.g., 0.4% Trypan Blue or 1% Naphthalene Black in D-PBSA or HBSS)..................... 1 mL
- ❑ Pasteur pipettes
- ❑ Microscope
- ❑ Tally counter

Protocol

1. Prepare a cell suspension at a high concentration ($\sim 1 \times 10^6$ cells/mL) by trypsinization or by centrifugation and resuspension.
2. Take a clean hemocytometer slide and fix the coverslip in place (*see* Protocol 22.1 and Fig. 21.1).
3. Mix one drop of cell suspension with one drop (Trypan Blue) or four drops (Naphthalene Black) of stain.
4. Load the counting chamber of the hemocytometer (*see* Protocol 22.1).
5. Leave the slide for 1–2 min (do not leave any longer, or viable cells will deteriorate and take up the stain).
6. Place the slide on the microscope, and use a 10× objective to look at the counting grid (*see* Fig. 21.1).
7. Count the total number of cells and the number of stained cells.
8. Wash the hemocytometer, and return it to its box.

Analysis. Calculate the percentage of unstained cells. This figure is the percentage viability by this method. If the respective volumes of cell suspension and stain are measured accurately at Step 3, then this method of viability determination can be incorporated into Protocol 22.1.

PROTOCOL 22.2. ESTIMATION OF VIABILITY BY DYE UPTAKE

Principle

Viable cells take up diacetyl fluorescein and hydrolyze it to fluorescein, to which the cell membrane of live cells is impermeable [Rotman & Papermaster, 1966]. Live cells fluoresce green; dead cells do not. Nonviable cells may be stained with propidium iodide and subsequently fluoresce red. Viability is expressed as the percentage of cells fluorescing green. This method may be applied to CCD analysis or flow cytometry (*see* Section 21.7.2).

Outline

Stain a cell suspension in a mixture of propidium iodide and diacetyl fluorescein, and examine the cells by fluorescence microscopy or flow cytometry.

Materials

Sterile or Aseptically Prepared:
- ❑ Single-cell suspension
- ❑ Fluorescein diacetate, 10 μg/mL, in HBSS
- ❑ Propidium iodide, 500 μg/mL

Nonsterile:
- ❑ Fluorescence microscope
- ❑ Filters:
  Fluorescein: excitation 450/590 nm, emission LP 515 nm
  Propidium iodide: excitation 488 nm, emission 615 nm

Protocol

1. Prepare the cell suspension as for dye exclusion (*see* Protocol 22.1), but in medium without phenol red.
2. Add the fluorescent dye mixture at a proportion of 1:10 to give a final concentration of 1 μg/mL of diacetyl fluorescein and 50 μg/mL of propidium iodide.
3. Incubate the cells at 37°C for 10 min.
4. Place a drop of the cells on a microscope slide, add a coverslip, and examine the cells by fluorescence microscopy.

Analysis. Cells that fluoresce green are viable, whereas those that fluoresce red are nonviable. Viability may be expressed as the percentage of cells that fluoresce green as a proportion of the total number of cells. The stained cell suspension can also be analyzed by flow cytometry (*see* Section 21.7.2).

Neutral red uptake. Living cells take up neutral red, 40 μg/mL in culture medium, and sequester it in the lysosomes. However, neutral red is not retained by nonviable cells. Uptake of neutral red is quantified by fixing the cells in formaldehyde and solubilizing the stain in acetic ethanol, allowing the plate to be read on an ELISA plate reader at 570 nm [Borenfreund et al., 1990; Babich & Borenfreund, 1990]. Neutral red tends to precipitate, so the medium with stain is usually incubated overnight and centrifuged before use. This assay does not measure the total number of cells, but it does show a reduction in the absorbance related to loss of viable cells and is readily automated.

22.3.2 Survival

Although short-term tests are convenient and usually are quick and easy to perform, they reveal only cells that are dead (i.e., permeable) at the time of the assay. Frequently,

however, cells that have been subjected to toxic influences (e.g., irradiation, antineoplastic drugs) show an effect several hours, or even days, later. The nature of the tests required to measure viability in these cases is necessarily different, because by the time the measurement is made the dead cells may have disappeared. Therefore, long-term tests are used to demonstrate survival rather than short-term toxicity, which may be reversible. Survival implies the retention of regenerative capacity and is usually measured by plating efficiency (*see* Protocol 21.10). Plating efficiency measures survival by demonstrating proliferative capacity for several cell generations, provided that the cells plate with a high-enough efficiency that the colonies can be considered representative of the entire cell population. Although not ideal, a plating efficiency of over 10% is usually acceptable.

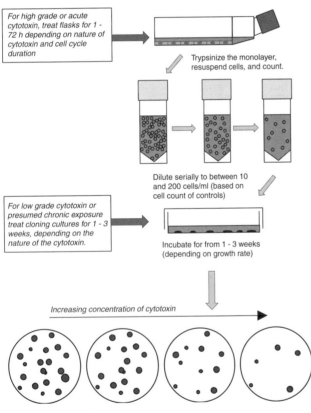

Fig. 22.1. Clonogenic Assay for Monolayer Cells. Cells are trypsinized, counted, and diluted as for monolayer dilution cloning (*see* Protocol 14.1). The test substance can be added before trypsinization or after cloning (*see* Section 22.3.2). The colonies are fixed and stained when they reach a reasonable size for counting by eye but before they overlap.

PROTOCOL 22.3. CLONOGENIC ASSAY FOR ATTACHED CELLS

Outline
Treat the cells with experimental agent at a range of concentrations for 24 h. Trypsinize the cells, seed them at a low cell density, and incubate them for 1–3 weeks. Stain the cells (*see* Plate 6a,e), and count the number of colonies (*see* Fig. 22.1).

Materials
Sterile:
❑ Growth medium
❑ D-PBSA
❑ Trypsin, 0.25%, crude
❑ Compound to be tested at 10× the maximum concentration to be used, dissolved in serum-free medium; check the pH and the osmolality of the test solution, and adjust if necessary
❑ Flasks, 25 cm²
❑ Petri dishes, 6 or 9 cm, labeled on the side of the base
Nonsterile:
❑ D-PBSA
❑ Methanol
❑ Crystal Violet, 1%
❑ Hemocytometer or electronic cell counter

Protocol
1. Prepare a series of cultures in 25-cm² flasks, three for each of six agent concentrations, and three controls. Seed the cells at 5×10^4 cells/mL in 4.5 mL of growth medium, and incubate them for 48 h, by which time the cultures will have progressed into the log phase (*see* Section 21.9.2).

2. Prepare a serial dilution of the compound to give 2 mL at 10× of each of the final concentrations required:
   (a) If you are testing a compound for the first time, use 5-fold dilutions over a range of 3–5 logs.
   (b) If you can predict the approximate toxic concentration, then select a narrower, arithmetic range over one or two decades.
3. Dilute each concentration 1:10 by adding 0.5 mL of 10× concentrate to each of three flasks for each concentration, starting with control medium (no compound added) and progressing from lowest to highest concentration.
4. Return the flasks to the incubator.
5. If the compound is slow acting or partially reversible, repeat Step 3 twice; that is, expose the cultures to the agent for 3 days, replacing the medium and compound daily by changing the medium. With fast-acting compounds, 1-h exposure may be sufficient.
6. Remove the medium from each group of three flasks in turn (working from lowest concentration of compound to highest), and trypsinize the cells.
7. Dilute and seed the cells into Petri dishes at the required density for clonal growth (*see* Protocol 14.1), diluting all of the cultures by the same amount as the control. Work from the flasks exposed to the highest concentration down to the lowest and then the control.
8. Incubate the cultures until colonies form.
9. Fix the cultures in absolute methanol, and stain them for 10 min in 1% Crystal Violet (*see* Protocol 16.3).
10. Wash the dishes in tap water, drain, and stand in an inverted position to dry.
11. Count the colonies with >50 cells (>5 generations).

Analysis of Survival Curve
12. Calculate the plating efficiency at each drug concentration.
13. Calculate the relative plating efficiency: the plating efficiency at each concentration as a fraction of the control. This is the *surviving fraction*.
14. Plot the surviving fraction on a log scale against the concentration on a linear or log scale, depending on the concentration range used (*see* Fig. 22.2).
15. Determine the IC_{50} or IC_{90}, which is the concentration of compound promoting 50% or 90% inhibition of colony formation, respectively. As this is a semilog plot, the IC_{90} is more appropriate, as it is more likely to fall on the

linear part of the curve, whereas the IC_{50} tends to fall on the knee of the curve, giving a less stable value.
16. Analyze the curve for differences in sensitivity:
    (a) *Slope of the curve and length of the knee.* A shallower slope and/or longer knee means reduced sensitivity; a steeper slope and/or shorter knee means increased sensitivity. Both the length of knee and the slope influence the IC_{50} and the IC_{90}, although a more significant difference can be observed in the IC_{90} (*see* Fig. 22.3) and the IC_{90} is often used as a simple derivative.
    (b) *Resistant fraction.* The fraction of resistant cells is indicated by a flattening of the lower end of the curve.
    (c) *Total resistance* is indicated by the lack of any gradient on the curve.
    (d) *Area under the curve:* Complex survival curves may be compared by calculating the area under the curve, but this is done for expediency and is not mathematically valid.

Variable parameters in survival assay
Concentration of agent. A wide range of concentrations in log increments (e.g., 1×10^{-6} M, 1×10^{-5} M, 1×10^{-4} M, 1×10^{-3} M, and control) should be used for the first attempt and a narrower range (log or linear), based on the results from the first range, for subsequent attempts.

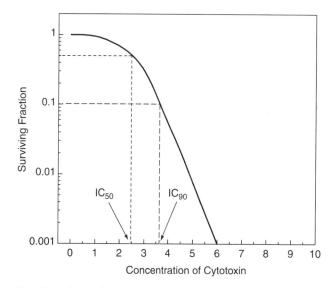

Fig. 22.2. Survival Curve. Semilog plot of the surviving fraction of cells (colonies forming from test cells/colonies forming from control cells) against the concentration of cytotoxin. Typically, the curve has a "knee", and the IC_{90} lies in the linear range of the curve. The IC_{50}, falling on the knee, is a less stable value.

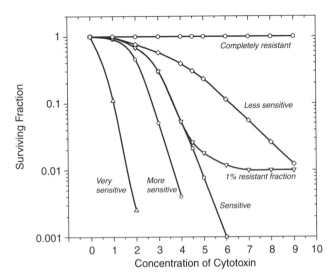

Fig. 22.3. Interpretation of Survival Curves. Semilog plot of cell survival against the concentration of cytotoxin. The slope increases with increasing sensitivity and decreases with reduced sensitivity until it becomes totally flat for complete resistance. Partial resistance can be expressed as the resistant fraction shown by the curve flattening out at the lower end.

Invariate agent concentrations. Some conditions that are tested cannot easily be varied—e.g., the quality of medium, water, or an insoluble plastic. In these cases, the serum concentrations can be varied. As serum may have a masking effect on low-level toxicity, an effect may only be seen in limiting serum.

Duration of exposure to agent. Some agents act rapidly, whereas others act more slowly. Exposure to ionizing radiation, for example, need last only a matter of minutes to achieve the required dose, whereas testing some cycle-dependent antimetabolic drugs may take several days to achieve a measurable effect. Duration of exposure (T) and drug concentration (C) are related, although $C \times T$ is not always a constant. Prolonging exposure can increase sensitivity beyond that predicted by $C \times T$, because of cell cycle effects and cumulative damage.

Time of exposure to agent. When the agent is soluble and expected to be toxic, the procedure in Protocol 22.3 should be followed, but when the quality of the agent is unknown, stimulation is expected, or only a minor effect is expected (e.g., 20% inhibition rather than severalfold), the agent may be incorporated during clonal growth rather than at preincubation. Confirmation of anticipated toxicity—e.g., for a cytotoxic drug—requires a conservative assay, with minimal drug exposure as compared to that *in vivo*, be applied before cloning. Confirmation of the lack of toxicity—e.g., for tap water or a nontoxic pharmaceutical—requires a more stringent assay, with prolonged exposure during cloning.

Cell density during exposure. The density of the cells during exposure to an agent can alter the response of the cells and the agent; for example, HeLa cells are less sensitive to the alkylating agent mustine at high cell densities [Freshney et al., 1975]. To determine the effects of cell density it should be varied in the preincubation phase, during exposure to the drug.

Cell density during cloning. The number of colonies may fall at high concentrations of a toxic agent, but it is possible to compensate for this effect by seeding more cells so that approximately the same number of colonies form at each concentration. This procedure removes the risk of a low clonal density influencing survival and improves statistical reliability, but is prone to the error that cells from higher drug concentrations are plated at a higher cell concentration, a factor that may also influence survival. It is preferable to plate cells on a preformed feeder layer, the density of which (5×10^3 cells/cm^2) greatly exceeds that of the cloning cells. This step ensures that the cell density is uniform regardless of clonal survival, which contributes little to the total cell density. Note that cloning on a feeder layer can sometimes reveal a resistant fraction of cells that is not apparent without the feeder layer (Fig. 22.4).

Colony size. Some agents are cytostatic (i.e., they inhibit cell proliferation) but not cytotoxic, and during continuous

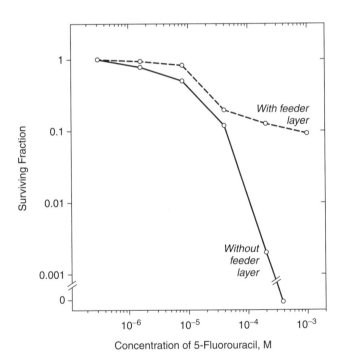

Fig. 22.4. Effect of Feeder Layer on Resistant Fraction. Human glioma cells were plated out in the presence (dashed line) and absence (solid line) of a feeder layer after treatment with various concentrations of 5-fluorouracil. A 10%-resistant fraction is apparent at 1×10^{-4} M drug only in the presence of a feeder layer. In the absence of the feeder layer, the small number of colonies making up the resistant fraction were unable to survive alone.

exposure they may reduce the size of colonies without reducing the number of colonies. In this case, the size of the colonies should be determined by densitometry [McKeehan et al., 1977], automatic colony counting, or visually counting the number of cells per colony.

For colony counting, the threshold number of cells per colony (e.g., 50 as in Protocol 21.9) is purely arbitrary, and it is assumed that most of the colonies are greatly in excess of this number. Colonies should be grown until they are quite large (>1 × 10³ cells), when the growth of larger colonies tends to slow down and smaller, but still viable, colonies tend to catch up with these larger colonies.

For *colony sizing*, stain the cultures earlier, before the growth rate of larger colonies has slowed down, and score all of the colonies.

Solvents. Some agents to be tested have low solubilities in aqueous media, and it may be necessary to use an organic solvent to dissolve them. Ethanol, propylene glycol, and dimethyl sulfoxide have been used for this purpose, but may themselves be toxic to cells. Hence, the minimum concentration of solvent should be used to obtain a solution. The agent may be made up at a high concentration in, for example, 100% ethanol, then diluted gradually with BSS and finally diluted into medium. The final concentration of solvent should be <0.5%, and a *solvent control* must be included (i.e., a control with the same final concentration of solvent but without the agent being tested).

Take care when using organic solvents with plastics or rubber. It is better to use glass with undiluted solvents and to use plastic only when the solvent concentration is <10%.

Although calculating the plating efficiency is one of the best methods for testing cell survival rates, it should be used only when the cloning efficiency is high enough for colonies that form to be representative of the whole cell population. Ideally, this means that controls should plate at 100% efficiency. In practice, however, this is seldom possible, and control plating efficiencies of 20% or less are often accepted.

22.3.3 Assays Based on Cell Proliferation

Cell counts after a few days in culture can also be used to determine the effect of various compounds on cell proliferation, but, at least in the early stages of testing, a complete growth curve is required (*see* Protocols 21.7, 21.8, and 21.9), because cell counts at a single point in time can be ambiguous (*see* Fig. 21.8, day 7). Growth curve analyses, using cell counting, are feasible only with relatively small numbers of samples, as they become cumbersome in a large screen. In cases for which there are many samples, a single point in time—e.g., the number of cells three to five days after exposure—can be used. The time should be selected as within the log phase, and preferably mid-log phase, of control cells. Any significant effect should be backed up with a complete growth curve over the whole growth cycle or by

an alternative assay, such as a survival curve by clonogenic assay (*see* Protocol 22.3) or MTT assay (*see* Protocol 22.4).

22.3.4 Metabolic Cytotoxicity Assays

Plating efficiency tests are labor intensive and time consuming to set up and analyze, particularly when a large number of samples is involved, and the duration of each experiment may be anywhere from 2 to 4 weeks. Furthermore, some cell lines have poor plating efficiencies, particularly freshly isolated normal cells, so a number of alternatives have been devised for assaying cells at higher densities, e.g., in microtitration plates (*see* Section 22.3.5). None of these tests measures survival directly, however. Instead, the net increase in the number of cells (i.e., the growth yield; *see* Section 22.3.3), the increase in the total amount of protein or DNA, or continued metabolic activity, such as the reduction of a tetrazolium salt to formazan or the synthesis protein or DNA, is determined. Survival in these cases is defined as the retention of metabolic or proliferative ability by the cell population as a whole some time after removal of the toxic influence. However, such assays cannot discriminate between a reduction in metabolic or proliferative activity per cell and a reduced number of cells, and therefore any novel or exceptional observation should be confirmed by clonogenic survival assay.

22.3.5 Microtitration Assays

The introduction of multiwell plates revolutionized the approach to replicate sampling in tissue culture. These plates are economical to use, lend themselves to automated handling, and can be of good optical quality. The most popular is the 96-well microtitration plate, each well having 28–32 mm² of growth area, 0.1 or 0.2 mL medium, and up to 1 × 10⁵ cells. Microtitration offers a method whereby large numbers of samples may be handled simultaneously, but with relatively few cells per sample. With this method, the whole population is exposed to the agent, and viability is determined subsequently, usually by measuring a metabolic parameter such as the ATP or NADH/NADPH concentration. A range of kits are available from Promega (*see* Appendix III).

The end point of a microtitration assay is usually an estimate of the number of cells. Although this result can be achieved directly by cell counts or by indirect methods, such as isotope incorporation, cell viability as measured by MTT reduction [Mosmann, 1983] is now widely chosen as the optimal end point [Cole, 1986; Alley et al., 1988]. MTT is a yellow water-soluble tetrazolium dye that is reduced by live, but not dead, cells to a purple formazan product that is insoluble in aqueous solutions. However, a number of factors can influence the reduction of MTT [Vistica et al., 1991]. The assay described in Protocol 22.4, provided by Jane Plumb of the Centre for Oncology and Applied Pharmacology, University of Glasgow, Scotland, United Kingdom, has been shown to give the same results as a standard clonogenic assay [Plumb et al., 1989]. It illustrates the use of microtitration in the assay of anticancer drugs, but would be applicable, with minor modifications, to any cytotoxicity assay.

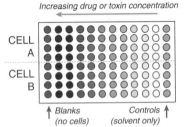

Blanks (no cells)

CELL A

CELL B

Set up microtitration plate and incubate for about two population doublings

Increasing drug or toxin concentration

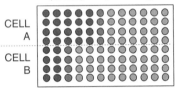

CELL A

CELL B

When cells are in exponential growth, add drug or toxin

↑ Blanks   Controls ↑
(no cells)   (solvent only)

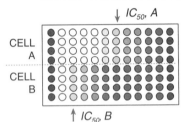

CELL A

CELL B

Remove drug and allow cells to recover in growth medium

After two or three more population doublings, remove medium and replace with MTT or XTT. Read on plate reader after 3-4 h. Use absorbance to plot inhibition curve and calculate IC_{50}

↓ IC_{50}, A

CELL A

CELL B

↑ IC_{50}, B

Fig. 22.5. *Microtitration Assay.* Stages in the assay of two different cell lines exposed to a range of concentrations of the same drug and then allowed to recover before the estimation of survival by the MTT reaction (*see* Protocol 22.4). The far left column has no cells and can be used as a blank to set the plate reader. This array is applicable when using plate sealers, when all wells are equivalent; however, with lids, there is a risk of an edge effect, probably due to evaporation, and it is better to leave the far left and far right columns blank (i.e., with medium only, as in Protocol 22.4) and some users leave the top and bottom rows blank as well.

PROTOCOL 22.4. MTT-BASED CYTOTOXICITY ASSAY

Principle
Cells in the exponential phase of growth are exposed to a cytotoxic drug. The duration of exposure is usually determined as the time required for maximal damage to occur, but is also influenced by the stability of the drug. After removal of the drug, the cells are allowed to proliferate for two to three population-doubling times (PDTs) in order to distinguish between cells that remain viable and are capable of proliferation and those that remain viable but cannot proliferate. The number of surviving cells

is then determined indirectly by MTT dye reduction. The amount of MTT-formazan produced can be determined spectrophotometrically once the MTT-formazan has been dissolved in a suitable solvent.

Outline
Incubate monolayer cultures in microtitration plates in a range of drug concentrations (Fig. 22.5). Remove the drug, and feed the plates daily for two to three PDTs; then feed the plates again, and add MTT to each well. Incubate the plates in the dark for 4 h, and then remove the medium and MTT. Dissolve the water-insoluble MTT-formazan crystals in DMSO, add a buffer to adjust the final pH, and record the absorbance in a plate reader.

Materials
Sterile:
❑ Growth medium
❑ Trypsin (0.25% + EDTA, 1 mM, in PBSA)
❑ MTT: 3-(4,5-dimethylthiazol-2-yl)-2,5-diphenyl-tetrazolium bromide (Sigma), 50 mg/mL, filter sterilized
❑ Sorensen's glycine buffer (0.1 M glycine, 0.1 M NaCl adjusted to pH 10.5 with 1 M NaOH)
❑ Microtitration plates (Iwaki)
❑ Pipettor tips, preferably in an autoclavable tip box
❑ Petri dishes (non-TC-treated), 5 cm and 9 cm or reservoir (Corning)
❑ Universal containers or tubes, 30 mL and 100 mL
Nonsterile:
❑ Plastic box (clear polystyrene, to hold plates)
❑ Multichannel pipettor
❑ Dimethyl sulfoxide (DMSO)
❑ DMSO dispenser (optional): Labsystems Microplate Dispenser (Cat No 5840 127, Thermo Electron)
❑ ELISA plate reader (Molecular Devices, with SOFTmax PRO; *see also* Appendix II)
❑ Plate carrier for centrifuge (for cells growing in suspension)

Protocol
Plating out cells:
1. Trypsinize a subconfluent monolayer culture, and collect the cells in growth medium containing serum.
2. Centrifuge the suspension (5 min at 200 g) to pellet the cells. Resuspend the cells in growth medium, and count them.
3. Dilute the cells to $2.5–50 \times 10^3$ cells/mL, depending on the growth rate of the cell line and allowing 20 mL of cell suspension per microtitration plate.
4. Transfer the cell suspension to a 9-cm Petri dish, and, with a multichannel pipette, add 200 μL of

the suspension into each well of the central 10 columns of a flat-bottomed 96-well plate (80 wells per plate), starting with column 2 and ending with column 11, thereby placing $0.5–10 \times 10^3$ cells into each well.

5. Add 200 μL of growth medium to the eight wells in columns 1 and 12. Column 1 will be used to blank the plate reader; column 12 helps to maintain the humidity for column 11 and minimizes the "edge effect."

6. Put the plates in a plastic lunch box, and incubate in a humidified atmosphere at 37°C for 1–3 days, such that the cells are in the exponential phase of growth at the time that drug is added.

7. For nonadherent cells, prepare a suspension in fresh growth medium. Dilute the cells to $5–100 \times 10^3$ cells/mL, and plate out only 100 μL of the suspension into round-bottomed 96-well plates. Add drug immediately to these plates.

Drug addition:

8. Prepare a serial fivefold dilution of the cytotoxic drug in growth medium to give eight concentrations. This set of concentrations should be chosen such that the highest concentration kills most of the cells and the lowest kills none of the cells. Once the toxicity of a drug is known, a smaller range of concentrations can be used. Normally, three plates are used for each drug to give triplicate determinations within one experiment.

9. For adherent cells, remove the medium from the wells in columns 2 to 11. This can be achieved with a hypodermic needle attached to a suction line.

10. Feed the cells in the eight wells in columns 2 and 11 with 200 μL of fresh growth medium; these cells are the controls.

11. Add the cytotoxic drug to the cells in columns 3 to 10. Only four wells are needed for each drug concentration, such that rows A–D can be used for one drug and rows E–H for a second drug.

12. Transfer the drug solutions to 5-cm Petri dishes, and add 200 μL to each group of four wells with a four-tip pipettor.

13. Return the plates to the plastic box, and incubate them for a defined exposure period. For nonadherent cells, prepare the drug dilution at twice the desired final concentration, and add 100 μL to the 100 μL of cells already in the wells.

Growth period:

14. At the end of the drug exposure period, remove the medium from all of the wells containing cells, and feed the cells with 200 μL of fresh medium. Centrifuge plates containing nonadherent cells (5 min at 200 *g*) to pellet the cells. Then remove the medium, using a fine-gauge needle to prevent disturbance of the cell pellet.

15. Feed the plates daily for 2–3 PDTs.

Estimation of surviving cell numbers:

16. Feed the plate with 200 μL of fresh medium at the end of the growth period, and add 50 μL of MTT to all of the wells in columns 1 to 11.

17. Wrap the plates in aluminum foil, and incubate them for 4 h in a humidified atmosphere at 37°C. Note that 4 h is a minimum incubation time, and plates can be left for up to 8 h.

18. Remove the medium and MTT from the wells (centrifuge for nonadherent cells), and dissolve the remaining MTT-formazan crystals by adding 200 μL of DMSO to all of the wells in columns 1 to 11.

19. Add glycine buffer (25 μL per well) to all of the wells containing DMSO.

20. Record absorbance at 570 nm immediately, because the product is unstable. Use the wells in column 1, which contain medium and MTT but no cells, to blank the plate reader.

Analysis of MTT Assay:

21. Plot a graph of the absorbance (*y*-axis) against the concentration of drug (*x*-axis).

22. Calculate the IC_{50} as the drug concentration that is required to reduce the absorbance to half that of the control. The mean absorbance reading from the wells in columns 2 and 11 is used as a control. The absorbance values in columns 2 and 11 should be the same. Occasionally, they are not, however, and this is taken to indicate uneven plating of cells across the plate.

23. The absolute value of the absorbance should be plotted so that control values may be compared, but the data can then be converted to a percentage-inhibition curve (Fig. 22.6), to normalize a series of curves.

Variations in MTT assay

Other applications. A similar assay has also been used to determine cellular radiosensitivity [Carmichael et al., 1987b]. MTT can be used to determine the number of cells after a variety of treatments other than cytotoxic drug exposure, such as growth factor stimulation. However, in each case, it is essential to ensure that the treatment itself does not affect the ability of the cell to reduce the dye.

Duration of exposure. As with clonogenic assays (*see* Protocol 22.3), some agents may act more quickly, and the exposure period and recovery may be shortened. The

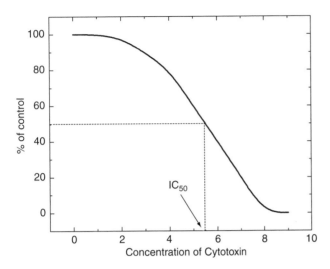

Fig. 22.6. Percentage Inhibition Curve. Test well values are calculated as a percentage of the controls and plotted against the concentration of cytotoxin. Typically, a sigmoid curve is obtained, and, ideally, the IC_{50} will lie in the center of the inflexion of the curve.

cells must remain in exponential growth throughout (*see* Sections 13.7.2 and 21.9.2), and the cell concentration at the end should still be within the linear range of the MTT spectrophotometric assay. When using a cell line for the first time, parallel plates should be set up for cell counts to generate a growth curve (*see* Protocol 21.8) and for MTT-formazan absorbance to ensure that absorbance is proportional to the number of cells. If the growth curve shows that the cells are moving into the stationary phase or the absorbance is

nonlinear when plotted against cell concentration, shorten the assay and proceed directly to Step 16 of Protocol 22.4.

Duration of exposure is related to the number of cell cycles that the cells have gone through during exposure and recovery. Not only will the cell density increase more rapidly during exposure, but in addition the response to cycle-dependent drugs will be quicker. Cell cycle time will influence the choice between a short-form and long-form assay (Figs. 22.7, 22.8). When first trying an assay, it may be desirable to sample on each day of drug exposure and recovery. If a stable IC_{50} is reached earlier, then the assay may be shortened.

End point. Sulforhodamine, a fluorescent dye that stains protein, can also be used to estimate the amount of protein

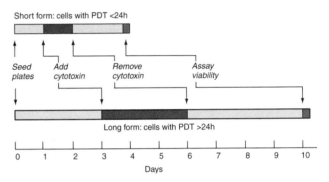

Fig. 22.7. Assay Duration. Pattern for short-form and long-form assays. The upper diagram represents an assay that is suitable for cell with a PDT <24 h, and the bottom diagram represents an assay that is suitable for cells with a PDT >24 h, although intermediate time scales are also possible.

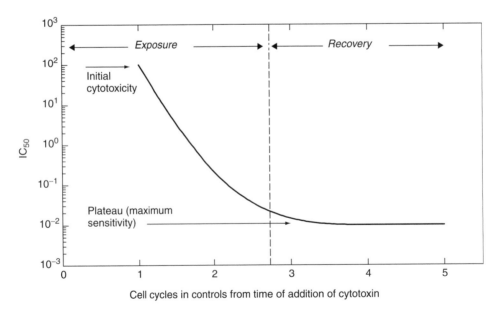

Fig. 22.8. Time Course of the Fall in IC_{50}. Idealized curve for an agent with a progressive increase in cytotoxicity with time, but eventually reaching a maximum effect after three cell cycles. Not all cytotoxic drugs will conform to this pattern [*see* Freshney et al., 1975].

(i.e., cells) per well on a plate reader with fluorescence detection [Boyd, 1989]. It stains all cells and does not discriminate between live and dead cells.

Labeling with [³H]thymidine (DNA synthesis), [³H]leucine [Freshney et al., 1975] or [³⁵S]methionine [Freshney & Morgan, 1978] (protein synthesis), or other isotopes can be substituted for MTT reduction. Quantitation is achieved by microtitration plate scintillation counting. Two types of scintillation counting are available: (1) The cellular contents may be aspirated onto filters by trypsinization and onto glass fiber filters by suction transfer, and the filters may be dried and counted in scintillant, or (2) the whole plate may be counted on a specially adapted scintillation counter (PerkinElmer).

In practice, it may not matter which criterion is used for determining viability or survival at the end of an assay; it is rather the design of the assay, e.g., duration of drug exposure and recovery, phase of the growth cycle (cell density, growth rate, etc.), that is more important. In a short assay with no or minimal recovery period, the end point must measure only viable cells (e.g., MTT), but in a longer assay the end point measures the difference between wells that have increased and those that have not, or have even decreased. In a monolayer assay, at least, nonviable cells will have been lost, and the increase or decrease relative to control wells is what is measured; whether by MTT, sulforhodamine, or isotope incorporation into DNA or protein becomes less important.

Handling. A variety of automated handling instruments are available, e.g., autodispensers, diluters, cell harvesters, and programmable plate readers (*see* Figs. 5.10–5.12), to reduce the handling time required per sample.

22.3.6 Comparison of Microtitration with Clonogenic Survival

The volume of medium required per sample for microtitration is less than one-fiftieth of that required for cloning, although the number of cells is approximately the same for both techniques. Microtitration assays are also shorter and more amenable to automated handling, data gathering, and analysis. Microtitration, however, is unable to distinguish between differential responses between cells within a population and the degree of response in each cell—e.g., a 50% inhibition of a metabolic parameter could mean that 50% of the cells respond or that each cell is inhibited by 50%—but this becomes less important in an assay with a prolonged recovery period, where the relative increase by cell proliferation becomes the major criterion of survival.

A comparison of the IC_{50} derived by microtitration and plating efficiency assays showed a good correlation between the two methods (Fig. 22.9) for the assay of antineoplastic drugs [Morgan et al., 1983]. The correlation for IC_{90} was not as tight.

A significant feature of microtitration assays, particularly with a photometric or radiometric end point, is the generation of large amounts of data, often in a format that is readily

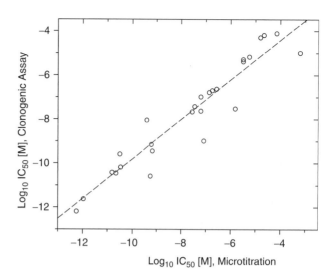

Fig. 22.9. Correlation Between Microtitration and Clonogenic Survival. Measurement of the IC_{50} values of a group of five cell lines from human glioma and six drugs (vincristine, bleomycin, VM-26 epidophyllotoxin, 5-fluorouracil, methyl CCNU, mithramycin). Most of the outlying points were derived from one cell line that later proved to be a mixture of cell types. The broken line is the regression, with the data points from the heterogeneous cell line omitted. Microtitration IC_{50} was derived by [³⁵S]methionine incorporation [Freshney & Morgan, 1978]. [After Freshney et al., 1982a].

analyzed by computer (*see* Section 21.8.2). It is important, however, to scan the raw data as well as the data-reduced end point because computer analysis may make different assumptions or corrections to deal with aberrant data points, which are not apparent unless the raw data is available for scrutiny.

22.3.7 Drug Interaction

The investigation of cytotoxicity often involves the study of the interaction of different drugs; drug interaction is readily determined by microtitration systems, in which several different ratios of interacting drugs can be examined simultaneously. Analysis of drug interaction can be performed by using an isobologram to interpret the data [Steel, 1979; Berenbaum, 1985]. A rectilinear plot implies an additive response, whereas a curvilinear plot implies synergy if the curve dips below the predicted line and antagonism if it goes above.

22.4 APPLICATIONS OF CYTOTOXICITY ASSAYS

22.4.1 Anticancer Drug Screening

Drug screening for the identification of new anticancer drugs can be a tedious and often inefficient method of discovering new active compounds. The trend is now more toward monitoring effects on specific molecular targets. However, there have been attempts to improve screening by adopting rapid, easily automated assays, like those based on the determination of the number of viable cells by staining the cells with MTT [Mosmann, 1983; Carmichael et al., 1987a; Plumb et al., 1989]. To further cut down on manipulations, the MTT incubation step may be omitted and the end point determined by measuring the amount of total protein with sulforhodamine B [Boyd, 1989]. Although this method is quicker and easier than the MTT assay, it should be remembered that nonviable, and certainly nonreplicating, cells will still stain, so the assay should be confirmed when activity is detected, using a more reliable indicator such as clonogenicity or MTT reduction.

22.4.2 Predictive Drug Testing for Tumors

The possibility has often been considered that measurement of the chemosensitivity of cells derived from a patient's tumor might be used in designing a chemotherapeutic regime for the patient [Freshney, 1978]. This technique has never been exhaustively tested, although the results of small-scale trials were encouraging [Hamburger & Salmon, 1977; Bateman et al., 1979; Hill, 1983; Thomas et al., 1985; Von Hoff et al., 1986]. What is required is the development of reliable and reproducible culture techniques for neoplastic cells from the most common tumors (e.g., breast, lung, colon), such that cultures of pure tumor cells capable of cell proliferation over several cell cycles may be prepared routinely. As many of the cells within a tumor have a limited life span, the main targets for chemotherapy are the clonogenic populations with infinite repopulation capacity [Al-Hajj et al., 2004; Jones et al., 2004]. Advances in stem cell recognition (see Section 23.7) may make isolation of tumor stem cells feasible. It was hoped that the soft agar clonogenic assay [Hamburger & Salmon, 1977] might isolate transformed stem cells for assay, but, although initially promising, isolated clones did not seem to have long-term regenerative capacity. Routine isolation of tumor cell populations with long-term repopulation efficiency has yet to be achieved, but when it is it may be possible to improve targeting and specificity of anticancer drugs, making predictive testing more meaningful. Assays might then be performed in a high proportion of cases, hopefully within 2 weeks of receipt of the biopsy.

The major problem, however, is one of logistics. The number of patients with tumors for which the correct target cells (1) will grow *in vitro* sufficiently to be tested, (2) can be expected to respond, and (3) will produce a response that can be followed up, is extremely small. Hence, it has proved difficult to use any *in vitro* test as a predictor of response or even to verify the reliability of the assay. The correlation of insensitivity *in vitro* with nonresponders is high, but few clinicians would withhold chemotherapy because of an *in vitro* test, particularly when the agent in question would probably not be used alone.

22.4.3 Testing Pharmaceuticals

A number of pharmaceutical companies maintain a program of *in vitro* testing on the assumption that it might prove more economical and ethically acceptable than animal testing. Legislation enforcing the use of animal tests is difficult to introduce as the complexity of the wide range of effects seen *in vivo* is still very difficult to model *in vitro*. However, there is considerable political pressure to introduce such legislation, and this is driving large-scale comparative surveys to determine whether any of the many existing tests may be acceptable [Knight & Breheny, 2002; Vanparys, 2002].

22.5 TRANSFORMATION AND MUTAGENESIS

Commonly used *in vitro* assays for transformation include anchorage independence (see Protocols 14.4, 14.5), reduced density limitation of cell proliferation (see Protocol 18.3), and evidence of mutagenesis. (See also Table 18.1.) Mutagenesis can be assayed by sister chromatid exchange; this procedure is described in Protocol 22.5, which was contributed by Maureen Illand and Robert Brown of the Centre for Oncology and Applied Pharmacology, University of Glasgow, Scotland, United Kingdom.

22.5.1 Mutagenesis Assay by Sister Chromatid Exchange

Sister chromatid exchanges (SCEs) are reciprocal exchanges of DNA segments between sister chromatids at identical loci during the S-phase of the cell cycle. As SCEs are more sensitive indicators of mutagenic activity than chromosome breaks, they have become a major tool in mutagenesis research [Latt, 1981].

With the development of the thymidine analog bromo-deoxyuridine (BUdR) and its subsequent use in DNA labeling experiments, the resolution of SCEs greatly improved in comparison to previous methods, which involved the incorporation of radioactive nucleotides into replicating DNA [Taylor, 1958]. Later, the fluorescence plus Giemsa (FPG) technique of Perry and Wolf [1974] for the scoring of SCEs was enhanced, and for the first time, permanent staining of SCEs was demonstrated. Previously, during the scoring process, rapid bleaching of the fluorescent stain occurred [Latt, 1981].

The FPG method involves two distinct steps: (1) Cells are labeled with BUdR for two complete cycles and then treated with colcemid to block the cells in metaphase. After BUdR exposure, the DNA of one chromatid of each chromosome contains bromouracil in one strand, while the DNA of its sister chromatid contains bromouracil in both strands. (2) Chromosomes are then prepared from these cells and stained with the fluorescent dye Hoechst 33258, and then the BUdR is photodegraded with ultraviolet light; this is followed by Giemsa staining. These final steps highlight the differential incorporation of bromouracil into the sister chromatids. DNA that contains bromouracil quenches the fluorescence of Hoechst-DNA complexes. Therefore, the chromatid containing bromouracil substituted in both strands fluoresces weakly and stains weakly with Giemsa, while the chromatid containing bromouracil in only one strand fluoresces more intensely, degrades the BUdR, and subsequently stains darkly with Giemsa (*see* Plate 17e,f). If any SCEs occur, this staining pattern produces what are called *harlequin chromosomes*.

PROTOCOL 22.5. SISTER CHROMATID EXCHANGE

Outline

Trypsinize metaphase-arrested cells that have been labeled with BUdR for two cell cycles, incubate the cells in hypotonic buffer, and then fix the cells. Prepare slides of the cells, after treating the cells with Hoechst 33258, and photodegrade the chromosome spreads. Stain the chromosomes with Giemsa, and visualize on a light microscope under oil immersion.

Materials

Sterile:

❑ D-PBSA

❑ PE: 10 mM EDTA in PBS

❑ Trypsin: 0.12% in PE

❑ BUdR (Sigma): Prepare 1 mM of stock in sterile UPW

❑ Karyomax: Colcemid, 10 μg/mL (Invitrogen)

❑ Growth medium

Nonsterile:

❑ SSC, 2×: 1:10 dilution of 20× SSC (*see* Appendix I)

❑ Hypotonic buffer: 0.075 M KCl

❑ Sorensen's buffer: phosphate buffer, 0.066 M, pH 6.8 (tablets from Merck)

❑ Methanol:acetic acid, 3:1, ice cold and freshly prepared

❑ Giemsa solution, 0.76%: Place 1 *g* of Giemsa powder (Merck) in 66 mL of glycerol and heat in a waterbath at 56–60°C for 11/2—2 h. Cool the solution, and add 66 mL of absolute alcohol.

❑ Giemsa, diluted to 3.5% in Sorensen's buffer, pH 6.8

❑ Hoechst 33258 (Sigma), 20 μg/mL, in UPW

❑ Latex photo-mountant or adhesive

❑ Xylene

❑ DPX mountant (Merck)

❑ Coverslips, 22 × 15 mm

❑ Coplin jar

❑ Slide rack (Thermo Electron)

❑ Short wave UV lamp in irradiation box

Δ **Safety Note.** Hoechst 33258 is carcinogenic; weigh it out and dissolve it in a fume hood.

Protocol

Pretreatment:

1. Seed the cells at the appropriate density (e.g., 1×10^6 cells per 75-cm² flask), and incubate for 2 days at 37°C.
2. Add BUdR to the growth medium at a final concentration of 10 μM.
3. Incubate the cells in the dark at 37°C for a further 48 h (∼2 cell cycles).
4. Add colcemid to the cells 1–6 h before harvesting, depending on the cycling time of the cells. For human cell lines, the final concentration of colcemid should be 0.01 μg/mL.

Harvesting cells:

5. Wash the cells with D-PBSA, and trypsinize them with 1 mL of 0.12% trypsin in PE.
6. Resuspend the cells in 10 mL of growth medium. Transfer the cell suspension to 50-mL centrifuge tubes.
7. Centrifuge the suspension at 1200 *g* for 5 min.
8. Remove the supernatant, leaving approximately 1.2 mL above the pellet. Flick the side of the tube to resuspend the pellet.
9. Slowly add 10 mL of hypotonic buffer (pre-warmed to 37°C), and incubate the cell suspension for 10–15 min at room temperature.
10. Spin the cells in a benchtop centrifuge at 1200 *g* for 5 min.
11. Remove the supernate, leaving 1.2 mL above the pellet. Flick the side of the tube to resuspend the pellet.
12. Add 10 mL of ice-cold fixative, initially drop by drop, mixing well after each addition. Leave the tube on ice for 10 min.
13. Repeat Steps 10–12 once more, letting the cells remain in fixative overnight at 4°C, to improve slide preparations.
14. Spin the fixed cells at 1200 *g* for 5 min, and resuspend the cells in 3–5 mL of methanol/acetic acid.
15. Store the cells at −20°C.

Slide preparation:

16. Slides should be clean and grease free before use, so wipe them with absolute alcohol.
17. Using a short glass Pasteur pipette, take up approximately 500 μL of the fixed cells.
18. Hold the slide at a downward angle, and, holding the Pasteur pipette at least 15 cm (6 in.) above the slide, drop 3 drops of the cell suspension onto the slide (*see* Protocol 16.7).
19. Air dry the slide in the dark.
20. Check the slide under phase contrast to ensure that there the metaphase spreads are

evenly distributed across the slide and that the chromosomes are well separated.

Harlequin staining:

21. Immerse the slides in a Coplin jar of Hoechst 33258 at a concentration of 20 μg/mL for 10 min. (Wear gloves, as Hoechst is toxic.)

22. Transfer the slides to a slide rack, and drop 500 μL of 2× SSC onto each slide.

23. Cover the slides with a 22 × 50-mm coverslip, and seal the edges with a temporary seal, such as latex photo-mountant, to prevent evaporation.

24. Place the slides in the slide rack, coverslips facing downwards, and place the slide rack on a shortwave UV box. Maintain a distance of approximately 4 cm between the slides and the UV source. The longer the slides are exposed to UV, the paler the pale chromatid will become; expose the slides for about 25–60 min.

25. Remove the coverslips from the slides, and wash the slides three times in UPW, 5 min per wash. Cover the slide holder with aluminum foil.

26. Air dry the slides in the dark.

27. Stain the slides in a Coplin jar containing 3.5% Giemsa solution in Sorensen's buffer, pH 6.8, for 3–5 min.

28. Carefully rinse the slides in tap water, and drain them with a paper tissue.

29. Air dry the slides on the bench for 1 h. Dip each slide into xylene, drop 4 drops of DPX mountant (Merck) onto the slide, and mount a 22 × 50-mm coverslip, expressing any air bubbles with tissue. (Carry out this final step in a fume hood, as xylene fumes are toxic. Also, wear gloves.)

30. Air dry the slides in a fume hood overnight.

Analysis:

31. Under the 40× objective of a light microscope, scan the slides for metaphase spreads.

32. Find an area on the slides where most of the metaphase spreads are located, and examine this area under oil immersion.

33. When no sister chromatid exchanges (SCEs) have occurred, each chromosome has one continuously staining pale chromatid and one continuously staining dark chromatid. One SCE has occurred when there is one area of dark staining and then light staining on one chromatid and on the sister chromatid one area of light staining and then dark staining (*see* Plate 17e,f). Each point of the discontinuity in staining is scored as one SCE.

34. Count the number of SCEs per cell and also the number of chromosomes per cell.

35. Larger chromosomes usually have a greater number of SCEs than smaller ones, and the

incidence of SCEs may vary from cell to cell. Therefore, scoring SCEs per chromosome is a more accurate measure of SCE rate. SCE score is calculated by the following formula:

$$\frac{\text{mean no. of SCEs cell}}{\text{mean no. of chromosomes/cell}}$$

Aim to score approximately fifty spreads per cell line being studied.

Variations. Pulse labeling of cells with BUdR and subsequent staining, as described previously, can detect differences in early and late replicating regions of chromosomes during the cell cycle. When cells are labeled with BUdR at the latter part of the cell cycle, DNA that replicates early will stain darkly with Giemsa, because of very little BUdR incorporation, and for regions of the chromosome that are pulsed at the earlier stages of the cell cycle, only those regions that replicate their DNA early will stain faintly with Giemsa [Latt, 1973].

Additionally, cells that have undergone only one cell cycle of continuous BUdR labeling show differential staining of chromatids only at certain bands (lateral asymmetry), due to the differences in thymine content of the DNA [Brito Babapulle, 1981]. After photodegradation of BUdR, the fluorescent dye acridine orange can also be used to stain SCEs. With this dye, green fluorescence is observed at regions that have double-stranded DNA and thus will have little BUdR incorporation, and red fluorescence is observed at regions that have single-stranded DNA, which will have incorporated the BUdR. Consequently, the red fluorescence is equivalent to lighter staining with Giemsa, and the green fluorescence is equivalent to the darker staining with Giemsa [Karenberg & Freelander, 1974].

22.5.2 Carcinogenicity

The potential for *in vitro* testing for carcinogenesis is considerable [Berky and Sherrod, 1977; Grafström, 1990a,b; Zhu et al., 1991], but this is one area in which *in vivo* testing is far from adequate; the models are poor, and the tests often take weeks, or even months, to perform. The development of a satisfactory *in vitro* test is hampered (1) by the lack of a universally acceptable criterion for malignant transformation *in vitro* and (2) by the inherent stability of human cells used as targets.

The most generally accepted tests so far assume that carcinogenesis, in most cases, is related to mutagenesis (*see* Section 18.3). This assumption is the basis of the Ames tests [Ames, 1980], wherein bacteria are used as targets and activation can be carried out with liver microsomal enzyme preparations. This test has a high predictive value, but nevertheless, dissimilarities in uptake, susceptibility, and type of cellular response have led to the introduction of alternative tests using mammalian and human cells as targets, e.g., sister chromatid exchange (*see* Section 22.5.1).

Some of these tests are also mutagenesis assays, using suspensions of L5178Y lymphoma cells as targets [Cole et al., 1990] and the induction of mutations or reversion, or cytological evidence of sister chromatid exchange (*see* Protocol 22.5), as evidence of mutagenesis. Others [Styles, 1977] have used transformation as an end point, assaying clonogenicity in suspension (*see* Section 18.5.1) as a criterion for transformation. Critics of these systems say that both use cells that are already partially transformed as targets; even the BHK21-C13 cell used by some workers is a continuous cell line and may not be regarded as completely normal. Furthermore, the bulk of the common cancers arise in epithelial tissues and not in connective tissue cells.

The demonstration of increased oncogene expression or amplification, or the presence of increased or altered oncogene products, may provide reliable criteria, in some cases functionally related to the carcinogen. This is now possible with microarray analysis. Likewise, the deletion or mutation of suppressor genes is open to molecular analysis, where deletions or mutations in the p53, Rb, p16, and L-CAM (E-cadherin) genes would cover a high proportion of malignant transformation events. It is now feasible to consider expression analysis and an alternative to mutagenesis in carcinogenicity testing [Nuwaysir et al., 1999; Desai et al., 2002; Vondracek et al., 2002].

22.6 INFLAMMATION

There is an increasing need for tissue culture testing to reveal the inflammatory responses that are likely to be induced by pharmaceuticals and cosmetics with topical application or by xenobiotics that may be inhaled or ingested and may be responsible for many forms of allergy. This is an area that is only at the early stages of development but bears great promise for the future. It is a sensitive topic in more ways than one. Animal rights groups are naturally incensed at the needless use of large numbers of animals to test new cosmetics that have little benefit except commercial advantage to the manufacturer, particularly when the testing of substances (such as shampoos) involves the Draize test, in which the compound is added to a rabbit's eye. More important, clinically, is the apparent increase in allergenic responses produced by pharmaceuticals and xenobiotics. These

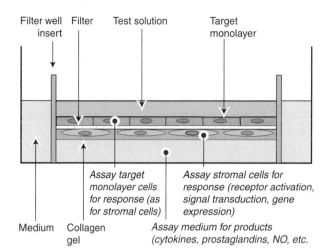

Fig. 22.10. Organotypic Assay. A hypothetical assay system for exposing one cell layer (e.g., epidermal keratinocytes) cocultured with another associated cell type (e.g., skin fibroblasts in collagen gel) to an irritant and measuring the response by cytokine release (*see also* Plate 21).

responses are little understood and poorly controlled, largely because of the absence of a simple reproducible *in vitro* test.

Since the advent of filter well technology, several models for skin and cornea have appeared [Braa and Triglia, 1991; Triglia et al., 1991; Fusenig, 1994b; Roguet et al., 1994; Kondo et al., 1997; Brinch & Elvig, 2001], utilizing the facility for coculture of different cell types that the filter well system provides. In these systems, the interaction of an allergen or irritant with a primary target (e.g., epidermis) is presumed to initiate a paracrine response, which triggers the release of a cytokine from a second, stromal component (e.g., dermis) (Fig. 22.10). This cytokine can then be measured by ELISA technology to monitor the degree of the response. Although still in the early stages of development, kits for the measurement of irritant responses are available [Epiderm (MatTek) [Koschier at al., 1997]; Episkin (Saduc) [Cohen et al., 1997]; SkinEthic [Brinch & Elvig, 2001; *see* Plate 21]. A protocol for corneal culture suitable for modeling irritant responses in the eye was provided by Carolyn Cahn (*see* Protocol 23.2).

It would seem that this type of system may be a major area of development, with the real prospect that allergen screening from patients' own skin in organotypic culture may become possible and that, ultimately, analysis of the GI tract will reveal allergens responsible for irritable bowel syndrome. In each case, and in many others, there is the possibility of specific mechanistic studies into the processes of abnormal cell interaction that typify many allergic and degenerative diseases.

CHAPTER 23

Culture of Specific Cell Types

For many people, cell culture is simply a substrate for producing enough cells for an experimental or production procedure. However, for others, there is increasing interest in specific cell types to examine specialized processes and their pathologies, or to use in cell-based therapies. The generation of a specialized cell culture can be achieved in several ways: (1) by isolating differentiated cells or tissue for short-term nonregenerative culture (*see* Chapter 25), (2) by isolating precursor cells, expanding them in selective culture conditions (*see* below and Sections 10.2.1, 14.6), and then inducing them to differentiate (*see* Section 17.7), or (3) by isolating stem cells and creating cultures that either regenerate the stem cell population or progress toward one of several possible differentiated phenotypes (*see* Section 23.7). In each case the technology is different, and this chapter sets out to introduce some of this technology for specific types of tissue.

23.1 CELL CULTURE OF SPECIALIZED CELLS

There are many reviews of culture techniques for specific cell types [Doyle et al., 1990–1999; Pollard & Walker, 1990; Freshney et al., 1994; Leigh & Watt, 1994; Freshney & Freshney, 1996; Ravid & Freshney, 1998; Haynes, 1999; Federoff & Richardson, 2001; Freshney & Freshney, 2002; Wise, 2002; Vunjak-Novakovic & Freshney, 2005; *see also* Table 3.1].

The development of techniques for cell line immortalization [Freshney & Freshney, 1996] (*see also* Section 18.4) has meant that it has been possible to generate continuous cell lines from a number of finite lines from untransformed tissue [Chang et al., 1982; Freshney et al., 1994; Klein et al., 1990; Steele et al., 1992; Wyllie et al., 1992]. In many cases the differentiated properties are lost, but by using a switchable promoter (e.g., temperature sensitivity), it may prove possible to recover the differentiated phenotype. The development of the transgenic mouse carrying the large T gene of SV40 has opened up a wide range of possibilities [Jat et al., 1991; Yanai et al., 1991; Morgan et al., 1994], because cells cultured from these animals are already immortalized but still retain some differentiated functions. As these cells carry a temperature-sensitive promoter for the immortalizing gene, it is possible to recover the normal phenotype at the nonpermissive temperature, 37°C.

Many specialized cells (e.g., epidermal keratinocytes, melanocytes, endothelial cells, smooth muscle cells, dermal fibroblasts, melanocytes, and mammary epithelium) are commercially available (*see* Appendix II). The cost is naturally very high, but as demand increases, the cost may fall. Skin cultures are also available for cytotoxicity and inflammation research from Episkin (Saduc), Epiderm (MatTek), and SkinEthic. These products are prepared in filter wells by combining keratinocytes with dermal fibroblasts and collagen supported by a nylon net in a so-called "skin equivalent." Other tissues—in particular, the cornea (SkinEthic)—have also been prepared in a similar way for toxicity studies (*see* Protocol 23.2; Plate 21).

A number of specialized procedures have now been devised, and some representative examples have been contributed by experts in various areas. The protocols in this chapter assume that the basic prerequisites of the cell biology laboratory, as specified in Chapter 4, will be available. Consequently, items such as laminar flow hoods inverted microscopes, bench centrifuges, and water baths will not be among the materials listed.

23.2 EPITHELIAL CELLS

Epithelial cells are responsible for the recognized functions of many organs (e.g., polarized transport in the kidney and gut, secretion in the liver and pancreas, gas exchange in the lung, and barrier protection in skin). Epithelial cells are also of interest as models of differentiation and stem cell kinetics (e.g., epidermal keratinocytes [Watt, 2001, 2002]) and are among the principal tissues in which common cancers arise. Consequently, the culture of various epithelial cells has been a focus of attention for many years. The major problem in the culture of pure epithelium has been the overgrowth of the culture by stromal cells, such as connective tissue fibroblasts and vascular endothelium. Most of the variations in technique are aimed at preventing such overgrowth by nutritional manipulation of the medium or alterations in the culture substrate (Table 23.1) that promote the growth of the undifferentiated epithelium and, preferably, the stem cells. Subsequent modifications may then be employed to enhance epithelial differentiation, although usually at the expense of proliferation.

Factors contained in serum—many of them derived from platelets—have a strong mitogenic effect on fibroblasts and tend to inhibit epithelial proliferation by inducing terminal differentiation. Consequently, one of the most significant events in the isolation and propagation of specialized cell cultures has been the development of selective, serum-free media (*see* Section 10.2.1), supplemented with specific growth factors (*see* Table 10.3).

The isolation of epithelial cells from donor tissue is often best performed with collagenase (*see* Protocol 12.8), which disperses the stromal cells but leaves the epithelial cells in small clusters, allowing their separation by settling and favoring their subsequent survival.

23.2.1 Epidermis
The epidermis represents one of the most interesting models of epithelial differentiation, partly because of its demonstration of developmental hierarchy by histology and partly because it was one of the first models to lend itself to 3T3 feeder layer-supported growth. Numerous attempts have been made to use cultured epidermal keratinocytes as models for differentiation, as ideal tissue constructs for organotypic culture and studies on cell interaction [Maas-Szabowski et al., 2002], and as grafts in burn or injury repair. Protocol 23.1 for the culture of epidermal keratinocytes was contributed by Norbert E. Fusenig and Hans Jürgen Stark, Division of Carcinogenesis and Differentiation, German Cancer Research Center, Im Neuenheimer Feld 280, 6900 Heidelberg, Germany.

Because of the progress in basic cell culture technology and in our understanding of the culture requirements of the various epithelial tissues, keratinocytes of most stratified epithelia can now be grown and studied in cell culture. Mostly, the squamous epithelia of the skin and their isolated epithelial cells, the keratinocytes, have been used to study their physiology and pathology *in vitro*. In addition, cells of the oral mucosa [Tomakidi et al., 1997; Grafström, 2002], as well as of skin appendages such as the

TABLE 23.1. Inhibition of Fibroblastic Overgrowth

Method	Agent	Tissue	Reference
Selective detachment	Trypsin	Fetal intestine, cardiac muscle, epidermis	Owens et al., 1974; Milo et al., 1980
	Collagenase	Breast carcinoma	Freshney, 1972; Lasfargues, 1973
Confluent feeder layers	Mouse 3T3 feeder cells	Epidermis	Rheinwald and Green, 1975
	Fetal human intestine feeder cells	Normal and malignant breast epithelium	Stampfer et al., 1980
		Colon carcinoma	Freshney et al., 1982b
Selective inhibitors	D-valine	Kidney epithelium	Gilbert and Migeon, 1975, 1977
	Cis-OH-proline	Cell lines	Kao and Prockop, 1977
	Ethylmercurithiosalicylate	Neonatal pancreas	Braaten et al., 1974
	Phenobarbitone	Liver	Fry and Bridges, 1979
	Antimesodermal antibody	Squamous carcinomas	Edwards et al., 1980
		Colonic adenoma	Paraskeva et al., 1985
	Geneticin	Melanocytes, melanoma	Levin et al., 1995
Selective media	MCDB 153	Epidermis	Boyce and Ham, 1983
	MCDB 170	Breast	Hammond et al., 1984
	Low Ca^{2+}	Epidermal melanocytes	Naeyaert et al., 1991

hair follicle [for a review, *see* Fusenig et al., 1994], have been isolated and cultured under various conditions, and reconstructed tissues have been formed in culture as well as in transplants [Limat et al., 1995; Maas-Szabowski et al., 2000, 2002].

Separation of the epithelial compartment from the underlying connective tissue is usually done by enzymatic digestion with trypsin [*see*, e.g., Smola et al., 1993], dispase type II (2.4 U/mL, Boehringer Mannheim [Tomakidi et al., 1997]), or thermolysin [Germain et al., 1993]. The isolated epithelium is further dispersed by additional incubation in trypsin or, mechanically, by pipetting, after which it is filtered through nylon gauze and propagated in a serum-free, low-calcium medium or on growth-arrested feeder cells by using different media formulations. Subpopulations of keratinocytes with stem cell characteristics can be isolated because of their selective attachment to basement membrane constituents [*see* Bickenbach and Chism, 1998].

In order to avoid overgrowth by mesenchymal cells, the epithelial compartment has to be separated from the connective tissue and dispersed into single cells, which are then cultured on different substrata by using different media formulations. The most commonly used and reliable method is the feeder layer technique first established by Rheinwald and Green [1975], who employed postmitotic (irradiated or mitomycin C-treated) 3T3 cells as mesenchymal feeders. Primary fibroblasts derived from dermis or submucosa and rendered postmitotic can be used instead [Limat et al., 1989; Tomakidi et al., 1997]. Feeder cells maintain a broad spectrum of physiological functions [Maas-Szabowski & Fusenig, 1996].

Alternatively, keratinocytes can be grown without feeder cells in serum-free media supplemented with pituitary extract or a variety of growth factors and other supplements, either at low (0.03–0.06 mM) Ca^{2+} concentrations to inhibit stratification or at normal (1.2–1.4 mM) Ca^{2+} concentrations to allow stratification and keratinization [for a review, *see* Holbrook & Hennings, 1983]. Although maintenance at high cell density in subcultures, in the presence of feeder cells (e.g., irradiated 3T3 cells), or culture in a low-calcium, serum-free medium helps to reduce fibroblast contamination, the mesenchymal cells are not completely eliminated by these procedures.

PROTOCOL 23.1. EPIDERMAL KERATINOCYTES

Outline

Separate the epidermis from the dermis enzymatically, disaggregate the keratinocytes, and seed them in a serum-free medium or on a growth-arrested feeder layer.

Materials
Sterile:
❑ Prepared feeder layers, 3 days old (*see* Section 14.2.3 and Protocol 23.4). Irradiate 3T3 at 30 Gy and human fibroblasts at 70 Gy, preferably in concentrated cell suspension.

❑ FAD: high-calcium medium for cultures with feeder layers; Ham's F12 and DMEM with additives, as formulated by Wu et al. [1982]: Ham's F12:DMEM, 1:3, supplemented with adenine (1.8×10^{-4} M) cholera toxin (1×10^{-10} M), EGF (1–10 ng/L), hydrocortisone (0.4 μg/mL, 1.1 μM), 5–10% FCS, penicillin (100 U/mL), and streptomycin (50 μg/mL)

❑ Keratinocyte growth medium (KGM): serum-free medium based on the original formulation of the medium MCDB 153 by Boyce and Ham [1983] (*see* Table 10.1 and Appendix II), supplemented with bovine pituitary extract with low Ca^{2+} (0.03–0.5 mM), which can be increased by adding $CaCl_2$

❑ Keratinocyte-defined medium (KDM): serum-free medium (*see* Appendix II) based on the original formulation of the medium MCDB 153 by Boyce and Ham [1983]; fully defined medium, tested as a fully competent medium in organotypic cultures [Stark et al., 1999]

❑ Supplemented KDM (SKDM):
(a) Keratinocyte-defined medium (KDM) without pituitary extract (Promocell, Clonetics)
(b) Supplement with: insulin (5 μg/mL), hydrocortisone (0.5 μg/mL), rhEGF (0.1 ng/mL), transferrin (20 μg/mL), 0.1% highly purified bovine serum albumin (free of endotoxins and fatty acids, e.g., A-8806, Sigma), and L-ascorbic acid (50 μg/mL).
(c) Adjust Ca^{2+} to 1.3 mM.
(d) Primary keratinocyte cultures can be maintained on uncoated plastic dishes in SKDM, although with low proliferative activity. Growth is improved when keratinocytes are plated on collagen (type 1)-coated dishes.

In organotypic cultures on lifted collagen gels populated with fibroblasts, this defined medium provides epidermal growth and differentiation indistinguishable from that obtained in FAD [Stark et al., 1999].

❑ D-PBSA
❑ EDTA (0.5%; ~15 mM)
❑ Glycerol (analytical grade)
❑ FAD medium with 20% FCS and 10% glycerol
❑ Enzymes:
(a) Thermolysin (T-7902, Sigma)
(b) Trypsin (1:250; 0.2 and 0.6%)

(c) Dispase (grade II, 2.4 U/mL, Boehringer Mannheim)
- Betadine solution (10% iodine)
- Cryopreservation ampoules
- Centrifuge tube, 50 mL
- Nylon gauze
- "Cell Strainer" (Becton Dickinson)
- Petri dishes, bacteriological grade, 10 cm
- Scalpels, curved forceps

Protocol

Specimens:

1. Obtain foreskin (neonatal as well as juvenile), the most frequent laboratory source for human skin, or trunk skin obtained from surgery or post mortem (up to 48 h). Keratinocytes derived from foreskin seem to attach and proliferate better than cells obtained from adult skin.
2. Incubate skin biopsies for up to 30 min in Betadine solution (10% iodine), to prevent infection. (This does not visibly decrease keratinocyte cell viability.)
3. Rinse twice in D-PBSA for 10 min each.

Epidermal Separation:

Split-thickness skin is optimal.

4. Slice full-thickness skin with a Castroviejo dermatome (Storz Instrument) set to 0.1–0.2 mm. Alternatively, subcutaneous tissue and part of the dermis can be eliminated with curved scissors.
5. Dissect skin into 1 × 2-cm pieces with scalpels.
6. Rinse the tissue two to five times in D-PBSA.
7. Incubate tissue with protease by one of the following steps:
   (a) Trypsin: Float skin samples on 0.6% trypsin in D-PBSA (pH 7.4) for 20–30 min at 37°C. Alternatively (and this works particularly well also with full-thickness skin):
   (b) Cold trypsin: Float the samples on ice-cold 0.2% trypsin at 4°C for 15–24 h. The pH of the trypsin solution has to be monitored with phenol red to avoid a shift leading to altered enzyme activity and cell viability.
   (c) Dispase (grade II): Incubate with 2.4 U/mL for 30 min at 37°C.
   (d) Thermolysin (0.5 mg/mL) [Germain et al., 1993]: Incubate for 12–16 h at 4°C.
8. Monitor the separation of the epidermis carefully. When the first detachment of the epidermis is visible at the cut edges of skin samples, place the pieces (dermis side down) in 10-cm plastic Petri dishes and irrigate with 5 mL complete culture medium including serum.

9. Peel off the epidermis with two fine curved forceps, and collect it in a 50-mL centrifuge tube containing 20 mL of complete culture medium.
   (a) When separation has been performed with trypsin, viable keratinocytes are detached from the epidermal parts by gently pipetting and sieving through nylon gauze of 100-μm mesh ("Cell Strainer," B-D Biosciences).
   (b) To obtain single cells from epidermis separated by dispase or thermolysin, additional incubation with trypsin (0.2%) and 1.3 mM EDTA (0.05%), for 10 min at 37°C is required.
10. After splitting the skin in trypsin, gently scrape off the loosely attached epidermal cells on the remaining dermal part, and add the cells to the epidermal suspension. This can lead to a higher cell yield if the purity of the keratinocyte population is not a major issue, but the amount of mesenchymal cell contamination in the resulting keratinocyte cultures is significantly higher.
11. Wash the isolated epidermal cells twice in culture medium by centrifugation at 100 g for 10 min, and count the total number of cells and the viable cells (those that do not absorb Trypan Blue; *see* Protocol 22.1).

Primary Culture:

12. Seed cells at 37°C in FAD or KGM medium at desired densities:
   (a) In FAD with feeder cells (dishes seeded 3 days previously with postmitotic 3T3 cells [Rheinwald & Green, 1975] (Protocol 14.3) or skin fibroblasts [Limat et al., 1989]), seed keratinocytes at $2-5 \times 10^4$ cells/cm^2.
   (b) For culture free of fibroblasts, seed primary keratinocytes at high density ($1-5 \times 10^5$ cells/cm^2) in FAD, and maintain them at high density in subcultures.
   (c) Alternatively, seed cells at low (1×10^3 cells/cm^2) and high (5×10^4 cells/cm^2) density in KGM in low Ca^{2+} (0.03–0.1 mM), and subculture in the same medium.
   (d) Cells will attach and grow, but at a lower rate, in SKDM, so this medium is advised for experimental conditions when low proliferative activities are required.
13. When cells have attached (after 1–3 days), rinse cultures extensively with medium to eliminate nonattached dead and differentiated cells, and continue cultivation in either FAD or KGM. Stratification and slowing down of growth can

be achieved by shifting the Ca^{2+} concentration in KGM.

Subculture:

14. Subculture as follows:
    (a) Cultures in FAD:
        i) Incubate in 1.3–2.7 mM (0.05–0.1%) EDTA for 5–15 min to initiate cell detachment, which is visible by the enlargement of intercellular spaces.
        ii) Incubate in 0.1% trypsin and 1.3 mM (0.05%) EDTA at 37°C for 5–10 min, followed by gentle pipetting, to completely detach the cells.
    (b) Cultures in KGM:
        i) EDTA pretreatment is not required, because of the low Ca^{2+} concentration.
        ii) Incubate in 0.1% trypsin with 1.3 mM EDTA, as with FAD cultures.

Cryopreservation:

15. Cryopreservation is often necessary to maintain large quantities of cells derived from the same tissue sample; the best results are reported when cells from preconfluent primary cultures are used.
    (a) Trypsinize cells as before, and centrifuge at 100 g for 10 min.
    (b) Resuspend cells in complete culture medium with serum, and count.
    (c) Dispense aliquots of 2×10^6 cells/mL in FAD medium with 20% FCS and 10% glycerol into cryopreservation tubes.
    (d) Equilibrate at 4°C for 1–2 h.
    (e) Freeze cells with a programmed freezing apparatus (Planer) at a cooling rate of 1°C per min. Alternatively, the cryopreservation tubes are inserted into cell freezing boxes filled with isopropanol (e.g., Cryo 1°C Freezing Container, cat. no. 5100-0001, Nalge Nunc), which allow an adequate cooling rate when put into a −80°C freezer.
    (f) To recover cells:
        i) Thaw cryotubes quickly in a 37°C water bath.
        ii) Dilute cells tenfold with medium.
        iii) Centrifuge cells and resuspend them at an appropriate concentration in the desired culture medium, and seed culture vessel.

Human and mouse cells can be grown in all three media for several months. Although mouse cells can be subcultured only once or twice, human cells can be passaged four and seven times in FAD and KGM, respectively.

Characterization of keratinocyte cultures. Cultured cells must be characterized for their epidermal (epithelial) phenotype to exclude contamination by mesenchymal cells. This is best achieved by using cytokeratin-specific antibodies for the epithelial cells. Contaminating endothelial cells can be identified by antibodies against CD31 or factor VIII-related antigen. Identifying fibroblasts unequivocally is difficult, because the use of antibodies against vimentin (the mesenchymal cytoskeletal element) is not specific; keratinocytes *in vitro* may initiate vimentin synthesis at frequencies that depend on culture conditions. As a practical assessment for mesenchymal cell contamination, cells should be plated at clonal densities ($1–5 \times 10^2$ cells/cm^2) on feeder cells, and clone morphology should be identified at low magnification after fixation and hematoxylin and eosin (H&E) staining of 10- to 14-day cultures. A more specific and highly sensitive method to identify contaminating fibroblasts is the analysis of expression of keratinocyte growth factor (KGF) by RT-PCR. Because this factor is produced in fibroblasts and not in keratinocytes, it represents a selective marker. Moreover, KGF expression is enhanced by cocultured keratinocytes, so a minority of contaminating fibroblasts will be detected by this assay [Maas-Szabowski et al., 1999]. Individual fibroblasts in a keratinocyte monolayer may be detected by *in situ* hybridization with a cDNA of KGF. Human fibroblasts can be identified and quantitated by the fibroblast-specific antibody AS02 (Dianova) by immunoperoxidase technique [Saalbach et al., 1997].

Variations. Keratinocytes can be obtained from the outer root sheaths (ORSs) of plucked scalp hair follicles by dissociating cells from the dissected follicle [Limat et al., 1989]. Up to 5×10^3 cells can be obtained from three hair follicles and plated on fibroblast feeder cells, resulting in about 1×10^6 cells within 15 days. Like interfollicular keratinocytes, ORS cells can be subcultured on feeder layer dishes and cryopreserved. These ORS-derived keratinocytes are similar in culture to interfollicular cells, form a regular neoepidermis when transplanted onto nude mice, and are used clinically to cover chronic wounds [Limat et al., 1996].

Keratinocytes can also be grown at clonal density to study clonal cell populations and their different proliferation potentials. For this purpose, keratinocytes are cocultured with X-irradiated 3T3 cells [Rheinwald & Green, 1975], at reduced Ca^{2+} concentrations with fibroblast-conditioned medium [Yuspa et al., 1981], or in defined serum-free medium [Boyce & Ham, 1983].

In order to provide more *in vivo*-like conditions and to study the regulation of skin physiology by epithelial–mesenchymal interactions, organotypic coculture systems have been developed by seeding the cells on collagen gels (populated with fibroblasts) or pieces of dermis and lifting the supports to the air–medium interface. [For a review of this procedure, *see* Fusenig, 1994a.] Under these improved growth conditions, keratinocytes express many aspects of growth and differentiation of the epidermis *in vivo*, including ultrastructural features and a complete basement membrane, features that are absent or less pronounced in submerged cultures on plastic [Smola et al., 1998]. Such organotypic cultures can now also be established and maintained for three weeks in defined SKDM medium, allowing the molecular interaction between epithelial and mesenchymal cells to be analyzed without disturbing influences from serum or other undefined tissue extracts [Stark et al., 1999].

Such organotypic keratinocyte cultures are also available commercially (*see* Fig. 22.10 and Section 22.6), either as cocultures with fibroblasts or as monocultures of keratinocytes growing on polycarbonate filter inserts (Falcon #3501), and are increasingly used for *in vitro* toxicity and biocompatibility testing of skin-related products [Fusenig, 1994b].

Optimal growth and differentiation of isolated keratinocytes are obtained under *in vivo* conditions when the cells are transplanted as intact cultures or in suspension onto nude mice [for review, *see* Fusenig 1994a]. This leads to an almost complete expression of cell, differentiation characteristics of normal epidermis, and all biochemical and ultrastructural features of a completely keratinized epithelium, including the formation of a basement membrane [Breitkreutz et al., 1997]. Combined cultures of epidermis and stroma, mounted on mesh filters, are now commercially available (Episkin, Epiderm, SkinEthic) and have been suggested as models for irritation and inflammation research.

23.2.2 Cornea

There have been a number of attempts to replace the rabbit's eye (Draize) test by using cultured corneal epithelium, so there has been considerable interest in culturing cornea [Dart, 2003]. The following introduction and Protocol 23.2 for the culture of normal human corneal cells in serum-free medium was provided by Carolyn Cahn, Gillette Medical Evaluation Laboratories, 401 Professional Drive, Gaithersburg, MD 20879.

The corneal epithelium contacts the external environment directly and is the first tissue compromised in ocular injury. Animal models are most frequently used to model the human ocular surface. The system to be described is being developed for *in vitro* toxicological investigations, which can complement *in vivo* studies.

The corneal epithelium *in vivo* is oxygenated by diffusion from the tear film; in addition, it undergoes frequent mitosis. These two characteristics, coupled with the availability of donor corneal tissue, combine to make the corneal epithelium a good starting material for the generation of primary cultures.

Among the sources for the acquisition of corneal tissue is a sophisticated eye banking system that provides donor corneas for transplantation. Eye bank tissue is downgraded and made available to researchers after three days of storage, because of the limited *in vitro* viability of the endothelial cell layer. Although the endothelium is labile, the epithelium retains its generative capacity for extended periods of time in storage at 4°C, making expired tissue suitable as a source of viable epithelium.

PROTOCOL 23.2. CORNEAL EPITHELIAL CELLS

Outline
Corneal tissue is placed on collagen and allowed to adhere, and the outgrowth is expanded and propagated in serum-free medium on fibronectin-collagen-coated surfaces.

Materials
Sterile or Aseptically Prepared:
- ❑ Keratinocyte serum-free medium (KGM, Clonetics) containing 0.15 mM of calcium, human epidermal growth factor (0.1 ng/mL), insulin (5 μg/mL), hydrocortisone (0.5 μg/mL), and bovine pituitary extract (30 μg/mL)
- ❑ Eagle's Minimal Essential Medium
- ❑ D-PBSA: Dulbecco's phosphate-buffered saline without Ca^{2+} and Mg^{2+}
- ❑ FBS
- ❑ Trypsin-EDTA: Trypsin, 0.05%, EDTA, 0.5 mM
- ❑ Fibronectin/collagen (FNC): fibronectin, 10 μg/mL, collagen, 35 μg/mL, with bovine serum albumin (BSA), 100 μg/mL, added as a stabilizer
- ❑ Biocoat six-well plate precoated with rat tail collagen, type I (B-D Biosciences)

Protocol
Primary Cultures:
1. Place donor corneas epithelial side up on a sterile surface, and cut them into 12 triangular-shaped wedges, using a single cut of the scalpel and avoiding any sawing motion. Careful handling of the cornea in this manner decreases damage to the collagen matrix of the stroma and minimizes the liberation of fibroblasts.
2. Turn each corneal segment epithelial side down, and place four segments in each well of a precoated six-well tray.
3. Press each segment down gently with forceps to ensure good contact between the tissue and the tissue culture surface. Allow the tissue to dry for 20 min.
4. Place one drop of KGM carefully upon each segment, and incubate the culture overnight at 37°C in 5% CO_2. Although the donor corneas received from the eye bank are stored in antibiotic-containing medium (either McCarey–Kauffman or Dexsol), all manipulations are performed under antibiotic-free conditions.

5. The next day, add 1 mL of KGM to each well. During the initial culture period, cells are observed to emigrate only from the limbal region of the cornea. No cells are observed to migrate away from the central cornea or the sclera. Fibroblast outgrowth is minimized by utilizing a serum-free medium that is low in calcium (0.15 mM) and that minimizes disruption of the collagen matrix.

6. Remove the tissue segment with forceps 5 days after the explantation of the donor cornea slice, and add 3 mL of medium. After removal of the donor tissue, adherent cells continue to proliferate, and within 2 weeks from the time of establishment of the culture, confluent monolayers form, displaying the typical cobblestone morphology associated with epithelia. The yield is approximately $1-5 \times 10^6$ cells/cornea.

Propagation:

7. Following the initial outgrowth period, feed the cultures twice per week.

8. At 70–80% confluence, rinse the cells in D-PBSA, and release with trypsin/EDTA for 4 min at 37°C.

9. Stop the reaction with 10% FBS in D-PBSA.

10. Wash the cells (centrifugation followed by resuspension in KGM), count them, and plate at 1×10^4 cells/cm² onto tissue culture surfaces coated with FNC.

11. Incubate the culture at 37°C in 95% air and 5% CO_2.

12. Change the culture medium to fresh medium 1 days after trypsinization and reseeding.

Immediately after passage, cells appear more spindle-shaped, are refractile, and are highly migratory. Within 7 days, control cultures become 70–80% confluent, continue to display a cobblestone morphology, and, if allowed to become postconfluent, retain the ability to stratify in discrete areas.

Although corneal epithelial cultures can be subcultured up to five times (approximately 9–10 population doublings), most of the proliferation occurs between the first and third passages. Approximate yields are $1-2 \times 10^6$ cells/cornea. Senescence always ensues by the fifth passage.

Development of continuous cell lines. Lines of human corneal epithelium (HCE) with an extended life span have been developed (*see* Protocol 18.1) to provide a larger supply of cells for experimental purposes.

Long-Term storage of cells. Cells can be stored frozen in liquid nitrogen (*see* Protocol 20.1).

Phenotypic development in vitro. Both primary cultures and HCE lines retain phenotypic characteristics of corneal epithelium *in situ*. They continue to synthesize collagenase and express EGF receptors and cornea-specific cytokeratins, although the level of expression under current culture conditions is less than that observed *in situ*. When corneal epithelium is cultured upon collagen membranes at air–liquid interfaces, its morphology and barrier function are fairly well preserved. Stratified membranes develop that are able to inhibit the diffusion of Na-fluorescein. Air–liquid interface cultures survive for 2 weeks *in vitro* and have been used to investigate injury and repair mechanisms. A more complete three-dimensional tissue model may be constructed by supplying corneal fibroblasts in a collagen gel (*see* Section 25.3.6).

Human corneal epithelial cells propagated *in vitro* may provide a suitable model for exploring basic cell biological mechanisms as well as toxicological phenomena. Although the primary cultures are adequate for *in vitro* studies, HCE lines with an extended life span provide a reliable source of material that can be shared among laboratories.

23.2.3 Breast

Milk [Buehring, 1972; Ceriani et al., 1979] and reduction mammoplasty [Stampfer et al., 2002] are suitable sources of normal ductal epithelium from the breast; the former gives purer cultures of epithelial cells. Disaggregation in collagenase [Speirs et al., 1996] is preferred for primary disaggregation, growth on confluent feeder layers of fetal human intestine [Stampfer et al., 1980; Freshney et al., 1982b] represses stromal contamination of both normal and malignant tissue (*see* Plates 6c,d, 15b), and optimization of the medium [Stampfer et al., 1980; Smith et al., 1981; Hammond et al., 1984] enables serial passage and cloning of the epithelial cells. Cultivation in collagen gel allows three-dimensional structures to form that correlate well with the histology of the original donor tissue [Berdichevsky et al., 1992; Gomm et al., 1997].

As with epidermis, cholera toxin [Taylor-Papadimitriou et al., 1980] and EGF [Osborne et al., 1980] stimulate the growth of epithelioid cells from normal breast tissue *in vitro*. The hormonal picture is more complex. Many epithelial cells survive better with insulin, (1–10 IU/mL) and hydrocortisone ($\sim 1 \times 10^{-8}$ M), added to the culture. The differentiation of acinar breast epithelium in organ culture requires hydrocortisone, insulin, and prolactin [Darcy et al., 1995], and estrogen, progesterone, and growth hormone have also been shown to be required in cell culture [Klevjer-Anderson & Buehring, 1980].

The following introduction and Protocol 23.3 for the culture of cells from human milk was contributed by Joyce Taylor-Papadimitriou, Cancer Research UK Breast Cancer Biology Group, Guy's Hospital, Thomas Guy House, London SE1 9RT, UK.

Milk from early lactation or after weaning gives the highest cell yield and contains clumps of epithelium that can proliferate in culture [Buehring, 1972; Taylor-Papadimitriou et al., 1980]. Primary cultures grown in a hormone-supplemented human serum-containing medium give cell lines that have limited life span but are clonogenic [Stoker et al., 1982]. These lines are eventually overtaken by nonepithelial "late milk" cells [McKay & Taylor-Papadimitriou, 1981].

PROTOCOL 23.3. MAMMARY EPITHELIUM

Outline
Cells centrifuged from early-lactation milk are grown in the presence of endogenous macrophages in an enriched medium and may be subcultured with a mixed protease chelating solution.

Materials
Sterile:
- ❑ Growth medium RPMI 1640
- ❑ Fetal bovine serum (FBS)
- ❑ Human serum (HuS; outdated pooled serum from blood banks; Australia antigen negative)
- ❑ Stock solutions:
  - (a) Insulin (Sigma), 1 mg/mL in 6 mM HCl
  - (b) Hydrocortisone, 0.5 mg/mL in physiological saline
  - (c) Cholera toxin (*see* Appendix II), 50 μg/mL in physiological saline

Note. Serum and stock solutions of insulin, hydrocortisone, cholera toxin, pancreatin, and trypsin should be kept at −20°C.

- ❑ Trypsinization solution (TEGPED):
  - (a) EGTA (Sigma, ethylene glycol-bis(β-aminoethylether)-*N*, *N*, *N′*,*N′*-tetraacetic acid), 13 mM in D-PBSA .... 10 mL
  - (b) EDTA (Sigma, diaminoethane tetraacetic acid), 7 mM in D-PBSA .... 4 mL
  - (c) Trypsin (Difco; B-D Biosciences), 0.2% in HBSS .... 4 mL
  - (d) Pancreatin (Difco, B-D Biosciences), 1.0% in HBSS .... 2 mL
- ❑ Growth medium: RPMI 1640 containing 15% FBS; 10% HuS; cholera toxin, 50 ng/mL; hydrocortisone, 0.5 μg/mL; insulin, 1 μg/mL
- ❑ Nunc plastic dishes, 5 cm, or 25-cm² flasks
- ❑ Universal containers or 20- to 50-mL centrifuge tubes

Protocol
Milk (2–7 days postpartum) can best be collected on hospital wards. The breast is swabbed with sterile

H_2O, and the milk is manually expressed into a sterile container. Five to 20 mL are usually obtained per patient. The milks are pooled and diluted 1:1 with RPMI 1640 medium to facilitate centrifugation.

Primary Cultures:
1. Spin diluted milk at 600–1,000 g for 20 min. Carefully remove the supernatant, leaving some liquid so as not to disturb the pellet.
2. Wash the pelleted cells two to four times with RPMI containing 5% FCS until the supernatant is not turbid.
3. Resuspend the packed cell volume in growth medium, and plate 50 μL of packed cells in 5-cm dishes (Nunc) in a 6-mL growth medium. Incubate the cultures at 37°C in 5% CO_2.
4. Change the medium after 3–5 days and thereafter twice weekly. Colonies appear around 6–8 days and expand to push off the milk macrophages, which initially act as feeders.

Subculture of Milk Cells:
5. Incubate the cells in TEGPED (1.5 mL per 5-cm plate) at 37°C for 5–15 min, depending on the age of the culture, to produce a single-cell suspension.
6. Centrifuge cells at 100 g for 5 min and resuspend in 6 mL of medium.
7. Divide the suspension, 2 mL into each of three fresh 5-cm dishes or 25-cm² flasks.
8. Dilute threefold by adding 4 mL of fresh medium.

Variations. It is convenient to use the macrophages that are already present in the milk as feeders, but they are gradually lost as the epithelial colonies expand. However, macrophages can be removed by absorption to glass, and in that case other feeders must be added. Irradiated or mitomycin-treated 3T6 cells (*see* Protocols 14.3, 23.4) show the best growth-promoting activity [Taylor-Papadimitriou et al., 1977]. Analogs of cyclic AMP can be used to replace the cholera toxin [Taylor-Papadimitriou et al., 1980], although this is not possible with macrophage feeders, which are killed by the analogs.

Uses and application of milk epithelial cell culture. Milk cultures provide cells from the fully functioning gland and allow the definition of phenotypes by immunological markers [Chang & Taylor-Papadimitriou, 1983]. Milk cells have been successfully transformed by SV40 virus [Chang et al., 1982] and used in genotoxicity studies [Martin et al., 2000].

23.2.4 Cervix
Cervical keratinocytes can be grown in serial culture at clonal density by using a modification [Stanley & Parkinson, 1979] of the method described for epidermal keratinocytes

[Rheinwald & Green, 1975]. Protocol 23.4 for the culture of epithelial cells from cervical biopsy samples was contributed by Margaret Stanley, Department of Pathology, University of Cambridge, UK.

PROTOCOL 23.4. CERVICAL EPITHELIUM

Outline
Single-cell suspensions from enzymatically disaggregated epithelium from punch or wedge cervical biopsies are inoculated into flasks or plates either with lethally inactivated Swiss 3T3 cells and grown in a medium supplemented with serum and growth factor (Protocol A) or grown in serum-free keratinocyte growth medium (KGM) without feeder support, by using modifications of the methods originally described for epidermal keratinocytes (Boyce & Ham, 1983). When the keratinocyte colonies that arise contain 1000 cells or more, the cultures can be trypsinized and passaged again with the use of fibroblast feeder support.

Materials
Sterile:
- Transport medium: Dulbecco's modification of Eagle's medium (DMEM), supplemented with 10% fetal calf serum, 100 μg/mL of gentamycin sulfate, and 10 μg/mL of amphotericin
- Trypsin EDTA solution for disaggregation of epithelial cells: 0.25% trypsin (v/v), 0.25 mM (0.01%) EDTA as the disodium salt with pH 7.4 in D-D-PBSA
- Medium for Protocol A: Keratinocyte culture medium (KCM): DMEM supplemented with 10% fetal bovine serum, 0.5 μg/mL of hydrocortisone, and 1×10^{-10} M cholera toxin:
  - (a) Fetal bovine serum: not all batches support growth adequately. Serum samples should be tested and a large batch of suitable quality bought.
  - (b) Epidermal growth factor (EGF): Sigma, 100 μg dissolved in 1 mL sterile UPW. Aliquot in 100-μL aliquots and store at −20°C. Prepare working stocks by diluting 100 μL of EGF at 100 μg/mL in 10 mL medium with serum. Store at 4°C. Dilute 1:100 in medium with serum for use at a final concentration of 10 ng/mL.
  - (c) Cholera toxin (CT): Sigma. Add 1.18 mL of sterile UPW to 1 mg of CT to give a 10^{-5} M solution. Store at +4°C. Working stocks should be 100 μL of 10^{-5} M CT in 10 mL of medium with serum. Filter sterilize the culture and keep

it at +4°C. Use at a final concentration of 10^{-10} M.
  - (d) Hydrocortisone: Sigma. Dissolve 1 mg of hydrocortisone in 1 mL of 50% ethanol/water (v/v). Dispense into 100-μL aliquots and keep at −20°C. Use at a final concentration of 0.5 μg/mL.
- Medium for Protocol B: Keratinocyte growth medium: KGM (Bio-Whittaker), or Keratinocyte-SFM (Invitrogen)
- Petri dishes for tissue culture, 6-cm and 9-cm diameter
- Forceps, rat toothed
- Curved iris scissors
- Disposable scalpels, No. 22 blade
- Pipettes
- Centrifuge tubes

Nonsterile:
- Hemocytometer
- Radiation source (e.g., X rays or ^{60}Co)

Swiss 3T3 Fibroblasts (Protocol A only):
- (a) A large master stock of cells should be prepared and frozen in individual ampoules of 1×10^6 cells. Cells should not be used for more than 20 passages.
  - i) Grow 3T3s in DMEM/10% calf serum in 175-cm² tissue culture flasks. Inoculate cells at 1.5×10^4 cells/cm². Change the medium after 2 days. Subculture every 4–5 days.
  - ii) To avoid low-level contamination, maintain one master flask of cells on antibiotic-free medium; these cells are then used at each passage to inoculate the flasks required for that week's feeder cells.
- (b) Feeder layers are inactivated by irradiation with 60 Gy (6000 rad), either from an X-ray or ^{60}Co source. Irradiated cells (XR-3T3) may be kept at 4°C for 3–4 days.
- (c) In the absence of a source of irradiation, inactivate feeder cells with mitomycin C.
  - i) Expose 3T3 cells growing in monolayer to 400 μg/mL of mitomycin C for 1 h at 37°C.
  - ii) Trypsinize the treated cells, resuspend and wash the cell pellet twice with fresh medium with serum, resuspend the cells at a suitable concentration in complete medium, and use.

Protocol A: with 3T3 Feeder Layers
Primary Culture:
1. Remove the cervical biopsy from the transport medium, and wash the cells two to three times with 5 mL of sterile D-PBSA containing gentamycin sulfate, 50 μg/mL, and amphotericin, 5 μg/mL.

2. Place the biopsy, epithelial surface down, on a sterile culture dish.

3. Using a disposable scalpel fitted with a No. 22 blade, cut and scrape away as much of the muscle and stroma as possible, leaving a thin, opaque epithelial strip.

4. Mince the epithelial strip finely with curved iris scissors.

5. Add 10 mL of trypsin/EDTA (prewarmed to 37°C) to the epithelial mince, and transfer the tissue to a sterile glass universal container containing a small plastic-coated magnetic stirrer bar.

6. Add a further 5–10 mL of trypsin/EDTA.

7. Place the universal container on a magnetic stirrer in an incubator or a hot room at 37°C, and stir slowly for 30–40 min.

8. Allow the suspension to stand at room temperature for 2–3 min.

9. Remove the supernate containing single cells, and filter it through a stainless steel or plastic mesh (e.g., Cell Strainer, BD Biosciences) into a 50-mL centrifuge tube (see Fig. 12.10).

10. Add 10 mL of complete medium to the filtrate.

11. Add a further 15 mL of warm trypsin/EDTA to the fragments in the universal container, and repeat Steps 5–9 twice. Combine the trypsin supernates and spin in a bench centrifuge at 1000 rpm (80 g) for 5 min.

12. Remove the supernate, add 10 mL of complete medium to the pellet, resuspend the cells vigorously to give a single-cell suspension, and count the cells with a hemocytometer. Assess cell viability with Trypan Blue exclusion (see Protocol 22.1).

13. Dilute the cervical cell suspension with KCM, and plate cells out at 2×10^4 cells/cm^2 together with 1×10^5 cells/cm^2 of lethally inactivated 3T3 cells (i.e., for 1×10^5 cervical cells, 5×10^5 XR-3T3/60-mm dish).

14. Incubate the cultures at 37°C in 5% CO_2.

15. Seventy-two hours after the initial plating, replace the medium with a complete medium supplemented with EGF at 10 ng/mL. Check the cultures microscopically to ensure that the feeder layer is adequate. Add further feeder cells if necessary.

16. Change the medium twice weekly; EGF should be present in the medium, except when the cells are initially plated. Keratinocyte colonies become visible on the microscope by days 8–12 and should be visible to the naked eye by days 14–16. Cultures should be passaged at this time.

Subculture:

17. Remove the medium from the cell layer, and remove the feeders by rinsing rapidly with 0.01% EDTA. Wash twice with D-PBSA.

18. To each culture dish, add enough prewarmed trypsin/EDTA to cover the cell sheet. Leave the cultures at 37°C until the keratinocytes have detached; check for detachment with a microscope. Do not leave the cells in trypsin for more than 20 min.

19. Remove the cell suspension from the plate and transfer it to a sterile centrifuge tube.

20. Rinse the growth surface with complete medium and add to the suspension. Mix and dispense the suspension with a 10-mL pipette.

21. Spin the cells at 80–100 g for 5 min.

22. Remove the supernate, add 10 mL of complete medium, and resuspend the cells vigorously with a 10-mL pipette to achieve a single-cell suspension.

23. Count the cells with a hemocytometer.

24. Cells may be replated on inactivated 3T3 cells and grown as just described or frozen for later recovery.

Protocol B: Serum-free Medium

Primary Culture:

1. Follow Steps 1–11 of Protocol A to obtain a single-cell suspension in medium supplemented with serum.

2. Spin the suspension at 1000 rpm (80 g) for 5 min. Remove the supernate and resuspend the cells in 10 mL of D-PBSA.

3. Spin the suspension again and wash the cells once more with D-PBSA. Resuspend the cells in 10 mL of keratinocyte growth medium (prepared as per the manufacturer's instructions), and seed into culture dishes or flasks at a density of 2×10^5 cells/cm^2 (10^6 cells/5-cm Petri dish, 4×10^6/9-cm Petri dish).

4. Incubate the cultures at 37°C in 5% CO_2.

5. Change the medium after 72 h and twice weekly thereafter. Cultures become confluent within 14–20 days and should be passaged at that time.

Subculture:

6. Follow steps 17–23 of Protocol A.

7. Spin the suspension at 80–100 g) for 5 min. Remove the supernate and resuspend the cells in 10 mL of D-PBSA.

8. Spin the suspension again and wash the cells once more with D-PBSA. Resuspend the cells in KGM and plate them onto culture dishes at 10^5 cells/cm^2 (5×10^5 cells/50-mm Petri dish, 2×10^6 cells/90-mm Petri dish).

9. Cells may also be frozen at this stage for recovery at a later date.

Cervical keratinocytes grown as just described can be used for a range of purposes, including investigations of papillomavirus carcinogenesis, differentiation studies, and examination of the response of these cells to mutagens and carcinogens.

23.2.5 Gastrointestinal Tract

The culture of normal epithelium from the gut lining has not been extensively reported, although there are numerous reports in the literature of continuous lines from human colon carcinoma (*see* Section 24.8.3). Owens et al. [1974] were able to culture cells from fetal human intestine as a finite cell line (FHS 74 Int), and extensive use has been made of the rat intestinal cell line IEC-6 [Rak et al., 1995; Bedrin et al., 1997]. Protocol 23.5 was condensed from Booth and O'Shea [2002].

The preparation of tissue specimens for cell culture is usually started within 1–2 h of removal from the patient. If this is impossible, fine cutting of the tissue into small pieces (1–2 mm) with scalpels and storage overnight at 4°C in washing medium (HBSS/PSG, *see* below) can also prove successful.

PROTOCOL 23.5. ISOLATION AND CULTURE OF COLONIC CRYPTS

Outline
Finely chopped tissue is disaggregated in collagenase and dispase, and the crypts are separated by sedimentation through sorbitol.

Materials
Sterile:
- HBSS/PSG: Hanks' balanced salt solution with penicillin, 100 U/mL, streptomycin, 30 μg/mL, and gentamycin, 25 μg/mL
- Digestion mix: collagenase Type XI, 150 U/mL, Dispase, 40 μg/mL, FBS, 1%, in DMEM
- Growth medium: DMEM with 25 mM glucose, 6.8 mM pyruvate, 0.25 U/mL insulin, EGF, 50 ng/mL, FBS, 2.5%, and antibiotics as in HBSS above, buffered to pH 7.4 with sodium bicarbonate under 7.5% CO_2
- S-DMEM: Growth medium with 2% sorbitol (0.11 M), sterilized by filtration
All media should be at 37°C.
- Petri dishes, 9 cm
- Universal containers or 30- to 50-mL centrifuge tubes

- Multiwell plates, 24 well, collagen coated:
  (a) Dilute stock Vitrogen tenfold.
  (b) Incubate plates with 0.5 mL/well for 3 h.
  (c) Remove excess Vitrogen and dry plates in laminar-flow hood.
  Can be stored wrapped in cling film (Saran Wrap) for up to 1 week at 4°C.
- Flasks, 25 cm²
- Scalpels, #4 holder, #22 blade
- Movette pipette (or similar wide-bore plastic pipette)

Protocol
Isolation of Crypts from Human Colon:
1. Tissue for culture should be placed in DMEM as soon as possible after excision from the patient. Preliminary studies indicate that as long as the tissue is quickly placed into DMEM it will withstand a short period (up to 2 h) at 4°C, although a longer period is not advisable.
2. Wash the intact tissue twice in 1× HBSS/PSG to remove any contaminating factors.
3. Transfer the tissue to a clean Petri dish with fresh HBSS/PSG.
4. Remove as much muscle as possible and chop the tissue into pieces with two opposed scalpels.
5. Wash the chopped tissue pieces by transferring them to a 25-cm² flask containing 20 mL HBSS/PSG and pipette up and down with a Movette pipette.
6. Allow contents of the flask to settle and remove the HBSS/PSG.
7. Repeat the process until the HBSS/PSG remains almost clear. This usually takes about 5 repeats of the washing process.
8. Transfer the tissue to a Petri dish and remove any excess HBSS/PSG.
9. Chop the tissue with the scalpels until it has a fairly smooth consistency. Transfer the chopped tissue to a 25-cm flask with a wide-bore pipette (ensure that the inside of the pipette is already wet with HBSS/PSG to prevent sticking).
10. Allow pieces to settle and remove excess HBSS/PSG.
11. Add 25–30 mL of the digestion mix.
12. Incubate at 37°C for 1 h.
13. Replace the digestion mix:
    (a) Allow the tissue to settle.
    (b) Carefully remove the digestion mix from the flask and discard it (being sure not to remove any undigested tissue in the process).
    (c) Add fresh digestion mix and continue incubation at 37°C for a further 1–2 h.

(d) Check the tissue every 15–30 min to ensure that crypts do not become overdigested. The isolation is ready when approximately 70–80% of the crypts have become liberated as individual free crypts.

N.B. Do not wait until all crypts are liberated as this may result in a reduction in yield caused by overdigestion.

Sedimentation of Human Colonic Crypts:
Released crypts are purified by sorbitol sedimentation. You may find that large aggregates of fatty/mesenchymal tissue rise to the top of the sedimentation tube; these should be discarded in the first steps because they may cause problems during the sedimentation process.

14. Divide the contents of the digestion flask between two 30-mL sterile universal containers and label "Set 1".
15. Top up to 25 mL with S-DMEM and remove any floating fatty or other mesenchymal tissue.
16. Allow the contents of the tubes to settle for a period of 1 min to allow any undigested matter to sediment.
17. Discard any fatty material floating on the surface and then gently remove the supernatant suspension from the tubes and transfer to two clean tubes; label "Set 2".
18. Let the solution settle for 30 s.
19. Transfer the contents to a further set of clean tubes; label "Set 3".
20. Top up with S-DMEM.
21. Centrifuge Set 3 tubes at 200–300 g for 4 min. (The slowest speed on a standard bench top centrifuge is usually about right.)
22. Discard the supernate.
23. Disperse the pellet by flicking/tapping the base of the tube.
24. Resuspend the crypts once more with S-DMEM and repeat the centrifugation step. This S-DMEM washing/centrifugation step must be repeated about 5 times, or until the supernate is clear.
25. The debris left behind in Set 1 and 2 tubes should be examined microscopically for the presence of crypts. If a high proportion of crypts remain, the material can be agitated and the initial sedimentation step repeated. If only undigested material remains, this can be discarded.
26. The crypts are ready to be plated out when the supernate appears clear. A critical plating density is required for good culture growth. Generally, the crypts appear to grow best when plated at a density of 800–1000 crypts/mL/well

in a 24-well plate. To determine the number of crypts isolated:
    (a) Resuspend the isolation in 10 mL of growth medium and agitate.
    (b) Remove 100 μL and add to 900 μL growth medium.
    (c) Take 100-μL sample from this and count the number of crypts present.
    No. of crypts/mL = No. of crypts counted ×100
27. Incubate the plates at 37°C in 7.5% CO_2 and allow the cells to attach for 2 days.
28. Remove the medium, containing all unattached material, from the plates and replace with fresh medium.
29. Supplement the medium at 5-day intervals with 0.5 mL fresh medium.

Crypts can be plated out on either collagen-coated plates or a 3T3 feeder layer. This method will allow routine culture for a period of about 1 wk. After this period the epithelial cells deteriorate and the contaminating fibroblasts become prevalent. The cultures do not subculture well. Various cell recovery agents have been utilized, but so far we have had limited success subculturing intestinal epithelial cells. Not surprisingly, fibroblasts can be quite easily subcultured and are a potential source of feeder layer cells.

23.2.6 Liver

Although cultures from adult liver do not express all the properties of liver parenchyma, there is little doubt that the correct lineage of cells may be cultured. So far, attempts at generating proliferating cell lines have not been particularly successful, but functional hepatocytes can be cultured under the correct conditions [Guguen-Guillouzo, 2002].

Some of the most useful continuous liver cell lines were derived from Reuber H35 [Pitot et al., 1964] and Morris [Granner et al., 1968] minimal-deviation hepatomas of the rat. Inducing tyrosine aminotransferase in these cell lines with dexamethasone proved to be a valuable model for studying the regulation of enzyme adaptation in mammalian cells [Granner et al., 1968; Reel & Kenney, 1968]. Cell lines such as Hep-G2 have also been generated from human hepatoma and retain some of the metabolizing properties of normal liver [Knowles et al., 1980].

Protocol 23.6 for the culture of isolated adult hepatocytes was contributed by Christiane Guguen-Guillouzo, Hôpital de Pontchaillou, INSERM U49, Rue Henri le Guilloux, 35033 Rennes, France.

When perfused into the liver through the vessels and capillaries at an adequate flow rate, proteolytic enzymes such as collagenase, which are relatively noncytotoxic, will disrupt intercellular junctions and will digest the connective framework within 15 min if the liver is previously cleared of blood and depleted of Ca^{2+} by washing with calcium-free buffer [Berry & Friend, 1969]. Hepatocytes are selected from the cell suspension by two or three differential centrifugations.

PROTOCOL 23.6. ISOLATION OF RAT HEPATOCYTES

Outline

Introduce a cannula in the portal vein or a portal branch, wash the liver with a calcium-free buffer (15 min), perfuse the liver with the enzymatic solution (15 min), collect and wash the cells, and count the viable hepatocytes [Guguen-Guillouzo & Guillouzo, 1986].

Materials

Sterile:

❑ L-15 Leibovitz medium
❑ Ham's F12 or Williams' E medium enriched with 0.2% bovine albumin (grade V, Sigma) and 10 μg/mL of bovine insulin (80–100 mL)
❑ Hepatocyte minimal medium (HMM): Eagle's MEM, 67.5%, medium 199, 22.5% (or William's E in place of both), bovine insulin, 5 μg/mL, BSA, 1 mg/mL, pyruvate 20 mM, penicillin, 100 U/mL, streptomycin, 100 μg/mL, and 20% FBS.
❑ HMM/SF: serum-free HMM with 1×10^{-5} M hydrocortisone hemisuccinate
❑ Calcium-free HEPES buffer pH 7.65: 160.8 mM NaCl; 3.15 mM KCl: 0.7 mM $Na_2HPO_4 \cdot 12H_2O$, 33 mM HEPES, sterilization by 0.22-μm micropore filter, and storage at 4°C (2 months)
❑ Collagenase solution: 0.025% collagenase (Sigma grade 1 or Roche #103578); 0.075% $CaCl_2 \cdot 2H_2O$ in calcium-free HEPES buffer pH 7.65; preparation and sterilization by filtration just before use
❑ Nembutal (Abbott, 5%)
❑ Heparin (Roche)
❑ Tygon tube (ID, 3.0 mm; OD, 5.0 mm)
❑ Disposable scalp vein infusion needles, 20G (Dubernard Hospital Laboratory, Bordeaux, France)
❑ Sewing thread for cannulation
❑ Graduated bottles and Petri dishes
❑ Surgical instruments (sharp, straight, and curved scissors and clips)
❑ Disposable syringes, 1 mL, 2

Nonsterile:

❑ Chronometer
❑ Peristaltic pump (10 to 200 rpm)
❑ Water bath

Protocol

1. Warm the washing HEPES buffer and collagenase solution in a water bath (usually approximately 38–39°C to achieve 37°C in the liver). Oxygenation is not necessary.
2. Set the pump flow rate at 30 mL/min.
3. Anesthetize the rat (180–200 g) by intraperitoneal injection of Nembutal (150 μL/100 g), and inject heparin into the femoral vein (1000 IU).
4. Swab the underside of the rat with 70% alcohol.
5. Open the abdomen, place a loosely tied ligature around the portal vein approximately 5 mm from the liver, insert the cannula up to the liver, and ligate.
6. Rapidly incise the subhepatic vessels to avoid excess pressure, and start the perfusion with 500 mL of calcium-free HEPES buffer at a flow rate of 30 mL/min; verify that the liver whitens within a few seconds.
7. Perfuse 300 mL of the collagenase solution at a flow rate of 15 mL/min for 20 min. The liver becomes swollen.
8. Remove the liver and wash it with HEPES buffer.
9. After disrupting the Glisson capsule, disperse the cells in 100 mL of L-15 Leibovitz medium.
10. Filter the suspension through two layers of gauze or 60- to 80-μm nylon mesh, allow the viable cells to sediment for 20 min (usually at room temperature), and discard the supernatant (60 mL) containing debris and dead cells.
11. Wash the cells three times by slow centrifugations (50 g for 40 s) to remove collagenase, damaged cells, and nonparenchymal cells.
12. Suspend isolated hepatocytes in HMM.
13. After 2–3 h, living cells will attach to plastic and begin to spread. Remove the dead cells with the supernatant medium.
14. After cell attachment, remove the FBS, and replace the medium with fresh HMM/SF. Renew every day thereafter.

When appropriate biomatrix components are used for coating the dish, e.g., fibronectin, collagen, or Matrigel (*see* Section 8.4.1), FBS addition can be avoided completely.

Analysis. Determine the cell yield and viability by the well-preserved refringent shape or the Trypan Blue exclusion test (0.2% w/v; usually 4–6×10^8 cells with a viability of more than 95%).

Variations

Isolation of hepatocytes from other species. The basic two-step perfusion procedure [Seglen, 1975] can be used for obtaining hepatocytes from various rodents, including mouse, rabbit, guinea pig, or woodchuck, by adapting the volume and the flow rate of the perfused solutions to the size of the liver. The technique has been adapted for the human liver [Guguen-Guillouzo et al., 1982] by perfusing a portion of the whole liver (usually with 1.5 L HEPES buffer and 1 L collagenase solution at 70 and 30 mL/min, respectively) or by taking biopsies (15–30 mL/min, depending on the size of the tissue sample). A complete isolation into a single-cell suspension can be obtained by an additional collagenase incubation at 37°C under gentle stirring for 10–20 min (especially for human liver) [Guguen-Guillouzo & Guillouzo, 1986]. Fish hepatocytes can be obtained by cannulating the intestinal vein and incising the heart to avoid excess pressure. Perfusion is performed at room temperature at a flow rate of 12 mL/min.

Maintenance and differentiation. Isolated parenchymal cells can be maintained in suspension for 4–6 h and used for short-term experiments. When seeded in nutrient medium supplemented with 1×10^{-6} M dexamethasone on plastic culture dishes (7×10^5 viable cells/mL), the cells survive for a few days. Survival for several weeks is obtained by seeding the cells onto a biomatrix [Rojkind et al., 1980]. However, the cells rapidly lose their specific differentiated functions. Higher stability (2 months) can be obtained by coculturing hepatocytes with rat liver epithelial cells presumed to derive from primitive biliary cells [Guguen-Guillouzo et al., 1983]. Hepatocytes are also more stable when seeded on Matrigel [Bissell et al., 1987] and are capable of undergoing from one to three rounds of cell division when they are cultured in a medium supplemented with EGF and pyruvate [McGowan, 1986] or with nicotinamide and EGF [Mitaka et al., 1991]. A combination of EGF and DMSO promotes differentiation of hepatocytes [Mitaka et al., 1993; Hino et al., 1999].

23.2.7 Pancreas

Both acinar [Bosco et al., 1994] and islet cells [Vonen et al., 1992; Cao et al., 1997] have been grown from pancreatic tissue, and lines have been established from the "immortomouse" [Blouin et al., 1997]. The conversion of adenoma to carcinoma has generated particular interest [Perl et al., 1998].

Protocol 23.7 for the culture of pancreatic acinar cells was contributed by Robert J. Hay and Maria das Gracas Miranda, American Type Culture Collection, 10801 University Boulevard, Manassas, VA 20110-2209. Fractionated populations of pancreatic acinar epithelia are inoculated onto collagen-coated culture dishes, and two-dimensional aggregate colonies are allowed to develop for subsequent study [Hay, 1979; Ruoff & Hay, 1979].

PROTOCOL 23.7. PANCREATIC EPITHELIUM

Outline

Dissociate guinea pig pancreatic tissue, filter the mixed-cell suspension through cheesecloth or nylon sieves, and layer the suspension over a BSA solution. After three sequential centrifugation and fractionation steps, dispense the cell pellet, and seed the cells in collagen-coated culture vessels.

Materials

Sterile:

- ☐ F12K tissue culture medium with 20% bovine calf serum (F12K-CS20)
- ☐ HBSS
- ☐ HBSS without Ca^{2+} and Mg^{2+} (HBSS-DVC)
- ☐ Dextrose (Difco, No. 0155–174, B-D Biosciences)
- ☐ Bovine serum albumin, fraction V (Sigma)
- ☐ Trypsin solution: trypsin 1:250 (Difco, B-D Biosciences), 0.25% in citrate buffer; 3 g/L of tri-sodium citrate; 6 g/L of NaCl, 5 g/L dextrose; 0.02 g/L of phenol red; pH 7.6.
- ☐ Collagenase solution: Dissolve collagenase (GIBCO) in 1× HBSS-DVC (pH 7.2) to give 1800 U/mL.
  Adjust the solution to pH 7.2, and dialyze, using a 12-KDa exclusion membrane for 4 h at 4°C, against 1× HBSS-DVC containing 0.2% glucose. The HBSS is discarded, and this step is repeated for 16 to 18 h with fresh HBSS-DVC. Filter, sterilize, and store the solution in 10- to 20-mL aliquots at −70°C or below.
- ☐ Conical flask, 25 mL (Bellco), siliconized
- ☐ Pipettes, 5 or 10 mL, wide bore (Bellco)
- ☐ Büchner funnel
- ☐ Dialysis tubing (Spectro-Por)
- ☐ Sidearm flask, 250 mL
- ☐ Polypropylene centrifuge tubes, 50 mL
- ☐ Cheesecloth
- ☐ Collagen-coated culture dishes (Biocoat, BD Biosciences)

Nonsterile:

- ☐ Agitating water bath (Backer model M5B#11-22-1)
- ☐ Centrifuge

Protocol

1. Make up the dissociation fluid by mixing 1 part of collagenase with 2 parts of trypsin solution, and warm to 37°C.
2. Aseptically remove the entire pancreas (0.5–1.0 g), and place the organ in F12K-CS20.
3. Trim away mesenteric membranes and other extraneous matter, and mince the pancreas into 1- to 3-mm fragments.

4. Transfer the fragments to a siliconized, 25-mL conical flask in 5 mL of prewarmed dissociation fluid. Agitate the solution at about 120 rpm for 15 min at 37°C in a shaker bath. Repeat this dissociation step two or three times with fresh fluid, until most of the tissue has been dispersed.

5. After each dissociation, allow large fragments to settle, and transfer the supernate to a 50-mL polypropylene centrifuge tube with approximately 12 mL of cold F12K-CS20 to neutralize the dissociation fluid.

6. Spin the suspension at 600 rpm for 5 min, and resuspend the pellet in 5 to 10 mL of cold F12K-CS20. Keep the cell suspension in ice.

7. Pool the cell suspensions and pass through several layers of sterile cheesecloth in a Büchner funnel. Apply light suction by inserting the funnel stem into a vacuum flask during filtration. Take an aliquot for cell quantitation.

8. Generally, $1–2 \times 10^5$ cells/mg of tissue with 90% to 95% viability are obtained at this step.

9. Layer 5×10^7 to 5×10^8 cells from the resulting fluid onto the surface of 2–4 columns consisting of 35 mL of 4% cold BSA in HBSS-DVC (pH 7.2) in polypropylene 50-mL centrifuge tubes. Centrifuge the tubes at 600 rpm for 5 min. This step is critical to achieving a good separation of acinar cells from islet, ductal, and stromal cells of the pancreas. Discard the supernate, and repeat this fractionation step twice more, pooling and resuspending the pellets in cold F12K-CS20 after each step.

10. Collect the cell pellets in 5–10 mL of cold F12K-CS20. Take an aliquot for counting. Yields of $2–5 \times 10^4$ cells/mg of tissue are obtained, with 80–95% of the total being acinar cells. Inoculation densities can be $3 \times 10^5/cm^2$. Cells adhere as aggregate colonies within 72 h.

Variations. After cells have adhered, F12K-C20 can be replaced by serum-free MEM with insulin (10 ng/mL), transferrin (5 ng/mL), selenium (9 ng/mL), and EGF (21 ng/mL). This medium supports pancreatic cell growth for at least 15 days without altering the cell morphology.

The preceding method has been applied to studies with guinea pig and human (transplant donor) tissues. The addition of irradiated human lung fibroblasts as a feeder layer produces a marked stimulation (up to 500%) in the incorporation of [³H]thymidine and prolongs survival by at least a factor of two.

23.2.8 Kidney

The kidney is a structurally complex organ in which the system of nephrons and collecting ducts is made up of numerous functionally and phenotypically distinct segments. This segmental heterogeneity is compounded by a cellular diversity that has yet to be fully characterized. Some tubular segments possess several morphologically distinct cell types. In addition, evidence points to rapid adaptive changes in cell ultrastructure that may correlate with changes in cell function [Stanton et al., 1981]. The structural and cellular heterogeneity presents a challenge to the cell culturist who is interested in isolating pure or highly enriched cell populations. The difficulty of the problem is further compounded in studies of the human kidney, with which form and access to the specimen may make some manipulations, such as vascular perfusion, difficult or impossible.

Several approaches have been used successfully to culture the cells of specific tubular segments. Density gradient methods are now commonly used to isolate enriched populations of enzyme-digested tubule segments and are particularly effective in establishing proximal tubule cell cultures from experimental animals [Taub et al., 1989]. Specific nephron or collecting duct segments can also be isolated by microdissection and then explanted to the culture substrate. This method, developed with the use of experimental animals [Horster, 1979], has been applied to the culture of human kidneys [Wilson et al., 1985] and the cyst wall epithelium of polycystic kidneys [Wilson et al., 1986]. Immunodissection [Smith & Garcia-Perez, 1985] and immunomagnetic separation [Pizzonia et al., 1991] methods have also been developed to isolate specific nephron cell types on the basis of their expression of cell type-specific ectoantigens. These elegant methods hold considerable promise for the study of specific kidney cell types in health and disease, but as yet have been applied almost exclusively to studies on experimental animals. The limited (and unscheduled) availability and inconsistent form (e.g., excised pieces, damaged vasculature, lengthy postnephrectomy period) of human donor kidneys make progressive enzymatic dissociation a more practical means for isolating human kidney cells for culture.

A number of methods for the primary culture of human kidney tissue have been reported [Detrisac et al., 1984; States et al., 1986; McAteer et al., 1991]. Protocol 23.8 and the preceding introduction were contributed by James McAteer, Stephen Kempson, and Andrew Evan, Indiana University School of Medicine, Indianapolis, IN 46202-5120.

The primary culture of tissue fragments excised from the outer cortex of the human kidney provides a means of isolating cells that express many of the functional characteristics of the proximal tubule [Kempson et al., 1989]. Progressive enzymatic dissociation and crude filtration yields single cells and small aggregates of cells that, when seeded at high density, give rise to a heterogeneous epithelium-enriched population. Large numbers of cells can be harvested, making it practical to establish multiple replicate primary cultures or to propagate cells for frozen storage. Experience with the method shows that the functional characteristics of such primary and subcultured cells are reproducible for kidneys from different donors and that, with proper handling of specimens, good cultures can be derived from kidneys after even a lengthy postnephrectomy period.

PROTOCOL 23.8. KIDNEY EPITHELIUM

Outline

Tissue fragments are excised from the outer cortex, minced, washed, and incubated (with agitation) in a collagenase-trypsin solution. The tissue is periodically pipette triturated, and the dissociated cells are collected and pooled. The cell suspension is filtered through a size-limiting screen to remove undigested fragments. The cells are then washed free of enzyme, suspended in medium supplemented with serum, and seeded on culture-grade plastic.

Materials

Sterile:
- Basal medium DMEM/F12: 50:50 mixture of DMEM and Ham's F12
- Fetal bovine serum (Sterile Systems, Hyclone)
- Complete culture medium: DMEM/F12 plus 10% fetal bovine serum
- BSS
- Collagenase (Worthington type IV), 0.1%, trypsin (1:250, Sigma), 0.1% solution in 0.15 M NaCl
- DNase, 0.5 mg/mL in saline
- Trypsin-EDTA solution: trypsin (1:250), 0.1%, EDTA (culture grade, Sigma) 1 mM
- Scalpels, tissue forceps
- Nitex screen (160 μm) (Tetko)
- Culture dishes, flasks
- Tubes, sterile, 50 mL

Nonsterile:
- Orbital shaker (Bellco)

Protocol

Tissue Dissociation:

1. Cut 5- to 10-mm-thick coronal slices of kidney, and wash the fragments in chilled basal medium.
2. Excise fragments from the outer cortex, and use crossed blades to mince the tissue into 1- to 2-mm pieces.
3. Transfer approximately 5 mL of tissue fragments to a 50-mL tube containing 20 mL of the warm collagenase-trypsin solution. Secure the tube to the platform of an orbital shaker within an incubator at 37°C.
4. Incubate the tissue with gentle agitation for 1 h.
5. Discard and replenish the enzyme solution.

6. Subsequently, collect the supernatant at 20-min intervals following the addition of 20 mL of basal medium containing 0.05 mg/mL DNase and gentle trituration through a 10-mL pipette (10 cycles).
7. Dilute the supernate that has been collected with an equal volume of complete culture medium, and hold the resulting suspension on ice.
8. Repeat this cell collection procedure five or more times, until the fragments are fully dispersed.
9. Pool the harvested supernatants and dispense them into aliquots in 50-mL tubes.
10. Centrifuge at 100 g for 15 min, and resuspend each pellet in 45 mL of complete medium containing DNase.
11. Filter the suspension through Nitex cloth (160 μm).
12. The isolation protocol yields cell aggregates as well as single cells. Hence, counting cells is unreliable. To propagate cells for subculture or cryopreservation, seed 15 mL into one 75-cm² flask.

Subculture and cryopreservation. Subculture NHK-C cells by routine methods. Dissociation with trypsin-EDTA produces a monodisperse suspension that is suitable for counting with a hemocytometer. Seed culture-grade dishes at approximately 1×10^5 cells per cm². Cryopreservation is performed by routine methods using complete medium containing 10% dimethyl sulfoxide (*see* Protocol 20.1).

Comments and safety precautions. A prominent variable in this method is the quality of the kidney specimen itself. Procedures for procuring tissue are impossible to standardize. Donor tissue commonly includes segments of kidney collected at surgical nephrectomy (e.g., for renal cell carcinoma) and intact kidney tissue that is judged unsuitable for transplantation (e.g., because it has an anomalous vasculature). These specimens are collected under different conditions (e.g., with or without perfusion, with varying periods of warm ischemia, and for different durations of the postnephrectomy period). In addition, differences in the age and health of the donor ensure that no two specimens are equivalent. Although not all kidneys yield satisfactory cultures, the foregoing protocol has been effective for a wide variety of specimens, including fresh surgical specimens and intact, perfused kidneys held on wet ice for nearly 100 h.

Δ **Safety Note.** Work involving any tissue or fluid specimen of human origin requires precautions to avoid unprotected contact with blood-borne pathogens (*see* Section 7.8).

Applications. NHK-C cells exhibit predominantly the functional characteristics of the renal proximal tubule, including a parathyroid hormone (PTH)-inhibitable Na^+-dependent inorganic phosphate transport system [Kempson et al., 1989]. They also show phlorizin-sensitive Na^+-dependent hexose transport, exhibit a proximal tubulelike pattern of hormonal stimulation of cAMP (PTH responsive, vasopressin insensitive), and express several proximal tubule brush border enzymes (maltase, leucine aminopeptidase, gamma-glutamyl transpetidase). In addition, NHK-C cells have been used to demonstrate the specificity of phosphonoformic acid, an inhibitor of Na^+/P_i cotransport [Yusufi et al., 1986], and they serve as a model of oxidant injury in the renal tubule [Andreoli & McAteer, 1990].

23.2.9 Bronchial and Tracheal Epithelium

Protocol 23.9 for the isolation and culture of normal human bronchial epithelial cells from autopsy tissue was contributed by Moira A. LaVeck and John F. Lechner, National Institutes of Health, Bethesda, MD. John Lechner's current address is Bayer Diagnostics, P. O. Box 2466, Berkeley, CA 94702, USA.

In the presence of serum, normal human bronchial epithelial (NHBE) cells cease to divide and, furthermore, terminally differentiate. The serum-free medium that has been optimized for NHBE cell growth does not support lung fibroblast cell replication, thus permitting the establishment of pure NHBE cell cultures [Lechner & LaVeck, 1985].

PROTOCOL 23.9. BRONCHIAL AND TRACHEAL EPITHELIUM

Outline

This procedure involves explanting fragments of large airway tissue in a serum-free medium (LHC-9) in order to initiate and subsequently propagate fibroblast-free outgrowths of NHBE cells; four subculturings and 30 population doublings are routine.

Materials

Sterile:

❏ Culture medium:
   (a) L-15 (Invitrogen)
   (b) LHC-9 [Lechner & LaVeck, 1985] (*see* Table 10.1)
   (c) HB [Lechner & LaVeck, 1985] (*see* Appendix I)
❏ Bronchial tissue from autopsy of noncancerous donors
❏ Plastic tissue culture dishes (6 and 10 cm)
❏ FN/V/BSA: Mixture of human fibronectin, 10 μg/mL (Collaborative Research, B-D Biosciences), collagen (Vitrogen 100, 30 μg/mL; Collagen Corp.), and crystallized bovine serum albumin (BSA), 10 μg/mL (Bayer) in LHC basal medium (Biosource International) [Lechner & LaVeck, 1985]

❏ Trypsin, 0.02%, EGTA (Sigma), 0.5 mM, and polyvinylpyrrolidine (USB), 1% solution [Lechner & LaVeck, 1985]
❏ Scalpels No. 1621 (B-D Biosciences)
❏ Surgical scissors
❏ Half-curved microdissecting forceps
❏ High-O_2 gas mixture (50% O_2, 45% N_2, 5% CO_2)

Nonsterile:

❏ Gloves (human tissue can be contaminated with biologically hazardous agents)
❏ Controlled-atmosphere chamber (Bellco No. 7741)
❏ Rocker platform (Bellco No. 7740)

Protocol

1. Coat a culture dish with 1 mL of the FN/V/BSA mixture per 6-cm dish, and incubate the dish in a humidified CO_2 incubator at 37°C for at least 2 h (not to exceed 48 h). Vacuum aspirate the mixture and fill the dish with 5 mL of culture medium.

2. Aseptically dissected lung tissue from noncancerous donors autopsied within the previous 12 h is placed into ice-cold L-15 medium for transport to the laboratory, where the bronchus is further dissected from the peripheral lung tissues.

3. Before culturing, scratch an area of one square centimeter at one edge of the surface of the 6-cm culture dishes with a scalpel blade.

4. Open the airways (submerged in the L-15 medium) with surgical scissors, and cut (slice, do not saw) the tissue with a scalpel into two pieces, 2 × 3 cm.

5. Using a scooping motion to prevent damage to the epithelium, pick up the moist fragments and place them epithelium side up onto the scratched area of the 6-cm dish. Remove the medium, and incubate the fragments at room temperature for 3–5 min to allow time for them to adhere to the scratched areas of the dishes.

6. Add 3 mL of HB medium to each dish and place them in a controlled-atmosphere chamber on a rocker platform.

7. Flush the chamber with a high-O_2 gas mixture.

8. Rock the chamber at 10 cycles per minute, causing the medium to flow intermittently over the epithelial surface.

9. Incubate rocking tissue fragments at 37°C, changing the medium and atmosphere after 1 day and at 2-day intervals for 6–8 days. This step improves subsequent explant cultures by reversing any ischemic damage to the epithelium that occurred from time of death of the donor until the tissue was placed in the ice-cold L-15 medium.

10. Before explanting, scratch seven areas of the surface of each 10-cm culture dish with a scalpel. Coat the surfaces of the scratched culture dishes with the FN/V/BSA mixture solution, and aspirate the surplus solution as before.

11. Cut the moist ischemia-reversed fragments into 7 × 7-mm pieces, and explant the pieces epithelium side up on the scratched areas.

12. Incubate the pieces at room temperature without medium for 3–5 min, as before.

13. Add 10 mL of LHC-9 medium to each dish, and incubate explants at 37°C in a humidified 5% air/CO_2 incubator.

14. Replace spent medium with fresh medium every 3 to 4 days.

15. After 8–11 days of incubation, when epithelial cell outgrowths radiate from the tissue explants more than 0.5 cm, transfer the explants to new culture dishes scratched and coated with FN/V/BSA to produce new outgrowths of epithelial cells. This step can be repeated up to seven times with high yields of NHBE cells.

16. Incubate the postexplant outgrowth cultures in LHC-9 medium for an additional 2–4 days before trypsinizing (with the trypsin/EGTA/PVP solution) for subculture or for experimental use.

23.2.10 Oral Epithelium

Oral keratinocytes have been cultured in medium similar to LHC-9 and MCDB153, again utilizing FN/C/BSA-coated culture dishes to promote attachment. Protocol 23.10 has been abridged from Grafström [2002].

PROTOCOL 23.10. ORAL KERATINOCYTES

Materials
Sterile:
- Transport medium: Leibowitz L-15 with 100 µg/mL gentamycin, 100 U/mL penicillin, 100 µg/mL, Fungizone, 1 µg/mL.
- Growth medium: modified MCDB 153 [Grafström, 2002]
- Antibiotic-supplemented growth medium: growth medium supplemented with gentamycin 100 µg/mL, penicillin-streptomycin 100 U/mL, and Fungizone, 1 µg/mL
- D-PBSA: Dulbecco's PBS lacking Ca^{2+} and Mg^{2+}
- Trypsin, 0.17% in D-PBSA
- PET: trypsin, 0.025%, EGTA, 0.5 mM, polyvinyl-pyrrolidone (PVP, 40 KDa), 1%, in D-PBSA
- Scalpels, #11 blade

- Scissors, fine
- Forceps, fine, 2 pairs
- Petri dishes, tissue-culture grade, 5 or 10 cm
- Petri dishes, non-tissue-culture grade for dissection, 3.5 cm, 10 cm
- Centrifuge tubes, 15 mL, conical
- Micropipette, e.g., Gilson, 100 µL
- FN/C/BSA-coated culture dishes, 5 cm (*see* Protocol 23.9)

Protocol
Primary culture:

1. Obtain the tissue from surgery or early autopsy, place in cold transport medium, and transfer to the laboratory as soon as possible.

2. Transfer the tissue to a 10-cm dish and rinse with phosphate-buffered saline (D-PBSA).

3. Remove as much connective tissue as possible, including parts containing blood. If the tissue specimen(s) have a surface area of ≥1 cm², divide them into smaller pieces.

4. Place the specimens in a 3.5-cm dish and add enough 0.17% trypsin solution for coverage (1–2 mL) and incubate overnight at 4°C.

5. Hold each specimen with forceps, then peel and scrape away the epithelium with another pair of forceps into the trypsin solution.

6. Triturate the suspension carefully a few times, to further disaggregate the cells, and transfer the cell suspension to a centrifuge tube.

7. Rinse the dish with D-PBSA and subsequently add this rinsing solution to the cell suspension. Use the same volume for rinsing as used for the trypsin solution used for tissue digestion in Step 4.

8. Remove an aliquot by micropipette for determination of cell yield.

9. Pellet the cells at 125 *g* at 4°C, preferably with a refrigerated centrifuge, and resuspend the cells in growth medium.

10. Dilute with additional growth medium as required and then seed the cells at 5 × 10³/cm² on FN/C/BSA-coated culture dishes. If cells are derived from autopsy material, use antibiotic-supplemented growth medium for the initial 3–4 days in culture.

11. Feed the cells with fresh medium after 24 h to remove possible cell debris and erythrocytes.

12. Feed the cells every second day.
Subculture:

13. Rinse the cells once with D-PBSA.

14. Add PET solution, covering the cells completely with the solution, e.g., use 3 mL/10-cm dish.

15. Incubate at room temperature until the cells begin to round up and/or detach from the dish. Follow the cell detachment under the microscope (the procedure generally takes 5–10 min). The rate of detachment can be enhanced by addition of trypsin at higher concentrations or by increasing the temperature to 37°C.
16. When most cells have detached, carefully tap on the side of dish and pipette the trypsin solution gently over the growth surface to mechanically enhance detachment.
17. When the cells have detached, add 5–10 mL of D-PBSA to the dish to inactivate the action of trypsin by dilution.
18. Transfer the cell suspension to centrifuge tubes and triturate gently to obtain an even distribution of the cells in the suspension.
19. Using a 1-mL pipette, take out a sample for a cell count and transfer to a hemocytometer chamber (see Protocol 21.1).
20. Determine the concentration of cells in the suspension.
21. Pellet the cells by centrifugation for 5 min at 125 g at 4°C.
22. Remove the supernate and tap the tube gently against the fingers until the pellet disperses.
23. Add growth medium as desired, disperse the cell suspension gently with a pipette, and add to each culture vessel.
24. Place all the vessels on a tray. Agitate the vessels gently by holding and moving the tray in your hands, alternating between different directions, to ensure even density of the cells in each vessel (simply swirling the dishes will focus the cells in the center of the dish). It is also important that the shelves are evenly fixed in the incubator where the vessels are placed, and that the incubator is free of vibration, to avoid an uneven distribution of cells during attachment.
25. Incubate the cells undisturbed for 4–24 h before changing the medium.

23.2.11 Prostate

A number of reports have used cells cultured from prostate tissue [Cronauer et al., 1997; Peehl, 2002] with a particular interest in regulating cell proliferation and differentiation by paracrine growth factors [Yan et al., 1992; Chopra et al., 1996; Thomson et al., 1997]. Protocol 23.11 for the primary culture of rat prostate epithelial cells was contributed by W. L. McKeehan, Texas A&M University, 2121 West Holcombe Boulevard, Houston, Texas 77030-3303, USA.

Isolated normal rat prostate epithelial cells are a valuable model for studying the cell biology of maintenance, growth, and functioning of normal prostate under investigator-controlled and -defined conditions [McKeehan et al., 1982, 1984; Chaproniere & McKeehan, 1986]. Normal cells serve as the control cell type out of which prostate adenocarcinoma cells arise. Conditions similar to those described here have also been useful for the primary culture of epithelial cells for transplantable rat tumors. Key features of this method are the specific support of epithelial cells by an improved nutrient medium and hormonelike growth factors [Yan et al., 1992], while concurrently inhibiting fibroblast outgrowth due to deficient or inhibitory properties of the medium.

PROTOCOL 23.11. PROSTATIC EPITHELIUM

Outline
Remove and prepare prostates for cell culture (30 min). Incubate prostatic tissue with collagenase (1 h), and collect single cells and small aggregates of cells (1 h). Inoculate and culture cells to confluent monolayer (7 days).

Materials
Sterile:
- Growth medium WAJC404 [McKeehan et al., 1984; Chaproniere & McKeehan, 1986] (see Table 10.1)
- Media salt solution (MSS) [McKeehan et al., 1984]:
  (a) NaCl .... 12 mM
  (b) KCl .... 2 mM
  (c) KH_2PO_4 .... 1 mM
  (d) HEPES, pH 7.6 .... 30 mM
  (e) Glucose .... 4 mM
  (f) Phenol red .... 3.3 µM
- Fetal bovine or horse serum
- Penicillin
- Kanamycin
- Cholera toxin
- Dexamethasone
- Epidermal growth factor (EGF)
- Ovine or rat prolactin
- Insulin
- Partially purified [McKeehan et al., 1984; Chaproniere & McKeehan, 1986] or purified [Crabb et al., 1986] prostatropin (prostate epithelial cell growth factor)
- Collagenase, type I (Sigma)
- Petri dishes (glass or plastic), 60 mm
- Scissors
- Syringe and 14-G cannula
- Erlenmeyer flasks, 25 mL
- Wire screen, 1-mm mesh, made to fit a 50-mL conical centrifuge tube
- Conical centrifuge tubes, plastic, 50 mL

- ❏ Nylon screen filters of 250, 150, 100, and 40 μm to fit 50-mL conical tubes
- ❏ 25-well plastic tissue culture dishes

Nonsterile:

- ❏ Shaking water bath at 37°C
- ❏ Centrifuge
- ❏ Hemocytometer or Coulter counter
- ❏ Rats, 10–12 weeks old, male

Protocol

1. Aseptically remove desired lobes of the prostate and place them in a sterile 60-mm Petri dish.
2. Trim fat from the lobes and weigh them.
3. Add 2 mL of collagenase at 675 U/mL in MSS containing 100 U/mL of penicillin and 100 μg/mL of kanamycin.
4. Mince the tissue with scissors to approximately 1-mm pieces, small enough to fit through a 14-G cannula.
5. Using a syringe and a 14-G cannula, transfer the minced tissue fragments to a sterile 25-mL Erlenmeyer flask. Add collagenase to 1 mL per 0.1 g of original wet tissue weight.
6. Incubate for 1 h at 37°C on a shaking water bath.
7. Aspirate the digested suspension three times through a 14-G cannula, and then pass the suspension through a coarse (1 mm) wire mesh screen fitted to 50-mL plastic conical tubes to remove debris and undigested material. Rinse the suspension with an equal volume of MSS containing 5% whole fetal calf serum or horse serum.
8. Collect cells by centrifugation at 100 *g* for 5 min at 4°C.
9. Resuspend the cell pellet in 5 mL of MSS plus 5% serum, and pass the suspension successively through nylon screen filters of mesh sizes 250, 150, 100, and 40 μm. Wash each screen with 5 mL of MSS plus 5% serum.
10. Collect cells by centrifugation, and resuspend them in 5 mL of nutrient medium WAJC404.
11. Count the cells and adjust the concentration to 4×10^6 cells/mL.
12. Inoculate 50 μL containing 2×10^5 cells into each well of a 24-well plate (area of well ~2 cm²) containing 1 mL of medium WAJC404, 10 ng/mL of cholera toxin, 1 μM dexamethasone, 10 ng/mL of EGF, 1 μg/mL of prolactin, 5 μg/mL of insulin, 10 ng/mL of prostatropin, 100 U/mL of penicillin, and 100 μg/mL of streptomycin. Partially purified sources of prostatropin can be substituted as described by McKeehan et al. [1984] and Chaproniere and McKeehan [1986].

13. Incubate the culture in a humidified atmosphere of 95% air and 5% CO_2 at 37°C. Change the medium at days 3 and 5. The cells should be near confluent by day 7.

Variations. The foregoing procedure can be applied to the culture of human prostate epithelial cells with modifications described by Chaproniere and McKeehan [1986].

The purified growth factor prostatropin is identical to heparin-binding fibroblast growth factor type one, previously called acidic FGF or HBGF-1 [Burgess & Maciag, 1989]. The new nomenclature, achieved by consensus, is now FGF-1. Molecular characterization of new members of the FGF ligand family, as well as the FGF receptor family in prostate cells, has revealed that prostate epithelial cells express a specific splice variant (FGF-R2IIIb) of one of the four FGF receptor genes. The specific receptor recognizes FGF-1 and stromal cell-derived FGF-7 (also called keratinocyte growth factor), but not FGF-2 (previously called bFGF or HBGF-2), described in Yan et al. [1992]. Prostate stromal cells express only the FGF-R2 gene, which recognizes FGF-1 and FGF-2, but not FGF-7. Because stromal cells respond to both FGF-1 and FGF-2, prostatropin/FGF-1 can be substituted for FGF-7 in the above procedure to provide an additional selection for epithelial cells. FGF-2 will not substitute for FGF-1 or FGF-7. If purified FGF-1 is used in the protocol, its activity can be potentiated by the addition of 10–50 μg per mL of heparin.

23.3 MESENCHYMAL CELLS

Included under this heading are those cells that are derived from the embryonic mesoderm, but excluding the hematopoietic system (*see* Section 23.5). Mesenchymal cells include the structural (connective tissue, muscle, and bone) and vascular (blood vessel endothelium and smooth muscle) cells. Whereas epithelial cells are classed as *parenchyma*, the main functional cells within an organ, the connective tissue cells are classed as *stroma*, or supporting tissue.

23.3.1 Connective Tissue

Connective tissue cells are generally regarded as the weeds of the tissue culturist's garden. They survive most mechanical and enzymatic explantation techniques and may be cultured in many simple media, particularly if serum is present.

Although cells loosely called fibroblasts have been isolated from many different tissues and assumed to be connective tissue cells, the precise identity of cells in this class is not always clear. Fibroblast lines (e.g., 3T3 from the mouse) produce types I and III collagen and release it into the medium [Goldberg, 1977]. Although collagen production is not restricted to fibroblasts, synthesis of type I in relatively large amounts is characteristic of connective tissue. However,

3T3 cells can also be induced to differentiate into adipose cells [Kuriharcuch & Green, 1978; Smyth et al., 1993] and $10T\frac{1}{2}$ cells can differentiate into chondrocytes and bone [Shea et al., 2003] in response to bone morphogenetic proteins (BMPs). It appears that mouse embryo fibroblastic cell lines, such as the 3T3 group of cell lines and $10T\frac{1}{2}$, are primitive mesodermal cells that may be induced to differentiate in more than one direction.

Human, hamster, and chick fibroblasts are morphologically distinct from mouse fibroblasts, as they assume a spindle-shaped morphology at confluence, producing characteristic parallel assays of cells distinct from the pavementlike appearance of mouse fibroblasts. The spindle-shaped cell may represent a more highly committed precursor and may be more correctly termed a fibroblast. NIH3T3 cells may become spindle shaped if allowed to remain at high cell density.

It has also been suggested that fibroblastic cell lines may be derived from vascular pericytes, connective-tissue-like cells in the blood vessels, but in the absence of the appropriate markers, this is difficult to confirm.

Clearly, cell lines loosely termed fibroblastic can be cultivated from embryonic and adult tissues, but these lines should not be regarded as identical or classed as fibroblasts without confirmation, by expression of the appropriate markers, that they can differentiate into fibrocytes. Collagen, type I, is one such marker. Thy I antigen has also been used [Raff et al., 1979; Saalbach et al., 1997], although it may also appear on some hematopoietic cells.

Cultures of fetal or adult fibroblasts can be prepared by conventional procedures such as primary explantation (*see* Protocol 12.4), warm (*see* Protocol 12.5) and cold (*see* Protocol 12.6) trypsinization, and collagenase digestion (*see* Protocol 12.8).

23.3.2 Adipose Tissue

Although it may be difficult to prepare cultures from mature fat cells, differentiation may be induced in cultures of mesenchymal cells (e.g., mouse 3T3-L1) by maintaining the cells at a high density for several days [Kuriharcuch & Green, 1978; Vierick et al., 1996]. An adipogenic factor in serum appears to be responsible for the induction. Primary cultures of fat cells can also be prepared from rat epididymis by collagenase digestion; Protocol 23.12 has been abridged from Quon [1998].

Adipose cells (adipocytes) are terminally differentiated, specialized cells whose primary physiological role has classically been described as an energy reservoir for the body. Adipocytes are a storage depot for triglycerides in times of energy excess and a source of energy in the form of free fatty acids released by lipolysis during times of energy need. Adipose cells have an important role as active regulators of carbohydrate and lipid metabolism [Flier, 1995; Reaven, 1995]. It is likely that specific abnormalities in adipose tissue can contribute directly to the pathogenesis of common diseases such as diabetes, hypertension, and obesity [Flier, 1995; Hotamisligil et al., 1995; Reaven, 1995; Walston et al., 1995; Zhang et al., 1994].

In Protocol 23.12 a modification of the isolation procedure originally described by Rodbell [1964] is optimized to obtain approximately 4 mL of packed adipose cells from the epididymal fat pads of four rats by flotation of disaggregated cells after centrifugation. This protocol can be scaled up or down as needed and has been used to provide insulin-responsive cells suitable for DNA transfer by electroporation.

PROTOCOL 23.12. PRIMARY CULTURE OF ADIPOSE CELLS

Materials

Sterile:

❑ DMEM-A: DMEM, pH 7.4, containing 25 mM of glucose, 2 mM of glutamine, 200 nM of (R)-N^6-(1-methyl-2-phenylethyl)adenosine (PIA, Sigma), 100 µg/mL of gentamycin, and 25 mM of HEPES. Prepare in 100-mL aliquots and warm to 37°C before use.

❑ DMEM-B: DMEM-A plus 7% BSA

❑ KRBH buffer: Krebs-Ringer medium, pH 7.4, containing 10 mM $NaHCO_3$, 30 mM HEPES (Sigma), 200 nM adenosine (Roche), and 1% (w/v) BSA (Intergen). To make 1 L of KRBH buffer, add 7 g of NaCl, 0.55 g of KH_2PO_4, 0.25 g of $MgSO_4 \cdot 7H_2O$, 0.84 g of $NaHCO_3$, 0.11 g of $CaCl_2$, 7.15 g of HEPES, 10 g of BSA, 1 mL of 200 µM adenosine, and UPW to a volume of 1 L. Dispense in aliquots of 100 mL. Warm to 37°C and adjust pH to 7.4 before use

❑ KRBH-A buffer: KRBH with 5% BSA

❑ Collagenase, 20 mg/mL in 2 mL of KRBH buffer, sterilized by passage through a 0.22-µm filter

❑ Tissue-culture Petri dish, 6-cm diameter (B-D Biosciences)

❑ Conical centrifuge tube, 50 mL (Corning)

❑ Polypropylene tubes, 17 mm × 100 mm

❑ Low-density polypropylene vial, 30 mL

❑ Nylon mesh filter, 250 µm, in holder or filter funnel

❑ Wide-bore pipette tips, 200 µL

❑ Scissors: two pairs of sharp dissecting scissors, 10 cm (4 in.); Perry forceps, 12.5 cm (5 in.)

Nonsterile:

❑ Male Sprague-Dawley rats (145–170 g, CD strain, Charles River Breeding Laboratories)

❑ Plastic box perfused with a gas mixture of 70% CO_2, 30% O_2

❏ Guillotine
❏ Pipettor for 200-μL tips

Protocol
Dissection of Epididymal Fat Pads:
1. Anesthetize rats in a plastic box with a gas mixture of 70% CO_2 and 30% O_2.
2. Decapitate rats with a guillotine and exsanguinate.
3. Soak the bodies briefly in 70% ethanol.
4. Remove the epididymal fat pads while maintaining the highest level of sterility possible.
   (a) Cut through the skin on the lower abdomen with one pair of scissors to expose the peritoneum.
   (b) Using a second pair of scissors, open the peritoneum and pull up the testes with a pair of forceps.
   (c) Trim fat pads from epididymides, taking care to leave the blood vessels behind.
5. Transport tissue to the culture laboratory.
Collagenase Digestion and Washing of Adipose Cells:
6. Add 4 g of fat pads (approximately equivalent to 8 fat pads) to a 30-mL low-density polypropylene vial containing 4 mL of KRBH buffer at 37°C.
7. Mince fat pads into pieces approximately 2 mm in diameter with scissors.
8. Add 1 mL of collagenase solution to the vial containing the minced fat pads, and incubate the pieces in a shaking water bath at 37°C for approximately 1 h, until the cell mixture takes on a creamy consistency.
9. After collagenase digestion, add 10 mL of KRBH buffer at 37°C to the vial.
10. Mix cells in the vial by swirling, and gently pass the cells through a 250-μm nylon mesh filter into a 50-mL conical tube.
11. Wash the cells by adding 30 mL of KRBH buffer at 37°C to the tube and centrifuging briefly at 200 g in a tabletop centrifuge. Remove infranate with a pipette. Note that adipose cells will be floating on top of the aqueous buffer.
12. Repeat the washing of cells by adding 40 mL of KRBH buffer, centrifuging, and removing infranate two additional times.
13. Wash cells twice with 40 mL of DMEM-A at 37°C.
14. After the final wash, resuspend the cells from the surface of the medium in DMEM-A at a cytocrit of approximately 40%.
Primary Culture of Adipose Cells:
15. Transfer 2 mL of the 40% cytocrit DMEM-A suspension, using 200-μL wide-bore pipette tips, into one 6-cm tissue culture dish.

16. Place the cells in a humid incubator at 37°C with 5% CO_2 for 1.5 h.
17. Add 5 mL of DMEM-B to each dish.

At this point, one typically performs a glucose uptake assay [Quon, 1998] on an aliquot of cells to check their viability.

23.3.3 Muscle

Smooth muscle. Smooth muscle cells have been grown from vascular tissue [Subramanian et al., 1991] and cocultured with vascular endothelium [Jinard et al., 1997; Klinger & Niklason, 2005]. Protocol 23.13 has been reproduced from Klinger and Niklason [2005].

PROTOCOL 23.13. ISOLATION AND CULTURE OF SMOOTH MUSCLE CELLS

Materials
Sterile:
❏ Smooth Muscle Cell Medium (SMCM): Dulbecco's Modified Eagle Medium (DMEM) supplemented with 20% fetal bovine serum, penicillin 100 U/mL, and streptomycin 100 μg/mL
❏ 0.25% trypsin/EDTA (*see* Protocol 12.1 and Appendix I)
❏ Petri dishes: 15 × 2.5 cm, 6 × 1.5 cm
❏ Scalpel and No. 10 surgical blade
❏ Dissection scissors
❏ Tissue forceps

Protocol
1. Obtain thoracic aorta from young calves.
2. Immerse aorta in Hanks' saline.
3. Place on ice until ready to isolate cells.
4. Place aorta into a 15 × 2.5-cm Petri dish for dissection.
5. Incise the aorta longitudinally with dissection scissors.
6. Gently scrape the luminal surface of the aorta with the scalpel blade to remove endothelial cells.
7. Dissect the medial layer of the aorta free from the intimal and adventitial layers.
8. Cut the medial layer into segments of approximately 1 cm^2.
9. Place medial segments intimal side down in 6 × 1.5-cm Petri dishes.
10. Allow the segments to adhere to the dishes for approximately 10 min.
11. Add 1 mL of SMCM directly onto the medial segment in each dish.

12. Place medial segments in humidified incubator at 37°C and 10% CO_2 overnight.

13. The following day, add 5 mL fresh SMCM to each dish, being careful not to detach the medial segment from the dish.

14. Maintain the medial segments in an incubator for an additional 10 days, after which time the SMCs will have migrated off the segments and become established in two-dimensional culture.

15. Passage cells at subconfluence, using 0.25% trypsin/EDTA.

Cardiac muscle and transplantation of skeletal myoblasts. Cardiac muscle can be isolated with the cold-trypsin method (Protocol 11.6) and will show contractions after about 3 days in culture [Deshpande et al., 1993]. Although human cardiac myocytes have also been grown [Goldman & Wurzel, 1992], transplantation of contractile and noncontractile cells from skeletal muscle into infarcted myocardium represents a new strategy for the treatment of acute myocardial infarctions; it has been performed in rabbit, rat, pig, dog, and sheep and shown to improve myocardial function *in vivo* [Léobon et al., 2003; Menasche, 2004; Taylor, 2001]. Because of these encouraging results, cell therapy, and especially cell transplantation, is evoked for several pathologies [Bhagavati & Xu, 2004].

Transplantation of autologous skeletal myoblasts into the postinfarction scar was recently applied to humans [Menasche et al., 2001]. These recent developments gave rise to a new interest in basic research on the molecular requirements for *in vitro* myogenic proliferation and differentiation [Neumann et al., 2003; Mal & Harter, 2003; Anderson & Wozniak, 2004; Li et al., 2004; Carvalho et al., 2004] together with a fast-developing interest in culture of stem cells giving myogenic cells [Bhagavati & Xu, 2004; Seruya et al., 2004; Bossolasco et al., 2004].

Myogenic cells from skeletal muscle. It is possible to culture myogenic cells from adult skeletal muscle of several species under conditions in which the cells continue to express at least some of their differentiated traits. Called *satellite cells,* the myogenic cells partially mimic the first steps of skeletal muscle differentiation. They proliferate and migrate randomly on the substratum and then align and finally undergo a fusion process to form multinucleated myotubes [Richler & Yaffe, 1970; Campion, 1984; Hartley & Yablonka-Reuveni, 1990]. Although three to four passages can be performed by means of trypsinization, subculture is no longer possible once differentiation (i.e., fusion) has taken place.

Two different protocols are described below. Protocol 23.13 is a traditional method (digestion and mechanical trituration followed by seeding of isolated cells); Protocol 23.14 is culture based on single myofibers. Both can be used

for animal cell culture, but the first can also be used for human cell culture. However, the second protocol gives a closer simulation of the myogenic process *in vivo*. The first method is easy to perform and gives good results in terms of cell yield and differentiation capacities, especially if you use enough starting material; however, the percentage of myogenic cells obtained per unit of muscle tissue is very low, <1% of all cells present [Rosenblatt et al., 1995], calling into question whether the resultant culture is truly representative of the originating tissue. This method is described for primary cultures from human healthy muscle biopsies, which are highly enriched in myogenic cells (at least 85% positive for desmin immunostaining at day 10 after seeding). It can also be used for animal muscles (e.g., rat, mouse, rabbit, pig, or dog).

For single myofiber culture, muscles are removed individually and incubated in collagenase to digest the connective tissue. Gentle mechanical trituration then frees intact individual myofibers. Myofibers, together with associated satellite cells, can be derived from individual muscles. This permits comparisons between the responses of different muscle groups to experimental procedures or between muscles of different ages or genetic origins. Furthermore, the myosin heavy chain type of individual myofibers can be determined by SDS gel electrophoresis, and so myofibers of different myosin heavy chain types can be compared [Rosenblatt et al., 1996].

Myogenic cultures can be initiated efficiently from single myofibers prepared from muscles of animals of any age, although a slight reduction is seen with increasing age. In animals suffering from chronic myopathies such as muscular dystrophy, there is some contamination by fibroblasts, probably due to an increase in connective tissue [Bockhold et al., 1998].

Protocol 23.14 and part of the preceding introduction were contributed by M. Malo, S. Bonavaud, Ph. Thibert, and G. Barlovatz-Meimon, Faculté des Sciences, Université Paris XII, Val de Marne, Avenue General de Gaulle, F-94010 Creteil, France.

PROTOCOL 23.14. CULTURE OF MYOBLASTS FROM ADULT SKELETAL MUSCLE

Outline
Primary cultures can be grown easily in Ham's F12 medium supplemented with 20% fetal bovine serum (FBS). Without modifying the culture conditions, these cells proliferate and differentiate by fusing to form multinucleated myotubes, confirming the myogenicity of the cultivated cells.

Materials

Sterile:

☐ Transport medium: Ham's F12 (Seromed 08101) without NaHCO₃ and with 20 mM HEPES (Serva 25245), 75 mL

☐ Growth medium: Ham's F12 with 20% fetal bovine serum (GIBCO 011–06290), 200 mL

☐ Pronase solution: 0.15% pronase (Sigma P-6911) and 0.03% EDTA (Merck 8418) with 20 IU/mL of penicillin, and 20 μg/mL of streptomycin (0.4% of Eurobio PES 3000), in Ham's F12 or D-PBS, 100 mL

☐ D-PBSA, 100 mL

☐ Nylon mesh,100 μm

☐ Scalpels

☐ Long, fine scissors

☐ Petri dishes for dissection, 9 cm

☐ Flasks, 25 cm², 4

☐ Centrifuge tubes, 20 mL, or universal containers, 5

Nonsterile:

☐ Hemocytometer

Protocol

Dissociation:

1. Trim off nonmuscle tissue from the biopsy with a scalpel, and rinse in D-PBSA.

2. Cut the biopsy into fragments parallel to the fibers and wash in D-PBSA before weighing the biopsy.

3. Place the fragments parallel to each other in the lid of a Petri dish, cut the fragments into thinner cylinders and then, finally, into 1-mm³ pieces, without crushing the tissue. The final cutting can be done in a tube with long scissors, again avoiding crushing.

4. Rinse with D-PBSA and let the pieces settle; discard the supernatant.

5. Digest the fragments for 1 h at 37°C in pronase solution. (Use 15 mL for a biopsy of 1–3 mg.) Shake the tube gently at 10- to 15-min intervals.

6. Triturate the culture with a pipette after incubation. The medium should become increasingly opaque as more and more cells are released.

7. Let the fragments settle to the bottom by gravity, forming pellet P1 and supernatant S1.

8. Filter S1 through a 100-μm nylon mesh into a 20-mL centrifuge tube. Shake or pipette the supernatant gently to resuspend the cells.

9. Centrifuge the tube 8–10 min at 350 g. Discard the supernatant by aspiration.

10. Resuspend the pellet very, very gently by means of a rubber bulb pipette in precisely 10 mL of growth medium, and count the cells with a hemocytometer.

11. Dilute the suspension in growth medium to seed culture flasks with about 1.5×10^4 cells/mL. About $1-2 \times 10^5$ cells/g are obtained from healthy donor biopsies.

12. Add 15 mL of digestion medium to P1, and incubate the fragments for 30 min in a water bath at 37°C, with periodic shaking.

13. Pipette the suspension to disaggregate the cells and then filter the suspension through nylon mesh. Rinse the filter with 20 mL of growth medium.

14. Centrifuge the suspension for 8–10 min at 350 g, count the cells, and seed as before.

15. Transfer the flasks to a 37°C humidified incubator with 5% CO₂.

Δ *Safety Note.* The rest of the biopsy and all tubes, pipettes, plates, etc., used in the procedure should be treated with hypochlorite before disposal (*see* Section 7.8.5).

Maintenance of Cultures:

16. Change the medium very gently 24 h after seeding and then every 3–4 days.

The development of these cultures is mainly toward differentiation. The timing of the three phases for human muscle cells is about 4–6 days for peak proliferation; then the cells align at about day 8, and around day 10 to 12 an increase in cell fusion and the formation of myotubes are observed. Nevertheless, one must keep in mind that some cells may differentiate earlier and that others will still proliferate when the majority of the culture is undergoing differentiation.

Subculture:

1. Add a small volume of EDTA gently to the cells and remove it immediately.

2. Add sufficient trypsin solution (0.25%) to form a thin layer over the cells.

3. When cells detach, add 5–10 mL of complete growth medium, pass the culture very gently in and out of a pipette, and then centrifuge the cells for 10 min at 350 *g*.

4. Count an aliquot and seed the cells at the chosen concentration.

Proliferation and differentiation indices. The growth curves of human myogenic cell cultures obtained in Ham's F12 medium supplemented with 20% FBS show the three traditional phases (*see* Section 21.9.2): the lag phase, the exponential phase, and the plateau, which corresponds to the onset of fusion. The last, evaluated in terms of the number of nuclei incorporated into myotubes or in terms of a fusion index (the percentage of nuclei incorporated into myotubes relative to the total number of nuclei), commences usually around day 8 after plating and rises dramatically around day 10. According to the sample, this chronology can gain or lose one day.

Hence, differentiation, expressed as the number of nuclei per myotube/cm², may be observed morphologically. But the differentiation process can also be monitored by the use of biochemical markers (such as the sarcomeric proteins), enzymes involved in differentiation (e.g., creatine phosphokinase and its time-dependent muscle-specific isoform shift), or the appearance of α-actin [Buckingham, 1992].

A family of genes, the best known of which is MyoD, was shown to activate muscle-specific gene expression in myogenic progenitors. [for review, *see* Buckingham, 1992.] Myogenic cell differentiation involves either the activation of a variety of other genes with concurrent changes in cell surface adhesive properties [Dodson et al., 1990] or requirement of cell surface plasminogen activator urokinase and its receptor [Quax et al., 1992].

Variations and applications. The explant-reexplant method has been developed by the group of Askanas [Askanas et al., 1990; Pegolo et al., 1990]; it has proved reliable for a long time, especially for studies on muscle defects.

When the aim is the study of the mechanism of differentiation, one has the choice between the explant and the enzymatic methods, followed by clonal or nonclonal culture or by three-dimensional cultivation. If the aim of the work is to obtain a large number of cells for biochemical or molecular studies, then the enzymatic method seems to be preferable.

Protocol 23.15 for single myofiber culture, and the associated introduction, above, was contributed by T. Partidge and L. Heslop, Muscle Cell Biology Group, MRC Clinical Sciences Centre, Hammersmith Hospital, Du Cane Road, London W12 ONN. This procedure is of value when it is necessary to extract all of the satellite cells or it is important to avoid contamination with nonmyogenic cells. The single myofiber isolation technique can be performed on any muscle that can be removed with a large proportion of its fibers undamaged, preferably tendon-to-tendon. In the contributors' laboratory they routinely use murine extensor digitorum longus (EDL), soleus, and tibialis anterior (TA) muscles.

PROTOCOL 23.15. SINGLE MYOFIBER CULTURE FROM SKELETAL MUSCLE

Outline

Single myofibers are released by trituration of collagenase-treated whole skeletal muscle, and the outgrowth is collected after culture on Matrigel.

Materials

Sterile:

- Matrigel™ (B-D Biosciences) 1 mg/mL in DMEM
- Dulbecco's modified Eagle's medium (DMEM, GIBCO, Invitrogen)
- L-glutamine, 200 mM (Sigma)
- Penicillin-streptomycin solution: penicillin 10^4 U/mL, streptomycin 10 mg/mL in 0.9% NaCl (Sigma)
- Chick embryo extract (MP Biomedicals)
- Horse serum (Invitrogen)
- Fetal bovine serum (FBS; PAA Laboratories)
- Plating medium: DMEM with horse serum, 10%, chick embryo extract 0.5%, L-glutamine, 4 mM, penicillin, 100 U/mL, and streptomycin, 100 µg/mL
- Proliferation medium: DMEM with 20% FBS, 10% horse serum, chick embryo extract, 1%, L-glutamine, 4 mM, penicillin, 100 U/mL, and streptomycin, 100 µg/mL.
- Collagenase type I (Sigma)
- Petri dishes, 9 cm
- Modified Pasteur pipettes: Pasteur pipettes are cut, using a diamond tipped pen to create apertures of varying sizes, and then fire-polished to prevent jagged edges damaging the myofibers.
- Multiwell plates, 24 well, coated with 1 mg/mL Matrigel™ (Primaria®, B-D Biosciences) in DMEM and incubated at 37°C for 30–60 min to allow the Matrigel™ to gel.

Nonsterile:

- Diamond-tipped pen
- Shaking water bath at 35°C
- Transilluminating stereo dissecting microscope
- Fetal bovine serum (PAA Laboratories)

Protocol

1. Dissect out the muscles immediately after killing the animal, taking care to handle the muscles by their tendons to minimize damage to the myofibers.
2. Place the muscles into 0.2% (w/v) type I collagenase in DMEM and incubate in a shaking water bath at 35°C for 60–90 min. Longer digestion times are required for larger muscles; as a general guide, an EDL muscle taken from a 6- to 8-week-old mouse requires 60 min.
3. Prepare Petri dishes to take myofibers by rinsing with horse serum to prevent myofibers from adhering to the surface of the dish, and add 20 mL DMEM.
4. After digestion, transfer the muscle to a Petri dish containing DMEM.

5. Under a transilluminating stereo dissecting microscope, liberate single muscle fibers by repeatedly triturating the muscle with a modified Pasteur pipette. Pipettes are rinsed in 10% horse serum in DMEM to prevent myofibers from adhering to the glass.

6. As myofibers are liberated, the diameter of the muscle decreases and Pasteur pipettes of progressively narrower aperture are used.

7. Once 20–30 myofibers have been liberated, the muscle bulk is transferred to a fresh Petri dish and more myofibers separated until sufficient numbers have been dissociated.

8. Larger muscles may require a second digestion step to free myofibers located in the core; after superficial myofibers have been dissociated, return the remaining muscle bulk to the collagenase for further digestion.

9. Plate myofibers on Matrigel-coated Primaria® tissue culture plates.

   (a) Place individual myofibers in the center of the well and allow them to attach to the Matrigel™. To ensure purity of the preparation, it is important only to plate undamaged fibers, free of regions of hypercontraction that obscure contaminating adherent cells often associated with fragments of microvessel.

   (b) Add plating medium slowly to the well, taking care not to disturb the plated myofiber.

10. Place plates in an incubator at 37°C and 5% CO_2.

Proliferation:

Satellite cells begin to migrate from the parent myofiber after 12–24 h in culture. By 3–4 days each myofiber will be surrounded by approximately 50–300 cells, the precise number of cells depends on a number of factors including the muscle from which the myofiber originated [Rosenblatt et al., 1996] and the age of the donor mouse [Bockhold et al., 1998].

11. When a number of cells have migrated from the myofiber, remove the myofiber from the culture with a Pasteur pipette pulled to a fine tip in a flame.

12. Continue culture of the remaining cells in original well.

13. Subculture (*see* Protocol 13.2), if required, and plate out in proliferation medium.

Differentiation. Proliferating myogenic cells can be induced to differentiate into myotubes by reducing the serum content of the medium to 5% horse serum in DMEM.

The myogenicity of the cells migrating from myofibers can be tested by immunostaining for desmin, a muscle-specific intermediate filament protein. In our hands all cells migrating from 98–99% of the single myofibers were found to express desmin.

Single myofibers can be prepared from the muscles of the $H-2K^b$-tsA58 transgenic mouse to obtain myogenic cells capable of many rounds of division *in vitro* [Morgan et al., 1994]. The $H-2K^b$-tsA58 transgenic mouse harbors the tsA58 temperature-sensitive mutant of the simian virus 40 (SV40) large T antigen under the control of an inducible promoter ($H-2K^b$) [Jat et al., 1991]. The tsA58 protein is active at 33°C, and transcription can be stimulated by addition of mouse γ-interferon. Myogenic cells derived from these mice proliferate in the presence of mouse γ-interferon (20 units/mL) at 33°C. When switched to 37–39°C in the absence of γ-interferon, either in culture or on transplantation *in vivo*, they differentiate to form myofibers in a rapid and synchronous manner.

23.3.4 Cartilage

Protocol 23.16 for culturing cells from articular cartilage was contributed by François Lemare, Sophie Thenet, and Sylvie Demignot of Laboratoire de Pharmacologie Cellulaire de L'Ecole Pratique des Hautes Etudes, Centre de Recherches Biomédicales des Cordeliers, 15 rue de L'Ecole de Médecine, 75006 Paris, France.

Chondrocytes are highly specialized cells of mesenchymal origin that are responsible for synthesis, maintenance, and degradation of the cartilage matrix. A great deal of research in the field of rheumatology has been focused on understanding the mechanisms that induce metabolic changes in articular chondrocytes during osteoarthritis and rheumatoid arthritis. Articular chondrocytes in culture are a very useful tool, but, cultured in a monolayer, they rapidly divide, become fibroblastic, and lose their biochemical characteristics [Benya et al., 1977; von der Mark, 1986]. As early as the first passage, there is a gradual shift from the synthesis of type II collagen to types I and III collagens and from aggrecan to low-molecular-weight proteoglycans. In parallel with the phenotypic modulation, the metabolic response to interleukin-1β is quantitatively affected by subculture [Blanco et al., 1995; Lemare et al., 1998].

Among the various methods explored for maintaining the phenotype of chondrocytes [for a review, *see* Adolphe & Benya, 1992], culture in alginate beads appears the most promising, because it has been shown that this culture system leads to the formation of a matrix similar to that of native articular cartilage [Häuselmann et al., 1996; Petit et al., 1996]. The system maintains the expression of the differentiated phenotype and is also able to restore it in dedifferentiated chondrocytes [Bonaventure et al., 1994; Lemare et al., 1998]. Another advantage over other three-dimensional methods is that cells can easily be recovered after the culture is completed, allowing protein and gene expression studies [Lemare et al., 1998].

PROTOCOL 23.16. CHONDROCYTES IN ALGINATE BEADS

Outline

Culture in alginate beads is based on the gelation in calcium chloride of a suspension of chondrocytes in an alginate solution.

Materials

Sterile:

- Cartilage dissection medium (CDM): Ham's F12 medium (Invitrogen), supplemented with netilmycin (20 µg/mL), vancomycin (10 µg/mL), and ceftazidim (30 µg/mL) (all from Sigma), is used for dissection of the joints and removal of cartilage, as well as for enzymatic digestion.
- Growth medium: DMEM (glucose, 1 g/L)/Ham's F12 (1:1, v/v), gentamicin, 4 µg/mL, supplemented with 10% FCS.

Note. Different batches of serum must be compared for their ability to support chondrocyte growth as well as the expression of the differentiated phenotype (assessed, for example, by immunolabeling of type II collagen). When one lot is selected, several months' supply should be purchased to avoid frequently changing batches of serum.

- Enzyme solutions for dissociation of cartilage:
  - **(a)** Trypsin, 0.05%, porcine pancreatic (Sigma) in Ham's F12
  - **(b)** Collagenase, 0.3%, type 1 (Worthington, CLS1) in Ham's F12
  - **(c)** Collagenase 0.06% in Ham's F12 + 10% FCS
  All solutions are sterilized with 0.22-µm filters.

Note. New batches of collagenase should be tested for efficient cartilage dissociation by comparison with the previous batch.

- Trypsin-EDTA solution (for subculture): 0.1% w/v porcine pancreatic trypsin in PBSA with 5 mM (0.02% w/v) of EDTA
- Sodium alginate solution: 1.25% w/v sodium alginate (Fluka) in 20 mM HEPES; 0.15 M NaCl

Preparation:

(a) Using a heated stirrer, first dissolve the HEPES in 0.15 M NaCl and heat to ~60°C.

(b) Add the sodium alginate and continue stirring until it dissolves completely (>2 h).

(c) Carefully adjust the solution to pH 7.4 with 1 M NaOH; do not exceed pH 7.4, as the addition of acid (HCl or H_2SO_4) results in irreversible precipitation.

(d) Dispense the solution into aliquots and sterilize it by autoclaving at 121°C for 10 min.

Alternative preparation: Sodium alginate dissolution in HEPES/NaCl can be achieved rapidly (~10 min) in a microwave oven, but care must be taken to avoid caramelization.

- Gelation solution: $CaCl_2$, 102 mM, HEPES, 5 mM, pH 7.4
- Solubilization solution: EDTA, 50 mM, HEPES, 10 mM, pH 7.4
- Sterile magnet

Protocol

Dissection:

1. Prepare cultures from knee, shoulder, and hip joints. Fetal or young donors are preferable to adults, as they provide higher quantities of cells and take longer to senesce. One-month-old Fauve de Bourgogne or New Zealand rabbits are ideal. However, hip joints obtained from surgical resection from human adults are also suitable sources of chondrocytes.

2. Wash tissue pieces thoroughly with CDM before dissection. Dissection should begin without delay. Remove skin, muscle, and tendons from joints. Carefully take cartilage fragments from articulations that are free of connective tissue.

Dissociation:

To prevent the cartilage drying, carry out all dissociation steps in 10-cm non-TC-treated dishes containing CDM. The quantities of enzyme solutions indicated in the rest of the protocol are suitable for tibiofemoral and scapulohumeral articulations from one rabbit, but they must be adapted to the quantity of cartilage to be dissociated.

3. Using crossed scalpels (*see* Protocol 12.4), mince cartilage slices into 1-mm^3 pieces.

4. Transfer cartilage fragments into a 30-mL flat-bottomed vial.

5. Add 10 mL of 0.05% trypsin solution, and incubate the fragments with moderate magnetic agitation for 25 min in a sealed vial at room temperature.

6. Remove the trypsin solution after the fragments sediment.

7. Add 10 mL of 0.3% collagenase solution, and incubate the culture with moderate magnetic agitation for 30 min in a sealed vial at room temperature.

8. Remove collagenase solution after the fragments sediment.

9. Add 10 mL of 0.06% collagenase solution.

10. Transfer the suspension into a 100-mm bacterial dish. Rinse the vial with another 10 mL of

0.06% collagenase solution, transfer the rinse into the dish, and incubate the cartilage fragments overnight at 37°C in an incubator with 5% CO_2.

11. Transfer the cell suspension into a 50-mL centrifuge tube and mix on a vortex mixer for a few seconds.

12. Remove residual material left after digestion by passing the digested material through a 70-μm nylon cell strainer (B-D Biosciences; *see* Fig. 12.10).

13. Centrifuge the filtrate at 400 *g* for 10 min.

14. Resuspend the cell pellet in 20 mL of CDM, and count the cells with a hemocytometer.

15. Centrifuge the cells at 400 *g* for 10 min.

Entrapment:

16. Resuspend the cell pellet in 1 mL of sodium alginate solution, and then dilute the suspension progressively in more sodium alginate solution, until a cellular density of 2×10^6 cells/mL is reached. This progressive dilution is necessary to obtain a homogeneous cell suspension in alginate.

17. Express the cell suspension in drops through a 21G needle into the gelation solution with moderate magnetic stirring, and allow the alginate to polymerize for 10 min to form beads.

18. Wash the beads 3 times in 5 vol NaCl, 0.15 M.

19. Distribute the beads, 5 mL (1×10^7 cells), into 75-cm^2 flasks containing 20 mL of growth medium.

20. Incubate the culture at 37°C in a humidified atmosphere of 5% CO_2 and 95% air.

Recovery:

21. Discard the medium.

22. Add 2 vol of solubilization solution to the beads.

23. Incubate the culture 15 min at 37°C.

24. Centrifuge the cells at 400 *g* for 10 min.

25. Resuspend the cell pellet in growth medium containing 0.06% collagenase.

26. Incubate the cells for 30 min at 37°C in a humidified atmosphere of 5% CO_2 and 95% air.

27. Centrifuge the cells at 400 *g* for 10 min.

28. Resuspend the cell pellet in serum-free DMEM/Ham's F12 medium, and count the cells with a hemocytometer.

29. Centrifuge the cells at 400 *g* for 10 min.

30. Repeat Steps 28 and 29, without counting the cells.

Analysis of chondrocyte phenotype

Collagen. Type II collagen represents 85–90% of the total collagen synthesized by chondrocytes in cartilage or primary culture. Type II collagen is highly specific to chondrocytes and is considered as their main differentiation marker. The expression of type II collagen can be analyzed by indirect immunofluorescence using specific antibodies. These must be very carefully checked, however, to make sure that there is no cross-reactivity with type I collagen, which is expressed by dedifferentiated cells. The immunocytochemical study of collagen expression in chondrocytes cultured in alginate beads necessitates sectioning the beads as described in Lemare et al. [1998].

For a more precise analysis of the collagen phenotype, SDS-PAGE and two-dimensional CNBr peptide maps of purified collagen after [³H]proline incorporation must be performed [Benya, 1981].

Concerning major collagens, SDS-PAGE analysis can reveal the presence of type III collagen and of the $\alpha2(I)$ chain of collagen I. However, it cannot separate the $\alpha1$ chain of type I and type II (i.e., it cannot make a distinction between type I and type II trimer collagen), which is why two-dimensional CNBr peptide mapping is necessary to assert type II collagen expression.

The expression of these genetically distinct collagens can also be analyzed at the transcriptional level by Northern blot using conventional transfer and hybridization techniques [Lemare et al., 1998] or by real-time PCR [Murphy & Polak, 2004].

Proteoglycans. Alcian Blue staining at acidic pH is widely used as an indicator of the presence of sulfated aggrecan [Lemare et al., 1998]. As with immunocytochemical studies, this staining necessitates sectioning the beads. A more precise analysis of the different types of proteoglycans synthesized can be performed after $^{35}SO_4$ incorporation [Verbruggen et al., 1990]. As for collagen genes, the analysis of aggrecan and link protein gene expression can be performed by Northern blot [Lemare et al., 1998] or real-time PCR [Murphy & Polak, 2004].

Variations and applications. Several studies have shown that chondrocytes cultured in alginate beads represent an attractive alternative chondrocyte model [Grandolfo et al., 1993; Häuselmann et al., 1994, 1996; Petit et al., 1996]. The main advantage of this technique over other three-dimensional methods is that cells can easily be recovered after culture, allowing protein and gene expression studies [Lemare et al., 1998].

The number of available cells often becomes a limiting factor in studies of chondrocyte metabolism, as articular cartilage is difficult to obtain in large quantities and has a low cellularity. In this case, the number of chondrocytes available can be amplified by first culturing the cells in a monolayer (2 passages) and then restoring the differentiated properties of chondrocytes by cultivating the cells in alginate beads for two weeks [Bonaventure et al., 1994; Lemare et al., 1998]. In parallel with this phenotypic restoration, the metabolic response to interleukin—1β of these cells was restored at a quantitative level similar to that of chondrocytes in a monolayer primary culture [Lemare et al., 1998]. This makes the culture in alginate beads a relevant model for the study of chondrocyte biology. Finally, scaffolds filled with chondrocytes embedded in alginate gel have been proposed for cartilage tissue engineering and graft [Cancedda et al., 2003; Marijnissen et al., 2002; Sah et al., 2005].

23.3.5 Bone

Although bone is mechanically difficult to handle, thin slices treated with EDTA and digested in collagenase [Bard et al., 1972] give rise to cultures of osteoblasts that have some functional characteristics of the tissue. Antiserum against collagen has been used to prevent fibroblastic overgrowth without inhibiting the osteoblasts [Duksin et al., 1975]. Propagated lines have been obtained from osteosarcoma [Smith et al., 1976; Weichselbaum et al., 1976], but not from normal osteoblasts. Mesenchymal stem cells (*see* Section 23.7.2) can be cultured and induced to form osteocytes, suitable for tissue engineering [Lennon & Caplan, 2005; Hofmann et al., 2005]

Protocol 23.17 for the culture of bone cells was contributed by Edith Schwartz, Departments of Orthopedic Surgery and Physiology, Tufts University School of Medicine, 136 Harrison Avenue, Boston, MA 02111.

Bone culture suffers from the inherent problem that the hard nature of the tissue makes manipulation difficult. However, conventional primary explant culture or digestion in collagenase and trypsin releases cells that may be passaged in the usual way.

PROTOCOL 23.17. OSTEOBLASTS

Outline
For explant culture, small fragments of tissue are allowed to adhere to the culture flask by incubation in a minimal amount of medium (*see* Protocol 12.4). The adherent explants are then flooded and the outgrowth is monitored. For disaggregated cell culture, trabecular bone is dissected down to 2- to 5-mm pieces and digested in collagenase and trypsin. Suspended cells are seeded into flasks in F12 medium.

Materials
Sterile:
❑ Ham's F12 medium with 12% fetal bovine serum, 2.3 mM Mg^{2+}, 100 U/mL of penicillin, and 100 μg/mL of streptomycin SO_4
❑ Trypsin solution to be used for cell passage: Dissolve 125 mg of trypsin (Sigma type XI) in 50 mL of Ca^{2+}, Mg^{2+}-free Tyrode's solution.

Adjust the resulting solution to pH 7.0 and filter sterilize it. Place aliquots of 5 and 10 mL in Pyrex tubes and store at $-20°C$.
❑ Digestion solution for the isolation of osteoblasts: Solution A: 8.0 g of NaCl, 0.2 g of KCl, 0.05 g of $NaH_2PO_4 \cdot H_2O$ in 100 mL of UPW
❑ Solution B (Collagenase-trypsin solution): Dissolve 137 mg of collagenase (type I, Worthington Biochemicals) and 50 mg of trypsin (Sigma, type III) in 10 mL of solution A. Adjust the combined solution to pH 7.2, and then bring the amount to 100 mL with UPW. Filter sterilize and distribute the solution into 10-mL aliquots that are stored at $-20°C$.
❑ Petri dishes, 9 cm
❑ Culture flasks 25 or 75 cm^2
❑ Scalpels
❑ Forceps

Protocol A: Explant Cultures
1. Obtain bone specimens from the operating room.
2. Rinse the tissue several times at room temperature with sterile saline.
3. If the bone cannot be used immediately, cover the specimen with sterile Ham's F12 medium containing 12% fetal calf serum, penicillin, and streptomycin sulfate. The bone may be stored overnight at 4°C.
4. The next morning, rinse the tissue with Tyrode's solution containing penicillin (100 U/mL) and streptomycin sulfate (100 μg/mL).
5. Place the bone in a Petri dish, and with the use of scalpel and forceps, remove the trabeculae and place them in a second Petri dish.
6. Add 10 mL of Tyrode's solution over the excised trabeculae. Rinse the trabeculae several times with Tyrode's solution until blood and fat cells are removed.
7. To initiate explant cultures, prepare a 25-cm^2 flask by preincubating it with 2 mL of complete medium for 20 min to equilibrate the medium with the gas phase. Adjust the pH as necessary with CO_2, 4.5% $NaHCO_3$, or HCl.
8. Cut the trabeculae into fragments of 1–3 mm.
9. Remove the preincubation medium from the flask, and add 2.5 mL of fresh Ham's F12 medium to the flask.
10. Transfer between 25 and 40 fragments of trabeculae to the flask.
11. With the flask in an upright position, slide the explant pieces along the base of the flask with the aid of an inoculating loop, and distribute the explant pieces evenly.

12. Permit the flask to remain upright for 15 min at 37°C.

13. Slowly restore the flask into a normal horizontal position. The explant pieces will stick to the bottom of the flask.

14. Leave the flask in the horizontal position at 37°C for 5 to 7 days. After this period, check for outgrowth, and replace the medium in the flask with fresh medium. To prevent the explant pieces from becoming detached, lift the flask slowly into the vertical position before carrying it from the incubator to the hood or to the microscope.

15. To maintain cultures, change the medium twice per week.

16. When confluence is reached, the explant pieces are removed, the cell layer is trypsinized, and the cells are isolated by centrifugation and seeded into flasks or wells.

Protocol B: Monolayer Cultures from Disaggregated Cells

1. Wash trabecular bone specimens repeatedly with Tyrode's solution to remove the fat and blood cells. The trabeculae are excised with scalpel and forceps under sterile conditions.

2. After collecting as much bone as possible, wash the remaining blood and fat cells away by rinsing the specimens three times with Tyrode's solution and cut into 2- to 5-mm fragments.

3. Wash the cut trabeculae with F12 medium containing fetal calf serum.

4. Place the pieces of bone in a small sterile bottle with a magnetic stirrer, and add 4 mL of digestion solution. (This amount should cover the bone specimens.)

5. Stir the solution containing bone fragments at room temperature for 45 min.

6. Remove the suspension of released cells and discard it, because these cells are most likely to contain fibroblasts.

7. Add a second aliquot of 4 mL of digestion solution to the bone fragments, and stir the mixture at room temperature for 30 min.

8. Collect the digestion solution from bone fragments, and centrifuge it for 2 min at 580 g at room temperature.

9. After removing the supernate, suspend the cells in 4 mL of Ham's F12 medium with 20% fetal calf serum, and count the cells.

10. Centrifuge the suspension at 580 g for 10 min, and resuspend the cells in 4 mL of complete medium. This suspension will become the inoculum.

11. Preincubate 75-cm^2 flasks for 20 min with 8 mL of complete F12 medium to equilibrate with the gas phase.

12. Remove the preincubation medium and add 2 mL of complete F12 medium.

13. Add 4 mL of medium containing the cell suspension. The inoculum should contain 6000–10,000 cells per cm^2 of surface area.

14. Finally, add another 6 mL of Ham's F12 medium, to give a total volume of 12 mL.

15. In the interim, add an additional 4 mL of digestion solution to the remaining pieces of bone, and repeat the digestion for 30 min. The released cells are harvested, and, if necessary, the digestion step is repeated several more times. With large amounts of bone, the digestion period can be increased to 1–3 h. Cell counts are performed after each digestion period, and the released cells are used to inoculate a different flask.

Passage of Cells in Culture:

1. Remove the pieces of explant.

2. Remove the medium and rinse the cell layer with D-PBSA, 0.2 mL/cm^2.

3. Add trypsin to the flask, 0.1 mL/cm^2, and incubate at 37°C until the cells have detached and separated from one another. Monitor cell detachment and separation on the microscope. In general, a 10-min incubation is sufficient.

4. Transfer the released cells to a centrifuge tube with an equal volume of Ham's F12 medium with 20% fetal calf serum (FCS).

5. Centrifuge the cells at 600 g for 5 min.

6. Discard the supernate, and resuspend the cells in complete medium by gentle, repeated pipetting.

7. Set one aliquot aside for the determination of cell concentration and another for DNA determination (see Protocol 21.3).

8. Inoculate the remaining cells into culture flasks or wells that have previously been equilibrated with medium. The cells should reattach within 24 h.

23.3.6 Endothelium

Much interest has been generated in endothelial cell culture because of the potential involvement of endothelial cells in vascular disease, the repair of blood vessels, and angiogenesis in cancer. Folkman and Haudenschild [1980] described the development of three-dimensional structures resembling capillary blood vessels derived from pure endothelial lines *in vitro*. Growth factors, including angiogenesis factor derived from Walker 256 cells *in vitro*, play an important part in maintaining cell proliferation and survival, so that secondary structures can be formed. Endothelial cultures are good models for contact inhibition and density limitation of growth, as cell proliferation is strongly inhibited after confluence is reached [Haudenschild et al., 1976]. Endothelial cell cultures are also available commercially (*see* Appendix II).

The following introduction and Protocol 23.18 for the culture of large-vessel endothelial cells was contributed by Dr. Charlotte Lawson, Royal Veterinary College, Royal College Street, London NW1 0UT, United Kingdom.

The inner surface of all blood vessels is composed of a single cell layer of endothelial cells. The endothelium provides a smooth anticoagulant surface, contributes to blood vessel contractility, and acts as a permeable barrier allowing the transport of nutrients and gases as well as influencing emigration of immune cells from the blood. Endothelial cells can be isolated from large blood vessels including human umbilical veins [Jaffe et al., 1973] and bovine [Booyse et al., 1975] or porcine [Bravery et al., 1995] aorta. Microvascular endothelial cells can be isolated from bovine adrenal gland [Folkman et al., 1979], rat brain [Bowman et al., 1981], human skin [Davison et al., 1983], human adipose tissue [Kern et al., 1983], as well as human [McDouall et al., 1996] and murine [Marelli-Berg et al., 2000; Lidington et al., 2002] heart.

PROTOCOL 23.18. ISOLATION AND CULTURE OF VASCULAR ENDOTHELIAL CELLS

Outline
Incubation of the blood vessel lumen with collagenase releases endothelial cells, which can be cultured for several population doublings on a gelatin-coated substrate in a suitable growth medium. Immunomagnetic separation (*see* Section 15.3.2; Figs. 15.6, 15.7) is used to isolate endothelial cells from disaggregated heart tissue.

A. Isolation of Endothelial Cells from Human Umbilical Vein (HUVECs)
[Adapted from the method of Jaffe et al., 1973]

Common Materials for All Endothelial Protocols
Sterile or aseptically prepared:
❏ D-PBSA on ice
❏ HBSS: Hanks' balanced salt solution
❏ HBSS/PSG containing 300 U/mL penicillin, 300 µg/mL streptomycin (Sigma), and 50 µg/mL gentamycin (Invitrogen)
❏ Trypsin, 500 µg/mL, EDTA, 5 mM (0.2%) (Sigma cat no. T4174)
❏ Culture flasks, 25 cm², coated with 0.5% gelatin
  (a) Dilute 2% solution of type B gelatin from bovine skin (Sigma, G1393) fourfold with D-PBSA

(b) Incubate in flask for 20 min at 37°C.
(c) Remove the gelatin just before addition of cells.
❏ Scalpel and No. 22 blade
❏ Needle
❏ Syringes, 20 mL
❏ Crocodile clips
❏ Artery clamps
❏ Forceps
❏ Sharp scissors
Nonsterile:
❏ Methanol, 70%
❏ Paper tissues
❏ Aluminum foil
❏ Clingfilm (or Saran Wrap)
❏ Cork board
❏ Extra strong thread

Materials for HUVECs
❏ Human umbilical cords in HBSS/PSG
❏ HUVEC growth medium: M199 medium with Hanks' salts, 2 mM L-glutamine, and 25 mM HEPES supplemented with 20% fetal bovine serum (FBS), 150 units/mL penicillin/streptomycin and endothelial cell growth supplement (ECGS) from bovine brain (Roche cat no 1033484, 75 mg; use at 20 µg/mL with heparin at 50 µg/mL or Sigma cat no E2759, 15 mg; use at 30 µg/mL with heparin at 10 µg/mL)
❏ Collagenase H: from *Clostridium histolyticum* (Roche cat no 1074059; use at 0.5 mg/mL in M199. Filter sterilize) .... 25 mL
❏ Newborn calf serum (NBS), cold .... 5 mL
❏ Adaptors (Luer female to Luer male adaptor Portex cat no 700/180/700) 2

Protocol
Isolation of endothelial cells:
1. Cover cork board with aluminum foil.
2. Place paper tissues on foil and soak with 70% methanol soaked tissues.
3. Squeeze down the length of the umbilical cord to expel any blood.
4. Cut 2 cm or more off the one end of the cord including any clamp marks.
5. Insert Luer adaptor all the way into the vein (Fig. 23.1).
6. Pin cord to cork board with needle near the vein but making sure not to pierce through to the lumen.
7. Using scalpel and forceps, strip back the arteries ~2 cm.
8. Secure the connector in place with thread, using two very tight opposing knots

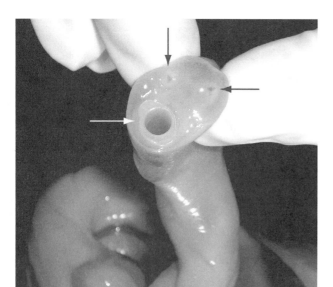

Fig. 23.1. View of Blood Vessels in Human Umbilical Cord. Black arrows, umbilical arteries. White arrow, umbilical vein with Luer adaptor inserted.

9. Clamping the other end of the cord shut, push up to 25 mL D-PBSA into the cord with a 20-mL syringe attached to the connector. With the vein distended under the pressure of the D-PBSA, check for any holes in the cord and clamp any holes with crocodile clips by nipping the exterior of the cord without closing the vein.

10. Empty cord of D-PBSA into waste pot.

11. Cut off ~2 cm from other end of cord and insert and secure connector as in Step 8.

12. Using another 20-mL syringe, push approximately 25 mL collagenase into the cord until it becomes distended.

13. Wrap the cord in methanol-sprayed Clingfilm and leave at 37°C for 8 min.

14. Squeeze collagenase out of cord and withdraw back into syringe(s).

15. Remove syringes and empty collagenase into tube containing 5 mL newborn calf serum.

16. Replace syringe on one end of cord and use the other syringe to push 25 mL D-PBSA into the cord.

17. Squeeze or tap cord along its length for a few moments, then push D-PBSA back into syringes.

18. Add to contents of tube from Step 15.

19. Centrifuge tube for 10 min, 210 *g*, 4°C.

20. Resuspend pellet in 5 mL HUVEC growth medium.

21. Remove excess gelatin from flask and add seed cells from Step 20.

22. The following day, remove the medium from the flask and discard.

23. Wash twice with D-PBSA and replace with fresh HUVEC growth medium.

Maintenance and culture:

24. Remove half the medium and replace with fresh three times per week.

25. Passage, 1:2 split, trypsin/EDTA into gelatin coated flasks approximately once per week.

26. Do not use beyond 6th passage.

B. Porcine Aortic Endothelial Cell (PAEC) Isolation

[Method adapted from the method of Bravery et al., 1995]

Materials for PAECs

☐ HBSS/PSGA: HBSS containing 300 units/mL penicillin/streptomycin, 50 µg/mL gentamycin, 2.5 µg/mL amphotericin B (GIBCO)

☐ Fresh porcine aorta (collected aseptically if possible), transported in HBSS/PSGA

☐ PAEC growth medium: M199 medium with Hanks' salts, 2 mM L-glutamine, and 25 mM HEPES, supplemented with 150 U/mL penicillin/streptomycin, 50 µg/mL gentamycin, 2.5 µg/mL amphotericin B, 10% FBS, 10% NBBS

☐ Collagenase H (as in HUVEC Materials) .... 50 mL

☐ FBS in 50-mL tube .... 5 mL

Protocol

1. Cover cork board with aluminum foil.

2. Lay aorta on methanol-sprayed tissue and pat dry.

3. On foil-covered board trim off excess tissue, leaving only branches—do not cut too short.

4. Tie off branches with thread and clamp thin end of aorta.

5. Pour in D-PBSA to fill; clamp top.

6. Clamp any leaks with sterile crocodile clips.

7. Discard D-PBSA.

8. Fill aorta with warm collagenase, reclamp, wrap in methanol-sprayed clingfilm and incubate 8 min at 37°C.

9. Gently massage aorta to loosen cells.

10. Pour collagenase from aorta into 50-mL tube containing 5 mL FBS

11. Fill aorta with ice-cold D-PBSA, clamp, massage, add to tube of cells in Step 10, and repeat this step.

12. Centrifuge cells at 500 *g* for 6 min at 4°C.

13. Resuspend cell pellet in 5 mL medium, place in 25-cm² flask and incubate overnight at 37°C.

14. The following day remove medium from flask and discard.

15. Wash twice with D-PBSA and replace with fresh medium.

Maintenance:

16. Replace medium three times per week; if cells are less than 50% confluent remove and replace only half the medium; if cells are 50–80% confluent remove and replace all the medium.
17. When cells are 80% confluent passage with 1:3 split ratio using trypsin/EDTA.
18. Do not use beyond 7th passage.

C. Murine Cardiac Endothelial Cell (MCEC) Isolation

[Adapted from the method of Marelli-Berg et al., 2000]

Materials for MCECs

❑ Mouse hearts, collected into HBSS/PSG (*see* Protocol 23.16A) containing 300 units/mL penicillin/streptomycin and 50 μg/mL gentamycin, 5
❑ MCEC growth medium: DMEM with 25 mM glucose and 4 mM L-glutamine, supplemented with 10% FBS, 150 U/mL penicillin, 150 μg/mL streptomycin, and endothelial cell growth supplement (ECGS) from bovine brain
❑ Medium for nonendothelial cells: DMEM (1000 mg/L glucose) containing 15% FCS
❑ HBSS/PS: HBSS with 150 U/mL penicillin, 150 μg/mL streptomycin
❑ HBSS/FB: HBSS with 1% FBS
❑ CMF: HBSS without Ca^{2+} and Mg^{2+} with 150 U/mL penicillin, 150 μg/mL streptomycin
❑ Trypsin/EDTA (e.g., Sigma catalog)
❑ FBS
❑ Collagenase A: from *Clostridium histolyticum* (Roche cat no 103586); use at 1 mg/mL in HBSS. Make up fresh and filter sterilize .... 10 mL
❑ Cell strainers, 70 μm mesh (B-D Biosciences)
❑ Petri dishes, 9 cm 2
❑ Irrelevant control antibody, e.g., OKT4 anti human CD4 (DAKO) .... 5 μg
❑ Rat-anti-mouse endoglin (CD105) antibody; clone MY7/18 (Pharmingen, B-D Biosciences) ... 5 μg
❑ Goat anti-rat Ig MicroBeads (Miltenyi cat no 130-048-501)
❑ MidiMacs separation unit (Miltenyi)
❑ LS MACS column (Miltenyi 130-042-401)

Protocol

1. Place hearts in Petri dish containing about ~5 mL HBSS/PS.
2. Remove aorta, pulmonary artery, and other appendages from heart.
3. Cut heart in half and remove any large blood clots and connective tissue.
4. Place hearts in fresh Petri dish with CMF, wash and replace with fresh CMF.
5. Dice up material in the dish with scalpels.
6. Wash with CMF by resuspension and settling.
7. Scrape and pick up tissue on side of scalpel blade and transfer into universal containing 10 mL warmed freshly prepared collagenase A at 1 mg/mL.
8. Incubate for 30 min at 37°C with agitation
9. Add CMF.
10. Disperse cells by shaking up tissue and allowing to settle. Aspirate off cell suspension and transfer to a fresh tube.
11. Repeat Step 10 several times.
12. Spin cell suspension at 300 *g* for 8 min.
13. Carefully remove supernate from the loose cell pellet.
14. Resuspend cell pellet in 5 mL trypsin/EDTA, try to break up any clumps of cells by pipetting, and incubate for 10 min at 37°C.
15. Add 10 mL HBSS/FB containing 1% FS to disaggregated cells digest and filter through 70 μm cell strainer.
16. Wash tissue in cell strainer with 20 mL cold HBSS/FB.
17. Centrifuge cells in filtrate for 8 min at 300 *g*.
18. Aspirate supernate carefully, discard, and retain the cell pellet.
19. Add ~5 μg irrelevant antibody (e.g., OKT4/OKT8) to the cell pellet, flick the tube to disperse the pellet, and incubate on ice for 20 min to block Fc receptors.
20. Add 5 μg rat anti-mouse endoglin to cell pellet, flick the tube (clone MY7/18), and incubate at 4°C for 20 min.
21. Add 30 mL HBSS and centrifuge at 300 *g* for 8 min at 4°C.
22. Add 50 μL goat anti-rat beads in 200 μL HBSS.
23. Incubate for 15 min at 4°C.
24. Add 30 mL HBSS to wash and centrifuge at 300 *g* for 8 min at 4°C.
25. Place MidiMacs column on magnet and wash with 500 μL HBSS/1% FCS according to manufacturer's recommendations.
26. Add 3 mL HBSS/1% FBCS to cell pellet and add to column.
27. Wash column 3× with 1 mL HBSS/FB followed by a final wash in 500 μL growth medium.
28. Remove column and place in 15-mL tube, add 4 mL MCEC growth medium, and put plunger in to collect bead bound cells. Move cell suspension to gelatin-coated 25-cm² flask

29. To keep nonendothelial cell population spin down washes from column and plate in 25-cm² flask in DMEM (1000 mg/L glucose) containing 15% FBS.

Maintenance:

30. Replace medium twice per week; if cells are less than 50% confluent remove and replace only half the medium; if cells are 50–80% confluent remove and replace all of the medium.

31. When cells are 80% confluent, passage with a split ratio of 1:3 with trypsin/EDTA.

32. Do not use beyond 3rd passage.

D. Human Cardiac Endothelial Cell (hCEC) Isolation

It is possible to isolate CEC from human left ventricle by following the method of McDouall et al. [1996].

Materials for HCEC

☐ Magnetic beads coated with antibodies to human MHC class II or human CD31 (Dynal)

☐ HCEC growth medium: M199 with 4 mM L-glutamine, supplemented with 10% FBS, 10% human male AB serum (Sigma), 150 U/mL penicillin, 150 μg/mL streptomycin, and ECGS as in Protocol A

☐ Collagenase II (Roche), 1 mg/mL

Protocol

1. Follow Protocol 23.18C.

2. Digest the tissue with collagenase II (Roche) at Step 7.

3. Use magnetic beads coated with antibodies to human MHC class II or human CD31 (Dynal) to select endothelial cells as at Step C20 and proceed as from Step C23.

4. Maintain hCEC in HCEC growth medium.

Identification of vascular endothelial cells. Vascular endothelial cells appear polygonal and have a characteristic "cobblestone" morphology when cultured under static conditions (Fig. 23.2). Fibroblast contamination should be easy to identify by their spindle-shaped morphology with a standard phase-contrast inverted microscope. Vascular endothelial cells can be identified by the production of factor VIII [Hoyer et al., 1973], angiotensin-converting enzyme [Del Vecchio & Smith, 1981], the uptake of acetylated low-density lipoprotein [Voyta et al., 1984], the presence of Weibel–Palade bodies [Weibel & Palade, 1964], and the expression of endothelium–specific cell surface antigens including PECAM-1 (CD31) [Parums et al., 1990; Albelda et al., 1990] and endoglin (CD105) [Gougos & Letarte, 1990]. Vascular endothelial cells will readily form tubes in Matrigel [McGuire & Orkin, 1987].

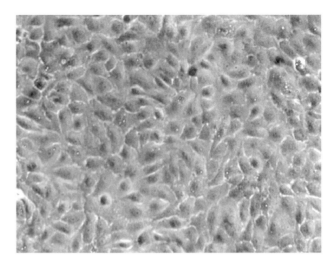

Fig. 23.2. Vascular Endothelial Cells in Culture. View of human aortic endothelial cell monolayer under phase contrast microscope. 10× magnification.

Variations. Large-vessel endothelial cells can be also isolated from human or bovine aorta, from muscular arteries including pulmonary, coronary, and internal mammary arteries, and from vein, with protocols similar to those outlined above. Microvascular endothelial cells can be isolated from skin [Davison et al., 1983], brain [Bowman et al., 1981], and lung [Magee et al., 1994] and from specialized vascular beds including liver [Shaw et al., 1984], kidney [Green et al., 1992], and lymph node [Ager, 1987].

23.4 NEUROECTODERMAL CELLS

23.4.1 Neurons

Nerve cells appear to be more fastidious in their choice of substrate than most other cells. They will not survive well on untreated glass or plastic, but will demonstrate neurite outgrowth on collagen and poly-D-lysine. Neurite outgrowth is encouraged by a polypeptide nerve growth factor [NGF; Levi-Montalcini, 1979] and other factors secreted by glial cells [Marchionni et al., 1993; Barnett, 2004; Miklic et al., 2004; Wu et al., 2004],. Cell proliferation has not been found in cultures of most neurons, even with cells from embryonic stages in which mitosis was apparent *in vivo*; however, recent studies with embryonal stem cells have shown that some neurons can be made to proliferate *in vitro* [Thomson et al., 1998] and recolonize *in vivo* [Snyder et al., 1992] and have even been recovered from human umbilical cord [Sanchez-Ramos, 2002; Newman et al., 2003].

Protocol 23.19 for the monolayer culture of cerebellar neurons was contributed by Bernt Engelsen and Rolf Bjerkvig, Department of Cell Biology and Anatomy, University of Bergen, Žrstadveien 19, N-5009 Bergen, Norway.

Cerebellar granule cells in culture provide a well-characterized neuronal cell population that is suited for morphological and biochemical studies [Drejer et al., 1983; Kingsbury et al., 1985]. The cells are obtained from the cerebella of 7- or 8-day-old rats, and nonneuronal cells are prevented from growing by the brief addition of cytosine arabinoside to the cultures.

PROTOCOL 23.19. CEREBELLAR GRANULE CELLS

Outline
The cerebella from four to eight neonatal rats are cut into small cubes and trypsinized for 15 min at 37°C in Hanks's BSS. The cell suspension is seeded in poly-L-lysine-coated culture wells or flasks.

Materials
Sterile:
- DMEM with
  (a) Glucose .... 30 mM
  (b) L-glutamine .... 2 mM
  (c) KCl 24.5 mM
  (d) Insulin (Sigma I-1882) .... 100 mU/L
  (e) p-Aminobenzoic acid
      (Sigma A-3659) .... 7 μM
  (f) Gentamycin .... 100 μg/mL
  (g) Fetal calf serum, heat
      inactivated 10%
- Hanks's BSS with 3 g/L BSA (HBSS)
- Poly-L-lysine, mol. wt. >300,000 (Sigma P-1524)
- Cytosine arabinoside (Cytostar, Upjohn; powder)
- Trypsin (type II): 0.025% in HBSS
- Silicone (Aquasil, Pierce 42799)
- Tissue culture Petri dishes, 3.5 cm
- Dissecting instruments: scalpels, scissors, and forceps
- Pasteur pipettes and 10-mL pipettes
- Test tubes, 12 and 50 mL
Nonsterile:
- Water bath
- Siliconization of Pasteur pipettes:
  (a) Dilute the Aquasil solution 0.1–1% in UPW.
  (b) Dip the pipettes into the solution or flush out the insides of the pipettes.
  (c) Air dry the pipettes for 24 h, or dry for several minutes at 100°C.
  (d) Sterilize the pipettes by dry heat.
- Poly-L-Lysine Treatment of Culture Dishes:
  (a) Dissolve the poly-L-lysine in UPW (10 mg/L).
  (b) Sterilize the mixture by filtration.
  (c) Add 1 mL of poly-L-lysine solution to each of the 35-mm Petri dishes.
  (d) Remove the poly-L-lysine solution after 10–15 min, and add 1–15 mL of culture medium.
  (e) Place the culture dishes in the incubator (minimum 2 h) until the cells are to be seeded.

Protocol
1. Dissect out the cerebella aseptically and place them in HBSS.
2. Mince the tissue with scalpels into small cubes approximately 0.5 mm³.
3. Transfer the minced tissue to test tubes (12 mL) and wash the tissue three times in HBSS. Allow the tissue to settle to the bottom of the tubes between each washing.
4. Add 10 mL of 0.025% trypsin (in HBSS) to the tissue, and incubate the tube in a water bath for 15 min at 37°C.
5. Transfer the trypsinized tissue to a 50-mL centrifuge tube, and add 20 mL of growth medium to stop the action of the trypsin.
6. Disaggregate the tissue by trituration through a siliconized Pasteur pipette, until a single-cell suspension is obtained.
7. Let the cell suspension stay in the centrifuge tube for 3–5 min, allowing small clumps of tissue to settle to the bottom of the tube. Remove these clumps with a Pasteur pipette.
8. Centrifuge the single-cell suspension at 200 g for 5 min, and aspirate off the supernate.
9. Resuspend the pellet in growth medium, and seed the cells at a concentration of 2.5–3.0 × 10⁶ cells/dish.
10. After 2–4 days (best results usually are obtained after 2 days), incubate the cultures with 5–10 μM cytosine arabinoside for 24 h.
11. Change to regular culture medium.

Characterization of neuronal cultures. Neurons can be identified immunologically by using neuron-specific enolase antibodies or by using tetanus toxin as a neuronal marker. Astrocyte contamination can be quantified by using glial fibrillary acidic protein as a marker.

Variations. A single-cell suspension can be obtained by mechanical sieving through nylon meshes of decreasing diameter (Protocol 12.9) or by sequential trypsinization (i.e., a 3- to 5-min trypsin treatment). Instead of HBSS, Puck's solution, Krebs, or other buffers with glucose can be used.

23.4.2 Glial Cells
Glial cells have been cultured from avian, rodent, and human brains. Human adult normal astroglial lines from brain lines express glial fibrillary acidic protein [GFAP; Burke et al., 1996], high-affinity γ-aminobutyric acid and glutamate uptake, and glutamine synthetase activity [Frame et al., 1980].

Oligodendrocytes do not readily survive subculture, but oligodendrocyte precursor cells from the optic nerve have been subcultured using PDGF and FGF2 as mitogens [Bögler et al., 1990]. Schwann cells have been subcultured using cholera toxin as a mitogen [Raff et al., 1978; Brockes et al., 1979], but now optimal growth has been found with the addition of glial growth factor and forskolin in medium containing 10% serum [Davis & Stroobant, 1990]. Cultured glial cells have had some success in transplant-mediated repair of spinal cord injury (*see* Section 25.3.8).

Protocol 23.20 for the purification of rat olfactory ensheathing cells from the olfactory bulb by fluorescence-activated cell sorting was contributed by Susan Barnett, Division of Clinical Neuroscience, University of Glasgow, CR UK Beatson Laboratories, Garscube Estate, Bearsden, G61 1BD, Glasgow, Scotland.

The ability to produce highly enriched populations of individual cell types is vital to the success of biological studies on the diverse elements that control cell growth, differentiation, and function. In this regard, the fluorescence-activated cell sorter (FACS) is a powerful tool for the enrichment of relatively small populations of cells from a heterogeneous cell population. Initially, the cell sorter was devised to characterize and sort cells of the hematopoietic system, for which a large number of cell-specific antibodies were available. With the increasing availability of antibodies that characterize both glia (Table 23.2) and neurons, it has now become possible to transfer this technique to the field of neurobiology [Abney et al., 1983; Williams et al., 1985; Trotter & Schachner, 1988; Barnett et al., 1993; Franceschini & Barnett, 1996].

The olfactory bulb is a tissue of considerable interest, because it is the only central nervous system (CNS) mammalian tissue that can support neuronal regeneration throughout life [Moulton, 1974; Graziadei & Monti Graziadei, 1979, 1980]. Once neuronal death occurs, either during the course of normal cell turnover or after injury, newly generated neurons are recruited from a pool of constantly proliferating putative stem cells in the olfactory

epithelium [Schwob, 2002]. These neurons are able to produce axons that reinnervate the olfactory bulb and reestablish olfaction. It is thought that a population of glial cells present in the olfactory nerve layer of the olfactory bulb, termed olfactory ensheathing cells (OECs), supports or encourages the growth of olfactory axons within the CNS [Doucette, 1984, 1990]. In fact, recent experiments have demonstrated that OECs are capable of remyelinating experimentally created demyelinated lesions in rat spinal cord [Franklin et al., 1996; Imaizumi et al., 1998] and that they support the regeneration of cut axons [Li et al., 1997, 2003a,b; Ramon-Cueto et al., 1998]. Many reviews have now been written regarding the use of these cells in CNS repair [Franklin & Barnett, 1997, 2000; Raisman, 2001; Wewetzer et al., 2002; Barnett & Riddell, 2004; Barnett & Chang, 2004].

To study the biology of OECs in detail, it is necessary to separate them from the other major cell types that make up the olfactory bulb: neurons, CNS glial cells, oligodendrocytes and astrocytes, endothelial cells, and meningeal cells [Barnett & Chang, 2004].

The rat olfactory bulb fulfills the criteria for successful cell separation by FACS in that (1) it readily forms a single-cell suspension after enzymatic treatment and (2) the mouse monoclonal antibody O4 [IgM; Sommer & Schachner, 1981], in conjunction with anti-galactocerebroside antibody [anti-GalC; Ranscht et al., 1982], can be used to distinguish OECs from other cell types that are present [Barnett et al., 1993; Franceschini & Barnett, 1996]. The O4 antibody recognizes sulfatide, seminolipid, and proligodendroblast antigen [Bansal et al., 1992] and labels many CNS and PNS glial cells, including Schwann cells, oligodendrocytes, and oligodendrocyte-type-2 astrocyte (O-2A) progenitor cells [Gard & Pfeiffer, 1990; Mirskey et al., 1990; Sommer & Schachner, 1981]. The olfactory bulb contains three cell populations that are labeled by means of the O4 antibody: OECs, oligodendrocytes, and O-2A progenitors [Barnett et al., 1993]. In contrast to OECs, oligodendrocytes and O-2A progenitors are also labeled by anti-GalC antibody, which can thereby be used as a negative selection marker. By using these two antibodies in conjunction, it has been found that 8–14% of the O4-positive cells are OECs and less than 1% of the total O4 positive cells are oligodendrocytes or O-2A progenitors. Moreover, the characteristic highly branched or bipolar morphology of oligodendrocytes and O-2A progenitors, respectively, can easily be distinguished from the flat fibroblast-like morphology of the OECs.

The O4-positive FACS-purified OECs can be classified into two types of cells—Schwann cell-like and astrocyte-like, which are distinguished by a panel of neural markers [Doucette, 1990; Franceschini & Barnett, 1996; Wewetzer et al., 2002]. These two cell types may have different functional roles *in vivo*. The growth factors that regulate their growth and differentiation are currently being investigated. In the presence of conditioned medium from type-1 astrocytes

TABLE 23.2. Characterization of Glial Cells by Immunostaining

Cell type	GFAP	Gal-C	O4	A2B5	NGF
O2-A	−	−	−	+	−
Type-1 astrocyte	+	−	−	−	−
Type-2 astrocyte	+	−	−	+	−
Oligodendrocyte	−	+	+[a]	+[a]	−
ONEC	+/−[c]	−	+[b]	−	+

[a]Mature oligodendrocytes lose A2B5 and become GalC[+] and O4[+] only.

[b]ONECs lose O4 in culture.

[c]GFAP may be fibrous or amorphous. ONECs gain GFAP expression in culture.

(ACM), OECs exhibit good viability and growth for at least 14 days. The method for growing astrocytes in culture can be found in Noble and Murrey [1984] and Wolswijk and Noble [1989]. Further experiments demonstrated that an isoform of neuregulin (neu–differentiation factor β) was identified as the factor in ACM [Pollock et al., 1999]. More recently, a combination of growth factors have been found to be potent OEC mitogens, which include a mix of ACM, FGF-2, forskolin, and neuregulin, in 5% fetal bovine serum [Alexander et al., 2002].

Protocol 23.20 describes a rapid method for purifying a minor population of glial cells from the olfactory bulb by using a cell surface antibody and FACS. These principles, however, can be applied to the purification of virtually any glial or neuronal population, provided that two main criteria are fulfilled: The researcher is able to produce a viable single-cell suspension, and suitable cell type-specific antibodies are available for identifying the cells of interest.

PROTOCOL 23.20. OLFACTORY ENSHEATHING CELLS

Materials
Sterile or Aseptically Prepared:
- Poly-L-lysine (Sigma), stock, 4 mg/mL, dilute 1:300 in sterile UPW for use at a final concentration of 13.3 µg/mL
- Collagenase type I (MP Biomedicals, cat no. 195109): stock, 2,000 U/mL in L-15
- Ca^{2+}- and Mg^{2+}-free Hanks's BSS (HBSS, Invitrogen)
- Leibowitz L-15 medium plus 25 µg/mL of gentamycin (MP Biomedicals)
- DMEM/FBS: DMEM (GIBCO) with 5% FBS
- DMEM/BS: DMEM (GIBCO) containing (all from Sigma) [Bottenstein & Sato, 1979]:
  - (a) Glucose .... 25 mM (4.5 g/L)
  - (b) Gentamycin (Invitrogen) ..... 25 µg/mL
  - (c) BSA Pathocyte (MP Biomedicals) .... 0.0286% (v/v)
  - (d) Glutamine .... 2 mM
  - (e) Bovine pancreas insulin ..... 10 µg/mL
  - (f) Human transferrin .... 100 µg/mL
  - (g) Progesterone .... 0.2 µM
  - (h) Putrescine .... 0.10 µM
  - (i) L-thyroxine .... 0.45 µM
  - (j) Selenium .... 0.224 µM
  - (k) 3,3',5-Triiodo-L-thyronine .... 0.49 µM
- Monoclonal antibodies:
  O4: MAB345 (Chemicon)
  Anti-galactocerebroside [Ranscht et al., 1982] (anti-GalC): MAB342 (Chemicon).

- Second antibodies: anti-mouse IgM-fluorescein and anti-mouse IgG3-phycoerythrein (Southern Biotechnology Associates or Cambridge Biosciences).
- Flasks, dishes and coverslips; coat all tissue culture plastics and glass coverslips with 13.3 mg/mL poly-L-lysine:
  - (a) Incubate for at least 1 h to a maximum of 24 h
  - (b) Wash the plastics and coverslips once with D-PBSA
  - (c) Air dry them before use.
- Petri dish
- Falcon 15-mL polypropylene conical tube with cap
- Forceps, curved and straight (size 7)
- Tubes, polystyrene, 12 × 75 mm, round bottom with caps [Falcon #2058, B-D Biosciences; these tubes fit both the sample tubing and the sort-collecting port of the FACS].
- Scissors, curved, dissecting
- Scalpel
Nonsterile:
- Sprague-Dawley rats, 6–8 days postpartum. A greater yield of OECs is obtained from neonatal rats rather than older animals, although it is possible to make preparations from animals of any age.
- Dissecting board and pins
- Alcohol, 70%, in a spray bottle
- Fluorescence-activated cell sorter (FACS; B-D Biosciences; see Section 15.4)

Protocol
Preparation of Dissociated Cell Suspension of OECs:
1. Kill the rats by decapitation. Use 15–20 animals for a FACS sort of 2–3 h to recover enough OECs for 10–15 coverslips (10,000 cells per coverslip).
2. Pin the head dorsal side up onto a dissecting board and spray with 70% ethanol.
3. Dip all dissecting instruments in 70% alcohol, before use, and shake the instruments dry.
4. Remove the skin from the head by using sharp curved scissors, and make a circular cut to remove the top of the skull, revealing the brain and the two olfactory bulbs at the anterior tip of the cortex near the nose (Fig. 23.3).
5. Using curved forceps, gently release the olfactory bulbs from the brain, and place them in a Petri dish containing a drop of L-15 plus gentamycin.
6. Using a sterile scalpel blade, chop the olfactory bulbs into small pieces.
7. Place the pieces into a small vial containing 1 mL of stock collagenase and 1 mL of L-15.
8. Incubate the pieces of olfactory bulbs at 37°C for 30–45 min.

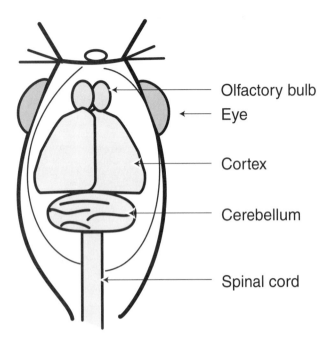

Fig. 23.3. Olfactory Bulb Dissection. Schematic diagram of the isolation of the olfactory bulbs from the brain of newborn rats. (Courtesy of Susan Barnett.)

9. Centrifuge the resultant suspension at 100 *g* for 5 min and remove the supernatant.
10. Resuspend the pelleted bulb tissue in 1 mL of Ca^{2+}- and Mg^{2+}-free HBSS.
11. Glial cells are fragile; therefore, to produce a single-cell suspension, the olfactory bulb tissue must be dissociated gently, taking care not to produce air bubbles. Dissociate the tissue gently through a 19G hypodermic needle followed by a 23G hypodermic needle.
12. Add 5 mL of Ca^{2+}- and Mg^{2+}-free HBSS, and spin the suspension again at 100 *g* for 10 min. Remove the supernate and resuspend the cells in DMEM-FBS at a concentration of 5×10^6 cells/mL.

FACS Sorting of OECs:
To remove as many of the $O4^+$ oligodendrocytes as possible from the sorted cell population, it is preferable to label them with a second antibody to galactocerebroside [anti-GalC; Ranscht et al., 1982; Barnett et al., 1993], thus allowing discrimination between oligodendrocytes and OECs. In this way, a dual-color sort can be carried out in which the unwanted cells, now labeled with both antibodies, can be ignored. On a FACSVantage the sort takes 1 hour.

13. After tissue dissociation, resuspend 5×10^6 cells in 500 μL of a mixture of O4 and anti-GalC antibody hybridoma supernates, and incubate at 4°C for 30–45 min in a 15-mL centrifugation tube. There will be more than 5×10^6 cells, so prepare enough tubes for the number of cells. If the hybridoma supernate does not contain high concentrations of antibody, then an increase in cell number will result in an apparent decrease in the percentage of $O4^+$ cells. A similar consideration applies to the anti-GalC hybridoma supernate. It is therefore always necessary to titrate antibodies by FACS analysis before use.
14. After 30–45 min, wash the cells twice in DMEM/FBS and resuspend in 500 μL of DMEM/FBS containing a 1:100 dilution of both goat anti-mouse IgM-fluorescein and goat anti-mouse IgG3-phycoerythrein. Incubate the cells for a further 30 min at 4°C.
15. Wash the cells twice in 15 mL of DMEM-FCS by centrifugation, and resuspend them at a concentration of 2×10^6 cells/mL for sorting in ice-cold DMEM/FBS media in cell-sorting tubes (B-D Biosciences 2085). Keep the cell suspension on ice. The FACS profile for antibody-labeled cells is stable for at least 3–4 h when the cells are kept on ice.
16. This protocol describes the method used for sorting cells on a FACStar cell sorter. However, any type of cell sorter should produce similar results. Align the laser to the 488-nm excitation line. Fluorescence compensation must be set to negate fluorescence spillover into the individual fluorescent channels. A control sample of cells labeled with the second antibodies alone should be analyzed before the labeled sample so that any nonspecific cell label can be accounted for and removed by decreasing the gain on the photomultiplier tube.
17. Set a cell sort window around the $O4^+$ $GalC^-$ cells. The average percentage of $O4^+$ olfactory bulb cells should be around $10 \pm 1\%$.
18. Sort the $O4^+$ $GalC^-$ cells into a polystyrene round-bottom tube containing 4 mL of DMEM/FBS.
19. After sorting, transfer the sorted cells to a 15-mL centrifugation tube, wash out the sort tube with medium, pool the washes, and fill the centrifuge tube with DMEM/FBS. Spin the centrifuge tube at 100 *g* for 15 min.
20. Remove the supernatant and resuspend the cells in DMEM/BS. Optimum cell viability and growth for the OECs may be obtained in a 1:1 dilution of DMEM/BS:ACM [Barnett et al., 1993]. ACM is collected from monolayers of purified cortical astrocytes [Noble & Murrey, 1984] incubated for

> 48 h in DMEM/BS. Cell can also be grown in ACM (1:5), FGF2 (10 ng/mL) forskolin (10 μM) diluted in DMEM/BS then diluted 1:1 in DMEM containing 10% FBS [Alexander et al., 2002].

This protocol produces a highly enriched population of O4+ olfactory bulb glial cells (>97%). These cells may be plated onto poly-L-lysine-coated coverslips, Petri dishes, or flasks and used for further investigations. Current evidence suggests that these cells do correspond to the OECs described in electron microscope studies [Doucette, 1984; Raisman, 1985], and research continues on their further characterization.

23.4.3 Endocrine Cells

The problems of culturing endocrine cells [O'Hare et al., 1978] are accentuated because the relative number of endocrine cells in a particular tissue relative to other parenchymal and stromal cells may be quite small. Sato and colleagues [Buonassisi et al., 1962; Sato & Yasumura, 1966] cultured functional adrenal and pituitary cells from rat tumors by mechanical disaggregation of the tumor [Zaroff et al., 1961] and regular monolayer culture. The functional integrity of the cells was retained by intermittent passage of the cells as tumors in rats [Buonassisi et al., 1962; Tashjian et al., 1968]. These lines are now fully adapted to culture and can be maintained without animal passage [Tashjian, 1979] and, in some cases, in fully defined media [Hayashi & Sato, 1976].

Fibroblasts have been reduced in cultures of pancreatic islet cells by treatment with ethylmercurithiosalicylate [Braaten et al., 1974], and islet cells have also been purified by density gradient centrifugation [Prince et al., 1978] and by centrifugal elutriation [Bretzel et al., 1990]. The islet cells apparently produce insulin, but not as propagated cell lines. Short-term cultures have also been generated from ductal progenitor cells, capable of differentiating toward insulin-secreting cells in serum-free medium [Gao et al., 2003].

Pituitary cells that produce pituitary hormones for several subcultures have been isolated from the mouse [de Vitry et al., 1974]. Although normal human pituitary cells do not survive well, and most pituitary adenoma cultures gradually lose the capacity to synthesize hormones, some cultures continue to produce growth hormone [see, e.g., Bossis et al., 2004] and three-dimensional cultures of pituitary adenomas continue to secrete prolactin [Guiraud et al., 1991].

23.4.4 Melanocytes

The high substrate dependency of cultured keratinocytes has been utilized to obtain preferential melanocyte attachment and growth in a hormone-supplemented medium [Naeyaert et al., 1991]. The system circumvents the problem

of keratinocyte contamination, while supporting good melanocyte proliferation with minimal supplementation of serum [Gilchrest et al., 1985; Wilkins et al., 1985] in the absence of chemotherapeutic agents or tumor promoters such as phorbol esters. These substances downregulate protein kinase C (PKC) and, hence, tyrosinase activity, curtailing the production of pigment [Park et al., 1993]. With conventional serum supplementation (5–20%), melanocyte growth is far better than reported in other systems, and fibroblast overgrowth can be prevented by reducing the Ca^{2+} concentration to 0.03 mM [Naeyaert et al., 1991]. The system described here can be further modified by removing cholera toxin (which otherwise promotes melanocyte growth by greatly increasing intracellular levels of cAMP) and substituting dibutyryl cAMP. This medium then provides physiological levels of both PKC and cAMP that permit studies of cAMP-modulating agents [Park et al., 1999] under conditions that are optimal for melanogenesis studies.

The preceding introduction and Protocol 23.21 for the culture of human melanocytes were contributed by Hee-Young Park, Mina Yaar, and Barbara A. Gilchrest, Department of Dermatology, Boston University School of Medicine, 80 East Concord St., Boston, MA 02118–2394.

PROTOCOL 23.21. MELANOCYTES

Outline
Epidermis, stripped from small fragments of skin after trypsinization, is dissociated in EDTA and cultured in serum-free medium supplemented with basic FGF (FGF-2).

Materials
Sterile:
- ❑ D-PBSA
- ❑ Fetal bovine serum
- ❑ Melanocyte hormone-supplemented medium (MHSM):

	Vol. stock	Final Conc.
(a) Medium 199 (GIBCO 400–1200)	93.1 mL	
(b) EGF (B-D Biosciences), 10 μg/mL stock	0.1 mL	10 ng/mL
(c) Transferrin (Sigma), 10 mg/mL stock	0.1 mL	10 μg/mL
(d) Insulin (Sigma), 10 mg/mL stock	0.1 mL	10 μg/mL
(e) Triiodothyronine (Sigma), 1 μM stock	0.1 mL	1 nM
(f) Hydrocortisone (Calbiochem), 0.28 mM stock	0.5 mL	1.4 μM

(g) Cholera toxin (Calbiochem),
  0.1 μM stock......................1.0 mL............1 nM
  Cholera toxin can be replaced by bovine pituitary extract (Clonetics), 35 μg/mL, stock 0.7 ng/mL final, and dibutyryl cAMP (Sigma), 1 mM stock, 100 μM final.
(h) Basic FGF (Amgen),
  200 ng/mL stock.................5.0 mL..........10 ng/mL

□ Trypsin/EDTA: Trypsin 0.25% in D-PBSA with 0.1% (3.5 mM) EDTA (GIBCO 610–5050)
□ Soybean trypsin inhibitor (Sigma), 10 μg/mL
□ Tissue culture dishes (Falcon, B-D Biosciences), 10, 6, and 3.5 cm in diameter
□ Pipettes
□ Centrifuge tubes, 15 and 50 mL
Nonsterile:
□ Humidified incubator (37°C, 7% CO_2)
□ Water bath
□ Electronic cell counter or hemocytometer

Protocol
Establishing the Primary Culture
Day 1:
1. Rinse the skin specimen in 70% alcohol and then twice in D-PBSA.
2. Transfer the tissue to a sterile 10-cm dish, epidermal side down.
3. With dissecting scissors, excise subcutaneous fat and deep dermis.
4. Cut the remaining tissue into 2 × 2-mm pieces with a scalpel, rolling the blade over the tissue. (Do not use a sawing action.)
5. Transfer the tissue fragments to cold 0.25% trypsin in a 15-mL centrifuge tube.
6. Incubate the tissue fragments 1–1½ h at room temperature. For some donors, incubation at 37°C for 1–2 h followed by incubation at 4°C for 18–24 h may give higher melanocyte yields.

Day 2:
7. Gently tap the centrifuge tube to dislodge fragments that have settled at the bottom, and then rapidly pour the contents of the tube into a 6-cm dish.
8. Using forceps, transfer the tissue fragments individually to a dry 10-cm dish, epidermal side down. Gently roll each tissue fragment against the dish. The epidermal sheet should adhere to the dish and allow a clean separation of the dermis with forceps. Discard the dermal fragments.
9. Transfer all epidermal sheets to a sterile 15-mL centrifuge tube containing 5 mL of 0.02% (0.7 mM) EDTA. Take care to place

each epidermal sheet in the EDTA solution, not on the plastic wall of the centrifuge tube.
10. Gently vortex the tube to disaggregate the epidermal sheets into a single-cell suspension.
11. Centrifuge the cells for 5 min at 350 g, and then aspirate the supernatant.
12. Resuspend the pellet in serum-free MHSM.
13. Count the cells with a hemocytometer, and inoculate 1 × 10⁶ cells (approximately 2–4 × 10⁴ melanocytes) per 3.5-cm dish in 2 mL of MHSM containing 5% FBS to facilitate attachment of the cells.
14. Refeed the culture with fresh MHSM containing growth factors and bovine pituitary extract twice weekly.

Days 3–30:
15. At 24 h, the cultures will contain primary keratinocytes with scattered melanocytes (Fig. 23.4a). Keratinocyte proliferation should cease with several days, and colonies should begin to detach during the second week. By the end of the third week, only melanocytes should remain (Fig. 22.4b,c). In most cases, cultures attain near confluence and are ready to passage within 2–4 weeks.

Subculture:
16. Gently rinse the culture dish twice with 0.02% (0.7 mM) EDTA.
17. Add 3 mL of 0.25% trypsin/0.1% (2.5 mM) EDTA, and incubate at 37°C. Examine the dish under phase microscopy every 5 min to detect cell detachment.
18. When most cells have detached, inactivate the trypsin with soybean trypsin inhibitor (10 μg/mL) or 1 mL of medium containing 10% calf serum. (Melanocytes maintained under serum-free conditions greatly benefit from a "serum kick" at the time of subculture.)
19. Pipette the contents of the dish to ensure complete melanocyte detachment.
20. Aspirate and centrifuge the cells for 5 min at 350 g.
21. Aspirate the supernatant, resuspend the cells in a calcium-free melanocyte medium containing 5% FBS without calcium, and replate at 2–4 × 10⁴ cells per 100-mm dish. FBS is required for good melanocyte attachment (>75%) to plastic dishes, but can be omitted if the dishes are coated with fibronectin or type I/III collagen [Gilchrest et al., 1985].
22. Refeed the culture twice a week with serum-free or serum-containing melanocyte medium. The serum-containing medium will give a greater

melanocyte yield, but encourages the growth of fibroblasts. Reducing the calcium concentration to 0.03 mM has no effect on the proliferation of melanocytes, but inhibits fibroblast overgrowth [Naeyaert et al., 1991].

23.4.5 Confirmation of Melanocytic Identity

Melanocyte cultures may be contaminated initially with keratinocytes and at any time by dermal fibroblasts. Both forms of contamination are rare in cultures established and maintained by an experienced technician or investigator but are common problems for the novice. The cultured cells can be confirmed to be melanocytes with moderate certainty by frequent examination of the culture under phase microscopy, assuming that the examiner is familiar with the respective cell morphologies (Fig. 23.4b,c). More definitive identification is provided by electron microscopic examination, DOPA staining, or immunofluorescent staining with Mel 5 antibody, directed against tyrosinase-related protein-1.

23.5 HEMATOPOIETIC CELLS

Hematopoietic cells have been grown in colony-forming assays, in long-term cultures from bone marrow, and as continuous cell lines [Freshney et al., 1994; Klug & Jordan, 2002]. Under the control of specific growth factors, hematopoietic stem cells, or CFU-A, can be recloned from agar cultures (*see* Protocol 23.23). MIP-1α appears to be the main factor that keeps the stem cells in the regenerative compartment. However, most suspension colonies, which contain cells of only one lineage, survive only as primary cultures that lose their repopulation efficiency and, hence, cannot be subcultured. Granulocytic colonies are the most common; but under the appropriate conditions, other lineages can be produced (*see* Protocol 23.23).

A number of myeloid cell lines have been developed from murine leukemias [Horibata & Harris, 1970] and, like some of the human lymphoblastoid lines [Collins et al., 1977], have been shown to make globulin chains and, in some cases, complete α- and γ-globulins (*see* Section 27.10). Some of these lines can be grown in serum-free medium [Iscove & Melchers, 1978]. T-cell lines require T-cell growth factors [e.g., interleukin IL-2; Gillis & Watson, 1981], and B-cell growth factors have also been described [Howard et al., 1981; Sredni et al., 1981; *see also* Freshney et al., 1994].

The first human lymphoblastoid cell lines were derived by culturing peripheral lymphocytes from blood at very high cell densities ($\sim 10^6$/mL), usually in deep culture (> 10 mm) [Moore et al., 1967]. Subsequently, immortalization was shown to be due to EBV, and reproducible techniques are now available for this procedure [Bolton & Spurr, 1996].

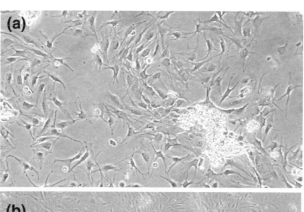

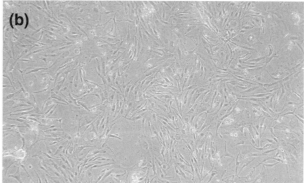

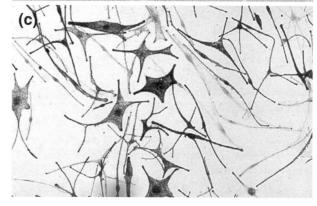

Fig. 23.4. Melanocyte Cultures. (a) Culture of newborn foreskin-derived melanocytes 1 week after inoculation. Note the keratinocyte colonies with central stratification and tightly apposed epithelial cells at the periphery. The melanocytes are the relatively small, dark dendritic cells, most of them in contact with the keratinocyte colonies by means of dendritic projections (seen via phase contrast). (b) Ten-day-old primary cultures of newborn epidermal melanocytes in medium lacking TPA. Many cells display branching dendrites, and other cells display bipolar to polygonal morphology (seen via phase contrast, ×320). (c) Secondary culture of TPA-treated epidermal melanocytes. Note the dendritic morphology of the cells, the pigmentation, and their slender spindle shape.

Rossi and Friend [1967] demonstrated that a mouse RNA virus (the "Friend virus") could cause splenomegaly and erythroblastosis in infected mice. Cell cultures taken from minced spleens of these animals could, in some cases, give rise to continuous cell lines of erythroleukemia cells. All of these lines are transformed by what is now recognized as a complex of defective and helper viruses derived from Moloney sarcoma virus [Ostertag & Pragnell, 1978, 1981]. Some cell lines can produce a virus that is infective *in vivo*, but not *in vitro*, and the cells can also be passaged as solid tumors or ascites tumors in DBA2 or BALB-c mice.

Treating cultures of Friend cells with a number of agents, including DMSO, sodium butyrate, isobutyric acid, and hexamethylene bisacetamide, promotes erythroid differentiation [Friend et al., 1971; Leder & Leder, 1975]. Untreated cells resemble undifferentiated proerythroblasts, whereas treated cells show nuclear condensation, a reduction in cell size, and an accumulation of hemoglobin (*see* Section 17.7) to the extent that centrifuged cell pellets are red in color. Evidence for differentiation can be demonstrated by staining the cells for hemoglobin with benzedine (*see* Plate 13a,b) and *in situ* hybridization of globin-specific messenger RNA (*see* Plate 13c,d). The human leukemic cell line K562 can also be induced to differentiate with sodium butyrate and hemin, though not with DMSO [Andersson et al., 1979c].

Macrophages may be isolated from many tissues by collecting the cells that attach during enzymatic disaggregation. The yield is rather low, however, and a number of techniques have been developed to obtain larger numbers of macrophages. Mineral oil or thioglycollate broth [Adams, 1979] may be injected into the peritoneum of a mouse, and 3 days later the peritoneal washings will contain a high proportion of macrophages. If necessary, macrophages may be purified by their ability to attach to the culture substrate in the presence of proteases (*see* Section 14.8.1). Macrophages can be subcultured only with difficulty, because of their insensitivity to trypsin. Methods have been developed using hydrophobic plastics (e.g., Petriperm dishes, Vivascience, Sartorius).

There are some reports of propagated lines of macrophages, mostly from murine neoplasia. Normal mature macrophages do not proliferate, although it may be possible to culture replicating precursor cells by the method of Dexter and Spooncer (*see* Protocol 23.22).

23.5.1 Long-Term Bone Marrow Cultures

The following introduction and Protocol 23.22 for the long-term culture of bone marrow was contributed by E. Spooncer, Department of Biomolecular Sciences, UMIST, Sackville Street, Manchester M60 1QD, U.K.

By culturing whole bone marrow, the relationship between the stroma and stem cells is maintained, and in the presence of the appropriate hematopoietic cell and stromal cell interactions, stem cells and specific progenitor cells can continue proliferating over several weeks [Dexter et al., 1984; Spooncer et al., 1992]. Progenitor cells from fresh marrow or long-term cultures may be assayed by clonogenic growth in soft agar [Heyworth & Spooncer, 1992] or in mice [Till & McCulloch, 1961].

PROTOCOL 23.22. LONG-TERM HEMATOPOIETIC CELL CULTURES FROM BONE MARROW

Outline

Marrow is aspirated into growth medium and maintained as an adherent cell multilayer for at least 12, and up to 30, weeks. Stem cells and maturing and mature myeloid cells are released from the adherent layer into the growth medium. Granulocyte/macrophage progenitor cells can be assayed in soft gels (*see* Protocol 23.23).

Materials

All reagents must be pretested to check their ability to support the growth of the cultures.

Sterile:

- Fischer's medium (Invitrogen) supplemented with 50 U/mL of penicillin and 50 μg/mL of streptomycin and containing 16 mM (1.32 g/L) of $NaHCO_3$
- Growth medium: Fischer's as above, 100-mL aliquots supplemented with 1 μM hydrocortisone, sodium succinate, and 20% horse serum (hydrocortisone sodium succinate made up as 1 mM stock in Fischer's medium and stored at −20°C)
- Syringes, 1 mL, with 21G needles
- Gauze, swabs, scissors, forceps
- Tissue culture flasks, 25 cm²

Nonsterile:

- Five mice: (C57Bl/6× DBA/2)F₁ bone marrow performs well in long-term culture, but marrow from some strains (e.g., CBA) does not [Greenberger, 1980]

Protocol

1. Kill the donor mice by cervical dislocation.
2. Wet the fur with 70% alcohol and remove both femurs. Collect 10 femurs in a Petri dish on ice containing Fischer's medium. One femur contains $1.5–2.0 \times 10^7$ nucleated cells.
3. In a laminar-flow hood:
   (a) Clean off any remaining muscle tissue with gauze swabs.
   (b) Hold the femur with forceps and cut off the knee end. The 21G needle should fit snugly into the bone cavity.
   (c) Cut off the other end of the femur as close to the end as possible.
   (d) Insert the tip of the bone into a 100-mL bottle of growth medium, and aspirate and depress the syringe plunger several times until all the bone marrow is flushed out of the femur.
   (e) Repeat Steps (a)–(d) with the other nine bones.

4. Disperse the marrow to a suspension by pipetting the large marrow cores through a 10-mL pipette. There is no need to disaggregate small clumps of cells.
5. Dispense 10-mL aliquots of the cell suspension into 25-cm^2 tissue culture flasks, swirling the suspension often to ensure an even distribution of the cells in the 10 cultures.
6. Gas the flasks with 5% CO_2 in air and tighten the caps.
7. Incubate the cultures horizontally at 33°C.
8. Feed the cultures weekly:
   (a) Agitate the flasks gently to suspend the loosely adherent cells.
   (b) Remove 5 mL of growth medium, including the suspension cells; take care not to touch the layer of adherent cells with the pipette.
   (c) Add 5 mL of fresh growth medium to each flask; to avoid damage, do not dispense the medium directly onto the adherent layer.
   (d) Gas the cultures and replace them in the incubator.

Analysis. Cells harvested during feeding can be investigated by a range of methods, including morphology, CFC assays (*see* Protocol 23.23), and the *in vivo* CFU-S assay for stem cells [Till & McCulloch, 1961].

Variations. Mouse erythroid [Dexter et al., 1981], B-lymphoid [Whitlock et al., 1984], and human long-term cultures [Gartner & Kaplan, 1980; Coutinho et al., 1992] have been grown.

23.5.2 Hematopoietic Colony-Forming Assays

Hematopoietic progenitor cells may be cloned in suspension in semisolid media in the presence of the appropriate growth factor(s) [Heyworth & Spooncer, 1992]. Pure or mixed colonies will be obtained, depending on the potency of the stem cells that are isolated. Assays for the detection of granulocyte and macrophage colony-forming cells (GM-CFC), erythroid burst-forming units (BFU-E), mixed colony-forming cells (CFC-mix), and granulocyte, erythrocyte, macrophage, and megakaryocyte colony-forming cells (CFC-GEMM) are described by Testa and Molineux [1993], and their place in routine hematopoietic cell culture technology is already well established [Metcalf, 1990]. The following introduction and Protocol 23.23 have been abridged from Freshney et al. [1994].

The efficiency of growth of colonies in these assays is increased by using Methocel instead of agar as the semisolid phase. This practice makes for a tighter colony that is easier to evaluate and count. Because Methocel is a high-viscosity liquid, and not a gel like agar, cells will sediment through

it, albeit slowly, and plate out on the plastic base of the dish. This route places them all in one focal plane for subsequent observation. The colonies will form if grown in an atmosphere of 5% CO_2 in air, but this may be at the expense of adequate hematopoiesis, and ideally, the gas phase should be 10% CO_2 and 5% O_2 in air [Bradley et al., 1978b]. If an incubator with this gas mixture is not available, a cylinder of mixed gases can be rented. Place the dishes in a plastic box with a lid with a hole in it. Seal the box with plastic tape, and gas via the hole before sealing it. A dish of water in the box will keep the atmosphere humid.

PROTOCOL 23.23. HEMATOPOIETIC COLONY FORMING ASSAYS

Outline
Suspend bone marrow cells in agar or Methocel, and seed the cells into dishes with the appropriate growth factors.

Materials
Sterile or Aseptically Prepared:
❑ Bone marrow cells (*see* Protocol 23.22, Steps 1–4 for preparation). Count the nucleated cells in a hemocytometer after staining them with methylene blue, or lyse the cells with Zapoglobin (Beckman Coulter) and count the nuclei on an electronic cell counter. Each femur will yield 1.0–1.5 × 10^7 cells.
❑ Methylcellulose, 4000 cP (Fluka)
❑ Noble agar (Difco, B-D Biosciences)
❑ Alpha MEM stock (Invitrogen)
❑ FBS
❑ Growth factors (Table 23.3), either recombinant (R&D) or from a conditioned medium
❑ BSA, 10% in D-PBSA
❑ Petri dishes, 3 cm
❑ Wehi-cell-conditioned medium (Wehi-CM): Wehi 3B is a mouse myelomonocytic cell line that when cultured, releases IL-3 (multi-CSF) into the medium [Bazill et al., 1983]. (*see* Protocol 14.2 for a method for making conditioned medium).

TABLE 23.3. Addition of Cells and Growth Factors for Colony-Forming Assays

Assay	IL-6	rMurGM[a]	rIL-3[b]	Epo	Cells	Incidence/10^5
BFU-E		0.1 ng	1 ng	2 U	5 × 10^4	40–80
CFC-mix			1 ng	2 U	5 × 10^4	100–180
CFU-GEMM	100 ng		1 ng	2 U	5 × 10^4	92–106

Growth Factors/mL

[a]10% AF1-19T CM can be substituted for GM-CSF [Pragnell et al., 1988].
[b]10% Wehi-CM can be substituted for rIL-3 [Bazill et al., 1983].

❑ Methylcellulose
  (a) Weigh out 7.2 g of Methocel, and add it to a 500-mL bottle containing a large magnetic stirrer bar.
  (b) Sterilize the Methocel by autoclaving.
  (c) Add 400 mL of sterile UPW heated to 90°C to wet the Methocel.
  (d) Stir the mixture at 4°C overnight to dissolve the Methocel. The solution is now Methocel 2×. It is more accurate to use a syringe (without a needle) than a pipette to dispense Methocel.
❑ Alpha medium stock solution:
  (a) Alpha medium, powder (GIBCO 10 L pack size)
  (b) MEM vitamin stock, 100×, 100 mL
  (c) Gentamycin sulfate, 200 mg. Stir the medium on a heated stirrer until it dissolves, and make to 3 L with UPW. (Do not allow the temperature to rise above 37°C.) Before final filtration through a 0.22-μm filter, it is advantageous to prefilter the medium through stacked filters of pore sizes 5, 1.2, 0.8, and 0.45 μm. Dispense the medium into 21-mL aliquots and store it in premeasured volumes at −20°C.
❑ Alpha medium, 2×:
  (a) Alpha medium stock solution, 21 mL
  (b) Fetal bovine serum (FBS), 25 mL
  (c) Glutamine (200 mM) 1 mL
  (d) NaHCO$_3$, 7.5%, 3 mL
  Mix the ingredients in a sterile bottle and equilibrate to 37°C.
Nonsterile:
❑ With agar, use two water baths, one at 37°C and the other at 55°C.
❑ Incubator: gas phase, 10% CO$_2$, 5% O$_2$, 85% N$_2$

Protocol A: BFU-E, CFC-Mix, and CFU-GEMM
1. Mix an equal volume of alpha 2× medium to which 1% BSA has been added, with 2× Methocel to make the required amount of medium for the experiment. Keep the mixture cold; Methocel is more liquid when it is cold.
2. Set up cultures in triplicate, but make enough mix for 4 dishes, as Methocel clings to the side of tubes and some is always lost.
3. Add the required concentration of growth factors to the tube (Table 23.3).
4. Add 5 × 10^4 cells.
5. Increase the total volume to 4.4 mL with the addition of 1× medium, and mix the tubes on a vortex mixer.
6. Using a syringe, plate out 1 mL of medium into 3-cm non-tissue-culture-grade dishes.

7. Incubate the culture at 37°C in a humid atmosphere of 10% CO$_2$ and 5% O$_2$ in air for 8–15 days.

Identification of the Colonies
BFU-E colonies can be either single colonies composed of very small cells or multicentric colonies (bursts), each with tightly packed very small pink or red cells. The incidence of these colonies is 40–80/10^5 bone marrow cells.

CFC-mix colonies can be single, compact colonies, usually with a halo of cells of widely varying size. They may be multicentric, but the cell population is obviously heterogeneous. The incidence of colonies in this assay is between 100 and 180/10^5 bone marrow cells, of which only about 10% will be mixed colonies containing erythroid cells. If the erythroid cells are not red, it can be very difficult for the inexperienced eye to identify the colonies accurately. The colonies should be photographed and then picked out, after which cytocentrifuge preparations should be made that can be fixed and stained with 10% Giemsa, and help sought with identification of the cells [Heyworth & Spooncer, 1992].

Protocol B: GM-CFC
1. Use 3 cm of non-tissue-culture-grade Petri dishes, lay out the dishes, and label them.
2. Make a 0.3% agar medium (*see* Protocol 14.4) and keep it at 37°C.
3. Prepare bone marrow cells (*see* Protocol 23.22). Count the nucleated cells after staining the cells with methylene blue, or lyse the cells with Zapoglobin (Beckman Coulter) and count the nuclei with an electronic cell counter. Each femur will yield 1.0–1.5 × 10^7 nucleated cells.
4. Add 0.1 ng/mL of rMurGM-CSF to each dish.
5. Add 1 mL of agar medium containing 7.5 × 10^4 cells and swirl gently to mix the cells and agar. Allow the agar to set at room temperature.
6. Alternatively, 0.8% Methocel medium can be used. This generally gives a tighter colony, which is easier to count.
7. Place the dishes in a clean plastic box.
8. Incubate the dishes for 6 days in a humidified incubator at 37°C in an atmosphere of 5% CO$_2$ in air.
9. Using an inverted microscope, count the colonies that contain more than 50 cells, and express the number as colonies/10^5 cells seeded.
10. The incidence of GM-CFC in normal bone marrow is 100–120/10^5 cells.

23.6 GONADS

Ovarian granulosa cells can be maintained and are apparently functional in primary culture [Orly et al., 1980], but specific functions are lost on subculture. A cell line started from Chinese hamster ovary [CHO-KI; Kao & Puck, 1968] has been in culture for many years, but its lineage still has not been identified. Although epithelioid at some stages of growth, it undergoes a fibroblast-like modification when it is cultured in dibutyryl cyclic AMP [Ilsie & Puck, 1971].

Cellular fractions from testis have been separated by velocity sedimentation at unit gravity, but no prolonged culture of the fractions has been reported. The TM4 is an epithelial line from mouse testis, although its differentiated features have not been reported. Sertoli cells have also been cultured from testis [Mather, 1979].

23.7 STEM CELLS

Culture of stem cells was developed first with hematopoietic cells, where the identification of lineage markers and the relative ease of disaggregation of hemopoietic tissues, such as bone marrow, made isolation feasible (*see* Protocol 23.22). Isolation and culture of stem cells from solid tissues has been much more problematical, but, as techniques develop, progress has been made in identifying and culturing cells, e.g., from mammary gland [Welm et al., 2003]. One problem has been the dearth of stem cell markers in solid tissues, but some progress has been made in this area [Zhou et al., 2001; Potten et al., 2003], opening up possibilities for cell sorting.

23.7.1 Embryonal Stem Cells

Embryonal stem cells from several species, but particularly mouse, have been used extensively to study differentiation [Martin & Evans, 1974; Martin, 1975, 1978; Rizzino, 2002; zur Nieden et al., 2003], because they may develop into a variety of different cell types (muscle, bone, nerve, etc.). Those that have been through animal passage form teratomas, analogous to spontaneous human teratomas. Cells grown on feeder layers of, for example, SCO mouse fibroblasts, will proliferate, but not differentiate, whereas, when the cells are grown on gelatin without a feeder layer or in nonadherent plastic dishes, nodules form that eventually differentiate. Leukemia inhibitory factor (LIF) appears to be one of the main regulatory factors that hold cells within the stem cell compartment, whereas retinoids, vitamin D3, and planar polar compounds induce lineage specific differentiation in mouse and human [Draper et al., 2002] cells.

Stem cell cultures have also been derived from human embryos and show differentiation down a number of different pathways, encouraging their use in tissue repair [Thomson et al., 1998; Rippon & Bishop, 2004].

23.7.2 Multipotent Stem Cells from the Adult

In addition, there are numerous reports of isolation of multipotent stem cells from a number of adult tissues, including bone marrow [Suva et al., 2004], liver [Deng et al., 2003], brain [Vescovi et al., 2002; Greco & Recht, 2003], muscle [Cao et al., 2003], and umbilical cord [Sanchez-Ramos, 2002; Newman et al., 2003]. The identification of multipotential stem cells in adult tissues opens up a wholly unexpected area in stem cell biology, previously locked into the concept that stem cell regeneration was tissue specific. It raises many exciting prospects for understanding commitment and differentiation in adult progenitor cells, but also raises a number of major questions. If stem cells exist in the brain, why are neurons not replaced? If there is a potential for circulating stem cells to repopulate other tissues, why do satellite cells from skeletal muscle not repair cardiac myocyte injury unless introduced artificially? What is the biological significance, from an evolutionary standpoint, of regenerative capacity and multipotentiality of stem cells in many tissues that is never used? The concept that stem cells can change their differentiative capacity, not just from one lineage to another (e.g., from astroglial to oligodendroglial, or from erythroid to myeloid) but from the derivative of one germ layer to the derivative of another (such as neuroectodermal to mesodermal [Wurmser et al., 2004] or endodermal to neuroectodermal [Deng et al., 2003]) conflicts so strongly with the established paradigm of lineage fidelity that many have claimed the atypical development of stem cells at an ectopic site is due to fusion of the incoming stem cell with a resident progenitor cell [Alvarez-Dolado et al., 2003; Greco & Recht, 2003]. However, although cell fusion probably does account for differentiation of some ectopic stem cells, the evidence for stem cell plasticity seems convincing [Wurmser et al., 2004]. In the light of evidence that precursor cells, previously thought to be lineage committed, appear to revert or convert to multipotent stem cells (*see* Section 17.1) it is possible that stem cells from one tissue can convert to a different stem cell with different potency at an ectopic site, perhaps via a totipotent cell.

23.7.3 Origin and Handling of Stem Cells

Mouse. Many of the original embryonal carcinoma lines, such as F9 and P19, were derived from mouse teratocarcinomas that had been serially transplanted under the kidney capsule [Martin & Evans, 1974]. Embryonal stem (ES) cell lines, such as ES-D3, J1 and R1, were derived from the inner cell mass of the preimplantation embryo [Evans & Kaufman, 1981; Martin, 1981], and embryonal germ (EG) cell lines were derived from primordial germ line cells [Matsui et al., 1991; Resnick et al., 1992]. They can be propagated in an undifferentiated state on feeder layers of STO cells or in the presence of LIF [Schamblott et al., 2002]. Differentiation occurs spontaneously in long-term culture in the absence of LIF or feeder layers, particularly if the cells are allowed to form aggregates. Differentiation can be promoted with retinoids, planar polar compounds such as DMSO or HMBA, dibutyryl cyclic AMP (dbcAMP), vitamin D3, and several cytokines.

Human. Human teratocarcinomas cells such as NTera-2 [Paquet-Durand et al., 2004] and ES cells [Thomson et al.,

1998] and human EG cells [Schamblott et al., 1998] can, similarly, be propagated in culture in the presence of LIF, FGF-2, and forskolin [Schamblott et al., 2002] with capacity to differentiate and significant potential for tissue engineering [Laslett et al., 2003; Rippon & Bishop, 2004]. Cultures derived from spontaneously generated embryoid bodies, isolated from EG cultures and disaggregated in collagenase and Dispase, give rise to EBD (embryoid body derived) cell lines that appear to be more easily maintained and have a longer life span in culture [Schamblott et al., 2002]. They are also capable of differentiation in a number of different directions and may have considerable potential for tissue engineering [Kerr et al., 2003].

Human adult stem cells can be isolated from a number of tissues with FACS or magnetic sorting of cells that lack lineage markers but possess stem cell markers, such as Sca-1 [Kawada et al., 2004; Oh et al., 2004], Kit [Das et al., 2004; Takahashi et al., 2004], c_Met [Suzuki et al., 2004], or efflux of Hoechst 33342 [Bhatt et al., 2003].

CHAPTER 24

Culture of Tumor Cells

24.1 PROBLEMS OF TUMOR CELL CULTURE

The culture of cells from tumors, particularly spontaneous human tumors, presents problems similar to those of the culture of specialized cells from normal tissue. The tumor cells must be separated from normal connective tissue cells, preferably by provision of a selective medium that will support tumor cells, but not normal cells. Although the development of selective media for normal cells has advanced considerably (*see* Section 10.2.1 and Chapter 23), progress in tumor culture has been limited by variation both among and within samples of tumor tissue, even from the same tumor type. It is often surprising to find that tumors that grow *in vivo*, largely as a result of their apparent autonomy from normal regulatory controls, fail to grow *in vitro*.

There are many possible reasons for the failure of some tumor cultures to survive. Their nutritional requirements may be different from those of the equivalent normal cells, or perhaps attempts to remove stroma may actually deprive the tumor cells of a matrix, nutrients, or signals necessary for survival. Alternatively, dilution of tumor cells to provide a sufficient amount of nutrients per cell may also dilute out autocrine growth factors produced by the cells. Strictly speaking, truly autocrine factors should be independent of dilution if they are secreted onto the surface of the cell and are active on the same cell, but it is possible that some so-called autocrine growth factors are in fact homocrine (*see* Section 3.5)—i.e., they act on adjacent similar cells and not only on the cell releasing them. Hence, a closely interacting population is required. Interaction with certain types of stromal cells may provide paracrine interaction if the

stroma are able to make the requisite growth factors, either spontaneously or in response to the tumor cells, necessary for their survival.

It may be incorrect to assume that the growth factor dependence of a tumor cell is similar to that of the normal cells of the tissue from which it was derived. Tumor cells may produce endogenous autocrine growth factors, such as TGF-α, and the provision of exogenous growth factors, such as EGF, may compete for the same receptor. Furthermore, the response of a tumor cell to a growth factor, or hormone, will depend on what other growth factors are present, some of which may be tumor cell derived, and on the status of the cell. A normal cell, capable of expressing growth suppressor and senescence genes, may respond differently from a cell in which one or more of these genes is inactive or mutated, and in which antagonistic, growth-promoting oncogenes are overexpressed.

So there are many possible reasons for a tumor cell population responding differently to the nutritional and mitogenic environment optimized for normal cells of the same lineage. To confirm this difference more information is required on the nutritional requirements of tumor cells, but, given the heterogeneity of tumors, the task is a daunting one. Because the potential therapeutic benefit to be derived from knowledge of the nutritional requirements of individual tumor cell lines is not likely to be great, greater emphasis has been placed on the response of tumor cells to growth factors and the differences in signal transduction.

It is probable that the bulk of the cells in a tumor have a limited life span, because of genetic aberration, terminal differentiation, apoptosis, or natural senescence, and only a few cells, analogous to a stem cell population in normal tissue, have the potential for continuous survival. Dilution into culture may reduce the number of these cells, as well as their interaction with other cells, such that survival is impossible. Cells from multicellular animals, unlike prokaryotes, do not survive readily in isolation. Even a tumor is still a multicellular organ and may require continuing cell interaction for survival. The lethality of the tumor to the host lies in its uncontrolled infiltration and colonial growth, but the origin of the bulk of the cell population may reside in a relatively small population of transformed stem cells [Jones et al., 2004; Al-Hajj et al., 2004]. This pool of stem cells may be so small that its dilution on explantation deprives it of some of its prerequisites for survival, particularly paracrine growth factors from stromal elements and homocrine interaction with other tumor subclones.

In sum, the goal is either to create the correct, defined nutritional and hormonal environment or, failing that, to provide a sustaining environment, as yet undefined but nevertheless able to permit the survival of an appropriate or representative population. There has been a continuing trend to use serum and feeder layers in order to get tumor cells to grow, and only a few tumors have responded to serum-free culture. As many transformed cells are not inhibited by TGF-β, there has not been the same need to eliminate serum, other than to repress fibroblastic growth, which remains a major problem. The adaptation of medium designed for equivalent normal cells is still the most logical approach to obtaining cell lines from tumors, even if supplemented with minimal amounts of serum or conditioned medium from other cells [Dairkee et al., 1995].

24.2 SAMPLING

In addition to preventing the overgrowth of connective tissue or vascular cells, both of which are stimulated to invade and proliferate by many tumors, tumor cell culture requires the separation of the transformed cells from the normal equivalent tissue cells, which may have similar characteristics. Furthermore, although any section of gastric epithelium may be regarded as representative of that particular zone of the gastric mucosa, tumor tissue, dependent as it is on genetic variation and natural selection for its development, is usually heterogeneous and composed of a series of often diverse subclones displaying considerable phenotypic diversity. Ensuring that cultures derived from this heterogeneous population are representative is difficult, and can never be guaranteed unless the whole tumor is used and survival is 100%. Because these conditions are practically impossible to achieve, the average tumor culture is a compromise. Assuming that representative subpopulations have been retained and are able to interact, the corporate identity may be similar to the original tumor.

The problem of selectivity is accentuated when sampling is carried out from secondary metastases, which often grow better, but may not be typical either of the primary tumor or of all other metastases. It is, however, interesting to speculate on the analogy between metastatic occurrences and ectopic development of stem cells (see Section 23.7.2).

In view of these practically overwhelming problems facing tumor culture, it is almost surprising that the field has produced any valid data whatsoever. In fact, it has, and this may result from (1) the aforementioned autonomy of tumor populations, which may have allowed the proliferation of tumor cells under conditions in which normal cells would not multiply; (2) the increased size of the proliferative pool in tumors, which is larger than that of most normal tissues; (3) the ability of tumor cells to give rise to tumors as xenografts in immune-deprived mice and increased success in deriving culture from the xenografts; and (4) the propensity of malignantly transformed cells to give rise to continuous immortalized cell lines more frequently than normal cells. This last feature, more than any other, has allowed extensive research to be carried out on tumor cell populations, even on apparently normal differentiation processes, despite the uncertainty of their relationship to the tumor from which they were derived.

The uncertainty of the status of continuous cell lines remains, but nevertheless they have provided a valuable source of human cell lines for molecular and virological research. The question of whether they represent advanced stages of progression of a tumor whose development has been accelerated in culture, a cryptic stem cell population, or a purely in vitro artifact is still to be resolved. They are certainly distinct from most early-passage tumor cultures but may still contain significant elements of the genotype of the parental cell from which they were derived. Their immortality is more likely to be due to the deletion or suppression of genes inducing senescence [Pereira-Smith & Smith, 1988; Goldstein et al., 1989; Holt et al., 1996; Sasaki et al., 1996] and to increased telomerase activity [Bryan & Reddel, 1997; Bodnar et al., 1998] than to overexpression of genes conferring malignancy per se.

24.3 DISAGGREGATION

Some tumors, such as human ovarian carcinoma, some gliomas, and many transplantable rodent tumors, are readily disaggregated by purely mechanical means, such as pipetting and sieving (see Section 12.3.8), which may also help to minimize stromal contamination, as stromal cells are often more tightly locked in fibrous connective tissue. Many of the common human carcinomas, however, are hard, or scirrhous, and the tumor cells are contained within large amounts of fibrous stroma, making mechanical disaggregation difficult, although scraping the cut surface of scirrhous tumors has been used successfully in the so-called spillage technique [Lasfargues, 1973; Oie et al., 1996], to release tumor cells from the fibrous stroma.

Enzymatic digestion has proved to be preferable to mechanical disaggregation in most cases. Although trypsin has often been used for this purpose, its effectiveness against fibrous connective tissue is limited, and it can reduce the seeding efficiency of the tumor cells [Lounis et al., 1994]; crude collagenase has been found to be more effective with several different types of tumor [Dairkee et al., 1997]. Enzymatic disaggregation also releases many stromal cells, requiring selective culture techniques for their elimination (*see* Section 24.7; Fig. 15.7d; Plate 23). Collagenase exposure may be carried out over several hours, or even days, in complete growth medium (*see* Protocol 12.8).

Extensive necrosis is also a problem of tumor tissue that is not usually encountered with normal tissue. Usually, the attachment of viable cells allows necrotic material to be removed on subsequent feeding, but if the amount of necrotic material is large and not easily removed at dissection, it may be advisable to use a Ficoll-paque separation (*see* Protocol 12.10) to remove necrotic cells.

24.4 PRIMARY CULTURE

Some cells—e.g., macrophages—attach to the substrate during collagenase digestion, but may be removed by transferring the disaggregated cell suspension to a fresh flask when the collagenase is removed. The adherent cells may be retained and cultured separately or irradiated or treated with mitomycin C and used as a feeder layer (*see* Section 14.2.3; Plates, 6d, 11a, 15b; Protocols 23.2, 23.4). The reseeded cells will contain many stromal cells (principally fibroblasts and endothelium), some of which may be removed by a second transfer to a fresh vessel in 2–4 h, because tumor cells, particularly clusters of malignant epithelium, often take longer to attach. This method of removal by serial transfer is generally only partially successful, however, and it will usually require selective culture conditions for the complete removal of the stromal cells. It may still be advantageous, however, to retain the stromal cell cultures to use as feeder layers.

Physical separation techniques have also been used to remove stromal contaminants [Csoka et al., 1995; Oie et al., 1996; *see also* Chapter 15], but, in general, these methods are suitable only if the cells are to be used immediately, as stromal overgrowth usually follows in the absence of selective conditions. However, magnetic sorting (*see* Section 15.3.2 and Plate 23) with anti-fibroblast microbeads (*see* Fig. 15.7d; Miltenyi Biotech) appears to be very effective.

Cloning as a method of purification has limitations, as tumor cells in primary culture often have poor plating efficiencies (<0.1%). Suspension cloning has been proposed in the past as a means of isolating not only tumor cells, but tumor stem cells [Hamburger & Salmon, 1977]. However, by the time that a clone has grown to sufficient numbers to be of potential analytical value, it may have changed considerably, and it may even have become heterogeneous itself, because

of genetic instability. Cloned isolates from a tumor should be studied collectively, and even in coculture, for a meaningful interpretation.

There has also been some difficulty in propagating cell lines from primary clones, particularly from clones isolated by the suspension method. It may be that although these cells are clonogenic, few of them really are stem cells, or, if they are, they mature spontaneously because of the suspension mode of growth and lose their regenerative capacity. Nevertheless, cell strains cloned directly from tumors would be valuable material for studying tumor clonal diversity and interaction, and they represent a key area of study for future investigation, particularly if selective conditions can be used to isolate tumor stem cells.

24.5 CHARACTERIZATION

The isolation of cells from tumors may give rise to several different types of cell line. Besides the neoplastic cells, connective-tissue fibroblasts, vascular endothelial and smooth muscle cells, infiltrating lymphocytes, granulocytes, and macrophages, as well as elements of the normal tissue in which the neoplasia arose, can all survive explantation. The hematopoietic components seldom form cell lines, although hematopoietic cell lines have been derived from small-cell carcinoma of the lung, causing serious confusion, because this carcinoma also tends to produce suspension cultures that can express myeloid markers [Ruff & Pert, 1984]. Macrophages and granulocytes are so strongly adherent and nonproliferative that they are generally lost at subculture. Smooth muscle does not propagate readily without the appropriate growth factors and selective medium, so the major potential contaminants of tumor cultures are fibroblasts, endothelial cells, and the normal equivalents of the neoplastic cells.

Of these contaminants, the major problem lies with the fibroblasts, which grow readily in culture and may also respond to tumor-derived mitogenic factors. Similarly, endothelial cells, particularly in the absence of fibroblasts, may respond to tumor-derived angiogenesis factors and proliferate readily. The role of normal equivalent cells is harder to define, as their similarity to the neoplastic cells has made the appropriate experiments difficult to analyze. Characterization criteria should be chosen to exclude non-tumor cells. For example, endothelial cells are factor VIII positive, contact inhibited, and sensitive to density limitation of growth; fibroblasts have a characteristic spindle-shaped morphology, are density limited for growth (though less so than endothelial cells), have a finite life span of 50 generations or so, make type I collagen, and are rigidly diploid.

In general, the normal cell component, phenotypically equivalent to the tumor cells, is harder to identify and eliminate. The cells will be diploid, although some tumor cells may be close to diploid. They are usually anchorage dependent and will have a finite life span, although, again, there are cases of normal epithelial cell lines becoming continuous [Boukamp et al., 1988], and, if the cells are epithelial, they are more likely to be inhibited by serum TGF-β. The tumor cells are more likely to show genetic aberrations, such as oncogene amplification, translocations, and suppressor gene deletions, identifiable by FISH (*see* Protocol 27.9). The tumor cells are also likely to be angiogenic, show a higher expression of urokinase-like plasminogen activator (uPA), and will be invasive (*see* Section 18.6).

From a behavioral aspect, the ability of neoplastic cells to grow on a preformed monolayer of the normal cells of same type is a good criterion for tumor cell identity and a potential model for separation. The normal cells also provide a feeder layer to sustain the tumor cells. Glioma, for example, will grow readily (better than on plastic in some cases) on a preformed monolayer of normal glial cells [MacDonald et al., 1985], but their normal counterparts will not, and the same may be true for hepatoma cells and skin carcinomas.

Normal cells tend to have a low growth fraction at saturation density (*see* Sections 18.5.2 and 21.11), whereas neoplastic cells continue to grow faster after reaching confluence. The maintenance of cultures at high density can sometimes provide conditions for overgrowth of the neoplastic cells (*see* Section 25.3.6 and Protocol 18.3).

24.6 DEVELOPMENT OF CELL LINES

Primary cultures of carcinoma cells do not always take readily to trypsin passage, and many of the cells in the primary culture may not be capable of propagation, because of genetic or phenotypic aberrations, terminal differentiation, or nutritional insufficiency. Nevertheless, some primary cultures from tumors can be subcultured, opening up major possibilities. Evidence for tumor cells in the subculture implies that they have not been overgrown, may even have a faster growth rate than contaminating normal cells, and may be available for cloning or other selective culture methods (*see* Sections 10.2.1, 14.6, 24.7.2).

One of the major advantages of subculture is amplification. Expanded cultures can be cryopreserved and replicate cultures prepared for characterization and assay of specific parameters such as genomic alterations, changes in gene expression, chemosensitivity, and invasiveness. Disadvantages of subculture include evolution away from the phenotype of the tumor, due to the inherent genetic instability of the cells and selective adaptation of the cell line to the culture environment.

24.6.1 Continuous Cell Lines

One major criterion for the neoplastic origin of a culture is its capacity to form a continuous cell line, which is usually aneuploid, heteroploid, insensitive to density limitation of growth, and anchorage independent, and often tumorigenic

(*see* Section 18.6.2). The relationship of this cell line to the primary culture and the parent tumor is still difficult to assess, however, as such cells are not always typical of the tumor population. The cells of a continuous cell line may represent (1) further transformation stimulated by adaptation to culture, made possible by the unstable genotypic characteristics of tumor cells, or (2) a specific subset or stem cell population of the tumor. Currently, the second possibility seems more likely [Petersen et al., 2003], as the emergence of a continuous cell line is often from colonies within the monolayer, suggesting that the continuous line arises from a minor immortalized subset of the tumor cell population, with cell culture merely providing the appropriate conditions for their expansion.

The capacity to form continuous cell lines is a useful criterion for a malignant origin, and some authors maintain that the characteristics of the cell lines (e.g., tumorigenicity, histology, chemosensitivity, etc.) still correlate with the tumor of origin [Tveit & Pihl, 1981; Minna et al., 1983]. In any event, these cell lines provide useful experimental material, although the time required for their evolution makes immediate clinical application difficult. Cancer cell lines are reviewed by Masters and Palsson in a series of books [Masters & Palsson, 1999–2000].

24.7 SELECTIVE CULTURE OF TUMOR CELLS

Three main approaches have been adopted to select tumor cells in primary culture: selective media (*see* Sections 10.2.1, 14.6), confluent feeder layers (*see* Section 24.7.2), and suspension cloning (*see* Section 14.3, 14.8.5).

24.7.1 Selective Media

There are only a few media that have been developed as selective agents for tumor cells, because of their inherent problems of variability and heterogeneity. HITES [Carney et al., 1981; *see* Table 10.2] is one such medium and may owe its success to the production of peptide growth factors by small-cell lung cancer, for which the medium was developed. A proportion, but not all, of small-cell lung cancer biopsies will grow in pure HITES; others will survive with a low-serum supplement (e.g., 2.5%). HITES medium is modified RPMI 1640 with hydrocortisone, insulin, transferrin, estradiol, and selenium. Of these constituents, selenium, insulin, and transferrin are probably the most important and are found in many serum-free formulations (*see* Table 10.2). The NCI group also produced a selective medium for adenocarcinoma; reputedly suitable for lung, colon, and, potentially, many other carcinomas [Brower et al., 1986]. It is also based on RPMI 1640, supplemented with selenium, insulin, and transferrin, with the addition of hydrocortisone, EGF, triiodothyronine, BSA, and sodium pyruvate (*see* Table 10.2). Other selective media have been used successfully with prostate [Uzgare et al., 2004], bladder [Messing et al., 1982], and mammary [Ethier et al., 1993] carcinoma.

Other types of selective media depend on the metabolic inhibition of fibroblastic growth and are not specifically optimized for any particular type of tumor (*see* Section 14.6). However, inhibitors have not been found to be generally effective, with the exception of the use of monoclonal antibodies against fibroblasts by Edwards et al. [1980] and Paraskeva et al. [1985]. These antibodies have proved useful in establishing cultures from laryngeal and colon cancer. Antibodies have also proved useful in either positively selecting epithelial cells or negatively sorting stromal cells from tumor cell suspensions by panning or magnetic sorting (*see* Sections 15.3.2, 24.4; Fig. 15.7d; Plate 23).

24.7.2 Confluent Feeder Layers

The use of confluent feeder layers (*see* Figs. 14.4 and 14.10; Plate 6c,d), perhaps more than any other method, has been applied successfully to many types of tumor. Smith and others [e.g., Lan et al., 1981] used confluent feeder layers of fetal human intestine, FHS74Int, to grow epithelial cells from mammary carcinoma, with media conditioned by other cell lines, although later reports suggest that selective culture in MCDB 170 is a more reproducible approach [Hammond et al., 1984]. Feeder layers of mouse 3T3 or STO embryonic fibroblasts were used successfully with breast, colon, and basal cell carcinoma [Rheinwald & Beckett, 1981; Leake et al., 1987].

Feeder layer techniques rely on the prevention of fibroblastic overgrowth by a preformed monolayer of other contact-inhibited cells. They are not selective against normal epithelium, as normal epidermis and normal breast epithelium both form colonies on confluent feeder layers. Results from glioma [MacDonald et al., 1985], however, suggest that selection against equivalent normal cells may be possible on a homologous feeder layer. Glioma grown on normal glial feeder layers should lose any normal glial contaminants. By the same argument, breast carcinoma seeded on growth arrested confluent cultures of normal breast epithelium—e.g., milk cells (*see* Protocol 23.3)—could become free of any contaminating normal epithelium.

PROTOCOL 24.1. GROWTH ON CONFLUENT FEEDER LAYERS

Outline

Treat feeder cells in the mid-exponential phase with mitomycin C, and reseed the cells to give a confluent monolayer. Seed tumor cells, dissociated from the biopsy by collagenase digestion, or from a primary culture with trypsin, onto the confluent monolayer (Fig. 24.1). Colonies from epithelial tumors may form in 3 weeks to 3 months. Fibrosarcoma and gliomas do not always form colonies, but may infiltrate the feeder layer and gradually overgrow.

Materials

Sterile:

❑ Feeder cells (e.g., 3T3, STO, 10T1/2, or FHS74Int)
❑ Mitomycin C (Sigma), 1 mg/mL

Note. It is advisable to do a dose-response curve with mitomycin C when using feeder cells for the first time, to confirm that this dose allows the feeder layer to survive for 2–3 weeks, but does not permit further replication in the feeder layer after about two doublings, at most. In this case, proceed as follows:

  (a) Treat the cells overnight (18 h) in 25-cm² flasks with 1–100 μg/mL of mitomycin C.
  (b) Trypsinize the cells, and reseed the entire contents of the flasks into a 75-cm² flask in 20 mL of fresh medium.
  (c) Grow the cells for 3 weeks, feeding them twice per week.
  (d) Stain the culture, and check for surviving colonies.

❑ Growth medium
❑ Collagenase, 2,000 U/mL, CLS grade (Worthington) or equivalent
❑ Trypsin, 0.25%, in PBSA
❑ Tumor biopsy or primary culture
❑ Forceps, fine curved
❑ Scalpels with #22 blades
❑ Petri dishes for dissection, as for primary culture

Protocol

1. Grow up the feeder cells to 80% confluence in six 75-cm² flasks.
2. Add mitomycin C to give the appropriate final concentration, usually around 5 μg/mL.
3. Incubate the cells overnight (~18 h) in mitomycin C.
4. Remove the medium with mitomycin C, and wash the monolayer with fresh medium.
5. Grow the cells for a further 24–48 h.
6. Trypsinize the cells, and reseed them in 25-cm² flasks at 5×10^5 cells/mL (1×10^5 cells/cm²). Incubate the cultures for 24 h.
7. If you are using biopsy material, the biopsy should be dissected and placed in collagenase during Step 2.
8. Seed a cell suspension from the biopsy, approximately 20–100 mg/flask, into two of the 25-cm² flasks, such that each flask holds 6 mL of suspension. Remove 1 mL of the suspension from each flask, and add it to 4 mL of medium in each of two more flasks. The third pair of flasks should be kept as controls to guard against feeder cells surviving the mitomycin C treatment.

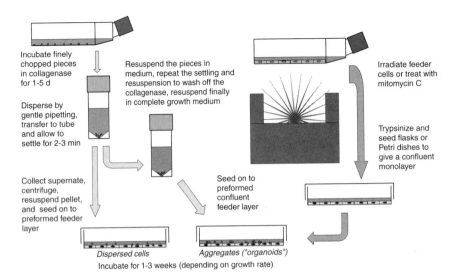

Fig. 24.1. Confluent Feeder Layers. Epithelial clusters from collagenase digestion will form colonies of epithelial cells on confluent feeder layers, such as fetal intestinal epithelium (FHS74Int), normal human glia, or irradiated 3T3 or STO cells. Dispersed cells, although containing more stromal cells, can also form colonies on confluent feeder layers with significant restriction of stromal overgrowth. Selection is against stromal components, but not normal epithelium.

If you are using a primary culture, trypsinize or dissociate the cells in collagenase, 200 U/mL final (*see* Protocol 12.8), and seed onto a feeder layer at 10^5 cells/mL in two flasks and 10^4 cells/mL in two flasks. If the cells are from a glioma (Plate 15b) or fibrosarcoma (Plate 1d), colonies may not appear as the tumor cells migrate freely among the feeder cells, and the surviving tumor will be confirmed only by subculturing the cells without a feeder layer (by which time contaminating normal cells should have been eliminated).

It is essential to confirm the species of origin of any cell line derived by this method, in order to guard against accidental contamination from resistant cells in the feeder layer. The species of origin can be confirmed by chromosomal analysis (*see* Protocol 16.7) and isoenzyme electrophoresis (*see* Protocol 16.10) if the feeder is of a different species from the primary culture. If feeder cells of the same species as the primary culture are used, it is necessary to DNA profile (*see* Protocol 16.9) or DNA fingerprint (*see* Protocol 16.8) both the feeder layer cells and any culture that is generated and compare the results with a portion of the biopsy or other tissue taken from the donor.

24.7.3 Suspension Cloning

The transformation of cells *in vitro* leads to an increase in their clonogenicity in agar (*see* Sections 14.3, 18.5.1); tumorigenicity has also been shown to correlate with cloning in Methocel [Freedman & Shin, 1974] (*see* Protocol 14.5). As cells may be cloned in suspension directly from disaggregated tumors [Hamburger & Salmon, 1977], or at least colonies may grow (they may not be clones) in preference to normal stromal cells, suspension cloning would seem to be a potentially selective technique. However, the colony-forming efficiency

is often very low (often <0.1%), and it is not easy to propagate cells isolated from the colonies. Although this method has not generated cell lines, it has been used for drug screening with tumor biopsies (*see* Section 22.4.1). FACS or immunomagnetic positive sorting of clonogenic cells by expression of stem cell markers, such as Sca-1, Kit [Takahashi et al., 2004], and the ABC transporter [Zhou et al., 2001] (*see also* Section 23.7.3) and negative sorting for more differentiated lineage markers may allow enrichment of a putative stem cell pool, which may make a better target for drug screening and molecular drug targeting.

24.7.4 Histotypic Culture

Apart from organ culture itself (Section 25.2), which is not fundamentally different for tumor tissue than for normal tissue, methods with particular application to tumor culture are spheroid culture (*see* Protocol 25.2) and filter well culture (*see* Protocol 25.4).

Spheroids. Normal stromal cells do not form spheroids or even become incorporated in tumor-derived spheroids. Hence, cultures from tumors allowed to form spheroids on nonadhesive substrates, like agarose (*see* Protocol 25.2), will tend to overgrow their stromal component. Some cultures from breast and small-cell lung carcinoma can generate spheroids or irregular cellular organoids that float off and may be collected from the supernatant medium, leaving the stroma behind. However, the spheroids or organoids do not always appear soon after culture and can sometimes take weeks, or even months, to form, suggesting derivation from a minority cell population in the tumor. In other cases, e.g., mammary carcinoma, the organoids may represent the normal epithelial component [Speirs, 2004].

Spheroid generation does not arise in all tumor cultures, but has been described in neuroblastoma, melanoma, and glioma (*see* Section 25.3.3). Its potential for generating cell lines has not been fully explored, however, as the bulk of attention has been given to forming either attached monolayers or suspension colonies in agar for assay purposes.

Filter well inserts. Filter well inserts are designed to recreate the cell and matrix interactions of the tissue from which the cells were derived. Hence, they provide an ideal model for the study of invasiveness (*see* Section 18.6.3), angiogenesis (*see* Section 18.6.4), and other abnormalities of cell interaction in cancer.

24.7.5 Xenografts

When cultures are derived from human tumors, the scarcity of material and the infrequency of rebiopsy make it difficult to make several attempts to culture the same tumor. The growth of some tumors in immune-deprived animals [Rofstad, 1994] provides an alternative approach that makes much greater amounts of tumor available. It has sometimes been found that cultures can be initiated more easily from xenografts than from the parent biopsy, but whether this is due to the availability of more tissue, enrichment of transformed cells, progression of the tumor, or modification of the tumor cells by the heterologous host (e.g., by murine retroviruses) is not clear.

Two main types of host are used: the genetically athymic nude mouse, which is T-cell deficient [Giovanella et al., 1974], and neonatally thymectomized animals that are subsequently irradiated and treated with cytosine arabinoside [Selby et al., 1980; Fergusson et al., 1980]. The first type of host is expensive to buy and difficult to rear, but maintains the tumor for longer. Thymectomized animals are more trouble to prepare, but cheaper and easier to provide in large numbers. They do, however, regain immune competence and ultimately reject the tumor after a few months. Take rates for tumors can be enhanced by using mice that are asplenic as well as athymic, genetically (e.g., *scid* mice) or by splenectomy, or by sublethally irradiating nude mice. Implantation with fibroblasts or Matrigel has also been reported to improve tumor take [Topley et al., 1993].

If access to a nude mouse colony is available, or facilities exist for neonatal thymectomy and irradiation, xenografting should be considered as a first step in generating a culture. Although only a small proportion of tumors may take, the resulting tumor will probably be easier to culture, and repeated attempts at culture may be made with subsequent passage of the tumor in mice. However, particular care must be taken, as with isolation from mouse feeder layers, to ensure that the cell line ultimately surviving is human, and not mouse, by proper characterization with isoenzyme (*see* Section 16.8) and chromosome analysis (*see* Section 16.5).

24.7.6 Characterization of Tumor Cell Cultures

General characteristics that may be used to identify tumor cells in culture are described in Chapter 18 (*see* Sections 18.3–18.6 and Table 18.1). Although these characteristics are often expressed in continuous cell lines, their detection in early-passage cultures may be more difficult because of greater heterogeneity. Detection of specific genetic abnormalities (polysomy, chromosomal deletions and translocations, oncogene overexpression, and tumor suppressor gene mutation or deletion) may be required, preferably by *in situ* analysis (*see* Section 27.8.1). Similarly, overexpressed oncogene products, such as mutant p53, or *erb*-B may be detected by immunostaining in cytological preparations or by flow cytometry.

24.7.7 Preservation of Tissue by Freezing

It is often difficult to take advantage of a large biopsy and utilize all of the valuable material that it provides. It is possible in these cases to preserve the tissue by freezing.

PROTOCOL 24.2. FREEZING BIOPSIES

Outline
Chop the tumor, expose the pieces to DMSO, and freeze aliquots in liquid nitrogen.

Materials
Sterile:
☐ Biopsy
☐ Plastic ampoules, 1.2 mL (Nalge Nunc)
☐ DBSS (*see* Appendix I)
☐ Collection medium (*see* Appendix I)
☐ DMSO (self-sterilizing if placed in a sterile container)
☐ Instruments (scalpels, forceps, dishes, etc., as for primary culture)

Protocol
1. After removing necrotic, fatty, and fibrous tissue, chop the tumor into about 1 to 2-mm pieces, and wash the pieces in DBSS, as for primary culture.
2. Place five to ten pieces in each ampoule.
3. Add 1 mL of growth medium containing 10% DMSO to the pieces, and leave them for 30 min at room temperature.
4. Freeze the ampoules at 1°C/min (*see* Protocol 20.1), and transfer them to a liquid nitrogen freezer (to avoid explosion risk, do not submerge in liquid nitrogen).
5. To thaw an ampoule, place it in 37°C water (with appropriate precautions; *see* Protocol 20.2).

6. Swab the ampoule thoroughly in alcohol, open it, allow the pieces to settle, and remove half of the medium.
7. Replace the medium slowly with fresh, DMSO-free medium. Mix by gentle shaking, and allow to stand for 5 min.
8. Gradually replace all of the medium with DMSO-free medium, transfer the pieces to a Petri dish, and proceed as for regular primary culture, but allowing twice as much material per flask.

24.8 SPECIFIC TUMOR TYPES

The general protocols described in Chapter 12 (*see* Protocols 12.3–12.10), together with selective culture (*see* Section 24.7 and Chapter 23), will provide a good starting point for culturing most tumor types. In general, a reasonable approach to tumor culture is to combine collagenase digestion (e.g., Protocol 12.8) with the tissue-specific approaches given in the protocols in Chapter 23, with or without the use of a feeder layer. Although serum-free conditions are often selective for cells from normal tissues, the nutritional and growth factor requirements of tumor cells may be more variable (*see* Section 24.7.1) and require the use of serum. Some specific examples of tumor culture are discussed briefly below.

24.8.1 Breast

Breast carcinoma can be cultured from collagenase digestion of biopsies [Leake et al., 1987; Dairkee et al., 1995, 1997; for protocols *see* Speirs, 2004] and propagated on feeder layers or in MCDB 170. However, many of the conditions used to derive cultures from normal breast (growth factor supplementation, collagen coating) may not be optimal for mammary carcinoma [Ethier et al., 1993], so a variety of conditions may need to be tested in preliminary attempts and the neoplastic identity of the cells cultured confirmed.

Identification of breast tumor cells, as distinct from normal breast cells, will require detection of specific genetic lesions including erbB-2, c-*myc*, and fibroblast growth factor receptor (FGFR) 2 [Ethier et al., 1993; Ray et al., 2004] and elevated cyclin E levels [Willmarth et al., 2004]. Lineage can be confirmed by cytokeratin, EMA [Heyderman et al., 1979], and anti-HMFG 1 and 2 [Burchell and Taylor-Papadimitriou, 1989]; it has been proposed that keratin 19 is more likely to be expressed in tumor cells than in normal mammary cells *in vitro* [Taylor-Papadimitriou et al., 1989].

24.8.2 Lung

Both small-cell lung carcinoma (SCLC) and non-small-cell lung carcinoma (NSCLC) have been cultured successfully [Oie et al., 1996; for protocols *see* Wu, 2004] with serum-free selective media—HITES for SCLC and ACL4 for NSCLC—mechanical spillage, and density gradient separation on Ficoll for isolating the cells. A substantial panel of these cell lines has been accumulated by the NCI, and some are available through the ATCC. The effects of matrix on oncogene and growth factor expression have also been studied [Pavelic et al., 1992] and have been used to facilitate culture of lung carcinoma cells from bone marrow micrometastases [Pantel et al., 1995]. The assay of chemotherapeutic drugs in brain metastases from lung assayed *in vitro* shows considerable heterogeneity of response [Marsh et al., 2004].

A number of markers are available for the identification of SCLC cells *in vitro*, including bombesin-like immunore-activity, DOPA-decarboxylase, N-*myc*, and creatine kinase BB isoenzyme overexpression [Carney et al., 1985; Pedersen et al., 2003], but variant SCLC and non-SCLC lack these markers. Squamous cell lung cancer has been shown to overexpress EGFR, erbB-2, and TGF-α [Piyathilake et al., 2002], and other NSCLC cells lines have been shown to express abnormal properties *in vitro*, including expression of HER2/neu, TP53, and K-*ras*, which correlate with in the *in vivo* phenotype of the tumor from which they were derived [Wistuba et al., 1999]

24.8.3 Stomach

Solid tumors from gastric cancer can be disaggregated by mincing with scissors and pipetting to free tumor cell aggregates from fibrous stroma. Ascites can be purified by centrifugation on Ficoll/metrizoate [for protocols *see* Park et al., 2004]. Enrichment for tumor cells is possible by harvesting tumor cell aggregates mechanically and subculturing, by scraping off fibroblasts, or by differential trypsinization. The morphology of the cell lines obtained range from well-flattened pavement-like monolayers, through more refractile attached cobblestone-like monolayers, to loosely attached aggregates (Fig. 24.2).

24.8.4 Colon

Serum-free conditions for the culture of some human colorectal cancer cell lines have been described [Fantini et al., 1987; Murakami & Masui, 1980], but these conditions are generally not suitable for newly isolated carcinoma cultures, which require serum. Colorectal carcinoma has been cultured from biopsies that have been taken from both primary tumors and metastases [Danielson et al., 1992; Paraskeva & Williams, 1992; Park & Gazdar, 1996]. As in lung carcinoma, some colorectal tumors occur with neuroendocrine properties, and some success has been reported on the use of HITES medium (*see* Table 10.2) with them [Lundqvist et al., 1991]. Density centrifugation on Percoll has been used to purify colonic carcinoma cells for primary culture in conventional medium (RPMI 1640 with 10% FB) [Csoka et al., 1995].

Protocols for culture of colon carcinoma are available in Whitehead [2004]. Protocol 24.3 has been taken from Paraskeva and Williams [1992].

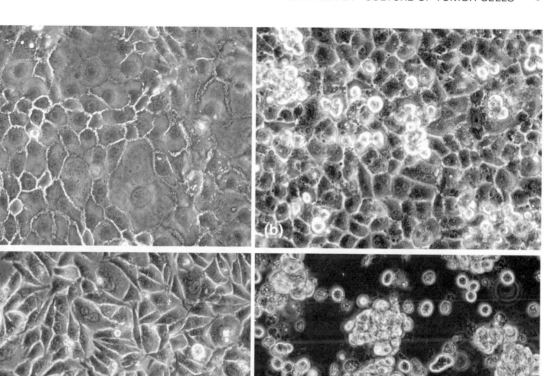

Fig. 24.2. Cell Lines from Gastric Carcinoma. (a) Well-flattened monolayer of SNU-216 gastric carcinoma cell line. (b) Pavementlike monolayer of SNU-484 gastric carcinoma cell line. (c) More refractile, cobblestone-like appearance of SNU-668 gastric carcinoma cell line. Cancer cells grow as adherent cultures, showing diffusely spreading growth of cultured tumor cells with fusiform or polygonal contours. (d) SNU-620 gastric carcinoma cell line. Cancer cells grow as both adherent and floating cell aggregates. Reproduced from Park et al. [2004].

PROTOCOL 24.3. CULTURE OF COLORECTAL TUMORS

Adenomas are usually digested enzymatically and carcinomas just chopped with surgical blades. If a well-differentiated colorectal cancer does not release tumor cells readily when the specimen is cut, it can be digested enzymatically.

Materials
Sterile:

❑ Growth medium: DMEM containing 2 mM glutamine and supplemented with 20% fetal bovine serum (batch selection is essential), hydrocortisone sodium succinate (1 µg/mL), insulin (0.2 U/mL), 2 mM glutamine, penicillin (100 U/mL), and streptomycin (100 µg/mL).

❑ Washing medium: growth medium with 5% FBS, 200 U/mL of penicillin, 200 µg/mL of streptomycin, and 50 µg/mL of gentamycin. Gentamycin is kept in the primary culture for at least the first week and then is removed.

❑ Digestion solution: DMEM and antibiotics, as described for the washing solution, with collagenase (1.5 mg/mL, Worthington type IV) hyaluronidase (0.25 mg/mL, Sigma type 1) and 2.5–5% FBS. Although we use Worthington collagenase, Sigma culture grades can also be tried.

❑ Dispase for subculture: Prepare Dispase (a neutral protease, Roche, grade 1) at 2 U/mL in DMEM containing 10% FBS, glutamine, penicillin, and streptomycin. Sterile filter the solution and store at −20°C. If a precipitate forms on thawing, the solution should be centrifuged and the active supernatant should be removed and used.

❑ Flasks, 25 cm², coated with collagen type IV, with Swiss 3T3 feeder cells at approximately 1×10^4 cells/cm² (*see* Protocol 23.4)

Protocol

Enzyme Digestion:

1. Wash tumor specimens four times in washing medium.
2. Mince in a small volume of the same medium, just enough to cover the tissue. (Do not allow the tissue to dry out!) Mince the tissues with crossed surgical blades or sharp scissors to fragments of approximately 1 mm³.
3. After cutting, wash the tissue again four times (the number of washings can be varied with experience, depending on whether contamination is a factor) by bench centrifugation (300 g for 3 min) and resuspension.
4. After washing the tumor fragments, put them into the digestion solution, and rotate at 37°C, usually overnight (approximately 12–16 h). The time the specimens are left in the solution is not critical, because digestion is a mild process, but it is important not to let the digestion medium become acid during the procedure. A low pH indicates that the specimens were left in the medium for too long or too much tissue was put in the volume of the digestion mixture. Approximately 1 cm³ of tumor tissue is put into 20–40 mL of digestion solution.

Nonenzymic Tissue Preparation:

Adenomas almost invariably need digestion with enzymes. However, with carcinomas, it is often found that during the cutting of the tumor with blades into 1-mm³ pieces, small clumps of tumor cells are released from the tumor tissue into the washing medium. In this case, the following procedure can be carried out.

1. Collect the washing medium containing released clumps of cells.
2. Separate into large and small clumps by allowing them to settle by gravity for a few minutes in a centrifuge tube.
3. Remove the supernatant phase.
4. Either put the remaining tissue pieces directly into culture or rotate the tissue gently for 30–60 min in washing medium to release more small clumps, which can then be collected, plated, and put into culture separately from the remaining larger pieces of tissue.
5. Wash all samples three times before putting them into culture.

Standard primary culture conditions:

1. Inoculate epithelial tubules and clumps of cells derived from tissue specimens into flasks in 4 mL of medium per collagen-coated 25-cm² flask of feeder cells.
2. Incubate at 37°C in a 5% CO_2 incubator.
3. Change the culture medium twice weekly.

The tubules and cells start to attach to the substratum, and epithelial cells migrate out within 1–2 days. Most of the tubules and small clumps of epithelium attach within 7 days, but the larger organoids can take up to 6 weeks to attach, although they will remain viable all that time.

The attachment of epithelium during primary culture and subculture is more reproducible and efficient when cells are inoculated onto collagen-coated flasks (Biocoat, B-D Biosciences; *see also* Protocols 23.2 and 23.9), and significantly better growth is obtained with 3T3 feeders than without. When the epithelial colonies expand to several hundred cells per colony, they become less dependent on 3T3 feeders, and no further addition of feeders is necessary.

Subculture and Propagation:

Most colorectal adenoma primary cultures and adenoma-derived cell lines cannot be passaged by routine trypsin/EDTA procedures [Paraskeva et al., 1984, 1985]. As disaggregation of the cultured adenoma cells to single cells with 0.1% trypsin in 0.25 mM (0.1%) EDTA results in extremely poor growth, Dispase is used instead.

1. Add Dispase to the cell monolayer, just enough to cover the cells (~2.5 mL/25-cm² flask), and leave the solution to stand for 40–60 min for primary cultures and 20–40 min for cell lines.
2. Once the epithelial layers begin to detach (they do so as sheets rather than single cells), pipette to help detachment and disaggregation into smaller clumps.
3. Wash and reseed the cells under standard culture conditions. It may take several days for clumps to attach, so replace the medium carefully when feeding.

Fibroblast contamination of colorectal tumor cultures. One or a combination of the following techniques can be employed to deal with fibroblast contamination:

(1) Physically remove well-isolated fibroblast colonies by scraping with a sterile blunt instrument (e.g., a cell scraper). Care has to be taken to wash the culture up to six times to remove any fibroblasts that have detached, in order to prevent them from reattaching.

(2) Differential trypsinization can be attempted with the carcinomas [Kirkland & Bailey, 1986].

(3) Dispase preferentially (but not exclusively) removes the epithelium during passaging and leaves behind most of the fibroblastic cells attached to the culture vessel [Paraskeva et al., 1984]. During subculture, cells that have been removed with dispase can be preincubated in plastic Petri dishes for 2–6 h to allow the preferential attachment of any fibroblasts that may have been removed together with the epithelium. Clumps of epithelial cells still floating can be transferred to new flasks under standard culture conditions. This technique takes advantage of the fact that fibroblasts in general attach much more quickly to plastic than do clumps of epithelial cells, so that a partial purification step is possible.

(4) Use a conjugate between anti-Thy-1 monoclonal antibody and the toxin ricin [Paraskeva et al., 1985]. Thy-1 antigen is present on colorectal fibroblasts, but not colorectal epithelial cells; therefore, the conjugate kills contaminating fibroblasts but shows no signs of toxicity toward the epithelium, whether derived from an adenoma or a carcinoma.

(5) Reduce the concentration of serum to about 2.5–5% if there are heavy concentrations of fibroblastic cells. It is worth remembering that normal fibroblasts have a finite growth span *in vitro* and that using any or all of the preceding techniques will eventually push the cells through so many divisions that any fibroblasts will senesce.

24.8.5 Pancreas

Cell lines from pancreatic primary tumors or metastases have been isolated and propagated in RPMI 1640 supplemented with fetal bovine serum. The cell lines were adapted to protein-free medium for the examination of cell products [Yamaguchi et al., 1990] (*see also* Protocol 23.7) Culture of pancreatic carcinoma have also been used to study the effect of genistein on angiogenesis [Buchler et al., 2004].

Pancreatic cell lines have been generated for studies in autocrine growth control by gastrin [Monstein et al., 2001]; no correlation was detected between the expression of gastrin and cholecystokinin (CCK) receptors, which normally bind gastrin. KCI-MOH1 is a cloned line of pancreatic adenocarcinoma isolated after passage in the SCID mouse [Mohammad et al., 1999]. Xenografted tumor was also used to establish the HPAC cell line for studies on glucocorticoid receptors [Gower et al., 1994]. Protocols for pancreatic carcinoma culture by digestion in collagenase and plating onto collagen-coated dishes are available in Iguchi et al. [2004], who derived cell lines from primary lesions, ascites, and liver metastases. Two lines from liver metastases were passaged through nude mice by splenic injection (Fig. 24.3).

24.8.6 Ovary

A number of cell lines have been established from ovarian epithelial tumors, e.g., OAW series [Wilson et al., 1996], OVCAR-3 [Hamilton et al., 1983], A2780, [Tsuruo et al.,

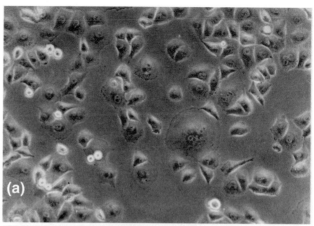

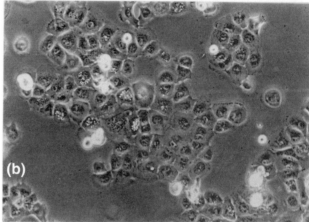

Fig. 24.3. Cell Lines from Pancreatic Cancer. Photomicrographs of cell lines KP-1N (a) and KP-3 (b) from liver metastases, xenografted in nude mice. Phase-contrast optics (×125). From Iguchi et al. [2004].

1986], some in serum-free medium [Jozan et al., 1992] and others in serum-containing medium, e.g., OSE medium, 50:50 M199:MCDB105, supplemented with 15% FBS used after collagenase digestion [Lounis et al., 1994]. Generation of cultures from ascites or pleural effusions appears to be easier than from solid tumor material [Verschraegen et al., 2003]. Density centrifugation on Percoll has also been used to purify ovarian carcinoma cells for primary culture, as for colonic carcinoma [Csoka et al., 1995]. Protocols for culture of ovarian carcinoma are available in Wilson [2004].

24.8.7 Prostate

The matrix-assisted method, described above for lung carcinoma, has also proved to be successful in isolating cell lines from prostate tumors [Pantel et al., 1995]. Serum-free culture has also been used for initiating cultures from normal prostate and from benign and malignant tumors [Chopra et al., 1996; Bright & Lewis, 2004] (*see also* Protocol 23.11). Long-term culture was possible after immortalization with a retroviral construct encoding the E6 and E7 transforming proteins of HPV16 [Bright & Lewis, 2004].

Development of prostatic epithelium is known to be under paracrine control from the stroma by the FGF family of growth factors, including KGF [Thomson et al., 1997]. When cultures were prepared from normal, benign prostatic hyperplasia (BPH) and carcinoma, it was shown that FGF-17 was elevated 2-fold in cultures from BPH [Polnaszek et al., 2004]. Tumor angiogenesis has also been studied with cultures from normal and neoplastic prostate utilizing nitric oxide [Wang et al., 2003] and VEGF signaling [Shih et al., 2003].

24.8.8 Bladder

Culture of bladder carcinoma has shown that superficial tumors tend to have a limited life span whereas cell lines from myoinvasive tumors often form continuous cell lines [Yeager et al., 1998]. Primary culture from transitional cell carcinomas of bladder have been used to study the role of FasL in immune protection of bladder cancer cells [Chopin et al., 2003]. Protocols for culture of bladder carcinoma can be found in Fu et al. [2004].

24.8.9 Skin

Melanoma. Pigment cells from skin can be cultured with the appropriate growth factors (*see* Protocol 23.21), and cultures can also be obtained from melanomas with a reasonable degree of success [Creasey et al., 1979; Mather & Sato, 1979a,b]. Primary melanomas are often contaminated with fibroblasts, but cloning on confluent feeder layers of normal cells [*see* Protocol 24.1; Creasey et al., 1979; Freshney et al., 1982b] may be possible. Cell cultures derived by mechanical spillage can be freed of fibroblasts by treatment with 100 μg/mL Geneticin (G418) [Halaban, 2004].

MCDB 153, supplemented with FGF-2, insulin, transferrin, α-tocopherol, bovine pituitary extract, hydrocortisone, and 5% serum, with catalase and PMA added for the first two passages, has been used to grow melanocytes from normal skin, dysplastic nevi, and melanotic metastases [Levin et al., 1995]. Protocols for melanoma culture are available in Halaban [2004].

Basal cell carcinoma. The 3T3 cell feeder layer technique has proved successful for culturing basal cell carcinoma of skin [Rheinwald & Beckett, 1981] (*see* Protocol 23.1). Oh et al. [2003] used primary cultures from basal cell carcinoma to study angiogenic potential and found a correlation with progression to aggressive disease. BCC cultures were also shown to be more sensitive to a cytotoxic effect of interferons than normal keratinocytes [Brysk et al., 1992].

Squamous cell carcinoma. SCC and erythroplakias have also been cultured on 3T3 feeder layers [for protocols *see* Edington et al., 2004] and have led to the establishment of valuable cell lines (the BICR series) representing different stages of malignancy and immortalization [Fitzsimmons et al., 2003; Gordon et al., 2003].

24.8.10 Cervix

Benign and malignant tumors may be established from cervical biopsies with the 3T3 feeder layer technique described for normal cervix [Stanley & Parkinson, 1979] (*see* Protocol 23.4). Cell cultures have been used to study chromosomal instability and the integration of human papillomaviruses, which are implicated in the development of cervical cancer [Koopman et al., 1999].

Protocols for cervical tumor culture are available in Stern et al. [2004].

24.8.11 Glioma

Cultures of human glioma can be prepared by mechanical disaggregation, trypsinization, or collagenase digestion [Westermark et al., 1973; Pontén, 1975; Freshney, 1980; for protocols *see* Darling, 2004; *see also* Fig. 16.2c and 24.4]. A number of gliomas have been cultured from rodents, among which the C6 deserves special mention [Benda et al., 1968]. This cell line expresses the astrocytic marker—glial fibrillary

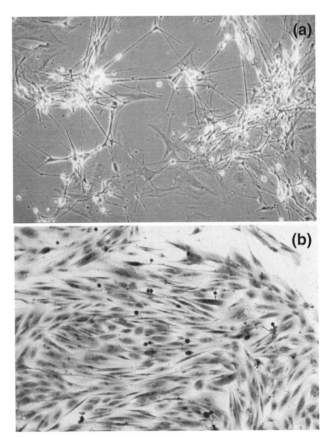

Fig. 24.4. Cultures from Human Glioma. Two cell cultures from human anaplastic astrocytoma. (a) Primary culture from collagenase digest, showing typical astrocytic cells, which may be differentiated tumor cells or reactive glia. (b) Continuous cell line MOG-G-UVW, showing one of several morphologies found in cell lines from glioma. This example shows a pleomorphic fibroblast-like morphology; there are no typical multipolar astrocytic cells, often lost in continuous cell lines. (*See also* Plates 7a,b,c, 9c, and 11b.)

acidic protein—in up to 98% of cells [Freshney et al., 1980a], but still carries the enzymes glycerol phosphate dehydrogenase and 2′,3′-cyclic nucleotide phosphorylase [Breen & De Vellis, 1974], both of which are oligodendrocytic markers. The line appears to be an interesting example of a precursor cell tumor that can mature along two distinct phenotype routes simultaneously.

Human glioma cultures have been used in predictive chemosensitivity testing [Thomas et al., 1985] and, more recently, in studies on retrovirus-mediated therapy [Rainov & Ren 2003] and targeting with brain tumor-selective peptide ligands fused to cytoxins [Liu et al., 2003].

24.8.12 Neuroblastoma

Several lines of neuroblastoma (e.g., SK-N-BE(2) [Biedler and Spengler, 1976], [Tumilowicz et al., 1970]) have been isolated and are of particular interest, because of their potential for differentiation [Dimitroulakos et al., 1994]. Mouse neuroblastomas have been found to be suitable substrates for

transmissible spongiform encephalopathies (TSEs) [Solassol et al., 2003].

24.8.13 Seminoma

Testicular seminomas have been cultured by using STO cells as a feeder layer and then have been supplemented with stem cell factor (SCF), LIF, and FGF-2, as used for embryonal stem cell cultures [Olie et al., 1995]. Although the cultures were heterogeneous, they did not give rise to primordial germ cell cultures.

24.8.14 Lymphoma and Leukemia

Routine generation of cell lines from lymphoma and leukemia has been difficult. Many of the continuous cell lines in existence were derived from Burkitt's lymphoma [Drexler & Minowada 2000] and shown to carry integrated EB viral genes. Protocols for lymphoma-leukemia culture can be found in Drexler [2004].

CHAPTER 25

Organotypic Culture

25.1 CELL INTERACTION AND PHENOTYPIC EXPRESSION

The historical divergence between maintenance of a fragment of explanted tissue and propagation of the cells that grew out from it led to the development of organ culture and cell culture (*see* Section 1.5), and it is cell culture that has become dominant. Now, although the potential uses of propagated cell lines are far from exhausted, many people are reverting to the notion that nutritional and hormonal supplementation are in themselves inadequate to recreate full structural and functional competence in a given cell population. The vital missing factor is cell interaction and the signaling capacity that it entails.

25.1.1 Effect of Cell Density

Cell–cell interaction manifests itself at the simplest level when a cell culture reaches confluence and the constituent cells begin to interact more strongly with each other because of contact-mediated signaling, formation of junctional complexes including gap junctions, and increased potential for exchange of homocrine factors (*see* Sections 3.5, 17.7.1). The first noticeable effect is cessation of cell motility and withdrawal from cell cycle (contact inhibition; *see* Section 18.5.2) in normal cells and reduced cell proliferation and increased apoptosis in transformed cells. Where cells have the capacity to differentiate, there is often an increase in the proportion of differentiated cells. C6 rat glioma cells show an increase in the percentage of GFAP-positive cells, (Fig. 25.1), secretory or absorptive epithelial cells form domes (*see* Fig. 16.1; Plate 12a,b), and skeletal myocytes fuse

to become multinucleated myotubes. Normal fibroblasts, although contact inhibited, will secrete more collagen after reaching confluence and will tend to multilayer, with the upper layer of cells smaller and darker staining, resembling fibrocytes (*see* Plate 10e).

25.1.2 Reciprocal Interactions

Interacting populations of different cells have a mutual effect on their respective phenotypes (*see* Section 17.7.1), and the resultant phenotypic changes lead to new interactions. Cell interaction is therefore not a single event, but instead a continuing cascade of events. Similarly, exogenous signals do not initiate a single event, as may be the case with homogeneous populations, but initiate a new cascade, as a result of the exogenously modified phenotype of one or both partners. For example, alveolar cells of the lung synthesize and release surfactant only in response to hormonal stimulation of adjacent fibroblasts [Post et al., 1984]; similarly, the response of prostate epithelium to stromal signals is in turn activated by androgen binding to the stroma [Thomson et al., 1997] and inducing the release of KGF. Linser and Moscona [1980] separated the Müller cells of the neural retina from pigmented retina and neurons and demonstrated that glucocorticoid-induced differentiation did not occur unless the Müller cells (astroglia) were recombined with neurons from the retina. Neurons from other regions of the brain were ineffective.

Epithelium differentiates in response to matrix constituents that are often determined jointly by the epithelium on one side and connective tissue on the other, as may be the case with the interaction between epidermis and dermis *in vitro* [Fusenig, 1994a; Limat et al., 1995]. Hence, the

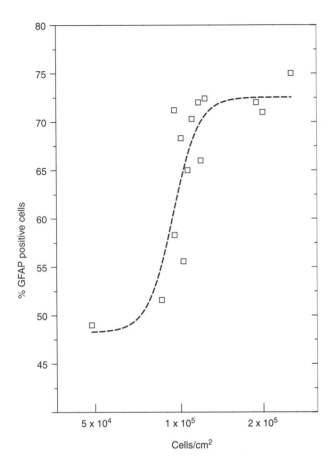

Fig. 25.1. Effect of Cell Density on Expression of GFAP in C6 Cells. Flask cultures were grown to different densities, fixed, and stained by immunoperoxidase for GFAP, and the percentage of stained cells was calculated from replicate counts in representative fields. (Courtesy of R. L. Shaw.)

whole integrated tissue may respond differently to simple ubiquitous signals, not because of the specificity of the signal or the receptor affinity, but because of the quality of the microenvironment encoded in the juxtaposition of one cell type with a specific correspondent. As in human society, the response of one individual to an exogenous stimulus is dictated as much by the spatial and temporal relationship of the individual with other individuals as by the endogenous makeup of the individual. Likewise, a primitive neural crest cell may become a neuron, an endocrine cell, or a melanocyte depending on its ultimate location, its interaction with adjacent cells, and its response, mediated by neighboring cells, to hormonal stimuli.

In essence, this preamble establishes that although some cell functions, such as cell proliferation, glycolysis, respiration, and gene transcription, proceed in isolation, their regulation as related to a functioning multicellular organism ultimately depends on the interaction among cells of the appropriate lineage, the appropriate stage in that lineage, and on the interaction among cells of different lineages occupying the same microenvironment. This concept suggests that if you

want to study the biology of isolated cells, or use the cells as a substrate, conventional monolayer or suspension cultures may be adequate, but if you want to learn something of the integrated function, or dysfunction, of cells *in vivo*, a histotypic or organotypic model will be required.

25.1.3 Choice of Models

There are two major ways to approach this goal. One is to accept the cellular distribution within the tissue, explant it, and maintain it as an organ culture. The second is to purify and propagate individual cell lineages, study them alone under conditions of homologous cell interaction, recombine them, and study their mutual interactions. These approaches have given rise to three main types of technique: (1) *organ culture*, in which whole organs, or representative parts, are maintained as small fragments in culture and retain their intrinsic distribution, numerical and spatial, of participating cells; (2) *histotypic culture*, in which propagated cell lines are grown alone to high density in a three-dimensional matrix; and (3) *organotypic culture*, in which cells of different lineages are recombined in experimentally determined ratios and spatial relationships to recreate a component of the organ under study (Fig. 25.2).

Organ culture seeks to retain the original structural relationship of cells of the same or different types, and hence their interactive function, in order to study the effect of exogenous stimuli on further development [Lasnitzki, 1992]. This relationship may be preserved by explanting the tissue intact or recreated by separating the constituents and recombining them, as in the now-classic experiments of Grobstein and Auerbach and others in organogenesis [Auerbach & Grobstein, 1958; Cooper, 1965; Wessells, 1977] (*see also* Section 17.7.1). Organotypic culture represents the synthetic approach, whereby a three-dimensional, high-density culture is regenerated from isolated (and, preferably, purified and characterized) lineages of cells that are then recombined, after which their interaction is studied and their response to exogenous stimuli is characterized. The exogenous stimuli may be regulatory hormones, nutritional conditions, or xenobiotics. In each case, the response is likely to be different from the responses of a pure cell type in isolation, grown at a low cell density.

25.2 ORGAN CULTURE

25.2.1 Gas and Nutrient Exchange

A major deficiency in tissue architecture in organ culture is the absence of a vascular system, limiting the size (by diffusion) and potentially the polarity of the cells within the organ culture. When cells are cultured as a solid mass of tissue, gaseous diffusion and the exchange of nutrients and metabolites is from the periphery and limits the size of the tissue. The dimensions of individual cells cultured in suspension or as a monolayer are such that diffusion is rapid,

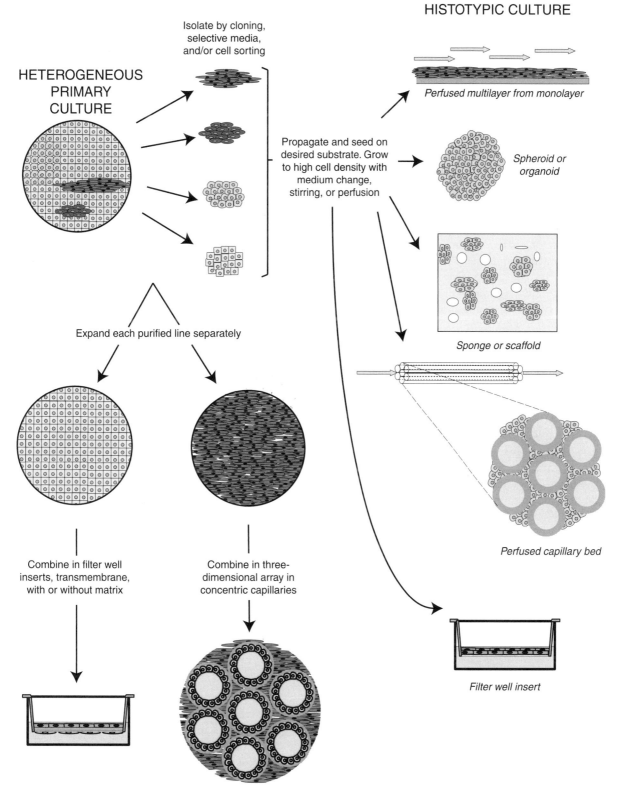

Fig. 25.2. Histotypic and Organotypic Culture.

but survival of cells in aggregates beyond about 250 µm in diameter (5000 cells) starts to become limited by diffusion, and at or above 1.0 mm in diameter ($\sim 2.5 \times 10^5$ cells) central necrosis is often apparent. To alleviate this problem, organ cultures are usually placed at the interface between the liquid and gaseous phases, to facilitate gas exchange while retaining access to nutrients. Most systems achieve this by positioning the explant in a filter well insert (*see* Figs. 25.7, 25.8) or on a raft or gel exposed to the air (Fig. 25.3), but explants anchored to a solid substrate can also be aerated by rocking the culture, exposing it alternately to a liquid medium and a gas phase [Nicosia et al., 1983; Lechner and LaVeck, 1985; *see* Protocol 23.9], or by using a roller bottle or tube (*see* Protocol 26.3).

Anchorage to a solid substrate can lead to the development of an outgrowth of cells from the explant and resultant alterations in geometry, although this effect can be minimized by using a nonwettable surface. One of the advantages of culture at the gas–liquid interface is that the explant retains a spherical geometry if the liquid is maintained at the correct level. If the liquid is too deep, gas exchange is impaired; if it is too shallow, surface tension will tend to flatten the explant and promote outgrowth.

Increased permeation of oxygen can also be achieved by using increasing O_2 concentrations up to pure oxygen or by using hyperbaric oxygen. Certain tissues—e.g., thyroid [de Ridder & Mareel, 1978], and prostate, trachea, and skin [Lasnitzki, 1992], particularly from a newborn or an adult—may benefit from elevated O_2 tension, but often this benefit is at the risk of O_2-induced toxicity. As increasing the O_2 tension will not facilitate CO_2 release or nutrient-metabolite exchange, the benefits of increased oxygen may be overridden by other limiting factors.

25.2.2 Structural Integrity

The maintenance of structural integrity, above other considerations, was and is the main reason for adopting organ culture as an *in vitro* technique in preference to cell culture. Whereas cell culture utilizes cells dissociated by mechanical or enzymic techniques or spontaneous migration, organ culture deliberately maintains the cellular associations found in the tissue. Initially, organ culture was selected to facilitate histological characterization, but ultimately it was discovered that certain elements of phenotypic expression were found only if cells were maintained in close association. The reasons for this are discussed above (*see* Section 25.1).

25.2.3 Growth and Differentiation

There is a relationship between growth and differentiation such that differentiated cells no longer proliferate (*see* Section 17.3). It is also possible that cessation of growth, irrespective of cell density, may in itself contribute to the induction of differentiation, if only by providing a permissive phenotypic state that is receptive to exogenous inducers of differentiation. Because of density limitation of cell proliferation and the physical restrictions imposed by organ culture geometry, most organ cultures do not grow, or, if they do, proliferation is limited to the outer cell layers. Hence, the status of the culture is permissive to differentiation and, given the appropriate cellular interactions and soluble inducers (*see* Section 17.7), should provide an ideal environment for differentiation to occur.

25.2.4 Limitations of Organ Culture

Experimental analysis of organ cultures depends largely on histological techniques, and they do not lend themselves readily to biochemical and molecular analyses. Biochemical monitoring requires reproducibility between samples, which is less easily achieved in organ culture than in propagated cell lines, because of sampling variation in preparing an organ culture, minor differences in handling and geometry, and variations in the ratios of cell types among cultures (*see* Table 1.4).

Organ cultures are also more difficult to prepare than replicate cultures from a propagated cell line and do not have the advantage of a characterized reference stock to which they may be related. Organ cultures cannot be propagated, and hence each experiment requires recourse to the original donor tissue, making the procedure labor intensive. Furthermore, as the population of reacting cells may be a minor component of the culture, attributing a molecular response to the correct cell type requires histological sectioning for autoradiographic, histochemical, or immunocytochemical *in situ* analysis.

Organ culture is essentially a technique for studying the behavior of integrated tissues rather than isolated cells. It is precisely in this area that a future understanding of the control of gene expression (and ultimately of cell behavior) in multicellular organisms may lie, but the limitations imposed by the organ culture system are such that recombinant systems between purified cell types may contribute more information at this particular stage. However, there is no doubt that organ culture has contributed a great deal to our understanding

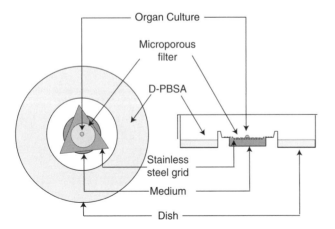

Fig. 25.3. Organ Culture. Small fragment of tissue on a filter laid on top of a stainless steel grid over the central well of an organ culture dish.

of developmental biology and tissue interactions and that it will continue to do so in the absence of adequate synthetic systems.

25.2.5 Types of Organ Culture

As techniques for organ culture have been dictated largely by the requirement to place the tissue at a location that allows optimal gas and nutrient exchange, most of these techniques put the tissue at the gas–liquid interface on semisolid gel substrates of agar [Wolff & Haffen, 1952] or clotted plasma [Fell & Robison, 1929] or on a raft of microporous filter, lens paper, or rayon supported on a stainless steel grid [Lasnitzki, 1992; Fig. 1.3] or adherent to a strip of Perspex or Plexiglas. This type of geometry is now most easily attained with filter well inserts (*see* Protocol 25.4) Protocol 25.1 uses organ primordia from chick embryo but is applicable to many other types of tissue.

PROTOCOL 25.1. ORGAN CULTURE

Outline
Dissect out the organ or tissue, reduce it to 1 mm^3, or to a thin membrane or rod, and place it on a support at the air–medium interface (e.g., filter well insert; *see* Figs. 25.7, 25.8). Incubate it in a humid CO_2 incubator, changing the medium as required.

Materials
Sterile or Aseptically Prepared:
- ❑ Instruments for dissection
- ❑ Medium (e.g., M199), with or without serum
- ❑ Filter well inserts, non-tissue-culture treated (e.g., Costar Transwells polycarbonate #3423, Corning)
- ❑ Multiwell plates, 12 well (Corning)
- ❑ *Nonsterile*:
- ❑ Fertile hen's eggs at 8 days of incubation

Protocol
1. Place the filter well inserts in the wells of a multiwell plate, and add sufficient medium to reach the level of the bottom of the filter (~1 mL).
2. Place the dishes in a humid CO_2 incubator to allow the pH of the medium to equilibrate at 37°C.
3. Prepare the tissue, or dissect out whole embryonic organs (e.g., 8-day femur or tibiotarsus of a chick embryo; *see* Protocols 12.2 and 12.7). The tissue must not be more than 1 mm thick, preferably less, in one dimension. (For example, 8-day embryonic tibiotarsus is perhaps 5 mm long, but only 0.5–0.8 mm in diameter. A fragment of skin might be 10 mm square but only 200 μm thick.

Tissue that must be chopped down to size, such as liver or kidney, should be no more than 1 mm^3).
4. For short dissections (<1 h), HBSS is sufficient, but for longer dissections, use 50% serum in HBSS buffered with HEPES to pH 7.4.
5. Take the dishes from the incubator, and transfer the tissue carefully to filters. A Pasteur pipette is usually best for this task and can be used to aspirate any surplus fluid transferred with the explant, although care should be taken not to puncture the filter. Wet the inside of the pipette with medium before aspiration, to prevent fragments of tissue from sticking to the pipette.
6. Check the level of medium, making sure that the tissue is wetted, but not totally submerged, and return the dishes to the incubator.
7. Check after 2–4 h to ensure that a film of medium remains over the filter and explant, but that it is not deep enough for the explant to float.
8. Incubate the dishes for 1–3 weeks, changing the medium every 2 or 3 days and sampling as required.

Variations. Most variations involve the following aspects:

(1) *Medium.* M199 or CMRL 1066 may be used with or without serum, and BGJ medium may be used [Biggers et al., 1961] for cartilage or bone.

(2) *Type of support.* Organ cultures may be supported by a filter (e.g., polycarbonate) lying on top of a stainless steel grid in a center-well organ culture dish (Falcon #3037, B-D Biosciences; Fig. 25.3). However, filter well inserts have a number of advantages in terms of handling, range of sizes, materials, and matrix coatings (*see* Fig. 25.7; Table 25.1). Different types of tissue may be combined on opposite sides of a filter to study their interaction (*see* Section 17.7.1). Furthermore, with small filter well inserts (e.g., 6.5 mm, as in Corning Costar #3423), the well formed on the top side of the filter assembly generates a meniscus of medium with a large surface area available for gas exchange. It is also possible to alter the configuration of the tissue by raising or lowering the level of medium in the dish, and therefore also in the well; deeper medium gives a spherical explant, and shallower medium flattens the explant.

(3) *O_2 tension.* Embryonic cultures are usually best kept in air, but late-stage embryo, newborn, and adult tissue are better kept in elevated oxygen [Trowell, 1959; de Ridder & Mareel, 1978; Zeltinger & Holbrook, 1997].

(4) *Stirred or static cultures.* Stirred cultures of small tissue fragments have been used for confrontational cultures for assay of invasion [Mareel et al., 1979; Bjerkvig et al., 1986a,b] (*see* Protocol 25.5).

(5) *Rocking or rotated cultures.* The tissue is anchored to a substrate and subjected alternately to liquid culture medium and the gas phase by placing the culture vessel on a rocking platform [*see* Protocol 23.9; Nicosia et al., 1983], or by anchoring the tissue to the wall of a rotating flask or tube (*see* Section 5.3.4 and Protocol 26.3).

25.3 HISTOTYPIC CULTURE

Histotypic culture, in this context, is defined as high-density cell culture with the cell density approaching that of the tissue *in vivo*. Various attempts have been made to regenerate tissuelike architecture from dispersed monolayer cultures. Green and Thomas [1978] showed that human epidermal keratinocytes will form dermatoglyphs (i.e., friction ridges) if they are kept for several weeks without transfer, and Folkman and Haudenschild [1980] were able to demonstrate the formation of capillary tubules in cultures of vascular endothelial cells cultured in the presence of endothelial growth factor and medium conditioned by tumor cells. As cells reach a high density, medium nutrients will become limiting. To avoid this, the ratio of medium volume to cell number should remain approximately as it was in low-density culture. This can be achieved by seeding cells on a small coverslip in the center of a large non-tissue-culture grade dish or by use of filter well inserts, which give the opportunity for the formation of both high-density polarized cultures and heterotypic combinations of cell types to create organotypic cultures (*see* Protocol 25.4). A high medium-to-cell ratio can also be maintained by perfusion (*see* Section 25.3.2).

25.3.1 Gel and Sponge Techniques

Leighton first demonstrated that both normal and malignant cells penetrate cellulose sponge [Leighton et al., 1968] facilitated by collagen coating; Gelfoam (a gelatin sponge matrix used in reconstructive surgery) can be used in place of cellulose [Sorour et al., 1975] and has been used in studies of the effect of mechanical strain on lung development [Liu et al., 1995]. These systems require histological analysis and are limited in dimensions, like organ cultures, by gaseous and nutrient diffusion. The use of three-dimensional sponges and gels has increased significantly with the development of tissue engineering (*see* Section 25.3.8).

Collagen gel. Collagen gel (native collagen, as distinct from denatured collagen coating) provides a matrix for the morphogenesis of primitive epithelial structures. Many different types of cell can be shown to penetrate such matrices and establish a tissuelike histology. Mammary epithelium forms rudimentary tubular and glandular structures when grown in collagen [Gomm et al., 1997], whereas breast carcinoma shows more disorganized growth [Berdichevsky et al., 1992]. The kidney epithelial cell line MDCK responds to paracrine stimulation from fibroblasts by producing tubular structures, but only in collagen gel [Kenworthy et al.,

1992]. Neurite outgrowth from sympathetic ganglia neurons growing on collagen gels follows the orientation of the collagen fibers in the gel [Ebendal, 1976].

Matrigel. Matrigel is a commercial product (B-D Biosciences), derived from the extracellular matrix of the Engelbreth−Holm−Swarm (EHS) mouse sarcoma, that has been used for coating plastic (*see* Section 8.4.1) but can also be used in gel form. It is composed of laminin, collagen, fibronectin, and proteoglycans with a number of bound growth factors, although it can be obtained in a growth factor-depleted form. It has been used as a substrate for epithelial morphogenesis [Larsen et al., 2004], formation of capillaries from endothelial cells [Jain et al., 1997; Vouret-Craviari et al., 2004], and in the study of malignant invasion [De Wever et al., 2004]. It is, however, a complex and not completely defined matrix and can inhibit some morphogenetic events, such as hepatocyte growth factor (HGF)-induced tubulogenesis of MDCK cells [Williams & Clark, 2003].

25.3.2 Hollow Fibers

Because medium supply and gas exchange become limiting at high cell densities, Knazek et al. [1972; Gullino and Knazek, 1979] developed a perfusion chamber from a bed of plastic capillary fibers, now available commercially (*see* Appendix II, Hollow fiber perfusion culture). The fibers are gas- and nutrient permeable and support cell growth on their outer surfaces. Medium, saturated with 5% CO_2 in air, is pumped through the centers of the capillaries, and cells are added to the outer chamber surrounding the bundle of fibers (*see* Fig. 8.11; *see also* Fig. 26.6). The cells attach and grow on the outside of the capillary fibers, fed by diffusion from the perfusate, and can reach tissuelike cell densities. Different plastics and ultrafiltration properties give molecular weight cut-off points at 10, 50, or 100 kDa, regulating the diffusion of macromolecules.

It is claimed that cells in this type of high-density culture behave as they would *in vivo*. For example, in such cultures, choriocarcinoma cells release more human chorionic gonadotrophin [Knazek et al., 1974] than they would in conventional monolayer culture and colonic carcinoma cells produce elevated levels of CEA [Rutzky et al., 1979; Quarles et al., 1980]. However, there are considerable technical difficulties in setting up the chambers, and they are costly. Furthermore, sampling cells from these chambers and determining the cell concentration are difficult. Overall, however, hollow fibers appear to present an ideal system for studying the synthesis and release of biopharmaceuticals and are now being exploited on a semi-industrial scale (*see* Section 26.2.6).

25.3.3 Spheroids

When dissociated cells are cultured in a gyratory shaker, they may reassociate into clusters. Dispersed cells from embryonic tissues will sort during reaggregation in a highly specific fashion [Linser & Moscona, 1980]. Cells in these heterotypic aggregates appear to be capable of sorting themselves into groups and forming tissuelike structures.

Homotypic reaggregation also occurs fairly readily, and spheroids generated in gyratory shakers or by growth on agar have been used as models for chemotherapy *in vitro* [Twentyman, 1980] and for the characterization of malignant invasion [Mareel et al., 1980]. As with organ cultures, the growth of spheroids is limited by diffusion, and a steady state may be reached in which cell proliferation in the outer layers is balanced by central necrosis (Fig. 25.4).

The following introduction and Protocol 25.2 for preparing multicellular tumor spheroids have been contributed by M. Boyd and R. J. Mairs, Center for Oncology and Applied Pharmacology, Glasgow University, Cancer Research UK Beatson Laboratories, Garscube Estate, Bearsden, Glasgow G61 1BD, Scotland.

Multicellular tumor spheroids provide a proliferating model for avascular micrometastases. The three-dimensional structure of spheroids allows the experimental study of aspects of drug penetration and resistance to radiation or chemotherapy that are dependent on intercellular contact. Spheroids are also well suited to the study of "bystander effects" in experimental targeted or gene therapy [Boyd et al., 2002, 2004]. Human tumor spheroids are more easily developed from established cell lines or from xenografts than from primary tumors [Sutherland, 1988].

From a single-cell suspension (trypsinized monolayer or disaggregated tumor), cells can be inoculated into magnetic stirrer vessels (Techne) and incubated to allow the formation of small aggregates over 3–5 days [Boyd et al., 2001]. This procedure is optimal; however, the majority of cell lines do not form spheroids in this manner. Alternatively, aggregates may be formed from cell suspensions in stationary flasks, previously base coated with agar. Aggregates may be left in the original flasks or transferred individually (by pipette) to multiwell plates, where continued growth over weeks will yield spheroids of maximum size, about 1000 μm [Yuhas et al., 1977; Sutherland, 1988].

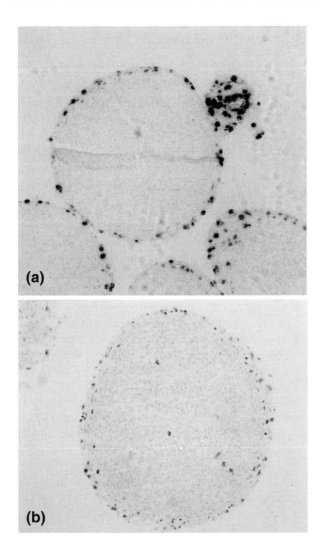

Fig. 25.4. Dividing Cells in Spheroids. Sections through mature spheroids of approximately 600- to 800-μm diameter. (a) Autoradiograph labeled with [^{125}I]iododeoxyuridine (IUdR), showing restriction of label to periphery. (b) Immunoperoxidase staining with anti-BUdR, showing similar restriction of label to periphery. (Courtesy of Ali Neshasterez.)

PROTOCOL 25.2. SPHEROIDS

Outline
Trypsinize monolayer cells, or disaggregate primary tissue, and seed the cells onto an agar-coated substrate. Transfer the aggregates to 24-well plates for analysis.

Materials
Sterile:
❑ Noble agar (Difco, B-D Biosciences)
❑ Growth medium
❑ Ultrapure water (UPW)
❑ Trypsin, 0.25%, in PBSA
❑ Flasks, 25 cm², or multiwell plates, 24 well
❑ Petri dishes, 9 cm

Note. When agar coating is used, all flasks, plates, and dishes should be sterile, but not necessarily tissue culture grade.

Nonsterile:
❑ Pi-pump (*see* Appendix II) or equivalent pipetting aid.

Protocol
Agar Coating In 25-cm² Flasks:
1. Add 1 g of Noble agar to 20 mL of UPW in a 100-mL borosilicate glass bottle with a loosely screwed-on cap.
2. Heat the agar in a water bath at 100°C for 10 min or until the agar has completely dissolved.
3. Add the contents of the bottle immediately to 60 mL of growth medium, previously heated to 37°C, and put 5-mL aliquots into each flask. Ensure that the agar is free from bubbles.

4. The agar will set at room temperature in ~5 min, giving a 1.25%-agar-coated flask.

Agar Coating In Multiwell Plates:

1. Add 0.5 g of agar to 10 mL of UPW, heat as in Step 2 for 25-cm^2 flasks, and then add 40 mL of UPW.

2. Place 0.5 mL of the resulting solution in each well of a 24-well plate to give a base coat of 1% agar. Accuracy and careful placement are important to ensure easy well-to-well focus of the microscope in subsequent viewing of spheroids.

Spheroid Initiation:

1. Trypsinize the confluent monolayer (for established lines; *see* Protocol 13.2) or disaggregate (for solid tumors; *see* Protocols 12.5, 12.6, and 12.8) to give a single-cell suspension.

2. Neutralize the trypsin with medium containing serum (if necessary).

3. Count the number of cells, using an electronic cell counter or a hemocytometer.

4. Place 5×10^5 cells in 5 mL of growth medium in each agar-coated 25-cm^2 flask, and incubate the cultures. If the cells are capable of spheroid formation, small aggregated clumps (about 100–300 μm in diameter) will form spontaneously in 3–5 days.

For subsequent growth, spheroids should be transferred to new 25-cm^2 flasks or 24-well plates.

Transfer to 25-cm^2 Flasks:

1. Transfer the contents of the original flasks to conical centrifuge tubes or universal containers.

2. Allow the spheroids to settle, and remove single cells with the supernate.

3. Resuspend the spheroids in fresh medium, and transfer the suspensions to new agar-coated flasks, where growth will proceed by division of cells in the outer layer.

Transfer to 24-Well Plates:

1. Transfer the contents of each 25-cm^2 flask into a 6-cm Petri dish.

2. Add 0.5 mL of medium to each agar-coated well of a 24-well plate.

3. Select individual spheroids of chosen dimensions under low-power magnification (×40), and, using a Pasteur pipette and a Pi-Pump or another pipetting aid with suitably fine control, transfer selected spheroids of similar diameter individually to the agar-coated wells of the 24-well plate.

4. Place the plate in a CO$_2$ incubator.

5. Replace the medium in the plate once or twice weekly (exchanging 0.5 mL each time), or add 0.5 mL of medium (without removing any medium) once or twice weekly, giving 2 mL/well after 2–4 weeks.

Analysis. Spheroid growth in wells or flasks may be quantified by regular (e.g., 2–3 times/week) measurement of the diameters of the spheroids, using a microscope eyepiece micrometer or graticule, or, preferably, by measurement of the cross-sectional area with an image analysis scanner. The most accurate growth curves are obtained when spheroids are grown in wells and are individually monitored.

Spheroids can also be utilized for clonogenic assay after treatment with test agents. Spheroids are collected in universal containers, washed with D-PBSA to remove residual medium, and then incubated with trypsin at 37°C for 5 min. Spheroids are then disaggregated to single-cell suspension by passage through a syringe needle and counted, and the number of viable cells is assessed by Trypan Blue exclusion (*see* Protocol 22.1). Cells can then be utilized for conventional clonogenic assay [Boyd et al., 2001, 2004].

Variations

Transfectant mosaic spheroids. Spheroids can be grown from populations of cells that have been transfected with different genes. The cells are first grown in monolayer and then are transfected with the transgene and subjected to selection for transgene-expressing cells. Mosaic spheroids are formed by the addition of both transfected and nontransfected monolayer cells in any desired proportions [Boyd et al., 2002, 2004]. The different cell populations are distributed throughout the resultant spheroids in approximately uniform mosaic patches, maintaining the same proportions of transfected to nontransfected cells as were added at the formation stage (Plate 12a)

Applications. Spheroids have wide applications in the modeling of avascular tumor growth [Ward & King, 1997], the role of three-dimensional spatial configurations in gene expression in cell populations [Waleh et al., 1995; Dangles et al., 1997], and the assessment of cytotoxic treatment. Treatment end points include growth delay, determination of the proportion of spheroids sterilized ("cured") by treatment, and colony formation in monolayer after disaggregation of treated spheroids [Freyer & Sutherland, 1980, Boyd et al., 1999, 2004]. An important area is the use of spheroids to study the penetration of cytotoxic drugs, antibodies, or other molecules used in targeted therapy [Sutherland, 1988; Carlsson & Nederman, 1989]. This category represents a special application that is not possible in single-cell suspensions or monolayer cultures. Spheroids have also proved useful in the study of cell killing by biologically targeted radionuclides [Mairs & Wheldon, 1996, Boyd et al., 2001, Fullerton et al., 2004]. Spheroid cultures have also been used in confrontation experiments to assess the invasiveness of spheroids derived from malignant cell populations that are grown in close proximity to normal cell cultures [de Ridder, 1997]; such techniques have also been used in nononcological studies of disease processes, such as studies of rheumatoid arthritis [Ermis et al., 1998]. Mosaic spheroids are a new variant form that has special applications in the assessment of bystander effects. For example, a current difficulty of gene therapy for cancer is the inefficiency of gene transfer procedures, leading to the requirement for bystander effects to eliminate cells in a tumor population that have not been transfected successfully. Mosaic spheroids mirror this situation *in vitro* and allow evaluation of different forms of the bystander effect, such as radiation cross fire when transfected cells are targeted with a radioactive agent [Boyd et al., 2002, 2004].

25.3.4 Rotating Chamber Systems

Miniperm bioreactor. Mixing and aeration can also be achieved by rolling the culture vessel, either in a conventional roller bottle (*see* Protocol 26.3) or in two-compartment chambers (Fig. 25.5). If the cell suspension is limited to one small compartment, then the cell concentration can be quite high, as the cells are not diluted by the bulk of the medium. The product concentration (e.g., antibody) accumulates in the cellular compartment, while nutrients and waste products diffuse across the semipermeable membrane to and from the medium compartment. An example of this kind of design is the MiniPERM™, a two compartment cylinder with cells in the smaller compartment and medium in the larger, separated by a semipermeable membrane (Vivascience, Sartorius; *see* Fig. 25.5). It is rotated to ensure mixing, and the medium can be sparged or replaced without disturbing the cells or product.

Although the major objective is bulk generation of product, e.g., monoclonal antibodies, the geometry of the chamber and the slow rotation tend to favor aggregate formation and this may enhance product formation.

Rotatory cell culture system (RCCS). Intrigued by the concept of growing cells in microgravity, in the 1980s NASA constructed a rotating chamber in which cells, growing in suspension, achieved simulated zero gravity with a slowly rotating chamber altering the sedimentation vector continuously (Fig. 25.6). The cells remain stationary, are subject to zero shear force, and tend to form three-dimensional aggregates, spheroidlike structures, which may

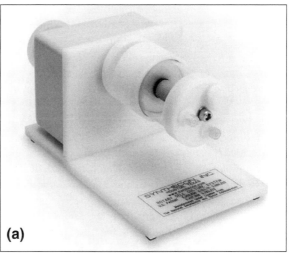

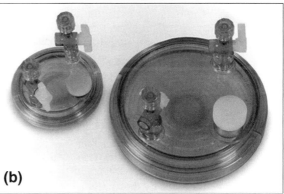

Fig. 25.6. Synthecon Rotatory Cell Culture System. In the Rotary Cell Culture System™, cells are maintained in suspension by adjusting the rotation speed of a cylindrical culture chamber. A gas permeable silicone membrane core gives the cells ample gas while disallowing shear causing bubbles. The NASA designed bioreactor is available from Synthecon (*see* Appendix III) and distributors world wide. (a) Reusable chamber, (b) Single use disposable chambers. (*See also* Plate 22c.)

be more differentiated with enhanced product formation [Proceedings—NASA bioreactors workshop on regulation of cell and tissue differentiation, 1997]. Gas exchange occurs from the cell-containing cylinder through a central silicone membrane. When the rotation stops, the aggregates sediment and the medium can be replaced. This bioreactor is available from Synthecon, Inc., as a disposable RCCS unit, or a reusable STLV (slow turning lateral vessel).

In addition to its original purpose, to determine the effects of microgravity on cells with application to the space program [Freed & Vunjak-Novakovic, 2002], this culture vessel has also provided a suitable bioreactor for bulk culture of tissue engineering constructs [Dutt et al., 2003; Vunjak-Novakovic, 2005].

25.3.5 Immobilization of Living Cells in Alginate

The technique of encapsulating living cells within alginate beads has been widely used in experimental research—e.g., hybridoma cells for monoclonal antibody production [Lang

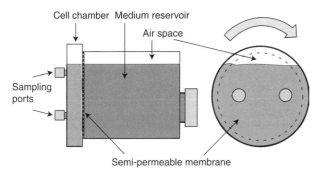

Fig. 25.5. Rotating Chamber System. Heraeus Miniperm. The concentrated cell suspension in the left chamber is separated from the medium chamber by a semipermeable membrane. High-molecular-weight products remain with cells and can be harvested from the sampling ports, while replenishment of the medium is carried out via the right-hand port. Mixing is achieved by rotating the chamber on a roller rack. (Miniperm: Vivascience, Sartorius AG.)

et al., 1995], hormone-producing cells used in animal models for the treatment of diabetes mellitus [Soon-Shiong et al., 1992] and chondrocytes (*see also* Protocol 23.16, Plate 18b).

Alginate is found primarily in brown seaweed *Laminaria, Macroaystis,* and *Ascophyllum* and consists mainly of two types of monosaccharides: L-guluronic acid (G) and D-mannuronic acid (M). It is composed of alternating molecules of M and G, and divalent cations bind strongly between separate G blocks and initiate the formation of an extended alginate gel network. Alginate gels can be formed into beads by dripping the alginate solution into a buffer containing divalent cations, such as Ca^{2+}. Mechanical strength, volume, stability, and porosity correlate with the G content such that alginate beads with a high G content have the largest pore sizes, ranging between 5 and 200 nm [Martinsen et al., 1989; Miura et al., 1986]. Pores of such sizes allow free diffusion of macromolecules out of, as well as into, the alginate. Host immune reactions to the alginate can be reduced substantially by using alginate with a high concentration of G and a low concentration of M [Otterlei et al., 1991].

At present, numerous cell types can be genetically engineered to produce specific proteins of choice. By encapsulating such cells in alginate, a valuable vehicle is obtained for delivering specific recombinant proteins to the organism. Thus, such alginate "bioreactors" may have an important therapeutic potential for the treatment of a number of diseases, in which the alginate may prevent the encapsulated cells from being destroyed by the immune system.

Protocol 25.3 and the preceding introduction to alginate encapsulation have been contributed by Tracy-Ann Read and Rolf Bjerkvig, Department of Anatomy and Cell Biology, University of Bergen, Norway.

PROTOCOL 25.3. ALGINATE ENCAPSULATION

Outline
Trypsinize producer cells from 75-cm² flask, count, and mix with sodium alginate solution. Drip from syringe into calcium chloride solution to form beads, and culture in suspension.

Materials
Sterile:
- ☐ Growth medium for selected cells
- ☐ D-PBSA
- ☐ Trypsin (concentration appropriate to subculture regime for cells)
- ☐ Saline solution containing:
    - **(a)** NaCl.........................8.0 g
    - **(b)** D-glucose..............................1.0 g
    - **(c)** Make up with UPW to....................1 L
    - **(d)** Adjust to pH 7.2–7.4 with HCl or NaOH
    - **(e)** Sterilize by autoclaving

- ☐ CaCl₂, 0.1 M, containing
    - **(a)** Saline solution.........................500 mL
    - **(b)** CaCl₂ · 2H₂O.........................7.35 g
    - **(c)** Adjust to pH 7.2–7.4 with HCl or NaOH
    - **(d)** Sterilize by autoclaving
- ☐ Sodium alginate (PRONOVA™ UP LVG): ultrapure, low viscosity, high guluronic acid content
- ☐ Sterilizing filters, 0.45 μm, nonpyrogenic (Millipore)

Protocol
1. Dissolve the alginate in saline solution to a concentration of 1.5%, and shake the resulting solution for a minimum 4 h at room temperature until the alginate has fully dissolved. Alginate dissolved in saline solution can be stored at 4°C for up to 3 days.
2. Filter the alginate 3 times with sterile nonpyrogenic 0.45-μm filters. The final filtration should be carried out immediately before the alginate is used.
3. Grow the cells to confluence.
4. Trypsinize the cells with 3 mL of trypsin, and count them.
5. Centrifuge the cells at 900 rpm for 4 min and remove the supernatant trypsin completely.
6. Mix the cells with the alginate to a concentration of 2×10^6 cells/mL by resuspending the cells gently in the alginate. At this point, it is crucial not to generate air bubbles in the alginate, as doing so will result in holes in the beads.
7. Transfer the alginate-suspended cells to a sterile syringe capped with a 27G needle.
8. Drip the cell-alginate suspension, applying a circular movement, into a beaker containing 0.1 M CaCl₂, which initiates the formation of alginate beads.
9. Allow the beads to gel for 10 min, and then wash them three times in D-PBSA.
10. Finally, wash them in growth medium.
11. Culture the beads in 175-cm² culture flasks containing growth medium at 37°C, 100% relative humidity, and 5% CO_2 in air.

25.3.6 Filter Well Inserts

Filter well inserts are a commercialization of a filter-based culture system the origins of which go back to the 1950s and used in various forms since then. A filter substrate provides an environment for studying cell interaction, stratification, polarization, and tissue modeling. Polarity and functional integrity can be established as in thyroid [Chambard et al., 1983], intestinal [Halleux & Schneider, 1994] and kidney [Mullin et al., 1997] epithelium (*see* Section 17.7.4). Filter cultures allow generation of stratified epidermis [Limat et al., 1995;

Kondo et al., 1997; Maas-Szabowski et al., 2000, 2002]. Others have used them to study invasion by granulocytes or malignant cells [McCall et al., 1981; Elvin et al., 1985; Repesh, 1989; Schlechte et al., 1990; Brunton et al., 1997].

One of the major advantages of filter well inserts is that they allow the recombination of cells at very high, tissuelike densities, with ready access to medium and gas exchange, but in a multireplicate form. Filter well inserts are now available from several suppliers (*see* Appendix II) in a variety of translucent or transparent materials, including polycarbonate, PTFE, and polyethylene teraphthalate, and ranging in size from 6.5 mm to 9 cm, suitable for 24-well, 12-well, and 6-well plates, or larger dishes (Figs. 24.5, 24.6; Plate 13; Table 25.1). Filters can be obtained precoated with collagen, laminin, fibronectin, or Matrigel.

PROTOCOL 25.4. FILTER WELL INSERTS

Outline
Seed cells into filter well inserts, and culture the cells in excess medium in multiwell plates.

Materials
Sterile:
- ❑ Approximately 0.5×10^6 cells per cm^2 of filter
- ❑ Growth medium, 1–20 mL per filter (depending on the vessel the housing filter)
- ❑ Filter well inserts
- ❑ Multiwell plates for filter inserts: 6, 12, or 24 well
- ❑ Forceps, curved

Nonsterile:
- ❑ Pipettor

Protocol
1. Place the filter wells in the plate or dish.
2. Add medium, tilting the dish to allow the medium to occupy the space below the filter and to displace the air with minimum entrapment. Add medium until it is level with the filter (2.5 mL for a 6-well plate; 1.0 mL for a 24-well plate).

---

3. Level the dish, and add 2×10^6 cells in 2 mL of medium to the top of the filter for a 25-mm-diameter filter, or 5×10^5 in 200 μL of medium for a 6.5-mm-diameter filter, taking care not to perforate the filter.
4. Place the dish in a humid CO_2 incubator in a protective box (*see* Section 6.6.2). It is critical to avoid shaking the box, and the cultures should not be moved in the incubator, to avoid spillage and resultant contamination.
5. Monolayers should become established in 3–5 days, although 5–10 days or longer may be required for histotypic differentiation (e.g., polarized transport) to become established.
6. Cultures may be maintained indefinitely, replacing the medium or transferring the insert to a fresh well or dish every 3–5 days.

Analysis.

(1) *Permeability.* Some epithelial cells (e.g., MDCK and Caco-2) and endothelial cells (e.g., from umbilical vein) form tight junctions several days after reaching confluence. This process is accelerated by precoating the membranes with collagen. Transepithelial permeability then becomes restricted to physiologically regulated transport through the cells, and pericellular transport falls to near zero. The process can be monitored by looking at dye (e.g., lucifer yellow), [^{14}C]methylcellulose, or [^{14}C]inulin transfer across the membrane, or by an increase in transepithelial electrical resistance (TEER).

(2) *Polarized transport.* The addition of labeled glucose or amino acid to the upper compartment of the filter well insert will show transport to the lower compartment, while the converse does not occur. If the cells possess P-glycoprotein or some other efflux transporter, cytotoxins (e.g., vinblastine) added to the lower compartment will be transported to the upper compartment, but not vice versa.

(3) *Penetration of cells through the filter.* Trypsinize and count each side of the filter in turn (trypsinized cells will not pass through even an 8-μm filter, as their spherical diameter

---

TABLE 25.1. Types of Filter Well Inserts

Make	Name	Material	Qualities	Transparency	Porosity (μm)
Millipore	Millicel	Nitrocellulose	Mesh	Opaque	0.45
		Polyolefin	Mesh	Transparent	0.45
Corning Costar	Transwells	Polycarbonate	Absolute	Transparent	5–8
				Translucent*	0.45–1.0
Falcon (B-D Biosciences)	Inserts	Polyethylene teraphthalate	Absolute	Transparent	0.45–3
Nunclon	Anocel	Ceramic	Sieve	Transparent	0.01
Earl-Clay	Ultraclone	Collagen	Mesh	Transparent	

*The higher the pore frequency, the lower the transparency. Low-porosity filters have a high pore frequency and are, consequently, less transparent.

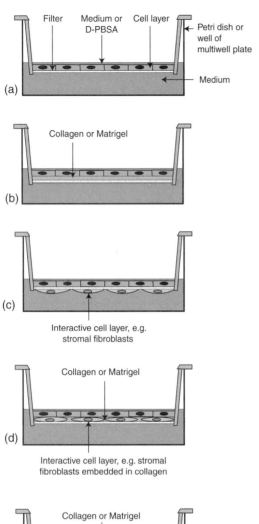

Fig. 25.7 labels, top diagram (a):
Filter — Medium or D-PBSA — Cell layer — Petri dish or well of multiwell plate — Medium

(b) Collagen or Matrigel

(c) Interactive cell layer, e.g. stromal fibroblasts

(d) Collagen or Matrigel — Interactive cell layer, e.g. stromal fibroblasts embedded in collagen

(e) Collagen or Matrigel — Interactive cell layer, e.g. endothelial cells below filter

Fig. 25.7. Filter Well Inserts. Sectional diagram of a hypothetical filter well insert. (a) Monolayer grown on top of the filter. (b) Monolayer grown on matrix on top of a filter. (c) Interactive cell layer added to the underside of a filter. (d) Interactive cell layer added to the matrix. (e) Interactive cell layer added to the underside of the filter with matrix coating above.

in suspension exceeds this size), or fix the filter, embed, and section and examine by electron microscope or conventional histology. Visualization is possible in whole mounts by mounting the fixed, stained (Giemsa) filter on a slide in DPX under a coverslip under pressure, to flatten the filter. Differential counting can then be performed by alternately focusing on each plane.

(4) *Detachment of cells from the filter to the bottom of the dish.* Count the cells by trypsinization or scanning.

Fig. 25.8. Transwells. Filter well inserts in a twelve-well plate, with a filter well insert alongside. (Corning.)

(5) *Partition above and below the filter.* Either count the cells as in (3), or prelabel the cells with rhodamine or fluorescein isothiocyanate (5 μg/mL for 30 min in a trypsinized suspension) and measure the fluorescence of solubilized cells (0.1% SDS in 0.3 N NaOH for 30 min) trypsinized from either side of the filter.

(6) *Cellular invasion* Precoat the filter with a cell layer (normal fibroblasts, MDCK, etc.), use microscopic examination to ensure that confluence is achieved, and then seed EDTA-dissociated test cells on top of a preformed layer (10^5–10^6 cells per filter). If the test cells are RITC- or FITC-labeled, then fluorescent measurements will reveal the appearance of the cells below the filter.

(7) *Matrix invasion* Coat the filter with Matrigel, apply the cells above the Matrigel, and monitor the appearance of the cells below the filter [Repesh, 1989; Schlechte et al., 1990]. Alternatively, seed the cells onto the lower surface of the filter, coat the upper surface of the filter with Matrigel, and monitor the invasion of the cells into the Matrigel by confocal microscopy [Brunton et al., 1997].

Variations

(1) *Depth of medium.* The depth of medium above the filter will regulate oxygen tension at the level of the cells. Keratinocytes or Type II pneumocytes from lung alveoli will require little medium and a high oxygen tension, whereas enterocytes, such as Caco-2, may be better off submerged and with a lower oxygen tension.

(2) *Filter porosity.* 1-μm filters allow cell interaction and contact without transit across the filter. 8-μm filters allow live cells to cross the filter. 0.2-μm filters probably do not allow cell contact. Low-porosity filters may be used to study cell interaction without permitting the cells to intermingle.

(3) *Transfilter combinations.* Invert the filter well insert, place upside down in a Petri dish, and load the underside (now

uppermost) with 0.5 mL, 2×10^6 cells/mL, of cell suspension. Place the lid on the top of the dish and touching the drop of cell suspension, before all of the medium drains through the filter. The depth of the dish will need to be about 0.5–1 mm higher than the height of the filter well insert so that a capillary space forms between the lid and the filter. Incubate the filter for 18 h. Capillarity will hold the medium and cells until the cells sediment onto the filter and attach [Brunton et al., 1997]. The next day, invert the filter and load the well with interacting cells or Matrigel, as described in Protocol 25.4.

25.3.7 Cultures of Neuronal Aggregates

Aggregating cultures of fetal brain cells have been extensively used to study neural cell differentiation [Seeds, 1971; Trapp et al., 1981; Bjerkvig et al., 1986a]. The aggregating cells follow the same developmental sequence as observed *in vivo*, leading to an organoid structure consisting of mature neurons, astrocytes, and oligodendrocytes. A prominent neuropil is also formed. In tumor biology, the aggregates can be used to study brain tumor cell invasion *in vitro* [Bjerkvig et al., 1986b].

The preceding introduction and Protocol 25.5 for aggregating cultures of brain cells have been contributed by Rolf Bjerkvig, Department of Cell Biology and Anatomy, University of Bergen, Žrstadveien 19, N–5009 Bergen, Norway.

PROTOCOL 25.5. NEURONAL AGGREGATES

Outline
Remove brains from fetal rats at day 17 or 18 of gestation (consult local Animal Ethics Committee, *see* Section 7.9.1), and prepare the brains as a single-cell suspension. Form brain aggregates by culturing in agar-coated wells in a multiwell plate. The cells in the aggregates form a mature organoid brain structure during a 20-day culture period.

Materials
Sterile:
- Dulbecco's modification of Eagle's medium, containing 10% heat-inactivated newborn calf serum; four times the prescribed concentration of nonessential amino acids; L-glutamine, 2 mM; penicillin, 100 U/mL; streptomycin, 100 μg/mL
- Phosphate-buffered saline (PBS) with Ca^{2+} and Mg^{2+}
- Trypsin type II (0.025% in D-PBSA)
- Agar (Difco)
- Multiwell tissue culture dishes (24-well plates; Nalge Nunc)
- Petri dishes, 10 cm
- Test tubes, 12 mL

- Scalpels, scissors, and surgical forceps
- Erlenmeyer flasks, 2 at 100 mL
- To coat the wells with agar-medium, use the following procedure:
  (a) Prepare a 3% stock solution (3 g of agar in 100 mL of D-PBSA) in an Erlenmeyer flask.
  (b) Heat the flask in boiling water until the agar is dissolved. Place an empty Erlenmeyer flask in boiling water, and add 10 mL of hot agar solution to it.
  (c) Slowly add warm complete growth medium to the flask until a medium-agar concentration of 0.75% is reached.
  (d) Add 0.5 mL of warm medium-agar solution to each well in the multiwell dish.
  (e) Allow agar to cool and gel.
  The multiwell dishes can be stored in a refrigerator for 1 week.

Nonsterile:
- Water bath

Protocol
1. Dissect out, aseptically, the whole brains from a litter of fetal rats at day 17 or 18 of gestation, and place the tissue in a 10-cm Petri dish containing D-PBSA.
2. Using scalpels, mince the tissue into small cubes, ~0.5 cm³.
3. Transfer the tissue to a test tube, and wash it three times in D-PBSA. Allow the tissue to settle to the bottom of the tube between each washing.
4. Add 5 mL of trypsin solution to the tissue, and incubate in a water bath for 5 min at 37°C.
5. Disaggregate the tissue by trituration through a Pasteur pipette approximately 20 times.
6. Allow the tissue to settle for 3 min, and transfer the clump-free milky cell suspension to a test tube containing 5 mL of growth medium.
7. Add 5 mL of fresh trypsin to the undissociated tissue, and repeat the trypsinization and dissociation procedure twice more.
8. Spin the cell suspension at 200 g for 5 min.
9. Aspirate and discard the supernate, resuspend the cells, and pool them in 10 mL of growth medium.
10. Count the cells, and add 3×10^6 cells in 1 mL to each agar-coated well.
11. Place the multiwell dish in a CO_2 incubator for 48 h.
12. Remove the aggregates to a sterile 10-cm Petri dish, and add 10 mL growth medium to the dish.
13. Transfer larger aggregates individually to new agar-coated wells by using a Pasteur pipette.
14. Change the medium every third day by carefully removing and adding new overlay medium.

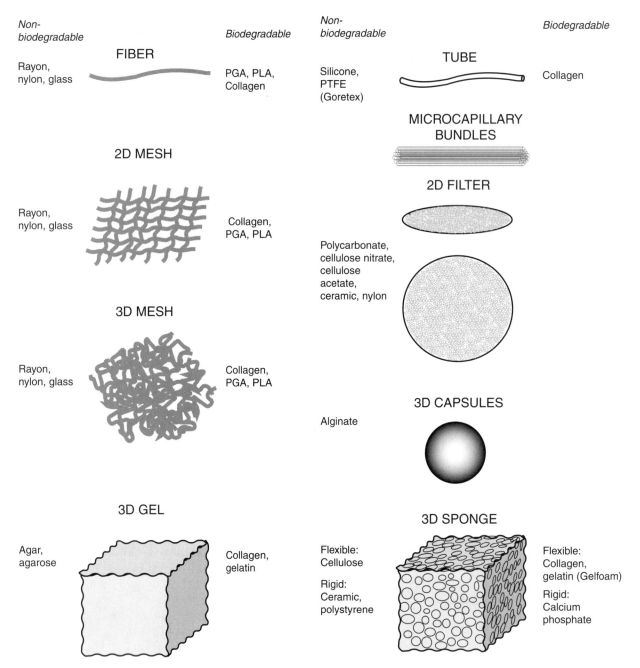

Fig. 25.9. Scaffolds and Matrices. Overview of types of scaffolds and matrices used in tissue engineering constructs. Many different geometries have been employed including linear fibers or tubes (top), two-dimensional mesh screens or filters (center), and three-dimensional cubes or spheres (bottom). Although nondegradable materials have been used (left-hand labels in each column) the trend is toward biodegradable scaffolds (right-hand labels in each column) such as collagen, gelatin, polyglycolic acid (PGA) or polylactic acid (PLA), and calcium phosphate.

During 20 days in culture, the aggregates will become spherical and develop into an organoid structure.

Analysis. Fix and embed in paraffin or epon for histological or electron microscopic evaluation. Oligodendrocytes, astrocytes, and neurons are identifiable by transmission electron microscopy or by immunohistochemical localization of myelin basic protein, glial fibrillary acidic protein, and neuron-specific enolase, respectively.

Variations. A single-cell suspension can be obtained by mechanical sieving through steel or nylon meshes [Trapp et al., 1981] (*see* Section 12.3.8). Reaggregation cultures can also be obtained by using a gyratory shaker. Select a speed

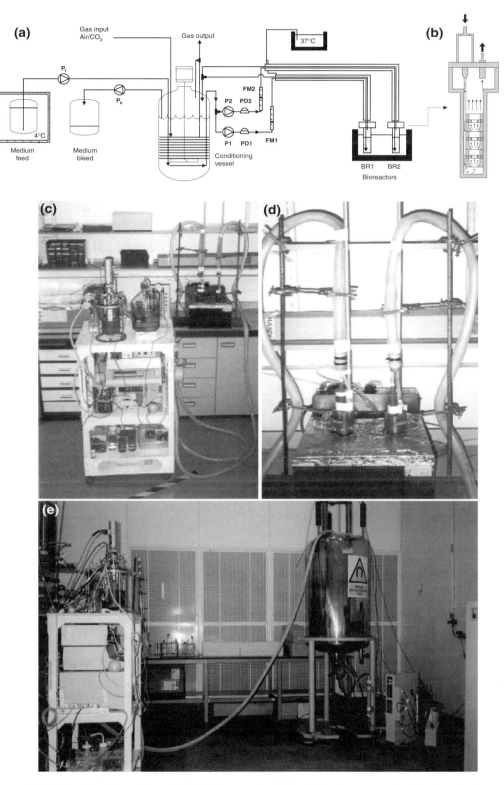

Fig. 25.10. Monitoring Cells within 3D Constructs by MRI. (a) Medium reservoir and gassing device supplying constructs of tissue-engineered cartilage in perfused bioreactors BR1 and BR2. (b) Enlarged view of bioreactor [*see also* Neves et al., 2003]. (c) Supply trolley with medium reservoirs and gassing facility, supplying the bioreactors with their temperature maintained in a water bath (d). (e) Taking a reading with the bioreactor, still connected to its medium supply, transferred to the NMR detector. The output can be visualized as a two- or three-dimensional MRI (*see* Fig. 25.11) or subjected to spectral analysis (*see* Fig. 26.18). (Courtesy of A. A. Neves & K. Brindle).

(about 70 rpm) such that the cells are brought into vortex, thereby greatly increasing the number of collisions between cells. This movement also prevents cell attachment to the culture flasks.

25.3.8 Tissue Equivalents and Tissue Engineering

The advent of filter well technology, boosted by its commercial availability, has produced a rapid expansion in the study of organotypic culture methods. Skin equivalents have been generated by coculturing dermis with epidermis (Mattek Epiderm, Episkin, SkinEthic), with an intervening layer of collagen, or with dermal fibroblasts incorporated into the collagen [Limat et al., 1995; Maas-Szabowski et al., 2000, 2002], and models for paracrine control of growth and differentiation have been developed with cells from lung [Speirs et al., 1991], prostate [Thomson et al., 1997], and breast [Van Roozendahl et al., 1992].

The opportunity for heterotypic cell interaction has also opened up numerous opportunities for studying inflammation and irritation *in vitro* (*see* Plate 21; *see also* Section 22.6) and for creating other models for tissue interaction with increased *in vivo* relevance [Emura et al., 1997; Gomm et al., 1997]. Construction of tissue equivalent cultures has also made tissue replacement therapy possible. Skin equivalent cultures have been used in burn repair [Hunziker & Limat 1999; Gobet et al., 1997; Wright et al., 1998], and now tissue engineering is being applied to the construction of tissue replacements for many different locations including cartilage, bone, ligament, cardiac and skeletal muscle, blood vessels, liver, and bladder [Atala & Lanza, 2002; Vunjak-Novakovic & Freshney, 2005].

Just as organotypic culture needs cell interaction, constructs for tissue engineering often require similar interactions, as in the interaction between endothelium and smooth muscle in blood vessel reconstruction [Klinger and Niklason, 2005]. In addition to biological interactions, some constructs also require physical forces; skeletal muscle needs tensile stress [Shansky et al., 1997; Powell et al., 2002; Shansky et al., 2005], bone [Mullender et al., 2004] and cartilage [Seidel et al., 2004] need compressive stress, and vascular endothelium in a blood vessel construct needs pulsatile flow [Niklason et al., 2001].

Engineering of tissue constructs depends on several components, depending on the tissue:

(1) Tissue cells of the correct lineage and at a proliferative, progenitor stage
(2) Interactive cells, e.g., dermal fibroblasts in skin, smooth muscle cells in blood vessels, or glial cells in neural constructs

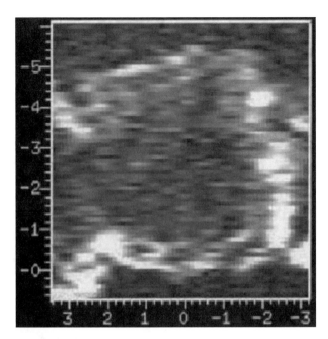

Fig. 25.11. MRI of Cartilage Construct. Brightness of image is proportional to cell density. Distances on grid are in mm. (Courtesy of Dr. Kevin Brindle; [from Thelwell et al., 2001]).

(3) A biodegradable scaffold to support the structure; e.g., polyglycolic acid (PGA); polylactic acid (PLA), or calcium phosphate (Fig. 25.9)
(4) Matrix, e.g., collagen, in place of the scaffold for soft tissues, or coating the scaffold to enhance cellular attachment
(5) Mechanical stress; tensile for muscle, compressive for cartilage and bone, and pulsatile for blood vessels

Neural cells have also been used for tissue reconstruction, but in these cases a cell suspension, rather than a construct, is injected into the injured site [Franklin & Barnett, 1997; Franklin & Barnett 2000; Wewetzer et al., 2002; Totoiu et al., 2004; Groves et al., 1993].

25.4 IMAGING CELLS IN 3D CONSTRUCTS

As microscopic observation becomes difficult when cells are incorporated into a scaffold in a 3D organotypic construct, alternative methods must be used to visualize the status of the cells within the construct. This can be done by NMR if the bioreactor housing the constructs is placed within an NMR detector (Fig. 25.10). The output can be displayed as an MRI (Fig. 25.11), and the emission spectrum can be analyzed (*see* Fig. 26.18).

CHAPTER 26

Scale-Up

The threshold between normal laboratory scale usage and large-scale or bulk culture is purely arbitrary, but is generally regarded as the level of cell production above which specialized apparatus and procedures will be required. Although $1 \times 10^9 - 1 \times 10^{10}$ cells can be produced in simple stirrer cultures of around 1- to 10-L capacity (*see* Section 13.7.5), larger-scale cultures of $1 \times 10^{11} - 1 \times 10^{12}$ cells will require apparatus ranging from a 100-L laboratory-scale fermentor to a semi-industrial pilot plant with capacities of from 100 to 1000 L. Full-scale industrial production uses 5000- to 20,000-L bioreactors, but these are beyond the scope of this book. The terms *stirrer culture* or *spinner culture* are synonymous and tend to be used for simple culture systems at the low end of the laboratory range of equipment. The terms *fermentor* and *bioreactor*, although not synonymous, have considerable overlap. The name *fermentor* derives from microbiological culture systems, designed originally for bacteria and yeast, and was used initially for laboratory equipment of around 50- to 100-L capacity, but the increase in scale in line with developments in the biotechnology industry, coupled with a greater diversity in design to cope with monolayer cultures as well as suspension, has led to the introduction of the name *bioreactor*. In tissue engineering, the term includes stirred culture of constructs in relatively small volumes.

The method employed to increase the scale of a culture depends on whether the cells proliferate in suspension or must be anchored to the substrate. Methods for suspension cells are generally simpler and will be dealt with first. Monolayer techniques, being dependent on increased provision of growth surface, tend to be more complex and will be dealt with later. Many of the monolayer techniques are also applicable to suspension cells.

26.1 SCALE-UP IN SUSPENSION

Scale-up of suspension cultures involves, primarily, an increase in the volume of the culture medium. Agitation of the medium is necessary when the depth exceeds 5 mm, and above $5-10$ cm (depending on the ratio of surface area to volume), sparging with CO_2 and air is required to maintain adequate gas exchange (Figs. 26.1, 26.2). Stirring of such cultures is best done slowly with a magnet encased in a glass pendulum (Techne, Integra) or with a large-surface-area paddle (Bellco). The stirring speed should be between 30 and 100 rpm, sufficient to prevent cell sedimentation, but not so fast as to create shear forces that would damage the cells. Antifoam (Dow Chemical Co.) or Pluronic F68 (Sigma), $0.01-0.1\%$, should be included when the serum concentration is above 2%, particularly if the medium is sparged. In the absence of serum, it may be necessary to increase the viscosity of the medium with carboxymethyl cellulose $(1-2\%)$ (molecular weight $\sim 10^5$).

PROTOCOL 26.1. STIRRED 4-LITER BATCH SUSPENSION CULTURE

The procedure for setting up a 4-L culture of suspended cells is as follows:

Culture of Animal Cells: A Manual of Basic Technique, Fifth Edition, by R. Ian Freshney
Copyright © 2005 John Wiley & Sons, Inc.

Fig. 26.1. Large Stirrer Flask. Techne 5-L stirrer flask on a magnetic stirrer. Note the offset pendulum that makes an excursion in the annular depression in the base of the flask. The side arms are for sampling or perfusion with CO_2 in air, which would be required with this volume of medium (*see* Figure 26.2).

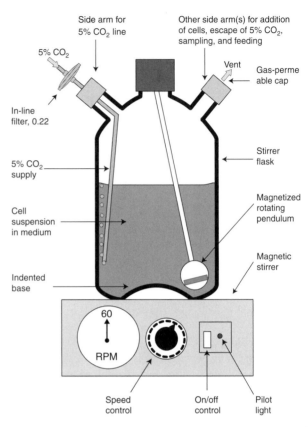

Fig. 26.2. Stirrer Culture. Diagram of a large stirrer flask suitable for volumes up to 8 L.

Outline

Grow a pilot culture of cells, and add it to a prewarmed aspirator of medium already equilibrated with 5% CO_2. Stir the cell suspension slowly, with sparging, until the required cell concentration is reached, and then harvest the cells.

Materials

Sterile or Aseptically Prepared:

☐ Starter culture vessel (small stirrer flask; *see* Figs. 7.6 and 12.7)
☐ Growth medium, warmed to 37°C
☐ Antifoam (silicone: Dow Corning, Merck; Pluronic F68: Sigma)
☐ Prepared large stirrer flask (Bellco, Techne; *see* Figs. 26.1 and 26.2) with:
  (a) A magnet enclosed in a glass pendulum, or a suitable alternative stirrer (*see* Section 26.1) suspended from the top cap (e.g., Techne).
  (b) One inlet port with a removable CO_2-permeable cap (i.e., a cap with an integral microporous membrane; *see* Section 8.2.3), or a cap with a silicone diaphragm pierced by a wide-bore needle with a Luer connection fixed to a sterile in-line filter (e.g., Millex).

  (c) A second inlet port carrying a tube for the gas line (e.g., Bellco #1965) with a 2- to 3-mm internal diameter, which reaches almost to the bottom of the vessel, but remains clear of the pendulum when it is stirring, and with the external entry terminating in a female Luer fitting guarded by aluminum foil.
  (d) Sterilize the stirrer culture vessel by autoclaving, at 100 kPa (15 lb/in.²) for 20 min, fully assembled, but with the CO_2-permeable cap removed and sterilized separately.
☐ An in-line sterile Luer-fitting micropore filter, 25-mm diameter, 0.2-μm porosity, (e.g., Millex), to attach to the gas entry line at the time of use

Nonsterile:

☐ Magnetic stirrer
☐ Supply of 5% CO_2, preferably from a metered supply at ~10–30 kPa (2–5 lb/in.²)
☐ D-PBSA for counting cells
☐ Electronic cell counter or hemocytometer

Protocol

1. Set up a starter culture (*see* Protocol 13.3 and Fig. 13.5) with 200 mL medium, and seed it with cells at 5×10^4–1×10^5 cells/mL. Place the culture on the magnetic stirrer, rotating at 60 rpm, and incubate the culture until a concentration of 5×10^5–1×10^6 cells/mL is reached. (*Note*: do not exceed 1×10^6 cells/mL as the cells may enter apoptosis.)

2. Set up the large stirrer flask as in Fig. 26.2.
3. Add 4 L of medium to the flask.
4. Add 0.4 mL of antifoam to the flask, using a disposable pipette or syringe.
5. Close the side arm with the CO_2-permeable cap.
6. Place the flask on the magnetic stirrer in a 37°C room or incubator.
7. Connect the 5%-CO_2 air line via the sterile, micropore in-line filter to the gas inlet Luer fitting.
8. Turn on the gas at a flow rate of approximately 10–15 mL/min.
9. Stir the culture at 60 rpm.
10. Incubate the large stirrer flask for about 2 h to allow the temperature and CO_2 tension to equilibrate.
11. Turn off the CO_2, and disconnect the flask from the CO_2 line.
12. Bring the large stirrer flask and the starter culture back to the laminar-flow hood.
13. Transfer the starter culture to the large stirrer flask by pouring in one single smooth action.
14. Return the large stirrer flask to the 37°C room or incubator.
15. Restart the stirrer at 60 rpm.
16. Reconnect the 5% CO_2 line, and adjust the gas flow to 10–15 mL/min. (Estimate the rate of flow by counting the bubbles if there is no flow meter on the line.)
17. Incubate the culture for 4–7 days, sampling every day to check the cell proliferation rate:
    (a) Bring the stirrer flask back to the laminar-flow hood.
    (b) Remove the side-arm cap.
    (c) Withdraw 5–10 mL of the cell suspension.
    (d) Count the cells, and check their viability by dye exclusion.
18. When the cell concentration reaches the desired level,
    (a) Turn off the gas and the magnetic stirrer;
    (b) Disconnect the stirrer flask from the gas supply;
    (c) Take the culture to the laboratory, and pour off the cells into centrifuge bottles.
    (d) Centrifuge the cells at 100 g for 10 min, and collect the cells (from the pellet) or the supernatant medium, as appropriate.

Analysis. Monitor the growth rate of the cells daily. For the best results, the cells should not show a lag period of more than 24 h and should still be in exponential growth when harvested. Plot the cell counts daily, and harvest the cells at approximately 1×10^6 cells/mL. Suspension cultures tend to enter apoptosis if this concentration is exceeded.

Variations
Adding medium. The following procedures are alternatives for adding medium to vessels larger than 5 L:

(1) Buy medium in media bags (*see* Appendix II) that can be hooked up alongside the stirrer flask, warmed to 37°C, and allowed to run in unattended.
(2) Sterilize the stirrer flask with a premeasured volume of UPW, and make up the medium from concentrates *in situ*. Mark the side of the flask to indicate the level of water, so that any water lost by evaporation during autoclaving can be made up.
(3) Use an autoclavable medium (*see* Appendix II), and autoclave it *in situ*.

Harvesting. Pouring from a large stirrer flask can be difficult, unless there is a bottom outlet. However, bottom outlets tend to clog with dead cells and are best avoided. Harvesting is best done by pumping or displacement:

(1) Remove the in-line filter from the gas line and add silicone tubing and either siphon off the cells or use a peristaltic pump, tilting the vessel when the fluid is near to the bottom.
(2) Disconnect the 5% CO_2 supply, replace the micropore filter with a flexible tube, attach the 5% CO_2 supply to the other port, and blow the cells out through the gas line.

26.1.1 Continuous Culture
If it is required that the cells be kept at a set concentration, the culture can be maintained in a *chemostat* or *biostat*. In this type of culture system, the cells are grown to the mid-log phase (monitored by daily cell counts); a measured volume of cells is removed each day and replaced with an equal volume of medium. Alternatively, the cells may be run off continuously, at a constant rate, at mid-log phase, and medium added at the same rate. The latter will require a stirrer vessel with four ports, two for CO_2 inlet and outlet, one for medium inlet, and one for spent medium outlet, collecting into a reservoir (Fig. 26.3). The flow rate of medium may be calculated from the growth rate of the culture [Griffiths, 2000] but is better determined experimentally by serial cell counting at different flow rates of the medium. Cell counts can be performed on the bioreactor by disconnecting the CO_2 inlet and withdrawing a sample by a syringe with a Luer double female adapter. The flow rate may be regulated by a variable peristaltic pump on the inlet line.

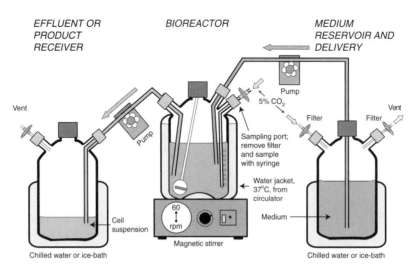

Fig. 26.3. Biostat. A modification of the suspension culture vessel in Fig. 26.1, with continuous matched input of fresh medium and output of cell suspension. The objective is to keep the culture conditions constant rather than to produce large numbers of cells (*see also* Figs. 26.17 and 26.18). Bulk culture, per se, is best performed in batches in the apparatus in Fig. 26.1 or Fig. 26.17.

Production of cells in bulk, in the 1- to 20-L range, is best done by the batch method described in Protocol 26.1. A steady state is required for monitoring metabolic changes related to cell density but is more expensive in medium and is more likely to lead to contamination. However, if the operation is in the 50- to 1000-L range, then more investment and time are spent in generating the culture, and the batch method becomes more costly in time, materials, and down time; thus, continuous culture may be better in such cases.

Suspension cultures can also be stirred by rotating bottles on a roller rack (*see* Fig. 26.11), as for monolayer cultures (*see* Protocol 26.3).

Adherent cells. Anchorage-dependent cells cannot be grown in liquid suspension, except on microcarriers (*see* Protocol 26.4), but transformed cells (e.g., virally transformed or spontaneously transformed continuous cell lines, such as HeLa-S$_3$) can. Because these cells are still capable of attachment, the culture vessels will need to be coated with a water-repellent silicone (e.g., Repelcote), and the calcium concentration may need to be reduced. S-MEM medium (Invitrogen, Sigma) is a variation of Eagle's MEM with no calcium in the formulation and has been used for the culture of HeLa-S$_3$ and other cells in suspension.

26.1.2   Scale and Complexity

Standard bench-top stirrer cultures operate satisfactorily up to around 10 L for the bulk production of cells or medium. If, however, more attention must be paid to process control, then a controlled fermentor or bioreactor should be used. These culture vessels have regulated input of medium and gas and provide the capability for data collection from oxygen, CO_2, and glucose electrodes in the culture vessel. They are regulated by a programmable control unit that records and

outputs data and can be used to regulate gas and liquid input and output, stirring speed, temperature, and so on (*see* Section 26.3).

26.1.3   Mixing and Aeration

Problems of increased scale in suspension cultures revolve around mixing and gas exchange. In contrast to simple stirrer cultures, most laboratory-scale fermentors agitate the medium with a rotating turbine or paddle. Designs of these fermentors vary, mostly to try to achieve maximum movement of liquid with minimum shear stress for the cells. Successful designs usually employ a slowly rotating large-bladed paddle with a relatively high surface area [Griffiths, 2000] and some additive designed to minimize the harmful effects of shear—e.g., Pluronic F68, carboxymethylcellulose (CMC), or polyvinylpyrrolidone (PVP) [Cherry & Papoutsakis, 1990].

Culture bags (see Appendix II). Plastic bags can be used to culture suspension cells. They are gas permeable and can be agitated by rocking on trays on a flat rocking platform.

Air lift fermentors. Large-scale fermentors frequently use the air lift principle (Fig. 26.4): 5% CO_2 in air is pumped into a porous steel ring at the base of the central cylinder, and bubbles stream up the center, carrying a flow of liquid with them, and are released at the top, while the medium is recycled to the bottom of the cylinder. This fairly simple type of fermentor is used extensively in the biotechnology industry, up to capacities of 20,000 L.

BelloCell aerator culture. This unusual device has a bellows medium compartment that alternately forces medium over cells anchored in porous matrices and withdraws it again, in a "breathing" motion of the bellows (Fig. 26.5). Optimum mixing and aeration is claimed

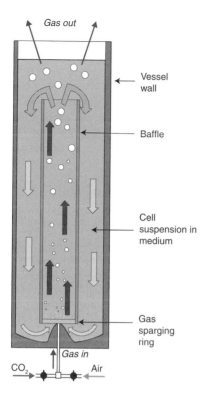

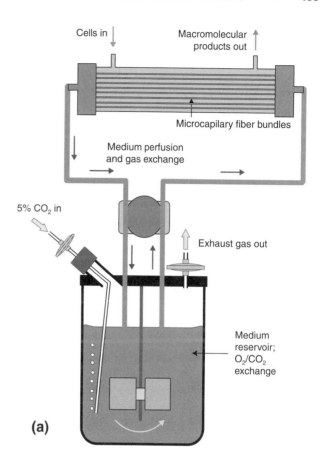

(a)

Fig. 26.4. Air Lift Fermentor. In an air lift fermentor, the bioreactor consists of two concentric cylinders, with the inner cylinder being shorter at both ends than the outer, thereby creating an outer and an inner chamber. The bottom of the inner chamber carries a sintered steel ring through which 5% CO_2 in air is bubbled. The bubbles rise, carrying the cell suspension with them. Air/CO_2 is vented from the top, and displacement ensures the return of the cell suspension down the outer chamber. (Modified from Griffiths [2000]).

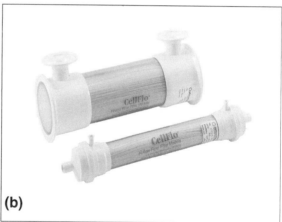

(b)

(c)

Fig. 26.5. BelloCell Aerator Culture. Cells are grown on matrix material that is alternately submerged in medium and exposed for gas exchange by pumping of the bellows compartment.

Fig. 26.6. Hollow Fiber Perfusion. (a) Sectional diagram of a hollow fiber perfusion system. Medium is circulated from a reservoir to the culture chamber by a peristaltic pump. Aeration and CO_2 exchange are regulated in the reservoir of medium. (b) Spectrapor hollow fiber cartridges. (c) Cartridge with removable ends. (b,c Courtesy of Spectrapor.)

with minimum shear (Cesco Bioengineering; Metabios; www.metabios.com/BelloCell.htm).

Perfused suspension culture. Hollow fiber and membrane perfusion systems also operate on the principle of compartmentalization. The cells are retained in a low-volume compartment at a very high concentration, and medium is perfused through hollow fibers within the cell compartment (Fig. 26.6), or through an adjacent membrane compartment (Membroferm, Polymun Scientific; Fig. 26.7). Regulation of gas exchange in the medium is external to the culture chamber. One version of the Membroferm allows for the product to be collected in a third membrane-bound compartment [Klement et al., 1987]. Like compartments can also be linked for serial perfusion. Hollow fibers are also available with double concentric spaces so that nutrient can be perfused down the center, product collected in the outer space, and cells grown in between the two concentric tubes.

Fluidized bed reactors for suspension cultures.
Although microcarriers were originally conceived for monolayer cultures (*see* Section 26.2.4), porous microcarriers can accommodate suspension cells within the interstices of the bead matrix. Because of the higher density of the microcarriers they can be perfused slowly from below, at such a rate that their sedimentation rate matches the flow rate. The beads therefore remain in stationary suspension, perfused by the medium, constantly replenishing nutrients and collecting the product into a downstream reservoir. Gas exchange is external to the reactor, and no mechanical mixing is required. Macroporous beads or fibrous matrices have also been used with entrapped suspension cells in fixed bed reactors with medium perfusion (CelliGen, New Brunswick Scientific).

26.2 SCALE-UP IN MONOLAYER

Regulations regarding the use of transformed cells for the production of biopharmaceuticals are now being relaxed, but a number of arguments for anchorage-dependent cells remain.

Although cells that grow in suspension are easier to manage, anchored cells, with their resultant potential for developing polarity, may yet represent a better model for posttranslational modification of proteins and the appropriate membrane flux, leading to their secretion from the cells in a bioactive form. For anchorage-dependent monolayer cultures, it is necessary to increase the surface area of the substrate in proportion to the number of cells and the volume of medium. This requirement has prompted a variety of different strategies, some simple and others complex.

26.2.1 Multisurface Propagators
Nunc Cell Factory. One of the simpler systems for scaling up monolayer cultures is the Nunclon Cell Factory (Fig. 26.8; *also* Table 7.1). This system is made up of rectangular Petri dish-like units, with a total surface area of 600–24,000 cm^2, interconnected at two adjacent corners by vertical tubes. Because of the positions of the apertures in the vertical tubes, medium can flow between compartments only when the unit is placed on end. When the unit is rotated and laid flat, the liquid in each compartment is isolated, although the apertures in the interconnecting tubes still allow connection of the gas phase. The cell factory has the advantage that it is not different in the geometry or the nature of its substrate from a conventional flask or Petri dish.

The recommended method of using the cell factory is described in Protocol 26.2 (*see* Fig. 26.9).

PROTOCOL 26.2. NUNC CELL FACTORY

Outline
Prepare a cell suspension in medium, and run the suspension into the chambers of the unit, lying on its long edge (*see* Fig. 26.9). Rotate through 90°, and then lay the unit flat and gas it with CO_2. Seal and incubate the unit.

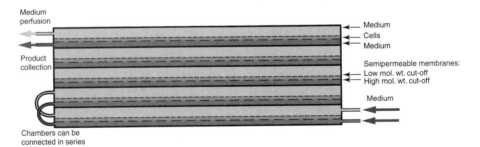

Fig. 26.7. Membroferm. Membroferm three-compartment membrane reactor [Klement et al., 1987]. A lamellate structure of semipermeable membranes, based on a triple unit wherein cells within a double membrane are fed from an upper compartment via a membrane with a low-molecular-weight cutoff. The cells secrete product into the lower compartment via a membrane that is microporous with a high-molecular-weight cutoff. Each base unit of three elements can be connected to create a perfusion system with the medium layer and product layers separate or combined.

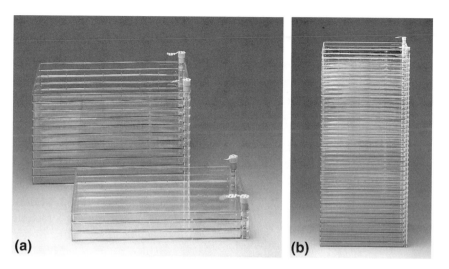

Fig. 26.8. Nunc Cell Factory. Stacked culture trays, each of 632 cm². (a) Two–chamber, 1264 cm², and 10-chamber, 6320 cm², Cell Factories. (b) 40-Chamber Cell Factory, 25,280 cm². (Courtesy of Nunc A/S).

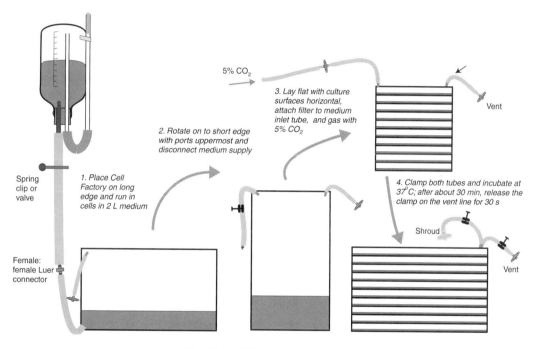

Fig. 26.9. Filling Nunc Cell Factory.

Materials

Sterile:

❑ Monolayer cells
❑ Growth medium
❑ Trypsin, 0.25% crude
❑ D-PBSA
❑ Nunc Cell Factory, 10-chamber
❑ Silicone tubing and connectors

Nonsterile:

❑ Hemocytometer or electronic cell counter and counting fluid

Protocol

1. Trypsinize the cells (*see* Protocol 13.2), resuspend them, count the cells (*see* Protocols 21.1, 21.2), and dilute the suspension to 2×10^4 cells/mL in 2 L of medium.

2. Place the chamber on a long edge, and connect the medium delivery tube to a female:female Luer connector in the supply tube to the bottom port (*see* Fig. 26.9).
3. Run the cells and medium in through the supply tube. Medium in all chambers will reach the same level.
4. Rotate the chamber through 90° in the plane of the monolayer, so that the unit lies on the short edge, with the entry ports and supply tube at the top.
5. Disconnect the medium delivery tube from the medium reservoir at the Luer connection.
6. Rotate the chamber through 90° perpendicular to the plane of the monolayer, so that it lies flat on its base, with the culture surfaces horizontal.
7. Purge the unit with 5% CO_2 in air for 5 min, and then clamp off both the supply and the outlet. (The chamber may be gassed continuously if desired.)
8. Transport it to the incubator, tipping the medium away from the supply and vent ports.
9. To change the medium (or collect the medium), follow Step 6 in reverse, remove the filter on the inlet tube, and then Step 4 in reverse.
10. Swab Luer connection, open the clamp, and drain off the medium.
11. Replace the medium as in Steps 2–8.
12. To collect the cells,
    (a) Remove the medium as in Steps 9 and 10.
    (b) Add 500 mL of D-PBSA and then remove it.
    (c) Add 500 mL of trypsin at 4°C, and remove it after 30 s.
    (d) Incubate the cells with the residual trypsin for 15 min.
    (e) Add medium to the cells, and rock the chamber to resuspend the cells.
    (f) Run the medium off the cells as in Steps 9 and 10.
13. The residue may be used to seed the next culture, although this method makes it difficult to control the seeding density. It is better to discard the chamber and start fresh.

Analysis. Monitoring cell growth in these chambers is difficult, so a single tray is supplied to act as a pilot culture. It is assumed that the single tray will behave as the multichamber unit. The supernatant medium can be collected repeatedly for virus or cell product purification. The collection of cells for analysis depends on the efficiency of trypsinization.

This technique has the advantage of simplicity, but can be expensive if the unit is discarded each time cells are collected. It was designed primarily for harvesting supernatant medium, but is also good for producing large numbers of cells ($1 \times 10^8 - 1 \times 10^{10}$).

26.2.2 Multiarray Disks, Spirals, and Tubes

Disks, spirals, and tubes have all been used to increase the surface area for monolayer growth, but few of these systems are now available on a commercial basis. Most matrix or multisurface propagators have now gone toward perfusion (*see* Section 26.2.6), with the emphasis on product recovery. Corning markets a multisurface perfusion system, called CellCube (Fig. 26.10), that is a hollow polystyrene cube with multiple inner lamellae, perfused with oxygenated, heated medium. The inner lamellae are capable of supporting monolayer growth on both surfaces.

26.2.3 Roller Culture

If cells are seeded into a round bottle or tube that is then rolled around its long axis on a roller rack (Fig. 26.11), the medium carrying the cells runs around the inside of the bottle. If the cells are nonadherent, they will be agitated by the rolling action, but will remain in the medium. If the cells are adhesive, they will gradually attach to the inner surface of the bottle and grow to form a monolayer (Fig. 26.12). This system has three major advantages over static monolayer culture: (1) the increase in utilizable surface area for a given size of bottle; (2) the constant, but gentle, agitation of the medium; and (3) the increased ratio of the medium's surface area to its volume, which allows gas exchange to take place at an increased rate through the thin film of medium over cells not actually submerged in the deep part of the medium.

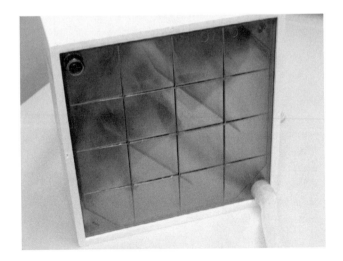

Fig. 26.10. *Corning CellCube.* Multisurface cell propagator with 6500-cm^2 available growth surface. (Courtesy of ATCC.)

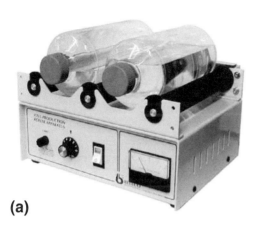

(a) **(b)**

Fig. 26.11. Roller Culture Bottles on Racks. (a) Small benchtop rack. (b) Large freestanding extendable rack. (Courtesy of Bellco.)

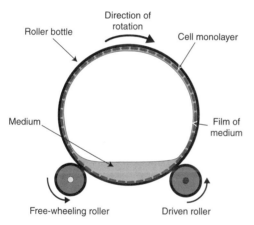

Fig. 26.12. Roller Bottle Culture. The cell monolayer (dotted line) is constantly bathed in liquid but is submerged only for about one-fourth of the cycle, enabling frequent replenishment of the medium and rapid gas exchange.

PROTOCOL 26.3. ROLLER BOTTLE CULTURE

Outline
Seed a cell suspension in medium into a round bottle, and rotate the bottle slowly on a roller rack.

Materials
Sterile or Aseptically Prepared:
❑ Medium and medium dispenser
❑ D-PBSA
❑ Crude trypsin, 0.25%
❑ Monolayer culture
❑ Roller bottles

Nonsterile:
❑ Hemocytometer, or electronic cell counter and counting fluid
❑ Supply of 5% CO_2
❑ Roller rack (*see* Fig. 26.11)

Protocol
1. Trypsinize the cells, and seed them at the usual density for these cells.

Note. The gas phase is large in a roller bottle, so, with media based on Hanks' salts and a gas phase of air (*see* Table 9.1), it may be necessary to blow a little 5% CO_2 into the bottle (e.g., for 2 s at 10 L/min). If the medium is CO_2/HCO_3-buffered, then the gas phase should be purged with 5% CO_2 (for 30 s–1 min at 20 L/min, depending on the size of the bottle; *see* Protocol 13.2).

2. Place the bottle on the roller rack, and rotate it at 20 rph until the cells attach (24–48 h).
3. Increase the rotational speed to 60–80 rph as the cell density increases.
4. To feed the cells or harvest the medium, take the bottles to a sterile work area, draw off the medium as usual, and replace it with fresh medium (*see* Protocol 13.1) A transfusion device (*see* Fig. 5.7) or media bag is useful for adding fresh medium, provided that the volume is not critical. If the volume of medium is critical, it may be dispensed by pipette or metered by a peristaltic pump (*see* Section 5.2.7).
5. To harvest the cells:
   (a) Remove the medium, rinse the cells with 50–100 mL of D-PBSA, and discard the D-PBSA.

(b) Add 50–100 mL of trypsin at 4°C, and roll the bottle for 15 s by hand or on a rack at 20 rpm.
(c) Draw off the trypsin, incubate the bottle for 5–15 min, and add medium to the bottle.
(d) Rock and rotate the bottle and wash off the cells by pipetting.

Analysis. Monitoring cells in roller bottles can be difficult, but it is usually possible to see the cells on an inverted microscope. However, with some microscopes, the condenser needs to be removed, and with others, the bottle may not fit on the stage. Therefore, choose a microscope with sufficient stage accommodation.

For repeated harvesting of large numbers of cells or for collecting supernatant medium, the roller bottle system is probably the most economical, although it is labor intensive and requires investment in a roller rack (*see* Fig. 26.11).

Variations

Aggregation. Some cells may tend to aggregate before they attach. This behavior is difficult to overcome, but may be reduced by lowering the initial rotational speed to 5, or even 2, rph, trying a different type or batch of serum, or precoating the surface of the roller bottle with fibronectin or polylysine (*see* Section 8.4).

Size. A range of bottles (around 500–1800 cm², both disposable and reusable, is available (*see* Table 8.1; Fig. 26.13; Appendix II: Roller bottles). Some bottles are provided with a ribbed inner surface to increase the surface area available for cell growth (Corning).

Volume. The volume of the medium may be varied. A low volume will give better gas exchange and may be better

Fig. 26.14. Roller Drum Apparatus. Roller drums are used for roller culture of large numbers of small bottles or tubes. (Courtesy of New Brunswick Scientific.)

for untransformed cells. Transformed cells, which are more anaerobic, grow faster, and produce more lactic acid, may be better in a larger volume of medium. The volumes given in Table 7.1 are mean values and may be halved or doubled as appropriate.

Mechanics. Roller racks are preferable for bottles (*see* Fig. 26.11b), as they are economical in space and allow easy observation of the bottles. Roller drums (Fig. 26.14), on the other hand, require more space for a given volume of culture bottle, but can be useful for large numbers of smaller bottles or tubes. Small bottles and tubes can also be rotated on a roller rack by enclosing them in a cylindrical holder.

26.2.4 Microcarriers

Monolayer cells can be grown on microbeads 90–300 μm in diameter and made of plastic, glass, gelatin, or collagen [Griffiths, 2000] (Table 26.1). Culturing monolayer cells on microbeads gives a maximum ratio of the surface area of the culture to volume of the medium, up to 90,000 cm²/L, depending on the size and density of the beads, and has the additional advantage that the cells may be treated as a suspension. Whereas the Nunc Cell Factory gives an increase in scale with conventional geometry (*see* Section 26.2.1), microcarriers require a significant departure from the usual substrate design. However, this difference has relatively little effect at the microscopic level (Plate 18c), as the cells are still growing on a smooth surface at the solid–liquid interface, although some cells are influenced by the radius of curvature and may prefer larger bead diameters or planar surfaces (*see* Plate 18d).

The major difference created by microcarrier systems is in the mechanics of handling [Griffiths, 1992]. Efficient stirring without grinding the beads is essential and can be achieved with a suspended rotating pendulum (Techne)

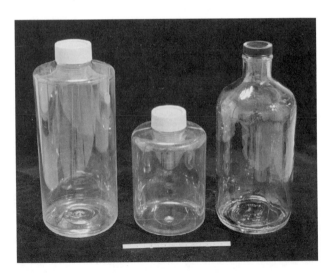

Fig. 26.13. Examples of Roller Culture Bottles. Center and left, disposable plastic (Falcon, Corning); right, glass.

TABLE 26.1. Microcarriers

Name	Supplier	Composition	Specific gravity	Diameter, μm
DEAE Dextran				
Cytodex 1	Amersham Pharmacia	DEAE-dextran	1.03	160–230
Cytodex 2	Amersham Pharmacia	DEAE-dextran	1.04	115–200
Microdex	Dextran Products	DEAE-dextran	1.03	150
Dormacell	JRH Biosciences	DEAE-dextran	1.05	140–240
Plastic				
Acrobeads	Galil	Coated polyacrolein	1.04	150
Biosilon	Nalge Nunc	TC-treated polystyrene	1.05	160–300
Biocarriers	Biorad	Polyacrylamide/DMAP	1.04	120–180
Bioplas	Whatman, Cellon	Polystyrene	1.04	150–210
Cytospheres	Lux	TC-treated polystyrene	1.04	160–230
BioSPEX PlastiSPEX	JRH Biosciences	Polystyrene	1.02, 1.03, 1.04	90–210
Rapidcell P	ICN	Polystyrene	1.02, 1.03	90–210
Gelatin				
Ventregel	Ventrex (Bayer)	Gelatin	1.03	150–250
Cytodex 3	Amersham Pharmacia	Gelatin-coated dextran	1.04	130–210
Cultisphere-G	Sigma	Gelatin	N/A	120–180
Cultisphere-GL	Sigma	Gelatin	N/A	150–330
Cultisphere-S	Sigma	Gelatin	N/A	120–180
Glass				
Bioglas	Whatman, Cellon	Glass-coated plastic	1.03	150–210
BioSPEX GlasSPEX	JRH Biosciences	Glass-coated polystyrene	1.02, 1.03, 1.04	90–210
Rapidcell G	ICN	Glass	1.02, 1.03	90–210
Ventreglas	Ventrex (Bayer)	Glass	1.03	90–210
Glass	Sigma	Glass, reusable	1.03, 1.04	95–210
Cellulose				
DE-52/53	Whatman	DEAE-Cellulose	1.03	Fibers
Collagen or collagen coated				
BioSPEX	JRH Biosciences	Collagen-coated polystyrene	1.2, 1.03, 1.04	90–210
Cytodex 3	Amersham Pharmacia	Collagen	1.04	130–210
Rapidcell C	ICN	Collagen coated	1.02, 1.03	90–210
Biospheres	Whatman, Cellon	Collagen-coated	1.02	150–210

(*see* Fig. 26.1; Fig. 26.2) or paddle (Bellco), as for suspension cell culture, rotating at 30 rpm. Technical literature is available from microcarrier suppliers to assist in setting up satisfactory cultures, and a number of protocols for microcarrier systems have been published [e.g., Griffiths, 1992].

PROTOCOL 26.4. MICROCARRIERS

Outline
Seed the cells at a high cell and bead concentration, dilute, stir, and sample the cells as required.

Materials
Sterile:
❑ Growth medium
❑ Microcarriers (*see* Table 26.1 and Appendix II)
❑ Starter culture

❑ Stirrer flask (*see* Appendix II)
Nonsterile:
❑ Magnetic stirrer (Techne, Bellco)

Protocol
1. Suspend the beads at 2–3 g/L, in one-third of the final volume of medium required.
2. Trypsinize and count the cells. Seed the cells at three to five times the normal seeding concentration into the bead suspension.
3. Stir the culture at 10–25 rpm for 8 h.
4. Add medium to reach the final volume, dictated by a bead concentration of 0.7–1 g/L.
5. Increase the stirring speed to approximately 60 rpm.
6. If the pH falls, feed the culture by switching off the stirrer for 5 min, allowing the beads to settle, and then replacing one-half to two-thirds of the medium.

7. To harvest the cells:
   (a) Remove the medium.
   (b) Wash the cells by settling.
   (c) Trypsinize the beads with trypsin/EDTA.
   (d) Allow the beads to settle.
   (e) Spin down the cells.
   (f) Wash the cells by resuspension and centrifugation.

Analysis. Cell counting on beads can be difficult, so the growth rate of the cells should be checked by determination of DNA (*see* Protocol 21.3); protein (*see* Protocol 21.4), if nonproteinaceous beads are used, or dehydrogenase activity, using the MTT assay (*see* Protocol 22.4) on a sample of the beads.

Variations. Most variations on this method arise from the choice of bead or design of the culture vessel and stirrer [Griffiths, 2000]. Bead density varies from 1.03 to 1.05 g/cc and influences the stirring speed, because a higher speed is required for higher-density beads. Composition, or coating, of the beads will also influence attachment, and more fastidious cells may prefer gelatin or collagen beads, which are also soluble when digested by protease activity. Glass beads are the easiest to recycle.

26.2.5 Macrocarriers

As well as porous microcarriers, which allow cells to grow within the interstices of the bead, there are also porous carriers that are larger with a macroscopic structure made up of a number of different materials, such as polylactic acid (PLA), polyglycolic acid (PGA), collagen, or gelatin (Gelfoam) in a variety of different geometries (*see* Fig. 25.9). These can be loaded with cells and stirred in a bioreactor or perfused in a fixed-bed or fluidized bed reactor (Fig. 26.15; Fig. 26.16). Fibracel is one such product designed for use in the Celligen bioreactor (New Brunswick Scientific; Fig. 26.15b) or a stirred bioreactor.

26.2.6 Perfused Monolayer Culture

Perfusion is frequently used to facilitate medium replacement and product recovery. The Cellcube, (Corning-Costar) is a perfused, multisurface, single-use propagator with growth-surface areas from 21,250 to 85,000 cm² (*see* Fig. 26.10), with associated pumps, oxygenator, and system controller. Other perfusion systems use cells anchored to macrocarriers or beads in fixed-bed reactors (*see* below).

Membrane perfusion. Many systems depend on filter membrane technology in which the culture bed is a flat, permeable sheet. Membroferm is compartmentalized in such a way that the cells, medium supply, and product occupy different membrane compartments (*see* Section 26.1.3; Fig. 26.7).

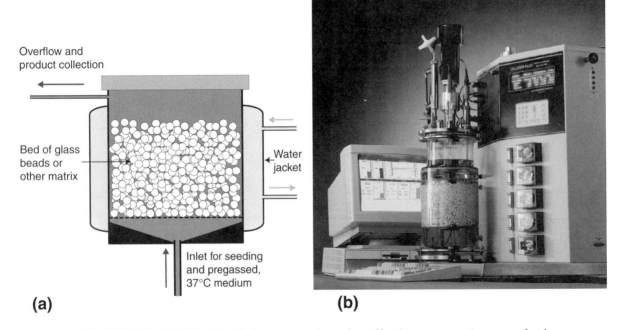

(a)   **(b)**

Fig. 26.15. Fixed-Bed Reactors. Cells grown on the surface of beads or macrocarriers are perfused with medium. The beads are usually glass and are settled in a dense bed resting on a perforated base at the bottom of the culture vessel, or, if of a lighter material, are restrained within a cage. Once the culture is established, the beads do not move, and medium percolates around them. (a) Hypothetical diagram. (b) Celligen reactor (Courtesy of New Brunswick Scientific).

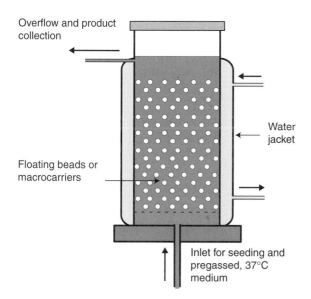

Fig. 26.16. Fluidized-Bed Reactors. In the fluidized-bed reactor low-density beads float on the perfusate, the flow rate of which is designed to be in equilibrium with the sedimentation rate of the beads.

Hollow fiber perfusion. There are a number of hollow fiber perfusion systems in which adherent cells grow on the outer surface of the perfused microcapillary bundles. High-molecular-weight products concentrate in the outer space with the cells, while nutrients are supplied and metabolites removed via the inner space (*see* Section 25.3.2 and Fig. 26.6). The potential of these systems for adherent cells lies in the re-creation of high, tissuelike cell densities, matrix interactions, and the establishment of cell polarity, all of which may be important for posttranslational processing of proteins and for exocytosis. Although originally designed for attached cells, hollow fiber systems are also used extensively for suspension cells, such as hybridomas (*see* Section 27.10).

Fixed-bed reactors. Systems have also been developed with glass or matrix beads (Fig. 26.15), with the medium being perfused upward through the bed or percolating downward by gravity. The product is collected with the spent medium in a reservoir. These systems are described in greater detail by Speir and Griffiths [1985–1990] and Griffiths [2000, 2001].

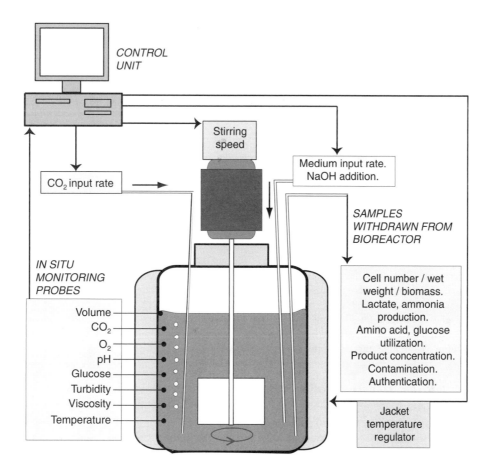

Fig. 26.17. Bioreactor Process Control. Schematic representation of a paddle-stirred bioreactor with direct-reading probes on the left, feeding to a control unit (top left) that stores the data and also regulates conditions within the bioreactor. A sampling port on the right withdraws the cell suspension from the bioreactor for analysis.

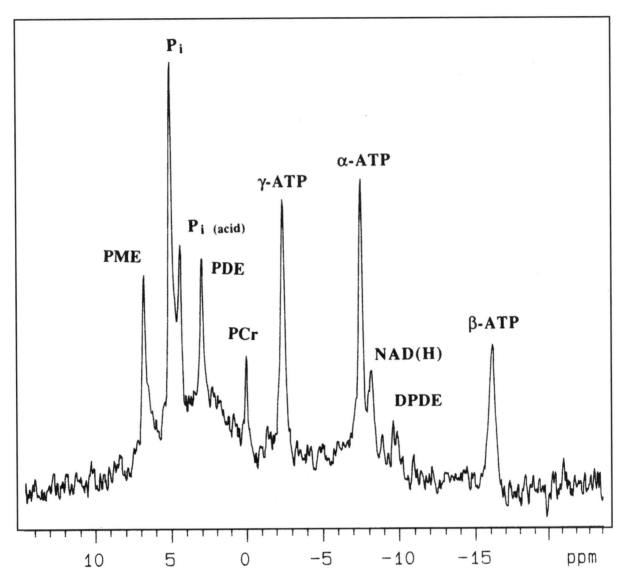

Fig. 26.18. Analysis by NMR. Cell growth in hollow fibers, analyzed by NMR. [^{31}P] NMR spectrum of CHO cells growing in a hollow fiber reactor. The cells had been grown on macroporous beads in the extracapillary space of a specially constructed hollow fiber cartridge that could be accommodated within a 25-mm-diameter NMR probe. The cells were present at a density of approximately 7×10^7 cells/mL. *Abbreviations:* PME, phosphomonoesters, including phosphocholine and phosphoethanolamine; P_i, extracellular inorganic phosphorus; P_i(acid), an acidic (pH 6.7) extracellular P_i pool within the bioreactor; PCr, phosphocreatine; γ-ATP, γ-phosphate of ATP; α-ATP, α-phosphate of ATP; PDE phosphodiesters; DPDE, diphosphodiesters, including UDP-glucose; β-ATP, β-phosphate of ATP NADH, reduced nicotinamide adenine dinucleotide. The chemical shift scale is referenced to phosphocreatine at 0.0 ppm. (Courtesy of Dr. Kevin Brindle.)

Fluidized–bed reactors for monolayer cultures. In fluidized-bed reactors, porous beads of a relatively low density—made of ceramics, or a mixture of ceramics and natural products, such as collagen—are suspended in an upward stream of medium when the flow rate of the medium matches the sedimentation rate of the beads (Fig. 26.16) or circulates (Cytopilot, Amersham Biosciences) in a pattern similar to an air-lift fermentor (*see* Fig. 26.4). While suspension cells lodge in the beads by entrapment,

monolayer cells attach to the outer surfaces as well as the interstices of the porous bead or macrocarrier.

Microencapsulation. Sodium alginate behaves as a sol or gel, depending on the concentration of divalent cations. It will gel as a hollow sphere around cells in suspension in a high concentration of divalent cations (*see* Protocols 25.3 and 23.16; Plate 18b). Because the alginate acts as a barrier to high-molecular-weight molecules, macromolecules secreted

by the cells are trapped within the vesicle, while nutrients, metabolites, and gas freely permeate the gel. The product and cells are recovered by reducing the concentration of divalent cations. These gels have a low immunoreactivity and can be implanted *in vivo*.

26.3 PROCESS CONTROL

The progress of suspension cultures is monitored via pH, oxygen, CO_2, and glucose electrodes that read from the culture *in situ*, and by assaying the utilization of nutrients, such as glucose and amino acids, or the buildup of metabolites, such as lactate and ammonia, and products, such as immunoglobulin from hybridomas (Fig. 26.17). The number of cells and other parameters, such as ATP, DNA, and protein, are determined in samples drawn from the culture and are used to calculate the total biomass. The temperature of the medium is regulated by preheating the input medium and by heating the surrounding water jacket regulated by feedback from the temperature probe. The flow rate of the medium is controlled and can match the output to the sample line, if the suspension is running as a biostat (*see* Section 26.1.1), and the stirring speed can be regulated, along with the viscosity of the medium, to reduce the shear stress.

There is a recurrent problem when monolayer cell cultures are scaled up, particularly in a fixed-bed or hollow fiber bioreactor: It is no longer possible to observe the cells directly, and both monitoring the progress of a culture by cell counting and determining the biomass become difficult. One approach to this problem [Brindle, 1998; Thelwall et al., 2001] uses nuclear magnetic resonance (NMR) to assay the contents of the culture chamber. By placing the cells in a perfused hollow fiber chamber within the magnetic field of an NMR spectroscope, a characteristic NMR spectrum is generated, enabling the identification and quantitation of specific metabolites (Fig. 26.18). NMR spectroscopes can also be used as imaging devices, producing a quasi-optical section through the chamber to reveal the distribution of the cells, and even distinguishing between proliferating and nonproliferating zones of the culture.

CHAPTER 27

Specialized Techniques

This chapter covers a number of techniques that are ancillary to or dependent on cell culture, but which are not an essential component of initiation of a culture or its regular maintenance. They represent specialist interests or applications that may be of direct value to some readers, or simply provide background on tissue culture applications for others.

27.1 LYMPHOCYTE PREPARATION

Flotation on a combination of Ficoll and sodium metrizoate (Ficoll-Hypaque) is the most widely used technique [Boyum, 1968a,b] for separating lymphocytes from plasma and erythrocytes. It is suitable for applications such as PHA stimulation (*see* Protocol 27.2) for chromosome analysis, enriching viable cells (*see* Protocol 12.10), and as a first step in the purification of lymphocyte subclasses.

PROTOCOL 27.1. PREPARATION OF LYMPHOCYTES

Outline
Layer whole citrated or heparinized blood or plasma, depleted in red blood cells by dextran-accelerated sedimentation, on top of a dense layer of Ficoll-Hypaque. After centrifugation, most of the lymphocytes are found at the interface between the Ficoll-Hypaque and the plasma.

Materials
Sterile:
- ❑ Blood sample in heparin or citrate (concentration determined by collection container, which will already contain citrate or heparin)
- ❑ Clear centrifuge tubes or universal containers
- ❑ D-PBSA
- ❑ Ficoll-Hypaque, adjusted to 1.077 g/cc (*see* Appendix II)
- ❑ Serum-free medium
- ❑ Syringe with blunt cannula, plastic Pasteur pipettes, or pipettor

Nonsterile:
- ❑ Hemocytometer or electronic cell counter
- ❑ Centrifuge

Protocol
1. Collect the blood sample in citrated or heparinized container and transport to laboratory.

△ *Safety Note.* Human blood may be infected with HIV, hepatitis, or other pathogens and should be handled with great care (*see* Section 7.8.3).

2. Dilute 1:1 with D-PBSA, and layer 9 mL onto 6 mL Ficoll-Hypaque. This should be done in a wide, transparent centrifuge tube with a cap such as the 25-mL Sterilin or Nunc universal container, or with double these volumes in a clear plastic Corning 50-mL tube.

---

3. Centrifuge the suspension for 15 min at 400 g (measured at the center of the interface).
4. Carefully remove the plasma/D-PBSA without disturbing the interface.
5. Collect the interface with a syringe (fitted with a blunt cannula), or a plastic Pasteur pipette, and dilute it to 20 mL in serum-free medium (e.g., RPMI 1640).

Δ *Safety Note.* Glass Pasteur pipettes and syringes with sharp needles should not be used with human blood. Instead, use a 1-mL pipettor or a syringe with a blunt cannula.

6. Centrifuge the diluted cell suspension from the interface at 70 g for 10 min.
7. Discard the supernate, and resuspend the pellet in 2 mL of serum-free medium. If several washes are required—e.g., to remove serum factors—repeat resuspension in 20 mL serum-free medium and centrifugation, two or three times more, before finally resuspending the pellet in 2 mL of serum-free medium.
8. Stain a sample of the cells with methylene blue, and count the nucleated cells on a hemocytometer. Alternatively, lyse a sample of the cells with Zapoglobin (Beckman Coulter), and count the nuclei on an electronic cell counter with a 70–100-μm orifice tube.

PROTOCOL 27.2. PHA STIMULATION OF LYMPHOCYTES

Materials
Sterile:
❏ Medium + 10% FBS or autologous serum
❏ Phytohemagglutinin (PHA), 50 μg/mL
❏ Test tubes or universal containers
❏ Microscope slides
❏ Colcemid, 0.01 μg/mL in BSS
❏ 0.075 M KCl

Protocol
1. Using the washed interface fraction from Step 7 of Protocol 27.1, incubate 2×10^6 cells/mL in medium, 1.5–2.0 cm deep, in HEPES or CO_2-buffered DMEM, CMRL 1066, or RPMI 1640 supplemented with 10% autologous serum or fetal bovine serum.
2. Add PHA, 5 μg/mL (final concentration), to stimulate mitosis from 24 to 72 h later.
3. Collect samples at 24, 36, 48, 60, and 72 h, and prepare smears or cytocentrifuge slides of the samples to determine the optimum incubation time (i.e., the peak mitotic index).
4. Add 0.001 μg/mL (final concentration) of colcemid for the 2 h during which the peak of mitosis is anticipated from observations made in Step 3 [Berger, 1979].
5. Centrifuge the cells after the colcemid treatment, resuspend the pellet in 0.075 M KCl for hypotonic swelling, and proceed as for chromosome preparation (*see* Protocols 16.7, 27.5).

Lymphocytes will be concentrated at the interface, along with some platelets and monocytes. Some granulocytes may be found in the interface, although most will be found in the Ficoll-Hypaque, and in the pellet created in Step 3. Monocytes and residual granulocytes can be removed from the interface fraction by taking advantage of their adherence to glass (beads or the surface of a flask) or to nylon mesh. Use a positive sort by MACS (*see* Section 15.3.2) or FACS (*see* Section 15.4) with specific lymphocyte subclass surface markers if purer preparations are required.

27.1.1 Blast Transformation

Lymphocytes in purified preparations or in whole blood may be stimulated with mitogens such as phytohemagglutinin (PHA), pokeweed mitogen (PWM), and antigen [Berger, 1979; Hume & Weidemann, 1980]. The resultant response may be used to quantify the immunocompetence of the cells. PHA stimulation is also used to produce mitosis for chromosomal analysis of peripheral blood [Rooney & Czepulkowski, 1986; Watt & Stephen, 1986] (*see* Protocol 16.7).

27.2 AUTORADIOGRAPHY

This section is intended to cover microautoradiography of any small molecular precursor into a cold acid-insoluble macromolecule, such as DNA, RNA, or protein. Other variations may be derived from this text or found in the literature [Rogers, 1979; Stein & Yanishevsky, 1979] (*see also* Protocol 27.8). Because autoradiography is used extensively to localize material in blots from electrophoresis as well as in microscope preparations, the two methods may be distinguished as macroautoradiography and microautoradiography, respectively (*see* Appendix IV: Glossary).

Isotopes suitable for microautoradiography are listed in Table 27.1. A low-energy emitter (e.g., ^3H or ^{55}Fe) in combination with a thin emulsion gives high intracellular resolution. Slightly higher-energy emitters (e.g., ^{14}C and

TABLE 27.1. Isotopes Suitable for Autoradiography

Isotope	Emission	Energy (mV) (mean)	$T_{1/2}$
^{3}H	β^-	0.018	12.3 yr
^{55}Fe	X-rays	0.0065	2.6 yr
^{125}I	X-rays	0.035, 0.033	60 days
^{14}C	β^-	0.155	5,570 yr
^{35}S	β^-	0.167	87 days
^{45}Ca	β^-	0.254	164 days

^{35}S) give localization at the cellular level. Still higher-energy isotopes (e.g., ^{131}I, ^{59}Fe, and ^{32}P) give poor resolution at the microscopic level but are used for macroautoradiographs of chromatograms and blots from DNA, RNA, and protein electrophoresis, for which the absorption of low-energy emitters would limit the detection of incorporation. Low concentrations of higher-energy isotopes (^{14}C and above), used in conjunction with thick nuclear emulsions, produce tracks that are useful in locating a few highly labeled particles (e.g., virus particles infecting a cell population or tissue).

Tritium is used most frequently for autoradiography at the cellular level, because the β particles released have a mean range of about 1 μm in aqueous media, giving very good resolution. Tritium-labeled compounds are usually less expensive than the ^{14}C- or ^{35}S-labeled equivalents and have a longer half-life. Because of their low energy of emission, however, it is important that the radiosensitive emulsion be positioned in close proximity to the specimen, with nothing between the cell and the emulsion. Even in this situation, only the incorporation in the top 1 μm of the specimen will be detected.

β-Particles entering the emulsion produce a latent image in the silver halide crystal lattice within the emulsion at the point where they stop and release their energy. The image may be visualized as metallic silver grains by treatment with an alkaline reducing agent (developer) and subsequent removal of the remaining unexposed silver halide by an acid fixer.

The latent image is more stable at low temperatures and in anhydrous conditions, so its sensitivity (signal relative to background) may be improved by exposure in a refrigerator over desiccant. This reduces the background silver grain formation by thermal activity.

PROTOCOL 27.3. MICROAUTORADIOGRAPHY

Outline
Incubate cultured cells with the appropriate isotopically labeled precursor (e.g., [^{3}H]thymidine to label DNA) and wash, fix, and dry the cells (Fig. 27.1). Perform any necessary extractions (e.g., to remove unincorporated precursors). Coat the specimen with emulsion in the dark and leave it to expose. After

---

development silver grains can be seen in overlying areas where radioisotope was incorporated (Fig. 27.2; *see* Plate 14d).

Materials

Setting up the culture:
Sterile:
❑ Cells
❑ D-PBSA
❑ Trypsin
❑ Growth medium
❑ Coverslips or slides, and Petri dishes (may be non-tissue-culture grade if coverslips or slides are used) or plastic bottles
Nonsterile:
❑ Hemocytometer, or electronic cell counter and counting fluid

Labeling with the Isotope:
Sterile:
❑ Isotope
❑ HBSS
Nonsterile:
❑ Protective gloves
❑ Containers for disposal of radioactive pipettes
❑ Container for radioactive liquid waste

Fixing and Processing the Cells:
Nonsterile:
❑ Acetic methanol (1:3, ice cold, freshly prepared)
❑ DPX
❑ TCA, 0.6 N
❑ Deionized water
❑ Gelatin: 0.2 g in 200 mL of UPW; microwave for 2 min, cool, and filter

Setting Up Autoradiographs:
Nonsterile:
❑ Safelight: Kodak Wratten II or equivalent
❑ Emulsion (Amersham Hypercoat Emulsion LM-1, RPN 40)

Note. It is convenient to melt the emulsion and disperse it into aliquots in dipping vessels (Amersham) suitable for the number of slides to be handled at one time. If the slides are sealed in a dark box, they may be stored at 4°C until required.

❑ Light-tight microscope slide boxes (Clay Adams, Raven)
❑ Silica gel (Fisher)
❑ Dark vinyl tape (e.g., electrical insulation tape)
❑ Black paper or polyethylene

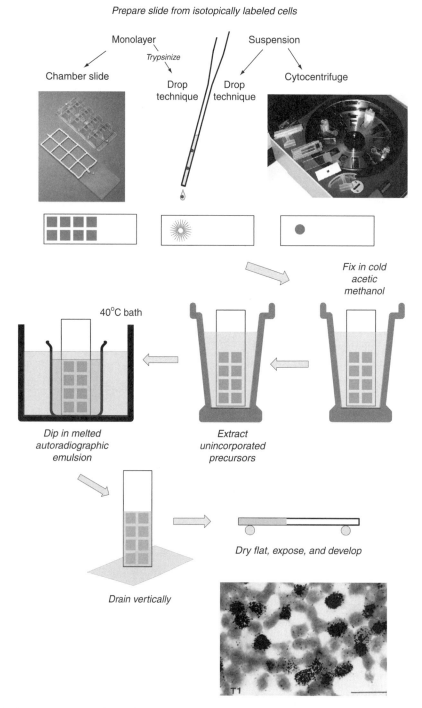

Fig. 27.1. Microautoradiography. Steps for preparing a microautoradiograph from a cell culture.

Note. All glassware must be carefully washed and free of isotopic contamination. Plastic coverslips should be used in preference to glass, to minimize radioactive background. Be particularly careful to prevent spillages, and immediately mop up any that do occur. Wear gloves, and change them regularly—e.g., when you move from incubation (a high level of isotope) to handling washed, fixed slides (a low level of isotope).

Processing Autoradiographs:
Nonsterile:
❑ Developer: Phenisol (Ilford), 20%
❑ Stop bath: 1% acetic acid
❑ Photographic fixer: 30% sodium thiosulfate
❑ Hypo clearing agent (Kodak)
❑ Coverslips (#00)

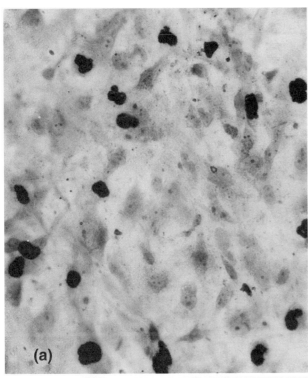

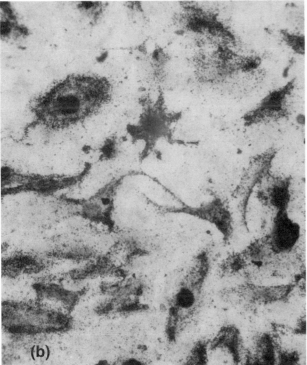

Fig. 27.2. Microautoradiograph. This pair of pictures is an example of [³H]thymidine incorporation into a cell monolayer. Normal glial cells were incubated with 0.1 μCi/mL (3.7 KBq/mL), 200 Ci/mmol (7.4 GBq/μmol), of [³H]thymidine for 24 h, washed, and processed as described in Protocol 27.3. (a) Typical densely labeled nuclei, suitable for determining the labeling index (*see* Protocol 21.11). (b) A similar culture, infected with mycoplasma showing [³H]thymidine incorporation in the cytoplasm.

Staining:

Nonsterile:
- ❏ Hematoxylin (filter before use) or Giemsa stain (*see* Protocol 16.2)
- ❏ Phosphate buffer, 0.01 M, pH 6.5
- ❏ Ethanol, 50, 70, and 100%
- ❏ Histoclear (Fisher)
- ❏ DPX (Merck)

Protocol

Setting up the Culture:
1. Set up replicate monolayer cultures on coverslips, slides (Nunc, Bellco), or Petri dishes.
2. Incubate the cultures at 37°C until the cells reach the appropriate stage for labeling.

Adding the Isotope:
3. Add the isotope, usually in the range of 0.1–10 μCi/mL (~4.0 KBq–0.4 MBq/mL), 100 Ci/mmol (~4 GBq/μmol), for 0.5–48 h as appropriate.

△ *Safety Note.* Follow local rules for handling radioisotopes. Because such rules vary, no special recommendations are made here, other than the following: Wear gloves; do not work in horizontal laminar flow; use a shallow tray with an absorbent liner to contain any accidental spillage; incubate the cultures in a box or tray labeled for use with radioactivity; and regulate the disposal of radioactive waste according to local limits (*see* Section 7.7.2).

4. Remove all medium containing isotope, and wash the cells carefully in BSS, discarding the medium and washes.

△ *Safety Note.* ³H-nucleosides are highly toxic, because of their ultimate localization in DNA (*see* Section 7.7.1).

Fixing and Processing the Cells:
5. Fix the cells in ice-cold acetic methanol for 10 min.
6. Prepare the slides:
   (a) Coverslips: Mount the coverslips on a slide with DPX or Permount, cells uppermost.
   (b) Cell suspensions (from growth in suspension or trypsinized): Centrifuge the cells onto a slide (*see* Protocol 16.4) or make drop preparations (*see* Protocol 16.7, Steps 4–12).
   (c) Prepare several extra control slides for use in determining the correct duration of exposure. All preparations are referred to as "slides" from now on.
7. Extract acid-soluble precursors (when labeling DNA, RNA, or protein) with ice-cold 0.6 N

TCA (3 × 10 min), and perform any other control extractions (e.g., with lipid solvent or enzymatic digestion).

8. Dip the slides in the gelatin solution for 2 min.

9. Drain the slides vertically, and let them dry. (The slides may be stored dry at 4°C).

Setting up the Autoradiographs:

10. Take the slides to the darkroom.

11. Under a dark-red safelight, melt the emulsion in a water bath at 46°C.

12. Mix the emulsion gently with a clean slide, taking care not to create bubbles.

13. Dip the slides:
    (a) Dip the slides in the emulsion for 5 s at 46°C, making sure that the cells are completely immersed. Note that the temperature is critical and will determine the thickness of the emulsion and, consequently, the resolution.
    (b) Withdraw the slides, and drain them for 5 s.
    (c) Dip the slides again in the emulsion for 5 s.
    (d) Withdraw the slides, and drain them vertically for 5 s.
    (e) Wipe the back of each slide with a paper tissue.
    (f) Allow the slides to dry flat on a tissue or piece of filter paper for 10 min.
    (g) Place the slides on a rack in a light-tight box, and allow them to dry completely. Do not force the slides to dry; dry them slowly in humid conditions to avoid crocking the emulsion.

14. When the slides are dry (2–3 h), transfer them to light-tight microscope slide boxes with a desiccant, such as silica gel. Make sure that you do not touch the slides and that they do not touch each other or the desiccant.

15. Seal the boxes with dark vinyl tape, wrap them in black paper or polythene, and place them in a refrigerator. Make sure that this refrigerator is not used for the storage of isotopes.

16. Leave the boxes at 4°C for 1–2 weeks. The exact time required will depend on the activity of the specimen and can be determined by processing one of the extra slides at intervals. Slides with prolonged exposure times have an increased background and are prone to latent image fade. It is better to increase the activity of the label than to increase the length of exposure.

Processing the Autoradiographs:

17. Return the boxes to the darkroom and, under a dark-red safelight, unseal the boxes.

18. Allow the slides to come to atmospheric temperature and humidity (~2 min).

19. Prepare the solutions for development at 20°C.

20. Place the slides in the developer for 2.5 min with gentle intermittent agitation.

21. Wash the slides briefly in UPW.

22. Transfer the slides to a photographic fixer for 5 min.

23. Rinse the slides in deionized water, and then place them in hypo clearing agent for 1 min.

24. Wash the slides in cold running water, or with five changes of water over 5 min.

25. Dry the slides, and examine them on the microscope. Phase contrast may be used by mounting a thin glass (#00) coverslip in water. Remove the coverslip when you are finished examining the slides and before the water dries out, or else the coverslip will stick to the emulsion.

Staining:

26. To stain with hematoxylin,
    (a) Stain the slides in freshly filtered hematoxylin for 45 s.
    (b) Rinse the slides in running water for 2 min.
    (c) Dehydrate the slides in a succession of 50, 70, and 100% EtOH.
    (d) Clear the slides in Histoclear; mount coverslips in DPX. If a coverslip is used, it must be #00 with a minimum of mountant, to allow sufficient working distance for a 100× objective.

27. To stain with Giemsa:
    (a) Immerse the dry slides in neat Giemsa stain for 1 min.
    (b) Dilute the stain *in situ* 1:10 in 0.01 M phosphate buffer (pH 6.5) for 10 min.
    (c) Remove the staining solution by upward displacement with water. The slides should not be withdrawn or the stain poured off, or else the scum that forms on top of the stain will adhere to the specimen.
    (d) Rinse the slides thoroughly under running tap water until the color is removed from emulsion, but not from the cells.

Analysis

Qualitative. Determine the specific localization of grains—e.g., over nuclei only or over one cell type rather than another.

Quantitative.

(1) *Grain counting.* Count the number of grains per cell, per nucleus, or localized elsewhere in the cells. This requires a low grain density—about 5–20 grains per nucleus, 10–50 grains per cell—no overlapping grains, and a low uniform background.

(2) *Labeling index.* Count the number of labeled cells as a proportion of the total number of cells (*see also* Protocol 21.11). The grain density should be higher for this assessment than for grain counting, to ease the recognition of labeled cells (*see* Plate 14d). If the grain density is high (e.g., ∼100 grains per nucleus), set the lower threshold at, say, 10 grains per nucleus or per cell; but remember that low levels of labeling, significantly over background, may be important, e.g., regarding DNA repair.

Microautoradiography is a useful tool for determining the distribution of isotope incorporation within a population, but it is less suited to total quantitation of isotope uptake or incorporation, for which scintillation counting is preferable.

Variations. Isotopes of two different energies—e.g., ^{3}H and ^{14}C—may be localized in one preparation by coating the slide first with a thin layer of emulsion, then coating that layer with gelatin alone, and finally coating the gelatin with a second layer of emulsion [Rogers, 1979]. The weaker β emission from the ^{3}H is stopped by the first emulsion and the gelatin overlay, while the higher energy β emission from the ^{14}C, having a longer mean path length, of around 20 μm, will penetrate the upper emulsion.

Adams [1980] described a method for autoradiographic preparations from Petri dishes or flasks such that liquid emulsion is poured directly onto fixed preparations without the need for trypsinization.

Soft β-emitters may also be detected in electron microscope preparations, using very thin films of emulsion or silver halide sublimed directly onto the section [Rogers, 1979].

Fluorescent and luminescent probes (Amersham Biosciences) are now being used in place of radioisotopically labeled probes (*see* Protocol 27.9). The resolution with this method is often superior, quantitation is possible by confocal microscopy (Bio-Rad) or a CCD camera (Dage-MTI), and disposal of reagents is environmentally friendly. However, the equipment for microscopic evaluation is expensive (∼\$50,000–\$200,000).

27.3 TIME-LAPSE RECORDING

Time-lapse recording was developed primarily as a cinematography technique by which naturally slow processes can be observed at a greatly accelerated rate. At its inception, the technique required a camera operated automatically by a signal from a timing device. Scientific cinematography started with the first cine camera itself when Marey used it to record animal movement in 1888. In the 1950s, Michael Abercrombie was the first to use the technique for behavior studies of tissue culture cells in a rigorously quantitative way [Abercrombie & Heaysman, 1954]. Video systems gradually replaced cine film, and "video microscopy" optimized their use in combination with light microscopes [Inoué & Spring, 1997].

The quality of video imaging progressively improved, and its digital form eventually superseded 16-mm film. The microscopy technique itself, however, remains the critical link in the whole process. Commercially available microscopy techniques usually require laborious manual, or at least semi-interactive, frame-by-frame analysis of the recordings. Automated analysis of cell behavior by digital image processing was achieved with highly refractile organisms [Wessels et al., 1996]. Fully automatic analysis with unstained vertebrate tissue culture cells requires special microscopy: digitally recorded interference microscopy with automated phase shifting [Dunn & Zicha, 1997]. Automatic analysis of cell behavior is an important aspect of time-lapse recording, because large amounts of data usually are required to investigate phenomena in this area with high intrinsic variability.

This background information and Protocol 27.4 for time-lapse recording has been contributed by Daniel Zicha, Cancer Research UK London Research Institute, Light Microscopy Laboratory, 44 Lincoln's Inn Fields, London, WC2A 3PX, England, UK.

PROTOCOL 27.4. TIME-LAPSE VIDEO RECORDING

Outline
Prepare cells in a culture chamber with suitable optical properties for the microscopy technique. The microscope is equipped with a video camera connected to a recording device, and additional equipment provides the desired temperature for the cell culture. During recording, individual exposures are triggered by a controller that also operates a shutter, eliminating unnecessary illumination of cells.

Materials
☐ Medium: MEM with Hanks' salts, 12 mM bicarbonate, 10% calf serum, and 4 mM L-glutamine
☐ Trypsin/EDTA: trypsin (Difco), 0.25% in EDTA, 0.5 mM
☐ Coverslips, 18 × 18 mm, No. 1.5 (Merck), washed with concentrated sulfuric acid and repeated boiling in UPW using Teflon coverslip holders (Eppendorf), and sterilized by autoclaving

❏ Petri dishes, plastic, 3.5 cm
❏ Filming chamber:
   (a) Glass slide, 1 mm thick, with a 15-mm-diameter hole in the center covered with a #3 coverslip, cemented to one face of the slide by UV-curing glass cement (Southern Watch and Clock Supplies).
   (b) Sterilize the chamber with 70% alcohol, followed by a blow with clean compressed air.
   (c) Wax mixture: beeswax (Fisher), soft yellow paraffin (Fisher), and paraffin wax (melting point of 46°C; Fisher), in a ratio of 1:1:1
❏ Microscope: Axiovert 135 TV (Zeiss)
❏ Perspex box, designed to encapsulate the stage of the microscope and equipped with a controlled heating system (custom made from controller 208 2739, T probe 219 4674, heater element 224 565, low-noise ac fan 583-325; RS Components, UK)
❏ Computer-controlled shutter and filter wheel (Ludl Electronic)
❏ Camera: RTE/CCD-1300-Y/HS (Princeton Instruments)
❏ Computer: Macintosh G3
❏ Imaging software IPlab (Scanalytics), which records microscope images in time-lapse mode

Protocol

1. Switch on all the equipment, especially the heater, well in advance, to allow the temperature within the cabinet to stabilize.
2. Release adherent culture cells into suspension, using trypsin/EDTA.
3. Seed around 20,000 cells on a coverslip in a Petri dish, and allow the cells to settle at 37°C in a humidified atmosphere with 3% CO_2 overnight.
4. Mount the coverslip on the filming chamber, which should be filled with culture medium, leaving a small bubble to ensure CO_2 equilibration with the Hanks' salts. Seal the coverslip in place with the hot wax mixture.
5. Use the appropriate microscopy technique—for example, phase contrast and fluorescence with LD Achroplan 32×/0.4 Ph2 objective. A higher-power objective with a short working distance will require an upright microscope to observe adherent cells moving over the ceiling of the chamber.
6. Prepare the computer for image acquisition:
   (a) Calibrate the image size, using the stage micrometer.
   (b) Choose the exposure time; 0.1 s is usually sufficient for phase contrast, whereas weak fluorescence requires a longer exposure of 0.4 s or more.
   (c) Select the time-lapse interval, the number of frames, and the file name. Analysis of cell translocation in slowly moving, fibroblast-like cells will give satisfactory results at 5-min lapse intervals. Observation of cell morphology that changes much faster will require lapse intervals of 1 min or even less. Also, translocation of fast cells, such as neutrophil leukocytes, will need 1 min or shorter lapse intervals.
   (d) Start the following script to perform the alternating acquisition of phase-contrast and fluorescence images:
      i) Set indexed file name for save;
      ii) Insert START label;
      iii) Set the filter wheel for fluorescence, and open the arc lamp shutter;
      iv) Full acquire for fluorescence image using the appropriate exposure time;
      v) Set the filter wheel for phase contrast, and close the arc lamp shutter;
      vi) Save the fluorescence image;
      vii) Open the halogen lamp shutter;
      viii) Full acquire for phase-contrast image using the appropriate exposure time;
      ix) Close the halogen map shutter;
      x) Save the phase-contrast image;
      xi) Pause for the time-lapse interval;
      xii) Close all windows;
      xiii) Loop to START label unless the specified number of frames have been recorded;
      xiv) Quit the program.
   Separate phase-contrast and fluorescence movie sequences can be viewed either independently or in combination with iPLab animation.
7. Store the image sequences, using a CD writer.

Δ **Safety Note.** (1) Take care when washing coverslips with concentrated acid. (2) The controlled electrical heater, built from Radio Spares (RS) components, must be properly insulated and earthed, and the fan must be mounted in a position that prevents the rotating blades from being touched accidentally.

Analysis. The imaging software IPLab can be used for the interactive measurement of positions, areas, and intensities in the images. Theory of moments provides a good basis for automatic analysis of cell behavior [Dunn & Brown, 1990]. It can be applied to fluorescence images for the evaluation of cell translocation using centroid positions, of cell spreading using cell area, of the intensity of the fluorescence signal, or of the shape factors in morphological studies.

Variations. The choice of medium depends on the requirements of the cells under observation. Medium with Hanks' salts has the advantage that its pH equilibrates automatically in a sealed culture chamber. An open chamber can be designed that contains a Petri dish positioned over a hole in the bottom of the chamber. This arrangement requires humidification and a supply of a mixture of CO_2 and air into the chamber. Hanks'-based medium with 12 mM bicarbonate will need around 3% CO_2, and Dulbecco's MEM will need 10% CO_2. An open chamber requires an inverted microscope, and high-power magnification can be achieved when glass-bottom, No. 1.5, uncoated, γ-irradiated 3.5-cm microwell dishes (MatTek) are used. Specialized chambers with local heating and a flow-through option are available (*see* Fig. 5.22; Intracell, Royston). It is no longer worth giving serious consideration to film-based methods, or even disk or tape time-lapse video recorders, because digital recording on a computer now provides greater convenience as well as other features—namely, image processing, including contrast enhancements, background noise subtraction, and support for analysis. An alternative time-lapse recording software for a PC is MetaMorph Imaging Software (Universal Imaging Corporation, PA, USA). Differential interference contrast (DIC) is a common alternative to phase contrast that improves the contrast of details at high resolution.

27.4 CONFOCAL MICROSCOPY

Time-lapse records, or real-time records made on videotape, can be made via the confocal microscope with a conventional high-resolution video camera or CCD. These systems allow the recording of events within the cell depicted by the distribution and relocation of, and changes in the staining intensity of, fluorescent probes. Fluorescent imaging can be used to localize cell organelles, such as the nucleus or Golgi complex [Lippincott-Schwartz et al., 1991]; measure fluctuations in intracellular calcium [Cobbold & Rink, 1987]; and follow the penetration and movement within the cell of a drug [Neyfakh, 1987; Bucana et al., 1990]. Measurements can be made in three dimensions, as the excitation and detection system is capable of visualizing optical sections through the cell, and over very short periods of time. Confocal microscopy can also visualize cells within a three-dimensional culture system, as in filter well invasion assays [Brunton et al., 1997].

27.5 CELL SYNCHRONY

In order to follow the progression of cells through the cell cycle, a number of techniques have been developed whereby a cell population may be fractionated or blocked metabolically so that on return to regular culture conditions, the cells will all be at the same phase and progress through the cycle in synchrony.

27.5.1 Cell Separation

Techniques for cell separation have been described in Chapter 15. Sedimentation at unit gravity can be used [Shall & McClelland, 1971; Shall, 1973], but centrifugal elutriation is preferable if a large number of cells ($>5 \times 10^7$) is required [Breder et al., 1996; Mikulits et al., 1997] (*see* Figs. 15.4 and 15.5). Fluorescence-activated cell sorting (*see* Figs. 15.8 and 15.9) can also be used, in conjunction with a nontoxic, reversible DNA stain, such as Hoechst 33342. The yield is lower for this method than for centrifugal elutriation ($\sim 10^7$ cells or less), but the purity of the fractions is higher.

One of the simplest techniques for separating synchronized cells is mitotic shake-off: monolayer cells tend to round up at metaphase and detach when the flask in which they are growing is shaken. This method works well with CHO cells [Tobey et al., 1967; Petersen et al., 1968] and some sublines of HeLa-S₃. Placing the cells at 4°C for 30 min to 1 h can also be used to synchronize cells in cycle [Rieder & Cole, 2002] and enhances the yield at mitotic shake-off [Miller et al., 1972].

27.5.2 Blockade

Two types of blocking have been used:

(1) *DNA synthesis inhibition (S phase).* The agents used to inhibit DNA synthesis include thymidine, hydroxyurea, cytosine arabinoside, and aminopterin [Stubblefield, 1968]. The effects of these agents are variable, because many are toxic and blocking cells in cycle at phases other than G_1 tends to lead to deterioration of the cells. Hence, the culture will contain nonviable cells, cells that have been blocked in the S phase but are viable, and cells that have escaped the block [Yoshida & Beppu, 1990].

(2) *Nutritional deprivation (G_1 phase).* In these cases, serum [Chang & Baserga, 1977] or isoleucine [Ley & Tobey, 1970] is removed from the medium for 24 h and then restored, whereupon transit through the cycle is resumed in synchrony [Yoshida & Beppu, 1990].

A high degree of synchrony (e.g., >80%) is achieved only in the first cycle; by the second cycle, the degree of synchrony may be <60%, and by the third cycle, cell cycle distribution may be close to random. A chemical blockade is often toxic to the cells, and nutritional deprivation does not work well in many transformed cells. Physical fractionation techniques are probably most effective and do less harm to the cells (*see* Sections 15.2, 15.4).

27.6 CULTURE OF AMNIOCYTES

The human fetal karyotype can be determined by culturing amniotic fluid cells obtained by amniocentesis. Amniocentesis can now be performed from 11 weeks of gestation onward, although most amniocentesis are still performed from 15 to 18 weeks of gestation.

Inborn errors of metabolism and other sex-linked or autosomal recessive and dominant conditions are diagnosed mainly on placental tissues obtained by chorionic villus sampling, using direct techniques (i.e., using uncultured material). In addition, methods are now available, using commercial probes (AneuVysion from Vysis), to detect the main trisomies and sex chromosome aneuploidy with uncultured amniocytes. Other approaches to detecting trisomies, using molecular polymorphisms, are also available. However, most prenatal diagnoses for Down syndrome are still based on a complete chromosome analysis, and a full chromosome analysis from amniotic fluid requires the culture of cells.

The coverslip method in Protocol 27.5 was originally submitted by Marie Ferguson-Smith, East Anglian Regional Genetics Service, Addenbrookes Hospital, Hills Road, Cambridge, CB2 2QQ England, UK. It has since been updated, and the alternative closed-tube method has been described by Mike Griffiths, Regional Genetics Laboratory, Birmingham Women's Hospital, Edgbaston, Birmingham, B15 2 TG, England, UK.

The principle of culturing amniotic fluid cells is based on separating the cells by centrifugation and setting up the cell suspension in a suitable culture vessel. A variety of approaches exist that use either closed or open systems to grow the cells in tubes, flasks, or slide chambers, or on coverslips in Petri dishes. Good-quality, rapid results can be obtained with any of these techniques, but a closed-tube system has the advantage of robustness, simplicity, and minimal use of resources.

PROTOCOL 27.5. CULTURE OF AMNIOCYTES

A. Closed-Tube Culture System:

Outline
Incubate cell cultures in a plastic Leighton tube in a standard incubator at 37°C, and harvest the cells with a trypsin suspension method.

Materials
Sterile or Aseptically Prepared:
- ☐ Amniotic fluid sample, 10–15 mL
- ☐ Complete medium: 100 mL of Ham's F10 medium with 20 mM HEPES buffer, 2 mM glutamine, 100 U/mL penicillin, and 100 µg/mL streptomycin with 10% fetal bovine serum or 2% Ultroser G (Invitrogen)
- ☐ D-PBSA
- ☐ Colcemid, 10 µg/mL: Add 1–9 mL of sterile D-PBSA for a working solution, and then add 0.1 mL of this for each 1 mL of medium in the culture, to reach a final concentration of 0.1 µg/mL
- ☐ Thymidine: Make a stock solution of 15 mg/mL in D-PBSA, and filter sterilize it. Add 0.1 mL

of the stock thymidine solution to 10 mL of medium. Change to the thymidine medium before harvesting cells. The final concentration in the medium should be 0.15 mg/mL.
- ☐ Bromodeoxyuridine, BUdR: A vial contains 250 mg; dissolve all of it in 25 mL of sterile D-PBSA, and filter sterilize the solution. Add 0.5 mL of the solution to 11.5 mL sterile of D-PBSA and 8 mL of diluted colcemid solution. Dispense the resultant solution into aliquots, and store them frozen. Add 0.1 mL of the BUdR-colcemid mix to each 1 mL of culture. The final concentrations should be 25 µg/mL of BUdR and 0.04 µg/mL of colcemid.
- ☐ Trypsin (Bacto, Difco): Reconstitute the contents of the vial with 10 mL of UPW, following manufacturer's instructions to give a 5% w/v solution.
- ☐ EDTA, 0.5 mM in isotonic saline (Invitrogen)
- ☐ Trypsin/EDTA (TE) solution for subculture and harvest: 2 mL of reconstituted Bacto trypsin in 100 mL of EDTA, giving 0.1% w/v trypsin
- ☐ Universal containers for collecting samples of amniotic fluid
- ☐ Plastic, disposable, 10-mL pipettes and a pipetting aid for setting up cultures
- ☐ Leighton tubes (flat sided, plastic) and caps (Nunc)
- ☐ Syringes, 20 mL; plastic and plastic mixing needles (Henley Medical) for feeding cultures
- ☐ Transfer pipettes/Pastettes, 1 mL and 3 mL, plastic, disposable
- ☐ Histopaque-1077 (Sigma); adapt the manufacturer's instructions

Nonsterile:
- ☐ Disinfectant [e.g., Virkon (Merck)]; follow the manufacturer's instructions
- ☐ Potassium chloride hypotonic solution, 0.075 M KCl in UPW; add 5.574 g of Analar potassium chloride to 1 L of UPW
- ☐ Fixative: Freshly mixed, 3 parts Analar methanol to 1 part Analar glacial acetic acid
- ☐ Transfer pipettes/pastettes, 1 mL, plastic, disposable
- ☐ Glass microscope slides, precleaned (Berliner Glas KG, from Skan, through Richardsons of Leicester)
- ☐ Coplin jars
- ☐ Hydrogen peroxide solution, 5% v/v: Dilute 1 part 30% hydrogen peroxide with 5 parts water
- ☐ Saline solution, 0.9% w/v NaCl in UPW
- ☐ Trypsin/saline solution for banding: Add 1.4 mL of reconstituted Bacto trypsin (Difco) to 50 mL of saline, giving 0.14% w/v trypsin.
- ☐ Buffer, pH 6.8 (Gurr, Merck): Add one buffer tablet to 1 L of UPW.

- Leishman stain (Gurr, Merck), usually diluted 1 part to 4 parts pH 6.8 buffer
- DPX slide mountant and coverslips
- Oven at 60°C
- Hot plate at 75°C
- Bright-field and phase-contrast upright microscope, to assess slide making and banding, and for chromosome analysis

Protocol

Δ Safety Notes

(a) Wear gloves and a laboratory coat. All samples should be handled under sterile conditions in a class II microbiological safety cabinet (*see* Section 7.8.2)

(b) Discarded media and hypotonic supernates (but not fixative) should be poured off into a disinfectant solution—e.g., Virkon or hypochlorite. Fixative should be discarded into sodium bicarbonate solution to neutralize the acid. After standing 2 h both types of waste may be discarded into normal drainage, with plenty of running water.

(c) Bromodeoxyuridine (BUdR) is a known mutagen, and its use is optional. An appropriate local safety assessment (e.g., COSHH) should be undertaken before proceeding with any work involving this chemical. Solutions and supernates containing significant amounts of the chemical may be discarded into sealable vessels containing vermiculite (or a similar absorbent material) and destroyed by incineration.

(d) Protocols involving the use of methanol and acetic acid for fixing and preparing slides should be carried out in a fume hood.

Setting up and Monitoring Tube Cultures:

1. Expect approximately 10–15 mL of amniotic fluid to be delivered to the laboratory in a plastic sterile universal container.
2. Use a sterile 20-mL syringe fitted with a sterile plastic mixing needle (or a 10-mL pipette) to divide the sample between three Leighton tubes. Label the tubes with patient and culture identification details. If the sample is small (a volume of 5 mL or less), set up only two cultures.
3. Centrifuge the tubes for 5 min at 150 g.
4. Pour the supernates back into their original container [for biochemical assay or immunoassay—e.g., α-fetoprotein (AFP) estimation], or discard them into disinfectant solution (Virkon).
5. Resuspend the cell pellets in 1 mL of culture medium per Leighton tube.
6. Incubate the tubes at 37°C.

7. Leave the cultures undisturbed for 5–7 days, to allow the cells to settle and establish colonies.
8. Assess the cultures. Depending on the degree of cell growth, either add 0.5 mL of fresh medium, or remove and discard the old medium and add approximately 1 mL of fresh medium.
9. Thereafter, reassess cell growth as necessary (every 2–4 days), changing the medium as appropriate until there is sufficient growth for a harvest, usually 7–12 days after the cultures were initiated. Change the medium on the day before harvesting, if possible.

Notes

(a) Rapid cell growth can be encouraged by spreading colonies using trypsin (*see Dispersing and Subculturing* below). Overgrown cultures can be recovered by subculturing into several additional Leighton tubes or flasks if large quantities of cells are required.

(b) Bloodstained amniotic fluid samples may not grow as well as clear samples. One approach to working with such samples is to separate some of the amniocytes from the contaminating red blood cells, using density gradient centrifugation (e.g., Histopaque; *see* Protocol 12.10).

(c) Assessment of heavily bloodstained samples at 6 to 7 days is usually not possible without removing the bloodstained medium first. The medium can be collected into supplementary tubes, which may be discarded if the original tubes show growth, or can be incubated further if necessary.

(d) Supplementary tubes may be established at the first change of medium in any case when discarding the original suspension is undesirable.

Dispersing and Subculturing:

10. Prewarm approximately 1.5 mL of sterile TE solution for each tube to be processed to 37°C.
11. Remove the medium from the culture tube, and rinse it once with 1 mL of sterile TE solution.
12. To disperse the culture,
    (a) Add 0.2 mL of TE to the culture, and incubate the culture at 37°C for 1–2 min to detach the cells.
    (b) Check the cells on an inverted microscope When the cells are in suspension, add 1 mL of culture medium, and return the tube to the incubator. (The serum in the medium will inactivate the TE solution.)
13. To subculture,
    (a) Add 0.5 mL of TE to the culture, and incubate it at 37°C for 1–2 min to detach the cells.
    (b) Check the culture on an inverted microscope. When the cells are in suspension, add

1–2 mL of culture medium, and divide the suspension between an appropriate number of subculture tubes. Two to eight subcultures may be seeded, depending on the initial cell density. It is usually worth varying the concentration of cells in each subculture tube, as doing so improves the chances of being able to harvest a tube at an optimal cell density.

(c) Top up each subculture with fresh medium to a final volume of 1 mL, and then incubate.

14. Check that the cells have resettled the next day, and change the medium.

Routine Tube Harvesting:

Harvests can be routinely carried out on primary cultures so long as other cultures are available as a backup. If only one culture remains, subculture it before harvesting the cells, or salvage it after the harvest.

15. When the cultures are ready to be harvested, add 0.1 mL of diluted colcemid to each culture.
16. Incubate the cultures for as long as is necessary to accumulate enough rounded-up mitotic cells. This is typically 2–3 h, but may be as little as half an hour for very active cultures or more than 4 h for slow-growing cultures.
17. Remove the medium into Virkon solution, and drain the tube briefly onto a paper towel.
18. Add 2 mL of TE, and incubate the cultures at 37°C for 3 min to detach the cells.
19. When the cells are in suspension, add 7 mL of KCl hypotonic solution, and leave the cultures at 37°C for 5 min.
20. Centrifuge the tubes at 150 g for 5 min.
21. Carefully remove the supernate by pouring it into Virkon solution. Drain the tube briefly onto a paper towel. Flick the tube gently to resuspend the cells in the small amount of remaining liquid.
22. Slowly fix the cells, using fresh fixative:
    (a) Flick the tube, and add the first 1–2 mL of fixative drop by drop, continually agitating the cells to avoid cell clumping.
    (b) Add a further 3–4 mL of fixative.
23. Centrifuge the tubes at 150 g for 5 min.
24. Remove the supernatant fixative by pouring it into sodium bicarbonate solution.
25. Gently resuspend the cell pellet, and add 5 mL of fixative to it.
26. Change the fixative by repeating the centrifugation, pour-off, and refixation steps (Steps 9–11).
27. Centrifuge the cells again, pour off the supernate, resuspend the cell pellet in residual fixative, and make slides of the cells (*see* later steps for slide preparation).

Notes

(a) Fixed-cell suspensions can be stored at −20°C at any of the fixed stages. Change the fix twice before making slides.

(b) Salvage harvests may be used as an alternative to subculture or when only a single culture remains. This method requires the addition of sterile colcemid and sterile TE in the initial stages of the harvest, followed by transfer of the cells to a separate centrifuge tube after the cells have detached, but before the addition of hypotonic solution. Fresh medium can be added to the original culture tube; the medium should then be changed the next day, after the cells have resettled, to remove residual traces of trypsin and colcemid.

(c) If a harvest produces an unacceptably low yield of metaphases, then alternative strategies such as the following can be used:
    (i) Subculture the cells into a flask that is supported at a slight angle to the horizontal while the cells settle. Tilting the flask in this manner ensures a leading edge of the cells that are always growing at an optimal rate across the whole width of the flask.
    (ii) Expose the cells to a reduced concentration of colcemid overnight.

(d) Longer chromosomes for higher-resolution analysis may be produced by using either thymidine synchronization or overnight exposure to colcemid in the presence of bromodeoxyuridine. However, both these approaches work best with cells in an exponential growth phase, requiring careful assessment of the growth.

Thymidine Synchronized Harvesting:

1. On the morning of the day before the cultures are to be harvested, change the medium in the cultures, using the thymidine-supplemented medium (to a final thymidine concentration of 0.15 mg/mL).
2. Early on the day of harvest, rinse out the thymidine, using prewarmed medium. Pour off the medium, and rinse the cells twice. Add 1 mL of medium (without thymidine), which releases the thymidine block. The time of release depends on the time you intend to harvest the culture.
3. Incubate the culture for 4 h, then add 0.1 mL of diluted colcemid.
4. Incubate the culture for a further 2 h, and then harvest as for routine tube harvests. (Start with Step 2 of *Routine Tube Harvesting*, earlier in this protocol).

Bromodeoxyuridine Overnight Colcemid Harvests:

1. Change the medium in the cultures to be harvested during the morning of the day before the harvest.
2. In the afternoon of the same day, add 0.1 mL of BUdR-colcemid solution for each 1 mL of culture medium (final concentrations in culture: BUdR, 25 µg/mL; colcemid, 0.04 µg/mL).
3. The next morning (after about 20 h of exposure), pour off the BUdR-colcemid medium into a sealable container filled with vermiculite. Add 2 mL of prewarmed TE to the culture tube, and incubate the tube for 3 min to detach the cells.
4. Check that the cells are in suspension, and then add 7 mL of 0.075 M KCl hypotonic solution. Incubate the culture for 5 min.
5. Centrifuge the culture at 150 g for 5 min.
6. Pour off the hypotonic supernate into the vermiculite container.
7. Gently resuspend the cell pellet, slowly fix the cells, and continue the harvest, starting with Step 6 of *Routine Tube Harvesting,* earlier in this protocol.

Slide Preparation:

1. Use cleaned slides. The slides may be purchased as precleaned slides; however, it may still be beneficial to add 1 drop of fresh fix to each slide and wipe the slide with a paper towel immediately before use. If it is possible to control the surrounding environment, make the slides at 20–25°C and a relative humidity of 40–50%.
2. Place one drop of cell suspension onto each slide by dropping the suspension from a height of 1–3 cm. Allow the drop to dry naturally.
3. Assess the quality of spreading with a phase-contrast microscope. If the spreading is acceptable, make the rest of the slides, adjusting the cell density by adding extra drops of fix to the suspension if necessary.
4. In some circumstances, the spreading may need to be improved, and the following suggestions may be helpful:
   (a) Breathe on the slide first, and then place one drop of cell suspension onto the slide from a height of 1–3 cm. Allow the slide to dry naturally.
   (b) Place one drop of cell suspension onto the slide, either with or without breathing first. Leave the slide for a few seconds, and then place a drop of fresh fix on top before the first drop has dried. Allow the slide to dry.
   (c) Place one drop of suspension onto the slide, either with or without breathing on the slide first. Leave the slide for several seconds, and then, as the drying surface becomes dimpled

or Newton rings become visible, place a drop of fresh fix on top. Allow the slide to dry.
   (d) If the quality of spreading is still unacceptable, place the fixed cell suspension in the freezer in fix overnight, and remake the slides the next day.

G-Banding with Trypsin:

1. Pretreatment of the slide may be carried out before trypsin exposure but, depending on the age of the slide, may not be essential. Hydrogen peroxide pretreatment works well with fresh slides. Concentrated Hanks' salt solution works well with older slides. Coplin jars are suitable staining vessels. Slides that are not pretreated may also be used.
2. Any of the following pretreatment methods may be used:
   (a) Immerse the fresh slides (after drying for 1 h) in 5% hydrogen peroxide for 20 s to 2 min.
   (b) Incubate the fresh slides in an oven at 60°C for a few hours or overnight, to age the slides. Immerse the slides in 5× Hanks' BSS for 5 min.
   (c) Immerse 1 day or older slides in pH-6.8 buffer for 60 s.
3. Rinse the slides well in pH 6.8 buffer.
4. Immerse the slides in 0.14% trypsin in saline solution for 3 s to 2 min.
5. Rinse the slides well in pH 6.8 buffer.
6. Stain the slides with Leishman stain, freshly diluted 1:4 with pH 6.8 buffer, for 4 min.
7. Wash the slides with tap water, and drain them to dry or blot them dry with care.
8. Assess banding under a bright-field microscope at 400× magnification. If the slides are underbanded, they may be destained in pH 6.8 buffer or methanol and the procedure repeated. If the slides are overbanded, start the procedure again with another slide and vary the exposure time to trypsin or change the pretreatment.
9. Leave the slides on a hot plate (75°C) for a few minutes to ensure that they are completely dry, and then mount them with a glass coverslip, using DPX.

Chromosome painting is dealt with later in this chapter (*see* Protocol 27.9).

B. Open-Coverslip Culture System:

Outline

Culture the cells on coverslips in a Petri dish in a 5%-CO_2-in-air, 95%-humidity incubator.

Materials

The materials for this procedure are mostly the same as for the closed-tube culture system, with the following variations:

Sterile:

❏ Sodium-bicarbonate-buffered Ham's F10 medium with L-glutamine, penicillin, streptomycin, fetal bovine serum, and Ultroser G, as for the closed-tube system
❏ Plastic Petri dishes, 3.5 cm, with vented lids
❏ Glass coverslips, 22 mm, dry heat sterilized
❏ Forceps

Nonsterile:

❏ Hypotonic solution, 0.05 M KCl in UPW
❏ Trypsin, 0.8% in saline solution, for banding (0.4 mL of reconstituted Difco trypsin in 50 mL of saline)

Protocol

Setting Up and Cell Culture:

1. Centrifuge the sample in two universal containers at 150 g for 10 min. If the sample arrives in only one universal container, split the sample into a second universal container before centrifuging, to avoid the risk of losing a sample in the centrifuge because of breakage.
2. Place sterile coverslips in two Petri dishes, and label the lids and dishes with the sample designation.
3. Remove the supernate with a pipette, leaving 1 mL of fluid above the cell pellet.
4. Resuspend the cell pellet(s), and transfer the suspension to the Petri dishes.
5. Add 3 mL of medium to each dish, and incubate the cultures at 38°C in a 5%-CO_2-in-air atmosphere in a humid incubator.
6. After 5–7 days inspect the cultures for cell growth, using an inverted microscope.
7. Partially change the medium by removing approximately 2 mL of the old medium with a sterile Pastette and replacing it with an equal volume of fresh medium.
8. Inspect the cultures and change the medium twice a week. Cells are ready to be harvested when a sufficient number of actively growing colonies of suitable size have developed; usually 10 to 14 days after the culture was initiated.

In situ Harvesting of Coverslips:

9. Transfer the coverslip to a new Petri dish (appropriately labeled) with medium up to 24 h before the intended harvest. The original dish should be kept as a reserve source of extra cells.

10. Add 0.3 mL of diluted colcemid to the Petri dish to give a final concentration of 0.1 µg/mL, and incubate the dish for 2–3 h.
11. Gently remove the medium with a Pastette, and replace it with 3 mL of 0.05 M KCl hypotonic solution, prewarmed to 37°C. Incubate the culture for 10 min.
12. Remove the Petri dish from the incubator, and place it on a suitable tray, with the labeled lid under the dish.
13. Add 6 drops of fixative gently down the side of the Petri dish. Allow the fixative to disperse in the hypotonic solution. Leave the dish for 1 min.
14. Add another 6 drops of fixative to the Petri dish and leave for 1 min.
15. Remove 2 mL of supernate from the Petri dish, and replace it with 2 mL of fresh fixative, adding the fixative gently down the side of the dish. Leave the dish for 2 min.
16. Remove all supernate from the Petri dish, and replace it with 3 mL of fresh fixative, adding the fixative gently down the side of the dish. Leave the dish for 10 min.
17. Remove all of the fixative, and allow the coverslip to air dry within the dish. After removal of the fixative, the tray can be raised along one side to encourage the residual fixative to drain off the coverslips.

G-Banding with Trypsin:

18. Age the coverslip (still in the Petri dish for identification) in an oven at 60°C overnight or on a hot plate at 75°C for 2 h.
19. Make up 0.04% trypsin in saline solution, and, immediately before banding, mix 1 part Leishman stain with 4 parts pH 6.8 buffer.
20. Using forceps, place the coverslip in the trypsin solution and agitate it gently for a few seconds (the exact amount of time varies according to the age of the material).
21. Rinse the coverslip in buffer, and return it to the Petri dish.
22. Immediately flood the Petri dish with diluted Leishman stain for 2–4 min (depending on the batch of stain).
23. Rinse the coverslip in buffer, stand it against the side of the Petri dish, on absorbent paper, to drain, and allow it to dry.
24. Coverslips may be examined for the quality of banding before being mounted.
25. Mount the coverslip on a labeled slide, using DPX.

27.7 CULTURE OF CELLS FROM POIKILOTHERMS

The approach to the culture of cells from cold-blooded animals (poikilotherms) has been similar to that employed for warm-blooded animals, largely because the bulk of present-day experience has been derived from culturing cells from birds and mammals. Thus, the dissociation techniques for primary culture use proteolytic enzymes, such as trypsin, with EDTA as a chelating agent. Fetal bovine serum appears to substitute well for homologous serum or hemolymph (and is more readily available), but modified media formulations may improve cell growth. A number of these media are available through commercial suppliers (*see* Appendix II), and the procedure for using these media is much the same as for mammalian cells: Try those media and sera that are currently available, assaying for growth, plating efficiency, and specialized functions (*see* Sections 9.6, 10.5). Because the development of media for many invertebrate cell lines is in its infancy, it may prove necessary to develop new formulations if an untried class of invertebrates or type of tissue is examined. Most of the accumulated experience so far relates to insects and mollusks.

Two reviews cover some of the early developments of the field [Vago, 1971, 1972; Maramorosch, 1976], and some recent exploitation in biotechnology is described in a publication of the European Society for Animal Cell Technology [Jain et al., 1991; Klöppinger et al., 1991; Speir et al., 1991]. The latter relates to the use of the baculovirus vector in insect cell lines, which has many of the advantages of posttranslational modification found in mammalian cells, without the regulatory problems related to the isolation of biopharmaceuticals from human and mammalian cells.

Culture of vertebrate cells other than bird and mammal has also followed procedures for warm-blooded vertebrates, and so far there has not been a major divergence in technique. Because this is a developmental area, certain basic parameters will still need to be considered to render culture conditions optimal, and if a new species is being investigated, optimal conditions for growth may need to be established—e.g., pH, osmolality (which will vary from species to species), nutrients, and mineral concentration. Temperature may be less vital, but it should be fixed within the appropriate environmental range and regulated within ±0.5°C; overheating is particularly damaging.

27.7.1 Fish Cells

Fish cell culture has become increasingly popular, because of the growing commercial interest in fish farming. Protocol 27.6 for culturing cells from zebrafish has been abridged from Collodi [1998].

The zebrafish possesses many favorable characteristics that make it a popular nonmammalian model for studies of vertebrate development and toxicology [Powers, 1989; Driever et al., 1994]. Zebrafish reach sexual maturity in approximately 3 months, and females produce 100 to 200 eggs each week throughout the year. Embryogenesis is completed outside of the mother in 3–4 days, and the large, transparent embryos are amenable to experimental manipulations involving cell labeling or ablation techniques [Westerfield, 1993]. *In vitro* approaches utilizing embryo cell cultures have also been employed for the study of zebrafish development. Cell lines and long-term primary cultures, initiated from blastula, gastrula, and late-stage embryos, have been established [Collodi et al., 1992; Ghosh & Collodi, 1994; Sun et al., 1995a,b; Peppelenbosch et al., 1995].

Cell lines derived from early-stage embryos. Because differentiation occurs during zebrafish gastrulation, the cells in earlier-stage embryos, such as the blastula, are pluripotent [Kane et al., 1992], and methods have been developed for the culture of cells from these early-stage embryos. The ZEM-2 cell line, initiated from mid-blastula-stage embryos, has been growing in culture for more than 300 generations in medium containing low concentrations of fetal bovine and trout sera, insulin, trout embryo extract, and medium conditioned by buffalo rat liver cells [Ghosh & Collodi, 1994].

A fibroblastic cell line, ZEF, has also been derived from early-stage embryos in medium supplemented with FBS and FGF. Once established, the line has been maintained in LDF medium (*see* Materials in Protocol 27.6) containing 10% FBS [Sun et al., 1995a], and ZEF cells have been utilized as a feeder layer for primary cultures of zebrafish embryo cells [Sun et al., 1995a, Bradford et al., 1994a].

Primary cultures derived from early-stage embryos. In addition to the continuously growing embryo cell lines that are available, methods have been developed for the initiation of primary cultures derived from early zebrafish embryos [Sun et al., 1995b, Bradford et al., 1994a,b]. Primary cultures, derived from early gastrula-stage embryos, maintained a diploid chromosome number and exhibited a morphology characteristic of pluripotent ES cells when derived on a feeder layer of ZEF fibroblasts in medium containing FBS, trout serum, fish embryo extract, insulin, and leukemia inhibitory factor [Sun et al., 1995b; Bradford et al., 1994a].

Cell lines derived from late-stage embryos. Late-stage zebrafish embryos (20–24 h postfertilization) have been used for the derivation of three fibroblastic cell lines: ZF29, ZF13 [Peppelenbosch et al., 1995], and ZF4 [Driever & Rangini, 1993]. The lines were derived in Leibowitz's L-15 (ZF29 and ZF13) or a mixture of Ham's F12 and Dulbecco's modified Eagle's media (ZF4) supplemented with FBS.

PROTOCOL 27.6. CELL CULTURES FROM ZEBRAFISH EMBRYOS

Materials

Sterile:

❏ LDF basal medium:
   Leibowitz's L-15 medium, 100 mL

DMEM, 70 mL
Ham's F-12, 30 mL
Sodium selenite 6 µM, 200 µL
Store refrigerated at 4°C.

❏ LDF primary medium: LDF basal medium, plus
FBS, 1%
Trout serum, 0.5%
Trout embryo extract, 40 µg of protein/mL
Insulin, 10 µg/mL
Leukemia inhibitory factor, human, recombinant, 10 ng/mL

❏ LDF maintenance medium: LDF basal medium, plus
FBS, 1%
Trout serum (Sea Grow, East Coast Biologicals), 0.5%
Trout embryo extract, 40 µg of protein/mL
Insulin, 10 µg/mL
BRL-conditioned medium, 50%

❏ D medium: DMEM/F12/10FB: 50/50 DMEM/Ham's F12 with 10% fetal bovine serum (FBS)

❏ Trout embryo extract:
  (a) Collect the embryos (Shasta Rainbow or other strains of trout, 28 days postfertilization, reared at 10°C; or zebrafish, three days postfertilization, reared at 28°C), and store them frozen at −80°C.
  (b) To prepare the extract, thaw the embryos (approximately 150 g) and homogenize them in 10 mL of LDF for 2 min on ice, using a Tissuemizer homogenizer (Tekmar).
  (c) Pass the homogenate through several layers of cheesecloth to remove the chorions, and then centrifuge the homogenate at (20,000 g for 30 min at 4°C).
  (d) After centrifugation, collect the supernate, leaving behind the bright-orange lipid layer present on the surface.
  (e) Transfer the supernatant to a new tube, and centrifuge it as before (in Step (c)).
  (f) Collect the supernate, leaving behind any remaining lipid, and then ultracentrifuge it at 100,000 g for 60 min at 4°C.
  (g) After ultracentrifugation, collect the supernate, leaving behind the lipid layer. Dilute the supernate with LDF (1:10), and filter sterilize.
  (h) The extract must be passed through a series of filters (1.2 µm, 0.45 µm, and 0.2 µm).
  (i) Store the extract frozen at −80°C in 0.5-mL aliquots.
  (j) To use the extract for cell culture, measure the concentration of protein (*see* Protocol 21.4)

and then dilute the extract with LDF to the desired working concentration.
  (k) Store the diluted extract refrigerated at 4°C for a maximum of two months.

❏ BRL cell-conditioned medium:
  (a) Culture BRL cells (ATCC) at 37°C in 75-cm² flasks in DMEM/F12/10FB.
  (b) When the cultures become confluent, replace the FD medium with LDF supplemented with 2% FBS, and incubate the cells at 37°C.
  (c) After 5 days, remove the LDF, filter it, and store it frozen at −20°C.
  (d) Add fresh LDF to the BRL cultures, and repeat the process 5 days later. Conditioned LDF medium can be collected up to three times from the same flask before the cells must be split and allowed to grow again to confluence.

❏ Holtfreter's buffer:
NaCl, 70 g
KCl, 1.0 g
NaHCO₃, 4.0 g
CaCl₂, 2.0 g
UPW, to 1000 mL
Store at 4°C. Prepare a working solution by diluting 1:20 with UPW.

❏ Pronase E, 0.5 mg/mL in Holtfreter's buffer

❏ Trypsin, 1%, EDTA, 1 mM, in D-PBSA

A. Fibroblast feeder layers

Outline
Prepare feeder layers of embryonic fibroblasts from gastrula-stage zebrafish [Sun et al., 1995a] by removing the chorion in pronase, culturing the cells in FGF-supplemented medium, and selecting the fibroblasts by differential trypsinization.

Materials for Feeder Layers
Sterile or aseptically prepared:
❏ Embryos, eight hours postfertilization
❏ Pronase
❏ Trypsin
❏ FBS
❏ LDF primary medium with 10 ng/mL bovine FGF ($a + b$ mixture)
❏ LDF basal medium with 10% FBS
❏ Flasks, 25 cm²

Protocol
1. Collect approximately 30 embryos (eight hours postfertilization).
2. Remove the chorion by pronase treatment [Sun et al., 1995a].
3. Incubate the embryos in trypsin (1 min) while gently pipetting to dissociate the cells.

4. Add FBS (10% final concentration) to stop the action of the trypsin.

5. Collect the cells into a pellet by centrifugation (at 500 g for 10 min).

6. Resuspend the cell pellet in 5 mL of LDF primary medium containing FGF, and transfer the cells to a 25-cm^2 flask.

7. Allow the cells to attach and grow to confluence at 26°C.

8. When the cells are confluent, passage the culture in the same medium.

9. After 2–3 passages, a mixed population of epithelial and fibroblastic cells will be present in the culture, and the cells can be maintained in LDF basal with 10% FBS. Select the fibroblasts for further culture, by differential trypsinization:

   (a) Treat the culture with trypsin for 1 min to remove most of the fibroblasts, and leave the epithelial cells attached to the plastic.

   (b) Transfer the fibroblasts to another flask, and repeat this process when the culture becomes confluent.

   (c) After two or three passages, the culture will consist of a homogeneous population of fibroblasts.

10. To prepare feeder layers of growth-arrested fibroblasts,

   (a) Add the cells to the appropriate culture dish or flask, and allow the fibroblasts to grow into a confluent monolayer.

   (b) Add mitomycin C, 10 μg/mL, to the cultures, and incubate the cultures for 3 h at 26°C.

   (c) After being rinsed three times with LDF, the growth-arrested fibroblasts can be used as feeder layers for zebrafish embryo cell cultures.

B. Primary Cultures

Outline

Collect embryos, remove the chorion, disaggregate the embryos in trypsin, and grow primary cultures derived from zebrafish blastula- and early gastrula-stage embryos on feeder layers of embryonic fibroblasts.

Materials for Primary Culture
Sterile:
☐ LDF primary medium
☐ Holtfreter's buffer
☐ Dilute bleach, 0.1% in UPW
☐ Pronase E solution
☐ Feeder layers of embryonic fibroblasts
☐ Human recombinant leukemia inhibitory factor (LIF), 1 μg/mL

Protocol

1. Harvest embryos at the midblastula or early gastrula stage, and rinse them several times with clean water.

2. After rinsing, transfer the embryos into 6-cm Petri dishes (50–100 embryos/dish), take them to a laminar-flow hood, and maintain them under aseptic conditions.

3. Soak the embryos for 2 min in dilute bleach, and rinse them several times in sterile Holtfreter's buffer.

4. Dechorionate the embryos by incubating them in 2 mL of Pronase E solution for ~10 min, and then gently swirl the embryos in the Petri dish to separate them from the partially digested chorion.

5. Tilt the dish to collect the embryos on one side, and gently remove 1.5 mL of the Pronase solution with a Pasteur pipette. To prevent the dechorionated embryos from adhering to the dish and rupturing, keep the dish tilted so that the embryos remain suspended in the remaining Pronase solution.

6. Gently rinse the embryos by adding 2 mL of Holtfreter's buffer and swirling gently.

7. Tilt the dish and remove most of the Holtfreter's buffer, leaving the embryos suspended in ~0.5 mL.

8. Repeat the rinse procedure two more times.

9. After the final rinse, leave the embryos suspended in 0.5 mL of Holtfreter's buffer, and add 2 mL of trypsin solution.

10. Incubate the embryos in the trypsin for 1 min and then dissociate the cells by gently pipetting 3–4 times.

11. Immediately transfer the cell suspension into a sterile polypropylene centrifuge tube, and add to the tube 200 μL of FBS to stop the trypsin.

12. Collect the cells by centrifugation (at 500 g, 5 min) and resuspend the pellet in LDF primary medium (without FBS or trout serum).

13. Seed the cells at 1×10^4 cells/cm^2 onto feeder layers of growth-arrested embryonic fibroblasts, contained in multiwell dishes or flasks.

14. Allow the cells to attach to the feeder layers (~15 min) before adding FBS and trout serum. Human recombinant leukemia inhibitory factor (10 ng/mL) is used in the medium in preference to BRL-conditioned medium [Sun et al., 1995a,b].

C. Cell lines

Materials for Cell Lines
Sterile or Aseptically Prepared:
☐ ZEM-2 cells (or equivalent)

❑ LDF maintenance medium

Protocol

1. Grow cultures derived from early zebrafish embryos, such as ZEM-2, in LDF maintenance medium to ~70% confluence.
2. Incubate the cultures at 26°C in ambient air.
3. Change the medium approximately every 5 days.
4. Subculture by trypsinization (*see* Protocol 13.2).

Trout serum and embryo extract have also been shown to stimulate the growth of embryo cells from other fish species [Collodi & Barnes, 1990].

27.7.2 Insect Cells

There has been considerable interest for some time in the culture of insect cells for studies of pest control and environmental toxicology. However, the greatest increase in the usage of insect cell culture has resulted from the use of baculovirus for gene cloning [Midgley et al., 1998]. Baculovirus is often grown in Sf9 cells, a continuous cell line from the fall armyworm *Spodoptera frugiperda*. Protocol 27.7, for the culture of Sf9 cells, has been abridged from Midgley et al. [1998]. In this method, Sf9 cells are kept growing continuously in a magnetic spinner culture flask (*see* Protocol 13.3) or a flat-bottom flask with a magnetic stirrer bar mixing at about 80 rpm, ensuring that the stirrer is not a source of heat. Ideally, the cells should be maintained at 27°C, but it is possible to grow them without an incubator in a room with constant temperature between 20 and 28°C. CO_2 is not required for these media. Cells can be maintained in standard plastic tissue culture flasks, but, because the cells attach to the surface of the flask, they must be detached for subculture by scraping or dislodging the cells with a jet of medium. However, this method will result in a lot of cell death, because the cells attach quite tightly to plastic when grown in the presence of serum. The cells should have a population-doubling time of <24 h.

PROTOCOL 27.7. PROPAGATION OF INSECT CELLS

Outline

Disperse the cells mechanically from the monolayer, and propagate them in suspension at 27°C.

Materials

Sterile or Aseptically Prepared:
❑ Cells: Sf9 [Smith et al., 1983] (ATCC #CRL-1711) or lines derived from the cabbage

looper *Trichoplusia ni* (Tn368, or BTI-TN-5B1-4; also known as "High Five," available from Invitrogen)
❑ Growth medium: EX-CELL 400 (JRH Biosciences) containing 2 mM L-glutamine, supplemented with 5% FBS, and 5 mL of penicillin/streptomycin solution (penicillin, 50 U/mL; streptomycin, 50 μmL). Store the medium at 4°C in the dark, and always warm it to room temperature before use. Sf9 cells are very sensitive to changes in growth medium, so for any change (e.g., to use serum-free EX-CELL 400), acclimatize the cells by gradually adding the new medium over a number of days.
❑ Dimethyl sulfoxide (DMSO), 10% in FBS
❑ Pluronic F68
❑ Culture flasks
❑ Spinner flask and magnetic stirrer
❑ Incubator at 27°C (CO_2 not required)

Protocol

Routine Maintenance:

1. Detach the cells from the flask culture by scraping or dispersing them with a jet of medium, or use cells grown in suspension in a spinner flask.
2. Count the cells by hemocytometer, and determine their viability by dye exclusion with Trypan Blue or Naphthalene Black (*see* Protocol 22.1).
3. Seed the spinner flask at $0.5-1 \times 10^6$ viable cells per mL (20–100 mL in a 500-mL spinner flask).
4. Incubate the cells at 20–28°C. (27°C is optimal.)
5. Dilute the cells to $0.5-1 \times 10^6$ cells per mL every 48–72 h, or when there are about $4-5 \times 10^6$ cells/mL.
6. Transfer the cells to a clean flask every 3–4 weeks.
7. If the cells clump, try stirring them slightly faster, and add the surfactant Pluronic F–68 (0.5–1.0% v/v) to reduce shearing.

Freezing Cells for Storage (see also Protocol 20.1):

1. Count the cells, and centrifuge at 1000 rpm (~200 *g*) for 5 min.
2. Resuspend the pellet at 1×10^7 cells/mL in 10% DMSO in FBS.
3. Dispense the cells into aliquots, place the aliquots into ampoules, and chill the ampoules on ice for 1 h.
4. Pack the tubes into a Styrofoam container.
5. Freeze the tubes slowly (~1°C/min) overnight at −70°C.
6. Transfer the tubes to a liquid nitrogen freezer.
7. To recover the frozen cells, thaw the ampoules rapidly at 37°C. (If the cells are stored in the liquid phase, take care to thaw them in a covered

vessel, to avoid risk of injury from explosion of the ampoule.)

8. Transfer the cells into a 25-cm^2 flask containing 5 mL of medium. Tip the flask to spread out the cells evenly.

9. Remove the medium after 2–3 h, when most of the cells should have attached, and add 5 mL of fresh medium.

10. Leave the cells 2–3 days to recover before detaching, and then transfer them to a stirrer flask or a larger plastic flask as described previously.

27.8 *IN SITU* MOLECULAR HYBRIDIZATION

Nucleic acid hybridization is used routinely for the detection of specific nucleotide sequences in DNA (Southern blotting) or RNA (Northern blotting). This technique can be applied as a cytological technique with fixed cells to detect nucleotide sequences *in situ* (*see* Plate 13c,d). Protocol 27.8, for *in situ* hybridization, has been contributed by W. Nicol Keith, Centre for Oncology and Applied Pharmacology, University of Glasgow, Garscube Estate, Switchback Rd., Glasgow G61 1BD, Scotland, UK.

27.8.1 Analysis of RNA Gene Expression by *In Situ* Hybridization

Molecular techniques can be roughly broken down into two groups: lysate analysis and *in situ* analysis. With lysate methods (Southern blot analysis and PCR), tissue biopsies are homogenized and the spatial relationships between the cells of the tissue are destroyed [Murphy et al., 1995]. This process leads to a loss of information on heterogeneity and small subpopulations, particularly in tumor biopsies, and presents an averaging of changes. However, quantitation can be simpler and more accurate than *in situ* approaches. In comparison, *in situ* techniques, such as RNA *in situ* hybridization, allow the visualization of gene expression in individual cells within their histological context [Soder et al., 1997, 1998; Keith, 2003].

The principle of *in situ* hybridization is based on the specific binding of a labeled nucleic acid probe to a complementary sequence in a tissue sample, followed by visualization of the probe. This process enables both detection and localization of the target sequence. A number of prerequisites for the success of this procedure include retention of the nucleic acid sequences in the sample and accessibility of these sequences to the probe. Specimens suitable for *in situ* hybridization (ISH) include cells from culture and tissue from samples of whole or biopsied organs.

PROTOCOL 27.8. AUTORADIOGRAPHIC *IN SITU* HYBRIDIZATION

Outline
Hybridize radiolabeled probes to fixed cells on microscope slides. Then visualize sites of hybridization by microautoradiography.

Materials
Nonsterile:
☐ Microcentrifuge tubes (e.g., Eppendorf)
☐ Microcentrifuge (Eppendorf)

Linearization of Plasmid DNA:
☐ RNA labeling kit (Amersham, RPN 3100)
☐ Diethylpyrocarbonate (DEPC; Sigma)
☐ DEPC-water: 1% DEPC in distilled water (dH$_2$O), autoclaved
☐ Phenol-chloroform isoamyl alcohol, pH 8 (Sigma)
☐ Sodium acetate (NaAc), 3 M, pH 8.0
☐ Absolute alcohol, analytical grade
☐ Glycogen, 20 mg/mL (Roche)
☐ Agarose, 1%, electrophoresis grade (Invitrogen)

Probe-Labeling Reagents and Solutions:
☐ Dithiothreitol (DTT), 0.2 M (Sigma)
☐ DTT, 50 mM: divided into aliquots and stored at −20°C
☐ [^{35}S]UTP (Amersham SJ 603)
☐ G50 Sephadex columns (Amersham Biosciences)
☐ Column buffer: 0.3 M NaAc, 1 mM EDTA, 1% SDS, autoclaved
☐ Phenol, pH 5.0 (Sigma)
☐ Chloroform isoamyl alcohol (Biogene)
☐ Scintillation fluid: Ecoscint A (National Diagnostics)

In Situ Hybridization Reagents and Solutions:
☐ Histoclear (Fisher)
☐ NaCl-DEPC: 0.85% NaCl, 1% diethyl pyrocarbonate (DEPC), autoclaved
☐ D-PBSA, 1% DEPC, autoclaved
☐ EDTA, 0.5 M: Dissolve in 1% DEPC in dH$_2$O; pH 7.5, autoclaved
☐ Proteinase K Buffer: 1 M Tris HCl, 0.5 M EDTA, 1% DEPC pH 7.5, autoclaved
☐ Proteinase K stock solution (Sigma): 20 mg/mL in DEPC H$_2$O; divide into aliquots and store at −20°C
☐ Formalin
☐ Triethanolamine, 0.1 M, with 1% DEPC, autoclaved
☐ Acetic anhydride (Sigma)

❑ DTT, 1 M: divided into aliquots and stored at −20°C
❑ Hybridmix, 60%:
  Formamide (Fluka), 6 mL
  Dextran sulfate, 50%, in 1% DEPC, 2 mL
  SSC, 20×, 1 mL
  Tris· HCl, 1 M, 100 µL
  Denhardt's solution, 50×, 200 µL
  SDS, 10%, 100 µL
  tRNA (10 mg/mL; Sigma), 400 µL
  Salmon DNA (10 mg/mL; Sigma), 200 µL
  Store at −20°C in 400-µL aliquots.

Washing Reagents and Solutions:
❑ SSC, 20× (see Appendix I; also from Invitrogen); dilute to 5× SSC, 2× SSC, and 0.1× SSC
❑ Formamide, 50% in 2× SSC
❑ β-Mercaptoethanol (Sigma)
❑ RNase buffer: 0.5 M NaCl, 0.5 M EDTA, 1 M Tris, pH 7.5
❑ RNaseA stock solution (Sigma): 10 mg/mL in DEPC H₂O; store at −20°C in 400-µL aliquots
❑ Gelatin: 0.2 g in 200 mL of dH₂O; microwave for 2 min, cool, and filter

Autoradiography Reagents and Solutions:
See Protocol 27.3.

Protocol

Handling RNA
All solutions involved in the preparation of the probe and up to the posthybridization wash steps must be free from RNase. Solutions should be treated with DEPC and autoclaved for 20 min at 121°C. This procedure removes the majority of RNases, but, because of the ubiquitous nature of RNases, it is not a substitute for care in handling the solutions, glassware, and pipettes. A set of pipettes dedicated for use only with RNA is worthwhile, and regular treatment of the pipettes with DEPC-water overnight or with a proprietary anti-RNase solution, such as RNase-Zap (Ambion), may be useful as well. All glassware should be wrapped in aluminum foil and autoclaved before use. Plastic Eppendorf tubes may be treated with DEPC-water or RNase-Zap before autoclaving.

Probe Preparation:

RNA probes (riboprobes)
Preparing RNA probes requires use of a DNA template of the target sequence, and generation of sense and antisense RNA probes, with radioactive nucleotides incorporated, is possible.

Single-stranded RNA probes are ideal if high sensitivity is required; probes of 200–1000 base pairs (bp) have been used, but probes of 150–200 bp are probably optimal, as tissue penetration can become reduced with longer probe size. Limited alkaline hydrolysis can be used to reduce probe size as required. The RNA-RNA or RNA-DNA hybrids are more stable than their oligonucleotide or DNA counterparts, rendering them the most popular probes.

Commercial probes and control probes
Commercially available (Ambion) DNA templates for Actin and GAPDH can be used to generate RNA probes for use as positive controls, because they are housekeeping genes and are ubiquitously expressed. Sense probes are commonly used as negative controls and are superior to the omission of a probe as a control. Well-characterized tumor samples with a range of RNA expression can be used as positive specimens during each run of slides. Tumor samples of a variety of tissues are also available commercially.

Linearization of Plasmid DNA:
1. Use 10–20 µg of DNA in a microcentrifuge tube.
2. Add 10 units of restriction enzyme per µg of DNA, and set up digestion as recommended by suppliers of the enzyme.
3. Leave the reaction at 37°C for 3 h or overnight.

Phenol Chloroform Extraction:
4. Add 400 µL of phenol-chloroform-isoamyl alcohol (pH 8.0), and vortex the mixture.
5. Spin the mixture for 3 min at 13,000 rpm in a microfuge at room temperature.
6. Collect the supernate, transfer it to a fresh tube, and add 10 µL of 3 M NaAc (pH 8) to it.
7. Add 250 µL of 100% ethanol (stored at −20°C).
8. Add 1 µL of glycogen to help precipitate the DNA.
9. Place the tube on dry ice for 1 h.
10. Spin for 15 min at 13,000 rpm.
11. Discard the supernate, and keep the pellet.
12. Add 400 µL of 70% ethanol (stored at −20°C) to wash the pellet.
13. Spin for 10 min at 13,000 rpm.
14. Remove the 70% ethanol, and air dry the pellet.
15. Resuspend the pellet in 10–20 µL of DEPC H₂O, depending on the amount of DNA used (see Step 1), aiming for a final concentration of 1 µg/µL.
16. Run 0.5 µL of this suspension on a 1% agarose gel.

RNA Labeling:

This part of the procedure involves the incorporation of radioactive nucleotides. Use the Amersham kit (RPN3100; Amersham Biosciences) according to the pack insert, with reference to the following method:

1. Combine the following in a microcentrifuge tube:
   (a) 5× transcription buffer .... 4 μL
   (b) DTT, 0.2 M .... 1 μL
   (c) HPR1 .... 1 μL
   (d) ATP, CTP, and GTP .... 0.5 μL
   (e) Linearized DNA template (1 μg/mL) ... 1 μL
   (f) [^{35}S]UTP .... 9.5 μL
   (g) RNA polymerase .... 2 μL
2. Mix the components, and place the solution at 37°C for 1.5 hours.

DNase Extraction of the DNA Template:

3. Add 10 U of DNase L.
4. Add 1 μL of RNase inhibitor.
5. Mix the solution, and place it at 37°C for 10 min.

Removal of Unincorporated Nucleotides:

6. Equilibrate a G50 Sephadex column with 2 mL of column buffer.
7. Add the probe to the column.
8. Add 400 μL of column buffer to the column, and allow the buffer to run through.
9. Add a further 400 μL of column buffer and collect the eluate in an Eppendorf tube.

Phenol-Chloroform Extraction:

10. Add 400 μL of phenol (pH 5.0) to the tube, vortex, and spin for 3 min at 13,000 rpm.
11. Collect the supernate, transfer it to a fresh microtube and add 400 μL chloroform-isoamyl alcohol to it. Vortex, and spin for 3 min at 13,000 rpm.
12. Collect the supernate and remove 1 μL of it for counting the incorporation.
13. Add 2.5 vol. of 100% ethanol (stored at −20°C) to the remaining supernate.
14. Add 1 μL of yeast glycogen to facilitate precipitation.
15. Place the tube on dry ice for 30 min.
16. Spin for 15 min at 13,000 rpm.
17. Remove the alcohol and leave the pellet undisturbed.
18. Wash the pellet with 70% ethanol, spin at 13,000 rpm for 10 min, and remove the ethanol.
19. Air dry the pellet and resuspend it in 50 mM DTT, calculating the volume of the 50 mM DTT as follows:

(a) Add the 1 μL of supernatant from Step 12 to 2–3 mL of scintillation fluid and determine counts per minute (CPM) on scintillation counter.

(b) Volume DTT $= \dfrac{\text{CPM} \times 400}{6 \times 10^5}$

where 400 is the volume after phenol/chloroform extraction and 6×10^5 is the required total CPM in the DTT solution.

(c) This is the volume of 50 mM DTT that the probe should be resuspended in.

20. Remove 1 μL and count again to confirm the activity of the probe.

In Situ Hybridization:

Preparation of the specimen:

The objective of this part of the procedure is to preserve the architecture and morphology of the tissue and to retain the RNA products. Rapid processing of the tissue sample, either by freezing or fixing in formalin, enables the RNA to be preserved. Crosslinking fixatives, such as 4% paraformaldehyde and 4% formaldehyde, are the fixatives of choice for the retention and/or accessibility of cellular RNA. The length of fixation will depend on the size of the specimen. Longer fixation times result in better tissue morphology, but may reduce access to the probe. Paraffin wax is the embedding medium of choice. It allows sectioning down to 1 μm in thickness and is easily removed before hybridization. As the sections will be processed through a number of solutions, coated slides are recommended to hold the specimen on the slide. Frozen samples should be chilled to −70°C, and, after cryosectioning, they should be placed on a coated slide and fixed.

Tissue preparation prehybridization:

This treatment of the tissue before hybridization attempts to increase the access of the probe to the target RNA sequence and to reduce nonspecific background binding. The specimen is subjected to protease treatment to increase the accessibility of the target nucleic acid to the probe, especially if the probe is greater than 100 bp. It is important to postfix the specimen in formaldehyde, to prevent disintegration of the tissue. Nonspecific binding to amino groups is reduced by acetylation with acetic anhydride. During tissue preparation, great care must be taken to protect the specimen from RNase. All glassware must be treated to remove any contamination, all solutions must be treated with DEPC, and gloves must be worn throughout the procedure. Handling of the tissue sections should be kept to a minimum.

Hybridization:

The hybridization temperature can be critical for some probe/target sequences. Formamide in the hybridization buffer, as a helix destabilizer, reduces the melting point of the hybrids and enables reduction of the temperature of hybridization. The lower temperature helps to preserve tissue architecture. A temperature of 52°C has been found to be the optimal. Dextran sulfate in the hybridization buffer, by volume exclusion, increases the concentration of the probe and reduces hybridization times. Although the hybridization reaction is almost complete after 5–6 h, it is convenient to leave the reaction overnight. The sodium ion concentration in the buffer serves to stabilize the hybrids.

Posthybridization washing:

The main objective of posthybridization washing is to remove unbound and nonspecifically bound probes by selection of the temperature, salt concentration, and formamide concentration. The use of RNase enables the digestion of single-stranded RNA, unbound to the target, but does not affect the bound RNA–RNA complexes.

Pretreatment of Paraffin Sections:

1. Dewax the paraffin sections with Histoclear, twice for 10 min.
2. Rehydrate through an ethanol series: 100%, 90%, 70%, 50%, and 30%, for 10 s each.
3. Rinse in 0.85% NaCl and D-PBSA solutions for 5 min each
4. Digest the section in Proteinase K, 400 μL of Proteinase K stock in 200 mL of Proteinase K buffer for 7.5 min
5. Rinse in D-PBSA; 3 min.
6. Postfix in 4% formalin or 4% paraformaldehyde.
7. Rinse in DEPC-water for 1 min.
8. Acetylate in 200 mL of 0.1 M triethanolamine with 500 μL of acetic anhydride for 10 min, stirring throughout in a fume hood.
9. Rinse in D-PBSA and 0.85% NaCl, for 5 min each.
10. Dehydrate through the ethanol series: 30%, 50%, 70%, 90%, and 100%, for 10 s each.
11. Air dry the section.

Preparation of the Probe and Hybridization:

12. For 20 paraffin sections, combine 16 μL of 1 M DTT, 344 μL of 60% Hybridmix, and 40 μL of the probe. Vortex and spin the solution briefly.
13. Denature the probe at 80°C for 3 min. Cool it on ice.
14. Apply 20 μL of the probe mix from Step 12 to each tissue section, and cover it with a glass coverslip.

15. Hybridize the solution at 52°C overnight in a humidified chamber.

Posthybridization Wash:

16. Preheat the solutions to the required temperature.
17. Wash the sections in 200 mL of 5× SSC with 250 μL of β-mercaptoethanol for 30 min at 50°C.
18. Wash in 200 mL of 50% formamide and 2× SSC with 1.4 mL of β-mercaptoethanol for 20 min at 65°C.
19. Wash in 200 mL of RNase buffer twice for 10 min, each time at 37°C.
20. Wash in 200 mL of RNase buffer with 400 μL of RNase A solution for 30 min at 37°C.
21. Repeat Step 19 but for 15 min each wash.
22. Repeat Step 18.
23. Wash in 200 mL of 5× SSC and 200 mL of 0.1× SSC, for 15 min each at 50°C.
24. Dehydrate in an ethanol series: 50%, 70%, and 100% for 1 min each.
25. Air dry the sections.
26. Dip the slides in the gelatin solution for 1 min, and then air dry them.

Autoradiography:

See Setting Up Autoradiographs in Protocol 27.3.

Analysis.

(1) Examine the sections, using light microscopy under bright- and dark-field illumination.
(2) Score the sections with reference to positive and negative controls.

27.8.2 Fluorescence *In Situ* Hybridization in the Analysis of Genes and Chromosomes

Protocol 27.9 for fluorescence *in situ* hybridization (FISH) has been provided by Nicol Keith, Centre for Oncology and Applied Pharmacology, University of Glasgow, Garscube Estate, Bearsden, Glasgow G61 1BD, Scotland, UK.

Fluorescence *in situ* hybridization is the most direct way of determining the linear order of genes on chromosomes. By using chromosome- and gene-specific probes, numerical and structural aberrations can also be analyzed within individual cells. These techniques have a wide variety of applications in the diagnosis of genetic disease and the identification of gene deletions, translocations, and amplifications during cancer development [Wiktor & Van Dyke, 2004; Krupp et al., 2004; Camps et al., 2004].

Nucleic acid probes are labeled nonisotopically by the incorporation of nucleotides modified with molecules such as biotin or digoxigenin. After hybridization of the labeled probes to the chromosomes, detection of the hybridized sequences is achieved by forming antibody complexes that recognize the biotin or digoxigenin within the probe. The hybridization is visualized by using antibodies conjugated to fluorochromes. The fluorescent signal can be detected in a number of ways. If the signal is strong enough, standard fluorescence microscopy can be used. However, data analysis and storage can be improved considerably by the use of digital imaging systems such as confocal laser scanning microscopy or cooled CCD camera. The major advantages of fluorescence *in situ* hybridization (FISH) are that it is nonisotopic, rapid, good for data storage and manipulation, and sensitive. It also shows accurate signal localization, allows simultaneous analysis of two or more fluorochromes, and provides a quantitative and spatial distribution of the signal.

PROTOCOL 27.9. FISH USING SINGLE-COPY GENOMIC PROBES AND CHROMOSOME PAINTING

Outline
Hybridize biotinylated or digoxigenin-labeled probes to denatured chromosomes and detect the probes by double-antibody fluorescent staining.

Materials
Nonsterile:
- SSC, 2× (*see* Appendix I for SSC, 20×)
- SSC, 1×
- D-PBSA
- Glycogen, 20 mg/mL (Boehringer Mannheim)
- EtOH, 70%
- EtOH, 100%
- Fixative: methanol:acetic acid, 3:1
- RNase, 100 μg/mL in 2× SSC
- Paraformaldehyde, 1%, in PBSA
- Formamide, 70%, in 2× SSC
- Hybridization buffer: formamide (50%), dextran sulfate (5%) 2× SSC, salmon sperm DNA (500 μg/mL)
- Labeled probe:
  Large cosmid clones are most suitable for probes to detect single-copy genes. However, cDNA probes can be used as well. DNA is labeled by nick translation, using commercial kits. The nick translation kit marketed by Roche (previously Boehringer Mannheim) can be used to incorporate either biotin or digoxigenin; follow the manufacturer's instructions.

Precipitation of a probe containing repetitive sequences:
Large cosmid probes often contain repetitive sequences that, if not suppressed before hybridization, result in high levels of nonspecific hybridization. The repeat sequences can be suppressed by competition with unlabeled human Cot1 DNA sequences that are enriched for repeat sequences. The Cot1 DNA can be included at the precipitation step (Step (b) in the following procedure).

- (a) For a 20-μL nick translation reaction, add to the probe 1 μL of 0.5 M EDTA, 2.5 μL of 4.0 M LiCl, 1 μL of glycogen, 100- to 1000-fold excess human Cot1 DNA (Invitrogen), and 100 μL of ethanol.
- (b) Place on dry ice for 30 min or at −20°C overnight to allow precipitate to form.
- (c) Spin in the microfuge for 15 min to pellet the DNA.
- (d) Wash the pellet in 70% ethanol.
- (e) Spin to repellet DNA and dry the pellet.
- (f) Resuspend DNA at 2–10 ng/μL in hybridization buffer.

Precipitation of probes without repetitive sequences: Follow the foregoing protocol, but leave out Cot1 DNA from the precipitation.

Probe denaturation:
- (a) For probes containing repetitive sequences that need to be suppressed using Cot1 DNA, heat the probe mix to 70°C for 10 min. Place the probe mix at 37°C for 1 h before application to a slide.
- (b) For probes without repetitive sequences, heat the probe mix to 70°C for 10 min. Chill the probe mix on ice for 10 min.
- Probe detection buffer, TBST: 0.05% Tween 20 in 0.1 M Tris, 0.15 M NaCl; pH 7.5
- Formamide, 50%, in 1× SSC
- Antibodies:
  - (a) First antibody, e.g., sheep polyclonal antiserum to digoxigenin or biotin; titration recommended by supplier (Roche)
  - (b) Second antibody, e.g., FITC-conjugated donkey anti-sheep IgG (Jackson Immunoresearch Inc.)
  - (c) Dilute the antibodies in 3% BSA in TBST
- Rubber latex adhesive
- Mountant: Vectashield; contains inhibitor of photobleaching (Vector Labs)

Protocol

Chromosome Preparation and Denaturation:
1. Prepare metaphase-arrested cells by the standard technique (*see* Protocols 16.7, 27.5), and drop the fixed cells onto slides. Mark the areas of spread with a diamond pencil.
2. Refix the cells for 1 h in fresh methanol: acetic acid, 3:1, and then air dry the slides.
3. Rinse the slides in 2× SSC for 2 min at room temperature.

4. Incubate in 100 μg/mL of RNase in 2× SSC at 37°C for 1 h.
5. Rinse in D-PBSA.
6. Refix in freshly prepared 1% paraformaldehyde in D-PBSA for 10 min at room temperature. (This step is optional.)
7. Rinse in D-PBSA.
8. Dehydrate for 2 min in 2 lots each of 70% ethanol and 100% ethanol. Air dry the slides.
9. Denature the chromosomes in 70% formamide in 2× SSC at 70°C for 2–4 min (determine the appropriate amount of time experimentally, starting at 2 min). Make sure that the 70% formamide is at 70°C before using it.
10. Wash the slides in several changes of ice-cold 70% ethanol.
11. Dehydrate as in Step 8, and air dry the slides.
12. The chromosomes are now ready for hybridization.

Hybridization:

13. Apply 10 μL of the denatured probe over the areas of spread, and cover the spreads with 22 × 22-mm coverslips.
14. Seal the coverslips around the edges with rubber latex adhesive.
15. Place the slides in a humidified box at 37°C overnight.

Probe Detection:

16. Remove the coverslips by immersing the slides in 2× SSC (at room temperature) and peeling off the adhesive. Place the slides in a Coplin jar.
17. Soak the slides, two times for 10 min each, in 50% formamide, 1× SSC, at 42°C.
18. Wash the slides, two times for 10 min each, in 2× SSC at 42°C.
19. Rinse in TBST.
20. Block nonspecific binding by incubating the slides with 3% bovine serum albumin (BSA) in TBST for 30–60 min at 37°C.
21. Add the first antibody to the slides. Use 100 μL of antibody per slide, and cover each slide with a Parafilm coverslip.
22. Incubate the slides for 1 h at 37°C.
23. Wash the slides in 500 mL of TBST for 10 min at room temperature.
24. Add the second, fluorochrome-conjugated, antibody, in 3% BSA/TBST, to the slides at a titration recommended by the supplier or determined by experiment.

Note. Be sure to use the correct antibody combinations—for example, sheep polyclonal antiserum to digoxigenin as a first antibody, followed by FITC-conjugated donkey anti-sheep IgG.

25. Incubate for 30 min at 37°C.
26. Wash for 30 min in 500 mL of TBST at room temperature.
27. Counterstain with 0.8 μg/mL of DAPI and/or 0.4 μg/mL of propidium iodide in TBST for 10 min. Mount coverslips in an antifade mountant.
28. View the slides, using fluorescence microscopy with appropriate filter combinations.

Variations

Chromosome painting. Chromosome paints are available commercially from a number of sources, including Invitrogen, Cambio, and Oncor. It is therefore no longer necessary to prepare your own paints. The hybridization and detection protocols vary with each commercial source. However, in general, section (a) (*Chromosome preparation and denaturation*) of the foregoing protocol can be used before hybridization. Hybridization and detection can then be carried out according to the supplier's instructions. If the paint is labeled with biotin, such as is the case for the Cambio paints, section (c) (*Probe detection*) can be followed, using the appropriate antibody combinations.

Recent advances. Classically, karyotypic analysis is carried out by chromosome banding, using dyes that differentially stain the chromosomes (*see* Protocol 27.5). Thus, each chromosome is identified by its banding pattern. However, traditional banding techniques cannot characterize many complex chromosomal aberrations. Recently, new karyotyping methods based on chromosome painting techniques—namely, spectral karyotyping (SKY) and multicolor fluorescence *in situ* hybridization (M-FISH)—have been developed. These techniques allow the simultaneous visualization of all 24 human chromosomes in different colors. Furthermore, visualization of the resulting fluorescence patterns by computer programs makes these techniques more sensitive than the human eye. These techniques are proving to be highly successful in the identification of new chromosomal alterations that were previously unresolved by traditional approaches [Wienberg & Stanyon, 1997; Ried et al., 1997; Macville et al., 1997; Nordgren, 2003; Lim et al., 2004].

27.9 SOMATIC CELL FUSION

Somatic cells fuse if cultured with inactivated Sendai virus or with polyethylene glycol (PEG) [Pontecorvo, 1975].

A proportion of the cells that fuse progress to nuclear fusion, and a proportion of these cells progress through mitosis, such that both sets of chromosomes replicate together and a hybrid is formed. In some interspecific hybrids—e.g., human–mouse—one set of chromosomes (the human) is gradually lost [Weiss & Green, 1967]. Thus, genetic recombination is possible *in vitro*, and, in some cases, segregation is possible as well.

Because the proportion of viable hybrids is low, selective media are required to favor the survival of the hybrids at the expense of the parental cells. TK⁻ and HGPRT⁻ mutants (*see* Section 27.9.1) of the two parental cell types are used, and the selection is carried out in HAT medium (hypoxanthine, aminopterin, and thymidine) (Fig. 27.3) [Littlefield, 1964a]. Only cells formed by the fusion of two different parental cells (heterokaryons) survive, because the parental cells and fusion products of the same parental cell type (homokaryons) are deficient in either thymidine kinase or hypoxanthine guanine phosphoribosyl transferase. The parental cells and homokaryons cannot, therefore, utilize thymidine or hypoxanthine from the medium, and because aminopterin blocks endogenous synthesis of purines and pyrimidines, they are unable to synthesize DNA.

Protocol 27.10 for somatic cell fusion has been contributed by Ivor Hickey, Science Department, St. Mary's University College, Belfast, Northern Ireland, UK.

Although many cell lines undergo spontaneous fusion, the frequency of such events is very low. To produce hybrids in significant numbers, cells are treated with the chemical fusogen polyethylene glycol (PEG) [Pontecorvo, 1975]. Selection systems that kill parental cells but not hybrids are then used to isolate clones of hybrid cells.

PROTOCOL 27.10. CELL HYBRIDIZATION

Outline
Bring the cells to be fused into close contact, either in suspension or in monolayers. Treat the cells with PEG briefly, to minimize cell killing. Usually, the cells are given a 24-h period to recover before selection for hybrids.

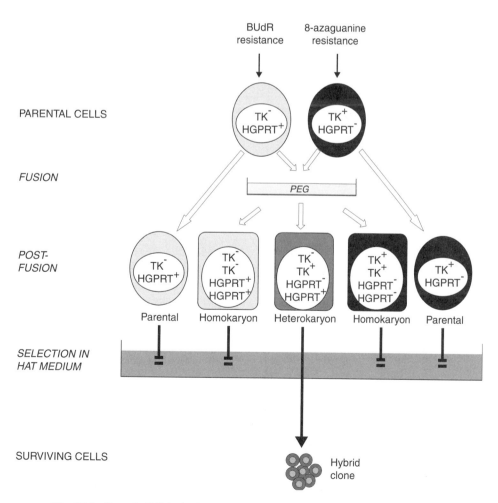

*Fig. 27.3. **Somatic Cell Hybridization.** Selection of hybrid cells after fusion (see text).*

Materials

Sterile:

☐ PEG 1000 (Merck):
  (a) Autoclave the PEG to liquefy and sterilize it.
  (b) Allow it to cool to 37°C, and then mix it with an equal volume of serum-free medium, prewarmed to 37°C.
  (c) Adjust the pH to approximately 7.6–7.9, using 1.0 M NaOH.
  (d) Store the solution at 4°C for up to 2 weeks.

☐ Complete growth medium
☐ Serum-free growth medium
☐ NaOH, 1.0 M
☐ Petri dishes, 5 cm
☐ Universal containers

Protocol

A. Monolayer Fusion:

1. Inoculate equal numbers of the two types of cells to be fused into 5-cm tissue culture dishes. Between 2.5×10^5 and 2.5×10^6 of each parental cell line per dish is usually sufficient.
2. Incubate the mixed culture overnight.
3. Warm the PEG solution to 37°C. It may be necessary at this point to readjust its pH, using NaOH.
4. Remove the medium thoroughly from the cultures and wash them once with serum-free medium.
5. Add 3.0 mL of the PEG solution and spread it over the monolayer of cells.
6. Remove the PEG solution after exactly 1.0 min, and rinse the monolayer three times with 10 mL of serum-free medium before returning the cells to complete medium.
7. Culture the cells overnight.
8. Add selection medium.

B. Suspension Fusion:

9. Centrifuge a mixture of 4×10^6 cells of each of the two parental cell lines at 150 g for 5 min at room temperature. Carry out centrifugation and subsequent fusion in 30-mL plastic universal containers or centrifuge tubes.
10. Resuspend the pellet in 15 mL of serum-free medium, and centrifuge again.
11. Aspirate off all of the medium.
12. Resuspend the cells in 1 mL of PEG solution by gently pipetting.
13. After 1.0 min, dilute the suspension with 9 mL of serum-free medium, and transfer half of the suspension to each of two universal containers or centrifuge tubes containing a further 15 mL of serum-free medium.

14. Centrifuge the suspensions at 150 g for 5 min. Remove the supernate, and resuspend the cells in complete medium.
15. Incubate the cells overnight at 37°C.
16. Clone the cells in selection medium.

Variations. A large number of variations of the PEG fusion technique have been reported. Although the procedure described here works well with a range of mouse, hamster, and human cells in interspecific and intraspecific fusions, it is unlikely to be optimal for all cell lines. Inclusion of 10% DMSO in the PEG solution has the advantage of reducing its viscosity and has been reported to improve fusion [Norwood et al., 1976]. Also, the molecular weight of the PEG used need not be 1000 Da. Preparations with molecular weights from 400 to 6000 Da have been successfully used to produce hybrids.

Selection of hybrid clones. The method of selection used in any particular instance depends on the species of origin of the two parental cell lines, the growth properties of the cell lines, and whether selectable genetic markers are present in either or both cell lines. Hybrids are most frequently selected with the HAT system: 10^{-4} M hypoxanthine, 6×10^{-7} M aminopterin, and 1.6×10^{-6} M thymidine [Littlefield, 1964a]. This system can be used to isolate hybrids made between pairs of mutant cell lines deficient in the enzymes thymidine kinase (TK$^-$) and hypoxanthine guanosine phosphoribosyl transferase (HGPRT$^-$), respectively. TK$^-$ cells are selected by exposure to BUdR and HGPRT$^-$ cells by exposure to thioguanine, following the procedures described by Biedler in Chapter 14 (*see* Protocol 14.9). When only one parent cell line carries such a mutation, HAT selection can still be applied if the other cell line does not grow, or grows poorly in culture (e.g., lymphocytes, senescing primary cultures).

Differential sensitivity to the cardiac glycoside ouabain is an important factor in the selection of hybrids between rodent cells and cells from a number of other species, including human. Rodent cells are resistant to concentrations of this antimetabolite up to 2.0 mM, whereas human cells are killed at 10 μM ouabain. The hybrids are much more resistant to ouabain than the human parental cells. If a rodent cell line that is HGPRT deficient is fused to unmarked human cells, then the hybrids can be selected in medium containing HAT and low concentrations of ouabain.

Although many other selection systems have been reported, few have been widely used. Exogenous dominant markers such as G418 resistance can be used successfully. It must be stressed that, whichever method is used to isolate clones of putative hybrid cells, confirmation of the hybrid nature of the cells must be obtained. This is usually done with cytogenetic (*see* Protocol 16.7) or molecular techniques.

In certain cases, comparing the number of hybrids with the frequency of revertants may be the only way of making this confirmation.

27.9.1 Nuclear Transfer

Genetic recombination experiments can also be carried out with isolated nuclei, but the major interest in this technique is related to cloning individual animals [Kono, 1997; Wolf et al., 1998]. Nuclei can be isolated by centrifuging cytochalasin B-treated cells and fusing the extracted nuclei to recipient whole cells or enucleated cytoplasts in the presence of PEG (Fig. 27.4). However, in animal cloning experiments micromanipulation techniques are used to remove the nucleus from one cell and inject it into a fertilized, preimplantation egg from which the existing nucleus has been removed.

27.10 PRODUCTION OF MONOCLONAL ANTIBODIES

Monoclonal antibodies have become indispensable tools in research, diagnostics, and therapeutics. Since hybridoma technology was first introduced by Kohler and Milstein in 1975 monoclonal antibodies have replaced polyclonal antibodies in many different applications. The following introduction and Protocol 27.11 were contributed by Janice Payne and Tina Kuus-Reichel of Hybritech Incorporated, a subsidiary of Beckman Coulter Inc., 7330 Carroll Road, San Diego, CA 92121.

Hybridomas are produced by fusing a nonsecreting myeloma cell with an antibody-producing B-lymphocyte in the presence of polyethylene glycol (Fig. 27.4). The myeloma cell is deficient in hypoxanthine-guanine phosphoribosyl transferase (HGPRT) or thymidine kinase (TK), necessary for DNA synthesis, and cannot survive in selection medium containing hypoxanthine, aminopterin, and thymidine. Any unfused B-lymphocytes from the spleen cannot survive in culture for more than a few days. Any B-cell-myeloma hybrids should contain the genetic information from both parent cells and are thus able to survive in the HAT selection medium. They can be cultured indefinitely and will produce unlimited quantities of antibody. Supernates from surviving hybridomas are screened for antibody by ELISA. Those hybridomas selected are then subcloned to ensure that they are producing antibody that is specific for a single epitope. Antibody production can be scaled up *in vivo* as ascites in mice or *in vitro* as a suspension culture. Hybridomas also grow very well in various hollow fiber and fermentation culture systems.

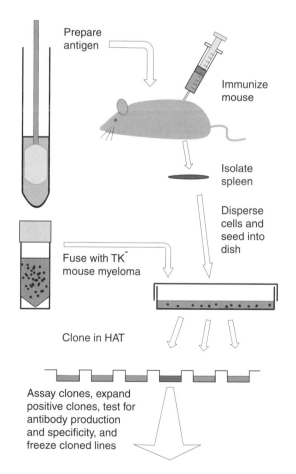

Prepare antigen

Immunize mouse

Isolate spleen

Disperse cells and seed into dish

Fuse with TK⁻ mouse myeloma

Clone in HAT

Assay clones, expand positive clones, test for antibody production and specificity, and freeze cloned lines

Fig. 27.4. Production of Hybridomas. Schematic diagram of the production of hybridoma clones capable of secreting monoclonal antibodies.

PROTOCOL 27.11. PRODUCTION OF MONOCLONAL ANTIBODIES

Outline

Using polyethylene glycol (PEG), fuse spleen cells from an immunized mouse with myeloma cells (P3.653). Select hybrid colonies (hybridomas) in HAT medium. Screen the supernates by ELISA 10–14 days after fusion (the ELISA screening protocol should be developed before the fusion), and expand, freeze, and subclone the desired hybridomas, to ensure monoclonality.

Materials
Sterile or Aseptically Prepared:
- ☐ Mice (Balb/c or A/J), 6–10 weeks old
- ☐ P3.653 myeloma: This myeloma cell line from ATCC does not secrete immunoglobulin and performs well in fusions.
- ☐ Antigen (125 μg per mouse is optimal; small antigens can be conjugated to keyhole limpet hemocyanin (KLH, Sigma) to increase antigenicity)
- ☐ Adjuvants (alum or Freund's adjuvant; complete and incomplete, Sigma): A detailed description of

these and other adjuvants can be found in Vogel and Powell [1995].

- ❏ TCD (T-Cell depletion) buffer: Hanks' balanced salt solution + 10 mM HEPES + 0.3% BSA
- ❏ NH_4Cl, 0.16 M
- ❏ Antimouse Thy 1.2 antibody (Accurate)
  For use, dilute 1:500 in TCD buffer, and filter sterilize
- ❏ Rabbit complement, Low Tox M (Accurate)
  Reconstitute in 1 mL of cold UPW, dilute 1:12 in TCD buffer, and filter sterilize
- ❏ D-PBSA
- ❏ SFM (serum-free medium): MEM that has been stored at room temperature, with the cap of the container loosened to allow the release of CO_2; the pH must be alkaline
- ❏ PEG (polyethylene glycol): Melt 10.5 mL of PEG 1450 (Sigma) in a 56°C water bath; add 19.5 mL of warm sterile MEM (pH 8.3 to 8.7); mix well; and allow the solution to equilibrate, with the cap of the container loosened, for 5–7 days before fusion.
- ❏ HAT stock (100×) (10 mM hypoxanthine, 40 μM aminopterin, 1.6 mM thymidine): 1.36 g of hypoxanthine, 729 μL of aminopterin (from 25 mg/mL stock), 0.387 g of thymidine, 0.022 g of glycine dissolved in 4 mL of 5 M NaOH + 26 mL UPW. Make up to 1 L with UPW. Filter sterilize.
- ❏ HAT medium: Basal medium, such as MEM or RPMI, + 10% FBS + 20% spleen-conditioned medium plus HAT stock (final dilution 1:100)
- ❏ SCM (spleen-conditioned medium): Tease 3–5 naive mouse spleens in D-PBSA. Count the cells, and resuspend them at 1×10^6 cells/mL in MEM + 10% horse serum. Transfer the suspension to a 500-mL spinner flask, and incubate it at 37°C in 5% CO_2 for 48 h. Remove the cells by centrifugation, and store the supernatant frozen.
- ❏ 8-Azaguanine stock, 10 mM (100×): 1.52 g of 8-azaguanine dissolved in 4 mL of 5 M NaOH + 21 mL of UPW. Make the solution up to 1 L with UPW. Filter sterilize.
- ❏ MEM, 10% fetal calf serum, 0.1 mM 8-azaguanine (for maintenance of P3.653 myeloma)
- ❏ HT stock (100×): 0.408 g of hypoxanthine, 0.1161 g of thymidine, 0.0067 g of glycine, dissolved in 2 mL of 5 M NaOH + 8 mL of UPW. Make the solution up to 300 mL with UPW. Filter sterilize.
- ❏ HT medium: HT stock diluted 100× in basal medium (as for HAT medium, above)
- ❏ Syringes, 1 mL with 25G needles
- ❏ Syringes, 1 mL, with 23G needles, ×2
- ❏ Dissecting instruments (scissors and forceps)
- ❏ Petri dishes, 60 × 15 mm
- ❏ Centrifuge tubes, 15 mL and 50 mL
- ❏ Multiwell plates, 24 well and 96 well
- ❏ Culture flasks
- ❏ 100-mL Nalgene bottle
- ❏ Pipette tips
- ❏ Reservoir for multipipettor (100 mL, Matrix Technologies; Corning Costar)

Nonsterile:
- ❏ Trypan Blue (Sigma)
- ❏ Multipipettor, 12 channel (Matrix Technologies)
- ❏ Inverted phase-contrast microscope
- ❏ Unopette microcollection system (BD Biosciences)
- ❏ Hemocytometer

Protocol

Immunization:
1. Bleed the mice on day 0 before the initial injection, and check the serum for background antigen reactivity.
2. Immunize the mice (A/J or Balb/c) with antigen emulsified in Freund's adjuvant or mixed with a 1/10 volume of alum and vortexed. Give three injections of antigen intraperitoneally, according to the following schedule:

Day	Amount of antigen	Adjuvant
0	50 μg	Alum or complete Freund's adjuvant
14	25 μg	Alum or incomplete Freund's adjuvant
28	25 μg	Alum or D-PBSA

3. Bleed the mice on day 35 and measure the serum titer of antibody by an ELISA assay.
4. Dilute the serum serially 1:4 after a 1:30 dilution, and up to 1:30,720.
5. Select mice with the highest ratio of serum titers to antigen for fusion.
6. Give the selected mice a final boost of 10 μg of antigen i.v. or 25 μg of antigen i.p. 3 days before fusion.

Myeloma:
1. It is convenient to perform fusions on a Thursday, with the mice receiving a final boost of antigen on a Monday.
2. Maintain the P3.653 myeloma cell line in MEM + 10% fetal calf serum + 8-azaguanine.
3. Dilute the P3.653 cells to 3.5×10^5 cells/mL each day for the three days before fusion.

T-Cell Depletion:
1. Bleed mice with appropriate serum titers, and sacrifice them by cervical dislocation.

2. Aseptically remove the spleens, and place them in a sterile Petri dish with 5 mL of sterile D-PBSA.

3. Gently tease the spleens with two 23G needles on 1-mL syringes. Teasing spleens roughly will result in a high concentration of fibroblasts.

4. Transfer the cells to a 15-mL conical tube, and allow clumps to settle.

5. Transfer spleen cells (without clumps) to a 50-mL conical tube, and, after a 1:100 dilution in a Unopette, count the cells with a hemocytometer.

6. Spin the cells at 200 g for 8 min.

7. To lyse the red blood cells, resuspend the resultant pellet in 0.84% NH_4Cl (10 mL/spleen), and incubate the suspension at 4°C for 15 min.

8. Underlayer the cell suspension with 14 mL of horse serum, and spin the solution at 450 g for 8 min.

9. Resuspend the resultant pellet in 50 mL of TCD buffer, and spin the suspension at 200 g for 8 min.

10. For T-cell depletion, resuspend the resultant cell pellet in diluted anti-Thy 1.2 at a final concentration of 1×10^7 cells/mL.

11. Incubate the suspension at 4°C for 45 min, and then spin it at 200 g for 8 min.

12. Resuspend the resultant pellet in diluted rabbit complement.

13. Incubate the suspension at 37°C for 45 min, and then spin it at 200 g for 8 min.

14. Count the cells by Trypan Blue exclusion on a hemocytometer (*see* Protocol 22.1). B-cell recovery should be 30–50%.

Fusion:

1. Mix the myeloma and B-cells in a 50-mL centrifuge tube. One fusion can be done on a maximum of 1.2×10^8 spleen cells. Mix the spleen cells with P3.653 myelomas at a ratio of 4:1; thus, the maximum number of P3.653 cells per fusion is 3×10^7 cells.

2. Centrifuge the suspension at 200 g for 8 min.

3. Break up the resultant pellet by tapping, and add 1 mL of PEG to the tube over 15 s.

4. Mix the suspension by gently swirling the tube for 75 s.

5. Add 1 mL of SFM over 15 s, and gently swirl the tube for 45 s.

6. Add 2 mL of SFM over 30 s, and swirl the tube for 90 s.

7. Add 4 mL of HAT medium over 30 s, and swirl the tube for 90 s.

8. Finally, add 8 mL of HAT medium over 30 s, and swirl tube for 90 s.

9. Add this volume (16 mL) to a sterile Nalgene bottle containing the calculated amount of HAT medium (125 mL if the maximum cell concentration has been used). 16 mL, containing 1.5×10^8 cells, from Step 1 in this section of the protocol plus 125 mL of HAT medium in the bottle makes 141 mL. With the wash in the next step (Step 10), the total volume is 150 mL and will result in a final concentration of 1×10^6 cells/mL.

10. Wash the 50-mL conical tube with 9 mL of HAT medium, and add this volume to the bottle.

11. Mix the contents of the bottle well, and transfer the cells to the sterile reservoir.

12. Using a 12-channel multipipettor, plate the cells at 200 µL/well into a sterile 96-well plate. The final concentration is then 2×10^5 cells/well.

Selection of Hybridomas:

1. Feed the fusion plates 5 days after fusion, by aspirating most of the culture media from the wells and replacing it with 150–200 µL/well of fresh HAT medium.

2. Feed the plates twice per week.

3. Screen the clones for selection of positive hybridomas, usually two weeks after fusion, using ELISA.

4. After a further 48 h, retest those clones that tested positive in the previous step.

5. Expand the most productive hybridomas by culturing them in two wells of a 96-well plate in media containing 10% FBS and HT.

6. Retest the clones, expand the positive hybridomas to a 24-well plate, and wean them off HT medium, at which time 2 mL of culture supernatant should be harvested for screening. At this step, enough volume is harvested to perform several selection assays to ensure that the antibody is directed only at the antigen of interest.

7. Expand the hybridomas to be kept to 4 wells of a 24-well plate, and cryopreserve them (*see* Protocol 20.1).

8. Perform a second cryopreservation after expanding the hybridoma to a 75-cm² flask.

Screening. Take care in developing the screening strategy to obtain a monoclonal antibody with the characteristics that you want. Hybridoma culture supernate should be screened as early as feasible for desired reactivity patterns. After initial selection by ELISA for reactivity to the immunogen, the expanded culture supernate should be tested in the application for which it was developed (e.g., Western blot, competitive immunoassay, flow cytometry, etc.). A more detailed discussion of ELISA and other immunoassays can be found in Knott et al. [1997].

Subcloning. To ensure monoclonality, subclone hybridomas of interest. This can be done by serially diluting cells and plating the equivalent of 1 cell per 3 wells in a 96-well plate (*see* Section 14.1) or by sorting with an automated cell deposition unit (ACDU) on a FACStarplus (BD Biosciences) and plating at one cell per well. Subcloning can be done on a mouse spleen feeder layer. Feeder layers are prepared by teasing a naive mouse spleen and resuspending the cells at 1×10^6 cells/mL. The cells are then plated in a 96-well microtitration plate at a final concentration of 2×10^5 cells/well. After subcloning, colonies can usually be seen at day 5 and must be checked visually for monoclonality. Plates are fed with fresh medium beginning on day 7. Screening for positive hybridomas is usually done between days 10 and 14. Those clones selected are then expanded and frozen in the same way as the parental hybridoma.

Antibody production. Concentrated antibody from clones of interest can be produced *in vivo* as ascites in IFA primed mice (Balb/c or nu/nu [Gillette, 1987]) or *in vitro* as a suspension culture. Several hollow fiber cell culture systems are also available (*see* Appendix II and Sections 25.3.2, 26.1.3, 26.2.5). When hybridomas are inoculated into a hollow fiber system, the cells are maintained in a compartment of the bioreactor, while fresh media and waste from the cells are recirculated. High concentrations of antibody are produced in the cell compartment, and culture supernate containing antibody can be harvested at multiple time points.

27.11 DNA TRANSFER

To study the function of individual genes and regulatory sequences, DNA fragments can be cloned and then transferred into host cells by a variety of techniques, such as transfection, lipofection, and retroviral infection (Fig. 27.5). Cloned DNA is often conveniently maintained as part of a bacterial plasmid. Many plasmids can attain a high copy number during bacterial growth, thus ensuring a plentiful stock of DNA for experimentation. Plasmid DNA is purified from the bacteria before use. Once the sequence of interest has been cloned, it can be manipulated further to isolate subclones containing, for example, promoter sequences. By genetic manipulation, promoter sequences can be linked to a reporter gene (e.g., β-gal or CAT) whose products (e.g., β-galactosidase, chloramphenicol acetyl transferase) can be readily assayed subsequent to transfection. Green fluorescent protein (GFP), from the Pacific jellyfish, *Aequoria victoria,* can be used as a reporter, will fluoresce in living cells irradiated with UV light on an inverted microscope, enables detection of transfected cells while still viable, and can be used to sort by FACS.

Transfections may be *transient* or *stable.* Transient transfections are short term and used shortly after transfection, and the efficiency of transfection is determined by reporter gene assays. DNA used for stable transfection contains a

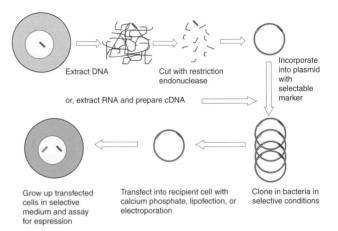

Fig. 27.5. DNA Transfer. Isolated DNA endonuclease fragments amplified by gene cloning techniques, added to whole cells, and incorporated by treatment with lipofection, electroporation, or coprecipitation with $Ca_3(PO_4)_2$.

selectable marker, such as *neo* or *hyg* B, that confers resistance to G418 (geneticin, an analog of neomycin) or hygromycin, respectively. Transfected cells are then selected by continued exposure to the selection agent, and resistant clones can be isolated (*see* Protocols 14.6 and 14.8).

Two protocols are presented below, one for stable transfection (using electroporation) and one for transient (using lipofection). Selection protocols for stable transfection can be added to the lipofection protocol, provided that the construct used for transfection contains the appropriate selectable marker. DNA transfer by calcium phosphate coprecipitation and viral transduction are described in Chapter 18 (*see* Protocols 18.1, 18.2).

27.11.1 Coprecipitation with Calcium Phosphate

The calcium phosphate technique for introducing genes into mammalian cells was first described by Graham and Van der Eb [1973] and is still widely used. In this method, exogenous DNA is mixed with calcium chloride and is then added to a solution containing phosphate ions. A calcium-phosphate-DNA coprecipitate is formed, which is taken up by mammalian cells in culture, resulting in expression of the exogenous gene. This method can be used to introduce any DNA into mammalian cells for transient expression assays or long-term transformation. This procedure has been described for fibroblast immortalization with SV40 T antigen (*see* Protocol 18.1).

27.11.2 Lipofection

The original method for cationic lipid-mediated DNA transfection into cultured cells [Felgner et al., 1987] was improved in 1993 by replacement of the monocationic lipid reagent with a polycationic one, Lipofectamine [Ciccarone et al., 1993; Hawley-Nelson et al., 1993]. The method is based on an ionic interaction of DNA and liposomes to form a complex, which can deliver functional DNA into cultured cells. Plasmid DNA is complexed, but not encapsulated, within unilamellar liposomes 600–1200 nm in size [Mahato et al., 1995a], formed by cationic lipids in water.

The advantages of cationic liposome-mediated transfection over other methods include generally higher efficiency; the ability to transfect successfully a wide variety (over 300 reported) of eukaryotic cell lines, many of which are refractive to other transfection procedures; and relatively low cell toxicity. Another advantage is that the basic procedure of DNA transfection can be adapted for transfection with RNA, synthetic oligonucleotides, proteins, and viruses. Finally, cationic liposomes can be used for the successful delivery of functional genes or viral genomes *in vivo* [Mahato et al., 1995b; Tagawa et al., 1996; Thorsell et al., 1996]. Its disadvantage is the relatively high cost of reagents, which practically precludes large-scale use.

Protocol 27.12 has been abridged from Bichko [1998].

PROTOCOL 27.12. TRANSIENT TRANSFECTION BY LIPOFECTION

Materials
Sterile or Aseptically Prepared:
❑ Cationic lipids
❑ Lipofectamine: 3:1 (w/w) liposome formulation of the polycationic lipid DOSPA (2,3-dioleyloxy-N[2(sperminecarboxamido)ethyl]-N,N-dimethyl-1-propanaminium trifluoroacetate) and the neutral lipid DOPE (dioleoylphosphatidylethanolamine) in water (Invitrogen)
❑ Lipofectin: 1:1 (w/w) liposome formulation of the cationic lipid DOTMA (N-[1-(2,3 dioleyloxy)-propyl]-N,N,N-trimethylammonium chloride) and the neutral lipid DOPE in water (Invitrogen)
❑ LipofectACE: 1:2.5 (w/w) liposome formulation of the cationic lipid DDAB (dimethyl dioctadecylammonium bromide) and the neutral lipid DOPE in water (Invitrogen)
❑ Cells for transfection
❑ DNA for transfection
❑ Growth medium: DMEM (1×) with 10% FBS, penicillin (100 U/mL), and streptomycin (100 μg/mL, or as appropriate to the cells being used)
❑ Reduced serum medium: OPTI-MEM 1 (contains 2% FBS; Invitrogen)
❑ Serum-free medium: DMEM without serum or antibiotics
❑ D-PBSA
❑ Buffered saline:
  NaCl, 0.15 M
  K_2HPO_4, 0.006 M
  KH_2PO_4 0.002 M

❑ Trypsin, 0.25%, in D-PBSA
❑ Multiwell plates, 6 well, or 3.5-cm Petri dishes

Protocol

A. Transfection of Adherent Cells
1. Seed approximately $1 \times 10^5 - 3 \times 10^5$ cells per well in 6-well plates in 3 mL of growth medium.
2. Incubate the cells at 37°C in a CO_2 incubator until the cells are 50% to 90% confluent. This step usually takes 18–24 h and should not take less then 16 h.
3. Before transfection, prepare the DNA and lipid solutions in sterile tubes:
   (a) For the DNA solution, dilute 1–2 μg of DNA into 0.5 mL of reduced-serum medium.
   (b) For the lipid solution, dilute 2–25 μL of cationic lipid reagent into 0.5 mL of serum-free medium.
   (c) Combine the two solutions, mix gently, and incubate at room temperature for 15–45 min, to allow the formation of DNA-lipid complexes.
4. Rinse the cells once with 2 mL of serum-free medium.
5. Overlay the DNA-lipid complex onto the cells. Antibacterial agents should be omitted during transfection.
6. Incubate the cells with the complexes at 37°C in a CO_2 incubator for 2–24 h; a period of 5 or 6 h is usually enough.
7. After incubation, remove the transfection mixture from the cells, and replace it with 3 mL of complete growth medium.
8. Replace the growth medium with fresh medium 18–24 h after the start of transfection.
9. Assay the growth medium or cells for transient gene activity as appropriate (*see* Protocols 27.14 and 27.15).

B. Transfection of Suspension Cells
1. Prepare the transfection mixture in sterile tubes as follows:
   (a) Dilute 2–5 μg of DNA in 0.5 mL of reduced-serum medium.
   (b) Dilute 2–20 μL of cationic lipid reagent in 0.5 mL of serum-free medium.
   (c) Combine the two solutions, mix the resultant solution gently, and incubate it at room temperature for 15–45 min, to allow the formation of DNA-lipid complexes.
2. Centrifuge a cell suspension containing approximately $1 \times 10^6 - 2 \times 10^6$ cells, and aspirate the supernatant medium.

3. Resuspend the cells in the transfection mixture, and transfer the suspension to a 3.5-cm dish.
4. Incubate the cells in a CO_2 incubator for 4–6 h.
5. To each dish, add 0.5 mL of growth medium, supplemented with 30% serum. (A large amount of serum is necessary to protect cells in suspension, which are more sensitive to the toxic effects of liposome reagents.)
6. Incubate the dishes in a CO_2 incubator overnight.
7. Add 2 mL of complete growth medium to each dish, and incubate the dishes in a CO_2 incubator.
8. At 24–72 h after the start of transfection, assay the cells or medium for gene activity as appropriate (see Protocols 27.14 and 27.15).

27.11.3 Electroporation

DNA can be introduced into cells by electroporation, when a high cell concentration is briefly exposed to a high-voltage electric field in the presence of the DNA to be transfected [Chu et al., 1987]. Small holes are generated transiently in the cell membrane [Zimmerman & Vienken, 1982], and the DNA is allowed to enter the cell and, in some of the cells, becomes incorporated into the genome. Equipment for electroporation is available commercially (Bio-Rad). Most cells refractory to chemical methods of gene transfer are successfully transfected by electroporation [Andreason & Evans, 1988; Chu et al., 1987].

Electroporation is usually performed at a constant capacitance setting (and therefore a constant pulse duration), with various field strengths (500–1500 kV/cm) for pilot investigations. For most cells, the settings at which approximately 20–50% of the cells remain viable after electroporation are sufficient for DNA transfer [Chu et al., 1987; Andreason & Evans, 1988]. Electroporation is usually performed at room temperature, and the cells are subsequently kept on ice, to extend the period of time that the membrane pores remain open [Andreason & Evans, 1989].

There is a linear relationship between DNA concentration, DNA uptake, and reporter gene expression [Chu et al., 1987]. It is believed that linearized DNA is more efficient for the production of stable transfectants than supercoiled DNA, presumably because of the increased efficiency with which linear DNA integrates into the genome DNA [Potter et al., 1984; Chu et al., 1987]. Electroporation results in the integration of DNA in low copy number [Boggs et al., 1986; Toneguzzo et al., 1988], although the copy number introduced can be adjusted by altering the concentration of DNA in the cell suspension. Chemical methods of transfection usually result in the integration of large concatamers, which may inherently interfere with cell function and obscure investigations involving specific gene overexpression [Robins et al., 1981; Kucherlapati & Skoultchi, 1984].

Suspension cells are more easily transfected by electroporation than adherent cells, because adherent cells must be detached from the culture vessel. The drawbacks of electroporation include its requirement for more cells and DNA than chemical methods of gene transfer and its variability in optimal parameters between cell types.

The above introduction and Protocol 27.13 have been abridged from Cataldo et al. [1998]. The protocol describes electroporation conditions for the stable transfection of a clone (Y10) of the murine megakaryocytic cell line L8057 [Ishida et al., 1993]. L8057-Y10 cells grow in suspension and are maintained in Ham's F12 medium supplemented with 10% heat-inactivated FBS, penicillin (50 U/mL), and streptomycin (50 µg/mL). The overall scheme presented here is for the stable transfection of L8057-Y10 cells.

PROTOCOL 27.13. STABLE TRANSFECTION BY ELECTROPORATION

Materials
Sterile or Aseptically Prepared:

Expression vectors and preparation of DNA:

❑ pSVTKGH [Selden et al., 1986], linearized. This vector contains the SV40 enhancer and the herpes virus thymidine kinase promoter sequences that drive the expression of human growth hormone (hGH). hGH is secreted directly into the tissue culture medium and is detected by radioimmunoassay [Ravid et al., 1991].

❑ pcDNA3 (Invitrogen) linearized. This expression vector contains the human cytomegalovirus enhancer-promoter sequences upstream from its multicloning site and the polyadenylation signal and transcription-termination sequences of the bovine growth hormone gene (bGH) downstream from the multicloning site. The pcDNA3 vector also contains the gene for neomycin resistance, alleviating the need to cotransfect an antibiotic resistance gene for the selection of stable transformants overexpressing a particular gene of interest. pcDNA3 may also be used as a selectable marker when cotransfected with reporter constructs, as presented in the protocol.

Tissue culture:

❑ L8057-Y10 cells
❑ Flasks, 75 cm²
❑ Plates, six well
❑ D-PBSA
❑ Medium F12/FB: Ham's F12 with glutamine (2 mM), penicillin (50 U/mL), and streptomycin

(5 μg/mL), supplemented with 10% FBS (heat inactivated; GIBCO #16140-014)
□ Geneticin (Invitrogen), 50 mg/mL in D-PBSA (adjust for potency—e.g., if the potency of a batch is 731 μg/mg, dissolve 64.4 mg of G418 per mL of D-PBSA)
□ Selective medium: F12/FB containing 600 μg/mL of G418

Electroporation:
□ Electroporation buffer [Ravid et al., 1991]:
NaCl, 30.8 mM
KCl, 120.7 mM
Na_2HPO_4, 8.1 mM
KH_2PO_4, 1.46 mM
$MgCl_2$, 5 mM
□ Gene Pulser Cuvettes
Nonsterile:
□ Gene Pulser II Apparatus (Bio-Rad)
□ Gene Pulser II Capacitance Extender Plus

Protocol
1. Seed L8057-Y10 cells at 5×10^5 cells/mL in a 75-cm² culture flask, and incubate the flask at 37°C in 5% CO_2 in F12/FB.
2. During the late log phase, collect the cells by centrifugation at 4°C at 380 g for 5 min.
3. Wash the resultant cell pellet in 10 mL of D-PBSA, and centrifuge at 4°C at 380 g for 5 min.
4. Wash the resultant cell pellet in 5 mL of electroporation buffer, and count the cells, using a hemocytometer.
5. Collect the cells by centrifugation at 4°C at 380 g for 5 min.
6. Resuspend the cells at 1×10^6 cells/0.8 mL of electroporation buffer.
7. Transfer 0.8 mL of cells into prechilled electroporation cuvettes.
8. Add 50 μg of linearized pSVTKGH, along with 5 μg of linearized pcDNA3, to the cell suspension. (When analyzing a specific gene promoter-reporter gene expression, a suitable negative control is to electroporate the cells in the absence of plasmid DNA.)
9. Mix the DNA/cell suspension by holding the sides of the cuvette and flicking the bottom. Incubate the suspension on ice for 10 min.
10. Electroporate the suspension at 400 V/500 μF, and record the duration of the shock.
11. Remove the cuvette from the shocking chamber, and place on ice for 10 min.
12. Transfer the electroporated cells into 10 mL of F12/FB, rinse the cuvette with medium to remove all of the cells, and collect the cells by centrifugation.
13. Resuspend the cells in 20 mL of F12/FB, and culture the suspension in a 75-cm² flask at 37°C in 5% CO_2 to allow expression of the neomycin selectable marker gene.
14. After 24–48 h, collect the cells by centrifugation, and then resuspend the cells in 20 mL of selective medium.
15. Change the selective medium every 2–4 days for at least 2 weeks, to remove the debris of dead cells and to permit resistant cells to grow.
16. Assay for hGH expression in the cell supernatant (*see* Section 27.11.5).

Variations
Geneticin. Cultured cell lines differ in their sensitivity to geneticin, and the most suitable concentration to use must be empirically determined by doing a kill curve (*see* Protocol 22.3) on the cell line that is being transfected. The cells should remain in selection medium throughout their growth period.

Picking colonies. See Protocols 14.6 and 14.8.

Transient transfection. Use circular plasmids, such as pCMVβ-gal, to replace pcDNA3. After Step 10, resuspend the cells in 2 mL of culture medium without selection agent, and culture them in a six-well plate for three to four days. Assay for transient hGH expression in the cell supernatant, and prepare a cell lysate to determine β-galactosidase activity.

Reporter gene expression. In the case of transient transfection, a plasmid containing the β-galactosidase reporter gene is cotransfected, in order to normalize hGH gene expression for overall electroporation efficiency. There are commercial kits available for the detection of both hGH secretion (Nichols Institute Diagnostics, #40-2205) and β-galactosidase activity (Promega, #E2000).

27.11.4 Other DNA Transfer Methods
Retroviral infection. Retroviruses have a high efficiency of gene transfer, are able to incorporate larger DNA fragments than plasmids, and infect host cells spontaneously [Ausubel et al., 1996; Hicks et al., 1998]. The introduced gene becomes permanent, and the process of insertion is achieved by normal cellular processes, is not harmful to the host cell, and does not cause any other genetic alterations (*see* Protocol 18.2).

Baculovirus. Inserting genomic sequences into baculoviruses, which are then propagated in insect cells (such as Sf9 cells), also allows large sequences (>100 kbp) to be cloned. The proteins produced have posttranslational modifications that are not available in prokaryotic systems, although there are differences in processing from mammalian cells.

Baculoviruses are not transmissible to mammalian cells, and Sf9 cells are unlikely to carry any risk of contamination with mammalian viruses, provided that any mammalian–derived supplements (e.g., FBS) are thoroughly screened before use. Sf9 can be subcultured without trypsin [*see* Protocol 26.9; Midgley et al., 1998].

Yeast artificial chromosomes. Yeast artificial chromosomes (YACs) [Strauss, 1998] also provide a genome that is capable of packaging larger sequences of DNA than bacterial plasmids, with downstream posttranslational processing such as found in eukaryotic cells, although this latter aspect is likely to have significant differences from that in mammalian cells. Propagation in yeast also gives a high-yield and stable culture system and is less difficult to maintain than large-scale insect or mammalian cell cultures.

Mammalian artificial chromosomes. There has been considerable success in applying the principle of YACs to mammalian systems [Bridger, 2004], in order to incorporate large mammalian sequences containing one or more structural and multiple regulatory genes into one construct. These constructs are known as mammalian artificial chromosomes (MACs) and are introduced into mammalian cells by monochromosomal transfer techniques (Newbold & Cuthbert, 1998). The technology has now progressed to the construction of human artificial chromosomes, opening up a whole new era for genetic therapy [Larin & Mejia, 2002; Katoh et al., 2004].

27.11.5 Reporter Genes

Assays of transfected cells for reporter genes, such as β-gal or CAT, confirm that the DNA construct has been incorporated and is expressed by the transfected population. β-gal staining can also be used to track cells in cell interaction studies [Bradley & Pitts, 1994]. Protocol 27.14 is abridged from Bichko [1998], based on Sanes et al. [1986], and modified by Invitrogen.

> **PROTOCOL 27.14. *IN SITU* STAINING FOR β-GALACTOSIDASE**
>
> **Materials**
> *Nonsterile:*
> ❑ D-PBSA
> ❑ Substrate: X-gal (Invitrogen), 20 mg/mL in dimethylformamide. Store at $-20°C$ in the dark, in a polypropylene tube, for up to 6 months.
> ❑ Fixative: Formaldehyde, 1.8%; glutaraldehyde, 0.05% in PBSA. Prepare the fixative by combining 85 mL of water, 10 mL of 10× D-PBSA, 5 mL of formalin (37% formaldehyde solution), and 0.2 mL of glutaraldehyde (25% solution). Store at 4°C.

> ❑ Stain solution: 5 mM potassium ferricyanide, 5 mM potassium ferrocyanide, 2 mM $MgCl_2$ in D-PBSA. Store at 4°C.
> ❑ Substrate/stain solution: 1 mg/mL of X-gal in stain solution. Prepare immediately before using.
> ❑ Formalin, 10%, in D-PBSA.
> ❑ In this case, the expression plasmid for β-gal would be used as a reporter gene for transfections (for example, pCMVβgal [MacGregor & Caskey, 1989]).
>
> *Protocol:*
> 1. Wash the cells (e.g., in 6-well plates) once with 2 mL of D-PBSA.
> 2. Fix the cells with 1 mL of fixative for 5 min at room temperature.
> 3. Wash the cells twice with 2 mL of D-PBSA.
> 4. Add 1 mL per well of substrate/stain solution to the cells, and incubate them 2 h to overnight at 37°C.
> 5. Rinse the cells in each well with 2 mL of D-PBSA. Observe the cells on an inverted microscope, and count the blue (β-gal positive) cells.
> 6. To store the plates, fix the cells in each well with 1 mL of 10% formalin in buffered saline for 10 min at room temperature, rinse the cells with buffered saline, and store in buffered saline at 4°C.

An expression plasmid for CAT (for example, pCMVCAT [Boshart et al., 1985]) can also be used as a reporter gene for transfection. Protocol 27.15 is abridged from Bichko [1998], based on Neumann et al. [1987], and modified by Invitrogen.

> **PROTOCOL 27.15. CHLORAMPHENICOL ACETYLTRANSFERASE (CAT) ASSAY**
>
> **Materials**
> *Sterile:*
> ❑ D-PBSA
> ❑ Multiwell plates: 6 well, 3.5 cm
> *Nonsterile:*
> ❑ Tris buffer: Tris·HCl, 0.1 M, pH 8.0
> ❑ Tris/Triton: Tris·HCl, 0.1 M, pH 8.0, 0.1% Triton X-100; store at 4°C
> ❑ CAT dilution buffer: Tris·HCl, 0.1 M, pH 8.0, 50% glycerol, 0.2% BSA
> ❑ Substrate: 250 mM chloramphenicol (GIBCO) in 100% ethanol; store in aliquots at $-70°C$.
> ❑ [^{14}C]-CoA: [^{14}C]butyryl Coenzyme A (10 μCi/mL; Amersham Biosciences)

- ❑ CAT enzyme standards (Invitrogen): Prepare standard solutions of 0.2, 1, 2, 4, and 10 U/mL in CAT dilution buffer. Store at 4°C.
- ❑ Liquid scintillation cocktail: Econofluor (Invitrogen)
- ❑ Deionized, distilled water (DDW)
- ❑ Polypropylene scintillation vials, 3.5 mL
- ❑ Microcentrifuge tubes
- ❑ Microcentrifuge
- ❑ Ice bath

Protocol

Cell Harvesting for 6-Well, 35-mm Plates:

1. Wash the cells once with D-PBSA 24–72 h after transfection.
2. Put the plates on ice, and add 1 mL of Tris/Triton per well.
3. Freeze the plates for 2 h at −70°C.
4. Thaw the plates at 37°C, and then put them on ice.
5. Transfer the cell lysates to microcentrifuge tubes, and spin the tubes for 5 min at maximum speed.
6. Collect the supernates, and heat them at 65°C for 10 min to inactivate the inhibitors of CAT.
7. Centrifuge at maximum speed for 3 min and collect the supernatant cell extract. Keep the supernatant at −70°C.

CAT Assay:

8. Put 5–150 μL of cell extract from each sample into a 3.5-mL polypropylene scintillation vial, and add sufficient 0.1 M Tris to it to reach a final volume of 150 μL.
9. For a negative control, use 150 μL of 0.1 M Tris.
10. For a positive control,
    (a) Add 150 μL of 0.1 M Tris to each of 5 vials.
    (b) Add 5 μL of each CAT standard solution to the vials, to give a standard curve of 1, 5, 10, 20, and 50 mU of CAT.
11. To each sample (including the controls), add 100 μL of the following mixture:
    (a) UPW .... 84 μL
    (b) Tris, 0.1 M .... 10 μL
    (c) Chloramphenicol .... 1 μL
    (d) [^{14}C]-CoA .... 5 μL (50 nCi)
12. Cap the samples, and incubate them at 37°C for 2 h.
13. Add 3 mL of Econofluor to all tubes, and then recap the tubes.
14. Mix the contents by inverting the tubes.
15. Incubate the tubes at room temperature for 2 h.
16. Count the samples for 30 s in a liquid scintillation counter.

CHAPTER 28

Problem Solving

No matter how well a laboratory is run, problems arise when new staff, new techniques, or any other new development destabilizes its normal routine. One solution to such problems is to make sure that procedures do not change—i.e., to define standard operating procedures (SOPs) and ensure that deviations from these procedures are made only after exhaustive testing of the possible repercussions. However, this is often difficult, particularly in a research environment, where progress demands change and new procedures are introduced continually.

The advice given in the preceding chapters has concentrated on practical, "how to do it" instructions, sometimes with indications of what might go wrong. This chapter attempts to summarize these potential problems under topic headings and adds a few more potential difficulties, queries, and, hopefully, solutions.

28.1 SLOW CELL GROWTH

28.1.1 Problems Restricted to your Own Stock

1. Check your cells for contamination.
   (a) If you are working without antibiotics (and you should be!), most contaminations will be obvious.
      i) Examine visually by naked eye and microscope (*see* Sections 28.6; 19.3.1).
      ii) Discard cells if contaminated.
   (b) Mycoplasma will not be obvious, and your cells should be checked regularly for it (*see* Protocols 19.2, 19.3).
   (c) Check potential routes or causes (*see* Section 28.6; 19.1).
2. Check the growth of your cells with other media and sera (i.e., a different batch or supplier). If you normally

use powder or 10×, buy in an alternative stock of 1×. If this change indicates that the problem is with the medium, see the subsection on medium later in this section.

3. If the problem is not with the medium, it may be with your cells.
   (a) Thaw out another ampoule from the freezer, and compare it with your current stock. Handle the two cultures separately, in case you have a contamination in the original stock. Do this before checking (or while checking) (b)–(d) below.
   (b) Count your cells at subculture, and set up a growth curve (*see* Protocols 21.7 and 21.8) to compare with your previous records for these cells. Check if the following factors apply to your cells:
      i) The seeding density was too low at transfer.
      ii) The cells were subcultured too frequently.
      iii) The cells were allowed to remain for too long in the plateau phase before subculture.
   (c) Was there a change in the batch of trypsin or another dissociation agent? Check the batch numbers and suppliers.
   (d) To assess the severity of dissociation, check to determine whether:
      i) The duration of exposure to trypsin (or other agents) was too long.
      ii) The dissociating agent was too concentrated or had specific activity that was too high.
      iii) The incubator used for trypsinization was too warm.
      iv) Pipetting during dissociation was too vigorous.

Culture of Animal Cells: A Manual of Basic Technique, Fifth Edition, by R. Ian Freshney
Copyright © 2005 John Wiley & Sons, Inc.

v) The cells were sensitive to EDTA (if EDTA was used).

vi) The wrong diluent was used for trypsin.

vii) A bad batch of trypsin diluent was used.

28.1.2 Problem is More General and Other People are having Difficulty

Check shared facilities and reagents
Hot room and incubators

(1) The temperature and stability are inadequately controlled

  a) The thermostats are faulty.

  b) The access to the incubator is too frequent.

  c) The hot room door is being left open.

  d) The circulating fan(s) has failed or is overheating.

Check incubators with portable recording thermometer (*see* Section 5.3.3).

(2) Humidity of CO_2 incubator is not being maintained

  a) The water tray is not filled.

  b) There is leakage around the doors.

  c) The access to the incubator is too frequent.

  d) Check evaporation rate by weighing Petri dish of PBSA daily.

(3) CO_2 concentration in CO_2 incubator is not properly controlled

  a) The access to the incubator is too frequent.

  b) There are problems with the CO_2 controller.

1. Check the pH *in situ* with a Petri dish of pretested medium.

2. Check incubator atmosphere with CO_2 tester (Carborite).

3. Recalibrate zero settings and CO_2 concentration on read-out with standard gas mixtures.

Medium and reagents. See Section 28.2.

28.1.3 Have any Changes Occurred in the Laboratory?

Staff

(1) Culture staff

(2) Preparation staff

(3) Preparation procedures

(4) Training

(5) Allocation of space; overcrowding

Materials
Chemical contamination. (*See also* Section 28.8.)

Media (see below)

(1) Supplier

(2) Batch

(3) Type

(4) Storage

Procedures

(1) Incubation times

(2) Speed of operations

(3) Sequence of operations

(4) Location

(5) Scale

Equipment

(1) Replacements: has any new equipment been added?

(2) Operational status: has any equipment become faulty?

(3) Location: Has position or location of equipment being used, or adjacent equipment, been changed?

(4) Operating temperature: Is the equipment working within its optimal temperature range and that of your samples within it?

28.2 MEDIUM

Are there problems either with your own stocks or with general stocks?
Age

1. Check the batch number and purchase date of the medium. It should not be more than 1 year old.

2. If the medium contains glutamine, then it is stable only for 1 month at $4°C$.

Storage conditions

1. Check that the temperature of the cold room or refrigerator is at or below $4°C$.

2. Check that the medium is stored in the dark, or at least in tungsten light. If not, it will degrade in fluorescent light, unless the light has a low UV output.

Adequacy of the medium

1. Check the medium against other media (*see* Section 9.6).

2. Buy in $1×$ medium, and compare it with your own.

3. If you already use $1×$ medium, try another supplier, and check individual additions to the medium (e.g., serum, growth factors, hormones, etc.).

Frequency of changing the medium

Check the cell concentration and pH. If the pH falls below 7.0 in under 48 h, the pH is too low at the start of the growth period, the cell concentration is too high, or there is a contamination.

Properties of medium
pH

Check that the pH is between 7.0 and 7.4 throughout the period of culture.

If there are pH fluctuations

1. Check the CO_2 supply to the incubator (*see Equipment*, above)

2. Check the CO_2 regulation of the incubator (*see Equipment*, above)

If the pH is too high

a) The CO_2 concentration in the incubator is too low.

b) The HCO_3^- concentration of the medium is too high.

c) If you are using DMEM, it should be used with a gas phase of 10% CO_2 if it has been made according to the original recipe.

Osmolality; the medium may be hyper- or hypotonic

(1) If the stock medium is not correctly formulated
(2) If one of the components is wrong
(3) If dilution procedure is wrong.
1. Check osmolality on an osmometer; it should be between 270 and 340 mOsmol/kg.
2. Check the preparation procedures.
   (a) Is the amount of water used to dilute a $10\times$ stock, or dissolve powder, correct?
   (b) Has a reagent been added that would alter osmolality?

Accidental omission of a component

1. Make up a fresh batch of medium (which is quicker than trying to decide which component is missing).
2. If a constituent of the medium has poor solubility or precipitates before filtration, check to ensure that all constituents are dissolved before filtering.
3. If there is precipitation on storage, check that the precipitate redissolves on diluting and/or heating to 37°C.
4. If the precipitate does not redissolve, discard the batch.
5. If precipitation recurs, complain to your supplier or change your supplier.

Defective component

1. Replace the components one at a time from an alternative source.
2. Keep a record of the batch numbers.

New batch of stock medium that appears to be faulty

1. Compare the batch with a previous batch, if one is still available.
2. Compare the batch with a batch of another supplier if no previous batches are available.

If medium is BSS-based, is BSS satisfactory?

1. Check with other users.
2. Try alternative sources of BSS or a different formulation.

If medium is water based, is ultrapure water satisfactory?

1. Check with other users
2. Check against fresh $1\times$ medium, bought in complete.

28.2.1   Selection

Has medium changed since the cells were acquired?

1. Revert to the previous type of medium or screen different media (*see* Section 11.6.3) if original unobtainable or inappropriate.
2. Obtain the medium from the same source as that of the supplier or originator of the cell line.
3. Compare the constituents of the medium with those of the supplier or originator of the cell line.

If starting a new culture:

1. Screen several media (*see* Section 11.6.3) for the following parameters:
   (a) Type of medium (*see* Section 9.6).
   (b) Type of serum (*see* Section 9.6).
   (c) Concentration of serum (*see* Protocols 21.7, 21.8, 21.9, 21.10, and 22.4)
2. Include serum-free media in screen (*see* Section 10.5).
   (a) Select by tissue-based criteria, e.g., MCDB 153 for epidermal keratinocytes.
   (b) Check growth factor supplementation (*see* Table 10.3).
3. Try feeder layer (*see* Sections 8.4.2, 14.2.3), conditioned medium (*see* Section 14.2.2), and/or matrix coating (*see* Section 8.4.1).

28.2.2   Unstable Reagents
Glutamine
1. Store it frozen at $-20°C$.
2. As it is reduced to 50% in 3–5 days at 37°C, replace medium after 3 days.
3. Use a stable alternative—e.g., dipeptide Glutamax—but test it first.

Serum
1. Store it frozen.
2. If it is partially thawed, thaw it completely, mix it, and refreeze it.
3. Test batches of serum before use.
4. Overlap batches, in case a deficiency is slow to appear. (It may take 2–3 subcultures).

Other Constituents
1. Store at 4°C.
2. Most constituents of media are stable for 1–2 weeks at 37°C.
3. Store supplements in aliquots that are used once only, and do not thaw and refreeze stock solutions repeatedly.

Trypsin
1. Store concentrated stock frozen in aliquots suitable for single use.
2. Make up with stock from concentrate, and store the diluted solution at 4°C for a maximum of 2 weeks.
3. Trypsin will degrade if left at room temperature for over 30 min.

28.3   PURITY OF CONSTITUENTS

28.3.1   Is the Water Purifier Working Correctly?
Quality of output
1. Test the conductivity of the water.
2. Test the total organic content (TOC; Millipore instrument) of the water.
3. Check medium made with the water against $1\times$ medium.

Maintenance regime

1. When was the deionizer last changed?
2. When was the reverse osmosis cartridge changed?
3. Is all connecting tubing clean? Are there any signs of algae, fungi, or bacteria?
4. Check the storage vessel for algal or fungal contamination.
5. Check for release of deionizer resin into output water.
6. Check for chemical contamination or residue in the glass boiler of the still
   (a) Dismantle the glass boiler.
   (b) Clean it in 1 M HCl and rinse thoroughly.
   (c) Discard first boiling from still.

Chemical traces in the plastic tubing.

1. Disconnect and clean tubing.
   (a) Soak in hot detergent.
   (b) Rinse and soak in 1 M HCl.
   (c) Rinse thoroughly in deionized water.
   (d) Reinstall and discard first batch of water.
2. Change the tubing to new, inert, washed, sterilized, plastic tubing (*see* Section 11.3.5).

28.3.2 Bicarbonate
Is the concentration correct?

1. Check the conductivity or osmolality against a reference standard solution.
2. Try another batch. (Make it up or buy it in.)
3. Check for signs of precipitate.

Wrong concentration

1. Check whether you are using the right amount for the medium (*see* Tables 9.1, 9.3, 10.1, and 10.2).
2. Check CO_2 concentration in incubator.

28.3.3 Antibiotics
Problems may arise because of:

(1) Frequency of use
(2) Concentration
(3) Batch number
(4) Combinations
(5) Fungicide (e.g., amphotericin B can be toxic)

28.3.4 Serum
Are you using a new batch?

1. Check the supplier's quality control.
2. Compare the batch with a previous batch or other batches.

Concentration
Too low? Too high?

1. Reconfirm the lack of toxicity, growth promotion, and plating efficiency
   (a) Create a growth curve (*see* Protocols 21.7, 21.8, 21.9).
   (b) Run a clonal growth assay (*see* Protocol 21.10).
2. Try replacing
   (a) With a new type of serum
   (b) With serum-free medium (*see* Section 10.5)
   (c) With serum substitute (*see* Section 10.5.2)

28.4 PLASTICS

Are you using a new make, type, or batch?

1. Check your batch against a previous batch.
2. Try an alternative supplier.

28.5 GLASSWARE

28.5.1 Washup
Are other cells showing symptoms of deterioration or impaired growth?
Are other users having trouble?
Is there trace contamination on the glass of storage bottles or pipettes?
(*See* Section 28.7.)
Have the caps not been properly rinsed? (Adding a few mL of BSS will show by a pH change if detergent has been left in the cap.)
Has any chemical glassware been mixed in with tissue culture glassware?

28.6 MICROBIAL CONTAMINATION

See also Table 18.1.

28.6.1 Confined to Single User
Sporadic
Single species

1. Check the aseptic technique of the operator (*see* Sections 6.3–6.5).
2. Are there any media or reagents that are unique to that user?
3. Check personal cleanliness of the operator.
   (a) Are hands washed before and after culture work?
   (b) Is the lab coat changed?
   (c) Is long hair tied back?
4. Get the operator to wear gloves.

Mycoplasma

1. Check with other users for evidence of mycoplasma contamination.
2. Ensure that the screening program is operative.
3. Imported cell lines are the most common source of infection. Quarantine them before screening for use in the main tissue culture area. Keep records on importation.
4. Check the natural-product reagents (serum, trypsin) with an indicator cell line (*see* Protocols 19.2, 19.3).
5. Restrict the use of antibiotics to primary culture and critical experiments.

Multispecific
May be due to lapses in sterile technique.

1. Ensure that all correct procedures have been maintained (*see* Protocol 6.1):

(a) Hood is uncluttered

(b) Materials have been swabbed before placing in hood

(c) All spillage is mopped up immediately

2. Check use of laboratory coats:

(a) Is the lab coat changed before commencing culture work?

(b) Is the lab coat buttoned? (Even people passing through with a flapping coat can disrupt the laminar air flow).

3. Check with other users of the same hood; the hood may be faulty and need to be serviced.

Repeated

Single species. The problem is usually a reagent or cell line.

1. Check for a unique reagent that no one else uses.

2. Is the cell line unique to that user?

Multispecific

1. Check sterile technique (*see* Sections 6.3–6.5).

2. Check for nonstandard procedures.

3. Check location: Is there too much equipment or traffic (*see* Section 6.2)?

4. Is the hood overcrowded?

5. Are nonsterile reagents being used?

6. Check use of laboratory coats (*see* Section 26.8.1).

Continuous
Single species
Contaminated solution

1. Mix medium (antibiotic free) 1:1 with nutrient broth (e.g., L-Broth), and incubate the solution.

2. Plate the medium out on blood agar, and incubate the solution, upside down, with blank controls.

Contaminated cell line

1. Check the cells by Hoechst staining (*see* Protocol 19.2).

2. Mix the cells with broth, and incubate.

Multispecific

1. Check sterile technique (*see* Sections 6.3–6.5).

2. Check for nonstandard procedures.

3. Check location: Is there too much equipment or traffic (*see* Section 6.2)?

4. Check how much is in the hood at any one time: Is the hood overcrowded?

5. Check whether nonsterile reagents are being used (inadvertently or deliberately).

28.6.2 Widespread
Sporadic
Single species

1. Check for an infrequently used reagent or medium that might be contaminated.

2. Check the incubators. Clean out if necessary (*see* Protocol 19.1).

3. Check for an increased spore count in the atmosphere by exposing bacteriological plates when room is quiet.

Multispecific
Sterilization failure

1. Check sterilizing ovens for:

(a) Overcrowding of contents preventing adequate air circulation

(b) Integrity of door seals and any other apertures

(c) Temperature and duration of sterilization cycle (160°C for 1 h)

2. Check autoclaves:

(a) For overcrowding of contents preventing adequate steam circulation

(b) For temperature and duration of sterilization cycle (121°C for 15–20 min) with probe in equivalent sample in center of load

(c) To ensure that all empty vessels are left open for steam circulation (*see* Fig. 11.3; Plate 22a).

New member of staff not following standard procedure

1. Check the training of the new staff with supervisor.

2. Update training procedures as necessary (*see* Sections 2.1, 6.3–6.5)

Contaminated storage (e.g., cold room or refrigerator)

1. Clean out the storage facility.

2. Make sure that sterilized items are not stored unsealed.

3. Check the turnover of sterile stocks.

Repeated, single species
Contaminated reagent or medium

1. Check the frequency of use of reagents among users to narrow the problem down to common reagents.

2. Test likely candidates by incubating them 1:1 in broth.

3. Check incubator for contamination and clean out as necessary (*see* Protocol 19.1).

4. Check hood for contamination.

(a) Swab work surface and sides of work area with phenolic disinfectant in 70% alcohol.

(b) Swab below work surface with phenolic disinfectant in 70% alcohol.

(c) Check other components of ductwork for leaks or contamination.

(d) Check integrity of filter with anemometer or nutrient agar plates (*see* Section 6.4).

(e) Check whether filters are dirty or blocked, e.g., pressure drop across filter (*see* Section 6.4).

(f) Check the cleaning schedules and update as necessary.

5. Check the maintenance schedules and update as necessary.

Repeated, multispecific
Sterilization failure

1. Check the autoclaves and sterilizing ovens for overcrowding.

2. Check the autoclaves and sterilizing ovens for electrical or mechanical failure.

3. Check the printouts and records of the sterilization cycles.
4. Check the integrity of the sterilization chamber of ovens.
5. Check whether new member of staff is not following standard procedure.
6. Check the training of new staff.
7. Check procedures with the supervisor.
8. Restrict the use of antibiotics.
9. Check for changes in procedures and/or requirements.
10. Redraft SOPs to suit any changed circumstances.

Contaminated storage (e.g., cold room or refrigerator)
1. Clean out the storage facility.
2. Make sure that sterilized items are not stored unsealed.

Contaminated room air
1. Check for contamination of the room's air by leaving blood agar plates open when the room is not in use, but when the ventilation is on.
2. Check the aseptic technique of operators.
3. Check the integrity of the hoods.
4. Check the air supply and air conditioners.
5. Check the quality of cleaning of incoming equipment.
6. Is there building work or other disturbance nearby? If so, try to improve isolation of culture laboratory (e.g., close doors, erect screens, exclude non-tissue culture staff, clean all equipment and materials entering the room).

28.6.3 Identification of Contamination
Bacterial, fungal
1. Check for bacterial or fungal contamination of cells or media by using high-power phase contrast.
2. Incubate the cells in broth.
3. Plate out the cells on blood agar, and incubate the culture.
4. Gram stain the cells, or consult a microbiologist.

Mycoplasma
1. Stain the culture with Hoechst 33258 (*see* Protocol 19.2).
2. Check for cytoplasmic DNA synthesis (incorporation of [^{3}H]thymidine), using autoradiography (*see* Protocol 27.3).
3. Get a commercial test done (*see* Appendix II—Mycoplasma Testing).

Viral
1. Test for viral contamination using:
   (a) Transmission or scanning electron microscopy
   (b) Immunostaining with a fluorescent or peroxidase-conjugated antibody (*see* Protocol 16.11)
   (c) ELISA assays
   (d) PCR
2. Obtain cell lines and biological reagents from virus free sources.

28.6.4 Decontamination
1. Discard stocks and do not attempt decontamination unless stocks are irreplaceable.

2. Decontaminate cells only if they are irreplaceable (*see* Protocol 19.4).
   (a) Should be carried out by an experienced member of staff.
   (b) Work in quarantine (*see* Section 19.1.8).

28.7 CHEMICAL CONTAMINATION
28.7.1 Glassware
Residue after cleaning
1. Keep tissue culture glassware separate from chemical glassware.
2. Ensure that there is no carry-over from the last rinse of a previous chemical wash in a washing machine.
3. Carry out spot checks by
   (a) Visual examination
   (b) Adding a small volume of BSS with phenol red and looking for pH change
   (c) Adding a small volume of UPW and checking conductivity
   (d) Cloning cells on the glass after sterilization by dry heat
4. Select detergent carefully (*see* Section 11.3.4).

Dust accumulation during storage
1. Foil cap all open vessels.
2. Store in dust-free area or container.

28.7.2 Pipettes
Residue or blockage after cleaning
1. Ensure that pipettes are washed and dried tip uppermost.
2. Collect pipettes into detergent, but rinse in water only.
3. Do not allow agar to be used in glass pipettes.
4. Check pipettes after washing and before sterilization by
   (a) Visual examination
   (b) Adding a small volume of BSS with phenol red and looking for pH change
   (c) Adding a small volume of UPW and checking conductivity.
5. Make sure cotton plugs are removed before washing.

28.7.3 Water Purification
See Sections 11.4.1, 28.3.1.

28.7.4 Other Reagents
Contamination of DMSO by dissolved plastic or rubber from container
1. Store DMSO in glass or polypropylene with glass or polypropylene cap.
2. Dispense with glass pipette or polypropylene pipettor tip.

Deterioration of glycerol on long-term storage
1. Buy in small amounts that will be used within 3–6 months.
2. Store in dark bottle.

28.7.5 Powders and Aerosols
1. Handle toxic chemicals that produce powders or aerosols in a fume hood (*see* Section 7.5.4).

2. Avoid drafts when weighing powders or dispensing liquids.
3. Control traffic of people and equipment into tissue culture laboratory and preparation areas.
4. Change to a clean laboratory coat before entering the tissue culture laboratory.

28.8 PRIMARY CULTURE

28.8.1 Poor Take in Primary Culture
Primary explants do not attach
1. Scratch the substrate through the explant (*see* Protocols 12.4 and 23.9).
2. Trap explant under a coverslip (*see* Protocol 12.4).
3. Embed in a plasma clot (*see* Protocol 12.4).

Disaggregation incomplete
1. Incubate the cells in protease for a longer amount of time.
2. Try cold pretreatment before incubation (*see* Protocol 12.6).
3. Use an alternative, or additional, protease (*see* Table 13.4; Section 12.3.6).

Complete disaggregation but poor attachment
Floating cells are viable
1. Treat the substrate
   (a) Coat the plastic with collagen, fibronectin, laminin, poly-D-lysine, or poly-L-lysine (*see* Section 8.4.1).
   (b) Use a feeder layer (*see* Section 8.4.2; Protocols 14.3, 23.4, and 24.1).
2. The culture may be nonadherent, so culture in suspension.

Floating cells are mostly nonviable
1. Adjust the concentration to a viable cell count.
2. Remove the nonviable cells (*see* Protocol 12.10).

Floating cells are nonviable and few cells have attached
Cell density
The cell density is too low; increase it to up to 1×10^6 cells/mL.

Cell adhesion is poor
Treat the substrate (*see* Sections 8.4.1, 8.4.2 and Protocols 14.3, 23.4, 24.1)

Enzymes used are too toxic
1. Change to a different protease (*see* Section 12.3.6 and Table 13.4).
2. Reduce the exposure time.
3. Try cold pretreatment with protease before incubation (*see* Protocol 12.6).

Medium is very acidic
1. Check the medium for contamination (*see* Section 19.3).
2. Reduce the cell concentration at seeding or 24 h later.
3. Add HEPES buffer, and vent the flask.

Wrong medium
1. Try a range of media (*see* Section 9.6; Tables 9.3, 9.6, 10.1, and 10.2; and 10.4).
2. Check the literature for media used with your cells (if you have not already done so).

Supplementation of medium
1. Use different types or batches of serum (*see* Section 9.6.2).
2. Replace the serum with serum-free medium (*see* Section 10.5).
3. Use different growth factors (*see* Section 10.4.3 and Table 10.3).
4. Use other mitogens (e.g., PMA; prostaglandins; hydrocortisone or other steroids; peptide hormones, such as insulin and transferrin; *see* Table 10.2).
5. Use conditioned medium (*see* Section 9.7.3; Protocol 14.2).

28.8.2 Wrong Cells Have Been Selected
Overgrowth by fibroblasts or endothelium
1. Use selective media (*see* Sections 10.2.1 and 14.6).
2. Use selective substrates (*see* Sections 8.4.1 and 14.8).
3. Use a selective feeder layer (*see* Protocol 24.1).
4. Check for cross-contamination from a feeder layer or a xenograft host (*see* Section 19.5; Protocols 16.8, 16.9, and 16.10).

Cross-contamination with another cell line
Check against other cell lines currently in culture (*see* Sections 7.10.1, 16.1, 16.3, 16.6.2, 19.5; Table 13.2).

28.8.3 Contamination
1. Check for contamination (*see* Section 19.3).
2. Pretreat the tissue with antibiotics (*see* Appendix I: Collection Medium and DBSS) or 70% ethanol (*see* Step 7 in Protocol 12.3).
3. Eradicate the contamination, but only if the material is irreplaceable (*see* Section 19.4).

28.9 DIFFERENTIATION

28.9.1 Cells Do Not Differentiate
1. Apply differentiation-inducing conditions (*see* Section 17.7).
2. Use a selective medium (*see* Section 10.2.1; Tables 10.1 and 10.2) at isolation and during propagation.
3. Coculture with appropriate feeder layer (*see* Sections 17.7.1, 17.7.3).

28.9.2 Loss of Product Formation
1. Check for the expression of the relevant gene, using RT-PCR or Northern blot [Ausubel et al., 1996, 2002] or *in situ* hybridization (*see* Section 27.8).
2. Check the reporter gene expression (using RT-PCR or Northern blot [Ausubel et al., 1996]; *see* Section 27.12.5).

3. Apply differentiation-inducing conditions (*see* Section 17.7).
4. Use the appropriate selective medium for the cell type (*see* Section 10.2.1; Tables 10.1 and 10.2) at isolation and during propagation.

28.10   FEEDING

28.10.1   Regular Monolayers
pH falls too quickly
1. Check for bacterial contamination (*see* Section 19.4.1).
2. Feed more frequently.
3. Seed at lower cell concentration.
4. Add HEPES buffer to medium (*see* Section 9.2.3) and vent flask (*see* Section 8.2.3).

pH rises after feeding
1. Reduce bicarbonate or other alkali in medium.
2. Increase CO_2 concentration.
3. Check for leakage in flasks incubated in air, or for film of liquid sealing caps or lids when in CO_2 incubator.

28.10.2   Clones
pH too high or too low
See Sections 28.12.1, 28.5.

Dishes get contaminated
1. Incubate dishes in box (*see* Section 6.6.2) and swab box before opening.
2. Discard lids of dishes or plates, particularly if they have medium on them. Swab outside of base with 70% alcohol.
3. Do not use fan in incubator.

Feeding required

Monolayer
Change medium (*see* Protocol 12.1) and check after 24 h.

Suspension
1. Add medium to agar cultures.
2. Add medium plus Methocel to Methocel clones; the nutrients will diffuse through the Methocel.

28.11   SUBCULTURE

This section discusses the problem of poor take or slow growth after subculture (*see also* Section 28.1).

28.11.1   Cell Cycle Phase at Subculture
1. Subculture from the exponential or late-exponential phase (*see* Protocol 13.2).
2. Ensure that the seeding concentration is correct—too low and the cells will have an extended lag period, too high and they will enter plateau before they are subcultured (*see* Section 13.7.3).

3. Do not subculture from plateau as plateau-phase cells tend to have a long lag phase. Some cultures, e.g., hybridomas, mouse leukemias, will deteriorate rapidly in the plateau phase.
4. Do not subculture more often than is necessary. Choose a cell concentration and subculture interval that are both convenient and give the most reproducible growth (*see* Section 13.7.2).

28.11.2   Senescence
1. Check the generation number; the cells may be approaching the end of a finite life span (*see* Sections 3.8.1, 13.7.2, 18.4.1).
2. Replace stocks routinely from freezer (*see* Section 20.4.2).
3. If necessary, try immortalization (*see* Protocols 18.1, 18.2).

28.11.3   Medium
As for slow cell growth (*see* Section 28.1).

28.11.4   Uneven Growth
Nonrandom distribution of cells
As for cloning (*see* Section 28.9.4).

Incorrect seeding
1. Do not cause the cell suspension to swirl when pipetting.
2. Do not pipette cells into the middle of dish without mixing.
3. Do not swirl the dish to mix the cells, but mix side-to-side and back-to-front, or pipette up and down gently without swirling.

Incubator vibration
1. Restrict the frequency of access to the incubator.
2. Check that the incubator is firmly seated.
3. Make sure that the shelves are level.
4. Place a cushion under the flasks, dishes, or plates.

Uneven heating (see Fig. 8.10)
1. Place flasks or dishes on an insulating tile or metal plate.
2. Check the air and temperature distribution.

28.12   CLONING

See also Slow Cell Growth in this chapter.

28.12.1   Poor Plating Efficiency
Too few colonies per dish
1. Increase the seeding concentration.
2. Use a feeder layer.
3. Improve the plating efficiency (*see* Section 14.2).

Colonies are too diffuse
1. Select a bigger Petri dish, to give more space for the colonies to spread out.
2. Use glucocorticoid (e.g., dexamethasone, 1×10^{-5}–1×10^{-6} M).

Mycoplasma contamination

1. Screen for mycoplasma (*see* Protocols 19.3.3–19.3.5)
2. Eradicate the mycoplasma, but only as a last resort, if the cell line is irreplaceable (*see* Sections 19.4.2; Protocol 19.4).

Poor handling

1. Have the cells have been out of the incubator too long?
2. Has a prolonged amount of time has been spent in dilution?
3. Has the medium evaporated during incubation?

Medium
Type

1. Choose a rich medium (e.g., Ham's F12).
2. If the medium is serum-free (e.g., MCDB 153 for keratinocytes; *see* Tables 10.1 and 10.2), try different serum-free media.
3. CO_2 is essential, so always incubate under a minimum of 2% CO_2.
4. Check for evaporation by weighing test dishes with medium, and under the same conditions, at start and during culture period.

If serum is essential

1. Use fetal bovine serum, which is usually better than calf or horse serum for cloning.
2. If fetal bovine serum is already being used, increase the concentration.
3. Select the batch based on the plating efficiency of the cells that you use (*see* Section 11.6.3, Protocol 21.10).

Poor substrate
Has there been a change in the culture plasticware supplier?

1. Check source of plasticware and change if necessary.
2. Try alternatively charged plastic, e.g., Primaria (BD Biosciences), CellBIND (Corning).
3. Coat the plastic with matrix (*see* Section 8.4.1, Protocol 8.1; 17.7.3; Protocol 23.9: FN/V/BSA).

28.12.2 Diffuse Colonies

1. Coat with matrix (*see* Section 8.4.1, Protocol 8.1; 17.7.3; Protocol 23.9: FN/V/BSA).
2. Use glucocorticoid (e.g., 1×10^{-7}–1×10^{-5} M dexamethasone) for the first 48–72 h.
3. Use a feeder layer (*see* Protocol 14.3).

28.12.3 Too Many Colonies per Dish

1. Reduce the seeding concentration (cells/mL).
2. Seed the same number of cells into a larger dish.

Overlapping colonies

1. Reduce the seeding density (cells/cm^2).
2. Grow the colonies for a shorter time (fewer cells/colony).
3. Seed the same number of cells on a larger dish.

28.12.4 Nonrandom Distribution

See also Section 28.12.4.

1. Add the cells to medium in a bottle, mix the cell suspension, and then seed the dishes.
2. Do not swirl the dishes to mix cells or when pipetting.
3. Ensure that the dishes are level.
4. Make sure that the medium covers all of the bottom of the dish evenly.
5. Ensure that the incubator is free from vibration.
6. Restrict access of other users to the incubator.
7. Label the box or tray containing the dishes with the phrase, "CLONING, DO NOT MOVE."
8. Place the box or tray at the back of the incubator.
9. Place the box or tray containing the dishes on a foam pad. (Wash the pad regularly.)

28.12.5 Nonadherent Cells
Anchorage-independent cells

1. Clone the cells in agar (*see* Protocol 14.4), agarose, or Methocel (*see* Protocol 14.5) over an agar underlay or in a non-tissue-culture-grade dish.
2. Add growth factors or supplements to underlay.
3. Use tissue culture grade dishes and seed feeder layer before pouring underlay.

Poorly attached cells
Clone the cells in Methocel in a tissue-culture-grade dish (*see* Protocol 14.5).

28.13 CROSS-CONTAMINATION

This section discusses the problem of cross-contamination with another cell line.

28.13.1 Symptoms of Cross-Contamination
Change in appearance
Morphology of cells changes. Cells pile up at high density in plateau when normally contact-inhibited.

Change in growth characteristics
Growth rate faster (i.e., shorter PDT).
Cells grow to a higher saturation density.

Cell line characteristics
Change in phenotypic characteristics (*see* Sections 16.1, 17.1, 18.3).

28.13.2 Preventing Cross-Contamination
See also Section 19.5.

1. Do not share your reagents with other users.
2. Do not share reagents among cell lines.
3. Do not handle more than one cell line at a time.
4. Do not return a pipette to medium after using it with cells.
5. Do not use pipettors for serial propagation unless filter tips are used.

6. Handle HeLa and other rapidly growing, high-plating-efficiency cells last.
7. Authenticate cell lines before freezing them (*see* Sections 16.1, 16.3).

28.13.3 Elimination of Cross-Contamination

1. Discard culture!
2. If detected early (i.e., there is evidence of a mixed culture), and no other stocks are available, clone the cells (*see* Protocol 14.1) and isolate likely colonies (*see* Protocols 14.6 and 14.7).
3. Do a positive or negative sort by FACS (*see* Section 15.4) or MACS (*see* Section 15.3.2).

28.14 CRYOPRESERVATION

28.14.1 Poor Recovery

Freezing rate

1. Change the cooling rate (*see* Section 20.3.4). although 1°C/min usually is optimal.
   (a) Change the wall thickness if using insulated container to freeze.
   (b) Use programmable freezer to change rate and shape of cooling curve.

Cell concentration
At freezing

Increase the cell concentration at freezing (optimum is usually from $1 \times 10^6 - 1 \times 10^7$/cells mL *see* Section 20.3.2; Protocol 20.1).

Thawing and reseeding

1. Dilute the culture more slowly after thawing (*see* Protocol 20.2) by gradually adding medium to the cells.
2. Reseed the cells at five times the normal seeding density (*see* Protocol 20.2).
3. Pool several ampoules (you will need to centrifuge to remove the preservative, *see* Protocol 20.2).
4. Remove the nonviable cells (*see* Protocol 12.10).

Preservative
DMSO

1. Check for chemical contamination in DMSO (from plastic or rubber).
   (a) Check color; should be colorless.
   (b) If colorless, do a spectrophotometric scan and compare with fresh DMSO.
2. Check if cells are susceptible to induction of differentiation by DMSO (*see* Section 17.7.2).
3. Centrifuge suspension-grown cells to remove the preservative.

Glycerol

Glycerol is toxic if stored for several months in light. (It gets converted to acrolein.)

1. Dispense glycerol in small volumes.
2. Store glycerol in the dark.

28.14.2 Changed Appearance after Cryopreservation

Mistaken identity

1. Check the labeling of the ampoule. If label illegible, discard the ampoule.
2. Check the records (*see* Section 20.3.6).
3. Check the authentication (*see* Section 16.3).

Changed conditions since the cells were last grown

(1) Are there new members of staff?
(2) Are there new procedures?
(3) Have the suppliers of media changed?
(4) Is a new serum batch being used?

28.14.3 Contamination

Leakage of ampoule

Check the seal: The cap should be tight, but the seal should not be distorted.

Contamination from water bath on thawing

1. Do not immerse ampoules; place in a rack only partially submerged.
2. Thaw the ampoules in a heating block.
3. Swab the ampoules carefully after thawing.

Δ *Safety Note.* Remember ampoules may explode if stored under liquid nitrogen, so place in rack and cover when thawing.

Contamination from liquid nitrogen

Do not submerge ampoules in liquid nitrogen. Use vapor-phase storage or a perfused jacket freezer.

28.14.4 Loss of Stock

User stock

Restock from the distribution stock.

Distribution stock

Restock from the seed stock.

Loss of seed stock

1. Check the security of the inventory control; it should be restricted to the curator only.
2. Restock from a reputable cell bank (*see* Appendix II: Cell Banks).

28.15 GRANULARITY OF CELLS

Intracellular
Are the cells unhealthy?

1. Check growth curve (*see* Protocols 20.7 and 20.8).
2. Check plating efficiency (*see* Protocol 20.9).

Is there a high rate of phagocytosis?
Check on high-power phase microscopy for uptake of neutral red or fluorescent dextran (FITC-dextran, Sigma).

Extracellular
Uniform particle size
Check for contamination (*see* Section 19.3).

Variable particle size
1. Check for precipitation from the medium (*see* Sections 28.2, 28.2.2).
2. Is there precipitation from serum (not usually harmful)?

28.16   CELL COUNTING

28.16.1   Hemocytometer
Variable counts
Sampling error
1. Mix the cell suspension thoroughly before sampling.
2. The cells should be singly suspended and not clumped (*see* Protocol 21.1).

Use of hemocytometer
1. Ensure that the coverslip is correctly attached. (Interference colors should be visible; *see* Section 21.1.1.)
2. Make sure that the counting chamber is not overfilled or underfilled.
3. A sufficient number of cells should be counted (>200).

Visibility of cells
1. The silvering on the counting chamber should be intact.
2. Use phase-contrast optics.
3. Use a noncentered light path if phase-contrast optics not available.
4. Try staining the cells before counting them.

28.16.2   Electronic Counting via Orifice by Resistance
Variable counts
Sampling error
1. Mix the cell suspension thoroughly before sampling.
2. The cells should be singly suspended and not clumped (*see* Protocol 21.2).

The count stops, will not start, or counts slowly (i.e., it takes longer than 25 s)

Orifice or cell channel blocked
1. Run the wash cycle.
2. Run the unblock cycle.
3. Rub the orifice with the tip of your finger or a fine brush.
4. Soak in detergent for 1–18 h and repeat unblock; repeat wash cycle three times.

The count is lower than expected, or the orifice blocks frequently
Cell suspension aggregated
1. Disperse the cells by pipetting the original sample vigorously, redilute the sample, and proceed.
2. Use different disaggregation technique (*see* Table 13.4).

Background is high but will not count
Electrode out of beaker or disconnected
1. Replace the electrode in the beaker.
2. Check that the electrode is secure.
3. Replace the electrode if the terminal plate is missing.

Count sequence will not start
Blocked orifice
1. Run the wash cycle.
2. Run the unblock cycles.
3. Rub the orifice with the tip of your finger or a fine brush.
4. Soak in detergent for 1–18 h, repeat unblock and repeat wash cycle three times.

Insufficient negative pressure
1. Check to ensure that the waste reservoir is not full.
2. Check the pump (see the manual for a diagnostic test) and connections.
3. Pump failure has occurred; call engineer.

Background is high
Line or radio interference from

There is a problem with electrical equipment (motors, fluorescent lights, incubators)
1. Check and eliminate potential problems by fitting suppressors to the equipment. Line filters are available, but are not always effective.
2. Check the grounding (earthing) of the counter, particularly the casing.

Particulate matter in counting fluid
1. Prepare fresh counting fluid.
2. Filter the counting fluid through a disposable Millex or equivalent filter.

28.17   VIABILITY

28.17.1   Morphological Appearance
Granularity and vacuolation (see also Section 28.15).
Intracellular granularity usually indicates that the cells are unhealthy.

Vacuolation usually indicates that the cells are unhealthy (*see* Fig. 13.1).

Loss of birefringence
Monolayer cells
If the cells are normally birefringent (edge of cell has a halo on phase contrast) and then lose birefringence,

then they usually have lost viability—e.g., by drying out during feeding.

Suspension cells

Suspension cells are normally clear and hyaline; granularity or opacity indicates dead or unhealthy cells.

Removing nonviable cells

1. Change the medium. Monolayer cells will float off if they are nonviable and thus do not need a special procedure for removal.
2. For suspension cells or primary disaggregates, enrich the viable cells by spinning them through Ficoll-Paque (*see* Protocol 12.10).

28.17.2　Testing Viability

If there is immediate cell loss following a manipulation, check by dye exclusion (*see* Protocol 22.1) or dye uptake (*see* Protocol 22.2 and try to eliminate cause).

(a) Primary cultures are usually 50–90% viable.
(b) Cell lines are usually 90–100% viable.
(c) Thawed cells are usually 50–80% viable.

28.17.3　Cytotoxicity

If deterioration is longer term, test for possible causes with clonogenic assay (*see* Protocol 22.3) or microtitration (*see* Protocol 22.4).

CHAPTER 29

In Conclusion

It has been my intention in the foregoing pages to describe the fundamentals of cell culture in sufficient detail that recourse to the literature is required only to extend your work beyond the basic procedures or to acquire some additional background detail. It is customary, when giving a lecture, to conclude with a summary that highlights the major points raised in the lecture, and that is how I would like to conclude this text.

There are certain requirements that are crucial to successful and reproducible cell culture, and they may be highlighted as follows:

- Ensure that your instruction, and the training of anyone who works with you, comes from an experienced source.
- Work in a clean, uncrowded, aseptic environment reserved for culture, and clear up when you have finished.
- To avoid the transfer of contamination, including cross-contamination, do not share media, reagents, cultures, or materials with others.
- Do not assume that your work is immune to mycoplasma because you have never seen it; test your cells regularly.
- Keep adequate records, particularly of changes in procedures, media, or reagents, and keep photographic records of the cell lines that you use.
- Work only on cell lines that have been obtained from a properly validated source, such as an international cell bank. Distrust any other cell line, even from the originator, unless you can prove that it is authentic.
- Become familiar with the cell lines that you use—their appearance, growth rate, and special characteristics—so that you can respond immediately to any change.
- Ensure that your work does not compromise your own safety or that of others working around you.
- Preserve cell line stocks in liquid nitrogen, and replace working stocks regularly.
- Protect seed stocks of valuable cell lines, and use other stocks for distribution.
- Try to work under conditions that are precisely defined, including minimal use of undefined media supplements, and do not change procedures for trivial reasons.
- Do not mix cell culture with other microbiological work.

To cover all of the fascinating aspects of cell and tissue culture would take many volumes and defeat the objective of this book. It has been more my intention to provide sufficient information to set up a laboratory and prepare the necessary materials with which to perform basic tissue culture, and to develop some of the more important techniques required for the characterization and understanding of your cell lines. This book may not be sufficient on its own, but with help and advice from colleagues and other laboratories, it may make your introduction to tissue culture easier and more satisfying and enjoyable than it otherwise might have been.

APPENDIX I

Preparation of Reagents

This appendix contains reagents that are used in several different protocols throughout the book. Specialized reagents used in only one protocol will generally be located within the Materials section of that protocol.

Note. Dilutions quoted as, for example, 1:10 or 1:100, are v/v and imply that the final volume is 10 or 100 parts, respectively.

Acetic/Methanol

Add 1 part glacial acetic acid to 3 parts methanol. Make up fresh each time used, and keep on ice.

Agar 2.5%

Agar	2.5 g
UPW	100 mL

(1) Boil to dissolve agar.
(2) Sterilize by autoclaving or boiling for 2 min.
(3) Store at room temperature.

Amido Black

See Naphthalene Black.

Amino Acids, Essential

See Section 9.4.1 and Table 9.3.
(Available as 50× concentrate in 0.1 M HCl from commercial suppliers such as Invitrogen, MP Biomedicals, and Sigma.)

(1) Make up tyrosine and tryptophan together at 50× in 0.1 M HCl and remaining amino acids at 100× in ultrapure water.
(2) Dilute for use as in Protocol 11.10.
(3) Sterilize by filtration.
(4) Store in the dark at 4°C.

Amino Acids—Nonessential

Ingredient	g/1 (100×)
L-Alanine	0.89
L-Asparagine H$_2$O	1.50
L-Aspartic acid	1.33
Glycine	0.75
L-Glutamic acid	1.47
L-Proline	1.15
L-Serine	1.05
Water	1000 mL

(1) Sterilize by filtration.
(2) Store at 4°C.
(3) Use at a concentration of 1:100.

Antibiotics

See under specific headings (e.g., penicillin, streptomycin sulfate, kanamycin sulfate, gentamycin, mycostatin).

Antifoam

e.g., RD emulsion 9964.40 (*see* Appendix II)

(1) Dispense into aliquots and autoclave to sterilize.
(2) Store at room temperature.
(3) Dilute 0.1 mL/liter (i.e., 1:10,000).

Bactopeptone, 5%

Difco Bactopeptone	5 g
Hanks' BSS	100 mL

(1) Stir to dissolve.
(2) Dispense in aliquots appropriate to use as a 1:50 dilution, and autoclave.
(3) Store at room temperature.
(4) Dilute 1/10 for use.

Culture of Animal Cells: A Manual of Basic Technique, Fifth Edition, by R. Ian Freshney
Copyright © 2005 John Wiley & Sons, Inc.

Balanced Salt Solutions (BSS)
(*See* Table 9.2.)

(1) Dissolve each constituent separately, adding $CaCl_2$ last.
(2) Make up to 1 L.
(3) Adjust pH to 6.5.
(4) Sterilize the solution by autoclaving or filtration. With autoclaving, the pH must be kept below 6.5 to prevent calcium phosphate from precipitating; alternatively, calcium may be omitted and added later. If glucose is included, the solution should be filtered to avoid caramelization of the glucose, or the glucose may be autoclaved separately (*see* Glucose, 20%, in this appendix) at a higher concentration (e.g., 20%) and added later.
(5) With autoclaving, mark the level of the liquid before autoclaving. Store the solution at room temperature, and if evaporation has occurred, make up to mark with sterile ultrapure water before use. If borosilicate glass is used, the bottle may be sealed before autoclaving and no evaporation will occur.

Broths
See manufacturers' instructions for preparation.
See also Bactopeptone and tryptose phosphate broth.
Sterilize by autoclaving.

Buffered glycerol mountant
See Mycoplasma

Carboxymethylcellulose (CMC)
(1) Weigh out 4 g of CMC and place it in a beaker.
(2) Add 90 mL of Hanks' BSS, and bring the mixture to boil in order to wet the CMC.
(3) Allow the solution to stand overnight at 4°C to clarify.
(4) Make volume up to 100 mL with Hanks' BSS.
(5) Sterilize the solution by autoclaving. The CMC will solidify again, but will redissolve at 4°C.
(6) For use (e.g., to increase the viscosity of the medium in suspension cultures), use 3 mL per 100 mL of growth medium.

Chick Embryo Extract [Paul, 1975]
(1) Remove embryos from eggs as described in Protocol 12.2, and place the embryos in 9-cm Petri dishes.
(2) Take out the eyes, using two pairs of sterile forceps.
(3) Transfer the embryos to flat- or round-bottomed 50-mL containers, two embryos to each container.
(4) Add an equal volume of Hanks' BSS to each container.
(5) Using a sterile glass rod that has been previously heated and flattened at one end, mash the embryos in the BSS until they have broken up.
(6) Let the mixture stand for 30 min at room temperature.
(7) Centrifuge the mixture for 15 min at 2000 g.
(8) Remove the supernate, and, after keeping a sample to check its sterility (*see* Section 11.6.2), dispense the solution into aliquots and store at −20°C.

Extracts of chick and other tissues may also be prepared by homogenization in a Potter homogenizer or Waring blender [Coon and Cahn, 1966].

(1) Homogenize chopped embryos with an equal volume of Hanks' BSS.

(2) Transfer the homogenate to centrifuge tubes, and spin at 1000 g for 10 min.
(3) Transfer the supernate to fresh tubes, and centrifuge for a further 20 min at 10,000 g.
(4) Check the sample for sterility (*see* Section 11.6.2).
(5) Dispense into aliquots.
(6) Store at −20°C.

Citric Acid/Crystal Violet
See Crystal Violet.

CMC
See Carboxymethylcellulose.

Colcemid, 100× Concentrate
Colcemid	100 mg
Hanks' BSS	100 mL

(1) Stir to dissolve.
(2) Sterilize by filtration.
(3) Dispense into aliquots and store at −20°C

Δ *Safety Note.* Colcemid is toxic; handle it with care by weighing in a fume cupboard and wearing gloves.

Collagenase
2,000 U/mL in Hanks' BSS

Worthington CLS-grade collagenase or the equivalent (specific activity 1500−2000 U/mg)	100,000 U
Hanks' BSS	50 mL

(1) To dissolve the mixture, stir at 37°C for 2 h or at 4°C overnight.
(2) Sterilize the solution by filtration, as with serum (*see* Protocol 11.15).
(3) Divide into aliquots, each suitable for 1−2 weeks of use.
(4) Store at −20°.

Collagenase−Trypsin−Chicken Serum (CTC) [Coon and Cahn, 1966]

Sterile	Volume	Final Concentration
Ca^{2+}- and Mg^{2+}-free saline [Moscona, 1952]	85 mL	
Trypsin stock, 2.5%, sterile	4 mL	0.1%
Collagenase stock, 1%, sterile	10 mL	0.1%
Chick serum	1 mL	1.0%
Dispense into aliquots and store at −20°C.		

Collection Medium (for Tissue Biopsies)
Growth medium	500 mL
Penicillin	125,000 units
Streptomycin	125 mg
Kanamycin	50 mg
or	
Gentamycin	25 mg
Amphotericin	1.25 mg

Store at 4°C for up to 3 weeks or at −20°C for longer periods.

Crystal Violet, 0.1%, in 0.1 M Citric Acid

Citric acid...21.0 g
Crystal violet.. 1.0 g

(1) Make up to 1000 mL with deionized water.
(2) Stir to dissolve.
(3) To clarify, filter the solution through Whatman No. 1 filter paper.

Crystal Violet 0.1% in Water

Crystal violet.. 100 mg
Water..100 mL

Filter through Whatman No. 1 paper before use.
Available ready made from Merck.

Dexamethasone (Merck)

1 mg/mL (100×)
This reagent comes already sterile in glass vials.

(1) To dissolve, add 5 mL water by syringe to the vial.
(2) Remove when dissolved
(3) Dilute to 1 mg/mL, approximately 2.5 mM.
(4) Divide the solution into aliquots and store at −20°C.
(5) For use, dilute the solution to give 10−50 nM (physiological concentration range), 0.1−1.0 μM (pharmacological dose range), or 25−100 μM (high dose range).

β-Methasone (Glaxo) and methylprednisolone (Sigma) may be prepared in the same way.

Dissection BSS (DBSS)

(1) To Hanks' BSS without bicarbonate, previously sterilized by autoclaving, add the following (all sterile):

Penicillin...250 U/mL
Streptomycin.. 250 μg/mL
Kanamycin... 100 μg/mL
or
Gentamycin...50 μg/mL
Amphotericin B...2.5 μg/mL

(2) Store at −20°C.

Dulbecco's PBS without Ca²⁺ and Mg²⁺ (D-PBSA)

(*See* Protocol 11.6)

EDTA (Versene)

(1) Prepare as a 10 mM concentrate, 0.374 g/L in D-PBSA.
(2) Sterilize by autoclaving or filtration.
(3) Dilute 1:10, or 1:5 for use at 1.0−2.0 mM, or, exceptionally, 1:2 for use at 5 mM, diluted in D-PBSA or trypsin in D-PBSA.

EGTA

As for EDTA, but EGTA may be used at higher concentrations because of its lower toxicity.

Ficoll, 20%

(1) Sprinkle 20 g of Ficoll (GE Heathcare) on the surface of 80 mL UPW

(2) Leave overnight for the Ficoll to settle and dissolve.
(3) Make up to 100 mL in UPW.
(4) Sterilize by autoclaving.
(5) Store at room temperature.

Fixative for Tissue Culture

See Acetic/methanol.
Alternatively, use pure anhydrous ethanol or methanol (*see* Protocol 16.2), 10% formalin, 1% glutaraldehyde, or 5% paraformaldehyde.

Gentamycin

Dilute to 50 μg/mL for use.

Gey's Balanced Salt Solution

	g
NaCl	7.00
KCl	0.37
CaCl₂	0.17
MgCl₂ · 6H₂O	0.21
MgSO₄ · 7H₂O	0.07
Na₂HPO₄ · 12H₂O	0.30
KH₂PO₄	0.03
NaHCO₃	2.27
Glucose	1.00
Water, up to	1000 mL
CO₂	5%

Giemsa Stain

Giemsa stain can be applied undiluted and then diluted with buffer or water (*see* Protocol 16.2), or diluted in buffer before use. The author has found the first method more successful for cultured cells.

(1) Prepare buffer:

$NaH_2PO_4 · 2H_2O$.....................0.01 M.......1.38 g/L
$Na_2HPO_4 · 7H_2O$.....................0.01 M.......2.68 g/L

Combine to give pH 6.5.

(2) Dilute prepared Giemsa concentrate 1:10 in 100 mL of buffer.
(3) Filter the solution through Whatman No. 1 filter paper to clarify.
(4) Make up a fresh solution each time, because the concentrate precipitates on storage.

Glucose, 20%

Glucose... 20 g
Hanks' BSS to..100 mL

(1) Sterilize by autoclaving.
(2) Store at room temperature.

Glutamine, 200 mM

L-glutamine...29.2 g
Hanks' BSS...1000 mL

(1) Dissolve the glutamine in BSS and sterilize by filtration (*see* Protocols 11.12, 11.13).
(2) Dispense the solution into aliquots and store at −20°C.

Glutathione

(1) Make 100× stock (i.e., 0.1 M in HBSS or D-PBSA), and dilute to 1 mM for use.
(2) Sterilize by filtration.
(3) Dispense into aliquots and store at −20°C.

Ham's F12

See Table 9.3.

Hanks' BSS

See Section 9.3 and Balanced Salt Solutions, in this appendix.

Hanks' BSS without phenol red:

Follow the preceding instructions, but omit phenol red.

HAT Medium

Drug	Concentration	Dissolve in	Molarity (100× final)
Hypoxanthine (H)....	136 mg/100 mL....	0.05 N HCl....	1×10^{-2} M
Aminopterin (A)......	1.76 mg/100 mL....	0.1 N NaOH...	4×10^{-5} M
Thymidine (T)........	38.7 mg/100 mL....	....HBSS....	1.6×10^{-3} M

(1) For use in the HAT selective medium, mix equal volumes of each, sterilize by filtration, and add the mixture to medium at 3% v/v.
(2) Store H and T at 4°C, A at −20°C.

HB Medium

Add the following to CMRL 1066 medium:

Insulin...	5 μg/mL
Hydrocortisone.......................................	0.36 μg/mL
β-Retinyl acetate.....................................	0.1 μg/mL
Glutamine...	1.17 mM
Penicillin..	50 U/mL
Streptomycin...	50 μg/mL
Gentamycin...	50 μg/mL
Fungizone...	1.0 μg/mL
Fetal bovine serum..................................	1%

HBSS

See Section 9.3 and Balanced Salt Solutions, in this appendix.

Hoechst 33258 [Chen, 1977]

2-[2-(4-Hydroxyphenol)-6-benzimidazoyl]-6-(1-methyl-4-piperazyl)benzimidazole trihydrochloride

(1) Make up 1 mg/mL stock in D-PBSA or HBSS without phenol red.
(2) Store the solution at −20°C. For use, dilute 1:20,000 (1.0 μL in 20 mL) in D-PBSA or HBSS without phenol red at pH 7.0.

△ *Safety Note.* Because this substance may be carcinogenic, handle it with extreme care. Weigh in a fume cupboard and wear gloves.

Kanamycin Sulfate (Kannasyn), 10 mg/mL

Kanamycin...	4, 1-g vials
Hanks' BSS..	400 mL

(1) Add 5 mL of HBSS from a 400-mL bottle of HBSS to each vial.

(2) Leave for a few minutes to dissolve.
(3) Remove the HBSS and kanamycin from the vials, and add them back to the HBSS bottle.
(4) Add another 5 mL of HBSS to each vial to rinse and return to the BSS bottle. Mix well.
(5) Dispense 20-mL aliquots of the solution into sterile containers and store at −20°C.
(6) Test for sterility: Add 2 mL of reagent to 10 mL of sterile medium, free of all other antibiotics, and incubate the solution at 37°C for 72 h.
(7) Use at 100 μg/mL.

Lactalbumin Hydrolysate 5% (10×)

Lactalbumin hydrolysate...	5 g
HBSS...	100 mL

(1) Heat to dissolve.
(2) Sterilize by autoclaving.
(3) Use at 0.5%.

McIlvaines Buffer, pH 5.5

	To make 20 mL	To make 100 mL
0.2 M Na$_2$HPO$_4$ (28.4 g/L)..........	11.37 mL........	56.85 mL
0.1 M citric acid (21.0 g/L)............	8.63 mL........	43.15 mL

Media

The constituents of some media in common use are listed in Chapters 9 and 10 (*see* Tables 9.3, 10.1, and 10.2), together with the recommended procedure for their preparation. For those media not described, *see* Morton [1970], or suppliers' catalogs (*see* Appendix II: Media).

MEM

See Section 9.4 and Table 9.3.

2-Mercaptoethanol (M.W. 78)

Stock solution, 5 mM......................................	4 μL
HBSS...	10 mL

(1) Sterilize by filtration in fume cupboard.
(2) Store at −20°C or make up a fresh solution each time.

Methocel

See Methylcellulose.

Methylcellulose (1.8%)

(1) Weigh out 7.2 g of Methocel, and add it to a 500-mL bottle containing a large magnetic stirrer bar.
(2) Sterilize by autoclaving with the cap loose for penetration of steam.
(3) Add 400 mL of sterile UPW heated to 90°C to wet the Methocel.
(4) Stir at 4°C overnight to dissolve. (The Methocel will form a solid gel if the magnet does not keep stirring.)
The resulting solution is now Methocel 2×, and for use, it should be diluted with an equal volume of 2× medium of your choice. It is more accurate to use a syringe (without a needle) than a pipette to dispense Methocel.

(5) For use, add a cell suspension in a small volume of growth medium (*see* Protocol 14.5)

Mitomycin C
Stock solution, 10 μg/mL (50×)

Mitomycin...2-mg vial

(1) Measure 20 mL of HBSS into a sterile container.
(2) Remove 2 mL of HBSS by syringe and add it to a vial of mitomycin.
(3) Allow the mixture to dissolve, withdraw the resulting solution, and add it back to the container.
(4) Store for 1 week only at 4°C in the dark. (Cover the container with aluminum foil.)
(5) For longer periods, store at −20°C.
(6) Dilute to 0.25 μg/mL for 18-h exposure or 20 μg/mL for 10-h exposure (*see* Protocols 14.3, 23.4)

Δ *Safety Note.* Because mitomycin is toxic, reconstitute it in the vial. Work in a fume hood when handling the substance in powder form.

MTT
(1) Dissolve 3-(4,5-dimethylthiazol-2-yl)-2,5-diphenyltetrazolium bromide (MTT) 50 mg/mL in D-PBSA
(2) Sterilize by filtration.

Δ *Safety Note.* MTT is toxic; weigh it in a fume cupboard and wear gloves.

Mycoplasma Reagents
Stain (*see* Hoechst 33258)
Mountant: Glycerol in McIlvaines Buffer pH 5.5

	To make 40 mL
0.4 M Na_2HPO_4 (56.8 g/L)	11.37 mL
0.2 M Citric acid (42.0 g/L)	8.63 mL
Glycerol	20.00 mL

(1) Add Vectashield (Vector) to reduce fluorescence fade (*see* manufacturer's instructions)
(2) Check the pH and adjust to 5.5.

Mycostatin (Nystatin)
(1) Prepare at 2 mg/mL (100×):

Mycostatin.. 200 mg
Hanks' BSS...100 mL

(2) Make up by same method as kanamycin.
(3) Final concentration, 20 μg/mL.

Naphthalene Black
(1) Prepare at 1% in Hanks' BSS

Naphthalene Black...1 g
Hanks' BSS... 100 mL

(2) Dissolve as much as possible of the stain in the HBSS.
(3) Filter the resulting saturated solution through Whatman No. 1 filter paper.

PBS
See Phosphate-Buffered Saline.

D-PBSA
See Phosphate-Buffered Saline.

PE
See Phosphate-Buffered Saline/EDTA.

Penicillin
(*e.g., Crystapen benzylpenicillin [sodium]*) *1,000,000 units per vial*

(1) Make up as for kanamycin, stock concentration 10,000 U/mL.

Crystapen..................................... 4 vials (4 × 10^6 U)
HBSS.......................................400 mL

(2) Store frozen at −20°C in aliquots of 5–10 mL.
(3) Use at 50–100 U/mL.

Percoll
(1) Ready made and sterile as purchased, Percoll should be diluted with medium or HBSS until the correct density is achieved.
(2) Check the osmolality. Adjusting it to 290 mOsm/Kg will require the diluent to be hypo- or hypertonic, so it is better to dilute a small sample first and check its osmolality, and then scale up.

Phosphate-Buffered Saline (Dulbecco Solution A; D-PBSA)
See Table 9.2 and Protocol 11.6.

Phosphate-Buffered Saline/EDTA, 10 mM (PE)
(1) Make up D-PBSA.
(2) Add EDTA disodium salt, 3.72 g/L, and stir.
(3) Dispense, autoclave, and store at room temperature.
(4) Dilute D-PBSA/EDTA 1:10 to give 1 mM for most applications or 1:2 (5 mM) for high chelating conditions (e.g., trypsinization of CaCo-2 cells).

Phytohemagglutinin (PHA)
(1) Prepare stock 500 μg/mL (100×) from lyophilized PHA by adding HBSS by syringe to an ampoule.
(2) Dispense into aliquots and store at −20°C.
(3) Dilute 1:100 for use.

SF12
Ham's F12 (*see* Table 9.3) with additional 2× Eagle's MEM essential amino acids and 1× nonessential amino acids but lacking thymidine, and with 10× folic acid concentration.

Sodium Citrate/Sodium Chloride See SSC
SSC 20× (Sodium Citrate/Sodium Chloride)
(1) Prepare concentrated SSC:

Trisodium citrate (dihydrate)	0.3 M	88.2 g
NaCl	3.0 M	175.3 g
Water		1000 mL

(2) Dilute to 1× or 2× as appropriate

Streptomycin Sulfate

(1) Take 2 mL from a bottle containing 100 mL of sterile HBSS, and add to a 1-g vial of streptomycin.

(2) When streptomycin has dissolved, return the 2 mL to the 98 mL of HBSS.

(3) Dilute 1:200 for use. The final concentration should be 50 μg/mL.

Trypsin

2.5% w/v in 0.85% (0.14 M) NaCl

Trypsin solutions can be bought commercially. Alternatively:

(1) Prepare a 2.5% solution:

Trypsin (e.g., Difco 1:250) .... 25 g
NaCl, 0.85% .... 1 L

(2) Stir trypsin for 1 h at room temperature or 10 h at 4°C. If the trypsin does not dissolve completely, clarify it by filtration through Whatman No. 1 filter paper.

(3) Sterilize by filtration.

(4) Dispense into 10- to 20-mL aliquots and store at −20°C.

(5) Thaw and dilute 1:10 in D-PBSA or PE for use.

(6) Store diluted trypsin at 4°C for a maximum of 3 weeks.

Note. Trypsin is available as a crude (e.g., Difco 1:250) or purified (e.g., Worthington or Sigma 3× recrystallized) preparation. Crude preparations contain several other proteases that may be important in cell dissociation, but may also be harmful to more sensitive cells. The usual practice is to use crude trypsin, unless the viability of the cells is diminished or reduced growth is observed, in which case purified trypsin may be used. Pure trypsin has a higher specific activity and should therefore be used at a proportionally lower concentration (e.g., 0.01 or 0.05%). Check for mycoplasma when preparing from raw trypsin.

Trypsin/EDTA

See Trypsin, Step (5)

Trypsin, Versene, Phosphate (TVP)

Trypsin (Difco 1:250) .... 25 mg (or 1 mL 2.5%)
Phosphate-buffered saline, D-PBSA .... 98 mL
Disodium EDTA (2H₂O) .... 37 mg
Chick serum (MP Biomedicals) .... 1 mL

(1) Mix D-PBSA and EDTA, autoclave the mixture, and store it at room temperature.

(2) Add chick serum and trypsin before use. If powdered trypsin is used, sterilize it by filtration before adding the serum.

(3) Dispense the solution into aliquots and store at −20°C.

Tryptose Phosphate Broth

(1) Prepare at 10% in HBSS

Tryptose phosphate (Difco) .... 100 g
Hanks' BSS .... 1000 mL

(2) Stir until dissolved.

(3) Dispense into aliquots of 100 mL and sterilize in the autoclave.

(4) Store at room temperature

(5) Dilute 1:100 (final concentration, 0.1%) for use.

Tyrode's Solution

	g
NaCl	8.00
KCl	0.20
CaCl₂	0.20
Mg₂Cl₂ · 6H₂O	0.10
NaH₂PO₄· H₂O	0.05
Glucose	1.00
UPW, up to	1 L
Gas phase	Air

Versene

See EDTA.

Viability Stain

See Naphthalene Black.
Trypan Blue is available from most tissue culture media suppliers (*see* Appendix II).

Vitamins

Detailed in media recipes (*see* Tables 9.3, 10.1, 10.2) and available commercially as 100× concentrates.

(1) Make up individually as 1000–10,000× stocks and combine as required to make up a 100× concentrate.

(2) Sterilize by filtration.

(3) Store at −20°C in the dark.

APPENDIX II

Sources of Equipment and Materials

The number of suppliers of reagents, equipment and materials used in cell culture is now so extensive that all possible sources are not given, but merely some examples of suppliers for each product. The suppliers' addresses are to be found in Appendix III. Additional suppliers are listed in the BiosupplyNet Source Book, Cold Spring Harbor Laboratory Press, and at *www.biosupplynet.com*, *www.biocompare.com*, *http://informagen.com/*, *http://www.martex.co.uk/laboratory-supplies/index.htm*, *http://www.biosciencetechnology.com/*, *www.cato.com/biotech/bio-prod.html*, *www.ispex.ca/naccbiologicals.html*, *www.biospace.com/service_and_supplier.cfm.*, *www.sciquest.com*, *http://www.coe.montana.edu/che/CompAlph.htm#A*.

Item	*Supplier*
ABI Prism 7000 real-time video system	Applied Biosystems
Accuspin tubes	Sigma Diagnostics
Accutase, Accumax	Sigma; TCS; Upstate Biotechnology
Acepromazine	Henry Schein
Acetonitrile	TAAB
Acetylcholine	Sigma
Actin, smooth muscle, alpha, antibody	DAKO
Activated charcoal	Sigma
Adenine (hydrochloride)	Sigma
Adipocytes	*See* Specialized cells
Agar	Invitrogen; BD Biosciences (Difco); Oxoid
Agar EM embedding kit	Plano; Polysciences
Agarose	Cambrex; FMC Bioproducts; Sigma
Agarose urea gel	Applied Biosystems
Agitating water bath	Baker; Grant; *see also* water baths
Air brush	Badger Air Brush Co.
Air velocity meters	*See* Anemometers
Albumin (BSA)	Sigma
Albumin antibody, rabbit anti-rat	MP Biomedicals
Albumin, rat	MP Biomedicals
Alcohol-resistant markers	Radleys
Alexa Fluor 488 goat anti-mouse IgG	Invitrogen; Molecular Probes
Alginate	ISP; NovaMatrix
Alkaline phosphatase assay materials	Sigma
Alkaline phosphatase reaction kit (Red)	Vector
Alpha medium	*See* Media
Amino acids	JT Baker; Merck; Sigma
Amino acids, essential	*See* MEM Essential Amino Acids
Aminoethanol	*See* ethanolamine
Aminopropyltrie-thoxysilane	Fluka
Aminopterin	Sigma
Amphotericin B (Fungizone)	Cambrex (BioWhittaker); Invitrogen; MP Biomedicals; PAA; Sigma
Ampoules, glass	Wheaton
Ampoules, plastic	Alpha Laboratories; CLP; Corning; Fisher Scientific; Greiner; Nalge Nunc
Anemometers	Technika; TSI; *see also* laminar-flow hoods

Anesthesia device, veterinary grade	SurgiVet
Angiocath (plastic catheter) 20 gauge	BD Biosciences
Anti-mouse IgG ABC alkaline phosphatase kit	Vector
Anti-rabbit IgG ABC peroxidase kit	Vector
Antibiotics	*See* Media
Antibodies	Amersham; BD Biosciences; Biopool; DAKO; Dianova; Invitrogen; MP Biomedicals; Peprotech; R&D Systems; Serotec; Roche Diagnostics; Santa Cruz Biotechnology; Sigma; Upstate Biotechnology; Vector
Antifoams	Bayer; Dow-Corning; Merck; Sigma
APAAP complex	DAKO; Vector
Apoptosis inhibitor	MP Biomedicals
Aprotinin (Trasylol)	Bayer; Serologicals; Sigma
Arginine HCl	*See* Chemicals
Ascorbic-2-phosphate	*See* Chemicals
Ascorbic acid	Sigma; Wako
Asparagine	*See* Chemicals
Aspartic acid	*See* Chemicals
Aspiration pipettes, unplugged	Corning Costar
Aspirators (reservoirs)	Bel-Art; Bibby Sterilin; Camlab; Corning; Integra; Kimble-Kontes; Techmate
Attachment factors	*See* Matrix
Authentication of cell lines	ATCC; Bioreliance; Cell Culture Characterization Services; Cellmark; DSMZ; ECACC; LGC Promochem; Orchid Cellmark; *see also* DNA profiling
Autoclavable bags and film	Altec; Applied Scientific; Bibby Sterilin; Elkay; Jencons; Lab Safety Supply; KNF; Portex
Autoclavable media	JRH Biosciences; Mediatech
Autoclavable nylon film and bags	Applied Scientific; KNF Corp.; Portex
Autoclaves	Ace; Astell; Bennet; Global; Integra; LTE; Precision Scientific; SP Industries; Steris; VWR
Automatic dispensers	Corning; Genetic Research Instr.; Gilson; Jencons; Matrix Technologies; Michael Smith; MP Biomedicals; Robbins Scientific
Automatic pipette	Alpha; BD Biosciences; Roche Diagnostics; Corning; Gilson; MP Biomedicals; Jencons; Labsystems; Rainin
Automatic pipette plugger	*See* Pipette plugger

Autoradiographic emulsion	*See* Emulsion for autoradiography
Autosequencer, ALFred	Amersham Biosciences
Avidin-biotinylated IgG kit	Dako Diagnostics
Bacteriological grade culture dish	BD Biosciences; Fisher
Bacteriocides for water baths	Guest Medical; Henry Schein
Bactopeptone	BD Biosciences (Difco); Invitrogen
Balb/c 3T3 A31 cells	ATCC
BALB/c nu/nu mice	Nippon CLEA
Basic fibroblast growth factor (bFGF)	*See* Growth factors
BEGM (bronchial epithelium growth medium)	Cambrex (Clonetics; BioWhittaker)
Benchcote	Whatman
Betadine	Bruce Medical
Bijou containers	Bibby Sterilin
Biochemicals	Calbiochem; Fluka; Merck; Pierce; Roche Diagnostics; Sigma; U.S. Biochemical
Biocoat multiwell plates	BD Biosciences
Bioelectrical monitoring	*See* Biosensors
Biohazard bags	*See* Autoclavable nylon film and bags
Biohazard safety cabinets	*See* Microbiological safety cabinets
Biopsy needles	Ranfac; Stille
Bioreactors	Alfa Laval; Bellco; Biotech Instruments; Biovest; Braun; Cellon; Charles River; Corning; Genetic Research Instrumentation; Global Medical; James Glass; KBI; New Brunswick Scientific; PerkinElmer; Sartorius; Sythecon; Vivascience; Wave Biotech
Biosensors	Analytical Technologies; Applied Biophysics; CellStat; Nova Biomedical; YSI
Biotain-MPS serum substitute	Cambrex (BioWhittaker)
Biotin	Sigma
Biotinylated anti-mouse IgG	Vector
Biotinylated IgG	Sigma; Vector
Blood urea nitrogen reagents	Stanbio Labs
BM-cycline	Roche Diagnostics
BMON software	Ingenieurbüro Jäckel
BMP-2, bone morphogenetic protein	Wyeth
Bone marrow	Cambrex
Bone morphogenic proteins (BMPs)	*See* Growth factors
Borate buffer	Sigma

Bottles	Applied Scientific; BD Biosciences; Bel-Art; Bellco; Bibby Sterilin; Caisson; Camlab; Corning; Fisher; Integra; Kimble Kontes; Nalge Nunc; Polytech; Radleys; Schott; Techmate; VWR
Bottle-top dispensers	Brand; Jencons; Fisher; Polytech
Bouin's fixative	Sigma
Bovine hypothalamus	Pel-Freez
Bovine pituitaries	Pel-Freez
Bovine pituitary extract	Cambrex (Clonetics); Cascade; Hammond Cell Technology; Invitrogen; PromoCell
Bovine serum (selected batch)	Gemini; Invitrogen; MP Biomedicals; Sigma; SeraLab
Bovine serum albumin	Amersham; Bayer; BioSource; BD Biosciences; Calbiochem; Cambrex (Clonetics); MP Biomedicals; Invitrogen; Intergen; Sigma
BrdU DNA cell labeling kit	Invitrogen (Invitrogen); Promega; Sigma
Bromodeoxyuridine	Sigma; *see also* BrdU cell labeling kit
Bronchial epithelial cells	*See* Specialized cells
Broths	*See* Nutrient broths
BSA	*See* Bovine serum albumin
BSS, Hanks', Earle's, etc.	*See* Media
BUdR	Sigma; *see also* BrdU cell labeling kit
Buffered formalin	Fisher
Butanedione monoxime (BDM)	Sigma
CaCl₂	*See* Chemicals
Cacodylate buffer	Plano; TAAB
Calcium assay kit	Sigma
Calcium binding reagent	Sigma
Calcium chloride	*See* Chemicals
Calcium/phosphorus combined standard	Sigma
Calf serum	*See* Serum
Calf thymus DNA Type I	Sigma
Cameras	Leica; Olympus; Nikon; Zeiss (*see also* CCD and Digital Cameras)
Cannulae	BD Biosciences; Roboz;
Capillary bed perfusion	*See* Hollow fiber perfusion culture
Carbodiimide, 1-ethyl-3-(dimethylaminopropyl)-EDC	Pierce
Carbon dioxide	Air Products; Cryoservice; Messer; Taylor-Wharton
Carboxymethyl-cellulose (CMC)	Fisher; Merck
Catheter extension kit, Interlink	Baxter HealthCare
Catheters	BD Biosciences
CCD cameras	BioWorld; Dage-MTI; Hamamatsu; Leica; Scanalytics; Sony. *See also* Microscopes

CD105 (555690), mouse anti-human	BD Biosciences (Pharmingen)
CD31-PE (555446) antibody	BD Biosciences (Pharmingen)
CD34-APC (555824) antibody	BD Biosciences (Pharmingen)
CD44-FITC (555478) antibody	BD Biosciences (Pharmingen)
CD45-APC (555485) antibody	BD Biosciences (Pharmingen)
cDNA libraries	Clontech
Cell banks	ATCC; Coriell Cell Repositories (CCR); ECACC; DSMZ; HSRRB; JCRB; Riken
Cell counters	Beckman Coulter; New Brunswick; Schärfe
Cell counting solutions	Beckman Coulter; Schärfe
Cell culture dishes, flasks, and plates	*See* Culture dishes, flasks, and plates
Cell culture inserts	*See* Filter well inserts
Cell dissociation agents	*See* individual enzymes and Trypsin replacements
Cell filter	BD Biosciences; Miltenyi
Cell line 5637	DSMZ
Cell line characterization	*See* Authentication of cell lines; DNA fingerprinting and profiling
Cell line databanks	ATCC; Coriell; DSMZ; ECACC; HSRRB; ICLC
Cell lines	*See* Cell banks
Cell proliferation assays	Lumitech (Cambrex); Promega; Upstate
Cell Quest acquisition software	BD Biosciences
Cell scraper	Applied Scientific; Bel-Art; BD Biosciences; Corning; Nalge Nunc; Techmate; Techno Plastic
Cell separation	Accurate Chemical & Scientific; Applied Immune Sciences; BD Biosciences; BioCarta; Biochrom; Dynal; Kendro; Miltenyi; PerkinElmer; Stem Cell Technologies
Cell sizing	Beckman Coulter; Schärfe
Cell strainers	BD Biosciences; Dynal; Miltenyi Biotec
CellFlo	Spectrum Laboratories
Cellgro	*See* Mediatech
Centrifugal elutriator	Beckman Coulter
Centrifuge tubes	Alpha; Bibby Sterilin; BD Biosciences (Falcon); Corning; Du Pont; Eppendorf; Greiner; Omnilab; Nalge Nunc; Techno Plastic Products
Centrifuges	Beckman; DuPont (Sorval); Fisher; Kendro; Thermo-IEC; Life Sciences International
Ceramic rods and scaffolds	Zimmer

Ceramics, coral-based	Interpore
Chamber slides	Applied Scientific; Bayer; BD Biosciences; Bibby Sterilin; Heraeus; Metachem; Nalge Nunc; Stem Cell Technologies
CHAPS buffer	Calbiochem
Chemicals	Calbiochem; Fisher; JT Baker; Merck; Pierce; Research Organics; Roche; Scientific & Chemical Supplies; Sigma; USB; Wako
Chick embryo extract	Accurate Chemical & Scientific; Invitrogen
Chicken plasma	Invitrogen; MP Biomedicals
Chicken serum	Invitrogen; MP Biomedicals; TCS Biologicals
Chloramphenicol	Sigma
Chloroform	*See* Chemicals
Chloroform isoamyl alcohol	Biogene
Chloros	Hays Chemical Distribution (*see also* Disinfectants)
Cholera toxin	EMD; List Biological; Merck; Sigma
Choline chloride	Sigma
Chondrocytes	*See* Specialized cells
Chondroitin sulfate	Sigma
Chromosome analysis	Cell Culture Characterization Services
Chromosome paints	Applied Imaging; Cambio; Vysis
Ciprofloxacin	Bayer
Clidox disinfectant	Tecniplast (Indulab)
Cloning disks, 3 mm	Sigma
Cloning rings/cylinders	BelArt; Bellco; Fisher; Scientific Laboratory Supplies
Clorox	Polyscience (*see also* Disinfectants)
Closed-circuit television (CCTV)	Dage-MTI; Hamamatsu; Leica (*see also* CCD cameras; microscopes)
CMC	*See* Carboxymethylcellulose
CMF: Ca^{2+}, Mg^{2+}-free EBSS (Earle's Balanced Salt Solution)	Invitrogen
CMF-PBS	*See* Media
CMRL-1066	*See* Media
CO_2	Air Products; Cryoservice; Taylor-Wharton
CO_2 automatic change-over unit for cylinders	Air Products and Chemicals Inc.; Gow-Mac; Lab Impex; Nuaire; Thermo-Shandon
CO_2 controllers	Air Products; Gow-Mac; Lab-Line; Lab Impex; Therma Electron (Forma; Hotpak)
CO_2 incubators	Barnstead-Thermolyne; Boro Labs; Camlab; Fisher; Heinecke; Kendro; Lab-Impex; Lab-Line; LEEC; Memmert; MP Biomedicals; Napco; New Brunswick Scientific; NuAire; Omnilab; Precision Scientific; Sanyo Gallenkamp; SP
	Industries; Thermo-Electron; Triple Red
CO_2-permeable caps	*See* Culture dishes, flasks, and plates
Colcemid	Sigma
Collagen	BD Biosciences; Biocolor; Biodesign; Biomedical Technologies; Biosource International; Cambridge Biosciences; Cellon; Chondrex; Cohesion Technologies; Collagen Corporation; CR Bard; Davol; Invitrogen; Inamed; Insmed; KeLab; Lab Vision; MP Biomedicals; Roche; Seikagaku; Sigma; Stratech; Universal Biologicals
Collagen sponge, Avitene Ultrafoam	CR Bard; Davol; McGhan Medical
Collagen type I antibody, rabbit anti-human	Biodesign
Collagenase	Chondrex; Invitrogen; Lorne; Roche; Serva; Sigma; Worthington
Collagen-coated culture dishes, flasks, plates	BD Biosciences; Corning; Iwaki Glass
Colony counters	BioWorld; BioDu Pont (NEN Life Sciences); Cole-Parmer; Don Whitley; Perceptive Instruments; PerkinElmer; Synbiosis
Colony ring marker	Nikon
Combi Ring Dish	Renner; Germany
Conductivity meters	Corning; Mettler Toledo; Quadrachem; Technika; Thermo Electron
Confocal microscope	Biotech Instruments; Leica; Molecular Dynamics; Nikon; Zeiss
Conical centrifuge tube	*See* Centrifuge tubes
Continuous roller pump	Cole-Parmer
Controlled-atmosphere chamber	Bellco; Vineland; NJ
Controlled-rate coolers for liquid N_2 freezing	Messer; Nalge Nunc; Planer; Thermo Electron; Statebourne; Taylor-Wharton
Coomassie Blue R	Sigma
Copper sulfate	*See* Chemicals
Cornwall syringe	Popper; *see also* Chempette from Cole-Parmer
Coverslip mounts	*See* Mountants
Coverslips, glass	Bayer; Corning; Fisher; Invitrogen; Wheaton
Coverslips, plastic	Bayer; Corning; MP Biomedicals; Lux; Invitrogen
CPSR serum substitute	Sigma
Cryofreezer	*See* Controlled-rate coolers *and* Freezers, liquid nitrogen
Cryogenic freezers	*See* Freezers, Liquid nitrogen
Cryogenic vials	*See* Ampoules

Cryolabels	Computer Imprintable Label Systems; GA International; Triple Red
Cryomarkers	GA International
Cryopreservation medium	MP Biomedicals; Invitrogen; JRH Biosciences; Sigma
Cryoprotective gloves	Taylor-Wharton; Jencons
Cryotubes	*See* Ampoules
Cryovials	*See* Ampoules
Crystal violet	Merck; Fisher
Culture bags	Cell Genix; Du Pont; Metabios; MP Biomedicals; PAW BioScience Products (*see also* Media Bags)
Culture dishes, flasks, and plates	BD Biosciences (Falcon); Bibby Sterilin; Corning (Costar); Fisher; Greiner; Invitrogen; Iwaki; MP Biomedicals (Lux); Nalge Nunc; Sarstedt; TPP
Culture media, salt solutions, etc.	*See* Media
Culture slides	*See* Chamber slides
Culture tubes	*See* Culture dishes, flasks, and plates.
Culture vessels	*See* Culture dishes, flasks, and plates.
Curved forceps	*See* Surgical instruments
Custom media	Caisson; *see also* Media
Cy5 fluorescent-labeled primers	Amersham Biosciences
Cyclic AMP	Sigma
Cyclosporine	Novartis; Sigma
Cyprofloxacin	*See* Ciprofloxacin
Cysteine	Sigma
Cysteine hydrochloride hydrate	*See* Chemicals
Cystine	*See* Chemicals
Cytobuckets	Thermo Electron Corporation
Cytocentrifuge	Bayer; CSP; Electron Microscopy Sciences; Thermo Electron; Sakura Finetek; Wescor
Cytofectin GSV	Glen Research
Cytogenetc analysis	Cell Culture Characterization Services
Cytokeratin antibodies	DAKO; Lab Vision; Santa Cruz Biotechnology; Zymed
Cytokines	*See* Growth factors
Cytometer	*See flow cytometer; scanning cytometer*
Cytoseal TM 60 fluorescence mountant	Microm; Stephens Scientific; VWR
Cytotoxicity assays	BioReliance (Invitrogen); MatTek; Promega; SkinEthic.
Cytotoxicity testing	CellStat; *see also* Cytotoxicity assays
Dacron vascular graft	Bard
Daigo's T medium	Wako
dbcAMP	Sigma
DEAE dextran	Amersham Biosciences; Bio-Rad

Decontamination (of equipment and facilities)	Anachem; Steris; *see also* Disinfectants
Deionizers	Barnstead Thermolyne; Bellco; Corning; Dow Corning; Elga; High-Q; Millipore; Purite; U.S. Filter; Vivendi Water Systems; VWR
Densitometers	Beckman Instruments; Pall Gelman Sciences; Gilford Instruments; Helena; Molecular Dynamics
Density marker beads	Amersham Biosciences
Density media	Amersham; Genetic Research Instr.; MP Biomedicals; Nycomed; Robbins Scientific; Sigma
Density meter	Mettler Toledo; Parr
Deoxyribonuclease	Sigma
Dermal puncher	Miltey
Dermatome	Stortz Instruments (*see also* Tissue slicers)
Desmin, mouse, antibody	Sigma
Detergents	Alconox; Calbiochem; Decon; MP Biomedicals; Pierce
Dexamethasone	Sigma; Merck
Dextran	Calbiochem; Fisher; Sigma
Dextrose	Fisher
Diacetyl fluorescein	Fisher; Sigma
Dialysis cassette	Pierce
Dialysis tubing	Cole-Parmer; Chemicon; Pierce; Serva; Spectrum
Diaminobenzidine	Sigma; Vector
Diethylpyrocarbonate (DEPC)	Sigma
Digital cameras	Canon; Kodak; Leica; Nikon; Olympus; Polaroid (*see also* CCD cameras; Microscopes)
Dimethylethylformamide	*See* Chemicals
Dimethylmethylene blue	Polysciences; Sigma
Dimethyl sulfoxide (DMSO)	ATCC; JT Baker; Merck; Sigma
Dipeptidyl peptidase IV (CD26)	Biosource; Neomarkers
Disk filter assembly for sterilization	*See* Filters
Dishes	*See* Culture dishes, flasks, and plates
Disinfectants	Anachem; BioMedical Products; DuPont Animal Health; Guest Medical; Hays Chemical Distribution; Johnson & Johnson; Lab Impex; Lab Safety Supply; Markson LabSales; MP Biomedicals; National Chemicals; Polyscience; Sigma; Steris; Tecniplast; Thomas; VWR Scientific
Disodium hydrogen orthophosphate	*See* Chemicals

Disodium hydrogen phosphate	*See* Chemicals	EHS matrix (Matrigel)	BD Biosciences
Dispase	BD Biosciences; Invitrogen; Roche	Elastase	Roche Diagnostics; Sigma
		Electronic cell counter	*See* cell counters
Dispensers, liquid	Accuramatic; Barnstead; Corning; Gilson; Jencons; Lawson Mardon Wheaton; Matrix Technologies; Mettler Toledo; Michael Smith; MP Biomedicals; Polytech; Popper; Robbins; Tecan; Thermo Electron; Zinsser	Electronic thermometers	Comark; Cole-Parmer; Fisher; Harvard; Grant; Labox; Omega; Pierce; Thermo Electron
		Electron microscopy	Electron Microscopy Services; TAAB; Structure Probe;
		Electrophoresis	Amersham Biosciences; Anachem; Bio-Rad; Invitrogen; Haake; Innovative Chemistry; Thermo Electron
Dissecting instruments	Cole-Parmer; EMS; Fine Scientific Tools; Fisher; Roboz; Swann-Morton; VWR	ELISA kit for collagen type I	Chondrex
Dissection microscope	Daigger; Olympus; Leica; Nikon; Zeiss	ELISA assays	Assay Designs; R & D Systems
		ELISA plate readers	*See* Plate readers
Dithiothreitol	Sigma	Emulsion for autoradiography	Amersham Biosciences
DMEM with stabilized glutamine	Invitrogen; Biochrom	Endofree Maxi Kit	Qiagen
DMEM/F12, 50/50	Biochrom; Cellgro-Mediatech; Invitrogen; Sigma	Endothelial cells	*See* Specialized cells
		Endothelin 1, human, porcine (ET-1)	Sigma
DNA fingerprinting & profiling	Anglia DNA Bioservices; ATCC; Cellmark UK; DNA Bioscience; DNA Solutions; ECACC; Laboratory Corporation of America; LGC Promochem; Orchid-Cellmark; Verilabs Biosciences; *see also* Authentication	Entellan	Merck
		Epidermal growth factor	*See* growth factors
		Epi-Life serum-free medium	Sigma
		Epinephrine	Sigma
		Epithelial membrane antigen (EMA) antibody	Dako
DNA Polymerase Premix	Takara Bio	Epon	Merck; LR-White; Roth
DNA preparation kit	Qiagen	Erlenmeyer flasks, tissue culture grade	*See* Glassware
DNA sequences	Geron		
DNA stains	Hoechst; Molecular Probes	Estradiol	Sigma
DNA templates	Ambion	Ethanol	*See* Chemicals
DNase	Lorne; Serologicals; Sigma; Worthington	Ethanolamine	Sigma
		Ethidium bromide	Sigma
Donkey anti-rabbit IgG	Jackson ImmunoResearch	Ethylene diamine tetraacetate disodium salt	*See* EDTA
Donor calf serum (DCS)	*See* Serum		
Donor horse serum	Gemini		
DPX, Permount	*See* Stains	Eukitt	Fluka
Dulbecco's modified Eagle's medium (DMEM)	*See* Media	Excell-900 serum substitute	JRH Biosciences
		Ex-cyte serum substitute	Bayer
		Extracellular matrix	*See* Matrix
Duran glass bottles	Camlab; Fisher; Schott	F12	*See* Media
ECGS	*See* Growth factors	F12:DMEM	*See* Media
ECM	*See* matrix	F12H medium	*See* Media
Ecoscint	BS & S; National Diagnostics; Perkin Elmer	FACS	BD Biosciences; Beckman-Coulter
EDC, 1-ethyl-3-(3-dimethylaminopropyl) carbodiimide	Pierce	FACSCalibur	BD Biosciences
		Factor VIII antibody	Roche Diagnostics
		Falcon tissue culture flasks	BD Biosciences
EDTA	JT Baker; Merck; Sigma	Fast Green	Sigma
EDTA, sterile	*See* Media	Fast red TR salt	SERVA
Edwards High Vacuum Grease	BOC Edwards	Fast Violet capsule	Sigma
		Fast-Track software Wavemaker32 ver.6.6	Instron
EGF	*See* Growth factors		
Egg incubators	G.Q.F. Manufacturing	FBS	*See* Serum
EGTA	Sigma	FCS	*See* Serum
Ehrlenmeyer flasks	*See* Glassware	$Fe_2SO_47H_2O$	*See* Chemicals

Fermentors	*See* Bioreactors
Ferrous sulfate	*See* Chemicals
Fetal bovine serum (FBS)	*See* Serum
FGF, human recombinant	*See* Growth factors
Fiber-optic illumination	Fischer Scientific
Fibrin sealant, Tisseel VH	Baxter
Fibroblast growth factor	*See* Growth factors
Fibroblasts	*See* Specialized cells
Fibronectin	Amersham; BD Biosciences; Biomedical Technologies; Biosource International; Invitrogen; Lab Vision; R & D; Sigma; Stratech; *See also* matrix
Fibronectin/collagen/BSA	Biosource International
Ficoll	Amersham Biosciences; MP Biomedicals; Sigma
Ficoll-Hypaque	Amersham Biosciences; Cambrex; MP Biomedicals; Nycomed; Sigma
Filter bottom multiwell plates	Nalge Nunc; Techmate; Whatman
Filter holders	*See* Filters
Filter well inserts	BD Biosciences; Bibby Sterilin; Corning; Integra; Millipore; Nalge Nunc
Filters, mesh or gauze	*See* Mesh filters
Filters, sterilization	BD Biosciences; Bibby Sterilin; Corning; Millipore; Nalge Nunc; Omnilab; Pall Gelman; Sartorius; Techno Plastic; VWR; Whatman.
Filtration equipment	*See* Filters
Fish serum	East Coast Biologics
FITC-avidin	Vector Laboratories
FITC-conjugated secondary antibody	*See* Antibodies
Fixed-bed bioreactors	New Brunswick
Flasks	*See* Culture dishes, flasks, and plates
FlexiPERM slide™ (8 chambers)	Sartorius
Flow cytometers	Agilent; BD Biosciences; Beckman Coulter; Genetic Research; Guava Technologies
Flow meters	*See* Gas flow meters
Fluidized-bed bioreactors	Amersham Biosciences
Fluorescence bleaching inhibitor	*See* Fluorescence fade retardant
Fluorescence fade retardant	Calbiochem; Citifluor; Vector
Fluorescence *in situ* hybridization (FISH)	Cell Culture Characterization Services (for reagents *see* Chromosome paints)
Fluorescence mountants	Biomedia; Citifluor; Microm; Stephens Scientific; Vector; VWR
Fluorescence-activated cell sorter (FACS)	BD Biosciences
FluorSave, fluorescent mounting medium	Calbiochem
Folic acid	Sigma
Forceps	*See* Dissecting instruments
Formaldehyde	*See* Chemicals
Formamide	Merck
Formic acid	*See* Chemicals
Formol saline, buffered	Sigma
Four-well plates	*See* Culture dishes, flasks, and plates
Fragment Manager software	Amersham Biosciences
Frame grabber card (time-lapse)	Scion
Freezer racks and canes	*See* Freezers, liquid nitrogen
Freezers, −20°C	Barnstead; Fisher; New Brunswick Scientific; local discount warehouses
Freezers, −70°C	Barnstead; Fisher; New Brunswick; Nuaire; Precision Scientific; Revco; Thermo Electron
Freezers, liquid nitrogen	Aire Liquide; Barnstead-Thermolyne; Boro Labs; Chart Biomed; Cryomed; Genetic Research; Jencons; Messer Cryotherm; Planer; Statebourne; Taylor-Wharton; Thermo Electron; VWR.
Freezing medium	*See* Cryopreservation medium
FuGENE 6	Roche
Fungicides	BioWhittaker; MP Biomedicals; Invitrogen; Sigma
Fungizone	*See* Amphotericin B
Gas blenders	Gow-Mac; Signal
Gas mixers	*See* Gas blenders
Gases	Air Products; British Oxygen; Cryoservice; Matheson Gas; Messer; Taylor-Wharton
Gas flow meters	Muis Controls; Omega Engineering; Titan Enterprises (*see also* Gas blendors)
Gas generators	Bioquell; Peak Scientific; Sartec; Texol
Gas-permeable caps	*See* Culture dishes, flasks, and plates
Gas-permeable tissue culture bags	*See* Media bags and Culture bags
Gauze	*See* Mesh filters and gauze
Gel/Mount water-based mounting medium	Biomeda
Geneticin	Invitrogen
Gentamicin	Gemini; Invitrogen; Sigma
German glass coverslips	Carolina Biological Supplies
Giemsa stain	Fisher; Merck; TCS Biologicals; Sigma
Gilson Pipetteman	Bellco; Corning; Gilson; Schott; Wheaton
Glass adhesive/cement, UV curing	Summers Optical

Glass coverslips	Chance Propper; Wheaton	Ham's F12	*See* Media
Glass fiber filters	Millipore; Pall; Whatman	Ham's F12:DMEM	*See* Media
Glass universal containers	Camlab	Hanks' balanced salt	*See* Media
Glassware	Bellco; Corning; Kimble-Kontes;	solution (HBSS)	
	Schott; Wheaton	Harris hematoxylin	Fisher
Glassware washing	Burge; Lancer; Miele; Scientek;	HARV bioreactor	Synthecon
machine	Scientific Instrument Centre;	HCl	*See* Chemicals
	Steris Corp.	Heat-inactivated fetal	*See* Serum
Gloves	Ansell Medical; Applied	bovine serum (HIFBS)	
	Scientific; Cryomed; Johnson	Heating blocks	Genetic Research Instr.; Robbins;
	& Johnson Medical; Lab Safety		VP Scientific
	Supply; Radleys; Renco Corp.;	Hematoxylin	Merck; Sigma
	Safeskin; Sentinel Laboratories;	Hemocytometer	Fisher; Omnilab; Thermo
	Stoelting; Surgicon; VWR		Electron
	Scientific	Hen's eggs	Charles River Laboratories
Gloves, nitrile	Ansell; Cole-Parmer; Lab Safety	Heparan sulfate	Sigma
	Supplies; Radleys; Sentinel;	Heparan sulfate	Sigma
	Stoelting	proteoglycans (HSPGs)	
Glucagon	Bedford Laboratories	Heparin	Roche; Sigma
Glucose	*See* Chemicals	Hepatocytes	*See* Specialized cells
Glucose, sterile	*See* Media	HEPES	*See* Media
Glutamic acid	*See* Chemicals	HepG2 cells, human	*See* Cell banks
Glutamine	*See* Chemicals	hepatoma cell line	
Glutamine, sterile	*See* Media	Hexafluoro-2-propanol	Aldrich
Glutaraldehyde	Fisher Scientific; Plano; Roth;	(HFIP)	
	Sigma	HGF/SF, human,	*See* Growth factors
Glutathione	Sigma	recombinant	
Glycerol	Fisher; Mallinckrodt; Merck;	High Aspect Ratio Vessel	Synthecon
	Sigma; VWR	(HARV)	
β-Glycerophosphate	Invitrogen; Sigma	Histidine HC·lH$_2$O	*See* Chemicals
Glycine	Sigma	Histoclear	Fisher
Glycine-arginine-glycine-	Sigma	HistoGel™	Lab Storage Systems Inc.
aspartate-serine		Histological stains	Merck; MTR Scientific; Sigma
(GRGDS) peptide		Histopaque	Sigma
GM-CSF	*See* Growth factors	Histostain-SP kit	Zymed Laboratories Inc.; San
Goat serum	Sigma	(Streptavidin-	Francisco; CA; U.S.A
GRGDS peptise	Sigma	peroxidase)	
Growth Factor Reduced	BD Biosciences; Fisher Scientific;	HLF cells	Coriell Institute for Medical
Matrigel			Research
Growth factors and	Amersham; Amgen; Austral	Hoechst 33258 fluorescent	MP Biomedicals (complete kit);
cytokines	Biologicals; BD Biosciences;	mycoplasma stain	Polysciences; Sigma
	Biodesign; Biosource;	Hoechst 33342 DNA stain	Sigma
	Cambrex; Cambridge	Hollow fiber perfusion	Argos; Bellco; Biovest; FiberCell;
	Bioscience; Collaborative	culture	JM Separations;
	Research (BD Biosciences);		Spectrum
	Genzyme; Invitrogen; MP	Hormones	Sigma
	Biomedicals; Paesel & Lorei;	Horse serum	*See* Serum
	Pepro-Tech; PerkinElmer;	HT Tuffryn filters	Pall-Gelman
	Promega; R&D Systems;	*Htrt* DNA	Geron
	Roche; Serologicals; Sigma;	Human bone marrow	Clonetic-Poietics
	Stratech; Universal Biologicals;	stromal cells	
	Upstate Biotechnology	Human epidermal growth	*See* Growth factors
Guanidine hydrochloride,	Pierce	factor (hEGF)	
8M		Human fibronectin	BD Biosciences (Collaborative
Gyratory shaker	Boeker Scientific; New		Research)
	Brunswick	Human transferrin	Sigma
Halothane	Henry Schein	Hyaluronidase	Sigma; Worthington
Ham's F-10	*See* Media	Hydrochloric Acid	*See* Chemicals
Ham's F-10 (10 μM	Invitrogen	Hydrocortisone	Amersham Biosciences; Merck;
tyrosine), powder			Sigma: Upjohn

Hydrocortisone and analogs	Biosource; Merck; Sigma
Hydrogels	Nektar
Hydrogen peroxide (30% solution)	*See* Chemicals
Hydroxylamine hydrochloride	Pierce
Hydroxysuccinimide (NHS)	Pierce
Hypaque-Ficoll media	*See* Ficoll-Hypaque
Hypochlorite disinfectant	*See* disinfectants
Hypoclearing agent	Kodak
Hypoxanthine	Sigma
IBMX, 3-isobutyl-1-methylxanthine	Sigma
Identity testing	*See* Authentication of cell lines and DNA fingerprinting & profiling
IgG, rat	Sigma
Image analysis	Applied Imaging; Bio-Rad; Carl Zeiss; Hamamatsu; Imaging Associates; Imaging Research; Leica; Molecular Dynamics; Nikon; Nonlinear Dynamics; Perceptive Instruments; PerkinElmer; Scanalytics; Syngene
Immunoanalyzers	Agilent; Guava (*see also* Flow cytometers)
Immunoglobulin antibodies, rabbit anti-mouse	Dako Diagnostics
Immunoglobulins, mouse anti-rabbit	Dako
Incubators	ATR; Barnstead; Bellco; Binder; Boro Labs; Camlab; Fisher; Genetic Research Instr.; Global Medical; Harvard; Kendro; Infors; Lab-Line; LEEC; LMS; LTE; Memmert; MP Biomedicals; Napco; New Brunswick Scientific; Nuaire; Omnilab; Precision Scientific; Robbins; RS Biotech; Sanyo Gallenkamp; Scientific Instrument Centre; SciGene; SP Industries; Thermo Electron; Triple Red; VWR.
Indomethacin	Sigma
Injection ports, Interlink system	BD Biosciences
Inorganic salts	Merck
Insect culture media	BD Biosciences; BioSource; Cambrex; Hyclone; Invitrogen; MP Biomedicals; Novagen; Sigma
Instruments, dissecting	Cole-Parmer; EMS; Fine Scientific Tools; Fisher; Roboz

Insulin	Cambrex (Clonetics); CP Pharmaceuticals; Eli Lilly; Intergen; Invitrogen; Sigma
Insulin/transferrin/selenium (ITS)	BD Biosciences; Fisher Scientific; Invitrogen; Roche Diagnostics; Sigma
Interleukins	*See* Growth factors
Inverted microscope	Leica; Nikon; Olympus; Zeiss
Iodonitrotetrazolium (INT) violet	Sigma
Irgacure 2959 photoinitiator for photopatterning	Ciba Specialty Chemicals
Iscove's modified Dulbecco's medium (IMDM)	*See* Media
Isoenzyme electrophoresis analysis	ATCC; Cell Culture Characterization Services; DSMZ; ECACC
Isoenzyme electrophoresis analysis kit	Innovative Chemistry
Isoleucine	*See* Chemicals
Isometric force transducer (for muscle culture)	Radnoti
Isoproterenol hydrochloride	Sigma
Isotonic counting solution	Beckman Coulter; Scharfe; *see also* Media for PBS
Isotopes	*See* Radioisotopes
ITS Premix serum substitute	MP Biomedicals
Kanamycin	Invitrogen; MP Biomedicals; Sigma
Karyotyping	Cell Culture Characterization Services
KCl	*See* Chemicals
KDM	*See* Keratinocyte defined medium
Keratin 19 (K19) antibody	Dako
Keratinocyte defined medium (KDM)	Cambrex; Cascade; Invitrogen; PromoCell; Sigma
Keratinocyte growth medium (KGM)	Cambrex; Cascade; Invitrogen; PromoCell; Sigma
Keratinocytes	*See* Specialized cells
Ketamine	Henry Schein; Sigma
Ketanest	Parke-Davis
KGF, human recombinant	*See* Growth factors
KGM	*See* Keratinocyte growth medium
KH$_2$PO$_4$	*See* Chemicals
Kodak X-OMAT film	Eastman Kodak
Krebs-Henseleit solution	Sigma
Labels	Triple Red; *See also* Cryolabels
Laboratory glassware washers	*See* Glassware washing machines
Laboratory suppliers, general	Camlab; Cole-Parmer; Fisher; Scientific Instrument Centre; Thermo Electron; Triple Red; VWR

Methanol	*See* Chemicals
Methionine	*See* Chemicals
Methocel	Dow Corning
Methylcellulose	Dow Corning; Fluka; Fisher; Sigma; Stem Cell Technologies
Metrizamide	Nycomed
MF-319 photopatterning/photetching developer	Microchem Corp.
Mg²⁺ and Ca²⁺-free Hanks' balanced salt solution (HBSS)	*See* Media
MgCl₂6H₂O	*See* Chemicals
MgSO₄7H₂O	*See* Chemicals
Microarray analysis	Affymetrix; Agilent; Bio-Rad; Brinkmann; Chemicon; Imaging Research; Nonlinear Dynamics; Schleicher & Schuell; Stratagene; Tecan
Microbiological safety cabinets	Plas Labs; *See also* Laminar-flow hoods
Microcaps, microcapillary tubes	(Drummond) Thermo-Shandon; Fisher
Microcarriers	Amersham Biosciences; Bibby Sterilin; Bio-Rad; Nalge Nunc; Invitrogen; MP Biomedicals; Sigma; SoloHill; TCS
Microcentrifuge filter tubes	Corning
Microdensitometry	*See* Image analysis; Scanning cytometer
Micromanipulators	Brinkmann; Eppendorf; Leica; Nikon; Steris; Stoelting
Micropipettes	Alpha; Anachem; Applied Scientific; Bibby Sterilin; Biohit; Brinkmann; Corning; Elkay; Eppendorf; Genetic Research Instr.; Gilson; Jencons; Lawson Mardon Wheaton; Oxford; Roche; Thermo Electron
Microplate reader	*See* Plate readers
Microscope incubation chambers	Buck Scientific; Imaging Associates;
Microscope slide containers	Corning; Raymond Lamb; Raven; Richardsons
Microscope slides	Bellco; Berliner Glas; H.. V. Skan; Laboratory Sales; Lab-Tek; Microm; Nalge Nunc; Propper; Raymond Lamb; Richardsons; VWR
Microscopes	Carl Zeiss; Leica; Olympus; Nikon; Prior
Microscopes, fluorescence	*See* Microscopes
Microtitration equipment	Anachem; Berthold; BioRad; Biospec; BMG Labtech; Brand; Camlab; Corning; Dynex; Elkay; ESA; Genetic Research Instr.; Gilson; Integra; Invitrogen; Molecular Devices; Nalge Nunc; Nonlinear Dynamics; PerkinElmer; R &

	D Systems; Robbins; Tecan; Techmate; Thermo Electron; VWR; Whatman; Zinsser
Microtitration plate centrifugation	Beckman Coulter; Thermo Electron
Microtitration plate sealers	*See* Plate sealers
Microtitration plates	Applied Biosystems; *See also* Culture dishes, flasks, and plates
Microtitration plate homogenizer	*See* MiniBead Beater
Microtitration plate readers	*See* Plate readers
Microtitration plate reagent dispensers	Thermo Electron
Microtitration plates with removable wells	Invitrogen
Millex-SV Syringe-driven filter unit, pore size 5 um	Millipore
Millicell-HA filter holder inserts	Millipore
MiniBead Beater microplate homogenizer	Biospec; Daintree
Minimal essential medium	*See* Media
Mitomycin C	Roche Diagnostics; Sigma
Mixed capillary bed perfusion	*See* Hollow fiber perfusion culture
Mountants	Biomeda; Citifluor; Merck; Microm; RA Lamb; Statlab; Stephens Scientific; Vector; VWR
Movette pipette	Invitrogen
MSCGM	Cambrex
MTT	Sigma
MUC-1 (mucin) antibody	Chemicon
Multichamber slides	*See* Chamber slides
Multipoint pipettors	*See* Micropipettes and pipettors
Multiwell filter plates	Whatman
Multiwell plates	*See* Culture dishes, flasks, and plates
Mycoplasma Removal Agent (MRA)	MP Biomedicals
Mycoplasma detection kits	ATCC; Gen-Probe; Irvine Scientific; Metachem; MP Biomedicals
Mycoplasma testing	ATCC; Bionique; ECACC; DMSZ; Mycoplasma Experience
Mycostatin, Nystatin	BioWhittaker; MP Biomedicals; Invitrogen; Sigma
Mylar plastic plate sealers	MP Biomedicals; Invitrogen
***myo*-Inositol**	Sigma
Na₂HPO₄7H₂O	*See* Chemicals
NaCl	*See* Chemicals
NaCl, 0.9%,	Baxter
NaHCO₃	*See* Chemicals
NaHCO₃, (7.5%) sterile	*See* Media
NaOH	*See* Chemicals
Naphthalene black	R.A. Lamb (*see also* Stains)
Naphthol AS-MX Alkaline Solution	Sigma

Naphthol-AS-MX-phosphate	Serva
NASA bioreactor	Cellon; Synthecon
Neck plug-controlled rate cooler	Taylor-Wharton
Needles (for syringes)	*See* syringes
Neutral protease (Dispase)	Roche Diagnostics
Neutral Red	TCS Biologicals
NGF	*See* growth factors
Niacinamide	Sigma
Nigrosin	Sigma; TCS Biologicals
Nile Red	Sigma
Nitex nylon filter, 100-μm	Tekmar-Dohrmann (*See also* Nylon mesh)
Nitrile gloves	Ansell; Cole-Parmer; Lab Safety Supplies; Radleys; Sentinel; Stoelting
Nitrogen freezers	*See* Freezers; liquid nitrogen
Nitrophenol standard solution	Sigma
Nitrophenyl phosphate	Sigma
Nonessential amino acids	Invitrogen; MP Biomedicals; Sigma
Normal goat serum	Vector
Nuclear Fast Red	Sigma
Nucleosides	Sigma
Nude mice	Charles River; Nippon CLEA
Nutridoma serum substitute	Roche Diagnostics
Nutrient agar plates	Oxoid
Nutrient broths	BD Biosciences (Difco); Invitrogen; MP Biomedicals; TCS
Nyaflo membranes	Gelman Sciences
Nylon film	Applied Scientific; Buck Scientific; KNF Corp.; Portex; Portland Plastics; Roth
Nylon mesh filters and gauze	*See* Mesh filters
Nystatin	Sigma; Invitrogen
OCG 825–835, 934, photopatterning/photoetching developers	Ciba Specialty Chemicals
Oncostatin M	*See* Growth factors
Opticell culture chamber	BioCrystal
Opti-MEM, reduced serum medium	Invitrogen
Orbital shaker	Bellco
Organ baths	Radnoti
Organ culture grids and dishes	BD Biosciences
Organotypic culture	ACM-Biotech; MatTek; Minucells; SkinEthic
ORIGEN™ DMSO freeze medium	Origen Biomedical
Osmium tetroxide	Plano; Roth; Sigma
Osmometer	Advanced Instruments; Analytical @ Technologies; Gonotec; Nova Biomedical; Wescor

Osteogenic protein 1 (OP-1; BMP-7)	*See* Growth factors
Ovens	Astell; Barnstead; Camlab; Dynalab; Harvard; Kendro; LEEC; LTE; Memmert; Precision Scientific; Scientific Instrument Centre; Thermo Electron
Oxygen probe	Microelectrodes
p53 antibodies: D0-1, PAb421, PAb240.	Oncogene Science
Packaging, sterile	*See* Sterile packaging
Packaging, transportation	*See* Sample containers for transportation
Pancreatic elastase	Sigma
Pancreatin	Invitrogen; Sigma
Panserin™ 401 serum free medium	CoaChrom
Pantothenic acid	Sigma
Papain	Worthington; Sigma
Paraformaldehyde	Electron Microscopy Sciences; Merck; Sigma
Pastette	Alpha Laboratories; Bibby Sterilin; Elkay; Polytech; Richardsons
PBS tablets	Oxoid; *See also* Media for powder or solution
Pen/Strep Fungizone mix	*See* Media
Penicillin	*See* Media
Penicillin/streptomycin	*See* Media
Pentobarbital	Abbott
Pepsin	Sigma
Peptidase, Dipeptidyl, IV (CD26)	Biosource; Neomarkers
Peptides: RGD, GRGDS, H-Gly-Arg-Gly-Asp-Ser-OH	Calbiochem
Percoll	Amersham Biosciences; Sigma
Peristaltic pumps	Altec; Amersham; Cole-Parmer; Gilson; Michael Smith; Watson Marlow
Perfusion cartridges, polycarbonate	Advanced Tissue Sciences
Perfusion culture	Amicon; Bioptechs; Cellco; Endotronics; Microgon
Permeable caps	*See* Gas-permeable caps
Permount mounting medium	Fisher
Peroxidase Blocking Reagent	Cytomation; Dako
Petri dishes	*See* Culture dishes, flasks, and plates
Petriperm dishes	Sartorius
PGA scaffolds	Albany International
Pharmed tubing	Cole-Parmer
Phenol Red	Sigma
Phenylalanine	*See* Chemicals
Phenylenediamine (OPD) tablets	Sigma
Phenylisothiocyanate	Pierce

Phosphocreatine	Sigma
Phosphoethanolamine	Sigma
Photographic chemicals, films, and papers	Agfa; Eastman Kodak; Fuji; Ilford (and local camera shops)
Photoinitiators for photopatterning, Irgacure 2959	Ciba Specialty Chemicals
Photopatterning/photo-etching	Microchem Corp; Ciba Specialty Chemicals
Photopatterning/photo-etching developer, MF-319	Microchem Corp.
Phytohemagglutinin	Amersham Biosciences; Sigma
PicoGreen	Molecular Probes
Pinning forceps	Fisher Scientific
Pipette tips, aerosol-resistant	*See* Pipettors and pipetting aids
Pipette cans	Bellco; Thermo Electron
Pipette cylinders or hods	Bel Art; Fisher; Nalge Nunc; Radleys
Pipette plugger	Bellco; Camlab; Volac
Pipette washer, drier	Bel Art; Thermo-Shandon; Radleys
Pipettes	Alpha Laboratories; BD Biosciences; Bellco; Corning; Fisher; Nalge Nunc; Sterilin
Pipettes, wide bore	Bellco
Pipettors and pipetting aids	Alpha Laboratories; Anachem; Applied Scientific; Barnstead; Bellco; Bibby Sterilin; Brand; Cole-Parmer; Corning; Daiger; Eppendorf; Fisher; Genetic Research Instr.; Gilson; Integra; Jencons; Lawson Mardon Wheaton; Matrix Technologies; Merck; Messer; Mettler-Toledo; MP Biomedicals; Polytech; Radleys; Rainin; Roche; Socorex; Thermo Electron; Whatman
Pi-Pump	Applied Scientific; Bel-art; Polytech; Radleys
Pituitary extract (PE)	Cambrex (Clonetics); Cascade; Invitrogen; PromoCell
Plasma, equine	MP Biomedicals
Plastic test-tubes	*See* Centrifuge tubes
Plasticware	Falcon (BD Biosciences)
Plate readers	Bio-Tek; Berthold; Bio-Rad; BMG; Camlab; Dynex; Fisher; Merck; Messer; Molecular Devices; Molecular Dynamics; PerkinElmer; Tecan; Thermo Electron.
Plate sealers	Anachem; Brandell; Elkay; Greiner; MP Biomedicals; Porvair
Plates	*See* Culture dishes, flasks, and plates
Platinum-cured silicone tubing, for air/CO_2 exchange	Cole-Parmer
Pleated cartridge filter	Pall Gelman
Pluronic F-68	Invitrogen; MP Biomedicals; Serva; Sigma
Polaron SC502 Sputter Coater	Fison
Poly (2-hydroxyethyl) methacrylate (Poly HEMA)	Sigma
Poly HEMA: poly (2-hydroxyethyl) methacrylate	Sigma
Poly(ethylene glycol) diacrylate (PEGDA), 3.4 kDa	Nektar; Shearwater
Polyacrylamide gels	Amersham Biosciences; Bio-Rad
Polycarbonate tubes	Nalge Nunc
Poly-D-lysine-coated dishes	BD Biosciences (Biocoat)
Polyethylene glycol-diacrylate (PEGDA)	Nektar Transforming Therapeutics; Shearwater
Polylysine, poly-D- and poly-L-	Biomedical Technologies; BD Biosciences; Sigma; Stratech
Polymyxin B	Invitrogen
Polyolefin heat shrink tubing	Appleton Electronics
Polypropylene jar, 30 ml	Nalge Nunc
Polystyrene flasks	*See* Culture dishes, flasks, and plates
Polyvinyl pyrrolidone	Calbiochem; Sigma; USB
Positive photoresists, OCG 825–835 St, Shipley 1813, or Shipley 1818	Shipley
Potassium chloride	*See* Chemicals
Potassium dihydrogen orthophosphate	*See* Chemicals
Potassium ferricyanide, $K_3Fe(CN)_6$	*See* Chemicals
Potassium ferrocyanide trihydrate, $K_4Fe_3(CN)_6 \cdot 3H_2O$	*See* Chemicals
Potassium phosphate, monobasic	*See* Chemicals
Precept tablets	Johnson & Johnson
Pressure cooker, bench-top autoclave	Astell Scientific; Harvard Apparatus; LTE; Napco; Valley Forge
Pressure monitor	Hewlett-Packard
Pressure transducers	Gould; Maxim Medica
Primaria® flasks and dishes	BD Biosciences (Falcon)
Primers	Applied Biosciences; Research Genetics; Riken
Progesterone in absolute ethanol	Sigma

Programmable freezer	Messer; Planer; Thermo Life Sciences (Cryomed); Statebourne
Proline	*See* Chemicals
Pronase	Calbiochem; Roche Diagnostics; Sigma
Propidium iodide	Sigma
Protease inhibitors	Intergen; Roche Diagnostics; Sigma
Protective clothing	Alexandria Workwear; Lab Safety Supply; Sigma
Proteinase K	Sigma
Protein assay	Pierce;
PTFE (Teflon) sheet (for disks) and tube (for cylinders)	Interplast; Plastim
Pump, bellows-style	Gorman-Rupp Industries
Pumps, peristaltic	Altec; Amersham; Cole-Parmer; Gilson; Michael Smith; Watson Marlow
Pumps, vacuum	BO; Cole-Parmer; Millipore; Pall Gelman; Varian
Putrescine 2HCl	Sigma-Aldrich
Pyridoxine HCl	Sigma-Aldrich
Qiagen Rneasy Mini Kit	Qiagen
QIAshredder	Qiagen
QuiAmp DNA mini kit	Qiagen
Quinacrine dihydrochloride	Sigma
Radioisotopes	Amersham; MP Biomedicals; PerkinElmer; Sigma
Random hexamers	Invitrogen
Rat serum	Pel-Freeze
Rat tail collagen	BD Biosciences
Rat tails	Biotrol
Rats	Charles River; Harlan Sera-Lab
RCCS (rotatory cell culture system)	Synthecon; Cellon
RDO decalcifying agent	Apex Engineering
Real-time monitoring	*See* Biosensors
Recombinant EGF	*See* Growth factors
Recording thermometer	Comark; Cole-Parmer; Fisher; Harvard; Grant; Omega; Pierce; Rustrack
Refractometers	Beckman; Cole-Parmer; Fisher; Technika.
Repeating pipettor	Brand; Cole-Parmer; Popper; Radleys
rEpidermal growth factor (rEGF)	*See* Growth factors
Reservoirs	*See* Aspirators
Retinoic acid	Sigma
Retinol acetate, all *trans*	Sigma
Reusable in-line filter assembly	*See* Filters
Reverse osmosis	Barnstead; Elga; Millipore; U.S. Filter
Reynolds' lead citrate	EMS; SPI Supplies
RGD: GRGDS: H-Gly-Arg-Gly-Asp-Ser-OH	Calbiochem
Rhodamine B	Sigma
Riboflavin	Sigma
Ring marker	Nikon
RNA preparation kit	Qiagen
*RNAimage*kits	GenHunter
RNAseZap®	Ambion
RNeasy mini kit	Qiagen
Robotics	Beckman Coulter; Bio-Rad; Cytogration; Gilson; Innovative Cell Technologies
Roccall	Henry Schein Rexodent
Rocker platform	Bellco
Roller bottle rack	Argos; Bellco; Genetic Research Instr.; Integra; Lawson Mardon Wheaton; New Brunswick Scientific; Robbins;
Roller bottles, glass	Bellco
Roller bottles, plastic	Applied Scientific; BD Biosciences; Caisson; Corning; Integra
Roller drum	Genetic Research Instr.; New Brunswick
Rompun, Xylazine hydrochloride	Bayer Vital
Rotameters	*See* Gas flow meters
Rotary shaker	New Brunswick
Rotatory Cell Culture System (RCCS)	Synthecon; Cellon
Rotatory Cell Culture Systems	Synthecon; Inc.
Round bottom test tubes with cap	Alpha Laboratories; BD Biosciences (Falcon); Bellco; Corning; Kimble-Kontes
RPMI 1640	*See* Media
Rubber pipette bulb	Bel Art; Bellco; Bibby Sterilin; Cole-Parmer; Fisher; Jencons Scientific; VWR
Safety cabinets	*See* Laminar-flow hoods
Safety products	Air Sea Atlanta; Altec; Bel-Art; Cellutech; Cin-Made; Jencons; Kimberly-Clark; Lab safety Supply; Saf-T-Pak
Safranin O	Sigma
Sample containers for transportation	Saf-T-Pak; Nalge Nunc; Air Sea Atlanta; Cellutech; Cin-Made Corp (*see also* www.uos.harvard. edu/ehs/bio_bio_shi.shtml and www.ehs.ucsf.edu/ Safety%20Updates/Bsu/ Bsu5.pdf)
SBTI (soybean trypsin inhibitor)	Sigma
Scaffolds, tissue engineering	Albany; Cook Biotech; EBI; Minucells; Zimmer
Scalpels	*See* Dissecting instruments
Scanning cytometer	Beckman Coulter; Compucyte (*see also* Microscopes)
Scanning electron microscopes (SEM)	ElectroScan; JEOL

SCID mice	Charles River
Scintillation fluid	Amersham Biosciences; BS & S; National Diagnostics; New England Nuclear (DuPont); PerkinElmer (Packard)
Scintillation vials, minivials	PerkinElmer (Packard)
Scion-Image software (time-lapse recording)	Scion
Scissors	*See* Dissecting instruments
Scrynel NYHC nylon gauze	Merck; *see also* Nylon mesh filters
Selenious acid	Merck; Sigma
SEM (scanning electron microscope)	ElectroScan; JEOL
Semipermeable nylon film	*See* Sterile packaging
Sequencing gel filter paper	Bio-Rad
Sequencing kit	Amersham Biosciences
Serine	*See* Chemicals
Serum	ATCC; Biochrom; Biosource; Caisson; Cambrex; Invitrogen; Gemini; Globepharm; Harlan Seralab; HyClone; Invitrogen; Irvine; Metachem; MP Biomedicals; JRH Biosciences; PAA; Perbio; PromoCell; Serologicals; Sigma; TCS Biologicals
Serum replacements	*See* Serum substitutes
Serum substitutes	Biosource; Irvine; Metachem; Protide; *see also specifics, below*
Serum substitutes, Biotain-MPS	Cambrex
Serum substitutes, CPSR	Sigma
Serum substitutes, Excell-900	JRH Biosciences
Serum substitutes, Ex-cyte	Bayer
Serum substitutes, ITS Premix	MP Biomedicals
Serum substitutes, Nutridoma	Roche Diagnostics
Serum substitutes, Serxtend	NEN (DuPont)
Serum substitutes, SIT	Sigma
Serum substitutes, TCM, TCH	Celox; MP Biomedicals
Serum substitutes, Ultroser	Invitrogen
Serum substitutes, Ventrex	JRH Biosciences
Serum-free media	BD Biosciences; Biosource; Cascade; CellGenix; CoaChrom; Roche Diagnostics; Biofluids; Cambrex (Clonetics); Cascade Biologicals; Hyclone; Hycor; Invitrogen; Irvine; JRH Biosciences; Mediatech; Metachem; MP Biomedicals; PAA; PromoCell; Roche; Sigma; Stratech; TCS VellWorks
Serxtend serum substitute	NEN (DuPont)
Shaking incubator	Camlab; New Brunswick; Radleys
Shipley 354, 1813, 1818; photopatterning developers	Microchem Corp.
Sieve with 200-mm mesh	BD Biosciences; Tekmar
Sieves	Markson; Retsch; *see also* Mesh filters and Gauze
Silica gel	Fisher; Merck
Silicone grease	BOC Edwards; British Oxygen; Dow Corning; Girovac
Silicone lubricant, inert	Dow Corning
Silicone rubber adhesive	Dow Corning; GE Silicones
Silicone rubber sheet	Nusil; Silicone Specialty Fabricators
Silicone rubber stoppers	Bibby Sterilin
Silicone tubing	Altec; Bibby Sterilin; Dow Corning; Cole-Parmer; Nusil; PAW BioScience Products; Thomas; Watson-Marlow
Silicones	Fisher; Merck; Serva; Sigma
Siliconizing solution (Sigmacote)	Sigma
Silver nitrate	*See* Chemicals
SIT serum substitute	Sigma
Skeletal muscle growth medium (SKGM)	Cambrex (Clonetics)
Slide boxes, light tight	Bel-Art; Cole-Parmer; Raven; Raymond Lamb
Slide boxes, light tight	BD Biosciences; Raven Scientific
Slide containers for emulsion (Cyto-Mailer)	Lab-Tek; Fisher; Statlab
Slide flasks	Nalge Nunc; *see also* Chamber slides
Slides, glass	*See* Microscope slides
Slow Turning Lateral Vessel (STLV)	Synthecon; Cellon
SLTV (Slow Turning Lateral Vessel)	Synthecon; Cellon
S-MEM	*See* Media
Smooth muscle alpha actin antibody	DAKO
Sodium alginate	ISP Alginate
Sodium azide	Sigma
Sodium butyrate (NaBt)	Sigma
Sodium carbonate	*See* Chemicals
Sodium chloride	*See* Chemicals
Sodium citrate	*See* Chemicals
Slide culture dishes	Vivascience; *see also* Culture dishes, flasks, and plates
Sodium deoxycholate	Sigma
Sodium formate	Fisher
Sodium hydroxide	Sigma
Sodium lauryl sulfate (SLS; sodium deoxycholate)	Sigma
Sodium orthovanadate, Na_3VO_4,	Sigma
Sodium phosphate, dibasic heptahydrate	*See* Chemicals
Sodium pyruvate	*See* Chemicals
Sodium pyruvate, sterile solution	*See* Media

Sodium selenate	Sigma
Sodium selenite	Sigma
Sodium thiosulfate	Sigma
SonicSeal slides, 4-well	Nalge Nunc
Sorbitol	Sigma
Soybean trypsin inhibitor (SBTI)	Sigma
Spatulas	Fisher; VWR
Specialized cell cultures	BD Biosciences; Cambrex; Cascade; Cell Systems; PromoCell; SkinEthic
Specimen containers	Alpha Laboratories; Corning; Bibby Sterilin; Nalge Nunc;VWR; see also Sample containers for transportation
Spectra Max 250 microplate spectrophotometer	Molecular Devices; see also Plate readers
Spinner flasks	See Stirrer flasks
Spring scissors	Fine Science Tools Inc.; VWR; see also Dissecting instruments
SPSS (version 10.0) statistical analysis software	SPSS
Sputter coater	Gatan; Fison
Stainless steel mesh	BD Biosciences; see also Mesh filters and gauze
Stains	Fisher; Merck; Molecular Probes; MTR; Sigma; TCS Biologicals
Stanzen Petri dishes	Greiner
Steam-permeable nylon film	See Autoclavable nylon film and bags
Stem cell factor (SCF)	See Growth factors
Stem cells (isolation, markers, preservation)	Chemicon; Geron; Metachem; Miltenyi; Origen; Roche; Universal Biologicals
Sterile filtration	See Filters, sterilization
Sterile indicating tape	See Sterility indicators
Sterile packaging: cartridge paper, semipermeable nylon film	Applied Scientific; Buck Scientific; KNF Corp; Portex; Portland Plastics; Roth
Sterility indicators	Appleton Woods; Applied Scientific; Bennett; Jencons; Popper; Raven; Roboz; SGM Biotech; Shamrock; Stoelting; Surgicon.
Sterilization bags	See Sterile packaging
Sterilization film	See Sterile packaging
Sterilizers	See Autoclaves; Ovens
Sterilizing and drying oven	See Ovens
Sterilizing filters	See Filters, sterilization
Sterilizing tape (indicator)	See Sterility indicators
Stills	Corning; Jencons; Steris
Stirrer flasks	Bellco; Corning; Genetic Research Instr.; Integra; Lawson Mardon Wheaton; Techne; see also Magnetic stirrers
Stirrers	See Magnetic stirrers
Streptavidin-FITC	Dako Diagnostics
Streptomycin	See Media
Sucrose	Fluka; Sigma
Sulfo-NHS, N-hydroxy-sulfosuccinimide	Pierce
Superscript amplification system	Invitrogen
Supplemented keratinocyte defined medium (SKDM)	See Keratinocyte growth medium
Surgical gauze	Johnson & Johnson; Kendall
Surgical Instruments	See Dissecting instruments
Surgilube	Fougera
Suture, Dacron®	Davis and Geck
Suture, Dexon®	Davis and Geck
Suture, silk	Harvard Apparatus
Swiss 3T3 cells	ATCC
Sylgard plastic	Dow Corning
Syringe filters	See Filters
Syringe needles, 22 g	See Syringes
Syringe pump	Harvard Apparatus
Syringes	Baxter; BD Biosciences; Popper; (and general laboratory suppliers)
Syringe-tip filters	Microgon; Millipore; Pall-Gelman; Sartorius; see also Filters
TCA: Trichloroacetic acid	See Chemicals
TCM, TCH serum substitutes	Celox; MP Biomedicals
TEER measurement	Applied Biophysics; BD Biosciences; World Precision Instruments
Temperature controllers	See Thermostats, proportional controllers
Temperature indicator strips	See Sterility indicators
Temperature recorders	See recording thermometers
Tension/compression system for muscle culture	Instron
Tetramethylbenzidine	Research Diagnostics
TGF-β1	R&D Systems; Research Diagnostics
Thermalog sterility indicator	Bennet; Popper
Thermanox	Nalge Nunc (see also Coverslips; plastic)
Thermometers, recording	Comark; Cole-Parmer; Fisher; Harvard; Grant; Omega; Pierce; Rustrack
Thermostats, proportional controllers	Controls & Automation; Fisher; Napco
Thiamine-HCl	Sigma
Thimerosal	Sigma
Thioctic acid	Sigma
Thioglycerol	Sigma
Three-way stop-cocks	Baxter Healthcare; Sherwood-Davis & Geck

Threonine	*See* Chemicals
Thymidine	Sigma
Time-lapse video	Dage-MTI; Hamamatsu; Imaging Associates; *See* CCD cameras; video camera; video recorder (*see also* Protocol 27.4)
Time-lapse observation chambers	Bioptechs; Buck Scientific; Carl Zeiss; Intracel
Tissue culture flasks	*See* Culture dishes, flasks, and plates
Tissue culture inserts	*See* Filter well inserts
Tissue culture media	*See* Media
Tissue culture plastic flasks and dishes	*See* Culture dishes, flasks, and plates
Tissue grinder: Pellet Pestle Mixer	Kimble-Kontes
Tissue sealant, fibrin, Tisseel VH	Baxter
Tissue slicers	Alabama Research & Development
Tissue Tek-O.T.C embedding compound	Sakura Finetek (Raymond Lamb in UK); Sakura Finetek Europe
Tocopherol	Sigma
Toluidine Blue	Sigma
Total organic carbon (TOC) meter	Millipore; Sartec; Thermo Electron
TPA, 12-O-tetradecanoyl phorbol-13-acetate	Sigma
Transepithelial electrical resistance (TEER)	Applied Biophysics; WPI
Transferrin	BD Biosciences; Sigma
Transmission electron microscope	Philips
Transplantation chamber (epidermis)	Greiner
Transportation containers	Air Packaging Technologies; Air Sea Atlanta; Altec; Bel-Art; Cellutech; Cin-Made; Jencons; Kimberly-Clark; Lab Safety Supply; Nalge Nunc; Saf-T-Pak
Transwell inserts	Corning; *see also* Filter well inserts
Trichlorotrifluoroethane	Sigma
Triethylamine	Fisher; Sigma
Triiodothyronine	Sigma
Tris	*See* Chemicals
Tritiated thymidine	NEN Life Sciences (DuPont)
Triton X-100	Sigma; Fisher
TRIzol	Invitrogen
Trypan blue viability stain	Merck; *See also* Media
Trypsin	Sigma; Worthington; *see also* Media
Trypsin inhibitors	Biosource; Cascade; Sigma; Serologicals
Trypsin/ETDA	*See* Media
Trypsin replacements/ substitutes	Hyclone; Innovative Cell Technologies; Invitrogen; Perbio; Sigma
Tryptophan	*See* Chemicals

Tryptose phosphate broth	BD Biosciences (Difco); Invitrogen; Oxoid
Trypzean	Sigma
Tube rotator	New Brunswick Scientific
Tuffryn filters	Pall-Gelman
Tween 20	Sigma
Tween 80	Sigma
Tyrode's salt solution (TBSS)	*See* Media
Tyrosinase (C-19) goat anti-human antibody	Santa Cruz Biotechnology
Tyrosine	*See* Chemicals
Ultrafoam collagen sponge	Davol
Ultra-low attachment plates, 35 mm	Corning (Costar)
Ultramicrotome	Leica
Ultra-TMB (3,3′,5,5′-tetramethylbenzidine)	Research Diagnostics
Ultroser G serum substitute	Invitrogen; *see also* Serum substitutes
Universal containers	Bibby Sterilin; Camlab; Corning; Elkay; Nalge Nunc
Uranyl acetate	Plano; SPI Supplies
Urea	*See* Chemicals
Urea nitrogen analysis	Stanbio Labs
UV-curing glass cement	Summers Optical
UV light sources	Cole-Parmer; UVP
Vacuum pump	*See* Pumps, vacuum
Validation of cell lines	ATCC; BioReliance; Cellmark; ECACC; LGC Promochem
Valine	*See* Chemicals
Vectashield	Vector Laboratories
Vectashield Mounting Medium with DAPI	Vector Laboratories
Vectastain Elite ABC Kit	Vector Laboratories
Vectastain Quick Kit	Vector Laboratories
Ventrex serum substitute	JRH Biosciences; *see also* Serum substitutes
Viability stains	*See* Media
Vials for freezing cells	*See* Ampoules
Vibratome slicer	Campden Instruments
Video camera	*See* CCD cameras; Digital cameras; Microscopes
Video recorder	*See* Time-lapse video
Villin	Novacastra Laboratories; Santa Cruz Biotechnology; Chemicon
Vimentin antibody	Sigma
Vinblastine	Sigma
Vinyl tape	3 M (general laboratory suppliers)
Virkon	Du Pont Animal Health Solutions; Scientific Laboratory Supplies
Vitamin B$_{12}$	Sigma
Vitamins	Sigma; *See also* Media
Vitamins, sterile solution	*See* media
Vitrogen	BD Biosciences; Cohesion Technology; Collagen Corp.; *see also* Collagen

von Willebrand factor (Factor VIII) antibody	Roche Diagnostics
Vortex mixer	Gallenkamp; Fisher; general laboratory suppliers
Wafer tweezers	Fluoroware
Water baths	Camlab; Cole-Parmer; Grant; Polyscience; SciGene; Thermo Electron
Water for tissue irrigation	Baxter
Water purification	Applied Biosciences; Barnstead; Elga; Genetic Research Instr.; High-Q; Millipore; Purite; Triple Red; U.S. Filter; Vivendi; Whatman
Wavemaker32, software for fatigue testing of ligament	Instron
Wescodyne	Steris
Weise buffer	Merck
Williams medium E	Invitrogen; Sigma
Winged infusion set syringes	Merck
WST-1 cytotoxicity indicator stain	Serva
X-gal: 5-Bromo-4-chloro-3-indolyl β-D-galactopyranoside	Roche
X-ray film	Eastman Kodak; Fuji
XTT	Sigma
Xylazine	NLS Animal Health; Sigma
Xylene	*See* Chemicals
$ZnSO_4 \cdot 7H_2O$	*See* Chemicals
Zyderm 2 Collagen Implant	Insmed; McGhan Medical

Suppliers and Other Resources

Telephone numbers come first and then fax numbers; two numbers separated by a semi-colon mean two lines are available. The international prefix is preceded by a "+" sign. which should be replaced by the appropriate international access dialing code. Products are listed last and are those products cited in, or most related to, the current context; this is not intended to be a comprehensive list of each supplier's products.

Abbott Laboratories. 100 Abbott Park Road, Abbott Park, North Chicago, IL 60064-3500, USA. 800-323-9100; 1-847-937-6100. *www.abbott.com.* Nembutal.

Accuramatic. 40–42 Windsor Road, Kings Lynn, Norfolk PE30 5PL, England, UK. 01553 777253. 01553 777253. *info@accuramatic.co.uk. www.accuramatic.co.uk.* Automatic dispenser; liquid handling; peristaltic filling machine.

Accurate Chemical & Scientific Corp. 300 Shames Dr., Westbury, NY 11590, USA. +1 516-333-2221; 800-645-6264. +1 516-997-4948. *info@accuratechemical.com. www.accuratechemical.com/.* Antibodies; cell separation; chick embryo extract; extracellular matrix assays; HeLa cells; microcentrifuges.

ACE—Autoclave Control Engineering. Unit E, Cavendish Courtyard, Sallow Road, Weldon North, Corby, Northants., NN1 5JX, England, UK. +44 (0)1536 206 200. +44 (0)1536 206 202. *sales@ace-autoclaves.co.uk. www.ace-autoclaves.co.uk* Large-capacity autoclaves; steam generators.

ACM-Biotech GmbH. Josef-Engert Strasse 9, D-93035 Regensburg, Germany. +49 (0)941 942 743. +49 (0)941 942 7444. *techservice@acm- biotech.com. www.acm-biotech.com.* Endothelial cells; Epiflow; hepatocytes; *in vitro* testing; keratinocytes; membrane perfusion; organotypic culture; selective media, serum-free.

Advanced Instruments Inc. Two Technology Way, Norwood, MA 02062, USA. 781 320 9000, 800 225 4034. 781 320 8181. *mail@aitests.com. www.aitests. com.* Osmometer.

AES Laboratoire. Rue Maryse Bastié, Ker Lann, CS 17219, F-35172 Bruz cedex, France. +33 (0)2 23 50 12 12. +33 (0)2 23 50 12 00. *aes@aeslaboratoire.com. www. aeslaboratoire.com.* Laminar-flow hoods; media; microbiological safety cabinets.

Affymetrix Inc. 888-362-2447. +1 408-731-5441. *support@affymetrix.com; sales@affymetrix.com. www. affymetrix.com.* DNA microarray analysis.

Affymetrix UK. +44 (0) 1628 552550. +44 (0) 1628 552598. *supporteurope@affymetrix.com; saleseurope @affymetrix.com. www.affymetrix.com.* DNA microarray analysis.

Agilent Technologies. Quantum Analytics, Inc., 363 Vintage Park Drive, Foster City, CA 94404, USA. 800-227-9770; 800-992-4199. *cag_enquiry@agilent.com. www. chem.agilent.com.* Fluorescence bioanalyzer; DNA microarrays.

Agilent Technologies UK Ltd. Life Sciences & Chemical Analysis Group, Lakeside, Cheadle Royal Business Park, Stockport, Cheshire SK8 3GR, England, UK. +44 (0)161 492 7500; 0845 712 5292. 0845 600 8356. *cag_enquiry@agilent.com. www.agilent.com/chem/uk.* Fluorescence bioanalyzer; DNA microarrays.

Culture of Animal Cells: A Manual of Basic Technique, Fifth Edition, by R. Ian Freshney
Copyright © 2005 John Wiley & Sons, Inc.

Air Packaging Technologies Inc. 25620 Rye Canyon Road, Valencia, CA 91355, USA. +1 661-294-2222. +1 661-294-0947. *llchappell@aol.com.* Airbox packaging; transport containers for cells.

Air Products & Chemicals, Inc. 7201 Hamilton Blvd, Allentown, PA 18195, USA. 610-481-4911; 800-654-4567. 800-880-5204. *info@apci.com. www.airproducts.com/.* CO_2, automatic change-over unit for cylinders, CO_2 controllers.

Air Products plc. 2 Millennium Gate, Westmere Drive, Crewe CW1 6AP, England, UK. 0800 389 0202. 01932 258 502. *info@apci.com. www.airproducts.co.uk.* CO_2, automatic change-over unit for cylinders, CO_2 controllers.

Air Sea Atlanta. Atlanta, GA. 1-404-351-8600;. *www.airseaatlanta.com.* Transportation containers; safety.

Aire Liquide. Division Materiel Cryogenique, Parc Gustave Eiffel, 8 rue Gutenberg, Bussy Saint Georges 77607, Marme-la-Valle, Cedex 3, France. +33 (0)1 64 76 15 00. +33 (0)1 64 76 16 99. *www.dmc.airliquide.com/.* Liquid nitrogen freezers; Dewars.

Aire Liquide America Corp. 2700 Post Oak Blvd., Suite 1800, Houston, Texas 77056, USA. +1 713 624 8000. +1 713 402 2149.*www.dmc.airliquide.com/.* Liquid nitrogen freezers; Dewars.

Aire Liquide (UK). *See* Borolabs. *www.dmc.airliquide. com/.* Liquid nitrogen freezers; Dewars.

Alabama Research and Development. A Division of Alabama Specialty Products, P.O. Box 739, Munford, AL 36268, USA. +1 (256) 358−0460. +1 (256) 358−4515. *www.alspi.com/Slicer.htm.* Krumdieck Tissue Slicer.

Albany International Research Co. Albany International Research Co., PO Box 9114, 777 West Street, Mansfield, MA 02048-9114, USA. +1 508 339 7300; 800 992 5017. *research@albint.com. www.airesco.com.* Tissue engineering scaffolds.

Alconox Inc. 30 Glenn Street, Suite 309, White Plains, NY 10603, USA. +1 914 948 4040. +1 914 948 4088. *cleaning@alconox.com. www.alconox.com.* Detergents.

Aldrich Chemical Co., Inc. 1001 W. St. Paul Ave., Milwaukee, WI 53233, USA. 800-325-3010. 414 273 4979. *www.sigma.sial.com/aldrich.* Chemicals; biochemicals; EDTA.

Alexis Corp. (UK) Ltd. P.O. Box 6757, Bingham, Nottingham NG13 8LS, England, UK. +44 (0)1949 836 111. +44 (0)1949 836 222. *alexix-uk@alexis-corp.com. www.alexis-corp.com.* Angiogenesis; antibodies: cytokines, interferons, receptors, signal transduction; VEGF.

Alexis Corp. (USA). 6181 Cornerstone Court East, Suite 103, San Diego, CA 92121, USA. +1 858-658-0065; 800-900-0065. +1 858-550-8825; 800-550-8825. *axxoraus@axxora.com. www.alexis-corp.com.* Angiogenesis; antibodies: cytokines, interferons, receptors, signal transduction; VEGF.

Alfa Laval. 7 Doman Road, Camberley, Surrey GU15 3DN, England, UK. +44 (0)1276 63383. +44 (0)1276 685035.

general.uk@alfalaval.com. www.alfalaval.co.uk. Fermentors, bioreactors, steam sterilization.

Alpha Laboratories Ltd. 40 Parham Drive, Eastleigh, Hants S05 4NU, England, UK. +44 (0) 23 8048 3000. +44 (0) 23 8064 3701. *info@alphalabs.co.uk. www.alphalabs. co.uk/.* Ampoules; automatic pipettes; centrifuge tubes; cryovials; dry block heaters; micropipettes; Pastettes; pipettes; pipette aids; Pasteur pipettes; pipette tips.

Alfa Medical. 59 Madison Ave., Hempstead, NY 11550, USA. 800-801-9934. +1 516-489-9364. *eMail@sterilizers. com. www.sterilizers.com/.* Sterility indicator strips.

Altec Products Ltd. Bude Business Centre, Bude, Cornwall, EX23 8QN, England, UK. 0845 359 9000; +44 1288 357820. 0845 359 9090; +44 1288 357822. *sales@altecweb.com. www.altecweb.com.* Autoclavable bags; Luer-Lok connectors; peristaltic pumps; safety products; silicone rubber tubing.

Ambion Inc. 2130 Woodward, Austin, TX 78744-1832, USA. 800 445−1161; +1 (512) 651-0200.+1 (512) 651-0201. *custom@ambion.com. www.ambion.com.* DNA templates; RNA isolation; RNAi; RNAseZap; siRNA; RT-PCR.

Ambion UK. Ermine Business Park, Spitfire Close, Huntingdon, Cambridgeshire, PE29 6XY, England, UK. +44 (0)1480 373020; 0800 138 1836. +44 (0)1480 373010. *custom@ambion.com. www.ambion.com.* DNA templates; RNA isolation; RNAi; RNAseZap; siRNA; RT-PCR.

American Association for Cancer Research. 615 Chestnut St., 17th Floor, Philadelphia, PA 19106-4404, USA. +1 (215) 440−9300. +1 (215) 440−9313. *membership@aacr.org. www.aacr.org.* Cancer society; journals; scientific meetings.

American Fluoroseal. 431-D East Diamond Avenue, Gaithersburg, MD 20877, USA.. 301.990.1407,. 301 990 1472. *info@americanfluoroseal.com. www.toafc.com/.* Media bags.

American Society for Cell Biology. 8120 Woodmont Avenue Suite 750, Bethesda, MD 20814-2762, USA. +1 301 347 9300. +1 301/347-9310. *ascbinfo@ascb.org. www.ascb.org/.* Journals; meetings; scientific society.

American Type Culture Collection. (*see* ATCC)

Amersham Biosciences. 800 Centennial Ave., P.O. Box 1327, Piscataway, NJ 08855-1327, USA. 800-526-3593. 877-295-8102. *cs-us@amersham.com. www4. amershambiosciences.com.* Antibodies; assays; chromatography media; cytokines; DEAE dextran; density media; DNA standards; electrophoresis; ELISA; epidermal growth factor; fibronectin; Ficoll-Hypaque; growth factors; Hybond N; lymphocyte preparation media; marker beads; microcarriers; Percoll; peristaltic pumps; phytohemagglutinin; power supplies; radioisotopes; receptors; signal transduction; sodium metrizoate; Ultraser-G; Verax.

Amersham Biosciences. Amersham Biosciences UK Ltd., Pollards Wood, Nightingales Lane, Chalfont

St.Giles, Bucks, HP8 4SP, England, UK. 0870 606 1921. +44 (0)1494 498 231; 0800 616 927. *orders. gb@amersham.com. www4.amershambiosciences.com*. For products *see* Amersham Biosciences (USA).

Amersham Pharmacia Biotech. *See* Amersham Biosciences..

Amgen, Inc. One Amgen Center Drive, Thousand Oaks, CA 91320-1799, USA. +1 805-447-1000. +1 805-447-1010. *newproducts@amgen.com. www.Amgen.com*. Growth factors; FGF.

Amicon, Inc. *See* Millipore.

Amphioxus Cell Technologies, Inc. 11222 Richmond Ave, Suite 180, Houston, TX 77082-2646, USA. 800-777-2707. 281-679-7910. *activox@amphioxus.com. www.amphioxus.com*. Human serum proteins; human liver cell line; hepatocytes; hollow fiber culture; *in vitro* toxicology assays.

Anachem Ltd. Anachem House, 20 Charles Street, Luton, Beds LU2 0EB, England, UK. +44 (0)1582 745 060. +44 (0)1582 745 115. *lifescience@anachem.co.uk; sales@anachem.co.uk. www.anachem-ltd.com*. Decontamination; digital imaging; disinfectant; electrophoresis; freezer storage; media; microtitration plates; microtubes; nucleic acid purification kits; pipettes; pipettors; plate sealers; racks; stirrer flasks; swabs; tips; tubes.

Analytical @ Technologies. Lynchford House, Lynchford Lane, Farnborough, Hampshire, GU14 6LT, England, UK. +44 (0)1252 514 711. +44 (0)1252 511 855. *info@analyticaltechnologies.co.uk.* www. *analyticaltechnologies.co.uk*. Biochemical analyzers; biosensors; colloid-osmometer; monitors: glucose, glutamine, glutamate, lactate; osmometers.

Anglia DNA Bioservices, Ltd. Norwich Research Park, Colney Lane, Norwich, NR4 7UH, Norfolk, UK. +44 (0)8454 565 365. *office@angliadna.co.uk. www.angliadna.co.uk*. Authentication; Genetic identity testing; DNA profiling.

Ansell Medical. Ansell Medical, 119 Ewell Road, Surbiton, Surrey KT6 6AL, England, UK. +1 (0)181 481 1804. +1 (0)181 481 1828. *www.ansell.com*. Nitrile gloves.

Ansell Healthcare. Ansell—Red Bank, 200 Schulz Drive, Red Bank, NJ 07701, USA. +1 732 345 5400. *www.ansell.com*. Nitrile gloves.

Anton Paar GmbH. Karnter Strasse 322, A-8054 Graz, Austria. +43 316 257 0. +43 316 257 257. *density@anton-paar.com; rheology@anton-paar.com. www.anton-paar.com/*. Density meter, rheometers, viscometers.

Anton Paar USA Inc. 10215 Timber Ridge Drive, Ashland, VA 23005, USA. (800)-722-7556; +1 804-550-1051. 804-550-1057. *info.us@anton-paar.com. www.anton-paar.com/*. Density meter, rheometers, viscometers.

Apex Engineering Products Corp. 1241 Shoreline Drive, Aurora, IL 60504, USA. +1 815 436 2200; 800 451 6291. +1 630-820-8886; 815 436 9418. *rdo@rdo-apex.com.. www.rdo-apex.com*. RDO decalcifying agent.

Appleton Electronics. 205 W. Wisconsin Ave., Appleton, WI 54911, USA. (800) 877−8919. 001 (920) 734−5172. *www.aedwis.com*. Electronic components; polyolefin heat shrink tubing.

Appleton Woods Ltd. Lindon House, Heeley Rd., Selly Oak, Birmingham B29 6EN, England, UK. +44 (0)121 472 7353. +44 (0)121 414 1075. *info@appletonwoods.co.uk. www.appletonwoods.co.uk*. Temperature indicator strips.

Applied BioPhysics, Inc. 1223 Peoples Ave., Troy, NY 12180, USA. +1 518 276 2165. +1 518 276 2907. *www.biophysics.com*. Cell adhesion; cell impedance; TEER; transepithelial resistance.

Applied Biosciences Corp. P.O. Box 520518 Salt Lake City, UT 84152, USA. 800-280-7852; 801-485-4988. 801-485-4987. *www.applied-biosciences.com*. Water purification.

Applied Biosystems. 850 Lincoln Centre Drive, Foster City, CA 94404, USA. 800 327 3002; +1 650 638 5800. +1 650 638 5884. *www.appliedbiosystems.com*. Primers; TaqMan 2X PCR Master Mix.

Applied Imaging. 120 Baytech Drive, San Jose CA 95134-2302, USA. +1 408 719 6400; 800 634 3622. +1 408 719 6401. *info@aicorp.com. www.aicorp.com*. Chromosome paints; image analysis; spot counting.

Applied Imaging International Ltd. BioScience Centre, Times Square, Newcastle upon Tyne, NE1 4EP, England, UK. +44 (0)191 202 3100. +44 (0)191 202 3101. *info@aii.co.uk. www.aicorp.co*. Chromosome paints; image analysis; spot counting.

Applied Immune Sciences. 5301 Patrick Henry Dr., Santa Clara, CA 95054, USA. +1 408-980-5812. Cell separation, immune panning.

Applied Medical Technology Ltd. 4 Orwell Furlong, Cowley Road, Cambridge CB4 0WY, England, UK. +44 (0)1223 420415. +44 (0)1223 420797. *support@applied-medical.co.uk. www.applied-medical. co.uk/*. Infusion pumps; infusion sets; Luer-Lok and QR connectors.

Applied Scientific. 154 W. Harris Ave., South San Francisco, CA 94080, USA. +1 650 244 9851. +1 650 244 9866. Antibiotics; autoclave bags; biohazard bags; cell scraper; cell strainer; chamber slides; cryovials; culture screening; flasks; gloves; medium; micropipettes; multiwell plates; needles; pipettes; pipettors; Pi-pump; roller bottles; sterility indicator tape; syringes.

Appropriate Technical Resources Inc. *See* ATR

Argos Technologies Inc. 1141 East Main Street, Suite 104, East Dundee, IL 60118, USA. +1 847 783 0456. +1 847 783 0380. *pkneisel@argos-tech.com. www.argos-tech.com*. Fluid handling; hollow fiber perfusion culture; roller culture; stirrer culture; vacuum pump.

Astell Scientific Ltd. Powerscroft Road, Sidcup, Kent, DA14 5DT, England, USA. +44 (0)20 8309 2023. +44 (0)20 8309 2036. *sales@astell.com. www.astell.com*.

Autoclaves; bench-top autoclaves; data loggers; sterilizers; ovens.

ATCC (American Type Culture Collection). 10801 University Boulevard, Manassas, VA 20110-2209, USA. +1 703-365-2700. +1 703-365-2701. Product info: *tech@atcc.org*; Database help: *help@atcc.org*. *www.atcc.org*. Cell lines; cell banking; feeder layers; hybridomas; media; mycoplasma detection service; PCR mycoplasma detection kit; serum; plasmids.

Atlas Clean Air Ltd. 176 Lomeshaye Business Village, Turner Road, Nelson, Lancashire BB9 7DR, England, UK. +44 (0)1282 447 666. +44 (0)1282 447 789. *info@atlascleanair.com. www.atlascleanair.com.* Laminar flow; microbiological safety cabinets.

ATR (Appropriate Technical Resources) Inc. P.O. Box 460, 9157 Whiskey Bottom Road, Laurel, MD 20723, USA. +1 301 470 2799. +1 410 792 2837. *inforeq@atrbiotech.com. www.atrbiotech.com.* Incubators; fermentors; shakers.

Austral Biologicals. 125 Ryan Industrial Ct., Suite 207, San Ramon, CA 94583, USA. +1 925 820 8390; 800 433 7105. +1 925 820 6843. *sales@australbio.com. www.australbiologicals.com.* Antibodies; cytokines; growth factors; receptors.

Autogen Bioclear. Holly Ditch Farm, Mile Elm, Calne, Wiltshire, SN11 0PY, England, UK. 0800 652 6744; +44 (0)1249 819008. +44 (0)1249 817266. *www.autogenbioclear.com.* Antibiotics; HEK 293 cells; MCDB 153; media; Primocin; Sebomed.

Axxora LLC. 6181 Cornerstone Court East, Suite 103, San Diego, CA 92121-4727, USA. 800 550 3066; +1 858 550 8824. +1 (858)550–8825; 1-800-550-8825. *www.biolog.de/usa.html.* Angiogenesis; antibodies: cytokines, receptors, signal transduction; VEGF.

Baker Co. Inc. P.O. Drawer E, 161 Gatehouse Road, Sanford, ME 04073, USA. +1 207 324 8773; 800 992 2537. +1 207 324 3869; 800-992-2537. *bakerco@bakerco.com. www.bakerco.com.* Laminar flow; microbiological safety cabinets.

Bard, C.R., Inc. *See* C. R. Bard Inc..

Barnstead Thermolyne Corp. 2555 Kerper Boulevard, P.O. Box 797, Dubuque, IA 52004-0797, USA. 800 446 6060; +1 319 589 0538. +1 319-589-0516. *mkt@barnsteadthermolyne.com. www.barnsteadthermolyne. com/.* CO_2 controllers; CO_2 incubators; dispensers; distillation reverse osmosis; freezers; heating mantles; heating tapes; hot plates; nitrogen freezers; ovens; pipettors; recorders; repipettors; sterilizers; stirrers; water purification.

Baxter HealthCare. One Baxter Parkway, Deerfield, IL 60015-4625, USA. 800-422-9837; +1 847-948-4770; +1 847-948-2000. +1 847-948-3642. *www.baxter.com.* Luer-lok valves, connectors; syringes; i.v. tubing sets.

Bayer Corp. Bayer Consumer Care Division, 36 Columbia Road, P.O. Box 1910, Morristown, NJ 07962-1910, USA. 800-331-4536. *www.bayermhc.com.* Antibiotics; ciprofloxacin; histology; slide processing; mycoplasma elimination; Thermanox coverslips.

Bayer plc. Strawberry Hill, RG 14 1JA Newbury, Berkshire, England, UK. +44 (0)1635 563 000. +44 1635 563 300. *medical.science@bayer.co.uk. www.bayer.co.uk/.* Antibiotics; cyprofloxacin; histology; slide processing; mycoplasma elimination; Thermanox coverslips.

BCL. *See* Roche Diagnostics..

BD Biosciences—Discovery Labware. 2 Oak Park, Bedford, MA 01730–9902, USA. 800 343 2035. 800 743 6200. *labware@bd.com. www.bd.com/.* Agar; automatic pipettes; Bactopeptone; bronchial epithelium; Caco-2 assay; cell culture inserts; cell scrapers; collagen; collagen-coated dishes; culture slides; fibronectin; filters; filter well inserts; FACS; flasks; flow cytometer; gas-permeable caps; growth factors; lactalbumin hydrolysate; laminin; Matrigel; microbiological media; multiwell plates; Natrigel; nutrient broths; organ culture dishes; organ culture grids; peptide hydrogel; Petri dishes; roller bottles; serum-free medium; TEER measurement; Tryptose; tubes; vitronectin.

BD Biosciences. Tullastrasse 8–12, 69126 Heidelberg, Germany. +49 6221 305 0. +49 6221 305 216. *mail@bdbiosciences.com. www.bd.com/.* For products *see* BD Biosciences—Discovery Labware.

BD Biosciences Immunocytometry Systems. 2350 Qume Dr., San Jose, CA 95131-1807, USA. 877 232 8995. +1 408-954-2347. *facservice@bdis.com. www.bd.com/.* Antibodies; cell separation; FACS; flow cytometers.

BD Biosciences UK Ltd. Between Towns Road, Cowley, Oxford, OX4 3LY, England, UK. +44 1 865 781 688. +44 1 865 781 578. *mail@bdbiosciences.com. www.bd.com/.* *See* BD Biosciences—Discovery Labware.

Beckman Coulter Inc. 4300 N. Harbor Blvd., Box 3100, Fullerton, CA 92834-3100, USA. 1–800 742 2345; +1 714 871 4848. 800 643 4366; +1 714 773 8283. *www.beckmancoulter.com/.* Cell counters; centrifugal elutriator; centrifuges; cell sizing; densitometers; DNA probes; flow cytometers; immuno-histochemistry reagents; latex particles (standard); liquid-handling, automated workstations; microtitration plate centrifugation; robotics systems for cell-based assays; scintillation counters; spectrophotometers (UV/VIS); Vi-CELL.

Beckman Coulter (UK) Ltd. Oakley Court, Kingsmead Business Park, High Wycombe, Bucks., HP11 1JU, England, UK. +44 (0)1494 441 181. +44 (0)1494 447 558. *beckmancoulter_uk@beckman.com. www.beckmancoulter.com.* For products *see* Beckman Coulter Inc.

Becton Dickinson. *See* BD Biosciences.

Bel-Art Products. 6 Industrial Road, Pequannock, NJ 07440-1992, USA. +1 973 694 0500; 800 4BEL-ART. +1 973 694 7199. *www.bel-art.com.* Aspirator bottles; biohazard bag holder; bottles; bottle racks; carboys; cell scraper;

cloning rings; culture boxes; culture chambers; culture flask racks; desiccator cabinets; flow meters; hods; magnetic cell scraper; pipette bulbs; pipette cylinders; pipettors; Pi-pump; racks; safety goggles; siphon pipette washer; slide boxes; stopcocks; tubing.

Bellco Glass Inc. 340 Edrudo Rd., Vineland, NJ 08360-3493, USA. 609 691 1075, 800 257 7043. 609 691 3247. *sales@bellcoglass.com. www.bellcoglass.com.* Bioreactors; borosilicate glass; cloning rings; deionizers; fermentors; flasks; incubators; magnetic stirrers; magnifying viewers; pipette cans; pipette plugger; pipettors; roller racks; shakers; trypsinization flask.

Bennett Scientific. Tor Green, Croft Road, East Ogwell, Newton Abbot, Devon, TQ12 6BA, England, UK. +44 (0)1626 369 990. +44 (0)1626 369 992. *sales@nbennett-scientific.com. www.bennett-scientific. com.* Autoclaves; magnetic stirrers; pumps: peristaltic, piston, vacuum; Thermalog sterility indicators.

Berthold Technologies. The Priory, High Street, Redburn, St Albans, AL3 7BR, England, UK. +44 (0)1582 79 1999. +44 (0)1582 79 1937. *James.Grand@Berthold.com. www.Bertholdtech.co.uk.* Absorbance; bioanalytical instruments; fluorescence; luminescence; microtitration plate readers; Mithras.

Berthold Technologies. Calmbacher Strasse 22, 75323 Bad Wildbad, Germany. +49 (0)7081-177 216. +49 (0)7081-177 301. *Winfried.Erdmann@Berthold.com. www.bertholdtech.com.* Absorbance; bioanalytical instruments; fluorescence; luminescence; microtitration plate readers; Mithras.

Bibby Sterilin Ltd. Beacon Road, Stone, Staffordshire, ST15 0SA, England, UK. +44 (0)1785 812 121. +44 (0)1785 815 066. *bsl@bibby-sterilin.com. www.bibby-sterilin.com.* Aspirators; biohazard bags; bottle cap liners; bottle top dispensers; butyl stoppers; chamber slides; cryovials; dishes; ESCO rubber; Fibra-Cel microcarriers; filters; flasks; glass-based dishes; magnetic stirrers; manual colony counter; media bottles; micropipettes; orbital shakers; pipettes; rocking racks; silicone tubing; silicone stoppers; Silescol; sterilization bags; universal containers.

Bigneat Ltd. 4 & 5 Piper's Wood Industrial Park, Waterbury Drive, Waterlooville, Hampshire, PO7 7XU, England, UK. +44 (0)23 92 266 400. +44 (0)23 92 263 373. *info@bigneat.com. www.bigneat.com.* Fume cabinets, ductless; laminar-flow hoods; safety cabinets.

Binder GmbH. Bergstrasse 14, D-78532 Tuttlingen, Germany. +49 (0) 7461 1792 0. +49 (0) 7461 1792 10. *info@binder-world.com. www.binder-world.com.* Incubators: anhydrous, CO_2, cooled.

BioCarta. 1-877-641-2355 x227. +1 760-804-1395. *info@biocarta.com. www.biocarta.com/.* Cell separation; cytokines; lymphocyte purification; magnetic cell isolation beads; PrepaCyte lymphocyte isolation reagent.

BioCarta Europe. +49 40 525 7030. +49 40 525 70377. *info.europe@biocarta.com. www.biocarta.com/.* Cell separation;

cytokines; lymphocyte purification; magnetic cell isolation beads; PrepaCyte lymphocyte isolation reagent.

Biochrom AG. POB 46 03 09, D-12213 Berlin, Germany. +49 30 77 99 06 0. +49-30 77 10 01 2; 49-30-77 99 06 66. *info@biochrom.de. www.biochrom.de.* Antibodies; Cell separation media; media; sera.

Biocolor Ltd. 10 Malone Road, Belfast BT9 5BN, Northern Ireland, UK. +44 (0) 1232 459 008. +44 (0) 1232 551 733. *info@biocolor.co.uk. www.biocolor.co.uk.* Collagen; collagen assay; Blyscan GAG assay; matrix assays.

Biocompare. *www.biocompare.com.* Directory of suppliers; products guide; sources of materials and equipment.

BioCrystal, Ltd. 575 McCorkle Blvd., Westerville, OH 43082-8888, USA. +1 (614) 818-0019.+1 (614) 818-1147. *sales@opticell.com. www.opticell.com.* OptiCell culture chamber.

Biodesign International. 60 Industrial Park Road, Saco, ME 04072, USA. 207-283-6500; 888-530-0140. 207-283-4800. *info@biodesign.com. www.biodesign.com.* antibodies; cardiac markers; CD markers; collagen; cytokines; growth factors; matrix; sera.

Biofluids. *See* Biosource International.

Biogene. BioGene House, 6 The Business Centre, Harvard Way, Kimbolton, Cambs PE28 0NJ, England, UK. +44 0845 1300 950. +44 0845 1300 960. *info@biogene.com. www.biogene.com/.* Chloroform isoamyl alcohol; molecular biology reagents.

Biohit, Inc. 3535 Route 66, Building 4, P.O. Box 308, Neptune, NJ 07754-0308, USA. +1 732 922 4900. +1 732 922 0557. *pipet@biohit.com. www.biohit.com.* Micropipettes; pipette tips; pipettors.

Biohit Oyj. Laippatie 1, 00880 Helsinki, Finland. +358-9-773 861. +358 9 773 86 200. *info@biohit.com. www.biohit.com.* Micropipettes; pipette tips; pipettors.

Biohit, Ltd. Unit 1, Barton Hill Way, Torquay, Devon TQ2 8JG, England, UK. +44 1803 315 900. +44 1803 315 530. *info@biohit.co.uk. www.biohit.com.* Micropipettes, pipette tips, pipettors.

Biomedical Technologies, Inc. 378 Page Street, Stoughton, MA 02072, USA. +1 781 344 9942. +1 781 341 1451. *info@btiinc.com. http://www.btiinc.com/.* Antibodies; attachment factors; matrix; collagen; fibronectin; laminin; medium additives; poly-D-lysine; growth factors.

Bionique Testing Laboratories, Inc. Bloomingdale Rd., RR1, Box 2, Saranac Lake, NY 12983, USA. +1 518 891 2356. +1 518 891 5753. *www.bionique.com/.* Mycoplasma testing.

Bio-Nobile Oy. P.O. Box 36, Tykistökatu 4 B, FIN-20521 Turku, Finland. +35 824 101 100. +35 824 101 123. *pickpen@bio-nobile.com. www.bio-nobile.com.* Magnetic separation; Pickpen colony counter.

Bioptechs, Inc. 3560 Beck Road, Butler, PA 16002, USA. 877 LIVE-CELL; +1 724 282 7145. +1 724 282 0745. *info@bioptechs.com. www.bioptechs.com.* Coverslip culture;

heated micro-observation dish; perfusion chambers; time-lapse observation.

Bioquell Medical Ltd. 29–31 Lynx Crescent, Weston-super-Mare, Somerset, BS24 9BR, England, UK. +44 (0)1934 410 500. *enquiries@bioquellmedical.co.uk.* *www.bioquellmedical.co.uk.* Hydrogen peroxide sterilization; laminar-flow hoods; safety cabinets; sterilizers.

Bioquell UK Ltd. 34 Walworth Road, Andover, Hants SP10 5AA, England, UK. 800 220 700; +44 (0)1264 835 835. +44 (0)1264 835 836. *enquiries@bioquell.com.* *www.bioquell.co.uk.* Clean rooms; isolators; hydrogen peroxide sterilization; gas generators; laminar-flow cabinets; microbiological safety cabinets.

Bio-Rad Laboratories. Life Science Group Div., 2000 Alfred Nobel Dr., Hercules, CA 94547, USA. 510-741-1000; 800 424 6723. 800-879-2289. *lsg_websupport@bio-rad.com.* *www.bio-rad.com.* Chromatography; DEAE dextran; DNA transfer; electrophoresis; gene transfection; image analysis; microcarriers; microtitration plate readers; microarray systems; microtitration robotics.

Bio-Rad Labs Ltd. Bio-Rad House, Maylands Avenue, Hemel Hempstead, Herts, HP2 7TD, England, UK. 0800 181134; +44 (0)20 8328 2000. +44 (0)20 8328 2550. *uk.lsg.marketing@bio-rad.com.* *www.bio-rad.com.* Chromatography; DEAE dextran; DNA transfer; electrophoresis; gene transfection; microcarriers; microtitration plate readers; microtitration robotics.

BioReliance. Invitrogen Bioservices, 14920 Broschart Rd., Rockville, MD 20850-3349, USA. +1 301 738 1000; 800 553 5372. +1 301 610 2590. *info@bioreliance.com.* *www.bioreliance.com.* *In vitro* toxicology; cytotoxicity assays; quality assurance assays; virus screening.

BioReliance (UK). Invitrogen Bioservices, Todd Campus, West of Scotland Science Park, Glasgow G20 OXA, Scotland, UK. +44 (0)141 946 9999. +44 (0)141 576 2412. *www.bioreliance.com.* *In vitro* toxicology; cytotoxicity assays; quality assurance assays; virus screening.

Biosource International. 542 Flynn Road, Camarillo, CA 93012, USA. 800 242 0607. 805 987 3385. *www.biosource.com.* Antibiotics; BSA coating; collagen; dipeptidyl peptidase IV (cd26) clone 202.36; fibronectin; FNC; gentamycin; growth factors; HEPES; hydrocortisone; keratinocytes; LHC medium; MCDB 402; medium; mycostatin; neural cell media; NR-1; selective medium; serum; serum-free medium; serum substitutes; penicillin; streptomycin; trypsin inhibitor; vitamins.

Biospec. Biospec Products, POB 788, Bartlesville, OK 74005, USA. +1-918-336-3363; 800-617-3363. 1-918-336-6060. *info@biospec.com.* *www.biospec.com.* Homogenization; microtitration plate sampling; MiniBeadBeater; RNA extraction.

BioSupplyNet. Cold Spring Harbor Laboratory Press. *www.biosupplynet.com.* Directory of suppliers; products guide; sources of materials and equipment.

Biotech Instruments, Ltd. Biotech House, 75a High Street, Kimpton, Hertfordshire SG4 8 PU, England, UK. +44 (0)1582 417 295. (+44 0)1582 457491. *www.biotinst.demon.co.uk/.* Confocal microscope; electrophoresis; fermentors.

Bio-Tek Instruments Inc. Highland Park, PO Box 998, Winooski, VT 05404-0998, USA. +1 802 655 4740; 888 451 5171. 802 655 7941. *labcsr@biotek.com.* *www.biotek.com/.* Plate readers.

Biotrol. Rat tails.

Biovest International. 8500 Evergreen Boulevard, Minneapolis, MN 55433, United States. +1 763 786 0302. +1 763 786 0915. *accountservices@biovest.com.* *www.biovest.com.* Bioreactors; hollow fiber perfusion culture; large scale culture; scale-up.

Biosero, Inc. 825 S. Primrose Ave, Suite G, Monrovia, CA 91016, USA. +1 626-303-0309. +1 626-256-6060. *info@bioseroinc.com.* *www.bioseroinc.com.* Laminar-flow cabinets; ductless fume cupboards; safety cabinets.

BMG Labtech GmbH. Hanns-Martin-Schleyer-Str. 10, D-77656 Offenburg, Germany. +49 781 969 68-0. +49 781 969 68-67. *germany@bmglabtech.com.* *www. bmglabtech.com.* Microtitration plate readers, absorbance, fluorescence, luminescence.

BMG Labtech Inc. 2415 Presidential Drive, Building 204, Suite 118, Durham, NC 27703, USA. +1 919 806 1735. +1 919 8068526. *usa@bmglabtech.com.* *www.bmglabtech.com.* Microtitration plate readers, fluorescence, luminescence, absorbance.

BMG Labtech Ltd. PO Box 73, Aylesbury, HP20 2QJ, England, UK. +44 (0)1296 336 650. +44 1296 336651. *uksales@bmglabtech.com.* *www.bmglabtech.com.* Microtitration plate readers, fluorescence, luminescence, absorbance.

BOC Edwards. Manor Royal, Crawley West, Sussex, RH10 9LW, UK. +(44) 1293 528844. +(44) 1293 533453. *www.bocedwards.com.* Pumps: diaphragm; PTFE, vacuum; silicone grease.

BOC Edwards USA. 301 Ballardvale Street, Wilmington, MA 01887, USA. 800 848 9800; +1 978 658 5410. +(1) 978 657 6546. *www.bocedwards.com.* Pumps: diaphragm; PTFE, vacuum; silicone grease.

Boehringer Mannheim. *See* Roche Applied Sciences.

Borolabs, Ltd. Unit F The Loddon Centre, Wade Road, Basingstoke, RG24 8FL, England, UK. +44 (0)870 300 1001. +44 (0)870 300 1004. *enquiries@borolabs.co.uk.* *www.borolabs.co.uk.* CO_2 incubators; data logger; nitrogen freezers; temperature recorders.

Brand GmbH. Postfach 11 55, D-97861, Germany. +49 (0) 9342 8080. +49 (0)9342 808 236. *info@brand.de.* *www.brand.de.* Bottle-top dispensers; glassware; microtitration plates; multipoint pipettors; pipettes; pipettors; pipetting aids; racks; repetitive pipettor; stepped pipettor; tubes.

BrandTech Scientific Inc. 11 Bokum Road, Essex, CT 06426-1506, USA. +1-860-767 2562. +1-860-767 2563. *mail@brandtech.com. www.brandtech.com. See* Brand GmbH.

Brandel Corp. 8561 Atlas Drive, Gaithersburg, MD 20877, USA. 800-948-6506. +1 301-869-5570. *sales@brandel. com. www.brandel.com.* Plate sealers.

Braun B Biotech, Inc. *See* Sartorius. Fermentors.

Brinkmann Instruments, Inc. 1 Cantiague Road, Westbury, NY 11590-0207, USA. 800 645 3050; +1 516 334 7500. +1 516 334 7506. *info@brinkmann.com. www.brinkmann.com.* Centrifuge tubes; gridded cover slips; liquid handling; microarray; microcapillaries; microcentrifuges; microcentrifuge tubes; microinjectors; micromanipulators; micropipettes; pipettor tips; pipettors.

British Association for Cancer Research (BACR). *Contact:* BACR Secretariat, c/o The Institute of Cancer Research, McElwain Laboratories, Cotswold Road, Sutton, Surrey SM2 5 NG, UK. +44 (0) 20 8722 4208. + 44 (0) 20 8770 1395. *bacr@icr.ac.uk. www.bacr.org.uk.* United Kingdom cancer society for scientists and clinicians.

British Oxygen Co., Ltd. The Priestly Centre, 10 Priestly Road, The Surrey Research Park, Guilford, Surrey GU2 5XY, England, UK. 800 111333; +44 (0)1483 579857. *www.boc.com.* Gases; gas regulators; CO_2; vacuum pumps.

British Society for Cell Biology. Contact M. Clements, Department of Zoology, Downing St., Cambridge CB2 3EJ, England, UK. +44 (0)1223 336 655. 00 44 (0) 1223 353 980. *bscb@bscb.org. http://www.kcl.ac.uk/kis/ schools/life_sciences/biomed/bscb/top.html.* Scientific society with special interest in cell and molecular biology.

Bruce Medical Supply. 411 Waverly Oaks Rd., Ste. 154, Waltham, MA 02452, USA. 800-225-8446; +1 781-894-6262. +1 781-894-9519. *sales@brucemedical.com. www.store.yahoo.com/brucemedical.* Antiseptics; Betadine; forceps; irrigation solutions; sample cups; scissors.

BS & S (Scotland), Ltd. 5/7 West Telferton, Portobello Industrial Estate, Edinburgh EH7 6UL, Scotland, UK. 0131 669 2282. 0131 657 4576. *bss@telferton.co.uk.* Ecoscint.

Buck Scientific, Inc. 58 Fort Point St., East Norwalk, CT 06855. 203-853-9444; 800-562-5566. 203-853-0569. *www.bucksci.com.* Microscope incubation chambers; time-lapse incubators.

Burge Scientific. 6 Langley Business Court, World's End, Beedon, Nr. Newbury, Berkshire RG20 8RY, England, UK. +44 (0)1635 248171. +44 (0)1635 248857. *sales@burge.co.uk. www.burge.co.uk/; http://www. arrowmight.co.uk/burge/index.htm.* Glassware washing machine.

Caisson Laboratories, Inc. 5 West Center, Sugar City, ID 83440, USA. +1 208 656 0880; 1877 840 0500. +1 208 656 0888. *custserv@caissonlabs.com. www.caissonlabs.com.* Custom media; fetal bovine serum; flasks; media; PBS; pipettes; roller bottles; trypsin; vented caps.

Calbiochem-Novabiochem Corp. *See* EMD Biosciences (North America), Merck Biosciences (elsewhere)..

Cambio, Ltd. The Irwin Centre, Scotland Road, Dry Drayton, Cambridge, CB3 8AR, England, UK. +44 (0)1954 210 200. +44 (0)1954 210 300. *support@cambio.co.uk. www.cambio.co.uk.* Chromosome paints.

Cambrex—Bioresearch Products. 8830 Biggs Ford Road, P.O. Box 127, Walkersville, MD 21793 0127, USA. +1 301-898-7025; 800-638-8174. 301-845-8338. *biotech-serv@cambrex.com. www.cambrex.com/bioproducts.* Agarose; antibiotics; penicillin, streptomycin; BSS: Hanks', Earle's; growth factors; media: selective, serum-free; specialized cells: astrocytes, B-cells, bone, hepatocytes, endothelial cells, epidermal keratinocytes, female reproductive cells, kidney epithelium, lung epithelium, mammary cells, melanocytes, neural progenitors, osteoblasts, prostate cells, renal cells, skin, smooth muscle cells, T-cells; vitamins.

Cambrex Bio Science Wokingham, Ltd. 1 Ashville Way, Wokingham, Berkshire RG41 2PL, England, UK. +44 (0)1189 795 234. +44 (0)1189 795 231. *sales.uk@cambrex.com. www.cambrex.com. See* Cambrex—Bioresearch Products.

Cambrex Bio Science Verviers, S.p.r.l. Parc Industriel de Petit Rechain, B-4800 Verviers, Belgium. +32-8-732-1609.+32-8-732-1634. *info.europe@cambrex.com. www. cambrex.com/bioproducts. See* Cambrex—Bioresearch Products.

Cambrex. Technical Sales Department, P.O. Box 127, 0245 Brown Deer Road, San Diego, CA 92121, USA. 800-852-5663. 858-824-0826. *www.clonetics.com..*

Cambridge BioScience Ltd. 24–25 Signet Court, Newmarket Road, Cambridge, CB5 8LA, England, UK. +44 (0)1223 316 855. +44 (0)1223 360 732. *Tech@cbio.co.uk. www.bioscience.co.uk.* Antibodies; apoptosis; baculovirus; cell cycle; collagen; cytokines; DNA assay; DNA transfer; ELISA; flow cytometry; growth factors; interleukin-2; molecular probes; oncogenes; signal transduction; tetracycline.

Camlab Ltd. Norman Way Industrial Estate, Over, Cambridge, CB4 5YE, England, UK. +44 (0)1954 233 100. +44 (0)1954 233 101. *mailbox@camlab.co.uk. www.camlab.co.uk.* Aspirators; automatic pipette plugger; carboys; centrifuge tubes; CO_2 incubators; containers; freezer inventory systems; glassware; glassware washing machines; glass universals; incubators; magnetic stirrers; medical flats; microtitration plate boxes and shakers; ovens; Parafilm; stirrers; pipette tips; pipettors; plasticware; plate readers; plate shakers; racks; reservoirs; shaking incubators; sterilizing and drying ovens; sterilizing filters; tubes; washup and sterilization; ultrasonic baths; vials; wash bottles; washing machines.

Campden Instruments. Leicester, UK. 0870 2403702. *UKsales@campdeninstruments.com. www.campden-inst. com/.* McIlwain tissue chopper.

Campden Instruments. Lafayette, IN, USA. +1 765 423 1505. *USsales@campdeninstruments.com. www.campden-inst.com/.* McIlwain tissue chopper.

Carl Zeiss AG. P.O.B. 4041, 37030 Gottingen, Germany. +49 (0) 551 5060 660. +49 (0) 551 5060 464. *mikro@zeiss.de. www.zeiss.de.* Apotome fluorescence imaging; confocal microscopy; image analysis; incubation chamber; microscopes: fluorescence, inverted, stereo, upright; optical sectioning; photomicroscopes; time-lapse video.

Carl Zeiss, Inc. Microscope Division, One Zeiss Dr., Thornwood, NY 10594, USA. +1 914 747 1800; 1800 233 2343. +1 914 681 7446. *micro@zeiss.com. www.zeiss.com/micro.* For products *see* Carl Zeiss AG.

Carl Zeiss Ltd. 15–20 Woodfield Road, Welwyn Garden City, Hertfordshire, AL7 1JQ, England, UK. +44 (0)1707 871200. +44 (0)1707 330237. *www.zeiss.co.uk.* For products *see* Carl Zeiss AG.

Cascade Biologics, Inc. 1341 Custer Drive, Portland, OR 97219, USA. 1 800 778 4770. +1 503 292 9521. *info@cascadebio.com. www.cascadebio.com.* Corneal epithelial cells; dermal fibroblasts; endothelial cells; epidermal keratinocytes; EpiLife growth supplements; MCDB 154; melanocytes; serum-free selective media; smooth muscle cells; specialized cell cultures; trypsin/EDTA; trypsin inhibitor.

Cascade Biologics, Inc., UK. Mansfield I-Centre, Oakham Business Park, Hamilton Way, Mansfield, Nottinghamshire NG18 5BR, England, UK. +44 (0) 1623 600 815. +44 (0) 1623 600 816. *ukinfo@cascadebio.com. www.cascadebio.com. See* Cascade Biologics, Inc., USA.

Cato Research. *www.cato.com/biotech/bio-prod.html.* Internet list of major biotechnology companies.

Cell Culture Characterization Services. 4581, Lapeer Road, Suite C, Orion Township, MI 48359. +1 248-656-2542; +1248-563-1493. +1 248-656-3899. *bhukku@cellcharacterization.com.* Authentication; chromosome analysis; chromosome painting; karyotyping; isoenzymes; FISH.

Cellex Biosciences Inc. *See* Biovest International.

CellGenix Technologie Transfer GmbH. Am Flughafen 16, D-79108 Freiburg. +49 761 888 89–330. +49 761 888 89–830. *info@cellgenix.com. www.cellgenix.com.* CellGro kit systems; cytokines; serum-free media; culture bags.

Cellgro. *See* Mediatech.

Cellmark Diagnostics. *See* Orchid Cellmark..

Cellmark UK. PO Box 265, Abingdon, Oxfordshire, OX14 1YX, England, UK. +44 (0)1235 528000. *cellmark@orchid.co.uk. www.cellmark.co.uk.* Cell line authentication; DNA profiling.

Cellon SA. CELLON S.A., 29 Am Bechler, L-7213 Bereldange, Luxembourg. +352 312 313. +352 311 052. *info@cellon.lu. www.cellon.lu/.* Bioreactors; collagen; culture bags; CulturSil; fermentors; matrix, media bags; NASA, Rotary Cell Culture System; scale-up.

Cellon UK Ltd. 61 Dublin St., Edinburgh, EN3 6NL, Scotland, UK. +44 (0)7714 332567. +44 (0)1472 852642. *info@cellon.lu. www.cellon.lu/.* Bioreactors; collagen; culture bags; CulturSil; fermentors; matrix, media bags; NASA, Rotary Cell Culture System; scale-up.

CellStat. 755 Page Mill Road, B-6, Palo Alto, CA 94304, USA. +1 650 802 0351. *www.cellstat.com.* Cytotoxicity testing; drug response and metabolic monitoring; real-time cell monitoring.

Cellular Products GmbH. Delitzscher Strasse 135, 04129 Leipzig, Sachsen, Germany. +49 (0)341 971 9758. +49 (0)341 972 5389. *info@cellular-products.de. www.cellular-products.de.* Human cytokines.

Cellutech. Watertown, NY, USA. 800-575-5945;. Transportation containers; spillage safety.

Celox. *See* Protide Pharmaceuticals.

Celtrix Laboratories. *See* Insmed.

Charles Austen Pumps Ltd. Royston Road, Byfleet, Surrey, KT14 7NY, England, UK. +44 (0)1932 355 277. +44 (0)1932 351 285. *info@charlesausten.com. www.charlesausten.com.* Pumps: diaphragm; PTFE, vacuum.

Charles River Laboratories. 251 Ballardvale St., Wilmington, MA 01887. +1 508 658 6000. 800-992-7329; 978 658 7132. *comments@criver.com. www.criver.com.* Endotoxin testing; laboratory animals.

Charles River (U.K.), Ltd. Manson Rd., Margate, Kent CT9 4LT, England. +44(0)1843 823575. +44(0)1843 823497. *crukcsd@uk.criver.com. www.criver.com/cruk.* Endotoxin testing; laboratory animals.

Chart Biomed. Domestic customer service. 800-482-2473; +1 770-257-12. 8-932-2473; 770-257-1300. *www.chartbiomed.com/.* Liquid nitrogen freezers.

Chart Biomed Europe. European Sales and Operation Director. +44 (0)1932 785570. +44 (0)1932 785 576. *www.chartbiomed.com/.* Liquid nitrogen freezers.

Chemap. *See* Alfa Laval.

Chemicon Europe. 9 Trident Way, Southall, UB18 7US, England, UK. +1 909 676 8080; 800 437 7500. +1 909 676 9209. *www.chemicon.com. See* Chemicon International.

Chemicon International Inc. 28820 Single Oak Drive, Temecula, CA 92590, USA. +1 909 676 8080; 800 437 7500. +1 909 676 9209. *www.chemicon.com.* Antibodies to cytokines; dialysis tubing; cytokine antibody arrays; GalC; EMA; GFAP; hypoxia detection; integrins; microarray technology; MUC-1; mucin neurotransporters; oncoproteins; signal transduction; stem cells; villin; viruses.

Chiron Corp. 4560 Horton Street, Emeryville, CA 94608-2916, USA. +1 510 923–2300. +1 510 923–3376. *communications@chiron.com. www.chiron.com.* Cytokines; pharmaceuticals; immunoassays.

Chondrex Inc. 2607 151st Place NE, Redmond, WA 98052, USA. (425) 702–6365 or (888) CHONDRE. *info@chondrex.com. www.chondrex.com.* Collagen; collagen antibodies; collagen staining; collagenase.

Ciba Specialty Chemicals Corp. 540 White Plains Road, P.O. Box 2005, Tarrytown, NY 10591-9005, USA. +1 914 785 2000. +1 914 785 2211. *kevin.bryla@cibasc.com. www.cibasc.com.* Photo and digital imaging.

Cin-Made Corp. 1780 Dreman Ave., Cincinnati, OH 45223, USA. 513-681-3600. 513-541-5945. *www.cinmadepackaginggroup.com.* Transportation containers; safety.

Clonetics. *See* Cambrex.

CoaChrom Diagnostica GmbH. Leo Mathauser Gasse 71, A1230 Vienna, Austria. +43 1699 97 97 0. +43 1699 18 97. *office@coachrom.com. www.coachrom.com/.* Antibodies; cytokines; endotoxin test; serum-free medium.

Cohesion Technologies. 2500 Faber Place, Palo Alto, CA 94303, USA;. +1 650-320-5500. +1 650 320−5522. *admin@cohesiontech.com. www.cohesiontech.com.* Cell Prime collagen coating; factor VIII antibody; Vitrogen.

Cole-Parmer Instrument Co. 625 E. Bunker Ct., Vernon Hills, IL 60061, USA. 847 549 7600, 800 323 4340. +1 847 247−2929. *techinfo@coleparmer.com. www.coleparmer.com/.* Data loggers; desiccator cabinets; electronic colony counters; meters; pH meters; pipettors; pumps; repeating syringe dispenser; slide boxes; stirring hot plate; thermometers; tubing; ultrasonic baths; water baths.

Cole-Parmer Instrument Co. Ltd. Unit 3, River Brent Business Park, Trumpers Way, Hanwell, London, W7 2QA, England, UK. +44 (0)208 574 7556. +44 (0)208 574 7543. *sales@coleparmer.co.uk. www.coleparmer.co.uk. See* Cole-Parmer Co., USA.

Collagen Corp. 2500 Faber Place, Palo Alto, CA 94303, USA. 415-856-0200. 415-856-2238. *www.collagen.com/.* Collagen, Vitrogen 100.

Collagen Corp. *See* Inamed Aesthetics.

Comark Ltd. Comark House, Gunnels Wood Park, Gunnels Wood Road, Stevenage, Hertfordshire, SG1 2TA, England, UK. +44 (0)1438 367 367. +44 (0)1438 367 400. *salesuk@comarkltd.com. www.comarkltd.com.* Electronic thermometers; recording thermometers.

Comark Instruments Inc. 9710 SW Sunshine Court, Beaverton, OR 97005, USA. 800 555 6658; +1 503 643 5204. +1 503 644 5859. *sales@comarkUSA.com. www.comarkltd.com/usa.* Electronic thermometers; recording thermometers.

CompuCyte Corp. 12 Emily Street, Cambridge, MA 02139, USA. 800-840-1303; +1 617-492-1300. +1 617 577 4501. *www.compucyte.com.* Fluorochromatic cytometer.

Computer Imprintable Label Systems Ltd. 2 Southdownview Way, Broadwater Business Park, Worthing, West Sussex, BN14 8BR, England, UK. +44 (0)1903 219 000. +44 (0)1903 219 111. *sales@cils-labels.com. www.cils-labels.com.* Autoclave labels; computer-printable labels; cryostorage labels; cryovial labels.

Contamination Control Products. 1 Third Ave, Neptune City, NJ 07753, USA. +1 877 553−2676. +1 732 869−2999. *info@ccpcleanroom.com. www.ccpcleanroom.com/.* Laminar flow hoods; microbiological safety cabinets.

Cook Biotech, Inc. 3055 Kent Avenue, West Lafayette, IN 47906, USA. +1 765 497 3355. +1 765 497 2361. *vivosis@cookbiotech.com. www.cookgroup.com.* ECM; extracellular matrix; tissue engineering scaffolds.

Coriell Cell Repository (CCR). 403 Haddon Avenue, Camden, NJ 08103, USA. 800-752-3805 (from US); 856-757-4848 (from abroad). 856-757-9737. *ccr@coriell.umdnj.edu.* *http://locus.umdnj.edu/ccr/.* Cell bank; Cell lines; NIGMS Human Genetic Mutant Cell Repository; NIA Aging Cell Repository.

Corning Inc., Life Sciences. 45 Nagog Park, Acton, MA 01720, USA. +1 978 635 2200; 800 492 1110. +1 978 635 2476. *clswebmail@corning.com. www.corning.com/Lifesciences.* (*See also* Sigma in Europe.) Automatic dispensers; bioreactors; bottles; Cellstack culture chambers; centrifuge tubes; conductivity meters; culture vessels; electrodes; filter well inserts; flasks; gas-permeable caps; glassware; magnetic stirrers; micropipettes; microscope slide containers; microtitration plates; multisurface propagators; multiwell plates; permeable caps; Petri dishes; pH meters; pipette aids; pipettes; roller bottles; spinner flasks; temperature controllers.

Corning B.V. Life Sciences. Koolhovenlaan 12, 1119 NE Schiphol-Rijk, The Netherlands. +31 (0) 20-659-60-51.+31 (0) 20-659-76-73. *cceurnl@corning.com. www.corning.com/lifesciences.* (*See also* Sigma.) For products *see* Corning Inc.

Costar. *See* Corning.

Coulter. *See* Beckman Coulter.

CP Pharmaceuticals Ltd. Ash Road North, Wrexham Industrial Estate, Wrexham LL13 9UF, Wales, UK;. *www.cppharma.co.uk/home.htm.* Insulin.

C.R. Bard Inc. Bard Urological Division, 8195 Industrial Boulevard, Covington, GA 30014, USA. +1 770-784-6100; 800-526-4455. *bob.mccall@crbard.com. www.bard-contigen.com.* Collagen Sponge.

Cryomed.. *See* Thermo Electron Corp., Thermo-Haake. *www.thermols.com.* Controlled-rate freezer; gloves; freezer racks; liquid nitrogen freezers.

Cryoservice Ltd. Cryoservice Limited, Prescott Drive, Warndon Business Park, Worcester, WR4 9RH, England, UK. +44 (0)1905 754 500. +44 (0)1905 754 060. *info@cryoservice.co.uk. www.cryoservice.co.uk.* CO_2; gases, liquid nitrogen.

CryoTrack. +1 650 305 9406; +1 650 941 6774. *info@cryotrack.com. www.cryotrack.com.* Cryogenic storage inventory control software.

Cytogration, Inc. 113 N. Washington Street, Suite 342, Rockville, MD 20850, USA. +1 301 330 4754; 877 330 4754. +1 301 519 9474. *info@cytogration.com. www.cytogration.com.* Robotics.

Dage-MTI, Inc. 701 N. Roeske Ave, Michigan City, IN 46360, USA. 219-872-5514. 219-872-5559. *info*

@dagemti.com. www.dagemti.com. CCD cameras; image intensifiers; image processors; monitors; time-lapse video.

Daigger. 620 Lakeview Parkway, Vernon Hills, IL 60061, USA. 1-800-621-7193. 1-800-320-7200;. *daigger @daigger.com. www.daigger.com.* Dishes; flasks; nitrile gloves; pipette aids; plasticware.

DakoCytomation Inc. 6392 Via Real, Carpinteria, CA 93013, USA. 805 566 6655, 800 235 5763. 805 566 6688. *customer.service@dakocytomation.com. www.dakocytomation.us.* Antibodies; apoptosis reagents; cytokeratins; flow cytometry reagents.

DakoCytomation Ltd. Marketing Dept., Denmark House, Angel Drove, Ely, Cambridge, CB2 4ET, England, UK. +44 (0)1353 669 911. +44 (0)1353 668 989. *www.dakoltd.co.uk www.dakocytomation.co.uk.* Antibodies; apoptosis reagents; cytokeratins; flow cytometry reagents.

DakoCytomation Denmark A/S. Produktionsvej 42, DK-2600 Glostrup, Denmark. 44 85 95 00. 44 85 95 95. *contact@dakocytomation.com. www.dakocytomation. com.* Antibodies; apoptosis reagents; cytokeratins; flow cytometry reagents.

Damon IEC. *See* Thermo Life Sciences.

Davol Inc. 100 Sockanossett Crossroads, Cranston, RI 02920, USA. 800 556 6756. +1 401 946 5379. *info @davol.com. www.davol.com.* Collagen; PTFE mesh.

Day-Impex. *See* Keison Products.

Decon Laboratories Ltd. Conway Street, Hove, East Sussex, BN3 3LY, England, UK. +44 (0)1273 739 241. +44 (0)1273 722 088. *mail@decon.co.uk. www.decon.co.uk.* Detergents; inks; marker pens, ultrasonic cleaning baths.

Deutsche Sammlung von Mikroorganismen und Zellkulturen (DSMZ). Mascheroder Weg 1B, D-38124 Braunschweig, Germany. *contact@dsmz.de. www.dsmz.de.* Cell bank; cell lines; cell line authentication; mycoplasma testing.

Dianova GmbH. Mittelweg 176, D-20148 Hamburg, Germany. +49 (0)40 450 670. +49 (0)40 450 67 490. *info@dianova.de. www.dianova.de.* Antibodies.

DiaSorin. 1951 Northwestern Avenue, P.O. Box 285, Stillwater, MN 55082-0285, USA. +1 612 439 9710, 800 328 1482. +1 651 351 5669. *www.diasorin.com/.* Calcitonin; estradiol; FSH; hGH; osteoclacin; prolactin; triiodothyronine; vitamin D.

DiaSorin S.p.A. Via Crescentino, 13040 Saluggia (VC), Italy. +39.0161.487093. +39.0161.487628. *www.diasorin.com/.* Calcitonin; estradiol; FSH; hGH; osteoclacin; prolactin; triiodothyronine; vitamin D.

Difco. *See* BD Biosciences. Agar; microbiological media; bactopeptone; lactalbumin hydrolysate; nutrient broths; tryptose phosphate broth.

DNA Solutions. (Branches worldwide.) *www.DNAsolutions. com.* Authentication; DNA profiling; forensics; identity testing.

Doity Engineering Ltd. P.O.Box 25, Isherwood Street, Rochdale, Lancashire OL11 1ZZ, England, UK. +44 (0)1706 646 971; +44 (0)1706 345 515. +44 (0)1706 640 454. *sales@doity.com. www.doity.com/.* Cold room shelving; hot room shelving; polypropylene-coated steel storage racks.

Don Whitley Scientific Ltd. 14 Otley Road, Shipley, West Yorkshire, BD17 7 SE, England, UK. +44 (0)1274 595728. +44 (0)1274 531197. *info@dwscientific.co.uk. www.dwscientific.co.uk.* Colony counters.

Dow Corning Corp. P.O. Box 0994, Midland, MI 48686-0994, USA. +1 517-496-4000. +1 517-496-4586. *www.dowcorning.com.* Antifoam; deionizers; Methocel; methylcellulose; silicone tubing.

Laboratory Corporation of America. 1447 York Court Burlington, NC 27215, USA. 800-334-5161; +1 336-584-5171. *www.labcorp.com/.* DNA profiling; paternity testing.

DSMZ. *See* Deutsche Sammlung von Mikroorganismen und Zellkulturen.

DuPont Animal Health Solutions. Chilton Industrial Estate, Sudbury, Suffolk CO10 2XD, England, UK. +44 1787 377305. +44 1787 310846. *biosecurity@gbr.dupont. com. www.antecint.com.* Virkon disinfectant.

Dynal Biotech (USA). 9099 N. Deerbrook Tr., Brown Deer, WI 53223, USA.. 1-800-558-4511. 414-357-4518. *uscustserv@dynalbiotech.com. /www.dynalbiotech.com.* Cell separation; MACS; magnetizable beads; immunoaffinity; cell separation; nylon mesh filters.

Dynal Biotech UK. 11 Bassendale Road, Croft Business Park, Bromborough, Wirral, CH62 3QL, UK. 0800 731 9037; +44 (0)151 346 1234. 0151 346 1223. *ukcustserv@dynalbiotech.com. /www.dynalbiotech.com.* *See* Dynal Biotech (USA).

Dynalab Corp. 350 Commerce Drive, Rochester NY 14623, USA. 800 828–6595. 716-334-9496. *labinfo @dyna-labware.com. www.dynalabcorp.com/.* Ovens; sample cups; storage.

Dynex Technologies Ltd. Columbia House, Columbia Drive, Worthing, England, UK. 01403 783381. 01403 784397. Microtitration: plate readers, auto samplers, diluters, dispensers, mixers, plate washers, microtitration software.

Dynex Technologies, Inc. 14340 Sullyfield Cir., Chantilly, VA 20151-1683, USA. +44 (0) 1903 267555. +44 (0) 1903 267722. *dynexuk@btconnect.com. www.dynextechnologies.com/.* Microtitration: plate readers, auto samplers, diluters, dispensers, mixers, plate washers, microtitration software.

East Coast Biologics, Inc. P.O. Box 489, North Berwick, ME 03906-0489, USA. +1 207-676-7639. +1 207-676-7658. *info@eastcoastbio.com. www.eastcoastbio.com.* Antibodies.

Eastman Kodak Co. Scientific Imaging Systems Div, 343 State St, Rochester, NY 14652-4115, USA. +1 203 786 5600; 800 225 5352. +1 203 786 5694. *www.kodak.com.* CCD; Digital cameras; digital imaging;

hypoclearing agent; microscope cameras; photographic developers; photographic fixatives; X-ray film.

EBI, L.P. 100 Interpace Parkway, Parsippany, NJ 07054, USA. 800-526-2579. *ebimedical_support@ebimed.com. www. ebimedical.com/.* Bone graft matrices; tissue engineering scaffolds.

ECACC (European Collection of Cell Cultures). CAMR, HPA Porton Down, Salisbury, Wiltshire, SP4 0JG, England, UK. +44 (0)1980 612512. +44 (0)1980 611315. *ecacc@hpa.org.uk. www.ecacc.org.uk.* Authentication; cell banking; cell lines, database; characterization; mycoplasma screening; validation.

Econo-med. Trafalgar House, Station Road, Long Sutton, Spalding, Lincolnshire, PE12 8BP, England, UK. +44 (0)1406 36 42 42. +44 (0)1406 36 46 76. *sales@econo-med.com. www.econo-med.com.* Gloves, nitrile, vinyl.

Electron Microscopy Services (EMS). 1560 Industry Road Hatfield, PA 19440, USA. +1 215-412-8400. +1 215-412-8450;. *sgkcck@aol.com. www.emsdiasum.com/ ems.* Cytotek cytocentrifuge; dissecting instruments; electron microscopy reagents; fixatives; lead citrate; uranyl acetate.

Elga LabWater, US Filter. 10 Technology Drive, Lowell, MA 01851, USA. +1 508 887 6300; 800 466 7873 ex 5000. +1 508 887 6266; 800 875 7873. *www.elgalabwater.com.* Filtration; ion exchange; photooxidation; reverse osmosis; water purification equipment.

Elga LabWater. High Street, Lane End, High Wycombe, Bucks HP14 3JH, England, UK. +44 1494 887 555. +44 1494 887 837. *sandra.welch@veoliawater.com. www.elgalabwater.co.uk.* Filtration; ion exchange; photooxidation; reverse osmosis; water purification equipment.

Elkay Laboratory Products. Unit 4, Marlborough Mews, Crockford Lane, Basingstoke, Hampshire, RG24 8NA, England, UK. +44 (0)1256 811118. +44 (0)1256 811116. *sales@elkay-uk.co.uk. www.elkay-uk.co.uk.* Biohazard bags; cryovials; disposable caps; filter tips; marker pens; media reservoirs; micropipettes; microtitration; pastettes; pipettes; pipette tips; pasteur pipettes; plate sealers; ripette; tips; transport tubes; tubes; tubing; vials; universal containers.

EMD Biosciences. 10394 Pacific Ctr. Ct., San Diego, CA 92121, USA. +1 858 450 9600; 800-854-3417. 800 776 0999; +1 858-453-3552. *customer. service@emdbiosciences.com. www.emdbiosciences.com.* Antibodies; biochemicals; buffers; chemicals; cholera toxin; detergents; dextran; EDTA; G418 sulfate; glycobiology reagents; hygromycin B; immunochemicals; enzymes.

EMS. *See* Electron Microscopy Services.

Endotronics. *See* Biovest International.

Envair Ltd. York Ave, Haslingden, Rossendale, Lancashire BB4 4HX, England, UK. +44 (0)1706 228416. +44 (0)1706 831957. *info@envair.co.uk. www.envair.co.uk.* Laminar flow hoods; CCTV monitoring; clean air rooms; microbiological safety cabinets.

Enzo Life Sciences, Inc. 60 Executive Blvd., Farmingdale, NY 11735, USA. +1 631 694 7070. +1 631 694 7501; 800 221 7705. *www.enzo.com.* Antibodies; m-chromogranin (clone PHE-5); microarray.

Eppendorf AG. Barkhausenweg 1, 22331 Hamburg, Germany. +49 405 380 10. +49 405 380 1556. *eppendorf@eppendorf.com. www.eppendorf.com.* Gridded cover slips; microcapillaries; microcentrifuges; microcentrifuge tubes; microinjectors; micromanipulators; micropipettes; pipettor tips; pipettors.

Eppendorf UK, Ltd. Eppendorf UK Limited, Endurance House, Chivers Way, Histon, Cambridge, CB4 9ZR, England, UK. +44 (0)1223 200 440. +44 (0)1223 200 441. *sales@eppendorf.co.uk. www.eppendorf.co.uk.* Gridded cover slips; microcapillaries; microcentrifuges; microcentrifuge tubes; microinjectors; micromanipulators; micropipettes; pipettor tips; pipettors.

Eppendorf USA. *See* Brinkmann.

Erlab DFS SA. Parc D'affaires Des Portes, BP 403, 27104 Val De Reuil Cedex, France. +33 2 32 09 55 80. +33 2 32 09 55 90. *Sales@erlab.net. www.erlab-dfs.com/.* Fume cupboards; laminar-flow hoods; protective enclosures.

Erlab DFS SA. St Thomas's House, St Thomas's Square, Salisbury, Wiltshire, SP1 1BA, England, UK. +44 (0)1722 341 940. +44 (0)1722 341 950. *SalesUK@erlab.net. www.captair.com.* Fume cupboards; laminar-flow hoods; protective enclosures.

ESA Inc. 22 Alpha Road, Chelmsford, MA 01824–4171, USA. 800 959–5095; +1 978 250 7000. +1 978 250–7090. *marketing@esainc.com. www.esainc.com.* Computer software; microtitration.

ESACT-UK. *Secretary:* Alison Stacey, ESACT-UK, PO Box 117, Saffron Walden, Essex, CB10 1XD, England, UK. *a.stacey@adpro.co.uk. www.esactuk.org.uk.* Julian Hanak, Chairman.

European Collection of Cell Cultures. *See* ECACC.

European Life Scientist Organisation. Contact: Ingeborg Fatscher, P.O. Box 1151, D-69199 Sandhausen, Germany. +49 6224 925613. +49 6224 925610,11,12. *contact@elso.org. www.elso.org.* Kai Simons, President.

European Society for Animal Cell Culture Technology (ESACT). *Office:* PO BOX 1723, 5 Bourne Gardens, Porton, Salisbury, Wilts SP4 0PL, England, UK. +44 1 980 610 405. +44 1 980 610 405. *contact@esact.org; office@esact.org. www.esact.org.* Otto-Wilhelm Merten, President.

European Society of Toxicology *In Vitro* (ESTIV). *ESTIV Secretary:* Dr Jan van der Valk, NCA, Dept. Animals, Science & Society, P.O. Box 80.166, NL-3508 TD Utrecht, The Netherlands. +31 30 253 2163. +31 30 253 9227. *secretary@estiv.org. www.estiv.org.* President: R. Combes (UK).

European Tissue Culture Society (ETCS). *Secretary:* Petra Boukamp, Deutsches Krebsforschungzentrum, Im Neuenheimer Feld 280, D-69120 Heidelberg, Germany. *www.etcs.info/etcsmain.html.* President: John Masters.

European Tissue Culture Society, UK Branch (ETCS-UK). *Contact:* David Lewis, ECACC, Heath Protection Agency, Porton Down, Salisbury, SP4 0JG, UK. +44 (0)1980 612 512. +44 (0)1980 611 315. *david.lewis@hpa.org.uk. www.ecacc.org.uk/..*

Falcon. *See* BD Biosciences.

FiberCell Systems, Inc. 905 West 7th Street # 334, Frederick, MD 21701, USA. +1 301-471-1269. +1 301. 865.6375. *gcadwell@adelphia.net. www.fibercellsystems. com.* Hollow fiber perfusion culture.

Fine Science Tools GmbH. Fahrtgasse 7–13, D-69117 Heidelberg, Germany. +49 (0)62 21/90 50 50. +49 (0)62 21/60 00 01. *europe@finescience.com. www.finescience.com/.* Dissecting instruments; Surgical instruments.

Fine Scientific Tools. 373-G Vintage Park Drive, Foster City, CA 94404-1139, USA. 800-521-2109; +1 650 349 1636. 800-523-2109; 650-349-3729. *info@finescience.com. www.finescience.com/.* Dissecting instruments; Surgical instruments.

Fisher Scientific Corp. 2000 Park Ln, Pittsburgh, PA 15275, USA. +1 412-490-8300; 800-766-7000. 800-926-1166. *www1.fishersci.com/.* Borosilicate glass bottles; carboxymethylcellulose (CMC); centrifuges; chemicals; cloning rings; CO_2 incubators; containers for nuclear emulsion; coverslips; crystal violet; diacetyl fluorescein; dissecting instruments; EDTA; Erlenmeyer flasks; electronic thermometers; freezers; Giemsa stain; glucose; glycerol; hemocytometers; magnetic stirrers; Microcaps (Drummond); Oxoid; pipettes; refractometer; sterilizing and drying oven; stirring hot plate; tape; temperature controllers; temperature recorders; vortex mixers.

Fisher Scientific U.K. Bishop Meadow Road, Loughborough, Leicestershire LE11 5RG, England, UK. +44 (0)1509 231166. +44 (0)1509 231893. *info@fisher.co.uk. www.fisher.co.uk.* For products *see* Fisher Scientific Corp.

Fison Instruments. *See* Fisher Scientific.

Fluka Riedel-de Haën. Industriestrasse 25, CH-9471 Buchs SG, Switzerland. +41 (0)81 755 25 11. +41(0)81 755 28 15. *Fluka@sial.com. www.sigmaaldrich.com/ Brands/Fluka.* Biochemicals; chemicals; EDTA; methylcellulose.

Fluka Chemical Corp. *See* Sigma-Aldrich. *www.sial.com.* Biochemicals; chemicals; EDTA; methylcellulose.

Fluorochem Ltd. Wesley Street, Old Glossop, Derbyshire SK13 9RY, England. 01457 868921. 01457 869360. *enquiries@fluorochem.co.uk. www.fluorochem.net/.* Density medium; flotation medium; fluorochemicals.

FMC BioProducts. *See* Cambrex. *http://www.cambrex. com/.* DNA sequencing; DNA separation; mutation detection; protein separation; PCR.

Forma. *See* Thermo Electron Corp. *www.thermo.com.* CO_2 incubators.

Fougera. Melville, NY. *www.fougera.com.* Sterile surgical lubricant; Surgilube.

Fresenius AG. D-61346 Bad Homburg v.d.H., Germany. +49 6172 608 0. *www.fresenius-ag.com/.* Oligonucleotide probes.

Fuji Medical Systems U.S.A., Inc. Corporate Office, 419 West Avenue, Stamford, CT 06902, USA. 800 431–1850; +1 203 324 2000; Customer Service: 800 872–3854; +1 203 602–3609. 203-327-6485. *ssg@fujimed.com. www.fujimed.com.* Photographic: film, developers, fixatives, emulsions; X-ray film.

Fuji Photo Film. PO Box 015, Leamington Spa, Warwickshire, CV 31 YA, England. +44 (0)1926 335 537. +44 (0)1926 887 793. *www.fujifilm-europe.com/.* Photographic: film, developers, fixatives, emulsions; X-ray film.

G.Q.F. Manufacturing Co. P.O. Box 1552, Savannah, GA 31402-1552, USA. +1 912 236–0651. +1 912 234 9978. *sales@gqfmfg.com. www.gqfmfg.com.* Egg incubators.

GA International. 965 Boul. Cure-Labelle, C.P.41534, Laval, Quebec, H7V 4A3, Canada. +1 450-973-9420, 1-800-518-0364. +1 450-973-6373. *mail@labtag.com. www.labtag.com.* cryolabels; markers.

GE Heathcare. *See* Amersham Biosciences.

GE Silicones. World Headquarters, Wilton, CT 06897. 800-255-8886. *www.gesilicones.com/.* Silicone rubber adhesive sealant.

Gemini Bio-Products. 1301 East Beamer Street, Woodland, CA 95776, USA. 800 543 6464; +1 530 668 3636. +1 530 668 3630. *www.gembio.com/.* Antibiotics; media; serum.

Genetic Research Instrumentation Ltd. Gene House, Queenborough Lane, Rayne, Braintree, Essex CM77 6TZ, England, UK. +44 (0)1376 332 900. +44 (0)1376 344 724. *gri@gri.co.uk. www.gri.co.uk.* Automatic dispensers; autoradiography film and cassettes; bioreactors; Buchler; controlled rate freezer; cryovials; density media; developers and fixers; fermentors; Guava flow cytometer; heating blocks; Hotpack incubators; Labconco; LEEC; lymphocyte preparation media; micropipettes; microtitration systems; nitrogen freezers; pipette tips; pipettors; refrigerated incubator; robotic liquid handling; roller rack incubators; roller drum; safety cabinets; stirrer flasks; water purification.

Gen-Probe, Inc. 10210 Genetic Center Drive, San Diego, CA 92121, USA. 619-546-8000; 800-523-5001. 800 288–3141; 800 342–7441. *customerservice@gen-probe. com. www.gen-probe.com/.* Mycoplasma detection kit.

Germfree Labs, Inc. 11 Aviator Way, Ormond Beach, FL 32174, USA. 800-888-5357; +1 386-677-7742. +1 386-677-1114. *info@germfree.com. www.germfree.com.* Microbiological safety cabinets; laminar-flow hoods; glove boxes; isolators.

Geron Corp.. 230 Constitution Drive, Menlo Park, CA 94025, USA. +1 650-473-7700. +1 650-473-7750. *info@geron.com. www.geron.com.* Human embryonic stem cell research; hESC directory.

GIBCO. *See* Invitrogen.

Gilson Inc. 3000 W. Beltline Hwy., PO Box 620027, Middleton, WI 53562, USA. 608 836 1551, 800 445 7661. 608 831 4451. *sales@gilson.com. www.gilson.com/.* Automatic dispensers; liquid handling; micropipettes; microtitration workstation; peristaltic pumps; pipettors; robotics; vacuum pumps.

Gilson S.A.S. 19, avenue des Entrepreneurs, BP 145, F-95400 Villiers le Bel, France. +33 01 34 29 50 00. +33 01 34 29 50 20. *www.gilson.com/.* Automatic dispensers; liquid handling; micropipettes; microtitration workstation; peristaltic pumps; pipettors; robotics; vacuum pumps.

Glen Research Corp. 22825 Davis Drive, Sterling, VA 20164, USA. +1 703 437 6191;. +1 703 435 9774. *support@glenres.com. www.glenres.com/.* Synthetic oligonucleotides.

Global Medical Instrumentation, Inc. (GMI). 6511 Bunker Lake Boulevard, Ramsey, MN 55303, UK. 763-712-8717. 763-497-0176. *richard@gmi-inc.com. www.gmi-inc.com.* Autoclaves; bioreactors; incubators; microbiological safety cabinets; sterilizers; Verax.

Gonotec GmbH. Eisenacher Strasse 56, 10823 Berlin, Germany. +49 (0)30 7809 588-0. +49 (0)30 7809 588-88. *contact@gonotec.com. www.gonotec.com. Osmometers.*

Gow-Mac Instrument Co. 277 Broadhead Rd., Bethlehem, PA 18017, USA. +1 610 954 9000. +1 610 954 0599. *sales@GOW-MAC.com. http://www.gow-mac.com/.* Gas analyzers; gas blenders; gas mixers.

Gow-Mac Instrument Co. (Ire) Ltd. Bay K 14a, Industrial Estate, Shannon, Cty. Clare, Ireland. +353-61-471 632. +353-61-471 042. *sales@GOW-MAC.ie. http://www.gow-mac.com/.* Gas blenders; Gas mixers; Gas analyzers.

Grace Bio-labs. PO Box 228, Bend, OR 97709, USA. +1 541 318 1208. +1 541-318-0242; 800-813-7339. *custservice@gracebio.com. www.gracebio.com.* Chamber slides; coverslip culture; multiwell slides.

Grant Instruments (Cambridge) Ltd. Shepreth, Cambridge, SG8 6 GB, England, UK. +44 (0)1763 260 811. +1 (0)1763 262 410. *info@grant.co.uk. www.grant.co.uk.* Electronic thermometers; recording thermometers; temperature recorders; water baths.

Greiner Bio-One GmbH. Maybachstrasse 2, D-72636 Frickenhausen, Germany. +49 (0)7022 948-0.+49 (0)7022 948-514. *info@de.gbo.com. www.greinerbiooneinc.com.* Dishes; filter well inserts; flasks; multiwell plates; plate sealers.

Greiner Bio-One Inc. 1205 Sarah St., Longwood, FL 32750, USA. +1 407-333-2800; 800 884 4703. +1 407-333-3001. *info@us.gbo.com. www.greinerbiooneinc.com.* Dishes; filter well inserts; flasks; multiwell plates; plate sealers.

Greiner Bio-One Ltd. Brunel Way, Stroudwater Business Park, Stonehouse, Glos., GL10 3SX, England, UK. +44 (0) 1453 825 255. +44 (0) 1453 827 277. *sales@uk.gbo.com. www.gbo.com.* Dishes; filter well inserts; flasks; multiwell plates; plate sealers.

Guava Technologies, Inc. 25801 Industrial Blvd., Hayward, CA 94545-2991, USA. +1 866 448 2827. +1 510-576-1500. *www.guavatechnologies.com.* Bench-top flow cytometer.

Guest Medical, Ltd. Enterprise Way, Edenbridge, Kent TN8 6EW, England, UK. 01732 876466. 01732 876476. *GuestMedical@BTInternet.com. www.guest-medical.co. uk.* Aquasan-water bath bacteriostat; biohazard spills kits; hypochlorite tablets; autoclavable disposal bags; glutaraldehyde.

Haake Buchler. *See* Thermo Electron Corp., Thermo-Haake. Electrophoresis; density gradient former; staining boxes.

Hamamatsu System Division. Div. of Hamamatsu Corp., 360 Foothill Rd., P.O. Box 6910, Bridgewater, NJ 08807, USA. 908-231-1116. 908-231-0852. *www.hamamatsu.com.* CCD cameras; closed-circuit TV; image analysis; videocameras; video recorders; time-lapse video.

Hamamatsu Photonics K.K. 325-6, Sunayama-cho, Hamamatsu City, 430 Japan. (81) 53 452 2141. (81) 53 456 7889. *www.hamamatsu.com.* CCD cameras; closed-circuit TV; image analysis; videocameras; video recorders; time-lapse video.

Hammond Cell Technology. PO Box 1424, Windsor, CA 95492. 707 473 0564. 707 473 0564. *hctculture@aol.com. www.hammondcelltech.com.* Bovine pituitary extract.

Hamo AG. CH-2542 Pieterlen, Switzerland. +41 (0)32 376 0200. +41 (0)32 376 0200. *info@hamo.com. www.hamo.com.* Laboratory glassware washers; washing machines.

Hamo UK Ltd. Spectra House, Boundary Way, Hemel Hempstead, HP2 7SH, England, UK. +44 (0) 1442 259 735. +44 (0) 1442 219 566. *hamo.uk@hamo.com. www.hamo.com.* Laboratory glassware washers; washing machines.

Hanna Instruments Inc. 584 Park East Dr., Woonsocket, RI 02895, USA. +1-401-765-7500.+1-401-762-5064. *sales@hannainst.com. www.hannainst.com.* Magnetic stirrers; meters: pH, conductivity.

Hanna Instruments. Eden Way, Pages Industrial Park, Leighton Buzzard, Beds, LU7 8TZ, England, UK. +44 (0)1525 850 855. +44 (0)1525 853 668. *salesteam@hannainst.co.uk. www.hannainst.com/.* Magnetic stirrers; meters: pH, conductivity.

Harlan Sera-Lab, Ltd. Hillcrest, Dodgeford Ln., Belton, Loughborough, Leicester LE12 9TE, England, UK. 01530 222 123. 01530 224 970. *hslcsd@harlanuk.co.uk. www.harlanseralab.co.uk.* Antibodies; serum.

Harvard Apparatus Inc. 84 October Hill Rd., Holliston, MA 01746, USA. +1 508-893-8999; 800-272-2775. +1 508-429-5732. *bioscience@harvardapparatus.com. www.harvardapparatus.com.* Bench-top autoclave; incubators; ovens; sterilizing oven.

Harvard Apparatus Ltd. Fircroft Way, Edenbridge, Kent TN8 6HE, England, UK. +44 (0)1732 864 001. +44 (0)1732 863 356. *sales@harvardapparatus.co.uk.*

www.harvardapparatus.co.uk. Bench-top autoclave; incubators; sterilizing oven.

Hays Chemical Distribution, Ltd. Hays House, Millmead, Guildford, GU2 5HJ, Surrey, England, UK. +44 (0)113 250 5811. +44 (0)113 250 5811. Chloros.

Hazelton. *See* Covance.

Health Science Research Resources Bank (HSRRB). Rinku-minamihama 2–11, Sennan-shi, Osaka 590-0535, Japan. + 81-724-80-1655. *www.jhsf.or.jp.* Cell banking; distribution of cell lines; DNA clones..

Helena Laboratories. PO Box 752, Beaumont, TX 77704-0752, USA. +1 409 842 3714; 800 231 5663. +1 409 842 6241. *www.helena.com/.* Densitometers.

Henry Schein Inc. 135 Duryea Road, Melville, NY 11747, USA. 800-711-6032; +1-631-843-5500×5117. 1-800-329-9109;. *custserv@henryschein.com. www.henryschein. com/.* Roccall.

Henry Schein Ltd. Centurion Close, Gillingham, Kent, ME8 0SB England, UK. 8700 102 043. 800 413 734. *sales@henryschein.co.uk. www.henryschein.co.uk.* Roccall.

Heraeus. Equipment, *see* Kendro. CO_2 incubators; centrifuges; laminar flow hoods; ovens.

Heraeus. Culture materials, *see* Vivascience AG. PetriPerm.

Heto-Holten. *See* Thermo Electron. Laminar flow hoods; microbiological safety cabinets.

High-Q Inc. P.O. Box 440, Wilmette, IL 60091, USA. 800-474-2674; +1-847-256-1231. 800-474-2672; +1-847-256-6114. *sales@high-q.com. www.high-q.com.* Distillation; water purification.

Hotpack. *See* SP Industries Co.

HRP, Inc. *See* Covance Research Products, Inc.

HSRRB. *See* Health Science Research Resources Bank.

H. V. Skan. Jeremy Drew, 425-433 Stratford Road, Shirley, Solihull, B90 4AE, England, UK. +44 (0)121 733 3003. 44 (0)121 733 1030. *info@skan.demon.co.uk. www.berlinerglas.com.* Microscope slides.

HyClone. 1725 South HyClone Road, Logan, UT 84321-6212, USA. 800-492-5663; 435-792-8000. 800-533-9450; +1 435-792-8001. *info@hyclone.com. www. hyclone.com.* Fetal bovine serum; medium; serum; serum-free medium; trypsin replacement.

Hyclone Europe S.A. Industriezone III, B-9320 Erembodegem-Aalst, Belgium. 53 83 44 04. 53 83 76 38. *www.hyclone.com.* Fetal bovine serum; medium; serum; serum-free medium; trypsin replacement.

Hycor Biomedicals, Inc. Garden Grove, CA 92841, USA. (800) 382–2527; (714) 933–3000. (714) 933–3220. ELISA; chemically-defined medium; serum-free medium.

ICLC Interlab Cell Line Collection. Istituto Nazionale per la Ricerca sul Cancro c/o CBA, Largo Rosanna Benzi, 10, Genova (GE), 16132 Italy. +39-0105737 474. +39-0105737 295. *iclc@ist.unige.it. www.biotech.ist.unige.it/cldb.* Cell bank; cell line databank.

ICN. *See* MP Biomedicals.

IEC. *See* Thermo Electron Corp. *www.thermo.com.* Centrifuges; CO_2 controllers; CO_2 incubators; cytocentrifuge adaptors; liquid handling.

Ilford Imaging UK Ltd. Town Lane, Mobberley, Knutsford, Cheshire, WA16 7JL, England, UK. +44 (0)1565 684 000. +44 (0)1565 873035. *www.ilford.com.* Photographic emulsions; developers; fixatives; film.

Ilford Imaging USA Inc. W. 70 Century Road, Paramus, NJ 07653, USA. +1 201-265-6000. *www.ilford.com.* Photographic emulsions; developers; fixatives; film.

Imaging Associates Ltd. 6 Avonbury Business Park, Howes Lane, Bicester, Oxon OX26 2UA, England, UK. +44 (0) 1869 356 240. +44 (0) 1869 356 241. *enquiries@imas.co.uk. www.imas.co.uk.* Image analysis; incubator chamber; microscope incubation chamber; microscopes: confocal, fluorescence, inverted, stereo, upright; photomicroscopes; time-lapse video.

Imaging Research Inc. Brock University, 500 Glenridge Avenue, St. Catherines, Ontario, L2S 3A1, Canada. 905 688 2040. 905 685 5861. Cytometer; image analysis; fluorescent probes; DNA and protein microarray analysis.

Inamed Aesthetics. 5540 Ekwill Street, Santa Barbara, CA 93111, USA. +1 805 683 6761. +1 805 967 5839. *www.inamedaesthetics.com.* Collagen implant; Zyderm 2.

Infors AG. CH-4103 Bottmingen, Basel, Switzerland. (0)61 425 7700. (0)61 425 7701. *Headoffice@infors-ht.com; service@infors-ht.com. www.infors-ht.ch.* Incubators; fermentors; shakers.

Infors U.K., Ltd. The Courtyard Business Centre, Dovers Fam, Lonesome Lane, Reigate, GB-RH2 7QT Surrey, England, UK. +44 (0)1737 22 31 00. +44 (0)1737 24 72 13. *infors.uk@infors-ht.com. www.infors.ch.* Incubators; fermentors; shakers.

Innovatis AG. Meisenstrasse 96, D-33607 Bielefeld, Germany. +49 (0)521 2997-300.+49 (0)521 2997-285. *info@innovatis.com. www.innovatis.com.* Cell counter; multiwell plate cell counting; viability determination.

Innovatis Inc. 333 Lancaster Avenue, Westgate V., 207, Frazer, PA 19355-1823, USA. +1-888-283-1564; +1- 610-889-7319.+1-888-283-1631; +1-610-889-7436. *info@innovatis.com. www.innovatis.com.* Cell counter; multiwell plate cell counting; viability determination.

Innovative Cell Technologies, Inc. 6790 Top Gun St. #1, San Diego, CA 92121, USA. +1 858 587 1716. +1 858.453–2117. *info@innovativecelltech.com. www.innovativecelltech.com.* Accumax; Accutase; cell dissociation agents; flow cytometry; robotics; trypsin replacement.

Innovative Chemistry, Inc. P.O. Box 90, Marshfield, MA 02050, USA. +1 781-837-6709. +1 781-834-7325. *www.innovativechem.com/.* Isoenzyme electrophoresis analysis kit.

Insight Biotechnology Ltd. P.O. Box 520, Wembley, Middlesex, HA9 7YN, England, UK. +44 (0) 20

8385 0303. +44 (0) 20 8385 0302. *info@insightbio.com. www.insightbio.com.* Antibodies; cytokines; ELISA.

Insmed Inc. 800 E. Leigh St., Richmond VA 23219, USA. +1 804-828-6893. +1 804-828-6894. *www. insmed.com/.* Collagen implant, Zyderm 2; Vitrogen 100.

Instron. 100 Royall Street, Canton, MA, 02021, USA. 1 800 564 8378; 1 781 828 2500. *www.instron.com.* Tension/compression system for muscle culture.

Integra Biosciences AG. Schonbuhlstr. 8, CH-7000, Chur, Switzerland. +41 (0)81 286 95 30. +41 (0)81 286 95 33. *info@integra-biosciences.com. www.integra-biosciences.com.* Aspiration pump; autoclaves; automatic pipettor; burner; culture flasks; dispensers; filter well inserts; filter flasks; filter caps; fluid handling; high-density culture flasks and chambers; liquid handling; magnetic stirrers; Maxisafe 2000; membrane flasks; microtitration; multipoint pipettors; perfused membrane culture system; pipettors; peristaltic pump pipettor; roller bottles; roller culture apparatus; six-well chambers; spinner culture; stirrer culture; Tecnomouse; vacuum pump.

Integra Biosciences UK. *See* Scientific Laboratory Supplies.

Integra Biosciences (USA). *See* Argos Technologies.

Intergen Co. *See* Serologicals.

International Association for Plant Tissue Culture (IAPTC). Abed A. Watad, Secretary-Treasurer, Department of Ornamental Horticulture, ARO, The Volcani Center, P.O.Box 6, Bet Dagan 50205 Israel. +972 3 9683500. +972 3 9660589. *vcwatad@volcani.agri.gov.il. http://indycc1.agri.ac.il/~tzvika/iaptc/ip-home.htm.*

Interplast Inc. 100 Connecticut Drive, Burlington, NJ 08016, USA. +1 609-386-4990; 1-800-220-1159. 609-386-9237. *www.interplastinc.com/.* PTFE; Teflon; tubes; sheets; rods.

Interpore Cross International. 181 Technology Drive, Irvine, CA 92618-2402, USA. +1 949 453–3200. +1 949 453–3225. *www.interpore.com/.* Coral-based ceramics.

Intracel, Ltd. Unit 4, Station Road, Shepreth, Royston, Herts SG86PZ, England, UK. 01763 262680. 01763 262676. Specialized heated chambers for time-lapse studies.

Invitrogen Corp. 1600 Faraday Avenue, Carlsbad, CA 92008, USA. 800-955-6288; (760) 603–7200. 800 331 2286; +1 760 602 6500. *catalog@invitrogen.com. www.invitrogen.com.* Agar; ampoules; antibodies; chick embryo extract; chick plasma; chicken serum; chromosome painting; DMEM; F12; gentamicin; glutamine; growth factors; Hanks' balanced salt solution (HBSS); HBSS without Ca^{2+} and Mg^{2+} (CMF); HEPES; insect medium; Iscove's modified Dulbecco's medium (IMDM); kanamycin; L-glutamine; McCoy's 5a medium; medium; microtitration plates; minimum essential medium–alpha (MEM-α) ; mycostatin; PBS; penicillin-streptomycin; Pluronic; recombinant epidermal growth factor (rEGF); RPMI 1640; serum; serum-free; trypsin; trypsin/EDTA.

Invitrogen Ltd. 3 Fountain Drive, Inchinnan Business Park, Paisley PA4 9RF, UK. +44 (0)141 814 6100; 0800 269 210 (sales); 0800 838 380 (tech. serv.). +44 (0)141 814 6260; 0800 243 485. *euroinfo@invitrogen.com. www.invitrogen.com. See* Invitrogen, Inc.

Irvine Scientific, Inc. 2511 Daimler St., Santa Ana, CA 94303, USA. 800 577 6097; +1 949 261 7800. +1 949 261 6522. *nucleus@irvinesci.com. www.irvinesci.com.* Medium; mycoplasma test kits; serum; serum-free medium; serum substitutes.

Irvine Scientific, UK. *See* Metachem Diagnostics; Smiths Medical International.

ISP Alginate. San Diego, CA, USA. *www.ispcorp.com/.* Sodium alginate.

J.T. Baker. Mallinckrodt Baker, Inc., 222 Red School Lane, Phillipsburg NJ 08865, USA. +1 314-530-2000; 800-354-2050. +1 314-530-2328. *infombi@mkg.com. www.jtbaker.com.* Bottle carrier; buffers; chemicals.

J.T. Baker UK. *See* Scientific & Chemical Supplies.

Jackson ImmunoResearch. P.O. Box 9, 872 West Baltimore Pike, West Grove, PA 19390, USA. +1 610 869 4024; 800 367 5296. +1 610 869 0171. *cuserv@jacksonimmuno.com. www.jacksonimmuno. com/.* Antibodies; FITC-conjugated donkey anti-sheep IgG.

Jackson ImmunoResearch (UK). *See* Stratech Scientific.

James Glass Co. 14 Hanover St., Hanover, MA 02339, USA. +1 781 829–0967. +1 781.829.2142. *sales@jamesglass.com. www.jamesglass.com.* Bioreactors; fermentors; glass blowing; laboratory glassware.

Japanese Cancer Cell Collection. *See* Health Science Research Resources Bank (HSRRB).

Japanese Collection of Research Bioresources. *See* JCRB and Health Science Research Resources Bank (HSRRB).

Japanese Tissue Culture Association (JTCA). +81-3-5814-5801.+81-3-5814-5820. *http://jtca.umin.jp.*

JCRB. Research Resource Division, National Institute of Biomedical Innovation, 7-6-8-Saito-Asagi, Ibaraki-shi, Osaka 567-0085, Japan. (For distribution of cell lines *see* HSRRB.) +81 (0)72 641 9811. *http://cellbank.nibio.go.jp.* Cell bank; acquisition od cell lines; DNA clones.

Jencons (Scientific) Ltd. Unit 6, Forest Row Business Park, Station Road, Forest Row, East Sussex, RH18 5DW, England, UK. +44 (0)1342 826836. +44 (0)1342 826771. *uksales@jencons.co.uk. www.jencons.co.uk.* Automatic dispensers; biohazard bags; CO_2 incubator; Dewars; liquid nitrogen freezers; magnetic stirrers; microbiological safety cabinets; micropipettes; neck plug controlled-rate freezer; pipettors; sterilizing tape; stirrers; Taylor Wharton.

Jencons Scientific Inc. 800 Bursca Drive, Suite 801, Bridgeville, PA 15017, USA. 800-846-9959; +1 412-257-8861. +1 412-257-8809. *info@jenconsusa.com.*

www.jenconsusa.com/. For products *see* Jencons (Scientific) Ltd.

JEOL. 11 Dearborn Road, Peabody, MA 01960, USA. +1 978 535 5900. +1 978 536 2205. *www.jeol.com*. Electron microscopes; scanning EM.

JEOL(UK) Ltd. JEOL House, Silvercourt Watchmead, Welwyn Garden City, AL7 1LT Herts, England, UK. +44 (0) 1707 37 71 17. +44 (0) 1707 37 32 54. *uk.sales@jeoleuro.com. www.jeoleuro.com/*. Electron microscopes; scanning EM.

JM Separations. P.O. Box 6268, 5600 HG Eindhoven, The Netherlands. +31 (0)40 219 7480. +31 (0)40 219 7499. *info@jmseparations.com. www.jmseparations.com*. Cellmax; hollow fiber perfusion culture; microcapillary perfusion; Triac concentric hollow fibers.

Johnson & Johnson, Medical, Ltd. Coronation Road, Ascot, Berks SL5 9EY, England, UK. +44 (0)1344 871 000. +44 (0)1344 872 599. Disinfectants; Precept tablets; hypochlorite; swabs; gloves.

Jouan. *See* Thermo Electron Corp. *www.thermo.com*. Centrifuges; incubators; ovens; microbiological safety cabinets; water baths.

Jouan Nordic A/S. *See* Thermo Electron.

JRH Biosciences. 13804 W. 107th Street, Lenexa, KS 66215 USA. +1 913-469-5580; 1 800-255-6032. +1 913-469-5584. *info-us@jrhbio.com. www.jrhbio.com*. Autoclavable medium; bulk culture; media bags; medium; serum; serum-free medium; Thermo-POW; Ventrex; vitamins.

JRH Biosciences Ltd. Smeaton Road, West Portway, Andover, Hampshire SP10 3LF, UK. +44 (0)1264-333311. +44 (0)1264-332412. *info-eu@jrhbio.com. www.jrhbio.com. See* JRH Biosciences (USA).

KBI BioPharma, Inc. P.O. Box 15579, 1101 Hamlin Road, Durham, NC 27704, USA. +1 919 479 9898. +1 919 620 7786. *sales@kbibiopharma.com. www.kbibiopharma.com*. Centrifugal BioReactor; high-density cell beds.

Keison Products. P.O. Box 2124, Chelmsford, Essex CM1 3UP, England, UK. +44 1245 600560. +44 1245 6000. *info@keison.co.uk. www.keison.co.uk/day-impex*. Dewars.

Keison Products, USA. 233 Rogue River Highway #425, Grants Pass, OR 97527-5429, USA. +1 541 956 9929. +1 541 610 1646. *info@keison.co.uk. www.keison.co.uk/day-impex*. Dewars.

KeLab. KeLab, Karl-Erik Ljung AB, Knipplagatan 10, 414 74 Göteborg, Sweden.. +46 (0)31-12 51 60. +46 (0)31-14 82 60. *goteborg@kelab-biochem.com. www.kelab-biochem.com/kontakt.html*. Collagen.

Kendro Laboratory Products. 308 Ridgefield Court, Asheville, NC 28806, USA. +1 828 658 2711. +1 828 645-3368. *info@kendro.spx.com. www.kendro.com*. Cell separator; centrifuges; CO$_2$ incubators, incubators, laminar-flow hoods; microbiological safety cabinets; microcentrifuges, ovens, freezers.

Kendro Laboratory Products. Robert-Bosch-Strasse 1, 63505 Langenselbold, Germany. 800-1-536 376 (Sales); 800-1-112 110 (Service). 800 1 112 114. *info.de@kendro.spx.com. www.kendro.com*. Cell separator; centrifuges, CO$_2$ incubators, incubators, laminar-flow hoods; microbiological safety cabinets; microcentrifuges, ovens, freezers.

Kendro Laboratory Products. Stortford Hall Park, Bishop's Stortford, Hertfordshire, CM23 5GZ, England, UK. +44 (0)1279 82 77 00. +44 (0)1279 82 77 50. *info.uk@kendro.spx.com. www.kendro.com*. Cell separator; centrifuges, CO$_2$ incubators, incubators, laminar-flow hoods; microbiological safety cabinets; microcentrifuges, ovens, freezers.

Kimberley-Clark. 12671 High Bluff Drive, San Diego, CA 92130, USA. 800 255 6401. *service@safeskin.com. www.kc-safety.com/*. Nitrile gloves.

Kimble-Kontes. Vineland, NJ, USA. +1 (888) 546–2531 Ex 1; (856) 692–3600 Ex 1. +1 (856) 794–9762. *cs@kimkon.com. www.kimble-kontes.com*. Plastic Pasteurs; glassware; centrifuge tubes.

Kinetic Biosystems, Inc. *See* KBI BioPharma.

KNF Corp. 734 West Penn Pike, Tamaqua, PA. 18252, USA. +1 (570) 386–3550. +1 (570) 386–3703. *www.knfCorp.com/*. Autoclavable nylon sheet.

Kodak Ltd. (*see also* Eastman Kodak). P.O. Box 66, Hemel Hempstead, Hertfordshire, HP1 1JU, UK. +44 (0)1442 261 122. +44 (0)1442 240 609. *www.kodak.co.uk*. CCD; digital cameras; digital imaging; hypoclearing agent; microscope cameras; photographic developers; photographic fixatives; X-ray film.

Kojair (UK) Ltd. 143 Reading Road, Wokingham, Berkshire, RG41 1LJ, England, UK. +44 (0)118 977 0957. +44 (0)118 977 5302. *info@kojair.co.uk. www.kojair.co.uk*. Laminar-flow hoods; microbiological safety cabinets.

Lab Safety Supply. PO Box 1368, Janesville, WI 53547-1368, USA. 800 356 0783; +1 608 754 7160. 800 543 9910; +1 608 754 3937. *custsvc@labsafety.com. www.labsafety.com/*. Biohazard bags; discard containers; disinfectants; eye wash bottles; face masks; nitrile gloves; labels; protective clothing; sharps disposal; sodium hypochlorite; tags.

Lab Vision Corp. 47790 Westinghouse Drive, Fremont, CA 94539, USA. 800-828-1628; +1 510-991-2800. +1 510-991-2826. *LabVision@LabVision.com. www.labvision.com*. Angiogenesis markers; antibodies: collagen, fibronectin, GFAP, laminin, mucin, apoptosis markers, cell cycle growth factors, oncogenes, heregulin, p21WAF1, MUC2, myelin, p16INK4a, BrdU, BUdR, cyclin, EGFR, E-cadherin, fibronectin, gastric mucin, cytokeratin, laminin receptor, nm23; concentrators; cytokines; extracellular matrix; extraction kits; invasion chambers.

Labcaire Systems, Ltd. 175 Kenn Road, Clevedon, North Somerset, BS21 6LH, England, UK. +44

(0)1275 793000. +44 (0)1275 341313. *info@labcaire.co.uk*. *www.labcaire.co.uk*. Class II microbiological safety cabinets; laminar-flow hoods.

Labco Limited (U.K.). Brow Works, Copyground Lane, High Wycombe, Buckinghamshire, HP12 3HE, United Kingdom. +44 (0)1494 459741. +44 (0)1494 465101. *sales@labco.co.uk*. *www.labco.co.uk*. Biohazard disposal bins.

Lab-Line. *See* Barnstead Thermolyne Corp.

Labmart (Cambridge), Ltd. 1 Pembroke Avenue, Waterbeach, Cambridge CB5, UK. +44 (0) 1223 441257. +44 (0) 1223 861990. Autoclaves; cryobiological storage systems; liquid nitrogen freezers; low-temperature freezers.

Laboratory Impex Systems Ltd. Impex House, 15 Riverside Park, Wimborne, Dorset, BH21 1QU, England, UK. +44 (0)1202 840685. +44 (0)1202 840701. *sales@lab-impex-systems.co.uk*. *www.lab-impex-systems. co.uk*. CO_2 controllers; disinfectants; fridges; freezers; freezer racks.

Laboratory Sales (UK) Ltd. Units 20−21, Transpennine Trading Estate, Rochdale OL11 2PX, England, UK. +44 (0)1706 356 444. +44 (0)1706 860 885. *sales@lg-uk.com*. *www.ls-uk.com*. Microscope slides and cover glasses; vials and closures.

Labox Ltd. The Oval, Hackney Road, London, E2 9DU, England, UK. +44 (0)207 613 2400. +44 (0)207 613 3420. *labox.uk@labox.biz*. *www.labox.biz*. Canisters; containers; electronic thermometers; plastics; pumps; siphons; tap siphon..

Labsales. *See* Camlab. Temperature data loggers; temperature recorders.

Lab Storage Systems Inc. P.O. Box 968, St. Peters, MO, 63376, USA. 800 345 4167; +1 636 928 8988. 800 345 4117; +1 636.922.4474. *www.labstore.com*. HistoGel embedding medium.

Lamb, R.A. *See* Raymond A. Lamb.

Lancer UK Ltd. 1 Pembroke Avenue, Waterbeach, Cambridge CB5 9QR, England, UK. +44 (0)1223 861 665. +44 (0)1223 861 990. *sales@lancer.com*. *www.lancer.co.uk*. Glassware washing machines.

Lancer USA Inc. 140 State Road 419, Winter Springs, FL 32708, USA. +1 407 327 8488. +1 407 327 1229. *sales@lancer.com*. *www.lancer.com*. Glassware washing machines.

Lawson Mardon Wheaton. 1501 N. Tenth St., Millville, NJ 08332-2092, USA. 609-825-1100; 800-225-1437. +1 856-825-1368; +1 856-825-4568. *www.wheatonsci.com*. Ampoules; borosilicate glass; coverslips; dispensers; Ehrlenmeyer flasks; Gilson Pipetteman; glassware; liquid handling; micropipettes; pipettors; plasticware; roller racks; stirrer flasks: multiuse, disposable.

LEEC Ltd. Private Road, No.7 Colwick Industrial Estate, Nottingham NG4 2AJ, England, UK. +1 (0)115 961 6222. +1 (0)115 961 6680. *sales@leec.co.uk*. *www.leec.co.uk*. CO_2 incubators; mechanical bottle brush; sterilizing ovens.

Leica Microsystems. 2345 Waukegan Road, Bannockburn, IL 60015, USA. +1 847 405 0123, 800 248 0123.

+1 847 405 0164. *www.leica-microsystems.com/*. Cameras; CCD cameras; CCTV; confocal microscope; image analysis; inverted microscope; karyotyping; micromanipulators; microscopes; refractometers.

Leica Mikrosysteme Vertrieb GmbH. Lilienthalstrasse 39−45, D-64625 Bensheim, Germany. +49 6251 136 0. +49 6251 136 155. *www.leica-microsystems.com/*. Cameras; CCD cameras; CCTV; confocal microscope; image analysis; inverted microscope; karyotyping; micromanipulators; microscopes; refractometers.

LGC Promochem. Queens Road, Teddington, Middlesex, TW11 0LY, England, UK. +44 (0)20 8943 8480. +44 (0)20 8943 7554. *uk@lgcpromochem.com*; *atcc@lgcpromochem.com*. *www.lgcpromochem.com*. Authentication; cell lines; DNA profiling; training courses.

Life Sciences International. *See* Thermo Electron Corp. *www.thermo.com*. Alarms; CO_2 incubators; controlled-rate freezer; dataloggers; freezer inventory systems; laminar flow hoods; liquid handling; liquid nitrogen freezers; low-temperature freezers; microbiological safety cabinets; micropipettes; microtitration: plates, plate readers, plate washers; magnetic sorting; nitrogen freezers; programmable freezer; radio data systems & alarms.

Life Technologies. *See* Invitrogen.

List Biological Laboratories. 501-B Vandell Way, Campbell, CA 95008, USA. 800 726−3213; +1 408 866 6363. +1 408 866 6364. *info@listlabs.com*. *www.listlabs.com*. Cholera toxin.

LMS Ltd. The Modern Forge, Riverhead, Sevenoaks, Kent, TN13 2EL, England, UK. +44 (0)1732 451 866. +44 (0)1732 450 127. *sales@lms.ltd.uk*. *www.lms.ltd.uk*. Benchtop autoclaves; cooled incubators; freezers; refrigerators.

Lorne Biochemicals. Reading, Berks., UK. +44 (0)118 934 2400. +44 (0)118 934 2788. *info@lornelabs.com*. *www.lornelabs.com*. Collagenase; DNase; hyaluronidase; trypsin.

LTE Scientific Ltd. Greenbridge Lane, Greenfield, Oldham, OL3 7EN, England, UK. +44 (0) 1457 876 221. +44 (0) 1457 870 131. *info@lte-scientific.co.uk*. *www.lte-scientific.co.uk*. Autoclaves; bench top autoclaves; incubators: CO_2, cooled; drying cabinets; laminar-flow cabinets; microbiological safety cabinets; ovens.

Ludl Electronic Products. 171 Brady Ave., Hawthorne, NY 10532, USA. +1 914 769−6111 x223. +1 914-769-4759. *Cust-Service@ludl.com*. *www.ludl.com/*. Electronics components; computer controlled shutter; filter wheel; time-lapse controls.

Lumitech Ltd. *See* Cambrex. Cytotoxicity kits; cell proliferation assays; mycoplasma assay; luciferase; adenylate nucleotide ratio assay.

MA Bioservices. *See* BioReliance.

Mallinckrodt-Baker. *See* J.T.Baker.

Markson LabSales Inc. 5285 N.E. Elam Young Pkwy., State A-400, Hillsboro, OR 97124, USA. +1 503 648 0762, 800 528 5114. +1 503 648 8118.

markson@teleport.com. www.markson.com. Chromatography reagents; disinfectants; filters; molecular weight sieves; refractometer.

Matrix Technologies Corp. 22 Friars Drive, Hudson, NH 03051, USA. +1 (0)603 595 0505, 800 345 0206. +1 (0)603 595 0106. *info@matrixtechcorp.com. www. matrixtechcorp.com.* Automatic dispensers; adjustable tip spacing multipoint pipettors, automatic pipettors; pipetting aids; robotic liquid handling.

Matrix Technologies Corp. Lower Meadow Road, Brooke Park, Handforth, Cheshire, SK9 3LW, England, UK. 0800 389 4431; +44 (0) 161 486 2110. +44 (0) 161 488 4560. *info@matrixtechcorp.eu.com. www. matrixtechcorp.com.* Automatic dispensers; adjustable tip spacing multipoint pipettors, automatic pipettors; pipetting aids; robotic liquid handling.

MatTek Corp. 200 Homer Avenue, Ashland, MA 01721, USA. 800-634-9018; +1 508-881-6771. +1 508-879-1532. *information@mattek.com. www.mattek.com/.* Cytotoxicity assays; organotypic cultures; tissue equivalents; transplants.

MBL International Corp. MBL International, 15 B Constitution Way, Woburn, MA 01801, USA. +1 781 939 6964; 1–800 200 5459. +1 781 939 6963. *info@mblintl.com. www.mblintl.com/.* Antibodies; apoptosis; BUdR; cell proliferation; PCNA; receptors; signal transduction.

MDH Ltd. *See* Bioquell.

Mediatech Inc. 13884 Park Center Road, Herndon, VA 20171, USA. 800 235 5476. *www.cellgro.com.* Autoclavable medium (MEM); glutamine; Ham's F12 powder; Mg^{2+} and Ca^{2+}-free Hanks' balanced salt solution (CMF-HBSS); insect culture media; Pluronic; serum-free media.

Medical & Biological Laboratories (MBL) Co Ltd. Sumitomoshoji Marunouchi Bldg. 5F, 5–10 Marunouchi 3 chome, Naka-ku, Nagoya 460-0002, Japan. +81 52 971 2081. +81 52 971 2337. *www.mbl.co.jp.* Antibodies; apoptosis; BUdR; cell proliferation; PCNA; receptors; signal transduction.

Medical Air Technology (MAT) Ltd. Airology Centre, Mars Street, Oldham OL9 6LY, England, UK. +44 (0)161 621 6200. +44 (0)161 624 7547. *Sales@medairtec.com. www.medicalairtechnology.com/.* Laminar-flow cabinets; laminar-flow canopies; microbiological safety cabinets.

Memmert GmbH. P.O. Box 1720, D-91107 Schwabach, Germany. +49 (0) 91 229 250. +49 (0) 91 221 4585. *info@memmert.com. www.memmert.com.* CO_2 incubators; cooled incubators; sterilizing ovens.

Merck Biosciences, Ltd. Padge Road, Beeston, Nottingham NG9 2JR, England, UK. 800 622 935; +44 (0)115 943 0840. +44 (0)115 9430951. *customer. service@merckbiosciences.co.uk. www.merckbiosciences. co.uk.* Antibodies; antifoam; biochemicals; buffers; carboxymethylcellulose; chemicals; cholera toxin; CMC;

crystal violet; Decon; detergents; dexamethasone; dextran; DMSO; EDTA; enzymes; G418 sulfate; Giemsa stain; glucose; glycerol; glycobiology reagents; hydrocortisone; hygromycin B; immunochemicals; mercaptoethanol; PEG; pipettors; plate readers; polyethylene glycol; roller culture; silica gel; stains; Trypan Blue.

Merck USA. *See* EMD Biosciences.

Merck KGaA. *See* VWR. *de.vwr.com.* For products *see* Merck Biosciences.

Messer Cryotherm. Euteneuen 4· D-57548 Kirchen/Sieg, Germany. +49 27 41/95 85-0. +49 27 41/69 00. *info@cryotherm.de. www.messer-cryotherm.de.* Cryofreezers; controlled-rate cooler; liquid nitrogen freezers.

Messer UK Ltd. Cedar House, 39 London Road, Reigate, Surrey RH2 9QE, England, UK. +44 (0)1737 241133. +44 (0)1737 241842. *www.messergroup.com.* Carbon dioxide; controlled-rate cooler; cryofreezing system; dry ice; gas mixtures; oxygen; nitrogen; liquid nitrogen freezers; calibration gas mixtures; gas regulators; pipettors; plate readers.

MetaBios Inc. 135-Innovation & Development Center, University of Victoria, R-Hut McKenzie Avenue, Victoria, British Columbia, V8W 3 W2, Canada. +1 (250) 472–4334. +1 (250) 721–6497. *info@metabios. com. www.metabios.com.* Bellows bioreactor; culture bags; media bags.

Metachem Diagnostics Ltd. 29 Forrest Rd., Piddington, Northampton, NN7 2DA, England, UK. +44-1604-870370.+44-1604-870194. *info@metachem.co.uk. www. metachem.co.uk.* Medium; mycoplasma test kits; serum; serum-free medium; serum substitutes.

Mettler-Toledo GmbH. PO Box VI-400, CH-8606 Greifensee, Switzerland. +41 1944 22 11. +41 1944 31 70. *www.mt.com.* Balances; conductivity meters; density meters; dispensers; pH meters; pipettors; refractometers.

Mettler-Toledo Inc. 1900 Polaris Parkway, Columbus, OH 43240. 800 METTLER; +1 614 438 4511. +1 614 438 4900. *us@mt-shop.com. www.mt.com.* Balances; conductivity meters; density meters; dispensers; pH meters; pipettors; refractometers.

Mettler-Toledo Ltd. 64 Boston Road, Beaumont Leys, Leicester, LE4 2ZW, England, UK. 070000 MTSHOP. 0116 236 5500. *uk@mt-shop.com. www.mt-shop.com.* Balances; conductivity meters; density meters; dispensers; pH meters; pipettors; refractometers.

Michael Smith Engineers Ltd. FREEPOST SCE 7470, Woking, Surrey, GU21 1BR, England, UK. 0800 316 7891. +44 (0)1483 723 110. *www.michael-smith-engineers.co.uk.* Automatic dispensers; metering and dispensing pumps; peristaltic pumps.

Microflow. *See* Bioquell.

MicroGeN2. *See* Texol Products.

Microgon Inc. *See* Spectrum. Filters; hollow fiber; sterilizing filters; syringe-tip filters.

Microm International GmbH. Robert-Bosch-Str. 49, D-69190 Walldorf, Germany. +49 6227-836 0. +49 6227-836 111. *www.microm.de.* Cytoseal 60; microscope slides; slide mountants.

Miele Inc. 9 Independence Way, Princeton, NJ 08540, USA. 609 419 9898; 800 843–7231. +1 609 419–4298. *www.labwashers.com/.* Glassware washing machines.

Miele Co., Ltd. Abingdon, Oxon, OX14 1TW, England, UK. 845 3303618. *www.mieleprofessional.co.uk.* Glassware washing machines.

Miles Inc. *See* Bayer Corp.

Miles Ltd. *See* Bayer plc.

Millipore Corp. 290 Concord Rd., Billerica, MA 01821, USA. +1 978 715 4321; (800) MILLIPORE. (781) 533–3110. *www.millipore.com/.* Carbon filtration; deionization; electrodeionization; filters; filter holders; filter sterilization; filter well inserts; molecular filtration; pumps; reverse osmosis; sterile filtration; Sterivex; ultrapure water; water purification.

Millipore (U.K.), Ltd. 3 & 5 The Courtyards, Hatters Lane, Waterford, Hertfordshire, WD18 8YH, England, UK. 870 900 46 45. 870 900 46 44. *www.millipore.com/purecommerce.* *See* Millipore Corp.

Miltenyi Biotec GmbH. Friedrich-Ebert-Strasse 68, 51429 Bergisch Gladbach, Germany. +49 2204 83060. +49 2204 85197. *macs@miltenyibiotec.de.* *www.MiltenyiBiotec.com.* Antibodies: Ac133, CD34; cell separation; cytokine secretion assays; epithelial cell separation; endothelial cell separation; fibroblast separation; MACS; magnetic microbeads; magnetic sorting; nylon mesh filters; stem cell marker.

Miltenyi Biotec Inc. 12740 Earhart Avenue, Auburn, CA 95602, USA. +1 530 888 8871; 800 FOR MACS. +1 530 888 8925. *macs@miltenyibiotec.com.* *www.MiltenyiBiotec.com.* *See* Miltenyi Biotec GmbH.

Miltenyi Biotec Ltd. Almac House, Church Lane, Bisley, Surrey GU24 9DR, England, UK. 01483 799 800. 01483 799 811. *macs@miltenyibiotec.co.uk.* *www.MiltenyiBiotec.com.* *See* Miltenyi Biotec GmbH.

Miltex Instruments Co., Inc. 700 Hicksville Road, Bethpage, NY 11714, USA. 800-645-8000; +1 717 840 9335. +1 717 840 9347. *customerservice@miltex.com.* *www.miltex.com.* Dissecting and surgical instruments.

Minucells and Minutissue Vertriebs-GmbH. Starenstrasse 2, D-93077 Bad Abbach, Germany. +49 (0)9405 962 440. +49 (0)9405 962 441. *minucells.minutissue@t-online.de.* *www.minucells.de.* Microscope chambers; organotypic culture; perfusion culture; scaffolds; tissue carriers; tissue culture courses.

Molecular Devices Corp. MaxLine Div., 1311 Orleans Dr., Sunnyvale, CA 94089, USA. 408-747-1700; 800-635-5577. 4+1 08-747-3602. *info@moldev.com.* *www.moleculardevices.com.* Microtitration: plate readers, fluorometers, software, spectrophotometers.

Molecular Devices Ltd. 135 Wharfedale Road, Winnersh Triangle, Winnersh, Wokingham, RG41 5RB, England, UK. 0118 944 8000. 0118 944 8001. *info@moldev.com.* *www.moleculardevices.com.* Microtitration: plate readers, fluorometers, software, spectrophotometers.

Molecular Dynamics. *See* Amersham Biosciences. *www1.amershambiosciences.com/.* Confocal microscopes; densitometers; plate readers.

Molecular Probes, Inc. 29851 Willow Creek Road, Eugene, OR 97402-0469, USA. 800 438–2209; +1 541 335 0338. +1 541 335-0305. *order@probes.com.* *www.probes.com.* DNA fluorochrome; enzyme substrates; fluorescent probes; PicoGreen; viability stains.

MP Biomedicals, Inc.. 15 Morgan, Irvine, CA 92618-2005, USA. 800.633.1352, +1 949.833.2500; Life sciences div: 800 854-0530 ; +1 330 562 1500. +1 949 859 5989. *custserv@mpbio.com.* *www.mpbio.com/.* Amino acid kits; antibiotics; antibodies; Apotame apoptosis inhibitor; automatic dispensers; chick plasma; chicken serum; CO_2 incubators; collagen; cryopreservation medium; detergents; disinfectants; Ficoll-Hypaque density media; glutamine; growth factors; HEPES; incubators; laminar-flow cabinets; lymphocyte preparation medium; Lymphoprep; medium; multisurface flasks; Mycoplasma Removal Agent (MRA); mycoplasma test kits; PBS; pipettors; plate sealers; Pluronic F68; penicillin; safety cabinets; serum-free media; streptomycin; trypsin; serum; Thermanox; Vitrogen.

MP Biomedicals NV/SA. Dornveld 10. 1731 Asse Relegem, Belgium. +32 2 466 00 00; 008000 7777 9999. +32 2 466 26 42. *www.mpbio.com/.* *See* MP Biomedicals, Inc.

MTR Scientific, LLC. 9639-122 Dr. Perry Road, Ijamsville, MD 21754, USA. +1 301-831-1377. +1 301-874-1899. *info@mtrscientific.com.* *www.mtrscientific. com.* DNA sequencing; human tissue arrays; stains.

Muis Controls Ltd. 10610-172 Street, Edmonton, Alberta, T5S 1H8, Canada. +1 780 486 2400. +1 780 486 2500. *www.MuisControls.com.* Gas flow meters; rotameters.

MVE Inc. *See* Chart Biomed.

Mycoplasma Experience. Reigate, UK. +44 (0)1737 226 662. +44 (0)1737 224 751. *mexp@mycoplasma-exp.com.* *www.mycoplasma-exp.com.* Mycoplasma testing.

Nalge Nunc (Europe) Ltd. Unit 1a, Thorn Business Park, Hereford HR2 6JT, England, UK. +44 (0) 1432 263933. +44 (0) 1432 376567. *sales@nalgenunc.co.uk.* *www.nalgenunc.com.* *See* Nalge Nunc International.

Nalge Nunc International. International Department, 75 Panorama Creek Drive, Rochester, NY 14625, USA. 800-625-4327; +1 716 264 3898. +1 585-586-8987; +1 716 264 3706. *nnics@nalgenunc.com.* *www.nalgenunc.com/.* Ampoules; canes; cell culture inserts; cell factories; cell scrapers; chamber slides; conical tubes; containers; coverslips; cryopens; cryotubes; cryovials; culture tubes; dishes; filter bottom microtitration plates; filters; filter sterilizer; filter well inserts; flasks; freezer canes; gas-permeable caps; marker pens; media bags; microcarriers; multisurface propagators; Thermanox coverslips.

Napco. 170 Marcel Dr., Winchester, VA 22602, USA. 800-621-8820. *info@napco2.com. www.napco2.com/.* CO_2 incubators; glassware washers.

Napco. *See* Precision Scientific/Napco. CO_2 incubators; glassware washers.

National Chemicals. 1259 Seaboard Industrial Blvd., Atlanta, GA 30318, USA. 800 237 0263. *www.natchem. com/.* Disinfectants; Iodophor.

National Diagnostics, Inc. 305 Patton Dr., S.W. Atlanta, GA 30336-1817, USA. +1 404 699–2121; 800-526-3867. +1 404-699-2077. *info@nationaldiagnostics.com. www.nationaldiagnostics.com.* Ecoscint scintillation fluid.

Nektar AL. 490 Discovery Drive, Huntsville, AL 35806, USA. 800 457 1806. +1 256 533 4805. *nektar@al.nektar.com. www.nektar.com.* Hydrogels; poly (ethylene glycol) diacrylate (PEGDA).

Nektar UK. 69 Listerhills Science Park, Campus Road, Bradford, West Yorkshire, BD7 1HR, England, UK. +44 (0)1274 305 540. +44 (0)1274 305 570. *nektar@uk.nektar.com. www.nektar.com.* Hydrogels; poly(ethylene glycol) diacrylate (PEGDA).

NEN Life Sciences. *See* PerkinElmer.

New Brunswick Scientific Co Inc. Box 4005, 44 Talmadge Road, Edison, NJ 08818-4005, USA. 800 631–5417; +1 732 287 1200. +1 732 287 4222. *bioinfo@nbsc.com. www.nbsc.com/.* Bioreactors; CO_2 incubators; drying oven; fermentors; gyratory shaker; incubators; magnifying viewers; rocker platforms; roller culture; roller drum; roller racks; scale-up; shaking incubators; sterilizing oven; ultralow-temperature freezers.

New Brunswick Scientific (U.K.), Ltd. 17 Alban Park, Hatfield Road, St Albans, Herts, AL4 0JJ, England, UK. 0800 581 331. +44 (0)1727 835 666. *bioinfo@nbsuk.co.uk. www.nbsuk.co.uk. See* New Brunswick Scientific Co Inc.

New England Biolabs. 32 Tozer Road, Beverly, MA 01915–5599, USA. 800-632-5227; +1 978-927-5054. 800-632-7440; +1978-922-7085. *customerservice@neb. com. www.neb.com.* Molecular biology reagents; vectors; signal transduction.

New England Biolabs (U.K.), Ltd. 73 Knowl Piece, Wilbury Way, Hitchin, Herts SG4 OTY, England, UK. +1 (0)1462 420 616; 800 318 486. +1 (0)1462 421 057; 800 435682. *info@uk.neb.com. www.neb.uk.com/.* Molecular biology, vectors, signal transduction.

Nichols Institute Diagnostics Div. 1311 Calle Batido, San Clemente, CA 92673, USA. +1 949-940-7200. +1 949-940-7271. *www.nicholsdiag.com/.* Immunoassays: bone, thyroid, adrenal function.

Nikon Europe B.V. P.O.Box 222, 1170 AE Badhoevedorp, The Netherlands. +31 20 4496 222. +31 20 4496 298. *www.nikon-instruments.jp/eng/. See* Nikon Inc.

Nikon Inc. 1300 Walt Whitman Road, Melville, NY 11747-3064, USA. +1 631 547 8500. +1 631 547 0306. *biosales@nikonincmail.com. www.nikonusa.com.* CCD cameras; colony ring marker; confocal microscope; digital cameras; fluorescence microscopes; image analysis; inverted microscopes; micromanipulators; microscopes; photomicrography; stereomicroscopes.

Nikon Instech Co., Ltd. Parale Mitsui Bldg., 8, Higashida-cho, Kawasaki-ku, Kawasaki, Kanagawa, 210-0005, Japan. +81-44-223-2167.+81-44-223-2182. *www.nikon-instruments.jp/eng/. See* Nikon Inc.

Nikon UK Ltd. Nikon House, 380 Richmond Road, Kingston upon Thames, Surrey KT2 5PR, England, UK. +1 (0)181 541 4440; +1 (0)208 481 6826. +1 (0)181 541 4584. *www.nikon-instruments.com. See* Nikon Inc.

NLS Animal Health. 800-638-8672.+1-888-568-2825. *webadmin@nlsanimalhealth.com. www.nlsanimalhealth. com/.* Marcaine anesthetic.

Nonlinear Dynamics. Cuthbert House, All Saints, Newcastle upon Tyne, NE1 2ET, England, UK. +44 (0)191 230 2121. +44 (0)191 230 2131. *www.nonlinear.com/.* Computer software; image analysis; microarray analysis; microtitration.

Nonlinear USA Inc. 4819 Emperor Blvd, Suite 400, Durham, NC 27703, USA. +1 919 313 4556. *www.nonlinear.com/.* Computer software; image analysis; microarray analysis; microtitration.

Nova Biomedical Corp. 200 Prospect Street, Waltham, MA 02454, USA. 800 458 5813; +1 781 894 0800. +1 781 893 6998. *info@novabio.com. www.novabiomedical. com.* Bioreactor analysers; biosensors: ammonia; electrolytes, glucose; glutamine; glutamate; lactate; osmometers; pH monitors.

Nova Biomedical. C3-5, Evans Business Centre, Deeside Industrial Park, Deeside, Flintshire, CH5 2JZ, Wales, UK. +44 1244 287 087. +44 1244 287080. *office@novabiomedical.co.uk. www.novabiomedical. com.* Bioreactor analysers; biosensors: ammonia; electrolytes, glucose; glutamine; glutamate; lactate; osmometers; pH monitors.

Novocastra Laboratories Ltd. Balliol Business Park West, Benton Lane, Newcastle upon Tyne, NE12 8EW, UK. +44 (0)191 215 0567. +44 (0)191 215 1152. *www.novocastra.co.uk.* Antibodies: NC-L-Villin.

NovaMatrix. FMC BioPolymer, Pharmaceutical, 1735 Market Street, Philadelphia, PA 19103, USA. +1 215 299 6420; +1 215 817 6571. +1 215 299 6669. *novamatrix_info@fmc.com. www.novamatrix.biz; www.fmc.com.* Alginate; chitosan salts and bases; encapsulation; matrix; sodium hyaluronates.

NovaMatrix. FMC BioPolymer, Gaustadalléen 21, N-0349 Oslo, Norway. +47 2295 8638. +47 2269 6470. *novamatrix_info@fmc.com. www.novamatrix.biz; www.fmc.com.* Alginate; chitosan salts and bases; encapsulation; matrix; sodium hyaluronates.

Novartis Pharmaceuticals UK Ltd. Frimley Business Park, GB- Frimley/Camberley, Surrey GU16 7SR, England, UK. +44 (0)1276 692 255. +44 (0)1276 692 508.. *www.novartis.co.uk.* Cyclosporin.

Novartis Institutes for BioMedical Research, Inc. 400 Technology Square, Cambridge, MA 02139, USA. +1 617 871 8000. +1 617 551 9540. *www.us.sandoz.com.* Cyclosporin.

Novo Nordisk A/S. Novo Nordisk A/S, Novo Allé, 2880 Bagsværd, Denmark. +45 4444 8888. +45 4449 0555. *webmaster@novonordisk.com. www.novonordisk.com.* Somatostatin.

NuAire Inc. 2100 Fernbrook Lane, Plymouth, MN 55447, USA. +1 763 553 1270; 1 800 328 3352. +1 763 553 0459. *nuaire@nuaire.com. www.nuaire.com.* CO_2 analyzers and alarms; CO_2 changeover valves; CO_2 incubators; laminar-flow cabinets; low-temperature freezers; safety cabinets.

Nuclepore Corp. *See* Corning.

Nunc. *See* Nalge-Nunc.

Nusil Silicone Technology. 1050 Cindy Lane, Carpinteria, CA 93013, USA. +1 805 684 8780. +1 805 566 9905. *steveb@nusil.com. www.nusil.com.* Adhesives; elastomers; gels; silicone rubber.

Nycomed Danmark ApS. Langebjerg 1, P.O. Box 88, DK-4000 Roskilde, Denmark. +45 46 77 11 11. +45 46 75 66 40. *www.nycomed.dk. www.nycomed.no.* Density medium; lymphocyte preparation medium; metrizamide; sodium metrizoate.

Nycomed UK Ltd. The Magdalen Centre, Oxford Science Park, Oxford OX4 4 GA, England, UK. +44 1865 784 500. +44 1865 784 501. *list@nycomed.com. www.nycomed.com/.* Density medium; lymphocyte preparation medium; Metrizamide; sodium metrizoate.

Olympus America Inc. 2 Corporate Center Drive P.O. Box 9058, Melville, NY 11747-9058, USA. 800-446-5967. +1 516 844 5112. *olympus@performark.com. www.olympusamerica.com/.* Confocal microscopy; digital cameras; inverted microscopes; photomicrography; research microscopes; stereomicroscopes.

Olympus Optical Co. (Europe) GmbH. Wendenstrasse 14–18, D-20097 Hamburg, Germany. +49 40 23 77 30. +49 40 233 765. *main@olympus.uk.com. www.olympus-europa.com. See* Olympus America, Inc.

Olympus U.K. Ltd. Dean Way, Gt. Western Industrial Park, Southall, Middlesex, UB2 4SB, England, UK. +44 (0)207 250 0179. +44 (0)207 250 4677. *microscope@olympus.uk.com. http://www.olympus.co.uk/microscopy/.* Confocal microscopy; digital cameras; inverted microscopes; photomicrography; research microscopes; stereomicroscopes.

Omega Engineering, Inc. One Omega Drive, Box 4047, Stamford, CT 06907-4047, USA. 800 848 4286; +1 203 359 1660. +1 203 359 7700. *info@omega.com. www.omega.com.* Conductivity meter; electronic thermometer; gas flow meters; rotameters; temperature indicator; pH meters.

Omnilab. Robert-Hooke-Str. 8, 28359 Bremen, Germany. +49 (0)421 175 99-0.+49 (0)421 175 99–300.

info@omnilab.de. www.omnilab.de. Centrifuge tubes; CO_2 incubators; filters; flasks; hemocytometers.

Oncogene Research Products. *See* EMB Biosciences (USA), Merck Biosciences (Calbiochem). Angiogenesis; apoptosis; ELISA kits; IGF; LIF; PCNA; TGF; TNF; VEGF.

Oncor, Inc. Marketing Division, 209 Perry Parkway, Gaithersburg, MD 20877, USA. +1 301 963 3500. +1 301 926 6129. *www.oncor.com.* Chromosome paints.

Orchid BioSciences, Inc. 4390 U.S. Route One, Princeton, NJ 08540, USA. +1 609 750 2200. +1 609-750-6400. *mail@orchid.com. www.orchid.com.* Authentication; DNA profiling; forensics; identity testing.

Orchid Cellmark. 20271 Goldenrod Lane, Suite 120, Germantown, MD 20876, USA. +1 301 428 4980, 800 872 5227. +1 301 428 4877. *www.orchidcellmark.com.* Authentication; DNA profiling; forensics; identity testing.

Origen Biomedical. 4020 S. Industrial Dr., Suite 160, Austin, TX 78744, USA. +1 512 474–7278. +1 512 617–1503. *sales@origen.com. www.origen.com/.* Cryostorage bags; freezing medium; stem cell preservation.

Orion Research Inc. *See* Thermo Electron Corp. *www.thermo.com.* Meters; dissolved oxygen, pH, temperature.

Oxford Instruments Inc. 130A Baker Ave. Extension, Concord, MA 01742, USA. +1 978 369 9933. +1 978 369 6616. *info@ma.oxinst.com. www.oxinst.com/.* Micropipettes.

Oxford Instruments plc. Old Station Way, Eynsham, Witney, Oxon OX29 4TL, UK. +44 (0) 1865 881 437. +44 (0) 1865 881 944. *info.oiplc@oxinst.co.uk. www.oxinst.com/.* Micropipettes.

Oxoid Ltd. Wade Road, Basingstoke, Hampshire, RG24 8PW, England, UK. +44 (0) 1256 841144. +44 (0) 1256 463388. *oxoid@oxoid.com. www.oxoid.com/uk.* Bacterial identification system; blood agar plates; microbiological agar plates; nutrient agar plates; nutrient broths; PBS tablets; Tryptone soya broth.

Oxoid Inc. 800 Proctor Avenue, Ogdensburg, New York 13669, USA. +1 613 226 1318. +1 613 226 3728. *www.oxoid.com/us. See* Oxoid Ltd.

PAA Laboratories GmbH. Haidmanweg 9, A-4061, Pasching, Austria. +43 7229 648 65. +43 7229 648 66. *info@paa.at. www.paa.at.* Antibiotics; attachment factors; bovine serum albumin; fetal bovine serum; media; selective serum-free media.

PAA Laboratories, Ltd. 1 Technine Guard Avenue, Houndstone Business Park, Yeovil, Somerset BA22 8YE, England, UK. +44 (0)1935 411 418. +44 (0)1935 411 480. *info@paalaboratories.co.uk. www.paa.at.* For products *see* PAA Laboratories GmbH.

Paar Scientific Ltd.(*see also*** Anton Paar).** 594 Kingston Rd., London SW20 8DN, England, UK. 0208 540 8553. 0208 543 8727. *paar@psl.anton-paar.co.uk. www.paar-scientific.com.* Density meters; viscometers.

Paesel & Lorei GmbH. & Co. Moselstr. 2b, D-63452 Hanau, Germany. +49 (0)6181 18 70-0.+49 (0)6181 18 70-70. *info@paesel-lorei.de. www.paesel-lorei.de.* Media; growth factors; antibodies.

Pall Corp. Bio Support Div., 25 Harbor Park Dr., Port Washington, NY 11746, USA. +1 516-484-3600; 800-289-7255.+1 (516) 484–3651. *custsvc@pall.com. www.pall.com/laboratory.* Filters; filter holders; filtration; pleated cartridge filter; pumps; vacuum pump; mycoplasma filtration; ultrafiltration; densitometers.

Pall Corp. (UK). Havant Street, Portsmouth, Hampshire, PO1 3PD, England, UK. +44 (0)23 9230 3303. +44 (0)23 9230 2509. *www.pall.com/laboratory. See* Pall Corp.

P & T Poultry Supplies and Equipment. Cleeton Cottage Farm, Cleeton Lane, Cleeton St Mary, Nr Cleobury Mortimer, Shropshire, DY14 0QU, England, UK. +44 (0)1584 890263. *info@pandtpoultry.co.uk. www. pandtpoultry.co.uk/.* Egg incubators; fertile eggs.

PAW BioScience Products, Inc. +1 732 842 3939. +1 732 842 6602. *admin@pawbioscience.com. www.pawbioscience.com.* Culture bags; hollow fiber perfusion bioreactor; silicone tubing.

Peak Scientific Instruments. 5424 Sea Edge Dr., Punta Gorda, FL 33950-8735, USA. +1 866 647 1649. +1 866 647 1649. *info@peakscientific.com. www. peakscientific.com/.* Gas generators; nitrogen generators.

Peak Scientific Instruments, Ltd. Fountain Crescent, Inchinnan Business Park, Inchinnan, Renfrewshire, PA4 9RE, Scotland, UK. +44 (0)141 812 8100. +44 (0)141 812 8200. *info@peakscientific.com. www. peakscientific.com/.* Gas generators; nitrogen generators.

PelFreeze. *See* Dynal Biotech.

Pepro Tech, Inc. Princeton Business Park. 5 Crescent Ave. Rocky Hill. NJ 08553-0275, USA.. (800) 436–9910; (609) 497-0253. (609) 497-0321. *info@peprotech.com; www.peprotech.com.* Antibodies; chemokines; cytokines; ELISA.

PeproTech EC Ltd. Peprotech House, 29 Margravine Road, London, W6 8LL, England, UK. +44 (0)20 7610 3062. +44 (0)20 7610 3430. *info@peprotech.co.uk. www.peprotech.com.* Antibodies; chemokines; cytokines; ELISA.

Perbio Science UK Ltd. Century House, High Street, Tattenhall, Cheshire CH3 9RJ, England, UK. +44 (0)1829 771 744. +44 (0)1829 771 644. *uk.info@perbio.com. www.perbio.com.* Culture bags; media; media bags; fetal bovine serum; trypsin replacement.

Perceptive Instruments.. Blois Meadow Business Centre, Steeple Bumpstead, Haverhill, Suffolk CB9 7BN, England, UK. 01440 730773. 01440 730630. *sales@perceptive.co.uk. www.perceptive.co.uk.* Colony counters; image analysis.

PerkinElmer LAS (Germany) GmbH. Ferdinand Porsche Ring 17, 63110 Rodgau-Jügesheim, Germany. 0800 1 81 00 32. 0800 1 81 00 31. *de.instruments.*

perkinelmer.com/. For products *see* PerkinElmer Life Sciences.

PerkinElmer LAS (UK) Ltd. P.O. Box 66, Hounslow TW5 9RT, England, UK. 0800 896046, 0800 891 715. 0800 891 714. *www.PerkinElmer.com.* For products *see* PerkinElmer Life Sciences.

PerkinElmer Life Sciences. 549 Albany Street, Boston, MA 02118, USA. 800-762-4000; +1 617 482 9595. 617 482 1380. *ProductInfo@perkinelmer.com. www. PerkinElmer.com.* Bioreactors; cell separation; centrifuges; colony counters; digital microscope camera; fermentors; fluoroimager; growth factors: NGF, bFGF (FGF-2), TGF-α, TGF-β; image analysis; luminometer; microtitration plate isotope counters; microtitration software; molecular probes; plate readers; radioisotopes; scintillation counters; scintillation fluids; serum replacement, SerXtend.

Pharmacia. *See* Amersham Biosiences.

Pharmingen. 10975 Torreyana Road, San Diego, CA 92121. +1 877 232 8995. +1 858 812 8888. *www. pharmingen.com.* Antibodies: signal transduction, cytokines, interleukin-2, receptors; flow cytometry reagents.

Phoretix International. *See* Nonlinear Dynamics.

Photometrics. 3440 East Britannia Drive, Tucson, AZ 85706, USA. +1 520 889 9933. *www.photomet.com/.* CCD cameras and accessories.

Pierce Chemical Co. 3747 N. Meridan Rd., P.O. Box 117, Rockford, IL 61105, USA. +1 815 968 0747; 800 874 3723. 800 842 5007; +1 815 968 8148. *cs@piercenet.com. www.piercenet.com.* Biochemicals; chemicals; detergents; matrix-coated plates; Pronectin; protein assay; sample vials.

Planer plc. Biomedical Division, Windmill Road, Sunbury, Middlesex TW16 7HD, England, UK. +44(0)1932 755 000. +44(0)1932 755 001. *www.planer.co.uk/.* Liquid nitrogen freezers; cell freezing; freezer control and data-logging software; inventory control; controlled-rate freezers.

Plas Labs Inc. 401 East North St., Lansing, MI 48906, USA. +1 517 372 7177. +1 517 372 2857. *PLI@Plas-Labs.com. www.Plas-Labs.com.* Biosafety cabinets; gloves boxes.

Plastim Ltd. Unit 100, Ashville Business Park, Commerce Road, Staverton, Glos, GL2 9QJ, England, UK. +44 (0)1452 857 733. +44 (0)1452 857 744. *www. plastim.co.uk.* PTFE products: rods, sheets, tubes.

Polaroid Corp. Polaroid Corp., North America Sales & Marketing, 1265 Main St. W2-2C, Waltham, MA 02451, USA. 800-343-5000. 617 386 6271. *www. polaroid.com/.* Instant cameras; closed-circuit TV; digital cameras; emulsion; instant film; microscope camera; photomicroscopy; Polaroid Land camera.

Polaroid (U.K.), Ltd. UK Sales & Marketing, 800 The Boulevard, Capability Green, Luton, LU1 3BA, Bedfordshire, England, UK. +44 (0)1582 409 800. +44 (0)1582 409 801. *www.polaroid.com. See* Polaroid Corp.

Polyfiltronics. *See* Whatman. Disk filters; filters; filtration equipment; filter holders; manifolds; microarray chip technology, microtitration; tubes; water purification.

Polymun Scientific. Nussdorfer Lande 11, 1190 Vienna, Austria. +43 1 36006 6202. +43 1 369 7615. *office@polymun.com.* *www.polymun.com.* Membroferm bioreactor.

PolyScience. 6600 West Touhy Avenue, P.O.Box 48312, Niles, IL 60714, USA. 800 229 7569; +1 847 647 0611. +1 847 647 1155. *customerservice@polyscience.com.* *www.polyscience.com/.* Analytical chemistry kits; chillers; circulators; Clorox; disinfectants; glutaraldehyde; water baths.

Polytech Scientific. 5 Haddonbrook Business Centre, Fallodan Road, Orton Southgate, Peterborough, Cambridgeshire, PE2 6YX, England, UK. +44 (0)1733 371 724. +44 (0)1733 371 645. *sales@polytechscientific.com.* *www.polytechscientific.com.* Bottle-top dispensers; bottles, plastic; cell counting cups; centrifuge tubes: conical and self-standing; cryovials; freezer inventory system; freezer storage; gas burners; pipette aids; pipette pump; Pi-pump; pipettors; plastic Pasteur pipettes; plastic stopcocks; sample cups; sterile indicating tape: autoclave and dry heat; test tube racks; universal container, 50 ml.

Popper & Sons, Inc. 300 Denton Ave., New Hyde Park, NY 11040, USA. 888 717 7677; +1 516 248 0300. 888 557 6773; +1 516 747 1188. *sales@popperandsons.com.* *www.popperandsons.com.* Cornwall syringe; needles; repeating syringe dispenser; sterility indicators; Thermalog; syringes; multipoint dispensers.

Porvair Sciences. Unit 6, Shepperton Business Park, Govett Avenue, Shepperton, Middlesex, TW17 8BA, England, UK. +44 (0)1932 240 255. +44 (0)1932 254 393. *www.porvair-sciences.com.* Plate sealers.

Precision Scientific/Napco. 170 Marcel Dr., Winchester VA, 22602 USA. (800) 621–8820; +1-540-869-9892. +1 540 869 0130. *www.precisionsci.com/lab/another/ napco2.* Autoclaves; centrifuges; CO_2 incubators, glassware washers; laminar-flow hoods; low-temperature freezers; microbiological safety cabinets; sterilizing and drying ovens; vacuum pumps..

Princeton Instruments. 3660 Quakerbridge Road, Trenton, NJ 08619, USA. +1 609-587-9797. +1 609-587-1970. *www.princetoninstruments.com/.* CCD cameras and accessories.

Prior Scientific Instruments Ltd. 3–4 Fielding Industrial Estate, Wilbraham Road, Fulbourn, Cambridge, CB1 5ET, England, UK. 44 (0)1223 881 711. 44 (0)1223 881 710. *uksales@prior.com.* *www.prior.com.* Microscopes: fluorescence, research, stereo; motorized stage; point counter.

Prior Scientific Inc. 80 Reservoir Park Drive, Rockland, ME 02370, USA. +1 781 878 8442. +1 781 878 8736. *info@prior.com.* Microscopes: fluorescence, research, stereo; motorized stage; point counter.

Priorclave Ltd. 129–131 Nathan Way, Woolwich, London, SE28 0AB, England, UK. +44 (0)208 316 6620. +44 (0)208 855 0616. *sales@priorclave.co.uk.* *www.priorclave.co.uk.* Autoclaves; sterilizers.

Promega Corp. 2800 Woods Hollow Rd., Madison, WI 53711, USA. +1 608 274 4330, 800 356 9526. 800 356 1970; +1 608 277 2601. *custserv@promega.com.* *www.promega.com.* Apoptosis assays; cell proliferation assays; cytotoxicity assays; DNA transfer; endothelial cells; gene transfection; growth factors; immortalization; molecular biology reagents.

Promega U.K. Delta House, Chilworth Research Centre, Southampton SO16 7 NS, England, UK. +44 (0)23 8076 0225; 800 378 994. +44 (0)1703 767014; 800 181037. *ukcustserve@promega.com.* *www.promega.com/uk/.* *See* Promega Corp.

PromoCell GmbH. Sickingenstrasse 63/65, D-69126 Heidelberg, Germany. +49 (0)6221 64934-0; 800 776 66 23. +49 (0)6221 64934-40; 800 100 83 06. *info@promocell. com.* *www.promocell.com/.* Adipocytes; chondrocytes; cytokines; dermal fibroblasts; ELISA assays; endothelial cells; epidermal keratinocytes; media; melanocytes; nasal epithelial cells; osteoblasts; perfusion chamber; selective media; serum; serum-free media; skeletal muscle cells; smooth muscle cells; specialized cell cultures; trypsin.

Protein Polymer Technologies Inc. 10655 Sorrento Valley Road, First Floor, San Diego, CA 92121, USA. 619 558 6064, 800 755 0407. 619 558 6477. *info@ppti.com.* *www.ppti.com.* Matrix; Pronectin; serum-free cell adhesion; tissue adhesives.

Protide Pharmaceuticals, Inc. 1311 Helmo Avenue, St. Paul, MN 55128, USA. +1 612 730 1500. +1 612 730 8900. *info@celox.com.* *www.protidepharma.com/.* BSS: Earle's; cryopreservatives; medium, serum-free; serum substitute; vitamins.

Purite Ltd. Bandet Way, Thame, Oxon OX9 3SJ, England, UK. +44 (0)1844 217 141. +44 (0)1844 218 098. *mail@purite.com.* *www.purite.com/.* Water purification.

Qiagen Inc. 27220 Turnberry Lane, Valencia, CA 91355, USA. 800 426 8157; 800-DNA-PREP (800 362 7737). 800 718 2056. *www.qiagen.com/.* DNA preparation; RNA preparation.

Qiagen Ltd. Qiagen House, Fleming Way, Crawley, West Sussex, RH10 9NQ, England, UK. +44 (0)1293 422 911. +44 (0)1293 422 911. *CustomerService-uk@qiagen.com.* *www.qiagen.com/.* DNA preparation; RNA preparation.

Qiagen, GmbH. QIAGEN GmbH., QIAGEN Strasse 1, 40724 Hilden, Germany. 800 79612; +49 (0)2103 29 12000. +49 (0)2103 29 22000. *orders-de@qiagen.com.* *www.qiagen.com/.* DNA preparation; RNA preparation.

QMX Laboratories Ltd. Bolford Street, Thaxted, Essex, CM6 2PY, England, UK. +44 (0)1371 831 611. +44 (0)1371 831 622. *sales@qmxlabs.com.* *www. qmxlabs.com.* Klarity™ syringe filters.

Quadrachem Laboratories Ltd. Kingfisher House, Forest Row Business Park, Forest Row, East Sussex, RH18 5DW, England, UK. +44 (0)1342 820 820. +44 (0)1342 820 825. *enquiries@qclscientific.com. www. qclscientific.com.* Dissolved oxygen meters; conductivity meters; pH meters.

R & D Systems. 614 McKinley Place, NE, Minneapolis, MN 55413, USA. 800 328 2400; +1 612 379 2956. 800 343 7475. *info@rndsystems.com. www.rndsystems.com.* Antibodies; cytokine assays; cytokines; eicosanoid assay kits; ELISA assays; epidermal growth factor; fibronectin; growth factors; interleukin-2; matrix; matrix metalloproteinases; microtitration; microtitration software; neutrophins.

R & D Systems Europe. 4–10 The Quadrant, Barton Lane., Abingdon, Oxon OX14 3YS, England, UK. 800 37 34 15; +44 (0)1235 529 449. +44 (0)1235 533 420. *info@RnDSystems.co.uk. www.rndsystems.com. See* R&D Systems (USA).

Radleys. Shire Hill, Saffron Walden, Essex CB11 3AZ, England, UK. +44 (0)1799 513 320. +44 (0)1799 513 283. *sales@radleys.co.uk. www.radleys.co.uk.* Alcohol-resistant markers; bottles; culture boxes; culture chambers; custom glassblowing; desiccators; flow meters; laboratory carts; nitrile disposable gloves; pipette bulbs; pipette cylinders; pipette hods; pipettors; Pi-pump; racks; repetitive dispenser; shaking incubator; siphon pipette washer; stepped pipettor; trolleys.

Radnoti Glass Technology Inc. 227 West Maple Avenue, Monrovia, CA91016, USA. 800 428 1416; +1 626 357 8827. +1 626 303 2998. *Desmond@radnoti.com. www. radnoti.com.* Organ and tissue baths.

Rainin Instrument Co., Inc. Mack Road, PO Box 4026, Woburn, MA 01888–4026, USA. +1 617 935 3050; 800 472 4646. +1 617 938 1152. *pipets@rainin.com. www.rainin.com/.* Automatic pipettes; pipettors.

Ranfac Corp. 30 Doherty Avenue, P.O. Box 635, Avon, MA 02322, USA. 800 272 6322; +1 508 588 4400. +1 508 584 8588. *info@ranfac.com. www.ranfac.com/.* Biopsy needles.

Raven Biological Laboratories. 8607 Park Dr., Omaha, NE 68106, USA. 800 728 5702; +1 402 593 0781. +1 402 593 0921; +1 402 593 0995. *info@ravenlabs.com. www.ravenlabs.com/.* Microbiological culture media; slide boxes; sterility indicators.

Raymond A. Lamb LLC. 7304 Vanclaybon Drive, Apex, NC 27502, USA.. +1 919 387 1237. +1 919 387 1736. *sales.na@ralamb.com. www.ralamb.net/information.html.* Coverglasses; microscope slides; slide trays; slide boxes.

Raymond A. Lamb Ltd. Units 4 & 5 Parkview Industrial Estate, Lottbridge Grove, Eastbourne, East Sussex BN23 6QE, England, UK. +44 (0)1323 737000. +44 (0)1323 733000. *sales@ralamb.com. www.ralamb.co.uk/..* Coverglasses; microscope slides; slide trays; slide boxes.

Research Diagnostics Inc. Research Diagnostics Inc., Pleasant Hill Road, Flanders, NJ 07836, USA. +1 973 584 7093; 800 631 9384. +1 973 584 0210;. *ResearchD@aol.com. www.researchd.com/.* Angiopoietin-1; antibodies; tetramethylbenzidine; TGF-β1; Ultra-TMB.

Research Organics Inc. 4353 E. 49th St., Cleveland, OH 44125-1083, USA.. 800 321 0570; +1 216 883 8025. *info@resorg.com. www.resorg.com.* Biochemicals; HEPES.

Retsch GmbH. & Co., KG. Rheinische Strasse 36, 42781 Haan, Germany. +49 21 29 55 61-0. +49 21 29 87 02. *info@retsch.de. www.retsch.de.* Sieves.

Revco Scientific, Inc. 308 Ridgefield Ct, Asheville, NC 28806, USA. 800-252-7100; (Canada) 800-447-3826; +1 828-658-2711. +1 828 645 3368. *sales@revco-sci.com. www.revco-sci.com.* Ultra deep freezers.

Richardsons of Leicester. 112A Milligan Road, Leicester, LE2 8FB, England, UK. +1 (0)116 283 8604. +1 (0)116 283 7109. *sales@richardsonsofleicester.co.uk. www.richardsonsofleicester.co.uk/.* Microscope slides (Berliner Glas KG); Pasteur pipettes; specimen containers.

Riken. 3-1-1 Koyadai, Tsukuba-shi, Ibaraki 305-0074, Japan. *www.brc.riken.go.jp/en/.* Cell bank; cell lines; DNA clones; primers.

Robbins Scientific (Europe) Ltd. The Exchange, 24 Haslucks Green Road, Shirley, Solihull, West Midlands, B90 2EL, England, UK. +44 (0)121 744 7445. +44 0121 744 0775. *sales@robsci.co.uk. www.robsci.co.uk.* Automatic dispensers; density media; heating blocks; lymphocyte preparation media; microtitration systems; robotic liquid handling; roller rack incubators.

Robbins Scientific (Europe) Ltd. *See also* Genetic Research Instrumentation.

Robbins Scientific, USA. *See* SciGene. *www.robsci.com.*

Roboz Surgical Instrument Co. Inc. PO Box 10710, Gaithersburg, MD 20898-0710, USA. 800-424-2984. +1 888 424 3121. *info@roboz.com. www.roboz.com.* Cannulas; instruments; sterilizers; syringes; sterilization indicator tape.

Roche Applied Science (Germany). Sandhofer Str 116, D-68305 Mannheim 31, Germany. 800-759-4152. Orders: +49 (0)621/759-4136. *www.roche-applied-science.com/.* Automatic pipettes; antibodies; apoptosis kits; biochemicals; BM cyline; bombesin; chromogranin; collagen; collagenase; DAPI; Dispase; DNA transfer; DOTAP; DOSPER; ELISA; EDTA; Fugene; growth factors; granulocyte-macrophage colony-stimulating factor (GM-CSF); heparin; interleukin-3 (IL-3); lipofection; micropipettes; medium; mycoplasma eradication; serum-free medium; stem cell factor (SCF); vitamins.

Roche Applied Science (UK). Bell Lane, Lewes, West Sussex BN7 1LG, England, UK. +44 (0)1273 480 444; 808 100 9998. 808 100 80 60. *www.roche-applied- science.com/. See* Roche Applied Science (Germany).

Roche Applied Science (USA). P.O. Box 50414, 9115 Hague Rd, P.O. Box 50414, Indianapolis, IN 46250-0414, USA. 800-428-5433. 800-428-2883. *www.roche-applied-science.com/. See* Roche Applied Science (Germany).

RS Biotech. Laboratory Equipment Division, 4 Mackintosh Place, South Newmoor, Irvine, Ayrshire, KA11 4JT,

Scotland, UK. +44 (0)1294 222770. +44 (0)1294 222248. *galaxy@rsbiotech.com. www.rsbiotech.com.* CO_2 incubators; Galaxy.

RS Components, Ltd. Birchington Road, Corby, Northants, NN17 9RS, England, UK. +44 (0)1536 444 222. *www.rs-components.com/index.htm.* Electronics; fans; heat controllers; heaters; thermistors.

Safelab Systems, Ltd. 2 Vines Industrial Centre, Nailsea, Bristol, BS48 1BG, England, UK. 44 (0)1275 855 292. +44 (0)1275 855 131. *sales@safelab.co.uk. www.safelab.co.uk.* Laminar-flow hoods; microbiological safety cabinets.

Safetec of America Inc. 887 Kensington Ave., Buffalo, NY 14215, USA. +1 716-895-1822; 800-456-7077. +1 716-895-2969. *www.safetec.com/.* Clean rooms; laminar-flow hoods; microbiological safety cabinets.

Saf-T-Pak. 10807-182 Street, Edmonton, Alberta, T5S 1J5, Canada. (780) 486-0211; 1-800-814-7484. (780) 486-0235; 1−888-814−7484. *www.saftpak.com/.* Transportation containers; safety.

Sakura Finetek Europe. *See* Bayer plc *and* Raymond A. Lamb. *www.sakuraeu.com/.* Cytocentrifuge; tissue embedding compound.

Sakura Finetek U.S.A., Inc. 1750 West 214th Street, Torrance, CA 90501, USA. 800-725-8723; +1 310-972-7800. +1 310-972-7888. *mail@corp.sakuraus.com. www.sakuraus.com.* Cytocentrifuge; tissue embedding compound.

Santa Cruz Biotechnology Inc. 2145 Delaware Ave., Santa Cruz, CA 95060-5706, USA. 800 457 3801; 831 457 3800. +1 831 457 3801. *scbt@netcom.com. www.scbt.com.* Antibodies: actin, integrins, cadherins, cell adhesion molecules, cytokeratin, desmin, fibronectin, GFAP, vimentin; flow cytometry reagents.

Sanyo Gallenkamp plc. Monarch Way, Belton Park, Loughborough, Leics LE11 5XG, England, UK. +44 (0)1509 265 265. +44 (0)1509 269 770. *sanyo@sanyobiomedical.co.uk. www.sanyogallenkamp.com/.* CO_2 incubators; freezers.

Sanyo Scientific. 1062 Thorndale Avenue, Bensenville, IL 60106, USA. 630-875-3530; 800-858-8442. 630-238-0074. *info@sss.sanyo.com. www.sanyobiomedical.com/.* Incubators; freezers.

Sarstedt AG & Co. Rommelsdorfer Strasse, Postfach 1220, 51582 Nümbrecht, Germany. +49 2293 305 0. +49 2293 305 1222. *info@sarstedt.com. www.sarstedt.com/.* CO_2-permeable caps; flasks.

Sarstedt Inc. 1025, St. James Church Road, P.O. Box 468, Newton NC 28658-0468, USA. +1 828 465 400003. +1 828 465 0718. *info@sarstedt.com. www.sarstedt.com/.* Tissue culture flasks, CO_2-permeable caps.

Sartec Group. Century Farm, Reading Street, Tenterden, Kent, TN30 7HS, England, UK. +44 (0)1233 758 157. +44 (0)1233 758 158. *sales@sartec.co.uk. www.sartec. co.uk.* Gas generators; TOC analyzers; TOC standards; total organic carbon analysis; water purity measurement.

Sartorius AG. Weender Landstrasse 94−108, D-37075 Gottingen, Germany. +49.551.308.00. +49.551.308. 3289. *www.sartorius.com/.* Balances; bioreactors; chromatography medium; disk filters; fermentors; filters; filter holders; filtration equipment; inverted microscopes; media bags; microscopes; Miniperm.

Sartorius Corp. 131 Heartland Blvd., Edgewood, NY 11717, USA. 516 254 4249, 800 368 7178. 516 254 4253. *www.sartorius.com/.* Balances; bioreactors; chromatography medium; disk filters; fermentors; filters; filter holders; filtration equipment; inverted microscopes; media bags; microscopes; Miniperm.

Sartorius Ltd. Blenheim Road, Longmead Industrial Estate, Epsom, Surrey KT19 9QN, England, UK. +44.1372.737100. +44.1372.720799. *sartorius.uk @sartorius.com. www.sartorius.co.uk.* Balances; bioreactors; chromatography medium; disk filters; fermentors; filters; filter holders; filtration equipment; inverted microscopes; media bags; microscopes; Miniperm.

Scanalytics, Inc. 8550 Lee Highway, Suite 400, Fairfax, VA 22031-1515, USA. +1 703-208-2230. +1 703-208-1960. *sales@scanalytics.com. www.scanalytics.com.* Imaging software; CCD cameras; image analysis.

Scharfe Systems. Krammerstrasse 22, D-72764 Reutlingen, Germany. +49 (0)7121 387 86-0.+49 (0)7121 387 86−99. *mail@CASY-Technology.com. www.CASY- Technology.com.* Cell counters; cell size analyzers; counting fluid; detergent; sample cups.

Schleicher & Schuell Bioscience. 10 Optical Avenue, Keene, NH 03431, USA. 800-245-4024. +1 603-357-7700. *custserv@schleicher-schuell.com. www.schleicher-schuell.com.* Blotting membranes; filter sterilization; microarray analysis.

Schleicher & Schuell BioScience GmbH. Hahnestrasse 3, D-37586 Dassel, Germany. +49 5561 791 463. +49 5561 791 583. *salesbio@schleicher-schuell.de. www. schleicher-schuell.com.* Blotting membranes; filter sterilization; microarray analysis.

Schleicher & Schuell, UK Ltd. Unit 11, Brunswick Park, Industrial Estate, London N11 1JL, England, UK. +44 (0) 208 361 3111. +44 (0) 208 361 6352. *Salesuk@schleicher-schuell.de. www.schleicher-schuell. com.* Blotting membranes; filter sterilization; microarray analysis.

Schott AG. Hattenbergstr. 10, 55122 Mainz, Germany. +49 (0)6131 66-0.+49 (0)6131 66−2000. *info@schott.com. www.schott.com.* Glassware; borosilicate glass; bottles; Duran glass bottles.

Schott North America, Inc. 555 Taxter Road, Elmsford, NY 10523, USA. +1 914 831 2200. +1 914 831 2201. *info@us.schott.com. www.schott.com.* Glassware; borosilicate glass; bottles; Duran glass bottles;.

Schott Corp. Drummond Road, Stafford, ST16 3EL, England, UK. +44 (0)1785 223 166. +44 (0)1785 223 522.

info.uk@schott.com. www.schott.com. Glassware; borosilicate glass; bottles; Duran glass bottles.

Scientek. 101–11151 Bridgeport Rd., Richmond, BC V6X 1T3, Canada. 1-866-321-3828;+1 604-273-9094. +1 604-273-1262. *sales@scientek.net. www.scientek.net/.* Glassware washing machine.

Scientific & Chemical Supplies. Carlton House, Livingstone Road, Bilston, West Midlands, England, UK. +44 (0)190–2402402. +44 (0)190 240 2343. *scs@scichem.co. uk. www.sci-chem.co.uk.* Chemicals.

Scientific Instrument Centre. Unit 34D Parham Drive, Boyatt Wood, Eastleigh, Hants, SO50 4NU, England, UK. +44 (0)23 8061 6821. +44 (0)23 8062 9700. *enquiries@sic.uk.com. www.sic.uk.com.* Incubators; detergents; glassware washing machines; nitrogen generators; ovens; ultrasonic baths.

Scientific Laboratory Supplies Ltd. Wilford Industrial Estate, Ruddington Lane, Wilford, Nottingham, NG11 7EP, England, UK. +44 (0)115 982 1111. +44 (0)115 982 5275. *sales@scientific-labs.com. www.scientific-labs.com.* Cloning rings; cryovials; dishes; flasks; multiwell plates.

SciGene. 530 Mercury Drive, Sunnyvale, CA 94085, USA. +1 408-733-7337. +1 408-733-7336. *sales@scigene.com. www.scigene.com/.* Incubators; water baths.

Scion Corp. Suite H, 82 Wormans Mill Ct, Frederick, MD 21701, USA. +1 301 695 7870. +1 (301) 695-0035. *info@scioncorp.com. www.scioncorp.com.* Frame grabber card (time-lapse).

Sefar Inc. Filtration Division, Moosstrasse 2, CH-8803 Rüschlikon, Switzerland. +41 1 724 65 11. +41 1 724 15 25. *www.scientific-labs.com/.* Gauze filters; mesh filters: Nitex, polyester, polyamine, polypropylene.

Sefar Filtration USA. 333 S. Highland Ave., Briarcliff Manor, NY 10510, USA.. +1 816 452 1520. +1 816 452 2183. *www.sefar.com.* Gauze filters; mesh filters: Nitex, polyester, polyamine, polypropylene.

Seikagaku America Inc. 124 Bernard St Jean Drive, East Falmouth, MA 02536–4445, USA.. 888 395 2221; 508 540 3444. +1 508-540-8680. *info@acciusa.com. www.seikagaku.com; www.acciusa.com.* Collagen Type I.

Sentinel Laboratories Ltd. Units 12–13 Lindfield Enterprise Park, Lewes Road, Lindfield, West Sussex, RH16 2LH, England, UK. +44 (0)1444 484044. +44 (0)1444 484045. *brian@sentinel-laboratories.com. www.sentinel-laboratories.com/.* Nitrile gloves, face masks.

Serologicals Corp. 5655 Spalding Drive, Norcross, GA 30092, USA. +1 678 728–2000. *info@serologicals.com. www.serologicals.com.* Antibodies; aprotinin; bovine serum albumin; chymotrypsin; cytokines; DNase; ELISA kits; growth factors; human serum; insulin; media supplements; molecular biology products; PCR; probes; protease inhibitors; serum; trypsin.

Serotec Inc. Serotec Inc., 3200 Atlantic Avenue, Suite 105, Raleigh, NC 27604, USA. 800-265 7376; +1 919 878 7978. +1 919 878 3751. *serotec @serotec-inc.com.*

www.serotec.com. Antibodies: CAMs, integrins, cytokines, EGFr, fibronectin, growth factors and cytokines, IL-2, IL-6, KGF, OSM, VEGF; ELISA.

Serotec Ltd. 22 Bankside, Station Approach, Kidlington, Oxford OX5 1JE, England, UK. +44 (0)1865 852700. +44 (0)1865 373 899. *serotec@serotec.co.uk. www.serotec.com.* Antibodies: CAMs, integrins, cytokines, EGFr, fibronectin, growth factors and cytokines, IL-2, IL-6, KGF, OSM, VEGF; ELISA.

SERVA Electrophoresis GmbH. Carl-Benz-Str. 7, P.O.B. 10 52 60, 69115 Heidelberg, Germany. 800 73 78 24 62; +49 (0)6221 13840-0.+49 (0)6221 13840-10. *info@serva.de. www.serva.de.* Albumins; antibiotics; collagenase; dextrans; dialysis tubing; electrophoresis reagents; HEPES; latex particles; Pluronic-F68; polyethylene glycol; silicones.

SGM Biotech, Inc. 10 Evergreen Drive Suite E, Bozeman, MT 59715, USA. +1 406 585–9535. +1 (406) 585–9219. *order@sgmbiotech.com. www.sgmbiotech.com.* Spore strips; sterility indicators.

Shamrock Scientific Speciality Systems, Inc. 34 Davis Dr., Bellwood, IL 60104, USA. +1 708 547 9005, 800 323 0249. (800) 248–1907; +1 708 547 9021. *sales@shamrocklabels.com. www.shamrocklabels.com.* Sterile-indicating tape; pressure-sensitive tapes; labels; labeling tapes; tags; temperature indicator strips.

Shandon-Scientific. *See* Thermo Electron Corp., ThermoShandon. *www.thermo.com.* Centrifuge cytospin; CO_2 automatic change-over unit for cylinders; cytocentrifuge; electrophoresis; hemocytometers; microcapillaries; pipette washer; pipette drier.

Sigma-Aldrich. PO Box 14508, St Louis, MO 63178, USA. 800 325 3010; +1 314 771 5765. +1 314 771 5757. *OC_DOM_HC@sial.com. www.sigmaaldrich.com.* Amino acids; antibodies; biochemicals; bovine serum albumin; chemicals; colcemid; collagen; collagenase; dexamethasone; dextran; diacetyl fluorescein; disinfectants; DMSO; DNase; ethidium bromide; EDTA; formamide; glutamine; glutathione; growth factors; media; MegaCell serum replacement; salts; serum-free; trypsin; Trypzean; vitamins.

Sigma-Aldrich Chemie GmbH. Eschenstrasse 5, 82024 Taufkirchen, Germany. 800 51 55 00. 800 64 90 000. *www.sigma-aldrich.com.* For products *see* Sigma-Aldrich (USA).

Sigma-Aldrich Co. Ltd. Fancy Road, Poole, Dorset BH17 7 NH, England, UK. + 44 1202 733114; 800 717 181. +44 1202 715460; 800 378 785. *ukcustsv@europe.sial.com. www.sigmaaldrich.com.* For products *see* Sigma-Aldrich (USA).

Signal Instrument Co Ltd. Standards House, 12 Doman Road, Camberley, Surrey, GU15 3DF, England, UK. +44 (0)1276 682 841. +44 (0)1276 691302. *instruments@signal-group.com. www.signal-group.com.* Gas mixers; gas blenders.

Signal USA. 355 North York Rd., Willow Grove, PA 19090, USA. 866-SIGNAL-U; +1 215 830 8882. +1

215 830 8922. *sales@k2bw.com. www.signal-group.com/.* Gas mixers; gas blenders.

Silicone Specialty Fabricators. 3077 Rollie Gates Drive, Paso Robles, CA 93446, USA. 800 394 4284; 805 239 4284. 805 239 0523. *www.ssfab.com/.* Silicone rubber sheet.

Skan. *See* H. V. Skan.

SkinEthic. 45 Rue St Philippe, 06000 Nice, France. +33 4 93 97 77 27. +33 4 93 97 77 28. *infos@skinethic.com. www.skinethic.com.* Alveolar epithelium; bladder epithelium; corneal cultures; cytotoxicity assays; irritancy assays; medium; oral epithelium; organotypic culture; reconstituted human epidermis; specialized cell cultures; tissue equivalents; vaginal epithelium.

Smiths Medical International Ltd. Military Road, Hythe, Kent, CT21 5BN, England, UK. +44 1303 260551. +44 1303 266761. *j.townsend@portex.com. www.smiths-medical.com/.* Semipermeable nylon film.

Society for *In Vitro* Biology (SIVB). SIVB Business Office, 13000-F York Road, #304 Charlotte, NC 28278, USA. (800) 761–7476; +1 704 588 1923. +1 704 588 5193. *sivb@sivb.org. www.sivb.org.* Biological society with interests in cell culture and related techniques..

Socorex ISBA SA. Ch. Champ-Colomb 7, P.O. Box 1024, Ecublens, Lausanne, Switzerland. +41 21 651 6000. +41 21 651 6001. *socorex@socorex.ch. www.socorex.ch.* Digital pipettors; multichannel pipettors; Stepper pipettor.

SoloHill Engineering Inc. 4220A Varsity Drive, Ann Arbor, MI 48108, USA. 866 807 3953; +1 734 973 2956. +1 734 973 3029. *solohill@ic.net. www.solohill.com.* Microcarriers.

Sorvall. *See* Kendro. *http://www.sorvall.com/..*

Southern Biotechnology Associates, Inc. P.O. Box 26221, Birmingham, AL 35260, USA. +1 205-945-1774; 800-722-2255. 205-945-8768. *info@SouthernBiotech. com. www.southernbiotech.com.* Antibodies, anti-mouse IgG3-phycoerythrin.

SP Industries Co. SP Industries Co., 935 Mearns Road, Warminster, PA, USA. +1 215-672-7800. +1 215 672 7807. *hotpack@hotpack.com. www.hotpack.com.* Autoclaves; CO_2 incubators; freezers; laminar flow hoods; safety cabinets.

Spectrum Europe B.V. P.O. Box 3262, 4800 Breda, The Netherlands. +31 76 5719 419. +31 76 5719 772. *info@spectrumeurope.nl. www.spectrumeurope.nl.* Cellmax; dialysis tubing; hollow fiber perfusion culture; microcapillary perfusion; concentric hollow fibers.

Spectrum Medical Inc. 18617 Broadwick Street, Rancho Dominguez, CA 90220, USA. +1 310 885 4600; 800 634 3300. +1 310 855 4666; 800 445 7330. *customerservice@spectrumlabs.com. www. spectrumlabs.com.* Cellmax; dialysis tubing; hollow fiber perfusion culture; microcapillary perfusion; concentric hollow fibers.

SPI Supplies. *See* Structure Probe, Inc.

SPSS Inc. 233 S. Wacker Drive, 11th Floor, Chicago, IL 60606. 312.651.3000. *www.spss.com/.* SPSS (version 10.0) statistical analysis software.

Staniar, J., & Co. 34 Stanley Road, Whitefield, Manchester, M45 8QX, England, UK. +44 (0)161 767 1500. +44 (0)161 767 1502. *sales@johnstaniar.co.uk. www.johnstaniar.co.uk/.* Gauze; nylon mesh; stainless steel mesh.

Statebourne Cryogenics. 18 Parsons Road, Parsons Industrial Estate, Washington, Tyne & Wear, NE37 1EZ, England, UK. +44 (0)191 416 4104. +44 (0)191 415 0369. *sales@statebourne.com. www.statebourne.com.* Controlled-rate freezer; dry shipper; liquid nitrogen freezers; storage Dewars.

Statlab Medical Products. 106 Hillside Driv, Lewisville, TX 75057, USA. +1 972 436–1010; 1–800 442-3573. +1 972 436–1369. *info@statlab.com. www.statlab.com/.* Coverslip mountants; slide container for autoradiographic emulsion.

StemCell Technologies France. 29 Chemin du Vieux Chene, Z.I.R.S.T, F-38240, Meylan, France. +33 (0)4 76 04 75 30. +33 (0)4 76 18 99 63. *info@stemcellfrance.com. www.stemcell.com. See* StemCell Technologies Inc.

StemCell Technologies Inc. 808-777 West Broadway, Vancouver, BC, Canada, V5Z 4J7. +1 604 877 0713; 800 667 0322. +1 604 877 0704; 800 567 2899. *infoweb@stemcell.com. www.stemcell.com.* Antibodies; cell separation; chamber slides; colony atlas; cytokines; dishes; hemopoietic progenitor culture; interleukin-2; methylcellulose media; software; spinner flasks; stem cell purification; videos.

Steris Corp. 5960 Heisley Rd., Mentor, OH 44060-1834, USA. +1 440-345-2600; 800-JIT-4-USE. +1 440-350-7081. *www.steris.com.* Autoclaves; decontamination; disinfectants; freeze drying; glassware washing machines; micromanipulators; microinjectors; stills; steam generators; Wescodyne.

Steris Corp. STERIS House, Jays Close, Viables, Basingstoke, Hampshire RG22 4AX, England, UK. +44 (0)1256 840 400. +44 (0)1276 685662/3. *www.steris.com.* Autoclaves; decontamination; disinfectants; freeze drying; glassware washing machines; micromanipulators; microinjectors; stills; steam generators; Wescodyne.

Stille AB. Gôrdsvõgen 14, Box 709, SE-169 27 Solna, Sweden. 0046 8 588 58 000. 0046 8 588 58 005;. *info@stille.se. www.stille.se.* Biopsy needles.

Stille-Sonesta, Inc. 2220 Canton Suite 209, P.O. Box 140957, Dallas, TX 75201, USA. (800) 665 1614; (214) 741 2464. (214) 741 2605;. *stille@airmail.net. www. stille.se/.* Biopsy needles.

Stoelting Co. 620 Wheat Lane, Wood Dale, IL 60191, USA. +1 630 860 9700. +1 630 860 9775. *physiology@stoeltingco.com. www.stoeltingco.com/physio.* Face masks; gloves, latex, nitrile; instrument sterilizer;

micromanipulators; Parafilm; sharps disposal containers; sterile indicator strips; surgical instruments; swabs.

Stratagene. 11011 N. Torrey Pines Rd., La Jolla, CA 92037-1073, USA. +1 512 321 3321; 800 894 1304. *Fax orders:* +1 512 321 3128; *Tech Services:* 858 535 0034. tech_services@stratagene.com. *www. stratagene.com.* Gene clone libraries; microarray analysis; PCR; vectors.

Stratech Scientific Ltd. Unit 4, Northfield Business Park, Northfield Road, Soham, Cambridgeshire, CB7 5UE, England, UK. +44 (0)1353 722 500. +44 (0)1353 727755. info@stratech.co.uk. www.stratech.co.uk. Antibodies: angiogenesis markers, collagen, fibronectin, GFAP, laminin, mucin, apoptosis markers, cell cycle, collagen I; concentrators; coverslip culture; cytokines; dishes; extracellular matrix; extraction kits; filter well inserts; flasks; growth factors; invasion chambers; matrix coating; multiwell slides; oncogenes; petri dishes; polylysine; serum-free media screening kit.

Structure Probe, Inc. 569 East Gay Street, West Chester, PA 19380, USA. 1-(800)-2424-SPI; 1-(610)-436-5400. 1-(610)-436-5755. spi3spi@2spi.com. www.2spi.com/. SPI-Chem; electron microscopy reagents; Epon substitute; light microscopy reagents.

Stuart Scientific. Holmethorpe Industrial Estate, Redhill, Surrey, RH1 2NB, England, UK. +44 (0)1737 766 431. +44 (0)1737 765 952. www.bibby-sterilin.com. Manual colony counter.

Summers Optical. Division of EMS Acquisition Corp., 1560 Industry Road, P.O. Box 380, Hatfield, PA 19440, USA. +1 215 412 8380. +1 215 41 8450; +1 215 412 8451. sgkcck@aol.com. www. emsdiasum.com/Summers/optical/cements/cements/uv. html. UV curing glass adhesive / cement.

Surgicon, Ltd. Wakefield Rd., Brighouse, W. Yorkshire HD6 1QL, England. +44 (0)1484 712 147. +44 (0)1484 400 106. Gloves; sterilization indicator tape (autoclave).

Swann-Morton Ltd. Owlerton Green, Sheffield, S6 2BJ, England, UK. +44 (0)114 234 4231. +44 (0)114 231 496. info@swann-morton.com. www.swann-morton.com. Surgical instruments; dissecting instruments; scalpels; blades.

Synbiosis. 97H Monocacy Blvd., Frederick, MD, 21701, USA. 603 465 3385. 603 465 2291. www.synbiosis.com/. Colony counters.

Synbiosis. Beacon House, Nuffield Road, Cambridge, CB4 1TF, England, UK. +44 (0)1223 727 125. +44 (0)1223 727 101. sales@synbiosis.com. www. synbiosis.com; www.acolytecounter.com. Acoltye colony counter; colony counters; colony sizing.

Syngene. Beacon House, Nuffield Road, Cambridge, CB4 1TF, England, UK. +44 (0)1223 727 123. +44 (0)1223 727 101. eurosales@syngene.com; intl-sales@syngene.com. www.syngene.com. Colony counting; gel analysis; image analysis.

Syngene USA. 5108 Pegasus Court, Suite M, Frederick, MD 21704, USA. +1 800-686-4407. +1 301-631-3977. ussales@syngene.com. www.syngene.com. Colony counting; gel analysis; image analysis.

Synthecon, Inc. 8054 El Rio, Houston, TX 77054, USA. +1 713 741 2582. +1 713 741 2588. rccs@synthecon.com. www.synthecon.com. NASA bioreactor; rotatory culture chamber.

TAAB Laboratories, Equipment, Ltd. 3 Minerva House, Calleva Park, Aldermaston, Berks, RG7 8NA, England, UK. +1 (0)118 9817775. +1 (0)118 9817881. sales@taab.co.uk. www.taab.co.uk. Acetonitrile; buffers; dissecting instruments; electron microscopy; fixatives; glutaraldehyde; immunocytochemistry; sodium cacodylate.

Taylor-Wharton Cryogenics. PO Box 568, Theodore, AL 36590-0568, USA. +1-251-443-8680; 800 898 2657. +1-251-443-2209. twsales@harsco.com. www.taylor-wharton.com/. CO_2; controlled-rate coolers; cryofreezers; Dewars; liquid nitrogen; liquid nitrogen freezers.

Taylor-Wharton GmbH. Postfach 14 70, D-25804 Huslum, Germany. 49 48 41 985 0. 49 48 41 985 30. www. taylor-wharton.com. CO_2; controlled-rate coolers; cryofreezers; Dewars; gases; liquid nitrogen freezers.

Taylor-Wharton UK. *See* Taylor-Wharton GmbH.

TCS Biologicals Ltd. Botolph Claydon, Buckingham MK18 2LR, England, UK. +44 (0)1296 714 222. +44 (0)1296 714 806. sales@tcsgroup.co.uk. www. tcsbio-sciences.co.uk/. Bacterial culture slides; blood products; chicken serum, plasma; crystal violet; Giemsa; hematoxylin; Leishman; microbiological stains; Neutral Red; Nigrosin; plasma; serum; storage vials;.

TCS CellWorks Ltd. Park Leys, Botolph Claydon, Buckinghamshire MK18 2LR, England, UK. +44 (0)1296 713 120. +44 (0)1296 713 122. office@tcscellworks.co.uk. www.tcscellworkscatalogue.co.uk/. Accutase; Accumax; angiogenesis kit; bronchial cells; cadherins; CAMs; chondrocytes; conditioned medium; corneal epithelium; endothelial cells; fibroblasts; hepatocytes; integrins; keratinocytes; leucocytes; mycoplasma treatment; nasal epithelial cells; osteoblasts; serum-free medium; smooth muscle cells; synovial cells.

Tebu-bio APS. Forskersparken CAT, DTU Bygning 347, 2800 Lyngby, Denmark. +45 45 25 64 01. +45 45 25 64 03. scandinavia@tebu-bio.com. www.tebu-bio.com. Matrix arrays; matrix coating.

Tebu-bio Ltd. Unit 7, Flag Business Exchange, Vicarage Farm Road, Peterborough, Cambs PE1 5TX, England, UK. +44 (0)1733 421 880. +44 (0)1733 421 882. uk@tebu-bio.com. www.tebu-bio.com. Matrix arrays; matrix coating.

Tecan Sales Switzerland AG. *See* strasse 103, CH-8708 Männedorf, Switzerland. +41 1(0) 922 89 22. +41 1 (0)922 89 23. tecan.sales.ch@tecan.com. www.tecan.com. Microarray scanners; microtitration: plate readers, dispensers, washers.

Tecan UK Ltd. Theale Court, 11–13 High Street, Theale, UK-Reading RG7 5 AH, England, UK. +44 (0)1189 300 300. +44 (0)1189 305 671. *helpdesk-uk@tecan.com. www.tecan-uk.co.uk.* Microtitration plate readers, dispensers, plate washers.

Tecan US Inc. P.O.Box 13953, Research Triangle Park, NC 27709, USA. +1 919 361 5200. +1 919 361 5201. *helpdesk-us@tecan.com. www.tecan.com.* Microtitration plate readers, dispensers, plate washers.

Techmate Ltd. 10 Bridgeturn Avenue, Old Wolverton, Milton Keynes, Bucks., MK12 5QL, England, UK. +44 (0)1908 322 222. +44 (0)1908 319 941. *sales@techmate.co.uk. www.techmate.co.uk.* Ampoules; aspirators; autoclavable baskets; bottles; cell culture inserts; cell scrapers; centrifuge tubes; chamber slides; conical tubes; containers; coverslips; cryotubes; cryotube markers; cryovials; culture tubes; desiccator cabinets; dishes; filter bottom microtitration plates; filters; filter sterilizer; filter well inserts; flasks; freezer storage racks; gas-permeable caps; hand vacuum pumps; media bags; pipette washers; plasticware.

Techne (Cambridge) Ltd. Duxford, Cambridge, CB2 4PZ, England, UK. 01223 832401. 01223 836838. *sales@techne.com. www.techne.com/.* Magnetic stirrers; stirrer flasks; suspension culture; microcarrier culture flasks.

Techne Inc. 3 Terri Lane, Suite 10, Burlington, N.J. 08016, USA. +1 609-589-2560; 800-225-9243. +1 609-589-2571. *www.techneusa.com/.* Magnetic stirrers; stirrer flasks; suspension culture; microcarrier culture flasks.

Technika. 4757 East Greenway Rd., Suite 107B PMB 177, Phoenix, AZ 85032, USA. +1 480 348 0279. *www.Technika.com.* Anemometers; conductivity meters; dissolved oxygen; refractometers.

Techno Plastic Products (TPP) AG. Zollstrasse 155, CH-8219 Trasadingen, Switzerland.. *info@tpp.ch. www.tpp.ch.* Cell scrapers; centrifuge tubes; dishes; filters; flasks; gas-permeable caps; Leighton tubes; multiwell plates; pipettes; scrapers.

Tecniplast UK Ltd. 2240 Parkway, Kettering Venture Park, Kettering Northants NN15 6XL, England, UK. *info@TecniplastUK.com. www.tecniplast.it.* Animal cages and equipment; disinfectants.

Tekmar-Dohrmann. *See* Teledyne Tekmar.

Tekto Inc. *See* Sefar Filtration USA.

Teledyne Tekmar. 4736 Socialville Foster Rd., Mason, OH 45040, USA. +1 513-247-7000; 800 874–2004. +1 513-247-7050. *sales@tekmar.com. www.tekmar.com.* Nylon mesh; Nitex; stirrers; TOC.

Texol Products Ltd. Myrekirk Road, Dundee, Scotland, DD2 4SX, UK. +44 (0) 1382 618400. +44 (0) 1382 618422. *info@texol.co.uk. www.gasgen.co.uk/.* nitrogen generator.

Thermo Electron Corp. 81 Wyman Street, Waltham, MA 02454–9046, USA. +1 781 622 1000. +1 781 622 1207. *Contact via product-directed search on website. www.thermo.com.*

Alarms; centrifuges; CO_2 controllers; CO_2 incubators; conductivity meters; controlled-rate freezer; cryopreservation equipment; Cytobuckets; cytocentrifuge; dataloggers; electronic thermometers; electrophoresis; freezer inventory systems; freezers; hemocytometers; incubators; laminar-flow hoods; liquid handling; liquid nitrogen freezers; low-temperature freezers; magnetic sorting; microbiological safety cabinets; Microcaps (Drummond); micropipettes; microtitration plates; microtitration plate reagent dispensers; nitrogen freezers; orbital shakers; ovens; pH meters; pipette drier; pipette washer; pipettors; plate readers; programmable freezer; radio data systems & alarms; safety cabinets; TOC meters; water baths.

Thermo Electron UK. Unit 5, The Ringway Centre, Edison Road, Basingstoke, Hampshire RG21 6YH, England, UK. +44 (0)870 609 0203. +44 (0)870 609 9202. *sales.btd.uk@thermo.com. www.thermo.com.* For products *see* Thermo Electron Corp.

Thermolyne. *See* Barnstead Thermolyne.

Thomas Scientific. 99 High Hill Rd., P.O. Box 99, Swedesboro, NJ 08085, USA. +1 856 467 2000; 800 345 2100. +1 856 467 3087. *value@thomassci.com. www.thomassci.com.* Disinfectants; dissecting instruments; iridectomy knives; magnetic stirrers; microsyringes; silicone tubing; surgical instruments.

Titan Enterprises Ltd. Unit 2, 5A Cold Harbour Business Park, Sherborne, Dorset, DT9 4JW, England, UK. +44 (0)1935 812 790. +44 (0)1935 812 890. *info@flowmeters.co.uk. http://www.flowmeters.co.uk/.* Gas flow meters; rotameters.

TPP. *See* Techo Plastic Products. *www.tpp.ch.*

Triple Red Laboratory Technology Ltd. Triple Red Ltd., Unit D4 Drakes Park, Long Crendon Ind Est., Long Crendon, Buckinghamshire, HP18 9BA, England, UK. +44 (0)1844 218 322. +44 (0)1844 218 332. *info@triplered.com. www.triplered.com.* CO_2 incubators; cryovial labels; label printers; labels; laminar flow hoods; water purification.

TSI Inc. 500 Cardigan Road, St. Paul, MN, 55126-3996, USA. 1-800-874-2811; +1 651-490-2811. +1 651-490-3824. *answers@tsi.com. www.tsi.com.* Air velocity meters; anemometers.

U.S. Filter (*see also* USF), Ltd. 40-004 Cook Street, Palm Desert, CA 92211, USA. 800.875.7873 ext. 5000; +1 760-340-0098. +1 760-341-9368. *labproducts@usfilter.com. www.usfilter.com.* Deionization; reverse osmosis; water purification.

Unibioscreen S.A. 40, Avenue Joseph Wybran, 1070 Brussels, Belgium. +32 (0)25 29 58 34; +32 (0)25 29 58 34. +32 (0)25 29 59 42. *support@unibioscreen.com. www.unibioscreen.com.* Cancer therapeutics; invasion; xenografts.

UniEquip. Fraunhoferstrasse 11, D-82152 Martinsried, Munich, Germany. +49 (0)89 857 52 00. +49 (0)89 856 13 04. *uniequip@t-online.de. www.uniequip.com.* Infra-red operated gas burner.

Universal Biologicals (Cambridge) Ltd. Passhouse Farmhouse, Papworth St Agnes, Cambridge, CB3 8QU, England, UK. +44 (0) 1480 839 015. +44 (0) 1480 831 912. *info@universalbiologicals.ltd.uk. www. universalbiologicals.ltd.uk.* Antibodies: actin, G proteins, neural stem cells, tubulin; collagen; cytometry reagents; growth factors; protein kinases.

Upstate Biotechnology. 1100 Winter Street, Suite 2300, Waltham, MA 02451, USA. 800-233-3991, 781-890-8845. 781-890-8845. *info@upstatebiotech.com. www.upstate. com.* Antibodies; apoptosis; assay kits; cell proliferation assays; cell signaling; enzymes; growth factors; neurobiology.

Upstate Biotechnology. Upstate House, Gemini Crescent, Technology Park, Dundee, DD2 1SW, Scotland, UK. +44 (0)1382 560812. 0800 0190 444. *ukinfo@upstate.com. www.upstate.com.* Antibodies; apoptosis; assay kits; cell proliferation assays; cell signaling; enzymes; growth factors; neurobiology.

USB Corp. 26111 Miles Rd., Cleveland, OH 44128, USA. +1 216 765 5000; 800 321 9322. 800 535 0898; +1 216 464 5075. *customerserv@usbweb.com. www.usbweb.com.* Biochemicals; polyvinylpyrrolidine.

UVP, Inc. 2066 W 11th St., Upland, CA 91786, USA. +1 909 946–3197; 800 452 6788. +1 909 946 3597. *info@uvp.com. www.uvp.com.* Mercury vapor lamps; fluorescence; UV lamps; UV transilluminators.

UVP, Ltd. Unit 1, Trinity Hall Farm Estate, Nuffield Road, Cambridge CB4 1 TG, England, UK. +44(0) 1223–420022. +44(0)1223–420561. *uvp@uvp.co.uk. www.uvp.com.* Mercury vapor lamps; fluorescence; UV lamps; UV transilluminators.

Varian. 3120 Hansen Way, Palo Alto, CA 94304-1030, USA. +1 650 213 8000. *custserv@varianinc.com. www.varianinc.com.* Vacuum pumps.

Vector Laboratories, Inc. 30 Ingold Road, Burlingame, CA 94010, USA. 800 227 6666; +1 650 697 3600. +1 650 697 0339. *vector@vectorlabs.com. www.vectorlabs.com.* ABC kit; alkaline phosphatase; antibodies; biotin/avidin; ELISAs; enzyme immunoassays; flow cytometry reagents; fluorescence bleaching retardant.; fluorescence fade retardant; fluorescence-quenching inhibitor; glucose oxidase; hybridoma screening; lectins; slide adhesive; UV quench inhibitor.

Vector Laboratories, Ltd. 3 Accent Park, Blakewell Road, Orton Southgate, Peterborough PE2 6XS, England, UK. +44 (0)1733 237 999. +44 (0)1733 237 119. *vector@vectorlabs.co.uk. www.vectorlabs.com.* For products *see* Vector Labs Inc.

Verilabs Europe. Postbus 1336, 2302 BH Leiden, Netherlands. +31 (0)900-3628378. *info@verilabs.nl.www. verilabs.nl/.* Authentication; DNA profiling; genetic identity testing.

Vision Biosystems. Balliol Business Park West, Benton Lane, Newcastle upon Tyne, NE12 8EW, England, UK. +44 (0)191 215 4242. +44 (0)191 2154227. *sales.eu@vision-bio.com. www.vision-bio.com.* Antibodies: NC–L-Villin.

Vision BioSystems Inc. 700 Longwater Drive, Norwell MA 02061, USA. +1 781 616 1190. +1 781 616 1193. *sales.usa@vision-bio.com. www.vision-bio.com/.* Antibodies: NC–L-Villin.

Vivascience AG. Feodor-Lynen-Strasse 21; 30625 Hannover, Germany. +49 (0)511 524 875-0.+49 (0)511 524 875-19. *info@vivascience.com. www.vivascience.de.* Bioreactors; chamber slides; culture dishes; Flexiperm; Miniperm; multiwell plates; Quadriperm; Petriperm; slide culture dishes.

Vivendi Water Systems. High Street, Lane End, High Wycombe, Bucks HP14 3JH, UK. +44 (0)1494 887 555. +44 1494 887 837. *sandra.welch@veoliawater.com. www.elgalabwater.com.* Filtration; ion exchange; photo-oxidation; reverse osmosis; water purification.

Volac. *See* Camlab; Cole-Parmer.

V&P Scientific Inc. 9823 Pacific Heights Boulevard, Suite T, San Diego, CA 92121, USA. 800 455 0644; +1 858 455 0643. +1 858 455 0703. *sales@vp-scientific.com. www.vp-scientific.com/.* Heating blocks.

VWR International. 1310 Goshen Pkwy, W. Chester, PA 19380, USA. +1 610 431 1700; 800 932 5000. +1 610 436 1761. *www.vwr.com.* Autoclaves; bottles; electrophoresis equipment; face masks; gloves: latex, nitrile, vinyl; incubators; lab supplies; magnetic stirrers; microsyringes; microtitration filter plates; nitrogen freezers; pH meters; pumps; slides and coverglasses; sterilizing filters; syringe filters; ultrasonic baths; water purifiers.

VWR International, Ltd. Hunter Boulevard, Magna Park, Lutterworth, Leics LE17 4XN, England, UK. 0800 22 33 44; +44 (0)1202 669 700. +44 (0)1455 558 586. *uk.sales@uk.vwr.com. www.vwr.com.* For products *see* VWR International, USA.

Vysis. 3100 Woodcreek Dr., Downers Grove, IL 60515. USA. 800-553-7042, Ex.1; +1 708 271 7000. +1 708 271 7008. *vysis_help@vysis.com. www.vysis.com.* Chromosome paints; molecular probes.

Wako Chemicals USA, Inc.. 1600 Bellwood Road, Richmond, VA 23237-1326, USA. +1 804 271 7791. +1 804 271 7791. *www.wakousa.com.* Antibodies; ascorbic acid; biochemicals; enzymes; Daigo's T medium.

Wallac. *See* PerkinElmer.

Watson-Marlow Bredel Pumps Inc. 37 Upton Technology Park, Wilmington, MA 01887-1018, USA. +1 978 658 6168; 800-282-8823. +1 978 658 0041. *support @wmbpumps.com. www.watson-marlow.com.* Peristaltic pumps; silicone tubing.

Watson-Marlow Bredel Pumps Ltd. Falmouth, Cornwall TR11 4BR, England, UK. +44 (0)1326 370 370; 800 0189 844. +44 (0)1326 376 009. *info@watson-marlow.co.uk. www.watson-marlow.co.uk.* Peristaltic pumps; Silicone tubing.

Wave Biotech AG. Ringstrasse 24, CH-8317 Tagelswangen, Switzerland. +41 (0)52 354 36 36. +41 (0)52 354 36 46. *Contact via web page. www.wavebiotech.ch.* Bioreactors; culture bags; rocker platform; wave-action agitation.

Webb Microtome Ltd. Dr Sim Webb, Webb Mini-Microtome, Brisimany, Walnut Hill, Surlingham, Norwich, NR14 7DQ, England, UK. +44 (0)1603 592 268. +44 (0)1603 592 250. *s.f.webb@eclipse.co.uk. www. microtome.co.uk.* Portable microtome.

Wescor, Inc. 459, South Main Street, Logan, Utah 84321, USA. 800 453–2725; +1 435 752 6011. 801 752 4127. *biomed@wescor.com. www.wescor.com/.* Cytocentrifuge; osmometers; slide stainer.

Whatman Inc. 9 Bridewell Pl., P.O. Box 1197, Clifton, NJ 07014, USA. +1 (617) 485–6590; 800-631-7290. +1 973 773 3991. *info@whatman.com. www.whatman.com.* Chromatography; filtered stoppers; filter paper; 24-Well filter plates; filters, glass fiber, paper; filter well plate; multichannel pipettors; microtitration plates; Nuclepore filters; pipettors; vacuum pumps.

Whatman plc. 27 Great West Road, Brentford, Middlesex, TW8 9BW, England, UK. +44 (0) 208 326 1740. +44 (0) 208 326 1741. *information@whatman.co.uk. www.whatman.plc.uk.* For products *see* Whatman Inc.

Wheaton Science Products. *See* Lawson Mardon Wheaton.

Wissenschaftlich-Technische Werkstatten GmbH. & Co. KG. Dr.-Karl-Slevogt-Strasse 1, D-82362 Weilheim, Germany. +49 (0)881 183-0.+49 (0)881 183–420. *info@WTW.com. www.WTW.com.* Colony counters; meters: conductivity, oxygen, pH, multi, thermometers; water analysis.

World Precision Instruments. 175 Sarasota Center Boulevard, Sarasota, FL 34240-9258, USA. +1 941 371 1003. +1 941 377 5248. *wpi@wpiinc.com. www.wpiinc.com.* Filter well inserts; resistance meter; TEER; transepithelial electrical resistance measurement.

World Precision Instruments. Astonbury Farm Business Centre, Aston, Stevenage, SG2 7EG, England, UK.

+44 (0)1438 880025. +44 (0)1438 880026. *wpiuk@wpi-europe.com. www.wpi-europe.com.* Filter well inserts; resistance meter; TEER; transepithelial electrical resistance measurement.

Worthington Biochemical Corp. 730 Vassar Avenue, Lakewood, NJ 08701, USA. 800-445-9603; +1 732 942 1660. 800 368–3108; +1 732 942 9270. *office@worthington-biochem.com. www.worthington- biochem.com.* Collagenase; DNase; hyaluronidase; trypsin.

WPI. *See* World Precision Instruments.

WTW Measurement Systems, Inc. 6E Gill St., Woburn, MA 01801, USA. 800 645 5999; +1 781 569 0095. +1 781-932-3198. *info@wtw-inc.com. www.wtw.com/.* Colony counters; meters: conductivity, oxygen, pH, multi, thermometers; water analysis.

YSI Inc. 1700/1725 Brannum Lane, Yellow Springs, Ohio 45387, USA. +1 937 767 7241; 800 659 8895. +1 937 767 8058. *Contact via web page. www.YSI.com.* Biochemical analyzers; biosensors; monitors: choline, glucose, glutamate, glutamine, lactate.

Zeiss. *See* Carl Zeiss.

Zimmer Corp. P.O. Box 708, 1800 West Center Street, Warsaw, IN 46581-0708, USA. 800-613-6131. +1 574-372-4988. *Contact via web page. www.zimmer.com.* Ceramic rods and scaffolds (tissue engineering).

Zinsser Analytic (U.K.), Ltd. Howarth Road, Stafferton Way, Maidenhead, Berks SL6 1AP, England. +44 (0)1628 773 202. +44 (0)1628 672 199. *officeuk @zinsser-analytic.com. www.zinsser-analytic.com/.* Automatic pipettes; vials; dispensers; liquid handling; microtitration; pumps: peristaltic, vacuum.

Zinsser Analytic GmbH.. Eschborner Landstrasse 135, D-60489 Frankfurt, Germany. +49 69 78 91 06-0. +49 69 78 91 06–80. *info@zinsser-analytic.com. www.zinsser-analytic.com.* Automatic pipettes; vials; dispensers; liquid handling; microtitration; pumps: peristaltic, vacuum.

Zymed Laboratories, Inc. 561 Eccles Avenue, South San Francisco, CA 94080, USA. 800-874-4494; 650-871-4494. +1 650 871 4499. *tech@zymed.com. www. zymed.com.* Antibodies: Ki-67, cytokeratin, breast cancer.

Glossary

[Modified after Schaeffer, 1990]

Adaptation. Induction or repression of synthesis of a macromolecule (usually a protein) in response to a stimulus; e.g., enzyme adaptation—an alteration in enzyme activity brought about by an inducer or repressor and involving an altered rate of enzyme synthesis or degradation.

Allograft. *See* Homograft.

Amniocentesis. Prenatal sampling of the amniotic cavity.

Anchorage dependent. Requiring attachment to a solid substrate for survival or growth.

Anemometer. An instrument for measuring flow rate of air.

Aneuploid. Not an exact multiple of the haploid chromosome number. (*See* Haploid.)

Apoptosis. Cell death by a biologically controlled intracellular process involving DNA cleavage and nuclear fragmentation.

Aseptic. Free of microbial infection.

Autocrine. Receptor-mediated response of a cell to a factor produced by the same cell.

Autograft. A graft from one individual transplanted back to the same individual.

Autoradiography. Localization of radioisotopes in cells and tissue sections (microautoradiography) and blots from electrophoresis preparations; achieved by exposure of a photographic emulsion placed in close proximity to the specimen.

Balanced salt solution. An isotonic solution of inorganic salts present in approximately the correct physiological concentrations; may also contain glucose, but is usually free of other organic nutrients.

Bioreactor. Culture vessel for large-scale production of cells, either anchored to a substrate or propagated in suspension.

Biostat. Culture vessel in which physical, physicochemical, and physiological conditions, as well as cell concentration, are kept constant, usually by perfusion, monitoring, and feedback.

Carcinoma. A tumor derived from epithelium, usually endodermally or ectodermally derived cells.

Cell concentration. Number of cells per mL of medium.

Cell culture. Growth of cells dissociated from the parent tissue by spontaneous migration or mechanical or enzymatic dispersal.

Cell density. Number of cells per cm^2 of substrate.

Cell fusion. Formation of a single cell body by the fusion of two other cells, either spontaneously or, more often, by induced fusion with inactivated Sendai virus or polyethylene glycol.

Cell hybridization. *See* Hybrid cell.

Cell line. A propagated culture after the first subculture.

Cell strain. A characterized cell line derived by selection or cloning.

Centipoise. Unit of viscosity; 1000 centipoises are equivalent to 1 Pascal-second.

Centromere. The point of adhesion between two *chromatids* in a *chromosome*; attaches to spindle during metaphase, telophase, and anaphase of cell division.

Chemically defined. Made entirely from pure defined constituents (said of a medium); distinct from "serum free," in which other poorly characterized constituents may be used to replace serum.

Chromatid. Paired constituent of a *chromosome* linked by a centromere.

Chromosome. A complex of DNA and nucleoproteins forming a defined morphological structure within the nucleus, visible at metaphase during cell division, made up of two morphologically identical *chromatids* joined at the *centromere*, and present in a defined number characteristic for each species.

Chromosome painting. Use of specific fluorescent probes to stain defined regions of the chromosome.

Clone. A population of cells derived from one cell.

Commitment. Irreversible progression from a stem cell to a particular defined lineage endowing the cell with the potential to express a limited repertoire of properties.

Confluent. A monolayer of cells in which all cells are in contact with other cells all around their periphery, and no available substrate is left uncovered.

Constitutive. Expressed by a cell in the absence of external regulation.

Contact inhibition. Inhibition of plasma membrane ruffling and cell motility when cells are in complete contact with other adjacent cells, as in a confluent culture; often precedes, but is not necessarily causally related to, cessation of cell proliferation.

Continuous cell line or cell strain. Cell line or strain having the capacity for infinite survival. Previously known as "established" and often referred to as "immortal."

Cyclic growth. Growth from a low cell density to a high cell density with a regular subculture interval; regular repetition of the growth cycle for maintenance purposes.

Cytokine. A factor, released by cells, that will induce a receptor-mediated effect on the proliferation, differentiation, or inflammation of other cells; usually a short-range paracrine, rather than systemic, effect.

Cytostasis. Cessation of cell proliferation.

Cytotoxicity. Cellular damage to one or more metabolic pathways, intracellular processes, or structures resulting in impaired function. Often, but not necessarily, linked to loss of viability.

Deadaptation. Reversible loss of a specific property due to the absence of the appropriate inducer (not always defined).

Dedifferentiation. Irreversible loss of the specialized properties that a cell would have expressed *in vivo*. As evidence accumulates that cultures dedifferentiate by a combination of the selection of undifferentiated stromal cells and deadaptation resulting from the absence of the appropriate inducers, the term is going out of favor. It is still correctly applied to mean the progressive loss of differentiated morphology in histological observations of, for example, tumor tissue.

Density limitation of growth. Mitotic inhibition correlated with an increase in cell density at confluence.

Diploid. Each chromosome represented as a pair, identical in the autosomes and female sex chromosomes and nonidentical in male sex chromosomes, and corresponding to the chromosome number and morphology of most somatic cells of the species from which the cells are derived.

Dome. A hemicystic or blisterlike structure in a confluent epithelial monolayer implying ion transport across the monolayer and resulting in the accumulation of water below the monolayer.

DNA fingerprinting. Binding of multilocus cDNA probes to hypervariable regions of satellite DNA cut by restriction endonucleases and visualized by autoradiography. Pattern specific to individual from whom, or from which, the DNA was derived.

DNA profiling. The generic term for assaying hypervariable regions of satellite DNA, now mainly used to detect the frequency of short tandem repeats (STRs) in microsatellite DNA with RT-PCR of single loci. More sensitive than multilocus *DNA fingerprinting* and readily quantifiable.

Ectoderm. The outer germ layer of the embryo, giving rise to the epithelium of the skin.

Embryonal stem cells. Totipotent stem cells isolated from the inner cell mass of an early embryo; can be propagated as cell lines with a wide range of differentiation capabilities.

Embryonic induction. The interaction (often reciprocal) of cells from two different germ layers, promoting differentiation.

Endocrine. Signaling factors, such as hormones, released by one tissue and having an effect on a distant tissue via the systemic vasculature.

Endoderm. The innermost germ layer of the embryo, giving rise to the epithelial component of organs such as the gut, liver, and lungs.

Endothelium. An epithelium-like cell layer lining spaces within mesodermally derived tissues, such as blood vessels, and derived from the mesoderm of the embryo.

Enzyme induction. An increase in synthesis of an enzyme produced by, for example, hormonal stimulation.

Epithelial. Describes cells derived from epithelium, but often used more loosely to describe any cells of a polygonal shape with clear, sharp boundaries between them. More correctly, the latter should be referred to as epithelium-like or epithelioid.

Epithelium. A covering or lining of cells, as in the surface of the skin or lining of the gut, usually derived from the embryonic endoderm or ectoderm, but sometimes derived from mesoderm, as with kidney tubules and mesothelium lining body cavities.

ES cells. *See* Embryonal stem cells.

Euploid. Exact multiple of the haploid chromosome set. The correct morphology characteristic of each chromosome pair in the species from which the cells are derived is not implicit in the definition, but is usually assumed to be the case; otherwise we should say "euploid, but with some chromosomal aberrations."

Explant. A fragment of tissue transplanted from its original site and maintained in an artificial medium.

Explantation. Isolation of tissue for maintenance *in vitro*, strictly as small fragments with accompanying outgrowth (*see* Primary explant), but often used as a generic term for the isolation of tissue for culture.

FACS. *See* Fluorescence-activated cell sorter.

Fermentor. Large-scale culture vessel, often used for cells in suspension; derived from same term applied to microbiological culture.

Fibroblast. A proliferating precursor cell of the mature differentiated fibrocyte.

Fibroblastic. Resembling fibroblasts [i.e., spindle shaped (bipolar) or stellate (multipolar)]; usually arranged in parallel arrays at confluence if contact is inhibited. Often, the term is used indiscriminately for undifferentiated mesodermal cells, regardless of their relationship to the fibrocyte lineage; implies a migratory type or cell with processes exceeding the nuclear diameter by threefold or more. More correctly, fibroblast-like or fibroblastoid.

Ficoll-paque. Density medium made up of Ficoll combined with a radiopaque iodinated substance, such as sodium metrizoate.

Finite cell line. A culture that has been propagated by subculture but is capable of only a limited number of cell generations *in vitro* before dying out.

FISH. *See* Fluorescence *in situ* hybridization.

Flow cytometer. An instrument providing quantitative and qualitative analysis of individual cells in a population by scanning a single cell stream with a laser, or with multiple lasers of different wavelengths, and recording the light that is scattered or the fluorescence that is emitted.

Fluorescence-activated cell sorter (FACS). A cell separation device based on electromagnetic sorting of a single-cell suspension by means of the scattering of light or the fluorescent properties of individual cells revealed by a laser scanning a single cell stream. (*See also* Flow cytometer.)

Fluorescence *in situ* hybridization (FISH). Binding of specific fluorescent probes to specific intracellular locations by *in situ hybridization*. (*See also* Chromosome painting).

Generation number. The number of population doublings (estimated from dilution at subculture) that a culture has undergone since explanation; necessarily contains an approximation of the number of generations in primary culture.

Generation time. The interval from one point in the cell division cycle to the same point in the cycle, one division later; distinct from population-doubling time, which is derived from the total cell count of a population and therefore averages different generation times, including the effect of nongrowing cells.

Genotype. The total genetic characteristics of a cell.

Glycocalyx. Glycosylated peptides, proteins, and lipids, and glycosaminoglycans attached to the surface of the cell.

Growth curve. A semilogarithmic plot of the cell number on a logarithmic scale against time on a linear scale, for a proliferating cell culture; usually divided into the lag phase (the phase before growth is initiated), the log phase (the period of exponential growth), and the plateau (a stable cell count achieved when the culture stops growing at a high cell density).

Growth cycle. Growth interval from subculture to the top of the log phase, ready for a further subculture.

Growth factor. A factor, released by cells, that induces proliferation in other cells; mostly paracrine in effect, but may be released into the blood by platelets or endothelium.

Growth medium. The medium used to propagate a particular cell line; usually a basal medium with additives such as serum or growth factors.

Haploid. That chromosome number wherein each chromosome is represented once; in most higher animals, the number present in the gametes and half the number found in most somatic cells.

Heterokaryon. Cell containing two or more genetically different nuclei; usually derived by cell fusion.

Heteroploid. A culture in which the cells have chromosome numbers other than diploid and differing from each other.

Histotypic. A culture resembling a tissue-like morphology *in vivo*. Usually, a three-dimensional culture re-created from a dispersed cell culture that attempts to regain, by cell proliferation and multilayering or by reaggregation, a tissuelike structure. Organ cultures cannot be propagated, whereas histotypic cultures can.

Holding medium. Medium, usually without serum and growth factors, or with minimal serum, designed to maintain cells in a viable state without proliferation (e.g., for collecting biopsies or maintaining cells at a plateau with no further cell proliferation).

Homeothermic. Able to maintain a constant body temperature despite environmental fluctuations.

Homograft (Allograft). A graft derived from a genetically different donor of the same species as the recipient.

Homokaryon. Cell containing two or more genetically identical nuclei; usually a product of cell fusion.

Hybrid cell. Mononucleate cell that results from the fusion of two different cells, leading to the formation of a synkaryon. (*See* Synkaryon.)

Ideogram. The arrangement of the chromosomes of a cell in order by size and morphology so that the karyotype may be studied and genetically analyzed.

Immortalization. The acquisition of an infinite life span. May be induced in finite cell lines by transfection with telomerase, oncogenes, or the large T-region of the SV40 genome, or by infection with SV40 (whole virus) or Epstein–Barr virus (EBV). Immortalization is not necessarily a malignant transformation, although it may be a component of malignant transformation.

Induction. An increase in effect produced by a given stimulus.

Infection (other than the commonplace definition). Transfer of genomic DNA with a retroviral construct containing the DNA sequence under investigation, usually packaged with a promoter sequence and a reporter gene, such as β-galactosidase; the product of an infection may be detected by staining with a chromogenic substrate.

In situ hybridization. Binding of specific complementary nucleic acid probes to intracellular locations; cDNA probes for the localization of mRNA sequences, or RNA probes for DNA localization (*see* Chromosome painting). Visualized by radioisotopic labeling of the probe and microautoradiography, or by using a fluorochrome bound to the probe.

Isograft (Syngraft). A graft derived from a genetically identical or nearly identical donor of the same species as the recipient.

In ovo. In the egg—usually, the hen's egg.

In vitro. Literally, "in glass," but used conventionally to mean cultured outside of the host as cell cultures, organ cultures, or short-term organ bath preparations; also used to indicate biochemical and molecular reactions carried out in a test tube, but these reactions are better referred to as *cell free*.

In vivo. In the living plant or animal.

Karyotype. The distinctive chromosomal complement of a cell.

Laminar flow. The flow of a fluid that closely follows the shape of a streamlined surface without turbulence; said of hoods or cabinets characterized by a stable flow of air over the work area so as to minimize turbulence.

Laminar-flow hood or cabinet. A workstation with filtered air flowing in a laminar (nonturbulent) manner parallel to or perpendicular to the work surface, to maintain the sterility of the work; the parallel flow is called *horizontal* laminar flow, the perpendicular flow *vertical* laminar flow.

Leukemia. Malignant disease of the hematopoietic system, evident as circulating blast cells.

LIF. Leukemia inhibitory factor, a cytokine of the inter-leukin-6 family; used to inhibit differentiation and maintain the stem cell phenotype in ES cells.

Lipofection. Transfection of DNA by fusion with lipid-encapsulated DNA.

Lymphoma. A solid tumor of lymphoid cells.

Log phase. *See* Growth curve.

Macroautoradiography. Localization of radioisotopes in whole body sections and blots from electrophoresis preparations, by exposure of a photographic emulsion placed in close proximity to the specimen, usually by placing the film with the blot in a cassette with an intensifier screen.

MACS (Magnetic-activated cell sorting). Sorting cells by the magnetic attraction of magnetizable antibody-coated ferritin beads that bind to specific cell surface antigens.

Malignant. Invasive or metastatic (i.e., colonizing other tissues); (said of a tumor). Usually progressive, leading to the destruction of host cells and, ultimately, death of the host.

Malignant transformation. The development of the ability to invade normal tissue without regulation in space or time; may also lead to metastatic growth (colonization of a distant site with subsequent unregulated invasive growth).

Manometer. A U-shaped tube containing liquid, the levels of which in each limb of the U reflect the pressure difference between the ends of the tube.

Medium. A mixture of inorganic salts and other nutrients capable of sustaining cell survival *in vitro* for 24 hours. *Growth medium*: A medium that is used in routine culture such that the cell number increases with time. *Maintenance medium*: A medium that will retain cell survival without cell proliferation (e.g., a low-serum or serum-free medium used with serum-dependent cells). The plural of medium is *media*.

Mesenchyme. Loose, often migratory embryonic tissue derived from the mesoderm, giving rise to connective tissue, cartilage, muscle, hemopoietic cells, etc., in the adult.

Mesenchymal stem cells (MSCs). Stem cells, usually derived from bone marrow, with multipotent differentiation capacity, e.g., cardiac muscle, neural cells, or hepatocytes, as well as hematopoietic lineages.

Mesoderm. A germ layer in the embryo arising between the ectoderm and endoderm and giving rise to mesenchyme, which, in turn, gives rise to connective tissue, etc. (*See* Mesenchyme.)

Microautoradiography. Localization of radioisotopes in cells and tissue sections, by exposure of a photographic emulsion placed in close proximity to the specimen, usually by dipping it in the melted emulsion. After development, the specimen may be viewed under a microscope.

Monoclonal. Derived from a single clone of cells. *Monoclonal antibody*: Antibody produced by a clone of lymphoid cells either *in vitro* or *in vivo*. *In vitro*, the clone is usually derived from a hybrid of a sensitized spleen cell and a continuously growing myeloma cell.

Morphogenesis. The development of form and structure of an organism.

Myeloma. A tumor derived from myeloid cells; used in monoclonal antibody production when the myeloma cell can produce immunoglobulin.

Neoplastic. A new, unnecessary proliferation of cells giving rise to a tumor.

Neoplastic transformation. The conversion of a non-tumorigenic cell into a tumorigenic cell.

Oncogene. A gene that, when transfected or infected into normal cells, induces malignant transformation; usually a positively acting gene coding for growth factors, receptors, signal transducers, or nuclear regulators.

Organ culture. The maintenance or growth of organ primordia or the whole or parts of an organ *in vitro* in a way that may allow differentiation and preservation of the architecture or function of the organ.

Organogenesis. The development of organs.

Organotypic. Histotypic culture involving more than one cell type to create a model of the cellular interactions characteristic of an organ *in vivo*. A reconstruction from dissociated cells or fragments of tissue is implied, as distinct from organ culture, in which the structural integrity of the explanted tissue is retained.

Osmolality. The concentration of osmotically active particles in an aqueous solution, expressed in osmoles/kg.

Osmolarity. The concentration of osmotically active particles in an aqueous solution, expressed in osmoles/L.

Osmole. The amount of a substance containing 1 mole of osmotically active particles.

Paracrine. An effect of one cell on another, adjacent cell mediated by a soluble factor without involvement of the systemic vasculature.

Parenchyma. That part of a tissue carrying out the major function of the tissue, e.g., the hepatocytes in liver; as distinct from the stroma, such as fibroblastic connective tissue, seen as supporting tissue.

Pascal. SI unit of pressure equivalent to 1 newton per square meter.

Passage. The transfer or subculture of cells from one culture vessel to another; usually, but not necessarily, involves the subdivision of a proliferating cell population, enabling the propagation of a cell line or cell strain.

Passage number. The number of times a culture has been subcultured.

Pavementlike. Cells in a regular monolayer or polygonal cells. More correctly, epithelioid or epithelium-like.

Phenotype. The aggregate of all the expressed properties of a cell; the product of the interaction of the genotype with the regulatory environment.

Plateau. *See* Growth curve.

Plating efficiency. The percentage of cells seeded at subculture that gives rise to colonies. If each colony can be said to be derived from one cell, plating efficiency is identical to cloning efficiency. Sometimes the plating efficiency is used loosely to describe the number of cells surviving after subculture, but this is better termed the *seeding efficiency*.

Ploidy. Relationship of chromosome number of a given type of cell to that found in normal somatic cells *in vivo*. (*See also* Haploid, Diploid, Euploid, Aneuploid, and Heteroploid).

Poikilothermic. Having a body temperature close to that of the environment and not regulated by metabolism.

Population density. The number of monolayer cells per unit area of substrate; for cells growing in suspension, the population density is identical to the cell concentration.

Population-doubling time. The interval required for a cell population to double at the middle of the logarithmic phase of growth.

Precursor cell. A cell at a stage in a cell differentiation pathway that is assumed to be committed to a particular type of differentiation. Early stages will be proliferative, late stages may not be.

Primary culture. A culture started from cells, tissues, or organs taken directly from an organism, and before the first subculture.

Primary explant. A fragment of tissue removed from the organism and placed in culture in such as way as to promote its survival and the outgrowth of viable cells.

Progenitor cells. Cells which are at an early stage of development, probably proliferating, and not yet expressing differentiated properties.

Pseudodiploid. Numerically diploid chromosome number, but with chromosomal aberrations.

Reagent. A substance (element, compound, or mixture) that participates in a chemical reaction.

Quasidiploid. *See* Pseudodiploid.

Sarcoma. A tumor derived from mesodermally derived cells [e.g., connective tissue, muscle (*myosarcoma*), or bone (*osteosarcoma*)].

Saturation density. Maximum number of cells attainable per cm^2 (in a monolayer culture) or per mL (in a suspension culture) under specified conditions.

Seeding efficiency. The percentage of the inoculum that attaches to the substrate within a stated period of time (implying viability, or survival, but not necessarily proliferative capacity).

Senescence. Biologically regulated loss of proliferative potential linked to shortening of the *telomeres* of the chromosomes.

Somatic cell genetics. The study of cell genetics by the recombination and segregation of genes in somatic cells, usually by fusion.

Split ratio. The divisor of the dilution ratio of a cell culture at subculture (e.g., one flask divided into four, or 100 mL up to 400 mL, would be a split ratio of 4).

Stem cell. The earliest detectable cell in a lineage with the capacity to generate all the cells in the lineage while maintaining its own population. A *unipotent* stem cell will only give rise to one differentiation pathway, *bipotent* to two, *multipotent* to more than two, *pluripotent* to several, and *totipotent* to all types of differentiated cell. (*See also* Embryonal stem cells and Mesenchymal stem cells.)

Stroma. That part of a tissue seen as having a purely supporting role, e.g., fibroblastic connective tissue and its vasculature.

Subconfluent. Less than confluent; not all of the available substrate is covered.

Subculture. *See* Passage.

Substrate. The matrix or solid underlay upon which a monolayer culture grows.

Superconfluent. Progressing beyond the state in which all the cells are attached to the substrate and multilayering occurs.

Suppressor gene. A gene that inhibits the transformed (malignant) phenotype, usually associated with dominant-negative regulation of cell proliferation or cell migration; often, suppressor genes are mutated or deleted in transformed cells and cancer.

Suspension culture. A culture in which cells will multiply when suspended in growth medium.

Synkaryon. A hybrid cell that results from the fusion of the nuclei it carries.

Telomeres. Terminal regions of the chromosomes that prevent recombination with other chromosomes and are able to maintain the proliferative capacity of the cell. Progressively shortened during senescence but maintained in stem cells and some tumor cells by telomerase.

Tetraploid. Twice the diploid (four times the haploid) number of chromosomes.

Tissue culture. Properly, the maintenance of fragments of tissue *in vitro*, but now commonly applied as a generic term denoting tissue explant culture, organ culture, and dispersed-cell culture, including the culture of propagated cell lines and cell strains.

Transdifferentiation. Cells from one lineage acquiring the ability to differentiate into cells of a different lineage.

Transfection. The transfer, by artificial means, of genetic material from one cell to another, when less than the whole nucleus of the donor cell is transferred. Transfection is usually achieved by transferring isolated chromosomes, DNA, or cloned genes.

Transformation. A permanent alteration of the cell phenotype, presumed to occur via an irreversible genetic change. May be spontaneous, as in the development of rapidly growing continuous cell lines from slow-growing early-passage rodent cell lines, or may be induced by chemical or viral action. Usually produces cell lines that have an increased growth rate, an infinite life span, a lower serum requirement, and a higher plating efficiency and that are often (but not necessarily) tumorigenic.

Validation (of cell lines). A process that includes authentication, characterization, and the demonstration of the lack of contamination of a cell line.

Variant. A cell line expressing a stable phenotype that is different from the parental culture from which it was derived.

Viral transformation. A permanent phenotypic change induced by the genetic and heritable effects of a transforming virus.

Xenograft. Transplantation of tissue to a species different from that from which it was derived; often used to describe the implantation of human tumors in athymic (nude), immune-deprived, or immune-suppressed mice.

APPENDIX V

General Textbooks and Relevant Journals

Alberts, B., Bray, D., Lewis, J., Raff, M., Roberts, K., & Watson, J. D. (2002). *The molecular biology of the cell*, 4th ed. New York, Garland.

Ausubel, F. M., Brent, R., Kingston, R. E., Moore, D. D., Seidman, J. G., Smith, J. A., Struhl, K. (Eds) (2002). Short Protocols in Molecular Biology, 5th Edition, Vol 1 & 2. Hoboken, NJ, John Wiley & Sons.

Barlovatz-Meimon, G., & Adolphe, M., (2003). *Culture de cellules animales; methodologies, applications*. Paris, Editions INSERM. *A collection of technique-oriented chapters on basic and advanced aspects of tissue culture.*

Butler, M. (ed.). (1991). *Mammalian cell biotechnology, a practical approach*. Oxford, U.K., IRL Press at Oxford University Press. *A useful introduction to basic biotechnology.*

Davis, J. M. (2002). *Basic cell culture, a practical approach*. Oxford, U.K., IRL Press at Oxford University Press.

Doyle, A., Griffiths, J. B., & Newell, D. G. (eds.). (1993). *Cell and tissue culture: Laboratory procedures*. Chichester, U.K., John Wiley & Sons. *A loose-leaf compendium of general and specialized techniques with regular updates. Very expensive, but a very good source for a wide variety of techniques.*

Doyle, A., Hay, R., & Kirsop, B. E. (eds.). (1990). *Living resources for biotechnology*. Cambridge, U.K., Cambridge University Press. *Useful information on databases and quality control.*

Freshney, R. I., & Freshney, M. G. (eds.). (2002). *Culture of epithelial cells*. New York, Wiley-Liss. *Invited chapters on specialized culture of epithelium; technique oriented.*

Freshney, R. I. (1999). *Freshney's Culture of animal cells, a multimedia guide*. New York, Wiley-Liss.

Freshney, R. I., & Freshney, M. G. (1996). *Culture of immortalized cells*. New York, Wiley-Liss.

Freshney, R. I., Pragnell, I. B., & Freshney, M. G. (eds.). (1994). *Culture of hematopoietic cells*. New York, Wiley-Liss. Second

in the series *"Culture of Specialized Cells."* Invited chapters on specialized techniques.

Haynes, L. W. (ed.) (1999). *The neuron in tissue culture*. Chichester, UK, John Wiley & Sons.

Leigh, I. M., Lane, E. B., & Watt, F. M. (eds.). (1994). *The keratinocyte handbook*. Cambridge, U.K., Cambridge University Press.

Leigh, I. M., & Watt, F. M. (eds.). (1994). *Keratinocyte methods*. Cambridge, U.K., Cambridge University Press.

Lodish, H., Berk, A., Matsudaira, P., Kaiser, C. A., Kreiger, M., Scott, M. P., Zipursky, S. L., & Darnell, J. (2004). *Molecular cell biology* 5th ed. New York, Scientific American Books, Freeman.

Masters, J. R. W., (ed.), *Animal cell culture, a practical approach*, 3rd ed. Oxford, U.K., IRL Press (2000).

Masters, J. R. W. (ed.). (1991). *Human cancer in primary culture*. London, Kluwer. *Product of a European Tissue Culture Society workshop.*

Masters, J. R. W., & Palsson, B. (1999). (eds.). *Human cell culture*. Dordrecht, The Netherlands, Kluwer.

Pfragner, R., & Freshney, R. I. (eds.) (2004). *Culture of human tumor cells*. Hoboken, NJ, Wiley-Liss. *Invited chapters with culture protocols for several types of human tumors.*

Pollack, R. (ed.). (1981). *Reading in mammalian cell culture*, 2nd ed. Cold Spring Harbor, NY, Cold Spring Harbor Laboratory Press. *A very good compilation of key papers in the field. Used as a tutorial, for general interest, and for teaching.*

Ravid, K., & Freshney, R. I. (eds.). (1998). *DNA transfer to cultured cells*. New York, Wiley-Liss. *Invited chapters with practical protocols on DNA transfer technology.*

Shahar, A., de Vellis, J., Vernadakis, A., & Haber, B. (1989). *A dissection and tissue culture manual of the nervous system*. New York, Wiley-Liss. *Useful short protocols; well illustrated.*

Vunjak-Novakovic, G., & Freshney, R. I. (eds.). (2005). *Culture of cells for tissue engineering*. Hoboken, NJ, Wiley-Liss (in press).

Invited chapters with practical protocols on preparation of cells and matrices for tissue engineering.

Useful Journals
Technique-Oriented Tissue Culture

Cell Preservation Technology
Cytotechnology (now incorporating *Methods in Cell Science*)
In Vitro Cell and Development Biology
Tissue Culture Research Communications (Japanese)

Cell Biology

Cell
Cell Biology, International Reports
Cellular Biology
Cell Growth & Differentiation

Current Opinion in Cell Biology
European Journal of Cell Biology
Experimental Cell Biology
Experimental Cell Research
Journal of Cell Biology
Journal of Cellular Physiology
Journal of Cell Science
Nature Biotechnology
Nature Cell Biology

Cancer

British Journal of Cancer
Cancer Research
European Journal of Cancer and Clinical Oncology
International Journal of Cancer
Journal of the National Cancer Institute

References

Aaronson, S. A., & Todaro, G. J. (1968). Development of 3T3-like lines from Balb/c mouse embryo cultures: Transformation susceptibility to SV40. *J. Cell Physiol.* **72**: 141–148.

Aaronson, S. A., Bottaro, D. P., Miki, T., Ron, D., Finch, P. W., Fleming, T. P., Ahn, J., Taylor, W. G., & Rubin, J. S. (1991). Keratinocyte growth factor: A fibroblast growth factor family member with unusual target cell specificity. *Ann. NY Acad. Sci.* **638**: 62–77.

Aaronson, S. A., Todaro, G. J., & Freeman, A. E. (1970). Human sarcoma cells in culture: Identification by colony-forming ability on monolayers of normal cells. *Exp. Cell. Res.* **61**: 1–5.

Abaza, N. A., Leighton, J., & Schultz, S. G. (1974). Effects of ouabain on the function and structure of a cell line (MDCK) derived from canine kidney; I: Light microscopic observations of monolayer growth. *In Vitro* **10**: 172–183.

Abbott, N. J., Hughes, C. C., Revest, P. A., & Greenwood, J. (1992). Development and characterisation of a rat brain capillary endothelial culture: Towards an *in vitro* blood-brain barrier. *J. Cell Sci.* **103** (Pt 1): 23–37.

Abercrombie, M., & Heaysman, J. E. M. (1954). Observations on the social behaviour of cells in tissue culture; II: "Monolayering" of fibroblasts. *Exp. Cell Res.* **6**: 293–306.

Abney, E. R., Williams, B. P., & Raff, M. C. (1983). Tracing the development of oligodendrocytes from precursor cells using monoclonal antibodies, fluorescence activated cell sorting and cell culture. *Dev. Biol.* **100**: 166–171.

Adams, D. O. (1979). Macrophages. In Jakoby, W. B., & Pastan, I. H. (eds.), *Methods in enzymology: vol. 57, cell culture.* New York, Academic Press, pp. 494–506.

Adams, R. L. P. (1980). In Work, T. S., & Burdon, R. H. (eds.): *Laboratory techniques in biochemistry and molecular biology: Cell culture for biochemists.* Amsterdam, Elsevier/North Holland Biomedical Press.

Adolphe, M. (1984). Multiplication and type II collagen production by rabbit articular chondrocytes cultivated in a defined medium. *Exp. Cell. Res.* **155**: 527–536.

Adolphe, M., & Benya, P. D. (1992). Different types of cultured chondrocytes: The *in vitro* approach to the study of biochemical regulation. In Adolphe, M., ed., *Biological regulation of the chondrocytes.* Boca Raton, FL, CRC Press, pp. 105–139.

Advisory Committee on Dangerous Pathogens (1995a). Categorisation of Biological Agents According to Hazard and Categories of Containment. The Stationery Office, P. O. Box 276, London SW8 5DT, England.

Advisory Committee on Dangerous Pathogens (1995b). Protection against blood-borne infections in the workplace: HIV and hepatitis. The Stationery Office, P. O. Box 276, London SW8 5DT, England.

Advisory Committee on Dangerous Pathogens (2003). Infection at work: Controlling the risks, Department of Health, PO Box 777, London SE1 6XH.

Ager, A. (1987). Isolation and culture of high endothelial cells from rat lymph nodes. *J Cell Sci* **87** (Pt 1): 133–144.

Agy, P. C., Shipley, G. D., & Ham, R. G. (1981). Protein-free medium for mouse neuroblastoma cells. *In Vitro* **17**: 671–680.

Ahmed, I., Collins, C. A., Lewis, M. P., Olsen, I., & Knowles, J. C. (2004). Processing, characterisation and biocompatibility of iron-phosphate glass fibres for tissue engineering. *Biomaterials* **25**(16): 3223–3232.

Aigner, J., Tegeler, J., Hutzler, P., Campoccia, D., Pavesio, A., Hammer, C., Kastenbauer, E., & Naumann, A. (1998). Cartilage tissue engineering with novel nonwoven structured biomaterial based on hyaluronic acid benzyl ester. *J. Biomed. Mater. Res.* **42**(2): 172–181.

Aitken, M. L., Villalon, M., Verdugo, P., & Nameroff, M. (1991). Enrichment of subpopulations of respiratory epithelial cells using flow cytometry. *Am. J. Resp. Cell. Mol. Biol.* **4**: 174–178.

Albelda, S. M., Oliver, P. D., Romer, L. H., & Buck, C. A. (1990). EndoCAM: a novel endothelial cell-cell adhesion molecule. *J. Cell. Biol.* **110**: 1227–1237.

Alberts, B., Bray, D., Johnson, A., Lewis, J., Raff, J., Roberts, K., & Walter, P. (1997). *Essential cell biology.* New York, Garland.

Alberts, B., Johnson, A., Lewis, J., Raff, M., Roberts, K., & Walter, P. (2002). *Molecular biology of the cell*, 4th ed. New York, Garland.

Albrecht, A. M., Biedler, J. L., & Hutchison, D. J. (1972). Two different species of dihydrofolate reductase in mammalian cells differentially resistant to amethopterin and methasquin. *Cancer Res.* **32**: 1539–1546.

Alexander, C. L., FitzGerald, U. F., & Barnett, S. C. (2002). Identification of growth factors that promote long-term proliferation of olfactory ensheathing cells and modulate their antigenic phenotype. *Glia* **37**: 349–364.

Al-Hajj, M., Becker, M. W., Wicha, M., Weissman, I., Clarke, M. F., (2004). Therapeutic implications of cancer stem cells. *Curr. Opin. Genet. Dev.* **14**: 43–47.

Ali, S., Muller, C. R., & Epplen, J. T. (1986). DNA finger printing by oligonucleotide probes specific for simple repeats. *Hum. Genet.* **74**: 239–243.

Alison, M. R., Vig, P., Russo, F., Bigger, B. W., Amofah, E., Themis, M., & Forbes, S. (2004). Hepatic stem cells: from inside and outside the liver? *Cell Prolif.* **37**: 1.

Alley, M. C., Scudiero, D. A., Monks, A., Hursey, M. L., Czerwiniski, M. J., Fine, D. L., Abbot, B. J., Mayo, J. G., Shoemaker, R. H., & Boyd, M. R. (1988). Feasibility of drug screening with panels of human tumour cell lines using a microculture tetrazolium assay. *Cancer Res.* **48**: 589–601.

Al-Mufti, R., Hambley, H., Farzaneh, F., & Nicolaides, K. H. (2004). Assessment of efficacy of cell separation techniques used in the enrichment of foetal erythroblasts from maternal blood: triple density gradient vs. single density gradient. *Clin. Lab. Haematol.* **26**(2): 123–128.

Al-Rubeai, M., & Singh, R. P. (1998). Apoptosis in cell culture. *Curr. Opin. Biotechnol.* **9**: 152–156.

Al-Rubeai, M., Welzenbach, K., Lloyd, D. R., & Emery, A. N. (1997). A rapid method for evaluation of cell number and viability by flow cytometry. *Cytotechnology* **24**: 161–168.

Alvarez-Dolado, M., Pardal, R., Garcia-Verdugo, J. M., Fike, J. R., Lee, H. O., Pfeffer, K., Lois, C.,. Morrison, S. J., & Alvarez-Buylla, A. (2003). Fusion of bone-marrow-derived cells with Purkinje neurons, cardiomyocytes and hepatocytes. *Nature* **425**: 968–973.

Ames, B. N. (1980). Identifying environmental chemicals causing mutations and cancer. *Science* **204**: 587–593.

Anderson, J. E., & Wozniak, A. C. (2004). Satellite cell activation on fibers: modeling events *in vivo*—an invited review. *Can. J. Physiol. Pharmacol.* **82**: 300–310.

Andersson, L. C., Jokinen, M., & Gahmberg, C. G. (1979c). Induction of erythroid differentiation in the human leukaemia cell line K562. *Nature* **278**: 364–365.

Andersson, L. C., Jokinen, M., Klein, G., & Nilsson, K. (1979b). Presence of erythrocytic components in the K562 cell line. *Int. J. Cancer.* **24**: 5–14.

Andersson, L. C., Nilsson, K., & Gahmberg, C. G. (1979a). K562—a human erythroleukemic cell line. *Int. J. Cancer* **23**: 143–147.

Andreason, G. L., & Evans, G. A. (1988). Introduction and expression of DNA molecules in eukaryotic cells by electroporation. *BioTechniques* **6**: 650–660.

Andreason, G. L., & Evans, G. A. (1989). Optimization of electroporation for transfection of mammalian cells. *Anal. Biochem.* **180**: 269–275.

Andreoli, S. P., & McAteer, J. A. (1990). Reactive oxygen molecule-mediated injury in endothelial and renal tubular epithelial cells *in vitro*. *Kidney Int.* **38**: 785–794.

Andzaparidze, O. G. (1968). Clinical experience with vaccines produced in the human diploid cell line WI-38. *Natl Cancer Inst Monogr.*, **29**: 477–84.

Antoniades, H. N., Scher, C. D., & Stiles, C. D. (1979). Purification of human platelet-derived growth factor. *Proc. Natl. Acad. Sci. USA* **76**: 1809.

Aoki, H., Motohashi, T., Yoshimura, N., Yamazaki, H., Yamane, T., Panthier, J. J., & Kunisada, T. (2005). Cooperative and indispensable roles of endothelin 3 and KIT signalings in melanocyte development. *Dev Dyn.* 2005 Mar 14. Online ahead of publication.

Armati, P. J., & Bonner, J. (1990). A technique for promoting Schwann cell growth from fresh and frozen biopsy nerve utilizing d-valine medium. *In Vitro Cell Dev. Biol.* **26**: 1116–1118.

Arrighi, F. E., & Hsu, T. C. (1974). Staining constitutive heterochromatin and Giemsa crossbands of mammalian chromosomes. In Yunis, J. (ed.), *Human chromosome methodology*, 2d ed. New York, Academic Press.

Artursson, P., & Magnusson, C. (1990). Epithelial transport of drugs in cell culture II: Effect of extracellular calcium concentration on the paracellular transport of drugs of different lipophilicities across monolayers of intestinal epithelial (Caco-2) cells. *J. Pharmaceut. Sci.* **79**: 595–600.

Askanas, V., Bornemann, A., & Engel, W. K. (1990). Immunocytochemical localization of desmin at human neuromuscular junctions. *Neurology* **40**: 949–953.

Atala, A., & Lanza, R. P. (2002). *Methods of Tissue Engineering.* San Diego, Academic Press.

Au, A. M.-J., & Varon, S. (1979). Neural cell sequestration on immunoaffinity columns. *Exp. Cell Res.* **120**: 269.

Auerbach, R., & Grobstein, C. (1958). Inductive interaction of embryonic tissues after dissociation and reaggregation. *Exp. Cell Res.* **15**: 384–397.

Ausubel, F. M., Brent, R., Kingston, R. E., Moore, D. D., Seidman, J. G., Smith, J. A., & Struhl, K. (eds.) (1996). *Current protocols in molecular biology*, New York, John Wiley & Sons.

Ausubel, F. M., Brent, R., Kingston, R. E., Moore, D. D., Seidman, J. G., Smith, J. A., & Struhl, K. (eds.) (2002). Short Protocols in Molecular Biology, 5th Edition, Vol 1 & 2. Hoboken, NJ, John Wiley & Sons.

Babich, H., & Borenfreund, E. (1990). Neutral red uptake. In Doyle, A., Griffiths, J. B., & Newall, D. G. (eds.), *Cell and tissue culture: Laboratory procedures.* Chichester, U.K., Wiley, Module 4B:7.

Bagley, R. G., Walter-Yohrling, J., Cao, X., Weber, W., Simons, B., Cook, B. P., Chartrand, S. D., Wang, C., Madden, S. L., & Teicher, B. A. (2003). Endothelial precursor cells as a model of tumor endothelium: characterization and comparison with mature endothelial cells. *Cancer Res.* **63**: 5866–5873.

Bakolitsa, C., Cohen, D. M., Bankston, L. A., Bobkov, A. A., Cadwell, G. W., Jennings, L., Critchley, D. R., Craig, S. W., & Liddington, R. C. (2004). Structural basis for vinculin activation at sites of cell adhesion. *Nature* **430**: 583–586.

Balin, A. K., Goodman, B. P., Rasmussen, H., & Cristofalo, V. J. (1976). The effect of oxygen tension on the growth and metabolism of WI-38 cells. *J. Cell Physiol.* **89**: 235–250.

Balkovetz, D. F., & Lipschutz, J. H. (1999). Hepatocyte growth factor and the kidney: It is not just for the liver. *Int. Rev. Cytol.* **186**: 225–260.

Ballard, P. L. (1979). Glucocorticoids and differentiation. *Glucocorticoid Horm. Action* **12**: 439–517.

Ballard, P. L., & Tomkins, G. M. (1969). Dexamethasone and cell adhesion. *Nature* **244**: 344–345.

Balmforth, A. J., Ball, S. G., Freshney, R. I., Graham, D. I., McNamee, B., & Vaughan, P. F. T. (1986). D-1 dopaminergic and beta-adrenergic stimulation of adenylate cyclase in a clone derived from the human astrocytoma cell line G-CCM. *J. Neurochem.* **47**: 715–719.

Baltimore, D. (2001). Our genome unveiled. *Nature,* **409**: 814–816.

Bansal, R., Stefansson, K., & Pfeiffer, S. E. (1992). Proligodendroblast antigen (POA), a developmental antigen expressed by A007/O4-positive oligodendrocyte progenitors prior to the appearance of sulfatide and galactocerebroside. *J. Neurochem.* **58**: 2221–2229.

Bard, D. R., Dickens, M. J., Smith, A. U., & Sarck, J. M. (1972). Isolation of living cells from mature mammalian bone. *Nature* **236**: 314–315.

Barnes, D., & Sato, G. (1980). Methods for growth of cultured cells in serum-free medium. *Anal. Biochem.* **102**: 255–270.

Barnes, W. D., Sirbasku, D. A., & Sato, G. H. (eds.). (1984a). *Cell culture methods for molecular and cell biology; Vol. 1: Methods for preparation of media, supplements, and substrata for serum-free animal cell culture.* New York, Alan R. Liss.

Barnes, W. D., Sirbasku, D. A., & Sato, G. H. (eds.). (1984b). *Cell culture methods for molecular and cell biology; Vol. 2: Methods for serum-free culture of cells of the endocrine system.* New York, Alan R. Liss.

Barnes, W. D., Sirbasku, D. A., & Sato, G. H. (eds.). (1984c). *Cell culture methods for molecular and cell biology; Vol. 3: Methods for serum-free culture of epithelial and fibroblastic cells.* New York, Alan R. Liss.

Barnes, W. D., Sirbasku, D. A., & Sato, G. H. (eds.). (1984d). *Cell culture methods for molecular and cell biology; Vol. 4: Methods for serum-free culture of neuronal and lymphoid cells.* New York, Alan R. Liss.

Barnett, S. C., & Chang, L. (2004). Olfactory ensheathing cells: Going solo or in need of a friend? *Trends Neurosci.* **27**(1): 54–60.

Barnett, S. C., & Riddell, J. S. (2004). Olfactory ensheathing cells (OECs) and the treatment of CNS injury; advantages and possible caveats. *J. Anat.* **24**: 57–67.

Barnett, S. C. (2004). Olfactory ensheathing cells: unique glial cell types? *J. Neurotrauma.* **21**: 375–382.

Barnett, S. C., Hutchins, A.-M., & Noble, M. (1993). Purification of olfactory nerve ensheathing cells of the olfactory bulb. *Dev. Biol.* **155**: 337–350.

Bateman, A. E., Peckham, M. J., & Steel, G. G. (1979). Assays of drug sensitivity for cells from human tumours: *In vitro* and *in vivo* tests on a xenografted tumour. *Br. J. Cancer* **40**: 81–88.

Battye, F. L., & Shortman, K. (1991). Flow cytometry and cell-separation procedures. *Curr. Opin. Immunol.* **3**: 238–241.

Bazill, G. W., Haynes, M., Garland, J., & Dexter, T. M. (1983). Characterisation and partial purification of a haemopoietic cell growth factor in WEHI-3 cell conditioned medium. *Biochem. J.* **210**: 747–759.

Beattie, G. M., Lappi, D. A., Baird, A., Hayek, A. (1990). Selective elimination of fibroblasts from pancreatic islet monolayers by basic fibroblast growth factor-saporin mitotoxin. *Diabetes.* **39**: 1002–5.

Beddington, R. (1992). Transgenic mutagenesis in the mouse. *Trends Genet.* **8**: 10.

Bedrin, M. S., Abolafia, C. M., & Thompson, J. F. (1997). Cytoskeletal association of epidermal growth factor receptor and associated signalling proteins is regulated by cell density in IEC-6 intestinal cells. *J. Cell. Physiol.* **172**: 126–136.

Bell, E., Sher, S., Hull, B., Merrill, C., Rosen, S., Chamson, A., Asselineau, D., Dubertret, L., Coulomb, B., Lapiere, C., Nusgens, B., & Neveux, Y. (1983). The reconstitution of living skin. *J. Invest. Dermatol.* **81**, no 1 Suppl: 2s–10s.

Benda, P., Lightbody, J., Sato, G., Levine, L., & Sweet, W. (1968). Differentiated rat glial cell strain in tissue culture. *Science* **161**: 370.

Benders, A. A. G. M., van Kuppevelt, T. H. M. S. M., Oosterhof, A., & Veerkamp, J. H. (1991). The biochemical and structural maturation of human skeletal muscle cells in culture: The effect of serum substitute, Ultroser. G. *Exp. Cell Res.* **195**: 284–294.

Benya, P. D. (1981). Two dimensional CNBr peptide patterns of collagen types I, II and III. *Coll. Relat. Res.* **1**: 17–26.

Benya, P. D., Padilla, S. R., & Nimmi, M. E. (1977). The progeny of rabbit articular chondrocytes synthesize collagen type I and III and I trimer, but not type II: Verification by cyanogen bromide peptide analysis. *Biochemistry* **16**: 865–872.

Berdichevsky, F., Gilbert, C., Shearer, M., & Taylor-Papadimitriou, J. (1992). Collagen-induced rapid morphogenesis of human mammary epithelial cells: The role of the alpha 2 beta 1 integrin. *J. Cell Sci.* **102**: 437–446.

Berenbaum, M. C. (1985). The expected effects of a combination of agents: The general solution. *J. Theor. Biol.* **114**: 413–432.

Berger, S. L. (1979). Lymphocytes as resting cells. In Jakoby, W. B., & Pastan, I. H. (eds.), *Methods in enzymology; vol. 57: Cell culture.* New York, Academic Press, pp. 486–494.

Berky, J. J., & Sherrod, P. C. (eds.). (1977). *Short term in vitro testing for carcinogenesis, mutagenesis and toxicity.* Philadelphia, Franklin Institute Press.

Bernstein, A. (1975). Differentiation of clonal lines of teratocarcinoma cells: Formation of embryoid bodies *in vitro*. *Proc. Natl. Acad. Sci. USA* **72**: 1441–1445.

Bernstine, E. G., Hooper, M. L., Grandchamp, S., & Ephrussi, B. (1973). Alkaline phosphatase activity in mouse teratoma. *Proc. Natl. Acad. Sci. USA* **70**: 3899–3903.

Berry, M. N., & Friend, D. S. (1969). High yield preparation of isolated rat liver parenchymal cells: A biochemical and fine structural study. *J. Cell Biol.* **43**: 506–520.

Bertheussen, K. (1993). Growth of cells in a new defined protein-free medium. *Cytotechnology* **11**: 219–231.

Bertoncello, I., Bradley, T. R., & Watt, S. M. (1991). An improved negative immunomagnetic selection strategy for the purification of primitive hemopoietic cells from normal bone marrow. *Exp. Hematol.* **19**: 95–100.

Bettger, W. J., Boyce, S. T., Walthall, B. J., & Ham, R. G. (1981). Rapid clonal growth and serial passage of human diploid fibroblasts in a lipid-enriched synthetic medium supplemented with EGF, insulin and dexamethasone. *Proc. Natl. Acad. Sci. USA* **78**: 5588–5592.

Bhagavati, S., & Xu, W. (2004). Isolation and enrichment of skeletal muscle progenitor cells from mouse bone marrow. *Biochem Biophys Res Commun.* **318**: 119–124.

Bhargava, M., Joseph, A., Knesel, J., Halaban, R., Li, Y., Pang, S., Golberg, I., Setter, E., Donovan, M. A., Zarnegar, R., Faletto, D., & Rosen, E. M. (1992). Scatter factor and hepatocyte growth factor activities, properties, and mechanism. *Cell Growth Differ.* **3**: 11–20.

Bhatt, R. I., Brown, M. D., Hart, C. A., Gilmore, P., Ramani, V. A., George, N. J., & Clarke, N. W. (2003). Novel method for the isolation and characterisation of the putative prostatic stem cell. *Cytometry* **54A**(2): 89–99.

Bichko, V. V. (1998). Cationic liposomes. In Ravid, K., & Freshney, R. I. (eds.), *DNA transfer to cultured cells*. New York, Wiley-Liss, pp. 193–212.

Bickenbach, J. R., & Chism, E. (1998). Selection and extended growth of murine epidermal stem cells in culture. *Exp. Cell Res.* **244**: 184–195.

Biedler, J. L. (1976). Chromosome abnormalities in human tumour cells in culture. In Fogh, J. (ed.), *Human Tumor Cells In Vitro*. New York, Academic Press.

Biedler, J. L., & Spengler, B. A. (1976). A novel chromosomal abnormality in human neuroblastoma and anti-folate resistant Chinese hamster cell lines in culture. *J. Natl. Cancer Inst.* **57**: 683–695.

Biedler, J. L., Albrecht, A. M., Hutchinson, D. J., & Spengler, B. A. (1972). Drug response, dihydrofolate reductase, and cytogenetics of amethopterin-resistant Chinese hamster cells *in vitro*. *Cancer Res.* **32**: 151–161.

Biggers, J. D., Gwatkin, R. B. C., & Heyner, S. (1961). Growth of embryonic avian and mammalian tibiae on a relatively simple chemically defined medium. *Exp. Cell Res.* **25**: 41.

Bignami, A., Dahl, D., & Rueger, D. G. (1980). Glial fibrillary acidic (GFA) protein in normal neural cells and in pathological conditions. In Federoff, S., & Hertz, L. (eds.), *Advances in cellular neurobiology*, vol. 1. New York, Academic Press.

Biosafety in Microbiological and Biomedical Laboratories (1984). Division of Safety, BG31, ICO2, NIH, Bethesda, MD.

Birch, J. R., & Pirt, S. J. (1970). Improvements in a chemically-defined medium for the growth of mouse cells (strain LS) in suspension. *J. Cell Sci.* **7**: 661–670.

Birch, J. R., & Pirt, S. J. (1971). The quantitative glucose and mineral nutrient requirements of mouse LS (suspension) cells in chemically-defined medium. *J. Cell Sci.* **8**: 693–700.

Birnie, G. D., & Simons, P. J. (1967). The incorporation of ^{3}H-thymidine and ^{3}H-uridine into chick and mouse embryo cells cultured on stainless steel. *Exp. Cell Res.* **46**: 355–366.

Bishop, A. E., Power, R. F., & Polak, J. M. (1988). Markers for neuroendocrine differentiation. *Pathol Res Pract.* 1988 Apr; **183**(2): 119–128.

Bishop, J. M. (1991). Molecular themes in oncogenesis. *Cell* **64**: 235–248.

Bissell, D. M., Arenson, D. M., Maher, J. J., & Roll, F. J. (1987). Support of cultured hepatocytes by a laminin-rich gel: Evidence for a functionally significant subendothelial matrix in normal liver. *J. Clin. Invest.* **79**: 801–812.

Bjerkvig, R., Laerum, O. D., & Mella, O. (1986a). Glioma cell interactions with fetal rat brain aggregates *in vitro*, and with brain tissue *in vivo*. *Cancer Res.* **46**: 4071–4079.

Bjerkvig, R., Steinsvag, S. K., & Laerum, O. D. (1986b). Reaggregation of fetal rat brain cells in a stationary culture system; I: Methodology and cell identification. *In Vitro* **22**: 180–192.

Blaker, G. J., Birch, J. R., & Pirt, S. J. (1971). The glucose, insulin and glutamine requirements of suspension cultures of HeLa cells in a defined culture medium. *J. Cell Sci.* **9**: 529–537.

Blanco, F. J., Geng, Y., & Lotz, M. (1995). Differentiation dependent effects of IL-1 and TGF-β on human articular chondrocyte proliferation are related to nitric oxide synthase expression. *J. Immunol.* **154**: 4018–4026.

Blouin, R., Grondin, G., Beaudoin, J., Arita, Y., Daigle, N., Talbot, B. G., Lebel, D., & Morisset, J. (1997). Establishment and immunocharacterization of an immortalized pancreatic cell line derived from the H-2 Kb-tsA58 transgenic mouse. *In Vitro Cell Dev. Biol. Anim.* **33**: 717–726.

Bobrow, M., Madan, J., & Pearson, P. L. (1972). Staining of some specific regions on human chromosomes, particularly the secondary constriction of no. 9. *Nature* **238**: 122–124.

Bochaton-Piallat, M. L., Gabbiani, F., Ropraz, P., & Gabbiani, G. (1992). Cultured aortic smooth muscle cells from newborn and adult rats show distinct cytoskeletal features. *Differentiation* **49**: 175–185.

Bockhold, K. J., Rosenblatt, J. D., & Partridge, T. A. (1998). Aging normal and dystrophic mouse muscle: analysis of myogenicity in cultures of living single fibres. *Muscle Nerve* **21**: 173–183.

Bodnar, A. G., Ouellette, M., Frolkis, M., Holt, S. E., Chiu, C.-P., Morin, G. B., Harley, C. B., Shay, J. W., Lichsteiner, S., & Wright, W. E. (1998). Extension of life-span by introduction of telomerase into normal human cells. *Science* **279**: 349–352.

Boggs, S. S., Gregg, R. G., Borenstein, N., & Smithies, O. (1986). Efficient transformation and frequent single-site, single-copy insertion of DNA can be obtained in mouse erythroleukemia cells transformed by electroporation. *Exp. Hematol.* **14**: 988–994.

Bögler, O., Wren, D., Barnett, S. C., Land, H., & Noble, M. (1990). Cooperation between two growth factors promotes extended self-renewal and inhibits differentiation of oligodendrocyte-type-2 astrocyte (O-2A) progenitor cells. *Proc. Natl. Acad. Sci. USA* **87**: 6368–6372.

Bolton, B. J., & Spurr, N. K. (1996). B-lymphocytes. In Freshney, R. I., & Freshney, M. G., (eds.), *Culture of immortalized cells*. New York, Wiley-Liss, pp. 283–298.

Bonaventure, J., Kadhom, N., Cohen-Solal, L., Ng, K. H., Bourguignon, J., Lasselin, C., & Freisinger, P. (1994). Reexpression of cartilage-specific genes by dedifferentiated human articular chondrocytes cultured in alginate beads. *Exp. Cell Res.* **212**: 97–104.

Booth, C., & O'Shea, J. A. (2002). Isolation and culture of intestinal epithelial cells. In Freshney, R. I. & Freshney, M. G. (eds.), *Culture of epithelial cells*, 2nd ed., Hoboken, NJ, Wiley-Liss, pp. 303–335.

Booyse, F. M., Sedlak, B. J., & Rafelson, M. E. (1975). Culture of arterial endothelial cells: Characterization and growth of bovine aortic cells. *Thromb. Diathes. Ahemorrh.* **34**: 825–839.

Borenfreund, E., Babich, H., & Martin-Alguacil, N. (1990). Rapid chemosensitivity assay with human normal and tumor cells *in vitro*. *In Vitro Cell Dev. Biol.* **26**: 1030–1034.

Bosco, D., Soriano, J. V., Chanson, M., & Meda, P. (1994). Heterogeneity and contact-dependent regulation of amylase release by individual acinar cells. *J. Cell Physiol.* **160**: 378–388.

Boshart, M., Weber, F., Jahn, G., Dorsch-Hasler, K., Fleckenstein, B., & Schaffner, W. (1985). A very strong enhancer is located upstream of an immediate early gene of human cytomegalovirus. *Cell* **41**: 521–530.

Bossis, I., Voutetakis, A., Matyakhina, L., Pack, S., Abu-Asab, M., Bourdeau, I., Griffin, K. J., Courcoutsakis, N., Stergiopoulos, S., Batista, D., Tsokos, M., & Stratakis, C. A. (2004). A pleiomorphic GH pituitary adenoma from a Carney complex patient displays universal allelic loss at the protein kinase A regulatory subunit 1A (PRKARIA) locus. *J. Med. Genet.* **41**: 596–600.

Bossolasco, P., Corti, S., Strazzer, S., Borsotti, C., Del Bo, R., Fortunato, F., Salani, S., Quirici, N., Bertolini, F., Gobbi, A., Deliliers, G. L., Pietro Comi, G., & Soligo, D. (2004). Skeletal muscle differentiation potential of human adult bone marrow cells. *Exp. Cell Res.* **295**: 66–78.

Bottenstein, J. E. (1984). Culture methods for growth of neuronal cells lines in defined media. In Barnes, D. W., Sirbasku, D. A., Sato, G. H., eds., *Methods for Serum-Free Culture of Neuronal and Lymphoid Cells*, New York, Alan R. Liss, pp 3–13.

Bottenstein, J. E., & Sato, G. (1979). Growth of a rat neuroblastoma cell line in serum free supplemented medium. *Proc. Natl. Acad. Sci. USA* **76**: 514–517.

Boukamp, P., Petrusevska, R. T., Breitkreutz, D., Hornung, J., & Markham, A. (1988). Normal keratinisation in a spontaneously immortalised, aneuploid human keratinocyte cell line. *J. Cell Biol.* **106**: 761–771.

Bourillot, P. Y., Waltzer, L., Sergeant, A., & Manet, E. (1998). Transcriptional repression by the Epstein-Barr virus EBNA3A protein tethered to DNA does not require RBP-Jkappa. *J Gen Virol.***79**: 363–370.

Bouzahzah, B., Nishikawa, Y., Simon, D., & Carr, B. I. (1995). Growth control and gene expression in a new hepatocellular carcinoma cell line, Hep40: Inhibitory actions of vitamin K. *J. Cell Physiol.* **165**: 459–467.

Bowman, P. D., Betz, A. L., Ar, D., Wolinsky, J. S., Penney, J. B., Shivers, R. R., & Goldstein, G. (1981). Primary culture of capillary endothelium from rat brain. *In Vitro* **17**: 353–362.

Boxberger, H. J., Meyer, T. F., Grausam, M. C., Reich, K., Becker, H. D., & Sessler, M. J. (1997). Isolating and maintaining highly polarized primary epithelial cells from normal human duodenum for growth as spheroid-like vesicles. *In Vitro Cell Dev. Biol. Anim.* **33**: 536–545.

Boxman, D. L. A., Quax, P. H. A., Lowick, C. W. G. M., Papapoulos, S. E., Verheijen, J., & Ponec, M. (1995). Differential regulation of plasminogen activation in normal keratinocytes and SCC-4 cells by fibroblasts. *J. Invest. Dermatol.* **104**: 374–378.

Boyce, S. T., & Ham, R. G. (1983). Calcium-regulated differentiation of normal human epidermal keratinocytes in chemically defined clonal culture and serum-free serial culture. *J. Invest. Dermatol.* **81**: 33s–40s.

Boyd, M., Mairs, R. J., Cunningham, S. H., Mairs, S. C., McCluskey, A. G., Livingstone, A., Stevenson, K., Brown, M. M., Wilson, L., Carlin, S., & Wheldon, T. E. (2001). A gene therapy/targeted radiotherapy strategy for radiation cell kill by [131I]MIBG. *J Gene Med* **3**: 165–172.

Boyd, M., Mairs, R. J., Keith, W. N., Ross, S. C., Welsh, P., Akabani, G., Owens, J., Vaidyanathan, G., Carruthers, R., Dorrens, J., & Zalutsky, M. R. (2004). An efficient targeted radiotherapy/gene therapy strategy utilising human telomerase promoters and radioastatine and harnassing radiation-mediated bystander effects. *J Gene Med* **6**: 937–947.

Boyd, M., Mairs, S. C., Stevenson, K., Livingstone, A., Clark, A. M., Ross, S. C., & Mairs, R. J. (2002). Transfectant mosaic spheroids: a new model for evaluation of tumour cell killing in targeted radiotherapy and experimental gene therapy. *J Gene Med* **4**: 1–10.

Boyd, M. R. (1989). Status of the NCI preclinical antitumor drug discovery screen. *Prin. Prac. Oncol.* **10**: 1–12.

Boyd, M., Cunningham, S. H., Brown, M. M., Mairs, R. J., & Wheldon, T. E. (1999). Noradrenaline transporter gene transfer for radiation cell kill by [131I] meta-iodobenzyl-guanidine. *Gene Ther* **6**: 1147–52.

Boyum, A. (1968a). Isolation of leucocytes from human blood: A two-phase system for removal of red cells with methylcellulose as erythrocyte aggregative agent. *Scand. J. Clin. Lab. Invest.* (Suppl 97) **21**: 9–29.

Boyum, A. (1968b). Isolation of leucocytes from human blood: Further observations—methylcellulose, dextran and Ficoll as erythrocyte aggregating agents. *Scand. J. Clin. Lab. Invest.* (Suppl 97) **31**: 50.

Braa, S. S., & Triglia, D. (1991). Predicting ocular irritation using three-dimensional human fibroblast cultures. *Cosmetics Toiletries* **106**: 55–58.

Braaten, J. T., Lee, M. J., Schewk, A., & Mintz, D. H. (1974). Removal of fibroblastoid cells from primary monolayer cultures of rat neonatal endocrine pancreas by sodium ethylmercurithiosalicylate. *Biochem. Biophys. Res. Commun.* **61**: 476–482.

Bradford, C. S., Sun, L., & Barnes, D. W. (1994a). Basic FGF stimulates proliferation and suppresses melanogenesis in cell cultures derived from early zebrafish embryos. *Mol. Mar. Biol. Biotech.* **3**: 78–86.

Bradford, C. S., Sun, L., Collodi, P., & Barnes, D. W. (1994b). Cell cultures from zebrafish embryos and adult tissues. *J. Tissue Cult. Methods* **16**: 99–107.

Bradford, M. (1976). A rapid and sensitive method for the quantitation of microgram quantities of protein utilizing the principle of protein-dye binding. *Anal. Biochem.* **72**: 248–254.

Bradley, C., & Pitts, J. (1994). The use of genetic marking to assess the interaction of sensitive and multidrug resistant cells in mixed culture. *Br. J. Cancer* **70**: 795–798.

Bradley, N. J., Bloom, H. J. G., Davies, A. J. S., & Swift, S. M. (1978a). Growth of human gliomas in immune-deficient mice: A possible model for pre-clinical therapy studies. *Br. J. Cancer* **38**: 263.

Bradley, T. R., Hodgson, G. S., & Rosendaal, M. (1978b). The effect of oxygen tension on haemopoietic and fibroblast cell proliferation *in vitro*. *J. Cell Physiol.* **97**(Suppl 1): 517–522.

Bravery, C. A., Batten, P., Yacoub, M. H., and Rose, M. L. (1995). Direct recognition of SLA- and HLA-like class II antigens on porcine endothelium by human T cells results in T cell activation and release of interleukin-2. *Transplantation* **60**: 1024–1033.

Breder, J., Ruller, S., Ruller, E., Schlaak, M., & van der Bosch, J. (1996). Induction of cell death by cytokines in cell cycle-synchronous tumor cell populations restricted to G1 and G2. *Exp. Cell Res.* **223**: 259–267.

Breen, G. A. M., & De Vellis, J. (1974). Regulation of glycerol phosphate dehydrogenase by hydrocortisone in dissociated rat cerebral cell cultures. *Dev. Biol.* **41**: 255–266.

Breitkreutz, D., Stark, H.-J., Mirancea, N., Tomakidi, P., Steinbauer, H., & Fusenig, N. E. (1997). Integrin and basement membrane normalization in mouse grafts of human keratinocytes: Implications for epidermal homeostasis. *Differentiation* **61**: 195–209.

Breitman, T. R., Kene, B. R., & Hemmi, H. (1984). Studies of growth and differentiation of human myelomonocytic leukaemia cell lines in serum-free medium. In Barnes, D. W., Sirbasku, D. A., & Sato, G. H. (eds.), *Methods for serum-free culture of neuronal and lymphoid cells*. New York, Alan R. Liss, pp. 215–236.

Bretzel, R. G., Bonath, K., & Federlin, K. (1990). The evaluation of neutral density separation utilizing Ficoll, sodium diatrizoate and Nycodenz and centrifugal elutriation in the purification of bovine and canine islet preparations. *Hormone Metab. Res.* (Suppl) **25**: 57–63.

Brewer, G. J. (1995). Serum-free B27/Neurobasal medium supports differentiated growth of neurons from the striatum, substantia nigra, septum, cerebral cortex, cerebellum, and dentate gyrus. *J. Neurosci. Res.* **42**: 674–683.

Bridger, J. M. (2004). Mammalian artificial chromosomes: modern day feats of engineering—Isambard Kingdom Brunel style. *Cytogenet. Genome Res.* **107**: 5–8.

Briggs, R., & King, T. J. (1960). Nuclear transplantation studies on the early gastrula (Rana pipiens). I. Nuclei of presumptive endoderm. *Dev Biol.* 1960 Jun; **2**: 252–70.

Bright, R. K., & Lewis, J. D. (2004). Long-term culture of normal and malignant human prostate epithelial cells. In Pfragner, R., & Freshney, R. I. (eds.), *Culture of human tumor cells*, Hoboken, NJ, Wiley-Liss, pp. 125–144.

Brinch, D. S., & Elvig, S. G. (2001). Evaluation of an *in vitro* human corneal model as alternative to the *in vivo* eye irritation testing of enzymes. *Toxicol. Lett.* **123**, suppl. 1: 22.

Brindle, K. M. (1998). Investigating the performance of intensive mammalian cell bioreactor systems using magnetic resonance imaging and spectroscopy. *Biotech. Genet. Eng. Rev.* **15**: 499–520.

British Standard BS5726 (1992). Microbiological Safety Cabinets, Parts 1–4. The Stationery Office, P. O. Box 276, London.

Brito Babapulle, V. (1981). Lateral asymmetry in human chromosomes 1, 3, 4, 15 and 16. *Cytogenet. Cell Genet.* **29**: 198–202.

Brockes, J. P., Fields, K. L., & Raff, M. C. (1979). Studies on cultured rat Schwann cells; I: Establishment of purified populations from cultures of peripheral nerve. *Brain Res.* **165**: 105–118.

Brouty-Boyé, D., Kolonias, D., Savaraj, N., & Lampidis, T. J. (1992). Alpha-smooth muscle actin expression in cultured cardiac fibroblasts of newborn rat. *In Vitro Cell Dev. Biol.* **28A**: 293–296.

Brower, M., Carney, D. N., Oie, H. K., Gazdar, A. F., & Minna, J. D. (1986). Growth of cell lines and clinical specimens

of human nonsmall cell lung cancer in a serum-free defined medium. *Cancer Res.* **46**: 798–806.

Brown, A. F., & Dunn, G. A. (1989). Microinterferometry of the movement of dry matter in fibroblasts. *J. Cell Sci.* **92**: 379–389.

Brunk, C. F., Jones, K. C., & James, T. W. (1979). Assay for nanogram quantities of DNA in cellular homogenates. *Anal. Biochem.* **92**: 497–500.

Brunton, V., Ozanne, B., Paraskeva, C., & Frame, M. (1997). A role for epidermal growth factor receptor, c-Src and focal adhesion kinase in an *in vitro* model for the progression of colon cancer. *Oncogene* **14**: 283–293.

Bruyneel, E. A., Debray, H., de Mets, M., Mareel, M. M., & Montreuil, J. (1990). Altered glycosylation in Madin-Darby canine kidney (MDCK) cells after transformation by murine sarcoma virus. *Clin. Exp. Metastasis* **8**: 241–253.

Bryan, D., Sexton, C. J., Williams, D., Leigh, I. M., & McKay, I. (1995). Oral keratinocytes immortalized with the early region of human papillomavirus type 16 show elevated expression of interleukin 6, which acts as an autocrine growth factor for the derived T103C cell line. *Cell Growth Differ.* **6**: 1245–1250.

Bryan, T. M., & Reddel, R. R. (1997). Telomere dynamics and telomerase activity in *in vitro* immortalised human cells. *Eur. J. Cancer* **33**: 767–773.

Brysk, M. M., Santschi, C. H., Bell, T., Wagner, R. F. Jr, Tyring, S. K., & Rajaraman, S. (1992). Culture of basal cell carcinoma. *J. Invest. Dermatol.* **98**: 45–49.

Bucana, C. D., Giavazzi, R., Nayar, R., O'Brian, C. A., Seid, C., Earnest, L. E., & Fan, D. (1990). Retention of vital dyes correlates inversely with the multidrug-resistant phenotype of adriamycin-selected murine fibrosarcoma variants. *Exp. Cell Res.* **190**: 69.

Buchanan, S. S., Gross, S. A., Acker, J. P., Toner, M., Carpenter, J. F., & Pyatt, D. W. (2004). Cryopreservation of stem cells using trehalose: evaluation of the method using a human hematopoietic cell line. *Stem Cells Dev.* **13**: 295–305.

Buchler, P., Reber, H. A., Buchler, M. W., Friess, H., Lavey, R. S., & Hines, O. J. (2004). Antiangiogenic activity of genistein in pancreatic carcinoma cells is mediated by the inhibition of hypoxia-inducible factor-1 and the down-regulation of VEGF gene expression. *Cancer* **100**: 201–210.

Buckingham, M. (1992). Making muscle in mammals. *Trends Genet.* **8**: 144–149.

Buehring, G. C. (1972). Culture of human mammary epithelial cells: Keeping abreast of a new method. *J. Natl. Cancer Inst.* **49**: 1433–1434.

Buick, R. N., Stanisic, T. H., Fry, S. E., Salmon, S. E., Trent, J. M., & Krosovich, P. (1979). Development of an agar-methyl cellulose clonogenic assay for cells of transitional cell carcinoma of the human bladder. *Cancer Res.* **39**: 5051–5056.

Buonassisi, V., Sato, G., & Cohen, A. I. (1962). Hormone-producing cultures of adrenal and pituitary tumor origin. *Proc. Natl. Acad. Sci. USA* **48**: 1184–1190.

Burchell, J., & Taylor-Papadimitriou, J. (1989). Antibodies to human milk fat globule molecules. *Cancer Invest.* **17**: 53–61.

Burchell, J., Durbin, H., & Taylor-Papadimitriou, J. (1983). Complexity of expression of antigenic determinants recognised by monoclonal antibodies HMFG 1 and HMFG 2 in normal and malignant human mammary epithelial cells. *J. Immunol.* **131**: 508–513.

Burchell, J., Gendler, S., Taylor-Papadimitriou, J., Girling, A., Lewis, A., Millis, R., & Lamport, D. (1987). Development and characterisation of breast cancer reactive monoclonal antibodies directed to the core protein of the human milk mucin. *Cancer Res*. **47**: 5476–5482.

Burgess, W. H., & Maciag, T. (1989). The heparin-binding fibroblast growth factor family of proteins. *Annu. Rev. Biochem.* **58**: 575–606.

Burke, J. F., Price, T. N. C., & Mayne, L. V. (1996). Immortalization of human astrocytes. In Freshney, R. I., & Freshney, M. G. (eds.), *Culture of immortalized cells*. New York, Wiley-Liss, pp. 299–314.

Burrows, M. T. (1912). Rhytmische Kontraktionen der isolierten Herzmuskelzelle Ausserhalb des Organismus. *Münch. Med. Wochenschr.* **LIX**: 1473.

Burwen, S. J., & Pitelka, D. R. (1980). Secretory function of lactating moose mammary epithelial cells cultured on collagen gels. *Exp. Cell Res*. **126**: 249–262.

Butler, M. (1991). *Mammalian cell biotechnology*. Oxford, IRL Press at Oxford University Press.

Butler, M., & Christie, A. (1994). Adaptation of mammalian cells to non-ammoniagenic media. *Cytotechnology* **15**: 87–94.

Cable, E. E., & Isom, H. C. (1997). Exposure of primary rat hepatocytes in long-term DMSO culture to selected transition metals induces hepatocyte proliferation and formation of duct-like structures. *Hepatology* **26**: 1444–1457.

Cai, J., Weiss, M. L., Rao, M. S. (2004). In search of "stemness". *Exp Hematol.*, **32**: 585–98.

Calder, C. J., Liversidge, J., & Dick, A. D. (2004). Murine respiratory tract dendritic cells: isolation, phenotyping and functional studies. *J. Immunol. Methods* **287**: 67–77.

Campion, D. G. (1984). The muscle satellite cell—a review. *Int. Rev. Cytol.* **87**: 225–251.

Camps, J., Morales, C., Prat, E., Ribas, M., Capella, G., Egozcue, J., Peinado, M. A., & Miro, R. (2004). Genetic evolution in colon cancer KM12 cells and metastatic derivates. *Int J Cancer*. **110**: 869–874.

Cancedda, R., Dozin, B., Giannoni, P. & Quarto, R. (2003). Tissue engineering and cell therapy of cartilage and bone. *Matrix Biol.* **22**: 81–91.

Cancela, M. L., Hu, B., & Price, P. A. (1997). Effect of cell density and growth factors on matrix GLA protein expression by normal rat kidney cells. *J. Cell Physiol.* **171**: 125–134.

Caniggia, I., Tseu, I., Han, R. N., Smith, B. T., Tanswell, K., & Post, M. (1991). Spatial and temporal differences in fibroblast behavior in fetal rat lung. *Am. J. Physiol. Lung Mol. Cell. Physiol.* **261**: L424–L433.

Cao, B., Zheng, B., Jankowski, R. J., Kimura, S., Ikezawa, M., Deasy, B., Cummins, J., Epperly, M., Qu-Petersen, Z., & Huard, J. (2003). Muscle stem cells differentiate into haematopoietic lineages but retain myogenic potential. *Nat. Cell Biol.* **5**: 640–646.

Cao, D., Lin, G., Westphale, E. M., Beyer, E. C., & Steinberg, T. H. (1997). Mechanisms for the coordination of intercellular calcium signaling in insulin-secreting cells. *J. Cell Sci.* **110**: 497–504.

Caplan, A. I. (1991). Mesenchymal stem cells. *J. Orthop. Res.* **9**: 641–650.

Caputo, J. L. (1996). Safety procedures. In Freshney, R. I., & Freshney, M. G. (eds.), *Culture of immortalized cells*. New York, Wiley-Liss, pp. 25–51.

Carlsson, J., & Nederman, T. (1989). Tumour spheroid technology in cancer therapy research. *Eur. J. Cancer Clin. Oncol.* **25**: 1127–1133.

Carmichael, J., DeGraff, W. G., Gazdar, A. F., Minna, J. D., & Mitchell, J. B. (1987a). Evaluation of a tetrazolium-based semi-automated colorimetric assay: Assessment of chemosensitivity testing. *Cancer Res.* **47**: 936–942.

Carmichael, J., DeGraff, W. G., Gazdar, A. F., Minna, J. D., & Mitchell, J. B. (1987b). Evaluation of a tetrazolium-based semiautomated colorimetric assay: Assessment of radiosensitivity. *Cancer Res.* **47**: 943–946.

Carney, D. N., Bunn, P. A., Gazdar, A. F., Pagan, J. A., & Minna, J. D. (1981). Selective growth in serum-free hormone-supplemented medium of tumor cells obtained by biopsy from patients with small cell carcinoma of lung. *Proc. Natl. Acad. Sci. USA* **78**: 3185–3189.

Carney, D. N., Gazdar, A. F., Bepler, G., Guccion, J. G., Marangos, P. J., Moodt, T. W., Zweig, M. H., & Minna, J. D. (1985): Establishment and identification of small cell lung cancer cell lines having classic and variant features. *Cancer Res.* **45**: 2913–2923.

Carpenter, G., & Cohen, S. (1977). Epidermal growth factor. In Acton, R. T., & Lynn, J. D. (eds.), *Cell culture and its application*. New York, Academic Press, pp. 83–105.

Carraway, K. L., Fregien, N., Carraway, K. L., III, & Carraway, C. A. (1992). Tumor sialomucin complexes as tumor antigens and modulators of cellular interactions and proliferation. *J. Cell Sci.* **103**: 299–307.

Carrel, A. (1912). On the permanent life of tissues outside the organism. *J. Exp. Med.* **15**: 516–528.

Carrel, A., & Ebeling, A. H. (1923). Survival and growth of fibroblasts *in vitro*. *J. Exp. Med.* **XXXVIII**: 487.

Cartwright, T., & Shah, G. P. (1994). Culture media. In Davis, J. M. (ed.), *Basic cell culture, a practical approach*. Oxford, U.K., IRL Press at Oxford University Press, pp. 58–91.

Carvalho, K. A., Guarita-Souza, L. C., Rebelatto, C. L., Senegaglia, A. C., Hansen, P., Mendonca, J. G., Cury, C. C., Francisco, J. C., & Brofman, P. R. (2004). Could the coculture of skeletal myoblasts and mesenchymal stem cells be a solution for postinfarction myocardial scar? *Transplant Proc.* **36**: 991–992.

Casanova, J. E. (2002). Epithelial cell cytoskeleton and intracellular trafficking. V. Confluence of membrane trafficking and motility in epithelial cell models. *Am. J. Physiol. Gastrointest. Liver. Physiol.* **283**(5): G1015–G1019.

Caspersson, T., Farber, S., Foley, C. E., Kudynowski, J., Modest, E. J., Simonsson, E., Wagh, U., & Zech, L. (1968). Chemical differentiation along metaphase chromosomes. *Exp. Cell Res.* **49**: 219–222.

Cataldo, L. M., Wang, Z., & Ravid, K. (1998). Electroporation of DNA into cultured cell lines. In Ravid, K., & Freshney, R. I. (eds.), *DNA transfer to cultured cells*. New York, Wiley-Liss, pp. 55–68.

Cavallaro, U., & Christofori, G. (2004). Cell adhesion and signalling by cadherins and Ig-CAMs in cancer. *Nat. Rev. Cancer* **4**: 118–132.

CDC Office of Health & Safety. (1999). Biosafety in Microbiological and Biomedical Laboratories (BMBL) 4th Edition. U.S. Department of Health and Human Services, Centers for Disease

Control and Prevention, and National Institutes of Health, Fourth Edition, May 1999. US Government Printing Office, Washington: 1999.

Centers for Disease Control. (1988). Update: Universal precautions for prevention of transmission of human immunodeficiency virus, hepatitis B virus, and other blood-borne pathogens in healthcare settings. *MMWR* **37**: 377–382, 387, 388.

Ceriani, R. L., Taylor-Papadimitriou, J., Peterson, J. A., & Brown, P. (1979). Characterization of cells cultured from early lactation milks. *In Vitro* **15**: 356–362.

Chambard, M., Mauchamp, J., & Chaband, O. (1987). Synthesis and apical and basolateral secretion of thyroglobulin by thyroid cell monolayers on permeable substrate: Modulation by thyrotropin. *J. Cell Physiol.* **133**: 37–45.

Chambard, M., Vemer, B., Gabrion, J., Mauchamp, J., Bugeia, J. C., Pelassy, C., & Mercier, B. (1983). Polarization of thyroid cells in culture: Evidence for the basolateral localization of the iodide "pump" and of the thyroid-stimulating hormone receptor–adenyl cyclase complex. *J. Cell Biol.* **96**: 1172–1177.

Chang, H., & Baserga, R. (1977). Time of replication of genes responsible for a temperature sensitive function in a cell cycle specific to mutant from a hamster cell line. *J. Cell Physiol.* **92**: 333–343.

Chang, R. S. (1954). Continuous subcultivation of epithelial-like cells from normal human tissues. *Proc. Soc. Exp. Biol. Med.* **87**: 440–443.

Chang, S. E., & Taylor-Papadimitriou, J. (1983). Modulation of phenotype in cultures of human milk epithelial cells and its relation to the expression of a membrane antigen. *Cell Differ.* **12**: 143–154.

Chang, S. E., Keen, J., Lane, E. B., & Taylor-Papadimitriou, J. (1982). Establishment and characterisation of SV40-transformed human breast epithelial cell lines. *Cancer Res.* **42**: 2040–2053.

Chaproniere, D. M., & McKeehan, W. L. (1986). Serial culture of adult human prostatic epithelial cells in serum-free medium containing low calcium and a new growth factor from bovine brain. *Cancer Res.* **46**: 819–824.

Chen, C.-S., Toda, K.-I., Maruguchi, Y., Matsuyoshi, N., Horiguchi, Y., & Imamura, S. (1997). Establishment and characterization of a novel *in vitro* angiogenesis model using a microvascular endothelial cell line, F-2C, cultured in chemically defined medium. *In Vitro Cell Dev. Biol. Anim.* **33**: 796–802.

Chen, T. C., Curthoys, N. P., Lagenaur, C. F., & Puschett, J. B. (1989). Characterization of primary cell cultures derived from rat renal proximal tubules. *In Vitro Cell Dev. Biol.* **25**: 714–722.

Chen, T. R. (1977). *In situ* detection of mycoplasm contamination in cell cultures by fluorescent Hoechst 33258 stain. *Exp. Cell Res.* **104**: 255.

Cherry, R. S., & Papoutsakis, E. T. (1990). Understanding and controlling fluid-mechanical injury of animal cells in bioreactors. In Spier, R. E., & Griffiths, J. B., *Animal Cell Biotechnology*, Vol. 4, London, Academic Press, pp. 71–121.

Cho, J.-K., & Bikle, D. D. (1997). Decrease of Ca-ATPase activity in human keratinocytes during calcium-induced differentiation. *J. Cell Physiol.* **172**: 146–154.

Chopin, D., Barei-Moniri, R., Maille, P., Le Frere-Belda, M. A., Muscatelli-Groux, B., Merendino, N., Lecerf, L., Stoppacciaro, A., & Velotti, F. (2003). Human urinary bladder transitional cell carcinomas acquire the functional Fas ligand during tumor progression. *Am. J. Pathol.* **162**: 1139–1149.

Chopra, D. P., Xue-Hu, I. C. (1993). Secretion of alpha-amylase in human parotid gland epithelial cell culture. *J Cell Physiol.*, **155**: 223–33.

Chopra, D. P., Grignon, D. J., Joiakim, A., Mathieu, P. A., Mohamed, A., Sakr, W. A., Powell, I. J., & Sarkar, F. H. (1996). Differential growth factor responses of epithelial cell cultures derived from normal human prostate, benign prostatic hyperplasia and primary prostate carcinoma. *J. Cell Physiol.* **169**: 269–280.

Christensen, B., Hansen, C., Kieler, J., & Schmidt, J. (1993). Identity of non-malignant human urothelial cell lines classified as transformation grade I (TGrI) and II (TGrII). *Anticancer Res.* **13**: 2187–2191.

Chu, G., Hayakawa, H., & Berg, P. (1987). Electroporation for the efficient transfection of mammalian cells with DNA. *Nucleic Acids Res.* **15**: 1311–1326.

Ciccarone, V., Hawley-Nelson, P., Gebeyehu, G., Jessee, J. (1993). Cationic liposome-mediated transfection of eukaryotic cells high-efficiency nucleic-acid delivery with lipofectin™, Lipofectace™, and Lipofectamine™ reagents. *FASEB J.* **7**: A1131.

Cioni, C., Filoni, S., Aquila, C., Bernardini, S., & Bosco, L. (1986). Transdifferentiation of eye tissue in anuran amphibians: Analysis of the transdifferentiation capacity of the iris of *Xenopus laevis* larvae. *Differentiation* **32**: 215–220.

Clarke, G. D., & Ryan, P. J. (1980). Tranquilizers can block mitogenesis in 3T3 cells and induce differentiation in Friend cells. *Nature* **287**: 160–161.

Clayton, R. M., Bower, D. J., Clayton, P. R., Patek, C. E., Randall, F. E., Sime, C., Wainwright, N. R., & Zehir, A. (1980). Cell culture in the investigation of normal and abnormal differentiation of eye tissues. In Richards, R. J., & Rajan, K. T. (eds.), *Tissue culture in medical research (II)*. Oxford, U.K., Pergamon Press.

Cobbold, P. H., & Rink, T. J. (1987). Fluorescence and bioluminescence measurement of cytoplasmic free Ca^{2+}. *Biochem. J.* **248**: 313.

Cohen, C., Roguet, R., Cottin, M., Olive, C., Leclaire, J., & Rougier, A. (1997). The use of Episkin, a reconstructed epidermis, in the evaluation of the protective effect of sunscreens against chemical lipoperoxidation induced by UVA. *J. Invest. Dermatol.* **108**: 77.

Cohen, S. (1962). Isolation of a mouse submaxillary gland protein accelerating incisor eruption and eyelid opening in the new-born animal. *J. Biol. Chem.* **237**: 1555–1562.

Cole, J., Fox, M., Garner, R. C., McGregor, D. B., & Thacker, J. (1990). Gene mutation assays in cultured mammalian cells. In Kirkland, D. J. (ed.), *Basic mutagenicity tests: UKEMS recommended procedures, Part 1 (revised)*. Cambridge, U.K., Cambridge University Press, pp. 87–114.

Cole, R. J., & Paul, J. (1966). The effects of erythropoietin on haem synthesis in mouse yolk sac and cultured foetal liver cells. *J. Embryol. Exp. Morphol.* **15**: 245–260.

Cole, S. P. C. (1986). Rapid chemosensitivity testing of human lung tumour cells using the MTT assay. *Cancer Chemother. Pharmacol.* **17**: 259–263.

Collen, D., Stassen, J. M., Marafino, B. J., Jr., Builder, S., De Cock, F., Ogez, J., Tajiri, D., Pennica, D., Bennett, W. F., Salwa, J., et al. (1984). Biological properties of human tissue-type plasminogen activator obtained by expression of recombinant

DNA in mammalian cells. *J Pharmacol Exp Ther.* **231**(1): 146–152.

Collins, C. H., & Kennedy, D, A. (1999). Evolution of microbiological safety cabinets. *Br. J. Biomed. Sci.* **56**: 161–169.

Collins, S. J., Gallo, R. C., & Gallagher, R. E. (1977). Continuous growth and differentiation of human myeloid leukaemic cells in suspension culture. *Nature* **270**: 347–349.

Collodi, P. (1998). DNA transfer to blastula-derived cultures from zebra fish. In Ravid, K., & Freshney, R. I. (eds.), *DNA transfer to cultured cells.* New York, Wiley-Liss, pp. 69–92.

Collodi, P., & Barnes, D. W. (1990). Mitogenic activity from trout embryos. *Proc. Natl. Acad. Sci. USA* **87**: 3498–3502.

Collodi, P., Kamei, Y., Ernst, T., Miranda, C., Buhler, D. R., & Barnes, D. W. (1992). Culture of cells from zebrafish (*Brachydanio rerio*) embryo and adult tissues. *Cell Biol. Toxicol.* **8**: 43–61.

Compton, C. C., Warland, G., Nakagawa, H., Opitz, O. G., & Rustgi, A. K. (1998). Cellular characterization and successful transfection of serially subcultured normal human esophageal keratinocytes. *J. Cell Physiol.* **177**: 274–281.

Coon, H. D. (1968). Clonal cultures of differentiated rat liver cells. *J. Cell Biol.* **39**: 29a.

Coon, H. G., & Cahn, R. D. (1966). Differentiation *in vitro*: Effects of Sephadex fractions of chick embryo extract. *Science* **153**: 1116–1119.

Cooper, G. W. (1965). Induction of somite chondrogenesis by cartilage and notochord: A correlation between inductive activity and specific stages of cytodifferentiation. *Dev. Biol.* **12**: 185–212.

Cooper, P. D., Burt, A. M., & Wilson, J. N. (1958). Critical effect of oxygen tension on rate of growth of animal cells in continuous suspended culture. *Nature* **182**: 1508–1509.

Cooper, S., & Broxmeyer, H. (1994). Purification of murine granulocyte-macrophage progenitor cells (CFC-GM) using counterflow centrifugal elutriation. In Freshney, R. I., Pragnell, I. B., & Freshney, M. G. (eds.), *Culture of haemopoietic cells.* New York, Wiley-Liss, pp. 223–234.

Coriell, L. L., Tall, M. G., & Gaskill, H. (1958). Common antigens in tissue culture cell lines. *Science* **128**: 198.

Cou, J. Y. (1978). Establishment of clonal human placental cells synthesizing human choriogonadotropin. *Proc. Natl. Acad. Sci. USA* **75**: 1854–1858.

Courtenay, V. D., Selby, P. I., Smith, I. E., Mills, J., & Peckham, M. J. (1978). Growth of human tumor cell colonies from biopsies using two soft-agar techniques. *Br. J. Cancer* **38**: 77–81.

Coutinho, L. H., Gilleece, M. H., de Wynter, E. A., Will, A., & Testa, N. G. (1992). Clonal and long-term cultures using human bone marrow. In Testa, N. G., & Molineux, G. (eds.), *Haemopoiesis: A practical approach.* Oxford, U.K., IRL Press at Oxford University Press, pp. 75–106.

Crabb, I. W., Armes, L. G., Johnson, C. M., & McKeehan, W. L. (1986). Characterization of multiple forms of prostatropin (prostate epithelial growth factor) from bovine brain. *Biochem. Biophys. Res. Commun.* **136**: 1155–1161.

Creasey, A. A., Smith, H. S., Hackett, A. I., Fukuyama, K., Epstein, W. L., & Madin, S. H. (1979). Biological properties of human melanoma cells in culture. *In Vitro* **15**: 342.

Croce, C. M. (1991). Genetic approaches to the study of the molecular basis of human cancer. *Cancer Res. (Suppl)* **51**: 5015s–5018s.

Cronauer, M. V., Eder, I. E., Hittmair, A., Sierek, G., Hobisch, A., Culig, Z., Thurnhur, M., Bartsch, G., & Klocker, H. (1997). A reliable system for the culture of human prostatic cells. *In Vitro Cell Dev. Biol. Anim.* **33**: 742–744.

Crouch, E. C., Stone, K. R., Bloch, M., & McDivitt, R. W. (1987). Heterogeneity in the production of collagens and fibronectin by morphologically distinct clones of a human tumor cell line: Evidence for intratumoral diversity in matrix protein biosynthesis. *Cancer Res.* **47**(22): 6086–6092.

Cruikshank, C. N. D., & Lowbury, E. J. L. (1952). Effect of antibiotics on tissue cultures of human skin. *Br. Med. J.* **2**: 1070.

Csoka, K., Nygren, P., Graf, W., Påhlman, L., Glimelius, B., & Larsson, R. (1995). Selective sensitivity of solid tumors to suramin in primary cultures of tumor cells from patients. *Int. J. Cancer* **63**: 356–360.

Cuttitta, F., Carney, D. N., Mulshine, J., Moody, T. W., Fedorko, J., Fischler, A., & Minna, J. D. (1985). Bombesin-like peptides can function as autocrine growth factors in human small-cell lung cancer. *Nature* **316**: 823.

Dairkee, S. H., Deng, G., Stampfer, M. R., Waldman, F. M., & Smith, H. S. (1995). Selective cell culture of primary breast carcinoma. *Cancer Res.* **55**: 2516–2519.

Dairkee, S. H., Paulo, S. C., Traquina, P., Moore, D. H., Ljung, B. M., & Smith, H. S. (1997). Partial enzymatic degradation of stroma allows enrichment and expansion of primary breast tumor cells. *Cancer Res.* **57**: 1590–1596.

Damon, D. H., Lobb, R. R., D'Amore, P. A., & Wagner, J. A. (1989). Heparin potentiates the action of acidic fibroblast growth factor by prolonging its biological half-life. *J. Cell Physiol.* **138**: 221–226.

Danen, E. H., & Yamada, K. M. (2001). Fibronectin, integrins, and growth control. *J. Cell. Physiol.* **189**: 1–13.

Dangles, V., Femenia, F., Laine, V., Berthelmy, M., LeRhun, D., Poupon, M. F., Levy, D., & Schwartz-Cornil, I. (1997). Two and three-dimensional cell structures govern epidermal growth factor survival function in human bladder carcinoma cell lines. *Cancer Res.* **57**: 3360–3364.

Danielson, K. G., McEldrew, D., Alston, J. T., Roling, D. B., Damjanov, A., Damjanov, I., Daska, I., & Spinner, N. (1992). Human colon carcinoma cell lines from the primary tumor and a lymph node metastasis. *In Vitro Cell Dev. Biol.* **28A**: 7–10.

Darcy, K. M., Shoemaker, S. F., Lee, P.-P. H., Vaughan, M. M., Black, J. D., & Ip, M. M. (1995). Prolactin and epidermal growth factor regulation of the proliferation, morphogenesis, and functional differentiation of normal rat mammary epithelial cells in three dimensional primary culture. *J. Cell Physiol.* **163**: 346–364.

Darling, J. L. (2004). *In vitro* culture of malignant brain tumors. In Pfragner, R., & Freshney, R. I. (eds.), *Culture of human tumor cells.* Hoboken, NJ, Wiley-Liss, pp. 349–372.

Darnell, J. E. (1982). Variety in the level of gene control in eukaryotic cells. *Nature* **297**: 365–371.

Dart, J. (2003). Corneal toxicity: the epithelium and stroma in iatrogenic and factitious disease. *Eye* **17**: 886–892.

Das, A. V., James, J., Zhao, X., Rahnenfuhrer, J., & Ahmad, I. (2004). Identification of c-Kit receptor as a regulator of adult neural stem cells in the mammalian eye: interactions with Notch signaling. *Dev Biol.* **273**(1): 87–105.

Davies, B., Brown, P. D., East, N., Crimmin, M. J., & Balkwill, F. R. (1993). A synthetic matrix metalloproteinase inhibitor decreases tumour burden and prolongs survival of mice bearing ovarian carcinoma xenografts. *Cancer Res.* **53**: 2087–2091.

Davis, J. B., & Stroobant, P. (1990) Platelet-derived growth factors and fibroblast growth factors are mitogens for rat Schwann cells. *J. Cell. Biol.* **110**: 1353–1360.

Davison, P. M., Bensch, R., & Karasek, M. A. (1983). Isolation and long-term serial cultivation of endothelial cells from microvessels of the adult human dermis. *In Vitro* **19**: 937–945.

Dawe, C. J., & Potter, M. (1957). Morphologic and biologic progression of a lymphoid neoplasm of the mouse *in vivo* and *in vitro*. *Am. J. Pathol.* **33**: 603.

Degos, L. (1997). Differentiation in acute promyelocytic leukemia: European experience. *J. Cell Physiol.* **173**: 285–287.

De Larco, J. E., & Todaro, G. J. (1978). Epithelioid and fibroblastoid rat kidney cell clones: Epidermal growth factor receptors and the effect of mouse sarcoma virus transformation. *J. Cell Physiol.* **94**: 335–342.

De Leij, L., Poppema, S., Nulend, I. K., Haar, A. T., Schwander, E., Ebbens, F., Postmus, P. E., & The, T. H. (1985). Neuroendocrine differentiation antigen on human lung carcinoma and Kulchitski cells. *Cancer Res.* **45**: 2192–2200.

Del Vecchio, P., & Smith, J. R. (1981). Expression of angiotensin converting enzyme activity in cultured pulmonary artery endothelial cells. *J. Cell Physiol.* **108**: 337–345.

Del Vecchio, S., Stoppelli, M. P., Carriero, M. V., Fonti, R., Massa, O., Li, P. Y., Botti, G., Cerra, M., d'Aiuto, G., Espositi, G., & Salvatore, M. (1993). Human urokinase receptor concentration in malignant and benign breast tumours by *in vitro* quantitative autoradiography: Comparison with urokinase levels. *Cancer Res.* **53**: 3198–3206.

DeMars, R. (1958). The inhibition of glutamine of glutamyl transferase formation in cultures of human cells. *Biochim. Biophys. Acta* **27**: 435–436.

Demers, G. W., Halbert, C. L., & Galloway, D. A. (1994). Elevated wild-type p53 protein levels in human epithelial cell lines immortalized by the human papillomavirus type 16 E7 gene. *Virology* **198**: 169–174.

Deng, J., Steindler, D. A., Laywell, E. D., & Petersen, B. E. (2003). Neural trans-differentiation potential of hepatic oval cells in the neonatal mouse brain. *Exp Neurol.* **182**: 373–382.

Dennis, C., (2003). Developmental biology: Synthetic sex cells. *Nature,* **424**: 364–6.

Dennis, C., Gallagher, R., & Campbell, P. (2001). The Humane Genome. *Nature* **209**: 813–858.

Department of Health and Social Security. (1986). *Good laboratory practice: The United Kingdom Compliance Programme.* London, HMSO.

De Ridder, L. (1997). Autologous confrontation of brain tumor derived spheroids with human dermal spheroids. *Anticancer Res.* **17**: 4119–4120.

De Ridder, L., & Calliauw, L. (1990). Invasion of human brain tumors *in vitro*: Relationship to clinical evolution. *J. Neurosurg.* **72**: 589–593.

De Ridder, L., & Mareel, M. (1978). Morphology and ^{125}I-concentration of embryonic chick thyroids cultured in an atmosphere of oxygen. *Cell Biol. Int. Rep.* **2**: 189–194.

Desai, K. V., Kavanaugh, C. J., Calvo, A., & Green, J. E. (2002). Chipping away at breast cancer: insights from microarray studies of human and mouse mammary cancer. *Endocr. Relat. Cancer.* **9**: 207–220.

Deshpande, A. K., Baig, M. A., Carleton, S., Wadgoankar, R., & Siddiqui, M. A. Q. (1993). Growth factors activate cardiogenic differentiation in avian mesodermal cells. *Mol. Cell Differ.* **1**: 269–284.

De Silva, R., Moy, E. L., & Reddel, R. R. (1996). Immortalization of human bronchial epithelial cells. In Freshney, R. I., & Freshney, M. G. (eds.), *Culture of immortalized cells.* New York, Wiley-Liss, pp. 121–144.

De Silva, R., Whitaker, N. J., Rogan, E. M., & Reddel, R. R. (1994). HPV-16 E6 and E7 genes, like SV40 early region genes, are insufficient for immortalization of human mesothelial and bronchial epithelial cells. *Exp. Cell Res.* **213**: 418–427.

Detrisac, C. J., Sens, M. A., Garvin, A. J., Spicer, S. S., & Sens, D. A. (1984). Tissue culture of human kidney epithelial cells of proximal tubule origin. *Kidney Int.* **25**: 383–390.

De Vitry, F., Camier, M., Czernichow, P., Benda, P., Cohen, P., & Tixier-Vidal, A. (1974). Establishment of a clone of mouse hypothalamic neurosecretory cells synthesizing neurophysin and vasopressin. *Proc. Natl. Acad. Sci. USA* **71**: 3575–3579.

De Vonne, T. L., & Mouray, H. (1978). Human α_2-macroglobulin and its antitrypsin and antithrombin activities in serum and plasma. *Clin. Chim. Acta* **90**: 83–85.

De Vries, J. E., Dinjens, W. N. M., De Bruyne, G. K., Verspaget, H. W., van der Linden, E. P. M., de Bruine, A. P., Mareel, M. M., Bosman, F. T., & ten Kate, J. (1995). *In vivo* and *in vitro* invasion in relation to phenotypic characteristics of human colorectal carcinoma cells. *Br. J. Cancer* **71**: 271–277.

De Wever, O., Westbroek, W., Verloes, A., Bloemen, N., Bracke, M., Gespach, C., Bruyneel, E., Mareel, M. (2004). Critical role of N-cadherin in myofibroblast invasion and migration *in vitro* stimulated by colon-cancer-cell-derived TGF-β or wounding. *J Cell Sci.* **117**: 4691–4703.

De Wynter, E., Allen, T., Coutinho, L., Flavell, D., Flavell, S. U., & Dexter, T. J. (1993). Localisation of granulocyte macrophage colony-stimulating factor in human long-term bone marrow cultures. *J. Cell Sci.* **106**: 761–769.

Dexter, T. J., Spooncer, E., Simmons, P., & Allen, T. D. (1984). Long-term marrow culture: An overview of technique and experience. In Wright, D. G., & Greenberger, J. S. (eds.), *Long-term bone marrow culture.* New York, Alan R. Liss, *Kroc Foundation Series 18*, pp. 57–96.

Dexter, T. M., Allen, T. D., Scott, D., & Teich, N. M. (1979). Isolation and characterisation of a bipotential haematopoietic cell line. *Nature* **277**: 417–474.

Dexter, T. M., Testa, N. G., Allen, T. D., Rutherford, S., & Scolnick, E. (1981). Molecular and cell biological aspects of erythropoiesis in long-term bone marrow cultures. *Blood* **58**: 699–707.

Dickson, M. A., Hahn, W. C., Ino, Y., Ronfard, V., Wu, J. Y., Weinberg, R. A., Louis, D. N., Li, F. P., & Rheinwald, J. G. (2000). Human keratinocytes that express hTERT and also bypass a p16(INK4a)-enforced mechanism that limits life span become immortal yet retain normal growth and differentiation characteristics. *Mol Cell Biol.* **20**: 1436–1447.

Dickson, J. D., Flanigan, T. P., & Kemshead, J. T. (1983). Monoclonal antibodies reacting specifically with the cell surface of human astrocytes in culture. *Biochem. Soc. Trans.* **11**: 208.

Dimitroulakos, J., Squire, J., Pawlin, G., & Yeger, H. (1994). NUB-7: A stable-1-type human neuroblastoma cell line inducible along N- and S-type cell lineages. *Cell Growth Differ.* **5**: 373–384.

Dinnen, R., & Ebisuzaki, K. (1990). Mitosis may be an obligatory route to terminal differentiation in the Friend erythroleukemia cell. *Exp. Cell Res.* **191**: 149–152.

Dirks, W. G., MacLeod, R. A. F., & Drexler, H. G. (1999). ECV3O4 (endothelial) is really T24 (bladder carcinoma): cell line cross-contamination at source. *In Vitro Cell. Dev. Biol. Anim.* **35**: 558–559.

Dobbs, L. G., & Gonzalez, R. F. (2002). Isolation and culture of pulmonary alveolar epithelial type II cells. In Freshney, R. I., & Freshney, M. G. (eds.), *Culture of epithelial cells*, 2nd ed., Hoboken, NJ, Wiley-Liss, pp. 278–301.

Dobbs, L. G., Pian, M., Dumars, S., Maglio, M., Allen, L. (1997). Maintenance of the differentiated type II cell phenotype by culture with an apical air surface. *Am. J. Physiol.*, **273**: L347–L354.

Dodson, M. V., Mathison, B. A., & Mathison, B. D. (1990). Effects of medium and substratum on ovine satellite cell attachment, proliferation and differentiation. *In Vitro Cell Differ. Dev.* **29**(1): 59–66.

Doering, C. B., Healey, J. F., Parker, E. T., Barrow, R. T., & Lollar, P. (2002). High level expression of recombinant porcine coagulation factor VIII. *J. Biol. Chem.* **277**: 38345–38349.

Doherty, P., Ashton, S. V., Moore, S. E., & Walsh, F. S. (1991). Morphoregulatory activities of NCAM and N-cadherin can be accounted for by G protein-dependent activation of L- and N-type neuronal Ca^{2+} channels. *Cell* **67**: 21–33.

Dolbeare, F., & Selden, J. R. (1994). Immunochemical quantitation of bromodeoxyuridine: Application to cell-cycle kinetics. *Method Cell Biol.* **41**: 297–316.

Dorazio, J. A., Cole, B. C., & Steinstreilein, J. (1996). Mycoplasma-arthritidis mitogen up-regulates human NK cell activity. *Infect. Immun.* **64**(2): 441–447.

Dotto, G. P., Parada, L. F., & Weinberg, R. A. (1985). Specific growth response of *ras*-transformed embryo fibroblasts to tumour promoters. *Nature* **318**: 472–475.

Doucette, J. R. (1984). The glial cells in the nerve fibre layer of the rat olfactory bulb. *Anat. Rec.* **210**: 285–391.

Doucette, J. R. (1990). Glial influences on axonal growth in the primary olfactory system. *Glia* **3**: 433–449.

Douglas, J. L., & Quinlan, M. P. (1994). Efficient nuclear localization of the Ad5 E1A 12S protein is necessary for immortalization but not cotransformation of primary epithelial cells. *Cell Growth Differ.* **5**: 475–483.

Douglas, W. H. J., McAteer, J. A., Dell'Orco, R. T., & Phelps, D. (1980). Visualization of cellular aggregates cultured on a three-dimensional collagen sponge matrix. *In Vitro* **16**: 306–312.

Dow, J., Lindsay, G., & Morrison, J. (1995). *Biochemistry: molecules, cells and the body*. Workingham, U.K., Addison Wesley.

Doyle, A., & Bolton, B. J. (1994). The quality control of cell lines and the prevention, detection and cure of contamination. In Davis, J. M. (ed.), *Basic cell culture, a practical approach*. Oxford, U.K., IRL Press at Oxford University Press, pp. 242–271.

Doyle, A., Griffiths, J. B., & Newall, D. G. (eds.). (1990–1999). *Cell and tissue culture: laboratory procedures*. Chichester, U.K., Wiley.

Doyle, A., Morris, C., & Mowles, J. M. (1990). Quality control. In Doyle, A., Hay, R., & Kirsop, B. E. (eds.), *Animal cells, living resources for biotechnology*. Cambridge, U.K., Cambridge University Press, pp. 81–100.

Draper, J. S., Pigott, C., Thomson, J. A., & Andrews, P. W. (2002). Surface antigens of human embryonic stem cells: changes upon differentiation in culture. *J Anat.* **200**: 249–258.

Drejer, J., Larsson, O. M., & Schousboe, A. (1983). Characterization of uptake and release processes for d- and l-aspartate in primary cultures of astrocytes and cerebellar granule cells. *Neurochem. Ref.* **8**: 231–243.

Drexler, H. G., Dirks, W. G., Matsuo, Y., & MacLeod, R. A. (2003). False leukemia-lymphoma cell lines: an update on over 500 cell lines. *Leukemia* **17**: 416–426.

Drexler, H. G., Gignac, S. M., Hu, Z.-B., Hopert, A., Fleckenstein, E., Viges, M., & Uphoff, C. C. (1994). Treatment of mycoplasma contamination in a large panel of cell cultures. *In Vitro Cell Dev. Biol.* **30A**: 344–347.

Drexler, H. G. (2004). Establishment and culture of human leukemia-lymphoma cell lines". In Pfragner, R., & Freshney. R. I. (eds.), *Culture of human tumor cells*. Hoboken, NJ, Wiley-Liss, pp. 319–348.

Drexler, H. G., & Minowada, J. (2000). Human leukemia-lymphoma cell lines: historical perspective, state of the art and future prospects. In Masters, J. R. W., & Palsson, B. O., (eds.), *Human cell culture, Vol. III Cancer cell lines, Part 3: Leukemia and lymphomas*, Dordrecht, The Netherlands, Kluwer, pp. 1–88.

Driever, W., & Rangini, Z. (1993). Characterization of a cell line derived from zebra-fish (*Brachydanio rerio*). *In Vitro Cell Dev. Biol. Anim.* **29A**: 749–754.

Driever, W., Stemple, D., Schier, A., & Solnica-Krezel, L. (1994). Zebrafish: Genetic tools for studying vertebrate development. *Trends Genet.* **10**: 152–159.

Duffy, M. J., Reilly, D., O'Sullivan, C., O'Higgins, N., Fennelly, J. J., & Andreasen, P. (1990). Urokinase-plasminogen activator, a new and independent prognostic market in breast cancer. *Cancer Res.* **50**: 6827–6829.

Duksin, D., Maoz, A., & Fuchs, S. (1975). Differential cytotoxic activity of anticollagen serum on rat osteoblasts and fibroblasts in tissue culture. *Cell* **5**: 83–86.

Dulbecco, R. (1952). Production of plaques in monolayers of tissue cultures by single particles of an animal virus. *Proc. Natl. Acad. Sci. USA* **38**: 747.

Dulbecco, R., & Elkington, J. (1973). Conditions limiting multiplication of fibroblastic and epithelial cells in dense cultures. *Nature* **246**: 197–199.

Dulbecco, R., & Freeman, G. (1959). Plaque formation by the polyoma virus. *Virology* **8**: 396–397.

Dulbecco, R., & Vogt, M. (1954). Plaque formation and isolation of pure cell lines with poliomyelitis viruses. *J. Exp. Med.* **199**: 167–182.

Duncan, E. L., De Silva, R., & Reddel, R. R. (1996). Immortalization of human mesothelial cells. In Freshney, R. I., & Freshney, M. G. (eds.), *Culture of immortalized cells*. New York, Wiley-Liss, pp. 239–258.

Dunham, L. J., & Stewart, H. L. (1953). A survey of transplantable and transmissible animal tumors. *J. Natl. Cancer Inst.* **13**: 1299–1377.

Dunn, G. A., & Brown, A. F. (1990). Quantifying cellular shape using moment invariants. In Alt, W., & Hoffman, G. (eds.),

Biological motion: Lecture notes in biomathematics 89. Berlin: Springer-Verlag, pp. 10–34.

Dunn, G. A., & Zicha, D. (1997). Using the DRIMAPS system of interference microscopy to study cell behavior. In Celis, J. E. (ed.), *Handbook of cell biology*, 2nd ed. New York, Academic Press, pp. 44–53.

Dunn, M. E., Schilling, K., & Mugnaini, E. (1998). Development and fine structure of murine Purkinje cells in dissociated cerebellar cultures: Neuronal polarity. *Anat. Embryol.* **197**: 9–29.

Dutt, K., Harris-Hooker, S., Ellerson, D., Layne, D., Kumar, R., and Hunt, R. (2003). Generation of 3D retina-like structures from a human retinal cell line in a NASA bioreactor. *Cell Transplant.* **12**: 717–731.

Eagle, H. (1955). The specific amino acid requirements of mammalian cells (stain L) in tissue culture. *J. Biol. Chem.* **214**: 839.

Eagle, H. (1959). Amino acid metabolism in mammalian cell cultures. *Science* **130**: 432.

Eagle, H. (1973). The effect of environmental pH on the growth of normal and malignant cells. *J. Cell Physiol.* **82**: 1–8.

Eagle, H., Foley, G. E., Koprowski, H., Lazarus, H., Levine, E. M., & Adams, R. A. (1970). Growth characteristics of virus-transformed cells. *J. Exp. Med.* **131**: 863–879.

Earle, W. R., Schilling, E. L., Stark, T. H., Straus, N. P., Brown, M. F., & Shelton, E. (1943). Production of malignancy *in vitro*; IV: The mouse fibroblast cultures and changes seen in the living cells. *J. Natl. Cancer Inst.* **4**: 165–212.

Easty, D. M., & Easty, G. C. (1974). Measurement of the ability of cells to infiltrate normal tissues *in vitro*. *Br. J. Cancer* **29**: 36–49.

Ebendal, T., (1976). The relative roles of contact inhibition and contact guidance in orientation of axons extending on aligned collagen fibrils *in vitro*. *Exp. Cell Res.* **98**: 159–169.

Echarti, C., & Maurer, H. R. (1989). Defined, serum-free culture conditions for the GM-microclonogenic assay using agar-capillaries. *Blut* **59**: 171–176.

Echarti, C., & Maurer, H. R. (1991). Lymphokine-activated killer cells: determination of their tumor cytolytic capacity by a clonogenic microassay using agar capillaries. *J. Immunol. Methods* **143**: 41–47.

Edelman, G. M. (1973). Nonenzymatic dissociations; B: Specific cell fractionation of chemically derivatized surfaces. In Kruse, P. F., Jr., & Patterson, M. K., Jr. (eds.), *Tissue culture methods and applications.* New York, Academic Press, pp. 29–36.

Edelson, J. D., Shannon, J. H., & Mason, R. J. (1988). Alkaline phosphatase: A marker of alveolar type II cell differentiation. *Am. Rev. Respir. Dis.* **138**: 1268–1275.

Edington, K. G., Berry, I. J., O'Prey, M., Burns, J. E., Clark, L. J., Mitchell, R., Robertson, G., Soutar, D., Coggins, L. W., & Parkinson, E. K. (2004). Multistage head and neck squamous cell carcinoma. In Pfragner, R., & Freshney. R. I. (eds.), *Culture of Human Tumor Cells.* Hoboken, NJ, Wiley-Liss, pp. 261–288.

Edwards, A. M., Silva, E., Jofre, B., Becker, M. I., & De Ioannes, A. E. (1994). Visible light effects on tumoral cells in a culture medium enriched with tryptophan and riboflavin. *J. Photochem. Photobiol. B* **24**: 179–186.

Edwards, R. G. (1996). The history of assisted human conception with especial reference to endocrinology. *Exp. Clin. Endocrinol. Diabetes* **104**(3): 183–204.

Edwards, P. A. W., Easty, D. M., & Foster, C. S. (1980). Selective culture of epithelioid cells from a human squamous carcinoma

using a monoclonal antibody to kill fibroblasts. *Cell Biol. Int. Rep.* **4**: 917–922.

Eisinger, M., Lee, J. S., Hefton, J. M., Darzykiewicz, A., Chiao, J. W., & Deharven, E. (1979). Human epidermal cell cultures—growth and differentiation in the absence of dermal components or medium supplements. *Proc. Natl. Acad. Sci. USA* **76**: 5340.

Elliget, K. A., & Lechner, J. F. (1992). Normal human bronchial epithelial cell cultures. In Freshney, R. I. (ed.), *Culture of epithelial cells.* New York, Wiley-Liss, pp. 181–196.

Elvin, P., Wong, V., & Evans, C. W. (1985). A study of the adhesive, locomotory and invasive behaviour of Walker 256 carcinosarcoma cells. *Exp. Cell Biol.* **53**: 9–18.

Emura, M., Katyal, S. L., Ochiai, A., Hirohashi, S., & Singh, G. (1997). *In vitro* reconstitution of human respiratory epithelium. *In Vitro Cell Dev. Biol. Anim.* **33**: 602–605.

Enders, J. F., Weller, T. J., & Robbins, F. C. (1949). Cultivation of Lansing strain of poliomyelitis in cultures of various human embryonic tissues. *Science* **109**: 85.

Eng, L. F., & Bigbee, J. W. (1979). Immunochemistry of nervous-system specific antigens. In Aprison, M. H. (ed.), *Advances in neurochemistry.* New York, Plenum Press, pp. 43–98.

Engel, L. W., Young, N. A., Tralka, T. S., Lippman, M. E., O'Brien, S. J., & Joyce, M. J. (1978). Establishment and characterization of three new continuous cell lines derived from breast carcinomas. *Cancer Res.* **38**: 3352–3364.

Eppig, J. J., & O'Brien, M. J. (1996). Development *in vitro* of mouse oocytes from primordial follicles. *Biol. Reprod.* **54**: 197–207.

Epstein, M. A., & Barr, Y. M. (1964). Cultivation *in vitro* of human lymphoblasts from Burkitt's malignant lymphoma. *Lancet* **1**: 252.

Ermis, A., Muller, B., Hopf, T., Hopf, C., Remberger, K., Justen, H. P., Welter, C., & Hanselmann, R. (1998). Invasion of human cartilage by cultured multicellular spheroids of rheumatoid synovial cells—a novel *in vitro* model system for rheumatoid arthritis. *J. Rheumatol.* **25**: 208–213.

Eroglu, A., Russo, M. J., Bieganski, R., Fowler, A., Cheley, S., Bayley, H., & Toner, M. (2000). Intracellular trehalose improves the survival of cryopreserved mammalian cells. *Nat. Biotechnol.* **18**: 163–167.

Espmark, J. A., & Ahlqvist-Roth, L. (1978). Tissue typing of cells in cultures; I: Distinction between cell lines by the various patterns produced in mixed haemabsorption with selected multiparous sera. *J. Immunol. Methods* **24**: 141–153.

Ethier, S. P., Mahacek, M. L., Gullick, W. J., Frank, T. S., & Weber, B. L. (1993). Differential isolation of normal luminal mammary epithelial cells and breast cancer cells from primary and metastatic sites using selective media. *Cancer Res.* **53**: 627–635.

European Committee for Standardisation. (1999). Biotechnology performance criteria for microbiological safety cabinets [WI91]. CEN/TC233 Biotechnology, Central Secretariat, European Committee for Standardisation, Rue de Stassart 36, B-1050 Brussels, Belgium.

European Pharmacopoeia. (1980). *Biological tests*, 2nd ed., Part 1, Vol. 2. Maisonneuve, S.A., France.

Evans, T., Rosenthal, E. T., Youngblom, J., Distel, D., & Hunt, T. (1983). Cyclin: a protein specified by maternal mRNA in sea urchin eggs that is destroyed at each cleavage division. *Cell* **33**: 389–396.

Evans, M. J., & Kaufman, M. H. (1981). Establishment in culture of pluripotential cells from mouse embryos. *Nature* **292**: 154–156.

Evans, V. J., & Bryant, J. C. (1965). Advances in tissue culture at the National Cancer Institute in the United States of America. In Ramakrishnan, C. V. (ed.), *Tissue culture*. The Hague, W. Junk, pp. 145–167.

Evans, V. J., Bryant, J. C., Fioramonti, M. C., McQuilkin, W. T., Sanford, K. K., & Earle, W. R. (1956). Studies of nutrient media for tissue C cells *in vitro*; I: A protein-free chemically defined medium for cultivation of strain L cells. *Cancer Res.* **16**: 77.

Evans, V. J., & Sanford, K. K. (1978). Development of defined media for studies on malignant transformation in culture. In Katsuta, H. (ed.), *Nutritional Requirements of Cultured Cells*, Baltimore, University Park Press, pp. 149–192.

Fantini, J., Galons, J. P., Abadie, B., Canioni, P., Cozzone, P. J., Marvali, J., & Tirard, A. (1987). Growth in serum-free medium of human colonic adenocarcinoma cell lines on microcarriers: A two-step method allowing optimal cell spreading and growth. *In Vitro* **23**: 641–646.

Fata, J. E., Werb, Z., & Bissell, M. J. (2004). Regulation of mammary gland branching morphogenesis by the extracellular matrix and its remodeling enzymes. *Breast Cancer Res.* **6**: 1–11.

Federoff, S. (1975). In Evans, V. J., Perry, V. P., & Vincent, M. M. (eds.), *Manual of the Tissue Culture Association* **1**: 53–57.

Federoff, S., & Richardson, A. (2001). *Protocols for neural cell culture*. Clifton, NJ, Humana Press.

Feinberg, A. P., & Vogelstein, B. (1983). A technique for radiolabeling DNA restriction fragments to high specific activity. *Anal. Biochem.* **132**: 6–13.

Felgner, P., Gadek, T., Holm, M., Roman, R., Chan, H. W., Wenz, M., Northrop, J. P., Ringold, G. M., & Danielsen, M. (1987). Lipofection: A highly efficient, lipid-mediated DNA-transfection procedure. *Proc. Natl. Acad. Sci. USA* **84**: 7413–7417.

Fell, H. B., & Robison, R. (1929). The growth, development and phosphatase activity of embryonic avian femora and limb buds cultivated *in vitro*. Biochem. J. **23**: 767–784.

Fenton, J. I., & Hord, N. G. (2004). Flavonoids promote cell migration in nontumorigenic colon epithelial cells differing in Apc genotype: Implications of matrix metalloproteinase activity. *Nutr Cancer.* **48**: 182–188.

Fergusson, R. J., Carmichael, J., & Smyth, J. F. (1980). Human tumour xenografts growing in immunodeficient mice: A useful model for assessing chemotherapeutic agents in bronchial carcinoma. *Thorax* **41**: 376–380.

Fernig, D. G., & Gallagher, J. T. (1994). Fibroblast growth factors and their receptors: An information network controlling tissue growth, morphogenesis and repair. *Prog. Growth Factor Res.* **5**: 353–377.

Finbow, M. E., & Pitts, J. D. (1981). Permeability of junctions between animal cells. *Exp. Cell Res.* **131**: 1–13.

Fiorino, A. S., Diehl, A. M., Lin, H. Z., Lemischka, I. R., & Reid, L. M. (1998). Maturation-dependent gene expression in a conditionally transformed liver progenitor cell line. *In Vitro Cell Dev Biol Anim.* **34**: 247–258.

Fischer, A. (1925). *Tissue culture. Studies in experimental morphology and general physiology of tissue cells in vitro*. London, Heinemann.

Fischer, G. A., & Sartorelli, A. C. (1964). Development, maintenance and assay of drug resistance. *Methods Med. Res.* **10**: 247.

Fisher, H. W., Puck, T. T., & Sato, G. (1958). Molecular growth requirements of single mammalian cells: The action of fetuin in promoting cell attachment of glass. *Proc. Natl. Acad. Sci. USA* **44**: 4–10.

Fitzsimmons, S. A., Ireland, H., Barr, N. I., Cuthbert, A. P., Going, J. J., Newbold, R. F., & Parkinson, E. K. (2003). Human squamous cell carcinomas lose a mortality gene from chromosome 6q14.3 to q15. *Oncogene* **22**: 1737–1746.

Flier, J. S. (1995). The adipocyte: storage depot or node on the energy information superhighway? *Cell* **80**: 15–18.

Flynn, D., Yang, J., Nandi, S. (1982). Growth and differentiation of primary cultures of mouse mammary epithelium embedded in collagen gel. *Differentiation*, **22**: 191–4.

Fogh, J. (1977). Absence of HeLa cell contamination in 169 cell lines derived from human tumors. *J. Natl. Cancer Inst.* **58**: 209–214.

Fogh, J., & Trempe, G. (1975). In Fogh, J. (ed.), *Human tumor cells in vitro*. New York, Academic Press, pp. 115–159.

Folkman, I., & Moscona, A. (1978). Role of cell shape in growth control. *Nature* **273**: 345–349.

Folkman, I., Haudenschild, C. C., & Zetter, B. R. (1979). Long-term culture of capillary endothelial cells. *Proc. Natl. Acad. Sci. USA* **76**: 5217–5221.

Folkman, J. (1992). The role of angiogenesis in tumor growth. *Sem. Cancer Biol.* **3**: 65–71.

Folkman, J., & D'Amore, P. A. (1996). Blood vessel formation: what is its molecular basis? Minireview. *Cell* **87**: 1153–1155.

Folkman, J., & Haudenschild, C. (1980). Angiogenesis *in vitro*. *Nature* **288**: 551–556.

Fontana, A., Hengarner, H., de Tribolet, N., & Weber, E. (1984). Glioblastoma cells release interleukin 1 and factors inhibiting interleukin 2-mediated effects. *J. Immunol.* **132**(4): 1837–1844.

Food and Drug Administration. (1992). Federal Register. *21 Code of Federal Regulations, Part 58*. Office of the Federal Register, National Archives and Records, Washington, DC.

Foreman, J., & Pegg, D. E. (1979). Cell preservation in a programmed cooling machine: The effect of variations in supercooling. *Cryobiology* **16**: 315–321.

Foster, R., & Martin, G. S. (1992). A mutation in the catalytic domain of pp60v-src is responsible for the host- and temperature-dependent phenotype of the Rous sarcoma virus mutant tsLA33-1. *Virology* **187**: 145–155.

Frame, M., Freshney, R. I., Shaw, R., & Graham, D. I. (1980). Markers of differentiation in glial cells. *Cell Biol. Int. Rep.* **4**: 732.

Franceschini, I. A., & Barnett, S. C. (1996). Low-affinity NGF-receptor and E-N-CAM expression define two types of olfactory nerve ensheathing cells that share a common lineage. *Dev. Biol.* **173**: 327–343.

Franklin, R. J. M., & Barnett, S. C. (1997). Do olfactory glia have advantages over Schwann cells for CNS repair? *J. Neurosci. Res.* **50**: 1–8.

Franklin, J. M., Gilson, J. M., Franceschini, I. A., & Barnett, S. C. (1996). Schwann cell-like myelination following transplantation of an olfactory-nerve-ensheathing-cell line into areas of demyelination in the adult CNS. *Glia* **17**: 217–224.

Franklin, R. J. M., & Barnett, S. C. (2000). Olfactory ensheathing cells and CNS regeneration—the sweet smell of success? *Neuron* **28**: 1–4.

Frazier, J. M. (1992). *In vitro toxicity testing*. New York, Marcel Dekker.

Fredin, B. L., Seiffert, S. C., & Gelehrter, T. D. (1979). Dexamethasone-induced adhesion in hepatoma cells: The role of plasminogen activator. *Nature* **277**: 312–313.

Freed, L. E., & Vunjak-Novakovic, G. (2002). Spaceflight bioreactor studies of cells and tissues. *Adv Space Biol Med.* **8**: 177–95.

Freedman, V. H., & Shin, S. (1974). Cellular tumorigenicity in nude mice: Correlation with cell growth in semi-solid medium. *Cell* **3**: 355–359.

Freshney, M. G. (1994). Colony-forming assays for CFC-GM, BFU-E, CFC-GEMM, and CFC-mix. In Freshney, R. I., Pragnell, I. B., & Freshney, M. G. (eds.), *Culture of hematopoietic cells.* New York, Wiley-Liss, pp. 265–268.

Freshney, R. I. (1972). Tumour cells disaggregated in collagenase. *Lancet* **2**: 488–489.

Freshney, R. I. (1976). Separation of cultured cells by isopycnic centrifugation in metrizamide gradients. In Rick-wood, D. (ed.), *Biological separations.* London and Washington, Information Retrieval, pp. 123–130.

Freshney, R. I. (1978). Use of tissue culture in predictive testing of drug sensitivity. *Cancer Top.* **1**: 5–7.

Freshney, R. I. (1980). Culture of glioma of the brain. In Thomas, D. G. T., & Graham, D. I. (eds.), *Brain tumours: Scientific basic, clinical investigation and current therapy.* London, Butterworths, pp. 21–50.

Freshney, R. I. (1985). Induction of differentiation in neoplastic cells. *Anticancer Res.,* **5**: 111–130.

Freshney, R. I. (ed.). (1992). *Culture of epithelial cells.* New York, Wiley-Liss.

Freshney, R. I., & Freshney, M. G. (eds.). (1996). *Culture of immortalized cells.* New York, Wiley-Liss.

Freshney, R. I., & Hart, E. (1982). Clonogenicity of human glia in suspension. *Br. J. Cancer* **46**: 463.

Freshney, R. I., & Morgan, D. (1978). Radioisotopic quantitation in microtitration plates by an autofluorographic method. *Cell Biol. Int. Rep.,* **2**: 375–380.

Freshney, R. I., Celik, F., & Morgan, D. (1982a). Analysis of cytotoxic and cytostatic effects. In Davis, W., Malvoni, C., & Tanneberger, St. (eds.), *The control of tumor growth and its biological base, Fortschritte in der Onkologie, Band 10.* Berlin, Akademie-Verlag, pp. 349–358.

Freshney, R. I., Hart, E., & Russell, J. M. (1982b). Isolation and purification of cell cultures from human tumours. In Reid, E., Cook, G. M. W., & Moore, D. J. (eds.), *Cancer cell organelles; methodological surveys (B): Biochemistry,* Vol. 2. Chichester, U.K., Horwood, pp. 97–110.

Freshney, R. I., Morgan, D., Hassanzadah, M., Shaw, R., & Frame, M. (1980a). Glucocorticoids, proliferation and the cell surface. In Richards, R. J., & Rajan, K. T. (eds.), *Tissue culture in medical research (II).* Oxford, U.K., Pergamon Press, pp. 125–132.

Freshney, R. I., Paul, J., & Kane, I. M. (1975). Assay of anticancer drugs in tissue culture: Conditions affecting their ability to incorporate ^{3}H-leucine after drug treatment. *Br. J. Cancer* **31**: 89–99.

Freshney, R. I., Pragnell, I. B., & Freshney, M. G., (eds.). (1994). *Culture of Haemopoietic Cells.* New York, Wiley-Liss.

Freshney, R. I., Sherry, A., Hassanzadah, M., Freshney, M., Crilly, P., & Morgan, D. (1980b). Control of cell proliferation in human glioma by glucocorticoids. *Br. J. Cancer* **41**: 857–866.

Freshney, R. I. (2002). Other epithelial cells. In Freshney, R. I. (ed.), *Culture of epithelial cells,* 2nd ed., Hoboken, NJ, Wiley-Liss, pp. 402–436.

Freshney, R. I., & Freshney, M. G. (2002). In Freshney, R. I. (ed.), *Culture of epithelial cells,* 2nd ed., Hoboken, NJ, Wiley-Liss, Chapters 6, 10, 11, 12, 13.

Freyer, J. P., & Sutherland, R. M. (1980). Selective dissociation and characterization of cells from different regions of multicell tumour spheroids. *Cancer Res.* **40**: 3956–3965.

Friend, C., Patuleia, M. C., & Nelson, J. B. (1966). Antibiotic effect of tylosine on a mycoplasma contaminant in a tissue culture leukemia cell line. *Proc. Soc. Exp. Biol. Med.* **121**: 1009.

Friend, C., Scher, W., Holland, J. G., & Sato, T. (1971). Hemoglobin synthesis in murine virus-induced leukemic cells *in vitro*; 2: Stimulation of erythroid differentiation by dimethyl sulfoxide. *Proc. Natl. Acad. Sci. USA* **68**: 378–382.

Friend, K. K., Dorman, B. P., Kucherlapati, R. S., & Ruddle, F. H. (1976). Detection of interspecific translocations in mouse–human hybrids by alkaline Giemsa staining. *Exp. Cell Res.* **99**: 31–36.

Froud, SJ. (1999). The development, benefits and disadvantages of serum-free media. *Dev. Biol. Stand.* **99**: 157–166.

Fry, J., & Bridges, J. W. (1979). The effect of phenobarbitone on adult rat liver cells and primary cell lines. *Toxicol. Lett.* **4**: 295–301.

Fu, V. X., Schwarze, S. R., Reznikoff, C. A., & Jarrard, D. F. (2004). The establishment and characterization of bladder cancer cultures *in vitro.* In Pfragner, R., & Freshney. R. I. (eds.), *Culture of human tumor cells,* Hoboken, NJ, Wiley-Liss, pp. 97–123.

Fujita, H., Asahina, A., Mitsui, H., & Tamaki, K. (2004). Langerhans cells exhibit low responsiveness to double-stranded RNA. *Biochem. Biophys. Res. Commun.* **319**: 832–839.

Fuller, B. B., & Meyskens, F. L. (1981). Endocrine responsiveness in human melanocytes and melanoma cells in culture. *J. Natl. Cancer Inst.* **66**: 799–802.

Fullerton, N. E., Boyd, M., Mairs, R. J., Keith, W. N., Alderwish, O., Brown, M. M., Livingstone, A., & Kirk, D. (2004). Combining a targeted radiotherapy and gene therapy approach for adenocarcinoma of prostate. *Prostate Cancer Prostatic Dis.* **7**: 355–63.

Furue, M., & Saito, S. (1997). Synergistic effect of hepatocyte growth factor and fibroblast growth factor-1 on the branching morphogenesis of rat submandibular gland epithelial cells. *Tissue Cult. Res. Commun.* **16**: 189–194.

Fusenig, N. E. (1994a). Epithelial–mesenchymal interactions regulate keratinocyte growth and differentiation *in vitro.* In Leigh, I. M., Lane, E. B., & Watt, F. M. (eds.), *Keratinocyte handbook,* Vol. I. Cambridge, U.K., Cambridge University Press, pp. 71–94.

Fusenig, N. E. (1994b). Cell culture models: reliable tools in pharmacotoxicology? In Fusenig, N. E., & Graf, H. (eds.), *Cell culture in pharmaceutical research.* Berlin & Heidelberg, Springer-Verlag, pp. 1–7.

Fusenig, N. E., Limat, A., Stark, H.-J., & Breitkreutz, D. (1994). Modulation of the differentiated phenotype of keratinocytes of the hair follicle and from epidermis. *J. Dermatol. Sci.* **7**: 142–151.

Gao, R., Ustinov, J., Pulkkinen, M. A., Lundin, K., Korsgren, O., & Otonkoski, T. (2003). Characterization of endocrine progenitor cells and critical factors for their differentiation in human adult pancreatic cell culture. *Diabetes* **52**: 2007–2015.

Gard, A. L., & Pfeiffer, S. E. (1990). Two proliferative stages of the oligodendrocyte lineage (A2B5+O4– and O4+GalC–) under different mitogenic control. *Neuron* **5**: 615–625.

Gardner, D. K., & Lane, M. (2003). Towards a single embryo transfer. *Reprod. Biomed. Online* 2003 Jun; **6**(4): 470–481.

Garrido, T., Riese, H. H., Aracil, M., & Perez-Aranda, A. (1995). Endothelial cell differentiation into capillary-like structures in response to tumour cell conditioned medium: A modified chemotaxis chamber assay. *Br. J. Cancer* **71**: 770–775.

Gartler, S. M. (1967). Genetic markers as tracers in cell culture. 2nd Decennial Review Conference on Cell, Tissue and Organ Culture. *NCI Monograph* **26**, pp. 167–195.

Gartner, S., & Kaplan, H. S. (1980). Long-term culture of human bone marrow cells. *Proc. Natl. Acad. Sci. USA* **77**: 4756–4759.

Gaudernack, T., Leivestad, T., Ugelstad, J., & Thorsby, E. (1986). Isolation of pure functionally active CD8+ T cells. Positive selection with monoclonal antibodies directly conjugated to monosized magnetic microspheres. *J. Immunol. Methods* **90**: 179–187.

Gaush, C. R., Hard, W. L., & Smith, T. F. (1966). Characterization of an established line of canine kidney cells (MDCK). *Proc. Soc. Exp. Biol. Med.* **122**: 931–933.

Gazdar, A. F., Carney, D. N., & Minna, J. D. (1983). The biology of nonsmall cell lung cancer. *Semin. Oncol.* **10**: 3–19.

Gazdar, A. F., Carney, D. N., Russell, E. K., Sims, H. L., Baylin, S. B., Bunn, P. A., Guccion, J. G., & Minna, J. D. (1980). Establishment of continuous, clonable cultures of small cell carcinoma of the lung which have amine precursor uptake and decarboxylation properties. *Cancer Res.* **40**: 3502–3507.

Gazdar, A. F., Zweig, M. H., Carney, D. N., Van Steirteghen, A. C., Baylin, S. B., & Minna, J. D. (1981). Levels of creatine kinase and its BB isoenzyme in lung cancer specimens and cultures. *Cancer Res.* **41**: 2773–2777.

Germain, L., Rouabhia, M., Guignard, R., Carrier, L., Bouvard, V., & Auger, F. A. (1993). Improvement of human keratinocyte isolation and culture using thermolysin. *Burns* **19**: 99–104.

Gerper, P., Whang-Peng, J., & Monroe, J. H. (1969). Transformation and chromosome changes induced by Epstein-Barr virus in normal human leukocyte cultures. *Proc. Natl. Acad. Sci. USA* **63**(3): 740–747.

Gey, G. O., Coffman, W. D., & Kubicek, M. T. (1952). Tissue culture studies of the proliferative capacity of cervical carcinoma and normal epithelium. *Cancer Res.* **12**: 364–365.

Ghigo, D., Priotto, C., Migliorino, D., Geromin, D., Franchino, C., Todde, R., Costamagna, C., Pescarmona, G., & Bosia, A. (1998). Retinoic acid-induced differentiation in a human neuroblastoma cell line is associated with an increase in nitric oxide synthesis. *J Cell Physiol.* **174**: 99–106.

Ghosh, C., & Collodi, P. (1994). Culture of cells from zebrafish (*Brachydanio rerio*) blastula-stage embryos. *Cytotechnology* **14**: 21–26.

Ghosh, D., Danielson, K. C., Alston, J. T., & Heyner, S. (1991). Functional differential of mouse uterine epithelial cells grown on collagen gels or reconstituted basement membranes. *In Vitro Cell Dev. Biol.* **27A**: 713–719.

Giard, D. J., Aaronson, S. A., Todaro, G. J., Arnstein, P., Kersey, J. H., Dosik, K., & Parks, W. P. (1972). *In vitro* cultivation of human tumors: Establishment of cell lines derived from a series of solid tumors. *J. Natl. Cancer Inst.* **51**: 1417.

Gignac, S. M., Steube, K., Schleithoff, L., Janssen, J. W., MacLeod, R. A., Quentmeier, H., & Drexler, H. G. (1993). Multiparameter approach in the identification of cross-contaminated leukemia cell lines. *Leukemia Lymphoma* **10**(4–5): 359–368.

Gilbert, S. F., & Migeon, B. R. (1975). d-Valine as a selective agent for normal human and rodent epithelial cells in culture. *Cell* **5**: 11–17.

Gilbert, S. F., & Migeon, B. R. (1977). Renal enzymes in kidney cells selected by d-valine medium. *J. Cell Physiol.* **92**: 161–168.

Gilchrest, B. A., Albert, L. S., Karassik, R. L., & Yaar, M. (1985). Substrate influences human epidermal melanocyte attachment and spreading in vitro. *In Vitro Cell Dev. Biol.,* **21**: 114.

Gillette, R. W. (1987). Alternatives to pristane priming for ascitic fluid and monoclonal antibody production. *J. Immunol. Methods* **99**: 21–23.

Gillis, S., & Watson, J. (1981). Interleukin-2 dependent culture of cytolytic T cell lines. *Immunol. Rev.* **54**: 81–109.

Gimbrone, M. A., Jr., Cotran, R. S., & Folkman, J. (1974). Human vascular endothelial cells in culture, growth and DNA synthesis. *J. Cell Biol.* **60**: 673–684.

Giovanella, B. C., Stehlin, J. S., & Williams, L. J. (1974). Heterotransplantation of human malignant tumors in "nude" mice; II: Malignant tumors induced by injection of cell cultures derived from human solid tumors. *J. Natl. Cancer Inst.* **52**: 921.

Giron, J. A., Lange, M., & Baseman, J. B. (1996). Adherence, fibronectin-binding and induction of cytoskeleton reorganization in cultured human-cells by *Mycoplasma penetrans*. *Infect. Immun.* **64**: 197–208.

Gjerset, R., Yu, A., & Haas, M. (1990). Establishment of continuous cultures of T-cell acute lymphoblastic leukemia cells at diagnosis. *Cancer Res.* **50**: 10–14.

Glavin, G. B., Szabo, S., Johnson, B. R., Xing, P. L., Morales, R. E., Plebani, M., & Nagy, L. (1996). Isolated rat gastric mucosal cells: Optimal conditions for cell harvesting, measures of viability and direct cytoprotection. *J. Pharmacol. Exp. Therapeut.* **276**: 1174–1179.

Gluzman, Y. (1981). SV40-transformed simian cells support the replication of early SV40 mutants. *Cell* **23**: 175–182.

Gobet, R., Raghunath, M., Altermatt, S., Meuli-Simmen, C., Benathan, M., Dietl, A., & Meuli, M. (1997). Efficacy of cultured epithelial autografts in pediatric burns and reconstructive surgery. *Surgery* **121**: 546–661.

Goding, C. R., & Fisher, D. E. (1997). Meeting review—regulation of melanocyte differentiation and growth. *Cell Growth Differ.* **8**: 935–940.

Goldberg, B. (1977). Collagen synthesis as a marker for cell type in mouse 3T3 lines. *Cell* **11**: 169–172.

Goldman, B. I., & Wurzel, J. (1992). Effects of subcultivation and culture medium on differentiation of human fetal cardiac myocytes. *In Vitro Cell Dev. Biol.* **28A**: 109–119.

Goldstein, S., Murano, S., Benes, H., Moerman, E. J., Jones, R. A., & Thweatt, R. (1989). Studies on the molecular genetic basis of replicative senescence in Werner syndrome and normal fibroblasts. *Exp. Gerontol.* **24**(5–6): 461–468.

Goldwasser, E. (1975). Erythropoietin and the differentiation of red blood cells. *Fed. Proc.* **34**: 2285–2292.

Gomm, J. J., Coope, R. C., Browne, P. J., and Coombes, R. C. (1997). Separated human breast epithelial and myoepithelial cells have different growth factor requirements *in vitro* but can reconstitute normal breast lobuloalveolar structure. *J. Cell Physiol.* **171**: 11–19.

Good, N. E., Winget, G. D., Winter, W., Connolly, T. N., Izawa, S., & Singh, R. M. M. (1966). Hydrogen ion buffers and biological research. *Biochemistry* **5**: 467–477.

Goodwin, G., Shaper, J. H., Abezoff, M. D., Mendelsohn, G., & Baylin, S. B. (1983). Analysis of cell surface proteins delineates a differentiation pathway linking endocrine and nonendocrine human lung cancers. *Proc. Natl. Acad. Sci. USA* **80**: 3807–3811.

Gordon, K. E., Ireland, H., Roberts, M., Steeghs, K., McCaul, J. A., MacDonald, D. G., & Parkinson, E. K. (2003). High levels of telomere dysfunction bestow a selective disadvantage during the progression of human oral squamous cell carcinoma. *Cancer Res.* **63**: 458–467.

Gospodarowicz, D., Rudland, P., Lindstrom, J., & Benirschke, K. (1975). Fibroblast growth factor: its localization, purification, mode of action, and physiological significance. *Adv. Metab. Disord.* **8**: 301–335.

Gospodarowicz, D. (1974). Localization of fibroblast growth factor and its effect alone and with hydrocortisone on 3T3 cell growth. *Nature* **249**: 123–127.

Gospodarowicz, D., & Moran, J. (1974). Growth factors in mammalian cell cultures. *Annu. Rev. Biochem.* **45**: 531–558.

Gospodarowicz, D., Delgado, D., & Vlodavsky, I. (1980). Permissive effect of the extracellular matrix on cell proliferation *in vitro*. *Proc. Natl. Acad. Sci. USA* **77**: 4094–4098.

Gospodarowicz, D., Greenburg, G., & Birdwell, C. R. (1978b). Determination of cell shape by the extracellular matrix and its correlation with the control of cellular growth. *Cancer Res.* **38**: 4155–4171.

Gospodarowicz, D., Greenburg, G., Bialecki, H., & Zetter, B. R. (1978a). Factors involved in the modulation of cell proliferation *in vivo* and *in vitro*: The role of fibroblast and epidermal growth factors in the proliferative response of mammalian cells. *In Vitro* **14**: 85–118.

Goto, S., Miyazaki, K., Funabiki, T., & Yasumitsu, H. (1999). Serum-free culture conditions for analysis of secretory proteinases during myogenic differentiation of mouse C2C12 myoblasts. *Anal Biochem.* **272**: 135–142.

Gougos, A., & Letarte, M. (1990) *J Biol Chem* **265**: 8361–8364.

Gower, W. R., Risch, R. M., Godellas, C. V., & Fabri, P. J. (1994). HPAC, a new human glucocorticoid sensitive pancreatic ductal adenocarcinoma cell line. *In Vitro Cell Dev. Biol.* **30A**: 151–161.

Grafström, R. C. (1990a). *In vitro* studies of aldehyde effects related to human respiratory carcinogenesis. *Mutation Res.* **238**: 175–184.

Grafström, R. C. (1990b). Carcinogenesis studies in human epithelial tissues and cells *in vitro*: Emphasis on serum-free culture conditions and transformation studies. *Acta Physiol. Scand.* **140**: 93–133.

Grafström, R. C. (2002). Human oral epithelium. In Freshney, R. I. & Freshney, M. G. (eds.), *Culture of epithelial cells*, 2nd ed. Hoboken, NJ, Wiley-Liss, pp. 195–255.

Graham, F. L., & Van der Eb, A. J. (1973). A new technique for the assay of infectivity of human adenovirus 5 DNA. *Virology* **52**: 456–461.

Graham, F. L., Smiley, J., Russell, W. C., & Nairn, R. (1977). Characteristics of a human cell line transformed by DNA from human adenovirus type. *J. Gen. Virol.* **36**: 59–72.

Graham, G. J. Freshney, M. G., Donaldson, D., & Pragnell, I. B. (1992). Purification and biochemical characterisation of human and murine stem cell inhibitors. *Growth Factors* **7**: 151–160.

Grampp, G. E., Sambanis, A., & Stephanopoulos, G. N. (1992). Use of regulated secretion in protein production from animal cells: An overview. *Adv. Biochem. Eng. Biotechnol.* **46**: 35–62.

Grandolfo, M., d'Andrea, P., Paoletti, S., Martina, M., Silvestrini, G., Bonucci, E., & Vittur, F. (1993). Culture and differentiation of chondrocytes entrapped in alginate beads. *Calcif. Tissue Int.* **52**: 131–138.

Granner, D. K., Hayashi, S., Thompson, E. B., & Tomkins, G. M. (1968). Stimulation of tyrosine aminotransferase synthesis by dexamethasone phosphate in cell culture. *J. Mol. Biol.* **35**: 291–301.

Graziadei, P. P. C., & Monti Graziadei, G. A. (1979). Neurogenesis and neuron regeneration in the olfactory system of mammals; I: Morphological aspects of differentiation and structural organisation of the olfactory sensory neurons. *J. Neurocytol.* **8**: 1–18.

Graziadei, P. P. C., & Monti Graziadei, G. A. (1980). Neurogenesis and neuron regeneration in the olfactory system of mammals; III: Deafferentation and reinnervation of the olfactory bulb following section of the *fila olfactoria* in rat. *J. Neurocytol.* **9**: 145–162.

Greco, B., & Recht, L. (2003). Somatic plasticity of neural stem cells: Fact or fancy? *J. Cell. Biochem.* **88**: 51–56.

Green, A. E., Athreya, B., Lehr, H. B., & Coriell, L. L. (1967). Viability of cell cultures following extended preservation in liquid nitrogen. *Proc. Soc. Exp. Biol. Med.* **124**: 1302–1307.

Green, D. F., Hwang, K. H., Ryan, U. S., & Bourgoignie, J. J. (1992). *Kidney Int* **41**: 1506–1516.

Green, H., & Kehinde, O. (1974). Sublines of mouse 3T3 cells that accumulate lipid. *Cell* **1**: 113–116.

Green, H., & Thomas, J. (1978). Pattern formation by cultured human epidermal cells: Development of curved ridges resembling dermatoglyphs. *Science* **200**: 1385–1388.

Green, H., Kehinde, O., & Thomas, J. (1979). Growth of cultured human epidermal cells into multiple epithelia suitable for grafting. *Proc. Natl. Acad. Sci. USA* **76**: 5665–5668.

Greenberger, J. S. (1980). Self-renewal of factor-dependent haemopoietic progenitor cell lines derived from long-term bone marrow cultures demonstrate significant mouse strain genotypic variation. *J. Supramol. Struct.* **13**: 501–511.

Greene, L. A., & Tischler, A. S. (1976). Establishment of a noradrenergic clonal line of rat adrenal pheochromocytoma cells which respond to nerve growth factor. *Proc Natl Acad Sci USA*, **73**: 2424–2428.

Greider, C. W., & Blackburn, E. H. (1996). Telomeres, telomerase, and cancer. *Sci.c Am.*, February 1996.

Griffiths, B. (2001). Scale-up of suspension and anchorage-dependent animal cells. *Mol. Biotechnol.* **17**: 225–238.

Griffiths, B. (2000). Scaling up of animal cell cultures. In Masters, J. R. W. (ed.), *Animal Cell Culture, a Practical Approach*, Oxford, U.K., Oxford University Press, pp. 19–67.

Griffiths, J. B., & Pirt, G. J. (1967). The uptake of amino acids by mouse cells (Strain LS) during growth in batch culture and chemostat culture: The influence of cell growth rate. *Proc. R., Soc. Biol.* **168**: 421–438.

Griffiths, J. B. (1991). Products from animal cells. In Butler, M. (ed.), *Mammalian cell biotechnology, a practical approach*, Oxford, U.K., IRL Press at Oxford University Press, pp. 207–235.

Grizzle, W. E., & Polt, S. H. (1988). Guidelines to avoid personal contamination by infective agents in research laboratories that use human tissue. *J. Tissue Cult. Methods* **11**: 191–200.

Gross, S. K., Lyerla, T. A., Williams, M. A., & McCluer, R. H. (1992). The accumulation and metabolism of glycosphingolipids in primary kidney cell cultures from beige mice. *Mol. Cell Biochem.* **118**: 61–66.

Groves, A., Barnett, S. C., Franklin, R. J. M., Crang, A. J., Mayer, M., Blakemore, W. & Nobel, M. (1993) Repair of demyelinated lesions by transplants of purified 0-2A progenitor cells. *Nature* **362**: 453–455.

Gugel, E. A., & Sanders, J. E. (1986). Needle-stick transmission of human colonic adenocarcinoma. *New Engl. J., Med.* **315**: 1487.

Guguen-Guillouzo, C. (2002). Isolation and culture of animal and human hepatocytes. In Freshney, R. I. and Freshney, M. G. (eds.), *Culture of epithelial cells*, 2nd ed., Hoboken, NJ, Wiley-Liss, pp. 337–379.

Guguen-Guillouzo, C., & Guillouzo, A. (1986). Methods for preparation of adult and fetal hepatocytes. In Guguen-Guillouzo, C., & Guillouzo, A. (eds.), *Isolated and cultured hepatocytes*. Paris, Les Éditions INSERM, John Libbey Eurotext, pp. 1–12.

Guguen-Guillouzo, C., Campion, J. P., Brissot, P., Glaise, D., Launois, B., Bourel, M., & Guillouzo, A. (1982). High yield preparation of isolated human adult hepatocytes by enzymatic profusion of the liver. *Cell Biol. Int. Rep.* **6**: 625–628.

Guguen-Guillouzo, C., Clement, B., Baffet, G., Beaument, C., Morel-Chany, E., Glaise, D., & Guillouzo, A. (1983). Maintenance and reversibility of active albumin secretion by adult rat hepatocytes co-cultured with another liver epithelial cell type. *Exp. Cell Res.* **143**: 47–54.

Guilbert, L. I., & Iscove, N. N. (1976). Partial replacement of serum by selenite, transferrin, albumin and lecithin in haemopoietic cell cultures. *Nature* **263**: 594–595.

Guillouzo, A. M. A. (1989). *Méthodes in vitro en pharmacotoxiocologie*. Les Éditions INSERM 170: 200.

Guiraud, J. M., Beuron, F., Sion, B., Brassier, G., Faivre, J., Thieulant, M. L., & Duval, J. (1991). Human prolactin producing pituitary adenomas in three dimensional culture. *In Vitro Cell Dev. Biol.* **27a**: 188–190.

Gullino, P. M., & Knazek, R. A. (1979). Tissue culture on artificial capillaries. In Jakoby, W. B., & Pastan, I. (eds.), *Methods in Enzymology, Vol. 58: Cell Culture*. New York, Academic Press, pp. 178–184.

Gumbiner, B. (1992). Epithelial morphogenesis. *Cell* **69**: 385–387.

Gumbiner, B. M. (1995). Signal transduction of betacatenin. *Curr. Opin. Cell Biol.* **7**: 634–640.

Gumbiner, B., & Simons, K. (1986). A functional assay for proteins involved in establishing an epithelial occluding barrier: Identification of a uvomorulin-like polypeptide. *J. Cell Biol.* **102**: 457–468.

Guner, M., Freshney, R. I., Morgan, D., Freshney, M. G., Thomas, D. G. T., & Graham, D. I. (1977). Effects of dexamethasone and betamethasone on *in vitro* cultures from human astrocytoma. *Br. J. Cancer* **35**: 439–447.

Gupta, K., Ramakrishnan, S., Browne, P. V., Solovey, A., & Hebbel, R. P. (1997). A novel technique for culture of human dermal microvascular endothelial cells under either serum-free or serum-supplemented conditions: Isolation by panning and

stimulation with vascular endothelial growth factor. *Exp. Cell Res.* **230**: 244–251.

Gustafson, C. J., Eldh, J., & Kratz, G. (1998). Culture of human urothelial cells on a cell-free dermis for autotransplantation. *Eur. Urol.* **33**: 503–506.

Hafny, B. E. L., Bourre, J.-M., & Roux, F. (1996). Synergistic stimulation of gamma-glutamyl transpeptidase and alkaline phosphatase activities by retinoic acid and astroglial factors in immortalised rat brain microvessel endothelial cells. *J. Cell. Physiol.* **167**: 451–460.

Halaban, R. (2004). Culture of melanocytes from normal, benign, and malignant lesions. In Pfragner, R., & Freshney, R. I. (eds.), *Culture of human tumor cells*. New York, Wiley-Liss, pp. 289–318.

Hallermeyer, K., & Hamprecht, B. (1984). Cellular heterogeneity in primary cultures of brain cells revealed by immunocytochemical localisation of glutamyl synthetase. *Brain Res.* **295**: 1–11.

Halleux, C., & Schneider, Y.-J. (1994). Iron absorption by Caco-2 cells cultivated in serum-free medium as *in vitro* model of the human intestinal epithelial barrier. *J. Cell. Physiol.* **158**: 17–28.

Ham, R. G. (1963). An improved nutrient solution for diploid Chinese hamster and human cell lines. *Exp. Cell Res.* **29**: 515.

Ham, R. G. (1965). Clonal growth of mammalian cells in a chemically defined synthetic medium. *Proc. Natl. Acad. Sci. USA* **53**: 288.

Ham, R. G. (1984). Growth of human fibroblasts in serum-free media. In Barnes, D. W., Sirbasku, D. A., & Sato, G. H. (eds.), *Cell culture methods for molecular and cell biology*, Vol. 3. New York, Alan R. Liss, pp. 249–264.

Ham, R. G., & McKeehan, W. L. (1978). Development of improved media and culture conditions for clonal growth of normal diploid cells. *In Vitro* **14**: 11–22.

Ham, R. G., & McKeehan, W. L. (1979). Media and growth requirements. In Jakoby, W. B., & Pastan, I. H. (eds.), *Methods in enzymology; vol. 58: Cell culture*. New York, Academic Press, pp. 44–93.

Hamburger, A. W., & Salmon, S. E. (1977). Primary bioassay of human tumor stem cells. *Science* **197**: 461–463.

Hamburger, A. W., Salmon, S. E., Kim, M. B., Trent, J. M., Soehnlen, B., Alberts, D. S., & Schmidt, H. J. (1978). Direct cloning of human ovarian carcinoma cells in agar. *Cancer Res.* **38**: 3438–3444.

Hames, B. D., & Glover, D. M. (1991). *Oncogenes*. Oxford, U.K., IRL Press at Oxford University Press.

Hamilton, T. C., Young, R. C., McKoy, W. M., Grotzinger, K. R., Green, J. A., Chu, E. W., Whang-Peng, J., Rogan, A. M., Green, W. R., & Ozols, R. F. (1983). Characterization of a human ovarian carcinoma cell line (NIH:OVCAR-3) with androgen and estrogen receptors. *Cancer Res.* **43**: 5379–5389.

Hamilton, W. G., & Ham, R. G. (1977). Clonal growth of Chinese hamster cell lines in protein-free media. *In Vitro* **13**: 537–547.

Hammond, S. L., Ham, R. G., & Stampfer, M. R. (1984). Serum free growth of human mammary epithelial cells: Rapid clonal growth in defined medium and extended serial passage with pituitary extract. *Proc. Natl. Acad. Sci. USA* **81**: 5435–5439.

Hanks, J. H., & Wallace, R. E. (1949). Relation of oxygen and temperature in the preservation of tissues by refrigeration. *Proc. Exp. Biol. Med.* **71**: 196.

Hansson, B., Rönnbäck, L., Persson, L. I., Lowenthal, A., Noppe, M., Alling, C., & Karlsson, B. (1984). Cellular composition of primary cultures from cerebral cortex, striatum, hippocampus, brain-stem and cerebellum. *Brain Res.* **300**: 9–18.

Hardman, P., Klement, B. J., & Spooner, B. S. (1990). Growth and morphogenesis of embryonic mouse organs on Biopore membrane. *In Vitro Cell Dev. Biol.* **26**: 119–120.

Harrington, W. N., & Godman, G. C. (1980). A selective inhibitor of cell proliferation from normal serum. *Proc. Natl. Acad. Sci. USA Biol. Sci.* **77**: 423–427.

Harris, H., & Watkins, J. F. (1965). Hybrid cells from mouse and man: Artificial heterokaryons of mammalian cells from different species. *Nature* **205**: 640–646.

Harris, L. W., & Griffiths, J. B. (1977). Relative effects of cooling and warming rates on mammalian cells during the freeze–thaw cycle. *Cryobiology* **14**: 662–669.

Harris, M. (1959). Essential growth factor in serum dialysate for chick skeletal muscle fibroblasts. *Proc. Soc. Exp. Biol. Med.* **102**: 468.

Harrison, R. G. (1907). Observations on the living developing nerve fiber. *Proc. Soc. Exp. Biol. Med.* **4**: 140–143.

Hart, I. R., & Fidler, I. J. (1978). An *in vitro* quantitative assay for tumor cell invasion. *Cancer Res.* **38**: 3218–3224.

Hartley, R. S., & Yablonka-Reuveni, Z. (1990). Long-term maintenance of primary myogenic cultures on a reconstituted basement membrane. *In Vitro Cell Dev. Biol.* **26**: 955–961.

Hashimoto, N. (2004). Stem cell systems in skeletal muscle. *Tanpakushitsu Kakusan Koso* **49**: 741–748.

Hass, R., Gunji, H., Hirano, M., Weichselbaum, R., & Kufe, D. (1993). Phorbol ester-induced monocytic differentiation is associated with G2 delay and down regulation of cdc25 expression. *Cell Growth Differ.* **4**: 159–166.

Hassell, J. R., Robey, P. G., Barrach, H. J., Wilczek, J., Rennard, S. I., & Martin, G. R. (1980). Isolation of a heparan sulfate-containing proteoglycan from basement membrane. *Proc. Natl. Acad. Sci. USA* **77**: 4494–4498.

Haudenschild, C. C., Zahniser, D., Folkman, J., & Klagsbrun, M. (1976). Human vascular endothelial cells in culture. *Exp. Cell Res.* **98**: 175–183.

Hauschka, S. D., & Konigsberg, I. R. (1966). The influence of collagen on the development of muscle clones. *Proc. Natl. Acad. Sci. USA* **55**: 119–126.

Häuselmann, H. J., Fernandes, R. J., Mok, S. S., Schmid, T. M., Block, J. A., Aydelotte, M. B., Kuettner, K. E., & Thonar, E. J. M. A. (1994). Phenotypic stability of bovine articular chondrocytes after long-term culture in alginate beads. *J. Cell Sci.* **107**: 17–27.

Häuselmann, H. J., Masuda, K., Hunziker, E. B., Neidhart, M., Mok, S. S., Michel, B. A., & Thonar, E. J.-M. A. (1996). Adult human chondrocytes cultured in alginate form a matrix similar to native human cartilage. *Am. J. Physiol. Cell Physiol.* **40**: C742–C752.

Hawley-Nelson, P., Ciccarone, V., Gebevehu, G., Jessee, J. (1993). A new polycationic liposome reagent with enhanced activity for transfection. *FASEB J.* **7**: A167.

Hay, E. D. (ed.). (1991). *Cell biology of extracellular matrix*. New York, Plenum Press.

Hay, R. J. (1979). Identification, separation and culture of mammalian tissue cells. In Reid, E. (ed.), *Methodological surveys in biochemistry; Vol. 8, Cell Populations*. London, Ellis Horwood, pp. 143–160.

Hay, R. J. (2000). Cell line preservation and characterization. In Masters, J. R. W. (ed.). *Animal cell culture, a practical approach*. Oxford, U.K., IRL Press at Oxford University Press, pp. 95–148.

Hay, R. J., & Cour, I. (1997). Testing for microbial contamination: Bacteria and fungi. In Doyle, A., Griffiths, J. B., & Newall, D. G. (eds.), *Cell and tissue culture: Laboratory procedures*. Chichester, U.K., Wiley, Module 7A: 2.

Hay, R. J., & Strehler, B. L. (1967). The limited growth span of cell strains isolated from the chick embryo. *Exp. Gerontol.* **2**: 123.

Hay, R. J., Caputo, J., & Macy, M. L. (1992). Establishing or verifying cell line identity. In *ATCC quality control methods for cell lines*, 2nd ed. Rockville, MD, American Type Culture Collection, pp. 52–66.

Hay, R. J., Miranda-Cleland, M., Durkin, S., & Reid, Y. A. (2000). Cell line preservation and authentication. In Masters, J. R. W. (ed.), *Animal cell culture*. Oxford University Press, Oxford, pp. 69–103.

Hayashi, I., & Sato, G. H. (1976). Replacement of serum by hormones permits growth of cells in a defined medium. *Nature* **259**: 132–134.

Hayflick, L. (1961). The establishment of a line (WISH) of human amnion cells in continuous cultivation. *Exp. Cell Res.* **23**: 14–20.

Hayflick, L., & Moorhead, P. S. (1961). The serial cultivation of human diploid cell strains. *Exp. Cell Res.* **25**: 585–621.

Haynes, L. W. (1999). *The neuron in tissue culture*. Chichester, U.K., John Wiley & Sons.

Heald, K. A., Hail, C. A., & Downing, R. (1991). Isolation of islets of Langerhans from the weanling pig. *Diabetes Res.* **17**: 7–12.

Health and Safety Commission (1985). Approved Code of Practice "The Protection of Persons Against Ionising Radiation Arising from An Work Activity." HMSO, London.

Health and Safety Commission (1991a). *Safe Working and the Prevention of Infection in Clinical Laboratories*. HMSO Publications, P. O. Box 276, London, SW8 5DT, England.

Health and Safety Commission (1991b). *Safe Working and the Prevention of Infection in Clinical Laboratories—Model Rules for Staff and Visitors*. HMSO Publications, P. O. Box 276, London, SW8 5DT, England.

Health and Safety Commission (1992). *Genetically Modified Organisms (Contained Use) Regulations*: SI 1992/3217, HMSO, ISBN 0-11-025332-9.

Health and Safety Commission (1999a). Carcinogens ACOP and Biological agents ACOP. *Control of Substances Hazardous to Health Regulations*, Approved Codes of Practice, L5, HSE Books, HMSO Publications Centre, P. O. Box 276, London SW8 5DT, England.

Health and Safety Commission (1999b). *Control of Substances Hazardous to Health Regulations*. Statutory Instrument No. 437, HSE Books, Sudbury, U.K.

Health Services Advisory Committee (1992). *Safe Disposal of Clinical Waste*. HSE Books, Sudbury, U.K.

Heffelfinger, S. C., Hawkins, H. H., Barrish, J., Taylor, L., & Darlington, G. (1992). SK HEP-1: A human cell line of endothelial origin. *In Vitro Cell Dev. Biol.* **28A**: 136–142.

Heldin, C. H., Westermark, B., & Wasteson, A. (1979). Platelet-derived growth factor: Purification and partial characterization. *Proc. Natl. Acad. Sci. USA* **76**: 3722–3726.

Hemmati-Brivaniou, A., Kelly, O. G., & Melton, D. A. (1994). Follistatin, an antagonist of activin, is expressed in the Spemann organizer and displays direct neuralizing activity. *Cell* **77**: 283–295.

Heyderman, E., Steele, K., & Ormerod, M. G. (1979). A new antigen on the epithelial membrane: Its immunoperoxidase localisation in normal and neoplastic tissue. *J. Clin. Pathol.* **32**: 35–39.

Heyworth, C. M., & Spooncer, E. (1992). *In vitro* clonal assays for murine multipotential and lineage restricted myeloid progenitor cells. In Testa, N. G., & Molineux, G. (eds.), *Haemopoiesis: A practical approach*. Oxford, U.K., IRL Press at Oxford University Press, pp. 37–54.

HHS Publication No. (CDC) 93–8395, 3rd Edition (1993). Biosafety in Microbiological and Biomedical Laboratories. U.S. Department of Health and Human Services, Public Health Service, Centers for Disease Control and Prevention, and National Institutes of Health. U.S. Government Printing Office, Washington.

Hicks, G. G., Chen, J., & Ruley, H. E. (1998). Production and use of retroviruses. In Ravid, K., & Freshney, R. I. (eds.), *DNA transfer to cultured cells*. New York, Wiley-Liss, pp. 1–26.

Higuchi, K. (1977). Cultivation of mammalian cell lines in serum-free chemically defined medium. *Methods Cell. Biol.* **14**: 131.

Hill, B. T. (1983). An overview of clonogenic assays for human tumour biopsies. In Dendy, P. P., & Hill, B. T. (eds.), *Human tumour drug sensitivity testing in vitro*. New York, Academic Press, pp. 91–102.

Hilwig, I., & Gropp, A. (1972). Staining of constitutive heterochromatin in mammalian chromosomes with a new fluorochrome. *Exp. Cell. Res.* **75**: 122–126.

Hince, T. A., & Roscoe, J. P. (1980). Differences in pattern and level of plasminogen activator production between a cloned cell line from an ethylnitrosourea-induced glioma and one from normal adult rat brain. *J. Cell Physiol.* **104**: 199–207.

Hino, H., Tateno, C., Sato, H., Yamasaki, C., Katayama, S., Kohashi, T., Aratani, A., Asahara, T., Dohi, K., & Yoshizato, K. (1999). A long-term culture of human hepatocytes which show a high growth potential and express their differentiated phenotypes. *Biochem. Biophys. Res. Commun.* **256**: 184–189.

Hirai, Y., Takebe, K., Takashina, M., Kobayashi, S., & Take-ichi, M. (1992). Epimorphin: A mesenchymal protein essential for epithelial morphogenesis. *Cell* **69**: 471–481.

Hlubinova, K., Feldsamova, A., & Prachar, J. (1994). Evaluation of two methods for elimination of mycoplasma. *In Vitro Cell Dev. Biol.* **30A**: 21–22.

Hofmann, S., Kaplan, D., Vunjak-Novakovic, G., Meinel, L. (2005). Tissue engineering of bone. In Vunjak-Novakovic, G., & Freshney. R. I., eds., *Culture of cells for tissue engineering*, Hoboken, NJ, Wiley-Liss.

Hoheisel, D., Nitz, T., Franke, H., Wegener, J., Hakvoort, A., Tilling, T., & Galla, H. J. (1998). Hydrocortisone reinforces the blood-brain properties in a serum free cell culture system. *Biochem. Biophys. Res. Commun.* **247**: 312–315.

Holbrook, K. A., & Hennings, H. (1983). Phenotypic expression of epidermal cells *in vitro*: A review. *J. Invest. Dermatol.* **81**: 11s–24s.

Hollenberg, M. D., & Cuatrecasas, P. (1973). Epidermal growth factor: Receptors in human fibroblasts and modulation of action by cholera toxin. *Proc. Natl. Acad. Sci. USA* **70**: 2964–2968.

Holley, R. W., Armour, R., & Baldwin, J. H. (1978). Density-dependent regulation of growth of BSC-1 cells in cell culture: Growth inhibitors formed by the cells. *Proc. Natl. Acad. Sci. USA* **75**: 1864–1866.

Holt, S. E., Shay, J. W., & Wright, W. E. (1996). Refining the telomere–telomerase hypothesis of ageing and cancer. *Nat. Biotechnol.* **14**: 836–839.

Hoober, J. K., Cohen, S. (1967). Epidermal growth factor. I. The stimulation of protein and ribonucleic acid synthesis in chick embryo epidermis. *Biochim. Biophys. Acta* **138**: 347–56.

Hopps, H., Bernheim, B. C., Nisalak, A., Tjio, J. H., & Smadel, J. E. (1963). Biologic characteristics of a continuous cell line derived from the African green monkey. *J. Immunol.* **91**: 416–424.

Horibata, K., & Harris, A. W. (1970). Mouse myelomas and lymphomas in culture. *Exp. Cell Res.* **60**: 61–77.

Horita, A., & Weber, L. J. (1964). Skin penetrating property of drugs dissolved in dimethylsulfoxide (DMSO) and other vehicles. *Life Sci.* **3**: 1389–1395.

Hornsby, P. J., Yang, L., Lala, D. S., Cheng, C. Y. & Salmons, B. (1992). A modified procedure for replica plating of mammalian cells allowing selection of clones based on gene expression. *Biotechniques* **12**: 244–251.

Horster, M. (1979). Primary culture of mammalian nephron epithelia: Requirements for cell outgrowth and proliferation from defined explanted nephron segments. *Pflugers Arch.* **382**: 209–215.

Hotamisligil, G. S., Arner, P., Caro, J. F., Atkinson, R. L., & Speigelman, B. M. (1995). Increased adipose tissue expression of tumor necrosis factor-α in human obesity and insulin resistance. *J. Clin. Invest.* **95**: 2409–2415.

Howard, M., Kessler, S., Chused, T., & Paul, W. E. (1981). Long term culture of normal mouse B lymphocytes. *Proc. Natl. Acad. Sci. USA* **78**: 5788–5792.

Howie Report. (1978). *Code of practice for prevention of infection in clinical laboratories and post-mortem rooms*. London, H.M. Stationery Office.

Hoyer, L. W., de los Santos, R. P., & Hoyer, J. R. (1973). Antihemophilic factor antigen: Localization in endothelial cells by immunofluorescence microscopy. *J. Clin. Invest.* **52**: 2737–2744.

HSE (1996). *A guide to the Genetically Modified Organisms (Contained Use) Regulations 1992 (as amended)*, HSE Books, ISBN 0 7176 1186 8.

Hsu, T. C., & Benirschke, K. (1967). *Atlas of mammalian chromosomes*, Vols. 1–4. New York, Springer.

Hu, G.-F., Riordan, J. F., & Vallee, B. L. (1997). A putative angiogenin receptor in angiogenin-responsive human endothelial cells. *Proc. Natl. Acad. Sci. USA* **94**: 204–209.

Hughes, S. E. (1996). Functional characterization of the spontaneously transformed human umbilical vein endothelial cell line ECV304: Use in an *in vitro* model of angiogenesis. *Exp. Cell Res.* **225**: 171–185.

Hull, R. N., Cherry, W. R., & Weaver, G. W. (1976). The origin and characteristics of a pig kidney cell strain, LLC-PKI. *In Vitro* **12**: 670–677.

Human Cytogenetic Nomenclature: *See* International System for Human Cytogenetic Nomenclature.

Hume, D. A., & Weidemann, M. J. (1980). *Mitogenic lymphocyte transformation*. Amsterdam, Elsevier/North Holland Biomedical Press.

Hunziker, T., & Limat, A. (1999). Cultured keratinocyte grafts. *Curr Probl Dermatol.*, **27**: 57–64.

Huschtscha, L. I., & Holliday, R. (1983). Limited and unlimited growth of SV40-transformed cells from human diploid MRC-5 fibroblasts. *J. Cell Sci.* **63**: 77–99.

Hynes, R. O. (1973). Alteration of cell-surface proteins by viral transformation and by proteolysis. *Proc. Natl. Acad. Sci. USA* **70**: 3170–3174.

Hynes, R. O. (1974). Role of cell surface alterations in cell transformation: The importance of proteases and cell surface proteins. *Cell* **1**: 147–156.

Hynes, R. O. (1992). Integrins: Versatility, modulation, and signaling in cell adhesion. *Cell* **69**: 11–25.

Hyvonen, T., Alakuijala, L., Andersson, L., Khomutov, A. R., Khomutov, R. M. & Eloranta, T. O. (1988). 1-Amino-oxy-3-aminopropane reversibly prevents the proliferation of cultured baby hamster kidney cells by interfering with polyamine synthesis. *J. Biol. Chem.* **263**: 1138–1144.

Ido, H., Harada, K., Futaki, S., Hayashi, Y., Nishiuchi, R., Natsuka, Y., Li, S., Wada, Y., Combs, A. C., Ervasti, J. M., & Sekiguchi, K. (2004). Molecular dissection of the alpha-dystroglycan- and integrin-binding sites within the globular domain of human laminin-10. *J. Biol. Chem.* **279**: 10946–10954.

Iguchi, H., Mizumoto, K., Shono, M., Kono, A., & Takiguchi, S. (2004). Pancreatic cancer-derived cultured cells: genetic alterations and application to an experimental model of pancreatic cancer metastasis. In Pfragner, R., & Freshney. R. I. (eds.), *Culture of human tumor cells*. Hoboken, NJ, Wiley-Liss, pp. 81–96.

Ikonomou, L., Schneider, Y. J., & Agathos, S. N. (2003). Insect cell culture for industrial production of recombinant proteins. *Appl. Microbiol. Biotechnol.* **62**: 1–20.

Illmensee, K., & Mintz, B. (1976). Totipotency and normal differentiation of single teratocarcinoma cells cloned by injection into blastocysts. *Proc. Natl. Acad. Sci. USA* **73**: 549–553.

Ilsie, A. W., & Puck, T. T. (1971). Morphological transformation of Chinese hamster cells by dibutyryl adenosine cycline 3' : 5'-monophosphate and testosterone. *Proc. Natl. Acad. Sci. USA* **2**: 358–361.

Imaizumi, T., Lankford, K. L., Waxman, S. G., Greer, C. A., and Kocsis, J. D. (1998). Transplanted olfactory ensheathing cells remyelinate and enhance axonal conduction in the demyelinated dorsal columns of the rat spinal cord. *J. Neurosci.* **18**: 6176–6185.

Inamatsu, M., Matsuzaki, T., Iwanari, H., & Yoshizato, K. (1998). Establishment of rat dermal papilla cell lines that sustain the potency to induce hair follicles from afollicular skin. *J. Invest. Dermatol.* **111**: 767–775.

Inokuchi, S., Handa, H., Imai, T., Makuuchi, H., Kidokoro, M., Tohya, H., Aizawa, S., Shimamura, K., Ueyama, Y., Mitomi, T., & Sawada, Y. (1995). Immortalisation of human oesophageal epithelial cells by a recombinant SV40 adenovirus vector. *Br. J. Cancer* **71**: 819–825.

Inoué, S., & Spring, K. R. (1997). *Video microscopy*, 2nd ed. New York and London, Plenum Press.

International Human Genome Sequencing Consortium (2001). Initial sequencing and analysis of the human genome. *Nature*, **409**: 860–921.

International System for Human Cytogenetic Nomenclature (1978). *Report of the Standing Committee on Human Cytogenetic Nomenclature*. Washington, DC, National Foundation of the March of Dimes.

Ireland, G. W., Dopping-Hepenstal, P. J., Jordan, P. W., & O'Neill, C. H. (1989). Limitation of substratum size alters cytoskeletal organization and behaviour of Swiss 3T3 fibroblasts. *Cell Biol. Int. Rep.* **13**: 781–790.

Iscove, N. N., Guilbert, L. W., & Weyman, C. (1980). Complete replacement of serum in primary cultures of erythropoietin-dependent red cell precursors (CFU-E) by albumin, transferrin, iron, unsaturated fatty acid, lecithin and cholesterol. *Exp. Cell Res.* **126**: 121–126.

Iscove, N., & Melchers, F. (1978). Complete replacement of serum by albumin, transferrin and soybean lipid in cultures of lipopolysaccharide-reactive B lymphocytes. *J. Exp. Med.* **147**: 923–933.

Ishida, Y., Levin, J., Baker, G., Stenberg, P., Yamada, Y., Sasaki, H., & Inoué, T. (1993). Biological and biochemical characteristics of murine megakaryoblastic cell line L8057. *Exp. Hematol.* **21**: 289–298.

Itagaki, A., & Kimura, G. (1974). TES and HEPES buffers in mammalian cell cultures and viral studies: Problems of carbon dioxide requirements. *Exp. Cell Res.* **83**: 351–360.

Iyer, V. R., Eisen, M. B., Ross, D. T., Schuler, G., Moore, T., Lee, J. C. F., Trent, J. M., Staudt, L. M., Hudson, J., Jr., Boguski, M. S., Lashkari, D., Shalon, D., Botstein, D., & Brown, P. O. (1999). The transcriptional program in the response of human fibroblasts to serum. *Science* **283**: 83–87.

Izutsu, K. T., Fatherazi, S., Belton, C. M., Oda, D., Cartwright, F. D., & Kenny, G. E. (1996). Mycoplasma orale infection affects K and Cl currents in the HSG salivary gland cell line. *In Vitro Cell Dev. Biol. Anim.* **32**: 361–365.

Jacobs, J. P. (1970). Characteristics of a human diploid cell designated MRC-5. *Nature* **227**: 168–170.

Jacobs, J. P., Garrett, A. J., & Meron, R. (1979). Characteristics of a serially propagated human diploid cell designated MRC-9. *J. Biol. Stand.* **7**: 113–122.

Jaffe, E. A., Nachman, R. L., Becker, G. C., & Ninick, C. R. (1973). Culture of human endothelial cells derived from umbilical veins. *J. Clin. Invest.* **52**: 2745–2756.

Jain, D., Ramasubramamanyan, K., Gould, S., Lenny, A., Candelore, M., Tota, M., Strader, C., Alves, K., Cuca, C., Tung, J. S., Hunt, G., Junker, B., Buckland, B. C., & Silberklang, M. (1991). In Speir, R. E., Griffiths, J. B., & Meignier, B. (eds.), *Production of biologicals from animal cells in culture*. Oxford, U.K., Butterworth-Heinemann, pp. 345–351.

Jain, R. K. Schlenger, K., Hockel, M., & Yuan, F. (1997). Quantitative angiogenesis assays: Progress and problems. *Nat. Med.* **3**: 1203–1208.

Janssen, M. I., van Leeuwen, M. B., Scholtmeijer, K., van Kooten, T. G., Dijkhuizen, L., & Wosten, H. A. (2003). Coating with genetic engineered hydrophobin promotes growth of fibroblasts on a hydrophobic solid. *Biomaterials* **23**: 4847–4854.

Jat, P. S., Noble, M. D., Ataliotis, P., Tanaka, Y., Yannoutsos, N., Larsen, L., & Kioussis, D. (1991). Direct derivation of conditionally immortal cell lines from an H-2 Kb-tsA58 transgenic mouse. *Proc. Natl. Acad. Sci. USA* **88**: 5096–5100.

Jeffreys, A. J., Wilson, V., & Thein, S. L. (1985). Individual specific "fingerprints" of human DNA. *Nature* **316**: 76–79.

Jeng, Y.-J., Watson, C. S., & Thomas, M. L. (1994). Identification of vitamin D-stimulated phosphatase in IEC-6 cells, a rat small intestine crypt cell line. *Exp. Cell Res.* **212**: 338–343.

Jenkins, N. (ed.). (1992). *Growth factors, a practical approach.* Oxford, U.K., IRL Press at Oxford University Press.

Jessell, T. M., & Melton, D. A. (1992). Diffusible factors in vertebrate embryonic induction. *Cell* **68**: 257–270.

Jin, D. I., Lee, S. H., Choi, J. H., Lee, J. S., Lee, J. E., Park, K. W., Seo, J. S. (2003). Targeting efficiency of a-1,3-galactosyl transferase gene in pig fetal fibroblast cells. *Exp Mol Med.* **35**: 572–7.

Jinard, F., Sergent-Engelen, T., Trouet, A., Remacle, C., & Schneider, Y.-J. (1997). Compartment coculture of porcine arterial endothelial and smooth muscle cells on a microporous membrane. *In Vitro Cell Dev. Biol. Anim.* **33**: 92–103.

Johnson, G. D. (1989). Immunofluorescence. In Catty, D. (ed.), *Antibodies; Vol. II: A practical approach.* Oxford, U.K., IRL Press at Oxford University Press, pp. 179–200.

Jones, R. J., Matsui, W. H., & Smith, B. D. (2004). Cancer stem cells: are we missing the target? *J. Natl. Cancer Inst.* **96**: 583–585.

Jones, E. L., & Gregory, J. (1989). Immunoperoxidase methods. In Catty, D. (ed.), *Antibodies; Vol. II: A practical approach.* Oxford, U.K., IRL Press at Oxford University Press, pp. 155–177.

Joukov, V., Kaipainen, A., Jeltsch, M., Pajusola, K., Olofsson, B., Kumar, V., Eriksson, U., & Alitalo, K. (1997). Vascular endothelial growth factors VEGF-B and VEGF-C. *J. Cell Physiol.* **173**: 211–215.

Jozan, S., Roche, H., Cheutin, F., Carton, M., & Salles, B. (1992). New human ovarian cell line OVCCR1/sf in serum-free medium. *In Vitro Cell Dev. Biol.* **28A**: 687–689.

Kaartinen, L., Nettesheim, P., Adler, K. B., & Randell, S. H. (1993). Rat tracheal epithelial cell differentiation *in vitro*. *In Vitro Cell Dev. Biol.* **29A**: 481–492.

Kahn, P., & Shin, S.-L. (1979). Cellular tumorigenicity in nude mice: Test of association among loss of cell-surface fibronectin, anchorage independence, and tumor-forming ability. *J. Cell Biol.* **82**: 1.

Kaltenbach, J. P., Kaltenbach, M. H., & Lyons, W. B. (1958). Nigrosin as a dye for differentiating live and dead ascites cells. *Exp. Cell Res.* **15**: 112–117.

Kaminska, B., Kaczmarek, L., & Grzelakowska-Sztabert, B. (1990). The regulation of G_0–S transition in mouse T lymphocytes by polyamines. *Exp. Cell Res.* **191**: 239–245.

Kane, D. A., Warga, R. M., & Kimmel, C. B. (1992). Mitotic domains in the early embryo of the zebrafish. *Nature* **360**: 735–737.

Kao, F. T., & Puck, T. T. (1968). Genetics of somatic mammalian cells; VII: Induction and isolation of nutritional mutants in Chinese hamster cells. *Proc. Natl. Acad. Sci. USA* **60**: 1275–1281.

Kao, W.-Y., & Prockop, D. I. (1977). Proline analogue removes fibroblasts from cultured mixed cell populations. *Nature* **266**: 63–64.

Karenberg, J. R., & Freelander, E. F. (1974). Giemsa technique for the detection of sister chromatid exchanges. *Chromosoma* **48**: 355–360.

Katdare, M., Osborne, M., Telang, N. T. (2004). Soy isoflavone genistein modulates cell cycle progression and induces apoptosis in HER-2/neu oncogene expressing human breast epithelial cells. *Int J Oncol.* **21**: 809–815.

Katoh, M., Ayabe, F., Norikane, S., Okada, T., Masumoto, H., Horike, S., Shirayoshi, Y., & Oshimura, M. (2004). Construction of a novel human artificial chromosome vector for gene delivery. *Biochem Biophys Res Commun.* **321**: 280–290.

Kawa, S., Kimura, S., Hakomori, S., & Igarashi, Y. (1997). Inhibition of chemotactic motility and trans-endothelial migration of human neutrophils by sphingosine 1-phosphate. *FEBS Lett.* **420**: 196–200.

Kawada, H., Fujita, J., Kinjo, K., Matsuzaki, Y., Tsuma, M., Miyatake, H., Muguruma, Y., Tsuboi, K., Itabashi, Y., Ikeda, Y., Ogawa, S., Okano, H., Hotta, T., Ando, K., & Fukuda, K. (2004). Nonhematopoietic mesenchymal stem cells can be mobilized and differentiate into cardiomyocytes after myocardial infarction. *Blood* **104**: 3581–3587.

Kawasaki, E. S. (2004). Microarrays and the gene expression profile of a single cell. *Ann N Y Acad Sci.,* **1020**: 92–100.

Kédinger, M., Simon-Assmann, P., Alexandre, E., & Haffen, K. (1987). Importance of a fibroblastic support for *in vitro* differentiation of intestinal endodermal cells and for their response to glucocorticoids. *Cell Differ.* **20**: 171–182.

Keen, M. J., & Rapson, N. T. (1995). Development of a serum-free culture medium for the large scale production of recombinant protein from a Chinese hamster ovary cell line. *Cytotechnology* **17**: 153–163.

Keilova, H. (1948). The effect of streptomycin on tissue cultures. *Experientia* **4**: 483.

Keith, W. N. (2003). *In situ* analysis of telomerase RNA gene expression as a marker for tumor progression. *Methods Mol. Med.* **75**: 163–176.

Keles, G. E., Berger, M. S., Lim, R., & Zaheer, A. (1992). Expression of glial fibrillary acidic protein in human medulloblastoma cells treated with recombinant glia maturation factor-beta. *Oncol. Res.* **4**: 431–437.

Kelley, D. S., Becker, J. E., & Potter, V. R. (1978). Effect of insulin, dexamethasone, and glucagon on the amino acid transport ability of four rat hepatoma cell lines and rat hepatocytes in culture. *Cancer Res.* **38**: 4591–4601.

Kempson, S. A., McAteer, J. A., Al-Mahrouq, H. A., Dousa, T. P., Dougherty, G. S., & Evan, A. P. (1989). Proximal tubule characteristics of cultured human renal cortex epithelium. *J. Lab. Clin. Med.* **113**: 285–296.

Kenworthy, P., Dowrick, P., Baillie-Johnson, H., McCann, B., Tsubouchi, H., Arakaki, N., Daikuhara, Y., & Warn, R. M. (1992). The presence of scatter factor in patients with metastatic spread to the pleura. *Br. J. Cancer* **66**: 243–247.

Kern, P. A., Knedler, A., & Eckel, R. H. (1983). Isolation and culture of microvascular endothelium from human adipose tissue. *J. Clin. Invest.* **71**: 1822–1829.

Kerr, D. A., Llado, J., Shamblott, M. J., Maragakis, N. J., Irani, D. N., Crawford, T. O., Krishnan, C., Dike, S., Gearhart, J. D., & Rothstein, J. D. (2003). Human embryonic germ cell derivatives facilitate motor recovery of rats with diffuse motor neuron injury. *J. Neurosci.* **23**: 5131–5140.

Khan, M. Z., Spandidos, D. A., Kerr, D. J., McNicol, A. M., Lang, J. C., de Ridder, L., & Freshney, R. I. (1991). Oncogene transfection of mink lung cells: effect on growth characteristics *in vitro* and *in vivo*. *Anticancer Res.* **11**: 1343–1348.

Kibbey, M. C., Royce, L. S., Dym, M., Baum, B. J., & Kleinman, H. K. (1992). Glandular-like morphogenesis of the human submandibular tumour cell line A253 on basement membrane components. *Exp. Cell Res.* **198**: 343–351.

Kiefer, J., Alexander, A., & Farach-Carson, M. C. (2004). Type I collagen-mediated changes in gene expression and function of prostate cancer cells. *Cancer Treat Res.* **118**: 101–124.

Kimhi, Y. H., Palfrey, C., & Spector, I. (1976). Maturation of neuroblastoma cells in the presence of dimethyl sulphoxide. *Proc. Natl. Acad. Sci. USA* **73**: 462–466.

Kingsbury, A., Gallo, V., Woodhams, P. L., & Balazs, R. (1985). Survival, morphology and adhesion properties of cerebellar interneurons cultured in chemically defined and serum-supplemented medium. *Dev. Brain Res.* **17**: 17–25.

Kinoshita, T., Miyajima, A. (2002). Cytokine regulation of liver development. *Biochim Biophys Acta.*, **1592**: 303–12..

Kinsella, J. L., Grant, D. S., Weeks, B. S., & Kleinman, H. K. (1992). Protein kinase C regulates endothelial cell tube formation on basement membrane matrix, Matrigel. *Exp. Cell Res.* **199**: 56–62.

Kirkland, S. C., & Bailey, I. G. (1986). Establishment and characterisation of six human colorectal adenocarcinoma cell lines. *Br. J. Cancer* **53**: 779–785.

Klagsbrun, M., & Baird, A. (1991). A dual receptor system is required for basic fibroblast growth factor activity. *Cell* **67**: 229–231.

Klein, B., Pastink, A., Odijk, H., Westerveld, A., & van der Eb, A. J. (1990). Transformation and immortalization of diploid xeroderma pigmentosum fibroblasts. *Exp. Cell Res.* **191**: 256–262.

Kleinman, H. K., Philp, D., & Hoffman, M. P. (2003). Role of the extracellular matrix in morphogenesis. *Curr Opin Biotechnol.* **14**: 526–532.

Kleinman, H. K., McGoodwin, E. B., Rennard, S. I., & Martin, G. R. (1979). Preparation of collagen substrates for cell attachment: Effect of collagen concentration and phosphate buffer. *Anal. Biochem.* **94**: 308–312.

Kleinsmith, L. J., & Pierce, G. B. (1964). *Cancer Res.* **24**: 1544–1551.

Klement, G., Scheirer, W., & Katinger, H. W. (1987). Construction of a large scale membrane reactor system with different compartments for cells, medium and product. *Dev Biol Stand.* **66**: 221–226.

Klevjer-Anderson, P., & Buehring, G. C. (1980). Effect of hormones on growth rates of malignant and nonmalignant human mammary epithelia in cell culture. *In Vitro* **16**: 491–501.

Klingel, S., Rothe, G., Kellermann, W., & Valet, G. (1994). Flow cytometric determination of cysteine and serine proteinase activities in living cells with rhodamine 110 substrates. *Methods Cell Biol.* **41**: 449–459.

Klinger, R. Y., & Niklason, L. E. (2005). Tissue engineered blood vessels. In Vunjak-Novakovic, G., & Freshney, R. I. (eds.), *Culture of cells for tissue engineering.*, Hoboken, NJ, Wiley-Liss (in press).

Klöppinger, M., Fertig, G., Fraune, E., & Miltenburger, H. G. (1991). High cell density perfusion culture of insect cells for production of baculovirus and recombinant protein. In Speir, R. E., Griffiths, J. B., & Meignier, B. (eds.), *Production of biologicals from animal cells in culture*. Oxford, U.K., Butterworth-Heinemann, pp. 470–474.

Klug, C. A., & Jordan, C. T. (2002). *Hematopoietic stem cell protocols.* Clifton, NJ, Humana Press.

Knazek, R. A., Gullino, P., Kohler, P. O., & Dedrick, R. (1972). Cell culture on artificial capillaries: An approach to tissue growth *in vitro*. *Science* **178**: 65–67.

Knazek, R. A., Kohler, P. O., & Gullino, P. M. (1974). Hormone production by cells grown *in vitro* on artificial capillaries. *Exp. Cell Res.* **84**: 251.

Knedler, A., & Ham, R. G. (1987). Optimized medium for clonal growth of human microvascular endothelial cells with minimal serum. *In Vitro* **23**(7): 481–491.

Kneuchel, R., & Masters, J. R. W. (1999). Bladder Cancer. In Masters, J. R. W., & Palsson, B. (eds.), *Human cell culture*, Vol. I, Dordrecht, Kluwer, pp. 213–230.

Knight, D. J., & Breheny, D. (2002). Alternatives to animal testing in the safety evaluation of products. *Altern. Lab. Anim.* **30**: 7–22.

Knott, C. L., Kuus-Reichel, K., Liu, R., & Wolfert, R. L. (1997). Development of antibodies for diagnostic assays. In Price, C., & Newman, D. (eds.), *Principles and practice of immunoassay*, 2nd ed. New York, Stockton Press, pp. 36–64.

Knowles, B. B., Howe, C. C., & Aden, D. P. (1980). Human hepatocellular carcinoma cell lines secrete the major plasma proteins and hepatitis B surface antigen. *Science* **209**: 497–499.

Kohler, G., & Milstein, C. (1975). Continuous cultures of fused cells secreting antibody of predefined specificity. *Nature* **256**: 495–497.

Kohlhepp, E. A., Condon, M. E., & Hamburger, A. W. (1987). Recombinant human interferon-α enhancement of retinoic acid induced differentiation of HL-60 cells. *Exp. Hematol.* **15**: 414–418.

Kondo, S., Kooshesh, F., & Sauder, D. N. (1997). Penetration of keratinocyte-derived cytokines into basement membrane. *J. Cell Physiol.* **171**: 190–195.

Kondo, T., & Raff, M. (2000). Oligodendrocyte precursor cells reprogrammed to become multipotential CNS stem cells. *Science*, **289**: 1754–7.

Kondo, T., & Raff, M. (2004). Chromatin remodeling and histone modification in the conversion of oligodendrocyte precursors to neural stem cells. *Genes Dev.*, **18**: 2963–72.

Kono, T. (1997). Nuclear transfer and reprogramming. *Rev. Reprod.* **2**: 74–80.

Koopman, L. A., Szuhai, K., van Eendenburg, J. D., Bezrookove, V., Kenter, G. G., Schuuring, E., Tanke, H., & Fleuren, G. J. (1999). Recurrent integration of human papillomaviruses 16, 45, and 67 near translocation breakpoints in new cervical cancer cell lines. *Cancer Res.* **59**: 5615–5624.

Kopper, L., & Hajdu, M. (2004). Tumor stem cells. *Pathol. Oncol. Res.* **10**: 69–73.

Koren, H. S., Handwerger, B. S., & Wunderlich, J. R. (1975). Identification of macrophage-like characteristics in a murine tumor cell line. *J. Immunol.* **114**: 894–897.

Korenberg, J. R., Chen, X. N., Adams, M. D., & Venter, J. C. (1995). Toward a cDNA map of the human genome. *Genomics* **29**: 364–370.

Koschier, F. J., Roth, R. N., Wallace, K. A., Curren, R. D., & Harbell, J. W. (1997). A comparison of three-dimensional human skin models to evaluate the dermal irritation of selected petroleum products. *In Vitro Toxicol.* **10**: 391–405.

Kreisberg, J. L., Sachs, G., Pretlow, T. G. E., & McGuire, R. A. (1977). Separation of proximal tubule cells from suspensions of rat kidney cell by free-flow electrophoresis. *J. Cell. Physiol.* **93**: 169–172.

Krotz, F., Sohn, H. Y., Gloe, T., Plank, C., & Pohl, U. (2003). Magnetofection potentiates gene delivery to cultured endothelial cells. *J Vasc Res.* **40**: 425–34.

Krupp, W., Geiger, K., Schobe, r R., Siegert, G., & Froster, U. G. (2004). Cytogenetic and molecular cytogenetic analyses in diffuse astrocytomas. *Cancer Genet. Cytogenet.* **153**: 32–38.

Kruse, R. H., Puckett, W. H., & Richardson, J. H. (1991). Biological safety cabinetry. *Clin. Microbiol. Rev.* **4**: 207–241.

Kruse, P. F., Jr., Keen, L. N., & Whittle, W. L. (1970). Some distinctive characteristics of high density perfusion cultures of diverse cell types. *In Vitro* **6**: 75–78.

Kucherlapati, R., & Skoultchi, A. (1984). Introduction of purified genes into mammalian cells. *CRC Crit.l Rev. Biochem.* **16**: 349–379.

Kujoth, G. C., & Fahl, W. E. (1997). c-sis/platelet-derived growth factor-B promoter requirements for induction during the 12 − O-tetradecanoylphorbol-13-acetate-mediated megakaryoblastic differentiation of K562 human erythroleukemia cells. *Cell Growth Differ.* **8**: 963–977.

Kuriharcuch, W., & Green, H. (1978). Adipose conversion of 3T3 cells depends on a serum factor. *Proc. Natl. Acad. Sci. USA* **75**: 6107–6110.

Kurtz, J. W., & Wells, W. W. (1979). Automated fluorometric analysis of DNA, protein, and enzyme activities: Application of methods in cell culture. *Anal. Biochem.* **94**: 166.

Labarca, C., & Paigen, K. (1980). A simple, rapid, and sensitive DNA assay procedure. *Anal. Biochem.* **102**: 344–352.

Laferte, S., & Loh, L. C. (1992). Characterization of a family of structurally related glycoproteins expressing beta 1-6-branched asparagine-linked oligosaccharides in human colon carcinoma cells. *Biochem. J.* **283**: 192–201.

Lag, M., Becher, R., Samuelsen, J. T., Wiger, R., Refsnes, M., Huitfeldt, H. S., & Schwarze, P. E. (1996). Expression of CYP2B1 in freshly isolated and proliferating cultures of epithelial rat lung cells. *Exp. Lung Res.* **22**: 627–649.

Lamb, R. F., Hennigan, R. F., Katsanakis, K. D., Turnbull, K., MacKenzie, E. D., Birnie, G. D., & Ozanne, B. W. (1997). AP-1-mediated invasion requires increased expression of the hyaluronan receptor, CD44. *Mol. Cell Biol.* **17**: 963–976.

Lan, S., Smith, H. S., & Stampfer, M. R. (1981). Clonal growth of normal and malignant human breast epithelia. *J. Surg. Oncol.* **18**: 317–322.

Lane, E. B. (1982). Monoclonal antibodies provide specific intramolecular markers for the study of tonofilament organisation. *J. Cell Biol.* **92**: 665–673.

Lang, M. S., Hovenkamp, E., Savelkoul, H. F. J., Knegt, P., & van Ewijk, W. (1995). Immunotherapy with monoclonal antibodies directed against the immunosuppressive domain of p15E inhibits tumour growth. *Clin Exp Immunol.* **102**: 468–475.

Langdon, S. P. (2004). Characterization and authentication of cancer cell lines: an overview. *Methods Mol. Med.* **88**: 33–42.

Lange, W., Brugger, W., Rosenthal, F. M., Kanz, L., & Lindemann, A. (1991). The role of cytokines in oncology. *Int. J. Cell Clon.* **9**: 252–273.

Larin, Z., & Mejia, J. E. (2002). Advances in human artificial chromosome technology. *Trends Genet.* **18**: 313–319.

Larsen, M. C., Brake, P. B., Pollenz, R. S., & Jefcoate, C. R. (2004). Linked expression of Ah receptor, ARNT, CYP1A1, and CYP1B1 in rat mammary epithelia, *in vitro*, is each substantially

elevated by specific extracellular matrix interactions that precede branching morphogenesis. *Toxicol. Sci.* **82**: 46–61.

Lasfargues, E. Y. (1973) Human mammary tumors. In Kruse, P., Patterson, M. K. (eds): *"Tissue Culture Methods and Applications."* New York, Academic Press, pp. 45–50.

Laslett, A. L., Filipczyk, A. A., & Pera, M. F. (2003). Characterization and culture of human embryonic stem cells. *Trends Cardiovasc. Med.* **13**: 295–301.

Lasnitzki, I. (1992). Organ culture. In Freshney, R. I. (ed.), *Animal cell culture, a practical approach.* Oxford, U.K., IRL Press at Oxford University Press, pp. 213–261.

Latt, S. A. (1973). Microfluorometric detection of DNA replication in human metaphase chromosomes. *Proc. Natl. Acad. Sci. USA* **70**: 3395–3399.

Latt, S. A. (1981). Sister chromatid exchange formation. *Annu. Rev. Genet.* **15**: 11–55.

Laug, W. E., Tokes, Z. A., Benedict, W. F., & Sorgente, N. (1980). Anchorage independent growth and plasminogen activator production by bovine endothelial cells. *J. Cell Biol.* **84**: 281–293.

Lavappa, K. S. (1978). Survey of ATCC stocks of human cell lines for HeLa contamination. *In Vitro* **14**(5): 469–475.

Law, L. W., Dunn, T. B., Boyle, P. J., & Miller, J. H. (1949). Observations on the effect of a folic acid antagonist on transplantable lymphoid leukemia in mice. *J. Natl. Cancer Inst.* **10**: 179–192.

Leake, R. E., Freshney, R. I., & Munir, I. (1987). Steroid responses *in vivo* and *in vitro*. In Green, B., & Leake, R. E., (eds.), *Steroid hormones, a practical approach*, Oxford, U.K., IRL Press at Oxford University Press, pp. 205–218.

Lebeau, M. M., & Rowley, J. D. (1984). Heritable fragile sites in cancer. *Nature* **308**: 607–608.

Lechardeur, D., Schwartz, B., Paulin, D., & Scherman, D. (1995). Induction of blood-brain barrier differentiation in a rat brain-derived endothelial cell line. *Exp. Cell Res.* **220**: 161–170.

Lechner, J. F., & LaVeck, M. A. (1985). A serum free method for culturing normal human bronchial epithelial cells at clonal density. *J. Tissue Cult. Methods* **9**: 43–48.

Lechner, J. F., Haugen, A., Autrup, H., McClendon, I. A., Trump, B. F., & Harris, C. C. (1981). Clonal growth of epithelial cells from normal adult human bronchus. *Cancer Res.* **41**: 2294–2304.

Leder, A., & Leder, P. (1975). Butyric acid, a potent inducer of erythroid differentiation in cultured erythroleukemic cells. *Cell* **5**: 319–322.

Le Douarin, N. M., Creuzet, S., Couly, G., Dupin, E. (2004). Neural crest cell plasticity and its limits. *Development* **131**: 4637–50.

Lee, M. W., Choi, J., Yang, M. S., Moon, Y. J., Park, J. S., Kim, H. C., Kim, Y. J. (2004). Mesenchymal stem cells from cryopreserved human umbilical cord blood. *Biochem Biophys Res Commun.*, **320**: 273–8.

Lee, D. R., Kaproth, M. T. & Parks, J. E. (2001). *In vitro* production of haploid germ cells from fresh or frozen-thawed testicular cells of neonatal bulls. *Biol. Reprod.* **65**: 873–878.

Lee, T. H., Baik, M. G., Im, W. B., Lee, C. S., Han, Y. M., Kim, S. J., Lee, K. K., & Choi, Y. J. (1996). Effects of EHS matrix on expression of transgenes in HC11 cells. *In Vitro Cell Dev. Biol. Anim.* **32**: 454–456.

Lehle, K., Buttstaedt, J., & Birnbaum, D. E. (2003). Expression of adhesion molecules and cytokines *in vitro* by endothelial cells seeded on various polymer surfaces coated with titaniumcarboxonitride. *J. Biomed. Mater. Res.* **65A**: 393–401.

Leibo, S. P., Mazur, P. (1971). The role of cooling rates in low-temperature preservation. *Cryobiology,* **8**: 447–52.

Leibovitz, A. (1963). The growth and maintenance of tissue cell cultures in free gas exchange with the atmosphere. *Am. J. Hyg.* **78**: 173–183.

Leigh, I. M., & Watt, F. M. (1994). *Keratinocyte methods.* Cambridge, U.K., Cambridge University Press.

Leighton, J. (1991). Radial histophysiologic gradient culture chamber rationale and preparation. *In Vitro Cell Dev. Biol.* **27A**: 786–790.

Leighton, J., Mark, R., & Rush, G. (1968). Patterns of three-dimensional growth in collagen coated cellulose sponge: Carcinomas and embryonic tissues. *Cancer Res.* **28**: 286–296.

Lemare, F., Steimberg, N., Le Griel, C., Demignot, S., and Adolphe, M. (1998). Dedifferentiated chondrocytes cultured in alginate beads: Restoration of the differentiated phenotype and of the metabolic response to interleukin−1β. *J. Cell Physiol.* **176**: 303–313.

Lennon, D., & Caplan, A. (2005). Mesenchymal stem cells. In Vunjak-Novakovic, G., & Freshney. R. I. (eds.), *Culture of cells for tissue engineering.* Hoboken, NJ, Wiley-Liss (In press).

Léobon, B., Garcin, I., Menasche, P., Vilquin, J. T., Audinat, E., & Charpak, S. (2003). Myoblasts transplanted into rat infarcted myocardium are functionally isolated from their host. *Proc. Natl. Acad. Sci. U S A* **100**: 7808–7811.

Le Poole, I. C., van den Berg, F. M., van den Wijn-gaard, R. M., Galloway, D. A., van Amstel, P. J., Buffing, A. A., Smiths, H. L., Westerhof, W., & Das, P. K. (1997). Generation of a human melanocyte cell line by introduction of HPV16 E6 and E7 genes. *In Vitro Cell Dev. Biol. Anim.* **33**: 42–49.

Le Roith, D., & Raizada, M. K. (eds.). (1989). *Molecular and cellular biology of insulin-like growth factors and their receptors.* New York, Plenum.

Lesuffleur, T., Barbat, A., Dussaulx, E., & Zweibaum, A. (1990). Growth adaptation to methotrexate of HT-29 human colon carcinoma cells is associated with their ability to differentiate into columnar absorptive and mucus-secreting cells. *Cancer Res.* **50**: 6334–6343.

Lever, J. (1986). Expression of differentiated functions in kidney epithelial cell lines. *Min. Elec. Metab.* **12**: 14–19.

Levi-Montalcini, R. (1966). The nerve growth factor: its mode of action on sensory and sympathetic nerve cells. *Harvey Lect.* **60**: 217–259.

Levi-Montalcini, R. C. P. (1979). The nerve-growth factor. *Sci. Am.* **240**: 68.

Levin, D. B., Wilson, K., Valadares de Amorim, G., Webber, J., Kenny, P., & Kusser, W. (1995). Detection of p53 mutations in benign and dysplastic nevi. *Cancer Res.* **55**: 4278–4282.

Levine, E. M., & Becker, B. G. (1977). Biochemical methods for detecting mycoplasma contamination. In McGarrity, G. T., Murphy, D. G., & Nichols, W. W. (eds.), *Mycoplasma infection of cell cultures.* New York, Plenum Press, pp. 87–104.

Ley, K. D., & Tobey, R. A. (1970). Regulation of initiation of DNA synthesis in Chinese hamster cells; II: Induction of DNA synthesis and cell division by isoleucine and glutamine in G_1-arrested cells in suspension culture. *J. Cell Biol.* **47**: 453–459.

Li, Y., Decherchi, P., & Raisman, G. (2003a) Transplantation of olfactory ensheathing cells into spinal cord lesions restores breathing and climbing. *J. Neurosci.* **23**: 727–731.

Li, Y., Foster, W., Deasy, B. M., Chan, Y., Prisk, V., Tang, Y., Cummins, J., & Huard, J. (2004). Transforming growth factor-beta1 induces the differentiation of myogenic cells into fibrotic cells in injured skeletal muscle: a key event in muscle fibrogenesis. *Am. J. Pathol.* **164**: 1007–1019.

Li, Y. M., Schacher, D. H., Liu, Q., Arkins, S., Rebeiz, N., McCusker, R. H., Jr., Dantzer, R., & Kelley, K. W. (1997). Regulation of myeloid growth and differentiation by the insulin-like growth factor I receptor. *Endocrinology.* **138**: 362–368.

Li, A. P., Roque, M. M., Beck, D. J., & Kaminski, D. L. (1992). Isolation and culturing of hepatocytes from human livers. *J. Tissue Cult. Methods* **14**: 139–146.

Li, Y. et al. (2003b) Transplanted olfactory ensheathing cells promote regeneration of cut adult rat optic nerve axons. *J. Neurosci.* **23**: 783–788.

Li, Y., Field, P. M., & Raisman, G. (1997). Repair of adult rat corticospinal tract by transplants of olfactory ensheathing cells. *Science* **277**: 2000–2002.

Lidington, E. A., Rao, R. M., Marelli-Berg, F. M., Jat, P. S., Haskard, D. O., & Mason, J. C. (2002) *Am. J. Physiol. Cell. Physiol.* **282**: C67–C74.

Liebermann, D., & Sachs, L. (1978). Nuclear control of neurite induction in neuroblastoma cells. *Exp. Cell Res.* **113**: 383–390.

Lillie, I. H., MacCallum, D. K., Jepsen, A. (1980) Fine structure of subcultivated stratified squamous epithelium grown on collagen rafts. *Exp Cell Res* **125**: 153–165.

Lim, G., Karaskova, J., Vukovic, B., Bayani, J., Beheshti, B., Bernardini, M., Squire, J. A., & Zielenska, M. (2004). Combined spectral karyotyping, multicolor banding, and microarray comparative genomic hybridization analysis provides a detailed characterization of complex structural chromosomal rearrangements associated with gene amplification in the osteosarcoma cell line MG-63. *Cancer Genet Cytogenet.* **153**: 158–164.

Limat, A., Breitkreutz, D., Thieköttter, G., Klein, E. C., Braathen, L. R., Hunziker, T., & Fusenig, N. E. (1995). Formation of a regular neo-epidermis by cultured human outer root sheath cells grafted on nude mice. *Transplantation* **59**: 1032–1038.

Limat, A., Hunziker, T., Boillat, C., Bayreuther, K., & Noser, F. (1989). Post-mitotic human dermal fibroblasts efficiently support the growth of human follicular keratinocytes. *J. Invest. Dermatol.* **92**: 758–762.

Limat, A., Mauri, D., & Hunziker, T. (1996). Successful treatment of chronic leg ulcers with epidermal equivalents generated from cultured autologous outer root sheath cells. *J. Invest. Dermatol.* **107**: 128–135.

Lin, C. C., & Uchida, I. A. (1973). Fluorescent banding of chromosomes (Q-bands). In Kruse, P. F., & Patterson, M. K. (eds.), *Tissue culture methods and applications.* New York, Academic Press, pp. 778–781.

Lin, M. A., Latt, S. A., & Davidson, R. L. (1974). Identification of human and mouse chromosomes in human−mouse hybrids by centromere fluorescence. *Exp. Cell Res.* **87**: 429–433.

Linser, P., & Moscona, A. A. (1980). Induction of glutamine synthetase in embryonic neural retina-localization in Muller fibers and dependence on cell interaction. *Proc. Natl. Acad. Sci. USA* **76**: 6476–6481.

Liotta, L. (1987). The role of cellular proteases and their inhibitors in invasion and metastasis: Introductionary overview. *Cancer Metastasis Rev.* **9**: 285–287.

Lippincott-Schwartz, J., Glickman, J., Donaldson, J. G., Robbins, J., Kreis, T. E., Seamon, K. B., Sheetz, M. P., & Klausner, R. D. (1991). Forskolin inhibits and reverses the effects of Brefeldin A on Golgi morphology by a cAMP-independent mechanism. *J. Cell Biol.* **112**: 567.

Littlefield, J. W. (1964a). Selection of hybrids from matings of fibroblasts *in vitro* and their presumed recombinants. *Science* **145**: 709–710.

Littlefield, J. W. (1964b). Three degrees of guanylic acid pyrophosphorylase deficiency in mouse fibroblasts. *Nature* **203**: 1142–1144.

Litwin, J. (1973). Titanium disks. In Kruse, P. F., & Patterson, M. K. (eds.), *Tissue culture methods and applications.* New York, Academic Press, pp. 383–387.

Liu, T. F., Cohen, K. A., Willingham, M. C., Tatter, S. B., Puri, R. K., & Frankel, A. E. (2003). Combination fusion protein therapy of refractory brain tumors: demonstration of efficacy in cell culture. *J. Neurooncol.* **65**: 77–85.

Liu, L., Delbe, J., Blat, C., Zapf, J., & Harel, L. (1992). Insulin like growth factor binding protein (IGFBP-3), an inhibitor of serum growth factors other than IGF-I and -II. *J. Cell Physiol.* **153**: 15–21.

Liu, M., Xu, J., Souza, P., Tanswell, B., Tanswell, A. K., & Post, M. (1995). The effect of mechanical strain on fetal rat lung cell proliferation: Comparison of two- and three-dimensional culture systems. *In Vitro Cell Dev. Biol. Anim.* **31**: 858–866.

Lopez-Casillas, F., Wrana, J. L., & Massague, J. (1993). Betaglycan presents ligand to the TGF-β signalling receptor. *Cell* **73**: 1435–1444.

Lotan, R., & Lotan, D. (1980). Stimulation of melanogenesis in a human melanoma cell line by retinoids. *Cancer Res.* **40**: 33–45.

Lounis, H., Provencher, D., Godbout, C., Fink, D., Milot, M.-J., & Mes-Masson, A.-M. (1994). Primary cultures of normal and tumoral ovarian epithelium: A powerful tool for basic molecular studies. *Exp. Cell Res.* **215**: 303–309.

Lovelock, J. E., & Bishop, M. W. H. (1959). Prevention of freezing damage to living cells by dimethyl sulphoxide. *Nature* **183**: 1394–1395.

Lowe, K. C., Davey, M. R., & Power, J. B. (1998). Perfluorochemicals: their applications and benefits to cell culture. *Trends Biotechnol.* **16**: 272–277.

Luikart, S. D., Maniglia, C. A., Furcht, L. T., McCarthy, J. B., & Oegama, T. R. (1990). A heparan sulphate-containing fraction of bone marrow stroma induces maturation of HL-60 cell *in vitro. Cancer Res.* **50**: 3781–3785.

Lundqvist, M., Mark, J., Funa, K., Heldin, N. E., Morstyn, G., Weddell, B., Layton, J., & Oberg, K. (1991). Characterisation of a cell line (LCC-18) from a cultured human neuroendocrine-differentiated colonic carcinoma. *Eur. J. Cancer* **12**: 1662–1668.

Lutz, M. P., Gaedicke, G., & Hartmann, W. (1992). Large-scale cell separation by centrifugal elutriation. *Anal. Biochem.* **200**: 376–380.

Maas-Szabowski, N., Stark, H. J., & Fusenig, N. E. (2000). Keratinocyte growth regulation in defined organotypic cultures through IL-1-induced KGF expression in resting fibroblasts. *J. Invest. Dermatol.* **114**: 1075–1084.

Maas-Szabowski, N., Stark, H. J., & Fusenig, N. E. (2002). Cell interaction and epithelial differentiation. In Freshney, R. I. & Freshney, M. G. (eds.), *Culture of epithelial cells,* 2nd ed. Hoboken, NJ, Wiley-Liss, pp. 31–63.

Maas-Szabowski, N., & Fusenig, N. E. (1996). Interleukin-1-induced growth factor expression in postmitotic and resting fibroblasts. *J. Invest. Dermatol.* **107**: 849–855.

Maas-Szabowski, N., Fusenig, N. E., & Shimotoyodome, A. (1999). Keratinocyte growth regulation in fibroblast co-cultures via a double paracrine mechanism. *J. Cell Sci.* **112**: 1843–1853.

MacDonald, C. M., Freshney, R. I., Hart, E., & Graham, D. I. (1985). Selective control of human glioma cell proliferation by specific cell interaction. *Exp. Cell Biol.* **53**: 130–137.

Macé, K., Gonzalez, F. J., McConnell, I. R., Garner, R. C., Avanti, O., Harris, C. C., & Pfeifer, A. M. A. (1994). Activation of promutagens in a human bronchial epithelial cell line stably expressing human cytochrome P450 1A2. *Mol. Carcinogen.* **11**: 65–73.

MacGregor, G., & Caskey, C. (1989). Construction of plasmids that express *E. coli* beta-galactosidase in mammalian cells. *Nucleic Acids Res.* **17**: 2363–2365.

Maciag, T., Cerondolo, J., Ilsley, S., Kelley, P. R., & Forand, R. (1979). Endothelial cell growth factor from bovine hypothalamus—identification and partial characterization. *Proc. Natl. Acad. Sci. USA* **76**: 5674–5678.

Macieira-Coelho, A. (1973). Cell cycle analysis; A: Mammalian cells. In Kruse, P. F., & Patterson, M. K. (eds.), *Tissue culture methods and applications.* New York, Academic Press, pp. 412–422.

Macklis, J. D., Sidman, R. L., & Shine, H. D. (1985). Cross-linked collagen surface for cell culture that is stable, uniform, and optically superior to conventional surfaces. *In Vitro* **21**: 189–194.

MacLeod, R. A., Dirks, W. G., Matsuo, Y., Kaufmann, M., Milch, H., Drexler, H. G. (1999). Widespread intraspecies cross-contamination of human tumor cell lines arising at source. *Int. J. Cancer,* **83**: 555–63.

Macpherson, I. (1973). Soft agar technique. In Kruse, P. F., & Patterson, M. K. (eds.), *Tissue culture methods and applications.* New York, Academic Press, pp. 276–280.

Macpherson, I., & Bryden, A. (1971). Mitomycin C treated cells as feeders. *Exp. Cell Res.* **69**: 240–241.

Macpherson, I., & Montagnier, L. (1964). Agar suspension culture for the selective assay of cells transformed by polyoma virus. *Virology* **23**: 291–294.

Macpherson, I., & Stoker, M. (1962). Polyoma transformation of hamster cell clones—an investigation of genetic factors affecting cell competence. *Virology* **16**: 147.

Macville, M., Veldman, T., Padilla-Nash, H., Wangsa, D., O'Brien, P., Schrock, E., & Ried, T. (1997). Spectral karyotyping, a 24-colour FISH technique for the identification of chromosomal rearrangements. *Histochemistry* **108**: 299–305.

Macy, M. (1978). Identification of cell line species by isoenzyme analysis. *Manual Am. Tissue Cult. Assoc.* **4**: 833–836.

Magee, J. C., Stone, A. E., Oldham, K. T., & Guice, K. S. (1994) *Am J Physiol Lung Mol. Cell. Physiol.* **267**: L433–L441.

Mahato, R. I., Kawabata, K., Nomura, T., Takakura, Y., & Hashida, M. (1995a). Physicochemical and pharmacokinetic characteristics of plasmid DNA/cationic liposome complexes. *J. Pharmaceut. Sci.* **84**: 1267–1271.

Mahato, R. I., Kawabata, K., Takakura, Y., & Hashida, M. (1995b). *In vivo* disposition characteristics of plasmid DNA complexed with cationic liposomes. *J. Drug Targeting* **3**: 149–157.

Mahdavi, V., & Hynes, R. O. (1979). Proteolytic enzymes in normal and transformed cells. *Biochim. Biophys. Acta* **583**: 167–178.

Mairs, R. J., & Wheldon, T. E. (1996). Experimental tumour therapy with targeted radionuclides in multicellular tumour spheroids. In Hagen, U., Jung, H., & Streffer, C. (eds.), *Radiation research, 1895–1995* (Proceedings of the 10th International Congress of Radiation Research, Wurzburg, Germany).

Mal, A., & Harter, M. L. (2003). MyoD is functionally linked to the silencing of a muscle-specific regulatory gene prior to skeletal myogenesis. *Proc. Natl. Acad. Sci. USA* **100**: 1735–1739.

Maltese, W. A., & Volpe, I. J. (1979). Induction of an oligoden-droglial enzyme in C-6 glioma cells maintained at high density or in serum-free medium. *J. Cell Physiol.* **101**: 459–470.

Management of Health and Safety at Work Directive 89/391/EEC.

Management of Health and Safety at Work Regulations. (1999). Statutory Instrument 1999 No. 3242. The Stationery Office Limited, ISBN 0 11 085625 2.

Mangi, A. A., Noiseux, N., Kong, D., He, H., Rezvani, M., Ingwall, J. S., & Dzau, V. J. (2003). Mesenchymal stem cells modified with Akt prevent remodeling and restore performance of infarcted hearts. *Nat. Med.* **9**: 1195–1201.

Maniatis, T., Hardison, R. C., Lacy, E., Lauer, J., O'Connell, C., Quon, D., Sim, G. K., & Efstradiadis, A. (1978). The isolation of structural genes from libraries of eukaryotic DNA. *Cell* **15**: 687–701.

Maramorosch, K. (1976). *Invertebrate tissue culture.* New York, Academic Press.

Marchionni, M. A., Goodearl, A. D., Chen, M. S., Bermingham-McDonogh, O., Kirk, C., Hendricks, M., Danehy, F., Misumi, D., Sudhalter, J., Kobayashi, K., Wroblewski, D., Lynch, C., Baldassare, M., Hiles, I., Davis, J. B., Hsuan, J. J., Totty, N. F., Otsu, M., McBurney, R. N., Waterfield, M. D., Stroobant, P., & Gwynne, D. (1993). Glial growth factors are alternatively spliced erbB2 ligands expressed in the nervous system. *Nature* **362**: 312–318.

Marcus, M., Lavi, U., Nattenberg, A., Ruttem, S., & Markowitz, O. (1980). Selective killing of mycoplasmas from contaminated cells in cell cultures. *Nature* **285**: 659–660.

Mardh, P. H. (1975). Elimination of mycoplasmas from cell cultures with sodium polyanethol sulphonate. *Nature* **254**: 515–516.

Mareel, M. M., Bruyneel, E., & Storme, G. (1980). Attachment of mouse fibrosarcoma cells to precultured fragments of embryonic chick heart. *Virchows Arch. B Cell Pathol.* **34**: 85–97.

Mareel, M., Kint, J., & Meyvisch, C. (1979). Methods of study of the invasion of malignant C3H-mouse fibroblasts into embryonic chick heart *in vitro*. *Virchows Arch. B Cell Pathol.* **30**: 95–111.

Marelli-Berg, F. M., Peek, E., Lidington, E. A., Stauss, H. J., & Lechler, R. I. (2000) *J. Immunol. Methods* **244**: 205–215.

Marh, J., Tres, L. L., Yamazaki, Y., Yanagimachi, R. & Kierszenbaum, A. L. (2003). Mouse round spermatids developed *in vitro* from preexisting spermatocytes can produce normal offspring by nuclear injection into *in vivo*-developed mature oocytes. *Biol. Reprod.* **69**: 169–176.

Marijnissen, W. J., van Osch, G. J., Aigner, J., van der Veen, S. W., Hollander, A. P., Verwoerd-Verhoef, H. L. & Verhaar, J. A. (2002). Alginate as a chondrocyte-delivery substance in combination with a non-woven scaffold for cartilage tissue engineering. *Biomaterials* **23**: 1511–1517.

Mark, J. (1971). Chromosomal characteristics of neurogenic tumours in adults. *Hereditas* **68**: 61–100.

Markovitz, D., Goff, S., & Bank, A. (1988). A safe packaging line for gene transfer: separating viral genes on two different plasmids. *J. Virol.* **62**: 1120–1124.

Marks, P. A., Richon, V. M., Kiyokawa, H., & Rifkind, R. A. (1994). Inducing differentiation of transformed cells with hybrid polar compounds: a cell cycle-dependent process. *Proc. Natl. Acad. Sci. USA* **91**: 10251–10254.

Markus, G., Takita, H., Camiolo, S. M., Corsanti, J., Evers, J. L., & Hobika, J. H. (1980). Content and characterization of plasminogen activators in human lung tumours and normal lung tissue. *Cancer Res.* **40**: 841–848.

Marsh, J. W., Donovan, M., Burholt, D. R., George, L. D., & Kornblith, P. L. (2004). Metastatic lung disease to the central nervous system: *in vitro* response to chemotherapeutic agents. *J. Neurooncol.* **66**: 81–90.

Marshall, C. J. (1991). Tumor suppressor genes. *Cell* **64**: 313–326.

Marte, B. M., Meyer, T., Stabel, S., Standke, G. J. R., Jaken, S., Fabbro, D., & Hynes, N. E. (1994). Protein kinase C and mammary cell differentiation: Involvement of protein kinase C alpha in the induction of beta-casein expression. *Cell Growth Differ.* **5**: 239–247.

Martin, F. L., Cole, K. J., Williams, J. A., Millar, B. C., Harvey, D., Weaver, G., Grover, P. L., & Phillips, D. H. (2000). Activation of genotoxins to DNA-damaging species in exfoliated breast milk cells. *Mutat. Res.* **470**: 115–124.

Martin, G. R. (1975). Teratocarcinomas as a model system for the study of embryogenesis and neoplasia. *Cell* **5**: 229–243.

Martin, G. R., & Evans, M. J. (1974). The morphology and growth of a pluripotent teratocarcinoma cell line and its derivatives in tissue culture. *Cell* **2**: 163–172.

Martin, G. R. (1978). Advantages and limitations of teratocarcinoma stem cells as models of development. In Johnson, M. H. (ed.), *Development in mammals*, Vol. 3. Amsterdam, North-Holland Publishing, p. 225.

Martin, G. R. (1981). Isolation of a pluripotent cell line from early mouse embryos cultured in medium conditioned by teratocarcinoma stem cells. *Proc. Natl. Acad. Sci. USA* **78**: 7634–7638.

Martinsen, A., Skjåk-Bræk, G., & Smidsrød, O. (1989). Alginate as immobilization material; 1: Correlation between chemical and physical properties of alginate gel beads. *Biotechnol. Bioeng.* **33**: 79–89.

Maskell, R., & Green, M. (1995). Applications of the comet assay technique. *Int. Micr. Lab.* **6**: 2–5.

Massague, J., Cheifetz, S., Laiho, M., Ralph, D. A., Weis, F. M., & Zentella, A. (1992). Transforming growth factor-beta. *Cancer Surveys* **12**: 81–103.

Masson, E. A., Atkin, S. L., & White, M. C. (1993). d-valine selective medium does not inhibit human fibroblast growth *in vitro*. *In Vitro Cell Dev. Biol. Anim.* **29A**: 912–913.

Masters, J. R. W., & Palsson, B. (eds.) (1999–2000). *Human cell culture*, Vol. I–IV, Dordrecht, Kluwer..

Masters, J. R., Bedford, P., Kearney, A., Povey, S., & Franks, L. M. (1988). Bladder cancer cell line cross-contamination: Identification using a locus-specific minisatellite probe. *Br. J. Cancer* **57**(3): 284–286.

Masters, J. R. W., Thomson, J. A., Daly-Burns, B., Reid, Y. A., Dirks, W. G., Packer, P., Toji, L. H., Ohno, T., Tanabe, H., Arlett, C. F., Kelland, L. R., Harrison, M., Virmani, A., Ward, T. H., Ayres, K. L., & Debenham, P. G. (2001). STR profiling provides an international reference standard for human cell lines. *Proc. Natl. Acad. Sci. USA* **98**: 8012–8017.

Masui, T., Lechner, J. F., Yoakum, G. H., Willey, J. C., & Harris, C. C. (1986b). Growth and differentiation of normal and transformed human bronchial epithelial cells. *J. Cell Physiol. (Suppl)* **4**: 73–81.

Masui, T., Wakefield, L. M., Lechner, J. F., LaVeck, M. A., Sporn, M. B., & Harris, C. C. (1986a). Type beta transforming growth factor is the primary differentiation-inducing serum factor for normal human bronchial epithelial cells. *Proc. Natl. Acad. Sci. USA* **83**: 2438–2442.

Mather, J. (1979). Testicular cells in defined medium. In Jakoby, W. B., & Pastan, I. H. (eds.), *Methods in enzymology; Vol. 57: Cell culture.* New York, Academic Press, p. 103.

Mather, J. P. (1998). Making informed choices: Medium, serum, and serum-free medium; how to choose the appropriate medium and culture system for the model you wish to create. *Methods Cell Biol.* **57**: 19–30.

Mather, J. P., & Sato, G. H. (1979a). The growth of mouse melanoma cells in hormone supplemented, serum-free medium. *Exp. Cell Res.* **120**: 191.

Mather, J. P., & Sato, G. H. (1979b). The use of hormone supplemented serum-free media in primary cultures. *Exp. Cell Res.* **124**: 215.

Matsui, A., Zsebo, K., & Hogan, B. L. M. (1992). Derivation of pluripotential embryonic stem cells from murine primordial germ cells in culture. *Cell* **70**: 841–847.

Matsui, Y., Toksoz, D., Nishikawa, S., Williams, D., Zsebo, K., Hogan, B. L. (1991). Effect of steel factor and leukemia inhibitory factor on murine primordial germ cells in culture. *Nature* **353**: 750–752.

Mayne, L. V., Price, T. N. C., Moorwood, K., & Burke, J. F. (1996). Development of immortal human fibroblast cell lines. In Freshney, R. I., & Freshney, M. G. (eds.), *Culture of immortalized cells.* New York, Wiley-Liss, pp. 77–93.

Mayne, L. V., Priestly, A., James, M. R., & Burke, J. F. (1986). Efficient immortalisation and morphological transformation of human fibroblasts with SV40 DNA linked to a dominant marker. *Exp. Cell Res.* **162**: 530–538.

McAteer, J. A., Kempson, S. A., & Evan, A. P. (1991). Culture of human renal cortex epithelial cells. *J. Tissue Cult. Methods* **13**: 143–148.

McCall, E., Povey, J., & Dumonde, D. C. (1981). The culture of vascular endothelial cells on microporous membranes. *Thromb. Res.* **24**: 417–431.

McCormack, S. A., Viar, M. J., Tague, L., & Johnston, L. R. (1996). Altered distribution of the nuclear receptor rar (*β*) accompanies proliferation and differentiation changes caused by retinoic acid in Caco-2 cells. *In Vitro Cell Dev. Biol. Anim.* **32**: 53–61.

McCormick, C., & Freshney, R. I. (2000). Activity of growth factors in the IL-6 group in the differentiation of human lung adenocarcinoma. *Br. J. Cancer* **82**: 881–890.

McCormick, C., Freshney, R. I., & Speirs, V. (1995). Activity of interferon alpha, interleukin 6 and insulin in the regulation of

differentiation in A549 alveolar carcinoma cells. *Br. J. Cancer* **71**: 232–239.

McDouall, R. M., Yacoub, M., and Rose, M. L. (1996). Isolation, culture, and characterisation of MHC class II-positive microvascular endothelial cells from the human heart. *Microvasc Res* **51**: 137–152.

McFarland, D. C., Liu, X., Velleman, S. G., Zeng, C., Coy, C. S., & Pesall, J. E. (2003). Variation in fibroblast growth factor response and heparan sulfate proteoglycan production in satellite cell populations. *Comp. Biochem. Physiol. C Toxicol. Pharmacol.* **134**: 341–351.

McGarrity, G. J. (1982). Detection of mycoplasmic infection of cell cultures. In Maramorosch, K. (ed.), *Advances in cell culture*, Vol. 2. New York, Academic Press, pp. 99–131.

McGowan, J. A. (1986). Hepatocyte proliferation in culture. In Guillouzo, A., & Guguen-Guillouzo, C. (eds.), *Isolated and cultured hepatocytes.* Paris, Les Editions Inserm, John Libbey Eurotext, pp. 13–38.

McGregor, D. B., Edwards, I., Riach, C. J., Cattenach, P., Martin, R., Mitchell, A. & Caspary, W. J. (1988). Studies of an S9 based metabolic activation system used in the mouse lymphoma L51768Y cell mutation assay. *Mutagenesis* **3**: 485–490.

McGuckin, C. P., Forraz, N., Allouard, Q., Pettengell, R. (2004). Umbilical cord blood stem cells can expand hematopoietic and neuroglial progenitors *in vitro. Exp Cell Res.*, **295**: 350–9.

McGuire, P. G., & Orkin, R. W. (1987) *Lab. Invest.* **57**: 94–105.

McIlwrath, A., Vasey, P., Ross, G., & Brown, R. (1994). Cell cycle arrests and radiosensitivity of human tumour cell lines: Dependence on wild-type p53 for radiosensitivity. *Cancer Res.* **54**: 3718–3722.

McKay, I., & Taylor-Papadimitriou, J. (1981). Junctional communication pattern of cells cultured from human milk. *Exp. Cell Res.* **134**: 465–470.

McKeehan, W. L. (1977). The effect of temperature during trypsin treatment on viability and multiplication potential of single normal human and chicken fibroblasts. *Cell Biol. Int. Rep.* **1**: 335–343.

McKeehan, W. L., & Ham, R. G. (1976a). Stimulation of clonal growth of normal fibroblasts with substrata coated with basic polymers. *J. Cell Biol.* **71**: 727–734.

McKeehan, W. L., & Ham, R. G. (1976b). Methods for reducing the serum requirement of growth *in vitro* of non-transformed diploid fibroblasts. *Dev. Biol. Standard.* **37**: 97–98.

McKeehan, W. L., & McKeehan, K. A. (1979). Oxocarboxylic acids, pyridine nucleotide-linked oxidoreductases and serum factors in regulation of cell proliferation. *J. Cell Physiol.* **101**: 9–16.

McKeehan, W. L., Adams, P. S., & Rosser, M. P. (1982). Modified nutrient medium MCDB 151 (WJAC401), defined growth factors, cholera toxin, pituitary factors, and horse serum support epithelial cell and suppress fibroblast proliferation in primary cultures of rat ventral prostate cells. *In Vitro* **18**: 87–91.

McKeehan, W. L., Adams, P. S., & Rosser, M. P. (1984). Direct mitogenic effects of insulin, epidermal growth factor, cholera toxin, unknown pituitary factors and possibly prolactin, but not androgen, on normal rat prostate epithelial cells in serum-free primary cell culture. *Cancer Res.* **44**: 1998–2010.

McKeehan, W. L., Hamilton, W. G., & Ham, R. G. (1976). Selenium is an essential trace nutrient for growth of WI-38 diploid human fibroblasts. *Proc. Natl. Acad. Sci. USA* **73**: 2023–2027.

McKeehan, W. L., McKeehan, K. A., Hammond, S. L., & Ham, R. G. (1977). Improved medium for clonal growth of human diploid cells at low concentrations of serum protein. *In Vitro* **13**: 399–416.

McLean, J. S., Frame, M. C., Freshney, R. I., Vaughan, P. F. T., & Mackie, A. E. (1986). Phenotypic modification of human glioma and non-small cell lung carcinoma by glucocorticoids and other agents. *Anticancer Res.* **6**: 1101–1106.

Medical Research Council (2002). Human Tissue and Biological Samples for Use in Research; Operational and Ethical Guidelines. Available from MRC External Communications +44 (0)2076 365 422 or online at www.mrc.ac.uk.

Meeker, A. K., & De Marzo, A. M. (2004). Recent advances in telomere biology: implications for human cancer. *Curr. Opin. Oncol.* **16**: 32–38.

Mege, R. M., Matsuzaki, F., Gallin, W. J., Goldberg, J. I., Cummingham, B. A., & Edelman, G. M. (1989). Construction of epithelioid sheets by transfection of mouse sarcoma cells with cDNAs for chicken cell adhesion molecules. *Proc. Natl. Acad. Sci. USA* **85**: 7274–7278.

Melera, P. W., Wolgemuth, D., Biedler, J. L., & Hession, C. (1980). Antifolate-resistant Chinese hamster cells: Evidence from independently derived sublines for the overproduction of two dihydrofolate reductases encoded by different mRNAs. *J. Biol. Chem.* **255**: 319–322.

Menasche, P., Hagege, A. A., Scorsin, M., Pouzet, B., Desnos, M., Duboc, D., Schwartz, K., Vilquin, J. T., & Marolleau, J. P. (2001). Myoblast transplantation for heart failure. *Lancet*, **357**: 279–280.

Menasche, P. (2004). Skeletal myoblast transplantation for cardiac repair. *Expert Rev. Cardiovasc. Ther.* **2**: 21–28.

Ment, L. R., Stewart, W. B., Scaramuzzino, D., & Madri, J. A. (1997). An *in vitro* three-dimensional coculture model of cerebral microvascular angiogenesis and differentiation. *In Vitro Cell Dev. Biol. Animal* **33**: 684–691.

Merten, O. W. (1999). Safety issues of animal products used in serum-free media. *Dev. Biol. Stand.* **99**: 167–180.

Messing, E. M., Fahey, I. L., deKernion, I. B., Bhuta, S. M., & Bubbers, I. E. (1982). Serum-free medium for the *in vitro* growth of normal and malignant urinary bladder epithelial cells. *Cancer Res.* **42**: 2392–2397.

Metcalf, D. (1969). Studies on colony formation *in vitro* by mouse bone marrow cells. I. Continuous cluster formation and relation of clusters to colonies. *J Cell Physiol.* **74**: 323–332.

Metcalf, D. (1990). The colony stimulating factors. *Cancer* **65**: 2185–2195.

Meyer, W., Latouche, G. N., Daniel, H. M., Thanos, M., Mitchell, T. G., Yarrow, D., Schonian, G., & Sorrell, T. C. (1997). Identification of pathogenic yeasts of the imperfect genus *Candida* by polymerase chain reaction fingerprinting. *Electrophoresis* **18**: 1548–1549.

Meyskens, F. L., & Fuller, B. B. (1980). Characterization of the effects of different retinoids on the growth and differentiation of a human melanoma cell line and selected subclones. *Cancer Res.* **40**: 2194–2196.

Michalopoulos, G., & Pitot, H. C. (1975). Primary culture of parenchymal liver cells on collagen membranes: Morphological and biochemical observations. *Exp. Cell Res.* **94**: 70–78.

Michler-Stuke, A., & Bottenstein, J. (1982). Proliferation of glial-derived cells in defined media. *J. Neurosci. Res.* **7**: 215–228.

Midgley, C. A., Craig, A. L., Hite, J. P., & Hupp, T. R. (1998). Baculovirus expression and the study of the regulation of the tumor suppressor protein p53. In Ravid, K., & Freshney, R. I. (eds.), *DNA transfer to cultured cells*. New York, Wiley-Liss, pp. 27–54.

Miklic, S., Juric, D. M., & Caman-Krzan, M. (2004). Differences in the regulation of BDNF and NGF synthesis in cultured neonatal rat astrocytes. *Int. J. Dev. Neurosci.* **22**: 119–130.

Mikulits, W., Dolznig, H., Edelmann, H., Sauer, T., Deiner, E. M., Ballou, L., Beug, H., & Mullner, E. W. (1997). Dynamics of cell cycle regulators: Artefact-free analysis by recultivation of cells synchronized by centrifugal elutriation. *DNA Cell Biol.* **16**: 849–859.

Milanesi, E., Ajmone-Marsan, P., Bignotti, E., Losio, M. N., Bernardi, J., Chegdani, F., Soncini, M., & Ferrari, M. (2003). Molecular detection of cell line cross-contaminations using amplified fragment length polymorphism DNA fingerprinting technology. *In Vitro Cell. Dev. Biol. Animal* **39**: 124–130.

Miller, A. D., Garcia, J. V., von Suhr, N., Lynch, C. M., Wilson, C., & Eiden, M. V. (1991). Construction and properties of retrovirus packaging cells based on gibbon abe leukemia virus. *J. Virol.* **65**: 2220–2224.

Miller, G., Lisco, H., Kohn, H. I., Stitt, D., & Enders, J. F.. (1971). Establishment of cell lines from normal adult human blood leukocytes by exposure to Epstein-Barr virus and neutralization by human sera with Epstein-Barr virus antibody. *Proc. Soc. Exp. Biol. Med.* **137**: 1459–1465.

Miller, G. G., Walker, G. W. R., & Giblack, R. E. (1972). A rapid method to determine the mammalian cell cycle. *Exp. Cell Res.* **72**: 533–538.

Miller, R. G., & Phillips, R. A. (1969). Separation of cells by velocity sedimentation. *J. Cell Physiol.* **73**: 191–201.

Mills, K. J., Volberg, T. M., Nervi, C., Grippo, J. F., Dawson, M. I., & Jetten, A. M. (1996). Regulation of retinoid-induced differentiation in embryonal carcinoma PCC4, aza1R cells: Effects of retinoid-receptor selective ligands. *Cell Growth Differ.* **7**: 327–337.

Milo, G. E., Ackerman, G. A., & Noyes, I. (1980). Growth and ultrastructural characterization of proliferating human keratinocytes *in vitro* without added extrinsic factors. *In Vitro* **16**: 20–30.

Minna, I. D., Carney, D. N., Cuttitta, F., & Gazdar, A. F. (1983). The biology of lung cancer. In Chabner, B. (ed.), *Rational basis for chemotherapy*. New York, Alan R. Liss.

Mirskey, R., Dubois, C., Morgan, L., & Jessen, K. R. (1990). O4 and A007 sufatide antibodies bind to embryonic Schwann cells prior to the appearance of galactocerebroside: Regulation by axon-Schwann cell signals and cyclic AMP. *Development* **109**: 105–116.

Mitaka, T., Norioka, K.-I., & Mochizuki, Y. (1993). Redifferentiation of proliferated rat hepatocytes cultures in L15 medium supplemented with EGF and DMSO. *In Vitro Cell Dev. Biol.* **29A**: 714–722.

Mitaka, T., Sattler, C. A., Sattler, G. L., Sargent, L. M., & Pitot, H. C. (1991). Multiple cell cycles occur in rat hepatocytes cultured in the presence of nicotinamide and epidermal growth factor. *Hepatology* **13**: 21–30.

Mitelman, F., (ed.) (1995). *ISCN. An international system for human cytogenetic nomenclature*, Basel, S. Karger.

Miura, Y., Akimoto, T., Kanazawa, H., & Yagi, K. (1986). Synthesis and secretion of protein by hepatocytes entrapped within calcium alginate. *Artif. Organs* **10**: 460–465.

Mohammad, R. M., Li, Y., Mohamed, A. N., Pettit, G. R., Adsay, V., Vaitkevicius, V. K., Al-Katib, A. M., & Sarkar, F. H. (1999). Clonal preservation of human pancreatic cell line derived from primary pancreatic adenocarcinoma. *Pancreas* **19**: 353–361.

Moll, R., Franke, W. W., & Schiller, D. L. (1982). The catalog of human cytokeratins: Patterns of expression in normal epithelia, tumours and cultured cells. *Cell* **31**: 11–24.

Monroy, B., Honiger, J., Darquy, S., & Reach, G. (1997). Use of polyethyleneglycol for porcine islet cryopreservation. *Cell Transplant.* **6**: 613–621.

Monstein, H. J., Ohlsson, B., & Axelson, J. (2001). Differential expression of gastrin, cholecystokinin-A and cholecystokinin-B receptor mRNA in human pancreatic cancer cell lines. *Scand. J. Gastroenterol.* **36**: 738–743.

Montagnier, L. (1968). Correlation entre la transformation des cellule BHK21 et leur resistance aux polysaccharides acides en milieu gélifié. *CR Acad. Sci. D* **267**: 921–924.

Montesano, R., Matsumoto, K., Nakamura, T., & Orci, L. (1991). Identification of a fibroblast-derived epithelial morphogen as hepatocyte growth factor. *Cell* **67**: 901–908.

Montesano, R., Soriano, J. V., Pepper, M. S., & Orci, L. (1997). Induction of epithelial branching tubulogenesis *in vitro*. *J. Cell Physiol.* **173**: 152–161.

Moore, G. E., Gerner, R. E., & Franklin, H. A. (1967). Culture of normal human leukocytes. *J. Am. Med. Assoc.* **199**: 519–524.

Moreno, R. F. (1990). Enhanced conditions for DNA fingerprinting with biotinylated M13 bacteriophage. *J. Forensic Sci.* **35**: 831–837.

Morgan, J. E., Beauchamp, J. R., Pagel, C. N., Peckham, M., Ataliotis, P., Jat, P. S., Noble, M. D., Farmer, K., & Partridge, T A.. (1994). Myogenic cell lines derived from transgenic mice carrying a thermolabile T antigen: a model system for the derivation of tissue-specific and mutation-specific cell lines. *Dev. Biol.* **162**: 486–498.

Morgan, D., Freshney, R. I., Darling, J. L., Thomas, D. G. T., & Celik, F. (1983). Assay of anticancer drugs in tissue culture: cell cultures of biopsies from human astrocytoma. *Br. J. Cancer* **47**: 205–214.

Morgan, J. E., Moore, S. E., Walsh, F. S., & Partridge, T. A. (1992). Formation of skeletal muscle *in vivo* from the mouse C2 cell line. *J. Cell Sci.* **102**: 779–787.

Morgan, J. G., Morton, H. J., & Parker, R. C. (1950). Nutrition of animal cells in tissue culture; I: Initial studies on a synthetic medium. *Proc. Soc. Exp. Biol. Med.* **73**: 1.

Morton, H. J. (1970). A survey of commercially available tissue culture media. *In Vitro* **6**: 89–108.

Moscona, A. A. (1952). Cell suspension from organ rudiments of chick embryos. *Exp. Cell Res.* **3**: 535.

Moscona, A. A., & Piddington, R. (1966). Stimulation by hydrocortisone of premature changes in the developmental pattern of glutamine synthetase in embryonic retina. *Biochim. Biophys. Acta* **121**: 409–411.

Mosmann, T. (1983). Rapid colorimetric assay for cellular growth and survival: Application to proliferation and cytotoxicity assays. *J. Immunol. Methods* **65**: 55–63.

Moulton, D. G. (1974). Dynamics of cell populations in the olfactory epithelium. *Ann. NY Acad. Sci.* **237**: 52–61.

Mowles, J. (1988). The use of ciprofloxacin for the elimination of mycoplasma from naturally infected cell lines. *Cytotechnology* **1**: 355–358.

Muirhead, E. E., Rightsel, W. A., Pitcock, J. A., & Inagami, T. (1990). Isolation and culture of juxtaglomerular and renomedullary interstitial cells. *Methods Enzymol.* **191**: 152–167.

Mullender, M., El Haj, A. J., Yang, Y., van Duin, M. A., Burger, E. H., & Klein-Nulend, J. (2004). Mechanotransduction of bone cells *in vitro*: mechanobiology of bone tissue. *Med. Biol. Eng. Comput.* **42**: 14–21.

Muller, U., Wang, D., Denda, S., Meneses, J. J., Pedersen, R. A., & Reichardt, L. F. (1997). Integrin alpha 8 beta 1 is critically important for epithelial–mesenchymal interactions during kidney morphogenesis. *Cell* **88**: 603–613.

Mullin, J. M., Marano, C. W., Laughlin, K. V., Nuciglio, M., Stevenson, B. R., & Soler, A. P. (1997). Different size limitations for increased transepithelial paracellular solute flux across phorbol ester and tumour necrosis factor treated epithelial cell sheets. *J. Cell Physiol.* **171**: 226–233.

Munthe-Kaas, A. C., & Seglen, P. O. (1974). The use of metrizamide as a gradient medium for isopycnic separation of rat liver cells. *FEBS Lett.* **43**: 252–256.

Murakami, H. (1984). Serum-free cultivation of plasmacytomas and hybridomas. In Barnes, D. W., Sirbasku, D. A., & Sato, G. H. (eds.), *Methods for serum-free culture of neuronal and lymphoid cells*. New York, Alan R. Liss, pp. 197–206.

Murakami, H., & Masui, H. (1980). Hormonal control of human colon carcinoma cell growth in serum-free medium. *Proc. Natl. Acad. Sci. USA* **77**: 3464–3468.

Murao, S., Gemmell, M. A., & Callaghan, M. F. (1983). Control of macrophage cell differentiation in human promyelocytic HL-60 leukemia cells by 1,25-dihydroxyvitamin D_3 and phorbol-12-myristate-13-acetate. *Cancer Res.* **43**: 4989–4996.

Murphy, C. L. & Polak, J. M. (2004). Control of human articular chondrocyte differentiation by reduced oxygen tension. *J. Cell. Physiol.* **199**: 451–459

Murphy, D. S., Hoare, S. F., Going, J. J., Mallon, E. E., George, W. D., Kaye, S. B., Brown, R., Black, D. M., & Keith, W. N. (1995). Characterization of extensive genetic alterations in ductal carcinoma *in situ* by fluorescence *in situ* hybridization and molecular analysis. *J. Natl. Cancer Inst.* **87**: 1694–1704.

Murphy, S. J., Watt, D. J., & Jones, G. E. (1992). An evaluation of cell separation techniques in a model mixed cell population. *J. Cell Sci.* **102**: 789–798.

Naeyaert, J. M., Eller, M., Gordon, P. R., Park, H.-Y., & Gilchrest, B. A. (1991). Pigment content of cultured human melanocytes does not correlate with tyrosinase message level. *Br. J. Dermatol.* **125**: 297–303.

Nardone, R. M., Todd, G., Gonzalez, P., & Gaffney, E. V. (1965). Nucleoside incorporation into strain L cells: Inhibition by pleuropneumonia-like organisms. *Science* **149**: 1100–1101.

National Research Council. (1989). *Biosafety in the laboratory: Prudent practices for the handling and disposal of infectious materials.* Washington, DC, National Academy Press.

National Sanitation Foundation. (1983). *Standard 49: Class II (laminar flow) biohazard cabinetry.* Ann Arbor, Michigan.

Nelson, J. B., ed. (1960). Biology of the pleuropneumonia-like organisms. *Ann. New York. Acad. Sci.* **79**: 305.

Nelson-Rees, W. A., Daniels, D., & Flandermeyer, R. R. (1981). Cross-contamination of cells in culture. *Science* **212**: 446–452.

Nelson-Rees, W., & Flandermeyer, R. R. (1977). Inter- and intraspecies contamination of human breast tumor cell lines HBC and BrCa5 and other cell cultures. *Science* **195**: 1343–1344.

Neugut, A. I., & Weinstein, I. B. (1979). Use of agarose in the determination of anchorage-independent growth. *In Vitro* **15**: 351.

Neumann, T., Hauschka, S. D., & Sanders, J. E. (2003). Tissue engineering of skeletal muscle using polymer fiber arrays. *Tissue Eng.* **9**: 995–1003.

Neumann, J. R., Morency, C. A., & Russian, K. O. (1987). A novel rapid assay for chloramphenicol acetyltransferase gene-expression. *Biotechniques* **5**: 444.

Neves, A. A., N., Medcalf, Brindle, K. (2003). Functional assessment of tissue-engineered meniscal cartilage by magnetic resonance imaging and spectroscopy. *Tissue Eng.* **9**: 51–62.

Newbold, R. F., & Cuthbert, A. P. (1998). Mapping human senescence genes using microcell-mediated chromosome transfer. In Ravid, K., & Freshney, R. I. (eds.), *DNA transfer to cultured cells.* New York, Wiley-Liss, pp. 237–264.

Newman, M. B., Davis, C. D., Kuzmin-Nichols, N., & Sanberg, P. R. (2003). Human umbilical cord blood (HUCB) cells for central nervous system repair. *Neurotox Res.* **5**: 355–68.

Neyfakh, A. A. (1987). Use of fluorescent dyes as molecular probes for the study of multidrug resistance. *Exp. Cell Res.* **174**: 168.

Nichols, W. W., Murphy, D. G., & Christofalo, V. J. (1977). Characterization of a new diploid human cell strain, IMR-90. *Science* **196**: 60–63.

Nicola, N. A. (1987). Hemopoietic growth factors and their interactions with specific receptors. *J. Cell Physiol. (Suppl.)* **5**: 9–14.

Nicolson, G. L. (1976). Trans-membrane control of the receptors on normal and tumor cells; II: Surface changes associated with transformation and malignancy. *Biochim. Biophys. Acta* **458**: 1–72.

Nicosia, R. F., & Ottinetti, A. (1990). Modulation of microvascular growth and morphogenesis by reconstituted basement membrane gel in three-dimensional cultures of rat aorta: A comparative study of angiogenesis in Matrigel, collagen, fibrin, and plasma clot. *In Vitro Cell Dev. Biol.* **26**: 119–128.

Nicosia, R. F., Tchao, R., & Leighton, J. (1983). Angiogenesis-dependent tumor spread in reinforced fibrin clot culture. *Cancer Res.* **43**: 2159–2166.

Nielsen, V. (1989). Vibration patterns in tissue culture vessels. *Nunc Bulletin 2* (May 1986; rev March 1989). Roskilde, Denmark, A/S Nunc.

Niklason, L. E., Abbott, W., Gao, J., Klagges, B., Hirschi, K. K., Ulubayram, K., Conroy, N., Jones, R., Vasanawala, A., Sanzgiri, S., & Langer, R. (2001). Morphological and mechanical characteristics of engineered bovine arteries. *J. Vasc. Surg.* **33**(3): 628–638.

Nilos, R. M., & Makarski, J. S. (1978). Control of melanogenesis in mouse melanoma cells of varying metastatic potential. *J. Natl. Cancer Inst.* **61**: 523–526.

Nims, R. W., Shoemaker, A. P., Bauernschub, M. A., Rec, L. J., & Harbell, J. W. (1998). Sensitivity of isoenzyme analysis for the detection of interspecies cell line cross-contamination. *In Vitro Cell Dev. Biol. Anim.* **34**: 35–39.

Noble, M., & Barnett, S. C. (1996). Production and growth of conditionally immortal primary glial cell cultures and cell lines. In Freshney, R. I., & Freshney, M. G. (eds.), *Culture of immortalized cells.* New York, Wiley-Liss, pp. 331–366.

Noble, M., & Murrey, K. (1984). Purified astrocytes promote the *in vitro* division of a bipotential glial progenitor cell. *EMBO J.* **3**: 2243–2247.

Nordgren, A. (2003). Hidden aberrations diagnosed by interphase fluorescence *in situ* hybridisation and spectral karyotyping in childhood acute lymphoblastic leukaemia. *Leuk. Lymphoma* **44**: 2039–2053.

Norwood, T. H., Zeigler, C. J., & Martin, G. M. (1976). Dimethyl sulphoxide enhances polyethylene glycol-mediated somatic cell fusion. *Somatic Cell Genet.* **2**: 263–270.

Nurse, P. (1990). Universal control mechanism regulating onset of M-phase. *Nature* **344**: 503–508.

Nuwaysir, E. F., Bittner, M., Trent, J., Barrett, J. C., & Afshari, C. A. (1999). Microarrays and toxicology: the advent of toxicogenomics. *Mol Carcinog.* **24**: 153–159.

Obata, Y., Kono, T. & Hatada, I. (2002). Gene silencing: maturation of mouse fetal germ cells *in vitro*. *Nature* **418**: 497.

O'Brien, S. J., Shannon, J. E., & Gail, M. H. (1980). Molecular approach to the identification and individualization of human and animal cells in culture: Isozyme and allozyme genetic signatures. *In Vitro* **16**: 119–135.

O'Farrell, P. H. (1975). High resolution two-dimensional electrophoresis of proteins. *J. Biol. Chem.* **250**: 4007–4021.

Office of Nuclear Regulatory Research, *see* U. S. Nuclear Regulatory Commission.

Oft, M., Akhurst, R. J., & Balmain, A. (2002). Metastasis is driven by sequential elevation of H-ras and Smad2 levels. *Nat. Cell Biol.* **4**: 487–494.

Oh, C. K., Kwon, Y. W., Kim, Y. S., Jang, H. S., Kwon, K. S. (2003). Expression of basic fibroblast growth factor, vascular endothelial growth factor, and thrombospondin-1 related to microvessel density in nonaggressive and aggressive basal cell carcinomas. *J Dermatol.*, **30**: 306–13.

Oh, H., Chi, X., Bradfute, S. B., Mishina, Y., Pocius, J., Michael, L. H., Behringer, R. R., Schwartz, R. J., Entman, M. L., & Schneider, M. D. (2004). Cardiac muscle plasticity in adult and embryo by heart-derived progenitor cells. *Ann. NY Acad. Sci.* **1015**: 182–189.

Oh, Y. S., Kim, E. J., Schaffer, B. S., Kang, Y. H., Binderup, L., MacDonald, R. G., & Park, J. H. (2001). Synthetic low-calcaemic vitamin D_3 analogues inhibit secretion of insulin-like growth factor II and stimulate production of insulin-like growth factor-binding protein-6 in conjunction with growth suppression of HT-29 colon cancer cells. *Mol. Cell Endocrinol.* **183**: 141–149.

O'Hare, M. J., Ellison, M. L., & Neville, A. M. (1978). Tissue culture in endocrine research: Perspectives, pitfalls, and potentials. *Curr. Top. Exp. Endocrinol.* **3**: 1–56.

Ohmichi, H., Koshimizu, U., Matsumoto, K., & Nakamura, T. (1998). Hepatocyte growth factor (HGF) acts as a mesenchyme-derived morphogenic factor during fetal lung development. *Development* **125**: 1315–1324.

Ohno, T., Saijo-Kurita, K., Miyamoto-Eimori, N., Kurose, T., Aoki, Y., & Yosimura, S. (1991). A simple method for *in situ* freezing of anchorage-dependent cells including rat liver parenchymal cells. *Cytotechnology* **5**: 273–277.

Oie, H. K., Russell, E. K., Carney, D. N., & Gazdar, A. F. (1996). Cell culture methods for the establishment of the NCI series of lung cancer cell lines. *J. Cell Biochem.* Suppl. 24: 24–31.

Olie, R. A., Looijenga, L. H. J., Dekker, M. C., de Jong, F. H., van Dissel-Emiliani, F. M. F., de Rooij, D. G., van der Holt, B., & Oosterhuis, J. W. (1995). Heterogeneity in the *in vitro* survival and proliferation of human seminoma cells. *Brit. J. Cancer* **71**: 13–17.

Olsson, I., & Ologsson, T. (1981). Induction of differentiation in a human promyelocytic leukemic cell line (HL-60). *Exp. Cell Res.* **131**: 225–230.

Operational Guidelines for Ethics Committees That Review Biomedical Research (2000). World Health Organization, Geneva, TDR/PRD/ETHICS/2000.1.

Orellana, S. A., Neff, C. D., Sweeney, W. E. & Avner, E. D. (1996). Novel Madin Darby canine kidney cell clones exhibit unique phenotypes in response to morphogens. *In Vitro Cell Dev. Biol. Animal* **32**: 329–339.

Organisation for Economic Co-operation and Development (2004). The application of the principles of GLP to *in vitro* studies. OECD Series on Principles of Good Laboratory Practice and Compliance Monitoring, No. 14; ENV/JM/MONO(2004)26.

Orly, J., Sato, G., & Erickson, G. F. (1980). Serum suppresses the expression of hormonally induced function in cultured granulosa cells. *Cell* **20**: 817–827.

Osborne, C. K., Hamilton, B., Tisus, G., & Livingston, R. B. (1980). Epidermal growth factor stimulation of human breast cancer cells in culture. *Cancer Res.* **40**: 2361–2366.

Osborne, H. B., Bakke, A. C., & Yu, J. (1982). Effect of dexamethasone on HMBA-induced Friend cell erythrodifferentiation. *Cancer Res.* **42**: 513–518.

Osborne, R., Durkin, T., Shannon, H., Dornan, E., & Hughes, C. (1999). Performance of open-fronted microbiological safety cabinets: the value of operator protection tests during routine servicing. *J. Appl. Microbiol.* **86**: 962–970.

Ostertag, W., & Pragnell, I. B. (1978). Changes in genome composition of the Friend virus complex in erythroleukaemia cells during the course of differentiation induced by DMSO. *Proc. Natl. Acad. Sci. USA* **75**: 3278–3282.

Ostertag, W., & Pragnell, I. B. (1981). Differentiation and viral involvement in differentiation of transformed mouse and rat erythroid cells. *Curr. Topics Microbiol. Immunol.* **94/95**: 143–208.

Otterlei, M., Østgaard, K., Skjåk-Bræk, G., Smidsrød, O., Soon Shiong, P., & Espevik, T. (1991). Induction of cytokine production from human monocytes stimulated with alginate. *J. Immunother.* **10**: 286–291.

Owens, R. B., Smith, H. S., & Hackett, A. J. (1974). Epithelial cell culture from normal glandular tissue of mice. Mouse epithelial cultures enriched by selective trypsinisation. *J. Natl. Cancer Inst.* **53**: 261–269.

Paddenberg, R., Wulf, S., Weber, A., Heimann, P., Beck, L., & Mannherz, H. G. (1996). Internucleosomal DNA fragmentation in cultured cells under conditions reported to induce apoptosis may be caused by mycoplasma endonucleases. *Europ. J. Cell Biol.* **71**: 105–119.

Pantel, K., Dickmanns, A., Zippelius, A., Klein, C., Shi, J., Hoechtlen-Vollmar, W., Schlimok, G., Weckermann, D., Oberneder, R., Fanning, E., & Rietmüller, G. (1995). Establishment of micrometastatic cell lines: A novel source of tumor cell vaccines. *J. Natl. Cancer Inst.* **87**: 1162–1168.

Paquet-Durand, F., Tan, S., & Bicker, G. (2004). Turning teratocarcinoma cells into neurons: rapid differentiation of NT-2 cells in floating spheres. *Brain Res. Dev. Brain Res.* **142**: 161–167.

Paquette, J. S., Tremblay, Bernier, P. V., Auger, F. A., Laviolette, M., Germain, L., Boutet, M., Boulet, L. P., & Goulet, F. (2003). Production of tissue-engineered three-dimensional human bronchial models. *In Vitro Cellular & Developmental Biology - Animal*, **39**: pp. 213–220.

Paraskeva, C., & Williams, A. C. (1992). The colon. In Freshney, R. I. (ed.), *Culture of epithelial cells*. New York, Wiley-Liss, pp. 82–105.

Paraskeva, C., Buckle, B. G., & Thorpe, P. E. (1985). Selective killing of contaminating human fibroblasts in epithelial cultures derived from colorectal tumors using an anti-Thy-1 antibody-ricin conjugate. *Br. J. Cancer* **51**: 131–134.

Paraskeva, C., Buckle, B. G., Sheer, D., & Wigley, C. B. (1984). The isolation and characterisation of colorectal epithelial cell lines at different stages in malignant transformation from familial polyposis coli patients. *Int. J. Cancer* **34**: 49–56.

Pardee, A. B., Cherington, P. V., & Medrano, E. E. (1984). On deciding which factors regulate cell growth. In Barnes, D. W., Sirbasku, D. A., & Sato, G. H. (eds.), *Methods for serum-free culture of epithelial and fibroblastic cells*. New York, Alan R. Liss, pp. 157–166.

Park, H. Y., Perez, J. M., Laursen, R., Hara, M., & Gilchrest, B. A. (1999). Protein kinase C-beta activates tyrosinase by phosphorylating serine residues in its cytoplasmic domain. *J Biol Chem.* **274**: 16470–8.

Park, H. Y., Russakovsky, V., Ohno, S., & Gilchrest, B. A. (1993). The β isoform of protein kinase C stimulates melanogenesis by activating tyrosinase in pigment cells. *J. Biol. Chem.* **268**: 11742–11749.

Park, J. G., & Gazdar, A. F. (1996). Biology of colorectal and gastric cancer cell lines. *J. Cell Biochem. Suppl.* **24**: 131–141.

Park, J.-G., Ku, J.-L., Kim, H.-S., Park, S.-Y., & Rutten, M. J. (2004). Culture of normal and malignant gastric epithelium. In Pfragner, R., & Freshney. R. I. (eds.), *Culture of human tumor cells*, Hoboken, NJ, Wiley-Liss, pp. 23–66.

Parker, R. C., Castor, L. N., & McCulloch, E. A. (1957). Altered cell strains in continuous culture. Special publications. *NY Acad. Sci.* **5**: 303–313.

Parker, R. C. (1961). *Methods of tissue culture*, 3rd ed., London, Pitman Medical, p 47.

Parker, R. C., Healy, G. M., & Fisher, D. C. (1954). Nutrition of animal cells in tissue culture. VII. Use of replicate cell cultures in the evaluation of synthetic media. *Canad. J. Biochem. Physiol.* **32**: 306.

Parkinson, E. K., & Yeudall, W. A. (2002). The epidermis. In Freshney, R. I. (ed.), *Culture of epithelial cells*, 2nd ed. New York, Wiley-Liss, pp. 65–94.

Parums, D. V., Cordell, J. L., Micklem, K., Heryet, A. R., Gatter, K. C., and Mason, D. Y. (1990) *J Clin Pathol* **43**: 752–757.

Pasch, J., Schiefer, A., Heschel, I., Dimoudis, N., & Rau, G. (2000). Variation of the HES concentration for the cryopreservation of keratinocytes in suspensions and in monolayers. *Cryobiology* **41**: 89–96.

Patel, K., Moore, S. E., Dickinson, G., Rossell, R. J., Beverley, P. C., Kemshead, J. T. & Walsh, F. S. (1989). Neural cell adhesion molecule (NCAM) is the antigen recognised by monoclonal antibodies of similar specificity in small-cell lung carcinoma and neuroblastoma. *Int. J. Cancer* **44**: 573–578.

Paul, J. (1975). *Cell and tissue culture.* Edinburgh, Churchill Livingstone, pp. 172–184.

Paul, J., Conkie, D., & Freshney, R. I. (1969). Erythropoietic cell population changes during the hepatic phase of erythropoiesis in the foetal mouse. *Cell Tissue Kinet.* **2**: 283–294.

Pavelic, K., Antonic, M., Pavelic, L., Pavelic, J., Pavelic, Z., & Spaventi, S. (1992). Human lung cancers growing on extracellular matrix: Expression of oncogenes and growth factors. *Anticancer Res.* **12**: 2191–2196.

Peat, N., Gendler, S. J., Lalani, N., Duhig, T., & Taylor-Papadimitriou, J. (1992). Tissue-specific expression of a human polymorphic epithelial mucin (MUC1) in transgenic mice. *Cancer Res.* **52**: 1954–1960.

Pedersen, N., Mortensen, S., Sorensen, S. B., Pedersen, M. W., Rieneck, K., Bovin, L. F., & Poulsen, H. S. (2003). Transcriptional gene expression profiling of small cell lung cancer cells. *Cancer Res.* **63**(8): 1943–1953.

Peehl, D. M., & Stanbridge, E. J. (1981). Anchorage-independent growth of normal human fibroblasts. *Proc. Natl. Acad. Sci. USA.* **78**: 3053–3057.

Peehl, D. M., & Ham, R. G. (1980). Clonal growth of human keratinocytes with small amounts of dialysed serum. *In Vitro* **16**: 526–540.

Peehl, D. M. (2002). Human prostatic epithelial cells. In Freshney, R. I. & Freshney, M. G., (eds.), *Culture of epithelial cells*, 2nd ed., Hoboken, NJ, Wiley-Liss, pp. 171–194.

Pegolo, G., Askanas, V., & Engel, W. K. (1990). Expression of muscle-specific isozymes of phosphorylase and creatine kinase in human muscle fibers cultured aneurally in serum-free, hormonally/chemically enriched medium. *Int. J. Dev. Neurosci.* **8**: 299–308.

Peppelenbosch, M. P., Tertoolen, L. G. J., DeLaat, S. W., & Zivkovic, D. (1995). Ionic responses to epidermal growth factor in zebrafish cells. *Exp. Cell Res.* **218**: 183–188.

Pereira, M. E. A., & Kabat, E. A. (1979). A versatile immunoadsorbent capable of binding lectins of various specificities and its use for the separation of cell populations. *J. Cell Biol.* **82**: 185–194.

Pereira-Smith, O., & Smith, J. (1988). Genetic analysis of indefinite division in human cells: Identification of four complementation groups. *Proc. Natl. Acad. Sci. USA* **85**: 6042–6046.

Perl, A.-K., Wilgenbus, P., Dahl, U., Semb, H., & Christofori, G. (1998). A causal role for E-cadherin in the transition from adenoma to carcinoma. *Nature* **392**: 190–193.

Perry, P., & Wolf, S. (1974). New Giemsa method for the differential staining of sister chromatids. *Nature* **251**: 156–158.

Pertoft, H., & Laurent, T. C. (1982). Sedimentation of cells in colloidal silica (Percoll). In Pretlow, T. G., & Pretlow, T. P. (eds.), *Cell separation, methods and selected applications*, Vol. 1. New York, Academic Press, pp. 115–152.

Peters, D. M., Dowd, N., Brandt, C., & Compton, T. (1996). Human papilloma virus E6/E7 genes can expand the lifespan of human corneal fibroblasts. *In Vitro Cell Dev. Biol. Animal* **32**: 279–284.

Petersen, O. W., Gudjonsson, T., Villadsen, R., Bissell, M. J., & Ronnov-Jessen, L. (2003). Epithelial progenitor cell lines as models of normal breast morphogenesis and neoplasia. *Cell Prolif.* **36** Suppl 1: 33–44.

Petersen, D. F., Anderson, E. C., & Tobey, R. A. (1968). Mitotic cells as a source of synchronized cultures. In Prescott, D. M. (ed.), *Methods in cell physiology*. New York, Academic Press, pp. 347–370.

Petit, B., Masuda, K., d'Souza, A., Otten, L., Pietryla, D., Hartmann, D. J., Morris, N. P., Uebelhart, D., Schmid, T. M., & Thonar, E. J.-M. A. (1996). Characterization of crosslinked collagens synthesized by mature articular chondrocytes cultures in alginate beads: Comparison of two distinct matrix compartments. *Exp. Cell Res.* **225**: 151–161.

Pevny, L., & Rao, M. S. (2003). The stem-cell menagerie. *Trends Neurosci.* **26**: 351–359.

Pfeffer, L. M., & Eisenkraft, B. L. (1991). The antiproliferative and antitumour effects of human alpha interferon on cultured renal carcinomas correlate with the expression of a kidney-associated differentiation antigen. *Interferons Cytokines* **17**: 30–31.

Pfragner, R., & Freshney, R. I. (eds.) (2004). *Culture of human tumor cells*, Hoboken, NJ, Wiley-Liss, Chapters 1, 2, 3, 5, 14.

Phillips, P. D., & Cristofalo, V. J. (1988). Classification system based on the functional equivalency of mitogens that regulate WI-38 cell proliferation. *Exp. Cell Res.* **175**: 396–403.

Pignata, S., Maggini, L., Zarrilli, R., Rea, A., & Acquaviva, A. M. (1994). The enterocyte-like differentiation of the Caco-2 tumour cell line strongly correlates with responsiveness to cAMP and activation of kinase A pathway. *Cell Growth Differ.* **5**: 967–973.

Pignatelli, M., & Bodmer, W. F. (1988). Genetics and biochemistry of collagen binding-triggered glandular differentiation in a human colon carcinoma cell line. *Proc. Natl. Acad. Sci. USA* **85**: 5561–5565.

Pipia, G. G., & Long, M. W. (1997). Human hematopoietic progenitor cell isolation based on galactose-specific cell surface binding. *Nat. Biotechnol.* **15**: 1007–1011.

Pitot, H., Periano, C., Morse, P., & Potter, V. R. (1964). Hepatomas in tissue culture compared with adapting liver *in vitro*. *Natl. Cancer Inst. Monogr.* **13**: 229–245.

Piyathilake, C. J., Frost, A. R., Manne, U., Weiss, H., Bell, W. C., Heimburger, D. C., & Grizzle, W. E. (2002). Differential expression of growth factors in squamous cell carcinoma and precancerous lesions of the lung. *Clin. Cancer Res.* **8**: 734–744.

Pizzonia, J. H., Gesek, F. A., Kennedy, S. M., Coutermarsh, B. A., Bacskal, B. J., & Friedman, P. A. (1991). Immunomagnetic separation, primary culture, and characterisation of cortical thick ascending limb plus distal convoluted tubule cells from mouse kidney. *In Vitro Cell Dev. Biol.* **27A**: 409–416.

Planas-Silva, M. D., & Weinberg, R. A. (1997). The restriction point and control of cell proliferation. *Curr. Opin. Cell Biol.* **9**: 768–772.

Planz, B., Wang, Q., Kirley, S. D., Lin, C. W., & McDougal, W. S. (1998). Androgen responsiveness of stromal cells of the human prostate: Regulation of cell proliferation and keratinocyte growth factor by androgen. *J. Urol.* **160**: 1850–1855.

Platsoucas, C. D., Good, R. A., & Gupta, S. (1979). Separation of human lymphocyte-T subpopulations (T-mu, T-gamma) by density gradient electrophoresis. *Proc. Natl. Acad. Sci. USA* **76**: 1972.

Plumb, J. A., Milroy, R., & Kaye, S. B. (1989). Effects of the pH dependence of 3-(4,5-dimethylthiazol-2-yl)-2,5-diphenyltetrazolium bromide-formazan absorption on chemosensitivity determined by a novel tetrazolium-based assay. *Cancer Res.* **49**: 4435–4440.

Pollack, M. S., Heagney, S. D., Livingston, P. O., & Fogh, J. (1981). HLA-A, B, C & DR alloantigen expression on forty-six cultured human tumor cell lines. *J. Natl. Cancer Inst.* **66**: 1003–1012.

Pollack, R. (1981). *Readings in mammalian cell culture*, 2nd ed. Cold Spring Harbor, NY Cold Spring Harbor Laboratory Press.

Pollard, T. D., & Borisy, G. G. (2003). Cellular motility driven by assembly and disassembly of actin filaments. *Cell* **112**: 453–465.

Pollard, J. W., & Walker, J. M. (eds.). (1990). *Animal cell culture: Methods in molecular biology*, 5. Clifton, NJ, Humana Press, pp. 83–97.

Pollock, G. S., Franceschini, I. A., Graham, G., & Barnett, S. C. (1999). Neuregulin is a mitogen and survival factor for olfactory bulb ensheathing cells and a related isoform is produced by astrocytes. *Eur. J. Neurosci.* **11**: 769–780.

Pollock, M. F., & Kenny, G. E. (1963). Mammalian cell cultures contaminated with pleuro-pneumonia-like organisms; III: Elimination of pleuro-pneumonia-like organisms with specific antiserum. *Proc. Soc. Exp. Biol. Med.* **112**: 176–181.

Polnaszek, N., Kwabi-Addo, B., Wang, J., & Ittmann, M. (2004). FGF17 is an autocrine prostatic epithelial growth factor and is upregulated in benign prostatic hyperplasia. *Prostate* **60**: 18–24.

Pontecorvo, G. (1975). Production of mammalian somatic cell hybrids by means of polyethylene glycol treatment. *Somat. Cell Genet.* **1**: 397–400.

Pontén, J. (1975). Neoplastic human glia cells in culture. In: Fogh, J. (ed.), *Human tumor cells in vitro*. New York, Plenum, pp. 175–185.

Pontén, J., & Westermark, B. (1980). Cell generation and aging of nontransformed glial cells from adult humans. In Fedoroff, S., & Hertz, L. (eds.), *Advances in cellular neurobiology*, Vol. 1. New York, Academic Press, pp. 209–227.

Poot, M., Hoehn, H., Kubbies, M., Grossmann, A., Chen, Y., & Rabinovitch, P. S. (1994). Cell-cycle analysis using continuous bromodeoxyuridine labeling and Hoechst 33358-ethidium bromide bivariate flow cytometry. *Methods Cell Biol.* **41**: 327–340.

Post, M., Floros, J., & Smith, B. T. (1984). Inhibition of lung maturation by monoclonal antibodies against fibroblast-pneumocyte factor. *Nature* **308**: 284–286.

Potten, C. S., Booth, C., Tudor, G. L., Booth, D., Brady, G., Hurley, P., Ashton, G., Clarke, R., Sakakibara, S. & Okano, H. (2003). Identification of a putative intestinal stem cell and early lineage marker; musashi-1. *Differentiation* **71**: 28–41.

Potter, H., Weir, L., & Leder, P. (1984). Enhancer-dependent expression of human k immunoglobulin genes introduced into mouse pre-B lymphocytes by electroporation. *Proc. Natl. Acad. Sci. USA* **81**: 7161–7165.

Powell, C. A., Smiley, B. L., Mills, J. & Vandenburgh, H. H., (2002). Mechanical stimulation improves tissue-engineered human skeletal muscle. *Am J Physiol Cell Physiol* **283**: C1557–C1565.

Powers, D. A. (1989). Fish as model systems. *Science* **246**: 352–357.

Pragnell, I. B., Wright, E. G., Lorimore, S. A., Adam, J., Rosendaal, M., deLamarter, J. F., Freshney, M., Eckmann, L., Sproul, A., & Wilkie, N. (1988). The effect of stem cell proliferation regulators demonstrated with an *in vitro* assay. *Blood* **72**: 196–201.

Prasad, K. N., & Edwards-Prasad, J., Ramanujam, S., & Sakamoto, A. (1980). Vitamin E increases the growth inhibitory and differentiating effects of tumour therapeutic agents on neuroblastoma and glioma cells in culture. *Proc. Soc. Exp. Biol. Med.* **164**: 158–163.

Preissmann, A., Wiesmann, R., Buchholz, R., Werner, R. G., & Noe, W. (1997). Investigations on oxygen limitations of adherent cells growing on macroporous microcarriers. *Cytotechnology* **24**: 121–134.

Pretlow, T. G., & Pretlow, T. P. (1989). Cell separation by gradient centrifugation methods. *Methods Enzymol.* **171**: 462–482.

Pretlow, T. G., Delmoro, C. M., Dilley, G. G., Spadafora, C. G., & Pretlow, T. P. (1991). Transplantation of human prostatic carcinoma into nude mice in Matrigel. *Cancer Res.* **51**: 3814–3817.

Prince, G. A., Jenson, A. B., Billups, L. C., & Notkins, A. L. (1978). Infection of human pancreatic beta cell cultures with mumps virus. *Nature* **271**: 158–161.

Proceedings—NASA bioreactors workshop on regulation of cell and tissue differentiation (1997). *In Vitro Cell. Dev. Biol. Anim.* **33**: 325–405.

Provision and use of work equipment (1995). Directive 89/355/EEC and 95/63/EC.

Provision and use of work equipment. (1992). *Health and Safety Executive*, Broad Lane, Sheffield S3 7HQ, England.

Puck, T. T., & Marcus, P. I. (1955). A rapid method for viable cell titration and clone production with HeLa cells in tissue culture: The use of X-irradiated cells to supply conditioning factors. *Proc. Natl. Acad. Sci. USA* **41**: 432–437.

Puck, T. T., Cieciura, S. J., & Robinson, A. (1958). Genetics of somatic mammalian cells; III: Long term cultivation of euploid cells from human and animal subjects. *J. Exp. Med.* **108**: 945–956.

Punchard, N., Watson, D., Thomson, R., & Shaw, M. (1996). Production of immortal human umbilical vein endothelial cells. In Freshney, R. I., & Freshney, M. G. (eds.), *Culture of immortalized cells*. New York, Wiley-Liss, pp. 203–238.

Quarles, J. M., Morris, N. G., & Leibovitz, A. (1980). Carcinoembryonic antigen production by human colorectal adenocarcinoma cells in matrix-perfusion culture. *In Vitro* **16**: 113–118.

Quax, P. H., Frisdal, E., Pedersen, N., Bonavaud, S., Thibert, P., Martelly, I., Verheijen, J. H., Blasi, F., & Barlovatz-Meimon, G. (1992). Modulation of activities and RNA level of the components of the plasminogen activation system during fusion of human myogenic satellite cells *in vitro*. *Dev. Biol.* **151**: 166–175.

Quintanilla, M., Brown, K., Ramsden, M., & Balmain, A. (1986). Carcinogen specific mutation and amplification of Ha-ras during mouse skin carcinogenesis. *Nature* **322**: 78–79.

Quon, M. J. (1998). Transfection of rat adipose cells by electroporation. In Ravid, K., & Freshney, R. I. (eds.), *DNA transfer to cultured cells*. New York, Wiley-Liss, pp. 93–110.

Rabito, C. A., Tchao, R., Valentich, J., & Leighton, J. (1980). Effect of cell substratum interaction of hemicyst formation by MDCK cells. *In Vitro* **16**: 461–468.

Radisky, D. C., Hirai, Y, Bissell, M. J. (2003). Delivering the message: epimorphin and mammary epithelial morphogenesis. *Trends Cell Biol.*, **13**: 426–34.

Raff, M. C. (1990). Glial cell diversification in the rat optic nerve. *Science* **243**: 1450–1455.

Raff, M. C., Abney, E., Brockes, J. P., & Hornby-Smith, A. (1978). Schwann cell growth factors. *Cell* **15**: 813–822.

Raff, M. C., Fields, K. L., Hakomori, S. L., Minsky, R., Pruss, R. M., & Winter, J. (1979). Cell-type-specific markers for distinguishing and studying neurons and the major classes of glial cells in culture. *Brain Res.* **174**: 283–309.

Rainov, N. G., & Ren, H. (2003). Clinical trials with retrovirus mediated gene therapy—what have we learned? *J. Neurooncol.* **65**: 227–36.

Raisman, G. (1985). Specialized neuroglial arrangement may explain the capacity of vomeronasal axons to reinnerveate central neurons. *Neuroscience* **14**: 237–254.

Raisman, G. (2001) Olfactory ensheathing cells—another miracle cure for spinal cord injury? *Nat. Rev. Neurosci.* **5**: 369–75.

Rak, J., Mitsuhashi, Y., Erdos, V., Huang, S.-N., Filmus, J., & Kerbel, R. S. (1995). Massive programmed cell death in intestinal epithelial cells induced by three-dimensional growth conditions: Suppression by mutant c-H-ras oncogene expression. *J. Cell Biol.* **131**: 1587–1598.

Ramaekers, F. C. S., Puts, J. J. G., Kant, A., Moesker, O., Jap, P. H. K., & Vooijs, G. P. (1982). Use of antibodies to intermediate filaments in the characterization of human tumors. *Cold Spring Harbor Symp. Quant. Biol.* **46**: 331–339.

Ramon-Cueto, A., Plant, G. W., Avila, J., & Bunge, M. B. (1998). Long-distance axonal regeneration in the transected adult rat spinal cord is promoted by olfactory ensheathing glia transplants. *J. Neurosci.* **18**: 3803–3815.

Ranscht, B., Clapshaw, P. A., Price, J., Noble, M., & Seifert, W. (1982). Development of oligodendrocytes and Schwann cells studied with monoclonal antibody against galactocerebroside. *Proc. Natl. Acad. Sci. USA* **79**: 2709–2713.

Rathjen, P. D., Lake, J., Whyatt, L. M., Bettess, M. D., & Rathjen, J. (1998). Properties and uses of embryonic stem cells: Prospects for application to human biology and gene therapy. *Reprod. Fertil. Dev.* **10**: 31–47.

Rattner, A., Sabido, O., Massoubre, C., Rascle, F., & Frey, J. (1997). Characterization of human osteoblastic cells: Influence of the culture conditions. *In Vitro Cell Dev. Biol. Anim.* **33**: 757–762.

Ravid, K., & Freshney, R. I. (eds.). (1998). *DNA transfer to cultured cells.* New York, Wiley-Liss.

Ravid, K., Doi, T., Beeler, D. L., Kuter, D. J., & Rosenberg, R. D. (1991). Transcriptional regulation of the rat platelet factor 4 gene: Interaction between an enhancer/silencer domain and the GATA site. *Mol. Cell. Biol.* **11**: 6116–6127.

Ray, M. E., Yang, Z. Q., Albertson, D., Kleer, C. G., Washburn, J. G., Macoska, J. A., & Ethier, S. P. (2004). Genomic and expression analysis of the 8p11–12 amplicon in human breast cancer cell lines. *Cancer Res.* **64**: 40–47.

Raz, A. (1982). B16 melanoma cell variants: Irreversible inhibition of growth and induction of morphologic differentiation by anthracycline antibiotics. *J. Natl. Cancer Inst.* **68**: 629–638.

Reaven, G. M. (1995). The fourth musketeer—from Alexandre Dumas to Claude Bernard. *Diabetologia* **38**: 3–13.

Reddel, R., Ke, Y., Gerwin, B. I., McMenamin, M. G., Lechner, J. F., Su, R. T., Brash, D. E., Park, J. B., Rhim, J. S., & Harris, C. C. (1988). Transformation of human bronchial epithelial cells by infection with SV40 or adenovirus-12 SV40 hybrid virus or transfection via strontium phosphate coprecipitation with a plasmid containing SV40 early region genes. *Cancer Res.* **48**: 1904–1909.

Reed, S. I. (2003). Ratchets and clocks: the cell cycle, ubiquitylation and protein turnover. *Nat. Rev. Mol. Cell Biol.* **4**(11): 855–864.

Reel, J. R., & Kenney, F. T. (1968). "Superinduction" of tyrosine transaminase in hepatoma cell cultures: Differential inhibition of synthesis and turnover by actinomycin D. *Proc. Natl. Acad. Sci. USA* **61**: 200–206.

Reeves, M. E. (1992). A metastatic tumour cell line has greatly reduced levels of a specific homotypic cell adhesion molecule activity. *Cancer Res.* **52**: 1546–1552.

Reid, L. M. (1990). Stem cell biology, hormone/matrix synergies and liver differentiation. *Curr. Opin. Cell. Biol.* **2**: 121–130.

Reid, L. M., & Rojkind, M. (1979). New techniques for culturing differentiated cells. Reconstituted basement membrane rafts. In Jakoby, W. B., & Pastan, I. H. (eds.) *Methods in enzymology, vol 57, cell culture.* New York, Academic Press, pp. 263–278.

Reitzer, L. J., Wice, B. M., & Kennel, D. (1979). Evidence that glutamine, not sugar, is the major energy source for cultured HeLa cells. *J. Biol. Chem.* **254**: 2669–2677.

Relou, I. A., Damen, C. A., van der Schaft, D. W., Groenewegen, G., Griffioen, A. W. (1998). Effect of culture conditions on endothelial cell growth and responsiveness. *Tissue Cell.* **30**: 525–30.

Repesh, L. A. (1989). A new *in vitro* assay for quantitating tumor cell invasion. *Invas. Metast.* **9**: 192–208.

Resnick, J. L., Bixler, L. S., & Donovan, P. J. (1992). Long term proliferation of mouse primordial germ cells in culture. *Nature* **359**: 550–551.

Rheinwald, J. G., & Beckett, M. A. (1981). Tumorigenic keratinocyte lines requiring anchorage and fibroblast support cultured from human squamous cell carcinomas. *Cancer Res.* **41**: 1657–1663.

Rheinwald, J. G., & Green, H. (1975). Serial cultivation of strains of human epidermal keratinocytes: The formation of keratinizing colonies from single cells. *Cell* **6**: 331–344.

Rheinwald, J. G., & Green, H. (1977). Epidermal growth factor and the multiplication of cultured human keratinocytes. *Nature* **265**: 421–424.

Richard, O., Duittoz, A. H., & Hevor, T. K. (1998). Early, middle, and late stages of neural cells from ovine embryo in primary cultures. *Neuroscience Res.* **31**: 61–68.

Richler, C., & Yaffe, D. (1970). The *in vitro* cultivation and differentiation capacities of myogenic cell lines. *Dev. Biol.* **23**: 1–22.

Richmond, A., Lawson, D. H., Nixon, D. W. & Chawla, R. K. (1985). Characterization of autostimulatory and transforming growth factors from human melanoma cells. *Cancer Res.* **45**: 6390–6394.

Rickwood, D., & Birnie, G. D. (1975). Metrizamide, a new density gradient medium. *FEBS Lett.* **50**: 102–110.

Ried, T., Liyanage, M., du Manoir, S., Heselmeyer, K., Auer, G., Macville, M., & Schrock, E. (1997). Tumor cytogenetics revisited: Comparative genomic hybridization and spectral karyotyping. *J. Mol. Med.* **75**: 801–814.

Rieder, C. L., & Cole, R. W. (2002). Cold-shock and the mammalian cell cycle. *Cell Cycle.* **1**: 169–175.

Rindler, M. J., Chuman, L. M., Shaffer, L., & Saier, M. H., Jr. (1979). Retention of differentiated properties in an established dog kidney epithelial cell line (MDCK). *J. Cell Biol.* **81**: 635–648.

Rippon, H. J., & Bishop, A. E. (2004). Embryonic stem cells. *Cell Prolif.* **37**: 23–34.

Rizzino, A. (2002). Embryonic stem cells provide a powerful and versatile model system. *Vitam Horm.* **64**: 1–42.

Robertson, K. M., & Robertson, C. N. (1995). Isolation and growth of human primary prostate epithelial cultures. *Methods Cell Sci.* **17**: 177–185.

Robins, D. M., Ripley, S., Henderson, A. S., & Axel, R. (1981). Transforming DNA integrates into the host chromosome. *Cell* **23**: 29–39.

Rodbell, M. (1964). Metabolism of isolated fat cells: Effects of hormones on glucose metabolism and lipolysis. *J. Biol. Chem.* **239**: 375–380.

Rodriguez, A. M., Elabd, C., Delteil, F., Astier, J., Vernochet, C., Saint-Marc, P., Guesnet, J., Guezennec, A., Amri, E. Z., Dani, C., & Ailhaud, G. (2004). Adipocyte differentiation of multipotent cells established from human adipose tissue. *Biochem. Biophys. Res. Commun.* **315**: 255–263.

Rofstad, E. K. (1994). Orthotopic human melanoma xenograft model systems for studies of tumor angiogenesis, pathophysiology, treatment sensitivity and metastatic pattern. *Br. J. Cancer* **70**: 804–812.

Rogers, A. W. (1979). *Techniques of autoradiography*, 3d ed. Amsterdam, Elsevier/North-Holland Biomedical Press.

Roguet, R., Cohen, C., & Rougier, A. (1994). A reconstituted human epidermis to assess cutaneous irritation, photoirritation and photoprotection *in vitro. In Vitro Skin Toxicol.* **10**: 141–149.

Rojkind, M., Gatmaitan, Z., Mackensen, S., Giambrone, M. A., Ponce, P., & Reid, L. M. (1980). Connective tissue biomatrix: Its isolation and utilization for long term cultures of normal rat hepatocytes. *J. Cell Biol.* **87**: 255–263.

Rooney, D. E., & Czepulkowski, B. H. (eds.). (1986). *Human cytogenetics, a practical approach.* Oxford, U.K., IRL Press at Oxford University Press.

Rooney, S. A., Young, S. L., & Mendelson, C. R. (1995). Molecular and cellular processing of lung surfactant. *FASEB J.* **8**: 957–967.

Rosenblatt, J. D., Lunt, A. I., Parry, D. J., & Partridge, T. A. (1995). Culturing satellite cells from living single muscle fiber explants. *In Vitro Cell. Dev. Biol.* **31**: 773–779.

Rosenblatt, J. D., Parry, D. J., & Partridge, T. A. (1996). Phenotype of adult mouse muscle myoblasts reflects their fiber type of origin. *Differentiation* **60**: 39–45.

Rosenfeld, M. A., Yoshimura, K., Trapnell, B. C., Yoneyama, K., Rosenthal, E. R., Dalemenas, W., Fukayama, M., Bargon, J., Stier, L. E., Stratford-Perricaudet, L., Perricaudet, M., Guggino, W. B., Pavirani, A., Lecocq, J.-P., & Crystal, R. G. (1992). *In vivo* transfer of the human cystic fibrosis transmembrane conductance regulator gene to the airway epithelium. *Cell* **68**: 143–155.

Rosenman, S. J., & Gallatin, W. M. (1991). Cell surface glycoconjugates in intercellular and cell-substratum interactions. *Semin. Cancer Biol.* **2**: 357–366.

Rossi, G. B., & Friend, C. (1967). Erythrocytic maturation of (Friend) virus-induced leukemic cells in spleen clones. *Proc. Natl. Acad. Sci. USA* **58**: 1373–1380.

Rothblat, G. H., & Morton, H. E. (1959). Detection and possible source of contaminating pleuropneumonia-like organisms (PPLO) in culture of tissue cells. *Proc. Soc. Exp. Biol. Med.* **100**: 87.

Rothfels, K. H., & Siminovitch, L. (1958). An air drying technique for flattening chromosomes in mammalian cells growth *in vitro. Stain Technol.* **33**: 73–77.

Rotman, B., & Papermaster, B. W. (1966). Membrane properties of living mammalian cells as studied by enzymatic hydrolysis of fluorogenic esters. *Proc. Natl. Acad. Sci. USA* **55**: 134–141.

Rous, P., & Jones, F. A. A. (1916). A method of obtaining suspensions of living cells from the fixed tissues, and for the plating out of individual cells. *J. Exp. Med.* **23**: 555.

Rowley, S. D., & Byrne, D. V. (1992). Low-temperature storage of bone marrow in nitrogen vapor-phase refrigerators: decreased temperature gradients with an aluminum racking system. *Transfusion* **32**: 750–754.

Rudland, P. S. (1992). Use of peanut lectin and rat mammary stem cell lines to identify a cellular differentiation pathway for the alveolar cell in the rat mammary gland. *J. Cell Physiol.* **153**: 157–168.

Ruff, M. R., & Pert, C. B. (1984). Small cell carcinoma of the lung: Macrophage-specific antigens suggest hematopoietic stem cell origin. *Science* **225**: 1034–1036.

Rules and guidance for pharmaceutical manufacturers and distributors. (1997). The Stationery Office, Ltd., London, U.K.

Rundlett, S. E., Gordon, D. A., & Miesfeld, R. L. (1992). Characterisation of a panel of rat ventral prostate epithelial cell lines immortalized in the presence or absence of androgens. *Exp. Cell Res.* **203**: 214–221.

Ruoff, N. M., & Hay, R. J. (1979). Metabolic and temporal studies on pancreatic exocrine cells in culture. *Cell Tissue Res.* **204**: 243–252.

Rush, L. J., Heinonen, K., Mrozek, K., Wolf, B. J., Abdel-Rahman, M., Szymanska, J., Peltomaki, P., Kapadia, F., Bloomfield, C. D., Caligiuri, M. A., & Plass, C. (2002). Comprehensive cytogenetic and molecular genetic characterization of the TI-1 acute myeloid leukemia cell line reveals cross-contamination with K-562 cell line. *Blood* **99**: 1874–1876.

Russo, J., Calaf, G., & Russo, I. H. (1993). A critical approach to the malignant transformation of human breast epithelial cells with chemical carcinogens. *Crit. Rev. Oncog.* **4**(4): 403–417.

Russo, A. A., Tong, L., Lee, J. O., Jeffrey, P. D., & Pavletich, N. P. (1998). Structural basis for inhibition of the cyclin-dependent kinase Cdk6 by the tumour suppressor p161NK4a. *Nature* **395**: 237–243.

Rutzky, L. P., Tomita, J. T., Calenoff, M. A., & Kahan, B. D. (1979). Human colon adenocarcinoma cells; III: *In vitro* organoid expression and carcino-embryonic antigen kinetics in hollow fiber culture. *J. Natl. Cancer Inst.* **63**: 893–902.

Rygaard, K., Moller, C., Bock, E., & Spang-Thomsen, M. (1992). Expression of cadherin and NCAM in human small cell lung cancer cell lines and xenografts. *Br. J. Cancer* **65**: 573–577.

Saalbach, A., Aust, G., Haustein, U. F., Herrmann, K., & Anderegg, U. (1997). The fibroblast-specific MAb AS02: a novel tool for detection and elimination of human fibroblasts. *Cell Tissue Res.* **290**: 593–599.

Sachs, L. (1978). Control of normal cell differentiation and the phenotypic reversion of malignancy in myeloid leukaemia. *Nature* **274**: 535–539.

Safe Use of Work Equipment. Provision and use of Work Equipment. Provision and Use of Work Equipment Regulations (1998). Approved Code of Practice and Guidance, L22, HSE Books, ISBN 0 7176 1626 6.

Safe working and the prevention of infection in clinical laboratories. (1991). Health and Safety Commission, HMSO Publication, P. O. Box 276, London, SW8 5DT, England.

Sager, R. (1992). Tumor suppressor genes in the cell cycle. *Curr. Opin. Cell Biol.* **4**: 155–160.

Sah, R., Thonar, E., & Masuda, K. (2005). Tissue engineering of articular cartilage. In Vunjak-Novakovic, G., Freshney. R. I. (eds.), *Culture of cells for tissue engineering.* Hoboken, NJ, Wiley-Liss (In press).

Saier, M. H. (1984). Hormonally defined, serum free medium for a proximal tubular kidney epithelial cell line, LLC-PKI. In Barnes, W. D. (ed.), *Methods for serum free culture of epithelial and fibroblastic cells.* New York, Alan R. Liss, pp. 25–31.

Sambrook, J., Fritsch, E. F., & Maniatis, T. (1989). *Molecular cloning: A laboratory manual,* 2d ed. Cold Spring Harbor, NY, Cold Spring Harbor Laboratory Press, 3 vols.

Sanchez-Ramos, J. R. (2002). Neural cells derived from adult bone marrow and umbilical cord blood. *J. Neurosci. Res.* **69**: 880–893.

Sandberg, A. A. (1982). Chromosomal changes in human cancers: Specificity and heterogeneity. In Owens, A. H., Coffey, D. S., & Baylin, S. B. (eds.), *Tumour cell heterogeneity.* New York, Academic Press, pp. 367–397.

Sanes, J. R., Rubenstein, J. L., & Nicolas, J. F. (1986). Use of recombinant retrovirus to study post-implantation cell lineage in mouse embryos. *EMBO J.* **5**: 3133–3142.

Sanford, K. K., Earle, W. R., & Likely, G. D. (1948). The growth *in vitro* of single isolated tissue cells. *J. Natl. Cancer Inst.* **9**: 229.

Sanford, K. K., Earle, W. R., Evans, V. J., Waltz, H. K., & Shannon, I. E. (1951). The measurement of proliferation in tissue cultures by enumeration of cell nuclei. *J. Natl. Cancer Inst.* **11**: 773.

Sanford, K. K., Westfall, B. B., Fioramonti, M. C., McQuilkin, W. T., Bryant, J. C., Peppers, E. V., & Earle, W. R. (1955). The effect of serum fractions on the proliferation of strain L mouse cells *in vitro. J. Natl. Cancer. Inst.* **16**: 789.

Sarang, Z., Haig, Y., Hansson, A., Vondracek, M., Wärngård, L., Grafström, R. C. (2003). Microarray assessment of fibronectin, collagen and integrin expression and the role of fibronectin-collagen coating in the growth of normal, SV40 T-antigen-immortalized and malignant human oral keratinocytes. *ATLA,* **31**: 575–585.

Sasaki, M., Honda, T., Yamada, H., Wake, N., & Barrett, J. C. (1996). Evidence for multiple pathways to cellular senescence. *Cancer Res.* **54**: 6090–6093.

Sato, G. H., & Yasumura, Y. (1966). Retention of differentiated function in dispersed cell culture. *Trans. NY Acad. Sci.* **28**: 1063–1079.

Sattler, G. A., Michalopoulos, G., Sattler, G. L., & Pitot, H. C. (1978). Ultrastructure of adult rat hepatocytes cultured on floating collagen membranes. *Cancer Res.* **38**: 1539–1549.

Saunders, N. A., Bernacki, S. H., Vollberg, T. M., & Jetten, A. M. (1993). Regulation of transglutaminase type-I expression in squamous differentiating rabbit tracheal epithelial cells and human epidermal keratinocytes—effects of retinoic acid and phorbol esters. *Mol. Endocrinol.* **7**: 387–398.

Scanlon, E. F., Hawkins, R. A., Fox, W. W., & Smith, W. S. (1965). Fatal homotransplanted melanoma (a case report). *Cancer* **18**: 782–789.

Schaeffer, W. I. (1990). Terminology associated with cell, tissue and organ culture, molecular biology and molecular genetics. *In Vitro Cell Dev. Biol.* **26**: 97–101.

Schaffer, K., Herrmuth, H., Mueller, J., Coy, D. H., Wong, H. C., Walsh, J. H., Classen, M., Schusdziarra, V., & Schepp, W. (1997). Bombesin-like peptides stimulate somatostatin release from rat fundic D cells in primary culture. *Am. J. Physiol. Gastrointest. Liver Physiol.* **273**: G686–G695.

Schamblott, M. J., Axelman, J., Sterneckert, J., Christoforou, N., Patterson, E. S., Siddiqi, M. A., Kahler, H., Ifeanyi, L. A., & Gearhart, J. D. (2002). Stem cell culture: pluripotent stem cells. In *Methods of tissue engineering,* Atala, A., & Lanza, R. P., Eds., San Diego, Academic Press, pp. 411–420.

Schamblott, M. J., Axelman, J., Wang, S., Bugg, E. M., Little-field, J. W., Donovan, P. J., Blumenthal, P. D., Huggins, G. R., & Gearhart, J. D. (1998). Derivation of pluripotent stem cells from cultured human primordial germ cells. *Proc. Natl. Acad. Sci. USA* **95**: 13726–13731.

Scher, W., Holland, J. G., & Friend, C. (1971). Hemoglobin synthesis in murine virus-induced leukemic cells *in vitro;* I: Partial purification and identification of hemoglobins. *Blood* **37**: 428–437.

Schimmelpfeng, L., Langenberg, U., & Peters, I. M. (1968). Macrophages overcome mycoplasma infections of cells *in vitro.* *Nature* **285**: 661.

Schlechte, W., Brattain, M., & Boyd, D. (1990). Invasion of extracellular matrix by cultured colon cancer cells: Dependence on urokinase receptor display. *Cancer Commun.* **2**: 173–179.

Schlessinger, J., Lax, I., & Lemmon, M. (1995). Regulation of growth factor activation by proteoglycans: What is the role of the low affinity receptors? *Cell* **83**: 357–360.

Schmidt, R., Reichert, U., Michel, S., Shrott, B., & Boullier, M. (1985). Plasma membrane transglutaminase and cornified envelope competence in cultured human keratinocytes. *FEBS Lett.* **186**: 204.

Schneider, E. L., & Stanbridge, E. I. (1975). A simple biochemical technique for the detection of mycoplasma contamination of cultured cells. *Methods Cell Biol.* **10**: 278–290.

Schoenlein, P. V., Shen, D.-W., Barrett, J. T., Pastan, I. T., & Gottesman, M. M. (1992). Double minute chromosomes carrying the human multidrug resistance 1 and 2 genes are generated from the dimerization of submicroscopic circular DNAs in colchicine-selected KB carcinoma cells. *Mol. Biol. Cell* **3**: 507–520.

Schor, S. L. (1994). Cytokine control of cell motility modulation and mediation by the extracellular matrix. *Prog. Growth Factor Res.* **5**: 223–248.

Schousboe, A., Thorbek, P., Hertz, L., & Krogsgaard-Larsen, P. (1979). Effects of GABA analogues of restricted conformation on GABA transport in astrocytes and brain cortex slices and on GABA receptor binding. *J. Neurochem.* **33**: 181–189.

Schulman, H. M. (1968). The fractionation of rabbit reticulocytes in dextran density gradients. *Biochim. Biophys. Acta* **148**: 251–255.

Schwartz Albiez, R., Heidtmann, H.-H., Wolf, D., Schirrma-cher, V., & Moldenhauer, G. (1991). Three types of human lung tumour cell lines can be distinguished according to surface expression of endogenous urokinase and their capacity to bind exogenous urokinase. *Br. J. Cancer* **65**: 51–57.

Schwob, J. (2002) Neuronal regeneration and the peripheral olfactory system. *Anat. Rec.* **269**: 33–49.

Scott, W. N., McCool, K., & Nelson, J. (2000) Improved method for the production of gold colloid monolayers for use in the phagokinetic track assay for cell motility. *Anal. Biochem.* **287**: 343–344.

Scotto, K. W., Biedler, I. L., & Melera, P. W. (1986). Amplification and expression of genes associated with multidrug resistance in mammalian cells. *Science* **232**: 751–755.

Seeds, N. W. (1971). Biochemical differentiation in reaggregating brain cell culture. *Proc. Natl. Acad. Sci. USA* **68**: 1858–1861.

Seglen, P. O. (1975). Preparation of isolated rat liver cells. *Methods Cell Biol.* **13**: 29–83.

Seidel, J. O., Pei, M., Gray, M. L., Langer, R., Freed, L. E., & Vunjak-Novakovic, G. (2004). Long-term culture of tissue engineered cartilage in a perfused chamber with mechanical stimulation. *Biorheology* **41**: 445–458.

Seifert, W., & Müller, H. W. (1984). Neuron-glia interaction in mammalian brain: Preparation and quantitative bioassay of a neurotropic factor (NTF) from primary astrocytes. In Barnes, D. W., Sirbasku, D. A., & Sato, G. H. (eds.), *Methods for serum-free culture of neuronal and lymphoid cells*. New York, Alan R. Liss, pp. 67–78.

Seigel, G. A. (1996). Establishment of an E1a-immortalised retinal cell culture. *In Vitro Cell Dev. Biol. Anim.* **32**: 66–68.

Selby, P. J., Thomas, M. J., Monaghan, P., Sloane, J., & Peckham, M. J. (1980). Human tumour xenografts established and serially transplanted in mice immunologically deprived by thymectomy, cytosine arabinoside and whole-body irradiation. *Br. J. Cancer* **41**: 52.

Selden, R. F., Howie, K. B., Rowe, M. E., Goodman, H. M., & Moore, D. (1986). Human growth hormone as a reporter gene in regulation studies employing transient gene expression. *Mol. Cell Biol.* **6**: 3173–3179.

Seruya, M., Shah, A., Pedrotty, D., du Laney, T., Melgiri, R., McKee, J. A., Young, H. E., & Niklason, L. E. (2004). Clonal population of adult stem cells: life span and differentiation potential. *Cell Transplant.* **13**: 93–101.

Shah, G. (1999). Why do we still use serum in the production of biopharmaceuticals? *Dev. Biol. Stand.* **99**: 17–22.

Shall, S. (1973). Sedimentation in sucrose and Ficoll gradients of cells grown in suspension culture. In Kruse, P. F., & Patterson, M. K. (eds.), *Tissue culture methods and applications*. New York, Academic Press, pp. 198–204.

Shall, S., & McClelland, A. J. (1971). Synchronization of mouse fibroblast LS cells grown in suspension culture. *Nat. New Biol.* **229**: 59–61.

Shansky, J., Chromiak, J., Del Tatto, M., & Vandenburgh, H. (1997). A simplified method for tissue engineering skeletal muscle organoids *in vitro*. *In Vitro Cell Dev. Biol. Anim.* **33**: 659–661.

Shansky, J., Ferland, P., McGuire, S., Powell, C., DelTatto, M., Nackman, M., Hennessey, J., & Vandenburgh, H. H. (2005). Tissue engineering human skeletal muscle for clinical applications. In Vunjak-Novakovic, G., & Freshney, R. I. (eds.), *Culture of cells for tissue engineering*. Hoboken, NJ, Wiley-Liss (in press).

Sharpe, P. T. (1988). *Methods of cell separation*. Amsterdam, Elsevier.

Shaw, R. G., Johnson, A. R., Schulz, W. W., Zahlten, R. N., & Combes, B. (1984) *Hepatology* **4**: 591–602.

Shay, J. W., & Wright, W. E. (1989). Quantitation of the frequency of immortalization of normal human diploid fibroblasts by SV40 large T antigen. *Exp. Cell Res.* **184**: 109–118.

Shay, J. W., Wright, W. E., Brasiskyte, D., & der Haegen, A. (1993). E6 of human papillomavirus type 16 can overcome the M1 stage of immortalization in human mammary epithelial cells but not in human fibroblasts. *Oncogene* **8**: 1407–1413.

Shea, C. M., Edgar, C. M., Teinhorn, T. A., Louis, C. & Gerstenfeld, L. C. (2003). BMP treatment of C3H10T1/2 mesenchymal stem cells induces both chondrogenesis and osteogenesis. *J. Cell. Biochem.* **90**: 1112–1127.

Shiga, M., Kapila, Y. L., Zhang, Q., Hayami, T., & Kapila, S. (2003). Ascorbic acid induces collagenase-1 in human periodontal ligament cells but not in MC3T3-E1 osteoblast-like cells: potential association between collagenase expression and changes in alkaline phosphatase phenotype. *J. Bone Miner. Res.* **18**: 67–77.

Shih, S. J., Dall'Era, M. A., Westphal, J. R., Yang, J., Sweep, C. G., Gandour-Edwards, R., & Evans, C. P. (2003). Elements regulating angiogenesis and correlative microvessel density in benign hyperplastic and malignant prostate tissue. *Prostate Cancer Prostatic Dis.* **6**: 131–137.

Shipley, G. D., & Ham, R. G. (1983). Multiplication of Swiss 3T3 cells in a serum-free medium. *Exp. Cell Res.* **146**: 249–260.

Simon-Assmann, P., Kédinger, M., & Haffen, K. (1986). Immunocytochemical localization of extracellular matrix proteins in relation to rat intestinal morphogenesis. *Differentiation* **32**: 59–66.

Simonsen, J. L., Rosada, C., Serakinci, N., Justesen, J., Stenderup, K., Rattan, S. I., Jensen, T. G., & Kassem, M. (2002). Telomerase expression extends the proliferative life-span and maintains the osteogenic potential of human bone marrow stromal cells. *Nat. Biotechnol.* **20**: 560–561.

Sinha, M. K., Buchanan, C., Raineri-Maldonado, C., Khazanie, P., Atkinson, S., DiMarchi, R., & Caro, J. F. (1990). IGF-II receptors and IGF-II-stimulated glucose transport in human fat cells. *Am. J. Physiol.* **258**: E534–E542.

Skehan, P., Storeng, R., Scudiero, D., Monks, A., McMahon, J., Vistica, D., Warren, J. T., Bokesch, H., Kenney, S., & Boyd, M. R. (1990). New colorimetric cytotoxicity assay for anticancer-drug screening. *J. Natl. Cancer Inst.* **82**: 1107–1112.

Skobe, M., & Fusenig, N. E. (1998). Tumorigenic conversion of immortal human keratinocytes through stromal cell activation. *Proc. Natl. Acad. Sci. USA* **95**: 1–6.

Smith, A. D., Datta, S. P., Smith, G. H., Campbell, P. N., Bentley, R., & McKenzie, H. A. (eds.). (1997). *Oxford dictionary of biochemistry and molecular biology*, Oxford, Oxford University Press.

Smith, A. G., Heath, J. K., Donaldson, D. D., Wong, G. G., Moreau, J., Stahl, M., & Rodgers, D. (1988). Inhibition of pluripotential stem cell differentiation by purified polypeptides. *Nature* **336**: 688–690.

Smith, H. S., Lan, S., Ceriani, R., Hackett, A. J., & Stampfer, M. R. (1981). Clonal proliferation of cultured non-malignant and malignant human breast epithelia. *Cancer Res.* **41**: 4637–4643.

Smith, H. S., Owens, R. B., Hiller, A. J., Nelson-Rees, W. A., & Johnston, J. O. (1976). The biology of human cells in tissue culture; I: Characterization of cells derived from osteogenic sarcomas. *Int. J. Cancer* **17**: 219–234.

Smith, M. D., Summers, M. D., & Frazer, M. J. (1983). Production of human beta interferon in insect cells infected with a baculovirus expression vector. *Mol. Cell Biol.* **3**: 2156–2165.

Smith, P. K., Krohn, R. I., Hermanson, G. T., Mallia, A. K., Gartner, F. H., Provenzano, M. D., Fujimoto, E. K., Goeke, N. M., Olson, B. J., & Klenk, D. C. (1985). Measurement of protein using bicinchoninic acid. *Anal. Biochem.* **150**: 235–239.

Smith, S., & de Lange, T. (1997). TRFI, a mammalian telomeric protein. *Trends Genet.* **113**: 21–26.

Smith, W. L., & Garcia-Perez, A. (1985). Immunodissection: Use of monoclonal antibodies to isolate specific types of renal cells. *Am. J. Physiol.* **248**: F1–F7.

Smola, H., Stark, H.-J., Thiekötter, G., Mirancea, N., Krieg, T., & Fusenig, N. E. (1998). Dynamics of basement membrane formation by keratinocyte–fibroblast interactions in organotypic skin culture. *Exp. Cell Res.* **239**: 399–410.

Smola, H., Thiekötter, G., & Fusenig, N. E. (1993). Mutual induction of growth factor gene expression by epidermal–dermal cell interaction. *J. Cell Biol.* **122**: 417–429.

Smyth, M. J., Rodney, L. Sparks, R. L., Wharton, W. (1993). Proadipocyte cell lines: models of cellular proliferation and differentiation. *J. Cell Sci.* **106**: 1–9.

Snyder, E. Y., Deitcher, D. L., Walsh, C., Arnold Aldea, S., Hartwieg, E. A., Cepko, C. L. (1992). Multipotent neural cell lines can engraft and participate in development of mouse cerebellum. *Cell* **68**, 33–51.

Soder, A. I., Going, J. J., Kaye, S. B., & Keith, W. N. (1998). Tumour specific regulation of telomerase RNA gene expression visualized by *in situ* hybridization. *Oncogene* **16**: 979–983.

Soder, A. I., Hoare, S. F., Muir, S., Going, J. J., Parkinson, E. K., & Keith, W. N. (1997). Amplification, increased dosage and *in situ* expression of the telomerase RNA gene in human cancer. *Oncogene* **14**: 1013–1021.

Solassol, J., Crozet, C., & Lehmann, S. (2003). Prion propagation in cultured cells. *Br. Med. Bull.* **66**: 87–97.

Sommer, I., & Schachner, M. (1981). Cells that are O4-antigen-positive and O1-antigen-negative differentiate into O1 antigen-positive oligodendrocytes. *Neurosci. Lett.* **29**: 183–188.

Soon-Shiong, P., Feldman, E., Nelson, R., Komtebedde, J., Smidsrød, O., Skjak-Bræk, G., Espevik, T., Heintz, R., & Lee, M. (1992). Successful reversal of spontaneous diabetes in dogs by intraperitoneal microencapsulated islets. *Transplantation* **54**: 769–774.

Sordillo, L. M., Oliver, S. P., & Akers, R. M. (1988). Culture of bovine mammary epithelial cells in d-valine modified medium: Selective removal of contaminating fibroblasts. *Cell Biol. Int. Rep.* **12**: 355–364.

Soriano, V., Pepper, M. S., Nakamura, T., Orci, L., & Montesano, R. (1995). Hepatocyte growth factor stimulates extensive development of branching duct-like structures by cloned mammary gland epithelial cells. *J. Cell Sci.* **108**: 413–430.

Sorieul, S., & Ephrussi, B. (1961). Karyological demonstration of hybridization of mammalian cells *in vitro*. *Nature* **190**: 653–654.

Sorour, O., Raafat, M., El-Bolkainy, N., & Mohamad, R. (1975). Infiltrative potentiality of brain tumors in organ culture. *J. Neurosurg.* **43**: 742–749.

Soule, H. D., Maloney, T. M., Wolman, S. R., Teterson, W. D., Brenz, R., McGrath, C. M., Russo, J., Pauley, R. J., Jones, R. F., & Brooks, S. C. (1990). Isolation and characterization of a spontaneously immortalized human breast epithelial cell line, MCF-10. *Cancer Res.* **50**: 6075–6086.

Soule, H. D., Vasquez, J., Long, A., Albert, S., & Brennan, M. (1973). A human cell line from a pleural effusion derived from a breast carcinoma. *J. Natl. Cancer Inst.* **51**: 1409–1416.

Southam, C. M. (1958). Homotransplantation of human cell lines. *Bull. NY Acad. Med.* **34**: 416–423.

Southern, P. J., & Berg, P. (1982). Transformation of mammalian cells to antibiotic resistance with a bacterial gene under control of the SV40 early region promoter. *J. Mol. App. Genet.* **1**: 327–341.

Southgate, J., Masters, J. W. R., & Trejdosiewicz, L. K. (2002). In Freshney, R. I., & Freshney, M. G., (eds), *Culture of epithelial cells*, 2nd ed. New York, Wiley-Liss, pp. 383–399.

Speir, R. E., Griffiths, J. B., & Meignier, B. (eds.). (1991). *Production of biologicals from animal cells in culture*. Oxford, U.K., Butterworth–Heinemann.

Speir, R., & Griffiths, J. B. (1985–1990). *Animal cell biotechnology*. London, Academic Press, 4 vols.

Speirs, V. (2004). Primary culture of human mammary tumor cells. In Pfragner, R., & Freshney. R. I. (eds.), *Culture of human tumor cells*. Hoboken, NJ, Wiley-Liss, pp. 205–219.

Speirs, V., Green, A. R., & White, M. C. (1996). Collagenase III: A superior enzyme for complete disaggregation and improved viability of normal and malignant human breast tissue. *In Vitro Cell Dev. Biol. Anim.* **32**: 72–74.

Speirs, V., Ray, K. P., & Freshney, R. I. (1991). Paracrine control of differentiation in the alveolar carcinoma, A549, by human foetal lung fibroblasts. *Br. J. Cancer* **64**: 693–699.

Spinelli, W., Sonnenfeld, K. H., & Ishii, N. (1982). Effects of phorbol ester tumor promoters and nerve growth factor on neurite outgrowth in cultured human neuroblastoma cells. *Cancer Res.* **42**: 5067–5073.

Splinter, T. A. W., Beudeker, M., & Beek, A. V. (1978). Changes in cell density induced by isopaque. *Exp. Cell Res.* **111**: 245–251.

Spooncer, E., Eliason, J., & Dexter, T. M. (1992). Long-term mouse bone marrow cultures. In Testa, N. G., & Molineux, G. (eds.), *Haemopoiesis: a practical approach*. Oxford, U.K., IRL Press at Oxford University Press, pp. 55–74.

Spremulli, E. N., & Dexter, D. L. (1984). Polar solvents: A novel class of antineoplastic agents. *J. Clin. Oncol.* **2**: 227–241.

Sredni, B., Sieckmann, D. G., Kumagai, S. H., Green, I., & Paul, W. E. (1981). Long term culture and cloning of non-transformed human B-lymphocytes. *J. Exp. Med.* **154**: 1500–1516.

Stacey, G. N., Bolton, B. J., & Doyle, A. (1993). Multilocus DNA fingerprinting used for definitive isolation of HeLa contamination in cell lines and determination of genetic diversity amongst HeLa cell clones. *In Vitro Cell Dev. Biol.* **29A**: 123A.

Stacey, G. N., Bolton, B. J., Morgan, D., Clark, S. A., & Doyle, A. (1992). Multilocus DNA fingerprint analysis of cell-banks: Stability studies and culture identification in human B-lymphoblastoid and mammalian cell lines. *Cytotechnology* **8**: 13–20.

Stacey, G. N., Hoelzl, H., Stehenson, J. R., & Doyle, A. (1997). Authentication of animal cell cultures by direct visualisation of repetitive DNA, aldolase gene PCR and isoenzyme analysis. *Biologicals* **25**: 75–85.

Stacey, G. N., Masters, J. R. M., Hay, R. J., Drexler, H. G., MacLeod, R. A. F., & Freshney, R. I. (2000). Cell contamination leads to inaccurate data: we must take action now. *Nature* **403**: 356.

Stampfer, M., Halcones, R. G., & Hackett, A. J. (1980). Growth of normal human mammary cells in culture. *In Vitro* **16**: 415–425.

Stampfer, M. R., Yaswen, P., & Taylor-Papadimitriou, J. (2002). Culture of human mammary epithelial cells. In Freshney, R. I. & Freshney, M. G. (eds.), *Culture of epithelial cells*, 2nd ed. Hoboken, NJ, Wiley-Liss, pp. 95–135.

Stanbridge, E. J., & Doersen, C.-J. (1978). Some effects that mycoplasmas have upon their injected host. In McGarrity, G. J., Murphy, D. G., & Nichols, W. W. (eds.), *Mycoplasma infection of cell cultures*. New York, Plenum Press, pp. 119–134.

Stanley, M. A., & Parkinson, E. (1979). Growth requirements of human cervical epithelial cells in culture. *Int. J. Cancer* **24**: 407–414.

Stanley, M. A. (2002). Culture of human cervical epithelial cells. In Freshney, R. I. & Freshney, M. G. (eds.), *Culture of epithelial cells*, 2nd ed. Hoboken, NJ, Wiley-Liss, pp. 138–169.

Stanners, C. P., Eliceri, G. L., & Green, H. (1971). Two types of ribosome in mouse–hamster hybrid cells. *Nat. New Biol.* **230**: 52–54.

Stanton, B. A., Biemesderfer, D., Wade, J. B., & Giebisch, G. (1981). Structural and functional study of the rat distal nephron: Effects of potassium adaptation and depletion. *Kidney Int.* **19**: 36–48.

Stark, H. J., Baur, M., Breitkreutz, D., Mirancea, N., & Fusenig, N. E. (1999). Organotypic keratinocyte cocultures in defined medium with regular epidermal morphogenesis and differentiation. *J. Invest. Dermatol.* **112**: 681–691.

States, B., Foreman, J., Lee, J., & Segal, S. (1986). Characteristics of cultured human renal cortical epithelia. *Biochem. Med. Metab. Biol.* **36**: 151–161.

Steel, G. G. (1979). Terminology in the description of drug-radiation interactions. *Int. J. Radiat. Oncol. Biol. Phys.* **5**: 1145–1150.

Steele, M. P., Levine, R. A., Joyce-Brady, M., & Brody, J. S. (1992). A rat alveolar type II cell line developed by adenovirus 12 SE1A gene transfer. *Am. J. Resp. Cell Mol. Biol.* **6**: 50–56.

Stein, H. G., & Yanishevsky, R. (1979). Autoradiography. In Jakoby, W. B., & Pastan, I. H. (eds.), *Methods in enzymology; vol. 57: Cell culture*. New York, Academic Press, pp. 279–292.

Steinberg, M. L. (1996). Immortalization of human epidermal keratinocytes by SV40. In Freshney, R. I., & Freshney, M. G. (eds.), *Culture of immortalized cells*. New York, Wiley-Liss, pp. 95–120.

Stern, P., West, C., & Burt, D. (2004). Culture of cervical carcinoma cell lines.. In Pfragner, R., & Freshney, R. I. (eds.), *Culture of human tumor cells*. Hoboken, NJ, Wiley-Liss, pp. 179–204.

Steube, K. G., Grunicke, D., Drexler, H. G. (1995). Isoenzyme analysis as a rapid method for the examination of the species identity of cell cultures. *In Vitro Cell Dev Biol Anim.*, **31**: 115–9.

Stewart, C. E. H., James, P. L., Fant, M. E., & Rotwein, P. (1996). Overexpression of insulin-like growth factor-II induces accelerated myoblast differentiation. *J. Cell. Physiol.* **169**: 23–32.

Stoker, M. G. P. (1973). Role of diffusion boundary layer in contact inhibition of growth. *Nature* **246**: 200–203.

Stoker, M. G. P., & Rubin, H. (1967). Density dependent inhibition of cell growth in culture. *Nature* **215**: 171–172.

Stoker, M., O'Neill, C., Berryman, S., & Waxman, B. (1968). Anchorage and growth regulation in normal and virus transformed cells. *Int. J. Cancer* **3**: 683–693.

Stoker, M., Perryman, M., & Eeles, R. (1982). Clonal analysis of morphological phenotype in cultured mammary epithelial cells from human milk. *Proc. R., Soc. Lond., Ser. B* **215**: 231–240.

Stoner, G. D., Katoh, Y., Foidart, J.-M., Trump, B. F., Steinert, P., & Harris, C. C. (1981). Cultured human bronchial epithelial cells: Blood group antigens, keratin, collagens and fibronectin. *In Vitro* **17**: 577–587.

Strange, R., Li, F., Fris, R. R., Reichmann, E., Haenni, B., & Burri, P. H. (1991). Mammary epithelial differentiation *in vitro*: Minimum requirements for a functional response to hormonal stimulation. *Cell Growth Differ.* **2**: 549–559.

Strangeways, T. S. P. & Fell, H. B. (1925). Experimental studies on the differentiation of embryonic tissues growing *in vivo* and *in vitro*. I. The development of the undifferentiated limb-bud (a) when subcutaneously grafted into the post-embryonic chick and (b) when cultivated *in vitro*. *Proc. Roy. Soc, London, ser. B*, **99**: 340–366.

Strangeways, T. S. P. & Fell, H. B. (1926). Experimental studies on the differentiation of embryonic tissues growing *in vivo* and *in vitro*. II. The development of the isolated early embryonic eye of the fowl when cultivated *in vitro*. *Proc. Roy. Soc, London, ser. B*, **100**: 273–283.

Strauss, W. M. (1998). Transfection of mammalian cells with yeast artificial chromosomes. In Ravid, K., & Freshney, R. I. (eds.), *DNA transfer to cultured cells*. New York, Wiley-Liss, pp. 213–236.

Strickland, S., & Beers, W. H. (1976). Studies on the role of plasminogen activator in ovulation: *In vitro* response of granulosa cells to gonadotropins, cyclic nucleotides, and prostaglandins. *J. Biol. Chem.* **251**: 5694–5702.

Stryer, L. (1995). *Biochemistry*, 4th ed. New York, W. H. Freeman, p. 505.

Stubblefield, E. (1968). Synchronization methods for mammalian cell cultures. In Prescott, D. M. (ed.), *Methods in cell physiology*. New York, Academic Press, pp. 25–43.

Styles, J. A. (1977). A method for detecting carcinogenic organic chemicals using mammalian cells in culture. *Br. J. Cancer* **36**: 558.

Su, X., Sorenson, C. M., & Sheibani, N. (2003). Isolation and characterization of murine retinal endothelial cells. *Mol. Vis.* **1**: 171–178.

Su, H. Y., Bos, T. J., Monteclaro, F. S., & Vogt, P. K. (1991). Jun inhibits myogenic differentiation. *Oncogene* **6**: 1759–1766.

Subramanian, M., Madden, J. A., & Harder, D. R. (1991). A method for the isolation of cells from arteries of various sizes. *J. Tissue Cult Methods* **13**: 13–20.

Subramanian, S. V., Fitzgerald, M. L., & Bernfield, M. (1997). Regulated shedding of syndecan- 1 and -4 ectodomains by thrombin and growth factor receptor activation. *J. Biol. Chem.* **272**: 14713–14720.

Suggs, J. E., Madden, M. C., Friedman, M., & Edgell, C.-J. S. (1986). Prostacyclin expression by a continuous human cell line derived from vascular endothelium. *Blood* **4**: 825–829.

Sugimoto, M., Tahara, H., Ide, T., & Furuichi, Y. (2004). Steps involved in immortalization and tumorigenesis in human B-lymphoblastoid cell lines transformed by Epstein-Barr virus. *Cancer Res.* **64**: 3361–3364.

Suh, H., Song, M. J., & Park, Y. N. (2003). Behavior of isolated rat oval cells in porous collagen scaffold. *Tissue Eng.* **9**: 411–20.

Sun, L., Bradford, C. S., & Barnes, D. W. (1995a). Feeder cell cultures for zebrafish embryonal cells *in vitro*. *Mol. Mar. Biol. Biotech.* **4**: 43–50.

Sun, L., Bradford, C. S., Ghosh, C., Collodi, P., & Barnes, D. W. (1995b). ES-like cell cultures derived from early zebrafish embryos. *Mol. Mar. Biol. Biotech.* **4**: 193–199.

Sundqvist, K., Liu, Y., Arvidson, K., Ormstad, K., Nilsson, L., Toftgård, R., & Grafström, R. C. (1991). Growth regulation of serum-free cultures of epithelial cells from normal human buccal mucosa. *In Vitro Cell Dev. Biol.* **27A**: 562–568.

Sutherland, R. M. (1988). Cell and micro environment interactions in tumour microregions: The multicell spheroid model. *Science* **240**: 117–184.

Suva, D., Garavaglia, G., Menetrey, J., Chapuis, B., Hoffmeyer, P., Bernheim, L., & Kindler, V. (2004). Non-hematopoietic human bone marrow contains long-lasting, pluripotential mesenchymal stem cells. *J. Cell. Physiol.* **198**: 110–118.

Suzuki, A., Nakauchi, H., & Taniguchi, H. (2004). Prospective isolation of multipotent pancreatic progenitors using flow-cytometric cell sorting. *Diabetes.* **53**: 2143–2152.

Suzuki, T., Saha, S., Sumantri, C., Takagi, M., & Boediono, A. (1995). The influence of polyvinylpyrrolidone on freezing of bovine IVF blastocysts following biopsy. *Cryobiology* **32**: 505–510.

Swope, V. B., Supp, A. P., Cornelius, J. R., Babcock, G. F., & Boyce, S. T. (1997). Regulation of pigmentation in cultured skin substitutes by cytometric sorting of melanocytes and keratinocytes. *J. Invest. Dermatol.* **109**: 289–295.

Sykes, J. A., Whitescarver, J., Briggs, L., & Anson, J. H. (1970). Separation of tumor cells from fibroblasts with use of discontinuous density gradients. *J. Natl. Cancer Inst.* **44**: 855–864.

Tagawa, M., Yokosuka, O., Imazeki, F., Ohto, M., & Omata, M. (1996). Gene expression and active virus replication in the liver after injection of duck hepatitis B virus DNA into the peripheral vein of ducklings. *J. Hepatol.* **24**: 328–334.

Takahashi, H., Yanagi, Y., Tamaki, Y., Muranaka, K., Usui, T., & Sata, M. (2004). Contribution of bone-marrow-derived cells to choroidal neovascularization. *Biochem. Biophys. Res. Commun.* **320**: 372–375.

Takahashi, K., & Okada, T. S. (1970). Analysis of the effect of "conditioned medium" upon the cell culture at low density. *Dev. Growth Differ.* **12**: 65–77.

Takahashi, K., Mitsui, K., & Yamanaka, S. (2003). Role of Eras in promoting tumour-like properties in mouse embryonic stem cells. *Nature* **423**: 541–545.

Takahashi, K., Suzuki, K., Kawahara, S., & Ono, T. (1991). Effects of lactogenic hormones on morphological development and growth of human breast epithelial cells cultivated in collagen gels. *Jpn. J. Cancer Res.* **82**: 553.

Takeda, K., Minowada, J., & Bloch, A. (1982). Kinetics of appearance of differentiation-associated characteristics in ML-1, a line of human myeloblastic leukaemia cells, after treatment with TPA, DMSO, or Ara-C. *Cancer Res.* **42**: 5152–5158.

Tarella, C., Ferrero, D., Gallo, E., Luyca Pagliardi, G., & Ruscetti, F. W. (1982). Induction of differentiation of HL-60 cells by dimethylsulphoxide: Evidence for a stochastic model not linked to the cell division cycle. *Cancer Res.* **42**: 445–449.

Tashjian, A. H., Jr. (1979). Clonal strains of hormone-producing pituitary cells. In Jakoby, W. B., & Pastan, I. H. (eds.), *Methods in enzymology; Vol. 57: Cell culture.* New York, Academic Press, pp. 527–535.

Tashjian, A. H., Yasamura, Y., Levine, L., Sato, G. H., & Parker, M. (1968). Establishment of clonal strains of rat pituitary tumor cells that secrete growth hormone. *Endocrinology* **82**: 342–352.

Taub, M.. (1984). Growth of primary and established kidney cell cultures in serum-free media. In Barnes, D. W., Sirbasku, D. A., Sato, G. H., eds., *Methods for Serum-Free Culture of Epithelial and Fibroblastic Cells*, New York, Alan R. Liss, pp 3–24.

Taub, M. L., Yang, S. I., & Wang, Y. (1989). Primary rabbit proximal tubule cell cultures maintain differentiated functions when cultured in a hormonally defined serum-free medium. *In Vitro Cell Dev. Biol.* **25**: 770–775.

Taylor, D. A. (2001). Cellular cardiomyoplasty with autologous skeletal myoblasts for ischemic heart disease and heart failure. *Curr. Control. Trials Cardiovasc. Med.* **2**: 208–210.

Taylor, J. H. (1958). Sister chromatid exchanges in tritium labeled chromosomes. *Genetics* **43**: 515–529.

Taylor-Papadimitriou, J., Purkiss, P., & Fentiman, I. S. (1980). Choleratoxin and analogues of cyclic AMP stimulate the growth of cultured human epithelial cells. *J. Cell Physiol.* **102**: 317–322.

Taylor-Papadimitriou, J., Shearer, M., & Stoker, M. G. P. (1977). Growth requirement of human mammary epithelial cells in culture. *Int. J. Cancer* **20**: 903–908.

Taylor-Papadimitriou, J., Stampfer, M., Bartek, J., Lewis, A., Boshell, M., Lane, E. B., & Leigh, I. M. (1989) Keratin expression in human mammary epithelial cells cultured from normal and malignant tissue: relation to *in vivo* phenotypes and influence of medium. *J. Cell Sci.* **94**: 403–413.

Tedder, R. S., Zuckerman, M. A., Goldstone, A. H., Hawkins, A. E., Fielding, A., Briggs, E. M., Irwin, D., Blair, S., Gorman, A. M., Patterson, K. G., Linch, D. C., Heptonstall, J., & Brink, N. S. (1995). Hepatitis B transmission from contaminated cryopreservation tank. *Lancet* **346**: 137–140.

Temin, H. M. (1966). Studies on carcinogenesis by avian sarcoma viruses; III: The differential effect of serum and polyanions on multiplication of uninfected and converted cells. *J. Natl. Cancer Inst.* **37**: 167–175.

Teofili, L., Rutella, S., Pierelli, L., Ortu la Barbera, E., di Mario, A., Menichella, G., Rumi, C., & Leone, G. (1996). Separation of chemotherapy plus G-CSF-mobilized peripheral blood mononuclear cells by counterflow centrifugal elutriation: *In vitro* characterization of two different CD34[+] cell populations. *Bone Marrow Transplant.* **18**: 421–425.

Terasaki, T., Kameya, T., Nakajima, T., Tsumuraya, M., Shimosato, Y., Kato, K., Ichinose, H., Nagatsu, T., & Hasegawa, T. (1984). Interconversion of biological characteristics of small cell lung cancer cells depending on the culture conditions. *Gann* **75**: 1689–1699.

Testa, N. G., & Molineux, G. (eds.). (1993). *Haemopoiesis.* Oxford, U.K., Oxford University Press.

Thacker, J., Webb, M. B., & Debenham, P. G. (1988). Fingerprinting cell lines: Use of human hypervariable DNA probes to characterise mammalian cell cultures. *Somat. Cell Mol. Genet.* **14**: 519–525.

Thelwall, P. E., Neves, A. A., & Brindle, K. M. (2001). Measurement of bioreactor perfusion using dynamic contrast agent-enhanced magnetic resonance imaging. *Biotechnol. Bioeng.* **75**: 682–690.

Thomas, D. G. T., Darling, J. L., Paul, E. A., Mott, T. C., Godlee, J. N., Tobias, J. S., Capra, L. G., Collins, C. D., Mooney, C., Bozek, T., Finn, G. P., Arigbabu, S. O., Bullard, D. E., Shannon, N., & Freshney, R. I. (1985). Assay of anti-cancer drugs in tissue culture: Relationship of relapse free interval (FRI) and *in vitro* chemosensitivity in patients with malignant cerebral glioma. *Br. J. Cancer* **51**: 525–532.

Thomas, S., Gray, E., & Robinson, C. J. (1997). Response of HUVEC and EAhy926 and fibroblast growth factors. *In Vitro Cell Dev. Biol. Anim.* **33**: 492–494.

Thompson, E. B., Tomkins, J. M., & Curran, J. F. (1966). Induction of tyrosine α-ketoglutarate transaminase by steroid hormones in a newly established tissue culture cell line. *Proc. Natl. Acad. Sci. USA* **56**: 296–303.

Thompson, L. H., & Baker, R. M. (1973). Isolation of mutants of cultured mammalian cells. In Prescott, D. (ed.), *Methods in cell biology*, Vol. 6. New York, Academic Press, pp. 209–281.

Thomson, A. A., Foster, B. A., & Cunha, G. R. (1997). Analysis of growth factor and receptor mRNA levels during development of the rat seminal vesicle and prostate. *Development* **124**: 2431–2439.

Thomson, A. W. (ed.). (1991). *Cytokine handbook*. London, Academic Press.

Thomson, J. A., Itskovitz-Eldor, J., Shapiro, S. S., Waknitz, M. A., & Swiergiel, J. J. (1998). Embryonic stem cell lines derived from human blastocysts. *Science* **282**: 1145–1147.

Thorsell, A., Blomqvist, A. G., & Heilig, M. (1996). Cationic lipid-mediated delivery and expression of prepro-neuro-peptide Y cDNA after intraventricular administration in rat: Feasibility and limitations. *Regul. Peptides* **61**: 205–211.

Till, J. E., & McCulloch, E. A. (1961). A direct measurement of the radiation sensitivity of normal mouse bone marrow cells. *Radiat. Res.* **14**: 213–222.

Tobey, R. A., Anderson, E. C., & Peterson, D. F. (1967). Effect of thymidine on duration of G1 in chinese hamster cells. *J. Cell Biol.* **35**: 53–67.

Todaro, G. J., & DeLarco, I. E. (1978). Growth factors produced by sarcoma virus-transformed cells. *Cancer Res.* **38**: 4147–4154.

Todaro, G. J., & Green, H. (1963). Quantitative studies of the growth of mouse embryo cells in culture and their development into established lines. *J. Cell Biol.* **17**: 299–313.

Tomakidi, P., Fusenig, N. E., Kohl, A., & Komposch, G. (1997). Histomorphological and biochemical differentiation capacity in organotypic co-cultures of primary gingival cells. *J. Periodont. Res.* **32**: 388–400.

Toneguzzo, F., Keating, A., Glynn, S., & McDonald, K. (1988). Electrical field mediated gene transfer: Characterization of DNA transfer and patterns of integration in lymphoid cells. *Nucleic Acids Res.* **16**: 5515–5532.

Topley, P., Jenkins, D. C., Jessup, E. A., & Stables, J. N. (1993). Effect of reconstituted basement membrane components on the growth of a panel of human tumour cell lines in nude mice. *Br. J. Cancer* **67**: 953–958.

Torday, J. S., & Kourembanas, S. (1990). Fetal rat lung fibroblasts produce a TGF-β homologue that blocks type II cell maturation. *Dev. Biol.* **13**: 35–41.

Totoiu, M. O., Nistor, G. I., Lane, T. E., & Keirstead, H. S. (2004). Remyelination, axonal sparing, and locomotor recovery following transplantation of glial-committed progenitor cells into the MHV model of multiple sclerosis. *Exp. Neurol.* **187**: 254–265.

Tozer, B. T., & Pirt, S. J. (1964). Suspension culture of mammalian cells and macromolecular growth promoting fractions of calf serum. *Nature* **201**: 375–378.

Traganos, F., Darzynkiewicz, Z., Sharpless, T., & Melamed, M. R. (1977). Nucleic acid content and cell cycle distribution of five human bladder cell lines analyzed by flow cytofluorometry. *Int. J. Cancer* **20**: 30–36.

Trapp, B. D., Honegger, P., Richelson, E., & Webster, H. de F. (1981). Morphological differentiation of mechanically dissociated fetal rat brain in aggregating cell cultures. *Brain Res.* **160**: 235–252.

Trickett, A. E., Ford, D. J., Lam-Po Tang, P. R. L., & Vowels, M. R. (1990). Comparison of magnetic particles for immunomagnetic bone marrow purging using an acute lymphoblastic leukaemia model. *Transpl. Proc.* **22**: 2177–2178.

Triglia, D., Braa, S. S., Yonan, C., & Naughton, G. K. (1991). Cytotoxicity testing using neutral red and MTT assays on a three-dimensional human skin substrate. *Toxic. In Vitro* **5**: 573–578.

Trotter, J., & Schachner, M. (1988). Cells positive for the O4 surface antigen isolated by cell sorting are able to differentiate into oligodendrocytes and type-2 astrocytes. *Dev. Brain Res.* **46**: 115–122.

Trowell, O. A. (1959). The culture of mature organs in a synthetic medium. *Exp. Cell Res.* **16**: 118–147.

Troyer, D. A., & Kreisberg, J. I. (1990). Isolation and study of glomerular cells. *Methods Enzymol.* **191**: 141–152.

Tsao, M. C., Walthall, B. I., & Ham, R. G. (1982). Clonal growth of normal human epidermal keratinocytes in a defined medium. *J. Cell Physiol.* **110**: 219–229.

Tsao, S.-W., Mok, S. C., Fey, E. G., Fletcher, J. A., Wan, T. S. K., Chew, E.-C., Muto, M. G., Knapp, R. C., & Berkowitz, R. S. (1995). Characterisation of human ovarian surface epithelial cells immortalized by human papilloma viral oncogenes (HPV-E6E7 ORFs). *Exp. Cell Res.* **218**: 499–507.

Tsuruo, T., Hamilton, T. C., Louis, K. G., Behrens, B. C., Young, R. C., & Ozols, R. F. (1986). Collateral susceptibility of adriamycin-, melphalan-, and cisplatin-resistant human ovarian tumor cells to bleomycin. *Jpn. J. Cancer Res.* **77**: 941–945.

Tumilowicz, J. J., Nichols, W. W., Cholon, J. J., & Greene, A. E. (1970). Definition of a continuous human cell line derived from neuroblastoma. *Cancer Res.* **30**: 2110–2118.

Turner, R. W. A., Siminovitch, L., McCulloch, E. A., & Till, J. E. (1967). Density gradient centrifugation of hemopoietic colony-forming cells. *J. Cell Physiol.* **69**: 73–81.

Tuszynski, M. H., Roberts, J., Senut, M. C., U, H. S., & Gage, F. H. (1996). Gene therapy in the adult primate brain: Intraparenchymal grafts of cells genetically modified to produce nerve growth factor prevent cholinergic neuronal degeneration. *Gene Therapy* **3**: 305–314.

Tveit, K. M., & Pihl, A. (1981). Do cells lines *in vitro* reflect the properties of the tumours of origin? A study of lines derived from human melanoma xenografts. *Br. J. Cancer* **44**: 775–786.

Twentyman, P. R. (1980). Response to chemotherapy of EMT6 spheroids as measured by growth delay and cell survival. *Eur. J. Cancer* **42**: 297–304.

U.S. Department of Health and Human Services (1993). *Biosafety in microbiological and biomedical laboratories*, 3d ed. Publication (CDC) 93–8395, Centers for Disease Control, US Govt. Printing Office, Washington, DC.

U.S. Nuclear Regulatory Commission (1997). Draft regulatory guide DG-0006. Guide for the preparation of applications for commercial nuclear pharmacy Licenses. Office of Nuclear Regulatory Research, U.S. Nuclear Regulatory Commission, Washington, DC 20555.

Uchida, I. A., & Lin, C. C. (1974). Quinacrine fluorescent patterns. In Yunis, J. (ed.), *Human chromosome methodology*, 2d ed. New York, Academic Press, pp. 47–58.

United States Pharmacopeia. (1985). *Sterility tests*, 21st revision. United States Pharmacopeial Convention, Inc., pp. 1156–1160.

Unkless, I., Dano, K., Kellerman, G., & Reich, E. (1974). Fibrinolysis associated with oncogenic transformation: Partial purification and characterization of cell factor, a plasminogen activator. *J. Biol. Chem.* **249**: 4295–4305.

Uphoff, C. C., & Drexler, H. G. (2002) Comparative PCR analysis for detection of mycoplasma infections in continuous cell lines. *In Vitro Cell. Dev. Biol. Animal* **38**: 79–85.

Uphoff, C. C., & Drexler, H. G. (2004). Elimination of Mycoplasma from infected cell lines using antibiotics. *Methods Mol. Med.* **88**: 327–334.

Ure, J. M., Fiering, S., & Smith, A. G. (1992). A rapid and efficient method for freezing and recovering clones of embryonic stem cells. *Trends Genet.* **8**: 6.

Uzgare, A. R., Xu, Y., & Isaacs, J. T. (2004). *In vitro* culturing and characteristics of transit amplifying epithelial cells from human prostate tissue. *J. Cell. Biochem.* **91**: 196–205.

Uzgare, A. R., Xu, Y., & Isaacs, J. T. (2004). *In vitro* culturing and characteristics of transit amplifying epithelial cells from human prostate tissue. *J. Cell. Biochem.* **91**: 196–205.

Vachon, P. H., Perreault, N., Magny, P., & Beallieu, J.-F. (1996). Uncoordinated transient mosaic patterns of intestinal hydrolase expression in differentiating human enterocytes. *J. Cell Physiol.* **166**: 198–207.

Vago, C. (ed.). (1971). *Invertebrate tissue culture*, Vol. 1. New York, Academic Press.

Vago, C. (ed.). (1972). *Invertebrate tissue culture*, Vol. 2. New York, Academic Press.

Vaheri, A., Ruoslahti, E., Westermark, B., & Pontén, J. (1976). A common cell-type specific surface antigen in cultured human glial cells and fibroblasts: Loss in malignant cells. *J. Exp. Med.* **143**: 64–72.

van Bokhoven, A., Varella-Garcia, M., Korch, C., & Miller, G. J. (2001a). TSU-Pr1 and JCA-1 cells are derivatives of T24 bladder carcinoma cells and are not of prostatic origin. *Cancer Res.* **61**: 6340–6344.

van Bokhoven, A., Varella-Garcia, M., Korch, C., Hessels, D., & Miller, G. J. (2001b). Widely used prostate carcinoma cell lines share common origins. *Prostate* **47**: 36–51.

Van Diggelen, O., Shin, S., & Phillips, D. (1977). Reduction in cellular tumorigenicity after mycoplasma infection and elimination of mycoplasma from infected cultures by passage in nude mice. *Cancer Res.* **37**: 2680–2687.

Van Helden, P. D., Wiid, I. J., Albrecht, C. F., Theron, E., Thornley, A. L., & Hoal-van Helden, E. G. (1988). Cross-contamination of human esophageal squamous carcinoma cell lines detected by DNA fingerprint analysis. *Cancer Res.* **48**: 5660–5662.

Vanparys, P. (2002). ECVAM and pharmaceuticals. *Altern Lab Anim.*, **30** Suppl 2: 221–3.

Van Roozendahl, C. E. P., van Ooijen, B., Klijn, J. G. M., Claasen, C., Eggermont, A. M. M., Henzen-Logmans, S. C., & Foekens, J. A. (1992). Stromal influences on breast cancer cell growth. *Br. J. Cancer* **65**: 77–81.

Varga Weisz, P. D., & Barnes, D. W. (1993). Characterization of human plasma growth inhibitory activity on serumfree mouse embryo cells. *In Vitro Cell Dev. Biol.* **29A**: 512–516.

Varon, S., & Manthorpe, M. (1980). Separation of neurons and glial cells by affinity methods. In Fedoroff, S., & Hertz, L. (eds.), *Advances in cellular neurobiology*, Vol. 1. New York, Academic Press, pp. 405–442.

Vaughan, A., & Milner, A. (1989). Fluorescence activated cell sorting. In Catty, D. (ed.), *Antibodies; Volume II: A practical approach*. Oxford, U.K., IRL Press at Oxford University Press, pp. 201–222.

Vaziri, H., & Benchimol, S. (1998). Reconstitution of telomerase activity in normal human cells leads to elongation of telomeres and extended replicative life-span. *Curr. Biology* **8**: 279–282.

Velcich, A., Palumbo, L., Jarry, A., Laboisse, C., Racevskis, J., & Augenlicht, L. (1995). Patterns of expression of lineage-specific markers during the *in vitro*-induced differentiation of HT29 colon carcinoma cells. *Cell Growth Differ.* **6**: 749–757.

Venitt, S. (1984). *Mutagenicity testing, a practical approach*. Oxford, U.K., IRL Press.

Venkatasubramanian, J., Sahi, J., Rao, M. C. (2000). Ion transport during growth and differentiation. *Ann N Y Acad Sci.* 2000;**915**: 357–72.

Verbruggen, G., Veys, E. M., Wieme, N., Malfait, A. M., Gijselbrecht, L., Nimmegeers, J., Almquist, K. F., & Broddelez, C. (1990). The synthesis and immobilization of cartilage-specific proteoglycan by human chondrocytes in different concentrations of agarose. *Clin. Exp. Rheumatol.* **8**: 371–378.

Verschraegen, C. F., Hu, W., Du, Y., Mendoza, J., Early, J., Deavers, M., Freedman, R. S., Bast, R. C. Jr., Kudelka, A. P., Kavanagh, J. J., & Giovanella, B. C. (2003). Establishment and characterization of cancer cell cultures and xenografts derived from primary or metastatic Mullerian cancers. *Clin. Cancer Res.* **9**: 845–852.

Vescovi, A., Gritti, A., Cossu, G., & Galli, R. (2002). Neural stem cells: plasticity and their transdifferentiation potential. *Cells Tissues Organs* **171**: 64–76.

Vierick, J. L., McNamara, P., & Dodson, M. V. (1996). Proliferation and differentiation of progeny of ovine unilocular fat cells (adipofibroblasts). *In Vitro Cell Dev. Biol. Anim.* **32**: 564–572.

Vilamitjana-Amedee, J., Bareile, R., Rouais, F., Caplan, A. I., & Harmand, M. F. (1993). Human bone marrow stromal cells express an osteoblastic phenotype in culture. *In Vitro Cell Dev. Biol.* **29A**: 699–707.

Visser, J. W., & De Vries, P. (1990). Identification and purification of murine hematopoietic stem cells by flow cytometry. *Methods Cell Biol.* **33**: 451–468.

Vistica, D. T., Skehan, P., Scudiero, D., Monks, A., Pittman, A., & Boyd, M. R. (1991). Tetrazolium-based assays for cellular viability: A critical examination of selected parameters affecting formazan production. *Cancer Res.* **51**: 2515–2520.

Vlodavsky, I., Lui, G. M., & Gospodarowicz, D. (1980). Morphological appearance, growth behavior and migratory activity of human tumor cells maintained on extracellular matrix versus plastic. *Cell* **19**: 607–617.

Vogel, F. R., & Powell, M. F. (1995). A compendium of vaccine adjuvants and excipients. In Powell, M. F., & Newman, M. (eds.), *Vaccine design: The subunit and adjuvant approach.* New York, Plenum Publishing, pp. 141–228.

von Briesen, H., Andreesen, R., Esser, R., Brugger, W., Meichsner, C., Becker, K., & Rubsamen-Waigmann, H. (1990). Infection of monocytes/macrophages by HIV *in vitro. Res Virol.* **141**: 225–231.

Von der Mark, K. (1986). Differentiation, modulation and dedifferentiation of chondrocytes. *Rheumatology* **10**: 272–315.

Vondracek, M., Weaver, D., Sarang, Z., Hedberg, J. J., Willey, J., Wärngård, L. and Grafström, R. C. (2002). Transcript profiling of enzymes involved in detoxification of xenobiotics and reactive oxygen in human normal and Simian virus 40 T antigen-immortalized oral keratinocytes. *Int. J. Cancer* **99**: 776–782.

Vonen, B., Bertheussen, K., Giaever, A. K., Florholmen, J., & Burhol, P. G. (1992). Effect of a new synthetic serum replacement on insulin and somatostatin secretion from isolated rat pancreatic islets in long term culture. *J. Tissue Cult. Methods* **14**: 45–50.

Von Hoff, D. D., Clark, G. M., Weis, G. R., Marshall, M. H., Buchok, J. B., Knight, W. A., & Lemaistre, C. F. (1986). Use of *in vitro* dose response effects to select antineoplastics for high dose or regional administration regimens. *J. Clin. Oncol.* **4**: 18–27.

Vouret-Craviari, V., Boulter, E., Grall, D., Matthews, C., & Van Obberghen-Schilling, E. (2004). ILK is required for the assembly of matrix-forming adhesions and capillary morphogenesis in endothelial cells. *J Cell Sci.* **117**: 4559–4569.

Voyta, J. C., Via, D. P., Butterfield, C. W., & Zetter, B. R. (1984). Identification and isolation of endothelial cells based on their increased uptake of acetylated–low density lipoprotein. *J. Cell Biol.* **99**: 2034–2040.

Vries, J. E., Benthem, M., & Rumke, P. (1973). Separation of viable from nonviable tumor cells by flotation on a Ficoll-triosil mixture. *Transplantation* **5**: 409–410.

Vunjak-Novakovic, G. (2005). Basic principles of tissue engineering. In Vunjak-Novakovic, G., & Freshney, R. I. (eds.), *Culture of cells for tissue engineering.* Hoboken, NJ, Wiley-Liss (in press).

Vunjak-Novakovic, G., & Freshney, R. I. (eds.) (2005). *Culture of cells for tissue engineering.* Hoboken, NJ, Wiley-Liss (in press).

Wada, T., Dacy, K. M., Guan, X.-P., & Ip, M. M. (1994). Phorbol 12-myristate 13 acetate stimulates proliferation and ductal morphogenesis and inhibits functional differentiation of normal rat mammary epithelial cells in primary culture. *J. Cell Physiol.* **158**: 97–109.

Waleh, N. S., Brody, M. D., Knapp, M. A., Mendonca, H. L., Lord, E. M., Koch, C. J., Laderoute, K. R., & Sutherland, R. M. (1995). Mapping of the vascular endothelial growth factor-producing hypoxic cells in multicellular tumour spheroids using a hypoxia-specific marker. *Cancer Res.* **55**: 6222–6226.

Walston, J., Silver, K., Bogardus, C., Knowler, W. C., Celi, F. S., Austin, S., Manning, B., Strosberg, A. D., Stern, M. P., Raben, N., Sorkin, J. D., Roth, J., & Shuldiner, A. R. (1995). Time of onset of non-insulin-dependent diabetes mellitus and genetic variation in the beta 3-adrenergic-receptor gene. *N. Engl. J., Med.* **333**: 343–347.

Walter, H. (1975). Partition of cells in two-polymer aqueous phases: A method for separating cells and for obtaining information on their surface properties. In Prescott, D. M. (ed.), *Methods in cell biology.* New York, Academic Press, pp. 25–50.

Walter, H. (1977). Partition of cells in two-polymer aqueous phases: A surface affinity method for cell separation. In Catsimpoolas, N. (ed.), *Methods of cell separation.* New York, Plenum Press, pp. 307–354.

Wang, J., Torbenson, M., Wang, Q., Ro, J. Y., & Becich, M. (2003). Expression of inducible nitric oxide synthase in paired neoplastic and non-neoplastic primary prostate cell cultures and prostatectomy specimen. *Urol. Oncol.* **21**: 117–122.

Wang, R. J., & Nixon, B. R. (1978). Identification of hydrogen peroxide as a photoproduct toxic to human cells in tissue-culture medium irradiated with "daylight" fluorescent light. *In Vitro.* **14**: 715–22.

Wang, H. C., & Fedoroff, S. (1972). Banding in human chromosomes treated with trypsin. *Nat. New Biol.* **235**: 52–53.

Wang, H. C., & Fedoroff, S. (1973). Karyology of cells in culture: Trypsin technique to reveal G-bands. In Kruse, P. F., & Patterson, M. J. (eds.), *Tissue culture methods and applications.* New York, Academic Press, pp. 782–787.

Ward, J. P., & King, J. R. (1997). Mathematical modelling of avascular tumour growth. *IMA J. Math. App. Med. Biol.* **14**: 39–69.

Watanabe, T., Kondo, K., & Oishi, M. (1991). Induction of *in vitro* differentiation of mouse erythroleukemia cells by genistein, an inhibitor of tyrosine kinases. *Cancer Res.* **51**: 764–768.

Watt, F. M. (2001). Stem cell fate and patterning in mammalian epidermis. *Curr Opin Genet Dev.* **11**: 410–417.

Watt, F. M. (2002). The stem cell compartment in human interfollicular epidermis. *J Dermatol Sci.* **28**: 173–180.

Watt, F. (1991). Annual Meeting of European Tissue Culture Society, Kraków, Poland.

Watt, J. L., & Stephen, G. S. (1986). Lymphocyte culture for chromosome analysis. In Rooney, D. E., & Czepulkowski, B. H. (eds.), *Human cytogenetics, a practical approach.* Oxford, U.K., IRL Press at Oxford University Press, pp. 39–56.

Waymouth, C. (1959). Rapid proliferation of sublines of NCTC clone 929 (Strain L) mouse cells in a simple chemically defined medium (MB752/1). *J. Natl. Cancer Inst.* **22**: 1003.

Waymouth, C. (1970). Osmolality of mammalian blood and of media for culture of mammalian cells. *In Vitro* **6**: 109–127.

Waymouth, C. (1974). To disaggregate or not to disaggregate: Injury and cell disaggregation, transient or permanent? *In Vitro* **10**: 97–111.

Waymouth, C. (1979). Autoclavable medium AM 77B. *J. Cell Physiol.* **100**: 548–550.

Waymouth, C. (1984). Preparation and use of serum-free culture media. In Barnes, W. D., Sirbasku, D. A., & Sato, G. H. (eds.), *Cell culture methods for molecular and cell biology; Vol. 1: Methods for preparation of media, supplements, and substrata for serum-free animal cell culture.* New York, Alan R. Liss, pp. 23–68.

Weibel, E. R., & Palade, G. E. (1964). New cytoplasmic components in arterial endothelia. *J. Cell Biol.* **23**: 101–102.

Weichselbaum, R., Epstein, I., & Little, J. B. (1976). A technique for developing established cell lines from human osteosarcomas. *In Vitro* **12**: 833–836.

Weinberg, R. A. (ed.). (1989). *Oncogenes and the molecular origins of cancer.* Cold Spring Harbor, NY, Cold Spring Harbor Laboratory Press.

Weiss, M. C., & Green, H. (1967). Human–mouse hybrid cell lines containing partial complements of human chromosomes and functioning human genes. *Proc. Natl. Acad. Sci. USA* **58**: 1104–1111.

Wells, D. L., Lipper, S. L., Hilliard, J. K., Stewart, J. A., Holmes, G. P., Herrmann, K. L., Kiley, M. P., & Schonberger, L. B. (1989). *Herpesvirus simiae* contamination of primary rhesus monkey kidney cell cultures: CDC recommendations to minimize risks to laboratory personnel. *Diagn. Microbiol. Infect. Dis.* **12**: 333–335.

Welm, B., Behbod, F., Goodell, M. A., & Rosen, J. M. (2003). Isolation and characterization of functional mammary gland stem cells. *Cell Prolif.* **36** (s1): 17–32.

Wessells, N. K. (1977). *Tissue interactions and development.* Menlo Park, CA, W. A. Benjamin.

Wessels, D., Titus, M., & Soll, D. R. (1996). A *Dictyostelium* myosin I plays a crucial role in regulating the frequency of pseudopods formed on the substratum. *Cell Motil. Cytoskeleton* **33**: 64–79.

Westerfield, M. (1993). *The zebrafish book: A guide for the laboratory use of zebrafish (Brachydanio rerio).* Eugene, OR, University of Oregon Press.

Westermark, B. (1974). The deficient density-dependent growth control of human malignant glioma cells and virus-transformed glialike cells in culture. *Int. J. Cancer* **12**: 438–451.

Westermark, B. (1978). Growth control in miniclones of human glial cells. *Exp. Cell. Res.* **111**: 295–299.

Westermark, B., Pontén, J., & Hugosson, R. (1973). Determinants for the establishment of permanent tissue culture lines from human gliomas. *Acta Pathol. Microbiol. Scand. A* **81**: 791–805.

Westermark, B., Wasteson, A. (1975). The response of cultured human normal glial cells to growth factors. *Adv Metab Disord.*, **8**: 85–100.

Westneat, D. F., Noon, W. A., Reeve, H. K., & Aquadro, C. F. (1988). Improved hybridisation conditions for DNA fingerprints probed with M13. *Nucleic Acids Res.* **16**: 4161.

Wewetzer, K., Verdu, E., Angelov, D. N., & Navarro, X. (2002) Olfactory ensheathing glia and Schwann cells: two of a kind? *Cell Tissue Res.* **309**: 337–345.

Whitehead, R. H. (2004). Establishment of cell lines from colon carcinoma. In Pfragner, R., & Freshney. R. I. (eds.), *Culture of human tumor cells.* Hoboken, NJ, Wiley-Liss, pp. 67–80.

Whitlock, C. A., Robertson, D., & Witte, O. N. (1984). Murine B cell lymphopoiesis in long term culture. *J. Immunol. Methods* **67**: 353–369.

Whur, P., Magudia, M., Boston, J., Lockwood, J., & Williams, D. C. (1980). Plasminogen activator in cultured Lewis lung carcinoma cells measured by chromogenic substrate assay. *Br. J. Cancer* **42**: 305–312.

Wienberg, J., & Stanyon, R. (1997). Comparative painting of mammalian chromosomes. *Curr. Opin. Genet. Dev.* **7**: 784–791.

Wiktor, A., & Van Dyke, D. L. (2004). Combined cytogenetic testing and fluorescence *in situ* hybridization analysis in the study of chronic lymphocytic leukemia and multiple myeloma. *Cancer Genet. Cytogenet.* **153**: 73–76.

Wiktor, T. J., Fernandes, M. V., Koprowski, H. (1964). Cultivation of Rabies Virus in Human Diploid Cell Strain WI-38. *J Immunol.* **93**: 353–66.

Wilkenheiser, K. A., Vorbroker, D. K., Rice, W. R., Clark, J. C., Bachurski, C. J., Oie, H. K., & Whitsett, J. E. (1991). Production of immortalized distal respiratory epithelial cell lines from surfactant protein C/simian virus 40 large tumor antigen transgenic mice. *Proc. Natl. Acad. Sci. USA* **90**: 11029–11033.

Wilkins, L. Gilchrest, B. A., Szabo, G., Weinstein, R., & Maciag, T. (1985). The stimulation of normal human melanocyte proliferation *in vitro* by melanocyte growth factor from bovine brain. *J. Cell Physiol.* **122**: 350.

Willey, J. C., Moser, C. E., Jr., Lechner, J. F., & Harris, C. C. (1984). Differential effects of 12-O-tetradecanoylphorbol-13-acetate on cultured normal and neoplastic human bronchial epithelial cells. *Cancer Res.* **44**: 5124–5126.

Williams, M. J., & Clark, P. (2003). Microscopic analysis of the cellular events during scatter factor/hepatocyte growth factor-induced epithelial tubulogenesis. *J. Anat.* **203**: 483–503.

Williams, B. P., Abney, E. R., & Raff, M. C. (1985). Macroglial cell development in embryonic rat brain: Studies using monoclonal antibodies, fluorescence-activated cell sorting and cell culture. *Dev. Biol.* **112**: 126–134.

Williams, G. M., & Gunn, J. M. (1974). Long-term cell culture of adult rat liver epithelial cells. *Exp. Cell Res.* **89**: 139–142.

Willmarth, N. E., Albertson, D. G., & Ethier, S. P. (2004). Chromosomal instability and lack of cyclin E regulation in hCdc4 mutant human breast cancer cells. *Breast Cancer Res.* **6**: R531–R539.

Wilson, A. P., Dent, M., Pejovic, T., Hubbold, L., & Rodford, H. (1996). Characterisation of seven human ovarian tumour cell lines. *Br. J. Cancer* **74**: 722–727.

Wilson, A. P. (2004). The development of human ovarian epithelial tumor cell lines from solid tumors and ascites. In *Culture of human tumor cells*, Ed. Pfragner, R., Freshney. R. I., Hoboken, NJ, Wiley-Liss, pp. 145–178.

Wilson, P. D., Dillingham, M. A., Breckon, R., & Anderson, R. J. (1985). Defined human renal tubular epithelia in culture: Growth, characterization, and hormonal response. *Am. J. Physiol.* **248**: F436–F443.

Wilson, P. D., Schrier, R. W., Breckon, R. D., & Gabow, P. A. (1986). A new method for studying human polycystic kidney disease epithelia in culture. *Kidney Int.* **30**: 371–378.

Wise, C. (2002). *Epithelial cell culture protocols.* Clifton, NJ, Humana Press.

Wistuba, I. I., Bryant, D., Behrens, C., Milchgrub, S., Virmani, A. K., Ashfaq, R., Minna, J. D., & Gazdar, A. F. (1999). Comparison of features of human lung cancer cell lines and their corresponding tumors. *Clin. Cancer Res.* **5**: 991–1000.

Witkowski, J. A. (1990). The inherited character of cancer—an historical survey. *Cancer Cells* **2**: 229–257.

Wolf, D. P., Meng, L., Ely, J. J., & Stouffer, R. L. (1998). Recent progress in mammalian cloning. *J. Assist. Reprod. Genet.* **15**: 235–239.

Wolff, E. T., & Haffen, K. (1952). Sur une méthode de culture d'organes embryonnaires *in vitro*. *Tex. Rep. Biol. Med.* **10**: 463–472.

Wolswijk, G., & Noble, M. (1989). Identification of an adult-specific glial progenitor cell. *Development* **105**: 387–400.

Wootton, M., Steeghs, K., Watt, D., Munro, J., Gordon, K., Ireland, H., Morrison, V., Behan, W., & Parkinson, E. K. (2003). Telomerase alone extends the replicative lifespan of human skeletal muscle cells without compromising genomic stability. *Hum. Gene Therapy* **14**: 1473–1487.

Work with Ionising Radiation. Ionising Radiation Regulations (1999). Approved Code of Practice and Guidance, L121, HSE Books, ISBN0 7176 1746 7.

Wright, K. A., Nadire, K. B., Busto, P., Tubo, R., McPherson, J. M., & Wentworth, B. M. (1998). Alternative delivery of keratinocytes using a polyurethane membrane and the implications for its use in the treatment of full-thickness burn injury. *Burns* **24**: 7–17.

Wu, D. K., & de Vellis, J. (1987). The expression of the intermediate filament-associated protein (NAPA-73) is associated with the stage of terminal differentiation of chick brain neurons. *Brain Res.* **421**: 186–93.

Wu, H., Friedman, W. J., & Dreyfus, C. F. (2004). Differential regulation of neurotrophin expression in basal forebrain astrocytes by neuronal signals. *J. Neurosci. Res.* **76**: 76–85.

Wu, R. (2004). Growth of human lung tumor cells in culture. In *Culture of human tumor cells*, Ed. Pfragner, R., Freshney. R. I., Hoboken, NJ, Wiley-Liss, pp. 1–21.

Wu, Y. J., Parker, L. M., Binder, N. E., Beckett, M. A., Sinard, J. H., Griffiths, C. T., & Rheinwald, J. G. (1982). The mesothelial keratins: A new family of cytoskeletal proteins identified in cultured mesothelial cells and nonkeratinizing epithelia. *Cell* **31**: 693–703.

Wuarin, L., Verity, M. A., & Sidell, N. (1991). Effects of interferon-gamma and its interaction with retinoic acid on human neuroblastoma differentiation. *Int. J. Cancer* **48**: 136–141.

Wurmser, A. E., Nakashima, K., Summers, R. G., Toni, N., D'amour, K. A., Lie, C. C., & Gage, F. H. (2004). Cell fusion-independent differentiation of neural stem cells to the endothelial lineage. *Nature* **430**: 350–356.

Wurster-Hill, D., Cannizzaro, L. A., Pettengill, O. S., Sorenson, G. D., Cate, C. C., & Maurer, L. H. (1984). Cytogenetics of small cell carcinoma of the lung. *Cancer Genet. Cytogenet.* **13**: 303–330.

Wyllie, F. S., Bond, J. A., Dawson, T., White, D., Davies, R., & Wynford-Thomas, D. (1992). A phenotypically and karyotypically stable human thyroid epithelial line conditionally immortalized by SV40 large T antigen. *Cancer Res.* **52**: 2938–2945.

Wysocki, L. J., & Sata, V. L. (1978). "Panning" for lymphocytes: A method for cell selection. *Proc. Natl. Acad. Sci. USA* **75**: 2844–2848.

Yaffe, D. (1968). Retention of differentiation potentialities during prolonged cultivation of myogenic cells. *Proc. Nat. Acad. Sci. USA* **61**: 477–483.

Yamada, K. M., & Geiger, B. (1997). Molecular interactions in cell adhesion complexes. *Curr. Opin. Cell Biol.* **9**: 76–85.

Yamada, T., Placzek, M., Tanaka, H., Dodd, J., & Jessell, T. M. (1991). Control of cell pattern in the developing nervous system: Polarizing activity of the floor plate and notochord. *Cell* **64**: 635–647.

Yamaguchi, N., Yamamura, Y., Koyama, K., Ohtsuji, E., Imanishi, J., & Ashihara, T. (1990). Characterization of new pancreatic cancer cell lines which propagate in a protein-free chemically defined medium. *Cancer Res.* **50**: 7008–7014.

Yan, G., Fukabori, Y., Nikolaropoulost, S., Wang, F., & McKeehan, W. L. (1992). Heparin binding keratinocyte growth factor is a candidate stromal to epithelial cell andromedin. *Mol. Endocrinol.* **6**: 2123–2128.

Yanai, N., Suzuki, M., & Obinata, M. (1991). Hepatocyte cell lines established from transgenic mice harboring temperature-sensitive simian virus 40 large T-antigen gene. *Exp. Cell Res.* **197**: 50–56.

Yang, J., Mani, S. A., Donaher, J. L., Ramaswamy, S., Itzykson, R. A., Come, C., Savagner, P., Gitelman, I., Richardson, A., & Weinberg, R. A. (2004). Twist, a master regulator of morphogenesis, plays an essential role in tumor metastasis. *Cell* **117**: 927–39.

Yasamura, Y., Tashjian, A. H., & Sato, G. (1966). Establishment of four functional clonal strains of animal cells in culture. *Science* **154**: 1186–1189.

Yeager. T. R., DeVries, S., Farrard, D. F., Kao, C., Nakada, S. Y., Monn, T. D., Bruskewitz, R., Stadler, W. M., Meisner, L. F., Gilchrest, K. W., Newyton, M. A., Waldman, F. M., & Reznikoff, C. A. (1998). Overcoming cellular senescence in human cancer pathogenesis. *Genes Dev.* **12**: 163–174.

Yen, A., Coles, M., & Varvayanis, S. (1993). 1,25-dihydroxy vitamin D_3 and $12 - O$-tetradecanoyl phorbol-13-acetate synergistically induce monocytic cell differentiation: FOS and RB expression. *J. Cell Physiol.* **156**: 198–203.

Yeoh, G. C. T., Hilliard, C., Fletcher, S., & Douglas, A. (1990). Gene expression in clonally derived cell lines produced by *in vitro* transformation of rat fetal hepatocytes: Isolation of cell lines which retain liver-specific markers. *Cancer Res.* **50**: 75–93.

Yerganian, G., & Leonard, M. J. (1961). Maintenance of normal *in situ* chromosomal features in long-term tissue cultures. *Science* **133**: 1600–1601.

Yevdokimova, N., & Freshney, R. I. (1997). Activation of paracrine growth factors by heparan sulphate induced by glucocorticoid in A549 lung carcinoma cells. *Brit. J. Cancer* **76**: 261–289.

Yoshida, M., & Beppu, T. (1990). In Doyle, A., Griffiths, J. B., & Newell, D. G., *Cell and tissue culture: Laboratory procedures*. Chichester, U.K., John Wiley & Sons, Module 4E2.

Yoshioka, M., Nakajima, Y., Ito, T., Mikami, O., Tanaka, S., Miyazaki, S., & Motoi, Y. (1997). Primary culture and expression of cytokine mRNAs by lipopolysaccharide in bovine Kupffer cells. *Vet. Immunol. Immunopathol.* **58**: 155–163.

Young, H. E., Duplaa, C., Romero-Ramos, M., Chesselet, M. F., Vourc'h, P., Yost, M. J., Ericson, K., Terracio, L., Asahara, T., Masuda, H., Tamura-Ninomiya, S., Detmer, K., Bray, R. A., Steele, T. A., Hixson, D., el-Kalay, M., Tobin, B. W., Russ, R. D., Horst, M. N., Floyd, J. A., Henson, N. L., Hawkins, K. C., Groom, J., Parikh, A., Blake, L., Bland, L. J., Thompson, A. J., Kirincich, A., Moreau, C., Hudson, J., Bowyer, F. P. 3rd, Lin, T. J., & Black, A. C. Jr. (2004). Adult reserve stem cells and their potential for tissue engineering. *Cell. Biochem. Biophys.* **40**: 1–80.

Yuhas, J. M., Li, A. P., Martinez, A. O., & Ladman, A. J. (1977). A simplified method for production and growth of multicellular tumour spheroids (MTS). *Cancer Res.* **37**: 3639–3643.

Yuspa, S. H., Koehler, B., Kulesz-Martin, M., & Hennings, H. (1981). Clonal growth of mouse epidermal cells in medium with reduced calcium concentration. *J. Invest. Dermatol.* **76**: 144–146.

Yusufi, A. N. K., Szczepanska-Konkel, M., Kempson, S. A., McAteer, J. A., & Dousa, T. P. (1986). Inhibition of human renal epithelial Na^+/Pi cotransport by phosphonoformic acid. *Biochem. Biophys. Res. Commun.* **139**: 679–686.

Zaroff, L., Sato, G. H., & Mills, S. E. (1961). Single-cell platings from freshly isolated mammalian tissue. *Exp. Cell Res.* **23**: 565–575.

Zeltinger, J., & Holbrook, K. A. (1997). A model system for long-term serum-free suspension organ culture of human fetal tissues: Experiments on digits and skin from multiple body regions. *Cell Tissue Res.* **290**: 51–60.

Zeng, C., Pesall, J. E., Gilkerson, K. K., & McFarland, D. C. (2002). The effect of hepatocyte growth factor on turkey satellite cell proliferation and differentiation. *Poult Sci.* **81**: 1191–1198.

Zhang, Y., Proenca, R., Maffei, M., Barone, M., Leopold, L., & Friedman, J. M. (1994). Positional cloning of the mouse obese gene and its human homologue. *Nature* **372**: 425–432.

Zhou, S., Schuetz, J. D., Bunting, K. D., Colapietro, A.-M., Sampath, J., Morris, J. J., Lagutina, I., Grosveld, G. C., Osawa, M., Nakauchi, H., & Sorrentino, B. P. (2001). The ABC transporter Bcrp1/ABCG2 is expressed in a wide variety of stem cells and is a molecular determinant of the side-population phenotype. *Nat. Med.* **7**: 1028–1034.

Zhu, C., & Joyce, N. C. (2004). Proliferative response of corneal endothelial cells from young and older donors. *Invest. Ophthalmol. Vis. Sci.* **45**: 1743–1751.

Zhu, S. Y., Cunningham, M. L., Gray, T. E., & Nettesheim, P. (1991). Cytotoxicity, genotoxicity and transforming activity of 4-(methylnitrosamino)-1-(3-pyridyl)-1-butanone (NNK) in rat trachea epithelial cells. *Mutation Res.* **261**(4): 249–259.

Zimmermann, H., Hillgartner, M., Manz, B., Feilen, P., Brunnenmeier, F., Leinfelder, U., Weber, M., Cramer, H., Schneider, S., Hendrich, C., Volke, F., & Zimmermann, U. (2003). Fabrication of homogeneously cross-linked, functional alginate microcapsules validated by NMR-, CLSM- and AFM-imaging. *Biomaterials* **24**: 2083–2096.

Zoubiane, G. S., Valentijn, A., Lowe, E. T., Akhtar, N., Bagley, S., Gilmore, A. P., & Streuli, C. H. (2003). A role for the cytoskeleton in prolactin-dependent mammary epithelial cell differentiation. *J. Cell Sci.* **117**: 271–280.

zur Nieden, N. I., Kempka, G. & Ahr, H. J. (2003). *In vitro* differentiation of embryonic stem cells into mineralized osteoblasts. *Differentiation* **71**: 18–27.

Zwain, I. H., Morris, P. L., & Cheng, C. Y. (1991). Identification of an inhibitory factor from a Sertoli clonal cell line (TM4) that modulates adult rat Leydig cell steroidogenesis. *Mol. Cell Endocrinol.* **80**: 115–126.

INDEX

NOTE: Page numbers followed by f refer to figures, page numbers followed by t refer to tables.

Culture of Animal Cells: A Manual of Basic Technique, Fifth Edition, by R. Ian Freshney
Copyright © 2005 John Wiley & Sons, Inc.